APG —   WS Phylogentic Group

Families of tracheophytes (other than angiosperms) covered in detail in Chapter 8.
See Table 8.1, p. 186–187, for a more detailed taxonomic presentation.

Family	Page	Family	Page	Family	Page	Family	Page
Araucariaceae	218	Dryopteridaceae	204	Ophioglossaceae	193	Selaginellaceae	189
Aspleniaceae	201	Ephedraceae	221	Osmundaceae	197	Taxaceae	219
Blechnaceae	203	Equisetaceae	193	Pinaceae	211	Thelypteridaceae	202
Cupressaceae	215	Ginkgoaceae	208	Podocarpaceae	217	"Woodsiaceae"	203
Cyatheaceae	199	Lycopodiaceae	188	Polypodiaceae	205	Zamiaceae	208
Cycadaceae	207	Marsileaceae	198	Psilotaceae	191		
Dennstaedtiaceae	200	Onocleaceae	204	Pteridaceae	201		

Families of angiosperms covered in detail in Chapter 9.
See Table 9.1, pp. 230–231, for a more detailed taxonomic presentation.

Family	Page	Family	Page	Family	Page	Family	Page
Acanthaceae	486	Casuarinaceae	406	Lauraceae	242	Polygonaceae	334
Adoxaceae	504	Celastraceae	351	Lecythidaceae	455	Pontederiaceae	283
Agavaceae	268	Ceratophyllaceae	248	Lentibulariaceae	488	"Portulacaceae"	328
Aizoaceae	327	Cistaceae	427	Liliaceae	257	Potamogetonaceae	256
Alismataceae	252	Clusiaceae	362	Loasaceae	443	Primulaceae	450
Alliaceae	270	Colchicaceae	258	Loranthaceae	336	Proteaceae	317
Altingiaceae	344	Combretaceae	416	Lythraceae	412	Ranunculaceae	309
Amaranthaceae	324	Commelinaceae	281	Magnoliaceae	237	Restionaceae	296
Amaryllidaceae	270	Convolvulaceae	462	Malpighiaceae	353	Rhamnaceae	388
Amborellaceae	232	Cornaceae	443	Malvaceae	424	Rhizophoraceae	364
Anacardiaceae	435	Crassulaceae	342	Marantaceae	304	Rosaceae	379
Annonaceae	240	Cucurbitaceae	396	Melanthiaceae	260	Rubiaceae	469
Apiaceae	495	Cyperaceae	294	Melastomataceae	418	Ruscaceae	266
Apocynaceae	471	Dioscoreaceae	275	Meliaceae	432	Rutaceae	429
Aquifoliaceae	494	Dipterocarpaceae	429	Menispermaceae	308	Salicaceae	367
Araceae	250	Droseraceae	332	Moraceae	392	Santalaceae	338
Araliaceae	499	Ebenaceae	449	Myricaceae	406	Sapindaceae	438
Arecaceae	278	Ericaceae	452	Myristicaceae	240	Sapotaceae	445
Aristolochiaceae	247	Eriocaulaceae	290	Myrtaceae	416	Sarraceniaceae	455
Asparagaceae	266	Euphorbiaceae	355	Nyctaginaceae	324	Saxifragaceae	338
Asphodelaceae	272	Fabaceae	371	Nymphaeaceae	233	Scrophulariaceae	484
Asteraceae	508	Fagaceae	401	Oleaceae	477	Simaroubaceae	435
Begoniaceae	398	Gentianaceae	471	Onagraceae	414	Smilacaceae	259
Berberidaceae	312	Geraniaceae	348	Orchidaceae	273	Solanaceae	459
Betulaceae	404	Gesneriaceae	481	Orobanchaceae	484	Theaceae	452
Bignoniaceae	486	Haemodoraceae	282	Oxalidaceae	351	Typhaceae	290
Boraginaceae	462	Hamamelidaceae	342	Papaveraceae	314	Ulmaceae	389
Brassicaceae	420	Hyacinthaceae	269	Passifloraceae	367	Urticaceae	393
Bromeliaceae	287	Hydrangeaceae	441	Phyllanthaceae	359	Verbenaceae	490
Burseraceae	437	Hydrocharitaceae	254	Phytolaccaceae	323	Violaceae	364
Cactaceae	330	Hypericaceae	362	Piperaceae	245	Vitaceae	346
Campanulaceae	508	Illiciaceae	235	Plantaginaceae	481	Winteraceae	244
Cannabaceae	391	Iridaceae	272	Platanaceae	316	Xyridaceae	292
Cannaceae	306	Juglandaceae	408	Poaceae	296	Zingiberaceae	302
Caprifoliaceae	501	Juncaceae	292	Polemoniaceae	457	Zygophyllaceae	350
Caryophyllaceae	320	Lamiaceae	492	Polygalaceae	377		

Handwritten annotations: Carrot / Queen Anne's Lace (Apiaceae); Sunflower (Asteraceae); Mustard (Brassicaceae); Honeysuckle / carnation (Caprifoliaceae / Caryophyllaceae); Bean / Oak (Fabaceae / Fagaceae); Iris (Iridaceae); Rush (Juncaceae); Mint (Lamiaceae); Lily (Liliaceae); Cotton (Malvaceae); Fig / Mulberry (Moraceae); Phil (Cornaceae); primrose (Onagraceae); last Monocot Grp (Orchidaceae); wood sorrel (Oxalidaceae); Rose (Rosaceae); coffee (Rubiaceae); potato / tomato / egg plant (Solanaceae); Elm (Ulmaceae); Walnut (Juglandaceae); grass / last dim. (Poaceae)

Plant Systematics

A PHYLOGENETIC APPROACH

THIRD EDITION

Plant THIRD EDITION
Systematics
A PHYLOGENETIC APPROACH

Walter S. Judd *University of Florida*

Christopher S. Campbell *University of Maine*

Elizabeth A. Kellogg *University of Missouri, St. Louis*

Peter F. Stevens *University of Missouri, St. Louis; Missouri Botanical Garden*

Michael J. Donoghue *Yale University*

 SINAUER ASSOCIATES, INC. *Publishers • Sunderland, Massachusetts USA*

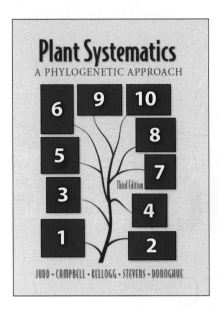

The Cover

(1) Lycophyte, *Lycopodium annotinum* (Lycopodiaceae). (2) Monilophyte (fern), *Woodwardia virginica* (Blechnaceae). (3) Conifer, *Chamaecyparis obtusa* (Cupressaceae). (4) Magnoliid, *Liriodendron tulipifera* (Magnoliaceae). (5, 6) Monocots: *Iris hexagona* (Iridaceae); *Carex verrucosa* (Cyperaceae). (7, 8) Rosids: *Meriania hernandii* (Melastomataceae); *Sorbus aucuparia* (Rosaceae). (9, 10) Asterids: *Lyonia lucida* (Ericaceae); *Hieracium aurantiacum* (Asteraceae/Compositae). Photographs of *Woodwardia virginica* by J. Richard Abbott, of *Meriania hernandii* by Darin S. Penneys, and of *Lyonia lucida* by Kurt M. Neubig; all others by Walter S. Judd.

***PLANT SYSTEMATICS: A PHYLOGENETIC APPROACH*, Third Edition**

Copyright © 2008 by Sinauer Associates, Inc. All rights reserved. No part of this book may be reprinted without written permission of the publisher.

Sinauer Associates, Inc.
23 Plumtree Road
Sunderland, MA 01375 U.S.A.
fax: 413-549-1118
email: orders@sinauer.com, publish@sinauer.com
www.sinauer.com

Library of Congress Cataloging-in-Publication Data

Plant systematics : a phylogenetic approach / Walter S. Judd ... [et al.]. — 3rd ed.
    p. cm.
  Includes bibliographical references and index.
  ISBN 978-0-87893-407-2 (casebound : alk. paper)
  1. Plants—Classification. I. Judd, Walter S.

QK95.P548 2008
580.1′2—dc22   2007019671

Printed in China

5 4 3 2 1

With appreciation, we dedicate this textbook to our mentor and friend,
Carroll E. Wood, Jr.,
whose kindness and knowledge of plant systematics
have helped many students and colleagues.

Contents in Brief

Contents

CHAPTER 3
Classification and System in Flowering Plants: Historical Background 39

CHAPTER 4
Taxonomic Evidence: Structural and Biochemical Characters 53

CHAPTER 5
Molecular Systematics 103

CHAPTER 6
The Evolution of Plant Diversity 119

CHAPTER 7
An Overview of Green Plant Phylogeny 153

CHAPTER 8
Lycophytes, Ferns, and Gymnosperms 185

CHAPTER 9
Phylogenetic Relationships of Angiosperms 225

APPENDIX ONE: Botanical Nomenclature 543

APPENDIX TWO: Specimen Preparation and Identification 553

Preface

We wrote this book because we believed that there was a need for an explanation of plant systematics (or taxonomy) that can be used in undergraduate courses and that incorporates current exciting developments in the field. In the past two decades abundant new systematic data, especially from DNA sequences, have come forth, and a rigorous phylogenetic approach has established novel ways of analyzing data and practicing systematics.

The first edition of this book was well received by many students and other botanists. It has been translated into French and Italian, and the authors of the first edition (Walter S. Judd, Christopher S. Campbell, Elizabeth A. Kellogg, and Peter F. Stevens) received the 1999 Engler Silver Award, an annual award of the International Association of Plant Systematists for the most outstanding book or monograph in plant systematics. Encouraged by this positive reception, we prepared a second edition (which was translated into Korean), and now this third edition, in order to keep the book as up-to-date as possible.

The basis for this book is the tree of life—the idea that all life is interrelated, like branches on a tree. We deal with the part of this tree occupied by the vascular plants or tracheophytes, and focus on the flowering plants, an important group that dominates most of Earth's terrestrial ecosystems and provides us with medicines, beautiful ornamentals, construction materials, fiber for paper and clothing, and most of our food. Historically, information about these and other attributes of vascular plants was essential to the development of human civilization. Survival rested upon knowing which plants were poisonous or were good to eat; which were potential food for animals; which were good for tools and weapons; which might cure ailments; and which might be useful in many other ways. Systematics as a science grew out of such interests in organisms and evolved from systems of information about the uses of different organisms to a broader understanding of biodiversity. To this end, plant systematists ask questions such as, How should the tremendous diversity of legumes be partitioned or classified into genera and then into species? What are the relationships among the major groups of grasses? How do we know whether or not a plant is a member of the orchid family? This text deals with how these and thousands of similar questions are answered; the final two chapters present some current answers to questions about evolutionary relationships among the major groups of tracheophytes.

Charles Darwin led the way in establishing the evolutionary perspective of modern biology. One impact of this view of life was to make it a major goal of systematics to uncover the evolutionary history—or phylogeny—of groups of organisms. Willi Hennig and W. H. Wagner, along with many others, developed explicit methods of hypothesizing and testing phylogenetic relationships and reflecting these in classifications. We follow a phylogenetic approach throughout this book.

Chapter 1 introduces and explains the concept of phylogeny, and outlines why systematics is important in biological investigations and to society. Chapter 2 explains how phylogenies are reconstructed from systematic evidence. The process of transforming evidence into a representation of phylogenetic relationships in an evolutionary tree is first presented using simple hypothetical data and then extended to complex real situations. This "primer of phylogeny" also incorporates the conversion of phylogenetic hypotheses into a classification.

The central position of phylogeny in classification has emerged only recently, but many groups of plants, such as legumes, grasses, and orchids, have long been recognized. Chapter 3 outlines how systematists perceived groups in the past, how higher taxa were formed, the historical background of phylogeny, and provides an explanation of plant groupings over the years. The information on which phylogenies and classifications are based, comes from a wide array of sources, including structural aspects of organisms, especially morphology, anatomy and secondary chemistry (Chapter 4), and DNA features, especially nucleotide sequences (Chapter 5). These chapters on phylogeny, classification, and taxonomic evidence provide essential information for understanding the vascular plant groups covered in this book.

Diversification and the evolution of plant species—how plant species form, how they interact, and how we define them, including a presentation of the effects of hybridization—are the fascinating and challenging topics of Chapter 6. Chapter 7 presents an overview of the phylogenetic history of green plants, especially tracheophytes, explains the origin of many of the important features used to identify plants, and lays the foundation for the final two chapters, which focus on vascular plant diversity.

Chapter 8 covers tracheophytes that do not have flowers, including lycophytes, monilophytes, and extant gymnosperms (conifers, cycads, ginkgo, and gnetophytes). Although these plants are certainly important, they are far surpassed in terms of number of species and ecological dominance by the flowering plants or angiosperms, the subject of Chapter 9. Chapters 8 and 9 present the diversity of these groups, with descriptions, illustrations, and discussions of phylogenetic relationships. Exceptionally high-quality illustrations from the Generic Flora of the Southeastern United States, which show taxonomically useful information for many families of plants, and a series of color photographic plates, are special assets of Chapters 8 and 9.

Chapters 8 and 9 also emphasize characterzation of major clades of vascular plants, and provide detailed coverage of 169 plant families. Two appendices provide additional information related to plant identification, including botanical nomenclature (the application of scientific names), collection of scientific plant specimens, an overview of plant identification, information about the plant systematics literature, and some useful sites on the World Wide Web.

The science of plant systematics has a large number of specialized terms. We have tried to minimize use of technical language, and to apply those terms used in a consistent and precise manner. The terms used in this book are not only included in the glossary, but they are also boldfaced and defined in the first six chapters and appendices (and included in the Subject Index).

Scientific names (and common names) are included in the Taxonomic Index.

Our knowledge of plant systematics is growing very rapidly, making this an exciting time to study plant systematics. We encourage students to get caught up in this excitement, to appreciate the beauty and importance of plants, which constitute one of the great branches of the Tree of Life, "which fills with its dead and broken branches the crust of the earth, and covers the surface with its ever-branching and beautiful ramifications" (Charles Darwin, *On the Origin of Species*, 1859, Chap. IV, page 163).

Media

An important tool for plant identification is the CD that accompanies this book. It contains over 3100 images of flowers, fruits, detailed dissections of flowers and fruits, and other parts of plants. Students can access these images through an alphabetical list of species, an alphabetical list of families (with the species listed alphabetically within each), or a list of orders and families following the arrangement adopted in this text. The CD was an effective study aid for students using the earlier editions of this text, and it is now greatly expanded in number of images, including those representing floral and fruit dissections or anatomical structures. An expanded illus-

trated glossary (a list of defined terms with links to particular images on the CD illustrating them) is included. The CD also contains an appendix, listing all the families of angiosperms according to the classification of the Angiosperm Phylogeny Group (which is followed in this text).

New for the Third Edition, the *Plant Systematics* Instructor's Resource CD contains all of the figures, tables, and plates from the textbook in both JPEG (high- and low-resolution) and PowerPoint™ formats. (Available to qualified adopters.)

Acknowledgments

This book is a group effort. Although each of us wrote certain chapters, we all benefited from numerous helpful comments from our coauthors. Walter S. Judd had primary responsibility for Chapters 4 and 9, and the two appendices. Christopher S. Campbell was responsible for Chapters 1, 6, and 8; he also contributed the treatment of Rosaceae in Chapter 9, the sections on pollination biology, embryology, chromosomes, and palynology in Chapter 4, and the section on the World Wide Web in Appendix 2. Elizabeth A. Kellogg wrote Chapter 5, a portion of Chapter 1, as well as the treatments for Poaceae, Juncaceae, Cyperaceae, and Santalales in Chapter 9, and the section on arguments against ranks in classifications in Appendix 1. Peter F. Stevens wrote Chapter 3 and Michael J. Donoghue wrote Chapter 7. Chapter 2 was co-written by Elizabeth and Walter.

The CD was the responsibility of Walter S. Judd, Daniel L. Nickrent, Kenneth R. Robertson, J. Richard Abbott, Christopher S. Campbell, Barbara S. Carlsward, Michael J. Donoghue, and Elizabeth A. Kellogg; this group also provided most of the photographs. Walter compiled the three appendices on the CD.

Carroll E. Wood, Jr. generously gave us permission to use numerous plates prepared for the Generic Flora of the Southeastern United States and assisted in their organization and electronic digitalization. These beautiful illustrations, which greatly enhance the value of this text, were drawn by a series of exceptionally talented artists:

IAS   Ihsan Al-Shehbaz

℘   Irene Brady

ADC   Arnold D. Clapman

SD   Sydney B. DeVore

DCJ   Diane C. Johnston

DHM   Dorothy H. Marsh

Sue Sargent (not initialed)

YS   Virginia Savage

KSV KS   Karen Stoutsenberger Velmure

LT.   LaVerne Trautz

mvm   Margaret van Montfrans

LAVorobik 1989   Linda A. Vorobik

RW   Rachel A. Wheeler

WBZ   Wendy B. Zomlefer

Carroll also provided editorial assistance, especially with the figure captions. Edward O. Wilson gave permission to use the previously unpublished plate of *Yucca filamentosa* and its pollinator; we thank Michael D. Frohlich for supplying materials for this plate and Kathy Horton for assistance in locating and transmitting it. Robert Dressler kindly allowed the use of his original illustration of *Encyclia cordigera*, and Wendy B. Zomlefer graciously allowed use of her beautiful illustration of *Schoenocaulon*. Robert K. Jansen generously gave permission to use the illustration of the chloroplast genome of *Vitis vinifera*. We thank Pamela and Douglas Soltis for the use of the *Tragopogon* diagram, and Sherwin Carlquist for the photo of *Argyroxiphium sandwichense*. H.-Dietmar Behnke prepared the informative plate illustrating various sieve-element plastid conditions; Y. Renea Taylor prepared the photo of epicuticular waxes; and Helmut Presser provided the photos of *Ophrys* pollination, Scott Hodges the photos of *Aquilegia*, and Rodney Barton the photo of *Iris fulva*. We are grateful for the numerous colleagues who contributed other illustrative materials; they are acknowledged in the figure captions.

Allison R. Minott assisted in the preparation of the captions for Chapter 9 and Reuben E. Judd helped take the photographs in Figure 1 of Appendix 1. Brian Moore and Mary Walsh helped with the figures for Chapter 7; Susan Donoghue edited the manuscript of this chapter. Alison E. Colwell, Paul Corogin, Steven P. Darwin, Peter K. Endress, Miguel A. García, Gretchen M. Ionta, Reuben E. Judd, Simon Malcomber, Litton J. Musselman, Robert F. C. Naczi, Kurt M. Neubig, Darin S. Penneys, Roger W. Sanders, David S. Seigler, J. Dan Skean Jr., Douglas E. Soltis, Margaret H. Stone, W. Mark Whitten, and Scott Zona provided some very useful slides for the CD that accompanies this text, and Jason Dirks and Christopher Small handled technical aspects relating to its production. We are deeply grateful for all the above-listed contributions.

We express our sincere appreciation to the following individuals, who read and commented on various sections of the book (and/or provided useful reprints or unpublished manuscripts): Pedro Acevedo, Victor A. Albert, Lawrence A. Alice, Ihsan Al-Shehbaz, Arne A. Anderberg, William R. Anderson, George W. Argus, Daniel F. Austin, David S. Barrington, David A. Baum, Paul E. Berry, Camilla P. Campbell, Lisa M. Campbell, Philip D. Cantino, Heather R. Carlisle, Mark W. Chase, Lynn G. Clark, David S. Conant, Garrett E. Crow, Steven P. Darwin, Claude W. de Pamphilis, Alison C. Dibble, James A. Doyle, Robert Dressler, Mary E. Endress, Peter K. Endress, Peter Goldblatt, Shirley A. Graham, Michael H. Grayum, Arthur D. Haines, Peter C. Hoch, Sara B. Hoot, Joachim W. Kadereit, Christine M. Kampny, Robert Kral, Kathleen A. Kron, Matthew Lavin, Steven R. Manchester, Paul S. Manos, Lucinda A. McDade, Alan W. Meerow, Laura C. Merrick, David R. Morgan, Cynthia M. Morton, Daniel L. Nickrent, Eliane M. Norman, Richard G. Olmstead, Clifford R. Parks, Gregory M. Plunkett, Robert A. Price, John F. Pruski, Kathleen M. Pryer, Susanne S. Renner, Karen S. Renzaglia, James L. Reveal, Kenneth R. Robertson, Edward E. Schilling, Alan R. Smith, Douglas E. Soltis, William L. Stern, Henk van der Werff, Paul van Rijckevorsel, Thomas F. Vining, Terrence Walters, Grady L. Webster, W. Mark Whitten, John H. Wiersema, Norris H. Williams, Wesley A. Wright, George Yatskievych, Wendy B. Zomlefer, and Scott A. Zona. We also thank the numerous individuals who sent corrections and/or suggestions relating to the earlier editions.

Walter thanks the University of Florida, College of Liberal Arts and Sciences, for providing a one-term sabbatical during which he was able to initiate work on this text.

We thank Andy Sinauer and all the staff at Sinauer Associates for their excellent advice and guidance, and we especially thank Carol Wigg, Laura Green, Jason Dirks, Christopher Small, Jefferson Johnson, Joanne Delphia, Norma Roche, and Marie Scavotto. Without their valuable work this project would never have reached completion.

Finally, Christopher, Michael, and Walter thank their wives, Margaret, Susan, and Beverly, respectively, for their emotional support and for their forbearance with their husband's preoccupation with plant systematics and this book. Elizabeth and Peter thank their son, Harry Stevens, for continuing to be patient with his parents.

We, the authors, assume all editorial responsibility for this book. Comments and corrections on the earlier editions have been helpful in preparing the third edition, and we would greatly appreciate any comments or corrections on this edition as well.

WALTER S. JUDD, *Gainesville, Florida*

CHRISTOPHER S. CAMPBELL, *Orono, Maine*

ELIZABETH A. KELLOGG, *St. Louis, Missouri*

PETER F. STEVENS, *St. Louis, Missouri*

MICHAEL J. DONOGHUE, *New Haven, Connecticut*

1

The Science of Plant Systematics

[T]he characters which naturalists consider as showing true affinity between any two or more species, are those which have been inherited from a common parent, all true classification being genealogical.

Charles Darwin 1859: 391

 What exactly is plant systematics? The question turns out to be more difficult than you may have imagined because both *plant* and *systematics* are rather hard to define. Considering these concepts in some detail will help us to define the science and clarify our aims.

What Do We Mean by *Plant*?

Most people have a commonsense notion of what a plant is: a plant is a living organism that is green and doesn't move around. For some, the notion of a plant encompasses the fungi, which are not green, and botany and plant biology departments in many colleges and universities include *mycologists*—people who study fungi. For some, the word *plant* is restricted to green organisms living on land and in water. Aquatic photosynthetic organisms, however, encompass a tremendous diversity of life forms, including green and non-green algae and related groups.

For the purposes of this book, we will consider the green plants, a major lineage that includes the so-called green algae and the land plants (Figure 1.1). Defined in this way, the green plants share a number of features, including: (1) the presence of the photosynthetic pigments chlorophylls a and b; (2) storage of carbohydrates, usually in the form of starch; and (3) the presence of two anterior whiplash flagella at some stage of the life cycle (often modified or sometimes lost).

FIGURE 1.1 Phylogeny of green plants (as pictured in a phylogenetic tree). Distinctive structural features that characterize various groups of green plants are indicated on the branches where the characteristics are thought to have evolved. Phylogenetic relationships among the liverworts, hornworts, and mosses are unclear; see also Figure 7.6.

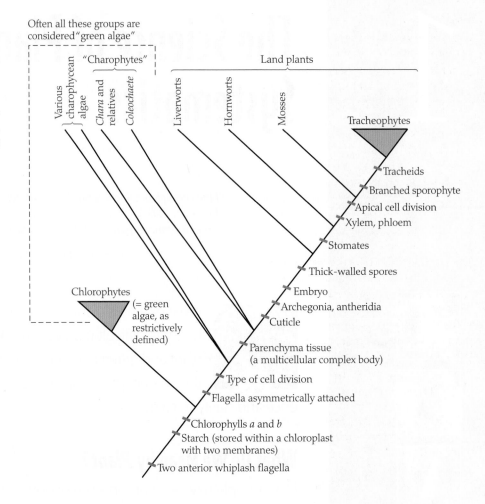

Within the green plants we will concentrate on the land plants—that is, the **embryophytes** (a few of which actually live in water), whose closest extant relatives are members of the "charophytes," a green algal group. Land plants have life histories involving alternation of two morphologically distinct bodies (a diploid sporophyte and a haploid gametophyte), thick-walled spores, an embryonic stage in the life cycle, specialized structures that protect the gametes (archegonia for eggs and antheridia for sperm), and a cuticle (a waxy protective layer over the epidermal cells). Along with these shared morphologies, numerous DNA characters also support the view that this group is **monophyletic**—that is, defined in this way, *plants* represent a single branch on the tree of life.

The land plants consist of three groups of rather small plants—liverworts, hornworts, and mosses—and the **tracheophytes**. *Tracheo* refers to the presence of tracheids—cells specialized for transport of liquids—and the Greek root *phyte* means plant. Tracheophytes, sometimes referred to as **vascular plants**, are by far the largest group of green plants, including about 260,000 species. They form the dominant vegetation over much of Earth's land surface and are the focus of this book.

The great majority of tracheophytes, all but about 12,000 species, are **flowering plants**, or **angiosperms**. In impor-

tance to the world's ecosystems and to human nutrition, medicine, and overall welfare, flowering plants far surpass the other tracheophytes, and most of this book is devoted to the angiosperms.

What Do We Mean by *Systematics*?

Even though this book focuses primarily on the angiosperm tracheophytes, the basic principles of systematics used here apply to all organisms. **Systematics** is the science of organismal diversity. It entails the discovery, description, and interpretation of biological diversity, as well as the synthesis of information on diversity in the form of predictive classification systems. According to paleontologist George Gaylord Simpson (1961: 7), "Systematics is the scientific study of the kinds and diversity of organisms and of any and all relationships among them." This view is so broad that it could encompass what we normally think of as ecology, and perhaps other disciplines as well, so it is necessary to consider in more detail the kinds of relationships that have specifically concerned systematists.

In our view, the fundamental aim of systematics is to discover all the branches of the evolutionary tree of life, to document changes that have occurred during the evolution of these branches, and to the greatest extent possible describe all species—the tips of the branches. Systematics is therefore the study of the biological diversity that exists on Earth today and its evolutionary history.

Systematists attempt to reconstruct the entire chronicle of evolutionary events, including the splitting of populations into separate lineages and any and all evolutionary modifications in the characteristics of organisms associated with these branching events, as well as with periods between branching. A secondary but critical aim of systematists is to convey their knowledge of the tree of life—of the terminal branches and their relationships to one another—in an unambiguous system of classification, which can then orient our understanding of life and the world around us. This is the *phylogenetic approach* to systematics.

We explicitly take the view that systematics is not just a descriptive science, but also aims to discover evolutionary relationships and real evolutionary entities that have resulted from the process of evolution. We take as a starting point the separation of a lineage into two or more lineages. We study the evolutionary modifications that have occurred (and will continue to occur) within these lineages. Our aim is to reconstruct the history of the separation of lineages and the history of their modifications as accurately as possible by bringing as much relevant information as possible to bear on the problem. Systematists continually put forward hypotheses about the existence of branches of the tree of life and test them with evidence derived from a wide variety of sources. Alternative hypotheses are evaluated, and some are provisionally chosen over others.

Some systematists see their work rather differently. They think of themselves as simply describing similarities and differences evident in the organisms around us without reference to any theory. They view branching diagrams such as Figures 1.1 and 1.2 and classifications (see Figure 1.5) only as efficient depictions of these similarities and differences. According to this approach, entities recognized by systematists are summaries of observed information and no more, whereas in our view, these entities are hypothesized branches of the evolutionary tree. Thus our approach extends beyond summation of the data at hand to statements about entities that we do not directly observe, but that we infer came into being through the evolutionary process.

This tension between theory-neutral and theory-grounded approaches pervades the history of science. There have always been those who consider theory-neutral observations to be both possible and desirable and who wish to define the basic terms of scientific discipline in terms of particular operations performed on the data. And there are those, like the authors of this book, whose concepts, definitions, and inferential procedures are explicitly grounded in theory and who wish to go beyond summarizing the evidence at hand to making claims about the world at large.

To some, the distinction we have just made may seem small, and in practice, it is true that systematists with different views of their activities all conduct their research in much the same way. We highlight the difference here because it may help interested readers comprehend some of the literature of systematics, and because it helps explain the orientation of our own treatment of plant systematics. Most importantly, throughout this book we will focus on how we interpret evidence of all kinds in relation to the fundamental aim of systematics stated here. Systematics occupies a central position in evolutionary biology and is playing an increasingly important role in many other disciplines, including ecology, molecular biology, developmental biology, anthropology, and even linguistics and philosophy.

The Phylogenetic Approach

Our view of the science of plant systematics explains an outlook that we hope will emerge again and again throughout this book—namely, that systematics is linked directly and centrally to the study of evolution in general, from the study of fossils to the study of genetic changes in local populations. The basic connection is extraordinarily simple: studies of the process of evolution benefit (usually enormously!) from knowledge of what we deduce has taken place during the evolution of life on Earth. For example, when one sets out a hypothesis about the evolution of a particular characteristic of an organism, it is assumed that the trait of interest did in fact originate within the group being studied. Furthermore, such hypotheses usually rely on some knowledge of the precursor condition from which the trait in question arose.

This kind of information on the sequence of evolutionary events is obtained by systematists who reconstruct the **phylogeny**—the evolutionary history—of a group of organisms. Similarly, studies of the rate of evolutionary change and of the ages and diversification patterns of lineages depend directly on knowledge of phylogenetic relationships.

How Do We Reconstruct Phylogeny?

A phylogeny consists of simple sets of statements of the following nature: groups A and B are more closely related to each other than either is to group C. Consider the simple example of three members of the rose family (Rosaceae): blackberry, cherry, and raspberry (Figure 1.2). Following the injunction "by their fruits ye shall know them," we can infer evolutionary relationships using only fruits as evidence. Blackberries and raspberries both have numerous small and fleshy fruits (drupes or stone fruits) that are clustered together (see Chapter 4 for a description of fruit types). Cherry fruits are also drupes, but they are borne singly and are larger than those of blackberries and raspberries.

With this information about the fruits of these three plants, we would judge that blackberries and raspberries are more closely related to each other than either is to cherries. This is equivalent to saying that blackberries and rasp-

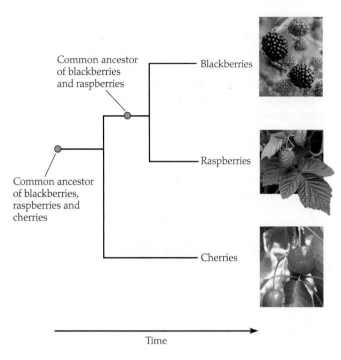

FIGURE 1.2 A simple phylogeny of three groups in the rose family.

berries share a more recent **common ancestor** than either does with cherries. Blackberries and raspberries are said to be **sister groups**, or closest relatives. An abundance of other evidence, from other structural features, chemical constituents, and DNA sequences, leads to the same conclusion about the relationships of these plants. We can present these phylogenetic relationships diagrammatically

as an **evolutionary tree** (also known as a **phylogenetic tree** or **cladogram**). This book contains scores of such trees; Figure 1.2 is one of the simplest.

More formally, a phylogenetic tree is a diagram that summarizes the relationships between ancestors and their descendants. Imagine a population of organisms that all look similar to each other. By some process, the population divides into two populations, and these two populations diverge from one another and evolve independently. In other words, two **lineages** (ancestor-descendant sequences of populations) are established. We know this has happened because members of the two new populations acquire, by the process of mutation, new characteristics in their genes, and possibly changes in their form, or **morphology**, that make members of one population look more similar to one another than to members of the other population or to the ancestral population. These characteristics are the evidence for evolution.

For example, a set of plants will produce offspring that are genetically related to their parents, as indicated by the lines in Figure 1.3. The offspring will produce more offspring, so we can view the population over several generations, with genetic connections indicated by lines.

If for some reason a population divides into two separate populations, each population will have its own set of genetic connections and will eventually acquire distinctive characteristics. For example, in the hypothetical populations shown in Figure 1.3, the population on the right develops red flowers, and the stems of the population on the left become woody. These changes are evidence that each of the populations constitutes a single lineage. The process can be repeated, and each of the new populations can divide again, with each of the newly formed popula-

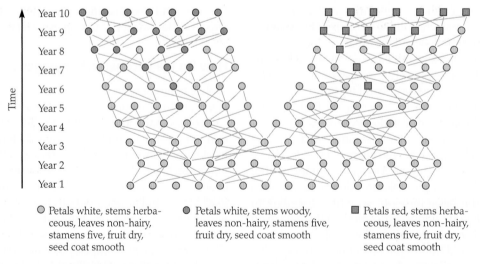

○ Petals white, stems herbaceous, leaves non-hairy, stamens five, fruit dry, seed coat smooth

● Petals white, stems woody, leaves non-hairy, stamens five, fruit dry, seed coat smooth

■ Petals red, stems herbaceous, leaves non-hairy, stamens five, fruit dry, seed coat smooth

FIGURE 1.3 Evolution of two hypothetical plant lineages. Each circle or square represents an individual plant. Lines extend upward from each plant to its descendants, and downward from each plant to its parents. In year 4, some process divides the population into two populations. A mutation in the population on the left causes a change from herbaceous to woody stems, which is transmitted to descendant plants. Over time, woody-stemmed plants gradually replace all the herbaceous ones in the population. A different mutation in the population on the right leads to a group of plants with red rather than white petals.

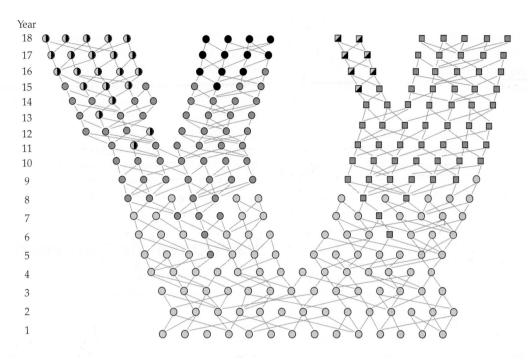

Year

○ Petals white, stems herbaceous,
leaves non-hairy, stamens five,
fruit dry, seed coat smooth

● Petals white, stems woody,
leaves non-hairy, stamens five,
fruit dry, seed coat spiny

■ Petals red, stems herbaceous,
leaves non-hairy, stamens five,
fruit dry, seed coat smooth

◪ Petals red, stems herbaceous,
leaves non-hairy, stamens four,
fruit dry, seed coat smooth

◔ Petals white, stems woody,
leaves non-hairy, stamens five,
fruit dry, seed coat smooth

◑ Petals white, stems woody,
leaves non-hairy, stamens five,
fruit fleshy, seed coat smooth

■ Petals red, stems herbaceous,
leaves hairy, stamens five,
fruit dry, seed coat smooth

FIGURE 1.4 The same hypothetical set of plants as in Figure 1.3 after 8 more years and two more divisions.

tions acquiring a new set of characteristics. Some of the woody plants now have fleshy fruits, and another group has a spiny seed coat. Meanwhile, some of the red-flowered plants now have only four stamens, and another set of red-flowered plants has hairy leaves (Figure 1.4).

The characteristics of plants, such as flower color or stem structure, are generally referred to as **characters**. Each character can have different values, or **character states**. In our example, the character "flower color" has two states: white and red. The character "stem structure" also has two states—woody and herbaceous—and so forth. All else being equal, plants with the same character state are more likely to be related to each other than are those with different character states.

The critical point in this example, however, is that character states such as red petals and woody stems are *new*: they are **derived** (or apomorphic) relative to the ancestral population, with its white flowers and herbaceous (nonwoody) stems. Only such derived character states tell us that a new lineage has been established; retention of the old character states (white flowers, herbaceous stems, nonhairy leaves, five stamens, dry fruit, smooth seed coat) does not tell us anything about what has happened.

A character state that is derived at one time may become ancestral at a later time. In Figure 1.4, woody stems are derived relative to the original population, but they are ancestral relative to the groups with fleshy fruits or spiny seed coats.

What Is Monophyly?

How does a systematist use a phylogeny to decide which groups of organisms, or **taxa** (singular **taxon**), to name in a classification? A phylogenetic approach demands that each taxon be a **monophyletic** group, defined as a group composed of an ancestor and all of its descendants (*mono*, single; *phylum*, lineage). The example in Figure 1.2 shows how one identifies monophyletic taxa. Assume for the sake of this example that the rose family contains only these three groups (it is, of course, much larger; see Chapter 9); that all three groups are themselves monophyletic; and that the figure shows the true phylogenetic relationships among the three groups. There are three possible subsets of two of the three taxa: (1) blackberries and raspberries, (2) blackberries and cherries, and (3) cherries and raspberries.

Which subsets of the three groups are monophyletic? Only subset 1 includes all the entities on a single branch of the phylogenetic tree as well as all the descendants of a common ancestor; only subset 1, therefore, is monophyletic. Monophyletic groups are also referred to as **clades**.

Another way to understand monophyly is with the simple rule that a monophyletic group is one that can be removed from the tree by one "cut" of the tree. Try this with Figure 1.2; removal of a nonmonophyletic group, such as blackberries and cherries, requires two cuts. Removal of a nonmonophyletic group from larger phylogenetic trees might require more than two cuts.

This particular definition of monophyly has been adopted only recently (see Chapter 3), and many traditionally recognized plant taxa are not monophyletic by this definition. A familiar example of a commonly recognized group that is not monophyletic is the "dicots." These angiosperms have characteristics, such as two cotyledons and flowers with multiples of four or five parts, that make them easily recognized. Nevertheless, they do not form a monophyletic group. The monocots, which apparently are monophyletic, are also descendants of the common ancestor of the "dicots," and the monocots are nested within the "dicots." Thus the "dicots" do not contain all the descendants of their common ancestor, and more than one cut is required to remove them from the tree of life.

Chapter 2 discusses monophyly in more detail, as well as how we interpret evidence for or against it. In keeping with our phylogenetic approach, we recognize only monophyletic groups in this book. For example we reject the nonmonophyletic "dicots" as a formal group, and place the names of this and other nonmonophyletic groups in quotation marks. A monophyletic group can be recognized as such by its shared derived characters (synapomorphies). **Synapomorphies** are character states that have arisen in the ancestor of the group and are present in all of its members (albeit sometimes in modified form). The concept of synapomorphies was first formalized by Hennig (1966) and Wagner (1980). In recent years, our ability to sequence DNA nucleotides has allowed us to compare the gene sequences of different organisms in our search for synapomorphies. Such studies are the stuff of **molecular systematics**, described in Chapter 5, and their results sometimes overturn long-held views about phylogenetic relationships.

In some cases, the evidence is not unambiguously for or against the monophyly of a group. For example, extant gymnosperms, which include familiar plants such as pines and redwoods, do not form a monophyletic group according to some data, while other data support their monophyly. Recent molecular data generally support the monophyly of extant gymnosperms, but further studies may well produce evidence to the contrary. We provisionally recognize extant gymnosperms as monophyletic in our classification. Further discussion of this difficult issue can be found in Chapters 7 and 8.

An important exception to the rule of monophyly in the recognition of taxa occurs at the level of species. The problem with monophyly at the species level has to do with the nature of relationships above and below the level of species. Above the level of species, the tree of life generally splits into separate branches, as in Figures 1.1 and 1.2. This is so because blackberries and cherries, for example, do not cross or hybridize with one another. Within species, however, branches join through matings between members of a species. Thus, during the separation of one species into two, matings may occur between members of the nascent lineages such that one cannot identify a common ancestor that is unique to either or both species. This problem and others concerning species are taken up in Chapter 6.

The Practice of Plant Systematics

Two important activities of systematists are classification and identification. **Classification** is the placement of an entity in a logically organized scheme of relationships. This scheme is usually hierarchical, consisting of large, inclusive groups of organisms (such as the green plant kingdom, consisting of all green plants), that contain less inclusive, progressively nested groups such as orders, families, genera, and species. The largest, most inclusive groups are the three great **domains** of life: the Bacteria, the Archaea (both of which consist of unicellular, prokaryotic organisms), and the Eukarya (all eukaryotic organisms, both unicellular and multicellular). The domain Eukarya is defined by several synapomorphies, including the presence of a cell nucleus.

Within Eukarya there are many mostly unicellular organisms, commonly lumped together as protists, and the three monophyletic, multicellular kingdoms: the animals, the fungi, and the green plants. Phylogenetic studies deal with groups ranging from domains to species. Figure 1.5 presents an example of the place of one plant species, *Solidago sempervirens* (seaside goldenrod), in a hierarchical classification scheme (see also Appendix 1). About 1.5 million species of organisms have been described and named, but Earth supports perhaps 10 to 20 times that number of species.

Systematics encompasses the discipline of **taxonomy**, a term tied to the word *taxon*. In taxonomy, groups of organisms are described and assigned scientific names. Names of taxa give us access to information about them, and thus it is valuable to have one name by which all can refer to a group of plants. This is especially true at the species level, which holds a special place in terms of usefulness and overall importance to humanity. The application of scientific names is the field of **biological nomenclature** (see Appendix 1).

Identification involves determining whether an unknown plant belongs to a known, named group of plants. In temperate regions, where the flora is generally well known, one can usually match a plant with a known species. An environmental consultant conducting an inventory of the plants of a temperate-zone salt marsh may find a plant and know it is a goldenrod, but be unable to identify the exact species; the numerous species belonging to the goldenrod

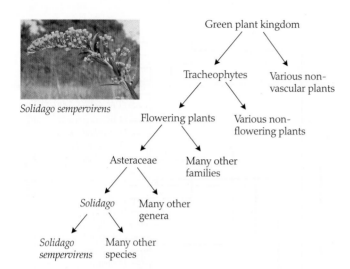

Green plant kingdom

Tracheophytes → Various non-vascular plants

Flowering plants → Various non-flowering plants

Asteraceae → Many other families

Solidago → Many other genera

Solidago sempervirens → Many other species

Solidago sempervirens

FIGURE 1.5 Portion of a hierarchical classification showing the placement of the species *Solidago sempervirens*, seaside goldenrod. Downward-pointing arrows indicate groups that are nested within the group above. In each case there are two arrows, one leading to the group containing the seaside goldenrod and the other leading to all other groups at the same level in the hierarchy. (Photo by David McIntyre.)

genus (scientific name *Solidago*) are sometimes difficult to distinguish from one another. The consultant needs to record information about the plant in order to identify it, but does not want to damage a plant of an unknown species because it might be rare or endangered. Therefore, she takes careful notes and photographs to document the plant's appearance. If circumstances render it prudent, it may be helpful to collect a single specimen, which can be preserved by pressing and drying (see Appendix 2) and used to identify a plant accurately.

The fastest way to identify a plant is to consult a professional botanist or well-trained naturalist who is familiar with the plants of a region. Literature can also be consulted; for most temperate regions, there are books devoted to naming and describing the plants; some covering all the plants, others focusing on a portion of the flora.

A third path to identification is to visit an **herbarium** (plural **herbaria**), a facility for the storage of scientific plant collections that is a standard component of universities and botanical institutions, and match the information, photographs, and/or specimen with a named specimen there. The World Wide Web, an increasingly useful resource for plant identification, provides images of plants and online keys (Farr 2006).

Plant identification in the tropics is far more challenging than in the temperate zone not only because tropical floras contain many more species than temperate floras, but also because tropical floras are far less well studied; vast numbers of tropical species have yet to be recognized, described, named, and collected for herbaria. Here the role of the specialist is critically important, yet there are fewer specialists in this area every year.

Why Is Systematics Important?

Systematics is essential to our understanding of and communication about the natural world. The basic activities of systematics—classification and naming—are ancient human methods of dealing with information about the natural world, and early in human cultural evolution they led to remarkably sophisticated classifications of important organisms. We depend on many species for food, shelter, fiber for clothing and paper, medicines, tools, dyes, and myriad other uses, and we can know and predict uses for these species in part because of our systematic understanding of the biota.

Although classification has always focused on describing and grouping organisms, it has only relatively recently been concerned with evolutionary phylogenetic relationships. Publication of Charles Darwin's *On the origin of species* in 1859 stimulated the incorporation of general evolutionary relationships into classification, an ongoing process that has yet to be fully realized (de Queiroz and Gauthier 1992). A critical step in this process has been the development of a phylogenetic perspective, to which Willi Hennig (a German entomologist, 1913–1976), Walter Zimmermann (a German botanist, 1892–1980), Warren H. Wagner, Jr. (a North American botanist, 1920–2000), and many others have contributed.

The more a classification reflects evolutionary history and phylogeny, the more useful it will be as a predictor. For example, the discovery of biochemical precursors of the drug cortisone in certain species of yams in the genus *Dioscorea* (in the family Dioscoreaceae; see Chapter 9) prompted a search for, and subsequent discovery of, higher concentrations of the drug in other *Dioscorea* species (Jeffrey 1982). The fact that these species are relatives of yams made it likely that they would share genetically controlled features, such as many chemical constituents, with the yams.

Indeed, our knowledge of systematics guides the search for plants of potential commercial importance. In the 1960s, while studying wild plants in the Peruvian Andes, botanist Hugh Iltis collected a species belonging to the potato genus, *Solanum*, which includes tomatoes. Iltis knew that wild relatives of the cultivated tomato could be important for the breeding of improved tomato crops. He sent some seeds to tomato geneticist Charles Rick in California, who identified the plant as a new species, *Solanum chmielewskii* (named in honor of the late Tadeusz Chmielewski, a Polish tomato geneticist). Rick crossed this wild relative with a cultivated tomato and thus introduced genes that markedly improved the taste of tomatoes (Rick 1982). Similar advances, by the hundreds, have improved yield, disease resistance, and other desirable traits in crops, commercial timber species, and horticultural varieties. Systematics is also critical in biological sciences involving diversity, such as conservation biology, ecology, and ethnobotany.

Systematics advances our knowledge of evolution because it establishes a historical context for understanding a wide variety of biological phenomena, such as ecological diversification and specialization, coevolutionary relationships of hosts and parasites and of plants and pollinators,

FIGURE 1.6 Habit of one of the Hawaiian silverswords, *Argyroxiphium sandwicense*, showing the basal cluster of sword-shaped leaves and the massive inflorescence, which can be up to 2 m tall. (Photo courtesy of Sherwin Carlquist and the Botanical Society of America.)

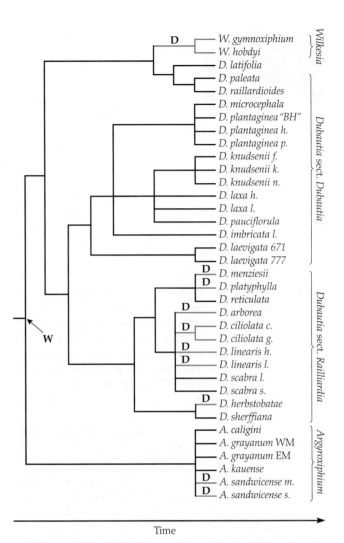

Time

FIGURE 1.7 Phylogenetic tree of the three genera (*Argyroxiphium*, *Dubautia*, and *Wilkesia*) and 28 species of Hawaiian silverswords, based on sequences from internal transcribed spacers in nuclear ribosomal DNA. The letters above the branches of the tree indicate the habitat (dry, D, or wet, W) of the lineage defined by the branch. Dry lineages are shown in blue. (After Baldwin and Robichaux 1995.)

biogeography, adaptation, speciation, and rates of evolution. Three examples will be presented here to show the importance of systematic approaches to evolutionary biology.

The Hawaiian silverswords of the daisy family (Asteraceae) are like many other groups on the Hawaiian Islands in exemplifying evolutionary radiations. The silversword alliance, a monophyletic group of 28 species in three genera (*Argyroxiphium*, *Dubautia*, and *Wilkesia*) that is endemic to the Hawaiian Islands, evolved from a single founding individual of a Californian ancestral species (Baldwin and Robichaux 1995). This group includes some of the most spectacular members of the Hawaiian flora. *Argyroxiphium* species, for example, have a basal rosette of green- or silvery-haired, swordlike leaves from which emerges a terminal inflorescence up to 2 m tall and with as many as 600 large heads (Figure 1.6).

In addition to rosette plants, members of the silversword alliance have diversified into many other growth forms, including trees, shrubs, subshrubs, cushion plants, and lianas. Silverswords occupy an exceptionally wide range of habitats, from 75 to 3750 m in elevation and from less than 400 to more than 12,300 mm in annual precipitation. A common pathway in the evolutionary radiation of the group apparently involves dispersal between islands followed by numerous major ecological shifts along moisture gradients and into bogs.

We can see this pattern of radiation in a phylogeny of the silversword alliance (Figure 1.7). When the habitat pref-

erences of silversword species are mapped onto a phylogenetic tree, it is clear that there have been numerous shifts from wet to dry habitats accompanying the evolution of the species. Diversification of the silversword alliance is estimated to have occurred over the past 5.2 million years, which is approximately the age of the oldest high Hawaiian island, Kaua'i. This diversification has been paced at a speciation rate of about 0.56 (±0.17) species per million years (Baldwin 2003), which is remarkably high compared with other groups of plants (see Speciation in Chapter 6).

The second example concerns evolutionary adaptations for pollination. The genus *Parkia* includes trees of the tropics, especially rain forests of the Amazon Basin. This genus belongs to the Fabaceae, the legume or bean family (see

Chapter 9), and some of its species produce seeds that are a popular vegetable in parts of the tropics. *Parkia* contains a large number of species whose flowers are pollinated by bats. Bat-pollinated *Parkia* flowers open at night, when bats are active, and produce an abundance of nectar to reward bats for pollinating them. Pollen is brushed onto a bat as it takes nectar from a flower, and the bat completes pollination by depositing pollen onto other flowers it visits.

Until a phylogeny of *Parkia* was established (Luckow and Hopkins 1995), it was not known whether bat pollination had evolved just once or multiple times in *Parkia*. Luckow and Hopkins identified a large clade within *Parkia* in which all the species for which the pollinator was known were bat-pollinated. Bat pollination is not known in *Parkia* outside this clade. This phylogeny was good evidence for a single origin of bat pollination. Several changes in the flowers occurred in the common ancestor of this bat-pollinated clade, the most striking of which was the evolution of specialized flowers that produce a large amount of nectar but no fruit (and which occur on a plant along with functional reproductive flowers). Luckow and Hopkins's phylogeny also suggested that the bat-pollinated species evolved from ancestors that were pollinated by nocturnal bees and thus had flowers that opened at night. Such flowers would have facilitated adaptation for bat pollination. Furthermore, Luckow and Hopkins's phylogeny indicated that the original evolution of bat pollination was followed by continued specialization for bat pollination. For example, the exterior surface of the pollen of some bat-pollinated *Parkia* is sculptured in a special way (Figure 1.8). This kind of pollen surface is called verrucate, which means wartlike. Members of the Fabaceae that are not closely related to *Parkia* but in which bat pollination has also evolved have similar verrucate pollen. The floral features associated with pollination by vertebrates and other aspects of pollination biology are dealt with in Chapter 4.

Our third example demonstrates the value of phylogeny to **biogeography**, the study of the geographic distributions of organisms. It concerns another group of important tropical trees, the baobabs, some of which are pollinated by bats. There are eight species of baobabs, all in the genus *Adansonia* of the Malvaceae (see Chapter 9). One species is native to Australia, six are confined to Madagascar, and *Adansonia digitata* grows in northeastern, central, and southern Africa. Indeed, *Adansonia digitata*, the African baobab, is emblematic of sub-Saharan Africa. It can live more than 1000 years and has an unusually massive trunk that can grow as large as 16 meters in diameter. The trunk has the capacity to store large volumes of water, which help the tree survive long periods of drought. The leaves, young sprouts, and seeds are edible and important in the diets of some Africans. The flowers of the African baobab are about 20 centimeters in diameter, open at night, and are pollinated by fruit bats. Bats also pollinate some of the Madagascan baobabs, although a nocturnal lemur also contributes significantly to the pollination of one of these species.

23.1 μm

FIGURE 1.8   A cluster of pollen grains of *Parkia sumatrana* var. *streptocarpa*, a bat-pollinated species. The surface of the pollen grains is described as verrucate (wartlike) and is thought to be a specialization for bat pollination. (From Luckow and Hopkins 1995.)

The three landmasses occupied by baobab species were once all part of Gondwana, the former Southern Hemisphere supercontinent that was separated by the forces of plate tectonics and continental drift starting about 120 million years ago. In this case, continental drift resulted in a series of separations, like the splitting of the lineages that make up a phylogeny. It is possible that the eight extant baobab species evolved simply because of the separation of Gondwana. Alternatively, the current geographic distribution of these species could be the result of seed dispersal between the landmasses. We could distinguish between these hypotheses if we had a good fossil record of baobabs, but we do not.

A phylogeny of *Adansonia*, together with estimates of the timing of lineage splitting within the genus, should provide a good test of our two biogeographic hypotheses. Baum et al. (1998) provided such a test with a phylogeny based on nuclear DNA sequences (Figure 1.9). Their phylogeny places the Australian *A. gibbosa* as sister to the remainder of the genus. Baum and colleagues then used the rate of evolution of the DNA sequences to estimate the time of divergence of *A. gibbosa* from the rest of the genus. If divergence is estimated to have begun before the time when Australia split away from Africa plus Madagascar, then divergence could simply have been the result of the isolation on landmasses separated by continental drift. Instead, the estimates obtained by Baum et al. indicated that this divergence was much more recent, suggesting seed dispersal across the ocean. Such long-distance seed dispersal is plausible, given that the fruits of many baobabs are woody and tough and are often dispersed by water.

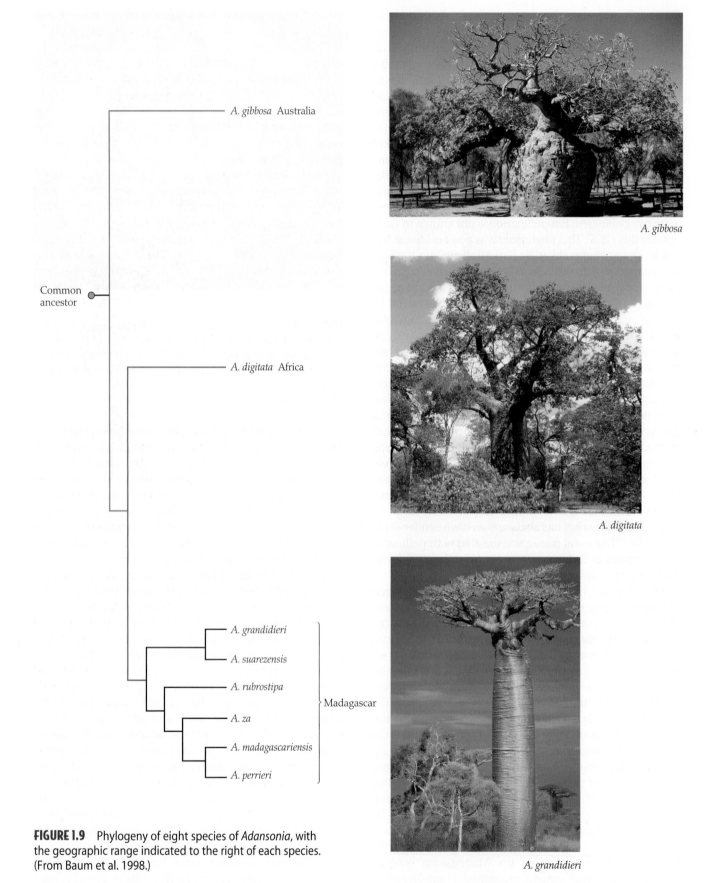

FIGURE 1.9 Phylogeny of eight species of *Adansonia*, with the geographic range indicated to the right of each species. (From Baum et al. 1998.)

Aims and Organization of This Book

This book takes a phylogenetic approach to plant systematics. Chapter 2 establishes the basic concepts and practices of phylogenetic systematics. To appreciate this approach, it is important to understand some of the history of plant systematics, the topic of Chapter 3.

We infer phylogenies from many sources of biological data, including both structural aspects of organisms (anatomy and morphology, or external features) and molecular features (biochemical constituents, such as proteins and flavonoids, and DNA). Chapter 4 is devoted to structural and non-DNA molecular systematic evidence; Chapter 5 focuses on evidence from DNA sequences.

Chapter 6 is about plant diversification and the evolution of plant species. How are species formed and maintained? How do we determine that two individuals belong to the same or different species? How do hybridization, polyploidy, and breeding systems shape diversification?

Chapter 7 presents an overview of the evolutionary history of plants, explaining the origin of many of the important features used to identify plants and creating a foundation for the final two chapters.

This book approaches identification by focusing on plant families, a good entry level for coverage of plant diversity. There are far too many plant species to learn effectively in an undergraduate course. Moreover, many plant families—such as the oak, pine, rose, grass, mustard, pea, and orchid families—are already familiar to many people and are relatively easy to recognize. Learning to recognize important families provides a classification within which one can then learn genera and species.

Chapters 8 and 9, which focus on plant diversity, contain numerous illustrations, descriptions, and **keys** (see Appendix 2). Keys organize information about a group (such as the families of conifers) in ways that facilitate identification. As noted earlier in this chapter, we follow the phylogenetic approach, and thus whenever possible have attempted to define families that are monophyletic. Chapter 8 covers the tracheophytes that are not angiosperms. More specifically, it introduces 26 families from six major groups: lycophytes (including club mosses and related plants), monilophytes (ferns, including whisk ferns and horsetails), cycads, ginkgo, conifers (pines, redwoods, et al.), and gnetophytes.

Angiosperm diversity, covered in Chapter 9, is very great; we describe more than 140 angiosperm families. The angiosperms encompass adaptations plants adopted for growth in almost every kind of habitat the world offers. They include all grain and major vegetable crops and the economically valuable broadleaved trees such as oaks and maples.

Two appendices deal with important components of the practice of systematics. In Appendix 1 we explain botanical nomenclature: the application of scientific names to plants. Appendix 2 covers the collection of scientific plant specimens, gives an overview of plant identification, and provides a brief guide to keeping abreast of future advances in plant systematics, using both the plant systematics literature and the World Wide Web.

The CD contains more than 3100 images of flowers, fruits, and other parts of plants, synapomorphies, and field identification characters that are useful for learning about the groups covered in this book. These features are also useful for identification, as are images that show detailed dissections of flowers and fruits. The CD also contains an illustrated glossary with links to one or more images that illustrate the defined feature. The images complement the family descriptions and botanical illustrations in the text.

The CD contains three appendices, each of which arranges the families covered in this text in one of three major plant classification systems: those of Cronquist (1981), Thorne (1992), and the Angiosperm Phylogeny Group (1998, 2003). The last is the classification system followed in the text.

It is important to appreciate that our knowledge of plant systematics is growing very rapidly. New phylogenetic hypotheses are emerging at a rapid pace, and we can expect major changes even within the next few years. Under these circumstances it is impossible for a textbook to remain current, and it will undoubtedly be necessary to augment the material presented here with additional information, perhaps obtained through computer databases (see Appendix 2 for a discussion of plant systematics on the World Wide Web).

Students may find it frustrating that, even in a discipline as old as plant systematics, our knowledge still requires frequent—indeed, constant—revision. We hope instead that this rapid change will be viewed positively, as a measure of the excitement and vitality of the field. As in any science, knowledge in plant systematics is forever provisional and must change to reflect new discoveries. Fortunately, recent developments in the logic and methods of inferring phylogenetic relationships, along with the availability of new sources of evidence, give us hope of achieving an increasingly accurate picture of evolutionary history. We will be satisfied if others join in the fun and help us achieve a better understanding of plant systematics.

LITERATURE CITED AND SUGGESTED READINGS

Items marked with an asterisk are especially recommended to those readers who are interested in further information on the topics discussed in this chapter.

Angiosperm Phylogeny Group. 1998. An ordinal classification for the families of flowering plants. *Ann. Missouri Bot. Gard.* 85: 531–553.

Angiosperm Phylogeny Group. 2003. An update of the Phylogeny Group classification for the orders and families of flowering plants: APGII. *Bot. J. Linnean Soc.* 141: 399–436.

Baldwin, B. G. 2003. A phylogenetic perspective on the origin and evolution of Madiinae. In *Tarweeds and silverswords: Evolution of the Madiinae (Asteraceae)*, S. Carlquist, B. G. Baldwin and G. D. Carr (eds.), 193–228. Missouri Botanical Garden Press, St. Louis.

Baldwin, B. G. and R. H. Robichaux. 1995. Historical biogeography and ecology of the Hawaiian silversword alliance (Asteraceae). New molecular phylogenetic perspectives. In *Hawaiian biogeography: Evolution on a hot spot archipelago*, W. L. Wagner and V. A. Funk (eds.), 259–287. Smithsonian Institution Press, Washington, DC.

Baum, D. A., R. L. Small and J. F. Wendel. 1998. Biogeography and floral evolution of baobabs (*Adansonia*, Bombicaceae) as inferred from multiple data sets. *Syst. Biol.* 47: 181–207.

Bremer, K. and H. Wanntrop. 1978. Phylogenetic systematics in botany. *Taxon* 27: 317–329.

*Briggs, D. and S. M. Walters. 1997. *Plant variation and evolution*, 3rd ed. Cambridge University Press, Cambridge.

Cronquist, A. 1981. *An integrated system of classification of flowering plants*. Columbia University Press, New York.

*Darwin, C. 1859. *On the origin of species*. Mentor edition, 1958. New American Library, New York.

*Davis, P. H. and V. H. Heywood. 1963. *Principles of angiosperm taxonomy*. Oliver & Boyd, Edinburgh, Scotland.

*de Queiroz, K. and J. Gauthier. 1992. Phylogenetic taxonomy. *Annu. Rev. Ecol. Syst.* 23: 449–480.

Donoghue, M. J. and J. W. Kadereit. 1992. Walter Zimmerman and the growth of phylogenetic theory. *Syst. Biol.* 41: 74–85.

Farr, D. F. 2006. On-line keys: More than just paper on the Web. *Taxon* 55: 589–596.

Graham, L. E., M. E. Cook and J. S. Busse. 2000. The origin of plants: Body plan changes contributing to a major evolutionary radiation. *Proc. Natl. Acad. Sci. USA* 97: 4535–4540.

Hennig, W. 1966. *Phylogenetic systematics*. University of Illinois Press, Urbana.

Hoch, P. C. and A. G. Stephenson. 1995. *Experimental and molecular approaches to plant biosystematics*. Missouri Botanical Garden, St. Louis.

Iltis, H. H. 1988. Serendipity in the exploration of biodiversity: What good are weedy tomatoes? In *Biodiversity*, E. O. Wilson and F. M. Peter (eds.), 98–105. National Academy Press, Washington, DC.

Jeffrey, C. 1982. *An introduction to plant taxonomy*. Cambridge University Press, Cambridge.

Lawrence, G. H. M. 1951. *The taxonomy of vascular plants*. Macmillan, New York.

Luckow, M. and H. C. F. Hopkins. 1995. A cladistic analysis of *Parkia* (Leguminosae: Mimosoideae). *Am. J. Bot.* 82: 1300–1320.

Niklas, K. J. 1997. *The evolutionary biology of plants*. University of Chicago Press, Chicago.

Radford, A. E., W. C. Dickison, J. R. Massey and C. R. Bell. 1974. *Vascular plant systematics*. Harper & Row, New York.

Renzaglia, K. S., R. J. Duff, D. L. Nickrent and D. J. Garbary. 2000. Vegetative and reproductive innovations of early land plants: Implications for a unified phylogeny. *Philos. Trans. R. Soc. London B* 355: 769–793.

Rick, C. M. 1982. The potential of exotic germplasm for tomato improvement. In *Plant improvement and somatic cell genetics*, I. K. Vasil, W. R. Scowocroft, and K. J. Frey (eds.), 1–28. Academic Press, New York.

Simpson, G. G. 1961. *Principles of animal taxonomy*. Columbia University Press, New York.

Stace, C. A. 1980. *Plant taxonomy and biosystematics*. University Park Press, Baltimore.

Stuessy, T. F. 1990. *Plant taxonomy: The systematic evaluation of comparative data*. Columbia University Press, New York.

Thorne, R. F. 1992. The classification and geography of the flowering plants. *Bot. Rev.* 58: 225–348.

Wagner, W. H., Jr. 1980. Origin and philosophy of the groundplan-divergence method of cladistics. *Syst. Bot.* 5: 173–193.

2

Methods and Principles of Biological Systematics

Biological systematics (or taxonomy) is the theory and practice of grouping individuals into species, arranging those species into larger groups, and giving those groups names, thus producing a **classification**. Classifications are used to organize information about plants, and keys can be constructed to identify plants.

There are many ways to construct a classification. For example, plants could be classified on the basis of their medicinal properties (as they are in some systems of herbal medicine) or on the basis of their preferred habitats (as they may be in some ecological classifications). A phylogeny-based classification, such as the one used in this book, attempts to arrange organisms into groups on the basis of their evolutionary relationships.

There are two main steps in producing such a classification. The first is determining the **phylogeny**, or evolutionary history, of a group of organisms. The second is basing the classification of the group on this history. These two steps can be, and often are, separated, such that every new theory of relationships does not lead automatically to a new classification. This chapter will outline how one goes about determining the history of a group and will then discuss briefly how one might construct a classification given that history.

How Are Phylogenies Constructed?

As described in Chapter 1, evolution is not simply descent with modification, but also involves the separation of lineages. This process can be visualized with diagrams such as those in Figures 1.3 and 1.4, but these are cumbersome to draw. Evolutionary history can be more conveniently summarized in a branching diagram (Figure 2.1A). (Some workers make distinctions between an evolutionary tree, a phylogeny, and a branching diagram or

FIGURE 2.1 (A) A simple way to redraw the pattern of changes shown in Figure 1.4. Full descriptions are provided for each of the ancestors and their descendants. (B) A simpler way to redraw Figure 2.1A, showing only the changes that have occurred in the various lineages.

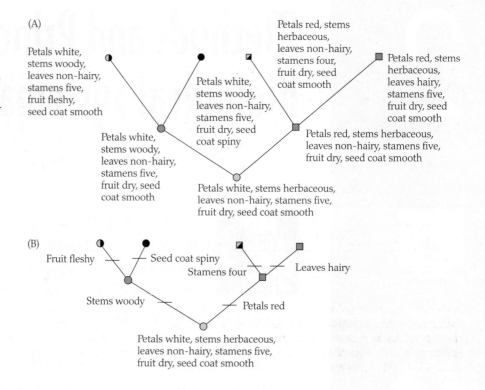

cladogram, but in this text the terms are used interchangeably.) To avoid repeating the ancestral character states retained in every group, systematists commonly note only the characters that have changed, and they place tick marks on the appropriate branches to indicate the relative order in which the character states originated (Figure 2.1B).

The shared derived character states in Figure 2.1B can be arranged in a hierarchy from more inclusive (e.g., stems woody or petals red) to less inclusive (e.g., leaves hairy, seed coat spiny). This arrangement then leads to the obvious conclusion that the plants themselves can be arranged in a hierarchical classification that is a reflection of their evolutionary history. The plants could be divided into two groups: one with the shared derived character state of red petals and the ancestral character state of herbaceous stems, the other with the shared derived character state of woody stems and the ancestral character state of white petals. Each of these groups could also be divided into two groups. Thus the classification could be derived directly from the phylogeny.

Note that the hierarchy is not changed by the order in which the branch tips are drawn. The shape, or **topology**, of the tree is determined only by the connections between the branches. We can tell the evolutionary "story" by starting at any point in the tree and working up or down. This means that the terms *higher* and *lower* are not really meaningful, but simply reflect how we have chosen to draw the evolutionary tree.

From this point of view, a plant systematics course could just as well begin by covering the Asteraceae, which some textbooks consider an "advanced" family, and then working out to other members of the asterid clade, as by starting

with the "primitive" families, such as Magnoliaceae and Nymphaeaceae. The latter families simply share a set of characters thought to be ancestral, but these are combined with a large set of derived characters as well.

Determining Evolutionary History

In the examples shown in Figures 1.3, 1.4, and 2.1, we described evolution as though we were watching it happen. This is rarely possible, of course, so part of the challenge of systematics is that we must *infer* what went on in the past. The first step in making such inferences is to examine extant species closely for characters that are believed to be heritable. A **heritable character** is any aspect of the plant's morphology that can be passed down genetically through evolutionary time and still be recognizable. For example, a flowering plant's petal color, inflorescence structure, and habit (general growth pattern) are all known to be under genetic control, and these characters are generally stably inherited from one generation to the next. Many examples of such heritable characters are described in Chapters 4 and 5.

Systematics entails the precise observation of organisms. Without careful descriptions of characters, phylogeny reconstruction and descriptions of evolutionary history are meaningless. Without accurate comparative morphology, classification of any sort is impossible. The assessment of similarity is the basis of comparative biology, and of systematics in particular. In making such assessments, it may be harder than you think to determine which structures of one plant can usefully be compared to structures of a differ-

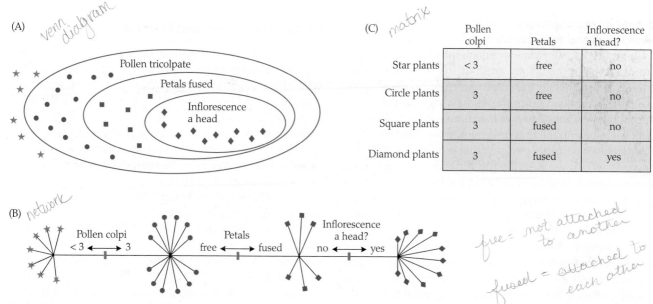

FIGURE 2.2 Each symbol represents a hypothetical pollen-producing plant species or group of species. A large subset of such plants have tricolpate pollen. Of that subset, a smaller group has fused petals, and of the plants with tricolpate pollen and fused petals, a subset has flowers arranged in a head. (A) The pattern described rendered as a Venn diagram of nested character states. (B) The pattern redrawn as an unrooted network; characters are indicated with bright green tick marks, with the different character states on either side. (C) The pattern redrawn as a matrix.

ent plant. Two structures may be deemed to be similar if (1) they are found in a similar position in both organisms; (2) they are similar in their cellular and histological structure; and/or (3) they are linked by intermediate forms of the structure (either by intermediates at different developmental stages of the same organism or by intermediates in different organisms). These three statements constitute **Remane's criteria** of similarity.

Remane (1952) actually called this list the "criteria of homology." In this book, however, we use the term *homology* in a more restricted sense, to mean **identity by descent**. In other words, if we say that a character is **homologous** among a group of species, we mean that all those species inherited that character from a common ancestor. Under this definition, observing similarity is only the first step in determining homology, since not every observed similarity is the result of homology.* (For example, structural similarities can evolve independently in unrelated plants that live in similar environments.) This text follows the viewpoint held by the many phylogenetic systematists who argue that homology can be determined only by constructing an evolutionary tree.

Characters, Character States, and Networks

From observations of heritable characters, plant groups that share particular character states can be identified. Suppose, for example, that pollen is observed to vary in the number

of grooves on its surface (a character), and that the pollen in a large number of plant species has three grooves (a character state). These grooves are in fact germination furrows called *colpi* (singular *colpus*), and three-grooved pollen is described as *tricolpate*. Within the large group of plant species with tricolpate pollen is a smaller group whose petals (character) are fused (character state), and within this fused-petal group is a still smaller group with flowers arranged in a head. These nested groups can be depicted as a set of concentric ovals in a **Venn diagram**, as shown in Figure 2.2A.

The information in the Venn diagram can also be drawn as a **network** (Figure 2.2B). Here the characters are shown as vertical green lines, or "tick marks" (a convention that is seen in illustrations throughout this text). Whereas the shapes (species) to the left of the "pollen" line have fewer than three colpi, those to the right of the line have tricolpate pollen. Likewise, the line for "petals" indicates a shift between the character states free and fused, and the inflorescence line indicates a shift between flowers clustered in a head and flowers borne separately. We can count the number of changes along the network to determine its *length*: proceeding from right to left, there is one change each in inflorescences, petal fusion, and pollen colpi, so the network can be described as having a length of 3.

The same information can be presented as a **matrix** in which the rows correspond to plants and the columns correspond to characters (Figure 2.2C). The character states are then used to fill in the matrix. The changes in character state are, or are hypothesized to be, genetic changes that potentially distinguish the groups of plants in the matrix. Thus the three changes in the network of Figure 2.2B represent three

---

*You should be aware that *homology* has several different meanings, and when reading the literature, it is worth checking what particular authors mean when they use the term.

changes in gene sequence (and thus in the resulting proteins) that altered the character states of some plants.

In Figure 2.2, all plants designated by the same shape are drawn as though they arose at the same time. This arrangement generally indicates ambiguity; for the purposes of this simplified example, we have not provided any information on their order of evolutionary origin. In addition, we have implied that determining the different character states is perfectly obvious. This is often not the case, however. When we describe the variation among similar morphological structures by dividing the character into character states, we are in fact putting forward a hypothesis of underlying genetic control, even though we rarely frame the assumption in these terms.

For example, if two species differ in the color of their flowers, we may score the character "petal color" as having two states, red and blue. By scoring it this way, we are hypothesizing that the genes underlying petal color have switched, over evolutionary time, to produce either red flowers from a blue-flowered ancestor or blue flowers from a red-flowered ancestor. In this instance, we know that there are in fact genes (e.g., components of the anthocyanin pathway) that control petal color, and thus the inference of two states controlled by a "genetic switch" is probably a reasonable one. In many cases, however, we have no idea of the genetic mechanisms that control the state of the structural characters observed. In proposing hypotheses about the nature of the underlying switches, often all we can do is be sure that the character states really are distinct. For quantitative characters such as leaf length or corolla tube width, this means graphing the quantitative data (i.e., the measurements) to be sure that the measurements of the species we are studying do not overlap.

For many characters, such measurements do overlap and also vary greatly, so much so that the assumption of underlying genetic switches—and therefore division into character states—is unsupported by any evidence. In these cases, the characters in question should be omitted from phylogenetic analysis (unless the overlap is caused by only a few individuals, in which case the character could be scored as polymorphic for that species and retained in the analysis). Even though such overlapping characters probably reflect genetic changes over evolutionary time, given our current state of knowledge, overlap makes it difficult to extract any reliable information on the underlying gene changes (although methods of dealing with plants with variable characters have been developed).

Variability and overlap in morphological characters are good reasons why many systematists have turned to molecular data in constructing phylogenies. With the emergence of nucleotide sequence data for many genes, the recognition of molecular character states (i.e., whether the nucleotide in a given position is A, T, G, or C) is often more precise. This may not be the case, however, if gene sequences are hard to align, or if restriction fragments are too similar in size. The use of molecular character states in plant systematics is detailed in Chapter 5.

Evolutionary Trees and Rooting

Figure 2.2 shows three different ways of recording and organizing observations about plants. Even though the network (Figure 2.2B) looks somewhat like a time line, it is not. It could be read from left to right, from right to left, or perhaps from the middle outward. To turn the network into an evolutionary tree, we must determine which changes are relatively more recent and which occurred further in the past. In other words, the tree must be **rooted**. Rooting *polarizes* the character changes, giving them a specific direction.

If you imagine that the network is a piece of string, you can keep the connections exactly the same, even when you pull down a root in different places. The network from Figure 2.2B is redrawn in Figure 2.3, but rooted in three different places. Notice that the length of each tree (or cladogram) is the same as the length of the original network—3—and that all the connections are the same, but that the order in which the character-change events occur differs considerably.

For example, in the rooting shown in Figure 2.3A, the ancestral plants had pollen with fewer than three colpi, petals not fused, and flowers not in heads, whereas we would conclude from Figure 2.3B that the ancestral plants had exactly the opposite character states. In Figure 2.3C, the tree is rooted in such a way that the ancestor had tricolpate pollen. The pollen later changed to having fewer than three colpi in one lineage, whereas the other lineage kept the pollen character state of three colpi and later acquired fused petals and flowers in heads.

Rooting a phylogenetic tree is critical for interpreting how plants evolved, and different rootings suggest different patterns of change (different character polarizations). There has been much discussion among systematists of how the position of the root should be determined. One frequent suggestion is that one should use fossils. But just because an extinct plant has been fossilized does not mean that its lineage *originated* earlier than those of plants now living; we know only that it died out earlier.

In determining evolutionary history, we are interested in determining when lineages diverged from one another (that is, when taxa originated). When taxa die out is interesting to know, but that fact by itself does not help in determining their origins. (Fossils are, of course, extremely useful when included as additional taxa in a phylogeny. They often have combinations of character states that no longer occur in extant taxa and can affect the overall structure of the tree, sometimes in surprising and informative ways.)

In general, evolutionary trees are rooted using a relative of the group under study: an **outgroup**. When selecting an outgroup, one must assume only that all ingroup members (members of the group under study) are more closely related to one another than to the outgroup; in other words, the outgroup must have separated from the ingroup lineage before the ingroup diversified. Often several outgroups are used. If an outgroup is added to a network, the point at which it attaches is determined as the root of the tree.

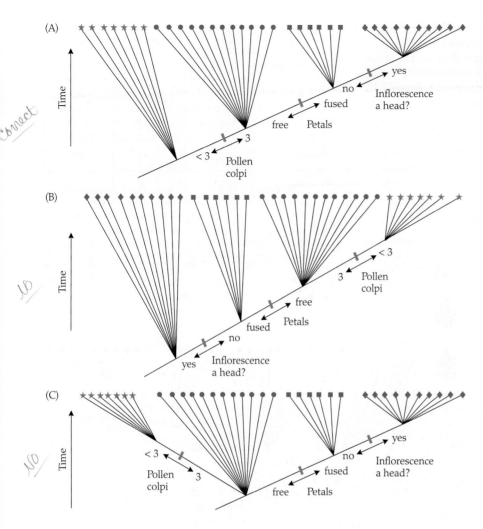

FIGURE 2.3 Three possible rootings of the network in Figure 2.2B. Note that in each case the number of evolutionary steps (character state changes) is the same as in the unrooted network.

In the case of Figures 2.2 and 2.3, the plants shown are all flowering plants (angiosperms), and their closest living relatives are the conifers, cycads, gnetophytes, or ginkgos, or a set of these (see Chapters 7 and 8). In Figure 2.4A, a conifer is added to the matrix from Figure 2.2C. (We could have used all gymnosperms as outgroups, but to keep the example simple we have chosen only one.)

Because conifers do not have petals or flowers, two of the characters must be scored as not applicable, but we do know that conifer pollen does not have three colpi. With this information, the conifer can be added to the network as an outgroup, as in Figure 2.4B. Because the conifer attaches among the star species, the tree can be rooted and redrawn as in Figure 2.4C. This tree corresponds to the rooted tree in Figure 2.3A and strengthens the hypothesis that Figure 2.3A accurately reflects evolutionary history.

Note that the tree can be drawn in different ways and still reflect the same evolutionary history. Comparing Figures 2.5A and B with Figure 2.4C shows that we can rotate the branches of the tree around any one of the branch points (**nodes**) without affecting the inferred order of events.

With a rooted tree (and only with a rooted tree), we can determine which groups are monophyletic (made up of an ancestor and *all* of its descendants). Therefore, in the example laid out in Figure 2.4C, the diamond plants are monophyletic (i.e., they form a clade). In fact, the flowering plants with fused petals and flowers arranged in a head are the family Asteraceae, which are known to form a monophyletic group. Thus having flowers in heads is a synapomorphy for (i.e., is a shared derived character for, or indicates the monophyly of) the Asteraceae, having fused petals is a shared derived character (synapomorphy) uniting the square species with the diamond species, and having tricolpate pollen indicates the monophyly of the circle plus square plus diamond species.

Notice how important rooting is for determining monophyly. If Figure 2.3B were the correct rooting of the flowering plant phylogeny, then fused petals and flowers in heads would be ancestral character states (usually called **symplesiomorphies**) rather than derived synapomorphies. In this case, the species indicated by diamonds and squares would not share any *derived* character and would not include *all* the descendants of their common ancestor; some of those descendants went on to become the circle and the star plants. Therefore, if Figure 2.3B were correct, the diamond plus square species would not be a monophyletic group (as they are with the rooting in Figure 2.3A). Instead, they

FIGURE 2.4   (A) The matrix from Figure 2.2C, but with character states added for a conifer. (B) The unrooted network from Figure 2.2B, but with the conifer attached according to the character states in Figure 2.4A. (C) The network of Figure 2.4B rooted with the conifer. Note that the evolutionary history is now the same as in Figure 2.3A.

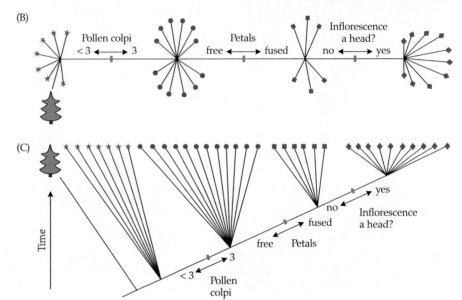

(A)	Pollen colpi	Petals	Inflorescence in heads?
Star plants	< 3	free	no
Circle plants	3	free	no
Square plants	3	fused	no
Diamond plants	3	fused	yes
Conifer	< 3	not applicable	not applicable

would constitute a **paraphyletic** group, one that includes a common ancestor and some, but not all, of its descendants.

As mentioned earlier, a character state that is derived (synapomorphic) at one time may become ancestral later. In Figure 2.4B, tricolpate pollen is a shared derived character of a large group of flowering plants. It is a synapomorphy and indicates monophyly of the group sometimes called the eudicots. For the group with fused petals, however, tricolpate pollen is an ancestral, or **plesiomorphic**, character state. It is something that all of the species in the group inherited from their common ancestor and thus indicates nothing about their relationships with one another. Plesiomorphic character states cannot show evolutionary relationships in the group being studied because they evolved earlier than any of the taxa being compared and have merely have been retained in the group's various lineages.

Sometimes monophyly of a group is indicated by the fact that its character states do not occur in any other organism. For example, all members of the grass family (Poaceae) have an embryo that is unlike the embryo of any other flowering plant. We can thus hypothesize that the grass embryo is uniquely derived in (is a synapomorphy for) the Poaceae and indicates that the family is monophyletic. This is the same as saying that any reasonable rooting of the phylogenetic tree will lead to the same conclusion.

It is often possible to find evidence that a group is monophyletic even without a large computer-assisted phylogenetic analysis. Indeed, most phylogenetic (sometimes called cladistic) analyses were done by hand until the mid-1980s. Characters are first divided into character states, as with any phylogenetic analysis. The character state in the outgroup (or outgroups) is then assumed to be ancestral (Stevens 1980; Watrous and Wheeler 1981; Maddison et al. 1984). In other words, each character is polarized, or given direction. The shared derived, or synapomorphic, character state can then be used as evidence of monophyly, and a cladogram can be constructed on the basis of synapomorphic character states (Box 2A). This kind of thinking is often useful in providing a first guess as to whether existing taxonomic groups are monophyletic and thus named appropriately.

Choosing Trees

As the preceding discussions have shown, determining the evolutionary history of a group of organisms is conceptually quite simple. First, characters are observed and divided into character states. Second, from the character states, a Venn diagram (see Figure 2.2A), a character × taxon matrix (see Figure 2.2C), or a branching network (see Figure 2.2B) can be constructed. Third, by inclusion of an outgroup, the net-

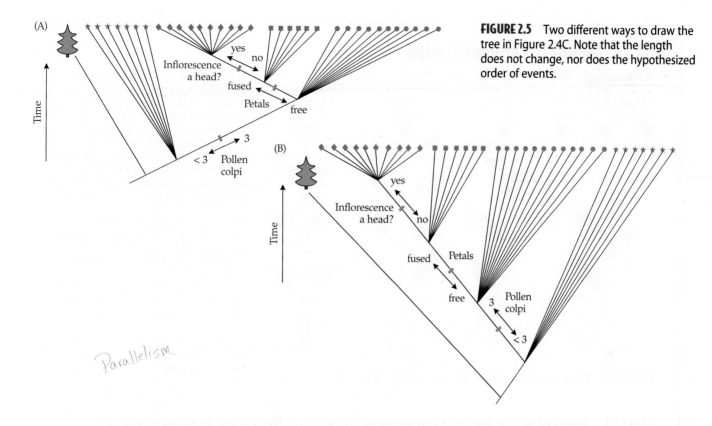

FIGURE 2.5 Two different ways to draw the tree in Figure 2.4C. Note that the length does not change, nor does the hypothesized order of events.

Parallelism

work can be rooted to produce an evolutionary tree, cladogram, or phylogeny.

Two phenomena, however, make it much harder in practice to determine evolutionary history: parallelism and reversal, which sometimes are referred to together as **homoplasy**. **Parallelism** is the appearance of similar character states in unrelated organisms. (Many authors make a distinction between parallelism and convergence, but for this discussion we will treat them as though they were the same.) A **reversal** occurs when a derived character state changes back to the ancestral state.

To provide a clear example, we will divide the group that we have called "star plants" into red star plants, gold star plants, and white star plants. Let us assume that the gold star and white star plants have only one cotyledon, whereas all the rest of the plants have more than one (including the conifer). Let us further assume that the white star plants have fused petals. We can add the character cotyledon number to the matrix in Figure 2.4A to create the matrix in Figure 2.8A, which gives the same information as the network in Figure 2.8B.

Now we see that, according to this network, there have been *two* parallel changes in petal fusion. Counting the number of changes on this network (its length), we find five: one each in pollen colpi, flowers in heads, and cotyledon number and two in petal fusion.

In this example, a group based on fused petals would be considered **polyphyletic**. Polyphyletic groups have two or more ancestral lineages in which the parallel character states evolved. (Although we distinguish here between paraphyletic and polyphyletic groups, many systematists

have observed that the difference is slight and simply call any para- or polyphyletic group nonmonophyletic.) Petal fusion in this case is nonhomologous because it fails the ultimate test of homology: congruence with other characters in a phylogenetic analysis.

Why not draw the network in such a way that petal fusion arises only once? Such a network is shown in Figure 2.8C. Now we have one change in petal fusion, but that requires two changes in cotyledon number and also two changes in number of pollen colpi, making the network six steps long.

Each of the networks can be converted into a phylogeny by rooting at the conifer, but they make different suggestions about how the plants have evolved. In Figure 2.8B, cotyledon number and number of pollen colpi have been stable over evolutionary time, whereas petal fusion has appeared twice, independently. In Figure 2.8C, we postulate that cotyledon number and number of pollen colpi have changed twice over evolutionary time, while petal fusion has evolved only once. By drawing either of these networks, we are proposing a hypothesis about how evolution has happened—about which genetic changes have occurred, at what frequency, and in which order.

As the two networks show, our two hypotheses differ. How do we determine which one is correct? There is no way to be certain. No one was there to watch the evolution of these plants. We can, however, make an educated guess, and some guesses seem more likely than others to be correct. One way to proceed is to ask, "What is the simplest explanation of the observations?" By asking this question, we apply a rule that is used throughout science, known as

BOX 2A Hennigian Argumentation

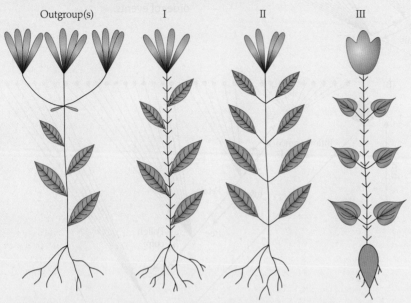

Outgroup(s)   I   II   III

FIGURE 2.6   Three imaginary species (I, II, and III) and an outgroup.

In the examples presented thus far, a network is constructed and is then polarized by determining where the outgroup attaches. Some systematists, however, prefer to polarize the characters first, using one or several outgroups, and then construct the phylogeny. This goes back to the original concept of phylogenetic analysis proposed by Willi Hennig (see Chapter 3).

Consider, for example, the imaginary plants presented in Figure 2.6. In this case the character states of the outgroup are assumed to be ancestral (plesiomorphic) and are represented by 0; derived states are represented by 1 or higher numerals (Table 2.1). These character states are then used to produce a character × taxon matrix (Table 2.2).

Next a phylogenetic tree (or cladogram) is constructed in which the taxa are grouped (placed on the same branch) according to evidence provided

TABLE 2.1   States of morphological characters used in cladistic analysis of the three imaginary species in Figure 2.6.

Morphological character	Character state[a]	
	Plesiomorphic	Apomorphic
1. Roots	Less than 1 mm thick (0)	Greater than 5 mm thick (1)
2. Stems	Glabrous (0)	Pubescent (1)
3. Leaves	Alternate (0)	Opposite (1)
4. Venation	Pinnate (0)	Palmate (1)
5. Petiole	Lacking (0)	Present (1)
6. Base of blade	Acute (0)	Cordate (1)
7. Perianth parts	4 (0)	3 (1)
8. Perianth parts	Separate (0)	Fused (1)
9. Flowers[b]	In a group of 3 (0)	Solitary (1)

[a]Character state codings are given in parentheses.
[b]Note that inflorescence condition (flowers solitary versus flowers in groups of 3) cannot be polarized unless additional outgroups are employed.

Occam's razor: Do not generate a hypothesis any more complex than is demanded by the data. Applying this principle of simplicity, or **parsimony**, leads us to prefer the shorter network. The fact that it is shorter does not make it correct, but it is the simplest explanation of the data.

In the example we have presented here, in which there are few characters and little homoplasy, it is easy to construct the shortest network that can link the organisms. In most real cases, however, many networks are possible, and it is not immediately obvious which one is the shortest. Fortunately, computer algorithms have been devised that

compare trees and calculate their lengths. Some of the most widely used programs are PHYLIP (Felsenstein 1989), NONA (Goloboff 1993), and PAUP*4.0 (Swofford 2000). These programs either evaluate data over all possible trees (an exhaustive search) or make reasonable guesses about the topology of the shortest trees (branch-and-bound searches or heuristic searches).

If the taxa are numerous, only heuristic algorithms can be used. These algorithms may not succeed in finding the shortest tree or trees because of the large number of possible trees. For example, the possible relationships of three

TABLE 2.2 Character × taxon matrix for the three hypothetical species of Figure 2.6, based on characters in Table 2.1.

Taxa	Characters								
	1	2	3	4	5	6	7	8	9
Species I	0	1	0	0	0	0	0	0	1
Species II	0	0	1	0	1	0	0	0	1
Species III	1	1	1	1	1	1	1	1	1
Outgroup(s)	0	0	0	0	0	0	0	0	0

by shared derived character states (synapomorphies). The presence of a derived character state (apomorphy) in two taxa suggests that they share a unique common ancestor in which the apomorphy first evolved; the two taxa are assumed to have inherited the apomorphy (or evolutionary novelty) from this ancestor. The cladogram, then, represents the simplest hypothesis that explains the pattern of derived character states, following the principle of parsimony.

A hypothesis of evolutionary relationships for species I, II, and III of Figure 2.6 is presented in Figure 2.7. Species II and III are hypothesized to share a unique common ancestor because they share the derived states of characters 3 and 5 (see Table 2.1). They both have opposite, petiolate leaves, which are hypothesized to have evolved in their common ancestor. Similarly, the shared possession of solitary flowers supports the recognition of a more inclusive monophyletic group containing species I, II, and III.

The presence of hairy stems in species I and III is homoplasious; that is, hairy stems are hypothesized to have evolved in parallel in these two species, so their similarity is not based on common ancestry. Note, however, that hairy

stems could have evolved in the most recent common ancestor of all three species and then been lost (a reversal) in species II.

The sharing of pinnate leaves with acute (forming an angle less than 90°) bases and flowers with four separate perianth parts by species I and II is symplesiomorphic; these are shared ancestral characters. Such characters do not indicate relationship. In contrast, palmate venation, cordate (heart-

shaped) leaf bases, and flowers with three fused perianth parts are derived character states unique to species III. These unique derived character states (autapomorphies) also tell us nothing about the phylogenetic relationships of species III.

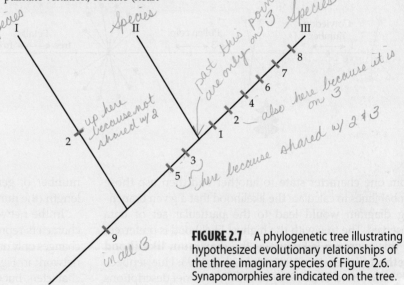

FIGURE 2.7 A phylogenetic tree illustrating hypothesized evolutionary relationships of the three imaginary species of Figure 2.6. Synapomorphies are indicated on the tree.

taxa can be expressed by only three rooted trees, [A(B,C)], [B(A,C)], and [C(A,B)]. But with larger numbers of taxa the number of potential trees expands rapidly; for example, four taxa yield 15 trees, five yield 105 trees, six yield 945 trees, and ten yield 34,459,425 trees!

The parsimony method is widely used, easily applicable to morphological changes, and possibly also the most intuitive of tree reconstruction methods. Parsimony works well when evolutionary rates are slow enough that chance similarities (due to the evolution of identical derived character states independently in two or more lineages) do not over-

whelm character states shared by the common ancestor. At higher rates of change, however, parsimony methods are susceptible to a phenomenon known as "long branch attraction" (Box 2B). Other methods of tree reconstruction use other criteria for choosing the preferred (optimal) tree. Instead of choosing the tree with the fewest evolutionary changes, one can convert the character × taxon matrix to a measure of similarity or dissimilarity among the plants, and then build a network that minimizes the dissimilarity; this is known as the **minimum-distance method**. Alternatively, one can develop theories about the probability of change

FIGURE 2.8 (A) A character × taxon matrix. (B) An unrooted network based on the matrix in 2.8A. Note that petal fusion appears to change twice. Network length is 5. (C) Another possible unrooted network based on the matrix in 2.8A. Unlike the network in 2.8B, petal fusion changes only once, but cotyledon number and pollen colpi change twice. Network length is 6.

(A)

	Pollen colpi	Petals	Inflorescence in heads?	Cotyledon number
Red star plants	< 3	free	no	2
Gold star plants	< 3	free	no	1
White star plants	< 3	fused	no	1
Circle plants	3	free	no	2
Square plants	3	fused	no	2
Diamond plants	3	fused	yes	2
Conifer	< 3	not applicable	not applicable	> 2

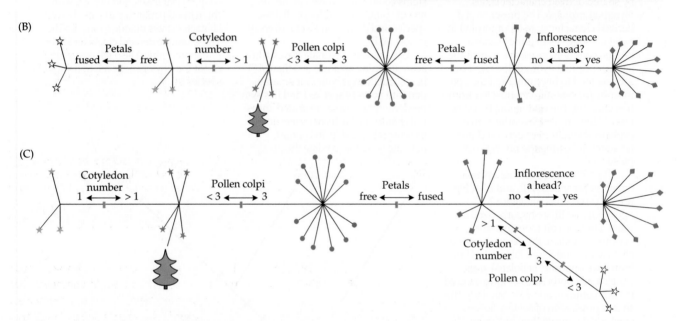

(B)

(C)

from one character state to another and then use those probabilities to calculate the likelihood that a given branching diagram would lead to the particular set of data observed. The tree with the highest likelihood is preferred, so this approach is known as the **maximum likelihood method** (Felsenstein 1981; Hillis et al. 1993; Huelsenbeck 1995; Swofford et al. 1996) (Box 2C). For brief descriptions of several current methods of phylogeny reconstruction, see Hall 2005.

Assessing Homoplasy

Parsimony analyses minimize the number of characters that change in parallel or reverse. If there are many such homoplasious characters, then the phylogenetic tree may be an artifact of the characters we have chosen, and a slight change in characters will lead to a different tree. The simplest, and most common, measure of homoplasy in a phylogenetic tree is the **consistency index (CI)**, which equals the minimum amount of possible evolutionary change (the

number of genetic switches) divided by the actual tree length (the number of actual genetic changes on the tree).

In the network shown in Figure 2.2B, each of the three characters represents a single genetic switch, and each one changes only once, so the consistency index is 3/3 = 1.0. In the network in Figure 2.8B, there are four binary (one-switch) characters, but one of those characters (petal fusion) changes twice on the tree, so the consistency index is 4/5 = 0.80.

Consistency indices may also be calculated for individual characters. In this case, the CI equals the minimum number of possible changes (one, for a binary character) divided by the actual number of changes on the tree. For example, the CI of petal fusion (see Figure 2.8B) is 1/2 = 0.50. For a given character × taxon matrix, the shortest network or tree will also have the highest consistency index. Lower consistency indices indicate the presence of many characters that contradict the evolutionary tree.

Comparing consistency indices across data sets is hazardous because the CI has some undesirable properties. For one thing, a character that changes once in only one taxon

BOX 2B Long Branch Attraction

Long branch attraction was identified originally by Felsenstein (1978) as a potential problem for phylogenetic analysis. If there are great differences in the rates of character evolution among lineages such that some lineages are evolving much more rapidly than others, and if characters have only a limited number of character states, then unusually long branches can be connected to each other in a tree whether or not they are actually closely related (Figure 2.9). This problem is particularly acute with DNA sequence data, for which each character has only four possible states, and for which mutation rates vary widely.

This phenomenon occurs because numerous random changes, some of which appear in parallel in the rapidly evolving lineages, outnumber the changes that provide information about common ancestry. The problem cannot be circumvented by adding more characters (base pairs, in DNA sequences); these merely add to the number of parallelisms linking the rapidly evolving lineages.

This situation can affect all methods of tree construction. With the correct model of evolution, however, maximum

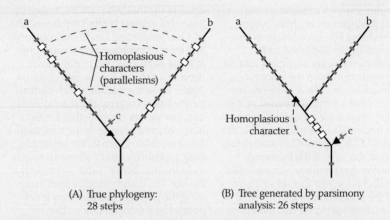

(A) True phylogeny: 28 steps

(B) Tree generated by parsimony analysis: 26 steps

FIGURE 2.9 Long branch attraction, a situation in which strongly unequal evolutionary rates cause parsimony to fail. (A) True phylogeny. Dashed lines show character states that have arisen in parallel in the lineages leading to a and b. (B) Phylogeny as reconstructed by parsimony. The number of parallelisms shared by a and b is greater than the number of characters linking a and c, so a and b appear to be sister taxa, with parallelisms (in the true phylogeny) treated as shared derived characters of a and b.

likelihood methods (see Box 2C) are less afflicted by this problem (although determining the correct model may be difficult). Long branch attraction is basically a sampling problem and may be alleviated by including taxa that are related to those terminating the long branches.

will have a consistency index of 1.0, but such a character tells us nothing about relationships. Such a uniquely derived character is sometimes called an **autapomorphy**. For example, if one of the red star plants in Figure 2.8B had hairy leaves while all the other plants in the network had hairless ones, leaf hairiness would not be of any help in indicating the relationship of the hairy-leaved plant to the other plants. In other words, the character would be **uninformative**. Because the uninformative character changes only once, however, it has a CI of 1.0. If we added many uninformative characters into the analysis, the overall CI would be inflated accordingly, and it would give a misleading impression that many characters supported the tree. Uninformative characters, therefore, are often omitted before the consistency index is calculated.

The consistency index is also sensitive to the number of taxa in an analysis (Sanderson and Donoghue 1989): analyses with many taxa tend to have lower CIs than analyses with fewer taxa. This relationship is true of both molecular and morphological data and of analyses of species, genera, or families.

The **retention index** (**RI**) circumvents the problems summarized in the previous two paragraphs, and also another limitation of the CI (Wiley et al. 1991; Forey et al.

1992). The CI should vary from near 0 (a character that changes many times on the tree) to 1.0 (a character that changes only once), but often the real range is much less. For example, in the matrix in Figure 2.8A, only two groups—the white star plants and the gold star plants—have a single cotyledon. If the one-cotyledon plants are all on a single branch of the network, as in Figure 2.8B, then the CI for cotyledon number is 1.0. If they are unrelated, as in Figure 2.8C, then the CI is 0.5 (1/2), which is the lowest possible value on the tree. Thus, instead of varying between 0 and 1, the CI varies between 0.5 and 1.0. The RI corrects for this narrower range of the CI by comparing the actual number of changes in the character to the maximum possible number of changes. The RI is computed by calculating the maximum possible tree length, which is the length that would occur if the derived character state originated independently in every taxon in which it appears (i.e., if all taxa with the derived character state were unrelated). The RI then equals the maximum length minus the actual length, divided by the maximum length minus the minimum length:

$$(L_{\text{max}} - L_{\text{actual}})/(L_{\text{max}} - L_{\text{min}})$$

In Figure 2.8B, then, the RI is $(9-5)/(9-4) = 4/5 = 0.80$.

BOX 2C Maximum Likelihood and Bayesian Methods

Parsimony analyses remain very common in phylogenetic analysis, but for analyses using DNA sequences as characters, maximum likelihood and Bayesian methods are becoming routine. These methods rely on the assumption that mutations in a DNA sequence are random. Over a particular period of evolutionary time, if the probability of a particular nucleotide base mutating is 1/100, then in a DNA sequence 100 bases long, we expect that one of the bases will mutate. We don't know which particular base will change, just that one of them will. If the period of evolutionary time is doubled, then we expect two mutations in our hypothetical sequence. In general, the expected number of changes will be

equal to the mutation rate multiplied by time; this formula is often symbolized by μt. Over longer and longer periods, more and more of the bases will change until at some point a second mutation will occur at a site that has already mutated. Again, we will not know which particular site has undergone the second mutation, but we can estimate that it must have occurred because of the total number of mutations seen in the sequence. Basic probability theory allows us to estimate the number of "extra" mutations at the site. The branch lengths used in creating the phylogenetic tree then incorporate these extra mutations that we infer must have occurred. All of our assumptions about the probabilities of

particular mutations together constitute a **model of evolution**. Both maximum likelihood and Bayesian methods are known as model-based methods because they incorporate ideas about the probability of change.

The theoretical statistical underpinnings of maximum likelihood and Bayesian approaches are very different. As a practical matter, however, a major distinction is simply computational speed. Maximum likelihood analyses take a long time to run, and bootstrap analyses require a high-performance computer. Bayesian methods estimate support for the tree at the same time as the tree is computed and so are faster.

Summarizing Evolutionary Trees

Parsimony analyses often find multiple trees, all with the same length but with different linkages among the taxa. Sometimes, too, different methods of analysis will find trees showing different topologies, and therefore different evolutionary histories, for the same taxa. In addition, studies using different kinds of characters (e.g., gene sequences, morphology) may find different trees. Rather than choosing among the trees in these cases, systematists may simply want to see what groups are found in all the shortest trees, or by all methods of analysis, or among different kinds of character matrices. The information in common in these trees can be summarized in a consensus tree.

Strict consensus trees contain only monophyletic groups that are common to all trees. For example, analyses of different sets of data have produced different ideas about the relationships among the early angiosperms. A study of the sequences of 4 genes led to the evolutionary tree shown in Figure 2.10A (which has been simplified for the purposes of this example) (Rydin et al. 2002). Adding more gene sequences, and analyzing them in a different way, led to the tree in Figure 2.10B (Burleigh and Mathews 2004). The trees both show that the angiosperms are sister to the gymnosperms and that the gymnosperms are monophyletic. Both trees also show that the Gnetales and the conifers (Pinaceae plus non-Pinaceae conifers) are closely related. The strict consensus of the two cladograms (Figure 2.10C) therefore shows the gymnosperms as monophyletic and the Gnetales plus conifers as a clade.

There are differences between the two evolutionary hypotheses, however. The 4-gene tree suggests that the Gnetales are sister to all conifers, but the 13-gene tree indicates that the Gnetales are sister only to the Pinaceae, which constitute a subset of the conifers. In the strict con-

sensus tree (Figure 2.10C), the Gnetales, Pinaceae, and the non-Pinaceae conifers appear as though they arose at the same time. This means that the available data cannot tell us whether they arose together or one after the other, nor can we determine the order in which they arose.

Having multiple lineages arising at the same apparent node in the diagram is usually an expression of ambiguity. The 13-gene tree suggests that the cycads are sister to all other gymnosperms, but in the 4-gene tree, the cycads, ginkgo, and the clade containing the rest of the gymnosperms all look as though they arose at the same time. The ambiguity in the 4-gene tree leads us to conclude that we really do not know which gymnosperm lineages appeared first. This uncertainty is reflected in the strict consensus tree by the fact that all those lineages are drawn as though they arose at the same time.

When many trees are compared, it is sometimes interesting to know whether a clade appears in most of the trees, even if it doesn't occur in all of them. A **majority-rule consensus tree** shows all groups that appear in 50% or more of the trees. If a particular clade is present in the majority of the most parsimonious trees, then this clade will be represented on the majority-rule consensus tree (along with an indication of the percentage of most parsimonious trees showing that clade). The majority-rule consensus tree will be inconsistent with some of the original trees and thus provides only a partial summary of the phylogenetic analyses.

A **semi-strict**, or **combinable-component**, **consensus tree** is often useful, particularly in comparisons of phylogenies with slightly different terminal taxa or constructed from different sources of characters. It is common, for example, to construct trees from two different sets of characters (e.g., a gene sequence and morphology) and to find that both sets of characters indicate monophyly of a particular group of species. Only one set of characters, however, may resolve

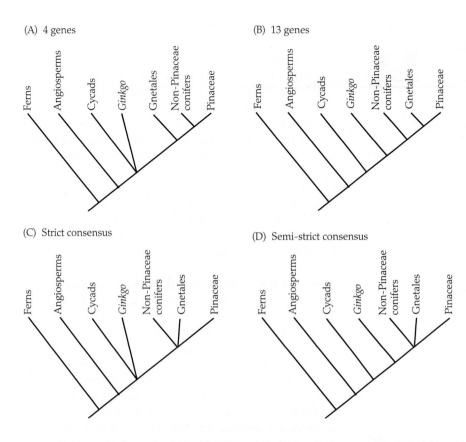

(A) 4 genes

(B) 13 genes

(C) Strict consensus

(D) Semi-strict consensus

FIGURE 2.10 (A) Phylogeny of seed plants based on DNA sequence data from 4 genes. (B) Phylogeny of seed plants based on DNA sequence data from 13 genes. (C) Strict consensus of the trees in A and B. (D) Semi-strict consensus of the trees in A and B. (A based on Rydin et al. 2002; D modified from Burleigh and Mathews 2004.)

relationships among the species. The semi-strict consensus tree then indicates all relationships supported by one tree or both trees and not contradicted by either.

For example, although the 4-gene tree (Figure 2.10A) does not give us any information about the order in which cycads, ginkgo, and the remaining gymnosperms originated, the 13-gene tree (Figure 2.10B) does. The two trees are not really conflicting; the 13-gene tree just provides more precise information. The semi-strict consensus tree thus follows the 13-gene arrangement of those three groups (Figure 2.10D).

The Probability of Evolutionary Change in Characters

In trying to infer the evolutionary history of a group, we depend on an implicit or explicit description (model) of the evolutionary process (see Box 2C). The more accurately the description reflects the underlying process, the more accurately we will be able to estimate the evolutionary history. This is particularly important for very divergent species in molecular phylogenies, for which parsimony methods often produce misleading results (see Box 2B). For nucleotides in a DNA sequence, mutation is assumed to be random, although this assumption is often modified to reflect hypothesized mechanisms of molecular evolution.

Developing a model is much more difficult for morphological characters because we usually have no idea how many genes are involved, nor do we know what kinds of changes in those genes lead to different character states. Nonetheless, certain assumptions must be made if one is to

proceed at all. (And, we note, *no* methods are entirely free of assumptions!) The major assumptions have to do with the likelihood of particular changes in character states and the likelihood of reversals and parallelisms.

Ordering character states The characters in Figure 2.8A have only two states. Such two-state (binary) characters are interpreted as representing a single genetic switch—"on" producing one state (e.g., tricolpate pollen), and "off" resulting in the other state (e.g., one-grooved, or monosulcate, pollen). Over evolutionary time, of course, such characters can continue to change. For example, tricolpate pollen is modified in some Caryophyllales so that it is spherical, with many pores evenly spaced around it (looking rather like a golf ball); this pollen is pantoporate.

If we were to include the character pollen colpi in a matrix containing some taxa with pantoporate pollen, that character would now have three states: monosulcate, tricolpate, and pantoporate. Pollen colpi would now be a multistate character, in contrast to the binary characters discussed previously. Multistate characters raise a difficult question: how many genetic switches are involved?

It is possible that monosulcate pollen changed to tricolpate pollen, which then changed to pantoporate pollen; this progression matches what we think happened in the angiosperms over evolutionary time (Figure 2.11A). (Recall that the outgroup does not have tricolpate pollen.) This scenario implies two genetic switches. It also implies that they must have occurred in order; that is, pantoporate pollen could arise only after tricolpate pollen did. If we accept this

(A)

(B)

(C)

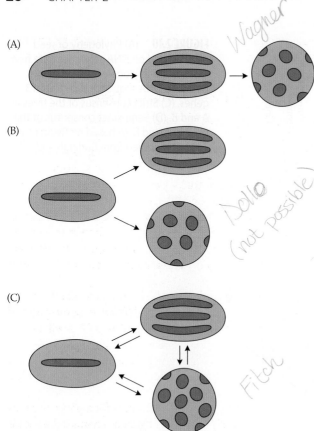

FIGURE 2.11 Three alternative hypotheses about the evolution of pollen morphology. (A) Monosulcate changed to tricolpate, which then changed to pantoporate. As drawn, the character is ordered and irreversible. (B) Monosulcate changed to tricolpate, and then independently changed to pantoporate. Again, the character is ordered and irreversible. If the arrows were drawn as double headed, the character would be interpreted as reversible. (C) Any pollen type can change to any other pollen type. The character is unordered and reversible.

series of events, the multistate character is considered **ordered**.

If we decide to allow for reversals of character states—that is, if we consider the possibility that pantoporate pollen might switch back to tricolpate and tricolpate to back monosulcate—the character is still ordered. It requires two evolutionary (genetic) steps to go from monosulcate to pantoporate pollen, and two steps to go from pantoporate to monosulcate pollen. A phylogenetic analysis in which all characters are treated as ordered is sometimes referred to in the literature as **Wagner parsimony**.

If we didn't know anything about the plants involved, we might want to consider the possibility that monosulcate pollen changed to tricolpate pollen, and in an independent event, monosulcate pollen changed to pantoporate pollen (Figure 2.11B). This sequence would suggest that there is a genetic switch that allows change from monosulcate to tricolpate pollen, as well as a switch that allows change from monosulcate to pantoporate pollen, but that a change from tricolpate to pantoporate pollen is impossible. The character

in this case is still ordered, but in a different way from what Figure 2.11A shows. If reversals are possible, then two steps are required to get from tricolpate to pantoporate pollen, and two from pantoporate to tricolpate pollen.

With morphological characters and character states, we are usually unsure of which switches are possible, so it is common to treat multistate characters as unordered (Figure 2.11C); this method is sometimes called **Fitch parsimony**. In the case of an unordered character, we postulate only one switch between any two states. DNA sequence characters are multistate characters with four states (adenine, thymine, guanine, cytosine). To treat these as ordered would be nonsensical; adenine does not need to change to cytosine before changing to guanine. DNA characters are therefore always treated as unordered and fully reversible.

Reversals, parallelisms, and character weighting   In the network in Figure 2.8B, we hypothesized that petal fusion arose twice, independently. To make the slightly longer network in Figure 2.8C, we had to let cotyledon number change from one to more than one and back to one again—that is, to reverse. In comparing the trees in Figures 2.8B and C, therefore, we are comparing two hypotheses: (1) that mutations in the genes leading to petal fusion have happened more than once and (2) that mutations in the genes controlling cotyledon number have happened and then their effects have been reversed. In deciding that the network in Figure 2.8B was shorter than the one in Figure 2.8C, we counted all the steps equally, whether they were parallelisms, reversals, or unique origins.

This approach may or may not be reasonable. Dollo's law, for example, suggests that for very complex characters, parallel origin is highly unlikely, whereas reversal may be quite easy (Mayr and Ashlock 1991). The assumption is that many genes must change in order for a morphological structure to be created, but only one of those genes needs to change for the structure to be lost.

We can build Dollo's law into the process of choosing a tree by making gains of structures count for more than losses; the process is then known as **Dollo parsimony**. (Defining the terms *gain* and *loss*, of course, requires a rooted tree; hence Dollo parsimony cannot be applied to an unrooted network.)

Certain characters are sometimes **weighted** in phylogenetic analyses. This weighting reflects the assumption that certain characters should be harder to modify than others. One might hypothesize, for example, that leaf anatomy is less likely to change than leaf hairiness (pubescence), and therefore a change in a leaf anatomical character could be counted as equivalent to two changes in pubescence for the purposes of counting steps in the tree.

Such weighting decisions can easily become subjective or arbitrary, and they risk biasing the outcome of the study toward finding particular groupings. (For example, the investigator might theorize, "My favorite species group has interesting leaf anatomy; therefore I think that leaf anatomy is phylogenetically important; therefore I will give it extra weight in the phylogenetic analysis." In this case it is no surprise when the favorite species group is shown to be monophyletic.)

Because of the possibility of bias, systematists generally attempt to base weighting decisions on an objective criterion. One approach is to do a preliminary phylogenetic analysis with all characters assigned equal weights. The results of this analysis will identify which characters have the least homoplasy on the shortest tree(s); the characters with less homoplasy can then be given more weight in subsequent analyses, a process known as **successive weighting**.

Another approach is to base weights on knowledge of the underlying genetic basis of characters. For example, in DNA sequence analyses, transversions (purine → pyrimidine or pyrimidine → purine changes) are weighted over transitions (purine → purine or pyrimidine → pyrimidine changes) because transitions are known to occur more frequently and to be easier to reverse. Restriction site gains may be weighted over restriction site losses because there are fewer ways to gain a restriction site than to lose one (see Chapter 5). And complex characters (presumably controlled by many genes) may be weighted over simple characters (presumably controlled by fewer genes), again because the latter are thought to be easier for selection to modify over evolutionary time.

The most common approach, used in most preliminary analyses, is to weight all characters equally. Although this approach sometimes is described as "unweighted," in fact it assumes that all characters are equally likely to change and weights them accordingly.

Underlying every discussion of weights is the assumption that all characters of organisms evolve independently. This assumption requires that change in one character not increase the probability of change in another character. Like the previous assumption, this one may be violated frequently; for example, a change in flower color might well lead to a shift in pollinators, which would then increase the probability that corolla shape would change. Violations of this assumption obviously affect char-

acter weighting, in that the likelihood of change for any two characters is not the same.

Do We Believe the Evolutionary Tree?

An evolutionary tree is simply a model or hypothesis, a best guess about the history of a group of plants. It follows that some guesses might be better, or at least more convincing, than others. Use of an optimality criterion is one way to evaluate the evolutionary tree; of all possible descriptions of history, we prefer the one that requires the fewest steps, or has the maximum likelihood, or the minimum distance. Trees can be evaluated more precisely, however. For the purposes of this discussion, we will continue to focus on phylogenies generated by parsimony methods (i.e., with the fewest evolutionary steps).

With parsimony methods, the shortest available tree is preferred over one that is longer. However, some parts of the tree may be more reliable than others. This will be the case if reversals and parallelisms (or simple misinterpretations of characters) affect some groups of plants more than others, or if there were very few evolutionary changes in the history of a particular group.

One simple way to evaluate support for a particular part of a tree is to note the number of genetic changes that occur on the branch leading to a particular group, along with the consistency indices of the characters. For example, a phylogeny of some members of the Ericaceae based on DNA sequence data (Figure 2.12; Kron and Judd 1997) found 18

FIGURE 2.12 The single most parsimonious tree found in analysis of *Lyonieae* (taxa in boldface type, lines in blue) using data from sequences of the *matK* gene. Branch lengths appear above lines; bootstrap values are in parentheses; decay indices (*d*) appear below lines. Length = 425, consistency index = 0.60. (From Kron and Judd 1997.)

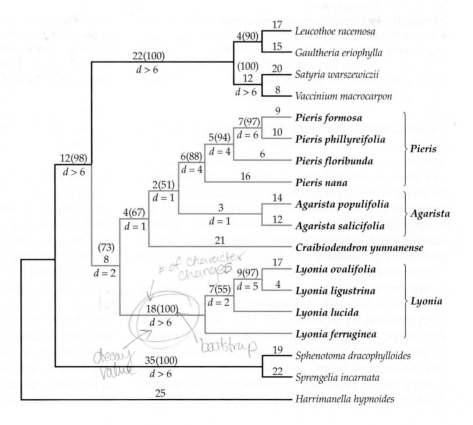

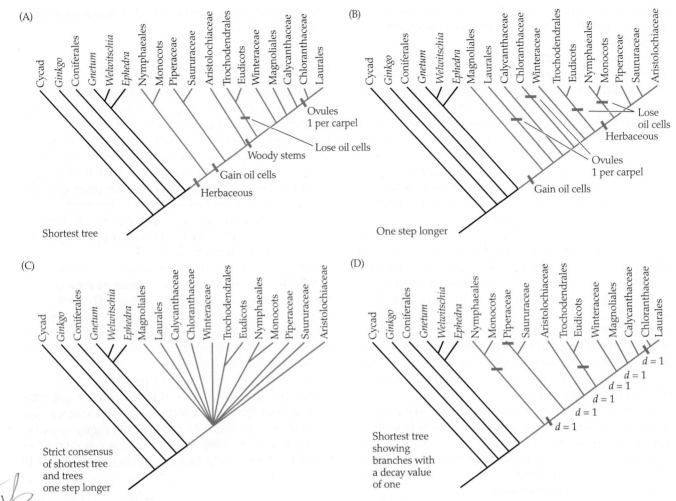

FIGURE 2.13 (A) Phylogeny of the angiosperms (blue lines), indicating patterns of change in the presence or absence of oil cells, ovule number per carpel, and plant habit (woody or herbaceous). (B) An alternative tree, only one step longer than the tree in A, showing patterns of change in the same characters. Note that herbaceousness now is hypothesized to have evolved only once, but loss of oil cells and reduction of ovule number occur twice. (C) Strict consensus of the shortest tree and trees one step longer (Figures 2.13A and B). (D) The same tree as in A, showing branches with a decay value of 1. (Data from Doyle et al. 1994.)

changes on the branch leading to the *Lyonia* clade. In an analysis of morphological characters for the same taxa, there were four characters that changed along the *Lyonia* branch and nowhere else on the tree. In other words, a number of the changes that occurred during the origin of the *Lyonia* clade produced novel characteristics, found nowhere else in the family. Groups like the *Lyonia* clade that share numerous characters that do not change elsewhere on the clado-gram are more believable than groups that share only a few highly homoplasious characters.

Another way to assess how well the data support the tree is to determine whether a group of interest occurs in other trees that are almost equally short. In other words, suppose we ask whether there are other ways to analyze the homoplasious characters that lead to trees that are one, two, or three steps longer.

For example, in a study of angiosperm diversification (Doyle et al. 1994), the shortest tree indicated that the earli-est-diverging lineages in the angiosperms were the mono-cots and the water lilies (Nymphaeaceae; see Chapter 9). This implied that the character herbaceous stems was gained once and then lost, whereas reduction in the number of ovules per carpel to one occurred only once, and oil cells were gained once and lost once (Figure 2.13A). On the other hand, trees one step longer, in which the earliest angiosperm lineages led to the magnolias, suggested that herbaceous stems evolved once, but reduction in ovule number occurred twice, and there were three changes in oil cells (gained once and lost twice or vice versa) (Figure 2.13B).

Thus, by looking at trees one step longer, we can hypothesize that some characters are less homoplasious, but some are more so. If we now take the strict consensus of all the trees, including the shortest ones and those one step longer, all early angiosperm lineages are drawn as though they radiate from a single point, indicating uncertainty about the order in which they evolved (Figure 2.13C).

You can see that many of the branches in the shortest trees do not appear in trees one step longer. Thus all those branches are not drawn in the strict consensus; in other words, they "collapse", or "decay." We can indicate this by placing a 1 next to each of the collapsing branches of the shortest tree (Figure 2.13D). This number is the **decay index**, sometimes called Bremer support, which represents how many extra steps are required to find trees that do not contain a particular group. It provides a relative measure of how much the homoplasy in the data affects support for a particular group.

The decay index is not statistical, which, depending on one's point of view, is either a virtue or a drawback. Because history happened only once and cannot be repeated, it is impossible to replicate the evolutionary experiment. It is certainly possible, however, to test whether character data are different from random expectations, although there are many possible ways to randomize systematic data. Many tests have been devised that use some sort of randomization technique. Probably the most widely used is bootstrap analysis.

Bootstrap analysis randomizes characters with respect to taxa. As an example, begin with the matrix in Figure 2.8A and randomize the columns while leaving the rows in place. Choose a column at random from the original matrix to become the first column of the new matrix. Then choose another column to become the second column, and so on until a new matrix is created with the same number of columns as the original. Because one returns to the original matrix each time to choose a new column, some characters may be represented several times in each new matrix, while others are omitted. This method is usually described as random sampling with replacement.

Figure 2.14 shows the matrix in Figure 2.8A randomly sampled with replacement; note that the first character from the original matrix (pollen colpi) has been selected twice, whereas the third character (inflorescence a head) has been missed by the random selection process. Multiple randomized matrices are constructed, and the most parsimonious trees are found for each new matrix. This process is used to create a set of at least 100 trees, which can be summarized by a consensus tree (see pages 24–25). In the bootstrap consensus tree, a clade with a bootstrap value of, say, 95% was present in 95% of the trees generated in the bootstrap analyses.

The phylogeny in Figure 2.12 shows both bootstrap and decay values, along with branch lengths. We see that bootstrap and decay values are high for the genus *Lyonia*, indicating that the data support monophyly of the genus, whereas the linkage of *Agarista* and *Pieris* is supported by only 51% of the bootstrap trees, and in trees only one step longer the two genera are not sisters, indicated by the decay value of 1.

Another excellent way to gain confidence in the groupings present in a tree is to compare phylogenies that have been based on different sets of characters. For example, phy-

	Pollen colpi	Petals	Pollen colpi	Cotyledon number
Red star plants	< 3	free	< 3	2
Gold star plants	< 3	free	< 3	1
White star plants	< 3	fused	< 3	1
Circle plants	3	free	3	2
Square plants	3	fused	3	2
Diamond plants	3	fused	3	2
Conifer	< 3	not applicable	< 3	> 2

FIGURE 2.14 The matrix from Figure 2.8A, sampled with replacement, as it would be for the first step of a bootstrap analysis. Note that in the sampling process, the character pollen colpi has been sampled twice, whereas the character inflorescence a head has been omitted.

logenies based on morphology, chloroplast DNA nucleotide sequences (cpDNA), and nuclear DNA nucleotide sequences could be (and often are) compared. If these phylogenies show similar groups, we can be more confident that they reflect the true order of events. For example, the monophyly of such families as the Poaceae, Onagraceae, Ericaceae, Asteraceae, and Orchidaceae has been supported by phylogenetic analysis of many kinds of data, including morphology, chloroplast gene sequences, and nuclear gene sequences.

Comparing trees is often particularly intriguing when the data come from different genes, as will be discussed in more detail in Chapter 5. It is also common to combine morphological and DNA characters in a single phylogenetic analysis, which often leads to more strongly supported phylogenies than either kind of data can produce alone. Morphological phylogenies assume that hybridization has not occurred, or at least is rare (Box 2D); multiple molecular trees can often test this assumption.

Describing Evolution: Mapping Characters on Trees

Phylogenies can be used to describe the evolutionary process and to develop hypotheses about adaptation, morphological and physiological change, or biogeography, among many other uses. If a phylogeny is to be used to describe evolutionary history, however, careful attention must be given to the characters and character states used in the description. In the discussion that follows we will focus on morphological characters, but many of the points apply to any kind of character.

Consider a group of plants for which the phylogenetic tree is known; a good example is the Ericaceae, for which much information is available (Figure 2.15). Assume for the purposes of this discussion that this tree is an accurate reflection of history and that each of the terminal genera really is monophyletic, as demonstrated by studies of multiple species of each. Then consider a study that is concerned with the gain or loss of fused petals, which are intimately connected with the evolution of pollination systems. This is the kind of study that systematists frequently engage in because the details of character evolution lead to hypotheses about how natural selection has worked. In addition, when constructing classifications, one frequently wants to know what morphological characters can be attributed to and distinguish a particular monophyletic group.

Figure 2.15 shows the observed character states for the genera. It seems trivially obvious from looking at the distribution of characters and character states that free petals must have evolved once in the lineage leading to *Ledum* (Labrador tea) and again in the lineage leading to *Vaccinium* sect. *Oxycoccum* (cranberries). Phrased another way, the ancestor of *Vaccinium* sect. *Oxycoccum* and all other vacciniums (blueberries) had fused petals, as did the ancestor of *Ledum* plus *Rhododendron* sect. 3.

Examine this "obvious" conclusion more closely. If we were studying only species of *Vaccinium*, we would have no way of knowing whether fused petals were ancestral or derived (Figure 2.16A). There must have been one genetic change, but it could have happened as easily in the lineage leading to the cranberries (sect. *Oxycoccum*) as in the lineage leading to the blueberries (other *Vaccinium*).

Only by reference to the outgroup *Epacris* can we determine when petal fusion was lost. Because *Epacris* has fused petals, free petals must have originated within *Vaccinium*; it

is simplest (most parsimonious) to assume just one genetic change, from fused to free (Figure 2.16B). This is the same as saying that the ancestor of blueberries plus cranberries had fused petals. If we were to postulate that the ancestor had free petals, we would need two changes to fused petals: one in *Epacris* and one in the blueberries. The same argument applies in the case of *Rhododendron* and *Ledum*.

Now suppose we were studying only species of *Vaccinium*, but this time, instead of using *Epacris* or other Ericaceae as outgroups, we used only *Ledum*. This could easily be the

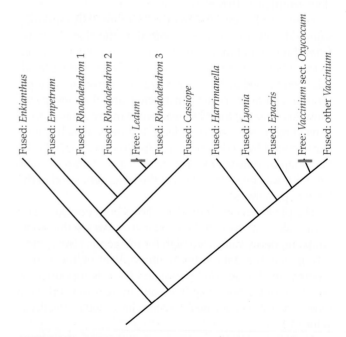

FIGURE 2.15 Phylogeny of a portion of the Ericaceae. The genus *Rhododendron* is paraphyletic and is represented by three separate lineages, numbered 1 to 3. Two changes to free petals are hypothesized. (Data from Stevens 1998.)

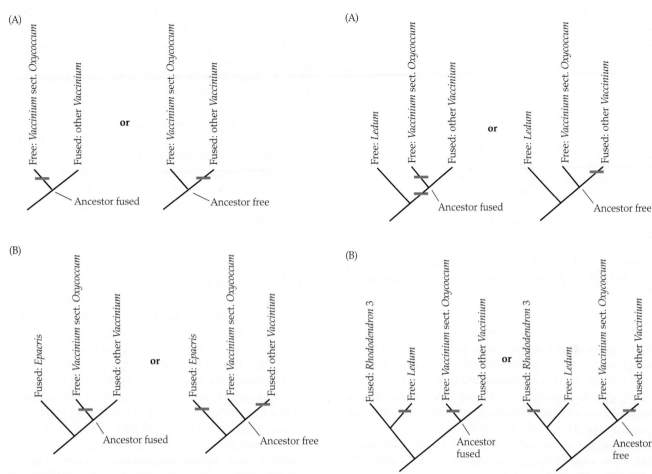

FIGURE 2.16   (A) Two *Vaccinium* taxa differ in character states. It is impossible to determine from this information alone what the character state of the ancestor was because either assumption involves one change in one descendant lineage. (B) The addition of an outgroup determines the character state of the ancestor. In this case, it is simpler (requires fewer steps) to assume that the ancestor had fused petals.

FIGURE 2.17   (A) Analysis of character state change in *Vaccinium* using a different outgroup. Note that the inference of the ancestral state is exactly the opposite of that reached when *Epacris* is used as an outgroup. (B) Analysis of character state change in *Vaccinium* using two outgroups that differ in state. It is now impossible to determine the character state of the ancestor.

case if material of the other genera were hard to obtain, or if those genera were extinct and we didn't even know they had existed. Now we would conclude that the ancestor of all vacciniums had free petals, and that in response to some unknown selective pressure there was a change to fused petals (Figure 2.17A). *This is exactly the opposite conclusion from the one reached in the previous example*, and the only difference is the genera included in the analysis.

One might try to improve the situation by using additional outgroups. For example, consider the same study of *Vaccinium*, but now use both *Ledum* and *Rhododendron* as outgroups. In this case the direction of change is completely ambiguous (Figure 2.17B). It is as simple to postulate that the ancestor of the group had fused petals and that there were two changes to free petals as it is to postulate that the ancestor had free petals and that there were two changes to fused.

These two choices are known as **equally parsimonious reconstructions**. For many characters on many trees, there are multiple equally parsimonious reconstructions. In other

words, there are multiple equally good hypotheses about the direction and timing of character state change. If you return to the example in Figure 2.13, you should be able to find equally parsimonious reconstructions that differ from the ones shown.

Ambiguity can also come from including taxa for which the character state is not known. Suppose, for example, two new taxa are discovered such that, on the basis of other characters, one is clearly sister to *Vaccinium* sect. *Oxycoccum* and the other is sister to the rest of *Vaccinium* (Figure 2.18). In addition, suppose that it is unclear whether the petals are fused or free. (This type of ambiguity is more common than you might think; it can occur when the original description is vague and/or illustrations are unclear, or when the original plant is known only from fruiting material.) In this example we do not know what the ancestral state was for *Vaccinium*, so we cannot make any hypothesis about the direction of evolutionary change. Nor can we be sure that the character "petals fused" is a synapomorphy for the genus.

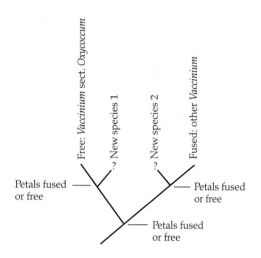

FIGURE 2.18 Addition of species for which the character state is unknown can prevent any inference about the ancestral state.

Various algorithms have been developed to assign character state changes to particular portions of trees (see Chapters 3 and 4 of Maddison and Maddison 2000 for a lucid and comprehensive discussion of these). Depending on the algorithm used, the character state changes can be biased in favor of parallelisms (the "delayed transformation," or DELTRAN, algorithm) or in favor of reversals ("accelerated transformation," or ACCTRAN). The results can have implications—sometimes major ones—for hypotheses about the evolutionary process, and they may also affect how organisms are described in a classification.

Constructing a Classification

The theory of classification is a topic with which systematists have been wrestling for centuries, and their struggles have led to a broad and frequently contentious literature (see Chapter 3). The principles of phylogenetic classification outlined here are commonly but not universally held. In general, however, classification has several goals. A classification is a common vocabulary designed to aid communication. Therefore a classification should be stable; names that are frequently changed become useless for communication. In addition, a classification should be predictive; that is, the name of a plant should help you to learn more about that plant and guide you to its literature.

Systematists generally agree about the goals of classification, but they may disagree profoundly about how to reach those goals. In this text we take a particular point of view, using phylogenetic classifications throughout. Thus, as far as possible, we recognize monophyletic and avoid paraphyletic or polyphyletic groups. In the few cases in which a nonmonophyletic family or order has not yet been divided into monophyletic units, we have placed the taxon name in quotation marks. The monophyly of many genera

of angiosperms is questionable, but relatively few phylogenetic analyses are available at this level, so generally we have not tried to indicate possible or probable paraphyly or polyphyly of genera.

The biological diversity on Earth is the result of genealogical descent with modification, and monophyletic groups owe their existence to this process. It is appropriate, therefore, to use monophyletic groups in biological classifications so that we may most accurately reflect this genealogical history. Classifications based on monophyletic groups are more predictive and of greater heuristic value then those based on overall similarity or on idiosyncratic weighting of particular characters (Farris 1979; Donoghue and Cantino 1988).

Phylogenetic classifications, because they reflect genealogy, will be the most useful in biological fields such as the study of plant distributions (phytogeography), host-parasite or plant-herbivore interactions, pollination biology, and fruit dispersal, or in answering questions related to the origin of adaptive characters (Nelson and Platnick 1981; Humphries and Parenti 1986; Brooks and McLennan 1991; Forey et al. 1992). Because of its predictive framework, a phylogenetic classification can direct the search for genes, biological products, biocontrol agents, and potential crop species. Phylogenetic information is also useful for making conservation decisions. Finally, phylogenetic classifications provide a framework for biological knowledge and a basis for comparative studies linking all fields of biology (Cracraft and Donoghue 2004).

Constructing a classification involves two steps. The first step is delimitation and naming of groups. In a phylogenetic classification this step is uncontroversial: named groups must be monophyletic. The second step involves ranking the groups and placing them in a hierarchy. This step remains problematic.

Grouping: Named Groups Are Monophyletic

A phylogenetic classification reflects evolutionary history and attempts to give names *only* to groups that are monophyletic—that is, composed of an ancestor and all its descendants. In the example in Figure 2.4C, we infer that the Asteraceae (the diamond plants) are monophyletic because they have flowers in heads. The square plants plus Asteraceae are also monophyletic because they share the derived character state of fused petals; this group also has a name, the Asteridae (or the asterid clade). Similarly, the entire group of plants with tricolpate pollen (circle plants plus Asteridae) is monophyletic and is known as the eudicots (or the tricolpate clade). This group could be given a formal Latin name, but it does not have one at the moment and may not actually need one.

In phylogenetic classification, paraphyletic groups are not named. In Figure 2.4C, a group made up of square plants plus circle plants would be paraphyletic. The most recent common ancestor shared by any square plant and a circle plant is also the most recent common ancestor of any circle plant and a diamond plant. In other words, the circle plants are as distantly related to square plants as any one of them is to diamond plants. Naming a group that included the square plus the circle plants would imply that the two plants are closely related even though they are not.

There are many examples in this book of named groups of plants that we now believe to be paraphyletic. One well-known example is "bryophytes," a group that traditionally includes the nonvascular land plants (liverworts, hornworts, and mosses; see Figure 1.1). But the liverworts, hornworts, and mosses are more distantly related to one another than the mosses are to the vascular plants (tracheophytes). Without quotation marks, the name *bryophytes* implies a closer relationship than actually exists.

Several traditionally recognized plant families, such as Apocynaceae and Capparaceae in the broad sense, are paraphyletic. In this text these families have been recircumscribed so as to recognize monophyletic groups: Apocynaceae has been combined with Asclepiadaceae, and Capparaceae has been combined with Brassicaceae (although some systematists divide it into Capparaceae s.s. and Cleomaceae).

Naming: Not All Groups Are Named

A phylogenetic classification attempts to name only monophyletic groups, but the fact that a group is monophyletic does not mean it needs to have a name. The reasons for this are practical. We could put every pair of species into its own genus, every pair of genera into its own family, every pair of families into its own superfamily, and so on. But such a classification would be cumbersome; in addition, it would not be stable because our view of sister species would change each time a new species was described, and our view of the entire classification would have to shift accordingly.

In practice, many monophyletic groups are not named. For example, the genus *Stenanthium* (Melanthiaceae) is monophyletic and contains four species (Zomlefer et al. 2001; Zomlefer and Judd 2002; Wofford 2006). Although it is clear that these four species fall into two monophyletic pairs, the two pairs of species are not named, and few systematists would consider doing so. In another example, over half of the genera of the grass family fall into a single large clade that contains four large, traditionally recognized subfamilies plus two smaller ones. Although agrostologists refer to this clade as the PACCAD clade (an acronym for Panicoideae-Arundinoideae-Centothecoideae-Chloridoideae-Aristidoideae-Danthonioideae), it has no formal Latin name.

How do systematists decide which monophyletic groups to name? There is no codified set of rules, but several criteria have been suggested, and some criteria are in common use despite not being fully articulated. A major criterion—perhaps *the* major criterion—is the strength of the evidence supporting a group. Ideally, only clades linked by many shared derived characters should be formally recognized and named in classifications. This makes sense if a classification is to function as a common vocabulary.

Names are most useful if they can be defined, and the more precise the definition the better. In other words, if a clade is to be named, it should have a set of characters by which it can be distinguished from other clades, or **diagnosed**. This criterion is also important to nomenclatural stability: if the meaning of a name shifts every time a new phylogeny is produced or a new character is examined, the name becomes effectively meaningless.

A second criterion is the presence of an obvious morphological character. Although systematists may not agree on the importance of this criterion, it is an important extension of the idea of a well-supported group and is also relevant to the use of classifications by nonsystematists for identification purposes. If, for example, the only way a field biologist can identify an organism is by knowing whether it has an alanine or a serine at position 281 in its ribulose 1,5-bisphosphate carboxylase/oxygenase molecule, she may not find the classification much help in making predictions about the organism. If, on the other hand, she knows that the organism is a grass with a particular spikelet structure, she can easily and reliably infer many aspects of its biology. (Lack of an obvious morphological synapomorphy is one of several reasons that the PACCAD clade of the grasses is not given a name.) The characters used for classification do not have to be those used for identification, but many systematists prefer to name clades that are easily recognized morphologically.

Another criterion is size of the group. Human memory is easily able to keep track of small numbers of items (in the range of 3 to 7) (Stevens 1998), but to organize and remember larger numbers of items requires additional mnemonic devices. (As an example, consider how many 9-digit zip codes you can remember compared with the 5-digit variety, or with 7-digit telephone numbers.) Dividing a large group into smaller groups is a way to organize one's thinking about large numbers of taxa. In the words of Davis and Heywood (1963: 83), "We must be able to place taxa in higher taxa so that we can find them again." The genus *Stenanthium* could be redefined to include only *Stenanthium gramineum* and *S. diffusum*, and a new genus could be described to include *S. densum* and *S. leimanthoides*. There seems little reason to do this, however, because four species is not a difficult number to keep track of. That said, there seems little reason to divide a large group if well-supported clades cannot be identified within it.

A fourth criterion is nomenclatural stability. A classification is ultimately a vocabulary, a means of communication. It cannot function this way if the meanings of the names continually change. Thus, given a set of well-supported, diagnosable, monophyletic groups, groups that have been named in the past can—and we would argue should—continue to be named. This is yet another argument against formally naming the PACCAD clade of the grasses, in that it would entail an unnecessary set of changes affecting long-standing taxonomic usage (Backlund and Bremer 1998; Stevens 1998).

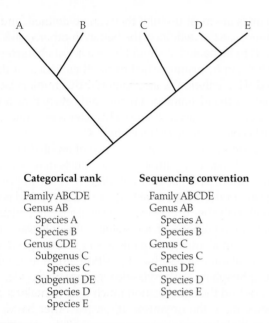

Categorical rank	Sequencing convention
Family ABCDE	Family ABCDE
Genus AB	Genus AB
Species A	Species A
Species B	Species B
Genus CDE	Genus C
Subgenus C	Species C
Species C	Genus DE
Subgenus DE	Species D
Species D	Species E
Species E	

FIGURE 2.19 Alternative classifications based on the phylogeny of a hypothetical group of taxa A, B, C, D, and E. The classification on the left uses four categorical ranks (family, genus, subgenus, and species); the one on the right uses only three ranks (family, genus, and species) plus a sequencing convention.

Ranking: Ranks Are Arbitrary

Having decided which monophyletic groups to name, we still have the question of exactly how to name them. The groups could, for example, be numbered, and a central index could list what is encompassed by each numbered group. This approach is similar to the system used by the telephone company to organize telephone numbers. The difficulty, of course, is that without a telephone book (a central index) and/or an excellent memory, the system is inaccessible.

Biological classification attempts to provide a working vocabulary that conveys phylogenetic information, yet can be learned by biologists who are not themselves primarily systematists. Because a phylogeny is similar in structure to a hierarchy, in which small groups are included in larger groups, which themselves are included in still larger groups, it makes sense for the classification to reflect it as a hierarchy.

Botanical classification uses a system developed in the eighteenth century in which taxa are assigned particular ranks, such as kingdom, phylum, class, order, family, genus, and species (i.e., Linnaean ranks) (see Chapter 3 and Appendix 1). A classification of named monophyletic groups should be logically consistent with the phylogenetic relationships hypothesized for the organisms being classified (as expressed in the sequence of branching points in the cladogram). That is, the categorical ranks of a Linnaean classification are used to express sister-group relationships.

Even though monophyletic taxa are considered to represent real groups that exist in nature as a result of the historical process of evolution, the categorical ranks themselves are only mental constructs. They have only relative (not absolute) meaning (Stevens 1998). In other words, the familial level is less inclusive than the ordinal level and more inclusive than the generic level, but no criteria are available to indicate that a particular taxon, such as the angiosperms, should be recognized at the level of phylum, class, or order.

In Figure 2.19, a cladogram of imaginary taxa A through E is first converted into a hierarchical classification according to Linnaean categorical ranks. Note that subgenus DE is nested within genus CDE, which is in turn nested within family ABCDE. (But we could have treated clade ABCDE as an order, clade CDE as a family, and clade DE as a genus.) Often, however, to fully express the sister-group relationships (in the cladogram), one needs more ranks than are available (in the taxonomic hierarchy), even after creating additional ranks by use of the prefixes *super-* and *sub-*.

One modification to the method of classification outlined here is the **sequencing convention**, which states that taxa that form an asymmetrical part of a cladogram may be placed at the same rank and arranged in their order of branching (Wiley 1979, 1981). Thus in Figure 2.19, AB, C, and DE could all be designated as genera. The sequence of names in the classification denotes the sequence of branching in the cladogram. Note that this is the same as saying that not all monophyletic groups are given names.

Even though ranking is arbitrary, the criteria described here for deciding which groups to name can also be applied to deciding the level at which to rank a group (see Stevens 1998 for full discussion). Nomenclatural stability again becomes important here. For example, it has recently been shown that the earliest-diverging lineage in the family Poaceae includes only two extant genera, *Anomochloa* and *Streptochaeta* (Figure 2.20). Thus one could, in principle, create a new family for *Anomochloa* and *Streptochaeta*; after all, it would be monophyletic and would leave the Poaceae as also monophyletic. For the purposes of stability, however, it makes sense to leave the two genera in Poaceae, where they have been given a subfamilial name: Anomochlooideae.

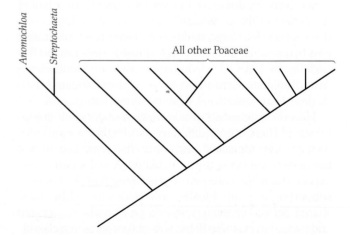

FIGURE 2.20 Phylogeny of Poaceae, showing the position of the genera *Anomochloa* and *Streptochaeta*.

Some systematists have proposed abandoning the Linnaean system altogether and replacing it with a "phylogenetic taxonomy." Full exploration of this possibility is beyond the scope of this text, but we will briefly address the arguments against the use of Linnaean ranks here.

Because rank is arbitrary, a genus (group of species) in one family may not be the same age as, encompass the same amount of variation as, or indeed have anything in common at all with a genus in another family—other than the fact that they are both monophyletic groups. Trained systematists are generally aware of this (Darwin was, for example) and realize that genera, families, and so on are not comparable units (Stevens 1997). Some scientists, however, frequently use such categories as though they were real. For example, it is common to measure plant diversity by listing the number of families represented by a local flora, even though the unit *family* does not mean anything in particular.

If rank is arbitrary, then one logical step would be to eliminate ranks altogether. Taxa would be placed in named groups, but the groups would not be designated as genus, family, order, or any other rank. Such categorization already exists informally, particularly among groups above the level of orders. The eudicots, for example, are widely recognized as monophyletic, but are not given a particular Linnaean rank. Similarly, few systematists worry about whether the angiosperms should be recognized as a division, class, subclass, superorder, or other rank; they are clearly monophyletic and designated by the non-Linnaean name *angiosperm*.

Eliminating ranks becomes more problematic among orders, families, and genera. Groups assigned to those ranks are familiar and their names are in common use, so an entirely new sort of nomenclature is unlikely to be accepted rapidly or without protest. Nonetheless, an alternative system of phylogenetic nomenclature, known as the *PhyloCode*, is being developed. The PhyloCode is designed entirely outside the rules of the International Code of Botanical Nomenclature (ICBN), which governs the use of Linnaean ranks and has long been used by all plant taxonomists (see Appendix 1). It is an alternative nomenclatural system rather than a revision of the existing system (see the PhyloCode Web site at www.ohiou.edu/phylocode).

Another result of phylogenetic studies is the observation that many phylogenies are only partially resolved, so that precise placement of taxa is impossible given the available data. This means that some species cannot be placed for certain in a genus, and some genera cannot be reliably assigned to a family. The current system allows uncertain placements above the rank of species to be reflected by the category *incertae sedis*—literally, "of uncertain position." An alternative would be a rank-free system, in which neither placement in a larger group nor naming all branches of a dichotomy or polytomy is necessary.

The authors of this text have been involved in reclassifications of genera, families, and orders on the basis of phylogenetic data and have found that—as long as the phylogeny is clear—use of the standard Linnaean hierarchy is

quite easy (especially when it is supplemented by unranked informal names). When the phylogeny is unclear, it is usually reasonable to wait for more data before modifying the classification.

For more discussion of the problems encountered in using the Linnaean system in phylogenetic classification, consult Wiley 1981; de Queiroz and Gauthier 1990, 1992; Wiley et al. 1991; Forey et al. 1992; and Hibbett and Donoghue 1998.

Comparing Phylogenetic Classifications with Those Derived Using Other Taxonomic Methods

Not all taxonomists use phylogenetic methods, although this is the majority approach. Some systematists have held the view that although evolution has occurred, parallelism and reversal have been so common that the details of evolutionary history can never be deciphered. This point of view led to a school of systematics known as **phenetics**. Pheneticists argued that because evolutionary history could never be unequivocally detected, organisms might best be classified according to overall similarity. Thus similar organisms were placed together in a group, while very different organisms were placed in different groups (Sneath and Sokal 1973).

One serious difficulty with the phenetic approach was that many systematists produced treelike diagrams that grouped organisms by overall similarity, but these diagrams were then interpreted as though they reflected evolutionary history. Sometimes this approach led to results similar to those produced by a phylogenetic analysis, but sometimes it led to the production of "groups" made up of organisms that shared only the fact that they were different from everything else, including one another. Such groups have since proved to be paraphyletic or polyphyletic.

The development of phenetic methods was an important prelude to the acceptance and use of phylogenetic approaches. A taxonomist constructing a phenetic classification first carefully observed as many characters as possible. These characters were divided into states, or the quantitative value of the character was recorded (e.g., a series of measurements of leaf length was taken and the mean recorded for each taxon). This information was arranged in a character × taxon matrix similar to that in Figure 2.8A. The matrix was converted to a similarity (taxon × taxon) matrix by the use of any of several mathematical measures of similarity (or dissimilarity; see Sneath and Sokal 1973; Abbot et al. 1985). The systematist then grouped the taxa that were most similar and illustrated the similarity relationships with either a maplike or a treelike diagram (a phenogram) (Figure 2.21). Phenograms were constructed using clustering algorithms, while maplike diagrams resulted from ordination studies employing multivariate statistical procedures (see Abbot et al. 1985).

(A) Map

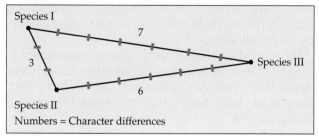

Numbers = Character differences

FIGURE 2.21 Two graphic means of expressing phenetic relationships. (A) Maplike diagram. (B) Phenogram.

(B) Phenogram

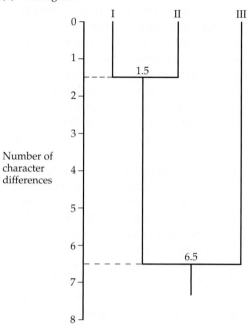

Number of character differences

Many of the classifications produced by phenetic methods are useful for identification and information retrieval. These classifications were not designed to retrieve evolutionary history, however, and are thus not appropriate for asking evolutionary questions. Phenetic systems do not distinguish between synapomorphy and convergent or parallel evolution.

Evolutionary taxonomy also differed from phylogenetic taxonomy in its approach to classification. The morphological similarity of a group was of utmost importance, and monophyly and paraphyly (in the strict cladistic senses of those words) were secondary. Thus a group could be recognized on the basis of some combination of derived and ancestral, unique and shared characters (Figure 2.22). Importance was

given to the recognition of "gaps" in the pattern of variation among phylogenetically adjacent groups (Simpson 1961; Ashlock 1979; Cronquist 1987; Mayr and Ashlock 1991). Characters considered to be evolutionarily (or ecologically) significant were stressed, and the expertise, authority, and intuition of individual systematists were central. Finally,

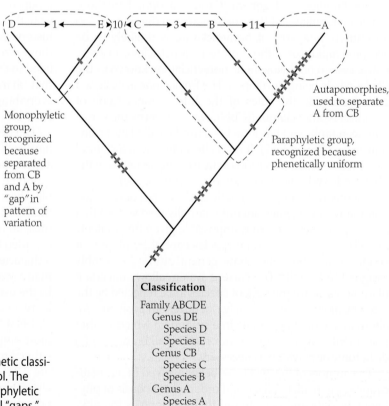

FIGURE 2.22 Phylogeny and a resulting nonphylogenetic classification produced according to the evolutionary school. The classification includes a mix of monophyletic and paraphyletic groups, separated from one another by morphological "gaps."

although evolutionary classifications usually referred to evolution, and the groups recognized in such classifications were often called monophyletic, the taxa were expected to be morphologically homogeneous and to be separated from one another by discrete gaps (Ashlock 1979; Stuessy 1983, 1990; Stevens 1986; Mayr and Ashlock 1991).

It has been said that systematics is as much an art as a science (although this statement begs the question of how one might define art and science), in part because so many aspects of the discipline seemed to have no objective basis. One fortunate result of phylogenetic systematics is that at least one major aspect of systematics—the delimitation of groups—has become formalized such that there is general agreement on how it should be done. Whereas phenetic and evolutionary classifications were ambiguous about grouping criteria, phylogenetic classifications are precise. A named group can be taken as monophyletic, including all descendants of a single common ancestor.

LITERATURE CITED AND SUGGESTED READINGS

Items marked with an asterisk are especially recommended to those readers who are interested in further information on the topics discussed in this chapter.

Abbot, L. A., F. A. Bisby and D. A. Rogers. 1985. *Taxonomic analysis in biology*. Columbia University Press, New York. [An introduction to various phenetic methods.]

Albert, V. A., A. Backlund, K. Bremer, M. W. Chase, J. R. Manhart, B. D. Mishler and K. C. Nixon. 1994. Functional constraints and *rbcL* evidence for land plant phylogeny. *Ann. Missouri Bot. Gard.* 81: 534–567.

Ashlock, P. D. 1979. An evolutionary systematist's view of classification. *Syst. Zool.* 28: 441–450. [Presentation of an easy-to-follow explicit method to construct an evolutionary taxonomic classification.]

*Backlund, A. and K. Bremer. 1998. To be or not to be: Principles of classification and monotypic plant families. *Taxon* 47: 391–400.

Bremer, K. 1983. Angiosperms and phylogenetic systematics. *Verh. Naturwiss. Verein Hamburg* 26: 343–354. [Discussion of reticulate evolution and cladistics.]

Bremer, K. 1988. The limits of amino acid sequence data in angiosperm phylogenetic reconstruction. *Evolution* 42: 795–803. [Decay indices.]

Bremer, K. and H.-E. Wanntorp. 1979. Hierarchy and reticulations in systematics. *Syst. Zool.* 28: 624–627.

*Brooks, D. R. and D. A. McLennan. 1991. *Phylogeny, ecology, and behavior*. University of Chicago Press, Chicago. [Excellent presentation of biological uses of phylogenetic hypotheses.]

Burleigh, J. G. and S. Mathews. 2004. Phylogenetic signal in nucleotide data from seed plants: Implications for resolving the seed plant tree of life. *Am. J. Bot.* 91: 1599–1613.

Cantino, P. D., H. N. Bryant, K. de Queiroz, M. J. Donoghue, T. Eriksson, D. M. Hillis and M. S. Y. Lee. 1999. Species names in phylogenetic nomenclature. *Syst. Biol.* 48: 790–807.

*Cracraft, J. and M. J. Donoghue. 2004. *Assembling the tree of life*. Oxford University Press, Oxford and New York.

Cronquist, A. 1987. A botanical critique of cladism. *Bot. Rev.* 53: 1–52.

*Dahlgren, R. and F. N. Rasmussen. 1983. Monocotyledon evolution: Characters and phylogenetic estimate. In *Evolutionary biol-

ogy*, vol. 16, M. K. Hecht, B. Wallace and G. T. Prance (eds.), 255–395. Plenum Press, New York. [Contains a simple introduction to cladistic methods.]

Davis, P. D. and V. H. Heywood. 1963. *Principles of angiosperm taxonomy*. Krieger, New York.

*de Queiroz, K. and J. Gauthier. 1990. Phylogeny as a central principle in taxonomy: Phylogenetic definitions of taxon names. *Syst. Zool.* 39: 307–322. [Proposal to abandon the Linnaean system.]

de Queiroz, K. and J. Gauthier. 1992. Phylogenetic taxonomy. *Annu. Rev. Ecol. Syst.* 23: 449–480.

de Queiroz, A., M. J. Donoghue and J. Kim. 1995. Separate versus combined analyses of phylogenetic evidence. *Annu. Rev. Ecol. Syst.* 26: 567–581.

*Donoghue, M. J. and P. D. Cantino. 1988. Paraphyly, ancestors, and the goals of taxonomy: A botanical defense of cladism. *Bot. Rev.* 54: 107–128.

Doyle, J. A., M. J. Donoghue and E. A. Zimmer. 1994. Integration of morphological and ribosomal RNA data on the origin of angiosperms. *Ann. Missouri Bot. Gard.* 81: 419–450.

*Eldredge, N. and J. Cracraft. 1980. *Phylogenetic patterns and the evolutionary process: Methods and theory in comparative biology*. Columbia University Press, New York.

*Farris, J. S. 1974. Formal definitions of paraphyly and polyphyly. *Syst. Zool.* 23: 548–554. [A polyphyletic group is defined as "a group in which the most recent common ancestor is assigned to some other group and not the group itself."]

Farris, J. S. 1979. The information content of the phylogenetic system. *Syst. Zool.* 28: 458–519.

Farris, J. S. 1989. *Hennig86*, Version 1.5. Port Jefferson Station, NY.

Felsenstein, J. 1978. Cases in which parsimony and compatibility methods will be positively misleading. *Syst. Zool.* 27: 401–410.

Felsenstein, J. 1981. Evolutionary trees from DNA sequences: A maximum likelihood approach. *J. Mol. Evol.* 17: 368–376.

Felsenstein, J. 1989. *PHYLIP 3.2 manual*. University of California Herbarium, Berkeley.

*Forey, P. L., C. J. Humphries, I. L. Kitching, R. W. Scotland, D. J. Siebert and D. M. Williams. 1992. *Cladistics: A practical course in systematics*. Oxford University Press, Oxford. [Summary of then-current cladistic methods.]

*Frohlich, M. W. 1987. Common-is-primitive: A partial validation by tree counting. *Syst. Bot.* 12: 217–237. [The principle that says that given an ingroup and an outgroup, both with states A and B of a homologous character, if state A is much more common than state B *in the outgroup*, then state A is likely to be ancestral *within the ingroup*.]

Funk, V. A. 1985. Phylogenetic patterns and hybridization. *Ann. Missouri Bot. Gard.* 72: 681–715.

Gift, N. and P. F. Stevens. 1997. Vagaries in the delimitation of character states in quantitative variation: An experimental study. *Syst. Biol.* 46: 112–125.

*Givnish, T. J. and K. J. Sytsma. 1997. Homoplasy in molecular vs. morphological data: The likelihood of correct phylogenetic inference. In *Molecular evolution and adaptive radiation*, T. J. Givnish and K. J. Sytsma (eds.), 55–101. Cambridge University Press, Cambridge.

Goloboff, P. A. 1993. *NONA*, Version 1.5.1. P. A. Goloboff, Tucumán, Argentina.

Hall, B. G. 2005. *Phylogenetic trees made easy*, 2nd ed. Sinauer Associates, Sunderland, MA.

*Hennig, W. 1966. *Phylogenetic systematics*. University of Illinois Press, Urbana.

Hibbett, D. and M. J. Donoghue. 1998. Integrating phylogenetic analysis and classification in fungi. *Mycologia* 90: 347–356.

Hillis, D. M., M. W. Allard and M. M. Miyamoto. 1993. Analysis of DNA sequence data: Phylogenetic inference. *Methods Enzymol.* 224: 456–487.

*Huelsenbeck, J. P. 1995. Performance of phylogenetic methods in simulation. *Syst. Biol.* 44: 17–48. [Comparison of performance in computer simulations of parsimony, maximum likelihood, and overall similarity methods of tree construction.]

*Huelsenbeck, J. P. and D. M. Hillis. 1993. Success of phylogenetic methods in the four-taxon case. *Syst. Biol.* 42: 247–264. [Several types of parsimony methods are relatively successful in reconstructing phylogeny.]

Humphries, C. J. and V. A. Funk. 1982. Cladistic methodology. In *Current concepts in plant taxonomy* (Systematics Association Special Volume, No. 25), V. H. Heywood and D. M. Moore (eds.), 323–362. Academic Press, London. [Introduction to basic cladistic methods.]

Humphries, C. J. and L. R. Parenti. 1986. *Cladistic biogeography*. Clarendon, Oxford.

Kellogg, E. A. 1989. Comments on genomic genera in the Triticeae. *Am. J. Bot.* 76: 796–805.

Kellogg, E. A., R. Appels and R. J. Mason-Gamer. 1996. When genes tell different stories: The diploid genera of Triticeae (Gramineae). *Syst. Bot.* 21: 321–347.

Kitching, I. J., P. L. Forey, C. J. Humphries and D. M. Williams. 1998. *Cladistics: The theory and practice of parsimony analysis*, 2nd ed. Oxford University Press, Oxford. [Summary of parsimony methods for constructing phylogenetic hypotheses, including issues dealing with characters.]

Kron, K. A. and W. S. Judd. 1997. Systematics of the *Lyonia* group (Andromedeae, Ericaceae) and the use of species as terminals in higher-level cladistic analyses. *Syst. Bot.* 22: 479–492.

*Maddison, D. R. and W. P. Maddison. 2000. *MacClade: Analysis of phylogeny and character evolution*, Version 4.0. Sinauer Associates, Sunderland, MA. [A useful program for exploring patterns of character change on a cladogram.]

Maddison, W. P., M. J. Donoghue and D. R. Maddison. 1984. Outgroup analysis and parsimony. *Syst. Zool.* 33: 83–103.

Mayr, E. and P. D. Ashlock. 1991. *Principles of systematic zoology*, 2nd ed. McGraw-Hill, New York.

*McDade, L. A. 1990. Hybrids and phylogenetic systematics. I. Patterns of character expression in hybrids and their implications for cladistic analyses. *Evolution* 44: 1685–1700.

*McDade, L. A. 1992. Hybrids and phylogenetic systematics. II. The impact of hybrids on cladistic analyses. *Evolution* 46: 1329–1346.

McDade, L. A. 1997. Hybrids and phylogenetic systematics. III. Comparison with distance methods. *Syst. Bot.* 22: 669–683.

Nelson, G. and N. Platnick. 1981. *Systematics and biogeography*. Columbia University Press, New York.

Pankhurst, R. J. 1991. *Practical taxonomic computing*. Cambridge University Press, Cambridge. [Use of computers in databases, identification, and phenetic and cladistic analyses.]

*Quicke, D. L. J. 1993. *Principles and techniques of contemporary taxonomy*. Blackwell, London.

Remane, A. 1952. *Die Grundlagen des naturlichen Systems, der vergleichenden Anatomie and der Phylogenetik* (The principles of the natural system, comparative anatomy, and phylogenetics). Geest & Portig, Leipzig.

Rydin, C., M. Källersjó and E. M. Friis. 2002. Seed plant relationships and the systematic position of Gnetales based on nuclear and chloroplast DNA. *Int. J. Plant Sci.* 163: 197–214.

*Sanderson, M. J. and M. J. Donoghue. 1989. Patterns of variation in levels of homoplasy. *Evolution* 43: 1781–1795.

Schuh, R. T. 2000. *Biological systematics: Principles and applications*. Cornell University Press, Ithaca, NY. [Detailed coverage of cladistic methods and application of cladistic results in classification, biogeography, ecological investigations, and conservation.]

Simpson, G. G. 1961. *Principles of animal taxonomy*. Columbia University Press, New York.

Sneath, P. H. A. and R. R. Sokal. 1973. *Numerical taxonomy*. Freeman, San Francisco.

Sokal, R. R. 1986. Phenetic taxonomy: Theory and methods. *Annu. Rev. Ecol. Syst.* 17: 423–442.

*Sokal, R. R. and F. J. Rohlf. 1980. An experiment in taxonomic judgment. *Syst. Bot.* 5: 341–365.

Stevens, P. F. 1980. Evolutionary polarity of character states. *Annu. Rev. Ecol. Syst.* 11: 333–358.

*Stevens, P. F. 1984. Homology and phylogeny: Morphology and systematics. *Syst. Bot.* 9: 395–409.

Stevens, P. F. 1986. Evolutionary classification in botany, 1960–1985. *J. Arnold Arbor.* 67: 313–339.

*Stevens, P. F. 1991. Character states, morphological variation, and phylogenetic analysis: A review. *Syst. Bot.* 16: 553–583.

Stevens, P. F. 1997. How to interpret botanical classifications: Suggestions from history. *BioScience* 47: 243–250.

Stevens, P. F. 1998. What kind of classification should the practising taxonomist use to be saved? In *Plant diversity in Malesia III: Proceedings of the 3rd International Flora Malesiana Symposium 1995*, J. Dransfield, M. J. E. Coode and D. A. Simpson (eds.), 295–319. Royal Botanical Gardens, Kew, London.

Stuessy, T. F. 1983. Phylogenetic trees in plant systematics. *Sida* 10: 1–13.

Stuessy, T. F. 1990. *Plant taxonomy*. Columbia University Press, New York.

*Swofford, D. L. 1993. *PAUP: Phylogenetic analysis using parsimony*, Version 3.1.1.

Distributed by the Illinois Natural History Survey, Champaign.

*Swofford, D. L. 2000. *PAUP*: Phylogenetic analysis using parsimony and other methods*, Version 4.0. [Beta-test edition, distributed by Sinauer Associates, Sunderland, MA.]

*Swofford, D. L., G. J. Olsen, P. J. Waddell and D. M. Hillis. 1996. Phylogenetic inference. In *Molecular systematics*, 2nd ed., D. M. Hillis, C. Moritz and B. K. Mable (eds.), 407–514. Sinauer Associates, Sunderland, MA. [Excellent summary of methods of tree construction.]

Sytsma, K. J. and J. C. Pires. 2001. Plant systematics in the next 50 years—Re-mapping the new frontier. *Taxon* 50: 713–732.

*Wagner, W. H., Jr. 1980. Origin and philosophy of the groundplan-divergence method of cladistics. *Syst. Bot.* 5: 173–193.

Wagner, W. H., Jr. 1983. Reticulistics: The recognition of hybrids and their role in cladistics and classification. In *Advances in cladistics: Proceedings of the second meeting of the Willi Hennig Society*, N. I. Platnick and V. A. Funk (eds.), 63–79. Columbia University Press, New York.

Watrous, L. E. and Q. D. Wheeler. 1981. The outgroup comparison method of character analysis. *Syst. Zool.* 30: 1–11.

Weins, J. J., ed. 2000. *Phylogenetic analysis of morphological data*. Smithsonian Institution Press, Washington, DC.

Wiley, E. O. 1979. An annotated Linnaean hierarchy, with comments on natural taxa and competing systems. *Syst. Zool.* 28: 308–337.

*Wiley, E. O. 1981. *Phylogenetics*. Wiley, New York. [Detailed discussion of cladistic principles.]

*Wiley, E. O., D. Siegel-Causey, D. R. Brooks and V. A. Funk. 1991. *The compleat cladist: A primer of phylogenetic procedures* (University of Kansas, Museum of Natural History, Special Publication, No. 19). University of Kansas, Museum of Natural History, Lawrence. [Summary of then-current cladistic methods.]

Wofford, B. E. 2006. A new species of *Stenanthium* (Melanthiaceae) from Tennessee, U.S.A. *Sida* 22: 447–459.

Zomlefer, W. B. and W. S. Judd. Resurrection of segregates of the polyphyletic genus *Zigadenus* s.l. (Liliales: Melanthiaceae) and resulting new combinations. *Novon* 12: 299–308.

Zomlefer, W. B., N. H. Williams, W. M. Whitten and W. S. Judd. 2001. Generic circumscription and relationships in the tribe Melanthieae (Liliales, Melanthiaceae), with emphasis on *Zigadenus*: Evidence from ITS and *trnL-F* sequence data. *Am. J. Bot.* 88: 1657–1669.

3

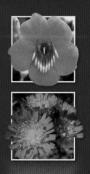

Classification and System in Flowering Plants: Historical Background

Throughout history, scientists have tried to determine the best way to classify living things. Their ideas on how to do this have changed considerably over time. In this chapter we discuss the principal ways that botanists have classified plants and some of the reasoning behind those classifications. The history of classification is only part of the history of systematics, however. In particular, the early collections on which so much of systematics is based are not neutral representations of nature, but are intimately connected with European colonial expansion, what price owners of private herbaria were prepared to pay for specimens, what kind of specimen they preferred, and the like. The complex relationships between professional and amateur botanists and the general public have also helped shape systematics. These relationships are an integral part of the historical background of our discipline, even if we know all too little about them (but see, for example, Allen 1976).

If you look at the phylogenetic trees in Chapters 7 through 9, you will see that it is possible to base classifications on them that capture precisely the clades in those phylogenies (see Chapter 2). Both phylogenies and classifications are hierarchical and are made up of groups nested within groups. However, some classifications in use today—in particular, evolutionary classifications—do not try to represent phylogenies in this way. Indeed, they are not strictly hierarchical. What they are trying to represent can be understood only in the context of a long history stretching back before anyone had any idea about evolution. Understanding classifications thus means that we need to understand their history.

In the past, makers of classifications had quite different ideas about nature and about the role of classification than they do today, yet we tend to assume that our ideas and theirs are the same. The problem is made worse because many terms have changed meaning over time. The term *system* is a prime example: it now refers to sets of relationships in genealogies (de Queiroz 1988), but in the late eighteenth century, *system* was used as a term to belittle classifications based on a single character—except in England, where it referred to what European botanists would have called *method*, classifications based on many characters! In addition, plant systematists, perhaps even more than other systematists, have long distrusted theory and have considered classification a theory-free, "empirical" operation (see Stevens 1986, 1990, 1994, 1998a, and Kornet 1991 for discussion). Theories, it was thought, should not affect a systematist's observations or classifications. For this reason, plant systematists have often been unwilling or unable to explain the reasons for their decisions regarding classification.

This chapter will describe some of the history of botanical classification in order to show how earlier, often non-phylogenetic, ideas about nature were incorporated into current classifications. First we will discuss the long-standing and continuing tension between the makers of classifications, most of whom want to understand relationships (although note that the term *relationship* has meant different things to different people), and many users of classifications, who simply want names to be stable. Then we will discuss how relationships are understood, how nature is visualized, and how higher taxa are delimited. We will also place the reasons for the changing circumscriptions of some of the major groups in their historical context. (Here we discuss only higher taxa—genera and above; for a discussion of species concepts, see Chapter 6 and Stevens 1992, 1997b.)

Classification, Nature, and Stability

For hundreds of years, botanists tried to develop classifications that were "natural." Until recently we have assumed that the history of systematics was in part the history of a single "natural" system that had been gradually developing over the centuries. Its principles were first outlined by Caesalpinus. Tournefort and Linnaeus described "natural" genera, and Linnaeus suggested a number of "natural" families, although he did not describe them. The "natural" method then received a major boost from A.-L. de Jussieu in his great *Genera plantarum* of 1789, in which he described both genera and families and placed the latter in classes. This Jussieuan foundation is the basis of our current classification system, and although new families have been added, the limits of existing families have been modified, and higher taxa such as orders have been added, nothing fundamental has changed.

Unfortunately, the word *natural* has no fixed meaning; rather, authors have used it to mean something that agrees with their own ideas about nature or about constructing classifications or systems. Eighteenth-century systematists had ideas about nature that were very different from ours—

they were certainly not evolutionary—and their systematic practice and classifications are best interpreted in terms of how they understood nature. Nineteenth-century systematists built on the work of their predecessors. Although they generally did not describe clearly their understanding of nature, which was changing, the way they discussed and depicted relationships was not much different from the practice of the preceding century. And many aspects of nineteenth-century classification practice persisted through the twentieth century.

Some historians of classification see a trend from analytic ways of grouping that were particularly common in the eighteenth century toward more synthetic procedures in the nineteenth century. In analytic grouping procedures, one or a few characters are used successively to define groups, so that organisms are divided up into smaller and smaller groups. The whole process is rather like using a key. In synthetic grouping procedures, many characters are used and groups are built up ("synthesized") (Mayr 1982).

The distinction between the two kinds of procedures is not always clear. Even in the twentieth century some botanists used single-character, analytic (divisive) systems (such as John Hutchinson, who divided dicotyledons into woody and herbaceous groups, a classification that even Linnaeus had dismissed as "lubricious"; see Hutchinson 1973). On the other hand, Jussieu's method for recognizing relationships, developed in the later part of the eighteenth century, is synthetic.

Classifications have been expected to do much more than reflect nature, however. They have also been expected to be (1) easy to use, (2) stable, (3) an aid to memory, (4) predictive, and (5) concise—a set of goals that are sometimes in conflict. These goals of classification were spelled out by Andreas Caesalpinus in 1583 (see Greene 1983, vol. 2: 815–817). Thus the systematist must not only describe nature (whatever he or she thinks nature is), but

ANDREAS CAESALPINUS
(1519–1603, Italian)

also must serve a community of users, many of whom will have no interest in the systematist's personal ideas about nature. Before the twentieth century, many of those users were medical personnel, but there was also a large group, that we can loosely call "amateurs," a group that included many women. By the end of the eighteenth century, *botany* had come to mean classification studies, while *philosophical botany* included what we now know as physiology and related fields.

During the nineteenth century, professional botanists were people who made classifications, enumerated the natural products of a country, and so forth; they were men. Amateur botanists, on the other hand, largely identified plants, for a surprisingly long time using the Linnaean system (see below). Many of these amateurs were women, and botany was a popular subject to teach to children. At the end of that century, as a laboratory-based "new" botany became popular in universities, the old or classificatory botany was effectively trivialized by portraying it as arbitrary classification, mere names, and more a pastime for women and children than a serious science (e.g., Coulter 1895). Systematic botany has since become rehabilitated, and its users now include a broad array of biologists; however, there is still a colloquial usage of the word *botany* that reflects these nineteenth-century tensions.

Stability of taxonomic names has been a perennial problem in systematic botany. Systematists have often preferred to leave the names of taxa unchanged—even if those names conflicted with their views of relationships among different species and genera—lest the users of their classifications become upset (Stevens 1994, Chapter 10; 1997a). Even George Bentham and J. D. Hooker, authors of the great three-volume *Genera plantarum* (1862–1883), circumscribed some taxa to reflect custom and convention; that is, they constructed some taxa that were not natural, even by their own definition of the word. Bentham, at least, ignored some of his own taxon circumscriptions when it came to discussing distributional patterns (Stevens 1997a). In this book, we have attempted to construct a classification that better reflects relationships, and it is highly likely that some readers will wish we had not changed so much. This point of view is, if nothing else, traditional! However, even if relationships are stable, it does not follow (unfortunately) that the Linnaean names we use will also be stable; we will discuss the reason for this later.

J. S. L. Gilmour (1940) promoted the idea that the best classification has maximum general utility. This idea, however, leaves open the definition of *utility*. The needs of the different groups of people who use classifications may change over the years, and differing needs may conflict (Stevens 1998b); the result may be classifications that conflict. Certainly Gilmour's ideas sharpened the conflict between those who wanted classifications to reflect evolutionary history and those who were not at all interested in phylogenetic relationships.

As early as 1778, Lamarck had suggested one solution to the problem: the characters used in the formal classification did not have to be those used for identification. The keys (see Appendix 2) that he promoted link users and experts. Easily visible characters could be used in keys, and these would not necessarily be the same as the sometimes inconspicuous characters used to distinguish the groups recognized in the formal classification. Keys made it easy to give plants their correct names, so groups in classifications did not—and do not—have to be easily recognizable.

A general respect for authority has also affected classifications. Some plant groups have been recognized for a long time, for example, Labiatae, Liliaceae, Cruciferae, and Compositae. Many of these names do not end in the conventional *-aceae*, which indicates that they are not based on particular genera, and these groups may even pre-date

"scientific" classification. The fact that such groups have always been recognized is sometimes used as evidence that they are "natural" groups. If they have been historically recognized by the acknowledged masters of the discipline, so the argument goes, then they must be correctly delimited. Systematists have been generally reluctant to modify such groups, although in our classification, Labiatae, and particularly Liliaceae, are circumscribed very differently from the way they were only 15 years ago.

These paradigmatic groups are generally ones that are obvious in the European flora, a fact that reflects the European origin of botanical systematics. Not only did the discipline originate in Europe, but it was dominated by Europeans for centuries. Not until Asa Gray (1810–1888) was there a North American botanist considered by Europeans to be fully their equal, and only with Charles Bessey (1845–1915) did North American botany become fully independent (Dupree 1959; Cuerrier et al. 1996). In other parts of the world, especially those that were then colonies of European nations, European domination persisted longer, and it was especially evident in the floras of countries that were written by European botanists and based on material held in European institutions.

ASA GRAY
(1810–1888, American)

Understanding Relationships

We mentioned in the preceding section that eighteenth- and nineteenth-century systematists saw nature quite differently from the way we do now. How can we know what they were thinking? The analogies they used when describing patterns of relationships they saw in nature and the diagrams they prepared to show these patterns are particularly good sources of information.

Although it is obvious to us that relationships can often be represented as treelike diagrams, the reason is that we share a common set of assumptions about how organisms came to be the way they are. Thus, throughout this book, we diagram relationships somewhat like pedigrees, with extant organisms linked by extinct ancestors (this kind of diagram is known as a **Steiner tree**).

Many of our predecessors did not share this view of nature. Picture yourself in the world before Darwin and before ideas of evolution. How would you imagine plants are related to one another? What would the word *relationship* mean? In fact, we find that many eighteenth- and nineteenth-century botanists thought of relationships as being like the relationships between countries on a map or stars in a constellation; that is, they were reticulating and even multidimensional.

**ANTOINE-LAURENT
DE JUSSIEU**
(1748–1836, French)

Antoine-Laurent de Jussieu described many of the families whose evolution we now attempt to study. His genera and families have been interpreted for two centuries as though they were more or less distinct groups, but this is not how he saw them. For Jussieu, relationships in nature formed continuous series that lacked any clear breaking points. Any divisions in these series were the work of man, not of nature. Jussieu emphasized that groups were linked, and his natural families, such as the Compositae, were natural for him precisely because they were examples of this continuous nature; not surprisingly, genera in such families were difficult to recognize.

For Jussieu, as for his colleague Lamarck, the taxonomic hierarchy was simply a set of words, each of which individually referred to a part of the continuum and together allowed the whole to be recalled to mind. Complications arose from Jussieu's descriptions of the groups he recognized. These groups were rarely wholly characterized by the features he listed for them, and a family description often referred only to characters of genera in the middle of the sequence in which he placed them (i.e., in the middle of the continuum).

A particularly interesting diagram from about this time is P. D. Giseke's "genealogical-geographical" map of 1792 (Figure 3.1). In this diagram, circles of various sizes, representing families, are placed at varying distances from one another. Giseke took pains to say that the relationships he showed were not those between "grandfather and grandson," but rather between "cousins or relatives by marriage." He noted whether or not there were intermediates between the families shown in the diagram, and he carefully distinguished between different kinds of relationships as he described the complex, two-dimensional spatial relationships between groups.

Much later, Bentham and Hooker's *Genera plantarum* (1862–1883) reflected the principles first outlined by Bentham in 1857. Both Bentham and Hooker thought that groups showed reticulating relationships and that their boundaries were sometimes, or even often, indistinct. All in all, their ideas were not too different from those of Giseke.

Through much of the nineteenth century, and even much of the twentieth, botanical relationships have been portrayed as highly complex and reticulating. Even when tree diagrams were used to show evolutionary relation-

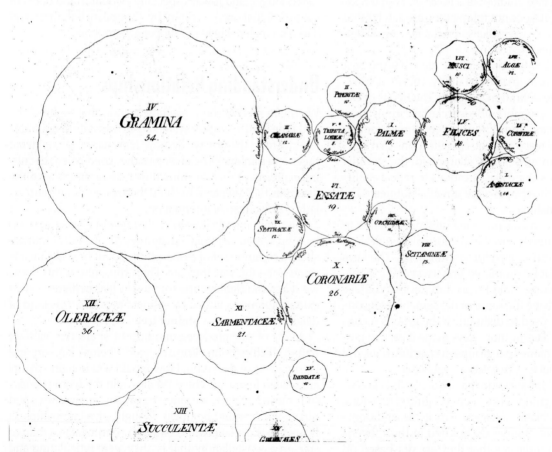

FIGURE 3.1   A portion of P. D. Giseke's "genealogical-geographical" map (1792).

ships, extant groups were linked directly to other extant groups. (These diagrams are known as **minimum spanning trees**.) Such diagrams imply that groups that exist in the world today are the ancestors of other groups that also currently exist, which doesn't make much sense in terms of evolutionary processes.

A more evolutionary view would be to say that two extant groups are descended from a single extinct ancestor. However, extant groups were linked directly in part because many botanists, from at least as early as Johann Georg Forster in 1786, have disliked talking about ancestors, whether because ancestor-descendant relationships could not be seen directly or, later, because the fossil record was simply too poor to detect them. There are a few examples of early trees with extinct ancestors, but such depictions are not common (for illustrations, see Lam 1936 and Voss 1952).

CHARLES EDWIN BESSEY
(1845–1915, American)

In addition to showing that groups were linked simultaneously to many other groups, the goal of these complex diagrams was often to indicate the relative "highness" or "lowness" of groups. This aim can be seen in the work of Charles Edwin Bessey, a major figure in North American botany at the end of the nineteenth century. He produced numerous diagrams showing relationships, and in the latter part of his career, these diagrams showed extant groups joined directly (Figure 3.2) (Cuerrier et al. 1996).

Bessey drew his diagrams to show major trends in advancement (and sometimes also reversals), and his classifications are to be read as sequences that in part reflect these diagrams. It is interesting that Bessey, in attempting to make systematics more philosophical, specifically dismissed the more usual maplike representations of nature. In fact, his freestanding "trees" are more like archipelagos or maps, even if they have been given an axis. They are conceptually similar to the representations of nature he dismissed, being quite like those of Linnaeus and Jussieu. Furthermore, although Bessey repeatedly emphasized that classifications should reflect phylogeny, the way he produced classifications made that goal very difficult to achieve.

ARTHUR CRONQUIST
(1919–1992, American)

Other major classification systems, such as that of Arthur Cronquist (1919–1992), are also minimum spanning trees that sometimes even allow reticulations between groups (Cronquist

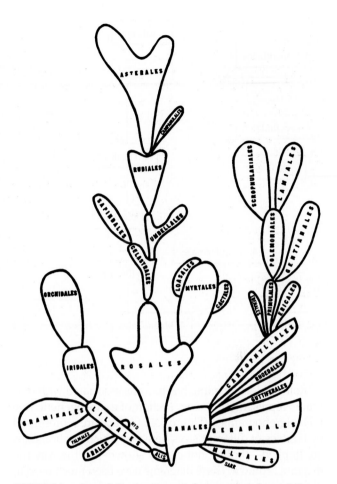

FIGURE 3.2 One of Charles Bessey's freestanding "trees." (From Bessey 1915.)

1981; Figure 3.3). They are certainly not readily interpretable in phylogenetic terms. An unwillingness to specify only historical connections between groups is associated with a tendency to emphasize parallel evolution, parallel tendencies, or even ideas of **orthogenesis** (directed evolution), as is particularly evident in Cronquist's earlier work. If two groups are not related directly, the argument goes, then the occurrence of the same characters in these groups must be explained by independent evolution.

H. F. Wernham, in an influential series of papers (1911–1912), asserted that the Sympetalae were polyphyletic—as were monocots, dicots, and even angiosperms as a whole. He felt that all important characters had evolved in parallel several times in closely related but independent lineages. In fact, suggestions that there were large-scale parallelisms in patterns of relationships are quite common from the late eighteenth century onward, and some nineteenth-century naturalists even claimed to see quite close parallelisms between the series into which animals and plants could be placed. The existence of such parallelisms was taken as evidence that the "real" pattern of relationships in nature had been discovered.

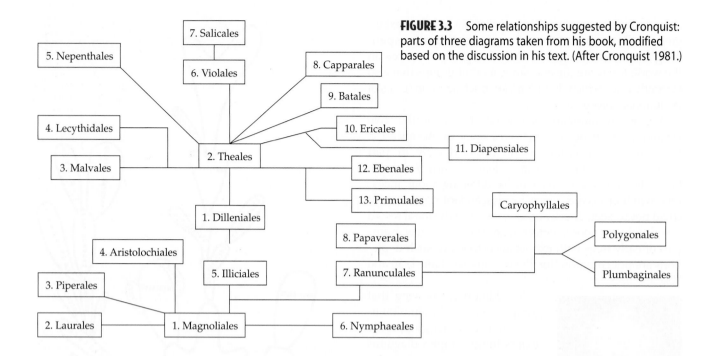

FIGURE 3.3 Some relationships suggested by Cronquist: parts of three diagrams taken from his book, modified based on the discussion in his text. (After Cronquist 1981.)

Rolf Dahlgren's name has become associated with diagrams representing a cross section of a phylogenetic tree ("Dahlgrenograms") (Figure 3.4). Groups are represented by bubbles of different sizes between which relationships are implied—though not clearly shown—by the way the diagram is drawn. These diagrams have been much used to display the results of broad surveys of variation in characters, such as the distribution of iridoids or the types of plastids in sieve tubes. Conceptually, Dahlgrenograms seem more closely related to Bessey's cactuslike diagram (see Figure 3.2) or Giseke's "genealogical-geographical" map (see Figure 3.1) than to the phylogenies used in this book. Indeed, for some authors, the fact that such diagrams did not have evolutionary implications was a virtue. They permitted one to think about general relationships without worrying about evolution (Heywood 1978). Note that in this context, again, the term *relationships* cannot mean evolutionary or, still less, phylogenetic relationships. However, as we will see, by the 1970s change was in the air, and Dahlgren himself was much interested in phylogeny.

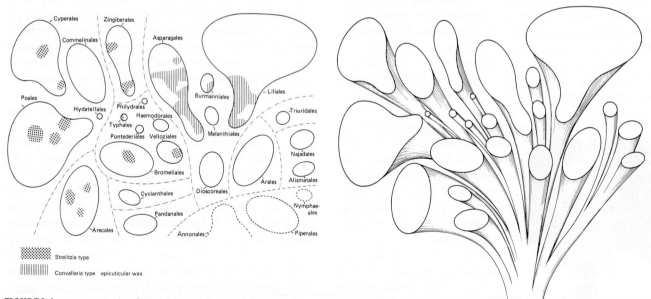

FIGURE 3.4 An example of a "Dahlgrenogram," a diagram representing a cross section of a phylogenetic tree. The diagram on the right is a three-dimensional representation of the diagram on the left. (From Dahlgren et al. 1985.)

Classifications and Memory

The use of classification as an aid to memory was critical in a time before computers existed and when even books were not common. A classification had to have a moderate number of families, and these families had to be divided into subgroups that were neither too small nor too large. Linnaeus emphasized the value of a system, grouping organisms in tens; this approach allowed him to place all of the fewer than 10,000 plant species that he thought existed into groups at just four hierarchical levels.

Jussieu, whose classification was imposed on what he saw as continuous natural variation, recognized only taxa that he thought were neither too small (there had to be at least 2 included members) nor too big (100 members may have been the uppermost limit). Thus he recognized no monogeneric families, and he divided the Compositae—which he thought were very "natural" but had well over 100 genera—into three families.

Similarly, Bentham and Hooker and some of their colleagues, including Asa Gray, had agreed before the monumental *Genera plantarum* was written that 200 was the upper limit of families to be recognized; otherwise there would be too many to memorize (201 was the final figure). Bentham and Hooker also agreed that the best size for taxa was 2 to 6—rarely, up to 12—included members. Yet some of the families they accepted had hundreds of genera.

Bentham and Hooker reconciled their intention of having a fixed, low number of often quite large families and the need to have small taxa at all hierarchical levels by interpolating formal or informal groupings wherever needed. As a result, all groups in their *Genera* above the level of genus have fewer than 14 included members. Such small groups are best suited for storage in and recall from memory, and the emphasis on recognizing large genera and families also minimized the number of names used in general and reduced that burden on memory (Stevens 1997a, 2002).

The Formation of Higher Taxa

The idea that almost all plants belong in genera with two or more species was suggested by Conrad Gessner around the middle of the sixteenth century (Morton 1981). However, similar groupings of plants (and animals) are evident in the classifications used by local peoples worldwide (Atran 1990; Berlin 1992). The recognition of such groupings is based on their salience, or obviousness and importance to the observer. This idea of salience is based on things such as the degree of similarity among members of a group, their commonness, and their utility for humans. The basic units in such classifications often are named with binomials, or, more generally, with a noun + adjective construction.

In herbals and other early botanical literature, plants are grouped in various ways (alphabetical arrangements are common), although why these groupings were recognized is often unclear. In 1694, Joseph Pitton de Tournefort provided clear guidelines for describing genera (see Dughi 1957). Generic characters should be recognizable in all members of the genus, he argued, and should be visible without the use of a microscope. When possible, these characters should be taken from features of the flower and fruit. Tournefort called groups based on these features primary genera. However, if

JOSEPH PITTON DE TOURNEFORT
(1656–1708, French)

these genera were too big, features from other parts of the plant could be used to characterize smaller genera (see also Walters 1986 and references therein). Tournefort called groups characterized by nonreproductive features secondary genera. He also suggested that it was important to keep the total number of genera to about 600 (Stevens 1998a). This number is in line with the basic units in folk taxonomies worldwide (see Berlin 1992). Tournefort's classification indeed has much in common with folk taxonomies.

Carolus Linnaeus focused on genera, and his descriptions were much more detailed than Tournefort's. Linnaeus believed that individual genera and species existed in nature, and that the ranks of genus and species were distinct ranks in the organization of nature, whereas larger groupings were matters of human convenience. He emphasized that features of the flower and fruit should be used to distinguish genera, and hence he combined most of Tournefort's secondary genera with the primary genera. He used vegetative features to distinguish species. But, as Linnaeus observed, *"Characterem non constituere Genus, sed Genus Characterem"* (Linnaeus 1751: 119)—loosely interpreted to mean that genera exist in nature independently of the features used to characterize them, certainly independently of any rigid application of "generic" characters.

CAROLUS LINNAEUS
(1707–1778, Swedish)

This and similar dicta, when coupled with the way Linnaeus went about recognizing and describing genera (for instance, he did not always change generic descriptions as he added new species to them), make his actual practice seem almost unprincipled at times (Stafleu 1971; Stevens 2002). Like Tournefort, Linnaeus considered large groups unwieldy, and he preferred small taxa at all hierarchical levels. Indeed, it has recently become common to think of Linnaeus as being some kind of essentialist, a person who believed that taxa have certain kinds of fixed characters without which they cannot be formally recognized (again, a loose interpretation). However, this is at best a great oversimplification (Winsor 2001; Stevens 2002).

Today's genera are ultimately built on a Linnaean foundation. Although many taxonomists, at least in theory, tolerate genera distinguished by characters other than those of flowers and fruits, in practice reproductive characters have served as a major source of generic-level differences. Nevertheless, many important nineteenth-century taxonomists did not believe that there was a rank of genus (or family, for example) in nature, and by the middle of the century this was true of the rank of species too (Stevens 2002).

By the 1870s it even seemed that most genera were known, but this state of affairs did not last long. Genera were seen as groups of species separated by morphological gaps of adequate size (see below). Hall and Clements (1923: 6) called for "experimental and statistical studies of the generic criteria in use" in a paper whose title indicated that they wanted to clarify how systematists detected phylogenies, but their aim was really to maintain the status quo, the conventional (broad) delimitation of genera (and species) because of "the significance of system, and of the mechanism of memory" (Hall and Clements 1923: 7). Moreover, they provided no new way of detecting phylogenies.

The idea that there were families of plants was specifically suggested by Pierre Magnol in 1689 (see Adanson 1763–1764, vol. 1: xxii–xxvii). Magnol used characters taken from all parts of the plant, or sometimes an "affinité sensible" that could not be expressed in words. He did not recognize all the families he might have because he wanted to keep their number small. He listed 76 families.

Linnaeus described classes and orders (= families) in his sexual system. Plants were assigned to these groups primarily on the basis of stamen number and arrangement and secondarily on the basis of ovary number (more exactly, on the number of stigmas or styles). For example, *Datura* and *Verbascum*, both of which have two carpels but only a single style, were placed in the Pentandria Monogyna.

Linnaeus also outlined a natural method, the goal of botany, in which genera were grouped into natural families (he recognized 67 in the year 1751, with a substantial residue of unplaced genera). He stressed the need for characters unique to a family and found in all its members, since without such characters the natural method would be like a bell without a clapper, as he graphically described the problem. However, Linnaeus was unable to provide such characters for even the most natural of families, such as the Umbelliferae (Apiaceae).

Although his largely artificial sexual system and his natural genera both relied almost exclusively on characters taken from the flower and fruit, when it came to natural families, Linnaeus (1751: 117) observed, *"Habitus occulte consulendus est"*—"the habit should secretly be consulted." *Habit*, for Linnaeus, comprised all parts of the plant other than flowers and fruit, including features such as leaf characteristics, and

could also be used to distinguish families. He regretfully noted that his natural method was incomplete because plants showed relationships in several directions, like territories on a map; because some plants were not yet discovered; and because the habits of plants were poorly known (Linnaeus 1751: 26–36, 137).

The need, then, was to find features that indicated higher-level relationships. Between 1763 and 1789, three authors outlined the issue in ways that defined the debate over the ensuing two centuries. In a series of tabulations by Michel Adanson (Adanson 1763–1764) showed that *every plant characteristic varied within natural groups*; Adanson thus concluded that no one character was essential for defining a group, and that groups could be defined only by combinations of characters. A classification could be produced only by an exhaustive comparison of all parts and properties of plants.

MICHEL ADANSON
(1727–1806, French)

Adanson did not state clearly how such a classification was to be carried out, but his contemporaries, such as Marie Jean Antoine Nicolas Caritat de Condorcet (1743–1793), noted that this might entail mechanization in the recording of characters (no easy task two centuries before the personal computer!). Other naturalists, presumably not wanting to have to look at every character in every plant, needed clear guidelines for deciding whether some characters were more important than others. In 1778 Jean-Baptiste de Lamarck came up with a numerical weighting scheme (the first one known in botany) that assigned similarity values to features that depended on how widely they were distributed in

JEAN-BAPTISTE-PIERRE-ANTOINE DE MONET DE LAMARCK
(1744–1829, French)

plants (e.g., how common a calyx was), although he took into account not simply presence or absence, but also the nature of the feature.

Jussieu (1789) built up "groups" by synthesis, successively forming species, genera, and families; ideas of general similarity seem to have guided this synthesis. He then showed how different features characterized groups of successively smaller circumscription. He described these features as if they were invariable at the level they characterized, and he strongly disagreed with Adanson's contention that there were no invariable, essential characters.

Augustin-Pyramus de Candolle saw a proper subordination of characters as being the third and final stage in

AUGUSTIN-PYRAMUS DE CANDOLLE
(1778–1841, Swiss)

detecting relationships, following the stages of "blind groping" and general comparison (Candolle 1813). This subordination of characters was similar to the way in which Jussieu had described the distributions of characters. (Candolle tended to give the same character equal weight, at least in related taxa.)

Note that Jussieu's emphasis on synthesis was compatible with his belief that there were no groups sharply separated from other such groups in nature; gradual synthesis produced the continuity in relationships that was a feature of continuous nature. Indeed, Jussieu confidently expected plants yet to be discovered to be intermediates that filled in the apparent gaps between groups. Candolle, on the other hand, tended to emphasize analysis. He asserted that there were distinct groups in nature, noting that botanical discoveries were not filling in the morphological gaps between groups, and he looked for features that characterized those groups.

During the nineteenth and twentieth centuries, the fundamental differences between Jussieu's and Candolle's understanding of nature were almost never discussed, and no accepted rationale for weighting characters was developed. Whether or how to weight remained a bone of contention for the next century and a half. Arguments between self-styled Jussieuans and Adansonians over weighting in the latter part of the nineteenth century centered more on what characters should be used, and how, and less on whether or not to use all characters.

A number of systematists, especially those in France and Germany, adopted concepts of **types**. These might be the common form in a group, or they might represent a "perfect" flower—a radially symmetrical form such as a *peloria* mutation in a bilaterally symmetrical group, or a bisexual flower in a monoecious or dioecious group. [The original *peloria* mutation was described by Linnaeus and so excited him that he initially wanted to name it as a new genus. His proposed "*Peloria*" in fact was a mutant of *Linaria* (Plantaginaceae) with five spurs radiating from the center of the flower instead of the normal single spur in the abaxial position.]

Such types could provide a way of understanding the diversity of form in a group and of relating one group to another. They were also in some ways an alternative to conventional weighting schemes. But typological thought, although widespread, never became systematized. Not only did the word *type* reflect a variety of very different ideas, but typological thought in general was equated by some (mostly English-speaking individuals) with speculation. Ideas of essences (i.e., essential characters, as described earlier) and types are often linked.

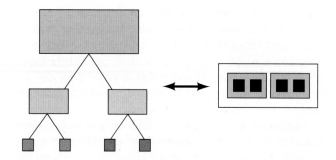

FIGURE 3.5 Although Charles Darwin's work did not provide any insights on how to rank the various taxa, he understood that some groups were subordinate to others.

Acceptance of evolution did not inspire any new ways of detecting relationships. For Charles Darwin (1809–1882), a perfect genealogy could be established only by fossils; all taxa were linked in a series in which individuals differed from one another only imperceptibly. Paradoxically, if the fossil record were perfect, classification would become impossible because classification, for Darwin (1859), depended on differences. Furthermore, Darwin provided no indication of how to rank taxa; indeed, he emphasized that rank was arbitrary. He did, however, describe relationships in terms of groups subordinate to other groups (Figure 3.5). Systematists such as George Bentham understood this to mean—when coupled with the idea of evolution—that the only difference between taxa occupying the highest and the lowest ranks of the hierarchy was one of degree (Bentham 1875). Taxa might be distinct, yet ranks were not fundamentally different; neither ranks nor individual taxa had essences. These ideas only compounded the problem of deciding at what rank to recognize a particular group, unless reference was made to previous taxonomic practice—that is, an appeal was made to established practice or convention.

GEORGE BENTHAM
(1800–1884, British)

Up until the middle of the twentieth century, systematists continued to delimit groups very much as they had at the beginning of the nineteenth century, although of course they knew much more about the basic morphology and anatomy of plants. Indeed, much descriptive work done between 1840 and 1920 remains invaluable; this work includes surveys of floral development by Jean-Baptiste Payer and of root anatomy by Philippe van Tieghem and Henri Douliot. However, although similarity in general morphology and anatomy had come to indicate closeness in evolutionary or phylogenetic relationships, there was still no way of deciding which particular characters indicated such relationships and which did not. Systematists sometimes tried to distinguish between characters that were adaptive, and therefore less valuable in assessing relationships, and those that were not adaptive and thus more valuable. Groups were circumscribed by morphological gaps, but there was no agreement as to when a gap was large enough for a group to be recognized at a particular rank.

In general, gap size has tended to be inversely proportional to the size of the groups involved (Davis and Heywood 1963). However, for about 170 years, it has been recognized that the application of criteria to evaluate closeness of relationship and to rank taxa has been inconsistent. So, for example, families in the Malvales have often been considered equivalent to tribes in the Rosaceae. It is not surprising that arguments over how broadly or narrowly taxa should be circumscribed have remained unresolved.

John Gilmour (1940; see also Winsor 1995, 2000) reopened discussion about the general issue of how to group organisms when he observed that the use of characters to establish evolutionary relationships tended to be circular: characters important in establishing evolutionary relationships were those important in delimiting groups, and vice versa. He suggested that groups in natural classifications were simply those that had many characters in common. Such groups had the property of being useful for a wide range of purposes; they were general-purpose classifications. In contrast, evolutionary classifications, by their very definition, would always be special-purpose classifications and so not of general use or interest. Gilmour believed that attributes (characters) of plants were sensory data, which, he thought, were facts. Classifications were "clips" that held these data together, so if different principles of classification were used, the clips—and so the groups recognized—would be different.

JOHN SCOTT LENNOX GILMOUR
(1906–1986, British)

What became known as phenetics or numerical taxonomy owes much to Gilmour's ideas. Pheneticists produced groups on the basis of overall similarity in hopes that this process would produce an objective, stable, and repeatable classification (Sokal and Sneath 1963; for more on this topic, see Chapter 2 and Vernon 1988). (Interestingly, Gilmour himself was not sympathetic to the use of computers that his approach encouraged.) The assumption that characters were observable facts soon proved to be a considerable oversimplification; what appeared to be basic characters could be subdivided. Furthermore, different numerical algorithms produced different phenograms (see Chapter 2), and hence could be the basis of different classifications, but it was often unclear why one algorithm should be preferred over another.

Phenetic theory and, in particular, practice had little effect on higher-level systematics in North America. It had rather more influence in England, where botanists were perhaps particularly distrustful of what they saw as being evolutionary speculation (Vernon 1993; Winsor 1995); good systematists, it was claimed, always had been more or less Gilmourian.

It has generally been conceded that genera, and particularly families, are less "natural" than species. What such comparisons might mean is unclear, however, because systematists have not been clear as to what is meant by *natural*. Is it that the rank of genus (for example) is a rank in nature, or that genera are discrete groups, or that members of a genus are more closely related to one another than they are to members of another genus?

In any event, there have always been dissenting voices. Linnaeus claimed that genera and species were equally natural. A.-L. de Candolle thought genera were more natural than species because genera were named as such by the common man (a similar idea was expressed by H. H. Bartlett). In a much-cited paper, Edgar Anderson (1940) reported on a survey he carried out to find out whether systematists thought genera or species were more natural. Some systematists who monographed groups, at least, were inclined to think that genera were more natural than species, despite the often-expressed views to the contrary. So were biogeographers such as Ronald Good, and it should not be forgotten that the genus is the basic unit of much of the biogeographic work that has focused on global patterns of diversity and relationships.

Early in the nineteenth century, Charles-François Brisseau de Mirbel (1776–1854) suggested that there were two main kinds of families (and genera). Whereas *familles en groupes* ("families in groups") were very natural and clearly circumscribed, *familles par enchaînement* ("families by chaining") were less natural and had less clear limits. All genera in the first kind of family tended to be united by one or more characters, but they were often difficult to distinguish from one another. In the second kind of family, individual genera were linked individually, forming a chain of similarities; these genera were often easily distinguished, even if the family was not. Similar distinctions, as between definable and indefinable families, have persisted (Davis and Heywood 1963: 107). It is perhaps ironic that some indefinable families, such as Rosaceae and Ranunculaceae, have turned out to be largely monophyletic, while definable families such as Lamiaceae or Liliaceae, as traditionally circumscribed, are strongly paraphyletic or polyphyletic.

Today most systematists realize that they need phylogenies. Before his untimely death in a car accident in 1989, Dahlgren had begun to elaborate relationships following more thoroughly phylogenetic principles. Such principles—especially the use of synapomorphies to diagnose monophyletic groups—were conceptualized by Willi Hennig (Hennig 1950, 1966) and Warren H. "Herb" Wagner (Wagner 1969, 1980); they are outlined in Chapter 2. These principles provide

EMIL HANS WILLI HENNIG
(1913–1976, German)

criteria for deciding which particular features indicate phylogenetic relationships.

For almost a quarter of a century, students of Wagner and others produced trees using Wagner parsimony, although most such studies involved only a few taxa. In the late 1970s, papers by Bremer, Wanntorp, and others popularized morphological Hennigian studies. A decade later, papers by Bremer (1987) and Jansen and Palmer (1987) suggested that both molecular and morphological data divided Asteraceae in a novel and exciting way. It was to be the combination of massive amounts of molecular and morphological data with methods of data analysis inspired ultimately by both Hennig and Sokal and Sneath that would transform systematists' ideas of higher-level relationships in land plants and, indeed, how systematists work (Stevens 2000a).

The style of systematists' work has also changed. In phylogenetic studies, at least, collaboration is the order of the day. No longer are systems "owned," as it were, by individuals, as phrases such as "Engler's system" or "Cronquist's system" might suggest. Collaboration is common in producing and analyzing the data as well as in suggesting possible classificatory interpretations (Angiosperm Phylogeny Group 1998, 2003; Grass Phylogeny Working Group 2001; see also Endersby 2001).

The arguments in systematics now focus on the use of particular statistical methods of evaluating the support for hypotheses of phylogenetic relationships and on the proper use of evolutionary models, as in maximum likelihood and Bayesian methods. Perhaps surprisingly, little progress has been made in understanding the relationship between morphological observations and hypotheses of phylogeny (Stevens 2000b). Aside from strictly paleobotanical studies, analyses of morphological characters alone are rather uncommon, yet the question as to whether the pattern of relationships among extant organisms can be understood without the incorporation of data from fossils remains contentious.

There is still disagreement over the relationship between phylogenies and the classifications that are based on them. Some systematists think that an important element shaping classifications should be how different one group looks from another; others think that classifications should be based primarily on phylogeny—that is, that all taxa should be monophyletic, whatever additional criteria are used to circumscribe them. The latter approach is the one taken in this book (see Chapter 2). This division is independent of the arguments for and against the PhyloCode, in which history has been invoked by all sides. Not only is there little evidence for some of the particular historical arguments advanced, however, but history is largely irrelevant to the discussion (Stevens 2006).

A final point bears on the weight or importance of particular kinds of characters in the detection of relationships. Whole suites of characters deemed to be important come into and go out of fashion over the years. Thus, in 1883, Ludwig Radlkofer proclaimed the following century to be that of the anatomical method in systematics. In 1924, Hermann Ziegenspeck produced the "Königsberger Stammbaum," a tree showing the serological relationships of all plants (with fossils placed in their appropriate positions). Anatomy has remained an important source of data in systematics, although generalized anatomical surveys such as those that Radlkofer favored became unfashionable well before the end of his "century of the anatomical method"; serological studies never really caught on. Features such as plant chemistry, chromosome number, and the morphology of sieve tube plastids have all had their moment of glory; indeed, the last in particular has turned out to be very valuable at higher levels. On the other hand, developmental studies have not been popular in systematics, despite Payer's remarkable work in the middle of the nineteenth century. This work was dismissed by systematists such as J. D. Hooker, but it has become widely cited in the comparative developmental studies that are adding so much to systematics today.

Because it was realized that single-character classifications were suspect, and because it was difficult to analyze and integrate all available data in one's head, botanists often restricted themselves to providing extensive surveys showing patterns of variation in individual characters. Indeed, despite claims that classifications did synthesize all available data (Lawrence 1951; Constance 1964), prior to the advent of the computer there was no way of analyzing the massive amounts of data that systematists had accumulated. (Despite this, the coverage of systematic data has often been sadly inadequate.) In any event, it was quite clear in the work of systematists such as Cronquist and G. L. Stebbins that single characters could be very important. Thus Stebbins (1974) was inclined to see direct relationships between *Paeonia* and gymnosperms because they all had in common embryos with a free nuclear stage.

There has been some tension between floras, with their proper emphasis on geographically circumscribed treatments focusing on characters that help in the identification of the plants (see Frodin 2000 and Appendix 2), and monographs that deal with taxa wherever they are found, which emphasize characters indicative of relationships, whether or not they can be used in identification. Major floras still often take more than 50 years to complete and are similar in their goals (and in the time it takes to finish them) to the colonial floras promoted by the directors of the Royal Botanic Gardens at Kew, England, in the later part of the nineteenth century. Accessible to a wide variety of scientists, they have helped stabilize usage of taxon names, although too little attention is paid to just how the contributors to these floras reach the taxonomic conclusions that they do (Stevens 1997a).

Plant Groupings over the Years

It is impossible to do more than mention a few of the main changes in ideas about relationships among plants in the years prior to the advent of phylogenetic methodology (see Lawrence 1951 for summaries). Some of the differences between what are now called monocotyledons and dicotyledons were already evident to Theophrastus in 300 B.C., but John Ray (1627–1705) was the first to make a major distinction between the two, although he subordinated it to his primary division into trees and herbs. Cotyledon number was the main character used by Jussieu in 1789 to divide plants, and it has almost always retained this prime position.

Jussieu placed the monocotyledons before the dicotyledons because they were simpler (monocots appeared to lack a corolla and had only a single cotyledon), and he began the dicots with such plants as Aristolochiaceae and many Caryophyllales (his nomenclature is modernized here); all, he thought, lacked a corolla (petals) and so were the simplest in the series of increasing complexity that his arrangement represented.

Many plants with catkins have flowers of different sexes, often on different individuals, and so they seemed to Jussieu to be by his criteria the most complex; these he placed near the end of the dicots, and thus near the end of his whole sequence. He placed the conifers (but not the cycads, which he included with the ferns) at the very end, probably in part because some genera have many cotyledons and so represented the culmination of the cotyledonary series. The distinctive nature of gymnospermy was demonstrated by Robert Brown in 1826, but gymnosperms (conifers and cycads) were not finally excluded from the angiosperms until much later in the century.

Two major arrangements adopted subsequently are associated with the names of A.-P. de Candolle and Adolf Engler. However, there has always been a plethora of alternative systems: in the twentieth century, Lam, Melville, Meeuse, and Hayata, to name just a few, all proposed their own sometimes very different systems.

Candolle (1813) began his system—which he said was not to be interpreted as being linear—with the Ranunculaceae on the grounds that one should place well-known organisms first. Simple plants tended to be least well known; the Ranunculaceae had, he thought, the most complex flowers and were well known. Candolle's series followed the sequence (1) Thalamiflorae (superior, sepals and petals distinct), (2) Calyciflorae (often a hypanthium, sepals and petals distinct), (3) Corolliflorae (sympetalous plants), (4) Monochlamydeae (only a single perianth series) (see Chapter 4 for discussion of these floral terms). Monocots followed the dicots, and gymnosperms, not named as such, straddled the end of the dicot series and the beginning of the monocots.

Although Bentham and Hooker largely followed the Candollean sequence, their classification delimits taxa that are quite often substantially different and hence implies different relationships. Furthermore, they noted that they adopted the dicot sequence Thalamiflorae-Gamopetalae-Monochlamydeae for convenience only, and that many Monochlamydeae, in particular, were probably related separately to Polypetalae (i.e., plants having distinct petals); they were not much happier with Gamopetalae. Their Gymnospermae was a fourth dicot group placed just before the monocots.

Engler's system is basically a modification of that of Adrien de Jussieu, who, like his father Antoine-Laurent, allowed that the basic sequence should be simple to complex (Jussieu 1843). However, he thought that monocots and dicots should be placed in parallel, not in series. Hence his dicot sequence began with plants that were absolutely simple, rather than those that Antoine-Laurent thought were most similar to monocots. Within the dicots, Adrien placed dioecious groups, divided into angiosperms and gymnosperms, first, and Amentiferae (i.e., species with reduced, wind-pollinated flowers borne in catkins or aments) were placed first within the angiosperms. The other three major groups of dicots that he recognized followed the morphological sequence (1) apetaly (petals lacking), (2) polypetaly (petals distinct), (3) monopetaly (petals fused).

HEINRICH GUSTAV ADOLF ENGLER
(1844–1930, German)

Engler excluded gymnosperms from the angiosperms and divided dicots into Archichlamydeae and Sympetalae. The angiosperms began with groups such as Piperaceae and Chloranthaceae before proceeding to Amentiferae and polypetalous plants. The basic arrangement is unchanged in recent editions of this system (Engler 1964), although Piperaceae have been moved. There is some debate whether Engler thought that the Amentiferae really were primitive, but some who used the Englerian sequence (or its precursors) certainly did think this to be the case.

Bessey's system combines features of both the Candollean and the Englerian ways of arranging plants (Bessey 1915). Bessey's dicta—guidelines for the production of phylogenies—have been particularly influential. Many of these dicta are specific evolutionary trends, and the identification of such trends long remained a major component of evolutionary thought. Recent systems, of which perhaps the most notable are those of Dahlgren (1983; Dahlgren et al. 1985), Thorne (1999, 2000), Takhtajan (1997), and Cronquist (1981), are largely variants of Besseyan and Englerian ideas combined (Cuerrier et al. 1996), although Thorne and particularly Dahlgren (as discussed earlier) paid more attention to phylogenetic principles.

Despite all these varying approaches, by the early 1980s something of a consensus about ideas of relationships seemed to be developing (Stevens 1986), and in North America Cronquist's system was much in use. Well documented and with descriptions that incorporated both anatomical and chemical information, it starts with families that are still considered to be members of "basal" angiosperm lineages, although they are now often placed in groups that are differently circumscribed. Elsewhere in Cronquist's system, particularly in groups such as the Rosidae, Dilleniidae, and Liliidae, there is little in common between the groups that he recognized and those that are recognized here.

Indeed, the consensus of the late twentieth century ignored those who still followed Englerian ideas, and it has not survived the effects of cladistic theory (see Chapter 2) and the recent spate of major molecular and morphological studies that utilize it (see Chapters 8 and 9). The changes have resulted from the existence of clear goals, large amounts of new data, and new methods of data analysis. These advances allow systematists to compare studies, to evaluate competing hypotheses of relationships, and to focus on taxa that are critical in understanding these rela-

tionships. The mindset of systematists is changing as we no longer describe relationships in nature, but instead propose hypotheses of phylogeny. Classifications are the work of humans, not of nature; we decide what groups we wish to talk about.

Although many important aspects of phylogenies remain unclear—for instance, the relationship of monocots to other angiosperms—the broad outlines of a new arrangement are evident (Angiosperm Phylogeny Group 1998, 2003; Chase 2004; Judd and Olmstead 2004; Soltis and Soltis 2004) and are reflected in the relationships discussed in Chapters 8 and 9 of this book and the sequence followed there. Indeed, with about ten major and for the most part not unexpected exceptions such as Saxifragaceae, Scrophulariaceae, Liliaceae, and Loganiaceae, family limits have changed little. It is the more inclusive clades that have shown more drastic changes, but again, the unsatisfactory nature of these higher groups was quite obvious to authors such as Davis and Heywood. Over the next decade we can expect to see substantial changes in many of our circumscriptions of lower-level clades such as genera.

LITERATURE CITED AND SUGGESTED READINGS

Items marked with an asterisk are especially recommended to those readers who are interested in further information on the topics discussed in this chapter.

Adanson, M. 1763–1764. *Familles des plantes*, 2 vols. Vincent, Paris.

Allen, D. E. 1976. *The naturalist in Britain: A social history*. A. Lane, London.

Anderson, E. 1940. The concept of the genus. II. A survey of modern opinion. *Bull. Torrey Bot. Club* 67: 363–369.

Angiosperm Phylogeny Group. 1998. An ordinal classification for the families of flowering plants. *Ann. Missouri Bot. Gard.* 85: 531–553.

Angiosperm Phylogeny Group. 2003. An update of the Angiosperm Phylogeny Group classification for the orders and families of flowering plants: APGII. *Bot. J. Linnean Soc.* 141: 399–436.

*Atran, S. 1990. *Cognitive foundations of natural history*. Cambridge University Press, Cambridge. [A challenging reinterpretation of early classifications.]

Bentham, G. 1857. Memorandum on the principles of generic nomenclature in botany as referred to in the previous paper. *J. Proc. Linnean Soc., Bot.* 2: 30–33.

Bentham, G. 1875. On the recent progress and present state of systematic botany. *Rep. Br. Assn. Adv. Sci.* (1874): 27–54.

Bentham, G. and J. D. Hooker. 1862–1883. *Genera plantarum*. 3 vols. Reeve & Co., London.

*Berlin, B. 1992. *Ethnobiological classification: Principles of categorization of plants and animals in traditional societies*. Princeton University Press, Princeton, NJ. [An excellent summary.]

Bessey, C. E. 1915. The phylogenetic taxonomy of flowering plants. *Ann. Missouri Bot. Gard.* 2: 109–164.

Bremer, K. 1987. Tribal interrelationships of Asteraceae. *Cladistics* 2: 210–253.

Brown, R. 1826. Character and description of *Kingia* … with observations … on the female flower of Cycadaceae and Coniferae. In *Narrative of a survey of the inter-tropical coasts of Western Australia …* , P. P. King (ed.), vol. 2, 538–565. Murray, London.

Candolle, A.-P. de. 1813. *Théorie élémentaire de la botanique*. Déterville, Paris.

Chase, M. W. 2004. Monocot relationships: An overview. *Am. J. Bot.* 91: 1645–1655.

*Constance, L. 1964. Systematic botany—An unending synthesis. *Taxon* 13: 257–273. [A clear statement of the goals of evolutionary systematics.]

*Coulter, J. M. 1895. *The botanical outlook*. Lincoln, Nebraska.

Cronquist, A. 1981. *An integrated system of classification of flowering plants*. Columbia University Press, New York.

*Cuerrier, A., R. Kiger and P. F. Stevens. 1996. Charles Bessey, evolution, classification, and the New Botany. *Huntia* 9: 179–213. [A study of the work of perhaps the most influential American systematist at the beginning of the twentieth century.]

Dahlgren, R. 1983. General aspects of angiosperm evolution and macrosystematics. *Nordic J. Bot.* 3: 119–149.

Dahlgren, R., H. T. Clifford and P. F. Yeo. 1985. *The families of monocotyledons*. Springer, Berlin.

Darwin, C. 1859. *On the origin of species by means of natural selection*. Reprinted in *On the origin of species: a facsimile of the first edition*, E. Mayr (ed.). 1964. Harvard University Press, Cambridge, MA.

Davis, P. H. and V. H. Heywood. 1963. *Principles of angiosperm taxonomy*. Edinburgh University Press, Edinburgh.

de Queiroz, K. 1988. Systematics and the Darwinian revolution. *Philos. Sci.* 55: 238–259.

Dughi, R. 1957. Tournefort dans l'histoire de la botanique. In *Tournefort*, R. Heim (ed.), 131–185. Muséum National d'Histoire Naturelle, Paris.

Dupree, H. 1959. *Asa Gray 1810–1888*. Belknap Press of Harvard University Press, Cambridge, MA.

Endersby, J. 2001. "The realm of hard evidence": Novelty, persuasion and collaboration in botanical cladistics. *Stud. Hist. Philos. Biol. Biomed. Sci.* 32: 343–360.

Engler, A. 1964. *Syllabus der Pflanzenfamilien*, H. Melchior (ed.), 12th ed., vol. 2. Borntraeger, Berlin.

Frodin, D. G. 2001. *Guide to standard floras of the world*, 2nd ed. Cambridge University Press, Cambridge.

*Gilmour, J. S. L. 1940. Taxonomy and philosophy. In *The new systematics*, J. Huxley (ed.), 461–474. Oxford University Press, Oxford. [A fascinating interpretation of the whys and wherefores of classification that remains worth reading.]

Giseke, P. D. 1792. *Praelectiones in ordines naturales plantarum*. Hoffmann, Hamburg, Germany.

Grass Phylogeny Working Group. 2001. Phylogeny and subfamilies of the grasses (Poaceae). *Ann. Missouri Bot. Gard.* 88: 373–457.

*Greene, E. L. 1983. *Landmarks of botanical history*, 2 vols., F. N. Egerton (ed.). Stanford University Press, Stanford, CA. [Written almost 100 years ago, but half of it published for the first time only in 1981 (but carefully edited then), this work covers botany up to the sixteenth century, although authors as late as Tournefort are also included.]

Hall, H. M. and F. E. Clements. 1923. *The phylogenetic method in taxonomy. The North American species of Artemisia, Chrysothamnus and Atriplex* (Carnegie Institute of Washington Publication No. 326). Carnegie Institution of Washington, Washington, DC.

*Haston, E., J. E. Richardson, P. F. Stevens, M. W. Chase and D. J. Harris. 2007. A linear sequence of Angiosperm Phylogeny Group II families. Taxon 56: 7–12.

Hennig, W. 1950. *Grundzüge einer Theorie der phylogenetischen Systematik*. Deutsche Zentralverlag, Berlin.

Hennig, W. 1966. *Phylogenetic systematics*. University of Illinois Press, Urbana.

Heywood, V. H. (ed.). 1978. *Flowering plants of the world*. Mayflower, New York.

Hutchinson, J. 1973. *The families of flowering plants; arranged according to a new system based on their probable phylogeny*. Clarendon Press, Oxford.

Jansen, R. K. and J. D. Palmer. 1987. A chloroplast DNA inversion marks an ancient evolutionary split in the sunflower family (Asteraceae). *Proc. Natl. Acad. Sci. USA* 84: 5818–5822.

Judd, W. S. and R. G. Olmstead. 2004. A survey of tricolpate (eudicot) phylogenetic relationships. *Am. J. Bot.* 91: 1627–1644.

Jussieu, A.-L. de. 1789. *Genera plantarum*. Hérissant and Barrois, Paris.

Jussieu, A.-L. de. 1843. *Cours élémentaire de histoire naturelle: Botanique*. Fortin Masson, Langlois and Leclerc, Paris.

Kornet, D. J. 1991. On specific and interspecific delimitation. In *The plant diversity of Malesia*, P. Baas, K. Kalkman and R. Geesink (eds.), 359–379. Kluwer, Dordrecht, Netherlands.

Lam, H. J. 1936. Phylogenetic symbols, past and present. *Acta Biotheor.* 2: 153–194.

Lamarck, J.-B.-P.-A. de M. de. 1778. *Flore française*, 3 vols. Imprimerie Royale, Paris.

Lawrence, G. H. L. 1951. *Taxonomy of vascular plants*. Macmillan, New York.

Linnaeus, C. 1751. *Philosophia botanica*. Kiesewetter, Stockholm.

*Mayr, E. 1982. *The growth of biological thought: Diversity, evolution and inheritance*. Belknap Press of Harvard University Press, Cambridge, MA. [History with a broad sweep, emphasizing animals and species.]

*Morton, A. G. 1981. *Outlines of botanical history*. Academic Press, London. [An invaluable account of all aspects of botanical knowledge up to the end of the nineteenth century.]

Sokal, R. R. and P. H. A. Sneath. 1963. *Principles of numerical taxonomy*. Freeman, San Francisco.

Soltis, P. S. and D. E. Soltis. 2004. The origin and diversification of angiosperms. *Am. J. Bot.* 91: 1614–1626.

*Stafleu, F. 1971. *Linnaeus and the Linnaeans*. Oosthoeck, Utrecht, Netherlands. [The classic treatment of the work of Linnaeus and his immediate successors.]

Stebbins, G. L. 1974. *Flowering plants: Evolution above the species level*. Belknap Press of Harvard University Press, Cambridge, Mass.

Stevens, P. F. 1986. Evolutionary classification in botany, 1860–1985. *J. Arnold Arbor.* 67: 313–339.

Stevens, P. F. 1990. Nomenclatural stability, taxonomic instinct, and flora writing—A recipe for disaster? In *The plant diversity of Malesia*, P. Baas, K. Kalkman and R. Geesink (eds.), 387–410. Kluwer, Dordrecht, Netherlands.

Stevens, P. F. 1992. Species: Historical perspectives. In *Keywords in evolutionary biology*, E. F. Keller and E. A. Lloyd (eds.), 302–311. Harvard University Press, Cambridge, MA.

*Stevens, P. F. 1994. *The development of biological systematics*. Columbia University Press, New York. [A discussion emphasizing that authors of the period 1780–1860 often did not produce classifications in the currently accepted sense of the word, and evaluating the history of systematics accordingly.]

*Stevens, P. F. 1997a. How to interpret botanical classifications—Suggestions from history. *BioScience* 47: 243–250. [A reinterpretation of Bentham and Hooker's *Genera Plantarum*, a major reference source for over a century.]

Stevens, P. F. 1997b. J. D. Hooker, George Bentham, Asa Gray and Ferdinand Mueller on species limits in theory and practice: A mid-nineteenth century debate and its repercussions. *Hist. Rec. Aust. Sci.* 11: 345–370.

Stevens, P. F. 1998a. Mind, memory and history: How classifications are shaped by and through time, and some consequences. *Zoologica Scripta* 26: 293–301.

Stevens, P. F. 1998b. What kind of classification should the practicing taxonomist use to be saved? In *Plant diversity in Malesia III*, J. Dransfield, M. J. E. Coode and D. A. Simpson (eds.), 295–319. Royal Botanical Gardens, Kew, London.

Stevens, P. F. 2000a. Botanical systematics 1950–2000: Change, progress, or both? *Taxon* 49: 635–659.

Stevens, P. F. 2000b. On characters and character states: Do overlapping and non-overlapping variation, molecules and morphology all yield data of the same value? In *Homology and systematics*, R. Scotland and R. T. Pennington (eds.), 81–104. Taylor & Francis, London.

*Stevens, P. F. 2002. Why do we name organisms? Some reminders from the past. *Taxon* 51: 11–26.

Stevens, P. F. 2006. An end to all things? Plants and their names. *Austral. Syst. Bot.* 19: 115–133.

Takhtajan, A. 1997. *Diversity and classification of flowering plants*. Columbia University Press, New York.

Thorne, R. F. 1999. The classification and geography of the monocotyledon subclasses Alismatidae, Liliidae and Commeliniidae. In *Plant systematics for the 21st century*, B. Nordenstam, G. El-Ghazaly and M. Kassas (eds.), 75–124. Portland Press, London.

Thorne, R. F. 2000. The classification and geography of the flowering plants: Dicotyledons of the class Angiospermae. *Bot. Rev.* 66: 441–647.

Tournefort, J. P. de. 1694. *Élémens de botanique*, 3 vols. Imprimerie Royale, Paris.

Vernon, K. 1988. The founding of numerical taxonomy. *Br. J. Hist. Sci.* 21: 143–159.

Vernon, K. 1993. Desperately seeking status: Evolutionary systematics and the taxonomists' search for respectability 1940–1960. *Br. J. Hist. Sci.* 26: 207–227.

*Voss, E. 1952. The history of keys and phylogenetic trees in systematic biology. *J. Sci. Lab. Denison Univ.* 43: 1–25. [A useful and well-illustrated survey.]

Wagner, W. H., Jr. 1969. The construction of a classification. In *Systematic biology: Proceedings of an International Conference* (National Research Council Publication 1692), 67–103. National Academy of Sciences, Washington, DC.

Wagner, W. H., Jr. 1980. Origin and philosophy of the groundplan-divergence method of cladistics. *Syst. Bot.* 5: 173–193.

*Walters, S. M. 1986. The name of the rose: A review of ideas on the European bias in angiosperm classification. *New Phytol.* 104: 527–546. [A valuable review with references to Walters's earlier publications.]

Wernham, H. F. 1911–1912. Floral evolution with particular regard to the sympetalous dicotyledons. *New Phytol.* 10: 73–83, 109–120, 145–159, 217–226, 203–235; 11: 145–166, 217–235, 290–305, 373–397.

*Winsor, M. P. 1995. The English debate on taxonomy and phylogeny, 1937–1940. *Hist. Philos. Life Sci.* 17: 105–130. [An illuminating account of a crucial period in systematics.]

Winsor, M. P. 2000. Species, demes and the omega taxonomy: Gilmour and the New Systematics. *Biol. Philos.* 15: 349–388.

Winsor, M. P. 2001. Cain on Linnaeus: The scientist-historian as an unanalysed entity. *Stud. Hist. Philos. Biol. Biomed. Sci.* 32C: 239–254.

4

Taxonomic Evidence: Structural and Biochemical Characters

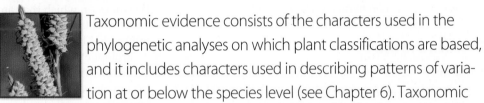 Taxonomic evidence consists of the characters used in the phylogenetic analyses on which plant classifications are based, and it includes characters used in describing patterns of variation at or below the species level (see Chapter 6). Taxonomic evidence can be gathered from a wide variety of sources from all parts of a plant during all stages of its development. In this chapter we summarize the use of characters from morphology, anatomy, embryology, chromosomes, palynology, secondary metabolites, and proteins. Nucleic acids (DNA and RNA) provide an increasingly important source of taxonomic characters; their use in plant taxonomy and the rapidly developing field of molecular systematics is discussed in detail in Chapter 5.

The practical discussion of plant characters in this chapter and the next provides a useful counterpart to the more theoretical discussion of characters in Chapter 2.

Morphology

Morphological characters are features of external form or appearance. They currently provide most of the characters used for practical plant identification and some of those used for hypothesizing phylogenetic relationships. These features have been used for a longer time than anatomical or molecular evidence, and they were the only source of taxonomic evidence in the beginnings of plant systematics. Morphological characters that are easily observed find practical use in keys and descriptions; those used in phylogeny reconstruction may not be so easily observed. Characters of both kinds are found in all parts of the plant, both vegetative and reproductive.

The vegetative parts of angiosperms are roots, stems, and leaves, and the reproductive parts are flowers, fruits, and seeds. The terms used in tracheophytes (vascular plants) to describe variation in these parts are outlined in the discussion that follows, although the reproductive terms strongly emphasize the angiosperms because they are the dominant group of vascular plants. (Specialized vegetative and especially reproductive terms relating to other groups of vascular plants are covered in Chapter 8.)

Many, if not all, of the terms outlined here should be considered merely convenient points along a continuum of variation in form. Although they are useful in communication, intermediate conditions will be encountered.

Duration and Habit

Duration is the life span of an individual plant. An **annual** plant lives for a single growing season. A **biennial** plant lives for two seasons, growing vegetatively during the first and flowering in the second. A **perennial** plant lives for three or more years and usually flowers and fruits repeatedly. Perennials may be herbaceous (lacking woody tissue), with only the underground portions living for several years, or woody, with a persistent aerial stem.

The general appearance, or **habit**, of plants varies greatly. Woody tissue is present in **trees** and **shrubs** but is lacking in **herbs**. Trees produce one main **trunk** (or **bole**); shrubs are usually shorter and produce several trunks. Climbing plants may be woody (**lianas**) or herbaceous (**vines**). **Suffrutescent** plants are intermediate between woody and herbaceous. Herbs or shrubs that grow upon another plant, which is used as a support, are called **epiphytes**.

The characteristic shape of a tree or shrub often relates to its pattern of growth or architecture, which is often of systematic value. Stems form the major plant axes and may be erect (**orthotropic**) or horizontal (**plagiotropic**). **Monopodial shoots** grow through the action of a single **apical meristem**: a single region of dividing, elongating,

and differentiating cells at the tip of the shoot. In other plants, **axillary branches** (i.e., branches that develop from buds associated with the leaves) take over the role of the main axis and provide for continuing growth, while the main axis slows its growth or dies; a series of such axillary branches constitutes a **sympodial shoot**.

Stems are provided with **buds**, which are small embryonic shoots, often protected by modified leaves (bud scales) or hairs. Buds may show a period of dormancy and, when they eventually grow out, may leave scars at the base of the new shoot. Such shoots are called **proleptic**. On the other hand, buds may develop and elongate at the same time as the shoot on which they are borne, in which case the new shoot usually lacks basal scars and has an elongated first internode. These shoots are called **sylleptic**.

Proleptic shoots are characteristic of temperate species, while sylleptic shoots are common in tropical plants. Some taxa, such as many Lauraceae, have both sylleptic and proleptic shoots. All shoots on a plant may be the same, or there may be two or more kinds of shoots. Thus *Ginkgo* and *Cercidiphyllum* both have some shoots with elongated internodes (**long shoots**) and others that produce only a few leaves and bud scales each year, all with only very short internodes (**short shoots**). These and other features are combined in various ways to produce an array of distinctive architectural growth patterns in trees and shrubs (see Hallé et al. 1978). Three examples are shown in Figure 4.1.

Axillary branches are initiated externally, in tissues immediately under the epidermis. Xylem and phloem are often arranged in a ring surrounding the **pith**, a central region with more or less isodiametric cells.

Roots

Roots usually branch irregularly. Lateral roots are initiated internally, in the endodermis and pericycle (the cell layers surrounding the conducting tissues) and erupt through the cortex, although lycophyte roots branch by a forking of the

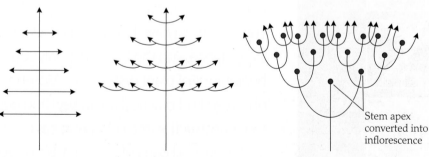

Examples: *Duabanga* (Lythraceae)   *Terminalia* and   *Rhus* (Anacardiaceae)
           Araucaria (Araucariaceae)   *Bucida* (Combretaceae)   *Pieris* (Ericaceae)

Stem apex converted into inflorescence

Main axis orthotropic   Main axis orthotropic   All stems similar,
and monopodial;   and monopodial;   orthotropic and
lateral branches   lateral branch systems   sympodial
plagiotropic and   plagiotropic and
monopodial   sympodial

FIGURE 4.1   Three architectural patterns of plant growth; branching is rhythmic in all three.

apical meristem. The xylem and phloem are situated in the central portion of the root, usually resulting in a lack of pith. Roots also lack the nodes and internodes that characterize stems (described below), and they are usually found underground.

The primary functions of roots are to hold the plant in place, to absorb water and minerals, and to store water and carbohydrates. Some roots are specialized for other functions, such as photosynthesis (as in some reduced epiphytic Orchidaceae), penetration of the tissues of a host species (as in parasitic species such as mistletoes, Santalaceae), constriction of the trunks of supporting trees (as in strangler figs, Moraceae), or aboveground support for the trunk or branches (as in banyan figs, Moraceae, and some mangroves, Rhizophoraceae).

Some plants, such as epiphytic aroids (Araceae), have dimorphic roots, with some functioning in water and mineral uptake and others providing attachment. Most roots grow downward, but exceptions occur, as in **pneumatophores**, which are specialized roots involved in gas exchange in some mangrove or swamp species.

Roots are quite uniform in appearance, and a plant usually cannot be identified without its aboveground parts. Roots are useful, however, in determining whether a plant is an annual or a perennial, and variation in the root system is sometimes taxonomically significant. Here are a few important terms relating to roots:

adventitious   developing from any plant part other than the embryonic root (radicle) or another root

aerial   growing above ground or water

fibrous   with all portions of the root system of more or less equal thickness, often well branched, and the primary root (taproot) absent or not obvious

fleshy   thick with water or carbohydrate storage tissue

haustorial   specialized for penetrating other plants and absorbing water and nutrients from them (as in parasites)

taproot   the major root, usually enlarged and growing downward

Stems

Stems—the axes of plants—consist of **nodes** (where leaves and axillary buds are produced) separated by **internodes** (Figure 4.2). They are frequently useful in identification and provide numerous systematically important characters.

Stems are usually elongated and function in exposing leaves to sunlight, flowers to pollinating agents, and fruits to dispersal agents. Some, however, are the primary photosynthetic organ of the plant (as in asparagus, Asparagaceae, and many cacti, Cactaceae), store water or carbohydrates (as in many cacti and other succulents), climb (as in hooked or twining stems of vines and lianas), or protect the plant (as in plants with thorns). Some important stem-related vocabulary is listed here:

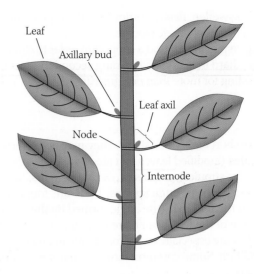

FIGURE 4.2   Generalized angiosperm stem showing nodes and internodes, leaves, leaf axils, and axillary buds.

acaulescent   having an inconspicuous stem

bulb   a short, erect, underground stem surrounded by thick, fleshy leaves or leaf bases

caulescent   having a distinct stem

corm   a short, erect, underground, more or less fleshy stem covered with thin, dry leaves or leaf bases

herbaceous   not woody; dying down at the end of the growing season

internode   the part of the stem between two adjacent nodes

lenticel   a wartlike protuberance on the stem surface involved in gas exchange

long shoot   a stem with long internodes; this term is applied only in plants in which internode length is clearly bimodal and both long and short shoots are present

node   region of the stem where the leaf and bud are borne

pith   soft tissue in the center of the stem, usually consisting of more or less isodiametric cells

rhizome   a horizontal stem, more or less underground, bearing scalelike leaves; often called a stolon (or runner) if above ground and having an elongated internode

scape   an erect, leafless stem bearing an inflorescence or flower at its apex; usually composed of a single elongated internode

scar   the remains of a point of attachment, as in leaf scar, stipule scar, bud scale scar

short shoot   a stem with short internodes in a plant in which other shoots have distinctly longer internodes; see long shoot

thorn   a reduced, sharp-pointed stem [In contrast, a reduced, sharp-pointed leaf or stipule or sharp-pointed marginal tooth is called a spine, and a sharp-pointed hair (involving epidermal tissue) or emergence (involving both epidermal and subepidermal tissues) is called a prickle.]

tuber   a swollen, fleshy portion of a rhizome involved in water or carbohydrate storage

twining   spiraling around a support in order to climb

woody   hard in texture, containing secondary xylem, and persisting for more than one growing season

Buds

Buds are short embryonic stems. They may be protected by **bud scales** (modified leaves that are sometimes represented only by stipules), a dense covering of hairs, and/or a sticky secretion. In angiosperms, buds are found at the nodes, in the **leaf axil** (the angle formed by the stem and the petiole of the leaf; see Figure 4.2), or at the end of the stem. They are especially useful for identifying twigs in winter condition. Some common terms pertaining to buds are listed here:

accessory bud   an extra bud (or buds) produced on either side of, or above or below, the main axillary bud

axillary bud   a bud located in the leaf axil

flower bud   a bud containing embryonic flowers

leaf bud   a bud containing embryonic leaves

mixed bud   a bud containing both embryonic flowers and embryonic leaves

naked   not covered by bud scales

pseudoterminal bud   an axillary bud that takes over the function of a terminal bud in sympodial shoots

superposed bud   a bud located above or below the axillary bud

terminal bud   a bud at the apex of a monopodial shoot

Leaves

Leaves are the major photosynthetic parts of most plants. They are borne at the nodes of a stem, usually below a bud (Figure 4.3; see also Figure 4.2). In contrast to stems, leaves usually do not continue to grow year after year. They are usually flat, and have one surface facing toward the stem axis (the **adaxial**, or upper, surface) and another surface facing away from the stem axis (the **abaxial**, or lower, surface). Most leaves are **bifacial**, having definite adaxial and abaxial surfaces, but sometimes they are **unifacial**, lacking such differentiation.

Leaves are homologous structures among the angiosperms, but not among vascular plants as a whole (see Chapters 7, 8, and 9). In addition to their obvious function in photosynthesis, leaves may be modified for protection, forming sharp-pointed **spines**; for water storage, as in many succulents; for climbing, as in vines or lianas with tendril leaves; for capturing insects, as in carnivorous plants; or for providing homes for ants or mites (domatia, described on page 61).

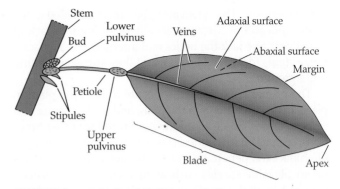

FIGURE 4.3   Generalized angiosperm leaf.

The major parts of a leaf are shown in Figure 4.3. The base of the **petiole** may have a narrow to broad point of attachment and may obscure the axillary bud. In monocots, the leaf is almost always broadly sheathing at the base, with the edges either fused or overlapping. In taxa such as grasses (Poaceae) and gingers (Zingiberaceae), there is an adaxial flap or **ligule** at the junction of the sheath and blade. A leaf that lacks a petiole is said to be **sessile**.

Pulvini (singular **pulvinus**), somewhat swollen and morphologically distinct parts of the petiole, are often present and involved in leaf movement. They may be at the leaf base; at the apex of the petiole, as in prayer plants (Marantaceae); in the middle of the petiole, as in a few Araceae; or on the petioles of leaflets (of compound leaves, as described on page 57).

Stipules are usually paired appendages located on either side of (or on) the petiole base. Stipules are sometimes single, in which case they are borne between the petiole and stem. They may be leaflike, scalelike, tendril-like, spinelike, glandular, very reduced, or completely lacking. They have various functions, but most often they help in protecting young leaves. Stipules are not always homologous.

Leaf arrangement   Leaves may be arranged in one of three major patterns (Figure 4.4). **Alternate** leaves are borne

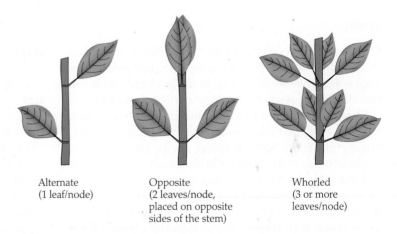

Alternate
(1 leaf/node)

Opposite
(2 leaves/node,
placed on opposite
sides of the stem)

Whorled
(3 or more
leaves/node)

FIGURE 4.4   Three major patterns of leaf arrangement.

singly and are usually arranged in a spiral pattern along the stem. Various kinds of spirals occur; they can be evaluated by determining the angle around the stem between the points of insertion of any two successive leaves or by following the spiral around the stem from any older, lower leaf to the first younger leaf directly in line above it. Alternate leaves are sometimes placed along just two sides of the stem (2-ranked, or distichous), or only three sides of the stem (3-ranked, or tristichous). [Two-ranked leaves that are flattened in the same plane with both surfaces identical, as in irises (Iridaceae), are called **equitant**.]

In contrast, **opposite** leaves are borne in pairs, the members of which are positioned on opposing sides of the stem. Opposite leaves may be spiraled, as in red mangroves (*Rhizophora*, Rhizophoraceae); 2-ranked, as in many Zygophyllaceae; or **decussate** (the leaves of adjacent nodes rotated 90°). The decussate arrangement is the most common condition among vascular plant species.

Finally, when three or more leaves are positioned at a node, they are considered to be **whorled**.

Leaf structure A leaf with a single blade is termed **simple**; a leaf with two or more blades, or **leaflets**, is said to be **compound**. The distinction between simple and compound leaves can be made by locating an axillary bud: an axillary bud is subtended by the entire leaf and never by individual leaflets. Leaflets may be arranged in various ways, as shown in Figure 4.5.

Leaf duration Leaves may function from a few days to many years, but most leaves function for only one or two growing seasons. **Deciduous** leaves fall (are abscised) at the end of the growing season; **evergreen** plants are leafy throughout the year. Some leaves, such as those of many members of Fagaceae, are **marcescent**; that is, they wither but do not fall off during the winter or dry season.

Venation types If there is one most prominent vein in a leaf, it is called the **midvein** or **primary vein**; branches from this vein are called **secondary veins**. **Tertiary veins** usually link the secondaries, forming a ladderlike (**percurrent** or **scalariform**) or netlike (**reticulate**) pattern (Figure 4.6).

There are three major patterns of organization of the major veins. The leaf may have a single primary vein with the secondary veins arising along its length like the teeth of a comb; this pattern is termed **pinnate**. Or the leaf may have several major veins radiating from the base (or near

Even-pinnate   Odd-pinnate

Palmate   Trifoliolate   Unifoliolate   Joint

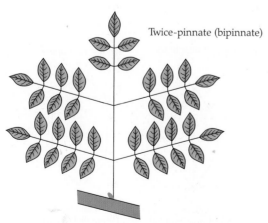

Twice-pinnate (bipinnate)

FIGURE 4.5 Arrangements of leaflets in compound leaves.

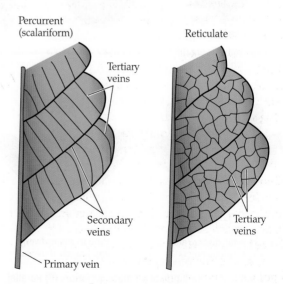

Percurrent (scalariform)   Reticulate

Tertiary veins

Secondary veins

Tertiary veins

Primary vein

FIGURE 4.6 Two patterns of tertiary veins.

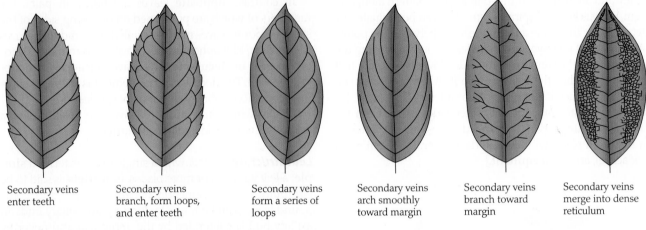

Secondary veins
enter teeth

Secondary veins
branch, form loops,
and enter teeth

Secondary veins
form a series of
loops

Secondary veins
arch smoothly
toward margin

Secondary veins
branch toward
margin

Secondary veins
merge into dense
reticulum

FIGURE 4.7   Some kinds of pinnate venation. (After Hickey 1973.)

the base) of the blade, like fingers from a palm; this pattern is called **palmate**. Many different kinds of pinnate (Figure 4.7) and palmate (Figure 4.8A) venation have been characterized (they are discussed in more detail in Hickey 1973 and Dilcher 1974). Finally, the leaf may have many parallel veins, a pattern termed **parallel** venation (Figure 4.8B).

Leaf shapes   A leaf may be considered to have one of four basic shapes (**ovate**, **obovate**, **elliptic**, **oblong**) depending on where the blade is the widest (Figure 4.9) (Hickey 1973).

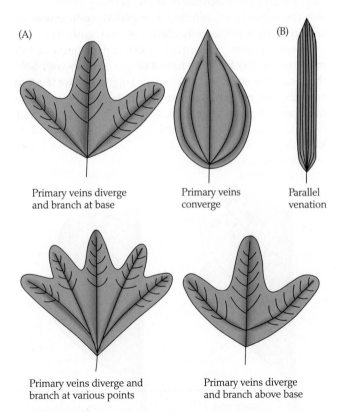

(A)

(B)

Primary veins diverge
and branch at base

Primary veins
converge

Parallel
venation

Primary veins diverge and
branch at various points

Primary veins diverge
and branch above base

FIGURE 4.8   (A) Four kinds of palmate venation. (B) Parallel venation. (After Hickey 1973.)

The meanings of these shape terms may be adjusted by the use of modifiers such as *broadly* or *narrowly*. If the petiole is attached away from the leaf margin, such that the leaf and its stalk form an "umbrella," the leaf is termed **peltate**, and such leaves may be any of a number of different shapes. A **linear** leaf, on the other hand, is very long and narrow. Various other specialized shape terms are sometimes employed, but their use is avoided as much as possible here. The blade of a leaf may be symmetrical or asymmetrical when viewed from above.

Very different leaf shapes may occur on the same plant, a condition known as **heterophylly**. Juvenile leaves may be quite different from adult leaves, but sometimes even an adult plant will bear several different kinds of leaves (as in *Sassafras*, Lauraceae).

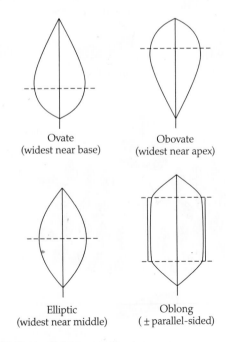

Ovate
(widest near base)

Obovate
(widest near apex)

Elliptic
(widest near middle)

Oblong
(± parallel-sided)

FIGURE 4.9   Leaf shapes.

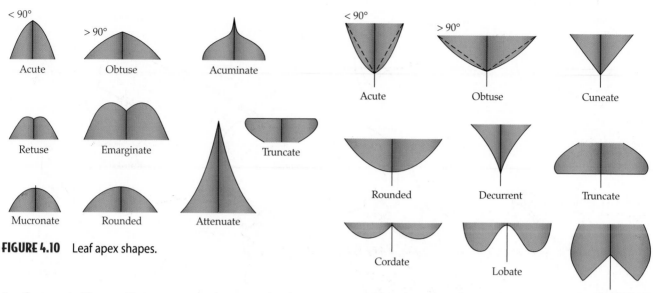

FIGURE 4.10 Leaf apex shapes.

FIGURE 4.11 Leaf base shapes.

Leaf apex and base Various terms relating to the shape of the leaf apex are illustrated in Figure 4.10. Terms relating to the shape of the leaf base are illustrated in Figure 4.11.

Leaf margin The leaf blade may have **lobed** or **unlobed** margins. These and other margin types are illustrated in Figure 4.12.

Various kinds of teeth may be defined according to anatomical features such as the pattern of the vein or veins entering the tooth, the shape of the tooth, and characters of the tooth apex such as glandularity. The more common tooth types are illustrated in Figure 4.13; others are defined where first encountered in Chapter 9 (see also Hickey and Wolfe 1975).

Leaf texture The leaf blade may be very thin (**membranous**), papery in texture (**chartaceous**), or very thick (**coriaceous**).

Ptyxis and vernation **Ptyxis** is the way in which an individual leaf is folded in the bud. A few ptyxis terms are illus-

trated in Figure 4.14 (see also Cullen 1978). **Vernation** is the way in which leaves are folded in the bud in relation to one another. Leaves that overlap in the bud are termed **imbricate**, while those with margins merely touching are called **valvate**. A few other vernation terms are defined in Chapter 9 (in the discussion of particular families).

Indumentum An **indumentum** (plural **indumenta**), or covering of hairs (**trichomes**), on the surface of an angiosperm gives that surface a particular texture. Most indumentum terms are ambiguous, and we will use only three in this text: **glabrous** (lacking hairs), **pubescent** (with various hairs), and **glaucous** (with a waxy covering, and thus often blue or white in appearance). A few more terms describing indumenta are listed here; we will not use them in this text, but you may encounter them, as well as many others, in botanical keys and descriptions:

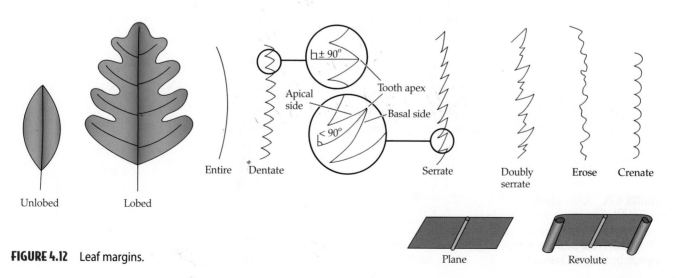

FIGURE 4.12 Leaf margins.

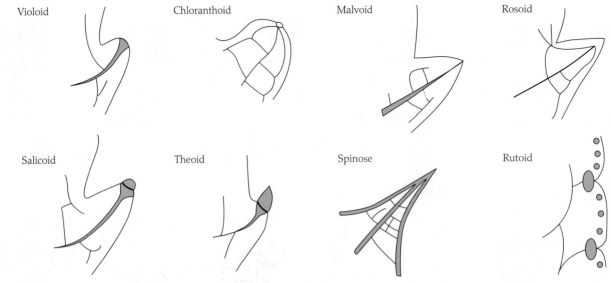

FIGURE 4.13   Some major tooth types.

arachnoid   having a cobwebby appearance
canescent   gray, with dense short hairs
hirsute   having long, often stiff, hairs
hispid   having stiff or rough hairs; bristly
lanate   woolly, with long, intertwined, somewhat matted
    hairs
pilose   having scattered, long, slender, soft hairs
puberulent   having minute, short hairs
scabrous   rough
sericeous   silky, with usually long, thin, appressed hairs
strigose   having stiff hairs, all pointing in one direction
tomentose   having densely matted soft hairs
velutinous   velvety
villous   covered with long, fine, soft hairs

We strongly recommend that the types of hairs occurring on a plant, along with their distribution and density, be carefully observed under a dissecting (or compound) microscope. Characters derived from such observations usually will be more useful (and consistently applicable) than the indumentum terms listed here.

Hairs may be **unicellular** or **multicellular, nonglandular** or **glandular**, and borne **singly** or in **tufts**, with surrounding cells of the epidermis modified or not. The shape of the individual hairs can be described in detail: Are they branched or simple? How are they branched; that is, are they **dendritic, stellate,** or **T-shaped**? Do they have a **flattened** or **globose** head, and is the stalk **uniseriate** (with one row of cells), **biseriate** (with two rows of cells), or **multiseriate** (with several rows of cells)? These terms are illustrated in Figure 4.15.

Some taxa have two or more types of hairs mixed together on their leaves or stems. For example, many species have nonglandular, unicellular hairs intermixed

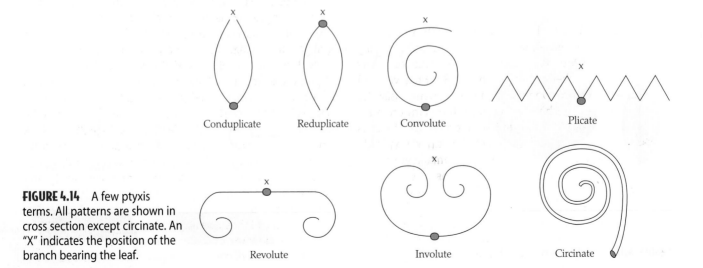

FIGURE 4.14   A few ptyxis terms. All patterns are shown in cross section except circinate. An "X" indicates the position of the branch bearing the leaf.

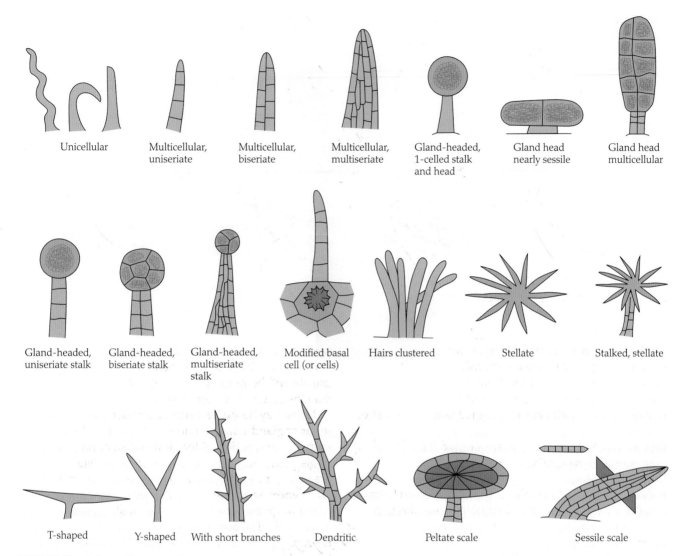

FIGURE 4.15 Selected features of hairs.

with gland-headed, multicellular hairs. The types of hairs, along with their density and distribution on the plant, are often of taxonomic value.

Domatia and glands Domatia (singular **domatium**)—literally, "tiny homes"—may contain organisms, usually mites or ants. Domatia occur on the leaves of many angiosperms (Pemberton and Turner 1989; Brouwer and Clifford 1990). Arthropod inhabitants of domatia assist the plant by deterring herbivory; in return, the plant provides them not only with a home but sometimes with food as well. Ant domatia are usually pouchlike and are typically found at the base of the leaf blade. Mite domatia are smaller and are usually found at vein junctions. They may be bowl-, volcano-, or pocket-shaped, formed by axillary hair tufts or by a revolute margin.

Various glandular structures may also occur on leaves. These structures usually secrete nectar and attract ants, which protect against herbivory (Bentley 1977).

Floral Morphology

The reproductive structures of angiosperms are called flowers. In our discussion of plant reproductive parts, we will focus on angiosperms; the specialized reproductive structures of the lycophytes ferns and their allies, conifers, cycads, ginkgos, and gnetophytes are described in Chapter 8.

A **flower** is a highly modified shoot bearing specialized appendages (modified leaves) (Figure 4.16). The modified shoot (or floral axis) is called the **receptacle**; the floral stalk is referred to as the **pedicel**. Flowers are usually borne in the axil of a more or less modified leaf, or **bract**; smaller, leaflike structures, the **bracteoles**, are often borne along the pedicel.

Flowers have up to three major parts: **perianth** (outer protective and/or colorful structures), **androecium** (plural **androecia**; pollen-producing structures), and **gynoecium** (plural **gynoecia**; ovule-producing structures). Flowers that have all three of these parts are said to be **complete**. If one or more of the three is lacking, the flower is **incomplete**. If at least the androecium and gynoecium are present, the

FIGURE 4.16 Parts of a generalized flower. Collective terms are in boldface.

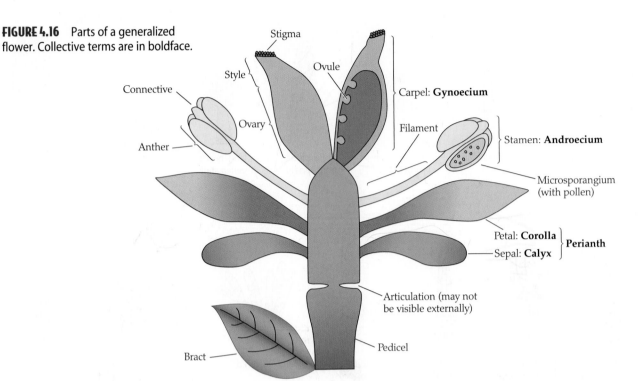

flower is termed **bisexual** (or **perfect**). If either is lacking, the flower is **unisexual** (or **imperfect**). It may be either **staminate**, if only the androecium is present, or **carpellate**, if only the gynoecium is present.

In **monoecious** species, both staminate and carpellate flowers are borne on a single individual; in **dioecious** species, the staminate and carpellate flowers are borne on separate individual plants. Various intermediate conditions exist, of course. **Polygamous** species have both bisexual and unisexual flowers (staminate and/or carpellate) on the same plant.

The perianth is always outermost in the flower, followed in nearly all flowers by the androecium, with the gynoecium in the center of the flower. The perianth parts may be undifferentiated, and the perianth composed merely of **tepals**. Alternatively, the perianth may be differentiated into two main parts, in which case it is composed of an outer whorl (or whorls or spirals) of **sepals**, collectively called the **calyx** (plural **calyces**), and an inner whorl (or whorls or spirals) of **petals**, collectively called the **corolla**.

The sepals typically protect the inner flower parts in the bud; the petals are usually colorful and assist in attracting pollinators (see the section on pollination biology on page 67). Corollas have evolved independently in various groups of angiosperms; in some families it is clear that the petals are showy, sterile stamens; in others the petals are modified sepals.

It is important to remember that although these perianth terms are useful in description and identification, they need to be used with caution in phylogenetic studies. Homology should not be assumed merely on the basis of a general similarity of form and function.

The androecium comprises all the **stamens** of the flower. Stamens are usually differentiated into an **anther** and a **filament**, although some are petal-like and are not differentiated into these two parts. Each anther usually contains four **pollen sacs**, or microsporangia, which are often confluent in two pairs. The pollen sacs are joined to each other and to the filament by a **connective**, which is occasionally expanded, forming various appendages or a conspicuous sterile tissue separating the pollen sacs.

Meiosis occurs within the pollen sacs, leading to the production of **pollen grains** (male gametophytes, or microgametophytes). The androecium is therefore often referred to as the "male part" of the flower. Of course, flowers, as part of the diploid plant (or sporophyte), cannot properly be said to be male (or female) because the sporophyte is involved only in spore production (associated with meiosis). Only the haploid plant (or gametophyte) is involved in gamete production (Figure 4.17).

Anthers open by various mechanisms, and pollen is usually released through longitudinal slits, although transverse slits, pores, and valves also occur. Anthers that open toward the center of the flower are said to be **introrse**; those that shed pollen toward the periphery are **extrorse**.

The gynoecium comprises all the **carpels** of the flower. The carpel is the site of pollination and fertilization. Carpels are typically composed of a **stigma**, which collects and facilitates the germination of the pollen carried to it by wind, water, or various animals; a **style**, a usually slender region specialized for pollen tube growth; and an **ovary**, an enlarged basal portion that surrounds and protects the **ovules**. The stigmatic surface may be variably papillate and either wet or dry.

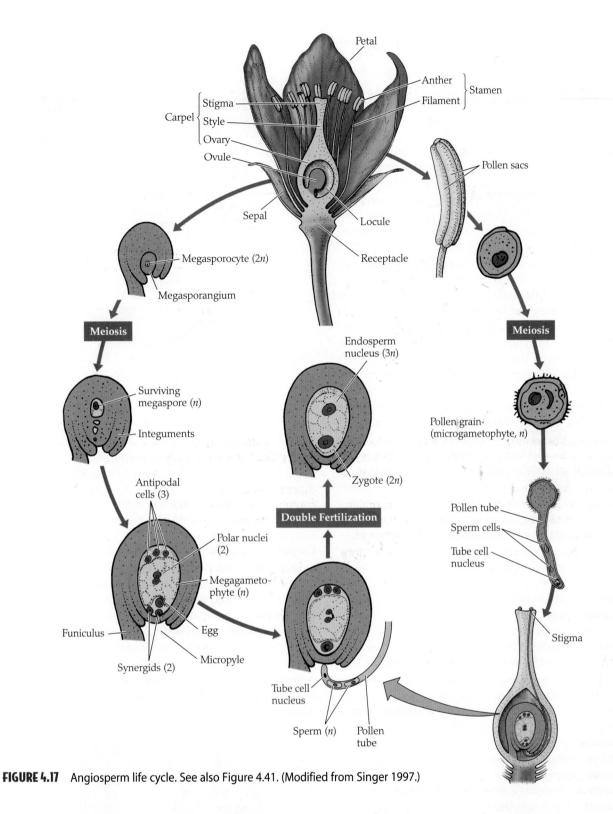

FIGURE 4.17 Angiosperm life cycle. See also Figure 4.41. (Modified from Singer 1997.)

Each ovule contains a megagametophyte (the female gametophyte, or embryo sac), which produces an egg and is usually provided with two protective layers called **integuments**. The ovule is attached to the ovary wall by a stalk called the **funiculus** (plural **funiculi**) or funicle. The gynoecium is often called the "female part" of the flower, although, as noted already, this is technically incorrect. As the ovule develops into a seed, the surrounding ovary develops into a fruit.

Various floral parts may be modified for the production of nectar or other pollinator attractants, such as oils or fragrances. **Nectaries** (nectar-producing glands) often form projections, lobes, or disklike structures. Nectaries are often produced near the base of the androecium and gynoecium or in nectar spurs formed by floral parts such as petals. Some flowers have an "extra" series of floral parts, often showy, called a **corona**. Coronal structures may be outgrowths of the perianth parts, stamens, or receptacle, and they are extremely diverse in form and function. (For a detailed discussion of the diversity of these and other floral structures, see Weberling 1989.)

The variation in floral features can be efficiently summarized by the use of floral formulas and diagrams (Box 4A).

Floral symmetry   The parts of some flowers are arranged so that two or more planes bisecting the flower through the center will produce symmetrical halves. Such flowers have **radial** symmetry, and they are also called **actinomorphic** or **regular** (Figure 4.18A). (A few radial flowers have only two planes of symmetry; these flowers are sometimes called **biradial**.)

The parts of other flowers are arranged so that they can be divided into symmetrical halves on only one plane. These flowers have **bilateral** symmetry, and are also called **zygomorphic** or **irregular** (Figure 4.18B).

A few flowers have no plane of symmetry and are **asymmetrical** (Figure 4.18C). In determining the symmetry of a flower, the position of the more conspicuous structures—that is, the perianth and/or androecium—is considered.

Fusion of floral parts   Floral parts may be fused together in various ways. Fusion of like parts (e.g., petals united to petals) is called **connation**; when like parts are not fused, they are said to be **distinct** (e.g., petals distinct from each other). Fusion of unlike parts (e.g., stamens united to petals) is called **adnation**; when unlike parts are not fused, they are said to be **free** (e.g., stamens free from petals). Fused structures may be united from the moment of origin onward, or they may grow together later in development.

Various other specialized terms are used for various types of connation and adnation:

apocarpous   carpels distinct
apopetalous   petals distinct
aposepalous   sepals distinct
apotepalous   tepals distinct
diadelphous   stamens connate by their filaments in two groups
epipetalous   stamens adnate to corolla
monadelphous   stamens connate by their filaments in a single group
sympetalous   petals connate
synandrous   stamens connate
syncarpous   carpels connate
syngenesious   stamens connate by their anthers
synsepalous   sepals connate
syntepalous   tepals connate

The corolla shape, especially in sympetalous flowers, may be valuable in identification. Specialized terms are applied to distinctive corolla shapes, including rotate (wheel- or disk-shaped), campanulate (bell-shaped), urceolate (urn-shaped), salverform (with slender tube and

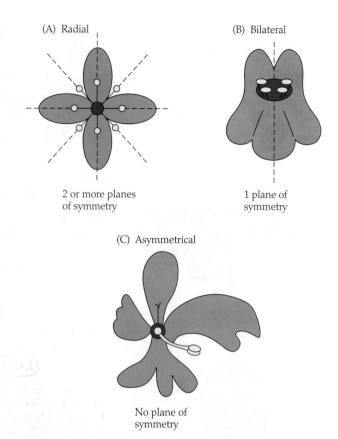

(A) Radial
2 or more planes of symmetry

(B) Bilateral
1 plane of symmetry

(C) Asymmetrical
No plane of symmetry

FIGURE 4.18   Patterns of floral symmetry.

abruptly flaring distal portion, or limb), funnelform (funnel-shaped), tubular, and bilabiate (2-lipped).

Carpel versus pistil   The term **pistil** is sometimes used for the structure(s) in the center of the flower that contain(s) the ovules. How does this term differ from *carpel*, the term introduced earlier and used throughout this book? Carpels are the basic units of the gynoecium; they may, of course, be distinct or connate. If they are distinct, then the term *pistil* is equivalent in meaning to the term *carpel*. If, however, the carpels are connate, then the terms are not equivalent because each carpel constitutes only one unit within a pistil, which is then considered to be *compound* (Figure 4.19).

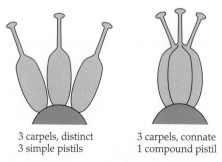

3 carpels, distinct
3 simple pistils

3 carpels, connate
1 compound pistil

FIGURE 4.19   The difference between the terms *carpel* and *pistil*.

BOX 4A Floral Formulas and Diagrams

A **floral formula** is a convenient short-hand method of recording floral symmetry, number of parts, connation and adnation, insertion, and ovary position. The formula consists of five symbols, as in the following example:

$$*, K5, C5, A\infty, G\underline{10}$$

The first symbol indicates either radial symmetry (*), bilateral symmetry (X), or asymmetry ($). The second item is the number of sepals, with "K" representing "calyx" (here K5, meaning a calyx of 5 sepals). The third item is the number of petals, with "C" representing "corolla" (here C5, meaning a corolla of 5 petals). The fourth, the androecial item, is the number of stamens, with "A" representing "androecium" (here numerous—the symbol for infinity is generally used when the number of stamens is more than about 12). The last item is the number of carpels, with "G" representing "gynoecium" (here G10, meaning a gynoecium of 10 carpels). The line below the carpel number indicates the position of the ovary with respect to other floral parts (here superior, and the flower hypogynous). If the ovary were inferior (and the flower epigynous), the line would have been drawn above the carpel number.

Connation is indicated by a circle around the number representing the parts involved. For example, in a flower with five stamens that are monadelphous (i.e., connate by their filaments), the androecial item of the floral formula would be indicated as

$$A\textcircled{5}$$

The plus symbol (+) may be used to indicate differentiation among the members of any floral part. For example, a flower with five large stamens alternating with five small ones would have the androecial item recorded as

$$A5 + 5$$

Adnation is indicated by a line connecting the numbers representing different floral parts. Thus, a flower that has a sympetalous corolla with epipetalous stamens—for example, two stamens adnate to the four connate petals—would have the numbers representing the corolla and androecial items indicated as

$$C\textcircled{4}, A\underline{2}$$

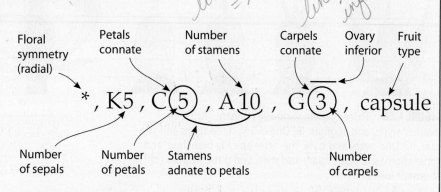

A sample floral formula

A sample floral formula

Sepals - Petals - Stamens - Carpels

The presence of a hypanthium (as in perigynous flowers) is indicated in the same fashion as adnation:

$$X, K\textcircled{5}, C\textcircled{5}, A\underline{10}, G5$$

Sterile stamens (staminodes) or sterile carpels (carpellodes or pistillodes) can be indicated by placing a dot next to the number of these sterile structures. Thus a flower with a syncarpous gynoecium composed of five fertile carpels and five sterile ones would be represented in the formula as

$$G\textcircled{5 + 5\bullet}$$

Variation in the number of floral parts within a taxon is indicated by using a dash (–) to separate the minimum and maximum numbers. For example, the formula

$$*, K4 - 5, C4 - 5, A8 - 10, G\textcircled{3}$$

would be representative of a taxon that has flowers with either 4 or 5 sepals and petals and from 8 to 10 stamens. Variation within a taxon in either connation or adnation is indicated by using a dashed (instead of a continuous) line:

$$*, K3, C\textcircled{3}, A\underline{6}, G\underline{1}$$

The lack of a particular floral part is indicated by placing a zero (0) in the appropriate position in the floral formula. For example, the floral formula

$$*, K3, C3, A0, G\textcircled{2}$$

represents a carpellate flower.

Flowers in which the perianth parts are not differentiated into a calyx and corolla (that is, flowers with a perianth of tepals) have formulas in which the second and third items (those representing sepals and petals) are combined into a single item (representing tepals). A hyphen (-) is placed before and after this item to indicate that the calyx and corolla categories have been combined, and the number is preceded by "T" indicating tepals. For example, an actinomorphic flower with 5 tepals, 10 stamens, and 3 connate carpels, with a superior ovary, would be represented as

$$*, T\text{-}5\text{-}, A10, G\textcircled{3}$$

The fruit type is often listed at the end of the floral formula:

$$*, T\text{-}5\text{-}, A10, G\textcircled{3}, capsule$$

A floral formula is by no means an end in itself; it is merely a convenient means of recording the information needed to identify a plant. Floral formulas also can be useful tools for remembering characteristics of the various angiosperm families. They are used extensively in this text (see Chapter 9). Their construction requires careful observation of individual flowers and of variation among the flowers of the same or different individuals.

Floral diagrams are stylized cross sections of flowers that represent the floral whorls as viewed from above. Rather like floral formulas, floral diagrams are used to show symmetry, numbers of parts, the relationships of the parts to one another, and degree of connation and/or adnation. Such diagrams cannot easily show ovary position. (For more information of floral diagrams, see Rendle 1925; Porter 1967; Correll and Correll 1982; Zomlefer 1994; Walters and Keil 1995.)

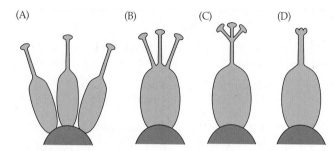

FIGURE 4.20 Three carpels, variously connate. (A) Three ovaries, styles, and stigmas. (B) One ovary, three styles and stigmas. (C) One ovary and style, the latter apically branched, and three stigmas. (D) One ovary and style, and three stigmas (or stigmatic lobes).

Number of parts Flowers differ in numbers of sepals, petals, stamens, and carpels. The number of parts is usually easily determined by counting, but extreme connation, especially of the carpels, may cause difficulties, and variation between different flowers of the same plant or between closely related species is common. Often it is possible to count fused carpels by counting the number of styles, stigmas, or stigmatic lobes (Figure 4.20). Placentation (which will be discussed shortly) may also be useful in determining carpel numbers.

Most flowers are based on a particular numerical plan—that is, on patterns of three, four, five, or various multiples of those numbers. The ending *-merous*, along with a numerical prefix, is used to indicate a flower's numerical plan. For example, a flower might have four sepals, four petals, eight stamens, and four carpels; such a flower would be described as 4-merous.

Insertion Attachment of floral parts is called **insertion**. Floral parts may be attached to the receptacle (or floral axis) in various ways. Three major insertion types are recognized: hypogynous, perigynous, and epigynous. The position of the ovary in relation to the attachment of floral parts also varies, from **superior** to **inferior** (Figure 4.21).

Flowers in which the perianth and androecium are inserted below the gynoecium are called **hypogynous**; the ovary of such flowers is said to be superior. Flowers in which a cuplike or tubular structure surrounds the gynoecium, but without being adnate to it, are called **perigynous**. In such flowers the perianth and androecium are attached to the rim of this structure, which is called the **hypanthium** (plural **hypanthia**; or **floral cup** or **floral tube**). The ovary of such flowers is also superior.

Hypanthia have evolved from various structures, such as from the fused basal portion of the perianth parts and stamens or from the receptacle. Flowers in which the perianth and stamens appear to be attached to the upper part of the ovary due to fusion of the hypanthium (or bases of floral and androecial parts) to the ovary are called **epigynous**. The ovary of such flowers is said to be inferior.

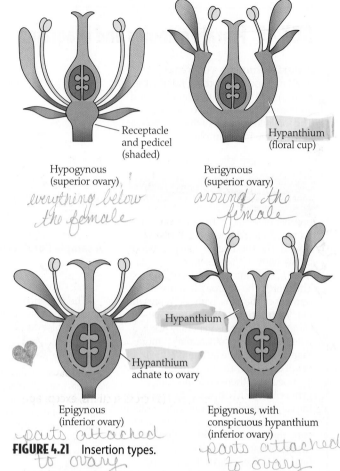

FIGURE 4.21 Insertion types.

In some epigynous flowers, the hypanthium may extend beyond the top of the ovary, forming a cup or tube around the style. If the hypanthium is fused only to the lower portion of the ovary, the latter is considered **half-inferior**. Insertion type and ovary position are best determined in a longitudinal section of the flower.

Floral parts making up adjacent whorls normally alternate with each other, so one would expect to find a petal, for example, inserted at the point between two adjacent sepals. An understanding of this common pattern can assist in interpreting the number of floral parts, especially when they are obscured by connation or adnation.

The gynoecium, or the androecium and gynoecium, occasionally are borne on stalks (the **androgynophore** and **gynophore**, respectively).

Placentation Ovules are arranged in various patterns within an ovary, allowing the recognition of various **placentation** types. Ovaries may contain from one to several chambers, or **locules**. The wall separating adjacent locules is called a **septum** (plural **septa**). The **placenta** (plural **placentae**) is the part of the ovary to which the ovules are attached. Major placentation types are illustrated in Figure 4.22. The number of ovules has no necessary correlation with the number of carpels, number of placentae, or placentation type.

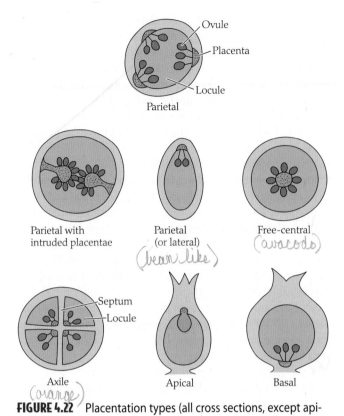

Ovule
Placenta
Locule
Parietal

Parietal with intruded placentae

Parietal (or lateral)
(bean like)

Free-central
(avacodo)

Septum
Locule

Axile
(orange)

Apical

Basal

FIGURE 4.22 Placentation types (all cross sections, except apical and basal).

Placentation type can be quite useful in determining the number of fused carpels in a flower. If placentation is **axile**, the number of locules is usually indicative of the number of carpels. In **parietal** placentation, the number of placentae usually equals the carpel number.

Miscellaneous floral terms The following list defines a few other floral terms commonly encountered in plant descriptions.

basifixed   referring to a structure, such as an anther, that is attached at its base

carpellode   a sterile carpel

centrifugal   developing first at the center and then gradually toward the periphery

centripetal   developing first at the periphery and then gradually toward the center

didynamous   having two long and two short stamens

exserted   sticking out, as in stamens extending beyond the corolla

included   hidden within, as in stamens not protruding from the corolla

pistillode   a sterile pistil

staminode   a sterile stamen

tetradynamous   having four long and two short stamens

versatile   referring to a structure, such as an anther, that is attached at its midpoint

Pollination Biology

Plants are stationary and thus depend on external forces to bring their gametes together. The sperm of ferns swim through water to reach the eggs. The sperm of seed plants are packaged in pollen grains for transport to the ovule or stigma—a process referred to as **pollination**. Conifer pollination occurs largely via wind. Most flowering plants are pollinated by animals, although wind pollination predominates in some large and ecologically dominant families and occurs sporadically in many others. Water pollination is rare.

Pollination has interested people since at least 1500 B.C., which is when Babylonians first noted that date palm (*Phoenix dactylifera*, Arecaceae) flowers produce a yellow powder (pollen). Through careful observation, they found that this powder must be applied to the flowers of fruit-bearing trees in order for the trees to produce fruit. This important discovery made it possible for them to increase the production of dates simply by spreading the yellow powder on date flowers by hand.

Pollination remains essential to human welfare today. Most human food comes from cereal grains (Poaceae) and beans (Fabaceae), all of which are the result of pollination. Nearly all the edible fruits, including apples, coconuts, figs, strawberries, and tomatoes, would not exist without pollination.

Flowers are adaptations for pollination. One can often infer the pollen vector from the morphology of a flower, as Darwin did when he predicted that there had to be a moth in Madagascar with a tongue long enough to reach the nectary in the 30-centimeter-long spur of the orchid *Angraecum sesquipedale* (see Chapter 6). The link between floral color, scent, time of flowering, structure, and rewards on the one hand and animal pollinator sensory capacity, behavior, and diet on the other is the basis of floral **pollination syndromes**. This link may be strong enough that a plant and a pollinator adapt to each other. We will examine an example of such **coevolution** involving the plant genus *Yucca* and its pollinator, the yucca moth, on page 69.

Pollination Syndromes

Wind and water pollination If you walk in a pine forest during pollen shed and pass by a pond, its surface may be covered by a yellow film of pine pollen. The pollen grains of wind-pollinated plants are often small, light, and have a smooth surface, but some species have larger grains with air spaces, which lower the density of the pollen grain and make it float better. The pollen-receptive surface of conifers and wind-pollinated angiosperms is either a sticky or a net-like trap for airborne pollen.

Wind-pollinated flowers are characterized by the production of a large amount of pollen that is readily transported by wind currents and by efficient means of trapping airborne pollen. Such flowers are small and lack much of a corolla (see, for example, illustrations of Betulaceae, Cyperaceae,

Fagaceae, Juglandaceae, and Poaceae in Chapter 9). In many wind-pollinated angiosperms, each flower has only a few ovules, and often only one of them produces a seed.

Wind pollination is unusual in the tropics, especially in lowland rain forests. Temperate forests, in contrast, are dominated by wind-pollinated trees: oaks (*Quercus*, Fagaceae), beeches (*Fagus*, Fagaceae), hickories (*Carya*, Juglandaceae), walnuts (*Juglans*, Juglandaceae), and birches (*Betula*, Betulaceae) in the Northern Hemisphere and southern beeches (*Nothofagus*, Nothofagaceae) in the Southern Hemisphere.

Pollen shed in wind-pollinated temperate species occurs at the beginning of the growing season, before or as leaves develop. A leafless forest has fewer obstacles to interrupt pollen flow. Pollen shed is also timed to avoid high humidity (which prevents the pollen from drying out and therefore reduces its buoyancy) and rain (which carries pollen down out of the air currents).

Corn (*Zea mays*, Poaceae) illustrates another common feature of wind pollination: unisexual flowers. The pollen is borne in staminate flowers in the corn "tassel," at the top of the plant, and the stigmas are the familiar "silk" that emerges, plumelike, from the top of the ear. Many people are unhappily aware of wind pollination because of hay fever, an allergic reaction to proteins in the outer pollen wall of some wind-pollinated species, such as ragweed (*Ambrosia*, Asteraceae).

Water pollination is limited to about 150 angiosperm species in 31 genera and 11 families (Cox 1988). Almost half of these species are marine or grow in brackish water, and 9 of the families are monocots. Pollen may be transported above, on, or below the water surface. Plants that pollinate under water often have filamentous, or eel-shaped, pollen borne in mucilaginous strands.

One of the longest-known and most fascinating examples of water pollination is *Vallisneria* (Hydrocharitaceae). In this genus, plants grow submersed. Their staminate flowers are released from the plant and float to the water surface, where they open and float about. At the same time, carpellate flowers rise to the surface on long peduncles and create a slight depression in the water surface, into which staminate flowers fall and are captured. Pollination follows this capture.

Animal pollination   Animal pollination is thought to be an important factor in the evolutionary success of angiosperms. Animals are often more efficient transporters of pollen than wind, and they can be found where there is little wind (such as within a dense tropical forest). They promote cross-pollination by moving between plants. Animal pollination has driven diversification in many groups, and evolution associated with pollination is nowhere more evident than in the Orchidaceae. Many species in this large family are apparently separated by floral structural reproductive isolation, as we will see in Chapter 6.

We are all familiar with bees buzzing around brightly colored flowers on warm summer days. In an apple orchard,

bees fly from flower to flower, collecting pollen in special pollen baskets on their legs and drinking the flowers' sugary nectar. Back at their hive, the bees convert the nectar into honey and feed their young the protein-rich pollen. In return for these rich rewards, the bees pollinate the apple flowers. As a bee drinks nectar and collects pollen, some pollen becomes stuck to its body hairs. That pollen is removed from its body by the stigmas of the next flower it visits. Without bees, there would be no or few apples. The same can be said for many other fruits.

In essence, bees and apple trees have a contract: the bees pollinate the apple trees, and the trees reward the bees with nectar and pollen. A plant's part of this contract consists of adaptations to attract pollinators, exploit their morphology and behavior to effect pollination, and ensure that they will return to its flowers by rewarding them. The animal's part of the contract is efficient pollination. Sometimes this contract is broken, by the plant or by the animal. The lack of reward in some cases of orchid pollination (which we will describe on page 71) represents a "breach of contract" on the part of the plant. On the part of the pollinator, some insects, called nectar robbers, chew into a flower to take the nectar without pollinating the flower.

The animals observed visiting flowers are not necessarily pollinators. For example, four species of *Marcgravia* (Marcgraviaceae) in Costa Rica are visited by small bats that take nectar from floral bracts modified into nectaries. The bats are also pollinators of these species, except for *M. nepenthoides*, which has much larger nectaries than the other three plants. The distance between nectary and flower in *M. nepenthoides* suggests that two species of opossums are more efficient pollinators of this species than the bats and are its primary pollinators (Tschapka and von Helversen 1999).

Many, though not all, flowers pollinated by one of the major types of animal pollinators have characteristic suites of floral adaptations (Table 4.1). Flowers attract their pollinators with colors and scents suited to those pollinators. Insects, for example, do not see the same spectrum of light that humans do; they are less sensitive to red but are able to see shorter wavelengths, down into the ultraviolet. Thus bee-pollinated flowers often have ultraviolet-reflecting or absorbing patterns that attract the insects to the flowers and then direct them to the nectar.

Flowers open and emit scents when their pollinators are active. Moths and bats, for example, are active at night. The flowers they visit are white and emit their scents at night, so that they are easier to locate in the darkness. Many floral fragrances are pleasant to humans; others are powerful and unpleasant. For example, flowers of *Weberocereus tunilla* (Cactaceae), an epiphytic rain forest cactus, give off a fragrance that has been described as similar to that of a corpse, and attract glossophagine bats (Tschapka et al. 1999).

Flowers that are pollinated by nonflying mammals (see Table 4.1) encompasses a broad range of floral characteristics because of the differences in morphology and foraging behavior of the three groups of pollinating mammals. Flowers pollinated by primates (such as lemurs and mon-

TABLE 4.1 **Floral pollination syndromes.**

| Pollinator | Floral characteristics | | | | |
	Color	Scent	Time of flowering	Corolla	Reward
Bee	Blue, yellow, purple	Fresh, strong	Day	Bilateral landing platform	Nectar and/or pollen
Butterfly	Bright; often red	Fresh, weak	Day	Landing platform; sometimes nectar spurs	Nectar only
Moth	White or pale	Sweet, strong	Night or dusk	Dissected; sometimes nectar spurs	Nectar only
Fly (reward)	Light	Faint	Day	Radial, shallow	Nectar and/or pollen
Fly (carrion)	Brownish, purplish	Rotten, strong	Day or night	Enclosed or open	None
Beetle	Often green or white	Various, strong	Day or night	Enclosed or open	Nectar and/or pollen
Bird	Bright; often red	None	Day	Tubular or pendant; ovary often inferior	Nectar only
Bat	Whitish	Musky, strong	Night	Showy flower or inflorescence	Nectar and/or pollen
Nonflying mammals	Dull-colored	Unscented to variously strong	Night	Robust, exserted styles and stamens	Copious nectar and/or pollen

keys) tend to be unscented and very large to accommodate the large size of the animals. Flowers pollinated by marsupials are usually located in the forest canopy, and rodent-pollinated flowers tend to be close to the ground and have a yeasty odor (Johnson et al. 2001). About 60 species of nonflying mammals have been documented as pollinators of about 100 plant species, especially in the Southern Hemisphere (Carthew and Goldingay 1997).

Flowers are built to fit their pollinators physically and to provide them with an appropriate reward. The corollas of many bee-pollinated flowers have nectar guides—lines or marks that direct bees to the nectar source. The corolla may form a landing platform that orients pollinators toward the nectar and/or pollen and forces them to perform the movements required for pollination. The pollen of many animal-pollinated flowers is covered with minute projections that stick to animal hair or feathers (see Figure 4.48E–H). Whereas bees consume both nectar and pollen, butterflies, moths, and birds are rewarded by nectar alone. In some cases pollinators are not rewarded, but rather are deceived into pollinating a flower.

Plant species may be specialized for one pollinator or may be served by a wide range of pollinators; the latter species are called **generalists**. *Acer saccharum* (sugar maple, Sapindaceae), for example, is both wind- and animal-pollinated. *Putoria calabrica* (Rubiaceae), a dwarf

shrub of the Mediterranean region, is another example of a generalist, being pollinated by various flies and butterflies (Ortiz et al. 2000). The syndromes we have listed in Table 4.1 are certainly not to be thought of as complexes of characters that are highly linked to the pollination of a particular flower type by a single pollinator, although this is sometimes the case.

Coevolution between Plant and Pollinator

One of the most fascinating cases of animal pollination is that of the yucca (*Yucca*, Agavaceae) and the yucca moth (*Tegiticula*). The yucca's white flowers open at night and attract the female yucca moth. Once inside the flower, the moth pollinates it. She carries pollen from another yucca plant in special tentacles under her head, and she stuffs this pollen into the stigmatic cavity (Figure 4.23). She then uses her ovipositor to penetrate the ovary wall and lay eggs among the ovules. Finally, just prior to leaving the flower, she climbs the filament, stabilizes herself by laying her tongue over the top of the filament, and collects pollen with her tentacles.

The moth takes no food from the flower because adult yucca moths do not eat; all consumption is by the larvae. The moth eggs hatch inside the fruit at about the time the seeds ripen, and the larvae consume some, but not all, of

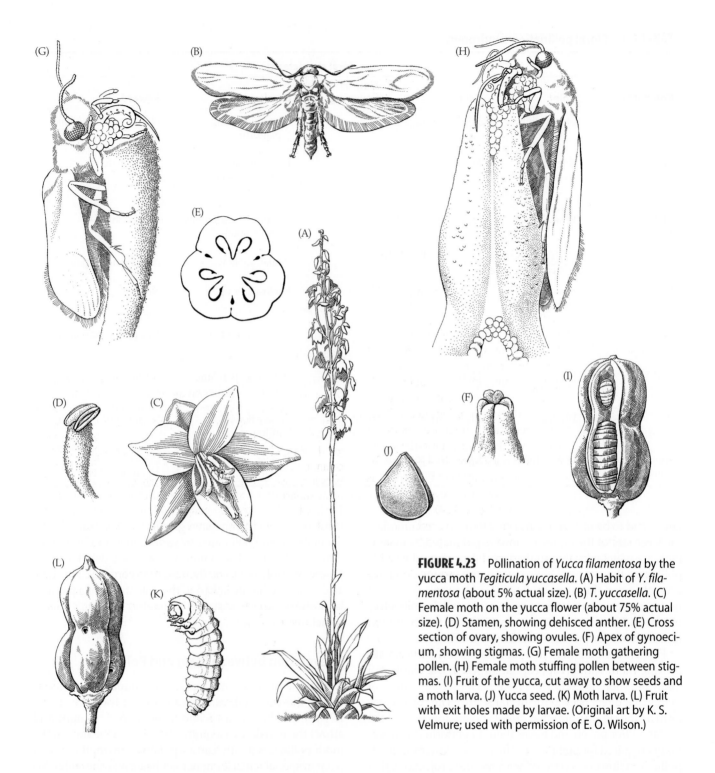

FIGURE 4.23 Pollination of *Yucca filamentosa* by the yucca moth *Tegiticula yuccasella*. (A) Habit of *Y. filamentosa* (about 5% actual size). (B) *T. yuccasella*. (C) Female moth on the yucca flower (about 75% actual size). (D) Stamen, showing dehisced anther. (E) Cross section of ovary, showing ovules. (F) Apex of gynoecium, showing stigmas. (G) Female moth gathering pollen. (H) Female moth stuffing pollen between stigmas. (I) Fruit of the yucca, cut away to show seeds and a moth larva. (J) Yucca seed. (K) Moth larva. (L) Fruit with exit holes made by larvae. (Original art by K. S. Velmure; used with permission of E. O. Wilson.)

the developing seeds. When they have eaten enough, the larvae bore out of the fruit, fall to the ground, and burrow into the soil to overwinter as pupae.

Coevolution is evident in the adaptations of yucca and moth that facilitate pollination. The plant produces stout, club-shaped filaments that the moth can easily climb to collect pollen. The anthers are positioned on top of the filament, where they are readily accessible to the moth. The moth's tentacles are an adaptation for collecting and trans-

porting pollen and placing it in stigmatic cavities. The moth's pollination behavior ensures a food supply for her young as well as reproduction by the plant that her species absolutely depends on.

Another tight coevolutionary relationship exists between figs (*Ficus*, Moraceae) and their fig wasp pollinators. Like the *Yucca* and its moth, figs and fig wasps are completely dependent on each other. There are about 750 species of figs, and most of these have their own species of fig wasp pollinator.

Deception and Nonnutritive Rewards in Orchid Pollination

Of the 19,500 species of orchids, about 8000 offer no food reward to their pollinators. Instead, they deceive their pollinators or provide them with nonnutritive rewards, such as fragrant chemicals. Some orchids, like some species in other families, bear purplish flowers with scents that resemble that of rotting flesh. These flowers, which are called carrion flowers (see Table 4.1), attract flies that lay their eggs in the flowers because they are tricked by the flowers' odor and color into perceiving the flowers as a source of food for their offspring. In the process of laying eggs, the fly moves within the flower and pollinates it.

Far more bizarre methods of ensuring pollination have evolved in some tropical orchids. We will see another example of orchid pollination deception in Chapter 6, in the flowers of *Ophrys* (Orchidaceae), which trigger mating behavior in male bees and wasps and provide no food reward.

Many New World tropical orchids are pollinated by male euglossine bees. After emerging from the nest, male euglossines fly through the forest, feeding on the nectar from various flowers and sleeping on the undersides of leaves or in tubular flowers. They are strongly attracted by the fragrant chemicals of certain orchids that do not produce nectar. They scratch at the source of the odor, and in the process of transferring the chemicals to their hind tibiae, involuntarily follow a path in the flower that leads to pollination. In some species of the orchid genera *Gongora* and *Stanhopea*, the bee is turned upside down and slides down the flower in such a way that it picks up the pollinium (Figure 4.24) (see the discussion of Orchidaceae in Chapter 9 for a description of the pollinium). The bee flies to another flower of the same species and eventually slides past the stigma, where the pollinium is deposited. Male euglossines use the floral fragrances they gather to attract females.

The Orchidaceae are an encyclopedia of pollination syndromes. In addition to pseudocopulating bees and wasps, carrion flies, and euglossine bees, pollinators of orchid species include beetles, bugs, moths, butterflies, mosquitoes, and birds. There are no records of nonflying mammal pollinators of orchids.

Modes of pollination are noted for nearly all of the families treated in Chapters 8 and 9.

Avoiding Self-Pollination

Pollination may occur within an individual (**self-pollination** or **selfing**) or between individuals (**cross-pollination**). It has long been known (Darwin 1876) that many plants avoid self-pollination and thus the possible harmful consequences of inbreeding depression. Such avoidance may be achieved by separating female and male gametes in space or in time.

Spatial separation may be accomplished through dioecy or monoecy. Temporal separation is achieved when the stigmas are receptive before the pollen is shed in a bisexual

FIGURE 4.24 *Stanhopea wardii* with *Eulaema* sp., a euglossine bee pollinator. The bee lands on the lip, which has a large brown spot at its base and spotted horn-like projections near its apex. The bee gathers fragrant compounds produced by the flower, and, in so doing, slides down the white, spotted column (on the right in this photo). When the bee reaches the tip of the column, it contacts the pollinia, which become attached to its back in a position that will bring them into contact with the stigmatic surface of the next *Stanhopea* flower visited by the bee. The bee in this photo apparently did not remove the pollinia as it exited the flower. Pollination is not always successful.

flower or monoecious individual (**protogyny**), or when the anthers shed pollen before the stigmas are receptive (**protandry**). The arrangement of stamens and stigmas and the movement of pollinators may also limit selfing.

Self-incompatibility Another mechanism for avoiding self-pollination is **self-incompatibility**, the inability of a bisexual plant to produce zygotes with its own pollen. Self-incompatibility is genetically controlled in many species by multiple alleles of a gene labeled simply *S*. Even though a diploid individual has only two of these alleles, there may be hundreds of other alleles in other individuals of the species. If a pollen grain has the same *S* allele as the carpellate plant, then mating will not be successful. Self-fertilization therefore will not occur, but with many different *S* genotypes possible, most matings with other individuals will be at least partially successful. Note that this mechanism will prevent not only selfing, but also matings between individuals that happen to have the same *S* alleles, regardless of how genetically different they are otherwise.

In **monomorphic self-incompatibility**, there are no morphological differences between the flowers of incompat-

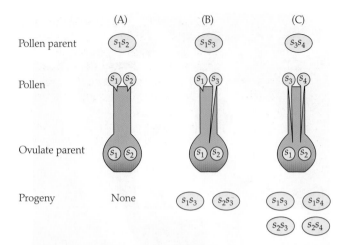

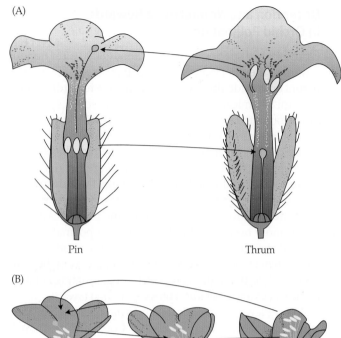

FIGURE 4.25 Consequences of a gametophytic self-incompatibility system on the production of progeny and the genetic variability of that progeny. (After Lewis 1949.)

ible individuals. Monomorphic self-incompatibility may be gametophytic or sporophytic. In **gametophytic self-incompatibility**, the genotype of the pollen determines what matings will be successful. If two individuals have the same two self-incompatibility alleles, then no pollinations between them will be successful (Figure 4.25). If one or both self-incompatibility alleles differ, however, then some or all matings will produce zygotes. Self-incompatibility between pollen parent and potential ovulate parent is most often recognized within the style, where a pollen tube with an S allele that is the same as one of the alleles of the ovulate plant simply stops growing.

In **sporophytic self-incompatibility**, the genotype of the anther (sporophyte) determines the fate of pollen grains. This form of self-incompatibility occurs in the Asteraceae and the Brassicaceae and a small number of other families; it is not as widespread as gametophytic self-incompatibility.

Heterostyly Heterostyly, or **heteromorphic self-incompatibility**, is a sporophytic self-incompatibility system that also involves floral morphological differences. The existence of two kinds of flowers in *Primula* (Primulaceae) was recognized long ago. One kind of flower has a long style and short stamens and is called a **pin** because of the resemblance of the long style to a pin. The other kind has a short style and long stamens that project out of the top of the corolla tube. This morph is called a **thrum**, referring to the resemblance of the stamens to pieces of thread coming out the end of a shirtsleeve (Figure 4.26A).

Darwin (1877) demonstrated that this heteromorphism is associated with self-incompatibility: only pollinations between different morphs are successful. The heteromorphism is governed by a supergene, a series of genes closely spaced or linked on a chromosome. Thrum individuals are heterozygous (Ss), and pins are homozygous recessive (ss), so the cross pin × thrum gives equal numbers of thrum and pin offspring.

FIGURE 4.26 Heterostyly. (A) Reciprocal anther and stigma positions in pin and thrum morphs of a distylous plant. (B) The three forms of a tristylous plant. Anthers and stigma are positioned at three levels. The arrows indicate the directions of compatible pollinations. (After Ganders 1979.)

Heterostyly is relatively infrequent. It is known from 24 families and is especially common in the Rubiaceae, where it has been documented in about 90 genera. It may be expressed as either two (distyly; Figure 4.26A) or three (tristyly; Figure 4.26B) floral morphs. Distyly, as in *Primula* and many Rubiaceae, is far more common than tristyly, which is known from only three families (Lythraceae, Oxalidaceae, and Pontederiaceae).

Inflorescences, Fruits, and Seeds

An **inflorescence** can be defined as "the shoot system which serves for the formation of flowers and which is modified accordingly" (Troll 1964, trans. Weberling 1989: 201). The arrangement of flowers on a plant (the inflorescence form and position) is important in routine identification as well as in the determination of phylogenetic relationships. Inflorescence categories have often been confused, however, because of the arbitrary separation of flower-bearing and vegetative regions of the plant (Figure 4.27).

Two quite different inflorescence types occur in angiosperms. In **determinate** (or **monotelic**) inflorescences, the

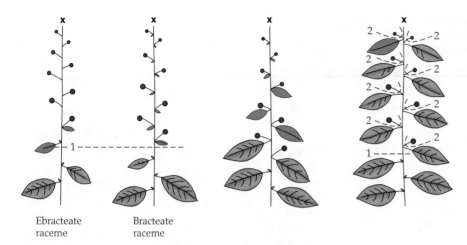

Ebracteate
raceme

Bracteate
raceme

FIGURE 4.27 Inflorescences (in this case, racemes) with various types of subtending leaves or bracts. Ebracteate and bracteate forms are frequently called terminal racemes because of the arbitrary delimitation of inflorescences from the vegetative region at 1. Leafy racemes (D) are often considered to have solitary, axillary flowers because of the arbitrary delimitation of inflorescences from the vegetative region at 2.

main axis of the inflorescence ends in a flower; in **indeterminate** (or **polytelic**) inflorescences, the growing point produces only lateral flowers or partial inflorescences (groups of flowers). Typical determinate and indeterminate inflorescences are shown in Figure 4.28 (see also Weberling 1989).

The flowering sequence of determinate inflorescences usually begins with the terminal flower at the top (or center) of the cluster of flowers. In indeterminate inflorescences, the flowering sequence usually starts at the base (or outside) of the cluster. Determinate inflorescences are generally ancestral to indeterminate ones, and transitional forms are known. Various kinds of determinate and indeterminate inflorescences, based on pattern of branching, have been described.

One of the more common types of determinate inflorescences is the **cyme** (or determinate thyrse), the lateral branches of which are composed of usually numerous three-flowered units, usually showing opposite branching (Figure 4.29). Cymes can be of many different shapes

because of differences in their branching patterns. If the inflorescence branches are initially monopodial—that is, producing several internodes before ending in a terminal flower—a panicle-like cyme results, and through reduction a racemelike cyme is formed.

The lateral branches (paraclades) of typical cymes or panicle-like cymes may be either alternately or oppositely arranged. **Scorpioid** and **helicoid** cymes are especially distinctive because of their coiled form, resulting from the abortion of one of the flowers within each three-flowered inflorescence unit.

The most common types of indeterminate inflorescences are racemes, spikes, corymbs, and panicles (Figure 4.30). A **raceme** is an inflorescence with a single axis bearing pedicellate flowers; a **spike** is similar, but the flowers are sessile (lacking a pedicel or stalk). In contrast, a **corymb** is a raceme with the pedicels of the lowermost flowers elongated, bringing all flowers to approximately the same

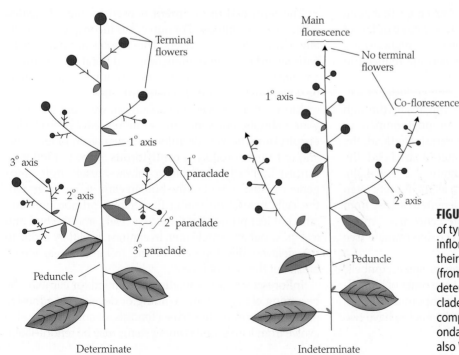

FIGURE 4.28 Diagrammatic representation of typical determinate and indeterminate inflorescences. The circles represent flowers; their size indicates the sequence of opening (from large to small). The individual units of determinate inflorescences are called paraclades; an indeterminate inflorescence is composed of the main florescence and secondary florescences (co-florescences). (See also Weberling 1989.)

FIGURE 4.29 Some common kinds of determinate inflorescences. The circles represent flowers; their size indicates the sequence of opening (from large to small).

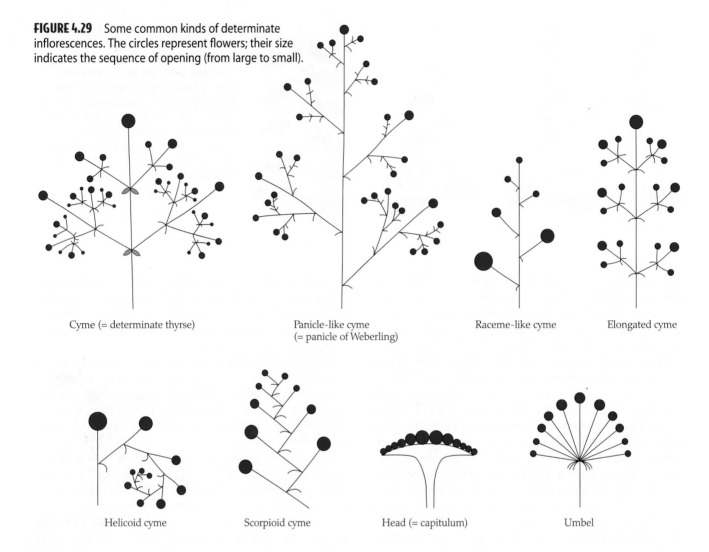

Cyme (= determinate thyrse)

Panicle-like cyme (= panicle of Weberling)

Raceme-like cyme

Elongated cyme

Helicoid cyme

Scorpioid cyme

Head (= capitulum)

Umbel

level. A **panicle** is merely a compound raceme—that is, an indeterminate inflorescence that has two or more orders of branching, with each axis bearing flowers or higher-order axes. Axillary racemes or cymes can become reduced in length, resulting in a fascicle.

A **head**, or **capitulum**, is a dense terminal cluster of sessile flowers. This inflorescence type can result through aggregation of the flowers of either an indeterminate or determinate inflorescence. In an indeterminate head, the peripheral flowers open first; in a determinate head, the central flowers open first (compare Figures 4.29 and 4.30). An **umbel** is an inflorescence in which all the flowers often have pedicels of approximately equal length that arise from a single region at the apex of the inflorescence axis. Umbels are often indeterminate, but may also be determinate (see Figures 4.29 and 4.30).

Simple inflorescences, such as racemes, spikes, umbels, and heads, have only a single axis (i.e., one order of branching). Compound inflorescences (e.g., compound racemes, compound umbels, cymes, thyrses, and panicles) have two or more orders of branching.

The term **catkin** or **ament** is used for any elongated inflorescence composed of numerous inconspicuous, usually wind-pollinated flowers. These terms are nonspecific with regard to order of branching and floral arrangement: aments may be either simple or compound, and they may be determinate or indeterminate structures.

Most inflorescences and solitary flowers are borne on young shoots, but some are borne on leaves (producing **epiphyllous** flowers or inflorescences) or on older stems and/or trunks (producing **cauliflorous** flowers or inflorescences). Epiphylly often (but not always) results from ontogenetic displacement of the bud; in early stages of growth the cells below the young flower bud primordium and adjacent leaf primordium divide actively, and the bud and leaf grow out as a single unit. In contrast, cauliflory is due to the delayed development of the inflorescences, which break out through old wood.

Inflorescences are sometimes modified for climbing by becoming elongated and twining or developing adhesive pads, thus forming **tendrils**. (Tendrils, of course, may also evolve from leaves, and twining stems may be tendril-like.)

FIGURE 4.30 Some common kinds of indeterminate inflorescences. The circles represent flowers; their size indicates the sequence of opening (from large to small).

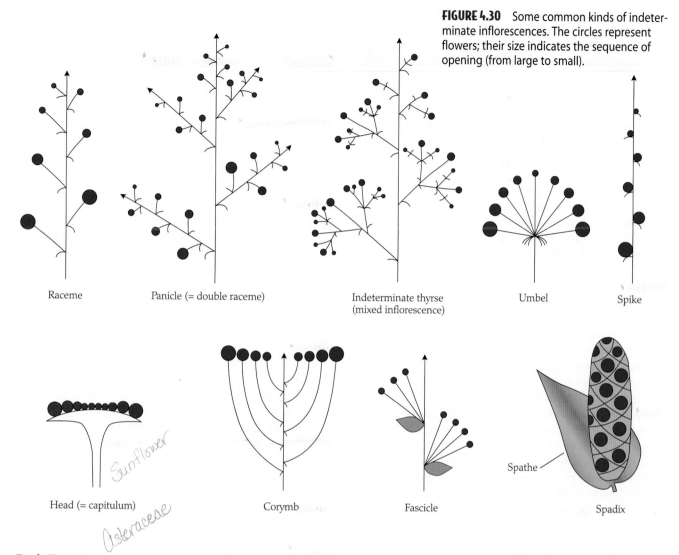

Raceme

Panicle (= double raceme)

Indeterminate thyrse (mixed inflorescence)

Umbel

Spike

Head (= capitulum)

Sunflower

Asteraceae

Corymb

Fascicle

Spathe

Spadix

Fruit Types

A **fruit** is a matured ovary along with fused accessory structures (hypanthium or perianth parts). The great diversity of size, form, texture, means of opening, and anatomy among fruits has long confounded plant systematists, and many different fruit types have been proposed.

All systems of fruit classification must deal with several difficulties. Foremost is the problem of the bewildering and often continuous variation in fruit structure; van der Pijl (1972: 17) concluded that "the fruit is too versatile and has too many aspects to be divided into strict categories." Second, additional complexities come from the extensive convergent/parallel evolution of fruiting structures; functionally similar fruits have arisen independently in different lineages of angiosperms from similar and different gynoecial conditions. Third, other parts of the flower (and even associated vegetative structures), in addition to the matured gynoecium, may form a functional part of the fruit. Examples of such **accessory structures** include the expanded fleshy receptacle of strawberries (*Fragaria*, Rosaceae), the fleshy perianth of seagrape fruits (*Coccoloba*, Polygonaceae),

the winglike calyx of dipterocarp fruits (*Dipterocarpus*, Dipterocarpaceae), and the fleshy inflorescence axis of figs (*Ficus*, Moraceae). Finally, tropical fruits have been neglected in many traditional fruit classifications.

In this text we employ an artificial system of descriptive fruit terms, based on the traditional fruit classification of Gray (1877). This system has been widely employed. It is based on the texture of the **pericarp**, or fruit wall (fleshy, dry, or hard), the pattern of **dehiscence** or **indehiscence** (type of fruit opening, or lack thereof), the shape and size of the fruit, and the carpel and ovule number.

Simple fruits (those resulting from a single flower) are divided into two categories: (1) those formed from a single carpel or several fused carpels, and (2) those that develop from several separate carpels of a single gynoecium (**aggregate fruits**). The individual units of an aggregate fruit may be any of the basic fruit types given in the list following this discussion. For example, the fruit of *Magnolia* (Magnoliaceae) is an aggregate of follicles, that of *Annona* (pawpaw, Annonaceae) an aggregate of berries, and that of *Rubus* (blackberry, Rosaceae) an aggregate of drupes.

If a fruit is the product of the gynoecia of several closely clustered flowers, it is termed a **multiple fruit**. As with aggregate fruits, the individual fruits composing the cluster may consist of any of the basic fruit types outlined in the list that follows this discussion. For example, the fruit of *Ananas* (pineapple, Bromeliaceae) is a multiple of berries, that of *Morus* (mulberry, Moraceae) a multiple of drupes, and that of *Platanus* (sycamore, Platanaceae) a multiple of achenes. The use of modifying terms (e.g., drupaceous schizocarp, winged or samaroid schizocarp, 1-seeded fleshy capsule) is encouraged.

This classification is presented by means of a key to a series of definitions. Although this system is admittedly arbitrary, the key and descriptions have proved useful in teaching and floristics. (For more information on fruit types, see Judd 1985; Weberling 1989; and Spjut 1994.)

achene a fairly small, indehiscent, dry fruit with a thin and close-fitting wall surrounding a single seed; includes the **cypsela**. Examples: *Bidens, Carex, Ceratophyllum, Clematis, Cyperus, Ficus, Fragaria, Helianthus, Medicago* (some), *Ostrya, Petiveria, Polygonium, Ranunculus, Rhynchospora, Rosa* (achenes enclosed in a fleshy hypanthium), *Rumex, Sagittaria, Taraxacum, Trifolium* (some), *Vernonia*. (See Figures 4.31A,B, 9.17, 9.58, 9.80, and 9.146.)

berry an indehiscent, fleshy fruit with one or a few to many seeds. The flesh may be ± homogeneous throughout, or the outer part may be hard, firm, or leathery; septa are present in some, and the seeds may be arillate (having a fleshy outgrowth of the funiculus) or with a fleshy testa (seed coat). Examples: *Actinidia, Annona, Averrhoa, Cananga, Citrus, Cucurbita, Eugenia, Litchi, Miconia, Musa* (some), *Opuntia, Passiflora, Phoenix, Punica, Sideroxylon, Smilax, Solanum, Tamarindus, Vaccinium, Vitis*. (See Figures 4.31G,H, 9.12, 9.48, 9.63, 9.52, 9.100, 9.108, and 9.121.)

capsule a dry to (rarely) fleshy fruit from a 2- to many-carpellate gynoecium that opens in various ways to release the seeds. Such fruits may have from one to many locules; if 2-locular, then the partition is not persistent. Examples: *Aesculus, Allium, Antirrhinum, Argemone, Aristolochia, Begonia, Blighia, Campsis, Clusia, Echinocystis, Epidendrum, Eucalyptus, Euonymus, Hibiscus, Hypericum, Ipomoea, Justicia, Lachnanthes, Lagerstroemia, Lecythis, Lyonia, Momordica, Oxalis, Papaver, Portulaca,* *Rhododendron, Swietenia, Triodanis, Viola*. (See Figures 4.31S, 9.32, 9.51, 9.55, 9.62, 9.71, 9.73, 9.109, 9.115, 9.116, 9.131, and 9.134.)

caryopsis (plural **caryopses**) (**grain**) a small, indehiscent, dry fruit with a thin wall surrounding and more or less fused to a single seed. Examples: most Poaceae. (See Figures 4.31C,D, 9.39 and 9.40.)

dehiscent drupe a fruit with a dry or fibrous to fleshy or leathery outer husk that early to tardily breaks apart (or opens), exposing one or more nutlike pits enclosing the seed(s). Examples: *Carya, Rhamnus* (some), *Sageretia*. (See Figure 9.96.)

drupe an indehiscent, fleshy fruit in which the outer part is more or less soft (to occasionally leathery or fibrous) and the center contains one or more hard pits (**pyrenes**) enclosing seeds. Examples: *Aegle, Arctostaphylos, Celtis, Clerodendrum, Cocos, Cordia, Cornus, Ilex, Juglans, Licania, Melia, Myrsine, Nectandra, Prunus, Psychotria, Roystonea, Rubus, Sabal, Scaevola, Syagrus, Terminalia, Toxicodendron*. (See Figures 4.31L,M, 9.13, 9.82, 9.95, 9.113, 9.130, and 9.137.)

follicle a dry to (rarely) fleshy fruit derived from a single carpel that opens along a single longitudinal suture; the seeds may be arillate or with a fleshy testa. Examples: *Akebia, Alstonia, Aquilegia, Asclepias, Caltha, Grevillea, Magnolia, Nerium, Paeonia, Sterculia, Zanthoxylum*. (See Figures 4.31O, 9.47, 9.61, 9.126, and 9.127.)

indehiscent pod an indehiscent, fairly dry fruit with few to many seeds. Examples: *Adansonia, Arachis, Bertholletia, Cassia* (some), *Crescentia, Kigelia, Medicago* (some), *Thespesia* (some).

legume a dry fruit derived from a single carpel that opens along ± two longitudinal sutures. Examples: many Fabaceae. (See Figures 4.31N, 9.74, 9.75, and 9.76.)

loment a dry fruit derived from a single carpel that breaks transversely into 1-seeded segments. Examples: *Aeschynomene, Desmodium, Sophora*.

nut a fairly large, indehiscent, dry fruit with a thick and bony wall surrounding a single seed. Examples: *Brasenia, Castanea, Corylus, Dipterocarpus, Fagus, Nelumbo, Quercus, Shorea*. (See Figures 4.31F, 9.91.)

pome an indehiscent, fleshy fruit in which the outer part is soft and the center contains papery or cartilaginous structures enclosing the seeds. Examples: most Rosaceae, tribe Pyreae. (See Figures 4.31J,K, 9.81.)

FIGURE 4.31 Some common kinds of fruits. (A) Achene with persistent bristles, *Scirpus tabernaemontani*. (B) The same, in longitudinal section. The fruit wall is indicated with radiating lines, the endosperm is stippled, and the embryo of the single seed is unshaded; the seed coat is too thin to show. (C) Caryopsis (grain), *Panicum clandestinum*. (D) The same, in longitudinal section, embryo to left, endosperm stippled. (E) Samara, *Ulmus rubra*. (F) Nut with cupule, containing a single seed, *Quercus alba*. (G) Berry, *Mosiera longipes*. (H) Same berry in cross section, showing numerous seeds. (I) *Poncirus trifoliata* berry in cross section, seeds embedded among pulp vesicles. (J) Pome of *Amelanchier laevis*. (K) The same, in longitudinal section; note inferior ovary and seed (at left). (L) Drupe with cupule, *Sassafras albidum*. (M) The same, in longitudinal section. Note embryo with large, fleshy cotyledons, pumule, and radicle. Endocarp is indicated with radiating lines. (N) Dehisced legume, *Vicia ludoviciana*. (O) Follicles from a 5-carpellate flower, *Caltha palustris*. (P) Schizocarp, *Croton linearis*. (Q) The same, in cross section, showing three seeds. (R) Silique, *Lepidium virginicum*. Note that it will split along the midline to release seeds. (S) Open capsule with escaping seeds, *Salix caroliniana*. (Compiled from various Generic Flora illustrations, courtesy of Carroll Wood.)

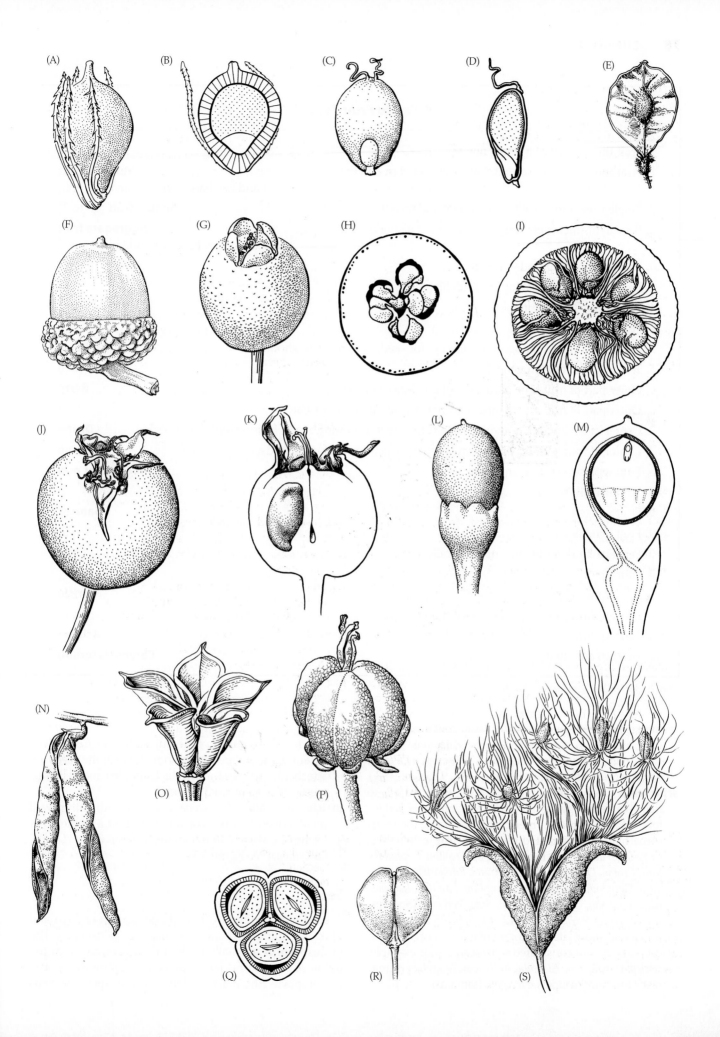

A Key to Fruit Types[a]

1. Fruit the product of a single flower 2
1. Fruit the product of several flowers clustered in one mass .... **Multiple fruit**
   [go to 3 and key based on an individual unit]

2. Single fruit (carpel solitary or several and fused) .... **Simple fruit** [go to 3]
2. Several distinct fruits (carpels several and distinct) .... **Aggregate fruit**
   [go to 3 and key based on individual units]

3. Fruit not opening (indehiscent) 4
3. Fruit opening or breaking apart (dehiscent) 13
4. Fruit fleshy (at least in part) 5
4. Fruit dry 8
5. Texture of fruit ± homogeneous (except for seeds), fleshy throughout .... **Berry**
5. Texture of fruit heterogeneous 6
6. Outer part of fruit firm, hard, or leathery; inner part softer .... **Berry**
6. Outer part of fruit ± soft; inner part papery, cartilaginous, or hard 7
7. Center of fruit with 1 or more hard pits (pyrenes) enclosing seeds; ovary inferior or superior .... **Drupe**
7. Center of fruit with papery or cartilaginous structures enclosing seeds; ovary inferior .... **Pome**
8. Fruit with several to many seeds .... **Indehiscent pod**
8. Fruit usually 1-seeded 9
9. Fruit winged .... **Samara**
9. Fruit wingless 10
10. Pericarp thick and bony; fruit generally large .... **Nut**
10. Pericarp thin; fruit smaller 11
11. Pericarp loose and free from seed .... **Utricle**
11. Pericarp firm, close-fitting or fused to seed 12
12. Pericarp firm, close-fitting, but free from seed .... **Achene**
12. Pericarp adnate (fused) to seed .... **Caryopsis (grain)**

[a] See Appendix 2 for explanation of dichotomous keys.

samara   a winged, indehiscent, dry fruit containing a single (rarely two) seed(s). Examples: *Ailanthus, Betula, Casuarina, Fraxinus, Liriodendron, Myroxylon, Ptelea, Stigmaphyllon, Ulmus.* (See Figures 4.31E, 9.85, and 9.93.)

schizocarp   a dry to rarely fleshy fruit derived from a 2- to many-carpellate gynoecium that splits into 1-seeded (or few-seeded) segments (**mericarps**). If desired, the mericarps may be designated as samara-like, achenelike, drupelike, and so on. Examples: *Acer, Apium, Cephalanthus, Croton, Daucus, Diodia, Erodium, Euphorbia, Glandularia, Gouania, Heliconia, Heliotropium, Lamium, Lycopus, Malva, Ochna, Oxypolis, Salvia, Sida, Verbena.* (See Figures 4.31P,Q, 9.67, 9.68, 9.112, 9.125, 9.135, and 9.139.) Fruits that show late developmental fusion of their apical parts are not considered schizocarps—for example, *Asclepias* (follicles), *Sterculia* (follicles), *Ailanthus* (samaras), *Simarouba* (drupes), *Pterygota* (samaras).

silique   a fruit derived from a 2-carpellate gynoecium in which the two halves of the fruit split away from a persistent partition (around the rim of which the seeds are attached); includes the **silicle**. Examples: many Brassicaceae. (See Figures 4.31R and 9.102.)

utricle   a small, indehiscent, dry fruit with a thin wall (bladderlike) that is loose and free from a single seed. Examples: *Amaranthus* (some), *Chenopodium, Lemna, Limonium.* (See Figure 9.54.)

Seeds

A **seed** is a matured ovule that contains an embryo and often its nutritive tissues (endosperm, perisperm). The **endosperm** is a usually triploid tissue derived from the union of the two nuclei in the central cell of the female gametophyte (the polar nuclei) with one sperm nucleus

13. Fruit from a single carpel 14

13. Fruit from a 2- to many-carpellate gynoecium 16

14. Fruit dehiscing along a single suture (slit) ....**Follicle**

14. Fruit dehiscing by two longitudinal sutures, or breaking up by transverse sutures 15

15. Sutures longitudinal ....**Legume**

15. Sutures transverse, the fruit breaking into 1-seeded segments ....**Loment**

16. Fruit with a dry/fibrous to leathery or fleshy outer husk that early to tardily breaks apart; center of fruit with hard pit(s) enclosing seed(s) ....**Dehiscent drupe**

16. Fruit lacking hard pit(s) enclosing seed(s); splitting open or into 1-seeded segments 17

17. Fruit splitting into 1- or few-seeded segments (mericarps) ....**Schizocarp**

17. Fruit splitting open and releasing seeds 18

18. Fruit 2-locular, the two valves splitting away from a persistent thin partition around the rim of which the seeds are attached ....**Silique**

18. Fruit 1- to several-locular, the partition not persistent if the fruit 2-locular ....**Capsule** [Go to 19]

19. Dehiscence circumscissile (splitting transversely), the top coming off like a lid ....**Circumscissile capsule (pyxis)**

19. Dehiscence not circumscissile 20

20. Fruit opening by pores, flaps, or teeth 21

20. Fruit opening longitudinally or irregularly 22

21. Fruit opening by a series of apical teeth ....**Denticidal capsule**

21. Fruit opening by pores or flaps (often near the top) ....**Poricidal capsule**

22. Fruit opening irregularly ....**Anomalicidal capsule**

22. Fruit opening longitudinally 23

23. Valves breaking away from the septa (partitions between the locules) ....**Septifragal capsule**

23. Valves remaining attached to the septa (at least in part) 24

24. Fruit splitting at the septa ....**Septicidal capsule**

24. Fruit splitting between the septa and into the locules of the ovary, or fruit 1-locular .....**Loculicidal capsule**

(see the section on embryology on page 87). Endosperm may be **homogeneous** (uniform in texture) or **ruminate** (dissected by partitions that grow inward from the seed coat). It may contain starch, oils, proteins, oligosaccharides, and/or hemicellulose, and it may be hard to soft and fleshy. The **perisperm** is a specialized diploid nutritive tissue derived from the megasporangium.

The seed is surrounded by a **seed coat**, which develops from the integument(s). The details of seed coat anatomy are quite variable. The **testa** (plural **testae**) develops from the outer integument and the **tegmen** (plural **tegmina**) from the inner integument. The prefixes *exo-*, *meso-*, and *endo-* refer to tissues developing from the outer epidermis, the middle portion, and the inner epidermis, respectively, of each of the two integuments. Seeds may be variously sized and shaped, and they may be associated with a wing or a tuft of hairs. The testa varies in surface texture due to the pattern and outgrowths of the individual cells composing its surface, and it is sometimes colorful and fleshy.

Some seeds are associated with a hard to soft, oily to fleshy, and often brightly colored structure called an **aril**. The aril is usually an outgrowth of the funiculus or the outer integument, although sometimes this term is restricted to structures derived from the funiculus, with those derived from the outer integument called **caruncles**. The seed bears a scar, called the **hilum** (plural **hila**), at the point where it was attached to the funiculus.

The embryo consists of an **epicotyl**, which will develop into the shoot; a **radicle**, which will develop into the primary root and usually gives rise to the root system; a **hypocotyl**, which connects the epicotyl and radicle; and usually one or two **cotyledons** (seedling leaves), which may be leaflike, fleshy, or modified as nutrient-absorptive structures.

Fruit and Seed Dispersal

Most fruit types can be dispersed by a variety of agents. Different parts of the fruit, seed, or associated structures (pedicel, perianth) may be modified for similar dispersal-related functions. For example, wind dispersal may be accomplished by (1) a tuft of hairs on the seeds, as in *Asclepias* (Apocynaceae), which has follicles that open to release hairy seeds; (2) wings on the fruits, as in *Fraxinus* (Oleaceae), which has samaras; (3) hair tufts on the fruits, as in *Anemone* (Ranunculaceae), which has an achene with a persistent style bearing elongated hairs; (4) a winglike perianth, as in *Dipterocarpus* (Dipterocarpaceae), which has nuts associated with elongated, winglike sepals; (5) association of the **infructescence** (mature inflorescence, with fruits) with an expanded, winglike bract, as in *Tilia* (Malvaceae), in which the fruits are nuts; or (6) a tumbleweed habit, as in *Cycloloma* (Amaranthaceae), in which the entire plant is blown across the landscape, dispersing its small fruits as it rolls.

Bird dispersal may be enhanced by (1) a colorful, fleshy seed coat, as in *Magnolia* (Magnoliaceae), which has an aggregate of follicles that open to reveal the fleshy seeds; (2) fleshy, indehiscent fruits, as in *Solanum* (Solanaceae), which has berries, *Prunus* (Rosaceae), which has drupes, or *Amelanchier* (Rosaceae), which has pomes; or (3) association of the fruit (or fruits) with fleshy accessory structures, as in *Coccoloba* (Polygonaceae), which has achenes surrounded by a fleshy perianth, *Fragaria* (Rosaceae), which has achenes borne on an expanded, colorful, fleshy receptacle, or *Hovenia* (Rhamnaceae), which has drupes associated with fleshy pedicels and inflorescence axes.

It is easy to see that similarly functioning fruiting structures may be derived from very different floral parts. Convergence (i.e., the derivation of similar structures from very different ancestral structures) is common in fruits, and similar fruits have evolved independently in many different angiosperm families. (For additional information on fruit dispersal, see van der Pijl 1972 and Weberling 1989.)

Some heavy fruits or seeds simply drop from the plant, land on the ground, and stay there. This means of dispersal is not very common, however, and may be characteristic only of species that have lost their primary dispersal agent, as may be the case for osage orange (*Maclura pomifera*, Moraceae).

Dispersal by the plant itself usually occurs through some kind of explosive discharge of seeds, fruits, or portions of fruits by means of swelling of seed mucilage, turgor pressure changes, or hygroscopic tissues. This category also includes passive movement of seed containers by wind, rain, or animals and creeping dispersal units whereby the fruit or seed moves itself by hygroscopic movements of bristles.

Adaptations for transport by wind include small, dust-like seeds; seeds with a balloonlike loose testa; inflated utricles, calyces, or bracts; or a pericarp with air spaces. Wind dispersal may be facilitated by a plume formed from a persistent style, long, hairy awns, a modified perianth (such as a pappus), placental outgrowths, outgrowths of the funiculus, elongation of the integument, a wing that splits apart, or hair tufts. Wings for wind dispersal may be present on fruits or seeds or developed from accessory parts (perianth, bracts). In tumbleweed dispersal, a large part of the plant or inflorescence breaks off and is blown around.

Dispersal by water occurs when seeds or fruits are washed away by rainfall or carried in water currents. Such seeds or fruits are often small, dry, and hard, and they may have spines or projections that act as anchoring structures (water burrs), a slimy covering, an unwettable surface layer, or low density and thus the ability to float.

Adaptations for transport on the outside surfaces of animals include small seeds or fruits with spines, hooks, or sticky hairs that are located at ground level and easily detach from the plant. This category also includes small and hard fruits or seeds that stick with mud to the feet of waterfowl, as well as trample burrs that become caught in the feet of large grazing mammals. Many sticky fruits attach to the feathers of birds.

Fruits and seeds may also be transported within an animal (after ingestion) or in its mouth. This category can be divided into subtypes by the kind of animal carrying the fruit or seed:

- *Fish* disperse some fleshy fruits or seeds of plants of riversides or inundated areas.
- Transport by *turtles* or *lizards* characterizes some fleshy fruits that have an odor. Such fruits are sometimes colored and often are borne near the ground or dropped from the plant at maturity. Some have a hard skin; others are hard but contain arillate seeds or seeds with fleshy testae.
- *Birds* may disperse nuts or seeds by carrying them in their bills or by hiding and burying them. Some viscid seeds stick to birds' bills. Bird-dispersed fruits or seeds often have an attractive edible part. The seeds of some fleshy fruits are protected from ingestion by a bony wall, bitter taste, or toxic compounds. When mature, these fruits have signaling colors that attract birds (often red contrasting with black, blue, or white). These fruits have no odor and no closed hard rind (or in hard fruits, the seeds are exposed or dangling), and they remain attached to the plant. Some have colorful hard seeds that mimic the colorful fleshy fruits of other bird-dispersed species.
- Transport by *nonflying mammals* is often associated with their stockpiling of fruits (especially nuts) or seeds. Mammal-dispersed fruits often have a high oil content and are frequently fleshy, with hard centers or leathery to hard skins that can be opened to reveal fleshy inner tissues, arillate seeds, or seeds with fleshy testae. The seeds may be toxic, bitter, or thick walled. Odor is very important in attracting mammals, but color is not essential. The fruits often drop from the plant.

- Fruits transported by *bats* share many of the characters listed in the previous entry for nonflying mammals, but they are usually borne in an exposed position (e.g., outside the dense crown of a tree). They have drab colors and a musty, sourish, or rancid odor. They are often large, fleshy, and easily digested, and they remain attached to the plant.
- Some seeds contain small, nutritious arils (or elaiosomes) and are dispersed by *ants*.

Modes of fruit and/or seed dispersal are noted for most of the families treated in Chapters 8 and 9.

Anatomy

Characteristics related to the internal structure of plants have been employed for systematic purposes for over 150 years, and they are useful in both practical identification and determination of phylogenetic relationships. **Anatomical characters** can be investigated by light microscope study, and many can be seen in the laboratory by use of the simplest of techniques. Characters observable with the transmission electron microscope (TEM) are frequently referred to as **ultrastructural**; those observable with the scanning electron microscope (SEM) are often called **micromorphological**. Some important characters of all three types are discussed briefly in this section.

Various distinctive cell types may be found in a variety of tissues. Sclereids and fibers are thick-walled and lignified cells. **Fibers** are elongated cells that frequently surround and protect the vascular tissue in the stem and the veins of the leaf. Such fibers vary in their pattern of arrangement and sometimes form girders connecting the vein to the adaxial and/or abaxial epidermis. **Sclereids** are thick-walled cells of various shapes. Secretory canals or cells, and cells containing various kinds of crystals (discussed on page 00), are often diagnostic for particular taxa. In **collenchyma**, the angles of the cell walls, in particular, are thickened with cellulose.

Secondary Xylem and Phloem

Wood (often called **secondary xylem**) is produced by a **vascular cambium** (plural **cambia**), a cylinder of actively dividing cells just inside the bark of a woody plant. Wood is a complex mixture of water-conducting cells (**tracheids** and/or **vessel elements**), cells with a support function (fibers), and living cells that run from the outside to the inside of the stem (**rays**) (Figure 4.32).

The water-conducting and support cells are both dead at maturity, which makes sense because cytoplasm would interfere with water transport. These cells have thick walls made up of both **cellulose** and **lignin**—the former a long chain of glucose molecules, the latter a complex polymeric substance composed of phenolic subunits. Whereas conifers and cycads have only tracheids for water conduction, gnetophytes and angiosperms usually have both tracheids and vessel elements. Angiosperms tend to have short, broad vessel elements with completely open ends that fit together like sewer pipes (forming a vessel) and conduct large amounts of water rapidly.

Many anatomists (e.g., Bailey 1944, 1951, 1957) have considered xylem types in terms of a priori defined trends

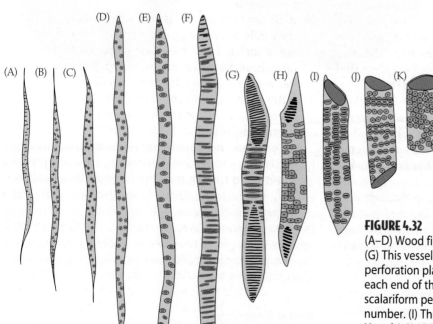

FIGURE 4.32 Some cell types of the secondary xylem. (A–D) Wood fibers. (E, F) Tracheids. (G–K) Vessel elements. (G) This vessel element type has many-barred scalariform perforation plates (the ladderlike pattern of openings at each end of the cell). (H) This vessel element also has scalariform perforation plates, but the bars are reduced in number. (I) This vessel element is an intermediate between H and J. (J, K) These short, broad, vessel elements have simple perforation plates. (Modified from Radford et al. 1974 and Bailey and Tupper 1918.)

FIGURE 4.33 A few examples of sieve tube element plastids. (A) S-type, with many starch grains (common in many angiosperms). (B) *Dracaena* (Convallariaceae); P-type, with numerous cuneate (wedge-shaped) crystalloids. (C) *Laurus* (Lauraceae); with both proteins and starch. (D) *Petiveria* (Petiveriaceae); P-type, with central protein crystalloid surrounded by protein filaments. (E) *Alternanthera* (Amaranthaceae); P-type, with protein filaments only. C, protein crystalloid; F, protein filaments; S, starch. (From Behnke 1975.)

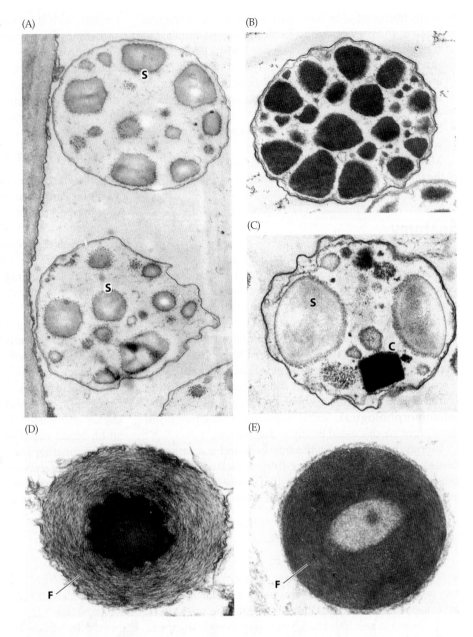

in vessel element evolution; that is, the progression from tracheids to long, narrow vessel elements with slanted, scalariform perforation plates to short, broad, vessel elements with simple perforation plates (Figure 4.32E–K). Carlquist (1988) has emphasized the linkage between xylem structure and its ecological function.

Other important aspects of wood anatomy include growth rings (bands or layers in the wood produced by seasonal variation in cambial activity) and the presence of various specialized cells containing crystals, resins, mucilage, or latex.

Secondary phloem provides fewer taxonomic characters than secondary xylem. There are two major types of carbohydrate-conducting cells in vascular plants: **sieve cells** and the more specialized **sieve tube elements**. The latter possess a distinctive sieve plate and companion cells. Stratified phloem consists of alternating bands of sieve tube elements

and companion cells and fibers; plants with such phloem often have fibrous bark.

Much emphasis has been placed on the structure of the plastids in sieve tube elements of angiosperms (Behnke 1972, 1975, 1977, 1981, 1991, 1994, 2000) because differences in their structure have been correlated with major clades, such as monocots, Caryophyllales, and Fabaceae. Behnke recognized two major categories of plastids in sieve tube elements: the S-type, which accumulates starch, and the P-type, which accumulates proteins (or proteins and starch) (Figure 4.33).

Nodal Anatomy

Nodal anatomy refers to the various patterns in the vascular connections between stem and leaf. The anatomy of the node is quite variable and often of systematic significance.

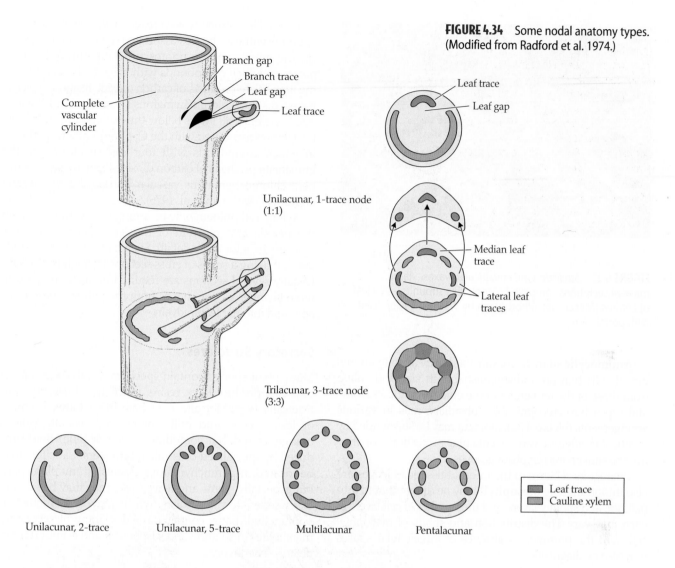

FIGURE 4.34 Some nodal anatomy types. (Modified from Radford et al. 1974.)

A key feature is the number of **leaf gaps**, or **parenchymatous interruptions**, left in the secondary vascular system of angiosperms by the departure of vascular bundles (**leaf traces**) to the leaves. The configuration of these leaf gaps is used as the basis of nodal types (Figure 4.34).

The nodal pattern can be expressed in terms of the number of traces (strands of vascular tissue) and the number of gaps, i.e., parenchymatous interruptions or **lacunae** (singular **lacuna**) (spaces between the vascular strands). For example, a unilacunar node with a single trace would be described as 1:1; a unilacunar node with two traces would be recorded as 2:1; and a trilacunar node with three traces would be 3:3.

Leaf Anatomy

Leaves are extremely varied anatomically and provide numerous systematically significant characters (Carlquist 1961; Dickison 1975; Stuessy 1990).

The **epidermis** (outer layer of the leaf) varies in the number of cell layers, the size and shape of individual cells, the thickness of cell walls, and the occurrence of **papillae** (singular **papilla**) (rounded bumps or projections of individual epidermal cells) or various kinds of hairs (trichomes; see page 00). Some leaves have a **hypodermis**, which is formed from one or more differentiated layers of cells beneath the epidermis. The **cuticle**, a waxy coating over the epidermis, varies in thickness and surface texture. Various wax deposits (**epicuticular waxes**) may also be deposited on top of the cuticle (Figure 4.35) (see also Behnke and Barthlott 1983; Barthlott 1990; Barthlott et al. 1998).

The epidermis contains pores, or **stomata** (singular **stoma**; also *stomates*), each surrounded by specialized **guard cells** that open or close the pore by means of changes in their internal water pressure. A variety of stomatal forms occur in vascular plants. Stomata are usually classified by the relationships of their **subsidiary cells** (epidermal cells associated with the stoma and morphologically distinguishable from the surrounding epidermal cells) to one another and to the guard cells. Note that the same configuration of subsidiary cells can result from different developmental pathways; studies of these pathways often aid in the interpretation of the phylogenetic significance of stomatal form.

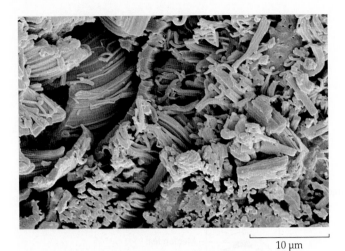

10 μm

FIGURE 4.35 *Strelitzia*-type epicuticular waxes—large, massive secretions composed of rodletlike subunits—in a leaf of *Latania* (Arecaceae). (Photograph by Y. Renea Taylor, used with permission.)

Anomocytic stomata are surrounded by a limited number of cells that are indistinguishable in size and shape from those of the remainder of the epidermis. Other stomatal types have recognizable subsidiary cells in various arrangements (Figure 4.36). Stomata may be surrounded by a cuticular ridge or various cuticular projections, or they may be sunken into crypts or grooves.

Characters relating to the internal tissues of leaves are also important. The **mesophyll** may be differentiated into palisade and spongy layers, and the number of cell layers in each may vary. The distribution and shape of mesophyll cells and the presence or absence of intercellular spaces may also be diagnostic.

Internal leaf structure is correlated with the biochemistry of photosynthesis. Leaves associated with C_3 photosynthesis, the most common photosynthetic pathway in green plants, in which compounds with three carbon atoms are the immediate products of carbon dioxide fixation, typically have a distinct chlorenchymatous layer of palisade cells in one to several cell layers below the leaf epidermis. In contrast, leaves associated with the C_4 photosynthetic pathway, in which compounds with four carbon atoms are the immediate products of carbon dioxide fixation, have prominent chlorenchymatous vascular bundle sheaths (**Kranz anatomy**) (Rathnam et al. 1976).

Xylem and phloem may be arranged in various ways in the petiole and midvein of the leaf (Figure 4.37) (see Howard 1974 for more detail). Such patterns are best studied by making a series of cross sections through the petiole. Usually several sections are required because the pattern often changes as one moves from the petiole's base to its apex and into the leaf midvein.

Secretory Structures

Many plant species contain specialized cells or groups of cells that produce latex, resins, mucilage, or essential oils (Metcalfe 1966; Metcalfe and Chalk 1950). **Latex** is a more or less opaque and milky or colored (usually yellow, orange, or red, but sometimes green or blue) fluid produced by specialized cells called **laticifers**. Laticifers may be located in parenchymatous tissues of any plant part, but especially stems and leaves, and they may be solitary or in series, forming tubes, which may be branched or unbranched. Latex contains a wide variety of secondary metabolites in solution and suspension and is important in deterring herbivory.

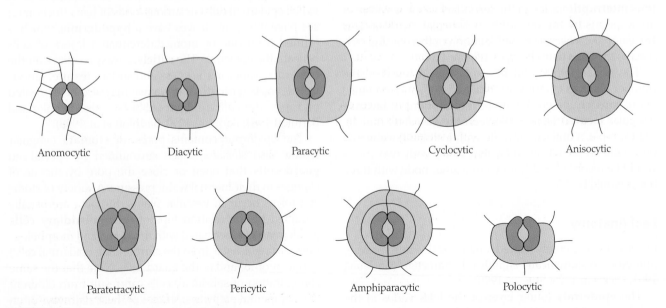

Anomocytic   Diacytic   Paracytic   Cyclocytic   Anisocytic

Paratetracytic   Pericytic   Amphiparacytic   Polocytic

FIGURE 4.36 Some important stomatal types.

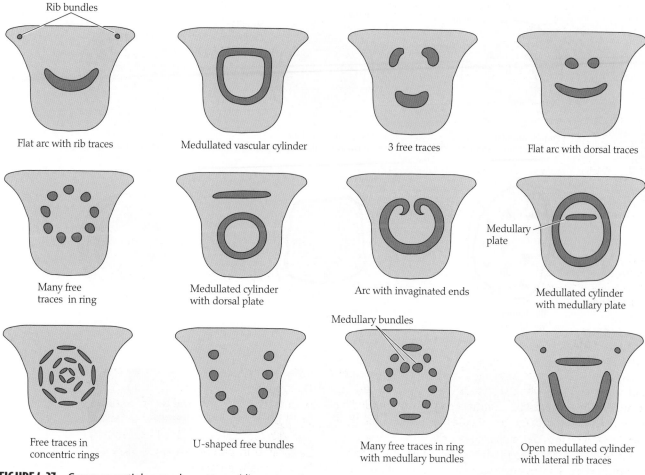

FIGURE 4.37 Common petiole vascular patterns (diagrammatic transverse sections at base of blade). (Modified from Radford et al. 1994.)

Some plant species produce clear **resins** (aromatic hydrocarbons that harden when oxidized) or **mucilage** (slimy fluids) in scattered cells, specialized cavities, or canals. **Essential oils** are highly volatile, aromatic organic compounds that are produced by specialized spherical cells scattered in the mesophyll or in cavities created by cellular breakdown or the separation of adjacent cells. Such cells or cavities in leaves give them a **pellucid dotted** appearance when viewed with transmitted light. The presence or absence of latex, resins, mucilage, and essential oils and the form and distribution of laticifers and secretory canals or cavities are often taxonomically significant.

Crystals

Crystals are common in vascular plants, usually located in cells, and variously shaped (Figure 4.38). They are usually composed of calcium oxalate, calcium carbonate, or silica. **Druses** (spherical groups of crystals), **raphides** (needlelike crystals), and **crystal sand** are the most common types. Silica bodies are often taxonomically important in monocots. Calcified bodies called **cystoliths** sometimes occur in spe-

cialized cells (called **lithocysts**); cystoliths often have systematic significance. The calcareous material in cystoliths is in the form of small amorphous particles.

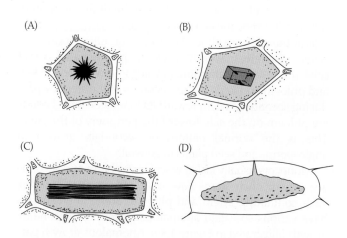

FIGURE 4.38 Crystal types. (A) Druse. (B) Prismatic form. (C) Raphides. (D) Cystolith.

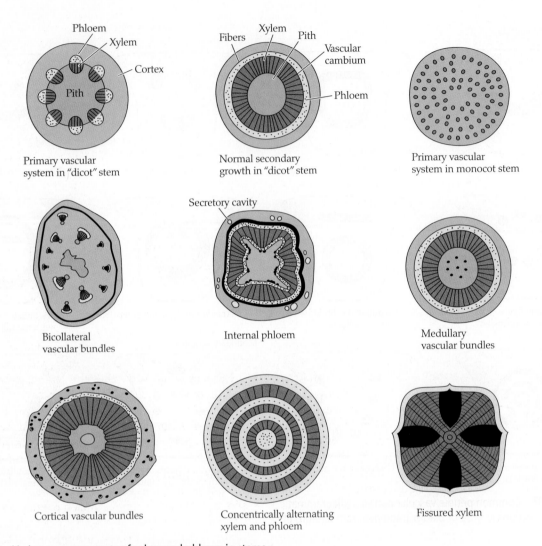

FIGURE 4.39   Various arrangements of xylem and phloem in stems.

Arrangement of Xylem and Phloem in the Stem

The stems of most seed plants contain a ring of primary xylem and phloem (as seen in cross section; Figure 4.39). Primary vascular tissues result from the differentiation of cells produced by the apical meristem, the zone of dividing cells at the apex of a shoot. The ring of bundles produced is called a **eustele**. In woody species, a meristematic layer of cells, the vascular cambium, develops between the xylem and phloem and adds to the thickness of the stem by producing secondary xylem toward its inner side and secondary phloem on the side toward the periphery of the stem. This is the normal pattern of secondary growth in angiosperms and seed plants in general.

Monocots usually lack a vascular cambium and secondary growth; their stems have scattered vascular bundles, each containing xylem and phloem. Some angiosperms have various so-called anomalous patterns of secondary growth (illustrated in Figure 4.39). Anomalous growth patterns are frequently encountered in succulents, where they allow for rapid increases in girth, and in lianas, where they resist damage due to twisting and bending. The following

terms describe different anatomical patterns, most relating to secondary growth:

axes more or less flattened or furrowed   the stem is flattened or furrowed due to unequally active cambium

concentrically alternating xylem and phloem   layers of phloem and xylem alternate in the stem because of the action of a series of vascular cambia

cortical vascular bundles   leaf trace bundles that run longitudinally in the cortex of the stem before joining the vascular system of the stem

fissured xylem   xylem broken up by the development of phloem or parenchyma tissue

included phloem   strands of phloem embedded in the secondary xylem

internal phloem   primary phloem in the form of strands or a continuous ring (as seen in a cross section of the stem) at the inner boundary of the xylem; the xylem is thus bounded by phloem on both its inner and outer surfaces. In species lacking secondary growth, development of internal phloem results in the presence of **bicol-**

lateral vascular bundles; that is, the vascular bundles of the stem have phloem on the inner as well as the normal outer side of the xylem

medullary vascular bundles   vascular bundles occurring in the pith of a stem

Floral Anatomy and Development

The pattern of vascular traces in flowers is often useful for understanding vestigial structures and homologies of parts in highly modified flowers. Extreme modification of parts is a particular problem in flowers that are greatly reduced and often densely clustered together in association with the evolution of wind pollination, such as those of the Betulaceae (see Chapter 9). The pattern of vascular traces in the gynoecium can usually be used to indicate the number of carpels, especially in gynoecia that have completely fused carpels with free-central, basal, or apical placentation.

The positioning of floral **primordia** (singular **primordium**) (floral parts or organs in their earliest condition) and their sequence of initiation are also of taxonomic significance (Evans and Dickinson 1996). Developmental studies are important for understanding homologies of various floral parts. For example, some flowers with many stamens have stamen primordia positioned in a spiral, while in other flowers they are clustered into five or ten groups. In addition, stamen primordia may be initiated **centripetally** (from periphery to center) or **centrifugally** (from center to periphery). Such variation suggests that multiple stamens have evolved many times among angiosperms.

Studies of corolla development have also been especially informative (Leins and Erbar 2003). In some sympetalous flowers the corolla develops from a ring primordium, which then develops lobes, while in others the corolla lobes are initiated first. Developmental studies have indicated that some flowers, such as those of Apiaceae and many Ericaceae, that appear to have separate petals are sympetalous early in development; they are therefore considered to have evolved from sympetalous ancestors. Studies have also shown that inferior and superior ovaries can be formed through various developmental pathways; for example, some flowers show a hypogynous developmental pathway with a convex floral apex, while others have an appendicular/epigynous developmental pathway with a concave apex (Soltis et al. 2003).

Embryology

Spores, gametophytes, and gametangia show marked evolutionary trends in tracheophytes (Table 4.2). Sporophytes produce spores in **sporangia** (singular **sporangium**). Most lycophytes and ferns (see Chapter 8) have only one kind of spore; these plants are **homosporous**. Their gametophytes are completely independent of the sporophyte (see Figure 8.1), bear both kinds of gametes (egg and sperm), and are either large (about 1 cm in diameter), green, and photosynthetic or curious subterranean structures that are saprophytic, acquiring nutrition from dead organisms with the aid of a fungus. Gametes are housed and protected in specialized structures within gametophytes called **gametangia** (singular **gametangium**): eggs in **archegonia** (singular **archegonium**) and sperm in **antheridia** (singular **antheridium**).

Heterosporous plants, which include a few genera of lycophytes and ferns and all seed plants, have two kinds of sporangia. **Megasporangia** contain **megaspores** that develop into **megagametophytes** (female gametophytes, or **embryo sacs**), and **microsporangia** (singular **microsporangium**) contain **microspores** that develop into **microgametophytes** (male gametophytes). Heterospory is correlated

TABLE 4.2   **General features of spores, gametophytes, and gametangia in major groups of tracheophytes.**

Group[a]	Heterospory	Gametophyte dependence on sporophyte	Gametophyte size (in cells)	Gametangia (antheridia and archegonia)
Lycopodiaceae, Equisetaceae, Psilotaceae, and most leptosporangiate ferns	No	None	Millions (macroscopic)	Present
Selaginellaceae, Isoetaceae, and aquatic leptosporangiate ferns (Marsileaceae and Salviniaceae)	Yes	Almost complete	Thousands (female); sometimes under 100 (male)	Present
Conifers, cycads, ginkgos, gnetopsids	Yes	Complete	1000s (female); a few (male)	Archegonia only
Angiosperms	Yes	Complete	About 7 (female); 3 or fewer (male)	Absent

[a] See Chapter 8 for more information about groups other than angiosperms and Chapter 9 for angiosperms.

with three gametophytic adaptations. First, gametophytes are small and develop inside the spore, unlike the relatively large, independent gametophytes of most tracheophytes that release their spores. Second, the two kinds of gametophytes are specialized: the megagametophyte is larger and is invested with a supply of nutrients, and the microgametophyte is small and dispersible. Third, the gametophytes are nutritionally dependent on the sporophyte. Reduction in gametophyte size continues in seed plants and is associated with the loss of recognizable antheridia in conifers and related plants and of antheridia and archegonia in angiosperms.

The embryological features of lycophytes, ferns and their allies, conifers, cycads, ginkgos, and gnetophytes are discussed further in Chapter 8. In this section we focus on angiosperm ovules, megagametophytes, embryos, and endosperm as well as on agamospermy.

Ovules and Megagametophytes

The **ovule** is a megasporangium surrounded by one or two protective layers (the integuments) and attached to the ovary wall by a stalk (the funiculus). The integuments nearly enclose the ovule, leaving only a small opening, the **micropyle**, through which pollen tubes usually enter the megasporangium. Ovules have been classified by their curvature. In an **orthotropous** ovule, the axis of the ovule and

funiclus are in a straight line; in an **anatropous** ovule, the ovule is inverted almost 180°. In a **campylotropous** ovule, the axis of the ovule is curved, and the ovule is held at about 90° (Figure 4.40; see also Figure 4.17). The ovule develops into a seed, and the integuments develop into a seed coat.

The megasporangium is the site of the meiosis that generates the megaspore. In at least 70% of angiosperms, meiosis yields four haploid megaspores, three of which degenerate. The fourth megaspore undergoes three mitotic divisions to produce a megagametophyte with eight nuclei in seven cells (Figure 4.41). Typically the egg and two other cells (the *synergids*) cluster near the micropyle, and a cell with two nuclei (the *polar nuclei*) ends up near the center. Three antipodal cells (whose functions are not clearly known) lie near the end of the megagametophyte away from the micropyle. However, in some basal angiosperms (e.g., Nymphaeaceae, Illiciaceae), the megagametophyte is composed of only the egg, synergids, and a single polar nucleus. The mature megagametophyte is surrounded by the sporangium wall, which in angiosperms also is called the **nucellus** (Figure 4.41). The nucellus may vary from a single layer to several layers of cells.

The pollen tube enters the megagametophyte through the micropyle and releases two sperm. One sperm fertilizes the egg to form the diploid **zygote**, the first cell of the next sporophytic generation. The other sperm usually fuses with both polar nuclei of the central cell to form the triploid primary endosperm nucleus, but in some basal angiosperms it fuses with the single polar nucleus, producing a diploid primary endosperm nucleus (Williams and Friedman 2004). This **double fertilization** process that gives rise to the endosperm is unique to angiosperms. The function of the other cells of the megagametophyte is unclear, although the synergids facilitate the fusion of sperm and egg in at least some species. Other, less common types of megagametophytes are restricted to certain genera or families.

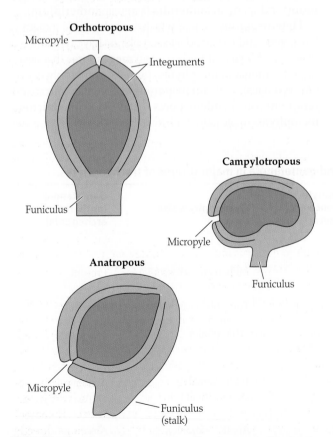

FIGURE 4.40 Three common types of ovule axes (in longitudinal section).

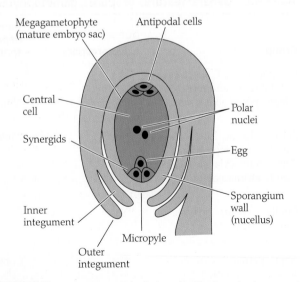

FIGURE 4.41 A mature ovule (in longitudinal section).

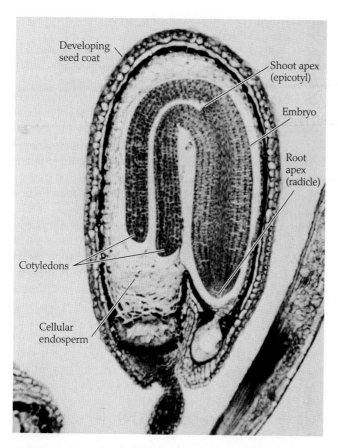

FIGURE 4.42 Longitudinal section of the seed of *Capsella* (Brassicaceae), showing embryo and endosperm. (From Gifford and Foster 1988. Copyright W. H. Freeman; used with permission.)

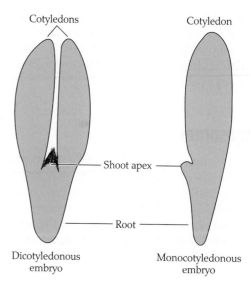

FIGURE 4.43 A comparison of "dicot" and monocot embryos. (Modified from Gifford and Foster 1988.)

Embryo and Endosperm

Embryo and endosperm are the two major components of angiosperm seeds. The development of embryos varies considerably within angiosperms. Mature embryos consist of an axis, one end of which is the root (radicle) and the other the shoot (epicotyl) (Figure 4.42).

When there are two cotyledons, the shoot apex lies between them, as in Figure 4.42. Flowering plants with such embryos are called **dicotyledons** and have long been considered a taxon, but they are actually a paraphyletic group. In **monocotyledons** the shoot apex is lateral, lying next to the solitary cotyledon (Figure 4.43). When the seed germinates and the embryo begins growing into a seedling, the cotyledon or cotyledons may enlarge and become green and photosynthetic. Alternatively, they may remain below ground or not develop beyond their embryonic state.

Endosperm is the specialized tissue of angiosperms that supplies nutrients to the developing embryo and in many cases to the seedling as well. The primary endosperm nucleus may divide many times without cell walls being formed. In most species with this **nuclear type** of endosperm development, cell walls do eventually form. The **cellular type** of endosperm development involves cell wall formation from the start. In the **helobial type** of endosperm development, which characterizes most monocots, the primary endosperm nucleus produces two cells; one divides little, while cells derived from the other make up most of the endosperm.

The endosperm may be completely absorbed by the cotyledons of the embryo before it matures, as in peas, beans, and walnuts (something you can easily confirm for yourself). The cotyledons of these **exalbuminous** seeds take over the role of the endosperm in nourishing the seedling. Examples of mature seeds with conspicuous endosperm are cereal grains—such as rice, wheat, corn, and oats—and palms. Corn endosperm is the source of popcorn, and the endosperm of coconuts (*Cocos nucifera*) is made up of the familiar coconut "milk" (nuclear endosperm before cell walls have formed) and "meat" (endosperm after cell walls have formed).

Agamospermy

Some plant species produce embryos without producing haploid gametes and without fertilization. This phenomenon is referred to as **agamospermy** (*a* = without; *gamo* = gametes; *sperm* = seed).

There are two major forms of agamospermy. In the first, a diploid cell functions as a megaspore to produce a gametophyte with the somatic chromosome number. Such megagametophytes are often similar in appearance to sexually derived gametophytes. The egg develops into an embryo without fertilization, a process called **parthenogenesis**. The second form of agamospermy, **adventitious embryony**, is best known in citrus plants (*Citrus*, Rutaceae) and is otherwise rare. Here the embryo develops directly from a somatic cell within the ovule without the formation of a gametophyte.

The best method of detecting agamospermy is to study megaspore development to determine whether the megaspores are mitotically or meiotically derived. The taxonomic distribution and systematic significance of agamospermy are discussed in Chapter 6.

Chromosomes

Chromosome number by itself may be a useful systematic character. Similar chromosome numbers may indicate close relationship; different chromosome numbers often create some reproductive isolation through reduced fertility of hybrids. Chromosome size, the position of the centromere, special banding patterns, and other features may also be systematically informative.

Chromosome Number

The lowest chromosome number in the somatic cells of a plant is its **diploid** number, often designated $2n$. Rice, for example, has a diploid number of 24. The diploid genome consists of two full chromosome sets or complements, one from the ovulate (maternal) parent and one from the pollen (paternal) parent. One full set, the **haploid** chromosome number, is borne by both spores and gametes (eggs and sperm). The haploid number, n, of rice is 12.

The hypothesized ancestral haploid number (often the lowest number) in a group of plants, such as a genus or family, is referred to as the **base number** and designated x. In the birches (*Betula*), for example, $x = 14$, and the hypothesized ancestral diploid number in that genus is therefore $2n = 28$ (Table 4.3).

The chromosome numbers of angiosperms range from $2n = 4$ (e.g., *Haplopappus gracilis*, Asteraceae) to $2n = 250$ (*Kalanchoe*, Crassulaceae). Ferns (see Chapter 8) show unusually high chromosome numbers, the highest known being $2n = 1440$ in *Ophioglossum reticulatum*. Knowledge of chromosome numbers is limited in many groups to a minority of species and often comes from a small sample of individuals within species.

The addition or loss of one or two whole chromosomes is referred to as **aneuploidy**. Examples of aneuploidy in *Clarkia* and *Lantana* are discussed on page 91. The presence of three or more whole sets of chromosomes in somatic cells is called **polyploidy**. A familiar polyploid is bread wheat (*Triticum aestivum*), which has six full sets of chromosomes. In *Triticum*, $x = 7$, and the chromosome number of bread wheat is therefore $2n = 6x = 42$. Polyploids are differentiated by the number of chromosome complements they contain. Triploids contain three chromosome complements, tetraploids four (hence *tetra-*), pentaploids five, and hexaploids six.

The addition of whole chromosome complements can occur in either somatic tissues or gametes. For example, if the nucleus of a cell that gives rise to a flowering branch fails to divide mitotically, then the chromosome number in that branch is automatically doubled. More commonly, production of chromosomally unreduced gametes leads to polyploidy. If, for example, an unreduced ($2n$) egg is fertilized by a reduced ($1n$) sperm, the zygote will be triploid ($3n$). If the plant that develops from this triploid zygote produces a triploid egg that is fertilized by a haploid sperm, the offspring will be tetraploid. Evolutionary and systematic aspects of polyploidy are discussed in Chapter 6.

There are two major forms of polyploidy. **Autopolyploidy** results from the union of three or more chromosome complements from the same (hence *auto-*) species; **allopolyploidy** results from the union of two or more different (hence *allo-*) genomes. Because they have diverged, chromosomes of different genomes often do not pair with each other. In diploids and allopolyploids there are two copies of each chromosome, and in meiosis the two homologous chromosomes pair to form a **bivalent**. Autopolyploids contain three or more homologues of each chromosome, and pairing of more than two chromosomes to form **multivalents** is thus a possibility. Multivalents lead to gametes with unbalanced chromosome numbers and sterility problems.

TABLE 4.3   **Some chromosome number variation in tracheophytes.**

Taxon	Family	2n
Asplenium trichomanes	Polypodiaceae	
ssp. *trichomanes*		72
ssp. *quadrivalens*		144
Lycopodium and related genera	Lycopodiaceae	34–36, 44–48, 68, 136, 260, 272
Betula	Betulaceae	28, 42, 56, 70, 84
Lantana	Verbenaceae	18, 22, 24, 27, 33, 44, 48, 72
Vicia	Fabaceae	10, 12, 14, 24, 28
Pyreae	Rosaceae	34, 51, 68
All genera	Pinaceae	24

TABLE 4.4 Examples of plant polyploids and their 2n chromosome numbers.

Species	Type of polyploid	2n
Taraxacum officinale, common dandelion	Allotriploid	21
Nicotiana tabacum, tobacco	Allotetraploid	48
Gossypium barbadense, cotton	Allotetraploid	52
Vaccinium corymbosum, highbush blueberry[a]	Allotetraploid	48
Betula papyrifera, white birch[a]	Allopentaploid	70
Triticum aestivum, bread wheat	Allohexaploid	42
Lythrum salicaria, purple loosestrife[a]	Autotetraploid	60
Phleum pratense, timothy grass[a]	Autohexaploid	42

[a]Other ploidy levels are known for these species.

Chromosomal pairing between genomes from various sources ranges from none (allopolyploids) to complete (autopolyploids), with a full range of intermediate levels corresponding to intermediate levels of genetic divergence. Hence it is best to consider autopolyploidy and allopolyploidy as extremes of a continuum. Some examples of polyploid plants are given in Table 4.4. Autopolyploidy is apparently not as common as allopolyploidy.

Chromosome number is generally constant within a species, although exceptions to this generality are fairly frequent. Chromosome number may also be constant within large groups. In Andropogoneae, the large grass tribe that includes *Zea*, *Sorghum*, and many important range grasses, $x = 10$ consistently; in the great majority of the approximately 1000 species of the tribe Pyreae, $x = 17$; and almost all members of Pinaceae are diploids ($2n = 24$).

In some species, chromosome number varies without correlated morphological variation. Autopolyploids, for example, may not differ morphologically from their diploid progenitors and are therefore often placed in the same species. Diploid ($2n = 14$) *Tolmiea menziesii* (Saxifragaceae), which grows in northern California and southern Oregon, and tetraploid ($2n = 28$) *T. menziesii*, which grows from central Oregon to southern Alaska, are morphologically very similar. In the spring beauty (*Claytonia virginica*, Portulacaceae) of eastern North America, aneuploidy is extensive: there are 50 different diploid numbers reported for this species, ranging from 12 to about 191.

Differences in chromosome number, when associated with morphological differences, may be recognized taxonomically, as in subspecies of *Asplenium trichomanes* (see Table 4.3). Species within many genera differ in ploidy level. The white-barked birches (*Betula*, Betulaceae) of North America, for example, include diploids [$2n = 28$: gray birch (*B. populifolia*) and mountain paper birch (*B. cordifolia*)], tetraploids [$2n = 56$: *B. cordifolia* and paper birch (*B. papyrifera*)], pentaploids ($2n = 70$: *B. papyrifera*), and hexaploids ($2n = 84$: *B. papyrifera*).

Differences in chromosome number between species often lead to reduced fertility in hybrids and the creation of a species boundary. *Clarkia biloba* (Onagraceae) is a widespread and variable Californian endemic with $n = 8$. *Clarkia lingulata* ($n = 9$) has an additional chromosome that is derived from chromosomes of *C. biloba*. The two species differ only in petal shape, but hybrids between them show very low fertility. Such aneuploid changes, however, do not always lead to speciation, as *Claytonia virginica* and other species with more than one chromosome number exemplify.

As an example of the taxonomic utility of chromosome number, consider *Lantana* (see Table 4.3). These tropical shrubs are taxonomically confusing due to aneuploidy, polyploidy, hybridization, and poorly resolved generic limits. There are two base numbers (11 and 12) and perhaps a third (9) in the genus. Diploid species, with $2n = 22$ or 24, are the foundations for polyploids, especially tetraploids ($2n = 44$ or 48). Triploids ($2n = 33$ or 36) result from crosses between diploids and tetraploids.

The base number 12 characterizes *Lantana* sect. *Callioreas*, and $x = 11$, found in *Lantana* sect. *Camara*, may have evolved via aneuploidy from $x = 12$. Several apparent synapomorphies support sect. *Camara* as monophyletic, but sect. *Callioreas* strongly resembles the related large genus *Lippia*. Additional studies are required to unravel the relationships among these groups.

At the species level, chromosome studies have augmented morphological studies in *Lantana*. Most triploids have some chromosomes that do not pair as bivalents in meiosis. Instead, they occur as **univalents** (solitary chromosomes for which there is no homologue) and multivalents. The presence of univalents corroborates the morphologically based hypothesis that many of the triploids are hybrids.

Cytology has played a key role in resolving relationships among Florida *Lantana* sect. *Camara* species. Morphologically based studies initially identified two species in the state: native *L. depressa* and the introduced tetraploid *L. camara*. Chromosome studies, coupled with other data, identified three diploid varieties of *L. depressa* in Florida (Figure 4.44), each of which hybridizes with *L. camara*. Chromosome numbers were critical data in resolving this systematic problem.

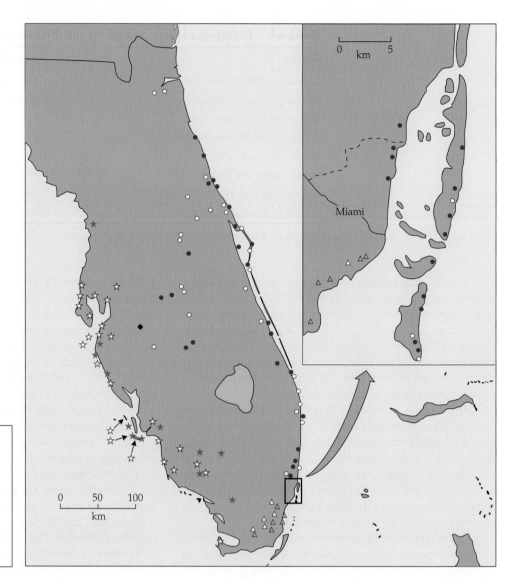

L. depressa
△ var. *depressa*
● var. *floridana*
★ var. *sanibelensis*
◆ intermed.?, vars.
   floridana & *sanibelensis*

△ Hybrids with *L. camara*
○ Hybrids with *L. camara*
☆ Hybrids with *L. camara*

Chromosome Structure

Chromosome number, size, and structural features make up what is called the **karyotype**, which may be useful in discriminating taxa. Chromosomes differ not only in overall length, but also in the length of the two chromosomal arms (Figure 4.45). The location of the **centromere**, the point on the chromosome where it is attached to the mechanism that separates chromosomes in cell division, determines whether the arms are more or less equal or unequal in length.

The combination of overall chromosome length and centromere location may allow discrimination of many of the chromosomes in a genome. Further distinctions are provided by specialized techniques for staining chromosomal bands (Nogueira et al. 1995). Genome mapping, an exciting approach that may soon have a major effect on systematic studies, is discussed in Chapter 5.

FIGURE 4.45   Chromosomes from *Callisia fragrans* (Commelinaceae), in which 2*n* = 12. Each object is a pair of homologous chromosomes. Some of the centromeres are indicated with arrows. (From Jones and Jopling 1972.)

(A)

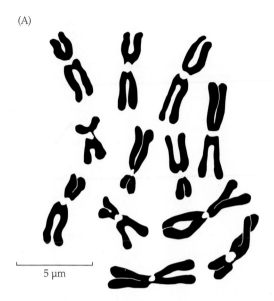

(B)

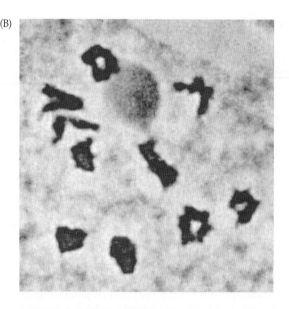

FIGURE 4.46 (A) Mitosis in *Rheo spathacea* (Commelinaceae), in which 2*n* = 12. Each structure consists of two duplicate DNA molecules (sister chromatids). (From Jones and Jopling 1972.) (B) Meiosis in *Andropogon gyrans* (Poaceae), in which *n* = 10 (× 940); each object is a pair (bivalent) of homologous chromosomes. The partial sphere just above the center is the nucleolus, where ribosomes are made. (From Campbell 1983.)

Methods of Chromosome Study

Determination of chromosome number and other karyotypic features is a routine component of plant systematics. Chromosome number may be studied in dividing cells undergoing either mitosis or meiosis (Figure 4.46). Mitosis is commonly examined in cells from actively growing root tips, but other tissues, such as expanding petals, may also be used. Meiosis is studied most often because it provides more information than mitosis about the relationships of genomes. **Microsporocytes**, the cells that give rise to pollen, are the cells of choice for meiotic study because they are easier to work with than megasporocytes and because they are easily removed from the anthers; there are also many more of them.

Protocols for chromosome study involve staining the cells with a chromosome-specific stain, such as carmine. Softening of the tissue facilitates squashing the cells so that the chromosomes separate from one another and are distinguishable for counting. Successful chromosome study may require considerable patience and skill because the chromosomes may be numerous and small, and because it may be difficult to collect material during the correct stage of meiosis. (Methods are outlined in Darlington and La Cour 1975 and Sessions 1990.)

There are now rapid procedures for estimating nuclear DNA content, which often strongly correlates with ploidy. One may therefore survey a large number of individuals for nuclear DNA content and, on the basis of direct knowledge of chromosome number in a smaller sample of individuals, at least infer ploidy (see, for example, Dickson et al. 1992; Cox et al. 1998; Obermayer et al. 1999; Talent and Dickinson 2005).

Palynology

Palynology is the study of pollen and spores. Pollen and spores are similar in size, but spores are the beginning of the gametophyte generation, whereas pollen grains are mature microgametophytes. The outer layers of pollen and spores are, however, equivalent. These outer layers often contain a special compound, sporopollenin, that resists degradation by various chemicals, bacteria, and fungi and contributes to the long persistence of pollen and spores in sediments. Pollen and spores, therefore, have been important in paleobotanical studies.

We will examine the spores of lycophytes and monilophytes (ferns and allies) in Chapter 8; here we will focus on pollen. But first we will briefly consider the development of the anther.

Development of the Anther

Most anthers consist of four microsporangia arranged in pairs. The anther wall is made up of several layers, and the innermost, the **tapetum** (plural **tapeta**), plays a key role in the development of the microspores and pollen. When the pollen matures and environmental conditions are appropriate, the anther opens to release the pollen.

The opening, or dehiscence, of most angiosperm anthers is accomplished by a longitudinal slit on each side of the anther between the paired microsporangia (**longitudinal**, or **slit**, **dehiscence**). In a few families, such as Ericaceae and Melastomataceae, pollen is shed through a small opening or pore at one end of the anther (**poricidal dehiscence**), while in others, such as Lauraceae, pollen is shed as the anther opens by flaps (**valvate dehiscence**).

Pollen Structure, Viability, and Methods of Study

Pollen grains may be released from the anthers singly or in clusters of two, four, or many. In many Apocynaceae (e.g., *Asclepias*) and Orchidaceae, pollen is aggregated into clusters called **pollinia** (singular **pollinium**). The smallest known pollen grains are about 10 μm in diameter, and the largest (in Annonaceae) are 350 μm in diameter. Pollen grains range in shape from spherical to rod-shaped (19 × 520 μm in some Acanthaceae).

The two most important structural features of pollen grains are the apertures and the outer wall. **Apertures** are areas in the pollen wall through which pollen tubes emerge during germination. Pollen grains are often described according to the shape of their aperture(s). **Colpate** apertures (also referred to as **sulcate** if they are positioned at the pole) are long and grooved (Figures 4.47A,C,E and 4.48A). **Porate** apertures are round and porelike (Figures 4.47B and 4.48C,D,E,G,H), and **zonate** apertures are ring-shaped or band-shaped . **Colporate** apertures combine the groove of colpate and the pore of porate apertures (Figures 4.47D,F and 4.48B). Apertures may be located at the pole or equator of the pollen grain (Figure 4.48A,D), or they may be more or less uniformly distributed over the grain surface (Figure 4.48E,G,H).

The nature and number of apertures is constant in many plant taxa. **Monosulcate** pollen grains (see Figures 4.47A

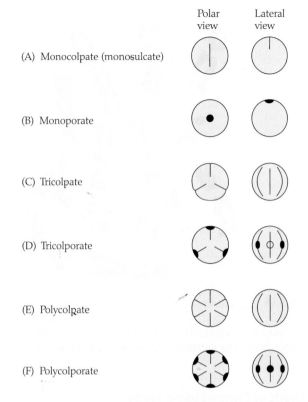

	Polar view	Lateral view
(A) Monocolpate (monosulcate)		
(B) Monoporate		
(C) Tricolpate		
(D) Tricolporate		
(E) Polycolpate		
(F) Polycolporate		

FIGURE 4.47 Some pollen aperture types. (After Gifford and Foster 1988; Faegri and Iverson 1950.)

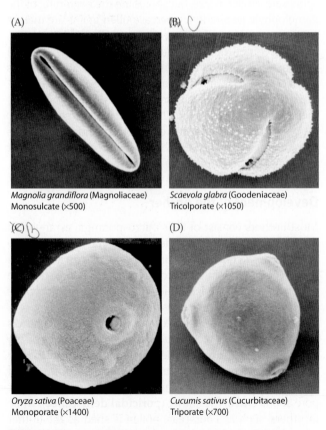

(A)

Magnolia grandiflora (Magnoliaceae)
Monosulcate (×500)

(B)

Scaevola glabra (Goodeniaceae)
Tricolporate (×1050)

(C)

Oryza sativa (Poaceae)
Monoporate (×1400)

(D)

Cucumis sativus (Cucurbitaceae)
Triporate (×700)

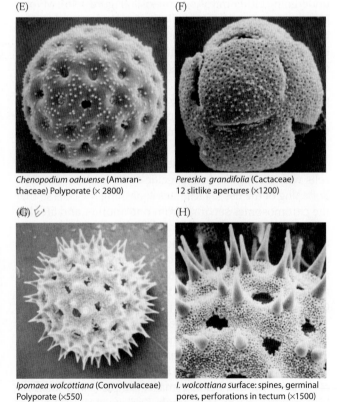

(E)

Chenopodium oahuense (Amaranthaceae) Polyporate (× 2800)

(F)

Pereskia grandifolia (Cactaceae)
12 slitlike apertures (×1200)

(G)

Ipomaea wolcottiana (Convolvulaceae)
Polyporate (×550)

(H)

I. wolcottiana surface: spines, germinal pores, perforations in tectum (×1500)

FIGURE 4.48 Scanning electron micrographs of representative angiosperm pollen grains, showing aperture types and surface features. (From Gifford and Foster 1988; original photos by J. Ward and D. Sunnell.)

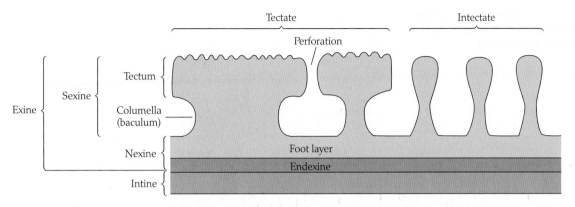

FIGURE 4.49 Cross section of a typical angiosperm pollen wall (exine and intine). (After Gifford and Foster 1988.)

and 4.48A) characterize many putatively basal woody angiosperms of the Magnoliales. Monocots too are basically a monosulcate group. In contrast, members of one large clade of angiosperms—the eudicots—bear **tricolpate** or tricolpate-derived pollen types (see Figures 4.47C,D and 4.48B,D).

The surface of the outer pollen wall, or **exine**, may be more or less smooth, as in many wind-pollinated species (see Figure 4.48C,D), or variously sculptured with spines, striations, reticulating ridges, knobs, and other features, as in most animal-pollinated species (see Figure 4.48E–H). These surface projections, which attach the pollen grain to animal pollinators, are a rich source of systematic characters. Systematists have also made use of internal exine features as characters at many taxonomic levels (Figure 4.49).

During the development of pollen, the microspore nucleus divides into a small **generative cell** and a much larger **vegetative cell**. The vegetative cell directs the growth of the pollen tube, and the generative cell usually divides into two sperm cells within the growing tube. In a minority of angiosperms, including both some tricolpates and some monocots, the generative cell divides into two sperm cells prior to anther dehiscence, and the pollen is shed in the 3-cell stage.

Pollen varies greatly in its ability to function (**viability**) after being shed from the anther. Viability is strongly affected by temperature and humidity, but these effects depend on the taxonomic group. For example, whereas grass pollen is short-lived, sometimes being viable for only minutes or hours, the pollen of many other species remains viable for up to several years if properly stored. One can evaluate viability by testing pollen for its capacity to germinate, for metabolic (enzymatic) activity, or for the presence of cytoplasm.

Features of the exine are obvious when a pollen grain is viewed by scanning electron microscopy (see Figure 4.48). In this procedure, the image is formed from electron beams. The internal structure of pollen, especially that of the exine (see Figure 4.49), is commonly examined with transmission electron microscopy.

Secondary Metabolites

The biochemical characters of plants have been employed taxonomically for some 100 years and indirectly—through the use of odors, tastes, and medicinal characteristics—for much longer. Chemical compounds have been used extensively in plant systematics, from analyses of infraspecific variation (see Adams 1977; Harborne and Turner 1984) to determination of phylogenetic relationships of families and other higher-level taxonomic groups (see Dahlgren 1975, 1983; Gershenzon and Mabry 1983).

Generally, emphasis has been placed on whether a particular compound is present or absent, but because many compounds can be formed by different synthetic pathways, much attention is now being placed on the elucidation of these pathways. Two major categories of systematically useful chemical compounds can be recognized: chemicals that perform metabolically nonessential functions in the plant, called **secondary metabolites**, and the information-containing molecules: proteins, DNA, and RNA. Proteins are discussed at the end of this chapter, and the taxonomic use of DNA and RNA is covered in detail in Chapter 5.

Most secondary metabolites function in defense against predators and pathogens, as allelopathic agents, or as attractants in pollination or fruit dispersal (Swain 1973; Levin 1976; Cronquist 1977). The major categories of secondary metabolites are discussed briefly in this section, and some aspects of their distribution among angiosperms are outlined. (For more information on the categories of secondary metabolites and their taxonomic use, see Gibbs 1974; Young and Seigler 1981; Gershenzon and Mabry 1983; Goodwin and Mercer 1983; Harborne 1984; Harborne and Turner 1984; Kubitzki 1984; Giannasi and Crawford 1986; Stuessy 1990.)

Alkaloids

Alkaloids are structurally diverse (Robinson 1981) and are derived from different amino acids or from mevalonic acid by various biosynthetic pathways. They are physiologically

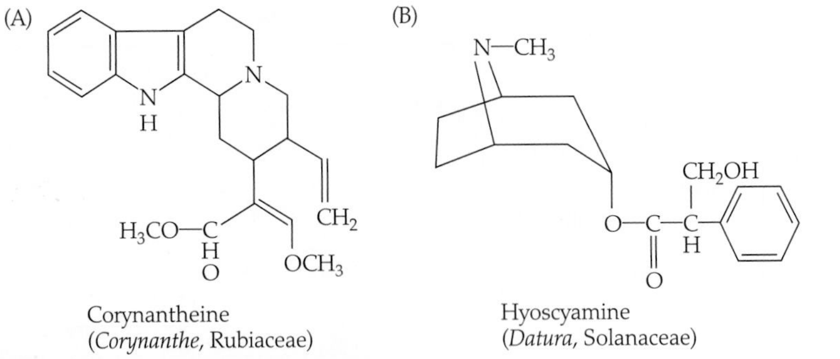

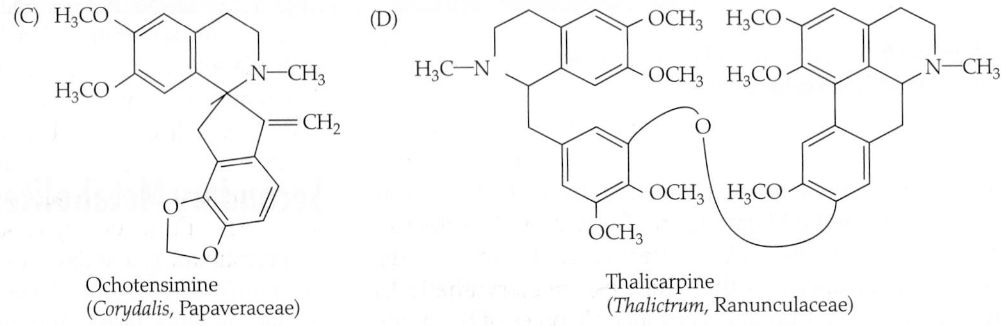

FIGURE 4.50 Structures of representative alkaloids. (A) A secologanin-type alkaloid. (B) A tropane alkaloid. (C, D) Benzylisoquinoline alkaloids.

(A) Corynantheine
(*Corynanthe*, Rubiaceae)

(B) Hyoscyamine
(*Datura*, Solanaceae)

(C) Ochotensimine
(*Corydalis*, Papaveraceae)

(D) Thalicarpine
(*Thalictrum*, Ranunculaceae)

active in animals, usually even at very low concentrations, and many are widely used in medicine (e.g., cocaine, morphine, atropine, colchicine, quinine, and strychnine). A few structural classes of alkaloids are shown in Figure 4.50 (see also Li and Willaman 1976).

Secologanin-type indole alkaloids (Figure 4.50A) are limited to the Apocynaceae, Gelsemiaceae, Loganiaceae, and Rubiaceae of the Gentianales. **Tropane alkaloids** (Figure 4.50B) occur in a wide array of families, but similar ones are characteristic of Solanaceae and Convolvulaceae (of the Solanales). **Benzylisoquinoline alkaloids** (Figure 4.50 C,D) occur in many members of Magnoliales, Laurales, and Ranunculales, as well as the Nelumbonaceae. Other groups, such as **isoprenoid alkaloids** and **pyrrolizidine alkaloids**, show a more scattered distribution among angiosperms and therefore are of less systematic interest.

Betalains and Anthocyanins

Betalains are nitrogenous red and yellow pigments (Figure 4.51A,B) that are restricted to Caryophyllales except for the Caryophyllaceae and Molluginaceae (Clement et al. 1994). In contrast, the red, yellow, blue, or purple pigments of most other plants are **anthocyanins** (a group of flavonoids) (Figure 4.51C,D).

Betalains and anthocyanins are mutually exclusive; they have never been found together in the same species. Such pigments, occurring in the perianth parts, are of course important in attracting pollinators, but they also occu r in young shoots, stems, leaves, fruits, and even roots, and they probably have additional functions, such as UV absorption and deterrence of herbivory.

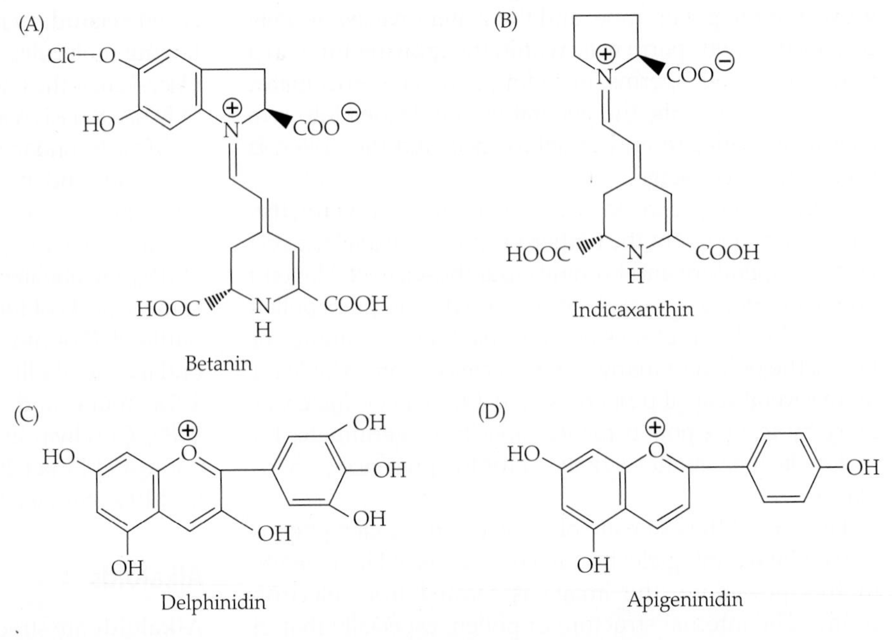

(A) Betanin

(B) Indicaxanthin

(C) Delphinidin

(D) Apigeninidin

FIGURE 4.51 Structures of representative betalains (A, B) and anthocyanins (C, D).

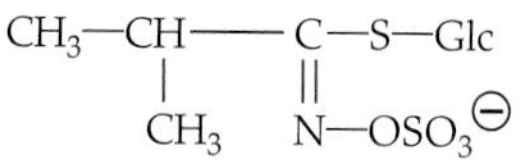

Isopropylglucosinolate
(*Tropaeolum*, Tropaeolaceae)

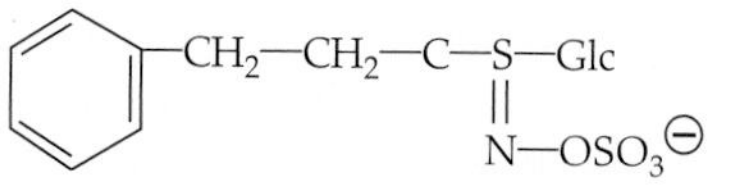

2-Phenylethylglucosinolate
(*Rorippa*, Brassicaceae)

FIGURE 4.52 Structures of representative glucosinolates.

Glucosinolates

The **glucosinolates** (Figure 4.52), also called mustard oil glucosides, are hydrolyzed by enzymes known as myrosinases to yield pungent hot mustard oils (Rodman 1981). Glucosinolates are synapomorphic for the Brassicales. The glucosinolates of the Brassicaceae, Resedaceae, and Tovariaceae, which are among the core Brassicales, are biosynthetically more complex than those of the other families of the order. Current evidence suggests that these compounds evolved only twice, in the common ancestor of Brassicales and in the common ancestor of the species of *Drypetes* (Putranjivaceae; see discussion of Euphorbiaceae) (Rodman et al. 1998).

Cyanogenic Glycosides

Cyanogenic glycosides (Figure 4.53) are defensive compounds that are hydrolyzed by various enzymes to release hydrogen cyanide (Hegnauer 1977). This process is called **cyanogenesis**. Cyanogenesis is widespread in angiosperms, and five different biosynthetic types of cyanogenic glycosides are known.

Some biosynthetic types have probably evolved numerous times, but others, such as the **cyclopentenoid cyanogenic glycosides**, are more restricted in their distribution (in this case to Achariaceae, Passifloraceae, Turneraceae, and Malesherbiaceae). Cyanogenic glycosides synthesized from leucine are common in the Rosaceae (*Prunus* and tribe Pyreae). Similar cyanogenic compounds are found in Fabaceae and Sapindaceae. Cyanogenic glycosides derived from tyrosine are common in several families of Magnoliales and Laurales.

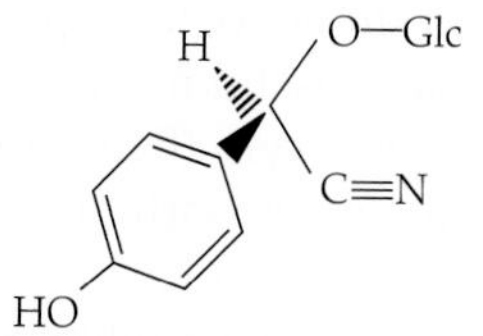

Taxiphyllin
(*Liriodendron*, Magnoliaceae)

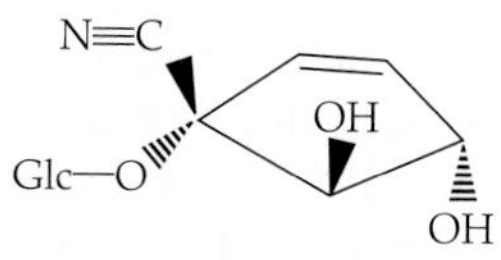

Gynocardin
(*Gynocardia*, Achariaceae)

FIGURE 4.53 Structures of representative cyanogenic glycosides.

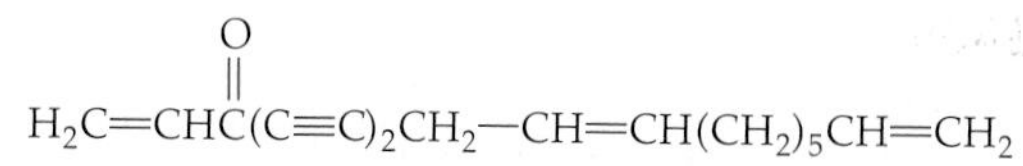

Dehydrofalcarinone
(*Artemisia*, Asteraceae)

FIGURE 4.54 Structure of a representative polyacetylene.

Polyacetylenes

Polyacetylenes (Figure 4.54) are a large group of nonnitrogenous secondary metabolites formed from the linking of acetate units via fatty acids. These compounds characterize a related group of asterid families, including Asteraceae, Apiaceae, Pittosporaceae, Campanulaceae, Goodeniaceae, and Caprifoliaceae. **Falcarinone polyacetylenes** are restricted to the Apiaceae, Araliaceae, and Pittosporaceae. These three families are also similar in their essential oils, oleanene- and ursene-type saponins, caffeic acid esters, furanocoumarins, and flavonoid profiles.

Terpenoids

Terpenoids are a large and structurally diverse group of secondary metabolites that are important in numerous biotic interactions (Goodwin 1971). They are formed by the union of 5-carbon isopentenyl diphosphate units formed in the mevalonic acid pathway. Terpenoids are very widely distributed, and many have primary physiological functions as components of membrane-bound steroids, carotenoid pigments, the phytyl side chain of chlorophyll, and the hormones gibberellic acid and abscisic acid.

The distribution of a few terpenoid types, however, is of taxonomic interest. Volatile **monoterpenoids** (Figure 4.55) and **sesquiterpenoids** (10-carbon and 15-carbon compounds, respectively) are the major components of **essential** (or **ethereal**) **oils**, which are characteristic of Magnoliales, Laurales, Austrobaileyales, and Piperales, as well as only distantly related clades, such as Myrtaceae, Rutaceae, Apiales, Lamiaceae, Verbenaceae, and Asteraceae. These

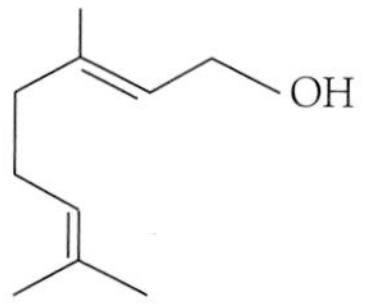

Geraniol

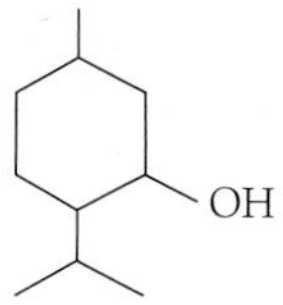

Menthol

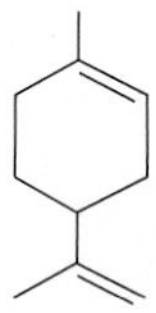

Limonene

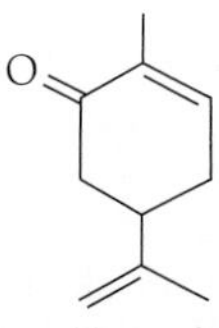

Carvone

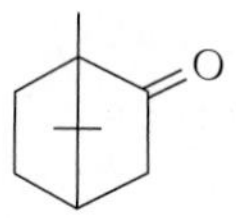

Camphor

FIGURE 4.55 Structures of representative monoterpenoids.

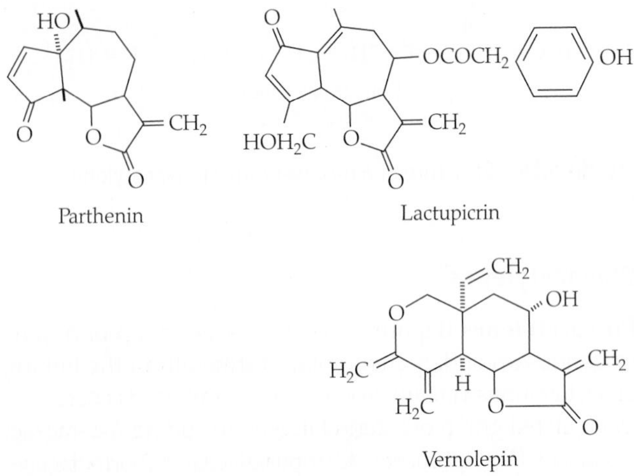

FIGURE 4.56 Structures of representative sesquiterpene lactones.

compounds occur not only in vegetative tissues in spherical cells or in various cavities or canals in parenchymatous tissues, but also in floral odor glands, where they are released and often function as floral attractants.

Sesquiterpene lactones (Figure 4.56) are known primarily from the Asteraceae (where they are diverse and taxonomically useful; Seaman 1982), but they also occur in a few other families, such as Apiaceae, Magnoliaceae, and Lauraceae.

Various **diterpenoids** (20-carbon), **triterpenoids** (30-carbon), and **steroids** (triterpenes based on the cyclopentane perhydro-phenanthrene ring system) are widely distributed and also have some systematic significance (Young and Seigler 1981). The triterpenoid **betulin** occurs in the bark of the white birches (*Betula papyrifera* and relatives); it is waterproof, highly flammable, and virtually unknown outside this group (O'Connell et al. 1988). **Triterpene saponins** occur in both Apiaceae and Pittosporaceae, supporting the hypothesized close phylogenetic relationship of these two families.

The triterpenoid derivatives **limonoids** and **quassinoids** (Figure 4.57), which are biosynthetically related, are limited to the Rutaceae, Meliaceae, and Simaroubaceae of the Sapindales; the bitter quassinoids constitute a distinctive

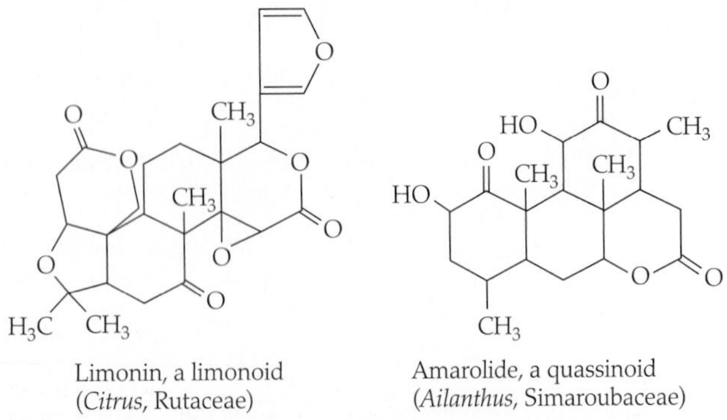

FIGURE 4.57 Structures of representative triterpenoid derivatives.

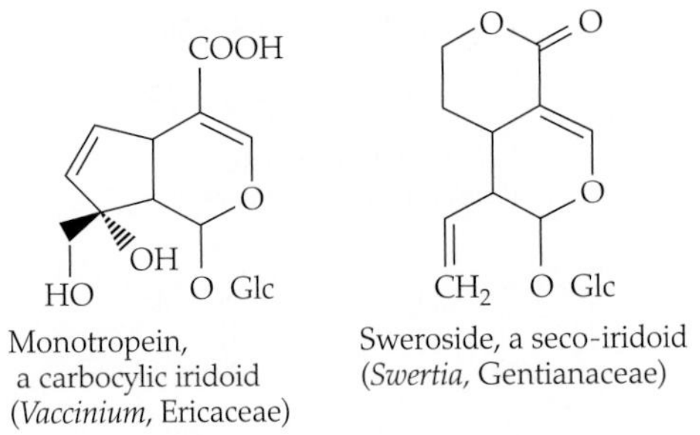

FIGURE 4.58 Structures of representative iridoids.

synapomorphy of the Simaroubaceae (including *Leitneria*). **Cardenolides** are highly poisonous glycosides of a type of 23-carbon steroid that occur in Ranunculaceae, Euphorbiaceae, Apocynaceae, Liliaceae, and Plantaginaceae.

Iridoids are 9- or 10-carbon monoterpenoid derivatives that usually occur as O-linked glycosides (Figure 4.58). Iridoid compounds are found in many families of the asterid clade, and iridoid types have been used to support relationships within this group (Jensen et al. 1975; Jensen 1992). For example, **seco-iridoids**, a chemically derived type of iridoid compound that lacks a carbocyclic ring, occur in Gentianales, Dipsacales, and many families of Cornales and Asterales. In contrast, **carbocyclic iridoids**, which have two rings, one composed entirely of carbon, are characteristic of the Lamiales, except for the Oleaceae, Tetrachondraceae, and Gesneriaceae. The presence of iridoids in the Ericales and Cornales provides evidence that these taxa actually belong to the asterid clade, even though they frequently have been excluded from that group (see Cronquist 1981).

Flavonoids

Flavonoids (Figure 4.59) are phenolic compounds that usually occur in a ring system derived through cyclization of an intermediate from a cinnamic acid derivative and three malonyl-CoA molecules. They probably function in defense against herbivores and in regulation of auxin transport. Flavonoids are employed extensively in plant systematics, probably because they can be fairly easily extracted and identified. They are found throughout the embryophytes (and are also known from the charophyte algae), and they have a diverse array of side groups attached to a common system of rings.

Although useful primarily in assessing relationships among closely related species (or in studies of infraspecific variation), flavonoids are occasionally useful in assessing phylogenetic relationships at higher levels (Bate-Smith 1968; Crawford 1978; Gornall et al. 1979; Harborne and Turner 1984). For example, the presence of certain 5-deoxyflavonoids in *Amphipterygium* (which has often been placed in its own family, the Julianiaceae; see Cronquist 1981) support its placement in Anacardiaceae.

Finally, flavonoid profiles have been shown to be quite useful in studies of interspecific hybridization (see Alston

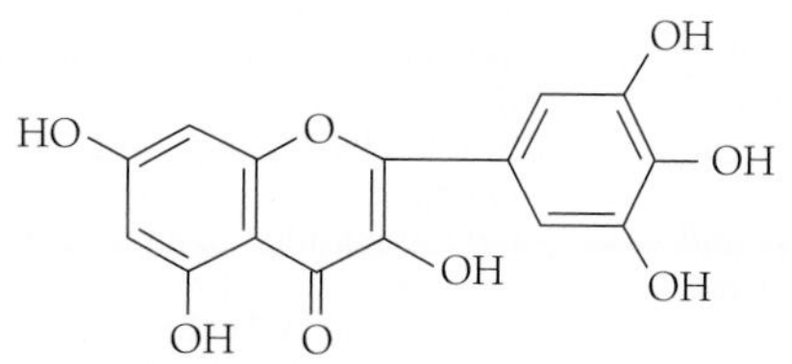

Myricetin
(*Limnanthes*, Limnanthaceae)

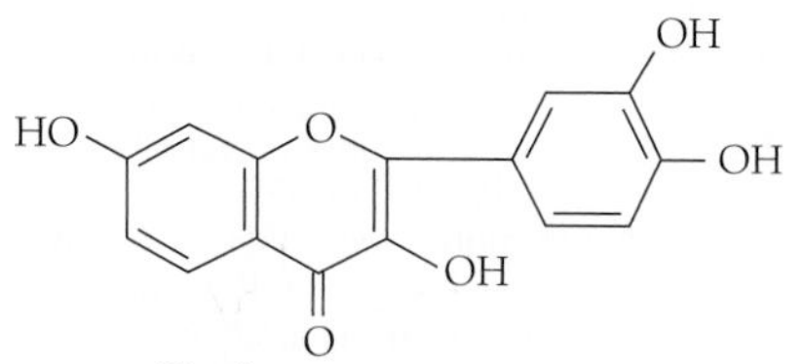

Fisetin
(*Amphipterygium*, Anacardiaceae)

FIGURE 4.59 Structures of representative flavonoids.

and Turner 1963; Smith and Levin 1963; Crawford and Giannasi 1982).

Proteins

Proteins are an extremely diverse class of molecules made up of amino acids linked by peptide bonds. The resulting chain of amino acids—a **polypeptide chain**—is three-dimensionally folded, resulting in a diversity of molecular shapes. Proteins function as enzymes, storage molecules, transport molecules, pigments, and structural materials. Proteins have been used systematically in several different ways, including amino acid sequencing and systematic serology (i.e., use of immunological reactions to assess levels of protein similarity between taxa), although in recent years DNA has largely replaced proteins as a source of systematic information. The use of DNA in systematics is detailed in the next chapter.

LITERATURE CITED AND SUGGESTED READINGS

Items marked with an asterisk are especially recommended to those readers who are interested in further information on the topics discussed in this chapter.

Morphology

Baumann-Bodenheim, M. G. 1954. Prinzipien eines Fruchtsystems der Angiospermen. *Bull. Soc. Bot. Suisse* 64: 94–112.

Bell, A. D. 1991. *Plant form: An illustrated guide to flowering plant morphology.* Oxford University Press, Oxford.

Bentley, B. L. 1977. Extrafloral nectaries and protection by pugnacious bodyguards. *Annu. Rev. Ecol. Syst.* 8: 407–427.

Brouwer, Y. M. and H. T. Clifford. 1990. An annotated list of domatia-bearing species. *Notes Jodrell Lab.* 12: 1–50.

Corner, E. J. H. 1976. *The seeds of dicotyledons.* 2 vols. Cambridge University Press, Cambridge.

Correll, D. S. and H. B. Correll. 1982. *Flora of the Bahama Archipelago.* J. Cramer, Vaduz, Germany. [Contains numerous floral diagrams.]

Cullen, J. 1978. A preliminary survey of ptyxis (vernation) in the angiosperms. *Notes R. Bot. Gard. Edinburgh* 37: 161–214.

D'Arcy, W. G. and R. C. Keating. 1996. *The anther: Form, function and phylogeny.* Cambridge University Press, Cambridge.

Davis, P. H. and V. H. Heywood. 1963. *Principles of angiosperm taxonomy.* Van Nostrand, Princeton, NJ. [Contains a detailed chapter on the use of morphological and anatomical characters in plant systematics.]

*Dilcher, D. 1974. Approaches to the identification of angiosperm leaf remains. *Bot. Rev.* 40: 1–157. [Very detailed summary of terms relating to angiosperm leaves.]

*Endress, P. K. 1994. *Diversity and evolution of tropical flowers.* Cambridge University Press, Cambridge. [Detailed presentation

of the structure and function of tropical flowers.]

Endress, P. K. 2003. Morphology and angiosperm systematics in the molecular era. *Bot. Rev.* 68: 545–570.

Endress, P. K., P. Baas and M. Gregory. 2000. Systematic plant morphology and anatomy—50 years of progress. *Taxon* 49: 401–434.

*Hallé, F., R. A. A. Oldeman and P. B. Tomlinson. 1978. *Tropical trees and forests: An architectural analysis.* Springer, Berlin.

*Harris, J. G. and M. W. Harris. 1994. *Plant identification terminology: An illustrated glossary.* Spring Lake Publishing, Spring Lake, UT.

Heslop-Harrison, Y. 1981. Stigma characteristics and angiosperm taxonomy. *Nordic J. Bot.* 1: 401–420.

*Hickey, L. J. 1973. Classification of the architecture of dicotyledonous leaves. *Am. J. Bot.* 60: 17–33. [Detailed summary of terms related to dicot leaves.]

Hickey, L. J. and J. A. Wolfe. 1975. The bases of angiosperm phylogeny: Vegetative morphology. *Ann. Missouri Bot. Gard.* 62: 538–589. [Note especially the careful treatment of tooth types.]

Jackson, B. D. 1928. *A glossary of botanic terms, with their derivation and accent,* 4th ed. Duckworth, London.

Keller, R. 1996. *Identification of tropical woody plants in the absence of flowers and fruits: A field guide.* Birkhauser, Basel, Switzerland.

*Kiger, R. W. and D. M. Porter. 2001. *Categorical glossary for the Flora of North America project.* Hunt Institute for Botanical Documentation, Carnegie Mellon University, Pittsburgh, PA.

Lawrence, G. H. M. 1951. *Taxonomy of vascular plants.* Macmillan, New York. [Appendix II is an illustrated glossary of taxonomic terms.]

Leaf Architecture Working Group. 1999. *Manual of leaf architecture—morphological description and categorization of dicotyledonous and net-veined monocotyledonous angiosperms.* Published by the authors, Washington, D.C.

Nettancourt, D. de. 1984. Incompatibility. In *Cellular interactions encyclopedia of plant physiology,* H. F. Linskens and J. Heslop-Harrison (eds.), new series, vol. 17, 624–639. Springer, Berlin.

O'Dowd, D. J. and R. W. Pemberton. 1998. Leaf domatia and foliar mite abundance in broadleaf deciduous forest of North Asia. *Am. J. Bot.* 85: 70–78.

Payne, W. W. 1978. A glossary of plant hair terminology. *Brittonia* 30: 239–255.

Pemberton, R. W. and C. E. Turner. 1989. Occurrence of predatory and fungivorous mites in leaf domatia. *Am. J. Bot.* 76: 105–112.

Philipson, W. R. 1977. Ovular morphology and the classification of dicotyledons. *Plant Syst. Evol.,* suppl. 1: 123–140.

Porter, C. L. 1967. *Taxonomy of flowering plants,* 2nd ed. Freeman, San Francisco. [Contains numerous floral diagrams.]

Rendle, A. B. 1925. *The classification of plants.* 2 vols. Cambridge University Press, Cambridge.

Richards, A. J. 1986. *Plant breeding systems.* Allen and Unwin, London.

Singer, S. R. 1997. Plant life cycles and angiosperm development. In *Embryology: Constructing the organism,* S. F. Gilbert and A. M. Raunio (eds.), 493–514. Sinauer Associates, Sunderland, MA.

Stearn, W. T. 1992. *Botanical Latin*, 4th ed. David and Charles, London.

Stuessy, T. S. 1990. *Plant taxonomy: The systematic evaluation of comparative data.* Columbia University Press, New York. [Contains a chapter on the use of morphological characters in plant systematics.]

Stuessy, T. S., V. Mayer and E. Hörandl (eds.). 2003. *Deep morphology: Toward a renaissance of morphology in plant systematics.* A. R. G. Gautner Verlag K. G. Ruggell, Liechtenstein.

Tryon, R. 1960. A glossary of some terms relating to the fern leaf. *Taxon* 9: 104–109.

van Balgooy, M. M. J. 1997. *Malesian seed plants.* Vol. 1: *Spot-characters. An aid for identification of families and genera.* Rijksherbarium/Hortus Botanicus, Leiden, Netherlands.

Walker-Larsen, J. and L. D. Harder. 2000. The evolution of staminodes in angiosperms: Patterns of stamen reduction, loss, and functional re-invention. *Am. J. Bot.* 87: 1367–1384.

Walters, D. R. and D. J. Keil. 1995. *Vascular plant taxonomy*, 4th ed. Kendall/Hunt, Dubuque, IA.

*Weberling, F. 1989. *Morphology of flowers and inflorescences* (R. J. Pankhurst, trans.). Cambridge University Press, Cambridge. [Includes a detailed presentation of variation in floral morphology and inflorescences, and associated terms.]

Willson, M. F. 1983. *Plant reproductive ecology.* Wiley, New York.

*Zomlefer, W. B. 1994. *Guide to flowering plant families.* University of North Carolina Press, Chapel Hill. [Contains a beautifully illustrated glossary of commonly used morphological terms.]

Pollination Biology

Arditti, J. 1992. *Fundamentals of orchid biology.* Wiley, New York.

Carthew, S. M. and R. L. Goldingay. 1997. Non-flying mammals as pollinators. *Trends Ecol. Evol.* 12: 104–108.

Cox, P. A. 1988. Hydrophyllous pollination. *Annu. Rev. Ecol. Syst.* 19: 261–280.

Darwin, C. 1862. *The various contrivances by which British and foreign orchids are fertilized by insects.* Murray, London.

Darwin, C. 1876. *The effect of cross- and self-fertilization in the vegetable kingdom.* Murray, London.

Darwin, C. 1877. *The different forms of flowers on plants of the same species.* Murray, London.

*Faegri, K. and L. van der Pijl. 1979. *The principles of pollination ecology*, 3rd ed. Pergamon, Oxford.

Ganders, F. R. 1979. The biology of heterostyly. *New Zealand J. Bot.* 17: 607–635. [Many other papers in this issue deal with reproduction in flowering plants.]

Gibernau, M., D. Barabé and D. Labat. 2000. Flowering and pollination of *Philodendron melinonii* (Araceae) in French Guiana. *Plant Biol.* 2: 331–334.

Johnson, S. D., A. Pauw and J. Midgley. 2001. Rodent pollination in the African lily *Massonia depressa* (Hyacinthaceae). *Am. J. Bot.* 88: 1768–1773.

Lewis, D. 1949. Incompatibility in flowering plants. *Biol. Rev.* 24: 472–496.

Nilsson, L. A. 1992. Orchid pollination biology. *Trends Ecol. Evol.* 7: 255–259.

Ortiz, P. L., M. Arista and S. Talavera. 2000. Pollination and breeding system of *Putoria calabrica* (Rubiaceae), a Mediterranean dwarf shrub. *Plant Biol.* 2: 325–330.

Pellmyr, O. 2003. Yuccas, yucca moths, and coevolution: A review. *Ann. Missouri Bot. Gard.* 90: 35–55.

Pellmyr, O., J. N. Thompson, J. M. Brown and R. G. Harrison. 1996. Evolution of pollination and mutualism in the yucca moth lineage. *Am. Nat.* 148: 827–847.

Proctor, M. and P. Yeo. 1972. *The pollination of flowers.* Taplinger, New York.

*Proctor, M., P. Yeo and A. Lack. 1996. *The natural history of pollination.* Timber Press, Portland, OR.

Real, L. (ed.) 1983. *Pollination biology.* Academic Press, Orlando, FL.

Tschapka, M. and O. von Helversen. 1999. Pollinators of syntopic *Marcgravia* species in Costa Rican lowland rain forest: Bats and opossums. *Plant Biol.* 1: 382–388.

Tschapka, M., O. von Helversen and W. Barthlott. 1999. Bat pollination of *Weberocereus tunilla*, an epiphytic rain forest cactus with functional flagelliflory. *Plant Biol.* 1: 554–559.

van der Pijl, L. and C. H. Dodson. 1966. *Orchid flowers: Their pollination and evolution.* University of Miami Press, Coral Gables, FL.

Inflorescences, Fruits, and Seeds

Gray, A. 1877. *Gray's lessons in botany and vegetable physiology.* Ivison, Blackman, Taylor, New York. [Contains an artificial classification of fruit types.]

Judd, W. S. 1985. A revised traditional/descriptive classification of fruits for use in floristics and teaching. *Phytologia* 58: 233–242.

Spjut, R. W. 1994. A systematic treatment of fruit types. *Mem. N.Y. Bot. Gard.* 70: 1–182.

Troll, W. 1964/1969. *Die Infloreszenzen, Typologie und Stellung im Aufbau des Vegetationskörpers.* 2 vols. Gustav Fischer, Jena, Germany.

*van der Pijl, L. 1972. *Principles of dispersal in higher plants.* McGraw-Hill, New York.

Weberling, F. 1965. Typology of inflorescences. *J. Linnean Soc. Bot.* 59: 215–221.

*Weberling, F. 1989. *Morphology of flowers and inflorescences* (R. J. Pankhurst, trans.). Cambridge University Press, Cambridge.

Anatomy

Ayensu, E. S. 1972. *Anatomy of the monocotyledons.* Vol. 6: *Dioscoreales.* Clarendon, Oxford.

Baas, P. 1982. *New perspectives in wood anatomy.* Nijhoff/Junk, The Hague, Netherlands.

Baas, P., E. Wheeler and M. W. Chase. 2000. Dicotyledonous wood anatomy and the APG system of angiosperm classification. *Bot. J. Linnean Soc.* 134: 3–17.

Bailey, I. W. 1933. The cambium and its derivative tissues. VIII. Structure, distribution and diagnostic significance of vestured pits in dicotyledons. *J. Arnold Arbor.* 14: 259–273.

Bailey, I. W. 1944. The development of vessels in angiosperms and its significance in morphological research. *Am. J. Bot.* 31: 421–428.

Bailey, I. W. 1951. The use and abuse of anatomical data in the study of phylogeny and classification. *Phytomorphology* 1: 67–69.

Bailey, I. W. 1957. The potentialities and limitations of wood anatomy in the phylogeny and classification of angiosperms. *J. Arnold Arbor.* 38: 243–254.

Bailey, I. W. and W. W. Tupper. 1918. Size variations in tracheary cells. I. A comparison between the secondary xylems of vascular cryptogams, gymnosperms, and angiosperms. *Proc. Am. Acad. Arts Sci.* 54: 149–204.

Barthlott, W. 1981. Epidermal and seed surface characters of plants: Systematic applicability and some evolutionary aspects. *Nordic J. Bot.* 1: 345–355.

Barthlott, W. 1990. Scanning electron microscopy of the epidermal surface in plants. In *Applications of the scanning EM in taxonomy and functional morphology*, D. Claugher (ed.), 69–94. Clarendon, Oxford.

*Barthlott, W., C. Neinhuis, D. Cutler, F. Ditsch, I. Meusel, I. Theisen and H. Wilhelmi. 1998. Classification and terminology of plant epicuticular waxes. *Bot. J. Linnean Soc.* 126: 237–260.

*Behnke, H.-D. 1972. Sieve-element plastids in relation to angiosperm systematics: An attempt towards a classification by ultrastructural analysis. *Bot. Rev.* 38: 155–197.

*Behnke, H.-D. 1975. The bases of angiosperm phylogeny: Ultrastructure. *Ann. Missouri Bot. Gard.* 62: 647–663.

Behnke, H.-D. 1977. Transmission electron microscopy and systematics of flowering plants. *Plant Syst. Evol.*, suppl. 1: 155–178.

Behnke, H.-D. 1981. Sieve-element characters. *Nordic J. Bot.* 1: 381–400.

Behnke, H.-D. 1991. Distribution and evolution of forms and types of sieve-element plastids in the dicotyledons. *Aliso* 13: 167–182.

Behnke, H.-D. 1994. Sieve-element plastids: Their significance for the evolution and systematics of the order. In *Caryophyllales*, H.-D. Behnke and T. J. Mabry (eds.), 87–121. Springer, Berlin.

Behnke, H.-D. 2000. Forms and sizes of sieve-element plastids and evolution of the monocotyledons. In *Monocots: Systematics and evolution*, K. L. Wilson and D. A. Morrison (eds.), 163–188. CSIRO, Collingwood, Australia.

Behnke, H.-D. and W. Barthlott. 1983. New evidence from ultrastructural and micromorphological fields in angiosperm classification. *Nordic J. Bot.* 3: 43–66.

*Carlquist, S. 1961. *Comparative plant anatomy: A guide to taxonomic and evolutionary applications of anatomical data in angiosperms.* Holt, Rinehart & Winston, New York.

Carlquist, S. 1988. *Comparative wood anatomy.* Springer Verlag, Berlin.

Cutler, D. F. 1969. *Anatomy of the monocotyledons.* Vol. 4: *Juncales.* Clarendon, Oxford.

Davis, P. H. and V. H. Heywood. 1973. *Principles of angiosperm taxonomy.* Krieger, New York. [Chapter on morphology, anatomy, palynology, and embryology.]

*Dickison, W. C. 1975. The bases of angiosperm phylogeny: Vegetative anatomy. *Ann. Missouri Bot. Gard.* 62: 590–620.

*Dickison, W. C. 2000. *Integrative plant anatomy*. Harcourt/Academic Press, San Diego, CA. [Includes chapters dealing with the usefulness of anatomical data in morphological and systematic studies.]

Esau, K. 1965. *Plant anatomy*, 2nd ed. Wiley, New York.

*Esau, K. 1977. *Anatomy of seed plants*, 2nd ed. Wiley, New York.

Evans, R. C. and T. A. Dickinson. 1996. North American black-fruited hawthorns. II. Floral development of 10- and 20-stamen morphotypes in *Crataegus* section *Douglasii* (Rosaceae: Maloideae). *Am. J. Bot.* 83: 961–978.

*Eyde, R. H. 1975. The bases of angiosperm phylogeny: Floral anatomy. *Ann. Missouri Bot. Gard.* 62: 521–537.

*Hickey, L. J. and J. A. Wolfe. 1975. The bases of angiosperm phylogeny: Vegetative morphology. *Ann. Missouri Bot. Gard.* 62: 538–589.

Howard, R. A. 1974. The stem-node-leaf continuum of the Dicotyledoneae. *J. Arnold Arbor.* 55: 125–181.

Jansen, S., P. Baas and E. Smets. 2001. Vestured pits: Their occurrence and systematic importance in eudicots. *Taxon* 50: 135–167.

Keating, R. C. 1984 [1985]. Leaf anatomy and its contribution to relationships in Myrtales. *Ann. Missouri Bot. Gard.* 71: 801–823.

Leins, P. 1964. Das zentripetale und zentrifugale Androeceum. *Berichte Deutsch. Bot. Gesellsch.* 77: 22–26.

Leins, P. and C. Erbar. 1997. Floral developmental studies: Some old and new questions. *Int. J. Plant Sci.* 158: 3–12.

Leins, P. and C. Erbar. 2003. Floral developmental features and molecular data in plant systematics. In *Deep morphology: Toward a renaissance of morphology in plant systematics*, T. F. Stuessy, V. Mayer and E. Hörland (eds.), 81–105. A. R. G. Gauter Verlag K. G. Ruggell, Lichtenstein.

Metcalfe, C. R. 1960. *Anatomy of the monocotyledons*. Vol. 1: *Gramineae*. Clarendon, Oxford.

Metcalfe, C. R. 1966. Distribution of latex in the plant kingdom. *Notes Jodrell Lab.* 3: 1–18.

Metcalfe, C. R. 1971. *Anatomy of the monocotyledons*. Vol. 5: *Cyperaceae*. Clarendon, Oxford.

Metcalfe, C. R. and L. Chalk. 1950. *Anatomy of the dicotyledons*. 2 vols. Clarendon, Oxford.

*Metcalfe, C. R. and L. Chalk. 1979. *Anatomy of the dicotyledons*, 2nd ed. Vol. 1: *Systematic anatomy of leaf and stem, with a brief history of the subject*. Clarendon, Oxford.

*Metcalfe, C. R. and L. Chalk. 1983. *Anatomy of the dicotyledons*, 2nd ed. Vol. 2: *Wood structure and conclusion of the general introduction*. Clarendon, Oxford.

Owens, S. A. 2000. Secondary and tertiary pulvini in the unifoliate leaf of *Cercis canadensis* L. (Fabaceae) with comparison to *Bauhinia purpurea* L. *Int. J. Plant Sci.* 161: 583–597.

Payne, W. W. 1979. Stomatal patterns in embryophytes: Their evolution, ontogeny and classification. *Taxon* 28: 117–132.

*Radford, A. E., W. C. Dickison, J. R. Massey and C. R. Bell. 1974. *Vascular plant systematics*. Harper and Row, New York. [Includes listings of numerous morphological and anatomical terms, with illustrations.]

Raghavendra, A. S. and V. S. Rama Das. 1978. The occurrence of C4 photosynthesis: A supplementary list of C4 plants reported during late 1944–mid 1977. *Photosynthetica* 12: 200–208.

Rasmussen, H. 1981. Terminology and classification of stomata and stomatal development—A critical survey. *Bot. J. Linnean Soc.* 83: 199–212.

Rathnam, C. K. M., A. S. Raghavendra and V. S. Rama Das. 1976. Diversity in the arrangements of mesophyll cells among leaves of certain C_4 dicotyledons in relation to C_4 physiology. *Z. Pflanzenphysiol.* 77: 283–291.

Soltis, D. E., M. Fishbein and R. K. Kuzoff. 2003. Reevaluating the evolution of epigyny: Data from phylogenetics and floral ontogeny. *Int. J. Plant Sci.* 164, suppl.: S251–S264.

*Stace, C. A. 1965. Cuticular studies as an aid to plant taxonomy. *Bull. Br. Mus. (Nat. Hist.) Bot.* 4: 1–78.

Stace, C. A. 1966. The use of epidermal characters in phylogenetic considerations. *New Phytol.* 65: 304–318.

Stace, C. A. 1989. *Plant taxonomy and biosystematics*, 2nd ed. Edward Arnold, London. [Chapter on structural information, including morphology and anatomy.]

Stuessy, T. F. 1990. *Plant taxonomy*. Columbia University Press, New York. [Includes a chapter on anatomy.]

Tomlinson, P. B. 1961. *Anatomy of the monocotyledons*. Vol. 2: *Palmae*. Clarendon, Oxford.

Tomlinson, P. B. 1969. *Anatomy of the monocotyledons*. Vol. 3: *Commelinales-Zingiberales*. Clarendon, Oxford.

Van Cotthem, W. R. J. 1970. A classification of stomatal types. *Bot. J. Linnean Soc.* 63: 235–246.

Wheeler, E. A., P. Baas and P. E. Gasson. 1989. I.A.W.A. list of microscopic features for hardwood identification. *IAWA Bull.* 10: 219–332.

Embryology

Asker, S. E. and L. Jerling. 1992. *Apomixis in plants*. CRC Press, Boca Raton, FL.

Dahlgren, G. 1991. Steps toward a rational system of the dicotyledons: Embryological characters. *Aliso* 13: 107–165.

Davis, G. L. 1966. *Systematic embryology of the angiosperms*. Wiley, New York.

Floyd, S. K. and W. E. Friedman. 2000. Evolution of endosperm developmental patterns among basal flowering plants. *Int. J. Plant Sci.* 161 (suppl. 6): 557–581.

Friedman, W. E. and S. K. Floyd. 2001. Perspective: The origin of flowering plants and their reproductive biology—A tail of two phylogenies. *Evolution* 55: 217–231.

Gifford, E. M. and A. S. Foster. 1988. *Morphology and evolution of vascular plants*, 3rd ed. Freeman, New York.

Herr, J. M., Jr. 1984. Embryology and taxonomy. In *Embryology of angiosperms*, B. M. Johri (ed.), 647–696. Springer Verlag, Berlin.

*Johri, B. M., K. B. Ambegaokar and P. S. Srivastra. 1992. *Comparative embryology of angiosperms*. 2 vols. Springer Verlag, New York.

Williams, J. H. and W. E. Friedman. 2004. The four-celled female gametophyte of *Illicium* (Illiciaceae; Austrobaileyales): Implications for understanding the origin and early evolution of monocots, eumagnoliids, and eudicots. *Am. J. Bot.* 91: 332–351.

Chromosomes

Campbell, C. S. 1983. Systematics of the *Andropogon virginicus* complex (Gramineae). *J. Arnold Arbor.* 64: 171–254.

Cox, A. V., G. J. Abdelnour, M. D. Bennett and I. J. Leitch. 1998. Genome size and karyotypic evolution in the slipper orchids (Cypripedioideae: Orchidaceae). *Am. J. Bot.* 85: 681–687.

Darlington, C. D. and L. F. La Cour. 1975. *The handling of chromosomes*, 6th ed. Wiley, New York.

Dickson, E. E., K. Arumuganthan, S. Kresovich and J. J. Doyle. 1992. Nuclear DNA content variation within Rosaceae. *Am. J. Bot.* 79: 1081–1086.

Flora of North America Editorial Committee. 1993. *Flora of North America north of Mexico*. Vol. 2: *Pteridophytes and gymnosperms*. Oxford University Press, New York.

Grant, W. F. (ed.). 1984. *Plant biosystematics*. Academic Press, Toronto, Ontario.

Jones, K. and C. Jopling. 1972. Chromosomes and the classification of the Commelinaceae. *Bot. J. Linnean Soc.* 65: 129–162.

Moore, D. M. 1976. *Plant cytogenetics* (Outline Series in Biology). Chapman & Hall, London.

Nogueira, C. Z., P. M. Ruas, C. F. Ruas and M. S. Ferrucci. 1995. Karyotype study of some species of *Serjania* and *Urvilliea* (Sapindaceae; Tribe Paullinieae). *Am. J. Bot.* 82: 646–654.

Obermayer, R. W., K. Üwiocicki and J. Greilhuber. 1999. Flow cyrometric determination of genome size in some Old World *Lupinus* species (Fabaceae). *Plant Biol.* 1: 403–407.

Qu, L., J. F. Hancock and J. H. Whallon. 1998. Evolution in an autopolyploid group displaying predominantly bivalent pairing at meiosis: Genomic similarity of diploid *Vaccinium darrowi* and autotetraploid *V. corymbosum*. *Am. J. Bot.* 85: 698–703.

*Raven, P. H. 1975. The bases of angiosperm phylogeny: Cytology. *Ann. Missouri Bot. Gard.* 62: 724–764.

Rieseberg, L. H., H. Choi, R. Chan and C. Spore. 1993. Genomic map of a diploid hybrid species. *Heredity* 70: 285–293.

Sanders, R. W. 1987a. Identity of *Lantana depressa* and *L. ovatifolia* (Verbenaceae) of Florida and the Bahamas. *Syst. Bot.* 12: 44–60.

Sanders, R. W. 1987b. Taxonomic significance of chromosome observations in Caribbean species of *Lantana* (Verbenaceae). *Am. J. Bot.* 74: 914–920.

*Sessions, S. K. 1990. Chromosomes: Molecular cytogenetics. In *Molecular systematics*, D. M. Hillis and C. Moritz (eds.), 156–203. Sinauer Associates, Sunderland, MA.

Soltis, D. E. and L. R. Rieseberg. 1986. Auto-polyploidy in *Tolmiea menziesii* (Saxifragaceae): Genetic insights from enzyme electrophoresis. *Am. J. Bot.* 73: 310–318.

Soltis, D. E. and P. S. Soltis. 1988. Are lycopods with high chromosome numbers ancient polyploids? *Am. J. Bot.* 75: 238–247.

Stace, C. A. 1984. *Plant taxonomy and biosystematics.* Edward Arnold, London.

Stace, C. A. 2000. Cytology and cytogenetics as a fundamental taxonomic resource for the 20th and 21st centuries. *Taxon* 49: 451–477.

Stebbins, G. L. 1971. *Chromosomal evolution in higher plants.* Addison-Wesley, Reading, MA.

Stuessy, T. F. 1990. *Plant taxonomy: The systematic evaluation of comparative data.* Columbia University Press, New York.

Talent, N. and T. A. Dickinson. 2005. Polyploid variation in *Crataegus* and *Mespilus* (Rosaceae, Maloideae): evolutionary inferences from flow cytometry of nuclear DNA amounts. *Can J. Bot.* 83:1268–1304.

Palynology

Blackmore, S. and I. K. Ferguson (eds.). 1986. *Pollen and spores: Form and function.* Academic Press, London.

*Erdtman, G. 1966. *Pollen morphology and plant taxonomy.* Hafner, New York.

Faegri, K. and J. Iversen. 1950. *Textbook of modern pollen analysis.* Munksgaard, Copenhagen.

Gifford, E. M. and A. S. Foster. 1988. *Morphology and evolution of vascular plants,* 3rd ed. Freeman, New York.

Graham, A., S. A. Graham, J. W. Nowicke, V. Patel and S. Lee. 1990. Palynology and systematics of the Lythraceae. III. Genera *Physocalymma* through *Woodfordia,* addenda, and conclusions. *Am. J. Bot.* 77: 159–177.

Nowicke, J. W. 1994. Pollen morphology and exine ultrastructure. In *Caryophyllales: Evolution and systematics,* H.-D. Behnke and T. J. Mabry (eds.), 167–221. Springer, Berlin.

Nowicke, J. W. and J. J. Skvarla. 1979. Pollen morphology: The potential influence in higher order systematics. *Ann. Missouri Bot. Gard.* 66: 633–700.

Stone, J. L., J. D. Thomson and S. J. Dent-Acosta. 1995. Assessment of pollen viability in hand-pollination experiments: A review. *Am. J. Bot.* 82: 1186–1197.

*Walker, J. W. and J. A. Doyle. 1975. The bases of angiosperm phylogeny: Palynology. *Ann. Missouri Bot. Gard.* 62: 664–723.

Secondary Metabolites

Adams, R. P. 1977. Chemosystematics: Analyses of populational differentiation and variability of ancestral and recent populations of *Juniperus ashei. Ann. Missouri Bot. Gard.* 64: 184–209.

Alston, R. E. and B. L. Turner. 1963. Natural hybridization among four species of *Baptisia* (Leguminosae). *Am. J. Bot.* 50: 159–173.

Bate-Smith, E. C. 1968. The phenolic constituents of plants and their taxonomic significance. *J. Linnean Soc. Bot.* 60: 325–383.

Bohlmann, F. 1971. Acetylenic compounds in the Umbelliferae. *Bot. J. Linnean Soc.* 64 (suppl. 1): 279–291.

Clement, J. S., T. J. Mabry, H. Wyler and A. S. Dreiding. 1994. Chemical review and evolutionary significance of the betalains. In *Caryophyllales,* H.-D. Behnke and T. J. Mabry (eds.), 247–261. Springer, Berlin.

Crawford, D. J. 1978. Flavonoid chemistry and angiosperm evolution. *Bot. Rev.* 44: 431–456.

Crawford, D. J. and D. E. Giannasi. 1982. Plant chemosystematics. *BioScience* 32: 114–118, 123–124.

Cronquist, A. 1977. On the taxonomic significance of secondary metabolites in angiosperms. *Plant Syst. Evol.,* suppl. 1: 179–189.

Cronquist, A. 1981. *An integrated system of classification of flowering plants.* Columbia University Press, New York.

Dahlgren, R. 1975. A system of classification of the angiosperms to be used to demonstrate the distribution of characters. *Bot. Notiser* 128: 119–147.

Dahlgren, R. 1983. General aspects of angiosperm evolution and macrosystematics. *Nordic J. Bot.* 3: 119–149.

*Gershenzon, J. and T. J. Mabry. 1983. Secondary metabolites and the higher classification of angiosperms. *Nordic J. Bot.* 3: 5–34.

Giannasi, D. E. and D. J. Crawford. 1986. Biochemical systematics. II. A reprise. *Evol. Biol.* 20: 25–248.

*Gibbs, R. D. 1974. *Chemotaxonomy of flowering plants.* 4 vols. McGill-Queen's University, Montreal, Quebec. [A detailed summary of secondary compounds occurring in various plant families and a discussion of their systematic significance.]

Goodwin, T. W. 1971. *Aspects of terpenoid chemistry and biochemistry.* Academic Press, London.

Goodwin, T. W. and E. I. Mercer. 1983. *Introduction to plant biochemistry,* 2nd ed. Pergamon, Oxford.

Gornall, R. J., B. A. Bohm and R. Dahlgren. 1979. The distribution of flavonoids in the angiosperms. *Bot. Notiser* 132: 1–30. [Includes discussions of the distribution of the various types of flavonoids and of their relative evolutionary advancement.]

Harborne, J. B. 1984. Chemical data in practical taxonomy. In *Current concepts in plant taxonomy,* V. H. Heywood and D. M. Moore (eds.), 237–261. Academic Press, London.

Harborne, J. B. and B. L. Turner. 1984. *Plant chemosystematics.* Academic Press, London.

Hegnauer, R. 1962–1996. *Chemotaxonomie der Pflanzen.* 11 vols. Birkhauser, Basel, Switzerland.

Hegnauer, R. 1977. Cyanogenic compounds as systematic markers in Tracheophyta. *Plant Syst. Evol.,* suppl. 1: 191–209.

Jensen, S. R. 1992. Systematic implications of the distribution of iridoids and other chemical compounds in the Loganiaceae and other families of the Asteridae. *Ann. Missouri Bot. Gard.* 79: 284–302.

Jensen, S. R., B. J. Nielsen and R. Dahlgren. 1975. Iridoid compounds, their occurrence and systematic importance in the angiosperms. *Bot. Notiser* 128: 148–180.

Kite, G. C., R. J. Grayer, P. J. Rudall and M. S. J. Simmonds. 2000. The potential for chemical characters in monocotyledon systematics. In *Monocots: Systematics and evolution,* K. L. Wilson and D. A. Morrison (eds.), 101—113. CSIRO, Collingwood, Australia.

Kubitzki, K. 1984. Phytochemistry in plant systematics and evolution. In *Current concepts in plant taxonomy,* V. H. Heywood and D. M. Moore (eds.), 263–277. Academic Press, London.

Levin, D. A. 1976. The chemical defenses of plants to pathogens and herbivores. *Annu. Rev. Ecol. Syst.* 7: 121–159.

Li, H. L. and J. J. Willaman. 1976. Distribution of alkaloids in angiosperm phylogeny. *Econ. Bot.* 22: 240–251.

O'Connell, M. M., M. D. Bentley, C. S. Campbell and B. J. W. Cole. 1988. Betulin and lupeol in bark from four white-barked birches. *Phytochemistry* 27: 2175–2176.

Robinson, T. 1981. *The biochemistry of alkaloids,* 2nd ed. Springer, New York.

Rodman, J. E. 1981. Divergence, convergence, and parallelism in phytochemical characters: The glucosinolate-myrosinase system. In *Phytochemistry and angiosperm phylogeny,* D. A. Young and D. S. Seigler (eds.), 43–79. Praeger, New York.

Rodman, J. E., P. S. Soltis, D. E. Soltis, K. J. Sytsma and K. G. Karol. 1998. Parallel evolution of glucosinolate biosynthesis inferred from congruent nuclear and plastid gene phylogenies. *Am. J. Bot.* 85: 997–1006.

Seaman, F. C. 1982. Sesquiterpene lactones as taxonomic characters in the Asteraceae. *Bot. Rev.* 48: 121–595.

Seigler, D. S. 1998. *Plant secondary metabolism.* Kluwer, Boston.

Smith, D. M. and D. A. Levin. 1963. A chromatographic study of reticulate evolution in the Appalachian *Asplenium* complex. *Am. J. Bot.* 50: 952–958.

Stace, C. A. 1989. *Plant taxonomy and biosystematics.* Edward Arnold, London. [Includes a chapter on chemical characters.]

Stuessy, T. F. 1990. *Plant taxonomy.* Columbia University Press, New York. [Includes a chapter on chemical characters.]

Swain, T. (ed.). 1973. *Chemistry in evolution and systematics.* Butterworth, London.

Swain, T. 1977. Secondary compounds as protective agents. *Annu. Rev. Plant Physiol.* 28: 479–501.

Young, D. A. and D. S. Seigler (eds.). 1981. *Phytochemistry and angiosperm phylogeny.* Praeger, New York.

5

Molecular Systematics

One of the most exciting and important developments of the past 25 years has been the application of nucleic acid data to problems in systematics. The term **molecular systematics** is used to refer to macromolecular systematics: the use of DNA and RNA sequences to infer evolutionary relationships among organisms. Although technically isozyme methods (which investigate variation among proteins) and flavonoid profiles (see Chapter 4) are also molecular, they have not had the overwhelming impact on the field that nucleic acid data have. In this chapter we focus on DNA sequence data, some of the common molecules and genomes in current use, and some aspects of data analysis that are unique to molecular data. We mention briefly several projects for sequencing whole genomes, which have accelerated data generation and have led to the development of many new computational tools that are increasingly applicable to systematics.

Molecular data have revolutionized our view of phylogenetic relationships, although not for the reasons initially suggested. Early proponents of molecular systematics claimed that molecular data were more likely than morphological data to reflect the true phylogeny, ostensibly because molecular data reflect gene-level changes, which were thought to be less subject to homoplasy than were morphological traits.

This early assurance now appears to be wrong. Molecular data are in fact subject to most of the same problems that morphological data are. The big difference is that there are simply many more molecular characters available, and their identity is generally easier to define: an adenine is an adenine, but compound leaves, for example, can form in quite different ways in different plants. As a result, DNA sequence data are now overwhelmingly the tool of choice for generating phylogenetic hypotheses.

In many cases, molecular data have supported the monophyly of groups that were recognized on morphological grounds (e.g., Poaceae, Fabaceae). More importantly, molecular data have often allowed systematists to choose among competing hypotheses of relationships (e.g., to decide what group is the sister group of the Asteraceae or the Poaceae). In other cases, molecular data have allowed the placement of taxa whose relationships were known to be problematic. For example, although Hydrangeaceae were traditionally placed in or near Saxifragaceae, it was clear that the two were unrelated. Only with molecular data, however, was there a strong alternative hypothesis for the placement of the Hydrangeaceae: in the order Cornales.

Molecular data have led to the recircumscription of many orders and genera, and they have pointed to some completely novel groupings, such as the monophyly of the glucosinolate clade (Brassicales) and the placement therein of the Limnanthaceae, and the recent placement of the parasitic Rafflesiaceae in Euphorbiaceae. They have also documented introgression between species that were apparently intersterile.

Sources of DNA Sequence Data

Good systematic work requires detailed knowledge of characters, their underlying biology, and the nature of their variation. For morphological characters, this research leads naturally into studies of developmental morphology. For molecular characters, it directs our attention to molecular biology and the structure and function of particular molecules. Each molecule has its own role in the cell, and its structure is constrained according to that role. Each molecule, like each set of morphological characters, has its own natural history, reflecting historical accident, developmental constraints, past and current adaptations (to both intra- and extracellular factors), and stochastic changes, whether fixed or transient.

This means that plant molecular systematists need to become as familiar with the structures and functions of the molecules they study as they are with the plants themselves. (At the same time, of course, they must be careful not to overlook the plants for the molecules!) Molecular genetics and biochemistry are becoming increasingly important as tools for understanding evolution, and any aspiring systematist should consider formal course work in plant molecular biology, molecular genetics, and/or biochemistry.

In this chapter we will describe the three plant genomes and some aspects of their molecular biology and evolution. We will then describe how DNA sequence data are generated, and mention a few of the major molecules used in systematic studies. We will conclude with brief descriptions of restriction site analysis, a historically important method that is used less often today, and an emerging technique, the mapping of nuclear genomes.

TABLE 5.1 Comparison of the three genomes in a plant cell.

	Genome size (kbp)	Inheritance
Chloroplast	135–160	Generally maternal (from the seed parent)
Mitochondrion	200–2500	Generally maternal (from the seed parent)
Nucleus	1.1×10^6 to 1.1×10^{11}	Biparental

Plant Genomes

The plant cell contains three different genomes: those of the chloroplast, the mitochondrion, and the nucleus (Table 5.1). Systematists use data from all three. The chloroplast and mitochondrion are generally inherited uniparentally (usually maternally in angiosperms); the nuclear genome is biparental. The three genomes differ dramatically in size, with the nuclear being by far the largest—measured in megabases of DNA. The mitochondrial genome includes several hundred kilobase pairs (kbp) of DNA (200–2500 kbp), which makes it small relative to the nuclear genome, but quite large relative to the mitochondrial genomes of animals (which tend to be about 16 kbp). The chloroplast genome is the smallest of the three plant genomes, in most plants ranging from 135 to 160 kbp.

Like the bacteria from which they are derived, mitochondria and chloroplasts have circular genomes. Large regions of noncoding DNA separate the genes in the mitochondrion, and their order in the genome is variable; in fact, their order changes so easily and frequently that many rearranged forms can occur even within the same cell. Rearrangements of the mitochondrial genome occur so often within individual plants that they do not characterize or differentiate species or groups of species and are thus not especially useful for inferring relationships.

The chloroplast, in contrast, is stable, both within cells and within species. The most obvious feature of the chloroplast genome is the presence of two regions that encode the same genes, but in opposite directions; these are known as **inverted repeats**. Between them are a small single-copy region and a large single-copy region (Figure 5.1).

Rearrangements of the chloroplast genome are rare enough in evolution that they can be used to demarcate major groups. For example, an early success of molecular systematics was the identification of the earliest-diverging members of Asteraceae by Jansen and Palmer (1987). They found that almost all members of the family have a unique order of genes in the large single-copy region of the chloroplast genome. This order could be explained if a large chunk of the DNA had been excised and reinserted in an

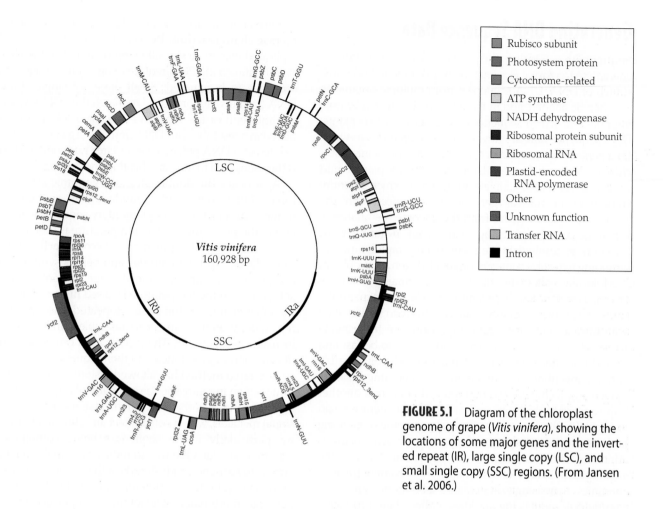

Rubisco subunit
Photosystem protein
Cytochrome-related
ATP synthase
NADH dehydrogenase
Ribosomal protein subunit
Ribosomal RNA
Plastid-encoded RNA polymerase
Other
Unknown function
Transfer RNA
Intron

FIGURE 5.1 Diagram of the chloroplast genome of grape (*Vitis vinifera*), showing the locations of some major genes and the inverted repeat (IR), large single copy (LSC), and small single copy (SSC) regions. (From Jansen et al. 2006.)

inverted orientation (an **inversion** of the DNA). All other angiosperms lack the inversion. The few Asteraceae that also lack the inversion are members of the subtribe Barnedisiinae, a South American group with bilabiate corollas. This finding strongly suggests that the Barnedisiinae (now treated as a subfamily, the Barnedisioideae) is the sister group to the rest of the enormous sunflower family, and that the latter is monophyletic.

Gains and losses of chloroplast genes, or their introns, are common enough to be worth looking for, but rare enough to be stable indicators of evolutionary change. (**Introns** are noncoding regions of genes that are interspersed between the coding **exons**.) Some groups, for instance, have lost one of the inverted repeats. This has occurred in a group of papilionoid legumes, in all conifers, and in *Euglena* (a flagellated photosynthetic eukaryote unrelated to green plants). In general, however, losses of a smaller piece of DNA, such as an intron or an entire gene, are more common than major rearrangements, and such losses may occur multiple times in evolution. For example, most angiosperms have an intron in the chloroplast gene *rpoC1*, but this intron has been lost in grasses, one subfamily of cacti (Cactoideae), at least two members of Goodeniaceae, some Aizoaceae, and some, but not all,

members of the genera *Passiflora* (Passifloraceae) and *Medicago* (Fabaceae) (Downie et al. 1996).

Although data are sparse for nuclear genomes, the order of their genes is presumed to be stable, at least within species, and may be stable across groups of species as well. Some information on nuclear gene order has been revealed by classic techniques of cytogenetics, but much more detailed information is now being provided by genome mapping, sequencing, and chromosome painting, particularly in Brassicaceae and Poaceae (Schranz et al. 2006; Kellogg and Bennetzen 2004). In the coming years these techniques could become important sources of systematic information.

DNA sequences change at a rate different from the rate of genomic rearrangement. Chloroplast genes tend to accumulate mutations more rapidly than do mitochondrial genes in plants. It is more difficult to generalize about nuclear genes, which is hardly surprising because there are so many of them. The frequency of mutation of a gene determines its utility for addressing particular phylogenetic problems. In general, a rapidly mutating gene is needed to assess relationships among closely related populations or species, whereas genes that mutate more slowly can be helpful in studies of older groups.

Generating DNA Sequence Data

Sequencing determines the precise order of nucleotides—adenine (A), cytosine (C), guanine (G), or thymine (T)—in a stretch of DNA. DNA sequences from multiple organisms can then be aligned, and mutations can be detected by noting points at which sequences differ between two plants. DNA sequence data are now being generated in two ways: (1) a gene-by-gene approach, in which a gene of interest is chosen, isolated from a large number of plants, and sequenced, and (2) a genomic approach, in which an entire chloroplast or nuclear genome is sequenced and the sequences of many genes from the genome are analyzed. While the gene-by-gene approach remains by far the most common, genomic data are accumulating rapidly and are beginning to resolve some important phylogenetic problems.

Molecular systematics has been and remains technique driven; as new methods become available, they expand the kinds and amounts of systematic data that can be extracted from nucleic acids. If useful comparisons are to be made across many taxa, the technique applied has to be fast and easy. This is why molecular systematics was barely possible until the invention of recombinant DNA, became easier as sequencing techniques were improved, and took another leap forward with the invention of the polymerase chain reaction (PCR) technique. As the major genome sequencing projects progress, the technology for gene sequencing is improving and becoming increasingly automated, and the techniques of molecular systematics are becoming those of genomics. Increasingly, systematics laboratories are using commercial sequencing facilities rather than producing sequences themselves. Sequences of both individual genes and whole genomes are available from GenBank, at the National Center for Biotechnology Information (http://www.ncbi.nlm.nih.gov/), which is the repository of a vast amount of publicly accessible data. We expect that in the future the attention of systematists will shift from sequence production itself to the more intellectually demanding work of analyzing sequences.

Gene-by-Gene Sequencing

Sequencing of genes, parts of genes, or noncoding regions is a common and fundamental aspect of systematic research. The central difficulty of sequencing has always been obtaining enough DNA to work with. The initial approach was to clone genes into bacteria and allow the bacteria to replicate the genes along with their own genomes. Genes were taken from genomic libraries, which researchers created by cutting all the DNA of an organism with restriction enzymes (described on pages 115–116) and then cloning all the resulting fragments into an appropriate plasmid, bacteriophage, or other vector. This method is quite slow, but it is reliable and avoids some of the possible artifacts of more efficient methods. It is also the only method available if only a few sequences of the gene of interest are known.

This laborious approach was later replaced by the **polymerase chain reaction (PCR)** technique, in which DNA is replicated enzymatically, allowing omission of the cloning step (Figure 5.2). PCR requires some knowledge of the sequence to be studied. Small pieces of single-stranded DNA (primers) are produced to match the DNA sequences at either end of the region of interest. These primers are placed in a test tube with DNA from the organismal genome of interest, a DNA polymerase, and free nucleotides. The mix is then subjected to repeated heating and cooling.

As it heats, the double-stranded genomic DNA denatures and becomes single-stranded. Then, as it cools, the primers bind to their complementary sequences at either end of the target region. The temperature is then raised to the point at which the polymerase becomes active. It binds to the DNA + primer complex and begins synthesizing a complementary strand using the free nucleotides in the solution. Then the temperature is raised higher to denature the DNA again, and the cycle is repeated. The DNA in the region between the primers is thus copied, and the amount increases exponentially. The PCR product can be sequenced directly, or it can be cloned and then sequenced.

This rapid method has allowed systematists to study the same region in many species of a particular group. A disadvantage of PCR is that the polymerase itself makes occasional mistakes, which could affect an estimate of phylogeny, particularly if the sequences being compared are extremely similar. One way to reduce potential sequencing error is to sequence both strands of the molecule, and some journals require this step before they will publish results; perhaps unfortunately, it is not universal practice. The decision of how accurately to sequence depends on the relative costs of an error versus the costs of repeatedly sequencing the same region. Systematists must often choose between highly accurate sequences from fewer taxa or less accurate sequences from more taxa.

Direct sequencing of the PCR product will not generally reveal minor variants of the sequence if they are present because the PCR product is actually a mixture of all the variants in the cell. This is often a problem with highly repetitive genes such as those encoding ribosomal RNA, for which the many copies often are not identical; the copies all end up jumbled together in the same test tube. Furthermore, direct sequencing cannot distinguish between alleles of the same gene. Imagine that two alleles differ from each other at two positions, such that one allele has an A at the first position and a T at the second, whereas the other has a T at the first position and an A at the second. Both positions will appear as A/T polymorphisms on a sequencing gel, and it is impossible to tell which allele has which base at which position. One can avoid this problem by cloning the PCR products. In this approach, each DNA molecule produced by PCR is inserted separately into a bacterium. The bacterium is then allowed to reproduce, making many copies of the single molecule. DNA sequencing is then carried out on a pure collection of identical molecules, rather than a mix.

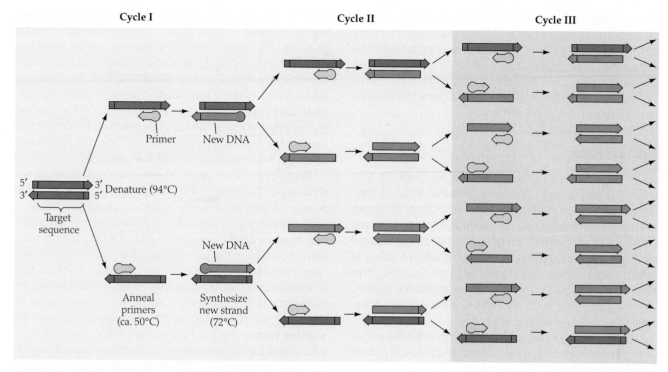

FIGURE 5.2 The polymerase chain reaction. The template DNA is shown in blue, the primers in yellow, and the newly synthesized DNA in red.

Whole-Genome Sequencing

As the cost of sequencing continues to fall, it is becoming cost-effective to sequence entire chloroplasts, or all the expressed genes in the genome, or even the entire nuclear genome. This has led to the field of genomics and to phylogenetic analyses of genomic data. One common method of sequencing a whole genome is to cut the genomic DNA with restriction enzymes and then clone large pieces of it into Bacterial Artificial Chromosomes (BACs), each of which can include a piece of DNA of more than 100 kbp. Each of the BACs is then sequenced. Powerful computer programs then compare the BAC sequences to look for ones that overlap. BACs with overlapping sequences are assumed to represent adjacent parts of the genomic sequence. By assembling many BAC sequences, the sequence of the entire genome is inferred.

Another approach is to sequence just the genes of the genome. In this method, the messenger RNAs (mRNAs) from a particular plant part are isolated as a group, and then each mRNA is cloned into a bacterial vector. These mRNAs are then sequenced, usually just from one end. These single-stranded sequences of parts of genes are known as Expressed Sequence Tags (**EST**s). Although they are not high-quality sequences, they are accumulating rapidly in public databases, and they are beginning to be used to address phylogenetic problems (e.g., De la Torre et al. 2006).

As this book goes to press, whole genomes have been sequenced for *Arabidopsis thaliana* (Brassicaceae), *Medicago truncatula* (Leguminosae), two subspecies of rice (*Oryza sativa*; Poaceae), and poplar (*Populus trichocarpa*; Salicaceae). Sequencing of sorghum (*Sorghum bicolor*; Poaceae), corn (*Zea mays*; Poaceae), *Lotus japonicus* (Leguminosae), potato (*Solanum tuberosum*; Solanaceae), tomato (*Solanum lycopersicon*; Solanaceae), and cassava (*Manihot esculenta*; Euphorbiaceae) is in progress. While these species currently represent a sparse sample of flowering plants, the numbers are likely to increase rapidly. In addition, EST data are available for dozens more plants (see http://www.ncbi.nlm.nih. gov/genomes/ PLANTS/PlantList.html#C_SEQ). There have already been efforts to mine these data for phylogenetic information (e.g., De la Torre et al. 2006; Sanderson et al. 2006) and to apply the analyses to major phylogenetic questions.

Analysis of DNA Sequence Data

There is a huge literature on the uses of DNA sequences in phylogeny reconstruction, and this literature has also converged with that on the exploding field of **bioinformatics** (the use of computers to manipulate and analyze biological data, particularly data on genes and genomes). Interested students are referred to one of the many excellent books available (e.g., Soltis et al. 1998; Page and Holmes 1998; Graur and Li 2000; Hall 2004; Mount 2001) or the extensive information available on the World Wide Web. Here we will discuss some of these uses and some examples that have affected our current view of phylogenetic relationships. The major issues to be addressed are mutation rate, alignment, analytic technique, and the relationship between the history of genes and the history of organisms (gene trees versus species trees).

Mutation Rates

Genes accumulate mutations at different rates, in part because gene products (RNAs or proteins) differ in how many changes they can tolerate and still function. Histones, for example, generally cease to work if many of their amino acids are replaced with different ones, whereas the internal transcribed spacer (ITS) of ribosomal RNA can still fold properly even if many of its nucleotides are changed. Thus genes for histones do not accumulate mutations rapidly, whereas genes for the ITS do, reflecting the different functional constraints on their gene products.

This simple observation has implications for the use of particular genes in phylogenetic reconstruction. If a gene is changing slowly, it will be difficult to find mutations from which a phylogeny can be constructed. At a very low mutation rate, the level of variation will approach the expected level of sequencing error (often estimated at about 3 in 10,000 bp for a double-stranded sequence), and inferences will become unreliable. Conversely, if a gene is changing too fast, parallelisms and reversals will accumulate to the point that all phylogenetic information is lost; the history of the sequence will be obliterated. The latter problem is particularly acute in work with noncoding sequences or remotely related taxa. Many systematists now do a preliminary study of multiple loci for their study group to determine which genes will have a level of variation appropriate for the question being asked.

Many of the methods used to analyze molecular data, and the limitations that apply to them, are similar to those for morphological data. Some methods, however, were developed specifically for use with molecular data (e.g., neighbor joining, maximum likelihood), and some problems, although present in all data sets, become more acute with molecular data.

Alignment of Sequences

Once sequences have been generated, they must be aligned. This is a critical step that determines which bases will be compared. It is the stage at which the scientist makes the initial assessment of similarity of nucleotide sites. Alignment is by far the most difficult part of using sequence data, and it is hard to automate. And clearly a poor alignment will lead to a meaningless phylogenetic tree.

Many computer programs will produce alignments, although in practice most systematists rely heavily on alignment "by eye." For many molecules currently used in plant systematics (e.g., protein-coding genes, noncoding regions in closely related plants), alignment is not a serious problem. For other molecules, such as genes encoding RNAs, alignment can be guided by models of the secondary structure of the gene product (the way the molecule folds). In this case the secondary structure is used as a template and the sequence is mapped on it. This method ensures that the proposed alignments maintain the structure of the molecule. (Methods for inferring secondary structure, however, have their own limitations.)

In protein-coding genes, alignments must consider the structure of the protein. The DNA sequence of such genes is read in groups of three bases, or **codons**, with each codon specifying a particular amino acid. Most commonly, insertions or deletions occur in sets of three bases as well, corresponding to the gain or loss of an amino acid; alignments need to incorporate this fact. Addition or subtraction of a single base (rather than a set of three) will change the entire structure of the protein encoded by the sequence by changing the start point of the subsequent codons (the **reading frame**). For example, the sequence AAATTGACT-TAC codes for the four amino acids lysine-leucine-threonine-tyrosine (K-L-T-Y). These are four sequential amino acids in the large subunit of Rubisco, shown in columns 31 to 42 in Figure 5.3. (Rubisco is an abbreviation for Ribulose 1,5-Bisphosphate Carboxylase/Oxygenase; the orthography varies in the literature, but is often either Rubisco, or RuBisCO.) If a single base were lost from the first lysine codon—leaving, for example, AATTGACTTAC—the protein would change. It would consist of an asparagine followed by a stop codon. The stop codon would prevent the remainder of the protein from being synthesized.

Nucleotide substitutions may or may not affect the protein produced. For example, at position 10 in the alignment shown in Figure 5.3 some taxa have an A and some have a C. This variation in the character state has an effect on the amino acid produced at that position; Rubisco in *Aristida, Stipagrostis, Eragrostis* and *Enneapogon* all have a glutamine (Q) at that position, whereas the other taxa have lysine (K). In contrast, the variation at position 24 (G vs. A) does not change the protein, because both AAG and AAA encode lysine. Note that both position 10 and position 24 provide potentially useful phylogenetic characters, even though only one has a biological effect on the protein.

Analytic Techniques

Some of the methods of phylogeny reconstruction that were described in Chapter 2 are particularly appropriate for use with DNA sequence data. These methods generally rely on statistical models of how DNA has changed over time. As noted in Chapter 2, for data with little homoplasy, virtually all methods will produce the same phylogenetic tree. In some cases, however, the choice of method will affect the result, and this is particularly true if rates of evolution are unequal in the lineages being compared.

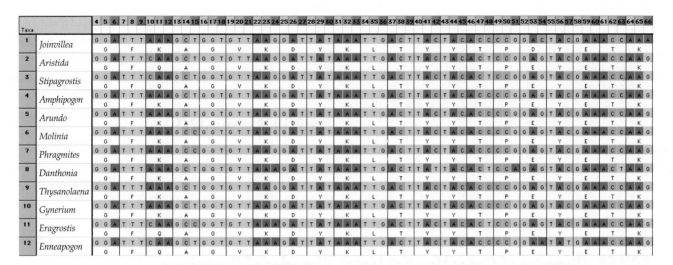

FIGURE 5.3 Alignment of part of the sequence for Rubisco. Each row in the alignment consists of a genus name followed by a set of nucleotides. Genera 2 through 12 are grasses (Poaceae); *Joinvillea* is an outgroup (family Joinvilleaceae). Each column (a taxonomic character) consists of nucleotides at corresponding positions (numbers at top) in the DNA molecules. Below each row of nucleotides is the amino acid translation, with the amino acid names abbreviated by a standard one-letter code. Thus the Rubisco protein in all taxa shown here has a glycine (G) followed by a phenylalanine (F). Variation at a given nucleotide position may or may not change the resulting protein. Note that an alignment is actually a character × taxon matrix similar to that in Table 2.2. (Alignment displayed in MacClade 4.0; Maddison and Maddison 2005.)

Homoplasy in molecular data creates particular problems. A reversal or convergence at a particular nucleotide is undetectable except via phylogenetic analysis; more detailed study of the character won't help. For example, an adenine at a particular position in a gene may have changed to guanine and then back to adenine; because only the adenine is visible in the sequence, there is no way to know that there had been a mutation to a guanine at that particular site. Thus, even though there may have been multiple mutations (sometimes called "multiple hits") at the same site, we only see the result of the most recent change (if any). Because of this, the actual number of mutations distinguishing two sequences, and thus of the divergence between them, may often be greater than the observed number of mutations. In this case a correction factor can be applied to estimate the actual evolutionary divergence. Which correction factor to use depends on estimates of the probability of particular types of mutations. (A full discussion of this kind of analysis can be found in Swofford et al. 1996; Page and Holmes 1998; or Graur and Li 2000.)

A high rate of mutation can lead to the problem of long branch attraction (see Box 2B), in which the wrong phylogenetic tree appears to be correct. If two unrelated sequences have very high rates of mutation, many sites will mutate multiple times. Because there are only four nucleotides, some of those mutations will lead to identical bases in both sequences, purely by chance. (In fact two random sequences will be 25% similar by chance alone.) In situations in which some sequences are mutating very rapidly and some quite slowly, accumulation of random mutations in the rapidly changing sequences will make them look alike, not because they are related, but because they are changing rapidly. The rapidly changing sequences can thus appear to be closely related even if they are not; in other words, the long branches "attract" each other. Long branch attraction could occur in principle with morphological data, but it is more likely with molecular data because the potential number of characters is so large and the available character states (A, C, G, T) are so few.

Gene Trees versus Species Trees

If a species has a single history, then we expect all parts of the plant to reflect that history. We also might expect any phylogeny based on any gene to reflect the history of the organisms bearing that gene, but in fact this is not always true. Nuclear genes may or may not track the history of the nucleus, and chloroplasts and mitochondria may or may not have a history different from that of the nucleus. There are three main reasons for these differences:

1. Mutation is a random process; therefore the phylogeny reconstructed for a particular gene may differ from that of other genes by chance alone.
2. Hybridization or introgression (described in Chapter 6) may transfer some DNA into a different lineage. This is particularly true in the case of chloroplasts and mitochondria, which are not linked to particular nuclear genomes.
3. Polymorphisms in an ancestral species can be lost in descendant species. By chance, this can result in a history of the genes that is actually different from the history of the organisms (Figure 5.4).

We now have multiple gene trees for many groups of organisms, and it is possible that none of those gene trees will be exactly the same as the species tree (Box 5A). For example, many plant taxa have now been shown to have the "wrong" chloroplast, presumably because of introgression. In one example, Soltis et al. (1991) found that individuals of *Tellima grandiflora* (Saxifragaceae) have two distinct chloroplast genomes, a "northern" type that occurs in plants from northern Oregon to Alaska, and a "southern" type found mostly in plants from northern Oregon south into California; there are also a few southern plants on Prince of Wales Island in the Alaska Panhandle and on the Olympic Peninsula in Washington. Because the southern chloroplast genome is most closely related to that of the genus *Mitella*, it is likely

that there was some ancestral introgression from *Mitella* into *T. grandiflora*. Hybridization and allopolyploidy can also lead to complex patterns of gene trees, as has been documented in the case of cotton (Box 5B).

Molecular Characters

In this section we will describe some of the major molecules used in systematic studies and what they do in the cell. The literature on biochemistry and molecular biology should be explored for each molecule used.

BOX 5A Histories of Genes May Not Match Histories of Organisms

The grass family has been the subject of many molecular systematic studies. Consider the subfamily Pooideae of the family Poaceae (Figure 5.5). This group was identified as monophyletic by cladistic studies of morphology, but several genera or small tribes, including the genus *Brachyelytrum* and the Stipeae, were sometimes placed with the pooids and sometimes placed in other subfamilies.

We now have five molecular phylogenies of the pooid clade, all of which

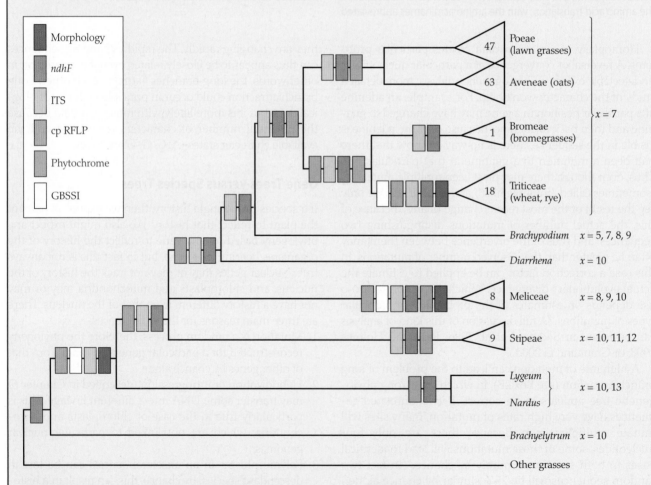

FIGURE 5.5 Phylogeny of the subfamily Pooideae. This semi-strict consensus tree shows clades supported by particular data sets (indicated by the colored rectangles) and not strongly contradicted by any other data set. Numbers in triangles are numbers of genera.

FIGURE 5.4   A comparison of gene trees and species trees. The history of a single gene (a gene tree) is shown by a black line, the history of the species as a whole (the species tree) by green shading. In the left-hand tree, the gene diverged at the same time as the species, such that the speciation event leading to B and C happened at the same time as divergence of the gene lineages that now exist in B and C. In the right-hand tree, a polymorphism appears in the lineage leading to species B and C. One of the two gene copies is more closely related to A. Sampling of this gene will lead to incorrect inferences about the species tree. (After Avise 1994.)

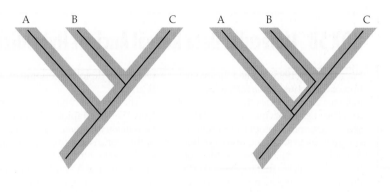

show the Stipeae as an early-diverging lineage. The morphological characters of the Stipeae are thus a mixture of synapomorphies linking them with the pooids and symplesiomorphies, which they share with many other grasses. Two of the studies that resulted in these phylogenies were based on chloroplast DNA, using restriction fragment length polymorphisms (cp RFLP; Davis and Soreng 1993) and sequences of *ndhF* (Catalán et al. 1997). We would expect these methods to give the same phylogeny because the chloroplast does not recombine and thus has a single history.

The other three studies were based on nuclear genes: those for the ITS (Hsiao et al. 1994), phytochrome *B* (Mathews and Sharrock 1996), and granule-bound starch synthase I (GBSSI) (Mason-Gamer et al. 1998). These studies support the same placement of the Stipeae. The fact that all data from both nuclear and chloroplast genomes suggest the same relationships indicates that the gene trees are probably good estimates of the organismal phylogeny. These data are also congruent with information on chromosome numbers.

A different result appears when we investigate relationships within the tribe Triticeae (Figure 5.6). For this group we have five molecular phylogenies, all dealing with the diploid genera. The two chloroplast phylogenies, based on RFLP analysis (Mason-Gamer and Kellogg 1996) and *rpoA* sequences (Petersen and Seberg 1997), suggest the

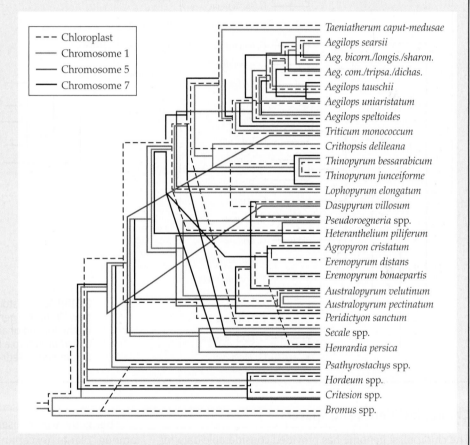

FIGURE 5.6   Phylogeny of the tribe Triticeae. Data for the chloroplast phylogeny is based on sequences of *rpoA* and RFLPs. The chromosome 1 and 5 histories are based on sequences of independent sets of 5S DNA spacers. The chromosome 7 history is based on the *GBSSI* gene.

same groupings, as expected. However, the three nuclear gene trees (based on sequences from three different chromosomes) are significantly different (Kellogg et al. 1996).

The explanation for this difference is not clear, but may involve a history of limited gene flow among the genera. The important point is that not all genes have identical histories. This means that one gene tree needs to be compared with a second one, preferably from a different genome, if we are to begin to infer organismal histories.

BOX 5B Molecular Data Reveal Ancient Hybridization

Wendel et al. (1995) studied the evolution of the genus *Gossypium*, which includes all the species that produce cotton. They used isozymes, nuclear ITS sequences, and chloroplast restriction site analysis to study the history of both diploid and tetraploid species. Most of their data indicate that the New World diploids (with a genome designated D; Figure 5.7) are monophyletic, as are the Old World diploids (genome groups A, B, and F).

The surprise came in analyzing the New World tetraploids, including *Gossypium hirsutum*, the source of most of the world's commercial cotton. These species were formed by allopolyploidization of A and D genomes. Wendel and his colleagues found that *G. hirsutum* has a chloroplast derived from one of the African species, and that it must have acquired that chloroplast only about 1–2 million years ago, well after the formation of the Atlantic Ocean. One other New World species has elements of an Old World ITS sequence as well. We do not know how such long-distance gene flow might have occurred.

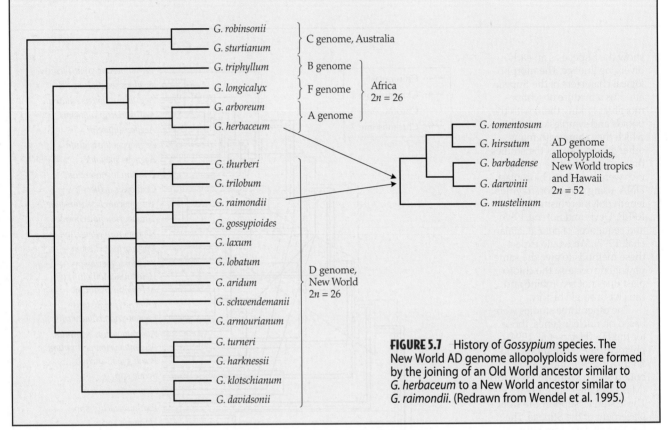

FIGURE 5.7 History of *Gossypium* species. The New World AD genome allopolyploids were formed by the joining of an Old World ancestor similar to *G. herbaceum* to a New World ancestor similar to *G. raimondii*. (Redrawn from Wendel et al. 1995.)

Chloroplast Genes and Spacers

The chloroplast genome has provided considerable data for phylogenetic study and continues to be widely used. DNA from the chloroplasts often constitutes over a fourth of the DNA in the cell, so it is abundant and easy to work with. This was particularly important in early molecular systematic studies, when many plant systematists were involved in a community-wide effort to generate a large database of sequences of the chloroplast gene *rbcL* (Chase et al. 1993). This gene encodes the large subunit of the photosynthetic enzyme Rubisco, which is the major carbon acceptor in all photosynthetic eukaryotes and cyanobacteria. The secondary structure of the protein is known (Figure 5.8), and amino acids can be assigned to particular structural components of the gene.

This gene was chosen because it is almost universal among plants (excepting only the parasites and some saprophytes), it is fairly long (1428 bp), and it presents no problems of alignment. The enthusiasm for sequencing the *rbcL* gene was aided by the generosity of Gerard Zurawski (then at the University of Georgia), who designed a set of near-universal PCR primers that he distributed freely to anyone who wanted them. The availability of these primers encouraged many plant systematists to generate *rbcL* sequences and has resulted in thousands of sequences, primarily of seed plants. The power of this broadly collaborative approach should not be underestimated.

The gene trees generated from these *rbcL* sequences have had an enormous influence on our view of relationships among angiosperm families, and they are referred to

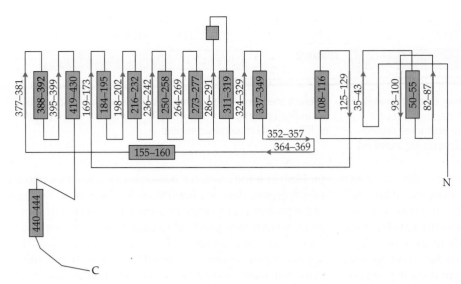

FIGURE 5.8 Secondary structure of the large subunit of Rubisco. The numbers refer to numbered amino acid residues. Green rectangles represent alpha helices; blue arrows represent beta sheets. (From Kellogg and Juliano 1997.)

throughout this book. In particular, several studies presented in a single issue of the *Annals of the Missouri Botanical Garden* in 1993 generated many hypotheses of relationship, which have been tested with other molecular and morphological data. These initial studies were remarkable for their tremendous heuristic value. Much attention focused on the work of Chase et al. (1993), who generated a phylogeny for all seed plants using 499 *rbcL* sequences; this phylogeny has been widely cited and is referred to often in this book.

The trees based on *rbcL* data support many ideas that were accepted on the basis of morphology, such as the monophyly of several well-known family groups (e.g., Asclepiadaceae/Apocynaceae, Brassicaceae/Capparaceae, Sapindaceae/Aceraceae/Hippocastanaceae) and the paraphyly (e.g., Caprifoliaceae) or polyphyly (e.g., Saxifragaceae) of others. In other cases, *rbcL* data helped resolve relationships that were previously ambiguous. The Ericaceae, for example, were once included in Sympetalae, but later were placed well outside it. Trees based on *rbcL* data, however, support placement of the Ericales in a larger clade with the Asteridae, reuniting much (but not all) of the Englerian Sympetalae (see Chapter 3).

Finally, in a few cases, *rbcL* data suggested something quite surprising. For example, they placed the nine families with nitrogen-fixing members in a single clade, along with a few families that do not fix nitrogen (Soltis et al. 1995). Because the nitrogen-fixing families had appeared to be completely unrelated, this finding suggests that these families may have more in common than previously believed.

One limitation of *rbcL* as a phylogenetic marker is its slow rate of change. The protein it encodes is a highly conserved molecule that is greatly constrained at the amino acid level. The *rbcL* gene is therefore not particularly useful for inferring relationships within or between closely related genera. Instead, other chloroplast genes have been used for such purposes, notably those that encode subunit F of NADP dehydrogenase (*ndhF*, in the small single-copy

region), those that encode the α and β″ subunits of RNA polymerase II (*rpoA* and *rpoC2*, in the large single-copy region), and a maturase gene (a gene encoding a protein that helps in intron removal) in the intron that separates the coding regions of *trnK* (*matK*). Because these sequences are all part of the same nonrecombining genome as *rbcL*, they all track the same (generally maternal) history.

The gene that encodes the β subunit of ATP synthase—*atpB*—has been used to address the same problems as *rbcL*. It appears to evolve at about the same rate and thus provides additional phylogenetically informative characters. Data from *atpB* have been combined with data from *rbcL* to refine the picture of angiosperm relationships (Qiu et al. 1999; Soltis et al. 1999, 2000). In addition, spacer regions (noncoding sequences between genes) and introns in the chloroplast have become popular for studies of closely related species, although some, like the widely used spacer between the transfer RNAs for leucine and phenylalanine (*trnL* and *trnF*, respectively), are often too short to produce a really definitive phylogeny.

Mitochondrial Genes

Relatively few studies have used plant mitochondrial genes for studying phylogeny. In general, these genes evolve slowly and are most useful for assessing ancient events such as the origin of the angiosperms (Qiu et al. 1999) or the phylogenies of large groups such as the seed plants (Soltis et al. 2002). The commonly used mitochondrial genes encode subunits of ATP synthase (*atp1*, *atpA*) and a maturase (*matR*).

Nuclear Genes

Ribosomal RNA genes Historically, the only nuclear genes with a copy number high enough for easy study were those encoding ribosomal RNA. These genes are arranged in tandem arrays of several hundred to several

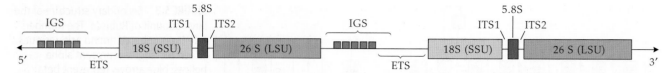

FIGURE 5.9 Structure of the ribosomal array. Coding regions of the 18S small subunit (SSU), 5.8S unit, and 26S large subunit (LSU) are shown as yellow, blue, and green rectangles, respectively; the transcription unit is bracketed. Spacers are indicated by black lines, and the short repeats in the intergenic spacer (IGS) are indicated by small red boxes. ETS, external transcribed spacer; ITS, internal transcribed spacer.

thousand copies. The general arrangement of these genes is shown in Figure 5.9. The genes encoding the small subunit (18S) and the large subunit (26S) of the ribosome are separated by a smaller (5.8S) gene, and the whole set of genes is transcribed as a single unit. There are short internal transcribed spacers (ITS) between the three genes. Each set of three genes is separated from the following set by a large spacer, the intergenic spacer (IGS). The middle portion of the spacer is not transcribed and is made up of variable numbers of short repeated sequences (about 100–300 bp each). These sequences are thought to play a role in gene regulation.

A completely separate rRNA array encodes only 5S RNA (not to be confused with 5.8S rRNA), a molecule that joins with the 26S RNA and a set of proteins to form the large ribosomal subunit. The 5S rRNA genes are in tandem arrays of several thousand copies and are separated by nontranscribed spacers.

Such highly repetitive sequences undergo homogenization processes known as **concerted evolution**. If a mutation occurs in one copy of a sequence, it is generally corrected to match the other copies. However, sometimes the nonmutated copies are "corrected" to match the mutated one, causing nucleotide changes to propagate throughout the array. In either case, the many copies of the sequence are generally more similar to one another than they are to copies in other species. Within-species variation does occur, however, so that some highly repetitive sequences can be used to assess variation within and among populations of the same species.

Sequences of the 18S and 26S genes have been used for studying relationships among large groups of plants. These genes are large (about 1800 and 3300 bp, respectively). They have some regions that are highly conserved, which helps in alignment, and others that are quite variable, which helps in distinguishing phylogenetic groups. A large cooperative effort, analogous to the *rbcL* study, generated a database of 18S sequences (Soltis et al. 1997), which were then combined with *rbcL* and *atpB* sequences to provide a picture of angiosperm evolution (Soltis et al. 1999).

The ITS region has become a common tool for determining relationships among species. In general, the ITS region has supported relationships inferred from the chloroplast or from morphology. In other cases, however, it has proved to be polymorphic within species, or even within individual plants, suggesting that concerted evolution has not completely homogenized the repeats. This polymorphism is a problem in some cases, but it also provides a tool for understanding gene flow and population-level variation. Because of the problem of incomplete homogenization of ITS genes, many systematists use PCR to amplify the ITS region, but then clone and sequence ITS copies individually. This approach provides an assessment of ITS variation within a plant and allows correction for ancestral polymorphisms. It has also led to the discovery that some copies of the ITS have accumulated so many mutations that the RNA transcribed from them would be expected not to fold properly. Such nonfunctional genes are known as **pseudogenes**.

Low copy number genes Nuclear genes with low copy numbers are being used more and more commonly as phylogenetic markers (Sang 2002; Small et al. 2004; Hughes et al. 2006). In order for a nuclear gene to be a useful phylogenetic indicator, it must not be easily confused with any other gene. Because many nuclear genes are duplicated or exist as part of a small set of genes (a **gene family**), some preliminary work is often required to be sure that all the sequences used are from genes that are related by descent (**orthologous genes**) and not simply recent duplicates (**paralogous genes**).

Population genetic theory suggests that allelic variation should not be misleading in studies of closely related species because alleles in one species should be more closely related to one another than they are to alleles in other species. Data on many nuclear genes support this expectation.

The phytochrome genes, which encode proteins that respond to light signals and control plant growth, development, and photosynthesis, have proved to be particularly illuminating. There are four major phytochrome genes in angiosperms, given the letter names *A*, *B*, *C*, and *E*. (Phytochrome *D* occurs in Brassicales and is a relatively recent duplicate of *B*.) All of these genes are descendants of an ancestral phytochrome gene that was duplicated once in the ancestor of the seed plants. One copy became the ancestor of phytochromes *A* and *C*, and the other the ancestor of *B* and *E*. The *A/C* duplication occurred before the origin of the angiosperms. Phytochromes *A* and *C* have each been used to produce a phylogeny of the early-diverging angiosperms (Mathews and Donoghue 1999), with results that are similar to those produced by chloroplast genes. In addition, individual phytochrome loci have been used to help resolve the phylogeny of *Aristolochia* (Ohi-Toma et al. 2006), Orobanchaceae (Bennett and Mathews 2006), and *Pereskia* and other related cacti (Edwards et al. 2005).

Introns of nuclear genes are often more variable than the ITS or the most variable chloroplast regions, and thus they are useful tools for assessing relationships among similar species or even among populations. For example, Olsen and Schaal (1999) used introns in the metabolic gene encoding glucose 6-phosphate dehydrogenase to determine where in Brazil cultivated cassava (*Manihot esculenta*) was domesticated from wild populations.

Malcomber (2002) used introns of the gene for triose phosphate isomerase and two different genes for phosphenolpyruvate carboxylase to assess relationships among species of *Gaertnera*, a member of the Rubiaceae. Introns of the gene that encodes granule-bound starch synthase I (*GBSSI*, orthologous to the gene known in corn [or maize, as it is called by those who work on this plant] as *waxy*) have been used to determine relationships among species of grasses, Rosaceae tribe Pyreae, potatoes, and morning glories (Mason-Gamer et al. 1998; Evans et al. 2000; Manos et al. 2001; Peralta and Spooner 2001).

When using nuclear introns as phylogenetic markers, it is often most productive to try several genes for a subset of the taxa to be studied. Because the choice of nuclear gene is not particularly critical (assuming that orthology can be demonstrated), it is most efficient to choose one that can be easily amplified through PCR, that can be readily aligned among the study taxa, and that provides sufficient variation for resolving relationships.

Combining data from many genes In many of the studies cited above, and many more cited in Chapters 8 and 9, data from several genes have been combined. As sequencing costs drop, the use of multiple genes is becoming more and more common. This practice often provides better support for particular relationships and clarifies patterns that are obscure when only a single gene is used.

High copy number noncoding nuclear sequences Unlike chloroplast, ribosomal RNA, and nuclear protein-coding genes, high copy number noncoding nuclear sequences evolve rapidly and are thus useful for addressing questions at the population level. The sequences used are generally short sequences that are repeated many times, often at many locations in the genome.

So-called **minisatellites** or **variable-number tandem repeats** (**VNTR**) are made up of repeated sequences that are generally tens of base pairs in length. In **microsatellites** the repeats are much shorter, consisting of only two or three nucleotides. Such repeated sequences are unstable and prone to errors in replication, usually deriving from **replication slippage** (although unequal crossing-over is also a possibility in some cases).

Replication slippage occurs as the DNA is being copied. The strands separate for replication but reanneal out of register, leading to a loop in the DNA. Mismatch repair mechanisms then either remove the loop (leading to the loss of a repeat unit) or insert extra bases on the opposite strand (leading to a duplication). Because of this instability, indi-

vidual organisms often vary in the number of repeats at a particular satellite locus. This variation can be used to determine a DNA "fingerprint" unique to a particular plant or closely related group of plants. Studies of population structure generally depend on accurate assessment of relationships among individual plants, and these markers are useful for such assessment.

Another method often used in studies at the population level is the **random amplified polymorphic DNA** (**RAPD**) method. In this technique, short (10 bp) PCR primers are designed with arbitrary sequences. These short random sequences will generally match one or more sequences somewhere in the genome of the plant, and the primers will bind to and amplify a fragment of DNA.

By performing many such PCRs with random primers, fragments can be found that distinguish individual plants or populations. These fragments allow rapid assessment of how many genotypes are present in a population and a rough estimate of how different those genotypes are. The technique is limited, however, because the identity of the fragments is not known. In other words, a fragment of 150 bp in one plant may not actually represent the same part of the genome as a 150 bp fragment in another plant, because the only criterion of similarity is fragment size. Verifying the identity of the fragments requires Southern blotting or restriction site analysis (see below), at which point the RAPD technique may become as laborious as a restriction site or sequencing study. Other techniques, such as amplified fragment length polymorphisms (AFLP), have been developed to circumvent the problems of RAPD, but a full discussion of these is beyond the scope of this book. Many studies employing high copy number genes are published in the journal *Molecular Ecology*, which is a rich source of information on systematic studies of closely related species.

Restriction Site Analysis

Most early molecular systematic studies employed **restriction site analysis**, the method of choice in the 1980s. This technique can be used to generate maps of individual genes or entire genomes. Much of what we now know about chloroplast and mitochondrial genome structure comes from such studies (see reviews by Olmstead and Palmer 1994 and Sytsma and Hahn 1997 and various chapters in Soltis et al. 1998).

In restriction site analysis, DNA is extracted from a plant and is then cut with **restriction enzymes**—enzymes that cut DNA at a particular sequence. The enzyme known as *Bam*HI, for example, cuts DNA everywhere it finds the sequence GGATCC, and *Eco*RI cuts at GAATTC.* A map is constructed by first cutting the DNA with one enzyme and

---

*The names of restriction enzymes are acronyms based on the first letter of the genus and first two letters of the species of the bacterium from which the enzyme was isolated. Thus *Bam*HI is from *Bacillus amyloliquefaciens*, *Eco*RI from *Escherichia coli*.

examining the resulting pattern of fragment sizes, then cutting it with a second enzyme, and finally cutting it with both enzymes together. This process creates a sort of puzzle from which the order of the restriction sites can be constructed by comparing the sizes of the fragments.

To piece together a restriction site map of the chloroplast genome, it is important to include only fragments of the chloroplast and not confuse them with pieces of DNA from the nucleus or mitochondrion. One way to do this is to grind up the plant and separate the chloroplasts from the rest of the tissue before isolating the DNA and cutting it. This approach is laborious, however, and was replaced by the technique of **Southern blotting** (named after E. M. Southern, who invented it), in which chloroplast, mitochondrial, and nuclear DNA are allowed to remain mixed and are cut simultaneously with the restriction enzyme. The cut DNA is then spread out by running it through an electrical gradient on a gel. The gel is then covered with a papery piece of nylon (called a membrane), and the DNA is transferred to the nylon in a process similar to making a silkscreen.

Next, a known piece of chloroplast DNA (the **probe**) is labeled with radioactive phosphorus and denatured to produce single-stranded DNA. This single-stranded DNA is then allowed to bind to the DNA on the membrane; it will bind only to matching (chloroplast) sequences. The membrane is then placed next to a piece of X-ray film. The bands of DNA to which the probe has bound appear as dark lines on the film. All the other DNA fragments (from the nucleus, mitochondrion, and other parts of the chloroplast) are present, but are invisible.

Restriction site analysis was initially used to determine the order of genes, particularly in the chloroplast genome. For example, the grass family (Poaceae) has three regions of the chloroplast genome that are inverted relative to those of most other angiosperms. One of these is unique to the Poaceae, one is shared with the Joinvilleaceae, and a third is shared with the Joinvilleaceae and Restionaceae (Doyle et al. 1992). The inversion that is unique to the Poaceae is hardly a surprise; the family is unquestionably monophyletic, a result that can be confirmed by almost any type of data. The inversions shared with the Joinvilleaceae and Restionaceae, however, helped clarify the morphological data, which suggested that either one might be sister to the grasses.

This relatively simple method has been turned into a powerful tool for systematics. It has been used most notably in studies of the chloroplast genome. Sequences (restriction sites) are scored as present or absent, and these scores are then used as characters in phylogenetic analysis. Presence or absence of restriction sites creates fragments of DNA of different lengths; these are known as **Restriction Fragment Length Polymorphisms**, or RFLPs. The methods used for this kind of study are the same as those used for mapping studies of the nuclear genome, which are now most commonly done with Southern blots. Yet another method, devised after PCR became widely available, is to amplify a particular piece of DNA and then cut it with restriction enzymes.

The advantage of using any restriction site approach is that it can potentially cover a large stretch of DNA and is thus thought to be less sensitive to local vagaries of selection or differences in mutation rate. This is also a disadvantage, of course; generally, the exact position of the restriction site is unknown (e.g., inside a gene or outside it, in the third position of a codon or not), so it is impossible to know whether a restriction site gain or loss is in exactly the same place among several taxa. Furthermore, with standard methods, estimates of the sizes of restriction fragments are accurate to only 50 or 100 bp, so that two sites very close to each other can easily be confused. Finally, a restriction site is a four- or six-base sequence of DNA that can be lost by a mutation in any one of its four or six bases. This means that different mutations will all look the same and cannot be distinguished.

Restriction site analysis is used less widely today than it was initially because sequencing has become so efficient. As a laboratory tool, however, it remains useful for screening clones of nuclear genes and for mapping the nuclear genome.

Nuclear Genome Mapping

Mapping of the nuclear genome is becoming increasingly common in evolutionary studies, particularly those of the close relatives of crop plants and the major model systems for molecular biology. Comparative studies using the nuclear genome have been done in the grasses, the Solanaceae, and the Brassicaceae and are appearing in other families as well. In addition, nuclear genome mapping studies are just beginning to address questions of speciation.

Generating a map of a nuclear genome requires a large commitment of time and effort. Two plants must be crossed and their F_1 offspring self-pollinated to produce a large number (100 or more) of F_2 plants. Then the genotypes of both parents and offspring are determined through the use of RFLP, RAPD, or AFLP markers; these markers must be polymorphic between the parents. This technique generally requires the use of sophisticated statistical programs to infer linkage relationships.

In a study of *Mimulus*, Bradshaw et al. (1995) found that the shift from bee pollination to bird pollination (see page 000) involved eight genes, which they were able to localize to linkage groups (groups of genes that are physically close together, i.e. "linked"; the largest possible linkage group is equivalent to a chromosome). In a similar study of *Helianthus*, Rieseberg and colleagues (1995, 1996) found that two species (*H. annuus* and *H. petiolaris*) differed by at least ten genomic rearrangements (three inversions and at least seven translocations), which affected genetic recombination and possibilities for introgression. The genome of their hybrid derivative, *H. anomalus*, has been rearranged relative to those of both parental species, so the species is partially reproductively isolated from both. Rieseberg's group then created new hybrids of *H. annuus* and *H. petiolaris* and found that the chromosomal rearrangements in the experimental hybrids were similar to those in the naturally occurring hybrid species, *H. anomalus*. They concluded that certain combinations of genes and gene rearrangements were selectively favored in the hybrid.

It is not clear whether the nuclear mapping approach will ever become simple enough to apply to the multiple species usually covered by a systematic study. Already, however, such studies are becoming extremely valuable for systematists interested in the mechanics of the speciation process.

Summary

Molecular techniques provide powerful tools for the study of evolution and phylogeny. Most data on relationships at the species level and above have so far come from the chloroplast genome and the highly repeated sequences of ribosomal RNA genes. Increasingly, however, low copy number nuclear genes are providing new insights. New tools are continually being developed for the study of variation within and among populations, including methods of genome mapping. As these tools come into more widespread use, they will provide new information on the processes of population-level differentiation.

No matter how powerful the molecular data, however, morphological data will remain critical for phylogenetic studies. The major questions in plant systematics are still morphological. Questions about the origin of species, the mechanisms of diversification, and the best way to classify diversity all require an understanding of morphology as well as phylogeny. We can now envision a time when robust phylogenies will have been constructed for all groups of plants, and the question of systematics will shift from "What is the phylogeny of my group?" to "How did the morphological diversity in my group arise?"

LITERATURE CITED AND SUGGESTED READINGS

Items marked with an asterisk are especially recommended to those readers who are interested in further information on the topics discussed in this chapter.

*Avise, J. C. 1994. *Molecular markers, natural history and evolution.* Chapman & Hall, New York.

Bennett, J. R. and S. Mathews. 2006. Phylogeny of the parasitic plant family Orobanchaceae inferred from phytochrome A. *Am. J. Bot.* 93: 1039–1051.

Bradshaw, H. D., S. M. Wilbert, K. B. Otto and D. W. Schemske. 1995. Genetic mapping of floral traits associated with reproductive isolation in monkeyflowers (*Mimulus*). *Nature* 376: 762–765.

Catalán, P., E. A. Kellogg and R. G. Olmstead. 1997. Phylogeny of Poaceae subfamily Pooideae based on chloroplast *ndhF* gene sequences. *Mol. Phylogenet. Evol.* 8: 150–166.

*Chase, M. W. and 41 others. 1993. Phylogenetics of seed plants: An analysis of nucleotide sequences from the plastid gene *rbcL*. *Ann. Missouri Bot. Gard.* 80: 528–580.

Davis, J. I. and R. J. Soreng. 1993. Phylogenetic structure in the grass family (Poaceae) as inferred from chloroplast DNA restriction site variation. *Am. J. Bot.* 80: 1444–1454.

De la Torre, J. E. B., M. G. Egan, M. S. Katari, E. D. Brenner, D. W. Stevenson, G. M. Coruzzi and R. DeSalle. 2006. ESTimating plant phylogeny: Lessons from partitioning. *BMC Evol. Biol.* 6: 48.

Downie, S. R., E. Llanas and D. S. Katz-Downie. 1996. Multiple independent losses of the *rpoC1* intron in angiosperm chloroplast DNAs. *Syst. Bot.* 21: 135–151.

Doyle, J. J., J. I. Davis, R. J. Soreng, D. Garvin and M. J. Anderson. 1992. Chloroplast DNA inversions and the origin of the grass family (Poaceae). *Proc. Natl. Acad. Sci. USA* 89: 7722–7726.

Edwards, E. J., R. Nyffeler and M. J. Donoghue. 2005. Basal cactus phylogeny: Implications of *Pereskia* (Cactaceae) paraphyly for the transition to the cactus life form. *Am. J. Bot.* 92: 1177–1188.

Evans, R. C., L. A. Alice, C. S. Campbell, E. A. Kellogg and T. A. Dickinson. 2000. Multiple putative *GBSSI* loci in the Rosaceae: Characterization and potential phylogenetic utility. *Mol. Phylogenet. Evol.* 11: 388–400.

*Graur, D. and W.-H. Li. 2000. *Fundamentals of molecular evolution*, 2nd ed. Sinauer Associates, Sunderland, MA.

Hall, B. G. 2004. *Phylogenetic trees made easy: A how-to manual*, 2nd ed. Sinauer Associates, Sunderland, MA.

Hsiao, C., N. J. Chatterton, K. H. Asay and K. B. Jensen. 1994. Molecular phylogeny of the Pooideae (Poaceae) based on nuclear rDNA (ITS) sequences. *Theor. Appl. Genet.* 90: 389–398.

Hughes, C. E., R. J. Eastwood and C. D. Bailey. 2006. From famine to feast? Selecting nuclear DNA sequence loci for plant species-level phylogeny reconstruction. *Philos. Trans. R. Soc. London* B 361: 211–225.

Jansen, R. K. and J. D. Palmer. 1987. A chloroplast DNA inversion marks an ancient evolutionary split in the sunflower family (Asteraceae). *Proc. Natl. Acad. Sci. USA* 84: 5818–5822.

Jansen, R. K., C. Kaittanis, S.-B. Lee, C. Saski, J. Tomkins, A. J Alverson and H. Daniell. 2006. Phylogenetic analyses of *Vitis* (Vitaceae) based on complete chloroplast genome sequences: effects of taxon sampling and phylogenetic methods on resolving relationships among rosids. *BMC Evolutionary Biology* 6:32–46.

Kellogg, E. A. and J. L. Bennetzen. 2004. The evolution of nuclear genome structure in seed plants. *Am. J. Bot.* 91: 1709–1725.

Kellogg, E. A. and N. D. Juliano. 1997. The structure and function of RuBisCO and their implications for systematic studies. *Am. J. Bot.* 84: 413–428.

Kellogg, E. A., R. Appels and R. J. Mason-Gamer. 1996. When genes tell different stories: The diploid genera of Triticeae (Gramineae). *Syst. Bot.* 21: 321–347.

Madison, D. and W. Maddison. 2005. MacClade 4.0. Sinauer Associates, Sunderland, MA.

Malcomber, S. T. 2002. Phylogeny of *Gaertnera* Lam. (Rubiaceae) based on multiple DNA markers: Evidence of rapid radiation in a widespread, morphologically diverse genus. *Evolution* 56: 42–57.

Manos, P. S., R. E. Miller and P. Wilkin. 2001. Phylogenetic analysis of *Ipomoea*, *Argyreia*, *Stictocardia*, and *Turbina* suggests a generalized model of morphological evolution in morning glories. *Syst. Bot.* 26: 585–602.

Mason-Gamer, R. J. and E. A. Kellogg. 1996. Chloroplast DNA analysis of the monogenomic Triticeae: Phylogenetic implications and genome-specific markers. In *Methods of genome analysis in plants: Their merits and pitfalls*, P. Jauhar (ed.), 301–325. CRC Press, Boca Raton, FL.

Mason-Gamer, R. J., C. F. Weil and E. A. Kellogg. 1998. Granule-bound starch synthase: Structure, function, and phylogenetic utility. *Mol. Biol. Evol.* 15: 1658–1673.

Mathews, S. and M. J. Donoghue. 1999. The root of angiosperm phylogeny inferred from duplicate phytochrome genes. *Science* 286: 947–950.

Mathews, S. and R. A. Sharrock. 1996. The phytochrome gene family in grasses (Poaceae): A phylogeny and evidence that grasses have a subset of the loci found in dicot angiosperms. *Mol. Biol. Evol.* 13: 1141–1150.

McMahon, M. M., and M. J. Sanderson. 2006. Phylogenetic supermatrix analysis of GenBank sequences for 2228 papilionoid legumes. *Syst. Biol.* 55: 818–386.

Mount, D. W. 2001. *Bioinformatics: Sequence and genome analysis.* Cold Spring Harbor Laboratory Press, Cold Spring Harbor, NY.

Ohi-Toma, T., T. Sugawara, H. Murata, S. Wanke, C. Neinhuis and J. Murata. 2006. Molecular phylogeny of *Aristolochia* sensu lato (Aristolochiaceae) based on sequences of *rbcL*, *matK*, and *phyA* genes, with special reference to differentiation of chromosome numbers. *Syst. Bot.* 31: 481–492.

*Olmstead, R. G. and J. D. Palmer. 1994. Chloroplast DNA systematics: A review of methods and data analysis. *Am. J. Bot.* 81: 1205–1224.

Olsen, K. M. and B. A. Schaal. 1999. Evidence on the origin of cassava: Phylogeography of *Manihot esculenta. Proc. Natl. Acad. Sci. USA* 96: 5586–5591.

*Page, R. D. M. and E. C. Holmes. 1998. *Molecular evolution: A phylogenetic approach.* Blackwell, Oxford.

Peralta, I. E. and D. M. Spooner. 2001. Granule-bound starch synthase (*GBSSI*) gene phylogeny of wild tomatoes (*Solanum* L. section *Lycopersicon* (Mill.) Wettst. subsection *Lycopersicon*). *Am. J. Bot.* 88: 1888–1902.

Petersen, G. and O. Seberg. 1997. Phylogenetic analysis of the Triticeae (Poaceae) based on *rpoA* sequence data. *Mol. Phylogenet. Evol.* 7: 217–230.

Qiu, Y.-L. and 9 others. 1999. The earliest angiosperms: Evidence from mitochondrial, plastid and nuclear genomes. *Nature* 402: 404–407.

Rieseberg, L. H., C. VanFossen and A. Desrochers. 1995. Genomic reorganization accompanies hybrid speciation in wild sunflowers. *Nature* 375: 313–316.

Rieseberg, L. H., B. Sinervo, C. R. Linder, M. Ungerer and D. M. Arias. 1996. Role of gene interactions in hybrid speciation: Evidence from ancient and experimental hybrids. *Science* 272: 741–745.

Sang, T. 2002. Utility of low-copy nuclear gene sequences in plant phylogenetics. *Critical Rev. Biochem. Mol. Biol.* 37: 121–147.

Schranz, M. E., M. Lysak and T. Mitchell-Olds. 2006. The ABCs of comparative genomics in the Brassicaceae: Building blocks at crucifer genomes. *Trends Plant Sci.* 30: 1–8.

Small, R. L., R. C. Cronn and J. F. Wendel. 2004. Use of nuclear genes for phylogeny reconstruction in plants. *Aust. Syst. Bot.* 17: 145–170.

Soltis, D. E., M. Mayer, P. S. Soltis and M. Edgerton. 1991. Chloroplast DNA variation in *Tellima grandiflora* (Saxifragaceae). *Am. J. Bot.* 78: 1379–1390.

Soltis, D. E., P. S. Soltis, D. R. Morgan, S. M. Swensen, B. C. Mullin, J. M. Dowd and P. G. Martin. 1995. Chloroplast gene sequence data suggest a single origin of the predisposition for symbiotic nitrogen fixation in angiosperms. *Proc. Natl. Acad. Sci. USA* 92: 2647–2651.

Soltis, D. E. and 15 others. 1997. Angiosperm phylogeny inferred from 18S ribosomal DNA sequences. *Ann. Missouri Bot. Gard.* 84: 1–49.

*Soltis, D. E., P. S. Soltis and J. J. Doyle (eds.). 1998. *Molecular systematics of plants II: DNA sequencing.* Kluwer, Boston.

Soltis, D. E. and 15 others. 2000. Angiosperm phylogeny inferred from 18S rDNA, *rbcL*, and *atpB* sequences. *Bot. J. Linnean Soc.* 133: 381–461.

Soltis, D. E., P. S. Soltis and M. J. Zanis. 2002. Phylogeny of seed plants based on evidence from eight genes. *Am. J. Bot.* 89: 1670–1681.

Soltis, P. S., D. E. Soltis and M. W. Chase. 1999. Angiosperm phylogeny inferred from multiple genes as a research tool for comparative biology. *Nature* 402: 402–404.

*Swofford, D. L., G. J. Olsen, P. J. Waddell and D. M. Hillis. 1996. Phylogenetic inference. In *Molecular systematics*, 2nd ed., D. M. Hillis, C. Moritz and B. K. Mable (eds.), 407–514. Sinauer Associates, Sunderland, MA.

*Sytsma, K. J. and W. J. Hahn. 1997. Molecular systematics: 1994–1995. *Prog. Bot.* 58: 470–499.

Wendel, J. F., A. Schnabel and T. Seelanan. 1995. An unusual ribosomal DNA sequence from *Gossypium gossypioides* reveals ancient, cryptic, intergenomic introgression. *Mol. Phylogenet. Evol.* 4: 298–313.

6

The Evolution of Plant Diversity

Planet Earth supports about 260,000 species of tracheophytes. They range from minute plants such as the floating aquatic duckweeds (*Lemna* and *Spirodela* of the Araceae), only millimeters in size and without leaves, to massive trees such as redwoods (*Sequoia*, Cupressaceae), over 100 meters tall and thousands of years old. As different as these extreme examples are, they are related through a common ancestor on the tree of life. Evolution—genetic change through time—is the source of this tremendous diversity and of the tree of life itself (Box 6A). The systematist's main concern is identifying and understanding the groups of organisms produced by evolution.

An essential ingredient of evolution is natural variation among individuals. Hence it is important to understand how this variation arises and is distributed geographically. The processes that create discrete units of variation (taxa) are of central interest to systematists, especially those that result in the formation of species (**speciation**). The role of **gene flow** (exchange of genes between populations) in creating and maintaining plant taxa is not clear. In order for one species to split into two new species, interruption of gene flow between the forming species is apparently necessary. Mating between individuals of different plant species may blur the boundary between species, but in many cases it has surprisingly little effect on morphological variation within a species.

Interspecific gene flow (**hybridization**, sometimes referred to as reticulation) plays a dual role in speciation. On the one hand, it may reduce diversity by merging species. On the other hand, it can be a powerful force leading to speciation, especially when coupled with polyploidy, an important source of genetic variation within plant species.

BOX 6A Theories of Organic Evolution

Evolution has most often been defined in two ways: (1) change in gene frequencies and (2) descent with modification. The latter definition is associated with Charles Robert Darwin (1809–1882) (Figure 6.1), who conceived the theory now called **Darwinism**. One of the most significant ideas in history, Darwinism has profoundly influenced not only the biological sciences, but also our perspective on our position in the world.

Others, most notably Jean-Baptiste Lamarck (1744–1829), also suggested that organisms change through time—an idea that was supported by the discovery of numerous fossils and of the great age of the Earth—but Lamarck believed in the inheritance of acquired characteristics. When Darwin began his work, the general view was that all organisms were the product of divine creation. Darwin's observations of natural history—both in England and during his travels as the naturalist aboard HMS *Beagle* on its round-the-world voyage from 1831 to 1836—and years of thought led him to his theory, which he presented in 1859 in his book *On the Origin of Species*. Darwin perceived that all organisms are related to one another in a branching tree of life, and he con-

FIGURE 6.1 Charles Darwin in his prime. (Courtesy of the American Philosophical Society.)

ceived of a process that could generate that pattern. He called that process **natural selection**. Its logic is based on a set of observations that Darwin brought together and two major inferences based on those observations (Figure 6.2).

Thomas Malthus (1766–1834), an influential English economist, observed that human populations are superfecund—that is, that they are capable of tremendous (geometric) growth in

numbers. Darwin extended this idea to the natural world. That natural populations do not experience explosive increases in their numbers, but instead generally persist in a steady state, could be explained by the constraint of limited resources. These first three facts—superfecundity, the steady state of populations, and limited resources—led to Darwin's first inference, that there is a struggle for existence. Only the best-adapted, or fittest, organisms—such as those plant individuals with flowers that most efficiently attract pollinators—will compete successfully against other individuals for the resources they need to survive and reproduce.

Darwin linked to this inference the fact that there is variation among individuals of a species. This concept was well understood by animal and plant breeders in Darwin's time, but it was otherwise not generally accepted. Since the ancient Greeks, a species, such as a particular species of oak, was considered the manifestation of a unique, unchangeable essence. Any deviations from this essence were regarded as unimportant. Second, from breeders and from his own experiments and observations, Darwin knew that offspring tend to resemble their parents.

Plant diversity is strongly shaped by breeding systems. Uniparental reproduction, by self-fertilization or by asexual means, tends to package variation in smaller, more numerous units than are found in groups with strictly biparental reproduction.

The final section of this chapter deals with species concepts, an issue that has been intensely debated by systematists and evolutionary biologists. Although many biologists define a species as a group of populations that does not exchange genes with other populations, gene flow fails as a criterion for species definition in plants when strictly applied because interspecific hybridization and uniparental reproduction are often common among them. What are commonly recognized as plant species frequently hybridize, and uniparental reproduction reduces a reproductive community to one individual or the progeny of one individual. Hence plant systematists by and large do not insist that species are equivalent to reproductive communities. Instead, they define species using a wide array of evidence that a population, or more usually a group of populations, forms an independent evolutionary lineage.

The characters most commonly used to define species are morphological: members of a given plant species are

morphologically more similar to one another than they are to members of other species. Molecular data, the use of which is becoming increasingly common, challenge and complement the delimitation of species on morphological grounds.

Plant Diversity Is the Result of Evolution

Support for evolution as the source of biodiversity comes from fossils, from the occurrence of features shared by groups of organisms, from observations of the geographic variation of organisms, and from environmental studies. Fossils of flowering plants (Figure 6.3) record the evolution of structures—flowers, fruits, leaves, and other parts—and the origin and extinction of species, genera, and families (Stewart and Rothwell 1993). Although there are many gaps in the fossil record, it overwhelmingly verifies morphological change in organisms (and therefore in their genes) over the millions of years of life on Earth.

The fact that living organisms function through common genetic and biochemical processes strongly supports a unique origin and common history of life on Earth. All

Although he did not understand the mechanism of heredity, he did see that it was essential to his second inference: natural selection.

When variation affects the outcome of the struggle for existence (as in the attractiveness of a flower to a pollinator), individuals with the most advantageous characteristics will survive and reproduce more often than others. The progeny of these successful individuals will be more frequent in future generations than those of the less successful individuals. Without heredity, by which successful characteristics are passed on to individuals of future generations, the population would not change through time. To summarize Darwinism, herita-

ble variation in fitness leads to natural selection and evolution.

Although natural selection is responsible for much evolutionary change, an alternative must operate some of the time. When the force of natural selection is weak, chance will govern the course of evolutionary change. A gene may be present in a population or species not because it confers higher fitness than another gene, but merely by chance. This process, known as random genetic drift, is described on pages 124–125 and is the basis of the **neutral theory of molecular evolution** (see Li 1997). The "neutral" part of this name refers to the notion that some variation does not affect (is neutral to) the survival of organisms.

Shortly after the publication of *On the Origin of Species*, an Austrian monk, Gregor Mendel (1822–1884), performed a series of elegant experiments demonstrating heredity in the common garden pea. These experiments were not widely appreciated until the beginning of the twentieth century, when they formed the basis of the field of genetics. Later, two British scientists (R. A. Fisher and J. B. S. Haldane) and an American (Sewall Wright) developed the field of population genetics and provided a quantitative and theoretical framework for studies of evolutionary change in populations. Transformed by population genetics, Darwinism was renamed **neo-Darwinism**.

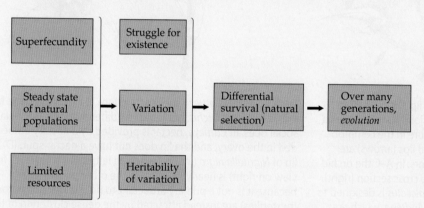

FIGURE 6.2 Darwin's logic. Three observations—that (1) organisms are capable of rapid increase in numbers of individuals (superfecundity), but (2) population size is generally stable and (3) resources are limited—led to the inference of a struggle for existence among individuals. Two additional observations, of natural variation among individuals and the heritability of this variation, led to the inference of differential reproduction (i.e., natural selection) and of evolution over many generations. (Modified from Mayr 1977.)

organisms perform certain essential functions via the same metabolic pathways. Glycolysis, for example, is a series of coordinated biochemical reactions that involves the partial breakdown of glucose to drive the synthesis of adenosine triphosphate (ATP) molecules in the cytoplasm of all living cells. The citric acid cycle is a continuation of glucose breakdown that generates more ATP and takes place in the mitochondria, which are found in all eukaryotic organisms.

DNA makes up the genotype in all organisms. Genetic information is encoded in the sequence of the nucleotides in DNA, forming a code that is nearly universal in living organisms. The **phenotype** (the observable traits of an organism) is the product of interactions between the genotype and the environment.

FIGURE 6.3 Reconstruction of leafy branch and flower of *Archaeanthus linnenbergeri*, an early angiosperm from the mid-Cretaceous. Fossils of such plants help us understand when angiosperms evolved and what some of the first members of this lineage may have looked like. Colors added to help distinguish structures. (From Dilcher and Crane 1984; original art by M. Rohn.)

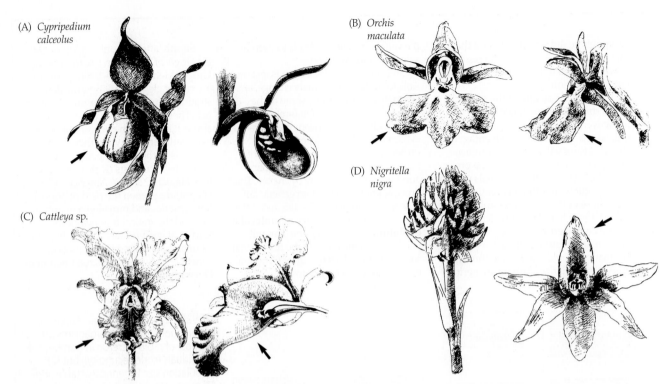

(A) *Cypripedium calceolus*

(B) *Orchis maculata*

(D) *Nigritella nigra*

(C) *Cattleya* sp.

FIGURE 6.4 Variations in the orchid floral lip. Despite differences in size, shape, and color, the orchid floral lip is thought to have derived from a single ancestral structure in the common ancestor of all orchids. Differences in orchid lips (arrows) are related to adaptations to different pollinators. In A–C the orchid flower is shown in front view (left) and side cross section (right). (A) The pouch-shaped lip of *Cypripedium calceolus* is designed to momentarily trap small bees of the genus *Andrena*, which are forced to pollinate the flower because the only possible way out of the lip takes them past the pollen and stigmatic surface. (B) The three-lobed lip of *Orchis maculata* attracts bees (and sometimes flies) and orients them toward the nectar in the spur of the lip; as pollinators take nectar, they pollinate the flower. (C) The large lip of this *Cattleya* species functions in much the same fashion as the lip of *Orchis*, but it attracts different pollinators—large social bees. In *Cattleya*, nectar is provided by a nectary embedded in the ovary, and the lip does not have a nectar spur. (D) The lip of *Nigritella nigra* (inflorescence on left; solitary flower in front view on right) is nearly identical to the other perianth parts because it is not a primary attractant to pollinators; pollinators (butterflies) are instead attracted by the dense clustering of the flowers. In addition, the lip is not in the lower position, as it is in the flowers in A–C; the lip of this orchid is specialized for butterflies because the opening to the small nectar spur is too small for the proboscis (tonguelike organ) of other insects. (From Faegri and van der Pijl 1979.)

Major genetic change is often recorded in phenotypes. The specialized lowermost petal (the lip, or labellum) of orchid flowers, for example, varies in size, color, and shape among species (Figure 6.4). Despite their differences, these structures trace back to a common ancestral lip; they are all modifications wrought by evolution over millions of years. Orchid lips have a fundamental similarity because of this common evolutionary history; in other words, they are homologous.

Many of the changes that make organisms different from one another affect their ability to survive and reproduce. This ability is referred to as **fitness** and is related to the concept of adaptation, the close correspondence between organisms and their environment. Variations in the color, size, and shape of flowers, for example, are adaptations that attract, manipulate, and reward dozens of species of birds and bats as well as thousands of species of bees, moths, and other insect pollinators (see Chapter 4). Flowers pollinated by bats, for example, are readily accessible to flying bats, are white

and easily located at night, and provide a plentiful nectar reward (Figure 6.5; see also Table 4.1).

Another remarkable example of adaptation is the nectar spur of the orchid *Angraecum sesquipedale* of Madagascar. The specific epithet refers to the great length of this spur: *sesqui* means "one-and-a-half" and *pedale* refers to "foot"; therefore, the literal meaning is "a foot and a half" (actually, the spur is only 30 cm long, not 45 cm). The pollinators of this species were not known in Darwin's time, but he hypothesized that there must be a moth with a tongue long enough to reach the nectary near the bottom of the spur. Later such a moth, a hawkmoth, was indeed discovered.

A final line of evidence for evolution is the intricacy of geographic variation within groups of organisms. Numerous studies document patterns of plant variation—morphological, ecological, phenological, and genetic—that are coincident with geographically varying physical or biotic features. Jeffrey pine (*Pinus jeffreyi*), a forest tree of montane regions from southwestern Oregon to Mexico, illustrates

FIGURE 6.5   The long-nosed bat (*Leptonycteris curasoae*) uses its highly extensile tongue to lick nectar from the cup-shaped corolla of the saguaro (*Carnegia gigantea*), which it pollinates in the process. The flower's size, white color, and nocturnal functioning are all adaptations for bat pollination. Bats are not capable of hovering, like hummingbirds, but instead drink briefly at a flower and then land, returning to the flower until the nectar is gone. Rarely, bats fatally impale themselves on the cactus's long spines.

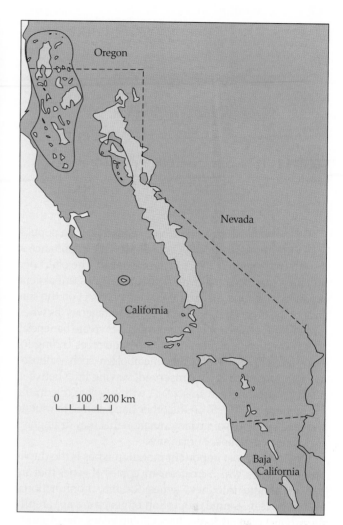

FIGURE 6.6   The global distribution of *Pinus jeffreyi*, with the distribution of serpentine soil ecotypes shown by the outlines in northern California and southwestern Oregon. (After Furnier and Adams 1986.)

such a pattern (Furnier and Adams 1986). The species maintains high genetic diversity, like most conifers, except in populations on serpentine soils (Figure 6.6). These soils, which are quite infertile, differ sufficiently from other, more fertile soils to induce the formation of **ecotypes** (populations specialized for certain ecological circumstances) adapted to serpentine soils. Different ecotypes have different genotypes. Serpentine ecotypes of Jeffrey pine harbor less genetic diversity than populations on more fertile soils, probably because of intense natural selection for genotypes tolerant of the rigorous soil conditions.

Evolution is frequently considered on two levels. Our discussion so far has focused on change within species. For example, an individual might bear a mutation (which is heritable) for enhanced chemical defense against herbivores. That individual might therefore produce more offspring than other individuals in the population, thereby increasing the frequency of the mutation within the population (and perhaps within other populations through gene flow). The accumulation of such changes within a lineage is sometimes referred to as **anagenesis**. Changes that lead to the forma-

tion of new species by the splitting of one lineage into two—**cladogenesis**—are of primary interest to systematics.

Variation in Plant Populations and Species

Sources of Variation

Mutation and genetic recombination are the primary sources of variation within plant populations and species, and they form the raw material for natural selection and random genetic drift. **Mutation** refers to alterations of DNA, ranging from changes in single bases (point mutations; Figure 6.7) to insertions, duplications, deletions, and inversions of parts of a chromosome, to gains or losses of whole chromosomes, and finally to changes in whole genomes (the full complement of chromosomes in the nucleus).

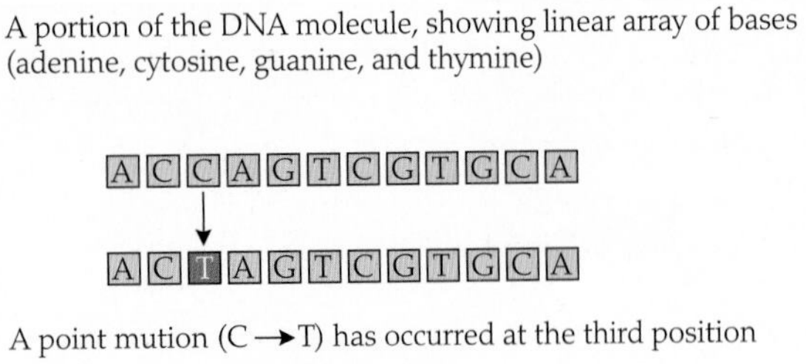

A portion of the DNA molecule, showing linear array of bases (adenine, cytosine, guanine, and thymine)

A C C A G T C G T G C A

↓

A C T A G T C G T G C A

A point mutation (C → T) has occurred at the third position

FIGURE 6.7 Point mutations in DNA are a source of genetic variation.

The extent to which a mutation spreads through a population or species varies greatly, depending on the importance of the affected region of DNA to the organism. The effect of a mutation may be lethal (when production of essential gene products is disrupted), neutral (having no effect on the survival of the organism), or selectively advantageous (as when new chromosomal arrangements of genes create beneficial, coordinated gene expression). DNA sequences coding for essential components of cellular metabolism, such as ribosomal subunits, are highly **conserved**, varying little between major groups of organisms. DNA with no clear role in the functioning of the cell, on the other hand, is free to mutate and regularly does so, creating variation that may distinguish closely related groups of organisms.

One of the most important mutation types is the duplication of genes, which creates extra copies of genes that are free to mutate into new genes (or into nonfunctional pseudogenes). A large proportion of genes are apparently the product of gene duplication. Globin genes, the products of which are essential for oxygen transport in our blood, have diversified following several duplications, and even in plants have duplicated and diverged. Leghemoglobin binds oxygen, playing an essential role in nitrogen fixation by legumes and other plants, and plastocyanin, another globin, is a photosynthetic pigment.

The largest mutations are aneuploidy and polyploidy (see Chapter 4). Aneuploidy is pronounced in some plant groups, even though the loss of essential genetic material or excesses of certain genes attending aneuploidy can be quite harmful. In the large genus of sedges (*Carex*, Cyperaceae), for example, the haploid chromosome number ranges from 6 to 56, and every number between 12 and 43 is found in at least one species. Bluegrasses (*Poa*, Poaceae) and willows (*Salix*, Salicaceae) also contain such extensive ranges of chromosome numbers or aneuploid series. Polyploidy leads to gene duplication and greater genetic diversity on which natural selection can act. Polyploidy is frequent and evolutionarily important in plants; it is discussed in more detail on pages 140–143.

Genetic recombination refers to the shuffling of genes that occurs primarily during meiosis, the special cell division that produces spores. Chromosomes occur in pairs, with one chromosome in each pair derived from the father and the other from the mother (Figure 6.8A)—each with the same genes in the same order. In meiosis, these homologous pairs line up together and routinely exchange chromosomal segments so that genes are rearranged. Rather than each of the homologues having only maternal or paternal genes, this reciprocal exchange, or **crossing-over**, gives them a mixture of parental genes (Figure 6.8B).

Genes are also recombined during meiosis by **independent assortment** of the chromosomes. After meiotic pairing, the homologues separate into spores (Figure 6.8C), and the maternal (or paternal) homologue of one pair may end up in a spore with the maternal or the paternal homologue of any other pair. An organism with 10 pairs of chromosomes, such as corn (*Zea mays*), can produce 210 (or 1024) genetically different spores and 10,242 (or 1,048,576) different zygotes just by independent assortment.

Recombination is a rich source of variation. It increases as the frequency of crossing-over and the number of chromosomes increase, and it is affected by population size, breeding system, seed and pollen dispersal, and other factors. Recombination is largely responsible for the great variation observed within natural populations and even among the progeny of a single mating.

Variation is also affected by gene flow, which may introduce new genetic material into a population, and by **random genetic drift**, the chance fixation of genes in small populations. To see how random genetic drift operates, consider a hypothetical population of 10 rose plants in which flower color is controlled by one gene that has two forms (alleles). One allele codes for red flowers and is designated by the letter *R*; the other allele codes for white flowers and is designated by the letter *W*. Each plant bears two alleles (one from each parent). Individuals with two *R* alleles (*RR* genotype) have red flowers; individuals with *WW* genotypes have white flowers; and individuals with *RW* genotypes have pink flowers. The fitness of the three flower colors is the same. Of the 10 plants in this population, 9 have red

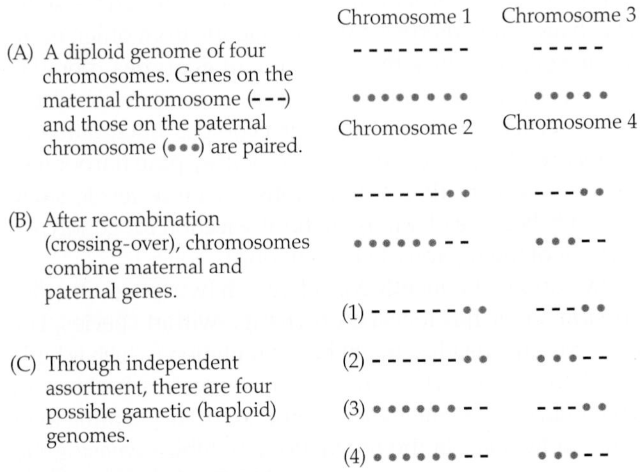

FIGURE 6.8 Genetic recombination between homologous chromosomes is a primary source of genetic variation.

flowers and 1 has pink flowers. There is a total of 20 alleles in this group of 10 plants; 19 of them are *R* alleles, and 1 is a *W* allele. The frequency of the *R* allele is therefore 95%, and that of the *W* allele only 5%.

Consider the possible outcomes of mating. The red roses produce gametophytes bearing gametes with the *R* allele; the one pink rose produces gametophytes bearing gametes with an *R* allele or a *W* allele. The *W* allele might not participate in any matings during one generation and therefore might be lost from the population. The *R* allele would then be the only allele in this population. The probability of the loss of the *W* allele is 95%, and that loss would represent a genetic change that is strictly the result of chance, not of natural selection.

Individual variation among plants, which is extensive in almost all plant groups, is discussed in detail by Briggs and Walters (1997). Next we will look at how this individual variation is distributed spatially and partitioned into taxonomic groups.

Local and Geographic Patterns of Variation

The starting point in studies of spatial variation is the population. A **population** is a group of individuals of a species occupying a more or less well defined geographic region and interacting with one another reproductively. Populations are difficult to characterize, however. They range in size from a single individual to millions of individuals, and they may persist for less than a year or thousands of years. They may consist of the offspring of one individual or be regularly stocked by immigrants. This variation in their history, coupled with different levels of genetic diversity, means that populations vary greatly at the genetic level.

Spatial variation may be more or less continuous, forming a **cline**, such as a continuous reduction in the height of trees from the bottom to the top of a mountain. Plants more commonly vary discontinuously in space. White fir (*Abies concolor*, Pinaceae), a conifer of mountainous regions in the western United States and northern Mexico, shows marked geographic discontinuities: northern California populations are unique in their hairy twigs and notched leaves; Baja California populations have uniquely short, thick leaves; a cluster of populations in Utah has even shorter leaves and a slightly different chemical profile than an otherwise similar set of populations in Colorado and northern New Mexico; and populations in southern New Mexico and Arizona are chemically like Colorado populations but morphologically like southern California populations.

Patterns of geographic variation covering large areas of the globe occur in many groups of plants. One source of some of these patterns is continental drift, which breaks up continents into pieces and separates taxa into geographically isolated groups that have the potential to diversify. A well-known example of this process, which is called **vicariance**, is provided by the southern beeches (the genus *Nothofagus* of the southern beech family, Nothofagaceae). There are about 35 species of southern beeches, and they now grow in southern South America, New Zealand, Australia, New Caledonia, and New Guinea. They are represented by fossils in these areas and Antarctica but nowhere else in the world. These areas are parts of the former supercontinent East Gondwana that the forces of continental drift separated starting about 80 million years ago. The genus consists of four clades, which correspond to subgenera. Subgenus *Brassospora*, which grows only in New Caledonia and New Guinea, is the sister group of subgen. *Nothofagus*, which is limited to South America. The geographic pattern of these subgenera represents a likely example of vicariance (Head 2006): the common ancestor of these two subgenera grew on East Gondwana, and, following the splitting up of this supercontinent, the descendents of the common ancestor evolved into the two subgenera in their separate ranges.

Careful studies of geographic variation often uncover patterns like that of white fir (or of Jeffrey pine, described earlier). Documenting patterns of geographic variation is an important step in understanding systematic diversity.

Speciation

Natural variation at and above the population level is usually not continuous, but occurs in discrete units, or taxa. Easily the most important taxonomic level is the species because it is often the smallest clearly recognizable and discrete set of populations. Understanding how species form and how to recognize them have been major challenges to systematists.

Speciation can be defined as the permanent severing of two or more sets of populations so that migrants from one population system would be at a disadvantage when entering another. This disadvantage could stem from a lack of mates for the migrants if the two systems were reproductively isolated. Alternatively, a migrant might not compete well with residents in withstanding pathogens, pests, and predators or in attracting pollinators and animals to disperse its fruits. Speciation may result either from adaptive changes or from changes that are due to chance.

One of the problems in studying speciation is that it may be a slow process. We see only moments in time and must infer a process from a pattern. For example, in *Gilia*, a genus of small herbs in the Polemoniaceae, some species in southwestern North America contain sets of populations that have been called geographical **races**. These races differ morphologically, but they often grow together, interbreed, and intergrade. Subspecies show less geographic overlap than races, but some intergradation and interbreeding still occurs. Species of *Gilia* generally are more strongly differentiated from one another and show less tendency to mate. From such patterns of races, subspecies, and species, the inference has been made that speciation is a gradual process of divergence and shutting off of gene flow.

Such gradual divergence through the intermediate steps of races, subspecies, and finally fully separate species

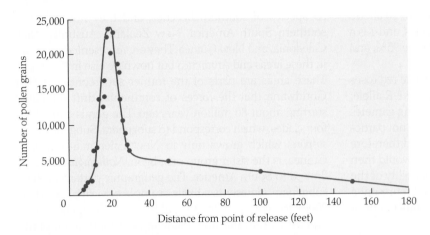

FIGURE 6.9 Distribution of pollen directly downwind from its source; most pollen grains fall near the parent plant. (Modified from Grant 1981.)

is the traditional view of speciation. According to this view, geographic isolation is required to prevent gene flow and allow divergence of the isolated sets of populations. This form of speciation is referred to as **allopatric**, or **geographic**, **speciation**.

Some scientists have questioned how races, especially ones that are geographically widespread, are transformed into coherent plant species (Levin 1993, 2000). The two most likely transforming mechanisms are gene flow and natural selection that is uniform in its effect over the range of a race. Genes are transported by pollen grains and fruits, most of which end up near the parent plant and rapidly diminish in numbers away from it (Figure 6.9). Thus gene flow is usually measured in meters (Figure 6.10), and it rarely extends to a kilometer. Overall, the diffusion of genes among populations of a race spread over a large region could require thousands of generations. On the other hand, occasionally pollen or fruit may travel a considerable distance, and gene flow in plants is apparently sufficient to allow the spread of strongly favorable alleles (Rieseberg and Burke 2001, Rieseberg et al. 2003).

Natural selection could direct different populations of a race toward a similar end point, but only if important elements of the environment are very similar in different parts of the range of the race. Given the complexity of biotic, physical, and climatic elements of the environment, such uniform change seems unlikely over large geographic regions. The likelihood that chance plays a part in speciation

further reduces the plausibility of allopatric speciation. Nevertheless, it is quite clear that allopatric speciation has occurred frequently (Coyne and Orr 2004).

Another possible mode of speciation does not require that gene flow be extensive or that selection be uniform over long distances. Instead, it postulates that speciation most often occurs in local populations or metapopulations (clusters of several local populations connected by occasional gene flow). This alternative is called **local speciation**, or the peripheral isolation model of speciation (Levin 1993).

Small populations at the margin of a species range are subject to random genetic drift, and environmental conditions at the margin may elicit adaptive change. These random and/or adaptive changes may be significant enough to create a new lineage (a **neospecies**). Such neospecies may fail to spread geographically and thus go extinct. If they do expand, their success may depend on their having unique ecological adaptations that allow them to avoid competition with their progenitor. Dispersal from the founding local population binds populations of the neospecies by common ancestry.

Both ecology and genomic changes can play major roles in speciation (Levin 2000). Ecology has been predominant in some cases, such as that of the Hawaiian silverswords. As discussed in Chapter 1, the evolutionary diversification of this group has apparently resulted from interisland dispersal

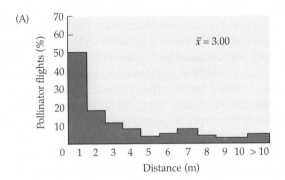

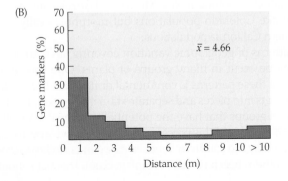

FIGURE 6.10 Gene flow in animal-pollinated *Phlox* (Polemoniaceae). (A) The distances (in meters) that pollinators traveled between plants. (B) The distances of gene flow. The mean distance ($\bar{x}$) is greater for gene flow than for pollinators because of carryover; that is, some pollen deposited by a pollinator is picked up by subsequent pollinators and "carried over" to other plants. (From Levin 1981.)

(A)

(B)

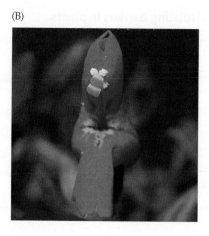

FIGURE 6.11   Flowers of *Mimulus lewisii* and *M. cardinalis*. The flowers of *M. lewisii* (A) are adapted for bumblebee pollination, whereas those of *M. cardinalis* (B) are adapted for hummingbird pollination. (From Schemske and Bradshaw 1999.)

followed by numerous, major ecological shifts along moisture gradients. Despite the morphological and ecological diversity in the silversword alliance, its genomes have not diverged in ways that prevent gene flow between species and even genera (Carr 1995; Caraway et al. 2001). More generally, the Hawaiian Islands and other oceanic archipelagos, such as the Canary Islands, provide many examples of evolutionary diversification illustrating the interplay of dispersal and ecological diversification (Baldwin et al. 1998).

Note that geographic separation does not automatically result in a loss of the capacity to interbreed. For example, the sycamores *Platanus occidentalis* (Platanaceae) of eastern North America and *P. orientalis* of the eastern Mediterranean region have been geographically separated for millions of years and have diverged from each other morphologically, and yet the artificial hybrid is vigorous, fertile, and frequently planted as a stately ornamental. Widely and long-separated populations may also be morphologically very similar.

Under special circumstances, speciation may occur without geographic isolation, although some form of strong barrier to gene flow is still necessary. This form of speciation, called **sympatric speciation**, occurs commonly in plants through polyploidy, which is described on pages 140–143. An example of sympatric speciation without polyploidy is provided by the genus *Stephanomeria* (Asteraceae). A new species of annual plant, *S. malheurensis*, is postulated to have arisen recently within a population of *S. exigua* ssp. *coronaria* in Oregon (Brauner and Gottlieb 1987). There were fewer than 250 individuals of this new species when it was discovered in the 1960s and early 1970s. The new species and its progenitor are similar genetically, but differ morphologically in quantitative characters (such as achene length). The progenitor species (*S. exigua* ssp. *coronaria*) cannot self-pollinate, but the derivative species (*S. malheurensis*) is predominantly self-pollinated, which reduces gene flow between it and the progenitor. This shift to self-pollination is governed by one gene, making the initiation of this speciation a relatively minor genetic event. Chromosomal structural differences between progenitor and derivative, however, lead to intersterility, a second barrier to gene flow.

Allopatric speciation has been inferred to require thousands or even millions of years to be completed (Coyne and Orr 2004). Speciation may also be rapid, either through polyploidy (discussed on pages 140–143) or without polyploidy. An example of rapid speciation without polyploidy involves two species in the monkeyflower genus *Mimulus* (Phrymaceae), in which a barrier to gene flow may have been caused by floral isolation (Bradshaw et al. 1995). *Mimulus lewisii* and *M. cardinalis* of the western United States differ strikingly in floral morphology (Figure 6.11). Flowers of *M. lewisii* are pollinated by bumblebees; their petals are pink with yellow nectar guides and landing platforms, and they contain low quantities of concentrated nectar. Flowers of *M. cardinalis* are adapted for hummingbird pollination; they are bright red, with a narrow corolla tube and copious nectar. Despite these major floral differences, experimental hybridizations between the two species yield vigorous and fertile offspring. Natural hybridization between the two species, however, has never been reported, even though they flower at the same time. The floral differences between the species are controlled by genes with relatively large effect. Hence reproductive isolation, and therefore speciation, could have been achieved quite rapidly.

Preservation of Diversity against Gene Flow

Barriers that prevent gene flow between closely related plant species are known as **reproductive isolating barriers**. As we consider these barriers, however, it is important to remember that the nature of gene flow between and within plant species is not used here as a criterion for their recognition. Gene flow does occur between individuals of many different plant species, and it does not occur between geographically distant populations of widely distributed species, such as the giant reed (*Phragmites australis*), which grows on several continents. Gene flow between plants from different continents must be very weak (or nonexistent), and yet the plants are globally very similar morphologically and are therefore considered members of the same species.

TABLE 6.1   **A classification of reproductive isolating barriers in plants.**

Premating	Postmating, prezygotic	Postmating, postzygotic

1. Temporal
    a. Seasonal
    b. Diurnal
	5. Incompatibility
	    a. Pollen-style
		5. Incompatibility
		    b. Seed

2. Habitat		6. Hybrid inviability

3. Floral
    a. Behavioral
    b. Structural
		7. Hybrid floral isolation

		8. Hybrid sterility

4. Reproductive mode
    a. Self-fertilization
    b. Agamospermy
		9. Hybrid breakdown

Source: After Levin 1971, 2000.

A Classification of Reproductive Isolating Barriers

Most reproductive isolating barriers act by preventing successful mating. Geographic separation is often considered a reproductive isolating mechanism, but it differs from the mechanisms in the classification scheme we present here, all of which act when two species are in proximity. Physical separation does isolate species, but one must be careful not to give it too much importance. Widely separated populations within a wide-ranging species, such as American beech (*Fagus grandifolia*, Fagaceae), which has a geographic range from New Brunswick to northern Florida, could be considered separate species if geographic isolation were the sole criterion for the definition of species. But beeches in Canada and Florida do not differ markedly, and they are recognized universally as members of the same species.

Often sets of populations within species are adapted to different climatic or soil conditions. Eastern red cedar (*Juniperus virginiana*, Cupressaceae) commonly grows in old fields and dry uplands throughout much of eastern North America. From North Carolina to central and northwestern Florida, however, this species grows on sand dunes and coastal river sandbanks. This ecotype is not recognized as a distinct species, just as the serpentine soil ecotypes of Jeffrey pine described earlier are not separate species because the ecological differences are not accompanied by morphological differences. Some species, however, are ecologically isolated from close relatives. Examples are provided in the section on habitat isolation below.

Reproductive isolating barriers can be classified according to the timing of their effect relative to reproduction (Table 6.1): premating (before pollination), postmating but prezygotic (after pollination and before fertilization), and postzygotic (after fertilization). Levin (2000) divides isolating barriers into an ecological group, corresponding to the premating group in Table 6.1, and a genomic group, corresponding

to the postmating group in Table 6.1. These groupings emphasizes that species interact with their environment (including resources such as pollinators) in different ways that restrict their capacity to mate, while genomic divergence is required for postmating isolation.

Temporal isolation   Flowering at different times of the year (Table 6.1: 1a) is common among many closely related species, such as several *Phlox* species of east-central Illinois and many species of oaks that grow together. Seasonal isolation is also effective among some species of willows (*Salix*). Among seven species from Ontario, there is an early-flowering group and a late-flowering group (Figure 6.12). Experimental hybridizations between the two groups are easy, but natural hybridization appears to be rare, even

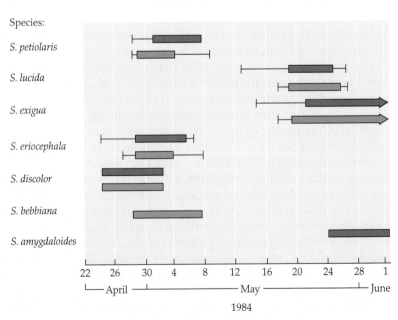

FIGURE 6.12   Flowering times for seven *Salix* species. The average first and last date of flowering is given by the thick bar; the thin line shows the range of flowering dates. For these dioecious species, red bars represent staminate plants, and blue bars represent carpellate plants. (From Mosseler and Papadopol 1989.)

though the species frequently grow together and are visited by the same kinds of pollinators. This observation suggests that seasonal isolation is effective, although molecular markers, which may be more sensitive to gene flow than morphological markers, have not been used to test for hybridization. Hybridization also appears to be rare among the species within each group, suggesting that postmating isolating barriers must exist as well.

Flowering at different times of day (diurnal isolation; Table 6.1: 1b) may also effectively isolate species that would otherwise hybridize. Creeping bentgrass (*Agrostis stolonifera*, Poaceae) flowers in the morning and another bentgrass species (*A. tenuis*) flowers in the afternoon. Unusual weather conditions may lead to simultaneous flowering and hybridization of these two species.

Habitat isolation Two species may be reproductively isolated if they occupy different habitats. The strength of this barrier depends upon the distance between the habitats relative to the distance over which pollination is effective. Some habitat isolation occurs between white lady's slippers (*Cypripedium*, Orchidaceae), which often grow on prairies, and yellow lady's slippers, which inhabit forests. Another example involves two species of *Lyonia* (Ericaceae) that grow in the southeastern United States. *Lyonia. ferruginea* (the rusty Lyonia or staggerbush) occurs in forests or scrub habitats on well-drained soils, and *L. fruticosa* (the coastal plain staggerbush) occurs in pine flatwoods where the soil is poorly drained because of a hard-pan of marine clay.

Floral isolation Floral adaptations to attract different pollinators limit or prevent gene exchange between many species. These adaptations may work through floral structure or through effects on pollinator behavior. Behavioral isolation (Table 6.1: 3a) reflects the capacity of pollinators to distinguish floral signals, such as color, shape, and fragrance (see the section on pollination biology in Chapter 4). Bees have long been known for their ability to recognize floral features; their sensitivity to the fragrances of orchids, for example, is the basis of many remarkable cases of reproductive isolation.

Some closely related orchids in the genus *Ophrys*, growing primarily in the Mediterranean region, produce different floral fragrances that attract males of different species of bees and wasps. Once the bee or wasp lands on the flower, continued stimulation by floral scents, along with the shape and texture of the flower, triggers the insect's mating instincts (Figure 6.13). Deceived into pseudocopulation with the flower, the insect pollinates the flower.

Another aspect of behavioral isolation is floral constancy, the restriction of a pollinator to one species even though other flowers are available. For example, bees attracted by the scent of one particular *Ophrys* species tend to visit the flowers of only that species during their forays, thus avoiding interspecific pollinations. Another example comes from the snapdragon species *Antirrhinum majus* and *A. glutinosum* (Plantaginaceae). Bees confine their visits to one species during a foray in mixed plantings, and seeds taken from these mixed plantings yield very few hybrids, even though the species are interfertile.

Two species of *Fuchsia* (Onagraceae) display a form of floral isolation known as structural isolation (Table 6.1: 3b). Where they are allopatric, *F. parviflora* and *F. enciliandra* ssp. *enciliandra* have flowers that are similar in size and are both pollinated by hummingbirds and bees. The petals of the former species are red, and those of the latter white or pink. Where the species' geographic ranges overlap, however, their flowers diverge dramatically: *F. parviflora* petals are

(A)

(B)

FIGURE 6.13 Pollination of orchids in the genus *Ophrys* by pseudocopulation. (A) A male *Andrena* bee attempting to mate with the lip of an *O. creticola* flower gets pollinia on the tip of its abdomen. (B) A male *Antophora erschowi* bee attempting to mate with the lip of an *O. elegans* flower gets pollinia on its head. (Photos courtesy of Helmut Presser.)

Character	*A. formosa*	*A. pubescens*
Flower position	Nodding	Erect
Flower color	Red	Pale
Spur length	10–17 mm	29–37 mm
Petal blade length	2–4 mm	9–12 mm

FIGURE 6.14 Flowers of *Aquilegia formosa* and *A. pubescens*. Whereas *A. formosa*'s nodding, red flowers with short spurs (10–17 mm) are adapted for hummingbird pollination, *A. pubescens*'s erect, pale flowers with long spurs (29–37 mm) are suited for hawkmoth pollination. (Modified from Arnold 1997; photos courtesy of Scott Hodges.)

white, and the hypanthium is shorter and broader, thus favoring bee pollination. Plants of *F. enciliandra* in the area of sympatry have red flowers with a longer, narrower hypanthium, making them more attractive to hummingbirds. Because these species are not known to hybridize, these floral changes are thought to be adaptations to reduce competition for pollinators. Such cases of divergence where species co-occur are called **character displacement**.

Hummingbirds are also involved in a well-studied case of structural floral isolation in *Aquilegia* (Ranunculaceae), the columbines. *Aquilegia formosa* flowers are red, nodding, with short nectar spurs (Figure 6.14), and they are pollinated by hummingbirds. *Aquilegia pubescens* flowers are pale yellow to white, upright in orientation, with long nectar spurs, and they are pollinated by hawkmoths. The two species grow mostly at different elevations, and when they do come together, some hybridization occurs, possibly because a pollinator other than hummingbirds or hawkmoths pollinates the flowers of both species. However, even though genes are exchanged between the species, the floral differences largely persist. These observations suggest that it is important for the species to maintain their floral syndromes and that floral isolation is the primary barrier to gene flow between them.

Milkweeds (*Asclepias*, Apocynaceae) exhibit another form of structural floral isolation. In this genus, pollen is packaged in pollinia that are delivered as a unit to the stigma. The stigmas lie within a slit, so precise orientation of the pollinium is required to effect pollination. The shapes of the pollinia of different species differ sufficiently that pollinia of one species are unlikely to fit the stigmatic slit of another species. Another example of floral structural isolation, involving the genus *Mimulus*, was discussed on page 127.

Reproductive mode   The shift from outcrossing to uniparental reproduction, either by self-fertilization (or selfing)

or agamospermy, has occurred in many plants and creates a barrier to mating. Self-fertilization requires that self-incompatibility (the mechanisms of which are described in Chapter 4) be replaced by **self-compatibility**. Self-fertilization is associated with certain changes in floral morphology, such as reduced amounts of pollen, small corollas and corolla lobes, small stigmatic lobes, and stigmas and anthers located close to each other. Self-compatible plants do not need to invest as much in pollen and corollas because they have no need to attract pollinators.

Self-compatibility has arisen multiple times within some lineages, as exemplified by a study in the genus *Linanthus* (Polemoniaceae). A molecular phylogeny based on ITS DNA sequences revealed that what had been called *L. bicolor* is actually represented by one clade occurring in southern California, a second clade in northern California and Oregon, and a third in Washington and British Columbia (Goodwillie 1999). Plants in these three clades are morphologically similar in floral features that are common in self-compatible plants and led to their inclusion in one species. However, their variation in less conspicuous features, such as calyx trichomes, is consistent with the molecular relationships of these three clades to other species, confirms the molecular perspective that *L. bicolor* is polyphyletic, and explains the similarity between the three clades as the result of convergent evolution.

One hypothesis for the evolution of self-compatibility is that it facilitates reproduction when cross-pollination is unlikely. Plants in the three clades of *L. bicolor* grow in dry habitats that contain relatively few pollinators at the times when the plants flower. The ancestors of these clades may have responded to a lack of pollinators when they migrated into these habitats by shifting to self-compatibility.

One consequence of a shift to selfing is reduction or elimination of gene flow. Loss of attractiveness of flowers to pollinators and early deposition of pollen onto the stigma of the same flower limit cross-pollination. A well-documented case of such reduction involves grasses that have developed tolerance for heavy metals, such as copper, that contaminate the soil around mines (Antonovics et al. 1971). A bentgrass species (*Agrostis tenuis*) and sweet vernal grass (*Anthoxanthum odoratum*) have subpopulations that are tolerant of metals and grow on contaminated soil. Selfing in these subpopulations limits gene flow from nearby outcrossing but nontolerant subpopulations and the consequent loss of tolerance.

The origin of *Stephanomeria malheurensis* by sympatric speciation (described on page 127) was mediated in part by a shift to self-compatibility and self-pollination. Many species eliminate cross-pollination altogether with self-pollination in unopened flowers (**cleistogamy**).

A shift in reproductive mode from sexuality to agamospermy (see Chapter 4) does not lead to complete isolation because most—perhaps all—agamospermous plants retain some sexuality. In addition, many agamosperms, even though they can produce seed without sex, produce viable pollen through meiosis and can mate with other species as a pollen parent. Box 6B examines hybridization between a sexual species, *Amelanchier bartramiana*, and an agamosperm, *A. laevis*, with the latter species serving as the pollen parent for the great majority of hybrids. Gene flow in this case is thus primarily from *A. laevis* to *A. bartramiana*, and agamospermy in *A. laevis* provides some reproductive isolation from *A. bartramiana*. Agamospermy, like self-fertilization, may provide reproductive assurance under conditions that are not favorable to cross-pollination.

Incompatibility   If pollen of one species reaches the stigma of another species, the stigma and style commonly will not permit the growth of the foreign pollen tube to the ovule (Table 6.1: 5a). The angiosperm stigma and style serve as effective sieves, allowing the maternal plant to accept or reject pollen. In some cases, the pollen of one species, such as corn, will not germinate on the stigmas of a close relative, such as *Tripsacum*, or grow in its styles, but the reciprocal cross is successful.

A well-documented case of the greater effectiveness of intraspecific pollen than interspecific pollen comes from the genus *Haplopappus* (Asteraceae). If pollen of *H. torreyi* is mixed with pollen of *H. graniticus* and applied to the stigma of *H. torreyi*, predominantly nonhybrid progeny are produced. If *H. graniticus* pollen is applied to *H. torreyi* stigmas and *H. torreyi* pollen is applied up to 24 minutes later, nonhybrid production is still favored. If the interval between application of interspecific pollen and intraspecific pollen exceeds 24 minutes, the offspring are mostly hybrids, but a delay of 1 to 2 hours is required to stop production of nonhybrid progeny altogether.

The success of intraspecific pollen over interspecific pollen is also striking in experimental pollinations of *Iris* (Iridaceae). Carney et al. (1994) applied a full range of mixtures of *I. fulva* and *I. hexagona* pollen to the stigmas of these species and then measured the proportion of hybrid and nonhybrid seeds. The results showed that intraspecific pollen contributes to far more seeds than interspecific pollen in all mixtures of pollen except for exclusively interspecific pollen (Figure 6.15). Intraspecific pollen may outcompete interspecific pollen because the pollen tubes of intraspecific pollen grow more quickly in styles of the same species. Hybridization does occur between these two *Iris* species and has been studied intensively (see Box 6D).

Even if a hybrid embryo forms, it may not develop into a viable seed because of incompatibility between the parental genomes within the embryo or between the hybrid embryo and the maternal endosperm (Table 6.1: 5b). The hybrid embryo resulting from the cross of *Primula elatior* and *P. veris* (Primulaceae) exemplifies this failure to mature.

Hybrid inviability   Hybrid inviability refers to the failure of hybrids to develop normally and reach reproductive maturity (Table 6.1: 6), as is the case for crosses between two species of the poppy genus, *Papaver dubium* × *P. rhoeas* (Papaveraceae). The habitat requirements of a hybrid are likely to be different from those of either parent, and hybrid inviability may be due to the lack of a suitable ecological niche. The habitat differ-

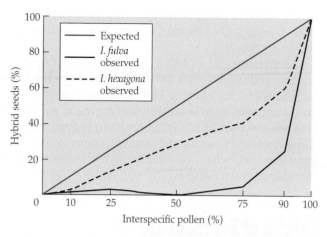

FIGURE 6.15 Percentage of hybrid seeds produced by different ratios of intraspecific and interspecific pollen applied to stigmas of *Iris fulva* and *I. hexagona* flowers. The solid red line connecting the corners of the graph represents the expected percentage of hybrid seeds, assuming no difference in the success of intraspecific and interspecific pollen. (From Carney et al. 1994.)

ences between white and yellow lady's slippers (see page 129) are strong enough that hybrids are not well adapted to any habitats and therefore have reduced fitness.

Hybrid floral isolation   The term *hybrid floral isolation* (Table 6.1: 7) refers to the absence of effective pollinators for a hybrid of two parental species that are adapted to very different pollinators.

Hybrid sterility   Hybrids may be perfectly viable but sterile (Table 6.1: 8) if their chromosomes fail to pair during meiosis because the chromosomes of the parental species differ in number or have diverged sufficiently to hinder pairing. Nonfunctional gametes are the result. The cross of broccoli (*Brassica oleracea*; this species of the Brassicaceae includes cabbage and related crops) with radish (*Raphanus sativus*) produces *Raphanobrassica*. The first-generation hybrids are vigorous, but the chromosomes from the two parental species do not pair with one another in meiosis, so functional gametes do not form.

Mooring (2001) found extensive hybrid sterility in some F_1 individuals of species in the *Eriophyllum lanatum* (Asteraceae) species complex. In these sterile F_1 individuals, chromosomes do not pair with one another successfully in meiosis, and the pollen has reduced viability. F_1 offspring of the Hawaiian silverswords similarly show irregular meiosis and low pollen viability (Carr 1995). In the silverswords, however, F_1 individuals retain enough fertility to mate with one of the parents, and the offspring of this mating may have much greater pollen fertility.

Hybrid breakdown   *Hybrid breakdown* refers to problems in later-generation hybrids (Table 6.1: 9). First-generation hybrids may be viable and fertile, but backcrosses or later-generation individuals may be inviable or sterile. For example, the hybrid of two grasses, *Festuca rubra* and *Vulpia fascic-ulata*, produces few offspring, and the F$_2$ plants are weak and do not flower.

Combinations of isolating mechanisms may operate together and reinforce one another. In jimsonweeds (*Datura*, Solanaceae), for example, both interspecific pollen-stigma incompatibility and hybrid inviability prevent gene flow between species. Gene flow between two species of avens (*Geum rivale* and *G. urbanum*, Rosaceae) is prevented by seasonal isolation, behavioral isolation, hybrid inviability, and hybrid breakdown (Table 6.1: 1a, 3a, 6, and 9).

Origins of Reproductive Isolating Barriers

There are three possible origins of reproductive isolation. First, as two species diverge from a common ancestor, accumulated genetic changes often diminish the likelihood of successful reproduction between them. Recall that speciation does not always result in reproductive isolation, however (see the sycamore example on page 127).

Second, premating isolation may be the result of natural selection against wastage of gametes on unsuccessful matings with other species. To understand how natural selection could select against hybridization, consider two individuals of the same species. The pollen and eggs of the first individual are committed exclusively to successful intraspecific matings and are never used for less successful interspecific matings, and the second individual uses some of its gametes in unsuccessful hybridizations. The first individual will have more offspring, and its mating behavior will become more common within the species.

Two observations demonstrate that selection against hybrids can increase reproductive isolation. First, removal of hybrids found in mixed plantings of two corn cultivars—white sweet and yellow flint—markedly decreased overlap in the flowering times of the two cultivars (resulting in temporal isolation).

Second, two species of *Phlox*—*P. cuspidata* and *P. drummondii*—illustrate a shift in floral traits contributing to reproductive isolation when in proximity to one another. These two species are fully cross-compatible, and their hybrids are vigorous but mostly sterile. Where the two species grow separately, both have pink flowers. Where they grow together, *P. drummondii* has undergone character displacement and has red flowers. When both pink- and red-flowered individuals of *P. drummondii* are experimentally introduced into a population of *P. cuspidata*, butterfly pollinators tend to visit flowers of one color on their forays (floral constancy). The result is that the offspring of red-flowered *P. drummondii* include fewer hybrids than those of pink-flowered individuals (Table 6.2). Reproductive isolation is further promoted by increased self-compatibility in sympatric populations of these two species.

TABLE 6.2 **Progeny of red- and pink-flowered *P. drummondii* grown with *P. cuspidata*.**

Flower color	Progeny	
	P. drummondii	Hybrid
Red	181 (87%)	27 (13%)
Pink	86 (62%)	53 (38%)

Source: From Levin 1985.

A third possible origin of reproductive isolation involves selection for premating isolation to reduce competition for animal pollinators. Closely related species often have similar flowers designed to attract the same or similar pollinators. If there are not enough pollinators to service two similarly adapted species, then seed set could be reduced. Flowering at different times and attracting different pollinators may increase the efficiency of the use of a limiting resource and thus elevate fitness.

Hybridization and Introgression

Reproductive isolating mechanisms are not always effective, and gene flow between plant taxa is common (Rieseberg and Morefield 1995; Ellstrand et al. 1996; Arnold 1997). Strictly speaking, hybridization is mating between unrelated individuals, but the term is most often used for matings between species. Such interspecific hybridization is critically important in plant evolution as a source of novel gene combinations and as a mechanism of speciation. Hybridization has also been very important in plant breeding as a means of moving desirable traits from one cultivated or even wild species into another cultivated species. The end result may be a new cultivar of agronomic (e.g., tomatoes and strawberries) or horticultural (e.g., roses) importance.

Hybridization is often associated with habitat disturbance (see Box 6B). The ecological adaptations that isolate two species may be broken down by natural disturbances (such as new pests or predators, blowdowns in a forest, fire, erosion, floods, or volcanic activity) that create a habitat suitable for hybrids. Reduced competition in the wake of disturbance also favors the growth of hybrids.

Human disturbance in Europe and North America is thought to have promoted extensive hybridization and subsequent morphological complexity in shadbush (*Amelanchier*, Rosaceae; also called serviceberry) (Figure 6.16), hawthorn (*Crataegus*, Rosaceae), blueberries and related plants (*Vaccinium*, Ericaceae), and blackberries and raspberries (*Rubus*, Rosaceae). Natural disturbance from volcanism and erosion favors hybridization in many Hawaiian groups (Carr 1995).

Hybridization between individuals belonging to different genera is unusual. It is important to bear in mind, however, that genera are human constructs; there is no clear "mark" for genera. Intergeneric hybridization does occur naturally be-

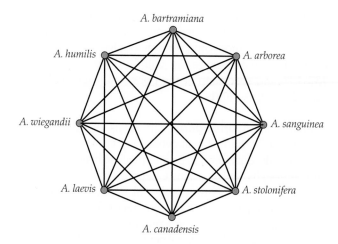

FIGURE 6.16 Reported natural hybridizations between members of eight species of *Amelanchier* in northeastern North America. (From Campbell and Wright 1996.)

TABLE 6.3   Examples of plant groups showing frequent hybridization.

Family	Genus
Asteraceae	Pussytoes (*Antennaria*), beggar's-tick (*Bidens*), sunflower (*Helianthus*)
Betulaceae	Birch (*Betula*)
Cupressaceae	Junipers (*Juniperus*)
Cyperaceae	Sedge (*Carex*), bulrush (*Scirpus, Schoenoplectus*)
Fagaceae	Oaks (*Quercus*)
Paeoniaceae	Peony (*Paeonia*)
Pinaceae	Pine (*Pinus*), spruce (*Picea*)
Poaceae	Bluegrass (*Poa*)
Aspleniaceae	Spleenwort (*Asplenium*)
Ranunculaceae	Columbine (*Aquilegia*)
Rosaceae	Shadbush (*Amelanchier*; Box 6B, Figure 6.16), hawthorn (*Crataegus*), avens (*Geum*), cinquefoils (*Potentilla*), roses (*Rosa*), blackberries and raspberries (*Rubus*), mountain ash (*Sorbus*)
Salicaeae	Poplars and aspens (*Populus*), willows (*Salix*)

tween genera of the Hawaiian silverswords (Carr 1995) and between several genera of the Pyreae (Rosaceae), such as *Amelanchier, Crataegus*, apple (*Malus*), pear (*Pyrus*), *Cotoneaster*, and *Sorbus*. In North America, for example, *Amelanchier* (with simple leaves) and *Sorbus* (with pinnately compound leaves) hybridize at times. The offspring of these morphologically divergent genera, named × *Amelasorbus*, amount to little more than curiosities. The intergeneric hybrid of wheat and rye has produced the commercially successful *Triticale*.

Frequency of hybridization   Hybridization is prevalent in plants. There are about 70,000 naturally occurring interspecific plant hybrids in the world, according to one estimate (Stace 1984), but not all plant groups hybridize. For example, locoweed (*Astragalus*, Fabaceae), a genus of flowering plants with about 2500 species, is not known for extensive hybridization. In contrast, hybridization is frequent between species of certain genera, some examples of which are presented in Table 6.3 (and see, for example, Figure 6.16).

Some large families, such as Apiaceae and Solanaceae, are not known to contain hybrids in certain well-studied parts of the world (Ellstrand et al. 1996). Genera within which hybridization is common tend to be perennials, outcrossers, pollinated by insects, and have some means of asexual reproduction. Hybridization is common enough to be an important creative force in plant diversity, to create complex patterns of variation, and in some cases to weaken the morphological distinctness of species.

Evolutionary consequences of hybridization   Hybridization has five potential consequences (Rieseberg and Ellstrand 1993; Rieseberg and Wendel 1993; Arnold 1994, 1997): (1) reinforcement of reproductive isolating mechanisms, (2) formation of a hybrid swarm through reproduction by hybrids at one site, (3) fusion of two species through interspecific gene flow (introgression), (4) creation of genet-

ic diversity and adaptation, and (5) evolution of new species. Hybridization can therefore maintain biodiversity (consequences 1 and 2), destroy it (consequence 3), or create it (consequences 4 and 5).

The observations on corn and *Phlox* described on page 132 show that selection against hybridization may strengthen reproductive isolating mechanisms. Species diversity in areas of species overlap may be preserved in this way. Numerous cases of apparent shifts in premating isolating mechanisms are also probably responses to selection against hybridization.

Crossing between two species followed by backcrossing of hybrids with one or both parental species may produce a **hybrid swarm** (Figure 6.17A) when the gene flow is limited to one or a few sites where the species grow together (see Boxes 6B and 6C). The formation of hybrid swarms is sometimes referred to as *local introgression*.

Hybrids may have higher fitness than the parental species—a well-known phenomenon called **hybrid vigor** (**heterosis**)—and may be adapted for new habitats. Heterosis is familiar in crop varieties; its importance to agriculture is exemplified by the high-yielding varieties of corn, tomatoes, cabbage, and many other species now widely available. It is not clear how important heterosis is in wild plants.

Introgression, the permanent incorporation of genes of one species into another species, has three potential consequences. First, it may lead to the merging of different species. *Gilia capitata* represents such a possible reversal of speciation

(A)

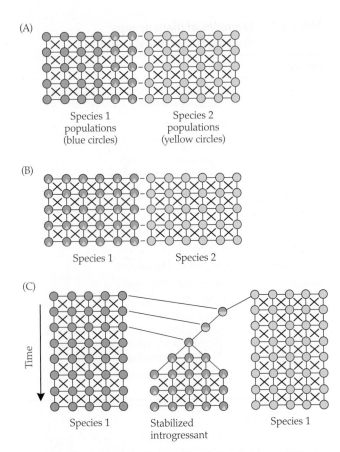

Species 1
populations
(blue circles)

Species 2
populations
(yellow circles)

(B)

Species 1

Species 2

(C)

Time

Species 1

Stabilized
introgressant

Species 1

FIGURE 6.17 Local introgression, dispersed introgression, and the origin of a stabilized introgressant. Black lines represent crosses between populations. (A) In a hybrid swarm (localized introgression), only some of the populations of either of the participating species have genes of the other species, as indicated here by the presence of some yellow (designating species 2 genes) in some populations of species 1. (B) Dispersed introgression from species 2 into species 1 is indicated by the presence of species 2 genes (some yellow) in all species 1 populations. (C) The origin of a stabilized introgressant from hybridization is followed first by backcrossing to one parental species, and then by reproduction strictly among hybrid populations. (From Rieseberg and Wendel 1993, by permission of Oxford University Press.)

by extensive gene flow (Grant 1963). This species consists of eight races found on the Pacific slope of North America. Three of these races are quite distinct and might be recognized as distinct taxa, were it not for the existence of intermediate races produced by hybridization among them. It is hypothesized that the three distinct races attained a maximum level of divergence in the Pliocene (roughly 2 to 5 million years ago). Later, gene flow among them created one species with continuously intergrading races.

Second, introgression may transfer genetic material from one species to another without merging the species (Figure 6.17B; see also Box 6D) and thus increase genetic diversity in the introgressed species.

Third, stabilized introgressants may form new species (Figure 6.17C). Hybrid speciation may occur at the diploid level, but it is far more commonly associated with an increase in ploidy. Diploid speciation will be discussed on pages 137–138, and speciation at the polyploid level will be discussed on pages 140–143.

Evidence for hybridization In terms of quantitatively inherited traits, F_1 (first filial generation; i.e., first-generation hybrids) individuals tend to be intermediate between the parents, although there are many exceptions to this general rule (Hardig et al. 2000; Schwarzbach et al. 2001). The dimensions of many plant parts—such as stem height, leaf length, petal length, and fruit diameter—are quantitatively (polygenically) inherited, controlled by several genes, each of which makes a partial contribution to phenotypic expression.

Consider a hypothetical example of hybridization between a shrub with stems under 2 m tall and a tree whose stems commonly exceed 15 m. The hypothetical hybrid would receive many genes coding for short stems from the shrubby parent and many genes for tall stems from the arboreal parent. The F_1 offspring would thus have stems of intermediate height. Box 6B describes an example of F_1 intermediacy in *Amelanchier*.

F_1 intermediacy is commonly expressed in morphology, chemistry, ecology, and in other ways, depending on which characters are examined. Characters governed by just one or a few genes do not show F_1 intermediacy (Rieseberg 1995). Instead, hybrids will have parental, novel, or extreme character states. For example, *Amelanchier bartramiana* is sexual, *A. laevis* produces seeds mostly asexually, and their hybrid (*A. × neglecta*), is strongly asexual. The form of asexual seed production (agamospermy) expressed in *Amelanchier* is thought to be controlled by one or a few, dominant genes.

Chemical and molecular characters are more likely than morphological characters to yield predictable expression in hybrids. In the case of nuclear DNA sequences, F_1 individuals show sequences from both parents in what is called an additive pattern. In contrast, molecular markers that are inherited from just one parent, such as chloroplast DNA, do not show additive patterns.

Morphological data have provided primary evidence for plant hybridization in hundreds of studies. Several factors other than hybridization can result in single characters showing patterns that resemble F_1 intermediacy, however, as we will see shortly. In addition, if there is considerable natural variation in a character, then the F_1 may not be intermediate between the parents.

The use of numerous morphological characters to study hybridization avoids these difficulties and increases the precision of analysis. With numerous characters, however, it becomes difficult to interpret all the variation simultaneously; indeed, one area of statistics—multivariate analysis— was developed by the geneticist R. A. Fisher for the purpose of analyzing hybridization in *Iris*. Quantitative morphological analysis of large data sets is described in Box 6C.

BOX 6B   Hybridization in Amelanchier

Amelanchier bartramiana (mountain shadbush) and *A. laevis* (smooth shadbush) commonly grow together and hybridize over much of northeastern North America. Mountain shadbush usually grows in cool, undisturbed forests or bogs, and smooth shadbush prefers early successional habitats such as roadsides, recently burned areas, cutover timberlands, and fields. The creation of a road or timber harvesting in the vicinity of a mountain shadbush population establishes conditions for invasion by smooth shadbush and its subsequent hybridization with mountain shadbush. The hybrid between these two species is named *A.* × *neglecta*; the multiplication symbol indicates that it is a hybrid.

Mountain shadbush and smooth shadbush differ conspicuously in their leaves and flowers. Mountain shadbush has short petioles, a tapering leaf base, and short petals. Smooth shadbush has longer petioles, a rounded or heart-shaped leaf base, and longer petals. *A.* × *neglecta* is intermediate between the two parental species in these characters (Figure 6.18).

A study of hybridization at one site (Weber and Campbell 1989) showed that most individuals of the hybrid are intermediate between the parental species for seven morphological characters that distinguish the two species (see

(A)

(B)

FIGURE 6.18   From left to right: *Amelanchier laevis*, *A.* × *neglecta*, and *A. bartramiana* flowers (A) and leaves (B). Note the intermediacy of *A.* × *neglecta* in petal length, petiole length, and the shape of the base of the leaf. Scale is in millimeters.

Figure 6.19A in Box 6C; Box 6C provides an explanation of the quantitative methods used). These characters include dimensions (e.g., petiole length) and quantified shapes (e.g., the angle between the base of the leaf blade and the petiole) that are likely to be quantitatively inherited.

Additional evidence that *A.* × *neglecta* is the hybrid of *A. bartramiana* and *A. laevis* comes from ribosomal DNA sequences and chloroplast DNA restriction sites (see Chapter 5). The two parental species differ at two nuclear ribosomal DNA (nrDNA) internal transcribed spacer (ITS) nucleotide sites, and the hybrid shows the nrDNA of both parents at both sites. Hybrids contain chloroplast DNA of one or the other parent because chloroplast DNA is inherited from the maternal parent only. Furthermore, hybridization is possible apparent-

ly because the two species flower at about the same time, are visited by the same pollinators, and produce viable, germinable seed when they are experimentally cross-pollinated.

At another site, where *A. laevis* grows with another species—*A.* "erecta" (purple shadbush; the specific epithet is placed in quotation marks because this species has not been formally named)—morphology suggests that hybridization has proceeded well beyond creation of the F₁ generation, and a hybrid swarm has formed (see Figure 6.20 in Box 6C) (Campbell and Wright 1996). Many plants at this site contain various combinations of parental character states and span the difference between the parental species. Hybridization between *A. laevis* and *A.* "erecta" is also interesting because both species are themselves probably of hybrid origin.

Careful documentation of the occurrence of F₁ individuals may include a determination of whether there is a statistically significant proportion of intermediate character states (Wilson 1992) for numerous characters and should use independent sets of characters (such as morphology, chemistry, and molecular data; see Boxes 6B and 6D). Care should be taken to discount characters that are correlated or genetically linked because they contain less information than those that are uncorrelated and unlinked.

Processes other than hybridization can generate a pattern resembling F₁ intermediacy. A character state could be intermediate because of mutation, phenotypic plasticity, or an evolutionary trend. For example, taxon B in Figure 6.19 is

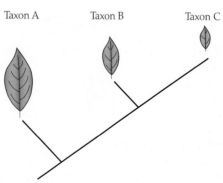

Taxon A   Taxon B   Taxon C

FIGURE 6.19   An evolutionary trend, such as the decrease in leaf size depicted here, may simulate hybridization.

BOX 6C Quantitative Morphological Analysis of Hybridization

Suppose we are testing the hypothesis that species A and B have hybridized to yield putative hybrid C. If we have identified one quantitatively inherited character that clearly separates A and B, we can ask whether C is intermediate between the parental species by arranging our samples of A, B, and C on a scale for our character.

For two or three characters, we can look at our sample in two or three dimensions. Take, for example, petal length and petiole length in *Amelanchier bartramiana* and *A. laevis*. It is easy to see the intermediacy of *A. × neglecta* when petal length is plotted against petiole length (Figure 6.20A). For more than three characters, however, we cannot readily visualize relationships among A, B, and C, and we must use quantitative morphological techniques.

Figure 6.20B and Figure 6.21 are based on **principal components analysis (PCA)** of seven and six characters, respectively. PCA is a statistical technique that reduces a data set of many characters to one, two, or three new characters. The first principal component is a composite character, the

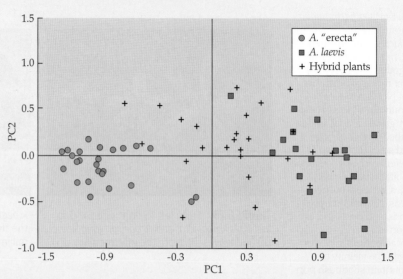

FIGURE 6.21 Principal components analysis of a hybrid swarm involving *Amelanchier* "erecta" (purple shadbush), *A. laevis* (smooth shadbush), and the putative hybrid of these two species. (After Campbell and Wright 1996.)

combination of all characters varying in a largely consistent way that captures maximal variation in the data set. Normally some variation is not explained by the first component because all characters are not perfectly consistent (correlated) with one another.

In Figure 6.21, for example, the first principal component explains 73% of the total variation in the data set. The second principal component explains the largest part of the variation not explained by the first principal component (17% of the total variation in Figure 6.21). The third and subsequent principal components (not shown in this figure) explain the

maximum variation not explained by the preceding principal components.

The principal components summarize the data and, when plotted against one another as in Figures 6.20B and 6.21, give a visual indication of the relationships of the taxa. In Figure 6.21, 90% of all the variation in the data set of six variables is expressed by the first two principal components. We have not lost much information by using this technique, but we have improved our ability to see what the data say. (For another example of quantitative morphological analysis of hybridization, see Hardig et al. 2000.)

FIGURE 6.20 Quantitative morphological study of hybridization. (A) Biplot of petiole length and petal length for individuals of *Amelanchier bartramiana*, *A. laevis*, and their putative hybrid, *A. × neglecta*. (B) Principal components analysis showing the first two principal components (PC1 and PC2) for a study of *Amelanchier* hybridization. (After Weber and Campbell 1989.)

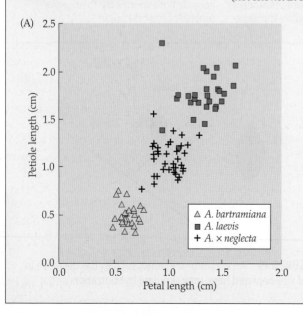

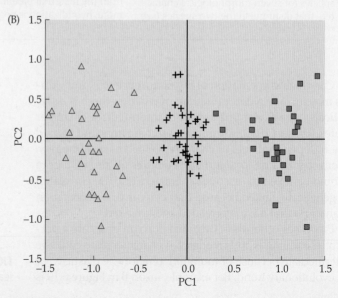

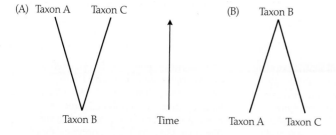

FIGURE 6.22 Evolutionary divergence versus hybridization. Similar patterns of intermediacy may result from (A) evolutionary divergence of taxa A and C from taxon B and (B) hybridization of taxa A and C to produce taxon B.

intermediate between taxa A and C in leaf length not because it is the hybrid of A and C, but because it is an intermediate step in an evolutionary trend toward smaller leaves. One distinguishes such a case of intermediacy from those generated by hybridization by considering other characters, at least some of which are likely to show intermediacy if hybridization has occurred.

A species may also become more or less intermediate between two other species through evolutionary convergence, thus simulating hybridization. The splitting of one species into two derivatives (evolutionary divergence) also simulates hybridization (Figure 6.22).

A combination of multiple data sets allows one to distinguish among hybridization, convergence, and divergence. Observations of the presence of both parental species at or near F_1 individuals in nature, of simultaneous flowering by the parents, and of pollinators that transfer pollen between the parents also support hypotheses of hybridization (see Box 6B). One can be sure that a certain plant is a hybrid, however, only when it has been experimentally produced. It is therefore best to refer to potential hybrids as putative hybrids.

Local reproduction by hybrids may produce a hybrid swarm (see Figure 6.17A and Box 6C). In the first *Amelanchier* case described in Box 6C, almost all individuals of *A.* × *neglecta* appear to be F_1 offspring (see Figure 6.20B); matings between F_1 individuals and either parental species (**backcrossing**) or among the F_1 offspring to produce an F_2 generation are apparently very uncommon. In the second case, however, a hybrid swarm has formed, and morphological features are being transferred from one species to another, primarily from purple shadbush to smooth shadbush (see Figure 6.21).

Hybrid swarms may confound identification; not only are the two parental species present, but the intermediates potentially span the gap between the parents. Pedicel length in purple shadbush averages 1.7 cm and ranges from 1.4 to 1.9 cm; smooth shadbush fruit pedicels average 2.9 cm in length and range from 2.7 to 3.5 cm. Putative hybrid individuals have pedicels intermediate between the parental species, averaging 2.4 cm, and their range from 1.5 to 3.1 cm spans nearly the full range of the two parental species.

Introgression   The irises of southern Louisiana have been models for understanding introgression (Box 6D; see Arnold 1994). Southern Louisiana supports considerable iris diversity. In the 1930s the numerous forms there were classified into over 80 species. Later, morphological, ecological, and genetic studies documented hybridization, and iris diversity was recast into four basic species plus numerous hybrids among them.

Two *Iris* species that grow together in southern Louisiana—*I. fulva* and *I. hexagona*—were central to Edgar Anderson's thesis of introgression, which he presented in his influential 1949 book *Introgressive Hybridization*. After the publication of that book, many plant systematists believed that introgression was relatively common and an important force in plant evolution. But *Iris* introgression was questioned in the 1960s, and few, if any, convincing cases of plant introgression had been documented by that time.

Recent molecular studies of Louisiana irises make clear that the first evidence for iris introgression was correctly interpreted. Introgression is common in plants, and there are now numerous well-documented cases and many additional possible examples (Rieseberg and Wendel 1993). The major reason for this recognition of the prevalence of introgression was the availability of molecular data. Low levels of introgression are difficult to detect with morphology, and it is difficult in some cases to refute alternatives to introgression.

Introgression may be bidirectional, in which case genes are exchanged between two species, or it may be unidirectional, with genes going exclusively from one species to another. The Hawaiian silverswords present a case of unidirectional introgression. At one site where two old lava flows are located near one another, *Dubautia ciliolata* is nearly restricted to the lava flow of 1855, and *D. scabra* is limited to the lava flow of 1935. RAPD (random amplified polymorphic DNA) markers show the occurrence of F_1 individuals, F_2 individuals, and backcrosses in which *D. ciliolata* mates with hybrids (Caraway et al. 2001). This unidirectional introgression may be providing *D. ciliolata* with the genetic material to move from the older to the newer lava flow.

Hybrid Speciation   Hybridization frequently produces new plant species. Although hybrid speciation is most often associated with polyploidy, there are a few well-documented cases of **diploid hybridization**, wherein a cross between two diploid species produces a diploid hybrid species. Considerable molecular evidence shows that two diploid sunflower species, *Helianthus annuus* and *H. petiolaris* (Asteraceae), hybridized in the past to produce three other diploid species: *H. anomalus*, *H. deserticola*, and *H. paradoxus* (reviewed in Rieseberg and Wendel 1993). These speciation events were mediated by chromosomal differences between the parental species that make the hybrids partially sterile. Recombination in hybrids restores some fertility, but they remain at least partially intersterile with the parents.

Another example, *Iris nelsonii*, is a diploid hybrid of three species: the two species discussed in Box 6D (*I. fulva* and *I. hexagona*) and *I. brevicaulis* (see Arnold 1992, 1994). The

BOX 6D   Introgression in Louisiana Irises

Iris fulva and *I. hexagona* hybridize in southern Louisiana where the Mississippi River bayou habitat of *I. fulva* mixes with the freshwater swamp and marsh habitat of *I. hexagona*, especially where the habitat has experienced human disturbance. Early studies used flower color—flowers are brick-red in *I. fulva* and blue in *I. hexagona*—and six other floral characters to differentiate the species and hybrids and support hypotheses of hybrid swarms and introgression.

Analyses of nuclear (ribosomal DNA, isozyme, and RAPDs) and chloroplast (cpDNA) data confirmed the hypotheses proposed on the basis of morphology (reviewed in Arnold 1992, 1994). Species-specific markers of each species for all four of these types of molecular data have been detected in the other species, both where the two species grow together and where they are allopatric. The evidence therefore clearly indicates that introgression has occurred in both directions between the species.

The two species differ in the presence of an insertion in the intergenic spacer (IGS) of rDNA. Whereas most allopatric populations of *I. hexagona* in Louisiana and Florida have this insertion in their rDNA repeats, most allopatric populations of *I. fulva* lack the insertion. Two populations are exceptional and exemplify dispersed introgression. One allopatric population is predominantly *I. fulva*, but 20 individuals were found to contain the insertion, although 4 also contained a minority of repeats lacking the insertion. Of the 12 individuals sampled from the second allopatric *I. fulva* population, 4 showed only the *I. fulva* type of rDNA, and the other 8 were predominantly *I. fulva*.

Patterns of gene flow in one area of overlap between *I. fulva* and *I. hexagona* differ strikingly for rDNA and cpDNA. Whereas nuclear genotypes indicate repeated backcrossing into *I. hexagona* (Figure 6.23A), cpDNA genotypes show considerably less gene flow (Figure 6.23B). This difference may be explained by a predominance of gene flow through pollen, which carries nuclear DNA but not maternally inherited cpDNA markers.

I. hexagona

I. fulva

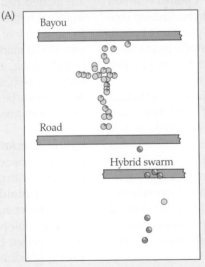

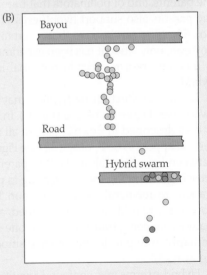

FIGURE 6.23 Nuclear and chloroplast DNA data for individual *Iris* plants from an area of overlap between *I. fulva* and *I. hexagona* near a road and bayou (the green bars) in Louisiana. Each circle represents genetic data derived from a single plant. (A) The relative proportion of *I. fulva* (blue portion of circles) and *I. hexagona* (yellow portion) nuclear markers in a sample of 37 plants. (B) The distribution of *I. fulva* (blue circles) and *I. hexagona* (yellow circles) chloroplast DNA markers for the same 37 plants and 3 additional plants. (After Arnold et al. 1992.) (Photo of *I. hexagona* by Walter Judd, of *I. fulva* courtesy of Rodney Barton.)

hybrid of two of these three species hybridized with the third; it is not known which two species formed the initial hybrid.

Past introgression has left its trace in many groups in the form of a conflict between phylogenies based on the uniparentally inherited chloroplast genome (cpDNA) and phylogenies based on nuclear markers (Rieseberg and Soltis 1991). *Helianthus annuus* is a weedy introduced species that has hybridized with several other sunflower species in the southwestern United States. The fate of chloroplast and nuclear genomes after these hybridizations has differed. In almost all cases, *H. annuus* nuclear DNA has been eliminated

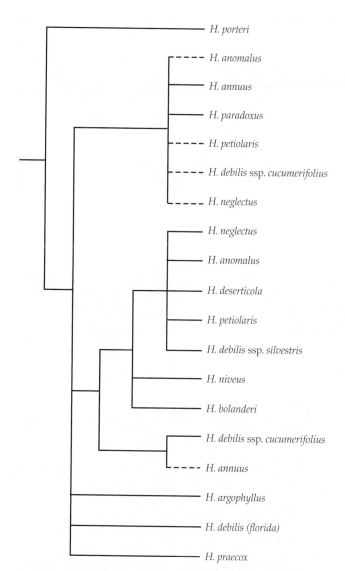

- *H. porteri*
- *H. anomalus*
- *H. annuus*
- *H. paradoxus*
- *H. petiolaris*
- *H. debilis* ssp. *cucumerifolius*
- *H. neglectus*
- *H. neglectus*
- *H. anomalus*
- *H. deserticola*
- *H. petiolaris*
- *H. debilis* ssp. *silvestris*
- *H. niveus*
- *H. bolanderi*
- *H. debilis* ssp. *cucumerifolius*
- *H. annuus*
- *H. argophyllus*
- *H. debilis* (*florida*)
- *H. praecox*

FIGURE 6.24 Single most parsimonious tree for *Helianthus* sect. *Helianthus* based on chloroplast DNA data. Dashed lines indicate discrepancies between morphological classification and cpDNA type that are thought to result from cytoplasmic introgression (chloroplast capture). (After Rieseberg et al. 1991.)

through backcrossing to the other parental species. *Helianthus annuus* cpDNA has been retained in the hybrids, however, and is now coupled with the nuclear genome of the other species (Figure 6.24).

This phenomenon, called **chloroplast capture** or *differential introgression*, has been documented in many groups (Rieseberg and Wendel 1993). Chloroplast DNA is more likely than nuclear DNA to cross species boundaries and leave a trace of past introgression (Rieseberg and Soltis 1991). The combination of cpDNA from one individual and nuclear DNA from another creates obvious conflicts in phylogeny reconstruction (Linder and Rieseberg 2004).

Hybridization and Phylogeny Reconstruction Hybrids pose problems for phylogenetic analyses because the reticu-late patterns of hybridization are fundamentally incompatible with the hierarchical patterns imposed by most current methods of phylogeny reconstruction (McDade 2000; Linder and Rieseberg 2004). Some systematists argue that hybrids might be detectable by the disruption they cause to phylogenies. Others recommend removing hybrids from data sets prior to phylogenetic study to avoid the potential chaos and poor resolution that hybrids are thought to bring to phylogeny.

McDade (1990, 1992a) produced 17 experimental hybrids from 12 parental species in the Central American genus *Aphelandra* (Acanthaceae) and assessed the effect of the hybrids' inclusion on resolution (number of trees), homoplasy (as measured by the consistency index, CI; see Chapter 2), and hypotheses of phylogenetic relationships among the parental species. The hybrids were added singly or in groups of two to five, randomly selected, to a data set containing only parental species. The hybrids did not reduce resolution, although they did significantly lower the CI. They disrupted the phylogeny only if they were very common in the ingroup and/or the parents of the hybrid were widely separated from one another on the tree (Figure 6.25).

Problems in phylogeny reconstruction due to frequent hybridization may be prevalent in groups such as those in Table 6.3, but there are few, if any, well-documented cases of hybridization between phylogenetically remote ancestors. One long-supported example of remote hybridization involved the Pyreae, the apple/pear tribe of the Rosaceae (Stebbins 1950). A leading hypothesis for the origin of this group was hybridization between ancestors of two other groups of the Rosaceae: tribe Amygdaleae (cherries and relatives) and tribe Spiraeeae (bridal wreaths). Molecular data, however, strongly refute this hypothesis and instead indicate that the Pyreae evolved from within one genus, *Gillenia*, of the southeastern United States (Evans and Campbell, 2002).

Similar studies of *Helianthus* parental species and three well-documented hybrids showed that inclusion of the hybrids did not reduce resolution, but did increase homoplasy in molecular data (summarized in Rieseberg and Ellstrand 1993). Homoplasy in morphological and chemical data did not change with the addition of sunflower hybrids. Finally, sunflower hybrids did not affect consensus tree topology.

Note that *Aphelandra* F_1 individuals do not behave differently from nonhybrid taxa in phylogenetic analyses; therefore, phylogenetic analyses cannot be used to detect them. If no F_1 individuals can be detected, hybridizations that occurred in the somewhat distant past certainly cannot be detected because evidence for hybridization has been subsequently more or less obscured.

Unless an investigator knows that hybrids are numerous in the ingroup or that there is great evolutionary distance between the parents of possible hybrids, it is generally acceptable practice to include potential hybrids in phylogenetic analyses. If a hypothesis of hybrid origin is strongly supported for a taxon, then it is not necessary to include it in phylogenetic analysis. Putative hybrids may be added to trees later (e.g., Sang et al. 1995; see also Chapter 2).

FIGURE 6.25 The effect of hybrids on topology and resolution in phylogenetic analyses. (A) Single most parsimonious tree for 12 species (each designated by the first two letters of the specific epithet) of Central American *Aphelandra* (Acanthaceae). (B) Effect of addition of hybrids on topology: addition of the hybrid LE × SI causes rearrangement of the entire tree relative to Figure 6.25A, with the loss of monophyly of clade SC-PA-GR-GO-TE-SI-ST and a shift in position of clade DA-HA-CA-LE-LI. (C) Consensus tree of three equally most parsimonious trees with inclusion of hybrids LE × GO and SC × PA, showing loss of resolution. (From McDade 1992a.)

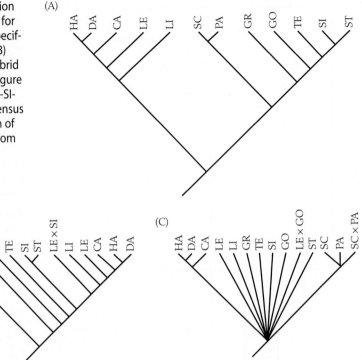

Polyploidy

Judging by its high incidence in many plant groups, polyploidy has played a major role in plant evolution. The addition of chromosome sets (see pages 90–91 in Chapter 4) provides redundant genetic material that may be free to mutate into new, adaptive genes. Polyploids thus often exhibit biochemical diversity, producing many kinds of gene products on which natural selection can operate. As a result, polyploids tend to be more widely distributed and found in more extreme habitats than their diploid relatives. The formation of a polyploid is often accompanied by radical changes in the genome (Soltis and Soltis 1999, 2000) and sometimes by loss of self-incompatibility and sexuality. Polyploidy is also a prime facilitator of the rapid speciation process called **polyploid speciation**.

Frequency of Polyploidy in Plants Our understanding of polyploidy has been greatly improved by molecular studies of individual genes and genomes (Soltis et al. 2003). These studies have shown that autopolyploidy, although probably not as common as allopolyploidy, is much more common than was traditionally thought (Soltis et al. 2003; see Chapter 4 for distinctions between these two types of polyploidy). It now appears that genome doubling is widespread in plants and many other groups of organisms, even those that have been considered diploid.

Polyploidy varies widely in its frequency among major groups of vascular plants. High chromosome numbers in hundreds of fern species have led to an estimate of 95% polyploidy in that group, although there is good evidence that some relatives of ferns with high chromosome num-

bers are actually diploid (see the discussion of Lycopodiaceae in Chapter 8). Only a few conifers, mostly in the Cupressaceae, are polyploid. Masterson (1994) attempted to estimate the frequency of polyploidy in flowering plants by determining ancestral base chromosome numbers. Using the correlation between cell size and DNA content, and thus chromosome number, she based estimates of ancient flowering plant chromosome numbers on the sizes of guard cells (leaf epidermal cells that control the opening and closing of stomata). The inferred chromosome number of primitive angiosperms was seven to nine, which means that approximately 70% of angiosperms have polyploidy in their ancestry.

Polyploid Speciation Polyploid speciation may occur after the formation of allopolyploids or autopolyploids. Allopolyploid speciation, which is thought to be more common than autopolyploid speciation, results from hybridization and subsequent doubling of the chromosome complement. A well-studied case of allopolyploid speciation involves three Eurasian goatsbeard species (*Tragopogon*, Asteraceae) that were introduced into North America in about 1900. Hybrids among these diploid species were first observed in 1949 in eastern Washington. Structural differences in the chromosomes of the three parental species prevent meiotic pairing in their diploid hybrids, making their gametes nonviable. Fertility is restored, however, by chromosome doubling. In the resulting allopolyploid *Tragopogon* hybrids, there is an exact duplicate with which each chromosome can pair so that meiosis is successful. A more detailed discussion of *Tragopogon* allopolyploidization is provided in Box 6E.

BOX 6E Allopolyploid Speciation in *Tragopogon*

Three diploid *Tragopogon* species were introduced into North America in about 1900: *T. dubius, T. porrifolius,* and *T. pratensis*. Hybrids among them were first observed in 1949 in eastern Washington State. *T. dubius* has hybridized with *T. porrifolius* to produce the allotetraploid *T. mirus*, and with *T. pratensis* to produce the allotetraploid *T. miscellus* (Figure 6.26). *T. mirus* and *T. miscellus* are morphologically distinct and reproductively isolated from the parental species because backcrosses are triploid and mostly sterile.

These hybridizations have been analyzed in great detail, using morphology, meiotic chromosome studies, genetic analysis of flower color, analysis of a gene controlling ligulate flower length, plant pigment chemistry, isozymes, and restriction site analysis of cpDNA and rDNA

(Soltis et al. 1995). Isozymes, cpDNA, and rDNA indicate that these hybridizations have occurred between 2 and 21 times for *T. miscellus* and between 5 and 9 times for *T. mirus*.

Chloroplast DNA is maternally inherited in this genus, so the direction of gene flow between the diploid parents can be determined by looking for parental cpDNA in the hybrid offspring. The maternal parent of *T. mirus* is *T. porrifolius* in all known origins. Both parental species have served as maternal parents for *T. miscellus*. Allopolyploids have arisen numerous times in other groups, such as *Dactylorhiza* (Orchidaceae; Hedren et al. 2001) and possibly *Spartina* (Ayers and Strong 2001), and it appears that single origins of polyploids are unusual (Soltis and Soltis 1999).

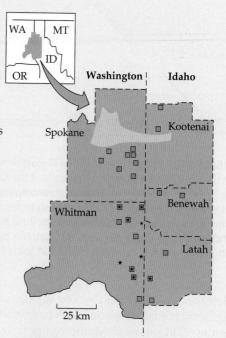

FIGURE 6.27 Distribution map of allotetraploid *Tragopogon* species within five counties of eastern Washington and adjacent Idaho. Populations of *T. mirus* are indicated by stars, populations of *T. miscellus* by the red squares. Localities with both species have both symbols. The yellow area indicates the continuous range of *T. miscellus* in the vicinity of Spokane. (From Novak et al. 1991.)

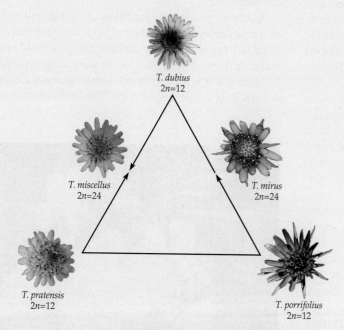

FIGURE 6.26 Allopolyploid speciation in *Tragopogon*. Diploid parental species (2*n*) are at the points of the triangle; tetraploids (4*n*) at the midpoints of lines between parental species. Arrows originate at the maternal parent. Both *T. pratensis* and *T. dubius* have been maternal parents in different originations of *T. miscellus*. (From Pires et al. 2004.)

What makes these allopolyploid speciation events especially interesting is that they are recent, and when and where they happened are known. They give us the opportunity to follow the ecology and dispersal of the new species as well as their molecular and morphological evolution. Both *Tragopogon* allopolyploids have increased in numbers and in geographic range in their region of origin in eastern Washington and northern Idaho (Figure 6.27). In fact, *T. miscellus* has become one of the most common weeds in the vicinity of Spokane, Washington.

An allopolyploidization in *Spartina* (Poaceae) resembles the *Tragopogon* example in being of recent occurrence following human introduction of a species into a new part of the world. *Spartina alterniflora* (with a diploid chromosome number, 2*n*, of 62)—salt marsh cordgrass—is an important component of salt marshes on the coasts of eastern North America. Individuals of this species from somewhere

between Boston, Massachusetts, and Newfoundland, Canada, were introduced into Southampton, United Kingdom, sometime prior to 1829 and became the seed parent in a hybridization with the native *S. maritima* (2*n* = 60), which resulted in the sterile F₁ *S. × townsendii* (2*n* = 62) by 1870.

The fertile allopolyploid *S. anglica* (2*n* = 120, 122, 124)— English cordgrass—had appeared by the end of the 1880s in

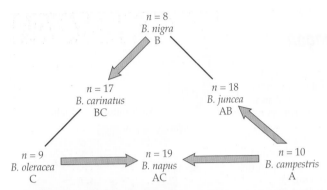

FIGURE 6.28   The triangle of U. The diploid parental species are at the points of the triangle, tetraploids at the midpoints of lines between parental species. Haploid chromosome number (*n*) and genome constitution (A, B, and C) are given for each species. Arrows originate at the maternal parent. Both *Brassica oleracea* and *B. campestris* have served as the maternal parent in different originations of *B. napus*. (After Soltis and Soltis 1993.)

Southampton and then spread rapidly along the coast, replacing the native species. Molecular markers (Ayers and Strong 2001) showed widespread genetic variation in *S. anglica*, which could be interpreted as a result of multiple origins. Alternatively, the variation could have developed from genetic recombination or loss of entire chromosomes in *S. anglica*. The genetic data indicated that the *S. alterniflora* component in all but one of the sampled *S. anglica* plants is identical to one *S. alterniflora* individual in Marchwood, the United Kingdom. The Marchwood plant may have been the seed parent progenitor of *S. anglica*. The common name "cordgrass" indicates that these plants produce rhizomes, which allow an asexual clone to cover a large area and persist for a long time through vegetative reproduction.

Allopolyploid speciation has played a major role in the evolutionary history of many crops, such as corn, sugarcane, cotton, and wheat, and has presumably facilitated the evolution of the traits that have made these plants useful to humans. One of the most interesting cases of hybridization involving crop species comes from *Brassica*. In 1935 the scientist U proposed that three *Brassica* species—*B. nigra* (black mustard), *B. oleracea* (the species from which brussels sprouts, broccoli, cabbage, cauliflower, kale, and kohlrabi have been developed), and *B. rapa* (= *B. campestris*, turnip)—hybridized to produce three other species. These relationships, which became known as the "triangle of U" (Figure 6.28), have been confirmed by numerous studies (reviewed by Song et al. 1988). *Brassica* allopolyploids share with other allopolyploids, such as *Tragopogon* (see Box 6E), a history of multiple origins and introgression (Soltis and Soltis 1993).

Brassica is involved in an even earlier and better-known case of allopolyploid speciation: the intergeneric hybrid *Raphanobrassica* (first mentioned on page 131). In the 1920s the Russian geneticist Georgi Karpechenko attempted to create a crop combining the edible leaves of cabbage and

the root of radish. Cabbage–radish hybrids are usually sterile, but Karpechenko found some individuals that were fertile because the gametes contributed by cabbage and radish were chromosomally unreduced. Cabbage and radish are both $2n = 9$, and the new fertile individuals were allotetraploids with $2n = 36$. Although this result was scientifically interesting, the new crop did not meet expectations: its roots were more like cabbage than radish, and its leaves resembled those of radish more than those of cabbage.

Recent studies show that polyploids are surprisingly dynamic (Soltis and Soltis 1999). In a brief period after their formation, polyploid genomes may undergo major restructuring, including exchanges of portions of chromosomes from the genomes of the parental species. The more divergent the parents of a polyploid are, the more drastic the genomic restructuring will be. Numerous studies have demonstrated that the same allopolyploid can arise multiple times (see Box 6E). If the parental species are genetically different, then repeated hybridizations between them will produce genetically distinct allopolyploids. Matings between these allopolyploids can, in turn, further expand genetic diversity.

Autopolyploids have traditionally not been recognized as species separate from their diploid ancestor, but Soltis et al. (2007) argue that some autopolyploids should be recognized as distinct species. One of the autopolyploids they support as a species occurs in *Chamerion angustifolium* (fireweed, Onagraceae; Figure 6.29). This herbaceous perennial

FIGURE 6.29   *Chamaerion angustifolium* (fireweed, Onagraceae). This large perennial herbaceous plant was photographed in northern Maine, well within the range of autotetraploid plants of this species. Soltis et al. (2007) favor recognition of the autotetraploids as a species separate from diploids. See text for more about polyploid speciation and this plant.

occurs over much of the Northern Hemisphere, mostly as diploids and autotetraploids. The geographic ranges of these two ploidy levels are largely separate, with diploids at higher latitudes than the autotetraploids, but the ranges do overlap in parts of North America. The two ploidy levels are morphologically distinct, the diploid having smaller flowers and shorter inflorescences. Diploids also flower earlier than tetraploids, which provides some reproductive isolation of ploidy levels. This temporal isolation of the ploidy levels plus some geographic isolation, floral constancy, self-fertilization, pollen-style incompatibility, and hybrid inviability were estimated to yield a total reproductive isolation of 99.7% (Husband and Sabara 2003). Hence, Soltis et al. (2007) concluded that both morphology and reproductive isolation strongly support species status for the autotetraploid.

Plant Breeding Systems

The primary means of reproduction in plants is biparental (Richards 1997). Uniparental reproduction via self-fertilization or asexual means is relatively common, however, and it is one of the most frequent evolutionary transitions in plants (Takebayashi and Morrell 2001). About 20% to 25% of plant taxa are predominantly selfing (Barrett and Eckert 1990). Because uniparental reproduction markedly restricts gene flow, it is often associated with complex patterns of morphological variation and with challenges in defining species.

As a starting point, consider a species, such as orchard grass (*Dactylis glomerata*), that forms large populations of individuals that always outcross. Such populations will contain considerable genetic diversity and will not be markedly differentiated from one another. In contrast, populations of plants derived from exclusively uniparental reproduction will tend to be genetically invariant and more or less differentiated from other populations of the species (Richards 1996).

Self-pollination and self-fertilization occur as a result of pollen transfer within a flower or among flowers of the same individual. Selfing affects genetic diversity within individuals, which is measured in terms of heterozygosity and homozygosity. **Heterozygosity** is the presence of two or more different alleles or other genetic elements at a locus within an individual. **Homozygosity** is the presence of only one allele within an individual. Selfing reduces heterozygosity and increases homozygosity.

Consider a gene that encodes an enzyme, such as malate dehydrogenase (MDH). A diploid heterozygous individual bears two alleles of the gene, and thus can make two forms of the enzyme. These two forms may differ in function, giving the heterozygote some biochemical flexibility that a homozygote would lack. We represent the genotype of such a heterozygote as *MDH-1/MDH-2*, indicating two alleles (1 and 2) of the *MDH* gene. Diploid heterozygotes generate two kinds of gametes: one with the *MDH-1* allele and the other with *MDH-2*.

		Pollen	
		MDH-1	*MDH-2*
Eggs	*MDH-1*	*MDH-1/MDH-1*	*MDH-1/MDH-2*
	MDH-2	*MDH-1/MDH-2*	*MDH-2/MDH-2*

FIGURE 6.30 Self-fertilization in a heterozygote for the malate dehydrogenase gene (*MDH*). Possible genotypes (based on alleles *MDH-1* and *MDH-2*) are shown for eggs and pollen. The offspring are ¼ homozygous for *MDH-1*, ¼ homozygous for *MDH-2*, and ½ heterozygous.

Self-fertilization by the heterozygote results in equal numbers of heterozygotes and homozygotes among the offspring (Figure 6.30). Each selfing thus halves the level of heterozygosity. In our example, we started with a heterozygote and a frequency of heterozygosity of 1. After one generation of selfing, heterozygote frequency is ½. If selfing continues in the four progeny in Figure 6.30, the two homozygotes will produce only homozygous offspring, and the heterozygotes will again produce equal numbers of heterozygotes and homozygotes. After this second generation of selfing, heterozygote frequency is $(\frac{1}{2})(\frac{1}{2}) = \frac{1}{4}$.

Thus repeated selfing reduces heterozygosity, as well as morphological variation among offspring, to a very low level. As a result, populations of selfers may be morphologically and genetically uniform. Selfing is the most extreme form of **inbreeding**, or mating with relatives. Inbreeders generally have higher levels of homozygosity and more uniform populations than outcrossers.

Many plants rely on a variety of forms of asexual vegetative reproduction (including stolons, rhizomes, buds, and fragmentation) to colonize an appropriate habitat and for limited dispersal. Asexual seed production, called agamospermy (see Chapter 4; an equivalent term is *apomixis*), differs from vegetative reproduction in retaining seed dispersal and some aspects of sexuality. Normal meiosis and full genetic recombination do not occur in agamosperms. Nevertheless, some genetic variation does develop in agamosperms, even though all sexuality has been lost (Mogie 1992).

Agamospermy occurs in about 34 families and 130 genera of flowering plants and is especially common in the Asteraceae (35 genera contain apomicts), Poaceae (37 genera), and Rosaceae (11) (Asker and Jerling 1992; Carman 1997; van Dijk and Vijverberg 2006). Genera with agamospermous species, such as *Amelanchier*, *Antennaria*, *Calamagrostis*, *Cotoneaster*, *Crataegus*, *Hieracium*, *Malus*, *Poa*, *Potentilla*, *Rubus*, *Sorbus*, and *Taraxacum*, have been as taxonomically difficult as any. If all sexuality is lost, agamospermy is obligate and is the only means of seed production. Usually, however, agamospermy is facultative: it occurs with sexuality in the same individual.

Species Concepts

How do we determine whether two plants are members of the same or different species? This question has been, and continues to be, intensely debated among systematists and evolutionary biologists. Despite the title of his major work, *On the Origin of Species*, Darwin said that species did not matter much and that they were whatever competent systematists said they were. Several recent books, however, focus on species concepts and speciation (King 1993; Lambert and Spencer 1995; Claridge et al. 1997; Howard and Berlocher 1998; Wilson 1999; Levin 2000; Wheeler and Meier 2000; Schilthuizen 2001; Coyne and Orr 2004). Systematists, evolutionary biologists, population biologists, conservation biologists, ecologists, agronomists, horticulturists, biogeographers, and many other scientists are more interested in species than in any other taxon.

As a starting point, consider the situation in animals, especially large vertebrates, in which the capacity to interbreed is usually the criterion used to distinguish species. Among most vertebrates, groups of interfertile individuals coincide with morphological, ecological, and geographic groups and units based on other criteria. Hence not only are species relatively easy to define, but one can also test them by determining whether they are interfertile. For example, horses and donkeys are morphologically distinct, and the hybrid of a male donkey and a female horse—the mule—is sterile. This definition of a species as a group of interfertile individuals is usually called the **biological species concept**, or **BSC** (Mayr 1963), although the other species concepts are "biological" as well. The BSC, which is also referred to as

the isolation species concept (ISC; Templeton 1989), has dominated the zoological literature (Mayr 1992, 2000; Coyne and Orr 2004) and, until recently, the botanical literature as well.

The appeal of the BSC lies in its simplicity, its agreement with the neo-Darwinian emphasis on gene flow and allopatric speciation, and its testability. A test of interfertility cannot be applied unambiguously in plants, however, because interfertility varies greatly among plants. The success of matings between members of different plant groups ranges from 0% to 100% (Figure 6.31), and species assignment in the case of intermediate levels of interfertility is ambiguous (Davis and Heywood 1963).

Because gene flow varies greatly among plant groups, reproductive communities range from one or a few individuals, as in selfing individuals and asexual clones, to morphologically heterogeneous assemblages when hybridization occurs between morphologically divergent groups. Strict application of the BSC would lead to the naming of potentially vast numbers of selfing individuals and asexual clones as species, many of which would differ little from other such species and would therefore be difficult to identify. Applying the BSC in the presence of frequent hybridization would create species that are broadly inclusive. This approach was actually advocated for the almost 1000 species included in the subtribe Pyrinae, some genera of which (i.e., *Amelanchier*, *Crataegus*, *Malus*, *Cotoneaster*, and *Sorbus*) occasionally interbreed.

Plant systematists have largely abandoned the BSC (Davis and Heywood 1963; Ehrlich and Raven 1969; Raven 1976; Mishler and Donoghue 1982; Donoghue 1985; Mish-

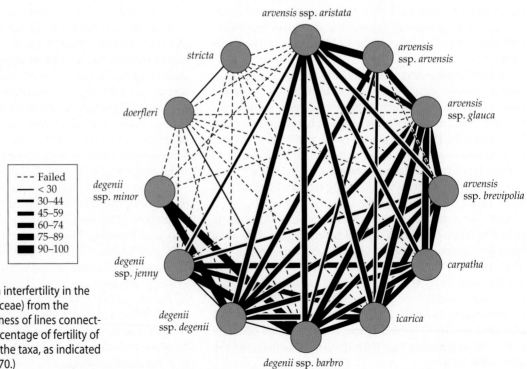

FIGURE 6.31 Variation in interfertility in the genus *Nigella* (Ranunculaceae) from the Aegean region. The thickness of lines connecting taxa indicates the percentage of fertility of hybrids formed between the taxa, as indicated in the key. (From Strid 1970.)

TABLE 6.4 Seven species concepts.

Concept	Criterion for species definition	Reference
Biological	Gap in interfertility between species	Mayr 1963
Recognition	Common fertilization system	Paterson 1985
Phenetic	Gap in the variation between species	Sokal and Crovello 1970
Evolutionary	Common evolutionary fate through time	Wiley 1978
Apomorphy	Monophyly	Donoghue 1985; Mishler 1985
Diagnosability	Unique combination of character states	Nixon and Wheeler 1990
Genealogical	Basal exclusivity	Baum and Shaw 1995

ler and Brandon 1987; Nixon and Wheeler 1990; Davis and Nixon 1992; Kornet 1993; Baum and Shaw 1995; McDade 1995; but see Schemske 2000). The occurrence of gene flow between plant species does not mean that they are not distinct lineages unless the hybridization is so pervasive that the species merge. Furthermore, forces other than gene flow must be responsible for the similarity of populations of cosmopolitan species, such as the giant reed (*Phragmites australis*), on different continents. For example, developmental constraints might have dictated a more or less uniform morphology throughout the geographic range. Rieseberg et al. (2006), however, argued against the common view that plant species are, unlike animal species, not reproductively isolated. Their comparison of numerous studies concerning plant and animal species showed that plant species are "actually more likely to represent reproductively isolated lineages."

Other species concepts have been put forth, including recognition, phenetic, evolutionary, and phylogenetic concepts (Table 6.4). The BSC and the **recognition species concept** (**RSC**) both focus on the role of gene flow, either as a diversifying force when there is a gap in gene flow (BSC) or as a cohesive force maintaining the similarity of individuals within a species (RSC). The **phenetic species concept** rests on the overall similarity of members of a species, which are separated from other species by a gap in variation. The **evolutionary species concept** centers on the recognition of evolutionary lineages, although it does not clearly prescribe how those lineages are to be identified. (These concepts and others are reviewed in Grant 1981; de Queiroz and Donoghue 1988; Templeton 1989; Baum 1992; Kornet 1993; Rieseberg 1994; Baum and Donoghue 1995; Baum and Shaw 1995; Davis 1997; Ghiselin 1997; Hull 1997; de Queiroz 1998; Mishler 1999; Levin 2000; Coyne and Orr 2004.)

The ascendancy of phylogeny as an organizing principle in systematics has motivated phylogenetic species concepts (PSC). A common feature of these concepts is that species are recognized as taxa that mark the boundary between reticulate relationships, such as those one finds among interbreeding members of a species, and divergent relationships, such as those of separate lineages between which there is no exchange of genes. Unfortunately, this boundary is often blurred. Because at least three quite different criteria have been advanced for a PSC, this term is ambiguous and therefore is not used in this book.

The criterion of monophyly stipulates that a species contains all the descendants of one ancestral population and is identifiable by autapomorphies (Donoghue 1985; Mishler 1985; Mishler and Brandon 1987; de Queiroz and Donoghue 1988). This concept can be referred to as the **apomorphy species concept**. Monophyly is an inappropriate criterion below the species level, however, because reticulate, nonhierarchical, and nondivergent relationships are incompatible with grouping by monophyly.

An alternative criterion for a PSC stresses diagnosability and defines a phylogenetic species as the "smallest aggregation of populations (sexual) or lineages (asexual) diagnosable by a unique combination of character states in comparable individuals" (Nixon and Wheeler 1990: 211; see also Davis and Nixon 1992). The character states must be fixed (invariant) within species.

Although this approach, which can be referred to as the **diagnosability species concept**, is simple and easily understood, there is uncertainty about what constitutes diagnosability. One small genetic characteristic unique to a series of populations would make that series diagnosable, but without other differences, most practicing systematists would not be willing to recognize it as a species. Insistence on fixation is also problematic; to be confident that a character state is actually invariant within a species requires much more extensive sampling than is commonly undertaken (Wiens and Servedio 2000).

Diagnosability is a character-based concept, as opposed to a historically based concept (Baum and Donoghue 1995). A third criterion for a PSC—basal exclusivity—tracks history. Exclusivity stipulates that members of a group must be more closely related to one another than to any organisms outside the group. Exclusive groups that contain no less inclusive groups form the basis of the **genealogical species concept (GSC)**.

Exclusivity may be determined by gene **coalescence**; individuals of a species are deemed more closely related to one another than to organisms from another species if their genes share more recent common ancestral genes (coalesce) than they do with individuals in the other species. To see how coalescence applies, consider that, overall, genes in siblings (brothers and sisters) share a more recent common ancestral gene than do genes in first cousins. Coalescence is a potentially flawed criterion, however, because different genes often give different patterns of coalescence (Doyle 1995); some genes link some sets of individuals, and others link different sets. Moreover, coalescence may never occur within a lineage in some genes because natural selection maintains polymorphism (Coyne and Orr 2004). Hence, it is unrealistic to require that all genes will coalesce within a species; but what proportion of genes should be required to achieve coalescence in the GSC?

From the preceding discussion it should be clear that there is no consensus about species concepts in plants. Part of the reason for this lack of consensus is that diversity is idiosyncratic. Each lineage has a unique history of genetic, morphological, and ecological changes; of interactions with other species and the physical world; and of migration, dispersal, and chance. Species differ to varying degrees in morphology, genes, ecology, geographic distribution, pollinators, breeding system, phenotypic plasticity, fruit dispersal, disease resistance, competitive ability, genome size, and numerous other aspects. Furthermore, speciation is a temporally extended process (de Queiroz 1998). We see in the world everything from the very beginning of the formation of new species and all subsequent stages of the speciation process up to species that have been established for a long time and are clearly differentiated from other species. Hence each species is unique.

Although species concepts should ideally be general, applicable, and theoretically significant, these three attributes often conflict (Hull 1997). For example, the BSC has theoretical significance, but as noted already, there are problems with its application. Moreover, it is not general (universal) because its requirement of sexuality excludes asexually reproducing organisms. Additionally, if one accepts the view that some gene flow does occur between species (Rieseberg et al. 2003; Coyne and Orr 2004), then application of the BSC requires choosing some level of gene flow. Such arbitrariness is not limited to the BSC, as noted by Coyne and Orr (2004, p. 34): "all species concepts require some subjective judgement." For the GSC, for example, the observation that some genes may not coalesce within a species requires a choice of a minimum fraction of genes to be recognized as a genealogical species.

The goal of systematics is to infer the products of the unique evolutionary histories of organisms and the relationships of those organisms. To this end, all information may

contribute. Because humans detect other organisms with our senses, especially sight, morphological markers are the primary evidence we use in species delimitation. Morphology is also the most accessible source of data about evolutionary relationships, and it remains the common way to recognize most species. Data from other sources, such as molecular markers, ecology, interfertility, geography, and pollination biology, are important as independent measures of the evolutionary reality of species.

Taxa are hypotheses, open to repeated testing as new data or methods of analysis become available. The basis for a taxonomic hypothesis is the systematist's knowledge of a plant group. This knowledge rests on field and laboratory study of morphology, ecology, breeding system, gene flow between closely related species, geographic distribution, and as many data as possible about structural and molecular variation.

Species are important units in disciplines other than systematics—notably agronomy, biogeography, conservation biology, ecology, genetics, horticulture, and physiology—and ideally should be recognizable by nonspecialists, people who have not dedicated months or years to the understanding of a group of plants. The use of morphologically significant variation is thus essential in practice. If species in a particular genus really are inherently difficult to distinguish, then systematists can assign those species to groups that the nonspecialist will be able to perceive without great difficulty. Some options are outlined in the following section.

Case Studies in Plant Species

Easily recognized species   An easily recognized species is characterized by outcrossing, unrestricted fertility among its members, and strong reproductive isolating mechanisms between its members and those of other species. Such species will not be internally fragmented and will not merge with other species through hybridization. Most vertebrate and vascular plant species fit this pattern. *Dactylis glomerata*, orchard grass, is such a species; it is easily recognized by field biologists and does not pose serious systematic problems. For these groups of plants, application of the different species concepts discussed in this chapter is likely to identify the same set of species.

Microspecies   **Microspecies** are minimally differentiated series of populations derived from uniparental reproduction. Members of the *Andropogon virginicus* species complex are abundant weeds of early succession in the eastern United States. Many of the taxa in this group differ from one another in minute, subtle ways and are difficult to identify. These taxa frequently grow together, and they flower at the same time, but they rarely hybridize (Campbell 1983). They have diverged little morphologically and yet are reproductively isolated. Some of the species in this complex are apparently outcrossing, and others are not only self-compatible but regularly self-pollinate because of cleistogamy.

Shifts from outcrossing to selfing have at least partially isolated some taxa in this group. In one case the shift from outcrossing to predominant selfing appears to be the result of precocious maturation of the flowers so that they self-pollinate before being exposed for outcrossing. Taxa that are reproductively isolated without much apparent phenotypic divergence are also referred to as **sibling species** or **cryptic species**.

Another example of cryptic species comes from *Asplenium nidus* (Aspleniaceae), an epiphytic fern of the Old World tropics. In a study of this species in Indonesia, three types of plants could be distinguished on the basis of *rbcL* DNA sequences (Yatabe et al. 2001). These three types all have the same chromosome number ($2n = 144$) and are indistinguishable morphologically, but they have different habitat preferences. Types A and B grow at the same elevations, but type A grows on the deeply shaded lower portions of the trunks of trees, whereas type B is found in partial shade on the upper parts of trees. Type C grows mostly at higher elevations than types A and B; it never grows with type B and seldom with type A. Controlled experimental hybridizations between types A and B and between types A and C failed to produce any hybrids between *rbcL* types. (Crosses between types B and C were not tested.) The authors concluded that the three types are cryptic species.

A solution to the microspecies problem is to reserve the species category for more strongly marked taxa, which can be broadly defined to encompass closely related but more obscure taxa. In the *Andropogon virginicus* complex, two categories below the species level have been used: varieties and variants. Varieties represent a level of divergence lower than that among species. Varieties are formal taxa represented by trinomials, such as *A. virginicus* var. *glaucus*. Variants are less distinct than varieties, corresponding in some cases to microspecies. They are informal taxa, with English names, such as *A. virginicus* var. *glaucus*, the wetlands variant. The section on nomenclature deals with taxa below the species level, such as varieties and subspecies.

Agamospecies **Agamospecies** are agamospermous microspecies. Many genera—for example, *Alchemilla*, *Amelanchier*, *Antennaria*, *Calamagrostis*, *Crataegus*, *Hieracium*, *Poa*, *Potentilla*, *Rubus*, *Sorbus*, and *Taraxacum*—contain one or more **agamic complexes**. Such groups include sexual taxa that hybridize and their hybrids that acquire agamospermy. Interaction between hybridization and agamospermy may create particularly intricate patterns of variation.

Agamospermy in these genera is often facultative, and some functional pollen is regularly produced. Hence agamosperms may serve occasionally as ovulate parents and regularly as pollen parents in matings with other facultative agamosperms or sexual species. The sterility problem that plagues sexual hybrids is not present in agamosperms. Species names by the hundreds, especially in *Crataegus* and *Rubus*, have been given to introgressants stabilized by agamospermy. A key element in understanding agamic complexes is identification of sexual taxa because these are the foundations of the complex (Bayer 1987).

Taraxacum, the dandelions, is probably the most extreme example of proliferation of agamospecies. Specialists in this genus recognize over 2000 species, most of them agamospecies (Kirschner and Stepánek 1994). Up to 100 microspecies have been documented from a 1-hectare area, creating significant difficulties for the nonspecialist in identifying the bewildering multitude of weakly differentiated taxa. For the most intractable plant species situations, all species concepts are difficult to apply. Use of the taxonomic rank of **section**, a group of species that are phenetically similar, has been advocated to simplify *Taraxacum* taxonomy.

Species that hybridize extensively Species of oaks (*Quercus*, Fagaceae) readily hybridize. This genus contains 500 to 600 species, many of which are common, dominant forest trees. Hybridization between closely related species makes the task of identifying oaks—already a challenge because of their large number of species, phenotypic plasticity, and juvenile versus adult character differences—especially difficult.

In oaks and other groups wherein hybridization occurs, a fundamental question is whether interspecific gene flow is merging the species. In a study of *Quercus gambelii* and *Q. grisea*, which hybridize frequently in the southwestern United States, Howard et al. (1997) discovered that "despite the occurrence of hybrids, both species remain distinct" (p. 747).

The consequences of hybridization in the genus range from infrequent interspecific gene flow to hybrid swarm formation to sets of hybridizing species referred to as **syngameons** to introgression. A syngameon is the most inclusive unit of interbreeding in a hybridizing species group (Figure 6.32). Adherents of the BSC use a term such as *semispecies* for each member of a syngameon.

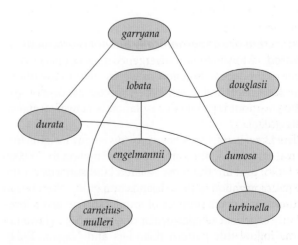

FIGURE 6.32 Semispecies in the California white oak syngameon. Lines connect species that produce fertile hybrids. (After Grant 1981.)

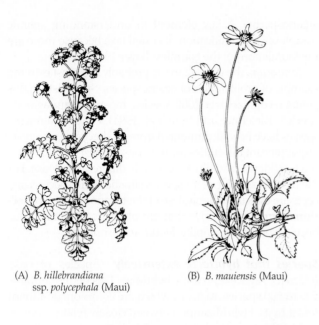

(A) *B. hillebrandiana*
ssp. *polycephala* (Maui)

(B) *B. mauiensis* (Maui)

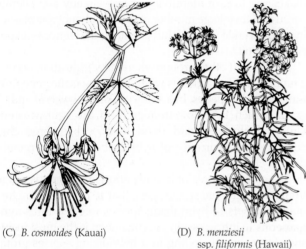

(C) *B. cosmoides* (Kauai)

(D) *B. menziesii*
ssp. *filiformis* (Hawaii)

FIGURE 6.33 Examples of four Hawaiian *Bidens* species, most of which are interfertile. (From Ganders and Nagata 1984.)

Species that are geographically but not reproductively isolated Morphological divergence does not necessarily ensure reproductive isolation, as we saw in the case of *Platanus occidentalis* and *P. orientalis*. A spectacular case of speciation without reproductive isolation is Hawaiian *Bidens* (Asteraceae).

The Hawaiian Islands have provided many insights into speciation because they are young (5.7 million to 700,000 years old) and are the most isolated land surfaces on the Earth. Colonization of these islands by a group often began with one or a small number of immigrants, which subsequently experienced spectacular diversification. Hawaiian *Bidens* follow this pattern (Ganders and Nagata 1984). Descended from one presumed ancestral immigrant, the genus now ranges from sea level to 2200 m in elevation and from semidesert habitats with less than 0.3 m rainfall per year to rain forests and montane bogs with more than 7 m of annual rainfall.

Taxonomists recognize 19 species and 8 subspecies of *Bidens* in Hawaii, which together exhibit a greater range of morphological and ecological diversity than the rest of the genus on five continents (Figure 6.33). The great morphological diversity within Hawaiian *Bidens* is not matched by genetic diversity, however; the different *Bidens* species are as genetically similar as populations of most other species. Morphological evolution has progressed much more rapidly than genetic evolution.

Almost all artificial crosses between Hawaiian *Bidens* species are successful; F_1 individuals produce good pollen and seed. Natural hybridization, however, is relatively rare; only about 5% of all possible hybridizations have been documented in nature. Hybridization does not occur in 85% of the possible cases because the species grow on different islands or different mountain ranges on the same island, and in an additional 8% because of ecological or seasonal isolation. Divergence without reproductive isolation is common in Hawaii (Carr 1995).

Guidelines for Recognizing Plant Species

The following list provides a few basic guidelines aimed at making it easier to recognize plant species:

1. *Get to know the plants in the field.* Extensive field studies throughout the growing season and throughout the geographic range provide invaluable information about morphology, local patterns of variation and discontinuities, breeding systems, flowering times, pollination, ecology, distribution, and the nature of any gene flow. Field studies are essential for rigorous sampling of structural, breeding system, and molecular variation.

2. *Collect data on morphological variation, molecular variation, breeding systems, flowering times, pollination, ecology, distribution, and gene flow.* Combinations of data, such as molecular and morphological, may complement one another. In studies of hybridization in *Salix* (Salicaceae) (Hardig et al. 2000) and altitudinal variation in *Eucalyptus* (Myrtaceae) (McGowen et al. 2001), for example, both molecular and morphological data provided useful insights. Taxonomic hypotheses are only as good as the data on which they are based, so it is important to establish rigorous sampling plans to adequately represent the existing variation. In temperate regions, considerable data for some plant groups may already exist. These extant data should not be accepted uncritically, but evaluated carefully and skeptically. Sampling by previous researchers may have been incomplete, data may have been misinterpreted or incorrectly scored, or important data may have been overlooked. A search for new systematic evidence may be very rewarding. In less studied regions, such as the tropics, extant data will often be minimal, necessitating more intensive field study and sampling.

3. *Analyze systematic data rigorously and display the results graphically to facilitate their interpretation.* Large amounts of data are not readily synthesized and understood simply by visual inspection. Commercially available software (e.g., NTSYS-pc; Rohlf 2005) includes a wide range of tools for statistical analyses of systematic data and graphic display of results. One of the most appropriate analytic tools is principal components analysis (see Box 6C). Principal components analysis and other statistical approaches have been used to assess the species issue by a number of researchers (Kellogg 1985; Leonard et al. 2005; Peirson et al. 2006; Sun et al. 2006).

4. *Hypothesize speciation scenarios and test them by observation and experimentation.*

Summary

Systematists study the results of evolution. We view the organic world as a tree of life with species at the branch tips that are connected to one another by common ancestry. We seek to understand how the raw material of evolution—heritable variation—is generated, broken up into discrete units such as species, and affected by gene flow. Our greatest challenge is selecting appropriate units to call species.

Darwinian natural selection and random genetic drift have driven the evolution of tremendous plant diversity. Inherited variation leads to differential reproduction via natural selection. Mutation and genetic recombination provide the variation required for evolutionary change. Mutation includes changes in DNA, from single-base changes to the addition of whole genomes. Recombination continually presents new arrangements of genes to be "tested" for their evolutionary fitness.

Speciation usually requires some ecological or geographic separation of population systems to prevent gene flow. Allopatric speciation involves the splitting of one species into two or more population systems that diverge in physical isolation. However, gene flow is not likely to be effective in genetically integrating geographically widespread races. Local speciation best explains the origin of divergence in small, marginal populations and the coherence of the species that result through common ancestry.

Some closely related species might well merge when they contacted one another if gene flow were unchecked. Many mechanisms of reproductive isolation retard interspecific gene flow. Natural selection may favor premating isolating mechanisms because they prevent hybridization or reduce competition for resources. Isolating mechanisms may also have evolved simply as a by-product of genetic divergence during speciation.

Hybridization, which is common in plants, can destroy, maintain, or create species diversity. Sometimes only the F_1 generation is formed because of factors such as hybrid sterility. If hybrids reproduce, they may backcross with parental species at the site of hybridization and create a hybrid swarm. If seeds or pollen of hybrid offspring disperse from the original site of hybridization, introgression may occur.

The basic incompatibility between the reticulating phylogenetic patterns of hybridization and the hierarchical patterns of most methods of phylogenetic analysis is a concern when hybrids are very common in the ingroup and/or the parents come from remote lineages in the ingroup. Hybridization is closely linked to polyploidy through allopolyploid speciation, which involves hybridization and subsequent doubling of the chromosome complement.

Polyploidy is believed to play a part in the ancestry of a majority of plant species and is therefore an important pathway in their diversification. Allopolyploids are highly dynamic, having multiple origins in the majority of studied cases, as well as the capacity for major and rapid genetic change through genome reorganization.

Plants differ markedly from animals in the diversity of their breeding systems. The prevalence of uniparental reproduction by selfing and agamospermy in plants fosters patterns of variation that do not readily fit into a taxonomic framework.

The species problem is heightened in plants by interspecific gene flow and by the compartmentalization of variation in uniparental reproducers. Consequently, plant systematists have largely abandoned the biological species concept. In defining species, they look for morphological, molecular, ecological, reproductive, geographic, and other evidence of independent and well-marked lineages. It is difficult to generalize about plant species because each one has a unique history.

LITERATURE CITED AND SUGGESTED READINGS

Items marked with an asterisk are especially recommended to those readers who are interested in further information on the topics discussed in this chapter.

Anderson, E. 1949. *Introgressive hybridization.* Wiley, New York.

Antonovics, J., A. D. Bradshaw and R. G. Turner. 1971. Heavy metal tolerance in plants. *Adv. Ecol. Res.* 7: 1–85.

Arnold, M. L. 1992. Natural hybridization as an evolutionary process. *Annu. Rev. Ecol. Syst.* 23: 237–261.

Arnold, M. L. 1994. Natural hybridization and Louisiana irises. *BioScience* 44: 141–147.

*Arnold, M. L. 1997. *Natural hybridization and evolution.* Oxford University Press, New York.

Arnold, M. L., J. J. Robinson, C. M. Buckner and B. D. Bennett. 1992. Pollen dispersal and interspecific gene flow in Louisiana irises. *Heredity* 68: 399–404.

*Asker, S. E. and L. Jerling. 1992. *Apomixis in plants.* CRC Press, Boca Raton, FL.

Ayers, D. R. and S. R. Strong. 2001. Origin and genetic diversity of *Spartina anglica* (Poaceae) using nuclear DNA markers. *Am. J. Bot.* 88: 1863–1867.

Baldwin, B. G. and R. H. Robichaux. 1995. Historical biogeography and ecology of the Hawaiian silversword alliance (Asteraceae): New molecular phylogenetic perspectives. In *Hawaiian biogeography: Evolution on a hot spot archipelago*, W. L. Wagner and V. A. Funk (eds.), 259–287.

Baldwin, B. G., D. J. Crawford, J. Francisco-Ortega, S.-C. Kim, T. Sang and T. F. Stuessy. 1998. Molecular phylogenetic insights on the origin and evolution of oceanic island plants. In *Molecular systematics of plants II: DNA sequencing*, D. E. Soltis, P. E. Soltis and J. J. Doyle (eds.), 410–441. Kluwer, Boston.

Barrett, S. C. H. and C. G. Eckert. 1990. Variation and evolution of mating systems in seed plants. In *Biological approaches and evolutionary trends in plants*, S. Kawano (ed.), 229–254. Academic Press, New York.

Baum, D. 1992. Phylogenetic species concepts. *Trends Ecol. Evol.* 7: 1–2.

*Baum, D. and M. J. Donoghue. 1995. Choosing among "phylogenetic" species concepts. *Syst. Bot.* 20: 560–573.

Baum, D. and K. L. Shaw. 1995. Genealogical perspectives on the species problem. In *Experimental approaches to plant systematics* (Monographs in Systematic Botany from the Missouri Botanical Garden, Vol. 53), P. C. Hoch and A. G. Stephenson (eds.), 289–303. Missouri Botanical Garden, St. Louis.

Bayer, R. J. 1987. Evolution and phylogenetic relationships of the *Antennaria* (Asteraceae: Inuleae) polyploid agamic complex. *Biol. Zentralblatt* 106: 683–698.

Bradshaw, H. D., Jr., S. M. Wilbert, K. G. Otto and D. W. Schemske. 1995. Genetic mapping of floral traits associated with reproductive isolation in monkeyflowers (*Mimulus*). *Nature* 376: 762–765.

Brauner, S. and L. D. Gottlieb. 1987. A self-compatible plant of *Stephanomeria exigua* var. *coronaria* (Asteraceae) and its relevance to the origin of its self-pollinating derivative *S. malheurensis*. *Syst. Bot.* 12: 299–304.

*Briggs, D. and S. M. Walters. 1997. *Plant variation and evolution*, 3rd ed. Cambridge University Press, Cambridge.

Campbell, C. S. 1983. Systematics of the *Andropogon virginicus* complex (Gramineae). *J. Arnold Arbor.* 64: 171–254.

Campbell, C. S. and W. A. Wright. 1996. Agamospermy, hybridization, and taxonomic complexity in northeastern North American *Amelanchier* (Rosaceae). *Folia Geobot. Phytotax.* 31: 345–354.

Caraway, V., G. D. Carr and C. W. Morden. 2001. Assessment of hybridization and introgression in lava-colonizing Hawaiian *Dubautia* (Asteraceae: Madiinae) using RAPD markers. *Am. J. Bot.* 88: 1688–1694.

Carman, J. G. 1997. Asynchronous expression of duplicate genes in angiosperms may cause apomixis, bispory, tetraspory, and polyembryony. *Biol. J. Linnean Soc.* 61: 51–94.

Carney, S. E., M. B. Cruzan and M. L. Arnold. 1994. Reproductive interactions between hybridizing irises: Analyses of pollen tube growth and fertilization success. *Am. J. Bot.* 81: 1169–1175.

Carr, G. D. 1995. A fully fertile intergeneric hybrid derivative from *Argyroxiphium sandwicense* ssp. *macrocephalum* × *Dubautia menziesii* (Asteraceae) and its relevance to plant evolution in the Hawaiian Islands. *Am. J. Bot.* 82: 1574–1581.

Claridge, M. F., H. A. Dawah and M. R. Wilson (eds.). 1997. *Species: The units of diversity*. Chapman & Hall, London.

Cockrum, E. L. and B. J. Hayward. 1962. Hummingbird bats. *Nat. Hist.* 71(8): 38–43.

*Coyne, J. A. and H. A. Orr. 2004. *Speciation*. Sinauer Assoc., Sunderland, Massachusetts.

*Darwin, C. 1859. *On the origin of species*. Murray, London.

Darwin, C. 1876. *The effects of cross- and self-fertilization in the vegetable kingdom*. Murray, London.

Darwin, C. 1877. *The various contrivances by which orchids are fertilised by insects*, 2nd ed. Appleton, New York.

Davis, J. I. 1997. Evolution, evidence, and the role of species concepts in phylogenetics. *Syst. Bot.* 22: 373–403.

Davis, J. I. and K. C. Nixon. 1992. Populations, genetic variation, and the delimitation of phylogenetic species. *Syst. Biol.* 41: 421–435.

*Davis, P. H. and V. H. Heywood. 1963. *Principles of angiosperm taxonomy*. Van Nostrand, Princeton, NJ.

de Queiroz, K. 1998. The general lineage concept of species, species criteria, and the process of speciation. In *Endless forms: Species and speciation*, D. J. Howard and S. H. Berlocher (eds.), 57–75. Oxford University Press, New York.

de Queiroz, K. and M. J. Donoghue. 1988. Phylogenetic systematics and the species problem. *Cladistics* 4: 317–338.

de Queiroz, K. and M. J. Donoghue. 1990. Phylogenetic systematics and species revisited. *Cladistics* 6: 83–90.

Dilcher, D. L. and P. R. Crane. 1984. *Archaeanthus*: An early angiosperm from the Cenomanian of the western interior of North America. *Ann. Missouri Bot. Gard.* 71: 351–383.

Donoghue, M. J. 1985. A critique of the biological species concept and recommendations for a phylogenetic alternative. *Bryologist* 88: 172–181.

*Donoghue, M. J. 1992. Homology. In *Keywords in evolutionary biology*, E. Fox Keller and E. Lloyd (eds.), 170–179. Harvard University Press, Cambridge, MA.

Doyle, J. J. 1995. The irrelevance of allele tree topologies for species delimitation and a non-topological alternative. *Syst. Bot.* 20: 574–588.

Ehrlich, P. R. and P. H. Raven. 1969. Differentiation of populations. *Science* 165: 1228–1232.

Ellstrand, N. C., R. Whitkus and L. H. Rieseberg. 1996. Distribution of spontaneous plant hybrids. *Proc. Natl. Acad. Sci. USA* 93: 5090–5093.

Evans, R. C. and C. S. Campbell. 2002. DNA sequences from the granule-bound starch synthase gene clarify the origin of the Maloideae (Rosaceae). *Am. J. Bot.* 89: 1478–1484.

Faegri, K. and L. van der Pijl. 1979. *The principles of pollination ecology*, 3rd ed. Pergamon, Oxford.

Furnier, G. R. and W. T. Adams. 1986. Geographic patterns of allozyme variation in Jeffrey pine. *Am. J. Bot.* 73: 1009–1015.

Ganders, F. R. and K. M. Nagata. 1984. The role of hybridization in the evolution of *Bidens* on the Hawaiian Islands. In *Plant biosystematics*, W. F. Grant (ed.), 179–194. Academic Press, Toronto, Ontario.

Ghiselin, M. T. 1997. *Metaphysics and the origin of species*. State University of New York Press, Albany.

Goodwillie, C. 1999. Multiple origins of self-compatibility in *Linanthus* section *Leptosiphon* (Polemnoniaceae): Phylogenetic evidence from internal transcribed spacer sequence DNA. *Evolution* 53: 1387–1395.

Grant, V. 1963. *The origin of adaptations*. Columbia University Press, New York.

*Grant, V. 1981. *Plant speciation*, 2nd ed. Columbia University Press, New York.

Hardig, T. M., S. J. Brunsfeld, R. S. Fritz, M. Morgan and C. M. Orians. 2000. Morphological and molecular evidence for hybridization and introgression in a willow (*Salix*) hybrid zone. *Mol. Ecol.* 9: 9–25.

Head, M. 2006. Panbiogeography of *Nothofagus* (Nothofagaceae): analysis of the main species massings. *J. Biogeogr.* 33:1066–1075.

Hedren, M., M. F. Fay and M. W. Chase. 2001. Amplified fragment length polymorphisms (AFLP) reveal details of polyploid evolution in *Dactylorhiza* (Orchidaceae). *Am. J. Bot.* 88: 1868–1880.

Howard, D. J. and S. H. Berlocher (eds.). 1998. *Endless forms: Species and speciation*. Oxford University Press, New York.

Howard, D.J., R.W. Preszler, J. Williams, S. Fenchel, and W.J. Boecklen. 1997. How discrete are oak species? Insights from a hybrid zone between *Quercus grisea* and *Q. gambelii*. *Evolution* 51: 747–755.

Hull, D. L. 1997. The ideal species concept—and why we can't get it. In *Species: The units of diversity*, M. F. Claridge, H. A. Dawah and M. R. Wilson (eds.), 357–380. Chapman & Hall, London.

Husband, B. C. and H. A. Sabara. 2003. Reproductive isolation between autotetraploids and their diploid progenitors in fireweed, *Chamerion angustifolium* (Onagraceae). *New Phytologist* 161: 703–713.

Kellogg, E. A. 1985. A biosystematic study of the *Poa secunda* complex. *J. Arnold Arbor.* 66: 201–242.

King, M. 1993. *Species evolution: The role of chromosome change*. Cambridge University Press, Cambridge.

Kirschner, J. and J. Stepánek. 1994. Clonality as a part of the evolutionary process in *Taraxacum*. *Folia Geobot. Phytotax.* 29: 265–275.

Kornet, D. J. 1993. *Reconstructing species: Demarcations in genealogical networks*. Unpublished doctoral dissertation, Institut voor Theoretische Biologie, Rijksuniversiteit, Leiden, Netherlands.

Kullenberg, B. and G. Bergstrom. 1976. The pollination of *Ophrys* orchids. *Bot. Notiser* 129: 11–19.

Lambert, D. M. and H. G. Spencer (eds.). 1995. *Speciation and the recognition concept: Theory and application*. Johns Hopkins University Press, Baltimore.

Leonard, M.R., R.E. Cook and J.C. Semple. 2005. A multivariate morphometric study of the aster genus *Sericocarpus* (Asteraceae: Astereae). *Sida* 21: 1471–1505.

*Levin, D. A. 1971. The origin of reproductive isolating mechanisms in flowering plants. *Taxon* 20: 91–113.

*Levin, D. A. 1981. Dispersal versus gene flow in plants. *Ann. Missouri Bot. Gard.* 68: 233–253.

Levin, D. A. 1985. Reproductive character displacement in *Phlox*. *Evolution* 39: 1275–1281.

*Levin, D. A. 1993. Local speciation in plants: The rule not the exception. *Syst. Bot.* 18: 197–208.

*Levin, D. A. 2000. *The origin, expansion, and demise of plant species*. Oxford University Press, New York.

Levin, D. A. 2001. The recurrent origin of plant races and species. *Syst. Bot.* 26: 197–204.

*Li, W.-H. 1997. *Molecular evolution*. Sinauer Associates, Sunderland, MA.

Linder, C.R. and L. H. Rieseberg. 2004. Reconstructing patterns of reticulate evolution in plants. *Am. J. Bot.* 91: 1700–1708.

Masterson, J. 1994. Stomatal size in fossil plants: Evidence for polyploidy in majority of angiosperms. *Science* 264: 421–424.

Mayr, E. 1963. *Animal species and evolution*. Harvard University Press/Belknap Press, Cambridge, MA.

*Mayr, E. 1977. Darwin and natural selection. *Am. Sci.* 65: 321–327.

Mayr, E. 1992. A local flora and the biological species concept. *Am. J. Bot.* 79: 222–238.

Mayr, E. 2000. Species concepts and phylogenetic theory. In *Species concepts and phylogenetic theory. A debate*. Q. D. Wheeler and R. Meier (eds.), 17–29. Columbia University Press, New York.

McDade, L. 1990. Hybrids and phylogenetic systematics. I. Patterns of character expression in hybrids and their implications for cladistic analysis. *Evolution* 44: 1685–1700.

McDade, L. 1992a. Hybrids and phylogenetic systematics. II. The impact of hybrids on cladistic analysis. *Evolution* 46: 1329–1346.

McDade, L. 1992b. Species concepts and problems in practice: Insight from botanical monographs. *Syst. Bot.* 20: 606–622.

McDade, L. 1995. Hybridization and phylogenetics. In *Experimental approaches to plant systematics* (Monographs in Systematic Botany from the Missouri Botanical Garden, Vol. 53), P. C. Hoch and A. G. Stephenson (eds.), 305–331. Missouri Botanical Garden, St. Louis.

*McDade, L. 2000. Hybridization and phylogenetics: Special insights from morphology. In *Phylogenetic analysis of morphological data*, J. J. Wiens (ed.), 146–164. Smithsonian Institution Press, Washington, DC.

McGowen, M. H., R. J. E. Wiltshire, B. M. Potts and R. E. Vaillancourt. 2001. The origin of *Eucalyptus vernicosa*, a unique shrub eucalypt. *Biol. J. Linnean Soc.* 74: 397–405.

Mishler, B. D. 1985. The morphological, developmental and phylogenetic basis of species concepts in bryophytes. *Bryologist* 88: 207–214.

Mishler, B. D. 1999. Getting rid of species? In *Species: New interdisciplinary essays*, R. Wilson (ed.), 307–315. MIT Press, Cambridge, MA.

Mishler, B. D. and R. N. Brandon. 1987. Individuality, pluralism, and the phylogenetic species concept. *Biol. Philos.* 2: 397–414.

Mishler, B. D. and M. J. Donoghue. 1982. Species concepts: A case for pluralism. *Syst. Zool.* 31: 491–503.

Mogie, M. 1992. *The evolution of asexual reproduction in plants*. Chapman & Hall, London.

Mooring, J. S. 2001. Barriers to interbreeding in the *Eriophyllum lanatum* (Asteraceae, Helenieae) species complex. *Am. J. Bot.* 88: 285–312.

Mosseler, A. and C. S. Papadopol. 1989. Seasonal isolation as a reproductive barrier among sympatric *Salix* species. *Can. J. Bot.* 67: 2563–2570.

Nixon, K. C. and Q. D. Wheeler. 1990. An amplification of the phylogenetic species concept. *Cladistics* 6: 211–223.

Novak, S. J., D. E. Soltis and P. S. Soltis. 1991. Owenby's *Tragopogons*: 40 years later. *Am. J. Bot.* 78: 1586–1600.

Paterson, H. E. H. 1985. The recognition species concept. In *Species and speciation* (Transvaal Museum Monograph, 4), E. E. Vrba (ed.), 21–39. Transvaal Museum, Pretoria, South Africa.

Peirson, J. A., P. D. Cantino, and H. E. Ballard. 2006. A taxonomic revision of *Collinsonia* (Lamiaceae) based on phenetic analyses of morphological variation. *Syst. Bot.* 31: 398–409.

Pires, J. C., K.Y. Lim, A. Kovarík, R. Matyásek, A. Boyd, A. R. Leitch, I. J. Leitch, M. D. Bennett, P. S. Soltis and D. E. Soltis. 2004. Molecular cytogenetic analysis of recently evolved *Tragopogon* (Asteraceae) allopolyploids reveal a karyotype that is additive of the diploid progenitors. *Am. J. Bot.* 91: 1022–1035.

Raven, P. H. 1976. Systematics and plant population biology. *Syst. Bot.* 1: 284–316.

Richards, A. J. 1996. Breeding systems in flowering plants and the control of variability. *Folia Geobot. Phytotax.* 36: 283–293.

*Richards, A. J. 1997. *Plant breeding systems*, 2nd ed. Chapman & Hall, London.

Rieseberg, L. R. 1994. Are many plant species paraphyletic? *Taxon* 43: 21–32.

Rieseberg, L. R. 1995. The role of hybridization in evolution: Old wine in new skins. *Am. J. Bot.* 82: 944–953.

Rieseberg, L. R. and J. M. Burke. 2001. The biological reality of species: Gene flow, selection and collective evolution. *Taxon* 50: 47–67.

Rieseberg, L. H., S. A. Church, and C. L. Morjan. 2003. Integration of populations and differentiation of species. *New Phytologist* 161: 59–69.

Rieseberg, L. R. and N. C. Ellstrand. 1993. What can molecular and morphological markers tell us about plant hybridization. *Crit. Rev. Plant Sci.* 12: 213–241.

*Rieseberg, L. R. and J. D. Morefield. 1995. Character expression, phylogenetic reconstruction, and the detection of reticulate evolution. In *Experimental approaches to plant systematics* (Monographs in Systematic Botany from the Missouri Botanical Garden, vol. 53), P. C. Hoch and A. G. Stephenson (eds.), 333–353. Missouri Botanical Garden, St. Louis.

Rieseberg, L. R. and D. E. Soltis. 1991. Phylogenetic consequences of cytoplasmic gene flow in plants. *Evol. Trends Plants* 5: 65–84.

Rieseberg, L. R. and J. F. Wendel. 1993. Introgression and its consequences in plants. In *Hybrid zones and the evolutionary process*, R. G. Harrison (ed.), 70–109. Oxford University Press, New York.

Rieseberg, L. R., S. M. Beckstrom-Sternberg, A. Liston and D. M. Arias. 1991. Phylogenetic and systematic inferences from chloroplast DNA and isozyme variation in *Helianthus* sect. *Helianthus*. *Syst. Bot.* 16: 50–76.

*Rieseberg, L. R., T. E. Wood and E. J. Baack. 2006. The nature of plant species. *Nature* 440: 524–527.

Rohlf, F. J. 2005. *NTSYS-pc: Numerical Taxonomy System*, vers. 2.2. Exeter Publishing, Ltd. Setauket, NY.

Sang, T., D. J. Crawford and T. F. Stuessy. 1995. Documentation of reticulate evolution in peonies (*Paeonia*) using internal transcribed spacer sequences of nuclear ribosomal DNA: Implications for biogeography and concerted evolution. *Proc. Natl. Acad. Sci. USA* 92: 6813–6817.

Schemske, D. W. 2000. Understanding the origin of species (Book review of D. J. Howard and S. H. Berlocher. 1989. *Speciation and its consequences*). *Evolution* 54: 1069–1073.

Schemske, D. W. and H. D. Bradshaw, Jr. 1999. Pollinator preference and the evolution of floral traits in monkeyflowers (*Mimulus*). *Proc. Natl. Acad. Sci. USA* 96: 11910–11915.

Schilthuizen, M. 2001. *Frogs, flies, and dandelions: Speciation—The evolution of new species*. Oxford University Press, Oxford.

Schwarzbach, A. E., L. A. Donovan and L. R. Rieseberg. 2001. Transgressive character expression in a hybrid sunflower species. *Am. J. Bot.* 88: 270–277.

Shaw, K. L. 1998. Species and the diversity of natural groups. In *Endless forms: Species and speciation*, D. J. Howard and S. H. Berlocher (eds.), 44–56. Oxford University Press, New York.

Sokal, R. R. and T. Crovello. 1970. The biological species concept: A critical evaluation. *Am. Nat.* 104: 127–153.

Soltis, D. E. and P. S. Soltis. 1993. Molecular data and the dynamic nature of polyploidy. *Crit. Rev. Plant Sci.* 12: 243–273.

Soltis, D. E. and P. S. Soltis. 1995. The dynamic nature of polyploid genomes. *Proc. Natl. Acad. Sci. USA* 92: 8089–8091.

*Soltis, D. E. and P. S. Soltis. 1999. Polyploidy: Recurrent formation and genome evolution. *Trends Ecol. Evol.* 14: 348–352.

Soltis, D. E., P. S. Soltis, D. W. Schemske, J. F. Hancock, J. N. Thompson, B. C. Husband, and W. S. Judd. 2007. Autopolyploidy in angiosperms: Have we grossly underestimated the number of species? *Taxon* 56: 13–30.

Soltis, D. E., P. S. Soltis, J. A. Tate. 2003. Advances in the study of polyploidy since *Plant speciation*. *New Phytologist* 161: 173–191.

Soltis, P. S., G. M. Pluncket, S. J. Novacek and D. E. Soltis. 1995. Genetic variation in *Tragopogon* species: Additional origins of the allotetraploids *T. mirus* and *T. miscellus*. *Am. J. Bot.* 82: 1329–1341.

Soltis, P. S. and D. E. Soltis. 2000. The role of genetic and genomic attributes in the success of polyploids. *Proc. Natl. Acad. Sci. USA* 97: 7051–7057.

Song, K. M., T. C. Osborn and P. H. Williams. 1988. *Brassica* taxonomy based on nuclear restriction fragment length polymorphisms. *Theor. Appl. Genet.* 5: 784–794.

Stace, C. A. 1984. *Plant taxonomy and biosystematics*. Edward Arnold, London.

*Stebbins, G. L. 1950. *Variation and evolution in plants*. Columbia University Press, New York.

Stevens, P. F. 1997. J. D. Hooker, George Bentham, Asa Gray and Ferdinand Mueller on species limits in theory and practice: A mid-nineteenth-century debate and its repercussions. *Hist. Rec. Aust. Sci.* 11: 345–370.

Stewart, W. N. and G. W. Rothwell. 1993. *Paleobotany and the evolution of plants*, 2nd ed. Cambridge University Press, Cambridge.

Strid, A. 1970. Studies in the Aegean flora, XVI. Biosystematics of the *Nigella arvensis* complex with special reference to non-adaptive radiation. *Opera Bot.* 28: 1–169.

Sun, F. -J., G. A. Levin, and S. R. Downie. 2006. A multivariate analysis of *Pseudocymopterus* (Apiaceae). *J. Torrey Bot. Club* 133: 499–512.

Takebayashi, N. and P. L. Morrell. 2001. Is self-fertilization an evolutionary dead end? Revisiting an old hypothesis with genetic theories and a macroevolutionary approach. *Am. J. Bot.* 88: 1143–1150.

Templeton, A. R. 1989. The meaning of species and speciation: A genetic perspective. In *Speciation and its consequences*, D. Otte and J. A. Endler (eds.), 3–27. Sinauer Associates, Sunderland, MA.

van Dijk, P. J. and K. Vijverberg. 2006. The significance of apomixis in the evolution of the angiosperms: A reappraisal. In *Plant species-level systematics: New perspectives on pattern and process* (Regnum Vegetabile, vol. 143), F. T. Bakker, L. W. Chatrou, B. Gravendeel and P. B. Peser (eds.), 101–116. A. R. G. Gantner Verlag, Ruggell, Liechtenstein.

Weber, J. E. and C. S. Campbell. 1989. Breeding system of a hybrid between a sexual and an apomictic species of *Amelanchier*, shadbush (Rosaceae, Maloideae). *Am. J. Bot.* 76: 341–347.

Wheeler, Q. D. and R. Meier (eds.). 2000. *Species concepts and phylogenetic theory: A debate*. Columbia University Press, New York.

Wiens, J. J. and M. R. Servideo. 2000. Species delimitation in systematics: Inferring diagnostic differences between species. *Proc. R. Soc. Lond. B.* 267: 631–636

Wiley, E. O. 1978. The evolutionary species concept reconsidered. *Syst. Zool.* 27: 17–26.

Wilson, P. 1992. On inferring hybridity from morphological intermediacy. *Taxon* 41: 11–23.

Wilson, R. A. (ed.) 1999. *Species: New interdisciplinary essays*. MIT Press, Cambridge, MA.

Yatabe, Y., S. Masyama, D. Darnaedi and N. Murakami. 2001. Molecular systematics of the *Asplenium nidus* complex from Mt. Halimun National Park, Indonesia: Evidence for reproductive isolation among three sympatric *rbcL* sequence types. *Am. J. Bot.* 88: 1517–1522.

7

An Overview of Green Plant Phylogeny

The word *plant* is commonly used to refer to any autotrophic eukaryotic organism capable of converting light energy into chemical energy via the process of photosynthesis. More specifically, plants produce carbohydrates from carbon dioxide and water in the presence of chlorophyll inside organelles called chloroplasts. Sometimes the term *plant* is extended to include autotrophic prokaryotic forms, especially the bacterial lineage known as the cyanobacteria (or blue-green algae). Many traditional botany textbooks even include the fungi, which differ dramatically from green plants in being heterotrophic eukaryotic organisms that enzymatically break down living or dead organic material and then absorb the simpler products of digestion. Fungi appear to be more closely related to animals, another lineage of heterotrophs characterized by relatively rapid movement and by eating other organisms and digesting them internally.

In this chapter we first briefly discuss the origin and evolution of several separately derived plant lineages, both to acquaint you with these important branches of the tree of life and to help put the green plant lineage in broad phylogenetic perspective. We then focus attention on the evolution of green plants, emphasizing several critical transitions. Specifically, we concentrate on the origin of the land plants (embryophytes), the vascular plants (tracheophytes), the seed plants (spermatophytes), and the flowering plants (angiosperms).

Although knowledge of fossil plants is critical to a deep understanding of each of these shifts, and although we will mention some key fossils, much of our discussion focuses on extant groups. In Chapter 8 you will find detailed descriptions of the major extant groups of vascular plants and seed plants, along with much more information on the biology of these plants. Likewise, Chapter 9 focuses on the attributes of flowering plant lineages and their phylogenetic relationships.

Our main aim in this chapter is to chronicle the evolutionary events leading up to the angiosperms. We therefore pay rather little attention to major branches such as the chlorophytes, the mosses, the lycophytes, and the ferns and their allies. From a phylogenetic standpoint, we could just as well "tell the story" of green plant evolution as leading up to the evolution of the mosses, the horsetails, or any other group (O'Hara 1992), but we follow the path leading to angiosperms simply because their diversity is the focus of this book.

Before we proceed, it is important to comment on the taxonomic names we will use in this chapter. Our knowledge of phylogenetic relationships among the major plant lineages has long been uncertain, and this is reflected in the existence of many contrasting classification systems. Sometimes the same name has been used to refer to different groups. For example, the name Chlorophyta is sometimes applied to the entire green plant clade, and sometimes to a branch within the green plants that includes most of the traditional "green algae." In other cases, different names have been used for the same group; for example, the green plants have been called Chlorophyta by some authors and Viridiplantae by others. To a large extent, these differences reflect the attempts of different authors to assign taxonomic ranks to groups in what they believe to be an internally consistent manner. However, as we have stressed elsewhere (see Chapter 2), the assignment of taxonomic ranks is basically arbitrary, and it typically reflects only the traditions of the relevant taxonomic community. Thus taxa assigned to a particular taxonomic rank (such as a class, order, or family) are not necessarily equivalent with respect to age, species diversity, or ecological breadth.

Other problems relate to changes in our knowledge of phylogeny. Progress in discerning relationships has quite often resulted in the realization that traditionally recognized groups are not, in fact, clades. For example, the name Bryophyta has long been applied to a group that includes the liverworts, mosses, and hornworts. In recent years, however, it has become clear that these groups probably do not form a clade; instead, "bryophytes" refers to a grade, or paraphyletic group, at the base of the embryophytes (land plants).

As we will emphasize, the same is true of several other traditional groups, including "green algae," "seedless vascular plants," "gymnosperms," and "dicotyledons." In some cases it is possible to abandon such names entirely, but in others it is tempting to retain them, either as common names for certain forms of organization (e.g., the "bryophytic" life cycle), or to refer to a clade (e.g., applying "gymnosperms" to a hypothesized clade containing just the extant "naked-seed plants").

In this chapter we will not refer to taxonomic ranks. Elsewhere in the text, major clades within the vascular plants are referred to orders and families, and we use the same names here. Likewise, standard genus and species names are used. However, whether a taxon is considered to be a class or an order by a particular author is not important for our discussion of green plant phylogeny.

In general, our choice of names reflects our sense of which ones are most commonly used in the literature and will therefore create the least confusion. Where possible, we have chosen names with rank-neutral endings, especially the ending -*phytes*, which means "plants." Efforts are under way to provide a new system of names for the major clades of vascular plants (Cantino et al., in press), and a number of small name changes have been made in this edition of the text for consistency with this treatment. Throughout, we have avoided using names that refer to nonmonophyletic groups, but when we do use such names (e.g., to clarify historical usage), we put them in quotation marks.

Endosymbiotic Events

The chloroplasts found in eukaryotes are endosymbiotic organelles derived ultimately from cyanobacteria. This view of the origin of plastids is now firmly established on the basis of structural evidence (e.g., the form and number of their membranes) and molecular studies establishing that the DNA in plastids is more closely related to that in free-living cyanobacteria than it is to DNA in the nucleus of the same cell.

Endosymbiosis entailed massive reductions in the size and gene content of the plastid genome relative to that of free-living cyanobacteria (Delwiche et al. 2004). For example, the free-living cyanobacterium *Nostoc* has a genome size of some 6400 kilobases and over 6500 genes, whereas a red algal chloroplast has only about 190 kilobases and 250 genes. Green algal chloroplasts are even smaller in most cases: about 120 kilobases and 120 genes. This reduction has involved the complete loss of some genes and the transfer of others from the chloroplast to the nucleus (e.g., Martin et al. 2002). There are many more proteins active within plastids (from 500 to 5000) than there are genes, which means that some of these proteins are products of genes that reside outside the plastids.

How many endosymbiotic events have there been? Recent phylogenetic evidence is consistent with just a single primary endosymbiotic event (Palmer 2003; Delwiche et al. 2004; Keeling 2004). For example, recent analyses of eukaryote phylogeny (see Baldauf et al. 2004) recover a clade containing viridophytes (green plants), rhodophytes (red algae), and glaucophytes, sometimes referred to as the archaeplastid clade or primoplantae (Figure 7.1). This result, combined with plastid gene order and composition and the presence of two membranes, suggests that a primary endosymbiotic event occurred in the common ancestor of this clade. In the glaucophytes, the cyanobacterial cell wall still surrounds the plastid, but the wall was lost in the lineage that includes red algae and green plants.

Plastids in red algae and in green plants differ significantly from each other (e.g., in structure and in light harvesting mechanisms), which makes it possible to distinguish with considerable confidence between a red plastid lineage and a green plastid lineage (Delwiche et al. 2004;

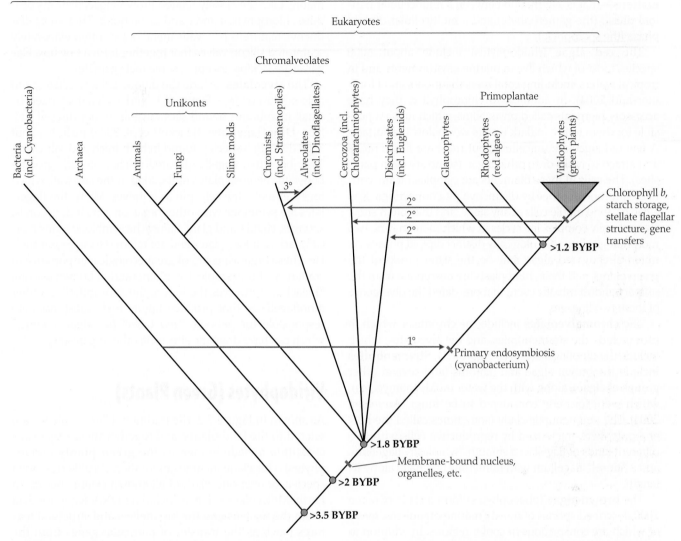

FIGURE 7.1 Phylogenetic tree of life, showing the positions of green plants (viridophytes) and various "algae" among the eukaryotes as well as characters marking several major clades. Red arrows represent primary, secondary, or tertiary endosymbi- otic events. One recent hypothesis for the eukaryotes places their root along the unikont branch, thus separating a clade that initial- ly had one cilium from a clade that initially had two cilia. BYBP, bil- lion years before present. (Adapted from Baldauf et al. 2004.)

Keeling 2004). This distinction helps us to identify instances in which plastids have been acquired by permanent incor- poration of either red or green eukaryotes (see Figure 7.1). It appears that red algal chloroplasts were acquired via such secondary endosymbiosis at the base of the chromalveolate clade, which includes a chromists line, with brown algae and diatoms, and an alveolate line, with dinoflagellates and apicomplexans (the later including *Plasmodium*, the malaria parasite, which contains remnant colorless plastids). Sec- ondary endosymbiotic events involving the uptake of green algae appear to account for the chloroplasts in euglenids (within Discicristates) and chlorarachniophytes (within Cercozoa). Dinoflagellates include a mixture of different types of plastids, and the chloroplasts in one subgroup may even have originated via a tertiary endosymbiotic event (Yoon et al. 2002).

Miscellaneous "Algae"

The term *algae* is applied to a wide variety of aquatic photo- synthetic organisms belonging to several lineages that are not directly related to one another. Before we provide brief descriptions of several of the major groups of "algae," we must briefly review life cycle diversity. In humans and other animals, the diploid phase of the life cycle is the dominant phase, and the only haploid cells are the gametes (pro- duced by meiosis). This kind of life cycle occurs in plants, but is very rare. Some plants have life cycles that are basi- cally the opposite of ours: a multicellular haploid organism is the dominant phase and gives rise to gametes by mitosis; syngamy (fusion of gametes) yields a diploid zygote that undergoes meiosis to yield haploid spores. Most auto- trophic life cycles lie somewhere between these two extremes and exhibit what is known as **alternation of generations**—that is, alternation between a multicellular haploid phase (the gametophyte) and a multicellular diploid phase (the sporophyte).

The **red algae** (rhodophytes) include about 6000 species, most of which live in marine environments and in tropical waters, including coral reefs (Saunders and Hommersand 2004). In addition to chlorophyll *a*, they have accessory pigments called phycobilins, which make it possible for them to live in dark waters well below the surface. A few red algae are unicellular, but most are filamentous and attach to rocks or to other algae (some are even parasites). The cells in these filaments are cytoplasmically connected to one another by distinctive pit connections. Red algae have no motile cells at any stage, and they often show exceptionally complex life cycles in which there may be two morphologically and ecologically distinct diploid phases. As noted already, red algae may be the sister group of the green plants, and their chloroplasts are descended from the primary endosymbiotic event that pre-dated the divergence of these two lineages.

The **chromalveolates** include the chromists, which in turn include the stramenopiles, and the alveolates, which include the dinoflagellates (see Figure 7.1). **Stramenopiles** include the brown algae and diatoms (and several other groups of algae), along with the water molds (oomycetes), which were formerly considered to be fungi (Andersen 2004). The stramenopile clade (sometimes called the heterokonts) is characterized by reproductive cells with two different kinds of flagellae: a smooth "whiplash" flagellum, and a "tinsel" flagellum with numerous fine hairs along its length.

The **brown algae** (phaeophytes) form a clade of some 2000 described species of mostly marine organisms, many of which are conspicuous in cooler regions. In addition to chlorophylls *a* and *c*, they have carotenoid pigments that account for their brown color. All brown algae are multicellular, but this condition presumably evolved within stramenopiles from a unicellular condition. Many brown algae are filamentous, but some are very large and show complex differentiation of the body into a holdfast, a stipe, a float, and one or more flat blades. Some of the larger forms show considerable anatomical differentiation, and some cells are even specialized for nutrient transport. Brown algal life cycles run the gamut from alternation between similar-looking diploid and haploid phases to extreme differentiation (usually with a dominant diploid phase). In *Fucus* and some related kelplike organisms, the multicellular haploid phase has been eliminated completely; in such cases the products of meiosis function directly as gametes, as they do in animals.

There are about 6000 living species of **diatoms** (bacillariophytes), and many more (perhaps as many as 40,000 species) are known from fossils. Owing to their still largely uncharted diversity (Norton et al. 2006), they may be "the insects of the microbial world." Diatoms are unicellular organisms (though they sometimes form loose filaments or clusters) found in both marine and freshwater environments. Like the closely related brown algae, diatoms produce chlorophylls *a* and *c* and carotenoids. Their most distinctive feature is cell walls made of two often elaborately sculptured silicon valves that together form a tiny box. Flagellae are lacking, except in some male gametes.

The **alveolates** include the dinoflagellates, ciliates, and apicomplexans (see Figure 7.1) and are characterized by small membrane-bound sacs (alveoli) under the cell surface. **Dinoflagellates** (Hackett et al. 2004) include about 3000 described species, found in both fresh and salt water. They have two flagellae located in characteristic grooves between cellulose plates embedded in the cell wall, which together make the cell spin as it moves. Many dinoflagellates are symbiotic with other organisms, including corals, sponges, squids, and giant clams. The symbiotic forms typically lack cellulose plates and are referred to as zooxanthellae. These organisms are of great ecological importance in coral reefs; for example, the phenomenon known as coral "bleaching" involves the loss of the zooxanthellae. Other dinoflagellates that produce highly toxic substances are responsible for periodic "red tides" or "algal blooms," which can have dramatic effects on other organisms.

Viridophytes (Green Plants)

As shown in Figure 7.2, the traditional "green algae" are related to the land plants, and together these organisms constitute a clade known as the **green plants** (viridophytes). This clade includes more than 300,000 described species, or over one-sixth of all known extant species on Earth. Molecular evidence, including DNA sequence data (from the nucleus and the organelles) and structural features (such as the transfer of particular genes from the chloroplast to the nucleus), strongly supports the monophyly of the green plants. This clade is also supported by numerous chemical and morphological features, including the loss of phycobilins (found in cyanobacteria, glaucophytes, and red algae) and the production of chlorophyll *b* (in addition to chlorophyll *a*). Green plants also store carbohydrates in the form of starch granules in their cells, and their motile cells have a characteristic stellate structure at the base of each of the usually two anterior whiplash flagellae.

Most phylogenetic analyses (e.g., Karol et al. 2001) have supported a basal split of green plants into a **chlorophyte** clade, containing most of the traditional "green algae," and a **streptophyte** clade, which includes the land plants and several other lineages formerly placed among the "green algae" (see Figure 7.2). Several lineages of unicellular organisms with distinctive scaly cell walls (so-called micromonads, or prasinophytes) are situated around the base of green plant phylogeny, and one of these, *Mesostigma*, has appeared as either the sister group of all other green plants (Turmel et al. 2002) or, more often, as the sister of the streptophyte line (Kim et al. 2006).

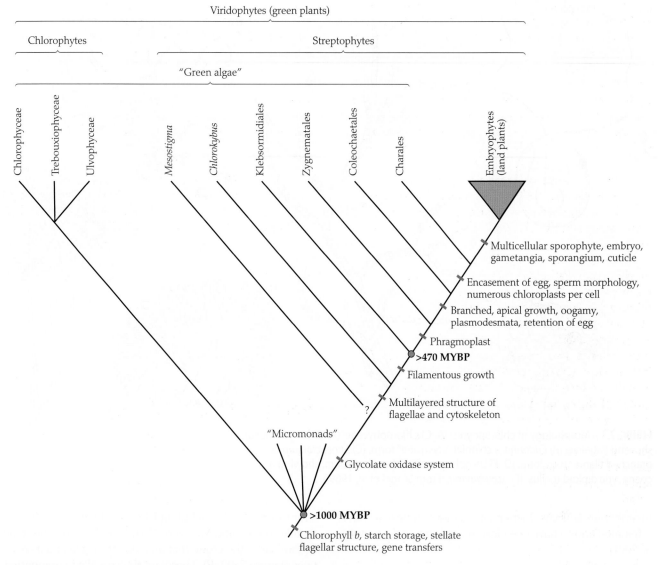

FIGURE 7.2   Green plant phylogeny, showing the separation of chlorophytes from streptophytes, the relationship of some former "green algae" to embryophytes, and characters marking major clades. MYBP, million years before present. (Adapted from Karol et al. 2001 and Delwiche et al. 2004.)

Chlorophytes

Within the chlorophytes there are three well-supported clades (see Figure 7.2): Chlorophyceae, Ulvophyceae, and Trebouxiophyceae (Lewis and McCourt 2004). Relationships among these clades remain unsettled, but gene order and other molecular characters suggest that the ulvophytes and the chlorophytes may be linked (Pombert et al. 2005).

The **Chlorophyceae** are marked by somewhat obscure ultrastructural features (such as clockwise rotation of the flagellar basal bodies), but they have been supported as a clade in most molecular studies. Included within this line is the so-called volvocine lineage, which encompasses progressively more complex colonies (from 4 cells in *Gonium* to as many as 500–50,000 cells in the hollow spherical

colonies of *Volvox*). These colonies were presumed to have been derived from unicells not unlike the model organism *Chlamydomonas* (Figure 7.3A,B). Recent studies indicate that the story is more complex, with several colonial lines derived independently, perhaps from within *Chlamydomonas* itself, which has hundreds of species.

The **Ulvophyceae** include many marine forms and are marked by the production of multinucleate cells (Figure 7.3D–F). In some, the entire body lacks walls between the nuclei except in the case of reproductive cells. Included in this group is the model organism *Acetabularia* (Figure 7.3F).

Finally, the **Trebouxiophyceae** contain forms with flagellate spores, but most are small round forms (apparently derived several times independently) that lack motile cells at any stage. Many of the nonmotile forms live in terrestrial habitats, often in association with lichen-forming fungi or

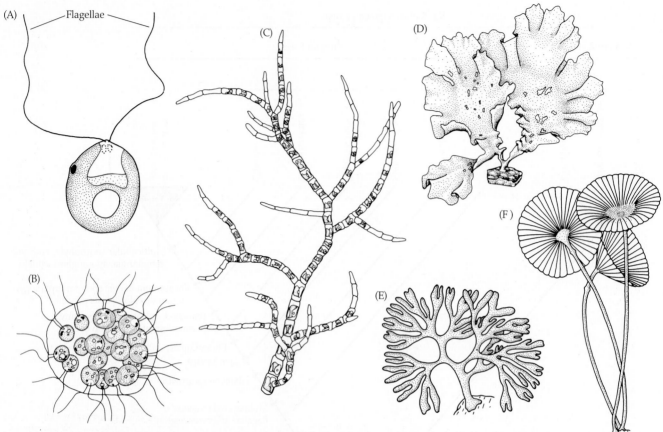

FIGURE 7.3 Morphology of chlorophytes. (A–C) Chlorophyceae: (A) *Chlamydomonas*, showing flagellae. (B) *Eudorina*, a colonial "volvocine" form. (C) *Stigeoclonium*, a branched filamentous form. (D–F) Ulvophyceae: (D) *Ulva*. (E) *Codium*, showing a coenocytic diploid thallus; (F) *Acetabularia*. (From Scagel et al. 1969.)

invertebrate animals. Lichen associations appear to have originated and to have been lost multiple times (Lutzoni et al. 2001).

Streptophytes

The discovery of the streptophyte lineage began in the late 1960s, when detailed ultrastructural studies of cell division first revealed a major difference in the orientation of the spindle microtubules among the organisms that had traditionally been classified as "green algae" (Pickett-Heaps 1979; Mattox and Stewart 1984). Some of these were found to have the phragmoplast orientation found in all land plants, in which the spindle is oriented perpendicular to the formation of the cell wall. A thorough survey showed that this phragmoplast occurred in the charophycean algae (or "charophytes"): the Coleochaetales and Charales. These plants show a range of different growth forms (including upright, branching forms, as in *Chara* and *Nitella*, and flattened forms, as in *Coleochaete*) and live in nearshore freshwater habitats (Figure 7.4A–C). As these organisms were studied in more detail, the idea emerged that they were actually more closely related to land plants than they were to other "green algae." It has since become clear that several other former green algal lineages also belong in the

streptophyte clade, including Klebsormidiales and Zygnematales (Lewis and McCourt 2004). The Zygnematales may be familiar as the group that includes *Spirogyra* and its relatives (Figure 7.4D–E). These are the so-called conjugating green algae, in reference to a form of sexual reproduction that involves the formation of a tubular connection between cells of adjacent filaments, passage of the protoplast from one cell to another, and the eventual fusion of nuclei to form a zygote.

The relationships among the streptophyte groups shown in Figure 7.2 have been confirmed by molecular data (Karol et al. 2001; Delwiche et al. 2004), including structural characters such as the movement of genes from the chloroplast to the nucleus. Coleochaetales and Charales possess some functionally important traits that are otherwise found only in land plants, such as flavonoids and the chemical precursors of a cuticle. Most important from the standpoint of the evolution of the land plant life cycle is the fact that they retain the egg and sometimes even the zygote (after fertilization) on the body of the haploid plant (Graham 1993).

These phylogenetic results have many important implications for our understanding of green plant evolution. For instance, they imply that there were several independent originations of multicellularity. As we have noted, the

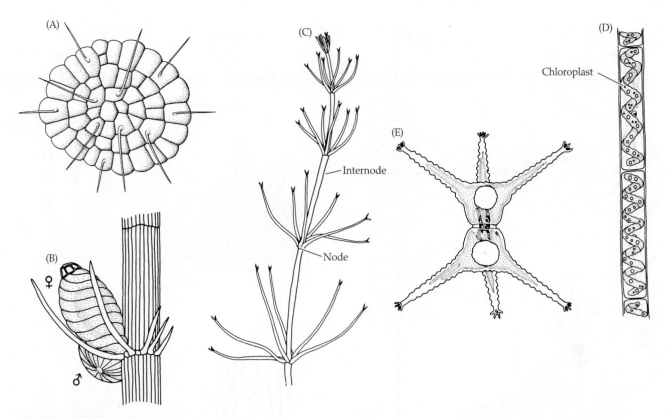

FIGURE 7.4 Morphology of basal streptophytes. (A) *Coleochaete*, showing a haploid discoidal thallus, with setae. (B, C) Charales: (B) *Chara*, showing a node with an egg-bearing structure (above) and a sperm-producing structure (below). (C) *Nitella* habit, showing node and internode construction. (D, E) Zygnematales: (D) *Spirogyra*, a filamentous form, showing helical chloroplasts. (E) *Staurastrum*, a unicellular desmid, forming two mirror-image semi-cells. (A from Taylor and Taylor 1993; B–E from Scagel et al. 1969.)

volvocine forms explored a lifestyle in which the cells became aggregated into colonies. The larger colonies show cytoplasmic interconnections and a division of labor, with some cells specialized for reproduction. Other chlorophytes formed filaments or membranous parenchymatous bodies of much larger size (such as the sea lettuce, *Ulva*, and its relatives), which show a more complex morphological integration and differentiation of cell functions. The Ulvophyceae followed a separate path involving multinucleate cells, sometimes forming filaments, and sometimes (as in *Codium*) forming a thallus by densely intertwining the filaments. Finally, multicellularity evolved separately in the streptophyte line. Many Zygnematales are filamentous, and parenchymatous forms (with plasmodesmata connecting adjacent cells) are found in the charophycean lineages plus the land plants.

Among the early-diverging lineages of green plants, we also encounter a wide variety of life cycles. Alternation of similar haploid gametophyte and diploid sporophyte generations (as in *Ulva*) is quite common. In contrast, *Codium* evolved a life cycle like that of humans, in which gametes are the only haploid cells. In the charophycean lineages, the plants are haploid, and the only diploid cell in the life cycle is the zygote, which results from fertilization of a large non-motile egg by a swimming sperm.

Embryophytes (Land Plants)

The land plants are depicted as stemming from a single common ancestor in Figure 7.2, a finding that is strongly supported by both molecular and morphological evidence (Kenrick and Crane 1997a,b; Karol et al. 2001; Wolf et al. 2005; Qiu et al. 2006). Land plants are also called embryophytes because they have a resting embryo stage early in the life of the sporophyte. *Embryophyte* is the preferable term because several algal lineages (e.g., some Trebouxiophyceae) have independently (though less conspicuously) made the transition to life on land. In addition to the embryo, embryophytes are characterized by the production of a multicellular sporophyte, multicellular reproductive structures (gametangia and sporangia), a cuticle, and thick-walled spores with characteristic trilete marks (see Figure 7.7A).

Traditionally, embryophytes have been classified as either bryophytes or vascular plants. There are three major lineages of bryophytes—liverworts, mosses, and hornworts—which we will characterize briefly in the next few paragraphs (see also Shaw and Goffinet 2000; Shaw and Renzaglia 2004). As we will see, it has become increasingly clear that the "bryophytes" are paraphyletic with respect to the vascular plants (see Figure 7.6).

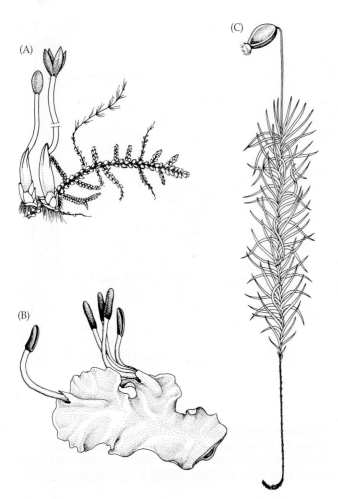
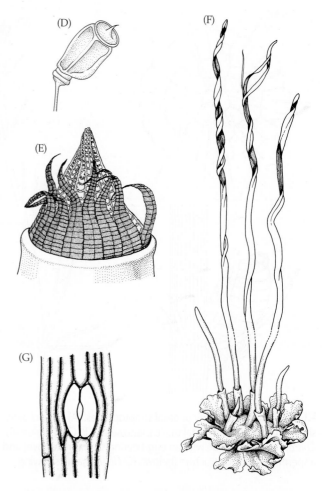

FIGURE 7.5 Morphology of basal embryophytes ("bryophytes"). (A, B) Liverworts: (A) A leafy liverwort, *Lepidozia reptans*, showing dehiscence of the sporangium by four valves. (B) Portion of a thalloid liverwort, *Monoclea forsteri*, showing sporangia with longitudinal dehiscence. (C–E) Mosses: (C) *Dawsonia superba* habit, showing upright, leafy gametophyte and unbranched sporophyte with terminal sporangium. (D) Sporangium (capsule) of a moss prior to dehiscence. (E) Apex of the dehiscing sporangium of a moss, *Fontinalis antipyretica*, showing the peristome teeth. (F, G) Hornworts: (F) *Phaeoceros laevis* habit, showing the thalloid gametophyte and dehiscing sporangia of the sporophyte. (G) Stoma, with guard cells, from the sporangium wall of *Anthoceros*. (A–B, D–G from Scagel et al. 1969; C from Barnes 1998.)

Liverworts

There are about 8000 species of **liverworts**, which come in a thalloid form or, more commonly, a derived leafy form (Figure 7.5A,B). Unlike mosses and hornworts, liverworts lack stomata, although some have epidermal pores without true guard cells. They also lack a characteristic columnar mass of sterile tissue (the columella) in the sporangium, which is present in mosses, hornworts, and early vascular plant lineages. These liverwort features have been interpreted as ancestral within land plants.

Sex in liverworts involves the production of sperm-producing antheridia and egg-containing archegonia. The sporophyte phase, with its terminal sporangium, is rather small and inconspicuous. The capsule typically opens through four valves, and sterile hygroscopic cells (elaters) among the spores may aid in dispersal.

Mosses

Mosses are probably the most familiar bryophytic plants, and with some 10,000 species, they are also the most diverse. The upright, leafy gametophyte is the dominant phase in the moss life cycle (Figure 7.5C). The sporophyte forms a single unbranched stalk terminated by a sporangium (or capsule) (Figure 7.5D). Haploid spores, produced via meiosis, are released from the sporangium; typically, dehiscence of the sporangium occurs by the detachment of a lid or operculum.

When a spore germinates, it forms a protonemal stage, which resembles a green algal filament. The protonema produces one or more upright, leafy gametophytes, which ultimately produce sperm and eggs in antheridia and archegonia, respectively. Fusion of the gametes yields the zygote, which develops through a series of mitotic divisions into the embryo and eventually into the mature sporophyte.

Analyses of relationships within mosses have generally supported the idea that *Sphagnum* (peat moss) is situated near the base of the tree and that *Andreaea* and a few close relatives also form an early branch (see Kenrick and Crane 1997a; Qiu et al. 2006). The enigmatic *Takakia*, which was considered to be a liverwort until the recent discovery of the sporophyte phase, now appears to be related to *Sphagnum*. The sporangium of *Andreaea* opens by four vertical slits, and that of *Takakia* by a single helical slit, in contrast to the lidlike operculum found in the vast majority of mosses. The operculum of most mosses is also characterized by a distinctive row of toothlike structures, which together make up the peristome (Figure 7.5E).

Hornworts

There are only about 100 species of hornworts (Figure 7.5F,G), which are encountered much more rarely than either mosses or liverworts. One presumably derived feature of this entirely thalloid group is the presence of a meristem in the sporophyte located at the base of the capsule. The activity of this meristem accounts for the contin-

ued upward growth of the capsule, which is quite extensive in some groups (e.g., *Anthoceros*).

Phylogenetic Relationships within Embryophytes

Phylogenetic analyses of land plants that have included an adequate sampling of species have found that the "bryophytes" are paraphyletic. However, the exact relationships are still controversial. Early morphological analyses supported a basal split between the liverwort lineage and everything else (Figure 7.6A), and placed the mosses as the sister group to the vascular plants (Mishler and Churchill 1985). Under this view, stomata are considered to be an innovation linking hornworts, mosses, and vascular plants, to the exclusion of liverworts. Likewise, specialized cells in the stems of mosses (in both the gametophyte and sporophyte of some species), called hydroids and leptoids, were interpreted as precursors of the water- and nutrient-conducting cells found in vascular plants. Mosses and vascular plants both have sporophytes that increase in height through cell divisions in an apical meristem, and the first vascular plants had upright gametophytes, as mosses do.

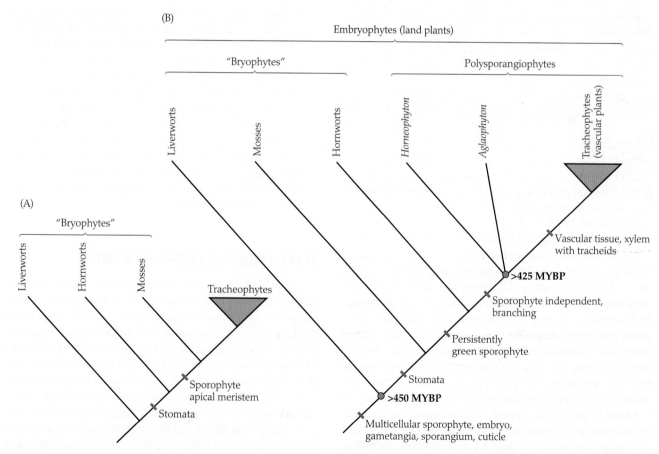

FIGURE 7.6 Phylogenetic relationships at the base of the embryophytes (land plants), showing characters that mark major clades under two hypotheses of how the bryophytic lineages (liv- erworts, mosses, and hornworts) are related to the vascular plants. MYBP, million years before present. (A adapted from Mishler and Churchill 1985; B adapted from Qiu et al. 2006.)

Several recent molecular studies, however—alone and in combination with a variety of ultrastructural characters (especially sperm ultrastructure)—have supported alternative hypotheses. In some trees, hornworts appeared as the sister group of all other extant land plants, and a clade containing mosses and liverworts was sister to the vascular plants (e.g., Renzaglia et al. 2000). The most recent analyses (e.g., Qiu et al. 2006) support the view depicted in Figure 7.6B, in which liverworts are sister to all other embryophytes and hornworts are sister to the vascular plants. This hypothesis remains consistent with a single origin of stomata, although hydroids and leptoids in mosses probably are not homologous with tracheids and sieve cells in vascular plants (Ligrone et al. 2000) and the stalk of the moss sporophyte may not be homologous with the stem in vascular plants (Kato and Akiyama 2005).

Transition to Land

This phylogenetic knowledge illuminates the origin of several key adaptations to life on land (Graham 1993; Waters 2003). Cuticle and sporopollenin (present in the thick spore wall) appear to be evolutionary responses to desiccation. Gas exchange is facilitated by small pores in the epidermis or by genuine stomata with guard cells that can open or close them depending on environmental conditions, thereby regulating water loss. Flavonoids help plants absorb damaging long-wavelength UV radiation. A glycolate oxidase system helps them cope with the inhibition of carbon dioxide fixation by oxygen, which is present in much higher concentrations in air than in water. The first land plants probably depended on symbiotic relationships with fungi to obtain nutrients from the soil, and such relationships have been documented in the major bryophytic lineages as well as in vascular plants (in which they are ubiquitous). The precursors of many of these adaptations can be found among the Coleochaetales and Charales, which therefore appear to have been preadapted to make the transition to land (Graham 1993).

Knowing that both the traditional "green algae" and the "bryophytes" are paraphyletic has also helped us understand the origin of the characteristic land plant life cycle, involving an alternation of multicellular gametophyte and sporophyte generations (Graham 1993). In Coleochaetales and Charales, the egg is retained on the haploid parent plant, and, in *Coleochaete*, the zygote (the only diploid stage) also remains on the parent plant until it undergoes meiosis to give rise to haploid spores. A key innovation in the line including the charophycean lineages and the embryophytes was the establishment of nutrient transport between haploid and diploid phases through a placental transfer tissue (Graham and Wilcox 2000). The land plant life cycle was probably derived from a charophyte-like ancestral condition by simple delay of meiosis and interpolation of a multicellular diploid phase via a series of mitotic divisions of the zygote.

In the embryophytes, the egg—and after fertilization, the embryo—is protected by a multicellular structure called an archegonium. Sperm are produced and protected by a multicellular structure called an antheridium. Initially, the gametophyte phase was dominant, as it is today in liverworts, mosses, and hornworts, and the sporophyte remained attached to, and was nutritionally dependent on, the gametophyte (though perhaps less so in hornworts; Qiu et al. 2006). In vascular plants, the sporophyte became dominant and nutritionally independent, and there was progressive reduction in gametophyte size (Kenrick and Crane 1997a,b).

These findings also help us interpret the absolute timing of events in embryophyte evolution (see Figures 7.1, 7.2, and 7.6). Green plants may be a billion or more years old, and some major green plant lineages existed in the Precambrian (Heckman et al. 2001). A variety of chlorophyte fossils have been found in the Cambrian (about 550 million years ago), such as well-preserved, lime-secreting Ulvophyceae, including relatives of *Acetabularia*. "Charophytes" (in the form of calcified Charales) do not appear in the fossil record until the mid-Silurian, but the wholesale occupation of land by green plants probably began in the mid-Ordovician, some 470 million years ago (Wellman et al. 2003; Sanderson 2003). Dispersed spores have been found from that time (and possibly even earlier, in the Cambrian), sometimes in envelope-enclosed tetrads or dyads (sets of four or two, respectively) resembling those seen today in some liverworts. Tiny bits of cuticle and tubular structures of plant origin also appear in the Ordovician, and individual spores with the characteristic trilete marks of land plants (Figure 7.7A) have been recovered from the early Silurian. It is probable, therefore, that liverworts, mosses, hornworts, and vascular plants were all in existence by the late Ordovician. Somewhat later, beginning in the mid-Silurian, well-preserved macrofossils representing the vascular plant lineage are found. The occupation of land was certainly in full swing by then.

Tracheophytes (Vascular Plants)

The first land plants were small and very simple in structure. In the case of the vascular plant lineage, the sporophyte was basically a dichotomously branching stem, about the height of a matchstick at first, with the sporangia (the site of meiosis yielding haploid spores) produced at the tips of the branches (Figure 7.7B,C). These plants had no leaves or roots. In some cases (e.g., *Aglaophyton*, formerly known as *Rhynia*, from the Rhynie chert in Scotland), the preservation of these plants is spectacular, and it is possible to discern many anatomical details, including stomata, spores, and vascular tissue inside the stem. These fossils revealed that the first **polysporangiophytes**—plants with branching sporophytes—did not actually produce specialized water-conducting cells (**tracheids**) in the xylem tissue and

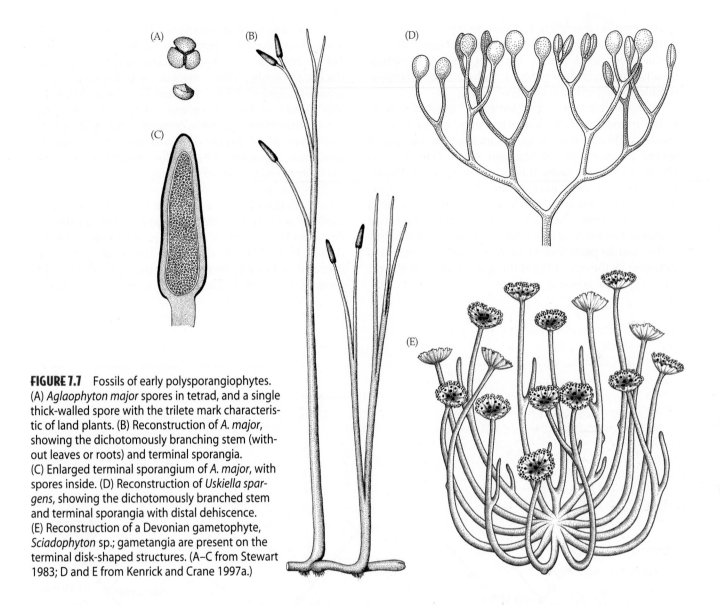

FIGURE 7.7 Fossils of early polysporangiophytes.
(A) *Aglaophyton major* spores in tetrad, and a single
thick-walled spore with the trilete mark characteristic of land plants. (B) Reconstruction of *A. major*,
showing the dichotomously branching stem (without leaves or roots) and terminal sporangia.
(C) Enlarged terminal sporangium of *A. major*, with
spores inside. (D) Reconstruction of *Uskiella spar-
gens*, showing the dichotomously branched stem
and terminal sporangia with distal dehiscence.
(E) Reconstruction of a Devonian gametophyte,
Sciadophyton sp.; gametangia are present on the
terminal disk-shaped structures. (A–C from Stewart
1983; D and E from Kenrick and Crane 1997a.)

must therefore have depended on turgor pressure to
remain upright (Kenrick and Crane 1997a). True water-conducting cells evolved somewhat later and characterize the
true vascular plants: the **tracheophytes** or **Tracheophyta**.

Tracheids are elongated cells with thickened walls that
are dead at maturity. Where one tracheid connects to the
next, there are characteristic openings, or pits, but a pit
membrane (primary cell wall) remains intact, and water
must pass through it as it moves from one cell to the next.
In the first tracheophytes, the tracheids were of a distinctive
type, in which the cell wall had only a very thin decay-
resistant layer (conferred by lignification of cellulose fibers).
Cell walls that are much more decay-resistant characterize
an included clade, which contains all extant vascular plants
(Kenrick and Crane 1997a). In these species, the strongly
lignified tracheids allow more efficient water conduction
and provide internal support, allowing the plants to grow
much taller.

In recent years, careful paleobotanical studies have
revealed that some early land plant fossils are actually haploid gametophytes, bearing antheridia and archegonia,
apparently on separate plants (Remy et al. 1993; Taylor et al.
2005). These fossils are remarkable because they are relatively large, upright, and branched, and in some ways
resemble the sporophyte phase of the life cycle (Figure
7.7E). This finding has led to the view that the first members of the vascular plant lineage exhibited alternation of
more or less similar generations. Thus, in comparison to the
bryophytic groups, it seems that both the gametophyte and
the sporophyte generations were initially elaborated.

This knowledge allows us to piece together a sequence
of events leading to the life cycle that we see in vascular
plants today. This life cycle includes a dramatic reduction in
the gametophyte phase and an equally impressive elaboration of the sporophyte phase. In the first vascular plants,
the gametophyte was nutritionally independent of the

sporophyte, a condition retained today in the "free-sporing" lineages such as ferns and lycophytes. With the evolution of seed plants, however, the gametophyte became much further reduced and eventually became completely dependent on the sporophyte.

Viewed in this context, the "bryophyte" groups (especially the mosses) and the vascular plants appear to have explored different mechanisms to increase the number of spores produced per fertilization event (Mishler and Churchill 1985). In mosses, this increase in spore production was accomplished by intercalation of a filamentous protonemal stage that could produce numerous unbranched leafy gametophytes, each bearing a single unbranched sporophyte terminated by a single sporangium. In contrast, in the vascular plant lineage the number of sporangia was increased by the branching of the sporophyte so that each branch tip could bear a sporangium.

What factors might have favored the elaboration of the sporophyte phase as opposed to the gametophyte phase (which became increasingly specialized for sexual reproduction)? One hypothesis invokes the buffering of diploid organisms against deleterious mutations. But an alternative hypothesis is that the sporophyte was free to become larger (which was advantageous in competing for light and may also have enhanced spore dispersal), whereas the gametophyte was dependent on water for fertilization as long as the sperm needed to swim to the egg.

Phylogenetic relationships among the major lines of extant vascular plants are shown in Figure 7.8. These conclusions are based on morphological and molecular evidence, and most of them are now quite strongly supported (Doyle 1998; Pryer et al. 2004a). The basal split, which occurred in the early to mid-Devonian (before 400 million years ago), separated a clade that includes the modern lyco-

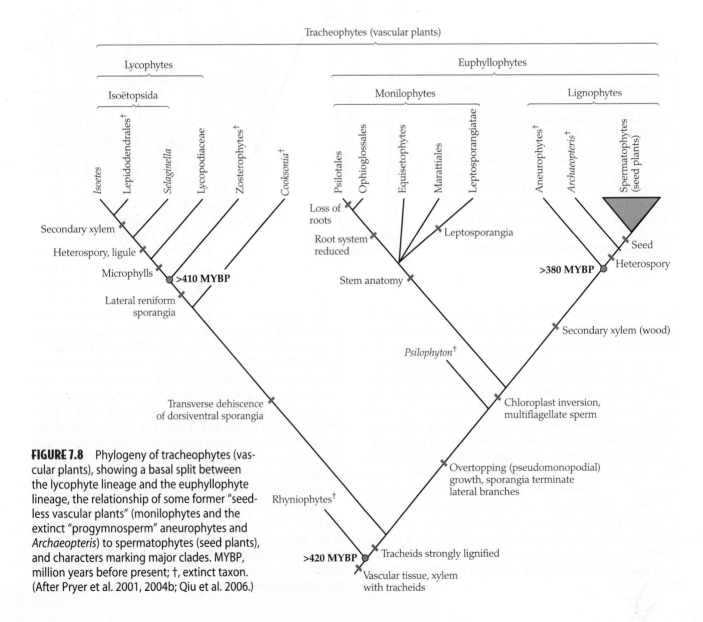

FIGURE 7.8 Phylogeny of tracheophytes (vascular plants), showing a basal split between the lycophyte lineage and the euphyllophyte lineage, the relationship of some former "seedless vascular plants" (monilophytes and the extinct "progymnosperm" aneurophytes and *Archaeopteris*) to spermatophytes (seed plants), and characters marking major clades. MYBP, million years before present; †, extinct taxon. (After Pryer et al. 2001, 2004b; Qiu et al. 2006.)

phyte lineage from another clade, known as the euphyllophytes, that contains all of the other extant vascular plant lineages. This split is marked by a variety of morphological features. One noteworthy feature is the presence of multiflagellate sperm in the euphyllophytes, as opposed to biflagellate sperm in the bryophytic lineages and in the lycophytes (except in *Isoetes* and *Phylloglossum*, in which multiflagellate sperm evolved independently). One compelling bit of molecular evidence is the derived presence in modern euphyllophytes of a 30-kilobase inversion in the chloroplast DNA (Raubeson and Jansen 1992; Wolf et al. 2005).

Lycophytes

The lineage that includes the modern **lycophytes**, or **Lycopodiophyta** (Lycophyta) (Figures 7.8 and 7.9A–C; see also Figures 8.1 and 8.2), appeared in the fossil record very soon after the first appearance of vascular plants. It is marked by the lateral position, reniform (kidneylike) shape, and transverse dehiscence of the sporangia. **Microphylls** (small leaves with a single vascular strand) evolved within this lineage (possibly through modification of lateral sporangia), as did distinctive dichotomously branching roots. During the Carboniferous period lycophytes were especially diverse and abundant, dominating coastal swamps of tropical lowlands (Bateman et al. 1998). The remains of these plants account for our major coal deposits.

Some lycophytes, such as *Lepidodendron*, became large trees, with secondary growth allowing an increase in girth (Figure 7.9D). The stems of these plants were covered by microphylls, which left the distinctive leaf bases seen in fossils (Figure 7.9E). These plants also evolved so-called stigmarian root systems; these are presumed to have been derived from rhizomes, in which case the spirally arranged rootlets may be modified leaves. Patterns of growth in these large plants are still poorly understood, but they may have grown very slowly in height at first (while the root system became established) and later elongated rapidly, and may have died after producing strobili (cone-like structures) at the tips of all the branches simultaneously (Phillips and DiMichele 1992; see Donoghue 2005).

Today there are some 1200 species of lycophytes belonging to several major lines (see Figure 7.8 and 7.9). Rhizomatous species of *Huperzia* and *Lycopodium* (club mosses) are commonly encountered in forests of the Northern Hemisphere. These plants and their tropical relatives are homosporous, meaning that they produce a single kind of spore, which gives rise to a bisexual gametophyte, producing both sperm and eggs.

The other living lycophytes (*Selaginella, Isoetes*) are heterosporous, producing microspores, which give rise to male gametophytes, and megaspores, which give rise to female gametophytes. The heterosporous taxa form a clade (Isoëtopsida; see Figure 7.8), which is also united by the association of a leaflike flap of tissue (the ligule) with the adaxial side of the leaf base.

Selaginella (spike mosses) (see Figure 7.9F–I), with over 700 species, is most diverse in the tropics, where many species grow as epiphytes. *Isoetes* (quillworts, with perhaps 150 species) is the only living remnant of the clade that included the giant lycopods of the Carboniferous, though it may have been derived from plants in this lineage that never attained the size of *Lepidodendron* and the other very large lycophyte trees. *Isoetes* has retained the cambium and some secondary growth, and it has rootlets that resemble those of the extinct trees (Figure 7.9J).

Euphyllophytes

The lineage that includes the modern **euphyllophytes**, or **Euphyllophyta** (see Figure 7.8), is marked by differentiation between a main axis and side branches (pseudomonopodial growth), a key development first seen in a variety of Devonian fossils known as trimerophytes (Figure 7.10A; see Donoghue 2005). According to the "telome theory" (Zimmermann 1965), **megaphylls** (the large leaves characteristic of the euphyllophytes) were derived from flattened lateral branch systems. This derivation involved planation (flattening) of the branch system and then webbing to form the leaf blade. It seems clear that leaves evolved independently, and by very different pathways, in the lycophyte and euphyllophyte lines. Even within the euphyllophytes, it appears that laminated megaphylls originated several times independently (e.g., in ferns, equisetophytes, and seed plants), in each case through the activity of a meristem located at the margin of the developing organ (Boyce and Knoll 2002).

Living euphyllophytes appear to belong to two major clades (see Figure 7.8): the seed plants (**spermatophytes** or **Spermatophyta**) and a clade that includes several "fern" lineages, the horsetails, and the whisk ferns (the **monilophytes** or **Monilophyta**). This new view of euphyllophyte relationships is supported by analyses of morphological characters and both chloroplast and nuclear genes (Pryer et al. 2004b; Rothwell and Nixon 2006). Within the monilophytes there are five major lineages, each discussed only briefly here: (1) leptosporangiate ferns (Leptosporangiatae), (2) Marattiales, (3) Ophioglossales, (4) Psilotales, and (5) equisetophytes (see also Chapter 8).

The common name *fern* is applied to the members of three of these major lineages: Leptosporangiatae, Marattiales, and Ophioglossales. These plants are superficially similar in usually having large (often highly dissected) frondlike leaves that unfold from a "fiddlehead" (so-called circinate vernation). These three lineages are usually divided into two groups on the basis of the structure and development of the sporangia. The Marattiales (Figure 7.10B,C) and the Ophioglossales are so-called eusporangiate ferns. These plants appear to have retained the ancestral condition, in which the sporangium develops from several initial cells and has a mature wall that is more than one cell layer thick. These eusporangia also tend to contain large num-

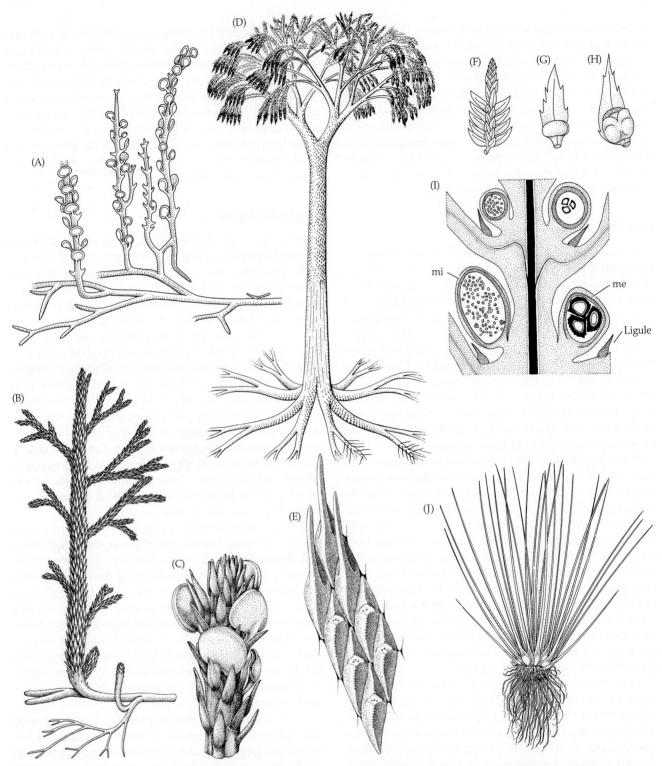

FIGURE 7.9 Morphology of lycophytes and their relatives. (A) Reconstruction of the extinct *Zosterophyllum deciduum*, showing prostrate rhizome bearing leafless upright axes with lateral reniform sporangia. (B) Reconstruction of the extinct *Asteroxylon mackiei*, showing upright dichotomously branching stems covered by microphylls, as well as rootlike axes. (C) *A. mackiei*, showing part of a fertile axis with reniform sporangia and transverse dehiscence. (D) Reconstruction of an extinct *Lepidodendron* sp., showing the dichotomously branching "root" system, the massive trunk with dichotomous branching above, and terminal strobili. (E) Portion of the surface of a stem of *Lepidodendron* sp., showing three attached microphylls and the scars left by the abscission of five others. (F) Tip of a branch of *Selaginella*, showing microphylls and a terminal strobilus. (G) Microsporangium of *Selaginella* in the axil of a microsporophyll. (H) Megasporangium of *Selaginella* in the axil of a megasporophyll. (I) Longitudinal section through a strobilus of *Selaginella harrisiana*, showing megasporangia (me) with four large megaspores, microsporangia (mi) with many tiny microspores, and ligules. (J) *Isoetes bolanderi* habit, showing leaves and roots. (A and J from Kenrick and Crane 1997a; B, C, and I from Stewart 1983; D and E from Gifford and Foster 1989; F–H from Barnes 1998.)

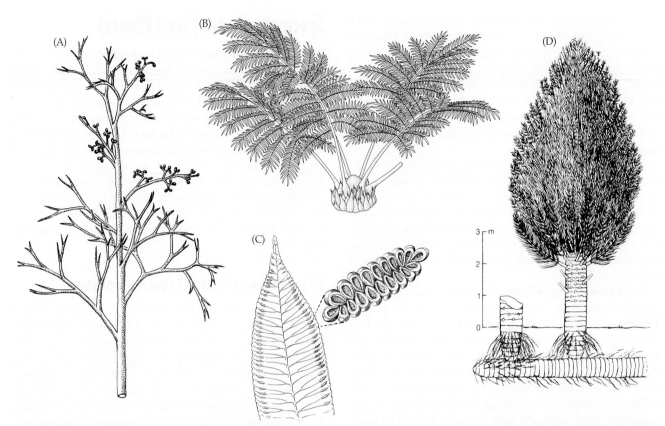

FIGURE 7.10   Morphology of various euphyllophytes.
(A) The extinct trimerophyte *Psilophyton forbesii*, showing
pseudomonopodial growth (differentiation between a main
trunk and side branches). (B) Large, arching leaves of *Angiopteris*
(Marattiales). (C) The lower (abaxial) surface of a fertile leaflet of
Angiopteris, showing a cluster of eusporangia. (D) Schematic rep-
resentation of an extinct treelike equisetophyte, *Calamites*, show-
ing the stout rhizome and tall, upright, branching shoot. (A from
Stewart 1983; B from Barnes 1998; C and D from Gifford and
Foster 1989.)

bers of haploid spores at maturity. In contrast, the leptosporangiates are characterized by a derived development in
which the sporangium arises from a single cell and has a
mature wall only one cell layer thick. These leptosporangia
are borne on a distinct stalk and have a characteristic structure called an **annulus**, consisting of a row of cells with
thickened inner walls and thin outer walls (see Figure 8.13).
The leptosporangia of most species contain a relatively
small and definite number of haploid spores (e.g., 16, 32,
64), which are ejected from the sporangia by a mechanism
driven by changes in moisture content in the annulus cells.

Probably the most familiar monilophytes are the **Leptosporangiatae**, of which there are more than 12,000 living
species (see Figures 8.4 and 8.8–8.19). Many of these plants
have highly dissected pinnate leaves, of the type we commonly associate with ferns, but leaf form is actually
extremely variable within this group, and some species
even have simple, undissected leaves. The sporangia are
typically produced in small clusters called **sori** (singular
sorus) on the undersides of the leaves. The sori are often
covered by a flap of tissue called an **indusium** (plural
indusia), though some are "naked." The structure and
position of the sori and the indusium vary enormously from
one fern group to another, and this variation has been

emphasized in taxonomic treatments (see Chapter 8). Fern
gametophytes are often small, heart-shaped structures,
with the archegonia present near the notch and antheridia
situated among the characteristic rhizoids. There is considerable variation, however, and in some ferns the gametophyte is even filamentous.

Within the leptosporangiate lineage, recent morphological and molecular studies have identified several noteworthy clades (Pryer et al. 2004b; Smith et al. 2006). As has
been long suggested on the basis of sporangium development (sporangia not in sori, rudimentary annulus, large
number of spores), Osmundaceae (cinnamon ferns) are
seen to be the sister group of the rest. One distinctive leptosporangiate clade includes the large tree ferns (Cyatheaceae), and another contains all the heterosporous aquatic
fern groups (placed in Marsileaceae and Salviniaceae).
Although the aquatic ferns are morphologically quite different from one another (e.g., *Salvinia* and *Azolla* with small
floating leaves versus *Marsilea* with leaves resembling
those of a four-leaf clover; see Figure 8.9), the existence of
fossil intermediates also supports the monophyly of the
group (Rothwell 1999).

The **Marattiales**, which are mainly plants of the wet
tropics, tend to have very large pinnate fronds with thick-

walled eusporangia in distinctive clusters (sometimes fused) on the lower surfaces (see Figure 7.10B,C). There are perhaps 150 living species in this clade, most of which belong to *Angiopteris* or *Marattia*, but it has a long fossil record, and extinct relatives (especially *Psaronius*) were important components of Carboniferous swamps. Consistent with their relative morphological stasis, these plants may also have a decelerated rate of molecular evolution (Soltis et al. 2002).

The **Ophioglossales** (with perhaps 80 species) are characterized by fronds that are divided into a flattened vegetative portion (or sterile segment) and a sporangium-bearing fertile segment (see Figure 8.6). This peculiar arrangement may have been derived from a dichotomous branch system. The gametophytes are subterranean, achlorophyllous, tuberlike structures that are associated with an endophytic fungus.

The **Psilotales**, or **psilophytes**, include about 15 species placed in *Psilotum* (the widespread whisk ferns) and *Tmesipteris* (from Australia and the South Pacific) (see Figure 8.5). Because the plant body consists of dichotomously branching stems, psilophytes have often been viewed as the last remnants of the first vascular plants. An alternative theory, based mainly on their subterranean gametophytes, which are associated with fungi, has been that they are reduced leptosporangiate ferns (possibly related to Gleicheniaceae; Bierhorst 1977). Recent molecular studies have established with considerable certainty that neither of these ideas is correct (Pryer et al. 2001, 2004b). Instead, it appears that the Psilotales are most closely related to the Ophioglossales, with which they share some similarities in gametophytes, in the reduction (or loss) of roots, and in the development and position of the sporangia. Under this view, the tiny leaves and the absence of true roots in the psilophytes are derived conditions.

Today there are only about 15 species of **equisetophytes**, or horsetails, all placed in *Equisetum* (see Figure 8.7). Equisetophytes have jointed, hollow stems, with distinct ridges where the epidermal cells deposit silica on their surfaces. The leaves are generally reduced to small scales and are borne in a whorl at each node. The haploid spores are produced in sporangia that are attached to the undersides of unusual peltate sporangiophores and clustered in strobili at the tips of the stems. Although the modern equisetophytes are homosporous, there is controversy over whether the gametophytes have separate sexes. Some gametophytes start out producing just antheridia and some only archegonia, but at least the female forms later become bisexual. Equisetophytes are well known as fossils, which can easily be identified by the characteristic stem architecture. Like the lycophytes, these plants were present in the Devonian but became much more abundant and diverse in the Carboniferous, when some of them also had much larger leaves, evolved heterospory, and became impressive trees (Figure 7.10D). The position of equisetophytes within the monilophytes remains uncertain (Pryer et al. 2004a).

Spermatophytes (Seed Plants)

The **Spermatophytes**, or **Spermatophyta**, are by far the most diverse lineage within the vascular plants, with about 270,000 living species. Most of this diversity is accounted for by just one subclade: the flowering plants, or angiosperms. Morphological evidence for the monophyly of seed plants includes the seed habit itself, but also the fact that the major extant seed plant lineages all share (at least ancestrally) the production of wood (secondary xylem) through the activity of a secondary meristem called the cambium. Another noteworthy vegetative characteristic of this clade is axillary branching, as compared with the unequal dichotomous branching that preceded it within euphyllophytes.

Major Characteristics of Spermatophytes

To understand the seed, it helps to think about how it evolved (see Figure 7.11C–E). Seed plants are nested well within a lineage characterized by homospory (one kind of spore, bisexual gametophytes). A critical step in the evolution of the seed was the evolution of heterospory: the production of two kinds of spores (microspores and megaspores), which produce two kinds of gametophytes (male or microgametophytes, which ultimately produce sperm; and female or megagametophytes, which produce one or more eggs).

Heterospory evolved several times within separate vascular plant lineages, including the lycophytes, the leptosporangiate ferns, the equisetophytes, and the line including the seed plants (Bateman and DiMichele 1994). In several of these cases, the evolution of heterospory was followed by a reduction in the number of functional megaspores. In the line leading to seed plants, the number was reduced to just one by abortion of all but one of the four haploid products of a single meiotic division. The single remaining megaspore was retained within the megasporangium and went on to produce a female gametophyte within the spore (endosporic development). Finally, the megasporangium became enveloped by sterile sporophyte tissue known as integument (see Figure 7.11D), except for a little hole left open at the apex, called the micropyle. In seed plants other than angiosperms, the micropyle serves as the entrance for one or more pollen grains, which are microspores within which the male gametophyte has developed.

It is also helpful to consider the developmental events leading to a mature seed in a plant such as a cycad or a pine tree. Within the ovule (young seed) a single meiotic division occurs within the megasporangium, three of the resulting haploid products disintegrate, and the female gametophyte develops within the remaining spore. Eventually the female gametophyte may contain thousands of cells, with one or more egg cells differentiated near the micropylar end of the seed. Microspores are produced in microsporangia, which may be borne elsewhere on the same plant (monoecy) or on separate plants (dioecy).

One or more pollen grains are transported to the vicinity of the micropyle—presumably by wind in the first seed plants. In many cases a drop of liquid (a pollen droplet) is exuded from the micropyle, which pulls adhering pollen grains inside when it retracts. A pollen grain germinates and sends out a tubular male gametophyte, which eventually delivers sperm to the vicinity of the egg. In modern cycads and ginkgos (discussed on page 171), the pollen tube is haustorial, ramifying slowly through the megasporangium wall, and two very large multiflagellate sperm are eventually produced. In the remaining modern seed plant lineages, a pair of nonmotile sperm are delivered directly to the female gametophyte by the pollen tube. Following fertilization, the diploid zygote develops into a new sporophyte embryo, and the female gametophyte serves as the nutritive tissue.

The second major characteristic of seed plants is the production of wood, or secondary xylem, which (along with a mechanism to regenerate the outer covering of the stem—the periderm) allows the development of a substantial trunk. Understanding how wood is produced requires some basic knowledge of how vascular plants develop. They grow in length through the activity of primary apical meristems at the tip of each shoot and of each root. These apical meristems are populated by undifferentiated cells that undergo mitotic cell divisions, leaving behind derivative cells that go on to differentiate into all of the different cell types and tissues in the plant body. Shoot apical meristems are also the site of initiation of new buds and leaves.

Some of the cells produced by the apical meristem differentiate within the stem into distinct strands of tissue that ultimately will function as vascular tissue. Within these strands, or vascular bundles, the differentiation of the first (primary) xylem, situated toward the inside of the plant axis, and of phloem, situated toward the outside, occurs. Between the xylem and the phloem there remains an undifferentiated layer of cells called the cambium. The cambium acts as a secondary meristem, giving rise to new cells toward both the inside and the outside of the stem, which then go on to differentiate into new xylem cells (such as tracheids) and new phloem cells (such as sieve cells).

The tissues that are produced through this process are referred to as secondary xylem and secondary phloem, respectively. Secondary xylem builds up over the years, forming wood, which is made up of dead, thick-walled cells that are quite sturdy and resistant to decay. Secondary phloem does not build up because phloem cells are not as thick-walled as xylem cells. Additionally, phloem cells have to be alive to carry out their function of transporting carbohydrates and nutrients upward and downward in the plant body.

It is interesting to note that in contrast to the bifacial cambium of seed plants, the giant lycophytes and most equisetophytes of the Carboniferous seem to have had unifacial cambia, producing secondary xylem toward the inside of the stem, but not secondary phloem. They also lacked the ability to substantially increase the size of the cambial ring, so wood production in these plants was actually quite limited (e.g., Cichan and Taylor 1990). The details of cambium function in these plants translated into a variety of highly unusual growth and life history strategies as compared with the familiar seed plants of today (Donoghue 2005).

Early Evolution of Spermatophytes

With this background on the seed and on wood, let us briefly consider the origin and early evolution of seed plants (Figures 7.11 and 7.12; see also Figure 7.8). Our knowledge of the relevant events relies heavily on well-preserved fossils from the late Devonian and early Carboniferous, which have been called "progymnosperms" and "seed ferns" (Figure 7.11A,B).

Recall that the differentiation of a main trunk and side branches had already evolved in the euphyllophyte lineage. One first sees the appearance of very large trunks, with wood rather similar in structural detail to that of modern conifers, in *Archaeopteris*, a "progymnosperm" of the late Devonian. Its trunk was connected to large, frondlike branch systems bearing many small leaves (Figure 7.11A). *Archaeopteris* was found to be heterosporous, yet without seeds.

The accurate reconstruction and phylogenetic placement of *Archaeopteris* and other "progymnosperms," such as *Aneurophyton* (Beck 1981, 1988), was fundamental in establishing that both heterospory and the production of wood pre-dated the evolution of the seed. The clade containing the seed plants plus "progymnosperms" has been called the **lignophytes** (Doyle and Donoghue 1986) or **Lignophyta**, in reference to the production of wood (see Figure 7.8).

The term *seed fern* is applied to a wide variety of early seed plants with large, frondlike leaves, resembling those seen in ferns today, but bearing bona fide seeds (Stewart and Rothwell 1993; Taylor and Taylor 1993) (Figure 7.11B). It is now clear that these plants are not all most closely related to one another and that a series of Paleozoic seed fern groups form a paraphyletic grade at the base of the seed plant radiation.

Careful analyses have revealed that the first seeds were situated in "cupules," and that each seed had an elaborate outgrowth of the sporangium wall that formed a specialized pollen chamber (e.g., Serbet and Rothwell 1992). This structure presumably functioned in pollen grain capture (Figure 7.11D). Integument tissue may have been derived from a series of sterilized sporangia, which initially formed lobes at the apex as opposed to a distinct micropyle (Figure 7.11E).

Through much of the last century, extant and extinct seed plant lineages were commonly divided into two major groups: the cycadophytes and the coniferophytes. The **cycadophytes**, including modern cycads, were distinguished by rather limited production of wood with wide rays (manoxylic wood) and by large, frondlike leaves and radially symmetrical seeds. In contrast, in **coniferophytes**, including the ginkgos and the conifers, the wood is well developed and dense (pycnoxylic), the leaves are simple and often needlelike, and the seeds are biradially symmetrical (platyspermic, or flattened). This distinction suggested

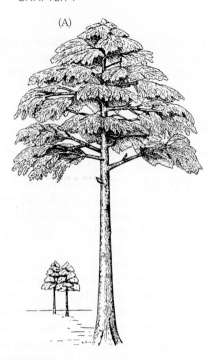

(A)

(B)

(C) Probable steps of seed evolution

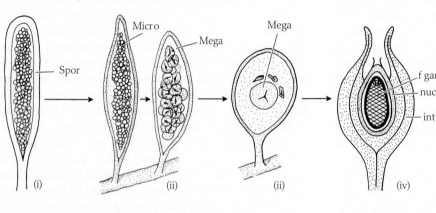

Micro

Mega

Mega

f gam

nuc

int

Spor

(i)   (ii)   (ii)   (ii)   (iv)

(F)

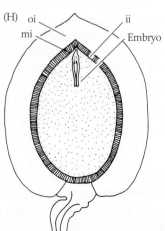

(D) Pollen-receiving structures

(i)   (ii)   (iii)

(E) Evolution of the integument

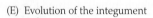

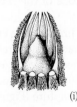

(i)   (ii)   (iii)   (iv)

(G)

(H)   oi   ii

mi   Embryo

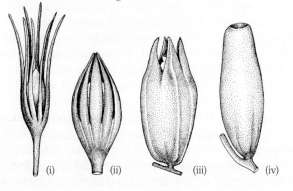

◀ **FIGURE 7.11** *Archaeopteris* and early seed plants. (A) Reconstruction of the habit of *Archaeopteris*, an extinct "progymnosperm" with a large trunk and flattened lateral branch systems. (B) Reconstruction of an extinct "seed fern," *Medullosa noei* (3.5–4.5 m high), showing the large compound leaves. (C) Probable steps in the evolution of the seed: (i) homospory in a distant ancestor; (ii) heterospory, with differentiation between sporangia that produce microspores and megaspores; (iii) reduction of the number of functional megaspores to one, and its development inside the megasporangium (endospory); (iv) envelopment of the megasporangium by integument tissue, leaving a micropyle at the apex. Spor, sporangium; Micro, microspores; Mega, megaspores; f gam, female gametophyte; int, integument; nuc, nucellus or megasporangium wall. (D) Pollen-receiving structures at the apex of the ovule in early seeds (all extinct): (i) *Physostoma elegans*; (ii) *P. ele-* gans, longitudinal section showing pollen chamber within; (iii) *Eurystoma angulare*, showing cup-shaped opening. (E) Stages in the evolution of the integument in early seeds (all extinct): (i) *Genomosperma kidstoni*, (ii) *G. latens*, (iii) *Eurystoma angulare*, (iv) *Stamnostoma huttonense*. (F) Portion of long shoot and spur shoot of the extant ginkgophyte, *Ginkgo biloba*, showing axillary microsporangiate strobili; detail of axis and four microsporangium-bearing structures at right. (G) Portion similar to that in F of an ovule-bearing plant of *G. biloba*, showing axillary stalks, each bearing a pair of ovules; detail of the tip of a stalk at right. (H) Longitudinal section of the seed of *G. biloba* with young embryo (ii, inner layer of integument; mi, middle layer of integument; oi, outer layer of integument). (A, F, and G from Bold 1967; B and D from Gifford and Foster 1989; C and H from Scagel et al. 1969; E from Stewart 1983.)

to some workers that seed plants actually originated twice. Under this view, the cycadophyte line was derived from a progymnospermous ancestor by the modification of flattened lateral branch systems into large, frondlike leaves. In coniferophytes, on the other hand, the individual leaves of a precursor like *Archaeopteris* might have been modified into needlelike leaves. This scenario implies that the seed itself evolved twice, corresponding to the two different symmetries.

However, phylogenetic analyses that have included the extant lineages along with representative fossils have generally supported the relationships shown in Figure 7.8 and Figure 7.12 (e.g., Doyle 1998, 2006). These studies imply that the seed evolved just once, and that the first seed plants were cycadophytic, at least in having large, dissected leaves and radially symmetrical seeds. Specifically, it appears that a series of Devonian-Carboniferous "seed ferns" (*Elkinsia*, *Lyginopteris*, and medullosans) are situated at the base of the seed plant phylogeny and that coniferophytes are nested well within the tree, in a platyspermic clade. This arrangement implies a later shift to small, needlelike leaves and to smaller, flattened seeds—both perhaps as adaptations to arid environments.

Extant Lineages of Spermatophytes

Today there are five major lineages of seed plants: cycads, ginkgos, conifers, gnetophytes, and flowering plants (angiosperms). The first four groups are often called gymnosperms, in reference to their naked seeds, as opposed to angiosperms, in which the seeds are enclosed inside a carpel. Despite many efforts to resolve the phylogenetic relationships among these lines using morphological and molecular data, they remain quite uncertain (see Figure 7.12).

Some recent molecular analyses have indicated that the extant groups of "naked-seed plants" form a clade, which is sister to the angiosperms. However, note that even if this were true, the gymnosperms as a whole would not be monophyletic. They are paraphyletic when one takes into

account the early-diverging fossil lineages already mentioned, as well as several other "seed fern" lineages from the later Permian and Mesozoic, some of which appear to be on the line leading to modern angiosperms (Doyle 2006). We will return to a discussion of these relationships following a brief introduction to each of the major groups (see also Chapter 8).

Cycads Cycads (**Cycadophyta** or Cycadales) were most abundant and diverse during the Mesozoic. Today there are perhaps 130 species left. Cycads generally produce squat trunks, with limited secondary xylem, and large compound leaves resembling those of ferns or palms (see Figure 8.21). They are dioecious, meaning that some plants bear strobili producing only seeds whereas others bear only pollen strobili. Both types of strobili are typically very large and in some cases brightly colored. Likewise, the seeds are generally large and usually have a fleshy and colorful seed coat, presumably to attract vertebrate dispersal agents.

Several cycad features may be ancestral within seed plants, such as haustorial male gametophytes and gigantic multiflagellate sperm. However, cycads are united by several apparently derived morphological characters, including the loss of axillary branching, the presence of "girdling" leaf traces, and the production of coralloid roots that house nitrogen-fixing cyanobacteria.

Within cycads, phylogenetic analyses indicate that the first split divides *Cycas* from the remaining groups (e.g., Rai et al. 2003). *Cycas* has retained the presumed ancestral condition (seen in some fossil relatives, such as *Taeniopteris*) of having several ovules borne on rather leaflike megasporophylls, which are not clustered into strobili. The derived condition, seen in the other line, is a reduction to two ovules borne on a peltate megasporophyll, with the ovules pointing inward toward the axis of the strobilus.

Ginkgos There is just one surviving species, *Ginkgo biloba*, within **ginkgophytes** (or Ginkgoales; Figure 7.11F–H). This species is hardly known in the wild, but it has been maintained for centuries around temples in China, and in

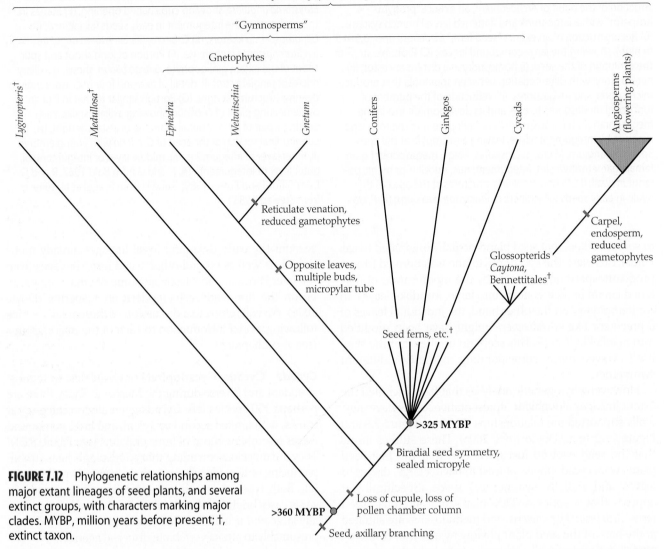

FIGURE 7.12 Phylogenetic relationships among major extant lineages of seed plants, and several extinct groups, with characters marking major clades. MYBP, million years before present; †, extinct taxon.

modern times it has been spread by humans as a street tree. Perhaps the most characteristic feature of the modern ginkgo is its production of deciduous, fan-shaped leaves with dichotomous venation. Ginkgophytes are well known in the fossil record, in which a greater diversity of leaf shapes is seen.

Like cycads, ginkgos are dioecious (Figure 7.11F,G). The ovules are borne in pairs on axillary stalks, thought to be reduced strobili. The integument tissue differentiates into a fleshy (and smelly) outer layer and a hard inner layer that encloses the female gametophyte (Figure 7.11H). Like cycads, ginkgos retain several ancestral characteristics, including haustorial male gametophytes and swimming sperm.

Conifers There are approximately 600 living species of **conifers** (**Coniferae** or Coniferales) (see Figures 8.24–8.27). These plants are shrubs or small trees with well-developed wood and often needlelike leaves. In most cases the leaves

are borne singly along the stem, but in pines (*Pinus*) they are clustered in short shoots. The needles often display additional adaptations to drought, such as sunken stomata. In some Southern Hemisphere conifers (e.g., *Podocarpus*, *Agathis*), however, the leaves are rather broad and flat, and in *Phyllocladus* the flattened branches resemble leaves.

Many conifers are monoecious, with both pollen-producing and seed-producing strobili borne on the same plant. Dioecy is found in other groups, such as the junipers (*Juniperus*), yews (*Taxus*), and podocarps (*Podocarpus*). In the pollen cones, microsporophylls bear microsporangia on the abaxial surface. The pollen grains often have a pair of saclike appendages, but these seem to have been lost in several lineages.

In the seed cones, receptive ovules are situated on the upper side of each cone scale. Meiosis occurs inside each ovule, and the one remaining haploid cell gives rise to the female gametophyte, which eventually produces one or more eggs at the micropylar end. A pollen tube grows through the wall of the megasporangium to deliver two sperm. The phenomenon of "polyembryony" is fairly common in conifers. Multiple embryos may be produced in an ovule either through separate fertilization events (depending on the number of eggs and pollen tubes) or, more commonly, by a characteristic subdivision of a single embryo into several genetically identical embryos early in development.

In modern conifers, the pollen strobili are said to be simple, whereas the seed cones are compound. The pollen strobilus is interpreted as a modified branch, and the microsporophylls as modified leaves. The seed cone, in contrast, was derived through modification of a branch that bore lateral branches in the axils of a series of leaves. This view is supported by fossils showing a series of steps in the reduction of a lateral branch bearing a number of seeds to the highly modified cone scale that we see in the modern groups (Figure 7.13A–E) (Florin 1954). It also follows from the observation that each cone scale is subtended by a bract, which represents the modified leaf. In a few conifers the subtending bract is noticeable, sticking out from between the cone scales. This is the case, for example, in the Douglas fir (*Pseudotsuga*), in which the cone scale is produced in the axil of a prominent three-pronged bract (Figure 7.13C). In many conifers, however, the bract is quite reduced. In Cupressaceae, such as *Taxodium* or *Cryptomeria*, the bract is fused to the cone scale, which still shows evidence of "leaves" (visible as small teeth or bumps).

Phylogenetic studies have yielded some important insights into the evolution of conifers (e.g., Stefanovic et al. 1998). Molecular data imply a basal split between the **Pinaceae** and a clade including all the other conifers, the **Cupressophyta** (Cantino et al., in press). The Pinaceae are distinguished by several features, including inversion of the ovules (with the micropyle facing the axis of the cone; Figure 7.13D) and the derivation of the wing of the seed from the cone scale. Within the Cupressophyta, the two major Southern Hemisphere groups—Podocarpaceae and Araucariaceae—form a clade, perhaps united by a shift to one ovule per cone scale. The Cupressaceae are marked by several potential apomorphies, such as fusion of the cone scale and the subtending bract. In turn, this group may be linked with the Taxaceae (the yews), which have highly reduced cones bearing just one terminal seed surrounded by a colorful fleshy aril. As noted on page 176, several recent molecular analyses have called into question the monophyly of the Coniferae, placing the gnetophytes within the conifers as the sister group of the Pinaceae (see Figure 7.15C).

Gnetophytes The fourth major extant lineage of seed plants is the **gnetophytes** (**Gnetophyta** or Gnetales) (Figure 7.13F–I; see also Figure 8.28). This group contains only about 75 living species, which belong to three quite distinct lineages. *Ephedra* (with about 40 species in deserts around the world) has very reduced scalelike leaves (see Figure 8.28). *Gnetum* (with about 35 species in tropical forests of

the Old and New Worlds) has broad leaves (Figure 7.13F–H), like those seen in most flowering plants. Finally, *Welwitschia* (with only one species, *W. mirabilis*, in southwestern Africa) produces just two (rarely four) functional leaves during its lifetime, which grow from the base and gradually fray out at the tips (Figure 7.13I).

Although these plants look very different from one another, they share some unusual features, such as opposite leaves, multiple axillary buds, vessels with circular openings between adjoining cells, compound pollen *and* seed strobili, and ancestrally ellipsoid pollen with characteristic striations running from tip to tip. The seeds also have two integumentary layers: the inner layer forms a micropylar tube that exudes the pollen droplet, and the outer layer is derived from a fused pair of bracts (Figure 7.13H). Molecular studies also strongly support the monophyly of this group.

Within the gnetophytes, *Gnetum* and *Welwitschia* form a well-supported clade. Morphological synapomorphies include reticulate leaf venation, further reduction of the male gametophyte, and aspects of female gametophyte structure (tetrasporic development, loss of archegonia, free nuclei functioning as eggs). The characteristic striated pollen found in *Ephedra* and *Welwitschia* was apparently lost along the line leading to *Gnetum* (which has spiny pollen grains with no apertures).

Aside from fossil pollen, the fossil record of the gnetophytes is rather limited (Crane 1996), with relatively few macrofossils described until recently (e.g., Rydin et al. 2004; reviewed in Won and Renner 2006). Although gnetophyte pollen grains are found as far back as the Triassic, it appears that the clade containing the modern groups diversified most significantly during the mid-Cretaceous, along with the angiosperms.

Like the angiosperms, the gnetophytes shortened the life cycle (and probably became herbaceous) and evolved insect pollination (found in some living species). In marked contrast to flowering plants, however, gnetophytes never became significant components of the vegetation at mid- and high paleolatitudes, and they underwent a dramatic decline during the late Cretaceous (Crane et al. 1995; Crane 1996).

Angiosperms (Flowering Plants)

With over 257,000 extant species, **flowering plants** (**Angiospermae**) account for most of green plant, land plant, and seed plant diversity. Strong evidence for the monophyly of angiosperms comes from molecular studies and from many shared derived morphological characters. Of these, some of the more obvious and important reproductive features are (1) seeds produced within a carpel with a stigmatic surface for pollen germination; (2) a very reduced female gametophyte, consisting in most cases of just eight nuclei in seven cells; and (3) double fertilization,

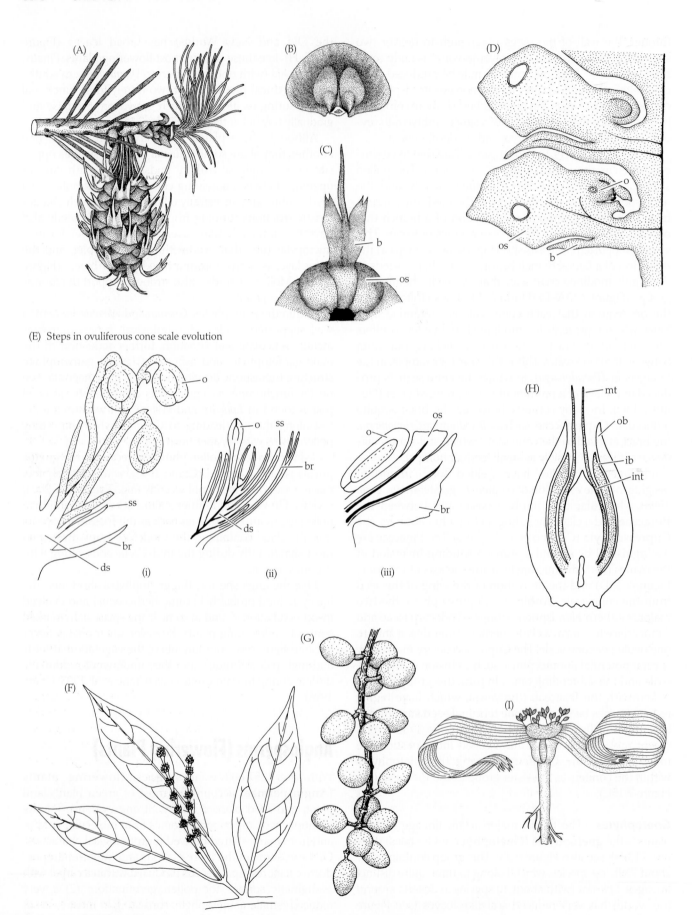

(E) Steps in ovuliferous cone scale evolution

◀ **FIGURE 7.13** Morphology of conifers and gnetophytes. (A) *Pseudotsuga*, showing a branch with a first-year seed cone. (B) Single ovuliferous cone scale of *Pseudotsuga*, showing two ovules on the upper surface. (C) A single bract-scale complex of *Pseudotsuga*, showing the exserted three-pronged bract (b) subtending the ovuliferous scale (os). (D) Longitudinal section through two bract-scale complexes in an ovulate cone of *Pinus strobus*, showing an ovule (o) with micropyle directed toward the cone axis, the ovuliferous scale (os), and the subtending bract (b). (E) Probable evolutionary steps in the origin of the ovuliferous cone scale of conifers: (i) the extinct *Cordaites*, with several ovules (o) and sterile scales (ss) attached to a dwarf shoot (ds) in the axil of a bract (br); (ii) the extinct *Lebachia*, in which the number of ovules is reduced; (iii) extant *Pinus*, with two ovules attached to the upper surface of the ovuliferous scale. (F) Leaves and compound microsporangiate strobili of *Gnetum*. (G) Mature seeds of *Gnetum*. (H) Longitudinal section through a young seed of *Gnetum*, showing the inner integument (int) extended into a micropylar tube (mt), surrounded by inner and outer bracteoles (ib, ob). (I) General habit of the gnetophyte *Welwitschia mirabilis*, showing the short woody stem with two large leaves, axillary position of the multiple strobili, and taproot. (A–D from Stewart 1983; E–H from Scagel et al. 1969; I from Barnes 1998.)

leading to the formation of a typically triploid nutritive tissue called endosperm.

Several derived vegetative characteristics are also noteworthy. Almost all angiosperms produce **vessels** in the xylem tissue, though this feature probably evolved within the group. Vessels differ from tracheids in that water can flow from one vessel element (an individual cell, evolutionarily derived from a tracheid) to the next without traversing a pit membrane (see Figure 4.33). Vessels are extremely efficient with respect to water transport but may be more prone to damage (especially through air embolisms) when subjected to drought stress. Angiosperm phloem differs from that of all other plants in that the sieve tube elements (living but enucleated cells functioning in the transport of carbohydrates) are accompanied by one or more companion cells that are derived from the same mother cell.

Flowers and the Angiosperm Life Cycle

The production of flowers is commonly considered the diagnostic feature of angiosperms, but the term *flower* is actually a bit nebulous. If flowers are short reproductive axes with closely aggregated sporophylls, then gnetophytes, for example, might also be said to have flowers. It is the particular construction and arrangement of the flower parts that sets the angiosperms apart from all other seed plants (see Figure 4.16). Most angiosperm stamens have a stalk portion (filament) and a tip portion (anther) bearing two pairs of microsporangia (pollen sacs). The angiosperm carpel is typically differentiated into a lower portion (ovary) that encloses the ovules and an elongated portion (style) that elevates a surface receptive to pollen (stigma). The angiosperm ovule is unusual in several ways (see Figures 4.41 and 4.42). It generally becomes curved over (anatropous) during development, so that the micropyle lies near the stalk of the ovule (in contrast to the orthotropous condition in other seed plants, in which the micropyle faces away from the stalk). In addition, whereas non-angiosperm seeds have one layer of integument tissue (sometimes differentiated into fleshy and hard layers), angiosperms typically have two distinct integuments (bitegmic ovules).

The angiosperm life cycle is also remarkably derived (see Figure 4.17). The male gametophyte has just three nuclei, or sometimes just two at the time the pollen is shed. A pollen grain that lands on a compatible stigma sends out a pollen tube that delivers the sperm directly to the female gametophyte inside the ovule. In the development of a typical angiosperm female gametophyte, meiosis is followed by the abortion of three products, and the remaining haploid nucleus undergoes a very small series of mitotic divisions (see Figure 4.42). Ultimately the egg is situated toward the micropylar end of the female gametophyte, along with two other cells (synergids) that appear to play a critical role in orienting the pollen tube and delivering the sperm nuclei. There are usually three cells (antipodals) at the opposite end, and two nuclei (polar nuclei) situated in a large cell in the middle. One of the two sperm nuclei fuses with the egg to give rise to the diploid zygote, and the other fuses with the two polar nuclei. This process is called double fertilization. The diploid zygote develops into an embryo, and the triploid product undergoes a series of mitotic divisions to produce endosperm, which serves as the nutritive tissue in the seed.

Time of Origin of Angiosperms

When did the flowering plants originate and radiate? It appears from the fossil record (which includes pollen, leaves, flowers, and fruits) that angiosperms underwent a major radiation starting in the early Cretaceous (Friis et al. 1987; Doyle and Donoghue 1993; Crane et al. 1995). The oldest unequivocal angiosperm fossils are pollen grains from about 135 million years ago. Extraordinarily complete macrofossils from China were first described as being from the late Jurassic (Sun et al. 2002), but are now interpreted as early Cretaceous. Many major angiosperm lineages can be recognized by the mid-Cretaceous (water lilies, Chloranthaceae, Winteraceae, and eudicots were present by 125 million years ago). Other Cretaceous fossils are difficult to assign to modern lineages (Friis et al. 2005). In any case, by the end of the Cretaceous, angiosperms had diversified extensively and were the dominant plants in many terrestrial environments (see Magallón and Sanderson 2001; Bell et al. 2005).

In discussing the age of the angiosperms (or any other group), it is important to distinguish clearly between the origin of the stem lineage—the line leading to the modern group (i.e., when this lineage split from its sister lineage that includes extant organisms)—and the origin of the crown clade—the least inclusive clade that contains all of the extant members. The clade that includes the angiosperm stem lineage has been referred to as the "angiophytes" (Doyle and Donoghue 1993), and more recently as the Pan-Angiospermae (Cantino et al., in press), to distinguish it from crown-clade angiosperms (Angiospermae).

It is possible that the angiophytes are quite ancient, whereas the crown angiosperms originated much more recently, perhaps not long before the radiation seen in the Cretaceous fossil record. That the Pan-Angiospermae may be quite old is suggested by the fact that all of the likely close relatives of angiosperms have fossil records going back at least to the Triassic. We might, therefore, expect to find stem-lineage fossils before the Cretaceous, though perhaps without the full complement of characters found in modern angiosperms. So far, however, putative angiosperm fossils from the Triassic and Jurassic have either turned out not to be related to the angiosperms or are equivocal on the basis of available material.

Estimates based on molecular data are faced with the problem of shifts in the rate of molecular evolution, possibly independently in different lineages. Early molecular clock studies yielded implausibly early ages for crown angiosperms. Progress has been made in "relaxing" the molecular clock assumption, however, and recent estimates place the origin of the angiosperm crown clade between 140 and 190 million years ago (Sanderson and Doyle 2001; Bell et al. 2005), somewhat before the unequivocal appearance of angiosperms in the fossil record.

Relationships of Angiosperms to Other Groups

Botanists have long puzzled over the relationships of angiosperms to other seed plants. This problem is complicated because, in addition to the other extant clades of seed plants (cycads, ginkgos, conifers, and gnetophytes), several extinct groups bear directly on the problem (see Beck 1988; Stewart and Rothwell 1993; Taylor and Taylor 1993). In particular, it has long been hypothesized that flowering plants are most closely related to some group of Mesozoic "seed ferns" (e.g., *Caytonia*, glossopterids), or perhaps to the Bennettitales (also known as "cycadeoids" because of their resemblance to cycads; Figure 7.14A). Bennettitales have been attractive candidates because some of them produced large, flowerlike reproductive structures, with pollen-producing organs surrounding a central stalk bearing naked seeds (Figure 7.14B).

Regarding the five extant lineages, ideas on relationships have shifted over the years (see Soltis et al. 2005). In the early 1900s (e.g., Arber and Parkin 1907), gnetophytes (along with the extinct Bennettitales) were widely believed to be related to angiosperms on the basis of several morphological similarities, such as vessels in the wood, net-

veined leaves in *Gnetum*, and flowerlike reproductive organs. These views had changed by the middle of the twentieth century with the reinterpretation of these characters. For example, vessel elements were interpreted as being derived independently in Gnetales (from tracheids with circular bordered pits) and in angiosperms (from tracheids with scalariform pits). This character, and several others, suggested instead that gnetophytes were related to conifers.

In the mid-1980s, several phylogenetic studies of seed plants were carried out using morphological characters (Crane 1985; Doyle and Donoghue 1986). These analyses concluded that angiosperms formed a clade with Bennettitales and Gnetales—a clade referred to as the "anthophytes" to highlight the flowerlike reproductive structures (Figure 7.15A). A number of independent morphological analyses yielded the same basic result, though in some the gnetophytes were paraphyletic with respect to angiosperms (Taylor and Hickey 1992; Nixon et al. 1994). The characters that appeared to unite the anthophytes varied among analyses, but they were mostly rather obscure and in some cases unknown in fossil groups—for example, lignin chemistry, the layering of cells in the apical meristem, and pollen and megaspore features (Donoghue and Doyle 2000).

In any case, the repeated recovery of the anthophyte clade favored a return to the view that gnetophytes and angiosperms are closely related. In turn, this conclusion influenced the interpretation of morphological evolution. Perhaps most notably, double fertilization (first reported for *Ephedra* in the early and mid-1900s) was interpreted as having evolved in the common ancestor of gnetophytes and angiosperms, with polyploid endosperm evolving later in the angiosperm line (see Friedman and Floyd 2001).

The first molecular phylogenetic studies of the problem yielded a variety of results and were viewed as at least consistent with the anthophyte hypothesis (see Donoghue and Doyle 2000). Starting in the late 1990s, however, a variety of molecular studies (especially those based on mitochondrial genes or on a combination of genes from different genomes) cast serious doubt on the existence of an anthophyte clade (e.g., Bowe et al. 2000; Chaw et al. 2000). These analyses suggested instead that the extant gymnosperm groups form a clade that is sister to the angiosperms, and that gnetophytes are related more directly to conifers (the gnetifer hypothesis; Figure 7.15B) or may even be nested within the conifers as the sister group of the Pinaceae (the gnepine hypothesis; Figure 7.15C). Detailed analyses of the molecular data sets (e.g., Graham and Olmstead 2000; Sanderson et al. 2000; Magallón and Sanderson 2002; Burleigh and Mathews 2004) have revealed several different signals, with some partitions of the data even favoring the placement of gnetophytes as sister to all other extant seed plant groups.

Unfortunately, these questions remain unresolved. It has become clear, however, that there are several potentially separate issues at stake. One important question is whether an anthophyte clade exists or whether, instead, gneto-

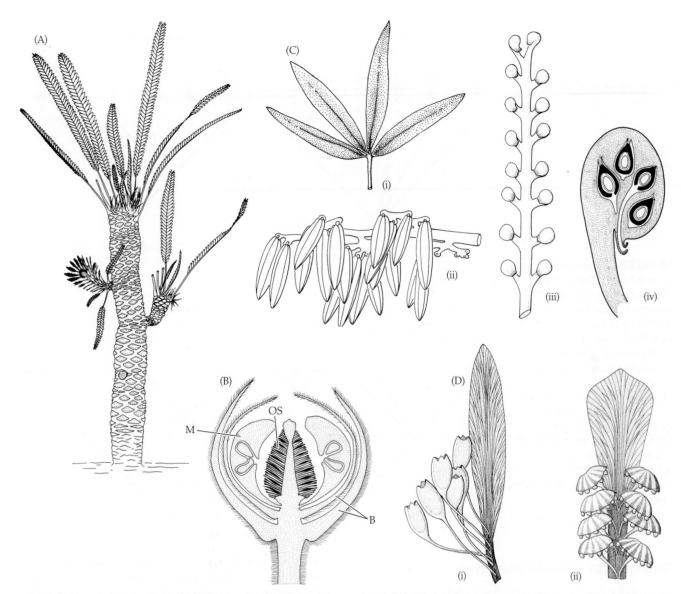

FIGURE 7.14 Reconstructions of Mesozoic fossils that may be closely related to angiosperms. (A, B) Bennettitales: (A) Habit of *Williamsonia sewardiana*, showing cycadlike trunk and compound leaves. (B) Longitudinal section of a flowerlike strobilus of *Williamsoniella*. B, bracts; M, microsporophyll with microsporangia; OS, stalked ovules and sterile scales borne on a central axis. (C) Caytoniales: (i) palmate leaf, *Sagenopteris phillipsi*; (ii) portion of a microsporophyll, *Caytonanthus kochi*; (iii) megasporophyll of *Caytonia nathorsti*, showing two rows of cupules; (iv) longitudinal section of a cupule of *Caytonia thomasi*, showing ovules within. (D) Glossopteridales: (i) ovulate portion of *Denkania indica*, showing six cupulelike structures attached to a leaf; (ii) *Lidettonia mucronata*, showing seeds attached on the lower surfaces of stalked disks borne on a leaf. (A from Taylor and Taylor 1993; B, C: ii–iv, and D from Gifford and Foster 1989; C: i from Stewart 1983.)

phytes are directly related to conifers. The bulk of the evidence now favors the latter view. A second issue is the rooting of the portion of the seed plant tree that includes the extant lineages. One possibility is a basal split into the angiosperms on the one hand and the extant gymnosperms on the other. But other possibilities are difficult to rule out on the basis of presently available data, such as placement of the root in the vicinity of cycads and ginkgos (Figure 7.15D). In any case, it is important to appreciate that "gymnosperms" (including Paleozoic and Mesozoic fossils) are paraphyletic with respect to angiosperms. To avoid con-

fusion, it will probably be best to apply a different name to the hypothesized clade including all extant groups of seed plants without carpels; Cantino et al. (in press) have proposed Acrogymnospermae for this purpose.

There is a distinct possibility that no living group of seed plants is very closely related to angiosperms. Recent results therefore accentuate the importance of fitting fossils into the picture, which will depend on more and better fossils and more attention to the phylogenetic analysis of morphological characters (Donoghue and Doyle 2000; Frohlich and Parker 2000; Doyle 2006).

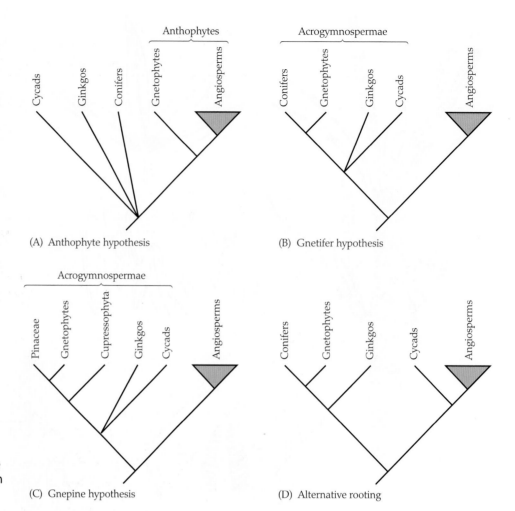

FIGURE 7.15 Alternative hypotheses of relationships among the five major extant lineages of seed plants. (A) According to the anthophyte hypothesis, gnetophytes are most closely related to angiosperms. (B) According to the gnetifer hypothesis, gnetophytes are most closely related to conifers. (C) According to the gnepine hypothesis, gnetophytes are most closely related to Pinaceae within the conifers. (D) An example of an alternative tree that is difficult to reject with current data.

Relationships within Angiosperms

Enormous progress has recently been made in understanding phylogenetic relationships at the base of the angiosperms themselves (Figure 7.16). Until quite recently, the problem of identifying the root of the angiosperms and the relationships among the basal branches looked intractable. Over the last decade, however, several different lines of evidence have converged on the same answer. These new findings are having a major effect on our interpretation of early angiosperm evolution and the factors that account for the enormous success of flowering plants (see Soltis et al. 2005).

Most students of angiosperm evolution have held that the first flowering plants were among the "Magnoliidae" (*sensu* Cronquist 1988; Takhtajan 1997)—a paraphyletic group including magnolias, avocados, water lilies, and black peppers, among others. Even if true, however, this conclusion is not very helpful in deriving an image of the first flowering plants because these plants display an impressive range of morphological forms. Some are woody plants and some are small herbs. Moreover, some, such as magnolias, have large flowers with many flower parts (stamens, carpels) spirally arranged on an elongated axis, while others, such as black peppers, have tiny flowers with few parts arranged in distinct whorls. Some early phylogenetic analyses suggested that the first flowering plants were woody with large flowers, while others implied that they were herbaceous with tiny flowers (see Doyle and Donoghue 1993).

Starting in 1999, a variety of molecular phylogenetic studies concluded that the first split within modern angiosperms was between a lineage that now includes a single species, *Amborella trichopoda* (and possibly also the water lilies, Nymphaeales), and all the rest of the extant angiosperm species (Mathews and Donoghue 1999; Qiu et al. 1999; Soltis et al. 1999; Parkinson et al. 1999; Barkman et al. 2000; Zanis et al. 2002). This conclusion has since been confirmed in all studies that have included a sufficient sample of taxa (e.g., Leebens-Mack et al. 2005). *Amborella trichopoda* is a shrubby plant from the island of New Caledonia with rather small flowers that have a limited number of spirally arranged parts (Endress and Igersheim 2000b). Pollen-producing flowers are borne on some plants and seed-producing flowers on others. The presence of staminodes in the carpellate flowers, however, implies that this species evolved from ancestors with bisexual (perfect) flowers. Unlike those in almost all other angiosperms, the water-conducting cells in the xylem of *Amborella* are tracheids (Feild et al. 2000), supporting the

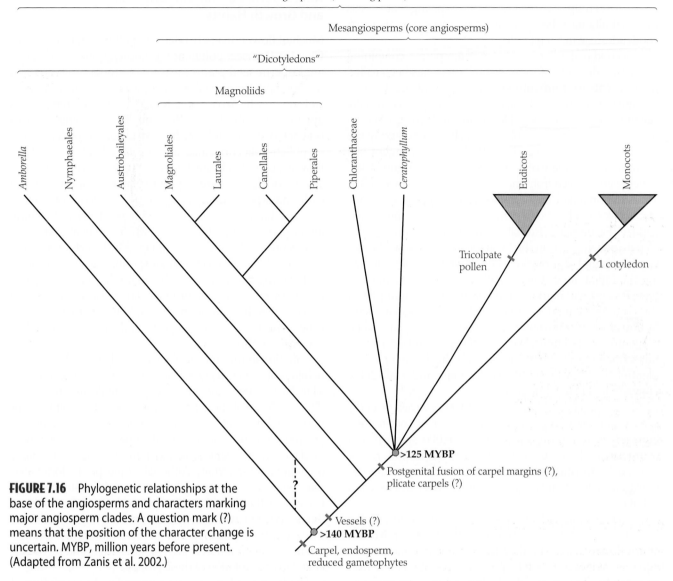

FIGURE 7.16 Phylogenetic relationships at the base of the angiosperms and characters marking major angiosperm clades. A question mark (?) means that the position of the character change is uncertain. MYBP, million years before present. (Adapted from Zanis et al. 2002.)

view that the first angiosperms lacked vessels (see Figure 7.16). *Amborella* female gametophytes are also highly unusual in having three, rather than two, synergid cells with the egg cell at the micropylar end (hence a total of nine nuclei in eight cells, as opposed to eight nuclei in seven cells as in most angiosperms; Friedman 2006).

The water lilies (Nymphaeales) form another very early branch of the angiosperm tree (Friis et al. 2001), as do Austrobaileyales (including Illiciaceae). Interestingly, the female gametophytes in these two lineages have just four cells and form diploid endosperm tissue (Friedman and Williams 2004). Along with *Amborella*, these two lineages subtend a well-supported **core angiosperm** clade that includes all the rest of the flowering plants, which Cantino et al. (in press) have named the **Mesangiospermae**.

Whereas in the basalmost lineages the carpels are typically sealed by a secretion, in members of the core angiosperm clade the carpels are usually sealed by post-

genital fusion of epidermal layers (Endress and Igersheim 2000a). In the three basal lineages, and also in Chloranthaceae (which may be at the base of the core angiosperms; Doyle and Endress 2000), the carpels are ascidiate, meaning that the primordium is U-shaped at first and then grows up like a tube, whereas in almost all Mesangiospermae the carpels are plicate, like a leaf folded down the middle. Although these observations help us to visualize the basal carpel condition in angiosperms, they leave open the controversial issue of whether the carpel was derived from a leaf or instead is a compound structure derived from a reduced branch and its subtending leaf (see Doyle 2006).

Relationships within the core angiosperm clade are still poorly resolved, with the placement of several enigmatic groups still uncertain, especially the Chloranthaceae and *Ceratophyllum* (Qiu et al. 2005). However, several major clades are rather well supported. First, a restricted **magno-liid** clade (**Magnoliidae**) includes the Magnoliales plus

Laurales and the Canellales plus Piperales. Winteraceae, a vessel-less group, is in the Canellales, implying that vessels may actually have been lost in some cases (see also Trochodendraceae in Chapter 9).

A second major lineage of core angiosperms, containing the remainder of the former dicotyledons, has been called the **eudicots** (or **Eudicotyledonae**). This lineage was first recognized in morphological analyses and was initially called the tricolpate clade (Donoghue and Doyle 1989), in reference to the main morphological character marking the group—namely, pollen grains with three colpi, or germinal furrows (and a variety of derivative forms; see Figure 4.48), which were derived from monosulcate forms (Doyle 2005). The additional germinal furrows may help to ensure contact between at least one germination site and the stigma surface (Furness and Rudall 2004). The appearance of tricolpate pollen grains in the fossil record at around 125 million years ago has provided a key calibration point for dating the radiation of flowering plants. Many eudicots also have flowers with parts in fours or fives, or in multiples of these numbers (Judd and Olmstead 2004). This major shift in flower organization within the eudicots appears to be correlated with duplications of genes encoding several transcription factors that play a key role in specifying organ identity and flower symmetry (Kramer and Hall 2005; Howarth and Donoghue 2006).

Altogether there are perhaps 160,000 species of eudicots. This huge group contains a number of species-rich lineages, including legumes (about 16,000 species) and composites (about 20,000 species) as well as buttercups, roses, oaks, mustards, tomatoes, mints, and snapdragons, to list only a few familiar groups among those discussed in detail in Chapter 9.

A third major clade, with some 65,000 species, corresponds to the traditional **monocotyledons** (or **Monocotyledonae**). Almost half of the species of monocots are either orchids (about 20,000 species) or grasses (about 9000 species), but this group also includes palms, bromeliads, bananas, aroids, lilies, irises, and many other familiar and important plants (see Chapter 9).

Many of the features traditionally cited in support of the monocots—such as flower parts in threes and monosulcate pollen—probably pre-dated the origin of this clade (Soltis et al. 2005). Other features may unite the monocots, such as scattered vascular bundles and loss of vascular cambium, parallel leaf venation, and development of the leaf blade from the basal part of the leaf primordium, but this will depend on exactly what their relatives turn out to be and on relationships within the monocot clade. In the end, the presence of one seed leaf, or cotyledon, may still be the morphological character that best distinguishes the monocotyledons (see Figure 4.44).

Note that the view of relationships we have just outlined is at odds with standard classifications in which flowering plants are divided into two major groups: the monocotyledons and the dicotyledons. Instead, the monocots make up a clade that is nested within the paraphyletic "dicots."

Angiosperm Pollination, Dispersal, and Growth Habits

Much of flower diversity relates to pollination biology (see Chapter 4). Insect pollination is known from several non-angiosperm seed plant lineages: the modern cycads and the gnetophytes, as well as the fossil Bennettitales and possibly some Mesozoic "seed ferns." Insect pollination was apparently established by the time the crown angiosperms originated. It was probably first carried out by pollen-eating or pollen-collecting insects, especially beetles and flies; flowers pollinated by nectar-collecting insects evolved later. These conclusions are supported by the morphology of early angiosperm fossils as well as by knowledge of pollination mechanisms in extant members of early-diverging angiosperm lineages (Friis et al. 1987; Thien et al. 2000).

It is unclear how much pollination by insects stimulated the early diversification of angiosperms, but the evolution of flowering plants apparently did not have a major effect on the origin of the major insect lineages, which evolved much earlier. It is abundantly clear, however, that diversification within some angiosperm and insect lineages has been causally linked.

Variation in fruit morphology is largely related to the use of different dispersal agents (see Chapter 4). Cretaceous fossil fruits and seeds are generally quite small, and there is no direct evidence of specialization for dispersal by mammals or birds (see Friis et al. 1987). Adaptations for dispersal by frugivorous and granivorous animals apparently did not appear until later in the Cretaceous, and in most lineages it probably originated in the Tertiary. Although angiosperm-dominated rain forest vegetation may have come into existence in the Cretaceous (Davis et al. 2005), fossil evidence indicates that it did not become widespread until the early Tertiary, at about the time when the radiation of modern birds and mammals occurred. The evolution of large, colorful fruits and seeds was linked to the evolution of these groups.

Finally, it is interesting to contemplate the evolution of growth form within the angiosperms and what effects it might have had on their diversification. Most recent studies position woody plant lineages near the base of the tree. *Amborella* and Austrobaileyales are mostly shrubs or small trees, though they show a tendency toward vinelike growth. Their modern representatives, at least, live in moist forest understory environments, and they show various adaptations to low-light environments. It has been argued that the first angiosperms grew in disturbed understory habitats or in shady streamside settings, and that movement out into more diverse environments might have stimulated diversification within the core angiosperms (Feild et al. 2004). A major exception among the early lineages is the water lily clade, whose members are herbaceous and live in high-light aquatic environments. The extinct *Archaefructus*, whose relationships remain poorly resolved, was also probably an aquatic plant (Sun et al. 2002; Friis et al. 2003).

The herbaceous habit evolved early in angiosperm evolution, and originated several times independently—for example, in Nymphaeales, Chloranthaceae, Piperales, and monocotyledons. In several cases, this development appears to be correlated with movement into aquatic habitats. Larger woody forms have reevolved from herbaceous plants on some occasions, though the evolution of "normal" wood was precluded in the monocots by the loss of the cambium. Within the monocots, large stature has been attained in several other ways—for example, through a specialized thickening mechanism in the apical meristem of palms; enlarged, stiffened leaf bases in the bananas and their relatives; and an anomalous form of cambial activity in the Ruscaceae, Agavaceae, and a few of their relatives (see Chapter 9).

Within the eudicots we see enormous variation in habit, but again, there have been many shifts from woody to herbaceous growth forms, some of these quite early in the evolution of the group. For example, herbaceous poppies (Papaveraceae) and buttercups (Ranunculaceae) may have evolved early, and independently, within one of the first major eudicot branches, the Ranunculales. *Nelumbo*, the water lotus, presents another early example involving a shift to the aquatic environment.

An important trend within eudicots has been the derivation of herbaceous lineages adapted to temperate climate zones from tropical woody plant lineages (Judd et al. 1994). These transitions often appear to be correlated with upward shifts in the rate of diversification (Judd et al. 1994; Magallón and Sanderson 2001), related perhaps in part to the geographic spread of many such lineages (e.g., around the Northern Hemisphere throughout the Tertiary; see Donoghue and Smith 2004). Taken together, all of these factors have had a profound effect on angiosperm diversity.

Summary

The tremendous progress made over the last few decades in establishing phylogenetic relationships is having a major impact on our understanding of green plant evolution. Recent phylogenetic analyses have shown that some traditionally recognized groups are not monophyletic. For example, we appreciate that "plants" (autotrophic eukaryotes) originated independently through several separate endosymbiotic events. Within the green plant clade, traditional "green algae" are paraphyletic with respect to land plants, as are "bryophytes" with respect to vascular plants, "seedless vascular plants" with respect to seed plants, "gymnosperms" with respect to flowering plants, and "dicotyledons" with respect to monocots. As such groups are dismantled, major new clades are being identified, such as the streptophytes (some "green algae" plus embryophytes) and euphyllophytes (some "seedless vascular plants" plus spermatophytes).

A variety of long-standing phylogenetic questions have also recently been answered with considerable confidence. For example, the whisk ferns (Psilotales) are not remnants of the first vascular plants, but instead are part of the monilophyte clade. Moreover, the very base of the angiosperm tree is finally being resolved, with the *Amborella* and water lily branches diverging before a core angiosperm clade that includes the eudicots and the monocots.

Although phylogenetic progress has been rapid, many key questions remain unresolved. For example, we are more uncertain today than we were a decade ago about relationships among the major seed plant lineages. Where do the gnetophytes really fit, and what really are the closest relatives of the flowering plants? And within the core angiosperms, what are the closest relatives of the monocots and the eudicots?

These important questions have been very difficult to resolve, but the successes of the last few decades suggest that the answers will eventually be forthcoming. Experience also implies that analyses integrating evidence from a wide variety of sources—molecular data, morphology, development, and the fossil record—stand the best chance of lasting success.

LITERATURE CITED AND SUGGESTED READINGS

References marked with an asterisk are recent reviews focused on phylogenetic relationships and are especially recommended for additional information and viewpoints. References marked with a dagger symbol are standard texts on comparative plant anatomy, morphology, and paleobotany that can be consulted for background information on specific characters and groups of organisms.

Andersen, R. A. 2004. Biology and systematics of heterokont and haptophyte algae. *Am. J. Bot.* 91: 1508–1522.

Arber, E. A. N. and J. Parkin. 1907. On the origin of angiosperms. *Bot. J. Linnean Soc.* 38: 29–80.

*Baldauf, S. L., D. Bhattacharya, J. Cockrill, P. Hugenholtz, J. Pawlowski and R. G. B. Simpson. 2004. The Tree of Life: An overview. In *Assembling the Tree of Life*, J. Cracraft and M. Donoghue (eds.), 43–75. Oxford University Press, New York.

Barkman, T. J., G. Chenery, J. R. McNeal, J. Lyons-Weiler and C. W. de Pamphilis. 2000. Independent and combined analyses of sequences from all three genomic compartments converge on the root of flowering plant phylogeny. *Proc. Natl. Acad. Sci. USA* 97: 13166–13171.

Barnes, R. S. K (ed.). 1998. *The diversity of living organisms*. Blackwell Publishing, Oxford, UK.

Bateman, R. M. and W. A. DiMichele. 1994. Heterospory: The most iterative key innovation in the evolutionary history of the plant kingdom. *Biol. Rev.* 69: 345–417.

*Bateman, R. M., P. R. Crane, W. A. DiMichele, P. R. Kenrick, N. P. Rowe, T. Speck and W. E. Stein. 1998. Early evolution of land plants: Phylogeny, physiology and ecology of the primary terrestrial radiation. *Annu. Rev. Ecol. Syst.* 29: 263–292.

Beck, C. B. 1981. *Archaeopteris* and its role in vascular plant evolution. In *Paleobotany, paleoecology, and evolution*, K. J. Niklas (ed.), vol. 1, 193–230. Praeger, New York.

Beck, C. B. (ed.). 1988. *Origin and evolution of gymnosperms*. Columbia University Press, New York.

Bell, C. D., D. E. Soltis and P. S. Soltis. 2005. The age of angiosperms: A molecular timescale without a clock. *Evolution* 59: 1245–1258.

Bierhorst, D. W. 1977. The systematic position of *Psilotum* and *Tmesipteris*. *Brittonia* 29: 3–13.

†Bold, H. C. 1967. *Morphology of plants*, 2nd ed. Harper & Row, New York.

†Bold, H. C. and M. J. Wynne. 1985. *Introduction to the algae: Structure and reproduction*. Prentice Hall, Englewood Cliffs, NJ.

Bowe, L. M., G. Coat and C. W. dePamphilis. 2000. Phylogeny of seed plants based on all three genomic compartments: Extant gymnosperms are monophyletic and Gnetales' closest relatives are conifers. *Proc. Natl. Acad. Sci. USA* 97: 4092–4097.

Boyce, C. K. and A. H. Knoll. 2002. Evolution of developmental potential and the multiple independent origins of leaves in Paleozoic vascular plants. *Paleobiology* 28: 70–100.

Burleigh, J. G. and S. Mathews. 2004. Phylogenetic signal in nucleotide data from seed plants: Implications for resolving the seed plant tree of life. *Am. J. Bot.* 91: 1599–1613.

Cantino, P. D., W. S. Judd, P. S. Soltis, D. E. Soltis, R. G. Olmstead, S. W. Graham and M. J. Donoghue. 2007. Towards a phylogenetic nomenclature of *Tracheophyta*. *Taxon* (in press).

Chaw, S. M., C. L. Parkinson, Y. C. Cheng, T. M. Vincent and J. D. Palmer. 2000. Seed plant phylogeny inferred from all three plant genomes: Monophyly of extant gymnosperms and origin of Gnetales from conifers. *Proc. Natl. Acad. Sci. USA* 97: 4086–4091.

Cichan, M. A. and T. N. Taylor. 1990. Evolution of cambium in geological time—A reappraisal. In *The vascular cambium*, M. Iqbal (ed.), 213–228. Wiley, Somerset, UK.

Crane, P. R. 1985. Phylogenetic analysis of seed plants and the origin of angiosperms. *Ann. Missouri Bot. Gard.* 72: 716–793.

Crane, P. R. 1996. The fossil history of the Gnetales. *Int. J. Plant Sci.* 157: S50–S57.

Crane, P. R., E. M. Friis and K. R. Pedersen. 1995. The origin and early diversification of angiosperms. *Nature* 374: 27–33.

Cronquist, A. 1988. *The evolution and classification of flowering plants*. New York Botanical Garden, New York.

Davis, C. C., C. O. Webb, K. J. Wurdack, C. A. Jaramillo and M. J. Donoghue. 2005. Explosive radiation of Malpighiales suggests a mid-Cretaceous origin of rain forests. *Am. Nat.* E36–E65.

*Delwiche, C. F., R. A. Andersen, D. Bhattacharya, B. D. Mishler and R. C. McCourt. 2004. Algal evolution and the early radiation of green plants. In *Assembling the Tree of Life*, J. Cracraft and M. Donoghue (eds.), 121–137. Oxford University Press, New York.

Donoghue, M. J. 2005. Key innovations, convergence, and success: Macroevolutionary lessons from plant phylogeny. *Paleobiology* 31(2): 77–93.

Donoghue, M. J. and J. A. Doyle. 1989. Phylogenetic analysis of angiosperms and the relationships of Hamamelidae. In *Evolution, systematics and fossil history of the Hamamelidae*, vol. 1, P. Crane and S. Blackmore (eds.), 17–45. Clarendon, Oxford.

Donoghue, M. J. and J. A. Doyle. 2000. Demise of the anthophyte hypothesis? *Curr. Biol.* 10: R106–R109.

Donoghue, M. J. and S. A. Smith. 2004. Patterns in the assembly of temperate forests around the Northern Hemisphere. *Philos. Trans. R. Soc. London* B 359: 1633–1644.

*Doyle, J. A. 1998. Phylogeny of the vascular plants. *Annu. Rev. Ecol. Syst.* 29: 567–599.

Doyle, J. A. 2005. Early evolution of angiosperm pollen as inferred from molecular and morphological phylogenetic analyses. *Grana* 44: 227–251.

Doyle, J. A. 2006. Seed ferns and the origin of angiosperms. *J. Torrey Bot. Soc.* 133: 169–209.

Doyle, J. A. and M. J. Donoghue. 1986. Seed plant phylogeny and the origin of angiosperms: An experimental cladistic approach. *Bot. Rev.* 52: 321–431.

Doyle, J. A. and M. J. Donoghue. 1993. Phylogenies and angiosperm diversification. *Paleobiology* 19: 141–167.

Doyle, J. A. and P. K. Endress. 2000. Morphological phylogenetic analysis of basal angiosperms: Comparison and combination with molecular data. *Int. J. Plant Sci.* 161: S121–S153.

Endress, P. K. and A. Igersheim. 2000a. Gynoecium structure and evolution in basal angiosperms. *Int. J. Plant Sci.* 161: S211–S223.

Endress, P. K. and A. Igersheim. 2000b. The reproductive structures of the basal angiosperm *Amborella trichopoda* (Amborellaceae). *Int. J. Plant Sci.* 161: S237–S248.

†Esau, K. 1977. *Anatomy of seed plants*. Wiley, New York.

Feild, T. S., M. A. Zweiniecki, T. Brodribb, T. Jaffre, M. J. Donoghue and N. M. Holbrook. 2000. Structure and function of tracheary elements in *Amborella trichopoda*. *Int. J. Plant Sci.* 161: 705–712.

Feild, T. S., N. C. Arens, J. A. Doyle, T. E. Dawson and M. J. Donoghue. 2004. Dark and disturbed: A new image of early angiosperm ecology. *Paleobiology* 30: 82–107.

Florin, R. 1954. The female reproductive organs of conifers and taxads. *Biol. Rev.* 29: 367–389.

Friedman, W. E. 2006. Embryological evidence for developmental lability during early angiosperm evolution. *Nature* 441: 337–340.

Friedman, W. E. and S. K. Floyd. 2001. Perspective: The origin of flowering plants and their reproductive biology—A tale of two phylogenies. *Evolution* 55: 217–231.

Friedman, W. E. and J. H. Williams. 2004. Developmental evolution of the sexual process in ancient flowering plant lineages. *Plant Cell* 16: S119–S132.

Friis, E. M., W. G. Chaloner and P. R. Crane. 1987. *The origins of angiosperms and their biological consequences*. Cambridge University Press, Cambridge.

Friis, E. M., K. R. Pederson and P. R. Crane. 2001. Fossil evidence of water lilies (Nymphaeales) in the early Cretaceous. *Nature* 410: 356–360.

Friis, E. M., J. A. Doyle, P. K. Endress and Q. Leng. 2003. *Archaefructus*—angiosperm precursor or specialized early angiosperm? *Trends Plant Sci.* 8: 369–373.

Friis, E. M., K. R. Pedersen and P. R. Crane. 2005. When Earth started blooming: Insights from the fossil record. *Curr. Opinion Plant Biol.* 8: 5–12.

Frohlich, M. W. and D. S. Parker. 2000. The mostly male theory of flower evolutionary origins: From genes to fossils. *Syst. Bot.* 25: 155–170.

Furness, C. A. and P. J. Rudall. 2004. Pollen aperture evolution—a crucial factor for eudicot success? *Trends Plant Sci.* 9: 154–158.

†Gifford, E. M. and A. S. Foster. 1989. *Morphology and evolution of vascular plants*, 3rd ed. Freeman, New York.

Graham, L. E. 1993. *Origin of the land plants*. Wiley, New York.

Graham, L. E. and L. W. Wilcox. 2000. The origin of alternation of generations in land plants: A focus on matrotrophy and hexose transport. *Philos. Trans. R. Soc. London* B 355: 757–767.

Graham, S. W. and R. G. Olmstead. 2000. Utility of 17 chloroplast genes for inferring the phylogeny of basal angiosperms. *Am. J. Bot.* 87: 1712–1730.

Hackett, J. D., D. M. Anderson, D. L. Erdner and D. Battacharya. 2004. Dinoflagellates: A remarkable evolutionary experiment. *Am. J. Bot.* 91: 1523–1534.

Heckman, D. S., D. M. Geiser, B. R. Eidell, R. L. Stauffer, N. L. Kardos and S. B. Hedges. 2001. Molecular evidence for the early colonization of land by fungi and plants. *Science* 293: 1129–1133.

Howarth, D. G. and M. J. Donoghue. 2006. Phylogenetic analyses of the "ECE " (CYC/TB1) clade reveal duplications that predate the core eudicots. *Proc. Nat. Acad. Sci. USA* 103: 9101–9106.

Judd, W. S. and R. G. Olmstead. 2004. A survey of tricolpate (eudicot) phylogenetic relationships. *Am. J. Bot.* 91: 1627–1644.

Judd, W. S., R. W. Sanders and M. J. Donoghue. 1994. Angiosperm family pairs—Preliminary phylogenetic analyses. *Harvard Pap. Bot.* 5: 1–51.

Karol, K. G., R. M. McCourt, M. T. Cimino and C. F. Delwiche. 2001. The closest living relatives of land plants. *Science* 294: 2351–2353.

Kato, M. and H. Akiyama. 2005. Interpolation hypothesis for origin of the vegetative sporophyte of land plants. *Taxon* 54: 443–450.

*Keeling, P. J. 2004. Diversity and evolutionary history of plastids and their hosts. *Am. J. Bot.* 91: 1481–1493.

Kenrick, P. and P. R. Crane. 1997a. *The origin and early diversification of land plants: A cladistic study*. Smithsonian Institution Press, Washington, DC.

Kenrick, P. and P. R. Crane. 1997b. The origin and early evolution of plants on land. *Nature* 389: 33–39.

Kim, E., L. W. Wilcox, M. W. Fawley and L. E. Graham. 2006. Phylogenetic position of the green flagellate *Mesostigma* based on α–tubulin and β–tubulin gene sequences. *Int. J. Plant Sci.* 167: 873–883.

Kramer, E. M. and J. C. Hall. 2005. Evolutionary dynamics of genes controlling floral development. *Curr. Opinion Plant Biol.* 8: 13–18.

Leebens-Mack, J. and 8 others. 2005. Identifying the basal angiosperm node in chloroplast genome phylogenies: Sampling one's way out of the Felsenstein zone. *Mol. Biol. Evol.* 22: 1948–1963.

*Lewis, L. A. and R. M. McCourt. 2004. Green algae and the origin of land plants. *Am. J. Bot.* 91: 1535–1556.

Ligrone, R., J. G. Duckett and J. S. Renzaglia. 2000. Conducting tissues and phyletic relationships of bryophytes. *Philos. Trans. R. Soc. London* B 355: 795–813.

Lutzoni, F., M. Pagel and V. Reeb. 2001. Major fungal lineages are derived from lichen symbiotic ancestors. *Nature* 411: 937–940.

Magallón, S. and M. J. Sanderson. 2001. Absolute diversification rates in angiosperm clades. *Evolution* 55: 1762–1780.

Magallón, S. and M. J. Sanderson. 2002. Relationships among seed plants inferred from highly conserved genes: Sorting conflicting phylogenetic signals among ancient lineages. *Am. J. Bot.* 89: 1991–2006.

Martin, W. and 9 others. 2002. Evolutionary analysis of *Arabidopsis*, cyanobacterial, and chloroplast genomes reveals plastid phylogeny and thousands of cyanobacterial genes in the nucleus. *Proc. Natl. Acad. Sci. USA* 99: 12246–12251.

Mathews, S. and M. J. Donoghue. 1999. The root of angiosperm phylogeny inferred from duplicate phytochrome genes. *Science* 286: 947–950.

Mattox, K. R. and K. D. Stewart. 1984. The classification of green algae: A concept based on comparative cytology. In *The systematics of the green algae*, D. Irvine and D. Johns (eds.), 29–72. Academic Press, London.

Mishler, B. D. and S. P. Churchill. 1985. Transition to a land flora: Phylogenetic relationships of the green algae and bryophytes. *Cladistics* 1: 305–328.

†Niklas, K. J. 1997. *The evolutionary biology of plants*. University of Chicago Press, Chicago.

Nixon, K. C., W. L. Crepet, D. Stevenson and E. M. Friis. 1994. A reevaluation of seed plant phylogeny. *Ann. Missouri Bot. Gard.* 81: 484–533.

Norton, T. A., M. Melkonian and R. A. Andersen. 1996. Algal biodiversity. *Phycologia* 35: 308–326.

O'Hara, R. J. 1992. Telling the tree: Narrative representation and the study of evolutionary history. *Biol. Philos.* 7: 135–160.

Palmer, J. D. 2003. The symbiotic birth and spread of plastids: How many times and whodunit? *J. Phycol.* 39: 4–11.

Parkinson, C. L., K. L. Adams and J. D. Palmer. 1999. Multigene analyses identify the three earliest lineages of extant flowering plants. *Curr. Biol.* 9: 1485–1488.

Phillips, T. L. and W. A. DiMichele. 1992. Comparative ecology and life-history biology of arborescent lycopsids in Late Carboniferous swamps of Euramerica. *Ann. Missouri Bot. Gard.* 79: 560–588.

Pickett-Heaps, J. D. 1979. Electron microscopy and the phylogeny of green algae and land plants. *Am. Zool.* 19: 545–554.

Pombert, J. F., C. Otis, C. Lemieux and M. Turmel. 2005. The chloroplast genome sequence of the green alga *Pseudendoclonium akinetum* (Ulvophyceae) reveals unusual structural features and new insights into the branching order of chlorophyte lineages. *Mol. Biol. Evol.* 22: 1903–1918.

Pryer, K. M., H. Schneider, A. R. Smith, R. Cranfill, P. G. Wolf, J. S. Hunt and S. D. Sipes. 2001. Horsetails and ferns are a monophyletic group and the closest living relatives to seed plants. *Nature* 409: 618–622.

*Pryer, K. M., H. Schneider and S. Magallón. 2004a. The radiation of vascular plants. In *Assembling the Tree of Life*, J. Cracraft and M. Donoghue (eds.), 138–153. Oxford University Press, New York.

Pryer, K. M., E. Schuettpelz, P. G. Wolf, H. Schneider, A. R. Smith and R. Cranfill. 2004b. Phylogeny and evolution of ferns (Monilophytes) with a focus on the early leptosporangiate divergences. *Am. J. Bot.* 91: 1582–1598.

Qiu, Y.-L. and 9 others. 1999. The earliest angiosperms: Evidence from mitochondrial, plastid and nuclear genomes. *Nature* 402: 404–407.

Qiu, Y.-L. and 19 others. 2005. Phylogenetic analyses of basal angiosperms based on nine plastid, mitochondrial, and nuclear genes. *Int. J. Plant Sci.* 166: 815–842.

Qiu, Y.-L. and 20 others. 2006. The deepest divergences in land plants inferred from phylogenomic evidence. *Proc. Natl. Acad. Sci. USA* 103: 15511–15516.

Rai, H. S., H. E. O'Brien, P. A. Reeves, R. G. Olmstead and S. W. Graham. 2003. Inference of higher-order relationships in the cycads from a large chloroplast data set. *Mol. Phyl. Evol.* 29: 350–359.

Raubeson, L. A. and R. K. Jansen. 1992. Chloroplast DNA evidence on the ancient evolutionary split in vascular plants. *Science* 255: 1697–1699.

†Raven, P. H., R. F. Evert and S. E. Eichhorn. 2005. *Biology of plants*, 7th ed. Worth, New York.

Remy, W., P. G. Gensel and H. Hass. 1993. The gametophyte generation of some early Devonian land plants. *Int. J. Plant Sci.* 154: 35–58.

Renzaglia, K. S., R. J. Duff, D. L. Nickrent and D. J. Garbary. 2000. Vegetative and reproductive innovations of early land plants: Implications for a unified phylogeny. *Philos. Trans. R. Soc. London* B 355: 769–793.

Rothwell, G. W. 1999. Fossils and ferns in the resolution of land plant phylogeny. *Bot. Rev.* 65: 188–218.

Rothwell, G. W. and K. C. Nixon. 2006. How does the inclusion of fossil data change our conclusions about the phylogenetic history of Euphyllophytes? *Int. J. Plant Sci.* 167: 737–749.

Rydin, C, K. R. Pedersen and E. M. Friis. 2004. On the evolutionary history of *Ephedra*: Cretaceous fossils and extant molecules. *Proc. Natl. Acad. Sci. USA* 101: 16571–16576.

Sanderson, M. J. 2003. Molecular data from 27 proteins do not support a Precambrian origin of land plants. *Am. J. Bot.* 90: 954–956.

Sanderson, M. J. and J. A. Doyle. 2001. Sources of error and confidence intervals in estimating the age of angiosperms from *rcbL* and 18S rDNA data. *Am. J. Bot.* 88: 1499–1516.

Sanderson, M. J., M. F. Wojciechowski, J.-M. Hu, T. Sher Khan and S. G. Brady. 2000. Error, bias, and long branch attraction in data for two chloroplast photosystem genes in seed plants. *Mol. Biol. Evol.* 17: 782–797.

Saunders, G. W. and M. H. Hommersand. 2004. Assessing red algal supraordinal diversity and taxonomy in the context of contemporary systematic data. *Am. J. Bot.* 91: 1494–1507.

†Scagel, R. F., R. J. Bandoni, G. E. Rouse, W. B. Schofield, J. R. Stein and T. M. C. Taylor. 1969. *Plant diversity: An evolutionary approach*. Wadsworth, Belmont, CA.

Serbet, R. and G. W. Rothwell. 1992. Characterizing the most primitive seed ferns. I. A reconstruction of *Elkinsia polymorpha*. *Int. J. Plant Sci.* 153: 602–621.

†Shaw, A. J. and B. Goffinet (eds.). 2000. *The biology of bryophytes*. Cambridge University Press, New York.

*Shaw, J. and K. Renzaglia. 2004. Phylogeny and diversification of bryophytes. *Am. J Bot.* 91: 1557–1581.

Smith, A. R., K. M. Pryer, E. Schuettpelz, P. Korall, H. Schneider and P. G. Wolf. 2006. A classification for extant ferns. *Taxon* 55: 705–731.

Soltis, P. S., D. E. Soltis and M. W. Chase. 1999. Angiosperm phylogeny inferred from multiple genes as a tool for comparative biology. *Nature* 402: 402–404.

Soltis, P. S., D. E. Soltis, V. Savolainen, P. R. Crane and T. G. Barraclough. 2002. Rate heterogeneity among lineages of tracheophytes: Integration of molecular and fossil data and evidence for molecular living fossils. *Proc. Natl. Acad. Sci. USA* 99: 4430–4435.

*Soltis, P. S., D. E. Soltis, M. W. Chase, P. K. Endress and P. R. Crane. 2004. The diversification of flowering plants. In *Assembling the Tree of Life*, J. Cracraft and M. Donoghue (eds.), 154–167. Oxford University Press, New York.

*Soltis, D. E., P. S. Soltis, P. K. Endress and M. W. Chase. 2005. *Phylogeny and evolution of angiosperms*. Sinauer Associates, Sunderland, MA.

Stefanovic, S., M. Jager, J. Deutsch, J. Broutin and M. Masselot. 1998. Phylogenetic relationships of conifers inferred from partial 28S rRNA gene sequences. *Am. J. Bot.* 85: 688–697.

Stewart, W. N. 1983. *Paleobotany and the evolution of plants*. Cambridge University Press, Cambridge.

†Stewart, W. N. and G. W. Rothwell. 1993. *Paleobotany and the evolution of plants*, 2nd ed. Cambridge University Press, New York.

Sun, G., Q. Ji, D. L. Dilcher, S. Zheng, K. C. Nixon and X. Wang. 2002. Archaefructaceae, a new basal angiosperm family. *Science* 296: 899–904.

Takhtajan, A. L. 1997. *Diversity and classification of flowering plants*. Columbia University Press, New York.

Taylor, D. W. and L. J. Hickey. 1992. Phylogenetic evidence for the herbaceous origin of angiosperms. *Plant Syst. Evol.* 180: 137–156.

†Taylor, T. N. and E. L. Taylor. 1993. *The biology and evolution of fossil plants*. Prentice Hall, Englewood Cliffs, NJ.

Taylor, T. N., H. Kerp and H. Hass. 2005. Life history biology of early land plants: Deciphering the gametophyte stage. *Proc. Natl. Acad. Sci. USA* 102: 5892–5897.

Thien, L. B., H. Azuma and S. Kawano. 2000. New perspectives on the pollination biology of basal angiosperms. *Int. J. Plant Sci.* 161: S225–S235.

Turmel, M., C. Otis and C. Lemieux. 2002. The complete mitochondrial DNA sequence of *Mesostigma viride* identifies this green alga as the earliest green plant divergence and predicts a highly compact mitochondrial genome in the ancestor of all green plants. *Mol. Biol. Evol.* 19: 24–38.

Waters, E. R. 2003. Molecular adaptation and the origin of land plants. *Mol. Phyl. Evol.* 29: 456–463.

Wellman, C. H., P. L. Osterloff and U. Mohiuddin. 2003. Fragments of the earliest land plants. *Nature* 425: 282–285.

Wolf, P. G. and 9 others. 2005. The first complete chloroplast genome sequence of a lycophyte, *Huperzia lucidula* (Lycopodiaceae). *Gene* 350: 117–128.

Won, H. and S. S. Renner. 2006. Dating dispersal and radiation in the gymnosperm *Gnetum* (Gnetales): Clock calibrations when outgroup relationships are uncertain. *Syst. Biol.* 55: 610–622.

Yoon, H. S., J. Hackett and D. Bhattacharya. 2002. A single origin of the peridinin- and fucoxanthin-containing plastids in dinoflagellates through tertiary endosymbiosis. *Proc. Natl. Acad. Sci. USA* 99: 11724–11729.

Zanis, M. J., D. E. Soltis, P. S. Soltis, S. Mathews and M. J. Donoghue. 2002. The root of the angiosperms revisited. *Proc. Natl. Acad. Sci. USA* 99: 6848–6853.

Zimmermann, W. 1965. *Die Telomtheorie*. Fischer, Stuttgart, Germany.

8

Lycophytes, Ferns, and Gymnosperms

 This chapter and Chapter 9 survey the diversity of living tracheophytes. The term *tracheo* refers to the presence of tracheids—cells specialized for transport of liquids—and the Greek root *phyte* means plant. Tracheophytes form a well-supported monophyletic group of generally large plants with branched sporophyte axes and well-developed tissues (with tracheids in the xylem and sieve cells in the phloem) for the transport of water and carbohydrates within the plant. As described in Chapter 7, the tracheophytes form a major clade within the embryophytes (land plants), nested within the paraphyletic "bryophytes" (see Figure 7.6). This implies the derivation of tracheophyte characteristics from those found in the bryophyte lineages, in which the small, unbranched sporophyte is nutritionally dependent on the dominant gametophyte phase of the life cycle.

There are two major lineages within the tracheophytes: the lycophytes and the euphyllophytes (see Figure 7.8). The euphyllophytes in turn comprise two major lineages of living plants: the monilophytes (the ferns, including Psilotaceae and Equisitaceae) and the spermatophytes, or seed plants. Finally, within the seed plants there are two major extant lineages: gymnosperms (conifers and others) and the angiosperms (flowering plants) (Table 8.1).

Gymnosperm means "naked seed," referring to the fact that the seeds are not enclosed in a protective structure, although they may sometimes be enclosed at maturity by fused cone scales or bracts, as in juniper "berries." In the angiosperms, seeds are enclosed in carpels (*angio* means "vessel," referring to the carpel).

Key to Major Groups of Tracheophytes

1. Sperm biflagellate; .... **Lycophytes**

1. Sperm multiflagellate (or sperm without flagella) .... 2 **(Euphyllophytes)**

2. Dispersal by spores; seeds absent .... 3 **(Monilophytes)**

2. Dispersal by seeds .... 6 **(Seed plants)**

3. Leaves generally less than 2 cm in length, with an unbranched vein or veinless, not divided into leaflets 4

3. Leaves generally more than 2 cm in length, with a branched vein, divided into leaflets or not 5

4. Leaves whorled, fused at base; branches, if present, whorled; internodes with conspicuous vertical ridges; sporangia clustered on peltate sporangiophores and aggregated in strobili; roots present and irregularly branching .... **Equisetaceae**

4. Leaves spiral, not fused at base; stem without vertical ridges; sporangia sometimes associated with forked or lobed leaves; roots lacking .... **Psilotaceae**

5. Leaves not coiled and unfolding lengthwise when enmerging; sporangia walls more than one cell in thickness, without an annulus, borne on or embedded in a special portion of the leaf (the **sporophore**) .... **Ophioglossaceae**

5. Leaves coiled and unfolding lengthwise when emerging; sporangia walls one-cell thick, mostly with an annulus, mostly clustered in sori on the abaxial leaf surface, less often scattered on the abaxial leaf surface or in sporocarps .... **Leptosporangiate ferns**

6. Carpels and endosperm present .... **Angiosperms**

6. Carpels and endosperm absent .... 7 **(Gymnosperms)**

7. Vessels present; leaves opposite, joined at base .... **Gnetales**

7. Vessels absent; leaves usually spiral, rarely opposite or joined at base 8

8. Sperm not flagellate, transported to ovule by pollen tube; leaves simple, small, and scale-like to larger and linear, with resin canals in most species .... **Conifers**

8. Sperm motile, swimming by means of flagella; leaves either pinnately compound or simple, large, and as broad or broader than long, with resin, if present, not in canals 9

9. Leaves pinnately or twice pinnately compound, persistent; stem mostly short and unbranched or dichotomously branching, sometimes subterranean; ovules borne on the margin of often peltate megasporophylls that are either loosely clustered at the shoot apex or grouped in strobili; outer layer of seeds mostly fleshy but not unpleasant-smelling; microsporangia clustered in sori on the abaxial surface of microsporophylls; mucilage canals present .... **Cycads**

9. Leaves simple, fan-shaped, deciduous; large, freely branching trees; ovules 2 at the end of a long stalk, often only one maturing; outer layer of seeds fleshy, unpleasant-smelling at maturity; microsporangia paired at the end of slender stalk; mucilage canals absent .... **Ginkgoaceae**

Phylogenetic relationships among the seed plant lineages are not yet well resolved, although studies, including fossils, do indicate that seed plant lineages with naked seeds are paraphyletic with respect to angiosperms (see Figure 7.12). One possibility is that Gnetales are more closely related to angiosperms than they are to the other extant gymnosperms (the "anthophyte hypothesis"). However, recent molecular studies support the view that the extant gymnosperm lineages form a clade that is sister to the angiosperms.

This chapter covers the major extant lineages of tracheophytes except for the angiosperms, which are covered in detail in Chapter 9. All together there are about 12,000 species of lycophytes, monilophytes, and gymnosperms, or about 5% of the total number of flowering plant species. In this chapter and the next, each family treatment includes a description in which useful identifying characters are indicated in italic print and synapomorphies (which may also be useful for identification) in boldface; a brief summary of distribution and ecology; the estimated number of genera

TABLE 8.1   Families of tracheophytes as classified in this book.[a]

LYCOPHYTES	Salviniales	SEED PLANTS
Lycopodiales	**Marsileaceae** (p. 198)	*Gymnosperms*
Lycopodiaceae (p. 188)	*Salviniaceae* (p. 198)	Cycadales (Cycads)
Selaginellaceae (p. 189)	Cyatheales	**Cycadaceae** (p. 207)
Isoetaceae (p. 187)	**Cyatheaceae** (p. 199)	**Zamiaceae** (p. 208)
	Cibotiaceae (p. 199)	Ginkgoales
MONILIPHYTES (Ferns)	*Culcitaceae* (p. 199)	**Ginkgoaceae** (p. 208)
Psilotales	*Dicksoniaceae* (p. 199)	Coniferales (Conifers)
Psilotaceae (p. 191)	*Loxomataceae* (p. 199)	**Pinaceae** (p. 211)
Ophioglossales	*Metaxyaceae* (p. 199)	**Cupressaceae** (including
Ophioglossaceae (p. 193)	*Plagiogyriaceae* (p. 199)	"Taxodiaceae") (p. 215)
Marattiales	*Thyrsopteridaceae* (p. 199)	*Sciadopityaceae* (p. 217)
Marattiaceae (p. 190)	Polypodiales	**Podocarpaceae** (p. 217)
Equisetales	**Dennstaedtiaceae** (p. 200)	**Araucariaceae** (p. 218)
Equisetaceae (p. 193)	**Pteridaceae** (p. 201)	**Taxaceae** (p. 219)
	Aspleniaceae (p. 201)	*Cephalotaxaceae* (p. 220)
Leptosporangiate ferns	**Thelypteridaceae** (p. 202)	Gnetales
Osmundales	**"Woodsiaceae"** (p. 203)	**Ephedraceae** (p. 221)
Osmundaceae (p. 197)	**Blechnaceae** (p. 203)	*Gnetaceae* (p. 221)
Hymenophyllales	**Onocleaceae** (p. 204)	*Welwitschiaceae* (p. 221)
Hymenophyllaceae (p. 197)	**Dryopteridaceae** (p. 204)	
Gleicheniales	**Polypodiaceae** (p. 205)	*Angiosperms: See Table 9.1*
Gleicheniaceae (p. 197)	*Davalliaceae* (p. 200)	
Dipteridaceae (p. 197)	*Lindsaeaceae* (p. 200)	
Matoniaceae (p. 197)	*Lomariopsidaceae* (p. 200)	
Schizaeales	*Oleandraceae* (p. 200)	
Schizaeaceae (p. 197)	*Saccolomataceae* (p. 200)	
	Tectariaceae (p. 200)	

[a]Families receiving full coverage in the text are indicated in **boldface**, while those only briefly characterized are in *italics*. Page numbers (in parentheses) indicate discussion of the family in this chapter.

and species (including a listing of major genera); a list of major economic plants and products; and a discussion. The family discussion includes information regarding characters supporting the group's monophyly, a brief overview of phylogenetic relationships within the family, information regarding pollination biology and seed dispersal (where relevant), and notes on other matters of biological interest. Finally, each family treatment includes a list of references that are useful sources of additional information.

References: Bowe et al. 2000; Burleigh and Mathews 2004; Chaw et al. 2000; Donoghue 1994; Doyle et al. 1994; Kenrick 2000; Kenrick and Crane 1997; Mishler et al. 1994; Nickrent et al. 2000; Nixon et al. 1994; Pryer et al. 2004; Renzaglia et al. 2000; Renzaglia et al. 2001; Rothwell 1999; Rydin et al. 2002; Stefanovic et al. 1998; Stewart and Rothwell 1993.

LYCOPHYTES

Arising at least 400 MYA, lycophytes were largest in size in the Carboniferous (345–290 MYA), when arborescent mem-

bers of this group such as *Lepidodendron*, some of them over 40 m tall and 2 m in diameter at the base, dominated forests. The remains of these ancient trees are the chief components of commercially important coal deposits of Europe and North America. Almost half of Carboniferous plant fossils are lycophytes.

The modern lycophytes consist of three families. One (Lycopodiaceae) is homosporous, and the other two (Selaginellaceae and Isoetaceae) are heterosporous. Gametophytes of Lycopodiaceae are bisexual, mycorrhizal, and usually subterranean. They may be partly green and on or near the surface of the ground or initially mycorrhizal and later photosynthetic. Gametophytes of Selaginellaceae are unisexual. The megagametophytes partly protrude from the megaspore wall, and microgametophytes develop wholly within the microspore, the wall rupturing to release the sperm. Isoetaceae, which are not treated in detail in this book, are sister to the Selaginellaceae. Isoetaceae are terrestrial or aquatic plants with erect, short cormlike stems and usually long leaves (see Figure 7.9J). Sporangia embedded in the adaxial face of the leaf bases are distinctive. The family has a nearly cosmopolitan distribution; the one genus

---

Key to Families of Lycophytes

1. Homosporous; leaves nonligulate .... **Lycopodiaceae**
1. Heterosporous; leaves ligulate, with a small projection on the adaxial surface 2
2. Sporangia borne singly in the axils of unmodified or modified leaves; leaves usually under 2 cm long .... **Selaginellaceae**
2. Sporangia at least initially embedded in adaxial face of leaf base; leaves 2–100 cm long .... Isoetaceae

---

(*Isoetes*) has about 150 species. Ironically, the small plants of *Isoetes* are the closest living relatives of the clade to which ancient lycophyte trees belonged.

Lycopodiales

This order contains three families, about five genera, and some 1280 species.

Lycopodiaceae Mirbel
(Club Moss Family)

Terrestrial or epiphytic plants usually about 5–20 cm tall (exceptionally up to about 2 m in pendant epiphytes), with *dichotomously branching stems*. Roots dichotomously branching. Leaves simple (microphylls), 0.2–2 cm long, with 1 unbranched vein, often densely covering the stem; linear and more or less spreading away from the stem or scale-like and appressed to the stem; spiral or opposite. Sporangia ± kidney-shaped, *opening by a transverse slit that divides the sporangium in two; solitary in leaf axils or borne on leaf bases*, sporophylls unmodified or modified and sometimes clustered into strobili. Homosporous; spores subglobose to tetrahedral, with a 3-branched scar. Gemmae (small structures of various sizes and shapes that detach from the plant for vegetative reproduction) present in some species of *Huperzia* (Figure 8.1).

Distribution and ecology: Cosmopolitan, rare in arid areas; most diverse in tropical montane and alpine habitats.

Genera/species: ca. 3/380. ***Major genera:*** *Huperzia* (300 spp.), *Lycopodiella* (40), and *Lycopodium* (40).

Economic plants and products: The family is not significant economically. Oily, highly flammable compounds in the spore wall ignite rapidly into a flash of light and were used by magicians and sorcerers in the Middle Ages, as a flash early in photography, and in the first (experimental) photocopying machines. The spores have been used as

(A)

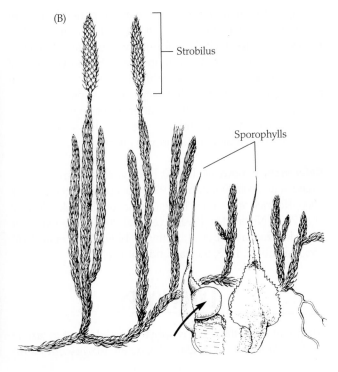

(B)

— Strobilus

Sporophylls

FIGURE 8.1 Lycopodiaceae. (A) *Lycopodium annotinum*: habit; (B) *L. clavatum*: habit (shown about two-thirds actual size) and sporophylls (× 7); the arrow points to a sporangium. (B from Øllgaard 1990.)

industrial lubricants and formerly were used to prevent rubber cohesion in condoms and surgical gloves.

Discussion: Taxonomic problems exist at generic and specific levels in the Lycopodiaceae. Traditionally treated as including just one large genus, *Lycopodium*, the family is here considered to contain three genera. Some of these genera contain monophyletic sections that some workers have recognized as genera. DNA sequences from *rbcL* divide the family into a *Huperzia* clade and a *Lycopodiella-Lycopodium* clade. A peculiar species, sometimes placed in a distinct genus (*Phylloglossum*), consists of plants only about 5 cm tall and highly reduced for arid environments of Australia, New Zealand, and Tasmania. Gametophyte morphology and *rbcL* sequences support inclusion of this species in *Huperzia*. Species of *Huperzia* are mostly tropical epiphytes with sporangia borne in leaf axils and not collected into strobili. *Huperzia* is divided into a neotropical and a paleotropical clade, each of which is thought to have diversified largely after continental drift separated South America and Africa about 80 MYA. Lycopodiaceae are an old family, extending back to about 380 MYA, but many of the species are relatively young. The formation of forests of broadleaved species shaded the forest floor and favored the evolution of epiphytism in the ancestors of *Huperzia*. The terrestrial habit has apparently evolved at least two times in neotropical mountains; one of these reversions is thought to have been in response to the formation of the Andes about 15 MYA. Species delimitation in *Huperzia* is poorly understood, at least in part due to frequent interspecific hybridization.

The *Lycopodiella-Lycopodium* clade is primarily temperate in distribution, and its sporangia are usually borne in conspicuous strobili. Some clades within these two genera have wider geographic distributions than the major neotropical and paleotropical clades in *Huperzia*, and some species, such as *Lycopodium clavatum*, are more or less cosmopolitan.

Members of this family may have somatic chromosome numbers up to about 275. These high numbers may be the result of repeated episodes of polyploidy, or these plants may simply be diploids with numerous chromosomes.

References: DiMichele and Skog 1992; Kenrick and Crane 1997; Lellinger 1985; Øllgaard 1990, 1992; Soltis and Soltis 1988; Tryon and Tryon 1982; Wagner and Beitel 1992, 1993; Wikström 1999; Wikström and Kenrick 1997; Wikström and Kenrick 2000; Wikström and Kenrick 2001.

Selaginellaceae Willk.

(Spike Moss Family)

Mostly terrestrial, herbaceous, perennial plants under 2 cm tall. Roots dichotomously branching; rhizophores usually produced from the stem, dichotomously branching. Stems erect or creeping. *Leaves about 0.5–1 cm long, spirally arranged and often 4-ranked on the secondary and ultimate branches*; ligulate; often dimorphic, those on upper side of stem small-

er than those on lower side; with a single, unbranched vein; with a ligule. Heterosporous; **sporangia borne in or near the axils of well-differentiated sporophylls, usually on 4-sided** (rarely cylindrical) **strobili**; strobili usually terminating branches, with either or both megasporangia and microsporangia; megasporangia usually with 4 megaspores 200–600 μm in diameter, with distinct ridges and conspicuously patterned with variously shaped projections; microsporangia with more than 100 microspores 20–60 μm in diameter (Figure 8.2; see also Figure 7.9F–I).

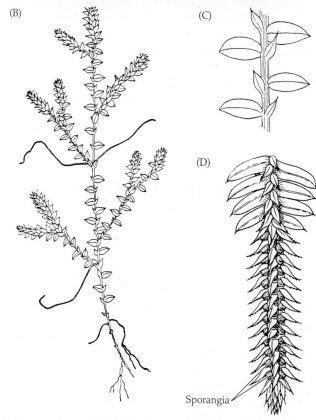

FIGURE 8.2 Selaginellaceae. (A) *Selaginella arenicola*: habit; (B, C) *S. apoda*: (B) habit (× 1.5); (C) sterile portion of plant (× 12). (D) *S. myosurus*: strobilus and vegetative branch bearing it (× 6). (B, C from Billington 1952; D from Alston 1932.)

Distribution and ecology: Mainly tropical, with a few species extending into arctic regions of both hemispheres; occupying a wide range of habitats.

Genera/species: 1 (*Selaginella*)/750.

Economic plants and products: The family is not significant economically, although a few species are ornamentals.

Discussion: Fossils of this family are known from tropical wetlands almost 350 MYA. Early in their history, members of Selaginellaceae evolved leaf dimorphism, with smaller leaves on the upper surface of the stem. Leaf dimorphism, which has been hypothesized to be an adaptation to poor light, may have been important for these early plants of Selaginellaceae and for modern forms as well, which mostly live on forest floors.

The majority of the species of the family occurs in tropical rainforests, but many species are capable of surviving long periods of drought because their small leaves are covered by thick cuticle and their branches curl into a ball. These plants can revive rapidly from apparent death by uncurling their branches, and they are therefore commonly called resurrection plants. The resurrection adaptation has evolved at least three times in the genus.

Although the small leaves and habit make many species of Lycopodiaceae and Selaginellaceae resemble one another, heterospory and ligulate leaves are synapomorphies linking Selaginellaceae and Isoetaceae.

References: Jermy 1990a,b; Kenrick and Crane 1997; Korall and Kenrick 2002 (their Figure 3 shows important morphological features); Lellinger 1985; Manhart 1995; Taylor et al. 1993; Tryon and Tryon 1982; Valdespino 1993; Webster 1992.

MONILOPHYTES (FERNS)

The monilophytes, or ferns, span a broad range of morphological diversity. The great majority of the members of this clade have subterranean stems bearing clumps of leaves that are compound or deeply lobed, many-veined, and comparatively large (up to 7 m or more in length). Such fern leaves are sometimes referred to as fronds and their petioles as stipes.

Ferns also include plants with leaves that are simple or resemble those of clovers; tree ferns with stems up to 20 m tall; and aquatics with filiform leaves. Two groups long referred to as "fern allies"—Psilotaceae and Equisetaceae—have, with few exceptions, rather small leaves with a single or no vein and superficially do not resemble ferns. For over 30 years, however, some workers have suggested that Psilotaceae and Equisetaceae are highly modified ferns, and recent DNA evidence clearly supports this view (see Chapter 7). Phylogenetic analyses indicate that ferns, as they are

defined in this text, are monophyletic. Ferns include about 300 genera and 9,000 species ranging in size from a few cm to 20 m tall.

There are two kinds of sporangia within ferns: **eusporangia** are plesiomorphic and the sporangium wall at dehiscence has two or more cell layers; in **leptosporangia**, the sporangium wall is composed of just one cell layer. The eusporangiate condition characterizes Psilotaceae, Ophioglossaceae, Equisetaceae, and Marattiaceae (as well as lycophytes). The Marattiaceae contain 4–7 genera and about 300 species, most found on the shaded floor of wet tropical forests; this family is not treated in detail in this book.

The great majority of ferns have leptosporangiate sporangia. In addition to sporangium wall thickness, the evolution of leptosporangiate ferns is associated with four other unique structural characteristics: the **annulus** (described later); a sporangial stalk with 4–6 cells in cross-section; a vertical first zygotic division; and primary xylem with scalariform bordered pits. Rearrangements of the chloroplast genome and DNA sequence data also mark the origin of leptosporangiate ferns.

DNA sequences from three chloroplast genes (*rbcL*, *atpB*, and *rps4*) and the nuclear 18S rDNA strongly link Ophioglossaceae and Psilotaceae (Figure 8.3). Relationships of Ophioglossaceae had long been a puzzle. Based on morphology, they appeared to be distantly related to other tracheophytes, or perhaps they had affinities with a long-extinct group of plants called progymnosperms. Relationships of Marattiaceae are also confused. The same DNA evidence that resolved a close relationship of Ophioglossaceae and Psilotaceae may place Equisetaceae and Marattiaceae in a clade that is sister to the leptosporangiate ferns (see Chapter 7). Because support for this clade is weak, we have not included it in Figure 8.3.

Spore shape, size, symmetry, the shape of a prominent scar—whether straight or three-branched—and the nature of the surface and layers of the spore wall have been taxonomically useful at all levels within the monilophytes. Spores of these plants are very small and are produced in great numbers; one estimate is that an individual tree fern in the genus *Cyathea* generates 1,250,000,000,000 spores in its lifetime. Spore dispersal is facilitated by adaptations such as elaters in *Equisetum* and the annulus in most ferns (see below).

Most ferns are homosporous, producing only one kind of spore. Their gametophytes are at least potentially bisexual, producing both gametes—eggs (in archegonia) and sperm, also called antherozoids (in antheridia; Figure 8.4). Hence individual fern gametophytes can produce zygotes and by themselves start a new population. Consequently, long-distance dispersal is more common in ferns than in flowering plants. Mating between gametophytes is promoted when gametophytes producing archegonia release compounds called antheridiogens that stimulate nearby gametophytes to produce antheridia. Marsileaceae are exceptional among the ferns treated in this book in being

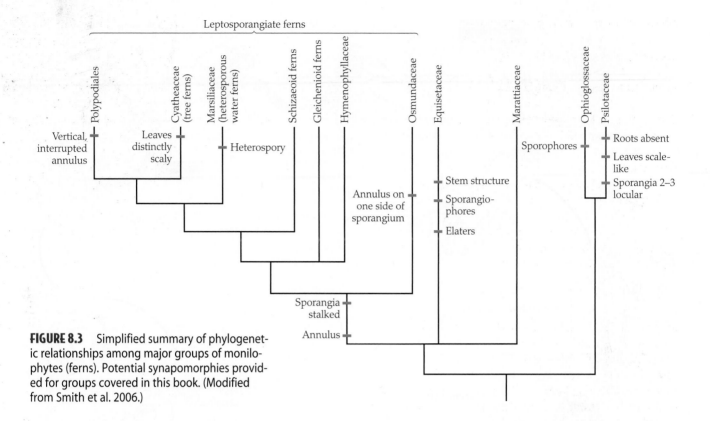

FIGURE 8.3 Simplified summary of phylogenetic relationships among major groups of monilophytes (ferns). Potential synapomorphies provided for groups covered in this book. (Modified from Smith et al. 2006.)

heterosporous and producing two kinds of spores (megaspores and microspores).

The gametophytes of ferns are generally less than 1 cm in any dimension. Gametophytes of Psilotaceae and Ophioglossaceae are subterranean, lack chlorophyll, and are mycotrophic, depending upon fungal partners for nutrition. Equisetaceae and most leptosporangiate ferns produce gametophytes that are photosynthetic and grow on the surface of the ground. Many leptosporangiate ferns have heart-shaped gametophytes (see Figure 8.4), although they may also be kidney-shaped, elongate, filamentous, or take various other forms. Gametophytes of some leptosporangiate ferns produce small buds (gemmae) that separate from the gametophyte for asexual reproduction. Marsileaceae are exceptional in having gametophytes that are endosporic—i.e., they are wholly or mostly retained within the spore wall. The megagametophytes protrude from the megaspores, and the microspores burst as the male gametophytes inside release sperm.

Fertilization in monilophytes usually requires water through which the flagellate sperm can swim from the antheridium to the archegonium. This requirement for water excludes these plants from some dry habitats. Many monilophytes can reproduce asexually by means of rhizomes, vegetative propagules on the gametophyte called **gemmae** in *Psilotum* and several leptosporangiate ferns, vegetative propagules on the sporophyte, direct outgrowths from gametophytic tissue (called apogamy), and spores that are produced via modified meiotic divisions.

References: Kenrick and Crane 1997; Pryer et al. 2004; Rothwell 1999; Rothwell and Nixon 2006; Smith et al. 2006; Sporne 1970; Tryon and Tryon 1982.

Psilotales

Psilotaceae Kanitz

(Whisk Fern Family)

Herbs, terrestrial but more *often epiphytic*. **Roots absent**, the plant anchored by subterranean stems that may bear gemmae. Aerial stems erect or hanging, glabrous, simple or dichotomously branched. Leaves spirally arranged, 2-ranked in some species; **scale-like or awl-shaped to lanceolate; simple or once-forked**, one-veined or veinless. **Sporangia 2- or 3-locular and -lobed**, large, sessile on or above the base of forked sporophylls. Homosporous; **spores bean-shaped**, pale in color (Figure 8.5).

Distribution and ecology: Pantropical and warm temperate except in dry areas, with the greatest number of species in Southeast Asia and the South Pacific; primarily at lower elevations, especially as epiphytes on tree fern and palm trunks.

Genera/species: 2/12. ***Major genera:*** *Tmesipteris* (10 spp.) and *Psilotum* (2).

Economic plants and products: The family is not significant economically.

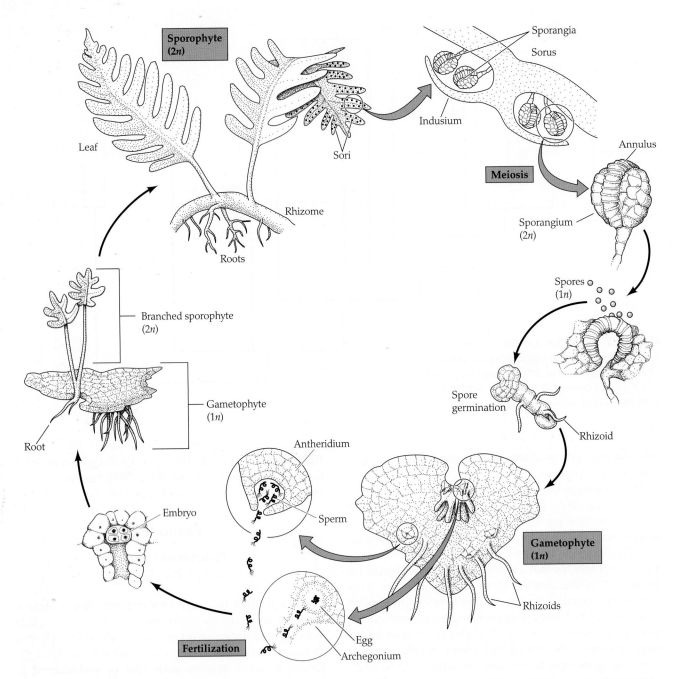

FIGURE 8.4 Life cycle of a fern of the genus *Polypodium*. The sporophyte generation is photosynthetic and is independent of the gametophyte. The cluster of sporangia plus a protective layer of cells called the indusium make up a sorus. Meiosis within the sporangia yields haploid spores that divide mitotically to produce a heart-shaped gametophyte that differentiates both archegonia and antheridia on one individual. The gametophyte is photosynthetic and independent, although it is reduced in size relative to the sporophyte. Fertilization takes place when sperm swim through water to the archegonia and fertilize the eggs. Unlike the gametophyte, the developing sporophyte has vascular tissue and roots. (From Singer 1997.)

Discussion: Although there is no extensive fossil record of Psilotaceae, this family has long been thought to be among the most primitive of the extant vascular plants, based on their similarities to ancient and simple fossil plants. Morphological, chemical, DNA sequence data, and detailed features of the architecture of sperm cells, how- ever, support a relationship of Psilotaceae to Ophioglos- saceae. The distinctive simplicity of these plants—part of the reason they do not resemble ferns—is probably reduc- tion associated with mycotrophy, epiphytism, and water stress. Growth of spores depends upon the presence of endophytic mycorrhizae.

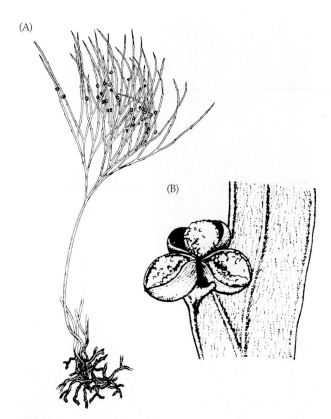

FIGURE 8.5 Psilotaceae. *Psilotum nudum*: (A) habit (about × 0.25); (B) portion of stem with open sporangium (× 15). (A from Knobloch and Correll 1962; original art by P. Horning; B from Brownlie 1977.)

References: Kramer 1990d; Kenrick and Crane 1997; Lellinger 1985; Manhart 1995; Pryer et al. 1995, 2001, 2004; Renzaglia et al. 2001; Smith et al. 2006; Thieret 1993b; Tryon and Tryon 1982; Wolf 1997; Wolf et al. 1998.

Ophioglossales

Ophioglossaceae C. Agardh

(Adder's-tongue Family)

Terrestrial, less often epiphytic. Stems unbranched, short, mostly subterranean. Leaves usually one at at time per stem, sometimes with a stalk that is expanded at the base into a sheath, usually **divided into a photosynthetic, sterile blade and one or more spore-bearing portions (the sporophore/s)**; the blade simple to more or less palmately compound to many times pinnately compound, from a few cm to about 50 cm long, rarely absent, conduplicate before unfolding, and nodding (but not circinate). Sporangia thick-walled (eusporangiate), not clustered in sori, separate or joined in synangia, exposed on branches of the sporophore or embedded in a spike-like sporophore; annulus absent. Homosporous; spores not green, thousands per sporangium (Figure 8.6).

Distribution and ecology: Worldwide, especially in temperate regions and the tropics. Members of this family may be associated with disturbance in pastures, old fields, and young second-growth forests, although many species grow in relatively undisturbed habitats.

Genera/species: 4/70–90. **Major genera:** *Botrychium* (45–60 spp.) and *Ophioglossum* (25–30).

Economic plants and products: None.

Discussion: The sporophore, which is assumed to be two basal leaf segments that have fused together, is the most unusual feature of the family and likely a synapomorphy. The family is morphologically distinct from other ferns, and the DNA sequence support for its relationship with Psilotaceae has no obvious support from morphology. Chromosome numbers in *Ophioglossum* are as high as $2n = 1400$, the highest of all tracheophytes.

References: Hasebe et al. 1995; Hauk et al. 2003; Pryer et al. 1995, 2001, 2004; Smith et al. 2006; Tryon and Tryon 1982; Wagner 1990; Wagner and Wagner 1993.

Equisetales

Equisetaceae Michx. ex DC.

(Horsetail Family)

Terrestrial to aquatic, rhizomatous perennials. Stems to 8 m in height, but under 1 m in most species; nodes swollen

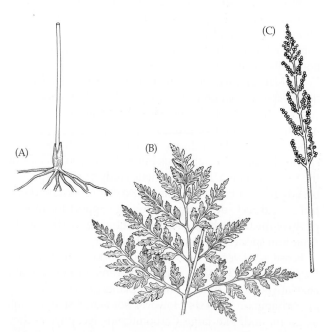

FIGURE 8.6 Ophioglossaceae. *Botrychium virginianum*: (A) fertile leaf (all × about 0.3); (B) portion of a sterile (photosynthetic) leaf; (C) base of plant. (From Wagner and Wagner 1993.)

(jointlike); **internodes with alternating vertical ridges and grooves externally and usually hollow, with a central canal and smaller canals under the ridges and valleys internally**; branches none or whorled and structurally similar to the main stem. **Leaves whorled, fused into sheaths at base**, usually much less than 2 cm long, the sheaths sometimes more or less swollen. Sporangia elongate, pendant beneath the expanded apex of sporangiophores; **sporangiophores peltate**, whorled in strobili terminating stems that are green or, in a few species, not green, unbranched, and either ephemeral or becoming green and branched when the spores are shed. Homosporous; spores spherical, green, and with **4–6 straplike elaters** wrapped around the spore that rapidly straighten and assist in spore dispersal (Figure 8.7).

Distribution and ecology: Cosmopolitan, except for Australia, New Zealand, and Antarctica, with the greatest number of species between 40° and 60° north latitude. These plants are primarily colonizers of unforested areas, lake margins, and wetlands.

Genera/species: 1 (*Equisetum*)/15.

Economic plants and products: The family is not significant economically. Silica in the stems made them useful to early European settlers in North America for cleaning cookware, a practice that is apparently the source of another common name, scouring rush.

Discussion: This family is morphologically distinct in its grooved and hollowed stem, whorled leaves, and peltate sporangiophores. Unitl recently the family was considered to be distinct from ferns and was referred to as a "fern ally." Molecular data and morphological features (such as the nature of the spermatozoids and root characters) show that horsetails are ferns (monilophytes).

Equisetum is divided into two subgenera: subgen. *Equisetum* (8 species; superficial stomates; stems branched) and subgen. *Hippochaete* (7 species; sunken stomates; stems unbranched). The two groups are recognized as distinct genera by some. Hybridization occurs frequently between species of the same subgenus but not between species of the two subgenera. A phylogenetic study of all 15 species, based on a combined analysis of two chloroplast markers, *trnL-F* and *rbcL*, demonstrates robust support for monophyly of the two subgenera. The relationships of the South American species, *E. bogotense*, are not clear.

Fossil relatives of Equisetaceae are known since the Devonian (408–360 MYA), and they became most abundant as relatively small (under 1 m tall) plants of the understory of Carboniferous forests. However, some had stems as tall as 20 m, and, like large Carboniferous lycophytes, they went extinct. The first fossils clearly assignable to *Equisetum* are from the Eocene (54–38 MYA), and the genus is thought to have diversified within the Cenozoic Era.

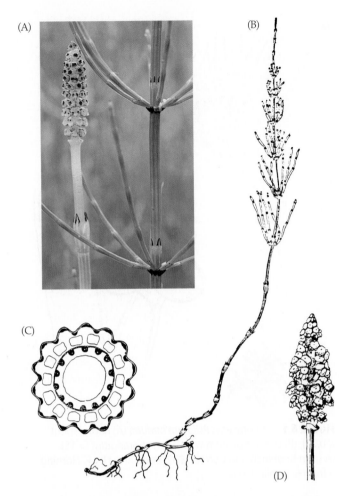

FIGURE 8.7 Equisetaceae. (A) *Equisetum palustre*: Fertile (left) and sterile (right) shoots. Sporangia (dark areas) on the fertile shoot are at the point of spore dispersal. (B–D) *E. arvense*: (B) habit of sterile plant (× 0.4); (C) cross-section through stem (× 9.5); (D) apex of fertile stem with strobilus (× 0.8). (B–C from Madalski 1954; D from Hauke 1990.)

References: Des Marais et al. 2003; Hauke 1990, 1993; Lellinger 1985; Pryer et al. 2001, 2004; Stewart and Rothwell 1993; Smith et al. 2006; Tryon and Tryon 1982.

Leptosporangiate Ferns

Terrestrial, epiphytic, or aquatic. **Primary xylem with scalariform bordered pits**. Stems rhizomatous to treelike and up to 20 m tall. Leaves (see Figures 8.8, 8.11–8.19) often much more than 2 cm long and with numerous and forking veins; mostly divided into separate lobes or separate leaflets, but sometimes simple and unlobed; spirally arranged, their bases often persistent and more or less covering the stem but sometimes cleanly abscising; with *circinate vernation* (see Figure 8.15)—forming a crozier, or "fiddlehead" (i.e., coiled and unfolding lengthwise when emerging); monomorphic or dimorphic (sterile and fertile leaves different). **Sporangia with stalk 3-6 cells in cross-section**, *mostly clustered in sori* on the abaxial leaf surface of normal leaves, or in specialized portions of leaves, or on completely separate fertile leaves (see Figure 8.17), less often scattered on the abaxial leaf surface or in sporocarps; usually with an **annulus** (a cluster or row of cells with thick walls that open the sporangium and catapult the spores into the air); the sporangium wall generally one cell layer thick at maturity. Homosporous, sometimes heterosporous; spores mostly green.

Discussion: Leptosporangiate ferns are distinguished by their usually large leaves that form croziers as they unfold and sporangia that often cluster in sori. The many species in

---

Key to Selected Families of Leptosporangiate Ferns

1. Plants floating in water or sometimes rooted in mud bordering water; leaves divided into 2 or 4 segments or threadlike and undivided; heterosporous, the sporangia lacking an annulus and of two kinds that are borne in the same specialized, hardened and bean- or pea-shaped sporocarp ....**Marsileaceae**

1. Plants mostly terrestrial; leaves mostly pinnately compound with numerous segments, at least pinnatifid, infrequently simple; homosporous, the sporangia annulate and usually borne in sori that are exposed on the leaf surface 2

2. Annulus merely a patch of cells on the side of the sporangium; sori absent; spores over 100 to several thousand per sporangium ....**Osmundaceae**

2. Annulus usually a much larger vertical or obliqueband of cells; sori usually present; spores usually 64 or fewer per sporangium 3

3. Stems usually massive and arborescent (delicate and creeping in a few species), reaching 20 m; annulus oblique, not interrupted by stalk of sporangium; sporangium stalk with 4-8 rows of cells ....**Cyatheaceae**

3. Stems usually creeping along the ground or subterranean, usually only tip of stem and leaves evident, or epiphytic; annulus vertical, interrupted by the sporangium stalk; sporangium stalk with 1–3 rows of cells ....4 **(Polypodiales)**

4. Leaves strongly dimorphic, the fertile leaves brown at maturity, not leaflike ....**Onocleaceae**

4. Leaves mostly monomorphic; occasionally somewhat dimorphic but leaflike 5

5. Sori mostly only on or near the leaf margin 6

5. Sori not limited to the leaf margin 7

6. Stems and petiole bases with hairs 1 cell wide and not scales 2 or more cells wide ....**Dennstaedtiaceae**

6. Stems and petiole bases with scales at least 2 cells wide ....**Pteridaceae**

7. Indusia absent 8

7. Indusia present 10

8. Leaves often with needlelike hairs ....**Thelypteridaceae**

8. Leaf hairs, if present, blunt 9

9. Petiole usually with scales persistent at the base ....**Dryopteridaceae**

9. Petiole lacking scales or if scaly, the scales not persistent ....**Polypodiaceae**

10. Sori often in chains or linear, parallel, and adjacent to midribs; indusia opening toward midribs ...**Blechnaceae**

10. Sori not in chains parallel to midribs; indusia not opening toward midribs 11

11. Indusia linear, laterally attached ....**Aspleniaceae**

11. Indusia round, kidney-shaped, or linear but not laterally attached 12

12. Leaves often with needlelike hairs ....**Thelypteridaceae**

12. Leaf hairs, if present, blunt 13

13. Petiole usually with scales persistent at the base ....**Dryopteridaceae**

13. Petiole lacking scales or if scaly, the scales not persistent ....**"Woodsiaceae"**

(A)

Polypodiales: Pteridaceae
Adiantum raddianum: leaf

(B)

Cyatheales: Cyatheaceae
Cyathea arborea: habit

(C)

Polypodiales: Dryopteridaceae
Cyrtomium falcatum: leaf with sori

(D)

Polypodiales: Onocleaceae
Matteuccia struthiopteris: crozier
(circinate leaf)

(E)

Salviniales: Marsileaceae
Marsilea sp.: habit

(F)

Osmundales: Osmundaceae
Osmunda cinnamomea: habit; fertile leaf

(G)

Polypodiales: Polypodiaceae
Pleopeltis polypodioides: leaves, sori on underside

PLATE 8.1 Leptosporangiate ferns

this group display tremendous leaf diversity, from the large, pinnately compound forms of the stereotypic fern to the clover-like leaves of the water-clover ferns, to the needlelike leaves of curly-grass ferns (Schizaeaceae, not treated here). Habit varies from aquatics (Marsileaceae), to epiphytes with ribbon-shaped leaves, to trees (Cyatheaceae) (Plate 8.1).

Currently about 33 families of leptosporangiate ferns are recognized. Here we treat 12 families that span the evolutionary diversity of this group as currently understood. Osmundaceae are sister to all other leptosporangiate ferns. Polypodiales plus the tree fern clade (including Cyatheaceae) and heterosporous aquatic ferns (including Marsileaceae) comprise what are sometimes called the core leptosporangiates. Schizaeoid ferns (not treated here) are likely the sister to the core leptosporangiate ferns (see Figure 8.3). Two other groups of leptosporangiate ferns, the gleichenoid ferns (Gleicheniaceae, Dipteridaceae, and Matoniaceae) and the filmy ferns (Hymenophyllaceae) are not treated in this text. Phylogenetic relationships among the leptosporangiate ferns are being clarified by recent papers (see Hasebe et al. 1995; Hennequin et al. 2006; Korall et al. 2006; Pryer 1999; Pryer et al. 1995, 2001, 2004; Schneider et al. 2004; Smith 1995; Smith and Cranfill 2002; Smith et al. 2006; Vangerow et al. 1999; Wolf 1997; Wolf et al. 1998, 1999).

The nature of the annulus is critical to the phylogeny of leptosporangiate ferns. In Osmundaceae it is merely a patch of cells on one side of the sporangium (Figure 8.8). Aquatic ferns, such as some species of *Ceratopteris* (Pteridaceae) and Marsileaceae have no annulus; sporangia of Marsileaceae are encased and dispersed in the sporocarp. The annulus of Cyatheaceae (see Figure 8.9) consists of a series of cells that are not vertical in orientation—it does not lie in the same plane as the sporangium stalk—and therefore is not interrupted by the stalk. In families of the Polypodiales, the annulus is a line of cells whose vertical orientation puts it in the plane of the stalk, but the annulus stops where it meets the stalk, and hence it is said to be "interrupted" by the stalk (see Figure 8.13).

Additional references: Flora of North America Editorial Committee 1993; Kramer and Green 1990; Lellinger 1985; Stein et al. 1992; Tryon and Tryon 1982; Wagner and Smith 1993.

Osmundales

Osmundaceae Bercht. & J. Presl
(Royal Fern Family)

Terrestrial. Stems branched, often covered by persistent leaf bases, horizontal to erect. Leaves about 0.5–2 m long, once- to thrice-pinnately compound, with an expanded petiolar base, circinate before unfolding; in some species wholly or partially dimorphic, with separate sporangium-bearing segments or whole leaves; often forming a vase-shaped clump. *Sporangia separate or in loose clusters, not clustered in sori, borne on wholly fertile portions of the leaf or on the abaxial surface of relatively unmodified portions of the leaf, with a short stalk of many rows of cells and a poorly differentiated* **annulus that consists of a group of thickened cells on the side of the sporangium**. Homosporous; spores green, 128–512 spores per sporangium (Figure 8.8).

Distribution and ecology: Worldwide except for very cold and dry climates and Pacific islands. *Osmunda*, the only genus in the Northern Hemisphere, is common in wetlands, such as swamps, swales, and lowland forests.

Genera/species: 3/18. ***Genera:*** *Osmunda* (10 spp.), *Leptopteris* (6), and *Todea* (2).

FIGURE 8.8 Osmundaceae. (A) *Osmunda regalis*: habit (about × 0.1), with an enlarged apical, partially fertile portion of the leaf to the left (about × 0.3). (B) *O. lancea*: sporangia, closed on left and open on right (× 315). Note thickened cell walls of annulus. (A from Taylor 1984; Milwaukee Public Museum, original art by P. Nelson; B from Hewitson 1962.)

Economic plants and products: Some species, such as cinnamon fern (*Osmunda cinnamomea*; Plate 8.1F) and royal fern (*O. regalis*), are grown as ornamentals.

Discussion: The numerous spores, rudimentary annulus, sporangium wall more than one cell thick, lack of a sorus, and *rbcL* DNA sequence data all support Osmundaceae as sister to all other leptosporangiate ferns. This conclusion is consistent with the long fossil record of the family dating to the Permian (286–245 MYA). *Osmunda claytoniana* has apparently lived since the late Triassic, about 220 MYA, and *O. cinnamomea* has grown in western North America more or less continuously for at least 70 million years.

References: Hasebe et al. 1994, 1995; Kramer 1990c; Lellinger 1985; Phipps et al. 1998; Pryer et al. 1995, 2004; Serbet and Rothwell 1999; Smith et al. 2006; Tryon and Tryon 1982; Whetstone and Atkinson 1993.

Salviniales

This order contains two families, five genera, and about 90 species of **heterosporous** ferns of wetland or aquatic habitats. These plants show a **differentiation of the sterile (i.e., photosynthetic) and fertile portions of their leaves**, and have blades with anastomosing veins. As in many aquatic groups, aerenchyma tissue is often present in the roots, stems, and petioles. Their sporangia **lack an annulus**. Salviniales are sister to Cyatheales plus Polypodiales.

Marsileaceae Mirbel
(Water-Clover Family)

Plants aquatic, with floating leaves or rooted in the mud bordering bodies of water. Stems slender, glabrous, and creeping, growing on soil surface or subterranean. *Leaves long-petioled, the blade divided into 2 or 4 leaflets or filiform and not expanded;* circinate before unfolding. Sori **with no indusial opening**, enclosed in bean- or pea-shaped *sporocarps* that are borne on short stalks near or at base of petioles; each sporocarp with at least 2 sori. Heterosporous. Sporangia without an annulus; megasporangia with one megaspore; microsporangia with 16–64 microspores (Figure 8.9).

Distribution and ecology: Nearly cosmopolitan in warm temperate and tropical areas; amphibious, growing in water or near water in very wet soil.

Genera/species: 3/76. ***Major genera:*** *Marsilea* (50–70 spp.) and *Pilularia* (6).

Economic plants and products: *Marsilea* is occasionally planted as a curiosity.

Discussion: The common name refers to the similarity between water-clover (*Marsilea*) leaves (Plate 8.1E) and

those of clovers (*Trifolium*, Fabaceae). *Pilularia*, which is widely distributed, has filiform leaves. Leaves of *Regnellidium*, a monotypic genus of southern Brazil and adjacent Argentina, have two leaflets. Recent studies show that *Pilularia* and *Regnellidium* form a clade, and *Marsilea* is the sister of that clade. The presence of one archegonium in megagametophytes is a possible synapomorphy of the family.

The sporocarps of Marsileaceae are remarkably durable. They protect spore viability for over 100 years in some species and may be an adaptation for growth in arid regions where rainfall is infrequent. Waterfowl consume the sporocarps, which pass through the digestive tract and thus disperse the spores.

The fossil record indicates that the genera of Marsileaceae diversified by the mid-Cretaceous.

Salviniaceae, which are not treated in detail in this book, are aquatic, cosmopolitan, and the two genera, *Azolla* and *Salvinia*, each have fewer than 10 species. These aquatics bear undivided, small (blades less than 15 mm long), floating leaves and flattened sporocarps. Salviniaceae are tropical or temperate in distribution. *Azolla* is an important economic plant, as an invasive and as a potential source of green manure, having a symbiotic cyanobacterium.

References: Johnson 1993a; Kramer 1990b; Lellinger 1985; Lupia et al. 2000; Pryer 1999; Pryer et al. 1995, 2004; Smith et al. 2006.

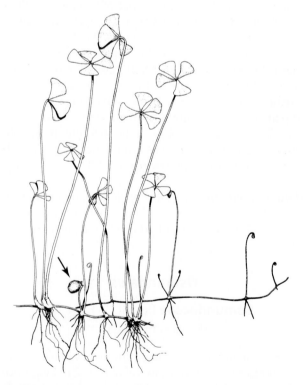

FIGURE 8.9 Marsileaceae. *Marsilea vestita*: habit (about × 0.7). Note mature, cloverlike leaves; numerous young, circinate leaves and single sporocarp (arrow). (From Taylor 1984; Milwaukee Public Museum, original art by P. Nelson.)

Cyatheales

This order contains 8 families, 13 genera, and at least 660 species. Most plants in this order are tree ferns, with trunk-like stems up to 20 m (Plate 8.1B), but there are also plants with small, creeping stems. The number of times that the tree habit has evolved and been lost in the order is uncertain. Although there are no clear morphological synapomorphies, DNA sequence data do indicate that the order is monophyletic. Molecular studies identify four major clades in the order, one of which is made up of the Cyatheaceae. Relationships of Cyatheaceae to other tree ferns are uncertain, and the second largest family of tree ferns, the "Dicksoniaceae," is clearly not monophyletic as it is usually circumscribed. The other families—Cibotiaceae, Culcitaceae, Loxomataceae, Metaxyaceae, Plagiogyriaceae, and Thyrsopteridaceae—are small. Pneumathodes (ventilation lines or patches on the leaf axis or petiole or on the rachis for gas exchange) and radial symmetry of the rhizome are potential synapomorphies of the tree fern clade. Molecular data tentatively support a sister-group relationship between this tree fern clade and the Polypodiales.

Cyatheaceae Kaulf.
(Scaly Tree Fern Family)

Stem usually a single, erect, arborescent trunk to 20 m tall, usually unbranched, but creeping along the ground and delicate in *Hymenophyllopsis*. **Leaves distinctly scaly**, *(0.5–) 2–3 (–5) m long*, most often pinnately or bipinnately compound, with leaflets usually deeply pinnately lobed, circinate before unfolding. Sporangia borne in sori on the abaxial leaf surface; the annulus continuous, not interrupted by the sporangium stalk. Indusium below the sorus or absent. Homosporous; spores not green, usually 64 (sometimes only 16) per sporangium (Figure 8.10).

Distribution and ecology: New and Old World tropical wet montane forests and cloud forests. Some species extend into south temperate areas (New Zealand and southern South America) and into north temperate India, China, and Japan. Some species are widely distributed, but local endemics are numerous on oceanic islands and tropical mountains. Many species are early successional on landslides.

Genera/species: about 5/at least 600. ***Major genera:*** *Alsophila* (230 spp.), *Sphaeropteris* (120), and *Cyathea* (110).

Economic plants and products: Members of this family are sometimes grown as ornamentals, and the fibrous rhizomes are used as a base for epiphytes in greenhouses. Human exploitation has led to extensive destruction of tree fern stands, but most species are now protected by international agreements barring trade in endangered species.

Discussion: Relationships among the three largest genera are uncertain. Moreover, monophyly of *Alsophila* and *Sphaeropteris* is questionable, and there is some molecular support for inclusion within *Cyathea* of *Hymenophyllopsis* (Hymenophyllopsidaceae). *Hymenophyllopsis*, with its small and delicate stems and leaves only 10–30 cm long, would seem to be unrelated to the large tree ferns of most of Cyatheaceae, but this genus does share with Cyatheaceae the presence of scales on the leaves. The name of the genus reflects the similarity of its delicate leaves to those of members of the Hymenophyllaceae, the filmy ferns, a family of about 600 species (not treated in this text). *Hymenophyllopsis*, which contains about eight species, is restricted to the Roraima geological formation in Venezuela, Guyana, and northernmost Brazil.

Groves of tree ferns, with the crown of large leaves atop the unbranched stems, make some of the most attractive displays in the natural world.

References: Conant et al. 1995; Conant and Stein 2001; Korall et al. 2006; Kramer 1990a; Pryer et al. 1995, 2004; Smith et al. 2006; Tryon and Tryon 1982; Wolfe et al. 1999.

Polypodiales

This order contains about 15 families, 220–260 genera, and over 7000 species; Polypodiales represent about 80% of the species diversity of leptosporangiate ferns. A brief description of the clade, which has often been treated as a broadly circumscribed family (Polypodiaceae), is provided below.

Terrestrial, sometimes growing on rocks, sometimes epiphytic or climbing, infrequently aquatic; perennial, rarely annual. Stems subterranean or barely evident at soil surface (except in epiphytes, climbers, and aquatics), sometimes horizontal (rhizomatous) and creeping, reproducing vegetatively; often with epidermal scales of various sizes, shapes, and textures. Leaves mostly pinnately lobed to once, twice, or several times pinnately compound (see Figures 8.11, 8.12, 8.14–8.19); less often palmately compound, simple, or ribbonlike; *circinate*, clustered or separated from

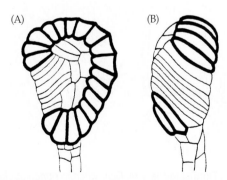

FIGURE 8.10 Cyatheaceae. *Cyathea capensis*: (A) sporangium, showing annulus with cells with thickened walls; (B) the same, different view (both × about 100). (From Holttum 1963.)

one another on an elongate stem; petiole usually well developed, rarely absent; sterile-fertile dimorphism well developed (see Figure 8.17) to absent; variously hairy or scaly. Sporangia with a well-developed **vertical annulus, interrupted by the sporangium stalk** (see Figure 8.13C); *usually borne in sori* that may be covered by an indusium. Sori variously positioned on the abaxial leaf surface, isolated from one another or covering the abaxial surface of the leaf; round (Figures 8.18 and 8.19) to elongate (Figure 8.16). Indusia round to elongate; sometimes absent, replaced by the recurving leaf margin, or combining with a recurving leaf margin. Homosporous; spores 64 (less often 32, 16, or 8) per sporangium, usually not green, monolete or trilete, the surface smooth or variously sculptured with ridges and spines.

Discussion: Polypodiales have long been recognized because of a distinctive synapomorphy, the interruption of the vertically oriented annulus by the stalk of the sporangium (Figure 8.13C). Some families within this order have been recognized for many decades, but the definition of other families has been controversial. Recent studies have largely resolved relationships within the order, leading to the recognition of 15 families, most of which are clearly monophyletic. Relationships among these 15 families are not completely resolved. The nine families of the order treated below represent almost 90% of all genera and over 90% of all species in the order. Other families in this order are Davalliaceae, Lindseaceae, Lomariopsidaceae, Oleandraceae, Saccolomataceae, and Tectariaceae.

Features common to many leptosporangiate ferns—large, clumped, pinnately compound leaves, and mesic forest habitat—hold for many species in this order. Nevertheless, as its description indicates, the order encompasses tremendous variation in habit, leaf morphology, and reproductive features. Characters of sori and indusia vary greatly as well. In some species, the sori and indusia are linear, and the indusium is attached beside and covers part or all the sorus, or the sori are round and more or less covered by an indusium. Sori of other species are close to the margin and partially covered by the recurving leaf margin, yet others are not covered at all. In addition to leaf morphology and these reproductive features, scales on the stem and leaves, leaf architecture, habit, number and arrangement of vascular bundles in the petiole, hairs, venation, spore morphology, and chromosome number are taxonomically informative.

Hybridization, polyploidy, and asexual spore production play prominent roles in the evolution and systematics of many of the larger genera, including *Asplenium, Athyrium, Ceratopteris, Cheilanthes, Cystopteris, Diplazium, Dryopteris, Gymnocarpium, Pellaea, Polypodium, Polystichum, Pteris,* and *Woodsia.*

References: Hasebe et al. 1995; Kramer and Green 1990; Lellinger 1985; Pryer et al. 1995, 2004; Schneider et al. 2004; Smith and Cranfill 2002; Smith et al. 2006; Tryon and Tryon 1982; Wagner and Smith 1993; Wolf 1997.

Dennstaedtiaceae Pic.Serm.
(Hay-Scented Fern Family)

Terrestrial or scrambling over other vegetation. Rhizomes long-creeping or rarely short-creeping, bearing jointed hairs. Leaves monomorphic; petiole pubescent or glabrous; blades often large (up to 7 m), 1–4-pinnate or more divided; indument of hairs, not scales. **Sori near the leaf margin, mostly linear, sometimes fused with a portion of the blade to form a cup or pouch or obscured in a recurved portion of the blade margin**. *Indusia linear or cuplike at blade margins, or reflexed over sori.* Sporangium stalk with 1–3 rows of cells. (Figure 8.11)

Distribution and ecology: Pantropical, with some groups in temperate zones and occupying moist forests, forest openings, pastures, rocky slopes, roadsides, and shaded banks. *Pteridium aquilinum* and *P. caudatum,* commonly called bracken ferns, are some of the most widely distributed plants. They grow worldwide except in regions that are extremely dry or cold, and are invasive weeds of fields and roadsides in many parts of the world. *Pteridium aquilinum* spreads effectively by long-creeping rhizomes,

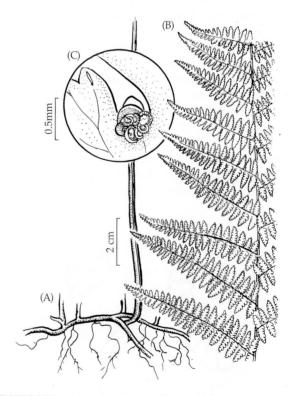

FIGURE 8.11 Dennstaedtiaceae. *Dennstaedtia punctilobula*: (A) rhizome and base of petiole; (B) portion of leaf; (C) enlarged view of a sorus in cup-shaped pocket formed by leaf margin. (From Nauman and Evans 1993.)

persisting effectively in stable habitats and producing spores only in highly disturbed sites.

Genera/species: 11/170. **Major genera:** *Dennstaedtia* (45 spp.).

Economic plants and products: Consumption of the unfolding leaves (crosiers or fiddleheads) of *Pteridium aquilinum*, which occurs in some temperate regions, is unhealthy because of the presence of carcinogenic compounds.

Discussion: The indusium in this family may consist of not only a true indusium but also a modified portion of the leaf blade that is fused with the true indusium. Some genera lack either the true indusium or the leaf modification.

The largest genus in the family, *Dennstaedtia*, is mostly tropical. Some of the tropical species have very large leaves, up to 3 m long. A temperate member of this genus, *D. punctilobula*, is common in much of eastern North America where, in the right conditions, it can form a dense patch that excludes almost all other vegetation. This species is called hay-scented fern because the foliage emits a fragrance reminiscent of hay. *Pteridium aquilinum* also emits a pleasant, somewhat sweet fragrance.

References: Nauman and Evans 1993; Smith et al. 2006.

Pteridaceae Ching
(Maidenhair Fern Family)

Terrestrial or epiphytic, rarely aquatic. Rhizomes long- to short-creeping, ascending to erect, scaly or less often hairy. Leaves monomorphic or dimorphic in a few genera; petiole usually with persistent scales near its base; blades simple to 1–6-pinnate; indumentum of hairs, glands, or scales. Sori near the margin and usually either forming a continuous band and protected by the reflexed leaf margin or arranged along all of the leaf veins. **Indusia absent** *(the reflexed margin that covers the sori in some genera is a false indusium)*. Sporangium stalk with 2–3 rows of cells (Figure 8.12; Plate 8.1A).

Distribution and ecology: Almost cosmopolitan, but most diverse in tropics. Many species of Pteridaceae are adapted to arid regions, an unusual habitat for ferns. Other species grow in tropical rainforests, temperate forests, ponds and other bodies of water, and many other habitats.

Genera/species: ca. 50/950. **Major genera:** *Pteris* (200 spp.), *Adiantum* (150), and *Cheilanthes* (150).

Economic plants and products: The so-called brakes (*Pteris*) and some members of the maidenhair genus (*Adiantum*) are grown as ornamentals.

Discussion: The morphological diversity of members of Pteridaceae matches the wide range of habitats they occu-

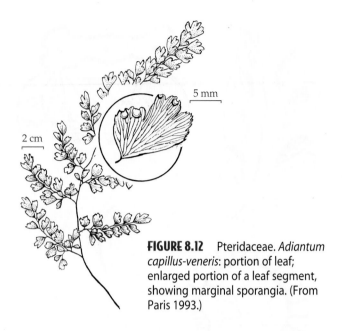

FIGURE 8.12 Pteridaceae. *Adiantum capillus-veneris*: portion of leaf; enlarged portion of a leaf segment, showing marginal sporangia. (From Paris 1993.)

py, from deserts to ponds to forest canopies to mangrove swamps. Pterid ferns that are adapted to dry habitats may have leathery leaves, and leathery leaves also characterize *Acrostichum*, a species adapted to tidally flooded mangrove communities. Epiphytes, such as *Vittaria* (the shoestring ferns), have long, linear leaves with indistinct petioles. Finally, the aquatics, *Ceratopteris* (the water ferns), may have inflated petioles for flotation, a weakly developed annulus, and distinctively ridged spore walls. Pteridaceae contains five clades that could be recognized as subfamilies. Some of the genera included in this family, such as *Cheilanthes*, are not monophyletic and are in need of extensive study.

References: Nauman and Evans 1993, Lloyd 1993; Smith et al. 2006; Windham 1993.

Aspleniaceae Newman
(Spleenwort Family)

Terrestrial or epiphytic. Rhizomes ascending or nearly erect, rarely creeping, scaly at the apex. Leaves monomorphic (rarely somewhat dimorphic); petiole base scaly; leaf blades simple to 5-pinnate, often with tiny glandular hairs and a few linear scales. Sori elongate along the veins, linear or curved. **Indusia linear**, *laterally attached*. Sporangium stalk with 1 row of cells (Figure 8.13).

Distribution and ecology: Almost cosmopolitan; most diverse in tropics. Most members of the family are epiphytic or grow on rocks (epipetric) in moist or wet forests, on the forest floor, in ravines, or along streams. Some species are adapted to drier sites, such as cliffs and lava beds. Other species occupy roadside banks, old walls, thickets, and other disturbed areas.

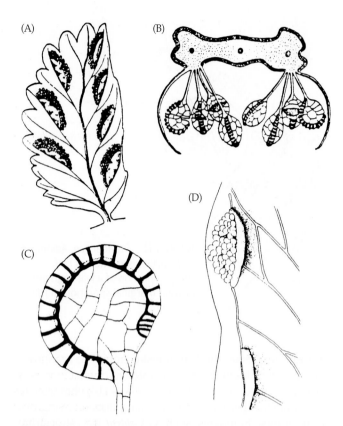

FIGURE 8.13 Aspleniaceae. (A, B) *Asplenium tripteropus*: (A) portion of underside of leaf, showing sori (× 4); (B) cross-section of fertile leaf, showing two sori and sporangia. (C) *A. nidus*: sporangium, showing interrupted annulus with cells with thickened walls (× 140). (D) *Asplenium* sp.: margin of leaf, with sori (× 8) (A, B from *Flora Tsinlingensis* 1974; C from Haider 1954; D from Pérez Arbeláez 1928.)

Genera/species: 1–10/700 or more. **Major genera:** *Asplenium* (700 spp.).

Economic plants and products: Some species of *Asplenium*, such as the hart's-tongue fern (*A. scolopendrium*), are grown as ornamentals.

Discussion: This family is part of a clade with Blechnaceae, Onocleaceae, Thelypteridaceae, and "Woodsiaceae." Members of Aspleniaceae display interesting morphologies. For example, *Asplenium rhizophyllum* of eastern North America is called the walking fern because its simple, elongate leaves may become elongated at the tip and root, thus "walking." This habit can lead to the formation of large clonal patches. The sister species of the walking fern, Ruprecht's walking fern (*A. ruprechtii*), is a native of eastern Asia and has the same "walking" habit. Both species inhabit shaded cliffs and mossy boulders.

Interspecific hybridization in *Asplenium* leads to the formation of sterile hybrids, fertile allopolyploids, and forms that produce spores without meiosis. Some of these hybrids have striking leaf forms and are thus popular with field naturalists and fern gardeners.

References: Wagner et al. 1993; Smith et al. 2006.

Thelypteridaceae Pic.Serm.
(Marsh Fern Family)

Terrestrial. Rhizomes creeping, ascending or erect, scaly at the apex. Leaves monomorphic or somewhat dimorphic; blade pinnate or pinnate-pinnatifid, rarely more divided or simple; indumentum of scales and transparent, often **needlelike hairs** or rarely absent. *Sori round to oblong, rarely elongate along the veins. Indusia kidney-shaped or absent.* Sporangium stalk with 3 rows of cells (Figure 8.14).

Distribution and ecology: Pantropical, especially in southeast Asia. Found in or at the edge of rainforests, and also swamps, stream margins, and disturbed sites such as landslides and roadside banks; only a few species in temperate regions.

Genera/species: 5–30/950. **Major genera:** *Cyclosorus* (600 spp.) and *Thelypteris* (280).

Economic plants and products: None.

Discussion: Thelypteridaceae in its current circumscription was first recognized almost 70 years ago. DNA sequence data divide the family into two major lineages, the phegopteroids and thelypteroids. Phegopteroids contain two genera with a total of about 30 species that are mostly tropical, and one genus (*Phegopteris*) with three species of the North Temperate Zone. Thelypteroids are mostly tropical and the number of genera varies widely depending upon classifi-

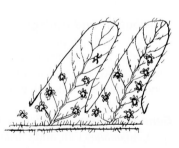

FIGURE 8.14 Thelypteridaceae. *Thelypteris noveboracensis*: portion of plant; enlarged portion of a leaf segment, showing sporangia. (From Smith 1993a.)

cation. Recent molecular phylogenetic work indicates that an intermediate number of genera should be accepted.

References: Smith 1993a; Smith and Cranfill 2002; Smith et al. 2006.

"Woodsiaceae" (A. Gray) Herter
(Lady Fern Family)

Mostly terrestrial. Rhizomes creeping, ascending or erect, bearing scales at the apex. Leaves monomorphic, rarely dimorphic; blades simple to 4-pinnate-pinnatifid. Sori round, U- to J-shaped, or linear. Indusia linear to kidney-shaped or rarely absent (Figure 8.15).

Distribution and ecology: Almost cosmopolitan, in rainforests, montane forests, thickets, open and shrubby areas, grassy and rocky places, and stream banks.

Genera/species: about 15/700. ***Major genera:*** *Diplazium* (400 spp.) and *Athyrium* (150).

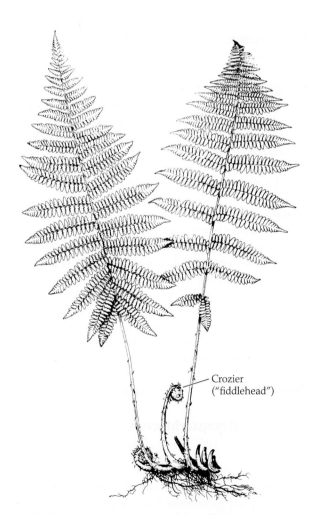

FIGURE 8.15 "Woodsiaceae." *Athyrium filix-femina*: habit (about one-third to one-quarter actual size). (From Taylor 1984; Milwaukee Public Museum, original art by P. Nelson.)

Economic plants and products: None.

Discussion: Support for monophyly of this family is lacking. The two largest genera, *Athyrium* and *Diplazium*, are very closely related and have been united in some classifications. Most species of *Athyrium* have a base chromosome number of 40, and the base chromosome number of most *Diplazium* is 41. *Athyrium filix-femina* is commonly called the lady fern and is widespread in North America, Central and South America, and the Old World. During the 1800s, when collecting and growing ferns was an avocation enthusiastically pursued by many, a variant of this species called 'Queen Victoria' was discovered. This cultivar has unusual leaves that fork near the base and have a criss-crossed appearance.

Reference: Smith et al. 2006.

Blechnaceae Bercht. & J. Presl
(Chain Fern Family)

Terrestrial. Rhizomes creeping, ascending or erect and forming a tree in some species, often bearing stolons, scaly at apex. Leaves monomorphic or often dimorphic, *often red when young,* petiole usually with persistent scales near its base; blades pinnatifid to 1-pinnate (rarely simple or 2-pinnate), large (often more than 1 m), glabrous or occasionally with scales or glands. *Sori often in chains or linear, often parallel and adjacent to midribs.* **Indusia linear and opening toward midribs.** Sporagium stalk with 3 rows of cells (Figure 8.16).

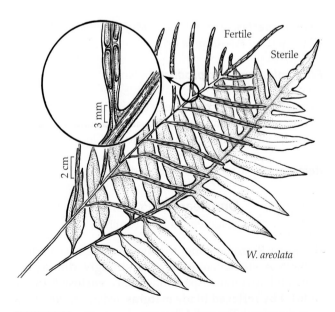

FIGURE 8.16 Blechnaceae. *Woodwardia areolata*: fertile and sterile leaves, about one-quarter size; enlargement of portion of fertile leaf showing sori in chains paralleling midrib. (From Cranfill 1993.)

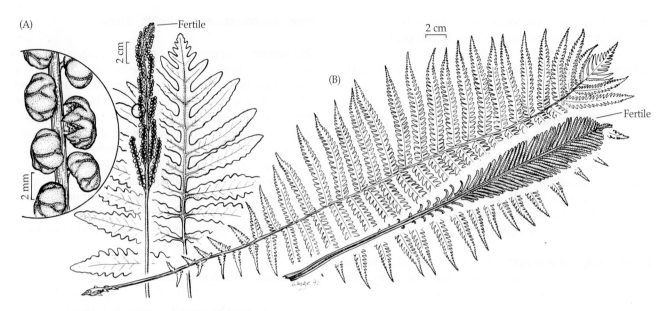

FIGURE 8.17 Onocleaceae. (A) *Onoclea sensibilis*: leaf; enlargement of portion of leaf showing indusia and sori. (B) *Matteucia struthiopteris*: fertile and sterile leaves. (From Johnson 1993b.)

Distribution and ecology: Widespread in tropical and especially south temperate regions. *Blechnum* is primarily in the southern hemisphere, but it is also important in the flora of Hawaii, and in the New World it is native from Alaska to the southernmost tip of South America. This genus grows mostly in wet forests, but also on the borders of forests and in swamps, thickets, and disturbed sites. *Woodwardia* is mostly North American and is found in forests, along streams, and moist meadows, especially in mountainous regions.

Genera/species: about 9/200. ***Major genera:*** *Blechnum* (175 spp.) and *Woodwardia* (14).

Economic plants and products: None.

Discussion: The occurrence of the sori in chains or parallel rows and the nature of the indusia are distinctive features of this family. Blechnaceae are sister to the Onocleaceae.

References: Cranfill 1993; Smith et al. 2006.

Onocleaceae Pic.Serm.
(Fiddlehead Fern Family)

Terrestrial. Rhizomes long- to short-creeping to ascending, sometimes stoloniferous. **Blades strongly dimorphic**, pinnatifid to pinnate-pinnatifid. **Sori enclosed (often tightly) by reflexed blade margins**. *Indusia membranous, often short-lived* (Figure 8.17).

Distribution and ecology: Almost entirely in north-temperate regions, especially in wetlands.

Genera/species: 4/5. ***Major genera:*** *Matteucia* (1 sp.) and *Onoclea* (1).

Economic plants and products: *Matteuccia struthiopteris*, a robust fern of alluvial woods in much of the Northern Hemisphere, is collected in the spring and its young leaves—often referred to as fiddleheads (Plate 8.1D)—are consumed fresh or canned; they are not known to be carcinogenic. This species is commonly called the fiddlehead fern and is often planted near houses.

Discussion: *Onoclea sensibilis* is a common wetlands plant over most of eastern North America and parts of eastern Asia. Its leaves are deeply pinnatifid, and the common name, sensitive fern, refers to the susceptibility of the leaves to frost. In contrast, the spore-bearing leaves of this species (and of *Matteuccia struthiopteris*) persist through the winter and release the spores in the spring. *Onoclea sensibilis* is known as a fossil from the Tertiary (about 60 MYA) not only from its current range, but also from Greenland, the British Isles, and western North America.

References: Johnson 1993b; Smith et al. 2006.

Dryopteridaceae Ching
(Wood Fern Family)

Terrestrial or epiphytic. Rhizomes creeping, ascending, or erect, with scales at the apex. Blades monomorphic, less often dimorphic; sometimes scaly or glandular, less often hairy; **petiole usually with scales persistent at the base**; blades simple to 1–5-pinnate or more divided. *Sori usually round, distinct or closely spaced and covering the leaf surface.*

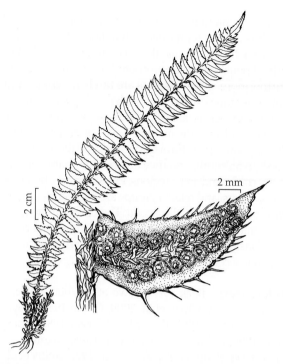

FIGURE 8.18 Dryopteridaceae. *Polystichum lonchitis*: fertile and sterile leaves; enlargement of portion of fertile leaf showing indusia and sori. (From D. H. Wagner 1993).

Indusia kidney-shaped, peltate, or absent. Sporangium stalk with 3 rows of cells (Figure 8.18).

Distribution and ecology: Almost cosmopolitan.

Genera/species: 40–45/1700. ***Major genera:*** *Elaphoglossum* (500 spp.), *Polystichum* (260), *Dryopteris* (225), and *Ctenitis* (150).

Economic plants and products: None.

Discussion: Hybridization and allopolyploidy are very frequent in many genera of this family. For example, in *Polystichum*, a complex genus of about 260 species found throughout most of the world, hybrids are common where two or more species grow together. Sterile hybrids have misshapen sporangia, which appear as small black dots rather than forming the normal sori (Plate 8.1C) with fertile sporangia.

References: Barrington et al. 1989; Smith et al. 2006, Wagner 1993.

Polypodiaceae Bercht. & J. Presl
(Polypody Family)

Mostly epiphytic or growing on rocks, a few terrestrial. Rhizomes long- to short-creeping, scaly. Leaves monomorphic or dimorphic; petiole lacking scales or sometimes scaly; blades simple, pinnatifid, 1-pinnate, or (uncommonly) more divided; hairs, scales, or glands present. *Sori round, oblong, elliptic, or elongate.* **Indusia absent**. Sporangium stalk with 1–3 rows of cells (Figure 8.19).

Distribution and ecology: Throughout the tropics with a few temperate representatives. Commonly epiphytic in the tropics in a wide range of forest types as well as other habitats. On rocks, including old stone walls and tile roofs, or soil in temperate regions.

Genera/species: ~56/1200. ***Major genera:*** *Grammitis* (400 spp.), *Polypodium* (150), *Pleopeltis* (50), and *Campyloneurum* (50).

Economic plants and products: Epiphytic *Platycerium*, *Aglaomorpha*, and *Phlebodium* are grown in greenhouses and tropical gardens for their interesting leaves (see discussion below).

Discussion: This family is part a clade with Dryopteridaceae and several families that are not treated here. Until recently, this family was not considered to include the grammitid ferns, but the green-spored *Grammitis* and related ferns nest within the Polypodiaceae clade in trees based on DNA sequence data. The boundaries between many of the genera are not clear.

The leaves of *Platycerium* are usually dimorphic, with so-called nest leaves that are closely attached to the substrate, vertically oriented, and up to 1 m long. Fertile leaves are erect to spreading or more often hanging and up to 3 m long. Plants can form massive clumps up to almost 2 m wide and heavy enough to break supporting branches. Some members of this genus are called staghorn ferns

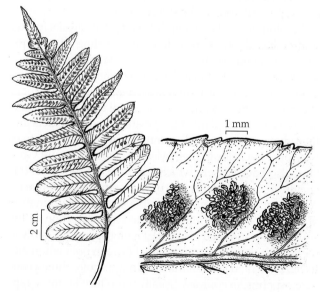

FIGURE 8.19 Polypodiaceae. *Polypodium californicum*: leaf; enlargement of portion of leaf showing sori. (From Haufler et al. 1993).

because of the similarity of the dichotomously branched and pubescent fertile leaves to developing antlers.

Pleopeltis polypodioides (resurrection fern; Plate 8.1G) is the most widespread epiphytic fern in temperate North America; the presence of peltate scales differentiates it from the similar *Polypodium virginianum*.

References: Smith 1993b; Smith et al. 2006; Tryon and Tryon 1982.

GYMNOSPERMS

Extant gymnosperms consist of cycads, Ginkgoaceae, one or two clades of conifers, and Gnetales. Monophyly of gymnosperms has been highly controversial. For many years they were not considered to form a monophyletic group (Crane 1988; Doyle 1998; Doyle et al. 1994; Nixon et al. 1994; Price 1996; Stefanovic et al. 1998). Certainly, when fossils such as the extinct seed ferns—which had naked seeds—are considered, groups with naked seeds do not represent a clade. The status of gymnosperms depends upon the location of the root of the seed plants. If, for example, Gnetales are sister to angiosperms, then gymnosperms are paraphyletic. However, most molecular datasets support monophyly of extant gymnosperms (Bowe et al. 2000; Burleigh and Mathews 2004; Chaw et al. 2000; Goremykin et al. 1996; Rydin et al. 2002; Soltis et al. 2002). Most of these data also suggest a close relationship of Gnetales and Pinaceae Under this so-called "gnepine" hypothesis, conifers themselves are not monophyletic, and Gnetales become extremely divergent conifers. A relationship between Gnetales and Pinaceae is unexpected given the many morphological differences between the two groups and a major genomic difference. Pinaceae and other conifers, but not Gnetales, have lost the inverted repeat from the plastid genome.

Resolution of relationships of seed plants awaits further studies based on more extensive sampling of gymnosperms, but here we treat conifers as monophyletic.

Together cycads, Ginkgoaceae, conifers, and Gnetales represent only about 15 families, 75–80 genera, and about 820 species. The evolutionary success of angiosperms relative to other seed plants may be attributed to vessels and reproductive features. All gymnosperms except Gnetales have only tracheids in their xylem. Angiosperms and Gnetales also have vessels, which are more efficient than tracheids in water transport under some circumstances. Angiosperm carpels make possible stigmatic germination of pollen and are variously adapted for protection of the young ovule and seed dispersal. Gymnosperms are slow to reproduce; up to a year may pass between pollination and fertilization, and seed maturation may require three years. Angiosperms, in contrast, usually reproduce far more rapidly, going from a seed to a seed in weeks in some annuals. With the exception of the cycads and some Gnetales, gymnosperms are pollinated by wind. Angiosperms have adapted in numerous ways to animal pollination, and they are thus able to reproduce in habitats where there is little wind, such as the forest floor. The highly specific nature of animal pollination may promote speciation (see Chapter 6). Furthermore, gymnosperms are rarely polyploid and have not undergone extensive allopolyploid speciation.

Gymnosperms are all woody—being trees, shrubs, or lianas—and include no true aquatics and few epiphytes. These plants grow throughout most of the world, from 72 north to 55 south, and they are the dominant vegetation in many colder and arctic regions. Pines, spruces, hemlocks, firs, yews, cedars, and related groups are familiar as ornamentals and supply high-quality wood. They include the tallest, the most massive, and the longest living individual plants. We will consider 8 of the 15 families of gymnosperms, which together account for the great majority of species.

References: Beck 1988; Bowe et al. 2000; Burleigh and Mathews 2004; Chaw et al. 2000; Crane 1988; Doyle et al. 1994; Friis et al. 1987; Gifford and Foster 1988; Goremykin et al. 1996; Kubitzki 1990; Nimsch 1995; Nixon et al. 1994; Price 1996; Rydin et al. 2002; Singh 1978; Soltis et al. 2002; Sporne 1974; Stefanovic et al. 1998; Stewart and Rothwell 1993; Taylor and Taylor 1993.

Cycadales (Cycads)

The cycads are an ancient group that has retained clearly primitive features, such as motile sperm. Cycads evolved in the Carboniferous or early Permian, about 280 million years ago and reached their peak abundance and diversity in the Mesozoic era. Now cycads are mostly Southern Hemisphere relicts, and many of the species are endangered or threatened by extinction.

The group is monophyletic, as judged by synapomorphies in structural features, such as girdling leaf traces, a specialized (omega-like) pattern of vascular bundles in the petiole, the presence of mucilage canals, and distinctive meristems, as well as poisonous compounds, cycasins. These and other toxic compounds may have been important in the evolution of cycads as defenses against bacteria and fungi. Cycad toxins have been responsible for partial or total paralysis of the hind limbs of livestock.

Another synapomorphy of cycads are special roots that are called **coralloid roots** because of their similarity in appearance to marine coral. These roots host cyanobacteria that carry out nitrogen fixation, like that of bacteria in legumes. The cyanobacteria convert gaseous atmospheric nitrogen, which cycads cannot use, into a form they can use, providing a source of nitrogen that promotes growth in the nutrient-poor soils cycads frequently occupy.

Cycads are often palmlike, with an unbranched upright stem up to 18–20 m tall and large compound leaves crowded toward the stem apex, or fernlike, with a subterranean stem and compound leaves. Most cycads bear **cataphylls**, scalelike leaves that occur among the normal leaves and

---

Key to Families of Cycads

1. Leaflets circinate when young, with a midvein and no lateral veins; megasporophylls leaflike, loosely clustered near stem apex and not forming a strobilus, pinnately lobed or toothed above the ovules, with 2–8 ovules attached laterally to the basal portion .... **Cycadaceae**

1. Leaflets flat or conduplicate when young, with or without a midvein but with numerous, ± parallel veins or midvein present, with numerous, dichotomously branched or simple lateral veins; megasporophylls greatly reduced, valvate or imbricate, forming a strobilus with 2 reflexed ovules .... **Zamiaceae**

---

often serve a protective function. Cycads are slow-growing, with stems reaching as little as 1 m in height in 500 years.

Cycad reproductive structures occur in strobili that consist of an axis and megasporophylls (ovule-bearing leaves) or microsporophylls (pollen-bearing leaves). These simple structures contrast with the complex seed cones of conifers (see below). All cycads have pollen strobili, and all except *Cycas* have ovulate strobili. Although cycads produce abundant, powdery pollen that suggests wind pollination, insects are the primary vectors of pollen. Beetles (and to a lesser extent bees) are either the sole pollen vector or move pollen to the ovule after wind has carried it to the ovulate strobilus from another plant. Pollination and fertilization may be separated by up to 7 months.

Cycad seeds often have a brightly colored (pink, orange, or red) and fleshy outer layer and are commonly dispersed by birds, bats, opossums, turtles, and many other animals. Brightly colored megasporophylls of some cycads also attract dispersers. Ocean currents transport cycads whose seeds are buoyant due to a spongy outer layer, while gravity disperses the seeds of others.

Dioecy characterizes all cycads and may be governed by sex chromosomes. Chromosome number varies considerably among, but usually not within, cycad genera, and is taxonomically useful.

The cycad clade consists of 2 families, 10–11 genera, and about 300 species. Cycadaceae contains only *Cycas*; Zamiaceae comprises the remaining 9–10 genera, including *Stangeria* (which has often been placed in the monogeneric Stangeriaceae).

References: Crane 1988; Johnson and Wilson 1990; Hemsen et al. 2006; Hill et al. 2003; Jones 1993; Landry 1993; Norstog and Nicholls 1997; Norstog and Fawcett 1989; Rai et al. 2003; Stevenson 1990, 1991, 1992.

Cycadaceae Pers.
(Cycad Family)

Palmlike plants, with an unbranched to sparsely branched, woody stem covered with old remnants of leaf bases and living foliage near the apex of the stem; or *fernlike* with a subterranean stem. Leaves persistent, spiral, *pinnately compound;* **leaflets circinate when young and unfolding, with one midvein and no lateral veins**, entire, lower leaflets often spinelike. Microsporophylls aggregated into compact strobili; pollen nonsaccate, with a single furrow. *Megasporophylls grouped at the stem apex, somewhat leaflike and not clustered into a strobilus;* ovules 2–8 on the megasporophyll margin. Seeds large, *slightly flattened,* and covered by a *brightly colored, fleshy outer layer.*

Distribution and ecology: Madagascar, possibly Africa, Southeast Asia, Malaysia, Australia, and Polynesia. In forests and savannas. Many species tolerate fire because the apical meristem is underground or protected by persistent leaf bases.

Genus/species: 1 (*Cycas*)/ca. 20.

Economic plants and products: Several species are popular as ornamentals in warm climates and as houseplants. The stem is a source of a starch called sago, especially in times of food shortage. Seeds may contain 20–30% of this starch, which is edible only after removal of toxins.

Discussion: This family is distinct because its megasporophylls have a well-developed and toothed to pinnately dissected blade and are not compactly clustered into a strobilus (Figure 8.20). *Cycas* is strongly supported by morphological and molecular data as the sister genus to the remainder of the cycads, and its fossil record extends back to the Permian, at least 250 million years ago. It is currently the most widely distributed genus of cycads.

Cyanobacteria in the coralloid roots of Cycadaceae produce BMAA (β-methylamino-L-alanine), a non-protein amino acid that is neurotoxic and that may increase in concentration as it passes up the food chain. In Guam, the Chamarro people, who suffer an extremely high rate of amyotrophic lateral sclerosis/Parkinsonism dementia complex, eat flying foxes (frugivorous bats), which consume cycad seeds. BMAA has been detected at high concentration in the brains of some Chamarros who died of this complex and may have contributed to their neurological disease.

Ovules

FIGURE 8.20 Cycadaceae. *Cycas circinalis*: ovule-bearing leaves of ovulate plant (× 0.5).

References: Cox et al. 2005; Hemsen et al. 2006; Hill et al. 2003; Johnson and Wilson 1990; Rai et al. 2003.

Zamiaceae Horianow
(Coontie Family)

Fernlike with subterranean stem or *palmlike* with aerial, unbranched stem to 18 m tall and large, pinnately compound leaves clustered near stem apex. Stem covered with persistent dead leaf bases or naked. Leaves pinnately (rarely twice pinnately) compound, spiral, persistent, leathery, with or without stout spines on the petiole and rachis; *leaflets flat when young and unfolding, with numerous ± parallel veins* (with a midvein and dichotomously branching secondary veins in *Stangeria*), entire, dentate, or with sharp spines. Microsporophylls aggregated into compact strobili, with numerous small microsporangia that are often clustered; pollen nonsaccate, with a single furrow. **Megasporangiate strobili** 1 to several per plant, more or less globose to ovoid or cylindrical, disintegrating at maturity; **megasporophylls** densely crowded, symmetrically to asymmetrically peltate, valvate or imbricate, **each with 2 ovules**. Seeds large (1–2 or more cm long), **± round in cross-section**, with an *often brightly colored and fleshy outer layer* and a hard inner layer; cotyledons 2 (Figure 8.21).

Distribution and ecology: Tropical to warm temperate regions of the New World, Africa, and Australia. From poor, dry soils of grasslands and woodlands to dense, tropical forests. The only cycad native to the United States is *Zamia floridana* in Florida and southern Georgia.

Genera/species: 9/111. **Major genera:** *Encephalartos* (35 spp.), *Zamia* (35), *Macrozamia* (14), *Ceratozamia* (10), and *Dioon* (10).

Economic plants and products: Many species are grown as ornamentals in warm climates and as houseplants. Like *Cycas,* the stem and seeds of many Zamiaceae species are a source of sago starch. Removal of the toxic glycosides cycasin and macrozamin by boiling and washing is required before consuming sago.

Discussion: The seed strobili of this family are among the heaviest and largest of seed plant reproductive structures, weighing up to 40 kg and measuring 60 cm in length and 30 cm in diameter. The seeds are also quite large, up to 4 cm long. *Zamia* is unusual in having species with several different chromosome numbers, and species delimitation is very difficult in this genus.

Bowenia, which has two or three species confined to tropical regions of northeastern Australia, is unique among cycads in its bipinnately compound leaves. This genus has been placed in its own family or united with *Stangeria*, which grows naturally only in southern Africa and contains only one species, in the Stangeriaceae. *Stangeria* is exceptional in Zamiaceae in its leaf venation (see family description). Molecular data analyzed to date clearly place these two genera within Zamiaceae but are not clear about the relationship between them and other members of the family.

Zamia is wide-ranging, morphologically diverse, and has numerous species. Pollen-eating beetles are responsible for pollinating *Zamia* in Florida and Mexico. Release of oil of wintergreen apparently draws the beetles to microsporangiate strobili, which provide an egg-laying site and food for adult and larval beetles. Adult beetles eat pollen and are covered by it. Pollination occurs when pollen-covered adults go to megasporangiate strobili where the beetles neither feed for lay eggs. Some molecular data suggest a relationship between *Zamia, Microcycas,* and *Ceratozamia,* and these three genera share valvate megasporophylls and articulate leaflets (likely synapomorphies).

References: Hill et al. 2003; Johnson and Wilson 1990; Landry 1993; Norstog and Fawcett 1989; Norstog and Nicholls 1997; Rai et al. 2003; Stevenson 1991, 1992.

Ginkgoales

Ginkgoaceae Engler
(Ginkgo or Maidenhair Tree Family)

Trees to 30 m, with a more or less asymmetrical crown and furrowed, gray bark. Resin canals absent. Leaves simple, spiral, and widely spaced on long shoots near the tips of branches, also *closely packed on stubby short shoots on older growth;* **fan-shaped**, bilobed or entire; **deciduous** and bright yellow in the fall; **dichotomously veined**. Dioecious.

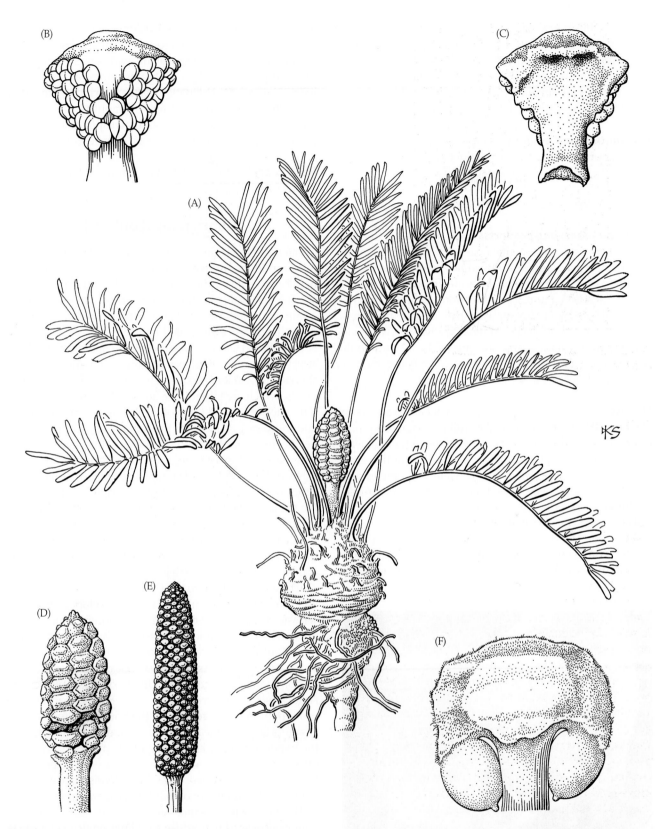

FIGURE 8.21 Zamiaceae. *Zamia floridana*: (A) habit of ovulate plant (some leaves removed) with strobilus at time of pollination; note large fleshy taproot with small lateral roots and coralloid roots (at right, near stem–root juncture) (× 0.75); (B) abaxial surface of microsporophyll, with microsporangia (× 4.5); (C) adaxial surface of microsporophyll (× 4.5); (D) ovulate strobilus at time of pollination; note separation of megasporophylls in lower portion of strobilus, allowing entry of pollinators and pollen (× 0.75); (E) microsporangiate strobilus during pollen shedding (× 0.5); (F) adaxial view of megasporophyll with two ovules, each with micropyle directed toward axis at bottom (× 4.5). (From Stevenson 1991, *J. Arnold Arbor.* Suppl. Series 1: pp. 367–384.)

Pollen strobili borne on spur shoots, long and pendant; pollen not winged. **Ovules paired on a long stalk from spur shoots**; seeds frequently 1 per stalk (the other ovule not maturing into a seed), ca. 2.5 cm in diameter, with a juicy, unpleasant-smelling outer coat and hard inner coat; cotyledons 2–3 (see Figure 7.11F–H).

Distribution and ecology: Limited to remote mountain valleys of China; possibly extinct in the wild. Little is known about the ecology of this species.

Genus/species: 1 (*Ginkgo*)/1 (*G. biloba*).

Economic plants and products: The maidenhair tree has been grown as an ornamental near religious institutions in eastern Asia for centuries. Individuals can live for over a thousand years, and these old trees are probably the source of plantings in many parts of the world, where *Ginkgo* is commonly used as ornamentals for shade and for their interesting and attractive foliage. Staminate individuals, which do not produce the unpleasant-smelling seeds, are commonly planted. The gametophyte and embryo, when boiled, fried, or roasted, are a delicacy in some Chinese dishes.

Discussion: The first representatives of *Ginkgo* appear in the late Triassic, more than 200 million years ago, and reproductive structures have changed little in their general appearance in about 120 million years. During the early Jurassic, extinct relatives of *Ginkgo* were widespread and diverse, consisting of perhaps three families. Now, ironically, although rare or possibly extinct in the wild, *Ginkgo* does well as a shade tree in urban situations.

The broad, deciduous leaves of *Ginkgo* are unlike those of nearly all other gymnosperms (Figure 8.22). Sperm motility, known elsewhere in seed plants only in the cycads, is clearly

FIGURE 8.22 Ginkgoaceae. Broad leaf of *Ginkgo biloba*. Comparison with a fossilized *Ginkgo* leaf from the Triassic (about 225 MYA) illustrates how little this genus has changed over evolutionary history.

a primitive feature, as is the lack of pollen tubes. Ginkgoaceae are not closely related to any extant groups.

Ginkgo is one of the few plants with sex chromosomes. Ovulate plants bear two X chromosomes, and staminate individuals are XY. Pollination occurs by wind in the spring, but fertilization is delayed for 4–7 months. The juiciness and odor of the seed suggest animal dispersal, but the dispersing taxa are unknown and may now be extinct.

References: Page 1990b; Whetstone 1993; Zhou and Zhengzoos.

Coniferales (Conifers)

Conifers are the largest and most ecologically and economically important group of gymnosperms. Pines, spruces, firs, hemlocks, cedars, cypresses, redwoods, and giant sequoias are familiar, valued, and revered trees. Members of this group are called conifers because most bear their seeds in specialized structures called cones. Cones protect ovules and seeds and also facilitate pollination and dispersal. These structures consist of an axis bearing highly modified short shoots, the ovuliferous scales. These scales are subtended by bracts, which are either large and conspicuous (as in some Pinaceae), very small (as in other Pinaceae), or small to large and more or less fused to the scale (as in Cupressaceae). Seeds are associated with the scales. Cone scales of most members of Pinaceae and Cupressaceae are woody or leathery (Plate 8.2A,F). The junipers (*Juniperus*) have cone scales that are more or less juicy and brightly colored, making the cones berrylike (see Figure 8.25H,P,Q) and animal-dispersed. In Podocarpaceae cones are often reduced, with highly modified, juicy, brightly colored scales and just one ovule (Plate 8.2B). Taxaceae bear solitary seeds partially or completely surrounded by a juicy aril (Plate 8.2D).

Conifers date back to the Carboniferous, some 300 million years ago. Many current families developed by the late Triassic or early Jurassic, and some contemporary genera appeared in the middle Jurassic. Today conifers remain important in colder regions, such as the boreal forests of North America and Asia, where species of pine, spruce, and fir dominate the vegetation. Other conifers—particularly Araucariaceae, Cupressaceae, and Podocarpaceae—are prominent in cooler regions of the Southern Hemisphere. Conifers are valuable ornamentals, and their wood is used for paper, building, and many other purposes. They are often referred to as "evergreens" because of the persistent foliage in most species, or as "softwoods" because their wood is softer than that of many angiosperm trees.

Pollination is by wind. Most conifers, like most non-angiosperm seed plants, use a pollination droplet, a sticky fluid extruded from the ovule at pollination, to catch airborne pollen. Pollen grains of most Pinaceae bear two **saccae**: small, winglike appendages that may serve to float the pollen grains upward in the pollination droplet toward the ovule or to orient the grains properly for germination.

Key to Selected Families of Conifers

1. Seeds in woody (fleshy only in *Juniperus*) cones, mostly hidden by cone scales, a few
to many per cone; plants highly resinous 2

1. Seeds partially to wholly enclosed or subtended by fleshy, often brightly colored structures,
usually solitary: plants slightly resinous 4

2. Seeds 1 per cone scale ....**Araucariaceae**

2. Seeds usually more than 1 per cone scale 3

3. Leaves scale-like or needle-like, spiral, opposite, or whorled, persistent on branches
after dying (but most branches shed with age); pollen nonsaccate; cone scales valvate or
imbricate (and then leaves scalelike and opposite), flat or peltate, fused to bract, juicy
in *Juniperus*; seeds with 2 or 3 lateral wings (infrequently wingless), 1–20 per scale ....**Cupressaceae**

3. Leaves linear to needle-like, spiral, shed from branches (or as short-shoots in *Pinus*);
pollen often saccate; cone scales imbricate, flat, distinct from bract; seeds terminally
winged (or rarely wingless), 2 per scale..**Pinaceae**

4. Seeds more or less surrounded by a specialized cone scale (the epimatium), not an aril,
and usually associated with colored, juicy bracts; pollen mostly saccate; pollen strobili
with 2 sporangia per microsporophyll ....**Podocarpaceae**

4. Seeds more or less surrounded by an aril, an outgrowth of the axis below the ovule;
pollen nonsaccate; pollen strobili with 2–9 sporangia per microsporophyll ....**Taxaceae**

Alternatively, the pollen may be trapped on more or less sticky structures in the vicinity of the ovule. The pollen then germinates and grows via a pollen tube to the ovule (the sperm lack flagella).

Conifer trees are often monopodial with a dominant central shoot or trunk. With age, the crown may branch irregularly. Branches are often whorled, at least when the plant is young.

The conifers comprise 7 families, 60–65 genera, and over 600 species. Relationships among the major groups of conifers are shown in Figure 8.23. The five families treated here include all but a few conifer species.

References: Brunsfeld et al. 1994; Eckenwalder 1976; Farjon 1990, 2005a; Hart 1987; Kelch 1997; Page 1990a,c–f; Price et al. 1987; Price and Lowenstein 1989; Quinn et al. 2002; Richardson 1998; Singh 1978; Sporne 1974; Stefanovic et al. 1998; Thieret 1993a; Watson and Eckenwalder 1993.

Pinaceae Adanson
(Pine Family)

Trees (occasionally shrubs), often emitting strong fragrances from bark and/or leaves; *resin canals present in wood and leaves*. Branches whorled or opposite (rarely alternate). Leaves simple, *linear to needlelike* (rarely narrowly ovate), spiral but often appearing 2-ranked by twisting of leaf base to bring most of the leaves into one plane, clustered or fascicled in groups of 2 to 5 in *Pinus*, sessile or short-petioled, on long shoots or tightly clustered on short shoots, persist-ent (deciduous in *Larix* and *Pseudolarix*). *Monoecious*. Microsporangiate strobili with spirally arranged, bilaterally symmetrical microsporophylls; microsporangia 2 on the abaxial microsporophyll surface; pollen grains with 2 saccae (saccae absent in *Larix, Pseudotsuga*, and all but two species of *Tsuga*). *Cones with spirally arranged, flattened bract-scale complexes*; scales persistent (deciduous in *Abies, Cedrus,* and *Pseudolarix*), *bracts free from the scale*, longer than the cone scale to much shorter than the cone scale; maturing in 2 (3) years; *ovules 2*, **inverted** (micropyle directed toward the cone axis), on the adaxial cone scale surface; archegonia few per ovule, not clustered. **Seeds with a long, terminal wing** derived from tissue of the cone scale (wing reduced or absent in some species of *Pinus*); embryo straight, cotyledons 2–18 (Figure 8.24; see also Figures 7.13A–E).

Distribution and ecology: Pinaceae are almost entirely limited to the Northern Hemisphere. Three or four genera grow only in eastern Asia; *Cedrus* is confined to North Africa, the Near East, Cyprus, and the Himalayas; and the remaining six genera (the major genera) all occur widely in the Northern Hemisphere. The family ranges from warm temperate climates to the limit of tree growth above the Arctic Circle, from permanently water-saturated soils to well-drained soils, and from sea level to alpine habitats up to 4800 m above sea level in eastern Tibet. The seeds of pines are primary components of the diets of many species of birds, squirrels, chipmunks, and other rodents. Members of the family provide cover for many wildlife species and are important in watershed protection.

(A)

Coniferales: Cupressaceae
Chamaecyparis obtusa: cones

(B)

Coniferales: Podocarpaceae
Podocarpus macrophyllus: mature ovulate cones

(C)

Coniferales: Pinaceae
Pinus palustris: habit

(D)

Coniferales: Taxaceae
Taxus baccata: aril

(E)

Coniferales: Araucariaceae
Araucaria heterophylla: habit

(F)

Coniferales: Pinaceae
Abies sp.: ovulate cone

(G)

Coniferales: Pinaceae
Picea torano: branch with leaves

PLATE 8.2 Gymnosperms: Coniferales

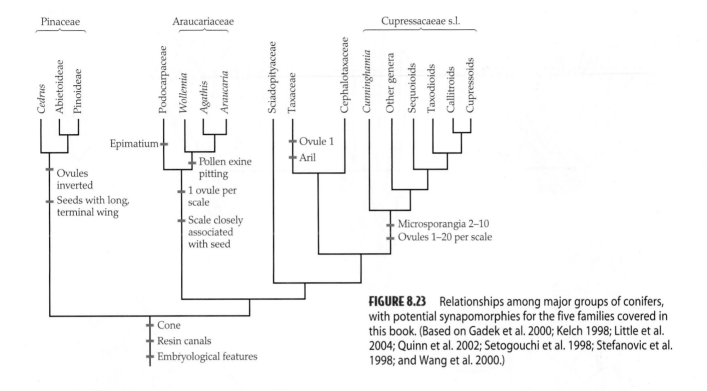

FIGURE 8.23 Relationships among major groups of conifers, with potential synapomorphies for the five families covered in this book. (Based on Gadek et al. 2000; Kelch 1998; Little et al. 2004; Quinn et al. 2002; Setogouchi et al. 1998; Stefanovic et al. 1998; and Wang et al. 2000.)

Genera/species: 10/220. ***Major genera:*** *Pinus* (100 spp.), *Abies* (40), *Picea* (40), *Larix* (10), *Tsuga* (10), and *Pseudotsuga* (ca. 5).

Economic plants and products: Pinaceae are probably the leading source of timber in the world. The wood of pines (*Pinus*), Douglas firs (*Pseudotsuga*), spruces (*Picea*), hemlocks (*Tsuga*), larches (*Larix*), firs (*Abies*), and cedars (*Cedrus*) is used extensively for construction, pulp for paper production, fence posts, telephone poles, furniture, interior trim for houses, woodenware, and numerous other purposes. Cedars, pines, and Douglas fir, in particular, have reputations for quality construction wood. The wood of spruces has long been preferred for the soundboards of stringed instruments, such as violins. Pines, spruces, hemlocks, cedars, Douglas firs, and firs are used extensively as ornamentals, and hundreds of cultivars have been developed in many of the species of these genera. Pine "nuts"—the more or less wingless seeds of the piñon pines of southwestern North America—were a staple in the diet of native North Americans. These seeds, and those of some Old World groups of pines, are now a gourmet food. Rosin and turpentine are extracted from various species of pines.

Discussion: Pinaceae are the largest and both economically and ecologically the most important family of conifers. Species in the three largest genera—*Abies*, *Picea*, and *Pinus*—are the primary components of many forests in cool and cold regions of the Northern Hemisphere. *Pinus* often dominates fire-maintained forests in warmer regions also, such as the southeastern United States (Plate 8.2C).

Numerous features—ovule inversion, prominent terminal seed wing, pattern of proembryogeny, protein-type sieve cell plastids, and the absence of biflavonoid compounds—establish the monophyly of Pinaceae. The family is not phylogenetically close to other extant conifer groups and is probably the sister group to the remaining conifers (see Figure 8.23).

Strongly congruent structural and seed protein immunological data divide Pinaceae into two subfamilies, Abietoideae and Pinoideae. Abietoideae include *Abies*, *Cedrus*, *Keteleeria*, *Pseudolarix*, and *Tsuga*, while *Cathaya*, *Larix*, *Picea*, *Pinus*, and *Pseudotsuga* make up Pinoideae. Phylogenetic analysis of three genes—*matK* from the chloroplast, mitochondrial *nad5*, and nuclear *4CL*—agree with this fundamental partition of the family except that *Cedrus* is sister to the remainder of the family. Pinoideae are supported by several synapomorphies (absence of resin canals in the seed coat, absence of a narrowed, pedicillate base of the cone scales, and presence of two resin canals in the vascular cylinder of the young taproot) and contain two clades, *Pseudotsuga* plus *Larix*, and *Cathaya*, *Picea*, plus *Pinus*. Abietoideae in the molecular phylogeny has two sets of sister taxa, *Abies* plus *Keteleeria* and *Pseudolarix* plus *Tsuga*. The monotypic genus *Nothotsuga* is very closely related to *Tsuga* and is best treated as a species of *Tsuga*.

Pinus is highly distinct in its leaves, which are clustered in groups of usually two to five, and its cone scales, which are apically thickened and often armed with a prickle. This genus also has the longest fossil record of extant Pinaceae, extending back to the Jurassic or early Cretaceous. By the late Cretaceous, two monophyletic subgenera—*Pinus* and

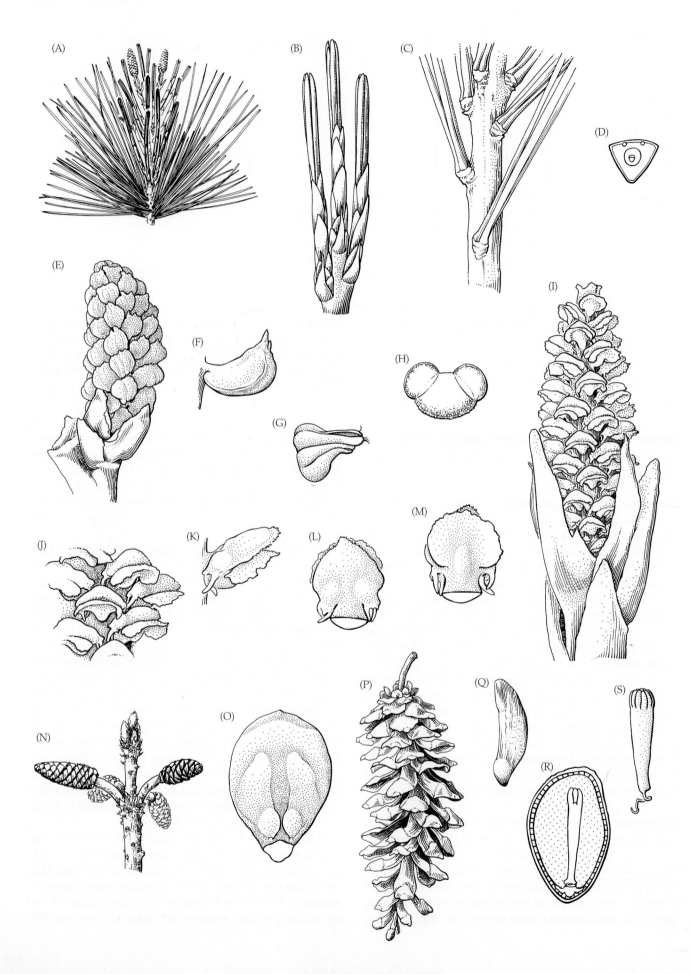

◀ **FIGURE 8.24** Pinaceae. *Pinus* subgen. *Strobus, P. strobus*: (A) tip of shoot with last season's leaves, new growth, and two ovulate cones at time of pollination (× 0.7); (B) short-shoots, each showing subtending bract, scale leaves, and five developing needle leaves (× 2.7); (C) detail of bases of mature short-shoots; the bracts and scale leaves have abscised (× 2.7); (D) diagrammatic cross-section of needle leaf, showing single fibrovascular bundle and two resin ducts (× 20); (E) microsporangiate strobilus just before dehiscence (× 5.3); (F,G) lateral and abaxial views of microsporophyll, showing the two abaxial sporangia and their dehiscence (× 13); (H) pollen grain, showing the two saccae (× 330); (I) ovulate cone at time of pollination (× 4); (J) detail of ovulate cone, showing scales with ovules and subtending bracts at time of pollination (× 13); (K) detail of lateral view of single cone scale and subtending bract at time of pollination; one ovule with two micropylar appendages visible (× 16); (L,M) abaxial and adaxial views of cone scale and bract at time of pollination, showing micropylar appendages on the two ovules (× 16); (N) four cones at the end of the first growing season, with all surrounding fascicles of leaves removed; lateral and terminal buds visible (× 0.7); (O) cone scale with two developing seeds at time of pollination of the next year's cones, showing remnants of the micropylar appendages and development of wings (× 4); (P) mature cone (× 0.7); (Q) mature seed, after wing has separated from cone scale (× 1.3); (R) longitudinal section of seed, wing removed, showing embryo surrounded by tissue of megagametophyte (stippled), micropyle facing base (× 7); (S) embryo, showing numerous cotyledons (× 8). (From Price 1989, *J. Arnold Arbor.* 70: pp. 247–305.)

Strobus—had differentiated. Members of subgenus *Pinus* are called hard pines because their wood is harder than that of subgenus *Strobus* species, which are commonly called soft pines. The two subgenera differ in other wood anatomical features and in the number of vascular bundles in each leaf, two in hard pines and one in soft pines. The leaf clusters are surrounded at the base by a sheath, which is persistent in hard pines and deciduous in soft pines. About 65 *Pinus* species are native to North America, with high concentrations of species in Mexico, California, and the southeastern United States.

Pinaceae include the longest-lived trees: intermountain bristlecone pine (*Pinus longaeva*), an alpine species of the southwestern United States, lives over 5000 years. While not matching Cupressaceae in height and massiveness, the pine family does have some huge trees. Douglas fir (*Pseudotsuga menziesii*), for instance, grows to over 80 m in height, and many pines and spruces exceed 60 m.

References: Farjon 2005c; Farjon and Styles 1997; Gernandt et al. 2005; Page 1990c; Price 1989; Richardson 1998; Stefanovic et al. 1998; Syring et al. 2005; Thieret 1993a; Wang et al. 2000; Wegst 2006.

Cupressaceae Gray
(Cypress or Redwood Family)

Trees or shrubs; wood and foliage often aromatic. Bark of trunks often fibrous, shredding in long strings on mature trees, or forming blocks. Leaves persistent (deciduous in three genera), simple, spiral or basally twisted to appear 2-ranked, opposite, or whorled, *scale-like, tightly appressed and as short as 1 mm to linear and up to about 3 cm long,* with resin canals, shed with the lateral branches; adult leaves appressed or spreading, sometimes spreading and linear on leading branches and appressed and scale-like on lateral branches; scale-like leaves often dimorphic, the lateral leaves keeled and folded around the branch and the leaves on the top and bottom of the branch flat. *Monoecious (dioe-*cious in *Juniperus*). Microsporangiate strobili with spirally arranged or opposite microsporophylls; **microsporangia 2–10** on the abaxial microsporophyll surface; pollen nonsaccate. Cone maturing in 1–3 years; scales peltate or basally attached and flattened, juicy in *Juniperus*, fused to bracts, persistent (deciduous in *Taxodium*); **ovules 1–20,** on adaxial scale surface, erect (micropyle facing away from the cone axis; in some the ovules may eventually be inverted); archegonia quite variable in number per ovule, clustered. Seeds with 2 (3) short lateral wings (wings absent in some genera); embryo straight, cotyledons 2–15 (Figure 8.25).

Distribution and ecology: This is a cosmopolitan family of warm to cold temperate climates. About three-quarters of the species occur in the Northern Hemisphere. About 16 genera contain only one species, and many of these have narrow distributions. Members of this family grow in diverse habitats, from wetlands to dry soils, and from sea level to high elevations in mountainous regions. The two species of *Taxodium* in the southeastern United States often grow in standing water.

Genera/species: 29–32/110–130.   ***Major genera:*** *Juniperus* (ca. 68 spp.), *Callitropsis* (18), *Callitris* (15), *Cupressus* (12), *Chamaecyparis* (7), *Thuja* (5), *Taxodium* (3), *Sequoia* (1), and *Sequoiadendron* (1).

Economic plants and products: The family produces highly valuable wood. *Cryptomeria, Chamaecyparis, Juniperus, Sequoia, Taxodium, Thuja,* and several other genera are suited for house construction, siding, decking, caskets, shingles, boat construction, paneling, wooden pencils, and many other purposes. Many woods from this family are naturally fragrant and have been used as a natural mothproofing for closets and chests and in the manufacture of perfumes. *Juniperus communis* cones are used to flavor gin. *Juniperus* pollen contains one of the most potent airborne allergens, and the tremendous amount of pollen produced by *Juniperus* species is highly correlated with nasal, sinus,

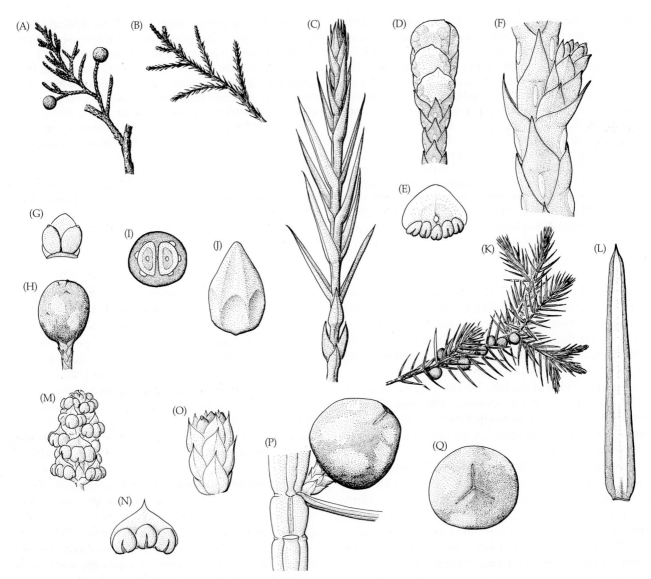

FIGURE 8.25 Cupressaceae. (A–J) *Juniperus virginiana*: (A) branchlets with only scale leaves, bearing mature ovulate cones (× 0.9); (B) branchlet with scale and needle leaves (× 0.9); (C) detail of branchlet with needle leaves, showing decurrent leaf bases (× 6.2); (D) microsporangiate strobilus before shedding of pollen, subtended by numerous scale leaves (× 6.2); (E) microsporophyll (abaxial view), showing dehisced sporangia (× 12); (F) branchlet with ovulate cone near time of pollination (× 9); (G) cone scale (adaxial view) with 2 erect ovules near time of pollination (× 12); (H) mature ovulate cone with fused cone scales (× 3.7); (I) cross-section of mature cone, only two seeds maturing (note resin vesicles outside seeds) (× 3.7); (J) seed, showing pits and ridges (× 6.2). (K–Q) *J. communis*: (K) branch, showing ternate leaves and axillary ovulate cones (× 0.9); (L) details of abscised portion of leaf in adaxial view, showing broad, white stomatal band (× 6.2); (M) microsporangiate strobilus after shedding of pollen (× 6.2); (N) microsporophyll, abaxial view; (O) axillary shoot with young ovulate cones at apex, showing three ovules near time of pollination (× 12); (P) portion of branchlet with mature ovulate cone; note remnant leaf bases fused to larger stem (× 3.7); (Q) apical views of ovulate cone, showing suture lines between three fused cone scales (× 3.7). (From Hart and Price 1990, *J. Arnold Arbor.* 71: pp. 275–322.)

and pulmonary allergies in humans and domestic animals. *Chamaecyparis, Cupressus, Juniperus, Platycladus, Thuja,* and other genera are grown extensively as ornamentals.

Discussion: This family was long split into Cupressaceae s.s. and Taxodiaceae on the basis of differences in the leaves. Cupressaceae s.s. leaves are either opposite and scale-like or whorled and linear, whereas those of Taxodiaceae are mostly spiral and linear. Leaves of *Metasequoia* (Taxodiaceae), however, are opposite, and those of *Athrotaxis* (Taxodiaceae) may be scale-like. There are numerous similarities (and potential synapomorphies) uniting these families: fusion of cone scale and bract; lateral wings on the seeds derived from the seed coat; microsporangia two or more per microsporophyll;

more than two seeds per cone scale; shedding of small branches; clustered archegonia; wingless pollen grains; peltate cone scales in many genera, and DNA sequence characters (Brunsfeld et al. 1994; Eckenwalder 1976; Hart 1987; Stefanovic et al. 1998; Tsumura et al. 1995; Watson and Eckenwalder 1993). Finally, Cupressaceae s.s. are monophyletic and probably arose out of a paraphyletic assemblage, the "Taxodiaceae." Hence the evidence comes down decisively on the side of merging the two families.

Cupressaceae s.s. divide into two well-supported clades: the cupressoid clade, with all Northern Hemisphere groups; and the callitroid clade, comprising all taxa from the Southern Hemisphere (see Figure 8.23). Cupressaceae s.s. are the sister group of the well-supported taxodioid clade of three genera: *Taxodium*, *Glyptostrobus*, and *Cryptomeria*. *Taxodium* grows in the eastern United States and Mexico and, like its sister genus *Glyptostrobus* (southern and central China), is deciduous. *Cryptomeria* is wide-ranging in China and Japan. Another well-supported trio of genera that diverged early in the evolution of Cupressaceae is the sequoioid clade of *Metasequoia*, *Sequoia*, and *Sequoiadendron*. *Metasequoia* was widely distributed and one of the most common genera of Cupressaceae in the Northern Hemisphere from the late Cretaceous to the Miocene. Its native range is now restricted to an isolated region of west-central China, and it was known outside its native range only as a fossil until 1944. Its deciduous habit must have evolved in parallel to that of *Taxodium* and *Glyptostrobus*. *Sequoia* and *Sequoiadendron*, like *Metasequoia* and several other genera, contain just one species each and are very geographically limited. *Sequoia* is restricted to coastal regions of northern California and southern Oregon and *Sequoiadendron* to the mountains of central California. *Cunninghamia*, a genus of about three Southeast Asian species, is the sister to the remainder of the family.

Juniperus is a member of the cupressoid clade and the second largest genus of conifers after *Pinus*. Junipers are mostly confined to the Northern Hemisphere, with centers of diversity in the deserts of Mexico and the southwestern United States, the Mediterranean, and central Asia and China. *Juniperus* species range from sea level to above treeline and from deserts to marshes. Some species in this genus are weedy and have invaded millions of acres of rangeland and farmland. The juicy cones of *Juniperus* are consumed by birds and small mammals, and long-distance dispersal by birds is suspected to have transported the genus to Atlantic islands such as the Azores, Bermuda, and the Canary Islands.

Juniperus is sister to the Old World genus *Cupressus*, which formerly included 16 species of western North and South America. These New World species have been transferred to *Callitropsis*. *Callitropsis nootkatensis*, the Nootka cypress (also called the Alaska cedar and yellow cedar), has been difficult to place taxonomically because of its distinctive morphology. This important timber species of northwestern North America had been placed in *Chamaecyparis*, *Cupres-*

sus, and *Xanthocyparis*. The closest relative of the Nootka cypress may be a recently described species from remnant forests of northern Vietnam, *Callitropsis vietnamensis*.

Sciadopitys, commonly called umbrella pine, was traditionally placed in "Taxodiaceae." The leaves in this genus appear to be fused in pairs but it is perhaps more likely that these paired structures are a kind of modified stem. Numerous morphological, molecular, and other differences argue for separate family status (Stefanovic et al. 1998), i.e., Sciadopityaceae.

Cupressaceae include the tallest (*Sequoia sempervirens*, redwood, almost 112 m tall and 6.7 m in diameter) and most massive (*Sequoiadendron giganteum*, giant sequoia, 106 m tall and 11.4 m in diameter) plants on earth. Some species live 2000–3500 years or more.

References: Adams 1993; Brunsfeld et al. 1994; Eckenwalder 1976; Farjon 2005b; Gadek et al. 2000; Hart 1987; Kusumi et al. 2000; Li and Xiang 2005; Little et al. 2004; Little 2006; Page 1990a,f; Price and Lowenstein 1989; Stefanovic et al. 1998; Tsumura et al. 1995; Watson and Eckenwalder 1993.

Podocarpaceae Endlicher
(Podocarp Family)

Shrubs or trees to 60 m tall, slightly resinous. Leaves simple, entire, varying greatly in shape (broadly linear and up to 30 cm long and 5 cm wide to scale-like), persistent, alternate. Dioecious (rarely monoecious). Microsporangiate strobili cylindrical, with numerous spirally arranged microsporophylls each with 2 microsporangia; pollen with 2 (0 or 3) saccae. Cones with 1 to many ovulate scales, each with 1 ovule and more or less reduced and fused to ovule, modified into a juicy structure (**epimatium**), and therefore *drupe-like*, rarely resembling a cone (Plate 8.2B). Cotyledons 2.

Distribution and ecology: Podocarpaceae are tropical and subtropical (less often cool temperate), especially in the Southern Hemisphere in the Old World. The family extends northward to Japan, Central America, and the Caribbean. Podocarps grow primarily in mesic forests.

Genera/species: 17/170 or more. **Major genera:** *Podocarpus* and *Dacrydium* (the number of species recognized in these genera, neither of which is monophyletic, depends upon how they are subdivided).

Economic plants and products: *Dacrydium*, *Podocarpus*, and other members of the family have valuable timber. *Podocarpus macrophyllus* is widely planted as an ornamental in mild climates.

Discussion: 28S rRNA gene sequences strongly support a sister-group relationship of Podocarpaceae and Araucariaceae (see Figure 8.23), sharing the following synapomorphies: one ovule per ovulate scale, ovulate scale closely asso-ciated with the seed, and possibly bract fused to scale. All but two genera of Podocarpaceae bear an epimatium, which is generally interpreted as a modified cone scale that partially folds around the ovule and is juicy at maturity. Podocarps also have an unusual (possibly synapomorphic) binucleate cellular stage early in embryogeny, and are unusual among conifer families in their diversity of cone structure and chromosome number. The family may have been long isolated from other conifers in the Southern Hemisphere.

DNA sequence data from *rbcL* support inclusion of the Southern Hemisphere genus *Phyllocladus* (Phyllocladaceae) in Podocarpaceae, as sister to the remainder of the family. Podocarpaceae also contains a bizarre shrub of remote forests of New Caledonia, *Parasitaxus usta*, that lacks roots and parasitizes another member of the Podocarpaceae, *Falcatifolium taxoides*.

There is considerable uncertainty about circumscription and relationships of genera of Podocapaceae. Morphology and 18S rDNA sequence data show that neither *Dacrydium* nor *Podocarpus* is monophyletic, and some workers recognize numerous segregate genera.

References: Axsmith et al. 1998; Conran et al. 2000; Kelch 1997, 1998; Page 1990d; Quinn et al. 2002; Stefanovic et al. 1998; Tomlinson 1992.

Araucariaceae Henkel & W. Hochst
(Norfolk Island Pine Family)

Long-lived trees to 65 m tall and 6 m in diameter at the base, highly resinous, usually very symmetrical and conical in growth form (Plate 8.2E). Leaves simple, entire, varying in shape (awl-like, scale-like, linear, oblong, or elliptic) sometimes on the same individual, persistent, sharp-point-ed in some species of *Araucaria*, spiral or opposite. Dioecious or monoecious. Microsporangiate strobili cylindrical, with numerous spirally arranged microsporophylls each with 4–20 microsporangia; pollen without saccae, the **exine pitted**. *Cones solitary, more or less erect, heavy*, maturing in 2–3 years and eventually disintegrating on the tree; *ovulate scales each with 1 ovule*, numerous, spirally arranged, flat-tened, linear to peltate, the bract more or less longer than and fused to the scale; *seeds large*, with or without marginal wings. Cotyledons 2, sometimes deeply divided and appearing like 4 (Figure 8.26).

Distribution and ecology: Araucariaceae are nearly restricted to the Southern Hemisphere, ranging from southeast Asia to Australia, New Zealand, and southern South America. Members of this family usually grow in tropical and subtropical rainforests as well as more temper-ate areas. The family is most diverse in New Caledonia, where 5 species of *Agathis* and 13 species of *Araucaria* are endemic. The New Caledonian species of each genus form a monophyletic subgroup according to *rbcL* analyses. The

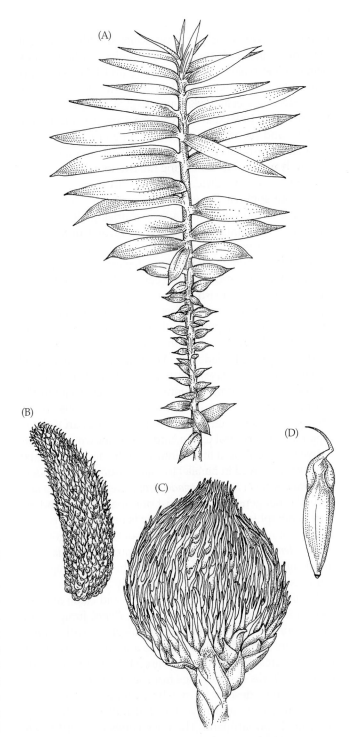

FIGURE 8.26 Araucariaceae. *Araucaria*: (A) *A. bidwilli* branch; (B-D) *A. araucana* microsporangiate strobilus, cone, and seed. (A from Page 1990e; B–D from Muñoz Pizarro 1959.)

small amount of genetic diversification of the species with-in each genus suggests a relatively recent radiation on the unusual (ultramafic) soils of New Caledonia. In some species of *Araucaria*, sharp-pointed leaves, ability to regen-erate branches, and protection of the growing apex by sur-rounding branches in young plants suggest adaptation against now-extinct herbivores.

Genera/species: 3/32.   ***Major genera:*** *Agathis* (13 spp.) and *Araucaria* (18).

Economic plants and products: Both of the major genera produce valuable timber. Larger individuals, such as those of *Agathis australis* of New Zealand, that reach 65 m in height and over 6 m in diameter, contain large amounts of high-quality wood. The Norfolk Island pine (*Araucaria hetrerophylla*) and the monkey puzzle tree (*A. araucana*) of Chile are prized ornamentals, either as landscape plants or house plants (Norfolk Island pine).

Discussion: Araucariaceae, like Podocarpaceae, are a distinct, almost exclusively Southern Hemisphere family. The fossil record of the Araucariaceae, and *Araucaria*, extends back into the Jurassic.

Agathis and *Araucaria* differ from one another strongly in foliar and reproductive structures. The leaves of *Agathis* are opposite and broad, whereas *Araucaria* leaves are spiral and linear to broad. *Agathis* is monoecious with ovules free from the cone scale and winged seeds. *Araucaria* is dioecious, with ovules that are fused to the cone scale and usually wingless seeds. DNA sequences from *rbcL* also support monophyly of the two genera. The *rbcL* data also agree with the division of the genus into four sections based on nonmolecular characters such as the number of cotyledons, position of microsporangiate cones, and cellular features of the leaf epidermis.

Wollemia nobilis, the Wollemi pine, was found in 1994 in the Wolllemi National Park north of Sydney, Australia by National Park and Wildlife Service officer David Noble. Known to science before 1994 only as a fossil extending back to 150 MYA, this is one of the world's rarest trees, with only about 43 adults in two populations 1.5 km apart. The trees, some of which are 500 to 1000 years old, have an unusual bark, described as "bubbling chocolate."

One of the largest and longest-lived trees in this family is *Agathis australis*, commonly called Kauri. One particular individual in northern New Zealand was 51.5 m tall and 13.8 m in circumference and about 2000 years old at the beginning of 2001. The Maori name for this individual is Tane Mahuta, which translates as "god of the forest."

References: Gilmore and Hill 1997; Page 1990e; Quinn et al. 2002; Setogouchi et al. 1998; Stefanovic et al. 1998; http://www.rbgsyd.gov.au/html/wollemi.html.

Taxaceae Bercht. & J. Presl
(Yew Family)

Small to moderately sized trees or shrubs, usually not resinous or only slightly resinous; fragrant or not. Wood without resin canals. Leaves simple, persistent for several years, shed singly, spiral (opposite in one species), often twisted so as to appear 2-ranked, *linear, flattened, entire, acute at apex,* with 0–1 resin canals. Dioecious (rarely monoecious). Microsporangiate strobili with 6–14 microsporophylls; microsporangia 2–9 per microsporophyll, radially arranged around the microsporophyll or limited to its abaxial surface; pollen nonsaccate. **Ovules solitary** *and cones lacking;* **seeds with a hard outer layer, associated with a fleshy, usually brightly colored aril** (Plate 8.2D); cotyledons 2 (occasionally 1 or 3) (Figure 8.27).

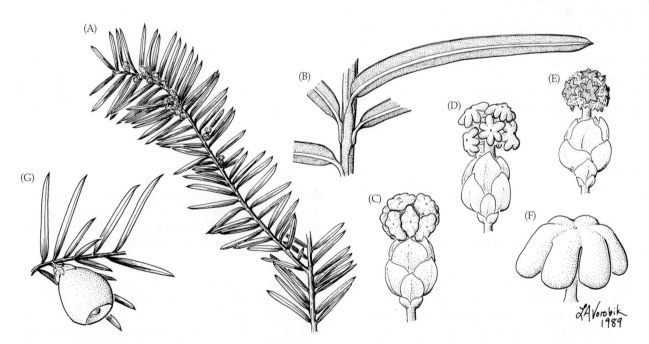

FIGURE 8.27   Taxaceae. *Taxus floridana:* (A) leafy shoot with microsporangiate strobili at time of pollen release (× 0.7); (B) detail of abaxial surface of leaf (× 3.3); (C–E) microsporangiate strobili before, during, and after shedding of pollen (× 6.5); (F) detail of microsporophyll (× 27); (G) shoot with arillate ovule (× 2). (From Price 1990, *J. Arnold Arbor.* 71: pp. 69–91.)

Distribution and ecology: Mostly Northern Hemisphere, extending south to Guatemala and Java, with one endemic genus of New Caledonia. Taxaceae tend to grow in damp valley bottom sites where leaf litter accumulates.

Genera/species: 5/20. ***Major genera:*** *Taxus* (10 spp.) and *Torreya* (4).

Economic plants and products: *Taxus* is widely grown as an ornamental and supplier of fine wood in North America and Europe. It is one of the finest coniferous woods, now used in high-grade furniture. *Torreya* is less important as an ornamental, but its wood, edible seed, and seed oil are valued in Asia. *Taxus* contains taxol, one of several highly poisonous alkaloids in the leaves, stems, and seeds. Taxol's potent antimitotic activity makes it of potential use as an anticancer chemotherapeutic compound.

Discussion: Taxaceae are unique among the conifers in that their solitary seed is not associated with cone scales. The aril is an outgrowth of the axis below the seed. Some systematists have removed Taxaceae from the conifers because of this lack of a cone, but embryology, wood anatomy, chemistry, and leaf and pollen morphology unquestionably tie this family to other conifers. The cone is thought to be lost, and the solitary seed with an aril is therefore a derived feature.

DNA sequences, morphology, anatomy, and alkaloid chemistry divide the family into two clades, one including *Taxus*, *Austrotaxus*, and *Pseudotaxus*, and the other comprising *Torreya* and *Amentotaxus*. The family is apparently most closely related to (and may be paraphyletic without) the Cephalotaxaceae, an East Asian monogeneric family having paired ovules associated with a small outgrowth considered to be a reduced cone scale and borne along the cone axis. Usually only one or two seeds mature, and these develop a juicy outer layer that resembles, but is not homologous with, the aril of Taxaceae. The solitary, drupelike seeds of many Podocarpaceae also resemble the arillate seeds of Taxaceae, but DNA sequence data indicate this fleshiness arose more than once (Stefanovic et al. 1998).

References: Cheng et al. 2000; Hils 1993; Li et al. 2001; Page 1990f; Price 1990; Quinn et al., 2002; Stefanovic et al. 1998.

Gnetales

The Gnetales are of particular interest in plant evolution because they show features both of conifers (seeds not enclosed in an ovary) and of angiosperms (vessels in the wood, somewhat flowerlike structures, and double fertilization). Together with angiosperms, Gnetales are sometimes referred to as anthophytes because of the presence of flowerlike structures or flowers, compound strobili with at least the rudiments of both megasporangia and microsporangia. Recent molecular phylogenetic studies (Bowe et al. 2000;

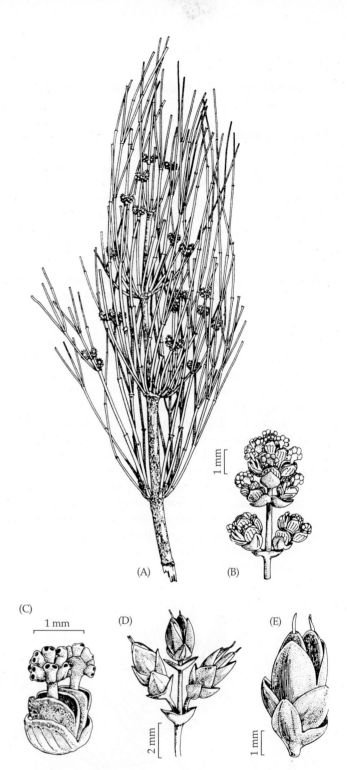

FIGURE 8.28 Ephedraceae. *Ephedra distachya*: (A) staminate branch; (B) microsporangiate strobilus; (C) microsporangiate reproductive structure; (D) ovulate strobilus; (E) ovulate reproductive structure. (From *Flora Ibérica* 1986.)

Chaw et al. 2000) do not support anthophytes as a clade and instead link Gnetales with conifers, but a link with angiosperms has not been ruled out by some workers (Rydin et al. 2002; Doyle 2006). Gnetales show several possible synapomorphies: presence of enveloping bracts around the ovules and microsporangia and a micropylar projection of the integument that produces a pollination droplet. Gnetales consist of three genera—*Gnetum*, *Welwitschia*, and *Ephedra*—

each of which is morphologically highly distinct. *Gnetum* (Gnetaceae) contains about 35 species of mostly tropical, dioecious lianas (less often trees or shrubs) with opposite, simple, broad leaves and seeds enclosed in a fleshy, brightly colored envelope (see Figure 7.13G–I). There is one species of *Welwitschia* (Welwitschiaceae), a bizarre plant of southwestern African deserts. The massive, short stem produces two enormous, straplike leaves (see Figure 7.13F) that wear away at the tips but continue to grow from the base for the entire life of the individual (up to 2000 years). *Ephedra* (Ephedraceae) is treated below.

Additional references: Kubitzki 1990; Price 1996; Dilcher et al. 2005.

Ephedraceae Dumort
(Mormon Tea or Joint Fir Family)

Mostly shrubs, less often clambering vines, and rarely small trees; often spreading by rhizomes. Wood with vessels. *Branches numerous, whorled or clustered, longitudinally grooved; usually green and photosynthetic.* Leaves opposite or whorled, **scale-like, fused basally into a sheath, often shed soon after developing**; resin canals absent. Mostly dioecious. Pollen strobili in whorls of 1–10, each consisting of 2–8 series of opposite or whorled bracts, the apical bracts each subtending a stalk with 2–10(–15) microsporangia. Pollen furrowed, without saccae. Ovulate strobili of 2–10 series of opposite or whorled bracts, those toward the apex subtending a pair of fused bracts forming a casing around the single ovule. Seeds 1–2(–3) per strobilus, yellow to dark brown; cotyledons 2 (Figure 8.28).

Distribution and ecology: Temperate regions worldwide except Australia. Ephedraceae are one of the few gymnosperms adapted to extremely arid regions; they often grow in dry, sunny habitats, such as deserts and steppes, and can occur as high as 4000 m in the Andes and Himalayas.

Genus/species: 1 (*Ephedra*)/ca. 60.

Economic plants and products: *Ephedra* has long been used for a variety of medicinal purposes, such as cough and circulatory weakness. Its primary use today is for the alkaloid ephedrine, which functions as a vessel constrictant.

Discussion: The photosynthetic stem and small leaves make these plants superficially resemble equisetophytes. Pollination is by wind, less often by insects that are attracted by nectar produced by the ovulate strobili. Dispersal is either by wind, promoted by keeled wings on the bracts of the seed strobilus, or by birds, which are attracted to the bright yellow, orange, or red, juicy, outer bracts. The genus likely evolved in the Old World.

References: Ickert-Bond and Wojciechowski 2004; Kubitzki 1990; Price 1996; Stevenson 1993.

LITERATURE CITED AND SUGGESTED READINGS

Items marked with an asterisk are especially recommended to those readers who are interested in further information on the topics discussed in this chapter.

Adams, R. P. 1993. *Juniperus*. In *Flora of North America*, vol. 2, 414–420.

Axsmith, B. J., T. N. Taylor and E. L. Taylor. 1998. Anatomically preserved leaves of the conifer *Notophytum krauselii* (Podocarpaceae) from the Triassic of Antarctica. *Am. J. Bot.* 85: 704–713.

Alston, A. H. G. 1932. Selaginellaceae. In *The pteridophyta of Madagascar* by C. Christiansen. *Dansk. Bot. Ark.* 7: 193–200.

Barrington, D. S., C. H. Haufler and C. R. Werth. 1989. Hybridization, reticulation, and species concepts in the ferns. *Am. Fern J.* 79: 55–64.

*Beck, C. B. (ed.). 1988. *Origin and evolution of gymnosperms*. Columbia University Press, New York.

Billington, C. 1952. Ferns of Madagascar. *Bull. Cranbrook Inst. Sci.* 32: 114.

Bowe, L. M., G. Coat and C. W. dePamphilis. 2000. Phylogeny of seed plants based on all three genomic compartments: Extant gymnosperms are monophyletic and Gnetales' closest relatives are conifers. *Proc. Nat. Acad. Sci. USA.* 97: 4092–4097.

Brownlie, G. 1977. *The pteridophyte flora of Fiji*. J. Cramer, Vaduz.

Brunsfeld, S. J., P. S. Soltis, D. E. Soltis, P. A. Gadek, C. J. Quinn, D. D. Strenge and T. A. Ranker. 1994. Phylogenetic relationships among the genera of Taxodiaceae and Cupressaceae: Evidence from *rbcL* sequences. *Syst. Bot.* 19: 253–262.

Burleigh, J. G. and S. Mathews. 2004. Phylogenetic signal in nucleotide data from seed plants: Implications for resolving the seed plant tree of life. *Am. J. Bot.* 91: 1599–1613.

Chaw, S.-M., C. L. Parkinson, Y. Cheng, T. M. Vincent and J. D. Palmer. 2000. Seed plant phylogeny inferred from all three genomes: Monophyly of extant gymnosperms and origin of Gnetales from conifers. *Proc. Nat. Acad. Sci. USA.* 97: 4086–4091.

Cheng, Y. C., R. G. Nicholson, K. Tripp and S. M. Chaw. 2000. Phylogeny of Taxaceae and Cephalotaxaceae genera inferred from chloroplast *matK* gene and nuclear DNA ITS region. *Mol. Phyl. Evol.* 14: 353–365.

Conant, D. S., L. A. Raubeson, D. K. Attwood and D. B. Stein. 1995. The relationships of Papuasian Cyatheaceae to New World tree ferns. *Am. Fern J.* 85: 328–340.

Conant, D. S. and D. B. Stein. 2001. Phylogenetic and geographic relationships of the tree ferns (Cyatheaceae) on Mount Kinabalu. *Sabah Parks Nature Journal* 4: 25–43.

Conran, J. G., G. M. Wood, P. G. Martin, J. M. Dowd, C. J. Quinn, P. A. Gadek and R. A. Price. 2000. Generic relationships within and between the gymnosperm families Podocarpaceae and Phyllocladaceae based on an analysis of the chloroplast gene *rbcL*. *Aust. J. Bot.* 48: 715–724.

Cox, P. A., S. A. Banack, S. J. Murch, U. Rasmussen, G. Tien, R. R. Bidigare, J. S. Metcalf, L. F. Morrison, G. A. Codd and B. Bergman. 2005. Diverse taxa of cyanobacteria produce β-N-methylamino-L-alanine, a neurotoxic amino acid. *Proc. Nat. Acad. Sci.* 102: 5074–5078.

Crane, P. R. 1988. Major clades and relationships in the higher gymnosperms. In *Origins and evolution of gymnosperms*, C. B. Beck. (ed.), 218–272. Columbia University Press, New York.

Cranfill, R. B. 1993. Blechnaceae. In *Flora of North America*, vol. 2, 223, 226–227.

Des Marais, D. L., A. R. Smith, D. M. Britton and K. M. Pryer. 2003. Phylogenetic relationships and evolution of extant horsetails (*Equisetum*) based on chloroplast DNA sequence data (*rbcL* and *trnL-F*). *Int. J. Plant Sci.* 164: 737–751.

Dilcher, D. L., M. E. Bernardes de Oliveira, D. Pons and T. A. Lott. 2005. Welwitschiaceae from the Lower Cretaceous of northeastern Brazil. *Am. J. Bot.* 92: 1294–1310.

*DiMichele, W. A. and J. E. Skog. 1992. The Lycopsida: A symposium. *Ann. Missouri Bot. Gard.* 79: 447–449.

Donoghue, M. J. 1994. Progress and prospects in reconstructing plant phylogeny. *Ann. Missouri Bot. Gard.* 81: 405–418.

Doyle, J. A. 1998. Phylogeny of vascular plants. *Ann. Rev. Ecol. Syst.* 29: 567–599.

Doyle, J. A. 2006. Seed ferns and the origin of angiosperms. *J. Torrey Bot. Club* 133: 169–209.

Doyle, J. A., M. J. Donoghue and E. A. Zimmer. 1994. Integration of morphological and ribosomal RNA data on the origin of angiosperms. *Ann. Missouri Bot. Gard.* 81: 419–450.

Eckenwalder, J. E. 1976. Reevaluation of Cupressaceae and Taxodiaceae: A proposed merger. *Madrono* 23: 237–256.

Farjon, A. 1990. *Pinaceae: Drawings and descriptions of the genera Abies, Cedrus, Pseudolarix, Keteleeria, Nothotsuga, Tsuga, Cathaya, Pseudotsuga, Larix, and Picea.* Koeltz Scientific Books, Konigstein, Germany.

Farjon, A. 2005a. *A bibliography of Conifers*, 2nd ed. Royal Botanic Gardens, Kew.

Farjon, A. 2005b. *A monograph of Cupressaceae and Sciadopitys.* Royal Botanic Gardens, Kew.

Farjon, A. 2005c. *Pinaceae, Drawings and Descriptions of the Genus Pinus*, 2nd ed. E. J. Brill, Leiden.

Farjon, A. and B. T. Styles. 1997. *Pinus* (Pinaceae). Monograph 75, *Flora Neotropica*. The New York Botanical Garden, New York.

Flora Ibérica. 1986. Jardín Botánico Real, Madrid.

*†Flora of North America Editorial Committee. 1993. *Flora of North America*, vol. 2, *Pteridophytes and gymnosperms.* Oxford University Press, New York.

Flora Tsinlingensis. 1974. Pteriodophyta. Academia Sinica, Beijing.

Friis, E. M., W. G. Falconer and P. R. Crane. 1987. *The origins of angiosperms and their biological consequences.* Cambridge University Press, Cambridge.

Gadek, P. A., D. L. Alpers, M. M. Helsewood and C. J. Quinn. 2000. Relationships within Cupressaceae *sensu lato*: A combined morphological and molecular approach. *Am. J. Bot.* 87: 1044–1057.

Gernandt, D. S., G. G. López, S. O. Garcia and A. Liston. 2005. Phylogeny and classification of *Pinus. Taxon* 54: 29–42.

*Gifford, E. M. and A. S. Foster. 1988. *Morphology and evolution of vascular plants*, 3rd ed. W. H. Freeman, New York.

Gilmore, S. and K. D. Hill. 1997. Relationships of Wollemi Pine (*Wollemia nobilis*) and a molecular phylogeny of the Araucariaceae. *Telopea* 7: 275–291.

Goremykin, V., V. Bobrova, J. Pahnke, A. Troitsky, A. Antonov and W. Martin. 1996. Noncoding sequences from the slowly evolving chloroplast inverted repeat in addition to *rbcL* data do not support gnetalean affinities of angiosperms. *Mol. Biol. Evol.* 13: 383–396.

Haider, K. 1954. Zur Morphologie und Physiologie der Sporangien leptosporangiater Farne. *Planta* 44: 370–411.

Hart, J. A. 1987. A cladistic analysis of conifers: Preliminary results. *J. Arnold Arbor.* 68: 269–307.

Hart, J. A. and R. A. Price. 1990. The genera of Cupressaceae (including Taxodiaceae) in the southeastern United States. *J. Arnold Arbor.* 71: 275–322.

*Hasebe, M., T. Omori, N. Nakazawa, T. Sano, M. Kato and K. Iwatsuki. 1994. *rbcL* gene sequences provide evidence for the evolutionary lineages of leptosporangiate ferns. *Proc. Nat. Acad. Sci., USA* 91: 7530–7534.

Hasebe, M., P. G. Wolf, K. M. Pryer, K. Ueda, M. Ito, R. Sano, G. J. Gastony, J. Yokoyama, J. R. Manhart, N. Murakami, E. H. Crane, C. H. Haufler and W. D. Hauk. 1995. Fern phylogeny based on *rbcL* nucleotide sequences. *Am Fern J.* 85: 134.

Haufler, C. H., M. D. Windham, F. A. Lang and S. A. Whitmore. 1993. *Polypodium.* In *Flora of North America*, vol. 2, 315–323.

Hauk, W. D., C. R. Parks and M. W. Chase. 2003. Phylogenetic studies of Ophioglossaceae: Evidence from *rbcL* and *trnL-F* plastid DNA sequences and morphology. *Mol. Phyl. Evol.* 28: 131–151.

Hauke, R. L. 1990. Equisetaceae. In *Families and genera of vascular plants*, vol. 1, K. U. Kramer and P. S. Green (volume eds.), 46–48.

Hauke, R. L. 1993. Equisetaceae. In *Flora of North America*, vol. 2, 76–84.

Hemsen, E. J., T. N. Taylor, E. L. Taylor and D. W. Stevenson. 2006. Cataphylls of the Middle Triassic cycad *Antarcticycas schopfii* and new insights into cycad evolution. *Am. J. Bot.* 93: 724–738.

Hennequin, S., A. Ebihara, M. Ito, K. Iwatsuki and J.-Y. Dubuisson. 2006. New insights into the phylogeny of the genus *Hymenophyllum* s.l. (Hymenophyllaceae): Revisiting the polyphyly of *Mecodium. Syst. Bot.* 31: 271–284.

Hewitson, W. 1962. Comparative morphology of Osmundaceae. *Ann. Missouri Bot. Gard.* 49: 57–93.

Hill, K. D., M. W. Chase, D. W. Stevenson, H. G. Hills and B. Schutzman. 2003. Families and genera of cycads: A molecular phylogenetic analysis of Cycadophyta based on nuclear and plastid DNA sequences. *Int. J. Plant Sci.* 164: 933–948.

Hils, M. H. 1993. Taxaceae. In *Flora of North America*, vol. 2, 423–427.

Holttum, R. E. 1963. *Flora Malesiana II*, vol. 1, *Pteridophyta: Morphology, key, Gleicheniaceae, Schizaeaceae.* 1–61.

Ickert-Bond, S. M. and M. F. Wojciechowski. 2004. Phylogenetic relationhships in *Ephedra* (Gnetales): Evidence from nuclear and chloroplast DNA sequences. *Syst. Bot.* 29: 834–849.

Jermy, A. C. 1990a. Isoetaceae. In *Families and genera of vascular plants*, vol. 1, K. U. Kramer and P. S. Green (volume eds.), 26–31.

Jermy, A. C. 1990b. Selaginellaceae. In *Families and genera of vascular plants*, vol. 1, K. U. Kramer and P. S. Green (volume eds.), 39–45.

Johnson, D. M. 1993a. Marsileaceae. In *Flora of North America*, vol. 2, 331–335.

Johnson, D. M. 1993b. *Matteuccia; Onoclea.* In *Flora of North America*, vol. 2, 249–251.

Johnson, L. A. S. and K. L. Wilson. 1990. Cycadales. In *Families and genera of vascular plants*, vol. 1, K. U. Kramer and P. S. Green (volume eds.), 363–377. Springer-Verlag, Berlin.

Jones, D. L. 1993. *Cycads of the world.* Smithsonian Institution Press, Washington, DC.

Kelch, D. G. 1997. The phylogeny of Podocarpaceae based on morphological evidence. *Syst. Bot.* 22: 113–131.

Kelch, D. G. 1998. Phylogeny of Podocarpaceae: Comparison of evidence from morphology and 18S rDNA. *Am. J. Bot.* 85: 986–996.

Kenrick, P. 2000. The relationships of vascular plants. *Phil. Trans. R. Soc. Lond. B.* 355: 847–855.

Kenrick, P. and P. R. Crane. 1997. *The origin and early diversification of land plants. A cladistic study.* Smithsonian Institution Press, Washington.

Knobloch, I. W. and D. S. Correll. 1962. *Ferns and fern allies of Chihuahua, Mexico.* Renner, TX.

Korall P. and P. Kenrick. 2002. Phylogenetic relationships in Selaginellaceae based on *rbcL* sequences. *Am. J. Bot.* 89: 506–517.

Korall, P., K. M. Pryer, J. S. Metzgar, H. Schneider and D. S. Conant. 2006. Tree ferns: Monophyletic groups and their relationships as revealed by four protein-coding plastid loci. *Mol. Phyl. Evol.* 39: 830–845.

Kramer, K. U. 1990a. Cyatheaceae. In *Families and genera of vascular plants*, vol. 1, K. U. Kramer and P. S. Green (volume eds.), 69–74.

Kramer, K. U. 1990b. Marsileaceae. In *Families and genera of vascular plants*, vol. 1, K. U. Kramer and P. S. Green (volume eds.), 180–183.

Kramer, K. U. 1990c. Osmundaceae. In *Families and genera of vascular plants*, vol. 1, K. U. Kramer and P. S. Green (volume eds.), 197–200.

Kramer, K. U. 1990d. Psilotaceae. In *Families and genera of vascular plants*, vol. 1, K. U. Kramer and P. S. Green (volume eds.), 22–25.

†Articles from this flora are cited throughout this bibliography. They are listed by author and cited as *Flora of North America*, vol. 2, with appropriate page ranges..

*§Kramer, K. U. and P. S. Green (volume eds.). 1990. *Pteridophytes and gymnosperms*. Vol. 1 in *Families and genera of vascular plants*, K. Kubitzki (ed.). Springer–Verlag, Berlin.

Kubitzki, K. 1990. Gnetatae. In *Families and genera of vascular plants*, vol. 1, K. U. Kramer and P. S. Green (volume eds.), 378–391.

Kusumi, J., Y. Tsumura, H. Yoshimaru and H. Tachida. 2000. Phylogenetic relationships in Taxodiaceae sensu stricto based on *matK* gene, *chlL* gene, *trnL-trnF* IGS region, and *trnL* intron sequences. *Am. J. Bot.* 87: 1480–1488.

Landry, G. P. 1993. Zamiaceae. In *Flora of North America*, vol. 2, 347–349.

*Lellinger, D. B. 1985. *A field manual of the ferns & fern–allies of the United States & Canada*. Smithsonian Institution Press, Washington, DC.

Li, J., C. C. Davis and M. J. Donoghue. 2001. Phylogeny and biogeography of *Taxus* (Taxaceae) inferred from sequences of the internal transcribed spacer region of nuclear ribosomal DNA. *Harvard Papers Bot.* 6: 267–274.

Li, J. and Q. Xiang. 2005. Phylogeny and biogeography of *Thuja* L. (Cupressaceae), an eastern Asian and North American disjunct genus. *J. Integ. Pl. Biol.* 47: 651–659.

Little, D. P., A. E. Schwarzbach, R. P. Adams and C.-F. Hsieh. 2004. The circumscription and phylogenetic relationships of *Callitropsis* and the newly described genus *Xanthcyparis* (Cupressaceae). *Am. J. Bot.* 91: 1872–1881.

Little, D. P. 2006. Evolution and circumscription of the true cypresses (Cupressaceae: *Cupressus*). *Syst. Bot.* 31: 461–480.

Lloyd, R. M. 1993 Parkeriaceae. In *Flora of North America*, vol. 2, 119–121.

Lupia, R., H. Schneider, G. M. Moeser, K. M. Pryer and P. R. Crane. 2000. Marsileaceae sporocarps and spores from the late Cretaceous of Georgia, U.S.A. *Int. J. Plant Sci.* 161: 975–988.

Madalski, J. 1954. *Atlas flory Polskiej i ziem osciennych 1.* Panstowowe Wydawnictwo Naukowe, Warsaw.

Manhart, J. R. 1995. Chloroplast 16S rDNA sequences and phylogenetic relationships of fern allies and ferns. *Am. Fern J.* 85: 182–192.

Mishler, B. D., L. A. Lewis, M. A. Buchheim, K. S. Renzaglia, D. J. Garbary, C. F. Delwiche, F. W. Zechman, T. S. Kantz and R. L. Chapman. 1994. Phylogenetic relationships of the "green algae" and "bryophytes." *Ann. Missouri Bot. Gard.* 81: 451–483.

Muñoz Pizarro, C. 1959. *Sinopsis de la Floa Chilena*. Universidad de Chile, Santiago.

Nauman, C. E. and A. M. Evans 1993. *Dennstaedtia*. In *Flora of North America*, vol. 2, 199–201.

Nickrent, D. L., C. L. Parkinson, J. D. Palmer and R. J. Duff. 2000. Multigene phylogeny of land plants with special reference to

bryophytes and the earliest land plants. *Mol. Biol. Evol.* 17: 1885–1895.

*Nimsch, H. 1995. *A reference guide to the gymnosperms of the world*. Koeltz Scientific Books, Champaign, IL.

Nixon, K. C., W. L. Crepet, D. Stevenson and E. M. Friis. 1994. A reevaluation of seed plant phylogeny. *Ann. Missouri Bot. Gard.* 81: 484–553.

Norstog, K. J. and P. K. S. Fawcett. 1989. Insect–cycad symbiosis and its relations to the pollination of *Zamia fufuracea* (Zamiaceae) by *Rhopalotria mollis* (Curculionidae). *Am. J. Bot.* 76: 1380–1394.

*Norstog, K. J. and T. J. Nichols. 1997. *The biology of cycads*. Cornell University Press, Ithaca, NY.

Øllgaard, B. 1990. Lycopodiaceae. In *Families and genera of vascular plants*, vol. 1, K. U. Kramer and P. S. Green (volume eds.), 31–39.

Øllgaard, B. 1992. Neotropical Lycopodiaceae: An overview. *Ann. Missouri Bot. Gard.* 79: 687–717.

Page, C. N. 1990a. Cupressaceae. In *Families and genera of vascular plants*, vol. 1, K. U. Kramer and P. S. Green (volume eds.), 302–318.

Page, C. N. 1990b. Ginkgoaceae. In *Families and genera of vascular plants*, vol. 1, K. U. Kramer and P. S. Green (volume eds.), 284–289.

Page, C. N. 1990c. Pinaceae. In *Families and genera of vascular plants*, vol. 1, K. U. Kramer and P. S. Green (volume eds.), 319–331.

Page, C. N. 1990d. Podocarpaceae. In *Families and genera of vascular plants*, vol. 1, K. U. Kramer and P. S. Green (volume eds.), 332–346.

Page, C. N. 1990e. Araucariaceae. In *Families and genera of vascular plants*, vol. 1, K. U. Kramer and P. S. Green (volume eds.), 294–299.

Page, C. N. 1990f. Taxaceae. In *Families and genera of vascular plants*, vol. 1, K. U. Kramer and P. S. Green (volume eds.), 348–353.

Page, C. N. 1990g. Taxodiaceae. In *Families and genera of vascular plants*, vol. 1, K. U. Kramer and P. S. Green (volume eds.), 353–361.

Paris, C. A. 1993. *Adiantum*. In *Flora of North America*, vol. 2, 125–130.

Pérez Arbeláez, E. 1928. Die natürlich Gruppe der Davalliaceen (Sm.). Kaulf. *Bot. Abh. Goebel* 14: 1–96.

Phipps, C. J., T. N. Taylor, E. L. Taylor, N. R. Cuneo, L. D. Boucher and X. Yao. 1998. *Osmunda* (Osmundaceae) from the Triassic of Antarctica: An example of evolutionary stasis. *Am. J. Bot.* 85: 888–895.

*Price, R. A. 1989. The genera of Pinaceae in the southeastern United States. *J. Arnold Arbor.* 70: 247–305.

*Price, R. A. 1990. The genera of Taxaceae in the southeastern United States. *J. Arnold Arbor.* 71: 69–91.

*Price, R. A. 1996. Systematics of Gnetales: A review of morphological and molecular evidence. *Intl. J. Plant Sci.* 157: S40–S49.

Price, R. A. and J. M. Lowenstein. 1989. An immunological comparison of the Sciadopityaceae, Taxodiaceae, and Cupressaceae. *Syst. Bot.* 14: 141–149.

Price, R. A., J. Olsen–Stojkovich and J. M. Lowenstein. 1987. Relationships among the genera of Pinaceae: An immunological comparison. *Syst. Bot.* 12: 91–97.

Pryer, K. M. 1999. Phylogeny of marsileaceous ferns and relationships of the fossil *Hydropteris pinnata* reconsidered. *Int. J. Plant Sci.* 160: 931–954.

Pryer, K. M., H. Schneider, A. R. Smith, R. Cranfill, P. G. Wolf, J. S. Hunt and S. D. Sipes. 2001. Horsetails and ferns are a monophyletic group and the closest living relatives to seed plants. *Nature* 409: 618–622.

*Pryer, K. M., E. Schuettpelz, P. G. Wolf, H. Schneider, A. R. Smith and R. Cranfill. 2004. Phylogeny and evolution of ferns (monilophytes) with a focus on early leptosporangiate divergences. *Am. J. Bot.* 91: 1582–1598.

Pryer, K. M., A. R. Smith and J. E. Skog. 1995. Phylogenetic relationships of extant ferns based on evidence from morphology and *rbcL* sequences. *Am. Fern J.* 85: 203–282.

Quinn, C. J., R. A. Price and P. A. Gadek. 2002. Familial concepts and relationships in the conifers based on *rbcL* and *matK* sequence comparisons. *Kew Bull.* 57: 513–531.

Rai, H. S., H. E. O'Brien, P. A Reeves, R. G. Olmstead and S. W. Graham. 2003. Inference of higher-order relationships in the cycads from a large chloroplast data set. *Mol. Phyl. Evol.* 29: 350–359.

Richardson, D. M. (ed.). 1998. *Ecology and biogeography of Pinus*. Cambridge University Press, Cambridge.

Renzaglia, K. S., R. J. Duff, D. L. Nickrent and D. J. Garbary. 2000. Vegetative and reproductive innovations of early land plants: Implications for a unified phylogeny. *Phil. Trans. R. Soc. Lond. B.* 355: 769–793.

Renzaglia, K. S., T. H. Johnson, H. D. Gates and D. P. Whittier. 2001. Architecture of the sperm cell of *Psilotum*. *Am. J. Bot.* 88: 1151–1163.

Rothwell, G. W. 1999. Fossils and ferns in the resolution of land plant phylogeny. *Bot. Rev.* 65: 188–218.

Rothwell, G. W. and K. C. Nixon. 2006. How does the inclusion of fossil data change our conclusions about the phylogenetic history of euphyllophytes? *Int. J. Plant Sci.* 167: 737–749.

Rydin, C., M. Kallersjo and E. M. Friis. 2002. Seed plant relationships and the systematic position of Gnetales based on nuclear and chloroplast DNA: Conflicting data, rooting problems, and the monophyly of conifers. *Int. J. Plant Sci.* 163: 197–214.

Schneider, H., A. R. Smith, T. J. Hildebrand, C. H. Hauffler and T. A. Ranker. 2004. Unraveling the phylolgeny of the polygrammoid ferns (Polypodiaceae and Grammitidaceae): Exploring aspects of the diversification of epiphytic plants. *Mol. Phyl. Evol.* 31: 1041–1063.

Serbet, R. and G. W. Rothwell. 1999. *Osmunda cinnamomea* (Osmundaceae) in the Upper Cretaceous of western North America: additional evidence for exceptional species longevity in filicalean ferns. *Int. J. Plant Sci.* 160: 425–433.

§Articles from this reference are cited throughout this bibliography. They are listed by author and cited as *Families and genera of vascular plants*, vol. 1, K. U. Kramer and P. S. Green (volume eds.), with appropriate page ranges.

Setogouchi, H. T. A. Osawa, J.-C. Pintaud, T. Jaffre and J.-M. Veillon. 1998. Phylogenetic relationships within Araucariaceae based on *rbcL* gene sequences. *Am. J. Bot.* 85: 1507–1516.

Singer, S. R. 1997. Plant life cycles and angiosperm development. In *Embryology: Constructing the organism*, S. F. Gilbert and A. M. Raunio (eds.), 493–514. Sinauer Associates, Sunderland, MA.

Singh, H. 1978. *Embryology of gymnosperms.* Gebruder Borntraeger, Berlin.

Smith, A. R. 1993a. Thleypteridaceae. In *Flora of North America*, vol. 2, 206–222.

Smith, A. R. 1993b. Polypodiaceae. In *Flora of North America*, vol. 2, 312–313.

Smith, A. R. 1995. Non–molecular phylogenetic hypotheses for ferns. *Am. Fern J.* 85: 104–122.

Smith, A. R. and R. B. Cranfill. 2002. Intrafamilial relationships of the thelypteroid ferns (Thelypteridaceae). *Am. Fern J.* 92: 131–149.

*Smith, A. R., K. M. Pryer, E. Schuettpelz, P. Korall, H. Schneider and P. G. Wolf. 2006. A classification of extant ferns. *Taxon* 555: 705–731.

Soltis, D. E. and P. S. Soltis. 1988. Are lycopods with high chromosome numbers ancient polyploids? *Am. J. Bot.* 75: 238–247.

Soltis, D. E., P. S. Soltis and M. J. Zanis. 2002. Phylogeny of seed plants based on evidence from eight genes. *Am. J. Bot.* 89: 1670–1681.

*Sporne, K. R. 1970. *The morphology of pteridophytes*, 3rd ed. Hutchinson University Library, London.

*Sporne, K. R. 1974. *The morphology of gymnosperms*, 2nd ed. Hutchinson University Library, London.

Stefanovic, S., M. Jager, J. Deutsch, J. Broutin and M. Masselot. 1998. Phylogenetic relationships of conifers inferred from partial 28S rRNA gene sequences. *Am. J. Bot.* 85: 688–697.

Stein, D. B., D. S. Conant, M. E. Ahearn, E. T. Jordan, S. A. Kirch, M. Hasebe, K. Iwatsuki, M. K. Tan and J. A. Thomson. 1992. Structural rearrangements of the chloroplast genome provide an important phylogenetic link in ferns. *Proc. Nat. Acad. Sci. USA* 89: 1856–1860.

*Stevenson, D. W. (ed.). 1990. *The biology, structure, and systematics of the Cycadales.* Memoirs of the New York Botanical Garden, Bronx, NY.

Stevenson, D. W. 1991. The Zamiaceae in the southeastern United States. *J. Arnold Arbor.*, Suppl. Series 1: 367–384.

Stevenson, D. W. 1992. A formal classification of the extant cycads. *Brittonia* 44: 220–223.

Stevenson, D. W. 1993. Ephedraceae. In *Flora of North America*, vol. 2, 428–434.

Stewart, W. N. and G. W. Rothwell. 1993. *Paleobotany and the evolution of plants*, 2nd ed. Cambridge University Press, Cambridge.

Syring, J., A. Willyard, R. Cronn and A. Liston. 2005. Evolutionary relationships among *Pinus* (Pinaceae) subsections inferredfrom multiple low-copy nuclear loci. *Am. J. Bot.* 92: 2086–2100.

Taylor, T. N. and E. L. Taylor. 1993. *The biology and evolution of fossil plants.* Prentice Hall, Englewood Cliffs, NJ.

Taylor, W. C. 1984. *Arkansas ferns and fern allies.* Milwaukee, WI.

Taylor, W. C., N. T. Luebke, D. M. Britton, R. J. Hickey and D. F. Brunton. 1993. Isoetaceae. In *Flora of North America*, vol. 2, 64–75.

Thieret, J. W. 1993a. Pinaceae. In *Flora of North America*, vol. 2, 352–398.

Thieret, J. W. 1993b. Psilotaceae. In *Flora of North America*, vol. 2, 16–17.

Tomlinson, P. B. 1992. Apsects of cone morphology and development in Podocarpaceae (Coniferales). *Int. J. Plant Sci.* 153: 572–588.

Tryon, A. F. and R. C. Moran. 1997. *The ferns and fern allies of New England.* Massachusetts Audubon Society, Lincoln, MA.

*Tryon, R. M. and A. F. Tryon. 1982. *Ferns and allied plants.* Springer–Verlag, New York.

Tsumura, Y., K. Yoshimura, N. Tomaru and K. Ohba. 1995. Molecular phylogeny of conifers using RFLP analysis of PCR-amplified specific chloroplast genes. *Theor. Appl. Genet.* 94: 1222–1236.

Valdespino, I. A. 1993. Selaginellaceae. In *Flora of North America*, vol. 2, 39–63.

Vangerow, S., T. Teerkorn and V. Knoop. 1999. Phylogenetic information in the mitochondrial *nad5* gene of pteridophytes: RNA editing and intron sequences. *Plant Biol.* 1: 235–243.

Wagner, D. H. 1993. *Poylstichum.* In *Flora of North America*, vol. 2, 290–299.

Wagner, W. H., Jr. 1990. Ophioglossaceae. In *Families and genera of vascular plants*, vol. 1, K. U. Kramer and P. S. Green (volume eds.), 193–197.

Wagner, W. H., Jr. and J. M. Beitel. 1992. Generic classification of modern North American Lycopodiaceae. *Ann. Missouri Bot. Gard.* 79: 676–686.

Wagner, W. H., Jr. and J. M. Beitel. 1993. Lycopodiaceae. In *Flora of North America*, vol. 2, 18–37.

Wagner, W. H., Jr. and A. R. Smith. 1993. Pteridophytes. In *Flora of North America*, vol. 2, 247–266.

Wagner, W. H., Jr. and F. S. Wagner 1993. Ophioglossaceae. In *Flora of North America*, vol. 2, 85–109.

Wagner, W. H., Jr.. R. C. Moran and C. R. Werth. 1993. Aspleniaceae. In *Flora of North America*, vol. 2, 228–245.

Wang, X.-Q., D. C. Tank and T. Sang. 2000. Phylogeny and divergence times in Pinaceae: Evidence from three genomes. *Mol. Phyl. Evol.* 17: 773–781.

Watson, F. D. and J. E. Eckenwalder. 1993. Cupressaceae. In *Flora of North America*, vol. 1, 399–422.

Webster, T. R. 1992. Developmental problems in *Selaginella* (Selaginellaceae) in an evolutionary context. *Ann. Missouri Bot. Gard.* 79: 632–647.

Wegst, U. G. K. 2006. Wood for sound. *Am. J. Bot.* 93: 1429–1448.

Whetstone, R. D. 1993. Ginkgoaceae. In *Flora of North America*, vol. 2, 350–351.

Whetstone, R. D. and T. A. Atkinson. 1993. Osmundaceae. In *Flora of North America*, vol. 2, 107–109.

Wikström, N. 1999. *Evolution of Lycopodiaceae (Lycopsida): Relationships and patterns of diversification.* Akademitryck, Edsbruk, Sweden.

Wikström, N. and P. Kenrick. 1997. Phylogeny of Lycopodiaceae (Lycopsida) and the relationships of *Phylloglossum drummondii* Kunze based on *rbcL* sequences. *Int. J. Plant Sci.* 156: 862–871.

Wikström, N. and P. Kenrick. 2000. Relationships of *Lycopodium* and *Lycopodiella* based on combined plastid *rbcL* gene and *trnL* intron sequence data. *Syst. Bot.* 25: 495–510.

Wikström, N. and P. Kenrick. 2001. Evolution of Lycopodiaceae (Lycopsida): Estimating divergence times from *rbcL* gene sequences by use of nonparametric rate smoothing. *Mol. Phyl. Evol.* 19: 177–186.

Windham, M. D. 1993. Pteridaceae. In *Flora of North America*, vol. 2, 122–124.

Wolf, P. G. 1997. Evaluation of *atpB* nucleotide sequences for phylogenetic studies of ferns and other pteridophytes. *Am. J. Bot.* 84: 1429–1440.

Wolf, P. G., K. M. Pryer, A. R. Smith and M. Hasebe. 1998. Phylogenetic studies of extant pteridophytes. In *Molecular systematics of plants II: DNA sequencing*, D. E. Soltis, P. S. Soltis and J. J. Doyle (eds.), 541–556. Kluwer Academic Publishers, Boston.

Wolf, P. G., S. D. Sipes, M. R. White, M. L. Martines, K. M. Pryer, A. R. Smith and K. Ueda. 1999. Phylogenetic relationships of the enigmatic fern families Hymenophyllopsidaceae and Lophosoriaceae: evidence from *rbcL* nucleotide sequences. *Plant Syst. Evol.* 219: 263–270.

Zhou, Z. and S. Zheng. 2003. The missing link in *Ginkgo* evolution. *Nature* 423: 821–822.

9

Phylogenetic Relationships of Angiosperms

The **angiosperms**, or flowering plants, are the dominant land plants. The monophyly of this group is strongly supported, as discussed in Chapter 7, and angiosperms are possibly sister to a group that includes all other extant seed plants (Bowe et al. 2000; Burleigh and Mathews 2004; Chaw et al. 1997; Soltis et al. 2002, 2005). The angiosperms have a long fossil record, going back to the earliest Cretaceous (ca. 135 million years ago); their fossils increase in abundance as one moves through the Cretaceous (Beck 1976; Crane et al. 1995). The group possibly originated during the Jurassic, more than 140 million years ago, but the lineage that gave rise to angiosperms may have been separate from other extant seed plants for twice that long.

Most angiosperm species fall into one of two great groups: the **monocots** (plants with a single cotyledon, and pollen grains usually monosulcate) or the **eudicots** (plants with two cotyledons, and pollen grains predominantly tricolpate or modifications thereof). Monophyly of the monocots is supported by many synapomorphies, including leaves with parallel venation, 3-merous flowers (with two whorls of perianth parts, two whorls of stamens, and a single whorl of carpels), embryos with a single cotyledon, sieve cell plastids with several cuneate protein crystals, stems with scattered vascular bundles, and an adventitious root system (although several of these characters are homoplasious; see page 249). Monophyly of monocots is also supported by nucleotide sequences of both chloroplast and nuclear genes and DNA regions (Bharathan and Zimmer 1995; Chase 2004; Chase et al. 1993, 1995a, 2000, 2006; Graham et al. 2006; Hilu et al. 2003; Källersjö et al. 1998; Qiu et al. 2005; Soltis et al. 1997, 1998, 2000, 2005). The eudicots or tricolpates are also monophyletic; synapomorphies include tricolpate pollen (or modifications of this basic pollen type) plus nuclear, chloroplast, and mitochondrial DNA sequences (Chase et al. 1993; Donoghue and Doyle 1989; Doyle et al. 1994;

Hilu et al. 2003; Hoot et al. 1999; Källersjö et al. 1998; Qiu et al. 2005; Savolainen et al. 2000a, b; Soltis et al. 1997, 2000, 2005; Zanis et al. 2003).

Data from nuclear, chloroplast, and mitochondrial DNA sequences and morphology show that the monocot and eudicot clades are derived from members of a morphologically disparate, paraphyletic group of families (Chase et al. 1993; Doyle 1996, 1998; Doyle et al. 1994; Donoghue and Doyle 1989; Graham and Olmstead 2000; Hilu et al. 2003; Mathews and Donoghue 1999; Qiu et al. 2005; Savolainen et al. 2000a; Soltis and Soltis 2004; Soltis et al. 2000, 2005; Zanis et al. 2003; Zimmer et al. 2000). Although members of this latter group share some features with eudicots—such as two cotyledons, a persistent radicle, stems with vascular bundles in a ring, secondary growth, and leaves with net venation—and thus are technically "dicots"—the shared characters are plesiomorphic within angiosperms; that is, these features evolved earlier in the phylogenetic history of tracheophytes.

Fortunately, relationships among the non-monocot, non-tricolpate angiosperms have been clarified by numerous recent phylogenetic analyses based on combined data sets. These analyses have used characters from two to several different genes (Graham and Olmstead 2000; Kim et al. 2004; Qiu et al 2000, 2005; Savolainen et al. 2000a; Soltis et al. 1998, 2000, 2005; Zanis et al. 2003), and sometimes also morphology (Doyle et al. 1994; Doyle 1998; Doyle and Endress 2000; Nandi et al. 1998). The cladograms in Figures 9.1 and 9.2 summarize the relationships among major clades as presented by Chase (2004), Judd and Olmstead (2004), Soltis and Soltis (2004), and Soltis et al. (2005). These are the relationships largely reflected in the classification of the Angiosperm Phylogeny Group, the APG (APG 1998, 2003). The non-monocot, non-tricolpate angiosperms represent an assemblage of lineages, and thus earlier attempts to place them in a single superorder (Annonanae; Thorne 1992) or subclass (Magnoliidae; Cronquist 1968, 1981, 1988; Takhtajan 1980, 1997) are misleading. Most analyses support the position of Amborellaceae, Nymphaeaceae, and Illiciaceae and their relatives in the ANITA grade, as the three lineages that first diverged from the remaining angiosperm groups. The ANITA grade may be characterized by carpel margins usually sealed by a secretion, whereas the carpel margins of most other angiosperms are closed by postgenital fusion of the epidermal layers (Doyle 1998; Endress 2004a,b; Endress and Igersheim 2000a,b; Soltis et al. 2005). Most members of these lineages may have 4-nucleate female gametophytes and diploid endosperm, but Amborellaceae has a 9-nucleate female gametophyte and triploid endosperm (Williams and Friedman 2004; Friedman 2006). (Most other angiosperms have an 8-nucleate female gametophyte and triploid endosperm; see Chapter 4.)

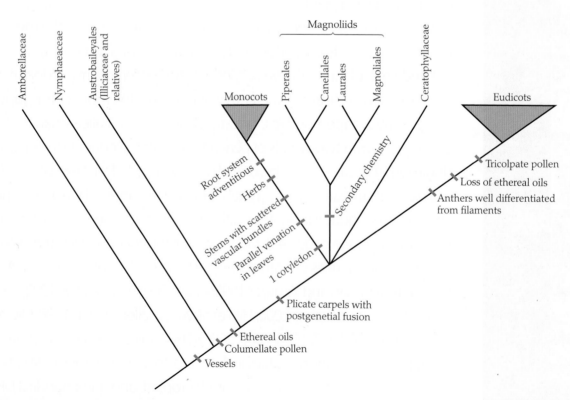

FIGURE 9.1 Cladogram of major angiosperm groups based on Soltis and Soltis (2004), showing selected morphological synapomorphies.

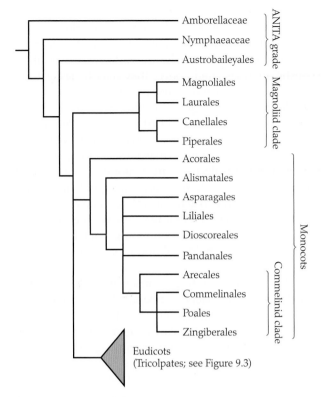

FIGURE 9.2 Major angiosperm clades based on the classification of the Angiosperm Phylogeny Group (APG 1998, 2003) and updated by recent phylogenetic analyses summarized in Chase (2004), Soltis and Soltis (2004), and Soltis et al. (2005).

The ANITA grade has a distinctive type of carpel development in which a meristematic cross-zone develops between the arms of the U-shaped primordium and the carpel grows up like a tube; most other angiosperms have plicate carpels—that is, the developing carpel is like a leaf folded down the middle, with the ovules in the plicate zone (Doyle and Endress 2000). In addition, *Amborella* (the only genus in Amborellaceae) lacks vessels, and most Nymphaeaceae either lack vessels or have vessels of an unusual, tracheid-like form (Schneider et al. 1995; Schneider and Carlquist 1995). Most other angiosperms have vessels, although these specialized water-conducting cells have been lost in a few groups (such as Winteraceae and some Chloranthaceae).

Illiciaceae and relatives have ethereal oils in specialized spherical cells in the leaf mesophyll, so their leaves have pellucid dots; such dots are also present in the leaves of other families, including Magnoliaceae, Annonaceae, Winteraceae, and Lauraceae of the magnoliids. But pellucid dots are absent from Nymphaeaceae and Amborellaceae; their lack is therefore likely to be the original condition within angiosperms.

Amborellaceae and Illiciaceae probably retain many plesiomorphic (ancestral) angiosperm characteristics that are also found in many magnoliids. Possible ancestral vegeta-

tive characters include the perennial, evergreen, woody habit with primary vascular bundles in a ring; unilacunar nodes with two vascular traces; and alternate, simple, more or less coriaceous, pinnately veined leaves with entire or sparsely toothed margins, and absence of stipules.

As judged by the distribution of characteristics in these families, the flowers of early angiosperms are likely to have been insect-pollinated, radially symmetrical, perfect and hypogynous, with the parts several to numerous, spirally arranged, and free and distinct. The perianth is characteristically of tepals (i.e., not differentiated into a calyx and corolla). In these families, stamens are generally poorly differentiated into an anther and filament; anthers have four microsporangia and open by longitudinal slits. The pollen grains are usually monosulcate, and well-developed columellae are often lacking. Their carpels have an elongated stigma and a style that is not well differentiated from the ovary; the ovules are attached to the locule walls (often lateral) and are typically anatropous, with two integuments and a thick megasporangium wall (nucellus). Fruits of ancestral angiosperms may have been indehiscent (and fleshy) or dehiscent (by a single slit, i.e., follicles), with medium-sized seeds that have abundant endosperm, and a minute embryo with two cotyledons. The Nymphaeaceae possess many specialized characters that likely evolved in association with their aquatic habit, including herbaceousness, stems with air canals and scattered vascular bundles, leaves with mucilage, elongate petioles, and dissected, peltate or nearly peltate blades, and flowers on long pedicels.

Most of the remaining non-monocot, non-eudicot angiosperms probably constitute a monophyletic group, which is here called the magnoliid clade, or simply the **magnoliids** (see Figure 9.2). The clade includes Magnoliales, Laurales, Canellales, and Piperales. The magnoliids are trees, shrubs, or lianas with alternate or opposite, usually pinnately veined, coriaceous leaves, and paracytic stomates. Flowers typically have several to numerous parts. Perianth parts are arranged spirally or in whorls of three, and stamens are often laminar. The filament is poorly differentiated from the anther, and the connective tissue is often well developed. The Piperales are a derived group of usually herbaceous (but sometimes secondarily soft-woody) plants. Their stems generally have swollen nodes and separated vascular bundles (or when wood develops, the rays are very broad). Leaves are alternate, often more or less palmately veined, soft-textured, and with anomocytic stomates. Members of the perianth and androecium are usually in whorls of three. The filament is well differentiated from the anther, and the connective is usually inconspicuous.

Note that the Nymphaeaceae and monocots share some of the above-listed characters of Piperales, including herbaceous habit and a tendency for 3-merous flowers. Given the cladogram shown in Figure 9.1, it is simplest to assume that these characters evolved independently in these three monophyletic groups. In addition, stems with scattered vascular bundles have evolved independently in Nymph-aeaceae, Piperaceae, and monocots. It is thus not surprising that some have considered Piperales and/or Nymph-aeaceae to be closely related to monocots (Chase et al. 1993; Donoghue and Doyle 1989; Doyle and Endress 2000; Doyle et al. 1994).

Magnoliids, along with the morphologically similar ANITA grade, traditionally have been considered to retain the greatest number of plesiomorphic features within the angiosperms (Cronquist 1968, 1981, 1988; Dahlgren 1977, 1983; Takhtajan 1969, 1980, 1997; Thorne 1974, 1992). This viewpoint has received some support from morphology-based cladistic analyses (Donoghue and Doyle 1989; Doyle et al. 1994; Loconte and Stevenson 1991), and receives at least partial support in recent molecular analyses because of the placement of Amborellaceae, Nymphaeaceae, and Illiciaceae (see, e.g., Graham and Olmstead 2000; Mathews and Donoghue 1999, 2000; Qiu et al. 2005; Soltis et al. 2000, 2005). Note, however, that magnoliids are more closely related cladistically to the monocots and eudicots than they are to the Amborellaceae, Nymphaeaceae, and Illiciaceae (see Figures 9.1 and 9.2).

Although the angiosperm cladogram is clearly rooted by the Amborellaceae (or Amborellaceae plus Nymph-aeaceae), the placement of a few groups is still equivocal. Ceratophyllaceae, especially, are variously placed in the cladistic analyses cited above; they may be sister to the monocots (Graham and Olmstead 2000; Zanis et al. 2003), sister to the eudicots (Qiu et al. 2005; Soltis et al. 2000), or sister to both together. The Chloranthaceae are another family of problematic placement. Recent evidence suggests that monocots and eudicots may be sister taxa, with mag-noliids, Ceratophyllaceae, and Chloranthaceae basal to them. The identity of the early divergent angiosperm clades will continue to generate interest and arguments. It is anticipated, however, that progress will be made most rapidly through analyses combining data from several sources, both morphological and molecular (Soltis et al. 1998, 2005).

A cladogram presenting a conservative estimate of our knowledge of phylogenetic relationships among angio-sperms is presented in Figures 9.2, 9.3, and 9.4. This tree is modified from that of the Angiosperm Phylogeny Group (1998, 2002), updated by recent phylogenetic analyses,

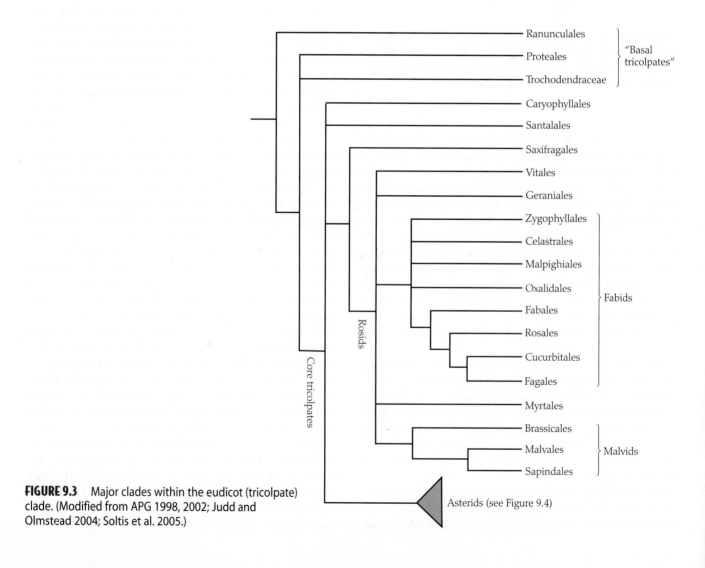

FIGURE 9.3 Major clades within the eudicot (tricolpate) clade. (Modified from APG 1998, 2002; Judd and Olmstead 2004; Soltis et al. 2005.)

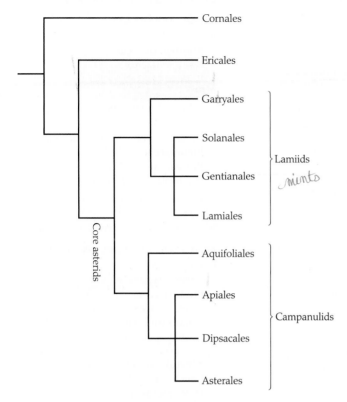

FIGURE 9.4 Major clades within the asterids. (Modified from APG 1998, 2002; Judd and Olmstead 2004; Soltis et al. 2005.)

summarized in Chase (2004), Judd and Olmstead (2004), Soltis and Soltis (2004), and Soltis et al. (2005). Hierarchial characterizations of all APG families may be found at http://www.mobot.org/MOBOT/research/APweb/.

The estimated 257,400 angiosperm species in 13,678 genera (Thorne 1992, 2001) occur in an extremely wide array of terrestrial habitats and exhibit an amazing diversity of morphological, anatomical, biochemical, and physiological characters. These plants are usually divided into about 450 families; Thorne (1992, 2001) recognized 490, Cronquist (1988) 387, Dahlgren (1983) 463, Takhtajan (1980) 589, and the Angiosperm Phylogeny Group (2003) 462.

This text treats 143 angiosperm families in detail, and an additional 97 families more briefly (Table 9.1). As in the presentation of non-angiosperm tracheophytes in Chapter 8, each family treatment includes (1) a description, in which useful identifying characters are indicated in *italic print* and presumed synapomorphies in **boldface**; (2) a brief summary of distribution, with indication of ecology when the family occurs in only a limited array of plant communities or ecological conditions; (3) an estimate of the number of genera and species (including a list of major genera); (4) a list of major economic plants and products; and (5) a discussion, with information regarding characters supporting the group's monophyly, a brief overview of phylogenetic relationships within the family, information regarding pollination biology and fruit dispersal, and notes on other mat-

ters of biological interest. Any useful sources not cited within the above elements are included in a list of additional references.

In addition to these elements, each angiosperm family presentation includes a **floral formula**—a graphic formula that summarizes floral symmetry and the number, fusion, and insertion of floral parts (see Box 4.1).

The family descriptions presented here are necessarily somewhat generalized, and exceptional conditions are not usually indicated. In these descriptions, anthers are assumed to be 4-locular (bithecate) and opening by longitudinal slits, and ovules are assumed to be anatropous, have two integuments, and a thick megasporangium, unless otherwise indicated. Endosperm is considered to be triploid and present in the seeds, and the embryo is considered to be straight, unless otherwise noted. The stem of any nonmonocot family is assumed to contain a ring of vascular bundles (eustele), while that of monocots is considered to have scattered bundles; therefore, only divergent conditions are described. Likewise, the embryos of any nonmonocot family are assumed to have two cotyledons, while those of monocots have only a single cotyledon, unless stated otherwise.

The drawings that illustrate many families were prepared in connection with the Generic Flora of the Southeastern United States Project unless otherwise indicated. Closely related families are treated within orders (i.e., names ending in -*ales*; see Appendix 1). **Ordinal treatments** include an outline of characters supporting the group's monophyly (ordinal synapomorphies) and a brief discussion of phylogenetic relationships of the families within the group. A **key** to all (or at least the most important) families constituting each order is included. Families with formal treatments in this text are indicated in **bold** in these keys. Circumscription of some families (and orders) has been altered from that traditionally used in order to render these groups monophyletic. Related orders are grouped into informal higher-level taxonomic groups. We largely follow the classification of the Angiosperm Phylogeny Group (APG 1998, 2003) because this classification is based on published cladistic analyses.

Families have been chosen for formal treatment on the basis of their number of genera and species, floristic dominance (especially in North America), economic importance, and phylogenetic interest. Table 9.1 lists the angiosperm families in an arrangement reflecting our understanding of their phylogenetic relationships, based on recent cladistic analyses and the classification proposed by the Angiosperm Phylogeny Group. Families receiving full coverage in the text are indicated in **boldface**, while those characterized only briefly are shown in *italics*. Because the phylogenetic relationships of some clades are still unclear, the sequence of orders employed in this chapter is somewhat arbitrary. Indeed any linear sequence is necessarily arbitrary because it cannot reflect the branching pattern expressed in a cladogram.

TABLE 9.1 Major families of angiosperms as classified in this book.[a]

ANITA GRADE

Amborellales
  Amborellaceae (p. 232)
Nymphaeales
  Nymphaeaceae (p. 233) (includes
    Barclayaceae, Cabombaceae)
Austrobaileyales
  Illiciaceae (p. 235)
  Schisandraceae (p. 235)
Chloranthaceae (p. 243)
  (placement uncertain)

MAGNOLIID COMPLEX

Magnoliales
  Magnoliaceae (p. 237)
  Annonaceae (p. 240)
  Myristicaceae (p. 240)
  Degeneriaceae (p. 237)
Laurales
  Lauraceae (p. 242)
  Calycanthaceae (p. 242)
  Hernandiaceae (p. 242)
  Monimiaceae (p. 242)
  Siparunaceae (p. 242)
Canellales
  Winteraceae (p. 244)
  Canellaceae (p. 244)
Piperales
  Piperaceae (p. 245)
  Aristolochiaceae (p. 247)
  Hydnoraceae (p 245)
  Lactoridaceae (p. 245)
  Saururaceae (p. 245)
Ceratophyllaceae (p. 248)
  (placement uncertain)

MONOCOTS

Acorales
  Acoraceae (p. 249)
Alismatales
  Araceae (p. 250) (includes
    Lemnaceae)
  Alismataceae (p. 252) (includes
    Limnocharitaceae)
  Hydrocharitaceae (p. 254) (includes
    Najadaceae)
  Potamogetonaceae (p. 256)
  Butomaceae (p. 250)
  Cymodoceaceae (p. 250)
  Posidoniaceae (p. 250)
  Ruppiaceae (p. 250)
  Tofieldiaceae (p. 262)
  Zannichelliaceae (p. 250)
  Zosteraceae (p. 250)
Liliales
  Liliaceae (p. 257) (includes
    Calochortaceae, Ulvulariaceae,
    in part)
  Colchicaceae (p. 258) (includes
    Ulvulariaceae, in part)
  Smilacaceae (p. 259)

Melanthiaceae (p. 260) (includes
  Trilliaceae)
Alstroemeriaceae (p. 257)
Asparagales
  Asparagaceae (p. 266)
  Ruscaceae (p. 266) (includes
    Convallariaceae, Nolinaceae,
    Dracaenaceae)
  Agavaceae (p. 268) (includes
    Hyacinthaceae subfam.
    Chlorogaloideae, Hostaceae)
  Hyacinthaceae (p. 269)
  Alliaceae (p. 270)
  Amaryllidaceae (p. 270)
  Asphodelaceae (p. 272)
  Iridaceae (p. 272)
  Orchidaceae (p. 273)
  Agapanthaceae (p. 270)
  Hemerocallidaceae (p. 265) (includes
    Phormiaceae, Johnsoniaceae)
  Hypoxidaceae (p. 265)
  Themidaceae (p. 270)
  Xanthorrhoeaceae (p. 265)
Dioscoreales
  Dioscoreaceae (p. 275)
  Burmanniaceae (p. 276)
  Nartheciaceae (p. 262)

COMMELINOID CLADE

Arecales
  Arecaceae (p. 278)
Commelinales
  Commelinaceae (p. 281)
  Haemodoraceae (p. 282)
  Pontederiaceae (p. 283)
  Philydraceae (p. 281)
Poales
  Bromeliaceae (p. 287)
  Typhaceae (p. 290) (includes
    Sparganiaceae)
  Eriocaulaceae (p. 290)
  Xyridaceae (p. 292)
  Juncaceae (p. 292)
  Cyperaceae (p. 294)
  Restionaceae (p. 296)
  Poaceae (p. 296)
  Flagellariaceae (p. 286)
  Joinvilleaceae (p. 286)
  Mayacaceae (p. 286)
Zingiberales
  Zingiberaceae (p. 302)
  Marantaceae (p. 304)
  Cannaceae (p. 306)
  Costaceae (p. 302)
  Heliconiaceae (p. 302)
  Musaceae (p. 302)
  Strelitziaceae (p. 302)

EUDICOTS (TRICOLPATES)

"BASAL TRICOLPATES"

Ranunculales
  Menispermaceae (p. 308)
  Ranunculaceae (p. 309)
  Berberidaceae (p. 312)
  Papaveraceae (p. 314)
    (includes Fumariaceae)
  Lardizabalaceae (p. 308)
Proteales and other "basal tricolpates"
  Platanaceae (p. 316)
  Proteaceae (p. 317)
  Nelumbonaceae (p. 316)
Trochodendraceae (p. 316) (includes Tet-
  racentraceae; placement uncertain)
Buxaceae (p. 316) (placement uncertain)

CORE EUDICOTS (CORE TRICOLPATES)

Caryophyllales
  Caryophyllaceae (p. 320)
  Phytolaccaceae (p. 323)
  Nyctaginaceae (p. 324)
  Amaranthaceae (p. 324) (includes
    Chenopodiaceae)
  Aizoaceae (p. 327)
  "Portulacaceae" (p. 328)
  Cactaceae (p. 330)
  Droseraceae (p. 332)
  Polygonaceae (p. 334)
  Nepenthaceae (p. 332)
  Petiveriaceae (p. 319)
  Plumbaginaceae (p. 332)
  Simmondsiaceae (p. 318)
Santalales
  Loranthaceae (p. 336)
  Santalaceae (p. 338) (includes
    Viscaceae)
  Misodendraceae (p. 336)
  "Olacaceae" (p. 336)
  Opiliaceae (p. 336)
  Schoepfiaceae (p. 336)
Saxifragales
  Saxifragaceae (p. 338)
  Crassulaceae (p. 342)
  Hamamelidaceae (p. 342)
  Altingiaceae (p. 344)
  Cercidiphyllaceae (p. 341)
  Grossulariaceae (p. 341)
  Haloragaceae (p. 341)
  Iteaceae (p. 341)
  Paeoniaceae (p. 338)

ROSID CLADE

Vitales
  Vitaceae (p. 346) (includes Leeaceae)
Geraniales
  Geraniaceae (p. 348)

Fabids (Eurosids I)

Zygophyllales
  Zygophyllaceae (p. 350)
  Krameriaceae (p. 351)

[a]Families receiving full coverage in the text are indicated in boldface, while those only briefly characterized are in italics and listed following those in boldface; thus the list does not necessarily place related families side by side. Page numbers (in parentheses) indicate the discussion of the family in this chapter.

TABLE 9.1 *(continued)*

Oxalidales
   Oxalidaceae (p. 351)
   Cephalotaceae (p. 351)
   Cunoniaceae (p. 351)
Celastrales
   Celastraceae (p. 351) (includes
     Hippocrateaceae)
   Parnassiaceae (p. 353)
Malpighiales
   Malpighiaceae (p. 353)
   Euphorbiaceae (p. 355)
   Phyllanthaceae (p. 359)
   Clusiaceae (p. 362)
   Hypericaceae (p. 362)
   Rhizophoraceae (p. 364)
   Violaceae (p. 364)
   Passifloraceae (p. 367)
   Salicaceae (p. 367; includes
     Flacourtiaceae, in part)
   Achariaceae (p. 355)
   Chrysobalanaceae (p. 354)
   Picrodendraceae (p. 353)
   Podostemaceae (p. 354)
   Putranjivaceae (p. 359)
   Rafflesiaceae (p. 353)
Fabales
   Fabaceae (p. 371)
   Polygalaceae (p. 377)
   Surianaceae (p. 372)
Rosales
   Rosaceae (p. 379)
   Rhamnaceae (p. 388)
   Ulmaceae (p. 389)
   Cannabaceae (p. 391) (includes
     Celtidaceae)
   Moraceae (p. 392)
   Urticaceae (p. 393) (includes
     Cecropiaceae)
Cucurbitales
   Cucurbitaceae (p. 396)
   Begoniaceae (p. 398)
   Datiscaceae (p. 396)
Fagales
   Fagaceae (p. 401)
   Betulaceae (p. 404)
   Casuarinaceae (p. 406)
   Myricaceae (p. 406)
   Juglandaceae (p. 408)
   Nothofagaceae (p. 400)
   Rhoipteleaceae (p. 401)
   Ticodendraceae (p. 404)
Myrtales (placement uncertain;
   possibly sister to Malvids)
   Lythraceae (p. 412) (includes Sonner-
     atiaceae, Trapaceae, Punicaceae)
   Onagraceae (p. 414)
   Combretaceae (p. 416)
   Myrtaceae (p. 416)
   Melastomataceae (p. 418)
   Memecylaceae (p. 412)
   Vochysiaceae (p. 412)

Malvids (Eurosids II)
Brassicales
   Brassicaceae (p. 420) (includes
     Capparaceae, Cleomaceae)
   Bataceae (p. 422)
   Caricaceae (p. 422)
   Moringaceae (p. 422)
   Resedaceae (p. 422)
Malvales
   Malvaceae (p. 424) (includes
     Tiliaceae, Sterculiaceae,
     Bombacaceae)
   Cistaceae (p. 427)
   Dipterocarpaceae (p. 429)
   Thymelaeaceae (p. 424)
Sapindales
   Rutaceae (p. 429)
   Meliaceae (p. 432)
   Simaroubaceae (p. 435)
   Anacardiaceae (p. 435) (includes
     Julianaceae)
   Burseraceae (p. 437)
   Sapindaceae (p. 438) (includes
     Aceraceae, Hippocastanaceae)

ASTERID CLADE (= SYMPETALAE)
Cornales
   Hydrangeaceae (p. 441)
   Loasaceae (p. 443)
   Cornaceae (p. 443) (includes
     Nyssaceae)
Ericales
   Sapotaceae (p. 445)
   Ebenaceae (p. 449)
   Primulaceae (p. 450) (includes
     Theophrastaceae, Maesaceae,
     Myrsinaceae)
   Theaceae (p. 452)
   Ericaceae (p. 452) (includes
     Pyrolaceae, Monotropaceae,
     Empetraceae, Epacridaceae)
   Sarraceniaceae (p. 455)
   Lecythidaceae (p. 455)
   Polemoniaceae (p. 457)
   Actinidiaceae (p. 445)
   Balsaminaceae (p. 445)
   Clethraceae (p. 445)
   Cyrillaceae (p. 445)
   Fouquieriaceae (p. 445)
   Pentaphragmataceae (p. 445) (includes
     Ternstroemiaceae)
   Styracaceae (p. 445)
   Symplocaceae (p. 445)

Lamiids (Euasterids I)
Garryales
   Garryaceae (p. 441)
Solanales
   Solanaceae (p. 459) (includes
     Nolanaceae)
   Convolvulaceae (p. 462) (includes
     Cuscutaceae)

   Boraginaceae (p. 462; includes
     Hydrophyllaceae, in part; Lenno-
     aceae) [placement uncertain]
   Hydroleaceae (p. 459f)
Gentianales
   Rubiaceae (p. 469)
   Gentianaceae (p. 471)
   Apocynaceae (p. 471) (includes
     Asclepiadaceae)
   Gelsemiaceae (p. 467)
   Loganiaceae (p. 467)
Lamiales
   Oleaceae (p. 477)
   Gesneriaceae (p. 481)
   Plantaginaceae (p. 481) (includes
     Callitrichaceae, Scrophulariaceae
     in part)
   Scrophulariaceae (p. 484) (includes
     Buddlejaceae, Myoporaceae)
   Orobanchaceae (p. 484) (includes
     former Scrophulariaceae,
     parasitic species)
   Bignoniaceae (p. 486)
   Acanthaceae (p. 486) (includes
     Avicenniaceae, Mendonciaceae)
   Lentibulariaceae (p. 488)
   Verbenaceae (p. 490)
   Lamiaceae (p. 492) (includes
     many genera formerly treated
     as Verbenaceae)
   Calceolariaceae (p. 477)
   Phrymaceae (p. 482) (includes
     Mimulus)
   Linderniaceae (p. 482)
   Tetrachondraceae (p. 475)

Campanulids (Euasterids II)
Aquifoliales
   Aquifoliaceae (p. 494)
   Helwingiaceae (p. 494)
   Phyllonomaceae (p. 494)
Apiales
   Apiaceae (p. 495) (includes
     Hydrocotylaceae, in part)
   Araliaceae (p. 499) (includes
     Hydrocotylaceae, in part)
   Myodocarpaceae (p. 497)
   Pittosporaceae (p. 497)
Dipsacales
   Caprifoliaceae (p. 501) (includes
     Dipsacaceae, Valerianaceae,
     Diervillaceae, Linnaeaceae)
   Adoxaceae (p. 504) (includes
     Sambucus, Viburnum)
Asterales
   Campanulaceae (p. 508) (includes
     Lobeliaceae)
   Asteraceae (p. 508)
   Calyceraceae (p. 506)
   Goodeniaceae (p. 506)
   Menyanthaceae (p. 506)
   Stylidiaceae (p. 506)

ANITA GRADE

Amborellales

Amborellaceae Pinchon
(Amborella Family)

Shrubs or small trees, lacking ethereal oils and pellucid dots; wood lacking vessels. Hairs simple. *Leaves alternate and 2-ranked, simple,* undulate, *obscurely serrate* (the teeth chloranthoid), with pinnate venation; stipules absent. Inflorescences determinate, axillary. **Flowers unisexual (plants dioecious)**, radial. *Tepals 5-11, ± distinct,* imbricate. *Stamens numerous,* poorly differentiated into an anther and filament, the latter short, represented by staminodes in carpellate flowers; **pollen grains monoaperturate (ulcerate) at distal pole, with poorly defined aperture margin.** *Carpels 5 or 6,* **borne on a slightly concave receptacle that tears with age**; ovaries superior, with lateral placentation; stigma extending down style on adaxial surface, with 2 flanges. Ovule 1. *Fruit an aggregate of* **drupes**, the pit with a pitted surface; seed with a minute embryo.

Floral formula:
Staminate: *T– 8–11– , A ∞ , G0
Carpellate: *T– 5–8– , A1–2•, G5–6, drupes

Distribution and ecology: Endemic to New Caledonia, in moist, shaded understory of montane forests.

Genus/species: 1 (*Amborella*)/1 (*A. trichopoda*).

Economic plants and products: None.

Discussion: *Amborella* was traditionally placed in the Laurales because its flowers have a more or less concave receptacle and its fruits are drupaceous (Plate 9.1B). It has

(A)

Nymphaeales: Nymphaeaceae
Nymphaea odorata: flower

(B)

A. trichopoda: drupes

Amborellales: Amborellaceae
Amborella trichopoda: flowers and leaves

(C)

I. cubense: flowers

Austrobailyales: Illiciaceae
Illicium parviflorum: fruit and leaves

(D)

Nymphaeales: Nymphaeaceae
Nuphar variegata: floating leaf blades, petioles, flower

PLATE 9.1 ANITA Grade

long been considered to be a primitive angiosperm, as suggested by its vessel-less wood, flowers with distinct tepals, numerous stamens with the anther and filaments more or less undifferentiated, several distinct carpels with stigmatic ridges unfused (and closed only by a secretion layer), and seeds with abundant endosperm and a minute embryo. The stomates are paracytic to anomocytic. These characters, along with the persistent, alternate, simple leaves, had suggested a placement within the magnoliids. Recent DNA-based phylogenetic analyses (Graham and Olmstead 2000; Hilu et al. 2003; Mathews and Donoghue 1999, 2000; Qiu et al. 2000; Soltis et al. 1999, 2000), however, distinguish *Amborella* from the magnoliids, and suggest that it may be the sister group of all other angiosperms (see Figures 9.1 and 9.2). However, this hypothesis requires additional study, as a few analyses have placed *Amborella* in a clade with Nymphaeaceae (Barkman et al. 2000). Especially distinctive features of Amborellaceae include the vessel-less wood; lack of ethereal oils; unisexual flowers with a shallow, cuplike receptacle; more or less inaperturate pollen; and aggregate of red drupes.

The flowers of *Amborella* are both insect- and wind-pollinated; a wide variety of insects visit them, but beetles are especially frequent. The drupes are dispersed by birds.

Additional references: Bailey and Swamy 1948; Bobrov et al. 2005; Buzgo et al. 2004; Endress 2001; Endress and Ingersheim 2000b; Feild et al. 2001; Hesse 2001; Philipson 1993; Ronse de Craene et al. 2003; Sampson 1993; Thien et al. 2003; Young 1982.

Nymphaeales

Nymphaeaceae Salisbury
(Water Lily Family)

Aquatic, **rhizomatous herbs**; *stem with vascular bundles usually scattered, with conspicuous air canals* and usually also **laticifers**; usually with distinct, stellate-branched sclereids projecting into the air canals; often with alkaloids (but not of the benzyl-isoquinoline type). Hairs simple, **usually producing mucilage (slime)**. *Leaves* alternate and spiral, opposite, or occasionally whorled, simple, peltate or nearly so, entire to toothed or dissected, short- to *long-petiolate, with blade submerged, floating, or emergent,* with palmate to pinnate venation; stipules present or absent. **Flowers** solitary, bisexual, radial, **with a long pedicel** *and usually floating or raised above the surface of the water*, **with girdling vascular bundles in receptacle**. Tepals 4–12, distinct to connate, imbricate, often petal-like. *Petals (petal-like staminodes) lacking or 8 to numerous,* inconspicuous to showy, often intergrading with stamens. *Stamens 3 to numerous,* the innermost sometimes represented by staminodes; filaments distinct, free or adnate to petaloid staminodes, slender and well differentiated from anthers to laminar and poorly differentiated from anthers; pollen grains usually monosulcate or lacking apertures. Carpels 3 to numerous,

distinct or connate; ovary/ovaries superior to inferior, *if connate then with several locules and* **placentation parietal (the ovules scattered on the partitions)**; *stigmas often elongate and radiating on an expanded circular to marginally lobed and grooved disk,* often surrounding the receptacle, which is visible as a knob or circular bump. Ovule 1 to numerous, anatropous to orthotropous. Nectaries lacking or sometimes present on the staminodes, although a sweet fluid may also be secreted by the stigma. *Fruit an aggregate of nuts or few-seeded indehiscent pods, a berry, or sometimes an irregularly dehiscent fleshy capsule;* seeds **usually operculate** (opening by a cap), often arillate; endosperm diploid, ± lacking, but with abundant starchy perisperm (Figure 9.5).

Floral formula:

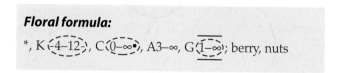

*, K $\widehat{4\text{–}12}$, C $\widehat{0\text{–}\infty}$, A3–∞, G $\overline{\widehat{1\text{–}\infty}}$; berry, nuts

Distribution and ecology: Widespread from tropical to cold temperate regions, occurring in rivers, ponds, lakes, and other freshwater wetlands.

Genera/species: 8/70. **Major genera:** *Nymphaea* (40 spp.) and *Nuphar* (15). These genera, along with *Cabomba* and *Brasenia*, occur in the continental United States and Canada.

Economic plants and products: Species of *Nymphaea* (water lily), *Nuphar* (yellow water lily, spatterdock), and *Victoria* (Amazon water lily) provide ornamental plants for pools. *Cabomba* is a popular aquarium plant.

Discussion: The family comprises two subfamilies: Cabomboideae, including *Cabomba* and *Brasenia*, and Nymphaeoideae, with the remaining genera; both are often recognized as families (Les et al. 1999). Cabomboideae are united by the presence of free-floating stems in addition to the rhizomes. They also have distinct carpels, nutlike fruits, and six-tepaled flowers that lack petal-like staminodes. Nymphaeoideae are separated from Cabomboideae by numerous synapomorphies: the presence of star-shaped sclerids, four or more sepals, spirally inserted stamens with laminar filaments, syncarpous fleshy fruits (berries), and laminar placentation (Les et al. 1999; Moseley et al. 1993). *Nuphar* likely is sister to the remaining genera of Nymphaeoideae. It has a superior ovary with separate stigmas and inconspicuous staminodes, while *Barclaya, Ondinea, Nymphaea, Victoria,* and *Euryale* have perigynous/epigynous flowers with a continuous stigmatic surface, showy staminodes, and zonosulcate pollen, all of which are likely synapomorphic. Within Nymphaeoideae the highest number of staminodial petals, stamens, and carpels occur in *Nymphaea* and the closely related *Victoria*, but even the flowers of *Nuphar* possess relatively numerous parts. The flowers of *Cabomba* have relatively few parts, and the numerous floral parts of Nymphaeoideae probably are

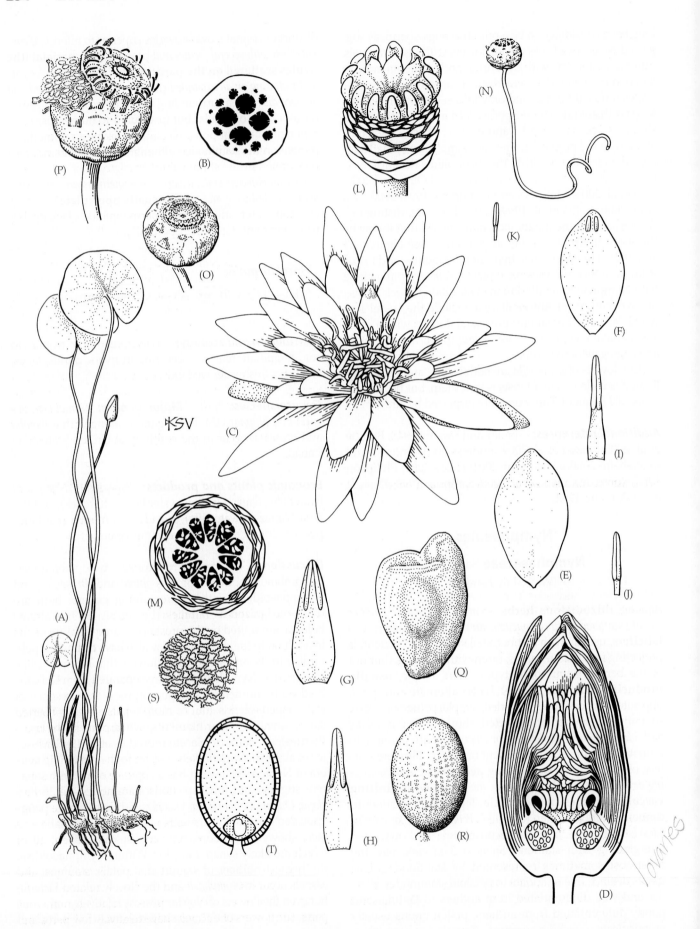

◀ **FIGURE 9.5** Nymphaeaceae. *Nymphaea odorata:* (A) habit (× 0.3); (B) petiole, in cross section (× 5); (C) flower (× 1); (D) flower, in longitudinal section (× 1.5); (E) petal (× 1.5); (F–H) petal-like stamens (× 1.5); (I–K) inner stamens (× 1.5); (L) gynoecium showing numerous carpels (× 2); (M) gynoecium in cross section (× 1.5); (N) fruit, with coiled floral peduncle (× 0.5); (O) fruit (× 1); (P) dehiscing fruit (× 1.5); (Q) seed, with aril (× 15); (R) seed (× 30); (S) seed coat (greatly magnified); (T) longitudinal seciton of seed, with embryo, endosperm, and perisperm. (× 30). (Original art prepared for the Generic Flora of the Southeast U.S. Project. Used with permission.)

the result of secondary increase. Nymphaeaceae, as here circumscribed, clearly are monophyletic (Chase et al. 1993; Donoghue and Doyle 1989; Doyle et al. 1994; Hamby and Zimmer 1992; Les et al. 1999; Qiu et al. 1993; Zimmer et al. 1989), although some botanists treat Cabomboideae at familial rank, as Cabombaceae.

Nelumbonaceae have often been placed in Nymphaeaceae, but all recent evidence places them within the tricolpate clade (Donoghue and Doyle 1989; Chase et al. 1993; Moseley et al. 1993; Qiu et al. 1993, 2000; Savolainen et al. 2000b; Soltis et al. 2000). Although superficially similar to Nymphaeaceae, Nelumbonaceae have tricolpate pollen, and numerous carpels that are sunken in pits within an enlarged, funnel-shaped, spongy receptacle.

The fragrant, showy flowers of Nymphaeaceae (Plate 9.1A) attract various insects (beetles, flies, and bees), which gather pollen or, less commonly, nectar. However, flowers of *Brasenia* lack nectar glands and are wind-pollinated, having numerous easily shaken anthers. Outcrossing is favored by protogyny. Flowers of *Victoria* and some species of *Nymphaea* attract beetles by providing food bodies (starch-filled carpel appendages) and producing heat along with a strong fruity odor. The flowers open and close daily and trap the beetles. In other species of *Nymphaea*, flies and small bees gather pollen from 2- or 3-day-old flowers. They are then attracted to a pool of sweet stigmatic fluid in young (1-day-old) flowers, in which they frequently drown. Pollen on their bodies becomes suspended in the stigmatic fluid and eventually germinates. The fleshy fruits of many Nymphaeaceae mature underwater and rupture irregularly due to swelling of the mucilaginous aril surrounding the seeds, which are water dispersed. In *Nuphar* the carpel segments separate and float away. Vegetative reproduction commonly occurs through production of rhizomes or specialized shoots/tubers.

Additional references: Endress 2001; Ito 1986, 1987; Les et al. 1991; Orgaard 1991; Osborn et al. 1991; Osborn and Schneider 1988; Meeuse and Schneider 1979; Padgett et al. 1999; Schneider and Carlquist 1995; Schneider and Jeter 1982; Schneider and Williamson 1993; Schneider et al. 1995; Taylor and Hickey 1996; Thorne 1974, 1992; Wiersema 1988; Wood 1959a.

Austrobaileyales

Illiciaceae A. C. Smith
(Star Anise Family)

Trees or shrubs; nodes unilacunar; with scattered spherical cells containing ethereal oils (aromatic terpenoids) and branched sclereids; often with tannins. Hairs simple. *Leaves alternate and spiral,* often clustered at the tips of the shoots, simple, *entire,* with pinnate venation, *blade with pellucid dots; stipules lacking. Inflorescences* determinate, with 1–3 flowers, *axillary.* Flowers bisexual, radial, with convex to shortly conical receptacle. *Tepals usually numerous,* distinct, the outer usually sepal-like, and the innermost sometimes minute, imbricate. *Stamens usually numerous,* distinct; *filaments short and thick, poorly differentiated from the anthers;* anthers with connective tissue extending between and beyond the apex of the pollen sacs; pollen grains tricolpate, but colpus position different from that of the eudicots. *Carpels usually 7 to numerous,* distinct, in a single whorl; ovaries superior, with ± basal placentation; stigma extending down style on adaxial surface. **Ovules 1 per carpel.** Nectar produced at base of stamens. **Fruit a starlike aggregate of 1-seeded follicles** (Plate 9.1C)**; seeds with smooth, hard coat**; embryo minute; *endosperm diploid, homogeneous* (Figure 9.6).

Floral formula: *, T-7–∞-, A7–∞, G7–∞; follicles

Distribution: Southeastern Asia, southeastern United States, Cuba, Hispaniola, and Mexico; mainly in moist forests.

Genus/species: 1/37. *Genus:* *Illicium.*

Economic plants and products: Anise oil is extracted from *Illicium verum* (star anise). Some species are used medicinally, and a few are used as ornamentals.

Discussion: Illiciaceae belongs within Austrobaileyales, an order of 4 families and about 100 species. Monophyly of Austrobaileyales is supported by DNA characters (Qiu et al. 2000; Renner 1999; Soltis et al. 2000) and possibly also by hardened mesotestal cells. Austrobaileyaceae and Trimeniaceae are early divergent clades, and the more specialized Illiciaceae and Schisandraceae (a small family of vines with unisexual flowers) form a clade (sometimes called Schizandraceae s.l.) united by the apomorphic characters of pollen grains with three (or six) furrows (colpi), a feature that is otherwise limited to the eudicot clade, and unilacunar nodes.

Flowers of *Illicium* (Plate 9.1C) are pollinated by a wide variety of small insects, particularly flies. The plants are self-incompatible. The follicles dehisce elastically, shooting out the smooth seeds. Both pollination and seed dispersal ap-

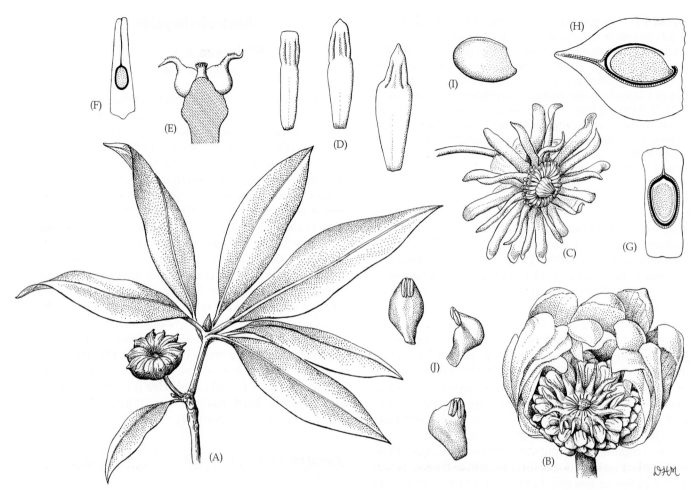

FIGURE 9.6 Illiciaceae. (A–I) *Illicium floridanum:* (A) fruiting branch (× 0.5); (B) opening flower bud with receptive carpels (× 4); (C) flower, later stage at shedding of pollen (× 1.5); (D) stamens, inner, outer, and an unusual subtepaloid form (× 7); (E) two carpels on receptacle (× 4); (F) carpel in cross section; note single ovule (× 15); (G) mature fruit, with single seed, endosperm stippled (× 3); (H) mature fruit in longitudinal section (× 3); (I) seed (× 3). (J) *I. parviflorum:* stamens (× 3). (From Wood 1958, *J. Arnold Arb.* 39: p. 317.)

pear to be quite local. Most species also form extensive clones due to the production of rhizomes.

Additional references: Endress 2001; Keng 1993; Roberts and Haynes 1983; Smith 1947; Thien et al. 1983; Thorne 1974; White and Thien 1985; Wood 1958.

MAGNOLIID CLADE

Magnoliales

Magnoliales are considered monophyletic on the basis of nuclear, chloroplast and mitochondrial sequence characters, along with their 2-ranked leaves, reduced fiber pit borders, stratified phloem, an adaxial plate of vascular tissue in the leaf midrib, and star-shaped sclerids in leaf mesophyll (see Donoghue and Doyle 1989; Hilu et al. 2003; Qiu et al. 1993, 2005; Sauquet et al. 2003; Zanis et al. 2003). Magnoliales are distinguished from Laurales by their trilacunar to multilacunar nodes, 2-ranked leaves, and frequently fleshy or arillate seeds. These families have retained numerous apparently plesiomorphic floral characters such as distinct and often numerous, spirally arranged stamens and carpels, superior ovaries, and seeds with a minute embryo and copious endosperm. Myristicaceae is undoubtedly sister to the remaining families (Doyle and Endress 2000; Doyle et al. 2004; Sauquet et al. 2003; Soltis et al. 2000). The order consists of 6 families and about 2840 species; noteworthy families include **Annonaceae**, **Magnoliaceae**, **Myristicaceae**, and Degeneriaceae.

References: Canright 1952, 1960; Cronquist 1981, 1988; Dahlgren 1983; Donoghue and Doyle 1989; Endress 1986b, 1994b; Igersheim and Endress 1997; Nandi et al. 1998; Takhtajan 1969, 1997; Thorne 1974, 1992; Weberling 1988b; Wood 1958.

Key to Major Families of Magnoliales

1. Stamens monadelphous, their filaments connate into a tube or column; seeds arillate ....**Myristicaceae**
1. Stamens distinct; seeds not arillate 2
2. Stipules present, sheathing the stem, and enveloping the apical bud; nodes multilacunar; endosperm homogeneous; perianth usually of tepals; receptacle elongate ....**Magnoliaceae**
2. Stipules lacking; nodes 3- to 5-lacunar; endosperm ruminate; perianth of sepals and petals; receptacle short to ± hemispherical 3
3. Carpel 1, the stigma running nearly its entire length; stamens laminar, with 3 veins, not packed into a tight ball; embryo with 3 or 4 cotyledons ....Degeneriaceae
3. Carpels more than 1, often numerous, the stigma ± elongate, but restricted to apical portion of each carpel; stamens short and stout, with an expanded connective, with 1 vein, and usually packed into a ball-like configuration; embryo with 2 cotyledons ....**Annonaceae**

Magnoliaceae A. L. de Jussieu
(Magnolia Family)

Trees or shrubs; **nodes multilacunar**; *with spherical cells containing ethereal oils (aromatic terpenoids)*; with alkaloids, usually of the benzyl-isoquinoline type. Hairs simple to stellate. *Leaves alternate*, spiral to 2-ranked, simple, sometimes lobed, *entire*, with pinnate venation, *blade with pellucid dots*; **stipules present, surrounding the terminal bud. Inflorescence a solitary, terminal flower** (Plate 9.2I), but sometimes appearing axillary (on a short shoot). **Flowers** usually bisexual, radial, **with an elongate receptacle** (Plate 9.2A). *Tepals 6 to numerous*, distinct, occasionally the outer 3 differentiated from the others and ± sepal-like, imbricate. *Stamens numerous*, distinct, often with 3 veins; *filaments short and thick, poorly differentiated from the anthers*; anthers with connective tissue often extending beyond the apex of the pollen sacs; pollen grains monosulcate. *Carpels usually numerous, distinct, on an elongate receptacle*; ovaries superior, with lateral placentation; stigma usually extending down style on adaxial surface, but sometimes reduced and terminal. Ovules usually 2 per carpel, sometimes several. Nectaries lacking. **Fruit an aggregate of follicles, which usually become closely appressed as they mature and open along the abaxial surface**, sometimes fleshy, with individual fruits becoming fused to one another as they mature, forming an indehiscent or irregularly dehiscent berrylike structure, or an aggregate of samaras; *seed with red to orange, fleshy coat* (except *Liriodendron*), usually dangling from a slender thread; embryo minute; *endosperm homogeneous* (Figure 9.7; see also Figure 4.47A).

Floral formula: *, T-6—∞-, A∞, G∞; follicles, samaras

Distribution and ecology: Temperate to tropical regions of eastern North America and eastern Asia, and tropical South America; mainly in moist forests.

Genera/species: 2/220. **Genera:** *Magnolia* (218 spp.) and *Liriodendron* (2).

Economic plants and products: *Liriodendron tulipifera* (tulip tree, tulip-poplar; Plate 9.2I) and several species of *Magnolia* are important ornamentals. Species of both genera are also used for timber.

Discussion: Cladistic analyses of *rbcL* sequences (Qiu et al. 1993) and morphological characters support monophyly of Magnoliaceae. Chloroplast genes (Azuma et al. 2001; Kim et al. 2001; Qiu et al. 1993, 1995) indicate that the family is composed of two clades: represented by *Liriodendron* and by *Magnolia* s. l. Recognition of *Talauma*, *Michelia*, and *Manglietia* leads to a paraphyletic *Magnolia*, and therefore, we circumscribe this genus broadly. *Liriodendron* has several apomorphies, including the striking lobed leaves, carpels with a restricted stigma, samaroid fruits, and seeds with a thin, more or less dry coat. The monophyly of *Magnolia* is supported by the follicles opening along the abaxial (or outer) surface. These can be described as backward-opening follicles because most follicular fruits open along the adaxial surface.

The showy flowers of Magnoliaceae are mainly pollinated by beetles, which may be trapped in the flower for a period of time, and often eat pollen and/or various floral tissues. *Liriodendron*, however, is bee-pollinated. Protogyny and self-incompatibility lead to outcrossing. The seeds of *Magnolia*, with bright red, pink, or orange, fleshy seed coats (Plate 9.2G), hang on thin springy threads—actually vessels with spiral thickenings—when the follicle opens and are dispersed by birds. The fleshy syncarps of some of the tropical species are also colorful and bird-dispersed. The samaras of *Liriodendron* are dispersed by wind.

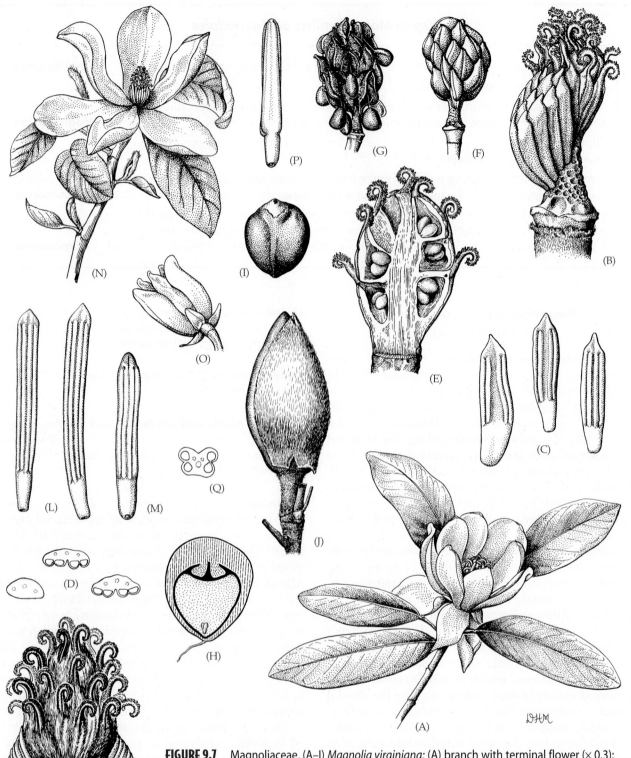

FIGURE 9.7 Magnoliaceae. (A–I) *Magnolia virginiana:* (A) branch with terminal flower (× 0.3);
(B) androecium (part removed) and gynoecium, on elongate receptacle (× 3.5); (C) stamens, adaxial surface (× 4); (D) stamens, in cross section (× 6); (E) gynoecium in longitudinal section, note two
ovules in each carpel (× 5); (F) nearly mature fruit (× 0.75); (G) mature fruit with pendulous seeds
(× 0.75); (H) seed in longitudinal section, note copious endosperm and minute embryo (× 2.5);
(I) seed, with fleshy outer seed coat removed (× 2.5). (J–L) *M. grandiflora:* (J) flower bud (× 0.75);
(K) floral receptacle with androecium (half of stamens removed) and gynoecium (× 2); (L) stamens,
adaxial surface (× 4). (M) *M. tripetala:* stamen, adaxial surface (× 4). (N–Q) *M. acuminata:* (N) branch
with terminal flower (× 0.3); (O) opening flower bud (× 0.3); (P) stamen, adaxial surface (× 4);
(Q) anther in cross section (× 6). (From Wood 1974, *A student's atlas of flowering plants*, p. 36.)

(A)

Magnoliales: Magnoliaceae
Magnolia × *soulangeana*:
flower (longitudinal section)

(B)

Canellales: Winteraceae
Drimys winteri: flower

(C)

Magnoliales: Annonaceae
Asimina incana:
branch with flowers

(E)

Piperales: Piperaceae
Piper aduncum: leaves and inflorescences

(D)

Piperales: Aristolochiaceae
Aristolochia gigantea: flower

(F)

Magnoliales: Annonaceae
Annona squamosa: branch with fruit

(G)

Magnoliales: Magnoliaceae
Magnolia virginiana:
fruits and seeds

(H)

Laurales: Lauraceae
Cinnamomum camphora:
leaves and fruits

(I)

Magnoliales: Magnoliaceae
Liriodendron tulipifera: leaves and flower

(J)

Laurales: Lauraceae
Persea americana: buds and flower

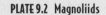

PLATE 9.2 Magnoliids

Additional references: Agababian 1972; Canright 1953; Endress 1994b; Gottsberger 1977, 1988; Nooteboom 1993; Thien 1974; Wood 1958.

Annonaceae A. L. de Jussieu
(Pawpaw or Annona Family)

Trees, shrubs, or lianas, with conspicuously fibrous bark; *nodes trilacunar;* **vessel elements with simple perforations**, and wood with broad rays; *with scattered spherical cells containing ethereal oils* (aromatic terpenoids), and often with scattered sclereids; usually with alkaloids of the benzyl-isoquinoline type; often with tannins. Hairs simple, sometimes stellate, or peltate scales. *Leaves alternate, 2-ranked,* simple, entire, *often short petioled,* with pinnate venation, blade *with pellucid dots; stipules lacking.* Inflorescences determinate, sometimes reduced to a single flower, terminal or axillary. *Flowers* usually bisexual, radial, **opening and then gradually increasing in size before flowering**, *with a short, flat to ± hemispherical receptacle. Sepals usually 3,* distinct or slightly connate, valvate or imbricate. *Petals usually 6,* distinct, *the outer 3 often larger and differentiated from the inner,* imbricate and/or valvate. *Stamens usually numerous, appearing peltate and packed into a ball-like or disklike configuration,* distinct, with 1 vein; *filaments short and thick, poorly differentiated from the anthers;* anthers with connective tissue extending beyond the apex of anther sacs; pollen grains various, some monosulcate, but often inaperturate, sometimes in tetrads or polyads. *Carpels (1-) 3 to numerous,* usually distinct, usually spirally arranged; ovaries superior, with lateral placentation; stigma extending down style on adaxial surface, or ± terminal. Ovules 1 to numerous per carpel. Nectaries or food tissues sometimes on inner petals. *Fruit an aggregate of berries,* these sometimes becoming connate as they develop; *seeds with raphe curving around the top,* often arillate; embryo minute; *endosperm ruminate* (Figure 9.8).

Floral formula: *, K3, C6, A∞, G(3–∞); berries

Distribution and ecology: Widely distributed in tropical and subtropical regions, and very characteristic of lowland wet forests.

Genera/species: 128/2300. **Major genera:** *Guatteria* (250 spp.), *Xylopia* (150), *Uvaria* (110), *Annona* (110), *Polyalthia* (100), *Artabotrys* (100), and *Rollinia* (65). The family is represented in the continental United States by *Asimina, Deeringothamnus,* and *Annona.*

Economic plants and products: Several species of *Annona* and *Rollinia* (cherimoya, guanabana, soursop, sugar apple, sweetsop) produce edible fruits (Plate 9.2F). Berries of *Asimina triloba* (pawpaw) are also edible. Flowers of *Cananga odorata* (ylang-ylang) are used in perfumes, and fruits of *Monodora myristica* are used as a substitute for

Myristica fragrans (nutmeg; Myristicaceae). *Annona, Cananga,* and *Polyalthia* are grown as ornamentals.

Discussion: Monophyly of Annonaceae is supported by DNA sequences (Qiu et al. 1993, 2000) and several morphological features (Doyle and Le Thomas 1994). A base chromosome number of eight probably is also synapomorphic (but reductions to *n* = 7 and increases to *n* = 9 have occurred). Tribes and genera have been distinguished by features such as indumentum, structure of the bud, sepal and petal aestivation, shape and texture of anthers, number and shape of carpels, extent of fruit connation, and number of seeds. Fruits in which the carpels become fused as they develop, forming a fleshy syncarp, as in *Annona* or *Rollinia,* are clearly derived in relation to persistently distinct berries, as in *Asimina* or *Cananga.* Connate carpels, as in *Monodora,* are also derived. *Anaxagorea* has follicle fruits, but most genera have fleshy, indehiscent fruits (berries). *Anaxagorea* also has laminar and elongated stamens, and endosperm with irregular ruminations, and it is sister to the rest of the family, which have more or less peltate stamens (packed into a characteristic ball-like or disk-like configuration) and spinose or thin-flat endosperm ruminations (extensions of seed coat into endosperm). A large group that includes *Xylopia, Monodora, Asimina, Annona,* and *Artabotrys* is characterized by pollen grains that lack germination apertures (Doyle and Le Thomas 1996; Doyle et al. 2004; Sauquet et al. 2003).

Flowers of most Annonaceae show various specializations leading to pollination by beetles, including closed flowers, fruity odors, feeding tissues, thick, fleshy petals, and structural protection of the reproductive organs; other pollinators include thrips, flies, and bees. Some species of *Annona* produce heat; beetles often stay in the flowers overnight, mate there, and frequently eat floral tissues. Many species of Annonaceae produce a sticky stigmatic fluid, which protects the carpels; the often hardened and more or less peltate anther connectives reduce consumption of pollen. Outcrossing in this family is favored by protogyny. The fleshy fruits are dispersed by birds, mammals, and turtles.

Additional references: Endress 1994b; Gottsberger 1988; van Heusden 1992; Kessler 1993a; Norman and Clayton 1986; Scharaschkin and Doyle 2006; Silberbauer-Gottberger et al. 2003; Thorne 1974; Vander Wyk and Canright 1956; Walker 1971; Wood 1958.

Myristicaceae R. Brown
(Nutmeg Family)

Trees, **with bark exuding a reddish sap when slashed**, *and usually with distinctive growth form, the trunk erect, the lateral branches appearing whorled and ± horizontal; with scattered spherical cells containing ethereal oils* (aromatic terpenoids); *with hallucinogenic phenolic compounds* (e.g., myristicin), tannins, and sometimes indole alkaloids. *Hairs simple, T-shaped, branched, or stellate. Leaves alternate, usually*

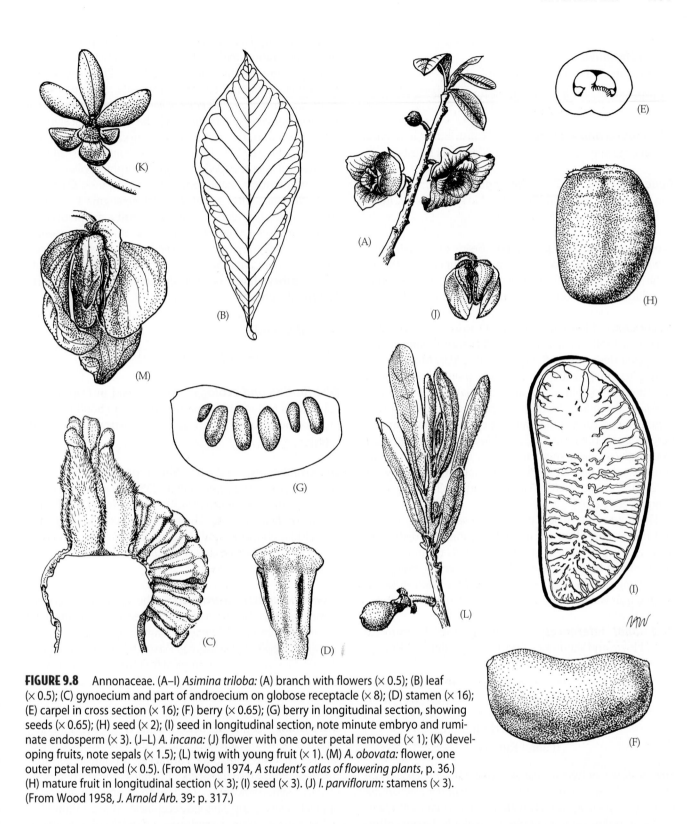

FIGURE 9.8 Annonaceae. (A–I) *Asimina triloba:* (A) branch with flowers (× 0.5); (B) leaf (× 0.5); (C) gynoecium and part of androecium on globose receptacle (× 8); (D) stamen (× 16); (E) carpel in cross section (× 16); (F) berry (× 0.65); (G) berry in longitudinal section, showing seeds (× 0.65); (H) seed (× 2); (I) seed in longitudinal section, note minute embryo and ruminate endosperm (× 3). (J–L) *A. incana:* (J) flower with one outer petal removed (× 1); (K) developing fruits, note sepals (× 1.5); (L) twig with young fruit (× 1). (M) *A. obovata:* flower, one outer petal removed (× 0.5). (From Wood 1974, *A student's atlas of flowering plants*, p. 36.) (H) mature fruit in longitudinal section (× 3); (I) seed (× 3). (J) *I. parviflorum:* stamens (× 3). (From Wood 1958, *J. Arnold Arb.* 39: p. 317.)

2-ranked, simple, entire, with pinnate venation, *blade with pellucid dots; stipules absent.* Inflorescences determinate, sometimes reduced and forming fascicles or heads, axillary. *Flowers inconspicuous, fragrant,* **unisexual (plants usually dioecious)**, radial. **Tepals** usually 3, **connate**, valvate. *Stamens 2-numerous;* **filaments connate into a ± solid column**; anthers distinct or also connate; pollen grains monosulcate to inaperturate. **Carpel 1**; ovary superior, *with ± basal placentation;* stigma 1, small, often bilobed. **Ovule 1.** Nectaries lacking. **Fruit ± leathery follicle, also dehiscing abaxially, exposing a single large seed and often colorful aril**; embryo minute; *endosperm ruminate.*

Floral formula:
Staminate: *, T–③–, A②–∞, G0
Carpellate: *, T–③–, A0, G1, fleshy capsule

Distribution and ecology: Widely distributed in tropical regions and very characteristic of lowland wet forests.

Genera/species: 17/370. **Major genera:** *Horsfieldia* (100 spp.), *Myristica* (70), and *Virola* (50). No species occur in the continental United States or Canada.

Economic plants and products: Seeds of *Myristica fragrans* (nutmeg tree) provide the spice nutmeg, while the aril is the source of mace. Several species of *Virola* are the source of hallucinogenic snuff. *Virola* and *Staudtia* provide timber.

Discussion: Monophyly of Myristicaceae is supported by molecular (Soltis et al. 2000) as well as morphological features (see description; Sauquet et al. 2003). Most genera belong to either a myristicoid clade (pollen with sulcate aperture, exine columellar, and reticulate; e.g., *Myristica*, *Horsfieldia*, *Virola*) or a mauloutchioid clade (pollen with spherical aperture, exine granular or mixed granular/columellate, and not reticulate; e.g., *Mauloutchia*, *Brochoneura*, *Doyleanthus*); many members of the latter clade have an androecium of numerous, partly separate anthers (Doyle et al. 2004; Sauquet 2003; Sauquet and Le Thomas 2003; Sauquet et al. 2003).

The flowers are insect pollinated, especially by beetles and thrips, and often are night blooming. The distinctive fruits are dispersed by birds and mammals (primates) that are attracted by the arillate seeds (with red to orange aril contrasting with the black seed coat) and the often strikingly colored fruit wall (inner surface can be red, pink, or white).

Additional references: Armstrong and Drummond 1986; Howe and Vande Kerckhove 1980; Kuhn and Kubitzki 1993; Wilson and Maculans 1967.

Laurales

Lauraceae A. L. de Jussieu
(Laurel Family)

Trees or shrubs, or twining, parasitic vine (*Cassytha*); *nodes unilacunar; with scattered spherical cells containing ethereal oils* (*aromatic terpenoids*); usually with tannins; usually with benzyl-isoquinoline and/or aporphine alkaloids. **Leaves alternate and spiral**, occasionally opposite, *but never 2-ranked*, simple, rarely lobed, **entire**, with usually pinnate venation, or sometimes the lowermost pair of secondary veins more prominent and arching toward apex and venation ± palmate, and all veins clearly visible, connected to adaxial and abaxial leaf surfaces by lignified tissue, blade *with pellucid dots; stipules lacking*. Inflorescences determinate to seemingly indeter-

minate, axillary. *Flowers* bisexual or unisexual (plants then ± dioecious), radial, *with distinctly concave receptacle*, usually small, pale green, white, or yellow (Plate 9.2J). *Tepals usually 6, distinct or slightly connate, imbricate. Stamens usually 3–12; filaments often with pairs of basal-lateral nectar- or odor-producing appendages* (staminodes), the 3 innermost stamens often also reduced to nectar- or odor-producing staminodes; *anthers opening by 2 or 4 flaps that curl from the base upward and pull out the sticky pollen*, often dimorphic; pollen grains without apertures, exine reduced to tiny spines. **Carpel 1**; *ovary superior, with ± apical placentation*; stigma 1, capitate, truncate, lobed, or elongate. *Ovule 1*. **Fruit a drupe** or occasionally a 1-seeded berry, *often associated with the persistent fleshy to woody receptacle (and sometimes also tepals) that often contrast in color with the fruit (i.e., fruit with a cupule; Plate 9.2H)*; **embryo large, with fleshy cotyledons; endosperm lacking** (Figure 9.9).

Floral formula:
*, T–⑥–, A3–12 + paired glands, G1; drupe

Distribution and ecology: Widespread in tropical and subtropical regions and especially diverse in Southeast Asia and northern South America; characteristic of tropical wet forests.

Genera/species: 50/2500. **Major genera:** *Litsea* (400 spp.), *Ocotea* (350), *Cinnamomum* (350), *Cryptocarya* (250), *Persea* (200), *Beilschmiedia* (150), *Nectandra* (120), *Phoebe* (100), and *Lindera* (100). The following occur in the continental United States and/or Canada: *Cassytha*, *Cinnamomum*, *Licaria*, *Lindera*, *Litsea*, *Nectandra*, *Persea*, *Sassafras*, and *Umbellularia*.

Economic plants and products: The family contains spice plants such as *Laurus nobilis* (bay leaves), *Cinnamomum verum* (cinnamon), *C. camphora* (camphor; Plate 9.2H), and *Sassafras albidum* (sassafras). *Persea americana* (avocado; Plate 9.2J) is an important tropical fruit tree. *Aniba rosaeodora* used in perfumes. *Beilschmiedia*, *Chlorocardium*, *Ocotea*, *Litsea*, and a few other genera contain species used for timber.

Discussion: Lauraceae belong to the large order Laurales, which consists of 7 families and about 3400 species; major families are Calycanthaceae, **Lauraceae**, Monimiaceae, Siparunaceae, and Hernandiaceae. Laurales are clearly monophyletic; synapomorphies include their unilacunar nodes, opposite leaves, cup-shaped receptacle, inner staminodes, pollen with sculptured apertures, and details of seed anatomy; some of these have been lost in many species. Monophyly of the order is supported in cladistic analyses based on both DNA sequences and morphology (Doyle and Endress 2000; Hilu et al. 2003; Qiu et al. 1993, 2000, 2005; Renner 1999; Soltis et al. 2000). Calycanthaceae are probably sister to all other Laurales (Doyle and Endress 2000; Renner and Chanderbali 2000). The other families are united by the addi-

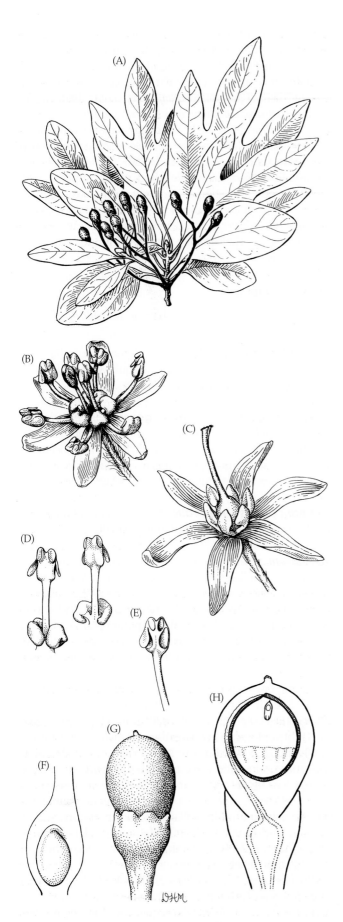

FIGURE 9.9 Lauraceae. *Sassafras albidum:* (A) fruiting branch (× 0.5); (B) staminate flower, androecium with nine stamens in three whorls of three (× 5); (C) carpellate flower, note staminodes (× 5); (D) two stamens of third whorl, each with two glands, note anthers opening by introrsely and laterally opening valves (× 6); (E) stamen of first whorl, note introrsely opening valves (two removed) (× 6); (F) ovary in longitudinal section, with single apical ovule (× 10); (G) mature drupe and cupule (× 3); (H) fruit and cupule in longitudinal section, note embryo with large, fleshy cotyledons (one cotyledon removed to show plumule and radicle) (× 4). (From Wood 1974, *A student's atlas of flowering plants*, p. 37.)

tional apomorphies of inaperturate pollen grains with a thin exine and spinules, stamens with paired appendages, anthers opening by flaps, and a single ovule per carpel (Donoghue and Doyle 1989; Doyle and Endress 2000; Renner 1999). Within this distinctive clade, Siparunaceae (and close relatives) form a clade distinct from a Lauraceae + Hernandiaceae + Monimiaceae clade. Lauraceae are likely sister to Hernandiaceae (Doyle and Endress 2000), although molecular studies (Renner 1999; Renner and Chanderbali 2000) possibly support a closer relationship with Monimiaceae. Lauraceae clearly are monophyletic (see description, and Renner and Chanderbali 2000). Lauraceae differ from Monimiaceae in their usually alternate and spiral, entire leaves (vs. opposite, toothed to entire leaves) and single carpel (vs. numerous carpels). Lauraceae differ from Hernandiaceae in having a superior ovary (vs. inferior) and drupaceous fruits (vs. fruits nuts, often associated with accessory structures). Siparunaceae, which often have been included within Monimiaceae, have several unusual apomorphic characters, e.g., bisporangiate anthers that open by a single slit, ovules with only a single integument, and flowers closed by a "roof" due to the extreme curvature of the cup-shaped receptacle (Renner et al. 1997).

Chloranthaceae have sometimes been placed in Laurales (Donoghue and Doyle 1989), but this placement is not supported by DNA-based cladistic analyses. Members of the family have reduced flowers, opposite and often toothed leaves, and swollen nodes with sheathing stipules. They may be a lineage diverging from remaining angiosperms after *Amborella*, Nymphaeaceae, and Austrobaileyales, or possibly are sister to magnoliids.

Lauraceae traditionally have been divided into two subfamilies. *Cassytha* is placed in the monotypic subfamily Cassythoideae on the basis of numerous specializations relating to its parasitic and viny habit, while all remaining genera are placed in the paraphyletic "Lauroideae." Rohwer and Rudolf (2005) suggest that *Hypodaphnis*, the *Cryptocarya* group, and *Cassytha* are successively sister to the rest of the family (see also Rohwer 2000, and Chanderbali 2001, for earlier studies).

Characters such as wood anatomy, stamen number, arrangement and number of anther locules, tepal persistence and form, fruit and cupule morphology, and inflorescence structure have been stressed in taxonomic groupings within "Lauroideae" (Burger 1988; Rohwer et al. 1991; van der Werff 1991; Rohwer 1993a, 1994; van der Werff and Richter 1996). Three tribes currently are recognized (see van der Werff and Richter 1996). Laureae have apparently racemose to umbellate inflorescences with involucral bracts and introrse anthers of the third whorl, and include *Litsea, Lindera, Laurus, Sassafras,* and *Umbellularia.* Perseeae, such as *Ocotea, Nectandra, Licaria, Persea, Phoebe,* and *Cinnamomum,* have cymose inflorescences lacking involucral bracts, and extrorse anthers of the third whorl. Recent molecular studies cast doubt on these tribal distinctions. Finally, Cryptocaryeae, which include *Beilschmiedia* and *Cryptocarya,* are similar to Perseeae, but their inflorescences have the lateral flowers of the three-flowered cymose units not quite opposite. The Cryptocaryeae are likely sister to the rest of the family (Chanderbali 2001; Rohwer 2000). Identification of genera and species is extremely difficult without flowering and fruiting material. Generic delimitation, which is often problematic, is discussed by van der Werff (1991) and Rohwer et al. (1991).

Flowers of Lauraceae are insect-pollinated, with flies and bees being the most common visitors. Modified staminodes paired at the base of some of the stamens produce odor and/or separate the stamens of the inner whorl and outer whorls; they sometimes also secrete nectar. In *Persea americana* outcrossing is enforced by a complicated system that involves two floral types, A and B flowers. The flowers have two periods of opening on successive days. The stigmas of A flowers are receptive in the morning of the first day, and anthers open in the afternoon of the second day. In B flowers, stigmas are receptive in the afternoon of the first day, while anthers shed their pollen during the morning of the second day. All flowers open on any particular tree will be in the same stage; thus trees of both A and B types are necessary for cross-pollination. Many other species of *Persea* (and some other genera) have similar systems. The drupes are dispersed mainly by birds, but dispersal by mammals also occurs. The color of the drupes often contrasts with that of the cupule, which is variously swollen pedicel, receptacle, and perianth, increasing the attractiveness of the fruits.

Additional references: Doyle et al. 1994; Endress and Igersheim 1997; Kubitzki and Kurz 1984; Thorne 1974; Weberling 1988b; Wood 1958.

Canellales

Winteraceae R. Brown ex Lindley
(Winter's Bark Family)

Trees or shrubs; nodes trilacunar; **vessels lacking**, with elongate, slender tracheids only; *with scattered spherical cells containing ethereal oils (aromatic terpenoids).* Hairs usually lacking. *Leaves alternate and spiral,* simple, *entire,* with pinnate venation, blade *with pellucid dots,* abaxial surface of stomates usually plugged by wax deposits; stipules lacking. Inflorescences determinate, sometimes reduced to a single flower, terminal or axillary. Flowers usually bisexual, radial, with short receptacle (Plate 9.2B). *Sepals usually 2–4,* distinct to connate, valvate, sometimes falling as a cap. *Petals usually 5 to many,* distinct, imbricate. *Stamens numerous,* distinct; *filaments ± flattened to laminar, usually poorly differentiated from the anthers;* anthers with connective tissue sometimes extending beyond the apex of the pollen sacs; **pollen grains uniporate, usually released in tetrads**. *Carpels 1 to numerous,* usually distinct, in a single whorl; ovaries superior, with lateral placentation; stigma extending down adaxial surface of style or ± capitate. Ovules 1 to several. Nectaries usually lacking. *Fruit a cluster of follicles or berries,* sometimes becoming connate as they mature; embryo minute; *endosperm homogeneous.*

Floral formula:

*, K $\widehat{(2-4)}$, C5–∞, A∞, G$\underline{1-∞}$; berries, follicles

Distribution and ecology: New Guinea, Australia, New Caledonia (and other islands of the southwestern Pacific), Madagascar, South America, and Mexico.

Genera/species: 5/90.   **Major genera:** *Tasmannia* (40 spp.) and *Bubbia* (30). The family is not represented in the continental United States or Canada, but *Drimys* occurs in Mexico. Most are understory species of moist forests, cool montane forests, or restricted to swampy habitats.

Economic plants and products: The bark of *Drimys winteri* (Winter's bark) has been used medicinally.

Discussion: Winteraceae and Canellaceae together constitute the Canellales. Canellaceae differ from Winteraceae in that their flowers have three sepals, 5 to 12 petals, 6 to numerous stamens that are fused into a tube, and two to six, connate carpels. Potential synapomorphies of Canellales include leaves with branched sclereids, secondary veins more or less irregular, carpels with well-differentiated pollen-tube transmission tissue, and ovules with outer integument of only 2–4 cell layers; the group's monophyly is supported in both molecular (Hilu et al. 2003; Qiu et al. 1999, 2000, 2005; Soltis et al. 2000; Zanis et al. 2002, 2003) and morphological investigations (Doyle and Endress 2000). The group is characterized by aporphine alkaloids, trilacunar nodes, and sieve tube plastids with starch and protein crystalloids and/or fibers. Canellales are most likely sister to Piperales, although some phylogenetic analyses place them sister to Magnoliales (Doyle and Endress 2000; Graham and Olmstead 2000; Mathews and Donoghue 2000; Qiu et al. 2000, 2005); the latter is currently thought to be the more likely position.

Phylogenetic relationships within Winteraceae have been assessed through cladistic analyses of morphology (Vink 1988) and nuclear DNA sequences (Karol et al. 2000;

Suh et al. 1993). Morphological and DNA characters agree on the close association of *Zygogynum, Exospermum, Belliolum,* and *Bubbia.* Vink (1988) treats this group as *Zygogynum* s.l. Suh et al. (1993), however, maintain *Bubbia,* which they consider to be sister to the *Zygogynum-Exospermum-Belliolum* clade. *Drimys* possesses the unusual autapomorphy of a reduced, more or less capitate stigma. *Tasmannia* may be sister to the above-mentioned genera, a hypothesis supported by ribosomal DNA sequences and by its low chromosome number. Finally, *Takhtajania* likely is sister to all other members of the family.

The lack of vessels in Winteraceae has often been considered a retained ancestral condition (see discussion in Bailey and Nast 1945; Cronquist 1981, 1988; Thorne 1974), a conclusion that is quite unparsimonious (Doyle and Endress 2000; Young 1981).

Flowers of Winteraceae, which are small to medium-sized, usually with a delicate, whitish corolla (Plate 9.2B), are pollinated by a variety of insects, especially small beetles, thrips, primitive moths, and flies. Pollen is the major pollinator reward, but in some species fluids produced by the stigmas or staminal glands provide additional rewards. Some species of *Tasmannia* are wind pollinated. Many species show self-incompatibility. Vertebrate dispersal characterizes the berry-fruited species.

Additional references: Doyle et al. 1990, 1994; Feild et al. 2000; Gottsberger 1988; Gottsberger et al. 1980; Keng 1993; Thien 1980; Thorne 1974; Vink 1993.

Piperales

Piperales are here circumscribed broadly, including 5 families; two, Piperaceae and Aristolochiaceae, are treated here. Monophyly of the order is strongly supported by phylogenetic analyses based on DNA sequence characters (Barkman et al. 2000; Graham and Olmstead 2000; Hilu et al. 2003; Mathews and Donoghue 2000; Nickrent et al. 2002; Qiu et al. 1999, 2000, 2005; Soltis et al. 2000; Zanis et al. 2003) and morphology (Doyle and Endress 2000). Possible morphological synapomorphies include two-ranked leaves, sheathing leaf base, and a single adaxial prophyll.

These plants are also characterized by stems with swollen nodes, distinct vascular bundles, often forming wood with broad rays, and vessel elements with simple perforations. They are herbs or soft-woody plants, with benzyl-isoquinoline and aporphine alkaloids, often with palmately veined leaves and more or less 3-merous flowers. Their 3-merous flowers and the monocot-like sieve tube plastids found in some Aristolochiaceae may suggest an affinity to the monocots (Dahlgren and Clifford 1982; Donoghue and Doyle 1989; Doyle and Endress 2000; Jaramillo et al. 2004). However, the order is here retained within the magnoliid clade.

Within Piperales, it is evident that Piperaceae are most closely related to Saururaceae (a small family of 4 genera and 6 species). The two families form a clade on the basis of tetracytic stomata, indeterminate, terminal, spicate inflorescences, rather obscure bilateral floral symmetry, minute pollen grains, orthotropous ovules, perisperm as nutritive tissue in the seed, and anatomical or developmental details of the flowers and seeds (Donoghue and Doyle 1989; Doyle and Endress 2000; Doyle et al. 1994; Igersheim and Endress 1998; Tucker et al. 1993). The Piperaceae + Saururaceae clade is also strongly supported by cladistic analyses based on DNA sequences (Doyle et al. 1994; Mathews and Donoghue 2000; Qiu et al. 1993, 2000; Soltis et al. 1997, 2000; Zimmer et al. 1989). Aristolochiaceae are most closely related to Lactoridaceae and Hydnoraceae (see discussion under Aristolochiaceae).

Additional reference: González and Rudall 2001.

Piperaceae Giseke
(Pepper Family)

Herbs to small trees, sometimes epiphytic; *nodes often ± swollen or jointed;* vessel elements with usually simple perforations; **stem with vascular bundles of more than 1 ring or ± scattered**; *with spherical cells containing ethereal oils;* often with alkaloids. Hairs simple. Leaves usually alternate and spiral, sometimes opposite, simple, entire, with palmate to pinnate venation, *with pellucid dots;* stipules lacking or adnate to petiole (and petiole sometimes sheathing stem). Inflores-

---

Key to Major Families of Piperales

1. Flowers minute, lacking a perianth, ± densely clustered in thick spikes; ovary superior, the placentation basal, with a single ovule; fruit a drupe; stems with primary vascular bundles in more than 1 ring, or ± scattered ....**Piperaceae**
1. Flowers conspicuous, with perianth of 3 connate, showy sepals, and sometimes also with 3 petals, not in spikes; ovary inferior or half-inferior, the placentation axile or parietal with intruded placentas; fruit a septicidal capsule or cluster of follicles; stem with primary vascular bundles in a single ring ....**Aristolochiaceae**

cences indeterminate, *of thick spikes,* **densely covered with minute flowers,** terminal or axillary, often displaced to a position opposite the leaf due to development of axillary shoot (Plate 9.2E). **Flowers** bisexual or unisexual (plants monoecious or dioecious), appearing to be radial, **inconspicuous, each with a broadly triangular to peltate bract.** *Perianth lacking.* Stamens 1–10, often 6; filaments usually distinct; pollen grains monosulcate or lacking apertures. **Carpels 1–4, connate;** ovary superior, *with basal placentation;* stigmas 1–4, capitate, lobed, or brushlike. **Ovule 1 per gynoecium,** orthotropous, with 1 or 2 integuments. Nectaries lacking. **Fruit usually a drupe;** endosperm scanty, supplemented by perisperm (Figure 9.10).

Floral formula: *, T-0-, A1–10, G$\underline{(1–4)}$; drupe

Distribution and ecology: Widely distributed in tropical and subtropical regions. Species of *Peperomia* are commonly epiphytes in moist broadleaved forests.

FIGURE 9.10   Piperaceae. (A) *Peperomia glabella:* leaf (× 0.75). (B–G) *P. humilis:* (B) flowering shoot (× 0.15); (C) portion of inflorescence (× 27); (D) flower with bract, showing gynoecium and two stamens (× 27); (E) gynoecium (× 27); (F) portion of spike with a mature drupe (× 18); (G) endocarp of drupe (× 27). (H–L) *P. obtusifolia:* (H) flowering shoot (× 0.75); (I) flower with bract, showing gynoecium and two stamens (× 27); (J) bract (× 27); (K) tip of spike with partly mature fruits, three removed to show immersion of base of fruit in tissue of axis, apex of spike with undeveloped flowers shown only in outline (× 9); (L) mature drupe, showing hooked apex and position of stigma (× 1991). (From Borstein 1991, *J. Arnold Arbor. Suppl. Ser.* 1, p. 359.)

Genera/species: 6/2020. **Major genera:** *Peperomia* (1000 spp.) and *Piper* (1000). Both occur within the continental United States.

Economic plants and products: Fruits of *Piper nigrum* provide black and white pepper, one of the oldest and most important spices. Leaves of *Piper betle* (betel pepper) are chewed (along with lime, various spices and fruits of *Areca catechu*, betel nut) and have a mildly stimulating effect. A few species of *Piper* are used medicinally. Several species of *Peperomia* are grown as ornamentals because of their attractive foliage.

Discussion: Piperaceae are clearly monophyletic (Tucker et al. 1993); in addition to the synapomorphies noted above, the female gametophyte is unusual in that it develops from all four products of megaspore meiosis. Members of the related family Saururaceae can be easily separated from Piperaceae by their 2 to 10 ovules per carpel and distinct styles (and often distinct carpels, both plesiomorphies), although Saururaceae probably are monophyletic (Tucker et al. 1993).

Within Piperaceae, *Zippelia* is the sister group to the remaining species, which form a clade on the basis of a reduction to three carpels and capitate stigmas (Tucker et al. 1993). Relationships among the remaining genera are problematic. *Piper* (circumscribed broadly, including *Macropiper*, *Pothomorphe*) is extremely diverse, varying in inflorescence position (terminal or axillary), structure (solitary or clustered spikes or racemes), sexuality (bisexual or unisexual, and the plants then dioecious), and number of stamens. Recent cladistic analysis based on nrDNA ITS sequences support the recognition of three major clades within the genus: South Pacific species, Asian species, and Neotropical species (Jaramillo and Manos 2001). *Peperomia* shows numerous apomorphies, such as a single carpel, 2 unilocular stamens, a 16-nucleate female gametophyte (versus 8 nuclei in other genera), ovules with a single integument, unusually small pollen grains that lack apertures, herbaceous habit, and succulent leaves.

The tiny flowers of Piperaceae probably are insect-pollinated, but more work on the pollination biology of the family is needed. Outcrossing may predominate due to protogyny. The drupes of *Piper* are mainly dispersed by birds and bats, while those of *Peperomia* are often sticky and may be externally transported by animals.

Additional references: Borstein 1991; Semple 1974; Tebbs 1993; Thorne 1974; Wood 1971.

Aristolochiaceae A. L. de Jussieu
(Dutchman's-Pipe Family)

Herbs, lianas, or occasionally shrubs; with spherical cells containing ethereal oils; **with aristolochic acids (bitter, yellow, nitrogenous compounds)** or alkaloids. Hairs simple. *Leaves alternate and spiral,* simple, sometimes lobed, entire, with palmate venation, with pellucid dots; stipules usually lacking. Inflorescences various. Flowers bisexual, radial to bilateral. *Sepals 3,* **connate,** *often bilateral, tubular, and S-shaped or pipe-shaped, with spreading 3-lobed to 1-lobed limb,* **showy,** dull red and mottled, valvate, and deciduous (distinct and green in *Saruma*). Petals usually lacking or vestigial, but in *Saruma* present and well developed, distinct, yellow, and imbricate. Stamens usually 6–12; filaments distinct, **slightly to strongly adnate to style;** pollen grains usually lacking apertures (monosulcate in *Saruma*). Carpels 4–6, connate (distinct in *Saruma*), often twisting in development; **ovary/ovaries half-inferior,** *inferior in the more derived taxa,* with axile placentation, or parietal with intruded placentas; stigmas 4–6, often lobed and spreading. **Ovules numerous.** Nectaries often patches of glandular hairs on calyx tube. *Fruit a septicidal capsule* (cluster of follicles in *Saruma*), often pendulous and opening from the base; seeds flattened, winged, or associated with fleshy tissue, **with the inner cells of testa with abundant oxalate crystals, and tegmen with 3 layers of fibers crossed at 90° angles** (Figure 9.11).

Floral formula:

* or X, K ③, C3 or 0, A6–12, G $\overline{(4-6)}$; capsule

Distribution and ecology: Widespread in tropical and temperate regions.

Genera/species: 7/460. **Major genera:** *Aristolochia* (370 spp.), *Asarum* (70). Both genera occur in the continental United States.

Economic plants and products: Many species of *Asarum* (wild ginger, incl. *Hexastylis*) and *Aristolochia* (Dutchman's-pipe; incl. *Isotrema, Pararistolchia*) are cultivated as ornamentals because of their unusual flowers or variegated leaves (Plate 9.2D). Some species of *Aristolochia* are used medicinally.

Discussion: Some DNA sequence studies indicate that the Aristolochiaceae may be paraphyletic, with *Lactoris* (Lactoridaceae) placed within it (see Qiu et al. 2000; Soltis et al. 2000), but the family here is considered monophyletic on the basis of its connate sepals, stamens more or less adnate to the gynoecium, usually 6-carpellate, more or less inferior ovary, distinctive chemistry and seed coat anatomy. The achlorophyllous root-parasitic Hydnoraceae, which have syntepalous flowers that arise endogenously from roots of the host, may also be closely related to the Aristolochiaceae (Nickrent et al. 2002).

Thorne (1992) and Huber (1993) divided Aristolochiaceae into Asaroideae, including *Asarum* and *Saruma*, which are perennial herbs with radial flowers in which the stamens are free or only slightly adnate to the gynoecium, and Aristolochioideae, including *Aristolochia*, which are subshrubs to lianas with usually bilateral flowers in which

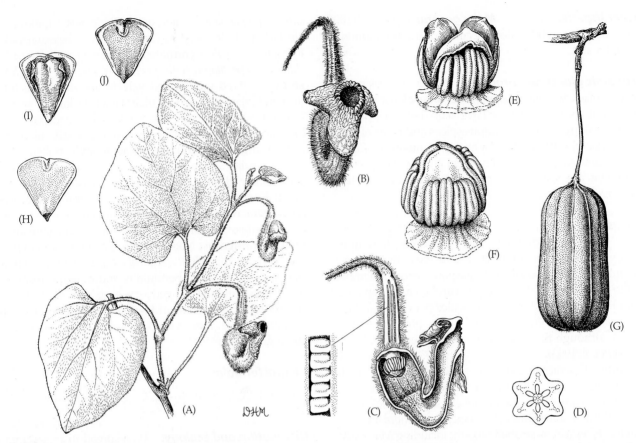

FIGURE 9.11 Aristolochiaceae. *Aristolochia tomentosa:* (A) branch with flower and bud 0.5; (B) flower (× 1); (C) flower with half of perianth and ovary removed, and detail of ovules in a single locule (× 1); (D) ovary in cross section (× 3); (E) receptive stigmas, at opening of flower (× 4); (F) stigmas at time of shedding of pollen (but adnate anthers shown as closed) now folded inward (× 4); (G) fruit (× 1); (H) seed, upper surface (× 1.5); (I) seed, lower surface, with corky funiculus (× 1.5); (J) seed, lower surface, corky funiculus removed (× 1.5). (From Wood 1974, *A student's atlas of flowering plants,* p. 21.)

the stamens are strongly adnate to the style. The subfamilies likely are monophyletic (Kelly and González 2003).

Flowers of Aristolochiaceae are mainly fly-pollinated. *Aristolochia* has elaborate trap flowers with a highly modified calyx (Plate 9.2D). Flies are attracted to these flowers by their dull red and often mottled coloration and by their odor, which may be fruity to fetid. Nectar, produced by glandular hairs on the calyx tube, functions as a pollinator reward. Pollen-bearing flies become trapped within an inflated portion of the calyx tube, which may be entered only through the constricted opening, and often is provided with downward-pointing hairs. During the first phase of the flower's life, pollen is deposited on the receptive stigmas that crown the gynoecium, projecting over the still unopened anthers. After pollination, the stigmas wither and become erect, thus exposing the dehiscing anthers. The flies become covered with pollen, and are able to leave the flower when the downward-pointing hairs and/or the calyx tube wither. Seeds of *Aristolochia* are typically flattened, and are dispersed by wind from the hanging, parachute-like capsules, although some species are water-dispersed or have sticky seeds that are externally transported

by animals; a few have fleshy fruits and are vertebrate-dispersed. Dispersal of seeds by ants is common in herbaceous taxa (e.g., *Asarum*); these seeds possess an aril-like structure.

Additional references: Faegri and van der Pijl 1980; Kelly 1997.

A CLADE OF UNCERTAIN POSITION

Ceratophyllales

Ceratophyllaceae S. F. Gray

(Hornwort Family)

Submersed, aquatic herbs; roots lacking, but often with colorless rootlike branches anchoring the plant; stems with a single vascular strand with central air canal surrounded by elongate starch-containing cells; with tannins. Leaves whorled, simple, often **dichoto-**

mously dissected, entire to serrate, **lacking stomates and a cuticle**; stipules lacking. **Inflorescences of solitary, axillary flowers. Flowers unisexual (plants monoecious)**, *radial, inconspicuous,* **with a whorl of 7 to numerous bracts** (possibly tepals). *Stamens 10 to numerous,* distinct; filaments not clearly differentiated from anthers; anthers with connective prolonged beyond pollen sacs and forming 2 prominent teeth; **pollen grains lacking apertures, with reduced exine, forming branched pollen tubes. Carpel 1**; ovary superior, **with ± apical placentation**; stigma elongated and extending along one side of style. Ovule 1 per carpel, **orthotropous, with 1 integument**. Nectaries lacking. **Fruit an achene, often with 2 or more projections, along with persistent style; seed with testa ± obliterated; endosperm lacking**.

> *Floral formula:*
> Staminate: *, T-7–∞-, A10–∞, G0
> Carpellate: *, T-7–∞-, A0, G1; achene

Distribution and ecology: Cosmopolitan; forming floating masses in freshwater habitats.

Genus/species: 1/6. ***Genus:*** *Ceratophyllum*.

Economic plants and products: *Ceratophyllum* is ecologically important in that it provides protection from predation for newly hatched fish. The foliage and fruits are important food items for migratory waterfowl; sometimes it is weedy, choking waterways.

Discussion: This family shows numerous adaptations to life as a submerged aquatic herb, has a fossil record extending back to the early Cretaceous, and is an ancient and peculiar taxon of uncertain affinity. In the Chase et al. (1993) cladistic analysis based on *rbcL* nucleotide sequences the Ceratophyllaceae is the sister group of all other angiosperms. However, in more recent molecular analyses the family is sister to the monocots (Graham and Olmstead 2000; Qiu et al. 2000), sister to the eudicots (Soltis et al. 2000; Hilu et al. 2003; Graham et al. 2006), or sister to all angiosperms except the ANITA grade (Matthews 2006).

The six species of *Ceratophyllum* are highly variable and taxonomically difficult. Relationships within the family are based on variation in the leaves and fruits. The inconspicuous flowers of Ceratophyllaceae are submerged and pollen is dispersed by water currents, as are the achenes, although dispersal by birds also occurs. The persistent style and variously developed appendages attach the small fruits to vegetation or sediments. Vegetative reproduction by fragmentation is common.

Additional references: Cronquist 1981; Dahlgren 1989; Dilcher 1989; Endress 1994a; Les 1988, 1989, 1993; Les et al. 1991; Thorne 1974; Wood 1959a.

MONOCOTS

The monocots are considered monophyletic based on their herbaceous habit, parallel-veined leaves with a sheathing base, embryo with a single cotyledon, sieve cell plastids with several cuneate protein crystals, stems with scattered vascular bundles, adventitious root system, and pentacyclic 3-merous flowers. Sieve cell plastids with several protein crystals also occur in some Aristolochiaceae (*Saruma* and *Asarum*), while scattered vascular bundles and adventitious roots also occur in Nymphaeaceae and some Piperaceae. Several monocots have pinnate to palmate leaves with obviously reticulate venation patterns (see Chase et al. 1995b; Dahlgren et al. 1985), but these are probably reversals associated with life in shaded forest understory habitats (Givnish et al. 2005). In addition, the leaves of most monocots, even those with a well-developed blade and petiole, are formed almost entirely from the basal end of the leaf primordium, while the leaves of nonmonocots are mainly derived from the apical end of the primordium. Monocots typically have monosulcate pollen, probably a retention of an ancestral angiosperm feature. It is noteworthy that monocots never have glandular-toothed leaves, any teeth being more or less spiny (as in Ceratophyllaceae).

Monophyly of the monocots is supported by DNA sequences and morphology (Bharathan and Zimmer 1995; Chase 2004; Chase et al. 1993, 1995a,b, 2000, 2006; Davis et al. 2004; Graham et al. 2006; Hilu et al. 2003; Savolainen et al. 2000a; Stevenson and Loconte 1995; Soltis et al. 1997, 2000, 2005). The taxonomic diversity of monocots is presented in detail by Kubitzki (1998a,b).

Within the monocots, Acoraceae (Bogner and Mayo 1998; Bogner and Nicolson 1991; Grayum 1987, 1990) are probably sister to the rest (Chase 2004; Chase et al. 1993, 1995b, 2006; Graham et al. 2006). They are wetland herbs with narrow, equitant leaves; small, bisexual flowers with six tepals, six stamens, and two or three fused carpels, which are borne on a thick, spicate inflorescence; and berry fruits. *Acorus* has a number of features that are plesiomorphic in monocots, perhaps including the ethereal oils (in specialized spherical cells) that give the group its common name, "sweet flag."

Alismatales

Cladistic analyses of both nuclear and chloroplast DNA sequences (Chase et al. 1993, 1995b, 2000, 2006; Duvall et al. 1993; Hilu et al. 2003; Källersjö et al. 1998; Soltis et al. 2000) support the monophyly of Alismatales, as do the possible morphological synapomorphies of stems with small scales or glandular hairs within the sheathing leaf bases at the nodes, extrorse anthers, and a large, green embryo (Dahlgren and Rasmussen 1983; Dahlgren et al. 1985; Stevenson and Loconte 1995). The Araceae are sister to the remaining families of the order, which share seeds lacking endosperm and root hair cells shorter than other epidermal cells. All members of this subclade occur in wetland or aquatic habitats.

Two major clades are recognized within this aquatic clade, which probably diverged very early in the evolution of monocots. The first clade contains Alismataceae, Hydrocharitaceae, and Butomaceae, and is supported by the apomorphies of perianth differentiated into sepals and petals, stamens more than six and/or carpels more than three (a secondary increase), and ovules scattered over the inner surface of the locules. Major families constituting the second clade are Potamogetonaceae, Ruppiaceae, Zosteraceae, Posidoniaceae, and Cymodoceaceae. This group is diagnosed on the basis of pollen that usually lacks apertures and more or less lacks an exine (Dahlgren and Rasmussen 1983; Cox and Humphries 1993). Plants of marine habitats evolved within Hydrocharitaceae, and more than twice in the sea grass families Zosteraceae, Cymodoceaceae, Ruppiaceae, and Posidoniaceae (Les et al. 1997a).

Alismatales contains about 14 families and about 3320 species; major families include **Araceae**, **Alismataceae**, **Hydrocharitaceae**, Butomaceae, **Potamogetonaceae**, Ruppiaceae, Zosteraceae, Posidoniaceae, Cymodoceaceae, and Tofieldiaceae.

Araceae A. L. de Jussieu
(Arum Family)

Terrestrial to aquatic herbs, often with rhizomes or corms, vines with aerial roots, epiphytes, or floating aquatics, the latter often very reduced with ± thalloid vegetative body; **grooved raphide crystals** *of calcium oxalate present in specialized cells, and associated chemicals causing irritation of mouth and throat if eaten;* cyanogenic compounds often present, and sometimes with alkaloids; often with laticifers, mucilage canals, or resin canals, and latex watery to milky. Hairs simple, but often lacking. *Leaves alternate and spiral or 2-ranked*, sometimes basal, usually simple, *blade often well developed*, sometimes strongly lobed, pinnately to palmately compound, usually entire, with parallel, pinnate, or palmate venation, sheathing at base; stipules lacking, but glandular hairs or small scales sometimes present at the node inside the leaf sheath. **Inflorescences indeterminate**, usually terminal, **forming a spike of numerous small flowers packed onto a fleshy axis (a spadix), which may be lacking flowers toward its apex, which is subtended by a large leaflike to petal-like bract (a spathe)**, but reduced in floating aquatic taxa. Flowers bisexual to unisexual (plants usually monoecious), radial, **lacking individual bracts**.

Key to Major Families of Alismatales

1. Flowers on a thick axis, the spadix, surrounded by or associated with a leaflike bract, the spathe; plants of various habitats .... **Araceae**
1. Flowers not on a spadix; plants aquatic or of wetlands 2
2. Perianth of sepals and petals 3
2. Perianth of tepals or lacking 5
3. Ovary inferior; carpels connate; fruits berries .... **Hydrocharitaceae**
3. Ovary superior; carpels ± distinct; fruits follicles or achenes 4
4. Pollen grains monosulcate; laticifers lacking and sap watery; fruits follicles; embryo straight .... Butomaceae
4. Pollen grains usually 4-multiporate or occasionally lacking apertures; laticifers present and sap milky; fruits achenes or follicles; embryo curved .... **Alismataceae**
5. Pollen grains globose or ellipsoid; plants of fresh, alkaline, or brackish water 6
5. Pollen grains threadlike; plants of marine environments 8
6. Flowers bisexual 7
6. Flowers unisexual .... Zannichelliaceae
7. Stamens 2, each with a small appendage; carpels on long stalks .... Ruppiaceae
7. Stamens 4, each with a large appendage; carpels sessile .... **Potamogetonaceae**
8. Flowers bisexual; stamens 3; carpel 1 .... Posidoniaceae
8. Flowers unisexual; stamens 1 or 2; carpels 2 9
9. Carpels distinct; stamens 2, connate .... Cymodoceaceae
9. Carpels connate; stamen 1 .... Zosteraceae

Tepals usually 4–6 or lacking, distinct to connate, inconspicuous and often fleshy, valvate or imbricate. Stamens 1–6 (–12); filaments distinct to connate; anthers sometimes opening by pores, distinct to connate; pollen grains various. Carpels usually 2–3, connate; ovary usually superior, placentation various; stigma 1, punctate or capitate. Ovules 1 to numerous, anatropous to orthotropous. Nectaries lacking. *Fruit usually a berry*, but occasionally a utricle, drupe, or nutlike; endosperm sometimes lacking (Figure 9.12).

Floral formula:

*, T-4-6- or -0-, A1–6, G$\underline{(1–3)}$; berry, utricle

Distribution and ecology: Cosmopolitan, but best developed in tropical and subtropical regions; very common in tropical forests and wetlands.

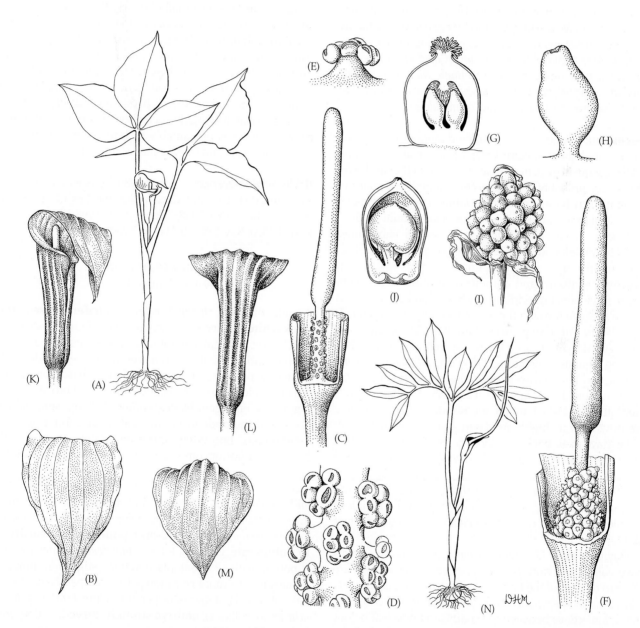

FIGURE 9.12 Araceae. (A–J) *Arisaema triphyllum:* (A) habit (× 0.2); (B) spathe, apical view (× 0.75); (C) staminate inflorescence, spathe removed (× 1.5); (D) portion of staminate spadix (× 9.75); (E) staminate flower (× 11.5); (F) carpellate inflorescence, spathe removed (× 1.5); (G) carpellate flower, longitudinal section, showing ovules (× 11.5); (H) ovule (× 30); (I) fruiting spadix (× 0.75); (J) fruit in longitudinal section, showing seed (× 3). (K–M) *A. triphyllum* var. *stewardsonii:* (K) inflorescence, lateral view (× 0.75); (L) inflorescence, back view (× 0.75); (M) spathe, apical view (× 0.75). (N) *A. dracontium:* habit (× 0.2). (From Wilson 1960a, *J. Arnold Arbor.* 41: p. 59.)

Genera/species: 109/2830. ***Major genera:*** *Anthurium* (900 spp.), *Philodendron* (500), *Arisaema* (150), *Homalomena* (140), *Amorphophallus* (100), *Schismatoglottis* (100), *Spathiphyllum* (60), *Monstera* (50), *Pothos* (50), *Xanthosoma* (40), *Dieffenbachia* (40), and *Syngonium* (30). Noteworthy genera of the continental United States and/or Canada are *Arisaema, Landoltia, Lemna, Lysichiton, Orontium, Peltandra, Pistia, Spirodela, Symplocarpus,* and *Wolffia.*

Economic plants and products: The starchy corms of *Alocasia, Colocasia* (taro), and *Xanthosoma* (yautia) are eaten (after proper treatment to remove the irritating chemicals). The berries of *Monstera* occasionally are eaten. The family contains numerous ornamentals, including *Philodendron, Zantedeschia* (calla lily), *Anthurium, Caladium, Colocasia, Dieffenbachia* (dumbcane), *Epipremnum, Monstera, Spathiphyllum, Syngonium, Aglaonema, Xanthosoma* (elephant's-ear), *Scindapsus, Spathicarpa,* and *Zamioculcus.*

Discussion: Araceae are considered monophyletic based on morphology (Grayum 1990; Mayo et al. 1995) and cpDNA sequences (Chase et al. 1993; French et al. 1995). The family probably is sister to the remaining families of Alismatales (Chase et al. 1995b; Dahlgren and Rasmussen 1983; Dahlgren et al. 1985; Stevenson and Loconte 1995).

Araceae have been divided into several subfamilies based on variation in habit, leaf arrangement and morphology, inflorescence structure, floral morphology, pollen structure, anatomy, and chromosome number (Grayum 1990). Phylogenetic relationships have also been assessed within Araceae through the use of *rbcL* sequences (French et al. 1995), *trnL-F* sequences (Tam et al. 2004), and morphology (Mayo et al. 1997). A few genera of very reduced floating aquatics, including *Spirodela, Landoltia, Lemna, Wolffia,* and *Wolfiella,* were once segregated as Lemnaceae (see Cronquist 1981; Dahlgren et al. 1985; den Hartog 1975; Landolt 1980, 1986; Landolt and Kandeler 1987), but they are now seen as highly modified Araceae (French et al. 1995; Mayo et al. 1995; Stockey et al. 1997; Tam et al. 2004). In *Lemna, Landoltia,* and *Spirodela* the spathe is represented by a membranous sheath, while it is completely lacking in *Wolffia* and *Wolfiella.* Although *rbcL* sequences place the much larger floating aquatic aroid *Pistia* in the Aroideae (see below), it is not close to *Lemna* and relatives (French et al. 1995).

Gymnostachys, Orontium, Symplocarpus, and relatives have condensed non-cormlike, thickened stems and are sister to the rest of the family. Most other Araceae are united by their leaves with expanded blades, major internode of the inflorescence between the spathe and next leaf below, the formation of a continuation shoot in the axil of the penultimate leaf below the spathe, and more or less basal placentation. The monoecious taxa comprise a large clade, the Aroideae, with 74 genera (*Zamioculca, Dieffenbachia, Spathicarpa, Philodendron, Caladium, Syngonium, Aglaonema, Zantedeschia, Amorphophallus, Peltandra, Asarum, Arum, Arisaema, Alocasia, Colocasia,* and *Pistia;* Mayo et al. 1997). A

second major clade, the Monsteroideae, is delimited on the basis of a undifferentiated spathe (i.e., lacking a tubular portion) that is soon deciduous or marcescent, with a distinct basal abscission zone. This group includes genera such as *Monstera, Scindapsus,* and *Epipremnum; Spathiphyllum* and relatives may be related (French et al. 1995). *Pothos* and relatives, and *Anthurium,* form the Pothoideae; these plants are characterized by fine leaf venation with the secondary and tertiary veins forming cross veins to the primaries (Mayo et al. 1997).

Inflorescences of Araceae are pollinated by several groups of insects, especially beetles, flies, and bees. The inflorescence usually produces a strong odor (sweet to noxious) and often heat. The gynoecium matures before the androecium, and when the flowers are unisexual, the carpellate flowers mature before the staminate, leading to outcrossing. In *Arisaema,* small (generally young) plants are staminate and larger (older) plants are carpellate, again leading to outcrossing. Dispersal of the green to brightly colored berries is presumably by birds or mammals. The utricles of *Lemna* and relatives are water-dispersed.

Additional references: Croat 1980; Igersheim et al. 2001; Keating 2004; Les and Crawford 1999; Les et al. 1997b, 2002; Maheshwari 1958; Mayo et al. 1998; Prychid and Rudall 2000; Ray 1987a,b; Wilson 1960a.

Alismataceae Ventenat
(Water Plantain Family)

Aquatic or wetland, rhizomatous herbs; **laticifers present, the latex white***;* tissues ± aerenchymatous. Hairs usually lacking. Leaves alternate, spiral or 2-ranked, usually ± basal, simple, entire, usually with a well-developed blade, with parallel or palmate venation, sheathing at base; sometimes polymorphic, with narrower submerged, and broader floating and/or emergent blades; stipules lacking; small scales present at the node inside the leaf sheath. Inflorescences determinate, but often appearing indeterminate, with branches or flowers often ± whorled, terminal, borne at the apex of a scape. Flowers bisexual or unisexual (plants then monoecious), radial, *with perianth differentiated into a calyx and corolla. Sepals 3,* distinct, imbricate. *Petals 3,* distinct, imbricate and crumpled, *usually white or pink. Stamens usually 6 to numerous;* filaments distinct; **pollen grains usually 2- to polyporate.** *Carpels* (3–) 6 to numerous, distinct, ovaries superior, with ± basal placentation; stigma 1, minute. Ovules few to more commonly 1 per carpel. Nectaries at base of carpels, stamens, or perianth parts. *Fruit a cluster of achenes* (or follicles); **embryo strongly curved**; endosperm lacking (Figure 9.13).

Floral formula: *, K3, C3, A6–∞, G6–∞; achenes

Distribution and ecology: Widely distributed; plants of freshwater marshes, swamps, lakes, rivers, and streams.

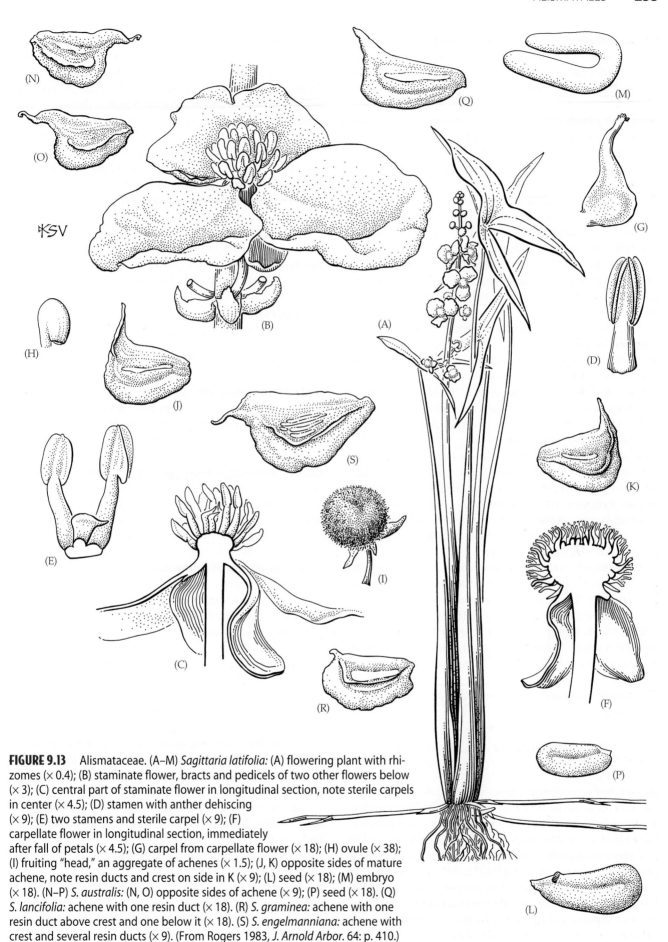

FIGURE 9.13   Alismataceae. (A–M) *Sagittaria latifolia:* (A) flowering plant with rhizomes (× 0.4); (B) staminate flower, bracts and pedicels of two other flowers below (× 3); (C) central part of staminate flower in longitudinal section, note sterile carpels in center (× 4.5); (D) stamen with anther dehiscing (× 9); (E) two stamens and sterile carpel (× 9); (F) carpellate flower in longitudinal section, immediately after fall of petals (× 4.5); (G) carpel from carpellate flower (× 18); (H) ovule (× 38); (I) fruiting "head," an aggregate of achenes (× 1.5); (J, K) opposite sides of mature achene, note resin ducts and crest on side in K (× 9); (L) seed (× 18); (M) embryo (× 18). (N–P) *S. australis:* (N, O) opposite sides of achene (× 9); (P) seed (× 18). (Q) *S. lancifolia:* achene with one resin duct (× 18). (R) *S. graminea:* achene with one resin duct above crest and one below it (× 18). (S) *S. engelmanniana:* achene with crest and several resin ducts (× 9). (From Rogers 1983, *J. Arnold Arbor.* 64: p. 410.)

Genera/species: 16/100. **Major genera:** *Echinodorus* (45 spp.) and *Sagittaria* (35). The family is represented in the continental United States and/or Canada by the above and *Alisma, Damasonium,* and *Limnocharis.*

Economic plants and products: *Sagittaria* (arrowhead), *Alisma* (water plantain), *Echinodorus* (burr-heads), and *Hydrocleis* (water poppy) provide pond and/or aquarium ornamentals. The rhizomes of *Sagittaria* may be eaten.

Discussion: Alismataceae are defined broadly (including the Limnocharitaceae; see Pichon 1946; Thorne 1992) and considered monophyletic on the basis of morphological and DNA characters (Dahlgren et al. 1985; Les et al. 1995; Soros and Les 2002). The genera with achenes and only a single basal ovule per carpel (e.g., *Alisma, Sagittaria,* and *Echinodorus*) may form a monophyletic subgroup (Chase et al. 1993, 1995b).

Species are often difficult to identify due to extensive variation in leaf morphology, which correlates with environmental parameters such as light intensity, water depth, water chemistry, and rate of flow (Adams and Godfrey 1961). Submerged leaves are usually linear, while floating or emergent leaves are petiolate with elliptic to ovate blade with an acute to sagittate base. Several different leaf forms may occur on the same plant.

Alismataceae (and Butomaceae) were once considered to represent primitive monocots (Cronquist 1981; Hutchinson 1973) due to their numerous, distinct stamens and carpels, in flowers that are superficially rather like those of Ranunculaceae. However, developmental and anatomical studies have indicated that these numerous stamens are actually due to secondary increase.

The showy flowers of Alismataceae are pollinated by various nectar-gathering insects (often bees and flies). In *Alisma* and *Echinodorus* the flowers are bisexual, while they are usually unisexual in *Sagittaria.* The achenes are often dispersed by water; they float due to the presence of spongy tissue and are resinous on the outer surface. They are also eaten (and dispersed) by waterfowl.

Additional references: Haynes et al. 1998a; Rogers 1983; Tomlinson 1982.

Hydrocharitaceae A. L. de Jussieu
(Frog's-Bit or Tape Grass Family)

Aquatic herbs, completely submerged to partly emergent, and rooted in the substrate or floating and unattached, in freshwater or marine habitats, often rhizomatous; tissues ± aerenchymatous. Hairs unicellular, thick-walled, prickle-like along leaf margins and/or veins. Leaves alternate and spiral, opposite, or whorled, along stem or in a basal rosette, simple, entire or serrate, sometimes with a well-developed blade, with parallel or palmate venation, or only midvein evident, sheathing at base; stipules lacking; small scales present at the node

inside the leaf sheath. Inflorescences determinate, sometimes reduced to a solitary flower, axillary, subtended by 2 often connate bracts. Flowers bisexual or unisexual (plants then monoecious or dioecious), usually radial, *with perianth differentiated into calyx and corolla. Sepals 3, distinct, valvate. Petals 3, distinct,* usually white, imbricate, *sometimes lacking.* Stamens 1, 2 or 3 to numerous; filaments distinct to connate; pollen grains usually lacking apertures, in *Thalassia* and *Halophila* united into threadlike chains. *Carpels usually 3–6, connate;* **ovary inferior**, with ovules scattered over surface of locules, the placentae often ± deeply intruded; styles often divided, appearing twice the number of carpels; stigmas elongate and papillose. Ovules numerous (or solitary and basal). Nectar often secreted from staminodes. **Fruit fleshy**, *either a berry or irregularly to valvately opening capsule;* embryo sometimes curved; endosperm lacking.

> **Floral formula:**
> Staminate: *, K3, C3, A2–∞, G0
> Carpellate: *, K3, C3, A0, G $\overline{(3-6)}$; berry, fleshy capsule

Distribution and ecology: Widely distributed, but most common in tropical and subtropical regions, in freshwater (most genera) and marine habitats (*Enhalus, Halophila, Thalassia*).

Genera/species: 18/140. **Major genera:** *Ottelia* (40 spp.), *Najas* (40) and *Elodea* (15). *Egeria, Elodea, Halophila, Hydrilla, Hydrocharis, Limnobium, Najas, Ottelia, Thalassia,* and *Vallisneria* occur in the continental United States and/or Canada.

Economic plants and products: Several genera, including *Hydrilla, Egeria,* and *Elodea* (waterweeds), *Vallisneria* (tape grass), and *Limnobium* (frog's-bit), are used as aquarium plants. Species of *Elodea, Hydrilla,* and *Lagarosiphon* are pernicious aquatic weeds.

Discussion: Hydrocharitaceae, although monophyletic (Dahlgren and Rasmussen 1983; Les et al. 2006), are morphologically heterogeneous, and have been divided into three to five subfamilies (Dahlgren et al. 1985). *Najas* has reduced flowers with a basal, erect ovule, but its placement within Hydrocharitaceae is supported by seed-coat anatomy and DNA sequences (Les 1993; Les and Haynes 1995; Les et al. 2006). *Zannichellia* (Zannichelliaceae; see key) probably also belongs here (Les et al. 1997a).

The family shows an interesting array of pollination mechanisms. Several species in *Egeria, Limnobium, Stratiotes,* and *Blyxa* have showy flowers that are held above the surface of the water and pollinated by various, usually nectar-gathering insects. In *Vallisneria, Enhalus,* and *Lagarosiphon,* the staminate flowers become detached and float on the water's surface, where they come into contact with the carpellate flowers (see Chapter 4). In *Elodea* the anthers of

(A)

Alismatales: Araceae
Monstera deliciosa: spathe and spadix

(B)

Alismatales: Potamogetonaceae
Potamogeton nodosus: fruits

(C)

Liliales: Smilacaceae
Smilax glauca: vine with tendrils

(D)

Liliales: Colchicaceae
Gloriosa superba: flowering plant

(E)

Liliales: Liliaceae
Lilium michiganense: flower

(F)

Alismatales: Alismataceae
Sagittaria lancifolia: staminate and
carpellate flowers

(G)

Liliales: Melanthiaceae
Trillium cuneatum: plant in bloom

(H)

Liliales: Melanthiaceae
Schoenocaulon dubium: flowers

(I)

Liliales: Liliaceae
Erythronium americanum: flower
(longitudinal section)

PLATE 9.3 Monocots
Alismatales and Liliales

the staminate flowers may explode, scattering pollen grains on the water's surface; the staminate flowers themselves are sometimes detached from the plant and float on the surface of the water to the stigma. In *Hydrilla* pollen transport may occur by wind or water. Finally, in *Thalassia* and *Halophila* pollination takes place underwater. Both outcrossing or selfing can occur. The fleshy fruits ripen below the water surface; fruits and/or seeds are either water- or animal-dispersed. Vegetative reproduction by fragmentation of rhizomes is common.

Additional references: Cook 1982, 1998; Cox and Humphries 1993; Cox and Tomlinson 1988; Haynes 1988; Haynes and Holm-Nielsen 2001; Kaul 1968, 1970; Tomlinson 1969b.

Potamogetonaceae Bercht. & J. Presl
(Pondweed Family)

Aquatic, rhizomatous herbs. Stems with reduced vascular bundles often in a ring, with air cavities; tannins often present. Hairs lacking. Leaves alternate and spiral, or opposite, blade sometimes well developed, simple, entire, with parallel venation or with only a single midvein, sheathing at base, the sheath open, **and ± separated from blade so that it appears to be a stipule,** the leaves sometimes heteromorphic, with submerged and floating forms; 2 to several small scales present at the node, inside the leaf sheaths. *Inflorescences* indeterminate, terminal and axillary, *spikelike* and elevated above or lying on water's surface. Flowers bisexual, radial, *not associated with bracts (at maturity).* Tepals lacking. **Stamens 4, with well-developed appendages at base of anther that form what appears to be a ± fleshy perianth**; pollen grains without an aperture, globose to ellipsoid. **Carpels usually 4,** distinct; *ovaries superior, with ± basal to apical placentation*; stigma 1, truncate to capitate. *Ovule 1, ± anatropous to orthotropous.* Nectaries lacking. *Fruit a cluster of achenes or drupes*; endosperm lacking.

Floral formula:

*, T-0-, A4 (appendaged), G$\underline{4}$; achenes, drupes

Distribution and ecology: Cosmopolitan; herbs of lakes, rivers, and other wetland habitats.

Genera/species: 4/100.   **Major genera:** *Potamogeton* (90 spp.) and *Stuckenia* (6); both occur in North America.

Economic plants and products: Although the family is of little direct economic importance, many species provide food for wildlife.

Discussion: *Ruppia,* a genus of alkaline, brackish, or occasionally salt water, is often placed here, but its inclusion makes the family biphyletic (Les et al. 1997a). Ruppiaceae

are characterized by flowers with two stamens that have minute appendages, pollen slightly elongated, and carpels with long stalks.

In *Potamogeton* and *Stuckenia* the flowers are usually raised above the surface of the water and wind-pollinated, while flowers of *Ruppia* are held at the water surface and are water-pollinated. The fruits are animal- or water-dispersed.

Additional references: Haynes 1978; Haynes et al. 1998b.

Liliales

Monophyly of Liliales is supported by cladistic analyses based on morphology and DNA sequences (Chase et al. 1995a,b, 2000, 2006; Davis et al. 2004; Fay et al. 2006; Goldblatt 1995; Graham et al. 2006; Hilu et al. 2003; Källersjö et al. 1998; Soltis et al. 2000; Stevenson and Loconte 1995; Vinnersten and Bremer 2001). Synapomorphies supporting this group include nectaries mostly on the base of tepals or filaments, septal nectaries nearly always lacking, extrorse anthers and the frequent presence of spots on the tepals. The outer epidermis of the seed coat has a cellular structure and lacks phytomelan (a black crust); the inner part of the seed coat also has cellular structure, both plesiomorphies (see also Stevenson et al. 2000; Rudall et al. 2000a). The group includes some species with the largest genomes in the flowering plants (Soltis et al. 2003a). Putative phylogenetic relationships within the order are shown in Figure 9.14.

Both the order Liliales and the family Liliaceae are here quite narrowly delimited, following Dahlgren et al. (1985) and recent cladistic analyses (see references cited above). Many of the families now treated within Dioscoreales, Asparagales, and Liliales were formerly considered within a more broadly circumscribed Liliales (Cronquist 1981; Thorne 1992) as the petaloid monocots, a "group" charac-

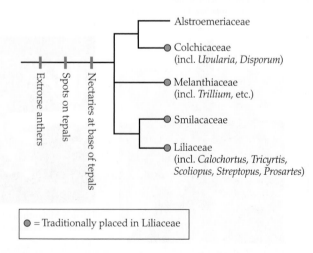

FIGURE 9.14   Cladogram showing hypothesized relationships within Liliales. (Modified from Soltis et al. 2005, Stevens 2001 and onward.)

terized by flowers with showy tepals and endosperm lacking starch. Cronquist (1981) placed most petaloid monocots with six-stamened flowers into a very broadly circumscribed—and we now know clearly polyphyletic—Liliaceae. Others have divided the petaloid monocots with six stamens into Liliaceae, including species with a superior ovary, and Amaryllidaceae, including species with an inferior ovary (Lawrence 1951). This separation is also artificial, separating clearly related genera such as *Agave* and *Yucca* (Agavaceae), and *Crinum* (Amaryllidaceae) and *Allium* (Alliaceae), as discussed in the family treatments (see Asparagales).

Although our knowledge of relationships within Liliales is now much improved, the delimitation of some families is still problematic (see especially discussions under Liliaceae, Melanthiaceae, and Colchicaceae). The Liliales include 11 families and ca. 1300 species; major families include Alstro-

emeriaceae, **Liliaceae, Colchicaceae, Smilacaceae,** and **Melanthiaceae**.

Liliaceae A. L. de Jussieu
(Lily Family)

Herbs with bulbs and contractile roots (or rhizomes); steroidal saponins often present. Hairs simple. *Leaves alternate and spiral, or whorled, along stem or in a basal rosette*, simple, entire, *with parallel venation*, and in *Prosartes* and *Tricyrtis* with clearly reticulate venation between the primary veins, often sheathing at base; stipules lacking. *Inflorescence* usually determinate, sometimes reduced to a single flower, *terminal. Flowers* bisexual, radial to slightly bilateral, ± *conspicuous. Tepals 6, distinct,* imbricate, *petaloid, often with spots or lines. Stamens 6;* filaments distinct; pollen grains usually monosulcate. *Carpels 3, connate; ovary superior, with*

Key to Major Families of Liliales

1. Vines, climbing by paired stipular tendrils at the base of the petiole .... **Smilacaceae**
1. Herbs, not climbing, and lacking tendrils 2
2. Ovary inferior; leaves usually twisted at base .... Alstroemeriaceae
2. Ovary superior; leaves not twisted at base 3
3. Bulbs present 4
3. Rhizomes or corms present 6
4. Capsule loculicidal; megagametophyte of *Fritillaria* type (egg, synergids, and one polar nucleus haploid, antipodals and second polar nucleus triploid) .... **Liliaceae** (subfam. Lilioideae)
4. Capsule septicidal or ventricidal; megagametophyte of *Polygonum* type (all cells haploid) 5
5. Capsules with carpel units splitting apart and their ventral margins separating; flowers usually small, tepals not spotted; anther locules confluent .... **Melanthiaceae**
5. Capsules septicidal; flowers large, tepals often with spots or lines; anther locules not confluent .... **Liliaceae** (*Calochortus*)
6. Corms present .... **Colchicaceae**
6. Rhizomes present 7
7. Leaves whorled, with parallel to palmate venation and pinnate secondary veins; perianth of sepals and petals .... **Melanthiaceae** (tribe Parideae)
7. Leaves alternate, with ± parallel venation; perianth of tepals 8
8. Inflorescences determinate, usually few-flowered and paniculate, with cymose branching, or reduced to paired or solitary flowers; flowers minute to conspicuous; style 1, but sometimes apically 3-branched 9
8. Inflorescences indeterminate, simple or compound racemes or spikes; flowers ± minute; styles 3 .... **Melanthiaceae**
9. Styles divided, thus stigmas 3; ovules ascending; leaves parallel-veined, but primaries often connected by evident reticulate veins .... **Liliaceae** (*Tricyrtis, Streptopus, Prosartes,* etc.)
9. Style not divided, thus stigma at most 3-lobed; ovules not ascending; leaves parallel-veined, not clearly reticulate .... **Colchicaceae** (*Uvularia, Disporum,* etc.)

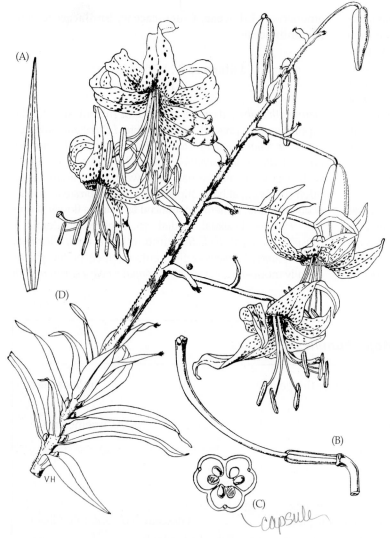

FIGURE 9.15 Liliaceae. *Lilium lancifolium:* (A) leaf (× 1); (B) gynoecium (× 1); (C) ovary, in cross-section (× 5); (D) flowering plant (× 0.3). (From Hutchinson 1973, *The families of flowering plants,* 3rd ed., p. 755.)

axile placentation; stigma 1, 3-lobed, or 3, ± elongated and extending along inner face of style branches. Ovules numerous, usually with 1 integument and a ± thin megasporangium; *megagametophyte often developing from 4 megaspores (Fritillaria type),* with some cells haploid and others triploid. *Nectar produced at base of tepals. Fruit a loculicidal or septicidal capsule,* occasionally a berry; seeds flat and disk-shaped or globose, *seed coat not black;* endosperm oily, its cells triploid *or pentaploid* (Figure 9.15).

Floral formula: *, T-6-, A6, G③; capsule, berry

Distribution and ecology: Widely distributed, mainly in temperate regions of the Northern Hemisphere; mainly spring-blooming plants of prairies, mountain meadows, and other open communities.

Genera/species: 16/635. **Major genera:** *Fritillaria* (100 spp.), *Gagea* (90), *Tulipa* (80), *Lilium* (80), and *Calochortus* (65). *Calochortus, Clintonia, Erythronium, Fritillaria, Medeola, Prosartes, Scoliopus, Streptopus,* and *Lilium* occur in the continental United States and/or Canada.

Economic plants and products: *Tulipa* (tulips), *Fritillaria* (fritillary), *Lilium* (lilies), *Calochortus* (mariposa lily) and *Erythronium* (trout lilies, adder's-tongue) are important ornamentals.

Discussion: Liliaceae, as here defined, are clearly monophyletic (Chase et al. 1995a,b), although difficult to diagnose morphologically (Tamura 1998b). *Calochortus, Prosartes, Scoliopus, Streptopus,* and *Tricyrtis* may form a clade, and are herbs with creeping rhizomes, apically divided styles, and the *Polygonum* type of megagametophyte development (i.e., megagametophyte developing from a single megaspore, with endosperm triploid). *Calochortus* has been placed in its own family, while the other genera have been placed in a heterogeneous Uvulariaceae (Dahlgren et al. 1985) or in an expanded Calochortaceae (Tamura 1998a). However, these members of Liliaceae are not closely related to the morphologically similar *Uvularia* and *Disporum* (Shinwari et al. 1994), and the latter here are placed in Colchicaceae. The remaining genera of Liliaceae constitute the Lilioideae, a large clade characterized by bulbs and contractile roots, and a megagametophyte developing from four megaspores (*Fritillaria* type). Monophyly of each of these two subclades of Liliaceae is supported by DNA sequence characters (Chase et al. 1995a, 2000).

The showy flowers of this family are insect-pollinated, especially by bees, wasps, butterflies, and moths; nectar and/or pollen are employed as pollinator rewards. The seeds are dispersed by wind or water; a few have aril-like structures and are dispersed by ants.

Colchicaceae A. P. de Candolle
(Colchicum Family)

Herbs with corms or elongated, corm-like structures, or creeping rhizomes; various alkaloids present, *including those of the colchicine-type (with a tropolone ring),* and steroidal saponins absent. Hairs simple. *Leaves alternate and spiral, along stem or in a basal rosette,* simple, entire, *with parallel venation,* usually sheathing at the base, occasionally ending in a tendril; stipules lacking. Inflorescence determinate or indeterminate, sometimes reduced to a single flower, terminal or axillary. *Flowers bisexual, radial, conspicuous. Tepals*

6, distinct to connate, **U-shaped and folded around each stamen in bud**, imbricate, petaloid, *often variegated, spotted, or with basal portion of a different color than the upper part. Stamens 6;* filaments distinct, sometimes adnate to tepals; pollen grains monosulcate to 2- or 4-sulcate or -foraminate. *Carpels 3, connate; ovary superior, with axile placentation;* stigmas 3, truncate to elongate. Ovules numerous, with 1 or 2 integuments and a ± thin megasporangium. *Nectar produced at base of tepals or stamens. Fruit a loculicidal or septicidal capsule* (or berry); seeds ± angular to globose, sometimes arillate, the seed coat not black; endosperm oily.

Floral formula: $*, T\underset{\sim}{\underline{6}}, A\underline{6}, \underline{G\textcircled{3}}$, capsule

Distribution and ecology: Widespread in temperate to tropical areas of North America, Africa, Europe, and Asia, to Australia and New Zealand; especially characteristic of winter rainfall ("Mediterranean") climates.

Genera/species: 18/225.   Major genera: *Colchicum* (90 spp.), *Wurmbea* (40), *Androcymbium* (40), *Iphigenia* (15), and *Disporum* (15). *Uvularia* is native, while *Colchicum* and *Gloriosa* are naturalized in the continental United States.

Economic plants and products: *Colchicum* (autumn crocus) and *Gloriosa* (glory lily, flame lily) are important ornamentals. Several genera are used medicinally because of their highly toxic alkaloids. The alkaloid colchicine is well known for its antimitotic effect; because it induces chromosome doubling, colchicine is much used in plant breeding to induce polyploidy.

Discussion: The monophyly of Colchicaceae, as here delimited, is supported by nucleotide sequences (Chase et al. 1995a, 2000; Rudall et al. 2000a) and morphology (U-shaped tepals, folded around each stamen in bud). The genera *Uvularia, Disporum, Tripladenia,* and *Schelhammera* were once placed in a non-monophyletic Uvulariaceae (see Dahlgren et al. 1985), but they are a paraphyletic assemblage with respect to the core Colchicaceae, which constitute a well-supported clade diagnosed by their corms or elongated, cormlike stems, and alkaloids of the colchicine type with a tropolone ring (Nordenstam 1998; Vinnersten and Reeves 2003). All members of Colchicaceae differ from Lilioideae in having the normal (*Polygonum*) type megagametophyte development.

The showy flowers of members of Colchicaceae are pollinated by various insects (bees, wasps, flies, butterflies, and moths); nectar and/or pollen are employed as pollinator rewards. The seeds often show no obvious dispersal adaptations; some have arils or aril-like structures and may be dispersed by ants or birds.

Additional reference: Wildman and Pursey 1968.

Smilacaceae Ventenat
(Catbrier Family)

Vines or occasionally erect herbs, *often with thick, tuberlike rhizomes;* steroidal saponins present. Hairs simple; prickles often present. **Leaves** alternate and spiral, simple, entire to spinose-serrate, **differentiated into a petiole and blade, with palmate venation, with primary veins converging, and clearly connected by a reticulum of higher-order veins, with a pair of tendrils near petiole base. Inflorescences** determinate, **umbellate**, terminal or axillary. **Flowers unisexual (plants dioecious)**, radial, and inconspicuous. *Tepals 6,* distinct to slightly connate, imbricate. Stamens usually 6; filaments distinct to slightly connate; **anthers** usually **unilocular** due to confluence of 2 locules; pollen grains monosulcate or ± lacking apertures, **exine with small spines**. Carpels 3, connate; ovary superior, usually with axile placentation; stigmas 3, ± elongate. **Ovules 1 or 2 in each locule**, anatropous to orthotropous. Nectar produced at base of tepals and stamens. **Fruit a 1–3-seeded berry**; seeds ± globose, not black (Figure 9.16).

Floral formula:
Staminate: $*, T\underset{\sim}{\underline{6}}, A\underline{6}, G0$

Carpellate: $*, T\underset{\sim}{\underline{6}}, A0, \underline{G\textcircled{3}}$; berry

Distribution: Widespread in tropical to temperate regions.

Genera/species: 1/310.   Genus: *Smilax* (310 spp.), which does occur in the continental United States and Canada.

Economic plants and products: Several species of *Smilax* are used medicinally; the genus is also the source of the flavoring sarsaparilla. Young stems, berries, and tubers are occasionally eaten.

Discussion: Monophyly of Smilacaceae, a distinctive family containing only *Smilax,* is supported by morphological and molecular data (Judd 1998; Cameron and Fu 2006). The family may be closely related to Rhipogonaceae, Philesiaceae, and Liliaceae (Chase et al. 1995a, 2006; Conran 1998; Fay et al. 2006; Rudall et al. 2000a). Although similar to Dioscoreaceae (Dahlgren et al. 1985), which are also vines with net-veined leaves, their morphology (Conran 1989) and DNA sequences (Chase et al. 1993, 1995a, 2000; Soltis et al. 2000) support placement of Smilacaceae in the Liliales. The family was once more broadly defined (Cronquist 1981), to include genera such as *Luzuriaga* and *Petermannia,* but inclusion of these genera would result in a paraphyletic Smilacaceae (Chase et al. 1995a, 2006; Fay et al. 2006; Rudall et al. 2000a). The non-climbing habit is derived and has evolved more than once within *Smilax* (Fu et al. 2005), and *S. aspera* is likely sister to the remaining species (Cameron and Fu 2006).

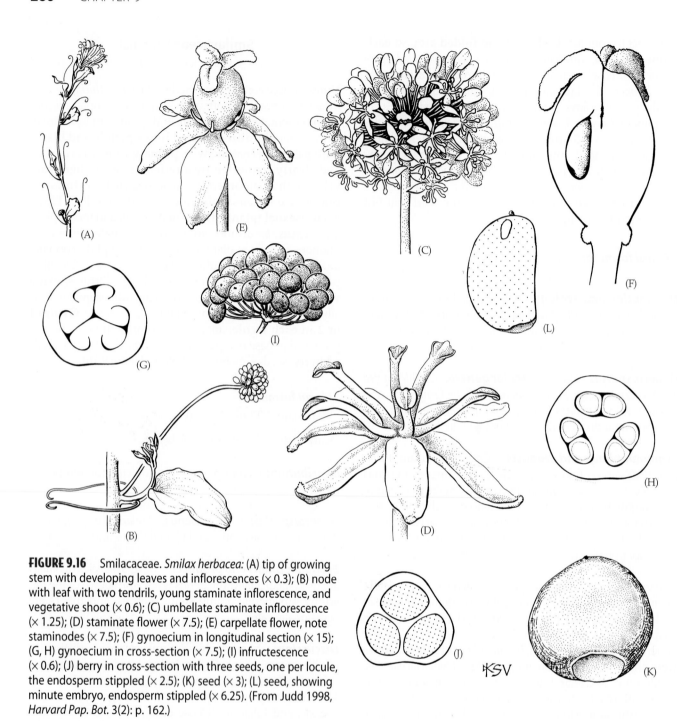

FIGURE 9.16 Smilacaceae. *Smilax herbacea:* (A) tip of growing stem with developing leaves and inflorescences (× 0.3); (B) node with leaf with two tendrils, young staminate inflorescence, and vegetative shoot (× 0.6); (C) umbellate staminate inflorescence (× 1.25); (D) staminate flower (× 7.5); (E) carpellate flower, note staminodes (× 7.5); (F) gynoecium in longitudinal section (× 15); (G, H) gynoecium in cross-section (× 7.5); (I) infructescence (× 0.6); (J) berry in cross-section with three seeds, one per locule, the endosperm stippled (× 2.5); (K) seed (× 3); (L) seed, showing minute embryo, endosperm stippled (× 6.25). (From Judd 1998, *Harvard Pap. Bot.* 3(2): p. 162.)

The small flowers of Smilacaceae are pollinated by insects (bees, flies). The fruits are bird-dispersed.

Melanthiaceae Batsch ex Borkh.
(Death Camas Family)

Herbs from short to elongate, slender to bulblike or tuber-like rhizomes; vascular bundles of stem sometimes in 3 rings (*Trillium* and relatives); roots sometimes contractile; steroidal saponins and *various toxic alkaloids often present.* Hairs simple. Leaves alternate and spiral, along stem or in a basal rosette (or whorled and usually the same number as the outer perianth whorl, in *Trillium* and relatives), simple, entire, with parallel venation (but ± palmate, with primary veins converging, secondary veins pinnate, and higher-order veins forming a distinct reticulum, in *Trillium* and relatives), sheathing at base; stipules lacking. *Inflorescences indeterminate,* terminal, (or reduced to a single flower in *Trillium* and relatives). *Flowers* usually bisexual, usually radial, small to large. *Tepals 6,* distinct to slightly connate, imbricate (but in *Trillium* differentiated into a calyx and corolla, with sepals and petals usually 3 or 4, distinct). Stamens 6 or 8 (rarely numerous); filaments distinct; anthers only 2-locular, *the locules often confluent,*

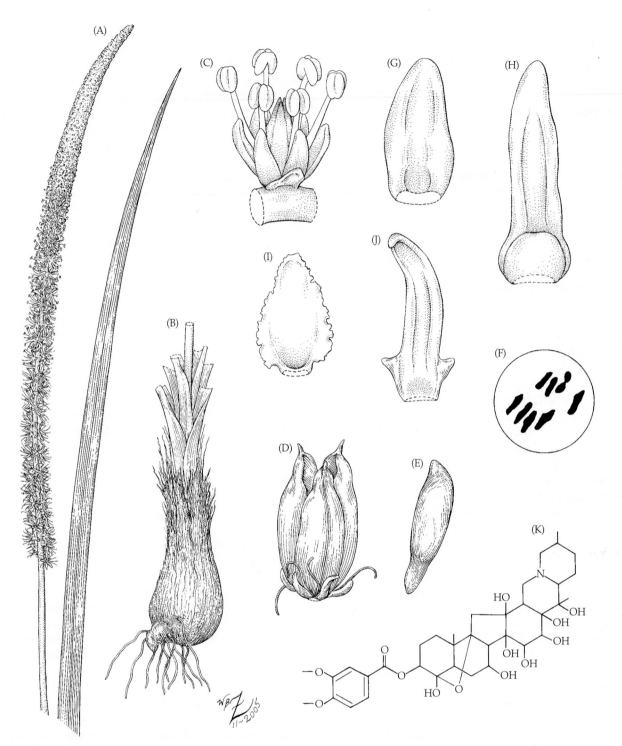

FIGURE 9.17 Melanthiaceae. *Schoenocaulon officinale*: (A) leaf and spicate inflorescence (× 0.5); (B) bulb with dark fibers (× 0.5). *S. dubium*: (C) sessile flower (× 7.5) (D) ventricidal capsule (× 4); (E) appended seed (× 6). (F) *S. texanum*: meiotic chromosomes, *n* = 8 (× 2000). (G–J) Outer tepals, adaxial view (× 12): (G) *S. dubium*; (H) *S. officinale*; (I) *S. yucatanense*; (J) *S. texanum*. (K) Structure of veratradine, a ceveratrum alkamine. (From Zomlefer et al. 2006b, Figure 2.)

opening by a single slit, resulting in a peltate appearance; pollen grains monosulcate or lacking apertures. Carpels 3–10, connate; ovary superior to slightly inferior, with axile placentation; **styles** usually **3, distinct**; stigmas 3, ± elongated. Ovules 2 to numerous in each locule; megagametophyte formed from one (*Polygonum* type) or two (*Allium* type) megaspore nuclei. Nectaries at base of perianth parts or lacking. *Fruit a ventricidal capsule* (carpels splitting apart and their ventral margins separating to release seeds), *loculicidal capsule, fleshy capsule, or berry;* seeds flattened to globose, sometimes winged or appendaged, not black (Figure 9.17).

Floral formula:

*, T⁻6⁻ [or K3-8, C3-8], A6-16, G③-8; capsule, berry

Distribution and ecology: Widely distributed in temperate and/or montane habitats; typically of herb-dominated communities.

Genera/species: 14/168. **Major genera:** *Veratrum* (45 spp.), *Trillium* (41), *Paris* (26), *Schoenocaulon* (24), *Toxicoscordion* (10), and *Anticlea* (10). All of the above genera (except *Paris*), along with *Amianthium, Chamaelirium, Helonias, Pseudotrillium, Stenanthium, Xerophyllum,* and *Zigadenus,* occur in the continental United States and/or Canada.

Economic plants and products: A few genera, especially *Trillium* and *Veratrum,* are used as ornamentals; several genera have various medicinal or insecticidal uses (due to their poisonous alkaloids). The leaves of *Xerophyllum* are used for making baskets.

Discussion: Monophyly of Melanthiaceae, when circumscribed to include *Trillium, Paris,* and *Pseudotrillium,* is well supported by DNA sequences (Chase et al. 1993, 1995a,b, 2000; Rudall et al. 2000a; Zomlefer et al. 2001, 2006a), although the family is not easily recognized (see description). In the past, the genera now placed in Tofieldiaceae and Nartheciaceae often were included in a morphologically heterogeneous "Melanthiaceae," while *Trillium* and relatives were excluded (as Trilliaceae; see Dahlgren et al. 1985; Zomlefer 1996), but both phenetic and cladistic studies confirm the heterogeneity of this traditional circumscription (Ambrose 1980; Goldblatt 1995). Thus, the family here excludes Nartheciaceae and Tofieldiaceae, which are not at all closely related to the remaining species (Zomlefer 1997a,b,c). Tofieldiaceae (e.g., *Tofieldia* and *Harperocallis,* in Alismatales) are distinct from Melanthiaceae (as here circumscribed) in their 2-ranked, equitant leaves, 2-sulcate pollen, presence of druses in parenchymatous tissues, and stalked ovary. Nartheciaceae (e.g., *Aletris, Narthecium, Lophiola,* in Dioscoreales) have unusual roots with air spaces in the cortex, a single style, and loculicidal capsules; some also have 2-ranked, equitant leaves. These two families also lack the *Veratrum* alkaloids and flattened to winged seeds characteristic of most species of Melanthiaceae.

Most members of the family belong to the clearly monophyletic and morphologically distinctive tribe Melanthieae, i.e., *Amianthium, Anticlea, Toxicoscordion, Schoenocaulon, Stenanthium, Veratrum,* and *Zigadenus,* a group characterized by inflorescences with bisexual or bisexual and staminate flowers, nectaries near the base of the tepals, anther locules apically confluent, opening by a single slit and resulting in a peltate appearance, a ventricidal capsule, and veratrum-alkaloids (Zomlefer et al. 2001). *Trillium, Paris,* and *Pseudotrillium* constitute the Parideae, another easily recog-

nizable tribe. These three genera stand out because of their whorled leaves with palmate, reticulate venation, solitary flowers with the perianth differentiated into a calyx and corolla, and fleshy capsules or berries—all likely synapomorphies. The haploid chromosome set in these genera consists of 5 large chromosomes, and this condition may be an additional synapomorphy. *Chionographis, Chamaelirium,* and *Xerophyllum* represent isolated lineages within the family.

Pseudotrillium nivale may be sister to the *Paris* + *Trillium* clade, and *Pseudotrillium* is distinctive in having spotted petals. *Paris* usually has 4- to 11-merous flowers with narrow petals, while *Trillium* has 3-merous flowers with broad petals (Farmer and Schilling 2002). Two large subgenera are traditionally recognized within *Trillium*: subg. *Trillium,* with pedicellate flowers, and subg. *Phyllantherum,* with sessile flowers. The pedicellate condition is plesiomorphic, calling the monophyly of subg. *Trillium* into question. Cladistic analyses based on DNA (Kato et al. 1995a,b) and morphology (Kawano and Kato 1995) also indicate that subg. *Trillium* is paraphyletic and subg. *Phyllantherum* is monophyletic.

The small flowers of Melanthiaceae are pollinated by beetles, bees, and flies (and possibly also by wind). The small seeds are probably wind dispersed, but dispersal by ants is characteristic in species of *Trillium* producing arillate seeds, while bird dispersal is likely in those with colorful berries.

Additional references: Tamura 1998c,d.

Asparagales

Monophyly of Asparagales is supported by cladistic analyses based on morphology, 18S rDNA, and several DNA sequences (Chase et al. 1995a,b, 2000, 2006; Conran 1989; Davis et al. 2004; Fay et al. 2000; Graham et al. 2006; Hilu et al. 2003; Källersjö et al. 1998; McPherson and Graham 2001; Pires et al. 2006; Rudall et al. 1997; Soltis et al. 2000; Stevenson et al. 2000). Supporting characters include their characteristic seeds, which have the outer epidermis of the coat obliterated (in most fleshy-fruited species) or present and with a black and carbonaceous phytomelan crust in many dry-fruited species. The inner part of the seed coat is usually completely collapsed. In contrast, seeds of the morphologically similar Liliales always have a well-developed outer epidermis, lack phytomelan, and usually retain a cellular structure in the inner portion of the coat. Asparagales also can be distinguished from Liliales by their usually nonspotted tepals, nectaries in the septa of the ovary (instead of at base of tepals or stamens), and sometimes anomalous secondary growth (versus lack of secondary growth).

The order consists of some 14–25 families and about 26,800 species; major families include **Orchidaceae**, Hypoxidaceae, **Iridaceae**, **Amaryllidaceae**, **Alliaceae**, **Hyacinthaceae**, Laxmanniaceae, **Agavaceae**, **Aspara-**

Asparagales: Amaryllidaceae
Galanthus nivalis: flower
(longitudinal section; ovary inferior)

Agave sp.: flower,
fruits, and seeds

Asparagales: Agavace
Agave americana: plant in bloom

Asparagales: Iridaceae
Gladiolus sp.: flowers
(3 stamens)

A. suffulta:
mucilaginous
interior of leaf

Asparagales: Aspholdelaceae
Aloe vera: leaves and inflorescence

Asparagales: Alliaceae
Allium christophii: flowers (ovaries superior)

Asparagales: Ruscaceae
Maianthemum canadense:
habit; fruiting plant

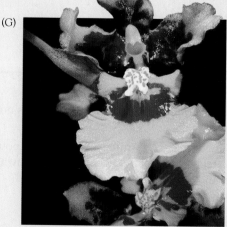

Asparagales: Orchidaceae
Oncidium sphacelatum: flowers

**PLATE 9.4 Monocots
Asparagales**

Asparagales: Ruscaceae
Sansevieria trifasciata:
plant in bloom

Key to Major Families of Asparagales

1. Ovary inferior 2

1. Ovary superior 6

2. Stamens adnate to style, usually only 1 or 2; one perianth member highly modified,
forming a lip; seeds lacking endosperm; placentation usually parietal ....**Orchidaceae**

2. Stamens not fused to style, usually 3 or 6; all perianth parts alike or 3 outer ±
differentiated from 3 inner; seeds with endosperm; placentation usually axile 3

3. Stamens 3; leaves equitant; seed coat not black, with cellular structure ....**Iridaceae**

3. Stamens 6; leaves not equitant; seed coat with outer epidermis obliterated, or present
and with black crust 4

4. Inflorescences scapose, umbellate (or reduced to a solitary flower); plants from a bulb,
with contractile roots ....**Amaryllidaceae**

4. Inflorescences paniculate, racemose, fasciculate, or reduced to a solitary flower, usually
not scapose; plants from a rhizome or corm, the roots contractile or not 5

5. Plants with rosettes of fleshy fibrous leaves; karyotype dimorphic, of 5 large and 25
small chromosomes; nectaries in the septa of the ovary; roots not contractile ....**Agavaceae** (Agavoideae)

5. Plants nonsucculent, leaves thin-textured and not strongly fibrous; karyotype not
dimorphic; nectaries lacking; roots usually contractile ....Hypoxidaceae

6. Fruit fleshy, a berry 7

6. Fruit dry, hard or leathery, usually a capsule, or triangular and nutlike 8

7. Leaves usually rudimentary, forming small scales and plants with cylindrical to
flattened green cladodes; seeds black ....**Asparagaceae**

7. Leaves usually ± large and photosynthetic; stems cylindrical, green to brown, but
not the major photosynthetic organ of the plant; seeds not black ....**Ruscaceae**
(various herbaceous genera, along
with ± woody plants, i.e., Dracaeneae)

8. Seeds not black; guard cells rich in oil; fruits dry, triangular, and nutlike ....**Ruscaceae** (Nolineae)

8. Seeds black (with phytomelan); guard cells lacking oil; fruits various, but usually
capsules and never as above 9

9. Plants from a bulb; inflorescence scapose (atop an elongate internode) 10

9. Plants from a rhizome; inflorescence usually not scapose 11

10. Inflorescence umbellate; plants usually with an onion or garlic smell ....**Alliaceae**

10. Inflorescence with an axis; plants without an onion or garlic smell ....**Hyacinthaceae**

11. Plants with woody trunk (or rhizome) and strongly fibrous leaves; anthers small
in relation to filaments; chromosomes dimorphic (5 large and 25 small) ....**Agavaceae** (Yuccoideae)

11. Plants with or without a woody trunk; leaves not strongly fibrous; anthers not small;
chromosomes more uniform in size 12

12. Pollen monosulcate; inflorescences indeterminate, forming simple or compound
racemes or spikes; seeds usually arillate; leaves usually succulent, often having
conspicuous gelatinous central zone, and often with colored sap produced
by specialized cells associated with vascular bundles of leaves; filaments
free from tepals ....**Asphodelaceae**

12. Pollen trichotomosulcate, with Y-shaped aperture; or if monosulcate then inflorescences
determinate, forming scorpioid cymes; seeds not arillate; leaves thin, lacking
gelatinous zone and colored sap; filaments ± adnate to tepals ....Hemerocallidaceae
[incl. Phormiaceae, Johnsoniaceae]

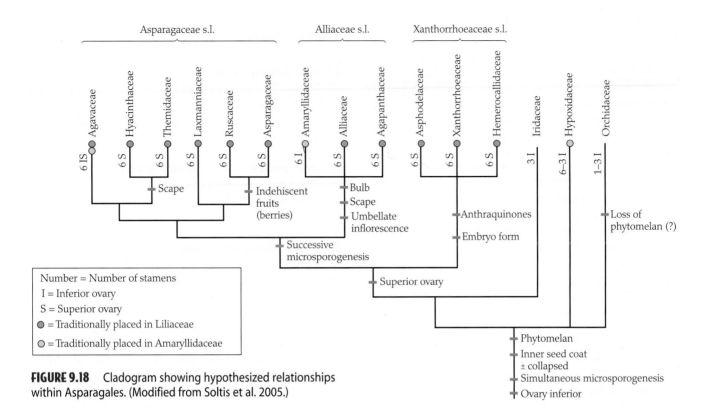

FIGURE 9.18 Cladogram showing hypothesized relationships within Asparagales. (Modified from Soltis et al. 2005.)

gaceae, **Convallariaceae**, **Asphodelaceae**, and Hemerocallidaceae.

Relationships within the Asparagales have been investigated by Dahlgren et al. (1985), Rudall and Cutler (1995), Chase et al. (1995a,b, 1996), Stevenson and Loconte (1995), Rudall et al. (1997a,b), Fay et al. (2000), McPherson and Graham (2001), Graham et al. (2006), and Pires et al. (2006). A paraphyletic assemblage including families such as Orchidaceae, Hypoxidaceae, Iridaceae, Asphodelaceae, and Hemerocallidaceae, is characterized by simultaneous microsporogenesis (the four microspores separate all at once from one another after both meiotic divisions have been completed). Within this assemblage, a clade including Alliaceae, Amaryllidaceae, Hyacinthaceae, Agavaceae, Asparagaceae, and Ruscaceae has successive microsporogenesis (Figure 9.18). In these families, one cell plate is laid down immediately after the first meiotic division and another is laid down in each daughter cell after the second meiotic division (Dahlgren and Clifford 1982). If the current phylogeny holds, we have an almost paradoxical—and evolutionarily fascinating—situation. Many of the "lower" Asparagales have a microsporogenesis pattern that is apomorphic (within the monocots), as well as an inferior ovary, also an apomorphic feature. The microsporogenesis pattern and ovary position of most "higher" Asparagales are like those of more basal monocots!

Orchidaceae and Hypoxidaceae are among the earliest branching clades in Asparagales. Hemerocallidaceae are close to Asphodelaceae and Xanthorrhoeaceae, and all three families have anthraquinones and a seedling with a nonphotosynthetic cotyledon. Alliaceae, Amaryllidaceae, and Agapanthaceae may form a clade, based on their bulbous rootstock, scapose umbellate inflorescences with spathelike bracts, and also cpDNA sequences (Fay et al. 2000). Ruscaceae and Asparagaceae may be related, perhaps as evidenced by their usually indehiscent, fleshy fruits; this relationship receives preliminary support from some analyses of DNA sequences (Chase et al. 1995a,b, 1996, 2000; Fay et al. 2000). Finally, Themidaceae, Hyacinthaceae, and Agavaceae form a clade (see Fay et al. 2000), although morphological support for this group is unclear. This clade likely is sister to the Ruscaceae + Asparagaceae + Laxmanniaceae clade.

Some familial limits are currently unclear, and certain botanists have proposed that several of the families recognized here be combined. Because criteria for determining absolute rank (see Chapter 3 and Appendix 1) do not exist, this is really an arbitrary taxonomic decision, although factors such as ease of recognition and level of support for the group's monophyly should be considered. The latter, of course, is vital. The recognition of nested monophyletic groups, and the determination of their diagnostic features, is the most important consideration. Exactly which of these groups are positioned at the rank of "family," which are recognized as "subfamilies," etc. is much less significant. Thus, some botanists treat Alliaceae, Amaryllidaceae, and Agapanthaceae, which together constitute a well-supported clade, as a single family, Alliaceae s.l. As discussed above (and see the family treatments), the fact that these three groups are closely related is beyond doubt, and it is an arbitrary decision whether they are considered as three closely related families or as three clades within a more broadly circumscribed family. Likewise, some combine Xanthorrhoeaceae, Asphodelaceae, and Hemerocallidaceae, recognizing these three clades within a broadly circumscribed Xanthorrhoeaceae. Finally, it has been suggested (see Angiosperm Phylogeny Group 2002) that Agavaceae, Asparagaceae, Hyacinthaceae, Ruscaceae, Themidaceae, and a few small families can be combined; the resulting broadly circumscribed family would be called Asparagaceae.

In using the "family" treatments provided in this text, remember that these named groups represent clades. In your systematics class you may be called upon to know the name and diagnostic characters of each of these groups, only some of them, or of groups within them, or your instructor may focus on more inclusive clades, such as the groups discussed above. The clades within Asparagales considered in this textbook (see the Key to Asparagales) may be treated as: (1) Asparagaceae, Ruscaceae, Asphodelaceae, Hemerocallidaceae, Agavaceae, Hyacinthaceae, Alliaceae, Amaryllidaceae, Iridaceae, Hypoxidaceae, and Orchidaceae (for a total of 11 families); or as (2) Asparagaceae s.l., Alliaceae s.l., Xanthorrhoeaceae s.l., Iridaceae, Hypoxidaceae, and Orchidaceae (for a total of 6 families). A similar taxonomic situation—the result of a current lack of consensus in the community of plant systematists regarding the number of "significant clades"—occurs in several other orders, especially Liliales, Malpighiales, Cornales, Lamiales, and Dipsacales.

Additional references: Huber 1977; Judd 2000; Tomlinson and Zimmerman 1969; Zomlefer 1998.

Asparagaceae A. L. de Jussieu
(Asparagus Family)

Rhizomatous herbs to shrubs, or scrambling to twining vines; *stems woody to withering annually, usually green, those associated with scale leaves forming leaflike phylloclades,* or stems reduced; steroidal saponins and ethereal oils present. Hairs simple. *Leaves alternate and spiral, simple, entire, usually reduced, ± scale-like, with a spiny basal spur,* the venation indistinct; stipules lacking. Inflorescences determinate, sometimes reduced to a single flower, axillary. Flowers bisexual or unisexual (plants monoecious or dioecious), radial, usually small. *Tepals 6, ± distinct, petaloid,* imbricate. *Stamens usually 6,* represented by staminodes in carpellate flowers; filaments distinct to connate, adnate to tepals; pollen grains monosulcate. *Carpels 3, connate, ovary superior, with axile placentation;* stigma usually 1, capitate to 3-lobed. Nectaries in septa of ovary. Ovules 1 to several in each locule, anatropous to ± orthotropous; **megagametophyte curved-asymmetric**. *Fruit usually a few-seeded berry;* seeds angular to ± globose, *seed coat with phytomelan* (black crust) and inner layers collapsed.

Floral formula: *, T–6–, A $\widehat{6}$, G $\underline{\textcircled{3}}$, capsule

Distribution and ecology: Widely distributed in the Old World from Europe and Africa to eastern Asia, Malesia, and Australia (*Asparagus*), and disjunct in Mexico (*Hemiphylacus*). The family is characteristic of regions of arid to Mediterranean climates.

Genera/species: 2/305.   ***Major genus:*** *Asparagus* (300 species). The family is represented in the continental United States and Canada only by four naturalized species of *Asparagus*.

Economic plants and products: The young shoots of *Asparagus officinalis* are eaten as a vegetable. Numerous species of *Asparagus* are used medicinally; several are popular ornamentals, e.g., *A. aethiopicus* and *A. setaceus*.

Discussion: Monophyly of Asparagaceae is supported by a few embryological features, i.e., the megagametophyte is curved-asymmetric and becomes even more so during development, and the ovule has a persistent nucellar epidermis of enlarged, cytoplasm-rich cells (Rudall et al. 1998). Most species belong to the large genus *Asparagus*, a clade diagnosed by several distinctive morphological specializations (leaves reduced to non-photosynthetic scales that bear in their axils green, solitary to clustered, terete to flattened phylloclades, and baccate fruits). DNA-based phylogenetic analyses (Chase et al. 1995a, 1996; Fay et al.2000; Rudall et al. 1997a,b) also support monophyly of Asparagaceae (including both *Asparagus* and *Hemiphylacus*).

The flattened photosynthetic organs of *Asparagus* have been the subject of much study and almost endless controversy (see Judd 2002; Kubitzki and Rudall 1998), however, most authors consider them to be flattened stems, a view accepted here. It is curious that similar photosynthetic branches have evolved independently in *Ruscus* and relatives (Ruscaceae). Certainly, Ruscaceae and Asparagaceae are close relatives.

The flowers of *Asparagus* are pollinated by various bees and beetles; the usually red to blue or black fruits are dispersed by birds.

Ruscaceae M. Roemer
(Butcher's Broom Family)

Rhizomatous herbs to trees; stems sometimes with anomalous secondary growth, sometimes with resin canals, occasionally flattened and photosynthetic (phylloclades); steroidal saponins present. Hairs simple. *Leaves* usually alternate and spiral, along stem or in a basal rosette, simple, entire, *with parallel venation,* occasionally petiolate, sheathing at base; stipules lacking. Inflorescences determinate, sometimes reduced to a single flower, terminal or axillary. *Flowers* bisexual, radial, *usually small. Tepals (4) 6, distinct, or*

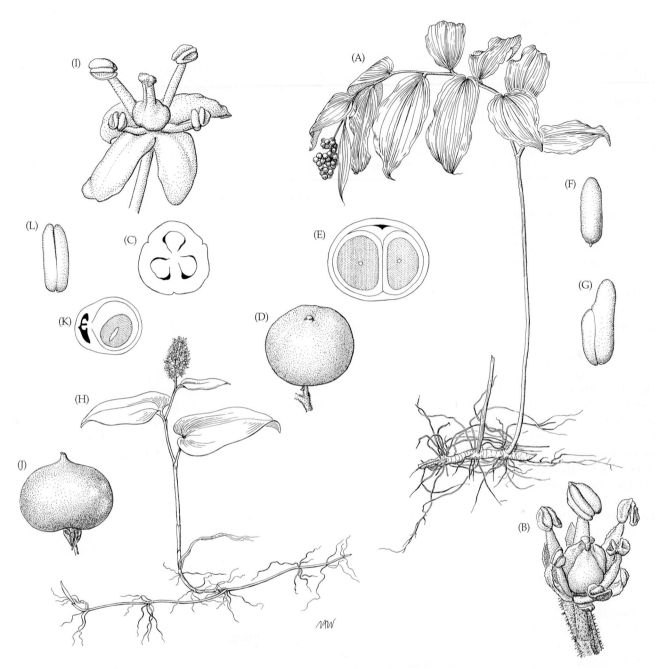

FIGURE 9.19 Ruscaceae. *Maianthemum racemosum:* (A) plant with rhizome (× 0.3); (B) flower (× 12); (C) ovary in cross-section (× 18); (D) berry (× 3.5); (E) berry in cross-section, embryo in center of endosperm (stippled) of each seed (× 3.5); (F, G) embryo and double embryo (× 12). *M. canadense:* (H) rhizome and flower-ing stem (× 6); (I) flower (× 12); (J) berry (× 3.5); (K) berry in cross-section, with two aborted ovules in left locule, single seed with embryo in endosperm (stippled) at right (× 3.5); (L) twin embryos from a single seed (× 12). (From Judd 2003, *Harvard Pap. Bot.* 7(2): p. 131.)

more commonly connate and perianth then urn-shaped to bell-shaped, or wheel-shaped, petaloid, *not spotted*, imbricate. *Stamens (4) 6*; filaments distinct or occasionally connate, often adnate to tepals; pollen grains monosulcate or lacking an aperture. *Carpels (2) 3, connate; ovary usually superior, with axile placentation*; stigma 1, capitate to 3-lobed. *Nectaries in septa of ovary.* Ovules 2 to few in each locule, anatropous to orthotropous. *Fruit usually a few-seeded berry*; seeds ± glo-bose, **seed coat** with outer epidermis lacking cellular struc-ture, and **lacking phytomelan** (black crust), and inner lay-ers collapsed (Figure 9.19).

Floral formula:

*, T-6-, A6, G③; berry, or dry 3-angled and nutlike

Distribution and ecology: Widely distributed from temperate to tropical regions; herbaceous species often understory herbs of moist forests and woody species often of arid regions.

Genera/species: 28/500. ***Major genera:*** *Dracaena* (80 spp.), *Polygonatum* (50), *Sansevieria* (50), *Maianthemum* (33), *Ophiopogon* (30), and *Nolina* (25). The family is represented in the continental United States and/or Canada by *Convallaria, Maianthemum, Polygonatum, Nolina, Dasylirion,* and *Sansevieria.*

Economic plants and products: Several genera, including *Aspidistra* (cast-iron plant), *Convallaria* (lily of the valley), *Dracaena* (dragon tree), *Liriope* (border grass), *Maianthemum* (false Solomon's seal; Plate 9.4F), *Polygonatum* (Solomon's seal), *Ophiopogon* (mondo grass), and *Sansevieria* (mother-in-law's tongue; Plate 9.4H) are used as ornamentals.

Discussion: Ruscaceae are broadly circumscribed, including Convallariaceae, Nolinaceae, and Dracaenaceae, as suggested by Chase et al. (1995a, 1996, 2000), Rudall et al. (2000b), Fay et al. (2000), and Judd (2003). Monophyly of the family is supported by DNA sequences (Chase et al. 1995a, 2000; Bogler and Simpson 1995, 1996; Rudall et al. 1997a,b, 2000b; Yamashita and Tamura 2000; Fay et al. 2000) and the lack of phytomelan in the seeds. Ruscaceae s. l. is sister to Asparagaceae (Bogler and Simpson 1996; Chase et al. 1995a, 1996, 2000; Fay et al. 2000).

Herbaceous members of Ruscaceae form a paraphyletic complex (Bogler and Simpson 1996; Conran 1989; Rudall et al. 2000b; Yamashita and Tamura 2000) from which are derived two woody clades, the Nolineae (*Nolina, Dasylirion, Calibanus,* and *Beaucarnea*) and the Dracaeneae (*Dracaena* and *Sansevieria*). Monophyly of the *Nolina* group (Nolineae) is supported by their dry, 3-angled, and nutlike fruits, and by leaves with minute longitudinal ridges and guard cells containing large amounts of oil. Monophyly of *Dracaena* and *Sansevieria* (Dracaeneae) is supported by resin canals in their leaves and bark, which thus frequently stain dark red or orange-red, hence the name "dragon's blood" used for the sap in some species of *Dracaena.*

Among the herbaceous genera, *Ruscus, Semele,* and *Danaë* (the Rusceae) are distinctive in having flattened photosynthetic stems, with the leaves reduced to scales. They are often confused with members of the Asparagaceae, but lack phytomelan in the testa of their seeds. Traditionally, the Ruscaceae has been restricted to just these three small, herbaceous genera.

Members of the Polygonateae, such as *Maianthemum* and *Polygonatum,* are easily recognized by their distinctive (and synapomorphic) habit (see Figure 9.19 and Plate 9.4F). These plants have broad leaves borne along aerial stems, which are connected by a sympodial rhizome system. *Liriope, Ophiopogon,* and relatives (the Ophiopogoneae) are noteworthy because their fruits rupture early to expose

seeds with a fleshy seed coat. These genera have spreading sympodial rhizomes with tufted clusters of often grasslike leaves; aerial stems are lacking. These leafy herbs, along with others, such as *Convallaria* and *Aspidistra,* have often been placed in the "Convallariaceae."

The small flowers of most Ruscaceae are insect-pollinated, especially by bees and wasps, which gather nectar or pollen. Their colorful berries usually are dispersed by birds. The dry, angular fruits of *Nolina* and relatives are wind-dispersed.

Additional references: Bogler 1998; Bos 1998; Conran and Tamura 1998.

Agavaceae Dumortier
(Agave Family)

Usually *large rosette herbs, trees, or shrubs,* rhizomatous; stems with *anomalous secondary growth;* raphidelike crystals of calcium oxalate present; steroidal saponins present. Hairs simple. *Leaves* alternate and spiral, *in rosettes at base or ends of branches,* simple, usually succulent, entire to spinose-serrate, and *usually with a sharp spine at apex,* with parallel venation, *the vascular bundles often associated with thick and tough fibers,* sheathing at base; stipules lacking. Inflorescences determinate, *usually paniculate,* terminal. Flowers usually bisexual, radial to slightly bilateral, often showy. *Tepals 6, distinct to connate,* and perianth then tubular to bell-shaped, imbricate, *petaloid, not spotted (and usually white to yellow).* Stamens 6; filaments distinct, sometimes adnate to the perianth; pollen grains monosulcate. *Carpels 3, connate; ovary superior or inferior,* with axile placentation; stigma minute, capitate to 3-lobed. Ovules ± numerous in each locule. *Nectaries in septa of ovary. Fruit a loculicidal capsule,* but sometimes fleshy and berrylike; *seeds flat, the seed coat with black crust (phytomelan;* Plate 9.4B) and inner layers ± collapsed; *karyotype usually of 5 large and 25 small chromosomes* (see Figure 4.23).

Floral formula: * or X, T$\underbrace{-6-}$, A6, G$\overline{\textcircled{3}}$; capsule

Distribution and ecology: Widely distributed in warm temperate to tropical regions of the New World, and especially diverse in Mexico; introduced in Old World; characteristic of arid and semiarid habitats.

Genera/species: 25/637. ***Major genera:*** *Agave* (300 spp.), *Chlorophytum* (150), *Anthericum* (65), *Yucca* (40), and *Hosta* (40). The family is represented in the continental United States and/or Canada by *Agave, Camassia, Chlorogalum, Furcraea, Hastingia, Hesperoaloe, Hesperocallis, Hesperoyucca, Manfreda, Schoenolirion,* and *Yucca.*

Economic plants and products: Several species of *Agave* (sisal hemp), *Furcraea,* and *Yucca* are used as fiber sources, and a few species of *Agave* are fermented to produce tequila and mescal. Both *Agave* and *Yucca* contain steroidal saponins that are used in the manufacture of oral contraceptives (due to their steroidal saponins). Several genera, including *Agave* (century plant; Plate 9.4B), *Hosta, Manfreda, Polianthes*, and *Yucca*, are used as ornamentals.

Discussion: Agavaceae are related to Hyacinthaceae and Themidaceae, and the phylogenetic position of some genera is problematic. However, both phenotypic and DNA characters support the family's monophyly (Bogler et al. 2006; Bogler and Simpson 1995, 1996; Chase et al. 2000; Pires et al. 2004). *Camassia, Hastingsia,* and *Chlorogalum* (usually treated in Hyacinthaceae, as subfam. Chlorogaloideae) and *Hosta* (usually placed in Hostaceae) are here included (see Bogler and Simpson 1995, 1996; Bogler et al. 2006; Chase et al. 1995a, 2000; Pfosser and Speta 1999). *Hosta* is a genus of rhizomatous herbs with broad-bladed leaves with prominent parallel veins, one-sided racemes of lilylike flowers, and black-seeded capsules. These four genera also have a distinctive bimodal karyotype. Molecular data (Chase et al. 1995a, 2000; Rudall et al. 1997b) also support inclusion of *Anthericum, Chlorophytum* and relatives (usually placed in Anthericaceae), a group of rhizomatous herbs with leaves in a basal rosette, leading to a broadly delimited Agavaceae, which are not easily characterized.

Woody Agavaceae usually are divided into the Yuccoideae (e.g., *Yucca, Hesperoaloe, Hesperoyucca*), with a superior ovary and tiny anthers, and the Agavoideae (e.g., *Agave, Furcraea, Manfreda, Polianthes*), with an inferior ovary and elongate anthers (Dahlgren et al. 1985); both are monophyletic (Bogler and Simpson 1995, 1996). The family was differently circumscribed by Cronquist (1981), who included genera that are here considered in Ruscaceae (*Nolina, Dasylirion, Beaucarnea, Dracaena,* and *Sansevieria*) and Laxmanniaceae (*Cordyline*). Such a broadly defined family is morphologically heterogeneous, unified only by the woody habit, and clearly polyphyletic (Dahlgren et al. 1985; Chase et al. 1995a,b; Bolger and Simpson 1995, 1996; Rudall et al. 1997).

The showy flowers of *Yucca* and *Hesperoyucca* are visited by small moths of the genus *Tegeticula* (see Figure 4.23). Other genera of Agavaceae are pollinated by birds (many species of *Beschornea*) or bats (several species of *Agave*). The black seeds are typically dispersed by wind, and fleshy-fruited species are dispersed by animals.

Additional references: Baker 1986; McKelvey and Sax 1933; Verhoek 1998.

Hyacinthaceae Batsch
(Hyacinth Family)

Herbs **with bulbs** *and contractile roots;* with steroidal saponins, *poisonous steroids* (bufodienolids and cardenolids), and mucilage cells or canals. Hairs simple. *Leaves alternate and spiral, in a basal rosette, simple, entire, with par-*

allel venation, sheathing at base; stipules lacking. *Inflorescence indeterminate (usually a raceme), on a scape.* Flowers bisexual, usually radial, often showy. *Tepals 6,* distinct to connate, and perianth then bell-shaped to tubular, imbricate, petaloid, not spotted. *Stamens 6;* filaments distinct to connate, sometimes adnate to tepals; pollen grains monosulcate. *Carpels 3, connate, ovary superior;* stigma 1, capitate to 3-lobed. Ovules 1 to numerous in each locule. Nectaries in septa of ovary. *Fruit a loculicidal capsule;* seeds globose to flattened, occasionally with aril-like structures, the seed coat usually with phytomelan and inner layers compressed or collapsed.

Floral formula: *, T$\underline{(6)}$, A(6), G$\underline{③}$, capsule

Distribution: Widespread from Europe and Africa to Asia, in temperate to tropical habitats, but most diverse in areas with a Mediterranean climate (pronounced summer dry season).

Genera/species: 63/850. **Major genera:** *Ornithogalum* (200 spp.), *Drimia* (100), *Albuca* (50), *Muscari* (50), and *Scilla* (30). A few species of *Muscari, Ornithogalum,* and *Scilla* are sparingly naturalized.

Economic plants and products: Several genera, e.g., *Urginea* and *Thuranthos*, are used medicinally because of their cardenolides. Species of genera such as *Scilla, Muscari, Hyacinthus, Puschkinia, Hyacinthoides, Eucomis,* and *Galtonia* are grown as ornamentals.

Discussion: Monophyly of Hyacinthaceae is supported by molecular data, which also supports exclusion of *Camassia, Chlorogalum,* and relatives, subfam. Chlorogaloideae (Chase et al. 1995a; Speta 1998). These genera are here considered as herbaceous members of Agavaceae, and are distinguished from Hyacinthaceae s.s. by their bimodal karyotype, and by the combination of more or less globose seeds with a firmly adhering testa, usually multinerved tepals, and distinctly 3-lobed stigmas. The exact placement of Hyacinthaceae within Asparagales is unclear, but they may be related to Themidaceae, and these two families may be associated with either the Agavaceae or (less likely) the Alliaceae + Amaryllidaceae clade (Chase et al. 1995a, 2000; Fay et al. 2000; Rudall et al. 1997a,b; Soltis et al. 2000). This uncertainty in placement makes determination of morphological synapomorphies problematic, although the presence of bulbs and poisonous steroids are noteworthy. The usually racemose or spicate, indeterminate inflorescence easily distinguishes the group from Themidaceae, Alliaceae, and Amaryllidaceae, all of which have umbellate inflorescences. The herbaceous habit, with a bulb, distinguishes the family from Agavaceae.

Hyacinthaceae are comprised of four monophyletic subfamilies; the largest are Ornithogaloideae (bracts large, protein crystals in nucleus; only *Ornithogalum*) and Hyacinthoideae (bracts usually small, protein crystals lacking; *Muscari, Scilla, Puschkinia, Massonia, Eucomis, Hyacinthus, Hyacinthoides*). Generic limits are controversial (Manning et al. 2004; Pfosser and Speta 1999; Wetsching and Pfosser 2003).

The showy flowers are pollinated by a wide range of insects (bees, wasps, flies, moths) as well as birds, and both nectar and pollen are offered as floral rewards. Asexual reproduction occasionally occurs (agamospermy or production of bulblets). Dispersal of seeds may occur by water, wind, or movement by ants.

Alliaceae Borkh.
(Onion Family)

Herbs with a bulb and contractile roots (or in *Tulbaghia*, a rhizome); stems reduced; **vessel elements with simple perforations; laticifers present (and latex ± clear)**; with steroidal saponins; **with onion- or garlic-scented sulfur compounds such as allyl-sulphides, propionaldehyde, propionthiol, and vinyl disulphide**. Hairs simple. Leaves alternate, usually 2-ranked, ± basal, simple, terete, angular, or flat, entire, with parallel venation, sheathing at base; stipules lacking. *Inflorescences determinate, composed of one or more contracted helicoid cymes and appearing to be an umbel, subtended by a few membranous spathelike bracts, terminal, at the end of a long scape.* Flowers bisexual, radial or bilateral, often showy; **individual flowers not associated with bracts**. *Tepals 6, distinct to connate*, and perianth then bell-shaped to tubular, imbricate, *petaloid, not spotted*; a corona (outgrowth of the perianth) sometimes present. *Stamens 6 (3)*; filaments distinct to connate, sometimes adnate to tepals, sometimes appendaged; pollen grains monosulcate. *Carpels 3, connate; ovary superior*, with axile placentation (Plate 9.4B); stigma 1, capitate to 3-lobed. Ovules 2 to numerous in each locule, anatropous to campylotropous. *Nectaries in septa of ovary. Fruit a loculicidal capsule; seeds globose to angular, the seed coat with phytomelan and inner layers compressed or collapsed; embryo ± curved.*

Floral formula: * or X, T⌣6⌣, A⌢6⌢, G③; capsule

Distribution and ecology: Widely distributed in temperate to tropical regions; frequently in semiarid habitats.

Genera/species: 13/645. **Major genera:** *Allium* (550 spp.), *Ipheion* (25), and *Tulbaghia* (24). The family is represented in the continental United States and/or Canada by *Allium* and *Nothoscordum*.

Economic plants and products: Several species of *Allium* (garlic, onions, shallots, chives, leeks, ramps) are important vegetables or flavorings. Their sap is mildly antiseptic, and several are used medicinally. A few genera, including *Allium, Gilliesia, Ipheion*, and *Tulbaghia* (society garlic), are used as ornamentals.

Discussion: Monophyly of Alliaceae is supported by morphology, chemistry, *rbcL* sequences (Fay and Chase 1996), and ITS sequences (Friesen et al. 2006). Alliaceae are closely related to Amaryllidaceae and Agapanthaceae. All three families are bulbous herbs with terminal umbellate inflorescences, which are subtended by spathaceous bracts and borne on a conspicuous scape; all of these features are probably synapomorphic. Cladistic analyses support the close relationship of these families (Chase et al. 1995a,b; Fay and Chase 1996; Fay et al. 2000; Pires and Sytsma 2002). Alliaceae and Agapanthaceae are often included within a broadened Amaryllidaceae (Hutchinson 1934, 1973), a decision that has merit from a practical standpoint since all these plants have many features in common.

Although *Agapanthus* (Agapanthaceae) has a superior ovary like Alliaceae, it lacks the sulfur-containing compounds, and cpDNA sequences do not support its placement within Alliaceae s.s. (Chase et al. 1995a; Fay et al. 2000). Themidaceae, which are herbs from corms and include *Dichelostemma, Triteleia*, and *Brodiaea*, often have been placed within Alliaceae (Dahlgren et al. 1998). Recent cladistic analyses have indicated that these genera are more closely related to Hyacinthaceae than to typical Alliaceae (Chase et al. 1995a; Fay and Chase 1996; Rudall et al. 1997a,b; Fay et al. 2000). Embryological characters do not support a close relationship between Alliaceae and Themidaceae (Berg 1996).

The showy flowers of Alliaceae are pollinated by a variety of insects (especially bees and wasps). The seeds are predominantly wind- or water-dispersed. A few species produce bulblets in the inflorescence.

Additional references: Mann 1959; Rahn 1998.

Amaryllidaceae J. St. Hilaire
(Amaryllis or Daffodil Family)

Herbs from a bulb with contractile roots; stems reduced; vessel elements with scalariform perforations; **characteristic "amaryllis" alkaloids present**. Hairs simple. Leaves alternate, usually 2-ranked, ± basal, simple, flat, entire, with parallel venation, sometimes differentiated into a blade and petiole, sheathing at base; stipules lacking. *Inflorescences determinate, composed of one or more contracted helicoid cymes, and appearing to be an umbel*, sometimes reduced to a single flower, *subtended by a few membranous spathelike bracts, terminal, on a long scape. Flowers bisexual, radial to bilateral, showy, each associated with a filiform bract. Tepals 6, distinct to connate*, imbricate, *petaloid, not spotted*; a corona (outgrowth of the perianth) sometimes present. *Stamens 6*; filaments distinct to connate, sometimes adnate to the perianth, sometimes appendaged (and forming a staminal corona); pollen grains monosulcate or bisulcate. *Carpels 3, connate;* **ovary inferior**, with axile placentation; stigma 1, minute to capitate or 3-lobed. Ovules ± numerous in each locule, sometimes with 1 integument. *Nectaries usually in septa of ovary. Fruit a loculicidal capsule* or occasionally a

(A)

(B)

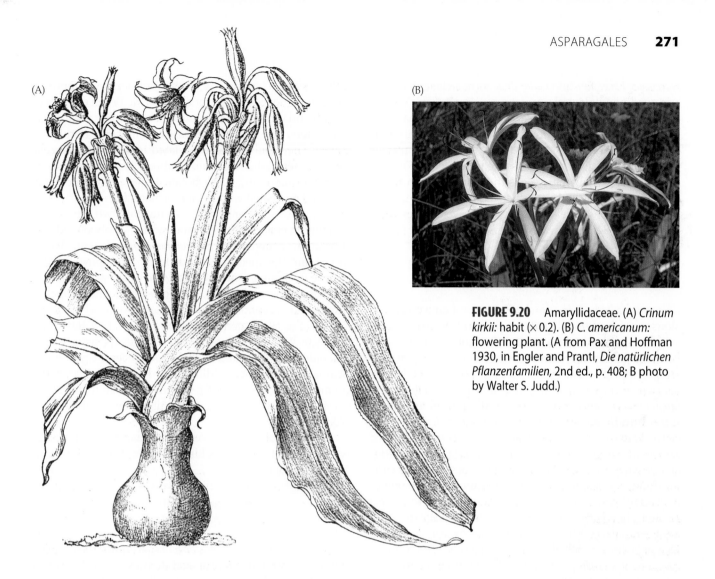

FIGURE 9.20　Amaryllidaceae. (A) *Crinum kirkii:* habit (× 0.2). (B) *C. americanum:* flowering plant. (A from Pax and Hoffman 1930, in Engler and Prantl, *Die natürlichen Pflanzenfamilien,* 2nd ed., p. 408; B photo by Walter S. Judd.)

berry; seeds dry to fleshy, flattened to globose, and sometimes winged, *the seed coat usually with black or blue crust,* but phytomelan sometimes lacking and outer epidermis then lacking cellular structure, inner layers also ± collapsed; embryo sometimes curved.

Floral formula:　* or X, T $\overline{(6-)}$, A $\widehat{(6)}$, G $\overline{③}$; capsule

Distribution:　Widely distributed from temperate to tropical regions, and especially diverse in South Africa, Andean South America, and the Mediterranean region.

Genera/species: 59/870.　**Major genera:** *Crinum* (130 spp.), *Hippeastrum* (70), *Zephyranthes* (60), *Hymenocallis* (50), *Cyrtanthus* (50), *Haemanthus* (40), and *Narcissus* (30). Common genera of the continental United States and/or Canada include *Crinum* (Figure 9.20), *Hymenocallis, Narcissus* (cultivated), and *Zephyranthes.*

Economic plants and products:　The family includes numerous ornamental genera: *Crinum* (crinum lily), *Eucharis* (Amazon lily), *Galanthus* (snowdrops), *Haemanthus* (blood lily), *Hippeastrum* (amaryllis), *Hymenocallis* (spider lily), *Narcissus* (daffodil, jonquil), *Zephyranthes* (rain lily,

zephyr lily), *Cyrtanthus* (kaffir lily), *Amaryllis* (Cape belladonna), and *Nerine* (Guernsey lily).

Discussion:　Monophyly of Amaryllidaceae is supported by secondary chemistry (amaryllid alkaloids), the inferior ovary (Plate 9.4A), and DNA sequences (Chase et al. 1995a; Meerow and Suijman 1998; Meerow et al. 1999a,b). Tribal characterizations within the family are provided by Meerow (1995), Meerow et al. (1999a,b, 2000a,b), and Meerow and Suijman (1998, 2006). The Amaryllideae (*Amaryllis, Nerine, Crinum,* and relatives) are sister to the remaining genera; Amaryllideae are easily recognizable by the apomorphies: presence of a sclerenchyma sheath in the scape, bisulcate pollen with a spiny exine, ovules with a single integument, non-dormant, water-rich seeds that lack phytomelan and have a chlorophyllous embryo. Among the remaining genera, Haemantheae (e.g., *Haemanthus, Clivia*) are diagnosed by berry fruits and occur in Africa, while the rest of the genera are American and Eurasian and form a clade. A large group of American genera (*Hymenocallis, Eucharis, Habranthus, Zephyranthes, Hippeastrum,* and *Griffinia*) probably form a monophyletic group. These may be recognized by their obvolute spathaceous bracts (i.e., two bracts overlapping in bud so that one half of each is external and the other half is internal). The Eurasian genera, such as *Lycoris,*

Narcissus, Leucojum, and *Galanthus,* form a clade that lacks apparent morphological support.

The showy flowers of Amaryllidaceae are pollinated by bees, wasps, butterflies, moths, and birds; most are outcrossing, but selfing also occurs. The seeds are usually wind- or water-dispersed, but some taxa, such as *Eucharis* subg. *Eucharis,* have blue seeds that contrast with bright orange capsules, leading to dispersal by birds. Large, fleshy seeds lacking phytomelan have evolved several times within the family. Some are water dispersed.

Asphodelaceae A. L. de Jussieu
(Aloe Family)

Rhizomatous herbs to trees or shrubs; stems often with anomalous secondary growth; *anthraquinones often present.* Hairs simple. *Leaves* alternate, spiral or 2– ranked, *in rosettes at base or ends of branches,* simple, *often succulent, entire to spinose-serrate,* with parallel venation; *often with vascular bundles (as seen in cross-section) arranged in a ring around central mucilaginous parenchyma tissue,* **with parenchymatous inner bundle sheath cells,** *these often modified, forming a cap of aloin cells at the phloem pole of most vascular bundles, containing colored secretions and generally accumulating anthraquinones,* not fibrous; sheathing at base; stipules lacking. Inflorescences indeterminate, terminal, but sometimes appearing lateral. Flowers usually bisexual, radial to bilateral, often showy. *Tepals 6, distinct to strongly connate* (perianth then ± tubular), imbricate, *petaloid, not spotted. Stamens 6;* filaments distinct; pollen grains monosulcate. *Carpels 3, connate; ovary superior,* with axile placentation; stigma 1, ± minute to discoid or slightly 3-lobed. Ovules 2 to numerous in each locule, anatropous to nearly orthotropous. *Nectaries in septa of ovary. Fruit a loculicidal capsule;* seeds **with dry aril** (occasionally lost) **that arises as an annular invagination at the distal end of the funiculus,** often flattened or winged, *the seed coat with phytomelan (a black crust)* and inner layers ± collapsed.

Floral formula: * or X, T$\widehat{6}$, A6, G$\underline{\textcircled{3}}$; capsule

Distribution and ecology: Distributed in temperate to tropical regions of the Old World, and especially diverse in southern Africa; usually in arid habitats.

Genera/species: 15/750. **Major genera:** *Aloe* (380 spp.), *Haworthia* (70), *Kniphofia* (70), and *Bulbine* (60). The family is represented in the continental United States only by two introduced taxa of *Aloe* (in southern Florida and California).

Economic plants and products: Several species of *Aloe* are used medicinally or in cosmetics; members of several genera, including *Aloe, Haworthia, Gasteria, Kniphofia,* and *Bulbine,* are used as ornamentals.

Discussion: Monophyly of Asphodelaceae is supported by *rbcL* sequences (Chase et al. 1995a; de Bruijn et al. 1995; Smith and Van Wyk 1998), arillate seeds, foliar anatomical characters, and possibly the base chromosome number of seven. The family is closely related to Xanthorrhoeaceae and Hemerocallidaceae and all three clades have anthraquinone pigments; some botanists have suggested that these all be treated within Xanthorrhoeaceae s. l.

Asphodelaceae are usually divided into two subfamilies (Dahlgren et al. 1985): the paraphyletic "Asphodeloideae" (including genera such as *Bulbine, Kniphofia,* and *Asphodelus*) and the specialized and clearly monophyletic Alooideae (including genera such as *Aloe, Gasteria,* and *Haworthia*). Synapomorphies for Alooideae include the distinctive leaves with a central gelatinous zone surrounded by vascular bundles associated with aloin cells (Plate 9.4D; Smith and Van Wyk 1991; Judd 1997a). *Bulbine* is sister to the Alooideae, as indicated by DNA sequences, its gelatinous leaves, and distinctive dimorphic karotype. *Aloe* is not monophyletic, having given rise to *Haworthia* (twice), *Gasteria,* and all other genera of the subfamily (Treutlein et al. 2003).

The colorful flowers of Asphodelaceae are pollinated by various insects as well as birds. The seeds are mainly dispersed by wind.

Iridaceae A. L. de Jussieu
(Iris Family)

Herbs with rhizomes, corms, or bulbs; **styloids (large prismatic crystals) of calcium oxalate present in sheaths of vascular bundles** (these occasionally lost); tannins and/or various terpenoids often present. Hairs simple. *Leaves alternate, 2-ranked,* **equitant (oriented edgewise to the stem), and with a unifacial blade,** along the stem to basal, simple, entire, with parallel venation, sheathing at base; stipules lacking. **Inflorescences determinate, a scorpioid cyme,** often highly modified, sometimes reduced to a single flower, terminal. Flowers bisexual, radial to bilateral, conspicuous, subtended individually by 1 or 2 bracts. *Tepals 6,* the outer sometimes differentiated from the inner, distinct or connate, imbricate, *petaloid, sometimes spotted.* **Stamens (2) 3** (Plate 9.4C); filaments distinct or connate, sometimes adnate to the perianth; anthers sometimes sticking to style branches; pollen grains usually monosulcate. *Carpels 3, connate; ovary usually inferior, with axile placentation;* style branches sometimes expanded and petaloid; stigmas (2) 3, terminal or on abaxial surface of style branches. Ovules few to numerous in each locule, anatropous or campylotropous. Nectaries in septa of ovary, on the tepals, or lacking. *Fruit a loculicidal capsule;* seeds sometimes arillate or with fleshy *seed coat,* the coat usually with cellular structure and ± *brown (black crust lacking).*

Floral formula: * or X, T$\widehat{6}$, A$\widehat{3}$, G$\overline{\textcircled{3}}$; capsule

Distribution: Widely distributed.

Genera/species: 67/1750. ***Major genera:*** *Gladiolus* (255 spp.), *Iris* (250), *Moraea* (125), *Sisyrinchium* (100), *Romulea* (90), *Crocus* (80), *Geissorhiza* (80), *Babiana* (65), and *Hesperantha* (65). Noteworthy genera occurring in the continental United States and/or Canada are *Alophia*, *Calydorea*, *Iris*, *Nemastylis*, and *Sisyrinchium*.

Economic plants and products: Stigmas of *Crocus sativus* are the source of the spice saffron. Numerous genera, including *Crocus*, *Tigridia* (tiger flower), *Freesia*, *Iris*, *Ixia* (corn lily), *Romulea*, *Neomarica*, *Moraea* (butterfly lily), *Nemastylis*, *Belamcanda*, *Sisyrinchium* (blue-eyed grass), *Gladiolus*, *Crocosmia*, and *Trimezia*, are used as ornamentals.

Discussion: Cladistic analyses based on morphology and cpDNA sequences support the monophyly of Iridaceae (Chase et al. 1995a, 2000; Goldblatt 1990; Rudall 1994; Reeves et al. 2001). Although morphological characters (Stevenson and Loconte 1995; Chase et al. 1995b) place the family within Liliales, DNA sequences place it within Asparagales (Chase et al. 1995a, 2000; Fay et al. 2000). Chloroplast DNA and morphology together placed the family within Asparagales, and therefore we consider the family to be a member of this order.

Three major clades (often recognized as subfamilies) are evident within Iridaceae (Goldblatt 1990; Goldblatt et al. 1998; Rudall 1994; Reeves et al. 2001). The Isophysidoideae, including only *Isophysis*, are distinguished by a superior ovary and are sister to the remaining groups. Iridoideae have flowers that often last only a single day; they can be recognized by the presence of nectaries on the tepals and by their very long, tubular style branches, divided below the level of the anthers, with each branch stigmatic only apically. The group also contains the free amino acids meta-carboxyphenylalanine and glycine. Several subclades are evident within Iridoideae, including Sisyrinchieae (style branches alternating with stamens vs. opposite them in other members of subfamily; *Sisyrinchium* and relatives) and a clade comprising Irideae, Mariceae, and Tigrideae (tepals differentiated into a limb and claw, upper apices of the style branches with petaloid appendages; e.g., *Iris*, *Belamcanda*, *Moraea*, *Nemastylis*, *Trimezia*, and *Tigridia*). Finally, Crocoideae (e.g., *Ixia*, *Crocosmia*, *Geissorhiza*, *Crocus*, *Romulea*, *Freesia*, *Gladiolus*, and *Hesperantha*) are hypothesized to be monophyletic based on their connate tepals, corms, sessile flowers, operculate pollen with micropunctate exine, and leaves with closed leaf sheaths and a pseudomidrib. Many members of this group retain the ancestral feature of septal nectaries.

The showy flowers of Iridaceae are mainly insect-pollinated (especially by beetles, bees, and flies), although some species are pollinated by birds. Nectar and/or pollen are the floral rewards. The seeds usually are dispersed by wind or water, but biotic dispersal also occurs.

Orchidaceae A. L. de Jussieu
(Orchid Family)

Terrestrial or epiphytic herbs, or occasionally vines, *with rhizomes, corms, or root-tubers,* occasionally mycoparasitic; *stems often basally thickened and forming pseudobulbs; roots strongly mycorrhizal, often with spongy, water-absorbing epidermis composed of dead cells (velamen).* Hairs various. Leaves usually alternate, spiral or 2-ranked, often plicate, basal or along the stem, sometimes reduced, simple, entire, with usually parallel venation, sheathing at base; stipules lacking. Inflorescences indeterminate, sometimes reduced to a single flower, terminal or axillary. Flowers usually bisexual, **bilateral, usually resupinate (twisted 180° during development),** often conspicuous. Outer tepals 3, distinct to connate, usually petaloid, imbricate. Inner tepals 3, distinct, sometimes spotted or variously colored, **the median one clearly differentiated from the 2 laterals, forming a lip (labellum),** often with fleshy bumps or ridges and of an unusual shape or color pattern. **Stamens 3 or fewer (usually only 1 or 2), adnate to style and stigma, forming a column**; pollen usually *grouped into soft or hard masses (pollinia).* Carpels 3, connate; *ovary inferior, usually with parietal placentation,* but occasionally axile; style and stigma highly modified, with portion of the latter usually nonreceptive (rostellum), a portion of which may form a sticky pad (viscidium) attached to the pollinia. Ovules numerous, with a thin megasporangium wall (Figure 9.21). Nectar produced in a lip-spur, by sepal apices, or in septal nectaries, but often lacking. *Fruit a capsule opening by (1–) 3 or 6 longitudinal slits; seeds minute,* the seed coat crustose or membranous, **lacking phytomelan,** with only the outer layer persisting, the inner tissues collapsed; **embryo very minute; endosperm lacking** (Figures 9.21 and 9.22).

Floral formula: X, T5+1, A1 or 2, G③; capsule

Distribution and ecology: Widely distributed, but most diverse in tropical regions (where frequently epiphytic).

Genera/species: 788/19,500. ***Major genera:*** *Pleurothallis* (1120 spp.), *Bulbophyllum* (1000), *Dendrobium* (900), *Epidendrum* (800), *Habenaria* (600), *Eria* (500), *Lepanthes* (460), *Maxillaria* (420), *Oncidium* (420), *Masdevallia* (380), *Stelis* (370), *Liparis* (350), *Malaxis* (300), *Oberonia* (300), *Encyclia* (235), *Eulophia* (200), *Angraecum* (200), *Taeniophyllum* (170), *Phreatia* (160), *Polystachya* (150), *Calanthe* (150), *Vanilla* (100), and *Catasetum* (100). The family is represented in the continental United States and/or Canada by numerous genera; some of the more noteworthy include *Bletia*, *Calopogon*, *Calypso*, *Cleistes*, *Corallorhiza*, *Cypripedium*, *Encyclia*, *Epidendrum*, *Goodyera*, *Habenaria*, *Harrisella*, *Hexalectris*, *Liparis*, *Listera*, *Malaxis*, *Oncidium*, *Orchis*, *Platanthera*, *Pogonia*, *Ponthieva*, *Spiranthes*, *Tipularia*, *Triphora*, and *Zeuxine*.

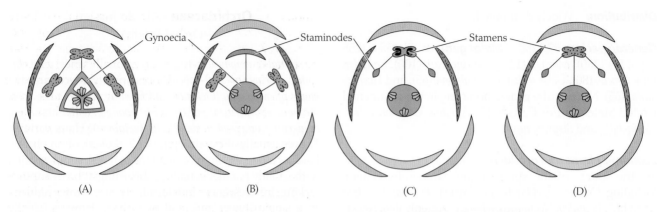

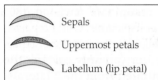

Sepals

Uppermost petals

Labellum (lip petal)

FIGURE 9.21 Floral diagrams of the major groups of Orchidaceae. (A) Apostasioideae. (B) Cypripedioideae. (C) Orchidoideae, Epidendroideae. (D) Vanilloideae. Within the four-lobed structures representing stamens, dots indicate pollen not in pollinia and solid black indicates pollen shed in pollinia. Gynoecia show placentation type. Adnation and connation are shown by lines connecting structures. (Modified from Endress 1994c and Dahlgren et al. 1985.)

Economic plants and products: Vanilla flavoring is extracted from the fruits of *Vanilla planifolia.* The family is economically important because of its numerous ornamentals, including *Cattleya* (corsage orchid), *Dendrobium, Epidendrum, Paphiopedilum* (slipper orchid), *Phalaenopsis, Vanda, Brassia, Cymbidium, Laelia, Miltonia, Oncidium, Encyclia,* and *Coelogyne.*

Discussion: Monophyly of Orchidaceae is supported by morphology and DNA sequences (Burns-Balogh and Funk 1986; Dressler 1981, 1993; Dressler and Chase 1995; Chase et al. 2000; Fay et al. 2000; Freudenstein et al. 2004; Judd et al. 1993). Phylogenetic relationships within the family have been addressed by several cladistic analyses of morphology and DNA sequences (Burns-Balogh and Funk 1986; Cameron 2006; Cameron and Chase 2000; Cameron et al. 1999; Dressler 1986, 1993; Dressler and Chase 1995; Freudenstein and Rasmussen 1999; Freudenstein et al. 2000, 2004; Judd et al. 1993; Molvray

FIGURE 9.22 Orchidaceae. *Encyclia cordigera:* (A) habit (× 0.75); (B) flower (× 2); (C) sepals and petals (× 1.5); (D) labellum (or lip) (× 4); (E) column, ventral view, note terminal anther and stigma (in darkened depression (× 4); (F) column, dorsal view, note terminal anther (× 4); (G) column, lateral view (× 4). (Original drawings by Robert Dressler, University of Florida, Gainesville.)

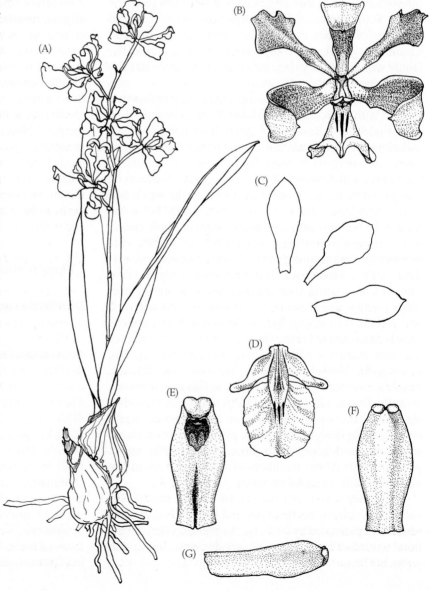

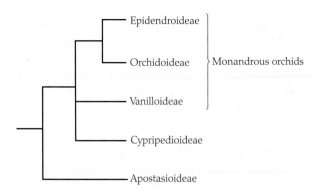

FIGURE 9.23 Phylogeny of Orchidaceae. (Modified from Cameron et al. 1999; Kocyan et al. 2004.)

et al. 2000; Van den Berg et al. 2005) although delimitation of some groups is still unclear. *Apostasia* and *Neuwiedia* (subfamily Apostasioideae) are sister to the remaining orchids (Cameron et al. 1999; Dressler 1993; Dressler and Chase 1995; Freudenstein and Rasmussen 1999; Judd et al. 1993; Neyland and Urbatsch 1996) (Figure 9.23). Monophyly of Apostasioideae is supported by their vessel elements with simple perforation plates and distinctive seeds. These two genera, and especially *Neuwiedia*, have retained many ancestral features, such as flowers with two (*Apostasia*) or three (*Neuwiedia*) stamens that are only slightly adnate to the style, axile placentation, pollen released as single, non-sticky grains, a symmetrical stigma, and all petals initiated simultaneously in development. The remaining orchids all have sticky (or fused) pollen and a broad, asymmetrical stigma with all lobes facing the center of the flower. Within this sticky-pollen clade, Cypripedioideae and Vanilloideae are sister to all the others. Cypripedioideae (e.g., *Cypripedium* and *Paphiopedilum*) are clearly monophyletic, as supported by their saccate (slipperlike) lip petal and median anther modified into a shieldlike staminode (see Figure 9.21); they have two functional stamens and lack pollinia. Vanilloideae (i.e., *Vanilla*, *Pogonia*, *Cleistes*) form a clade based on *rbcL* sequence data (Dressler and Chase 1995). This group is distinctive in that its flowers have only a single functional stamen and lack pollinia. Many Vanilloideae are vines with net-veined leaves; their ovaries are sometimes 3-locular. All other orchids have pollinia and a completely fused filament and style (Burns-Balogh and Funk 1986; Dahlgren et al. 1985; Judd et al. 1993; see Figure 9.21). This pollinia-forming clade, like Vanilloideae, has flowers with only a single functional stamen (monandry; the two lateral stamens are represented by slender staminodes or are entirely lacking). Morphological analyses (Freudenstein and Rasmussen 1999) and some molecular analyses (Cameron et al. 1999; Molvray et al. 2000) support the hypothesis that monandrous orchids are monophyletic, but other molecular analyses (Cameron 2006; Cameron and Chase 2000; Freudenstein et al. 2004) support the view that the reduc-

tion to a single functional stamen has occurred twice. Among monandrous orchids with pollinia, two large subfamilies, Epidendroideae and Orchidoideae (incl. Spiranthoideae), are recognized here (see also Figure 9.23; compare with Dressler 1981, 1986, 1993). Epidendroideae share the apomorphies of a beaked and incumbent anther (i.e., anther bent over apex of column), while Orchidoideae share the apomorphies of acute anther apex, soft stems, leaves convolute but not plicate, and lack of silica bodies (see also Dressler 1993; Stern et al. 1993). Epidendroideae contain numerous tropical epiphytes; representative genera include *Bulbophyllum, Catasetum, Dendrobium, Epidendrum, Encyclia, Maxillaria, Oncidium* (Plate 9.4G) *Pleurothallis,* and *Vanda*. Generic delimitation in this group is notoriously problematic, and most large genera are non-monophyletic. Representative members of the Orchidoideae include *Cynorkis, Diuris, Goodyera, Habenaria, Orchis, Platanthera, Spiranthes,* and *Zeuxine*.

Orchid flowers are extremely varied in form and attract a wide array of insects (bees, wasps, moths, butterflies, flies) as well as birds (see Chapter 4). Some attract generalist visitors, but many are quite specialized, attracting only one or a few species as pollinators. Pollen, nectar, or floral fragrances may be employed as pollinator rewards, although some flowers (e.g., *Cypripedium*) manipulate their pollinators and provide no reward, and some species of *Ophrys* and *Cryptostylis* mimic the form and scent of female bees, wasps, or flies, and are pollinated when the males attempt to mate with the flower (pseudocopulation; see Figure 6.12). Generally, the lip functions as a landing platform and provides visual or tactile cues orienting the pollinator. The pollinia become attached to the pollinator's body, and often are deposited in the stigma (usually a depression in the underside of the column) of the next flower visited. In some species pollination is a fairly uncommon event, and flowers may remain functional and showy for many days, with wilting of the perianth occurring rapidly after fertilization. Most species are outcrossing, but selfing is known to occur. The tiny, dustlike seeds are dispersed by wind and require nutrients supplied by a mycorrhizal fungus for germination.

Additional references: Cox et al. 1997; Kocyan and Endress 2001; Kocyan et al. 2004; van der Pijl and Dodson 1966.

Dioscoreales

Dioscoreaceae R. Brown

(Yam Family)

Twining vines *with thick rhizomes or a large tuberlike swelling; stem with vascular bundles in 1 or 2 rings;* steroidal sapogenins and alkaloids commonly present. Hairs simple to stellate; prickles sometimes present. *Leaves* usually alternate and spiral, simple, but sometimes palmately lobed or compound, entire, **differentiated into a petiole and**

blade, with palmate venation, the major veins converging and connected by a network of higher-order veins; *the petiole usually with an upper and lower pulvinus*, sometimes with stipule-like flanges, not sheathing; bulbils sometimes present in leaf axils. Inflorescences determinate, but sometimes appearing indeterminate, axillary. *Flowers usually unisexual* (plants dioecious), radial. *Tepals 6*, distinct to slightly connate, imbricate. *Stamens 6* (3); filaments distinct to slightly connate, adnate to the base of the tepals; **microsporogenesis simultaneous**; pollen grains monosulcate to variously porate. *Carpels 3, connate; ovary inferior, with axile placentation;* stigmas 3, minute to slightly bilobed. Ovules 2 to numerous in each locule. Nectaries in septa of ovary or base of tepals. *Fruit usually a triangular and 3-winged, loculicidal capsule*, but sometimes a berry or samara; *seeds usually flattened or winged*, the coat with yellow-brown to red pigments, crystals, **and thickened endotesta**; embryo with 1 or occasionally 2 cotyledons (Figure 9.24).

Floral formula:

Staminate: *, T-6-, A(6), G0

Carpellate: *, T-6-, A0, G③; winged loculicidal capsule, samara, berry

Distribution: Widely distributed in the tropics and subtropics, with a few in temperate regions.

Genera/species: 4/434.   **Major genera:** *Dioscorea* (400 spp.) and *Tacca* (30); *Dioscorea* and *Stenomeris* occur in the continental United States.

Economic plants and products: The starchy "tubers" of many species of *Dioscorea* (yams) are edible; these "tubers" should not to be confused with the roots of *Ipomoea batatas* (Convolvulaceae), which are also called yams. Other species are medicinally valuable due to the presence of alkaloids or steroidal sapogenins; the latter are used in anti-inflammatory medications and oral contraceptives.

Discussion: Dioscoreaceae are placed in Dioscoreales, an order with many vines with net-veined leaves, the mycoparasitic Burmanniaceae, and autotrophic Nartheciaceae. Dioscoreales, as here circumscribed, are monophyletic (Caddick et al. 2002a,b; Chase et al. 1995a, b, 2000, 2006; Soltis et al. 2000). Possible morphological synapomorphies include stems with vascular bundles in rings, the perianth persistent in fruit, and seeds with a short embryo. Dioscoreaceae may be distinguished from the phenetically similar Stemonaceae by 3-merous (vs. 2- or 5-merous) flowers and the consistently inferior ovary. They are easily separated from Burmanniaceae, a family of mycoparasitic herbs with scale-like leaves. Smilacaceae are also vines with net-veined leaves, but can easily be distinguished by their superior ovaries, few-seeded berries, and leaves with paired

stipular tendrils. As circumscribed by Dahlgren et al. (1985), Dioscoreales also contained Trilliaceae, Stemonaceae, Taccaceae, Smilacaceae, and a few other families, but the order as so circumscribed, is not monophyletic (Chase et al. 1996b, Conran 1989; Stevenson and Loconte 1995). In this text Trilliaceae and Smilacaceae are referred to Liliales and Stemonaceae to Pandanales. Dioscoreales had been considered as primitive monocots (Dahlgren et al. 1985; Stevenson and Loconte 1995), a position not supported by cladistic analyses based on DNA sequences.

Monophyly of Dioscoreaceae is supported by DNA sequence and morphological data (Caddick et al. 2000, 2002a; see description). The small genus *Stenomeris* may be sister to the remaining taxa, which form a clade supported by the putative synapomorphies of root-tubers, ridged or winged fruits, and a multilayered seed coat with a crystalline inner layer. *Tacca* and *Trichopus* may have diverged next, and both have perfect flowers. *Tacca* is morphologically distinctive and often has been treated in its own family (Taccaceae); it is easily distinguished from other members of the family by its acaulescent habit, parietal placentation, and unusual scapose and umbellate inflorescences with filamentous bracts. The remaining species of Dioscoreaceae are dioecious and belong to the large genus *Dioscorea*. Phylogenetic relationships within this dioecious clade have been investigated on the basis of DNA sequences, and morphology (Caddick et al. 2000, 2002a; Wilkin et al. 2005) and recognition of genera such as *Tamus* and *Rajania* make *Dioscorea* paraphyletic.

The inconspicuous flowers of Dioscoreaceae are insect-pollinated (mainly by flies). Dispersal is usually by wind, as indicated by the specialized fruits: 3-winged capsules with flattened and/or winged seeds or samaras (as in species often segregated as *Rajania*).

Additional references: Al-Shehbaz and Schubert 1989; Bouman 1995.

COMMELINOID MONOCOTS

The commelinoid monocots constitute a monophyletic group supported by *rbcL* sequences (Chase et al. 1993, 1995b; Duvall et al. 1993), *rbcL* and *atpB* sequences (Davis et al. 2004), sequences of multiple DNA regions (Chase et al. 2000, 2006; Graham et al. 2006; Soltis et al. 2000), and morphology (Dahlgren and Rasmussen 1983; Stevenson et al. 2000). Putative synapomorphies include *Strelitzia*-type epicuticular wax (see Figure 4.34), starchy pollen, and UV-fluorescent compounds—ferulic and coumaric acids—in the cell walls (Barthlott and Fröhlich 1983; Dahlgren et al. 1985; Harley and Ferguson 1990; Harris and Hartley 1980; Zona 2001). Most have starchy endosperm, but the endosperm of Arecaceae (Palmae) is nonstarchy; Dahlgren et al. (1985) suggested that this may represent an evolutionary loss. Some analyses using DNA sequences (Chase

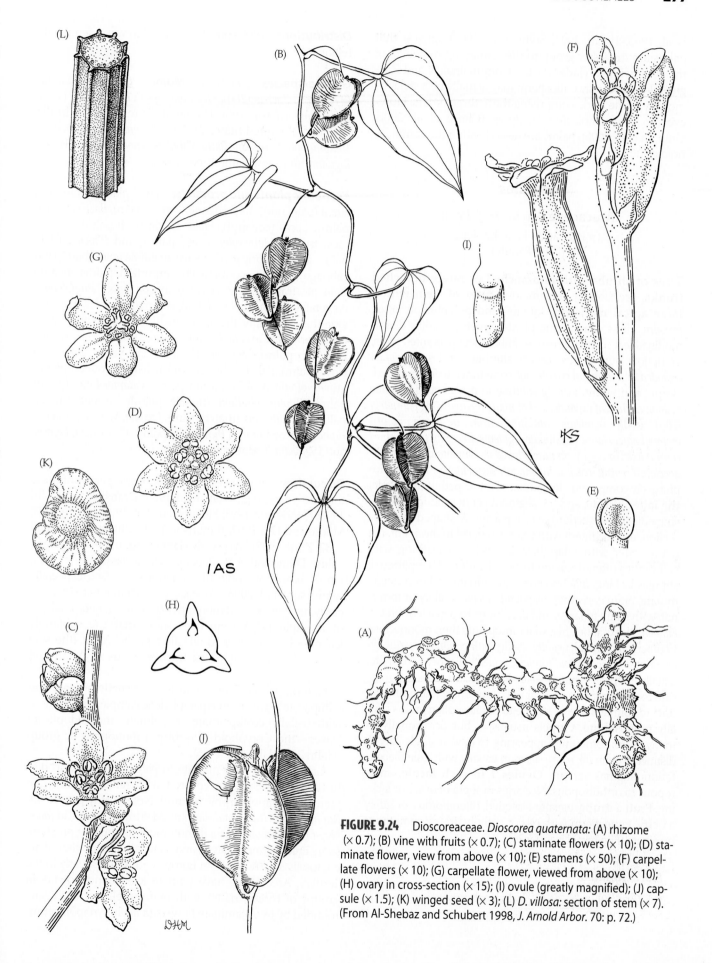

FIGURE 9.24   Dioscoreaceae. *Dioscorea quaternata:* (A) rhizome (× 0.7); (B) vine with fruits (× 0.7); (C) staminate flowers (× 10); (D) staminate flower, view from above (× 10); (E) stamens (× 50); (F) carpellate flowers (× 10); (G) carpellate flower, viewed from above (× 10); (H) ovary in cross-section (× 15); (I) ovule (greatly magnified); (J) capsule (× 1.5); (K) winged seed (× 3); (L) *D. villosa:* section of stem (× 7). (From Al-Shebaz and Schubert 1998, *J. Arnold Arbor.* 70: p. 72.)

et al. 1993, 1995b, 2000; Soltis et al. 2000) suggest that palms are the sister group to the remaining members of the commelinoid clade: Poales, Commelinales, and Zingiberales, so starchy endosperm may actually be a synapomorphy of the clade comprising these three orders. However, in other molecular analyses (Chase et al. 2006; Graham et al. 2006), palms are nested within the commelinoid clade.

Arecales

Arecaceae Bercht. &. J. Presl
(= Palmae A. L. de Jussieu)
(Palm Family)

Trees or shrubs with unbranched or rarely branched trunks, occasionally rhizomatous; **apex of stem with a large apical meristem**; tannins and polyphenols often present. Hairs various, and plants sometimes spiny due to modified leaf segments, exposed fibers, sharp-pointed roots, or petiole outgrowths. *Leaves alternate and spiral, often crowded in a terminal crown*, but sometimes well separated, simple and entire, *usually splitting in a pinnate to palmate fashion as the leaf expands, and at maturity appearing palmately lobed (with segments radiating from a single point), costapalmately lobed (± palmate segments diverging from a short central axis, or costa), pinnately lobed or compound (with well-developed central axis bearing pinnate segments)*, or rarely twice pinnately compound, differentiated into a petiole and blade, **the latter plicate**, and the segments either induplicate (V-shaped in cross-section) or reduplicate (Λ-shaped in cross-section; each segment with veins ± parallel to divergent, the petiole often with a flap (hastula), its base sheathing, with soft tissues often decaying to reveal various fiber patterns; stipules lacking. *Inflorescences* determinate and/or indeterminate, often *appearing compound-spicate*, axillary or terminal, with small to large, and deciduous to persistent bracts. *Flowers* bisexual or unisexual (and plants then monoecious to dioecious), radial, *usually sessile, with perianth usually differentiated into calyx and corolla*. Sepals 3, distinct to connate, usually imbricate. Petals usually 3, distinct to connate, imbricate to valvate. Stamens 3 or 6 to numerous; filaments distinct to connate, free or adnate to petals; pollen grains usually monosulcate. Carpels usually 3, but occasionally as many as 10, sometimes appearing to have a single carpel, distinct to connate; *ovary superior, usually with axile placentation*; stigmas various. **Ovules 1 in each locule**, anatropous to orthotropous. Nectaries in septa of ovary or lacking. **Fruit a drupe**, usually 1-seeded, often fibrous, or rarely a berry; endosperm with oils or carbohydrates, sometimes ruminate (Figures 9.25 and 9.26).

Floral formula:

*, K $\underline{\textcircled{3}}$, C $\underline{\textcircled{3}}$, A $\underline{\textcircled{6-\infty}}$, G $\textcircled{3}$; drupe, berry

Distribution: Widespread in tropical to warm temperate regions.

Genera/species: 200/2780. **Major genera:** *Calamus* (370 spp.), *Bactris* (200), *Daemonorops* (115), *Licuala* (100), and *Chamaedorea* (100). The family is represented in the continental United States by *Coccothrinax, Pseudophoenix, Rhaphidophyllum, Roystonia, Sabal, Serenoa, Thrinax, Washingtonia*, and a few naturalized genera.

Economic plants and products: Food plants come from *Areca* (betel nut), *Attalea* (American oil palm), *Bactris* (peach palm), *Cocos* (coconut), *Elaeis* (African oil palm), *Euterpe* (acai fruits), *Metroxylon* (sago palm), and *Phoenix* (date palm); a great many genera have an edible apical bud (palm cabbage). Other economically important palms include *Calamus* (rattan), *Copernicia* (carnuba wax), *Phytelephas* (vegetable ivory), *Raphia* (raffia), and many genera that provide thatch. Finally, the family includes a large number of ornamentals: *Caryota* (fishtail palm), *Chamaerops* (European fan palm), *Livistona* (Chinese fan palm), *Roystonea* (royal palm), *Sabal* (cabbage palm), *Syagrus* (queen palm), *Washingtonia* (California fan palm), *Chamaedorea* (parlor palm), *Raphidophyllum* (needle palm), *Thrinax* (thatch palm), *Coccothrinax* (thatch palm), *Licuala, Veitchia, Acoelorraphe* (paurotis palm), *Butia* (jelly palm), *Copernicia, Dypsis*, and *Wodyetia* (foxtail palm).

Discussion: A great deal of work over a period of more than 30 years (e.g., Dransfield 1986; Dransfield and Uhl 1998; Henderson 1995; Henderson et al. 1995; Moore 1973; Moore and Uhl 1982; Tomlinson 1990; Uhl and Dransfield 1987; Zona 1997) has greatly clarified our understanding of palms. Dransfield et al. (2005) present a classification of the family based on molecular relationships (see especially Asmussen et al. 2006). Arecaceae (or Palmae) are distinct, easily recognized, and monophyletic. Phylogenetic analyses of multiple DNA sequences show that the subfamily Calamoideae is sister to all other palms. *Nypa* (the sole genus of Nypoideae), a distinctive genus of Asian and western Pacific mangrove communities, is then sister to the remaining palms. Those genera with usually pinnate and reduplicate leaves form a paraphyletic complex, while those with usually costapalmate or palmate and induplicate leaves—the Coryphoideae—form a monophyletic group (Hahn 2002; Uhl et al. 1995).

Calamoideae have pinnate to palmate leaves and distinctive fruits that are covered with reflexed, imbricate scales (a synapomorphic feature). Noteworthy genera are *Raphia, Mauritia, Lepidocaryum, Metroxylon*, and *Calamus*. *Nypa* (Nypoideae) has a dichotomous-branching prostrate stem, erect, pinnate leaves, and its tepals are undifferentiated; fossils are known from Europe and America early in the Tertiary. Arecoideae have pinnate leaves and flowers in groups of three (triads), with one carpellate flower surrounded by two staminate flowers (a likely synapomorphy,

FIGURE 9.25 Arecaceae (Palmae, Coryphoideae).
Acoelorraphe wrightii: (A) habit (greatly reduced); (B) junction of leaf blade and petiole, showing hastula (× 0.5); (C) portion of inflorescence (× 0.5); (D) portion of inflorescence axis with flowers (× 7); (E) calyx (× 14); (F) spread-out corolla and androecium (× 14); (G) petal; (H) gynoecium (× 23); (I) drupe (× 3.5); (J) seed (× 4.5). (From Zona 1997, *Harvard Pap. Bot.* 2(1): p. 97.)

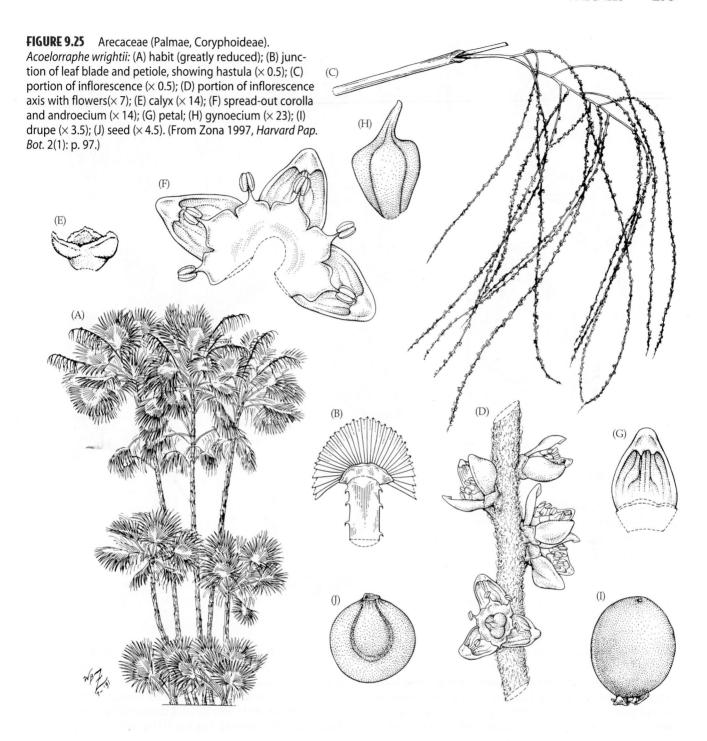

but lost in some). Within Arecoideae, a few clearly defined monophyletic groups are evident. Hyophorbeae (e.g., *Chamaedorea*, *Hyophorbe*) have imperfect flowers in lines. Cocoseae have the inflorescence associated with a large, persistent, woody bract, and fruits with bony, three-pored endocarps, and include genera such as *Elaeis*, *Cocos*, *Syagrus*, *Attalea*, *Bactris*, *Desmoncus*, and *Jubaea*. Iriarteae (e.g., *Iriartea*, *Socratea*) have stilt roots, and leaf segments with blunt apices and diverging veins. Most Arecoideae are placed within a heterogeneous Areceae (Baker et al. 2006); representative genera include *Areca*, *Dypsis*, *Wodyetia*,

Veitchia, *Ptychosperma*, and *Dictyosperma*. These palms sometimes have a crownshaft, a structure formed from a series of broad, overlapping leaf bases, which appears to be a vertical extension of the stem.

Coryphoideae include the monogeneric Phoeniceae (*Phoenix*, the date palms) with distinctive pinnate leaves in which the basal segments are spinelike. Borasseae (e.g., *Latania*, *Borassus*, *Lodoicea*, and *Hyphaene*) have staminate flowers embedded in thickened inflorescence axes. Corypheae include *Chamaerops*, *Rhaphis*, *Licuala*, *Copernicia*, *Corypha*, *Washingtonia*, *Serenoa*, *Livistona*, *Rhapidophyllum*, and *Acoe*-

FIGURE 9.26 Arecaceae (Palmae, Arecoideae). *Roystonia regia*: (A) habit (greatly reduced), note crownshaft and inflorescences; (B) portion of inflorescence axis, showing triads of flower buds, two staminate and one carpellate (× 4); (C) staminate flower bud with petal and three stamens removed (× 7); (D) open staminate flower, three stamens removed to show central pistillodium (× 7); (E, F) stamens, before and after dehiscence of anther (× 7); (G) nearly mature carpellate bud in partial section, showing nectariferous staminodial cup (hatched) and gynoecium (× 6.2); (H) corolla and staminodial cup of carpellate flower (× 6.2); (I) drupe (× 2.8); (J) endocarp (× 2.8); (K) seed, showing hilar scar (× 2.3). (From Zona 1997, *Harvard Pap. Bot.* 2(1): p. 101.)

lorraphe, and are difficult to characterize. Genera such as *Sabal*, *Thrinax*, and *Coccothrinax* are phenetically similar, and have been included here; their inclusion would make the tribe non-monophyletic. Caryoteae (fishtail palms; e.g., *Caryota* and *Arenga*) form a distinctive clade within the Coryphoideae (Asmussen et al. 2000; Asmussen and Chase 2001; Hahn 2002) because of their floral triads (evolved in parallel with those of the Arecoideae). This group has induplicate, blunt leaf segments with diverging veins.

Flowers of palms usually are insect-pollinated, especially by beetles, bees, and flies; nectar is often employed as a pollinator reward (Henderson 1986). Palm fruits are usually fleshy and dispersed by a wide variety of mammals and birds, although some (e.g., *Nypa* and *Cocos*) are water-dispersed and float in ocean currents (Zona and Henderson 1989).

Commelinales

Haemodoraceae, **Pontederiaceae**, Philydraceae, **Commelinaceae**, and Hanguanaceae form the Commelinales. These families have many-flowered helicoid cymes and an amoeboid tapetum—the innermost layer of the anther tapetum shows early breakdown of inner and radial cell walls, with the nuclei and cytoplasm moving into the anther cavity (Dahlgren and Clifford 1982), derived characters that they share with the likely related order Zingiberales. Monophyly of Commelinales is supported by DNA sequences (Chase et al. 1995a, 2000, 2006; Davis et al. 2004; Linder and Kellogg 1995; Soltis et al. 2000), but morphological synapomorphies are ambiguous. Pontederiaceae share with Haemodoraceae nontectate-columellate exine structure, while Philydraceae, like the Haemodoraceae, have harrow, unifacial leaves (Simpson 1990; Dahlgren et al. 1985). All three families have tannin cells in the perianth and sclereids in the placentae.

The placement of Commelinaceae has been disputed. Morphology places this family with Eriocaulaceae and Mayacaceae, in the group here called Poales (Stevenson and Loconte 1995), while DNA sequences place it with the Haemodoraceae, Pontederiaceae, and Philydraceae (Linder and Kellogg 1995; Chase et al. 2000). Flowers with sepals and petals and moniliform hairs, the characters used to support a relationship with some members of the Poales, probably have evolved independently in Commelinaceae. As considered here, Commelinales comprise 5 families and about 780 species.

Commelinaceae Mirbel
(Spiderwort Family)

Herbs, sometimes succulent, with well-developed stems that are ± swollen at the nodes, or stems sometimes short; often with mucilage cells or canals containing raphides. Hairs simple, uniseriate or unicellular. Leaves alternate, 2-ranked or spiral, scattered along stem, simple, narrow or somewhat expanded, flat to sharply folded (V-shaped in cross-section) *and often with the opposite halves rolled separately against the midrib in bud,* entire, with parallel venation, but midvein often prominent, **with closed basal sheath; stomata tetracytic**; stipules lacking. Inflorescences deter-

minate, composed of few to several helicoid cymes, sometimes reduced to a solitary flower, terminal or axillary, often subtended by a folded, leafy bract. *Flowers* usually bisexual, radial to bilateral, *with perianth differentiated into a calyx and corolla. Sepals 3,* usually distinct, imbricate or with open aestivation. *Petals 3,* distinct and usually clawed to connate, and corolla then with a short to elongate tube and flaring lobes, *quickly self-digesting,* 1 petal sometimes differently colored and/or reduced, imbricate and crumpled in bud. Stamens 6, or 3 and then often with 3 staminodes; *filaments* slender, distinct to slightly connate, sometimes adnate to petals, *often with conspicuous moniliform (beadlike) hairs*; anthers occasionally with apical pores; pollen grains usually monosulcate. Carpels 3, connate; *ovary superior, with axile placentation*; stigma 1, capitate, fringed, or 3-lobed. Ovules 1 to several in each locule, anatropous to orthotropous. Nectaries lacking. *Fruit usually a loculicidal capsule* (occasionally a berry); **seeds with a conspicuous conical cap** (Figure 9.27).

Floral formula: *, K3, C$\widehat{3}$, A3 or 6, G$\underline{\widehat{3}}$; capsule

Distribution: Widespread in tropical to temperate regions.

Genera/species: 40/650. **Major genera:** *Commelina* (230 spp.), *Tradescantia* (60), *Aneilema* (60), *Murdannia* (45), and *Callisia* (20). *Callisia, Commelina, Gibasis, Murdannia,* and *Tradescantia* occur in the United States and/or Canada.

Discussion: Monophyly of Commelinaceae is supported by both the morphological features described above and by and molecular data (Evans et al. 2000; Linder and Kellogg 1995). The genus *Cartonema* may be sister to the remaining members of the family; it has yellow, radially symmetrical flowers, and lacks both the glandular microhairs and raphide canals present in other taxa (Evans et al. 2000, 2003; Faden 1998).

Endemic to Asia / Eurasia

Key to Families of Commelinales

1. Perianth of sepals and petals; leaves with the opposite halves rolled separately against the midrib in bud; seed with a conical cap .... **Commelinaceae**
1. Perianth of tepals; leaves not as above; seeds lacking a conical cap 2
2. Leaves with expanded, bifacial blade; petiole present .... **Pontederiaceae**
2. Leaves equitant, unifacial; petiole lacking 3
3. Stamen solitary; tepals 4 and distinct; flowers lacking septal nectaries; arylphenalenones absent, the roots never reddish .... Philydraceae
3. Stamens 3 or 6; tepals 6 and distinct to connate; flowers with septal nectaries, but these sometimes poorly developed; arylphenalenones (often reddish polyphenolics) present, often giving an orange-red to purple color to the rhizomes and roots .... **Haemodoraceae**

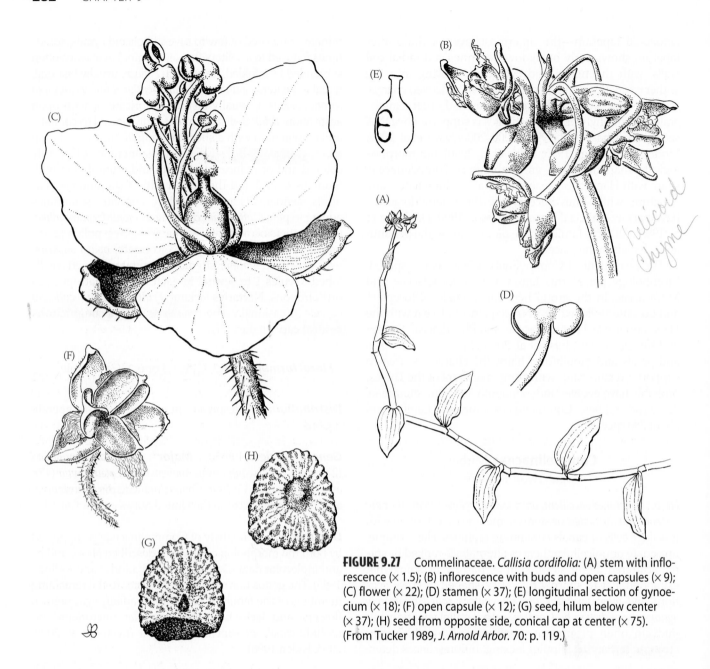

FIGURE 9.27 Commelinaceae. *Callisia cordifolia:* (A) stem with inflorescence (× 1.5); (B) inflorescence with buds and open capsules (× 9); (C) flower (× 22); (D) stamen (× 37); (E) longitudinal section of gynoecium (× 18); (F) open capsule (× 12); (G) seed, hilum below center (× 37); (H) seed from opposite side, conical cap at center (× 75). (From Tucker 1989, *J. Arnold Arbor.* 70: p. 119.)

Most genera of Commelinaceae belong to one of two large tribes (Faden and Hunt 1991; Tucker 1989): Tradescantieae (25 genera, e.g., *Callisia, Tradescantia,* and *Gibasis*) and Commelineae (13 genera, e.g., *Commelina, Murdannia,* and *Aneilema*). The former group is characterized by spineless pollen grains, medium to large chromosomes, radial flowers, and filament hairs (when present) moniliform; the latter has spiny pollen, small chromosomes, radial to bilateral flowers, and filament hairs (when present) usually not moniliform. There has been much convergence in floral characters within the family due to strong pollinator selection, and anatomical features (e.g., stomatal structure) may be quite useful in diagnosing major clades (Evans et al. 2000, 2003).

Flowers of Commelinaceae function for only a day at most. Pollination is usually accomplished by pollen-gathering bees or wasps. Interestingly, the staminodes are often more conspicuous than the stamens, and bilateral flowers are held so that the odd sepal is adaxial. The moniliform hairs often present on the filaments of Tradescantieae may delude bees into scraping them, as if gathering pollen. Self-pollination is common in some species.

Haemodoraceae R. Brown
(Bloodwort or Blood Lily Family)

Herbs, with rhizomes, corms, or bulbs, and roots often with orange-red pigment, **containing various phenalones, i.e., polyphenolic compounds.** *Hairs simple to dendritic or stellate, and usually densely covering inflorescence axes, bracts, and outer perianth parts. Leaves* alternate, *2-ranked, equitant, unifacial,* those of upper portion of stem reduced, simple, entire, with parallel venation, sheathing at base; stipules lacking. *Inflorescences determinate, consisting of a series of helicoid cymes,* but sometimes appearing indeterminate, terminal. Flowers bisexual, radial to bilateral. *Tepals 6,* showy, distinct to connate, the perianth tube (when present), sometimes with a slit along upper surface, imbricate or valvate. Stamens 3 or 6, sometimes reduced to 1, occasionally dimorphic; filaments free or adnate to the tepals; pollen grains monosulcate or 2–7-porate. *Carpels 3, connate;* ovary superior or inferior, with axile placentation; stigma 1, capitate to 3-lobed. Ovules 1 to numerous in each locule, anatropous to orthotropous. Nectaries in septa of ovary, but often poorly developed. *Fruit a loculicidal capsule;* seeds often winged.

Floral formula: * or X, T-6-, A 6 or 3, G③; capsule

Distribution and ecology: Widespread in Australia, southern Africa, and northern South America, but a few in North America; often wetland plants.

Genera/species: 13/100. **Major genera:** *Conostylis* (30 spp.), *Haemodorum* (20), *Anigozanthos* (11). The family is represented in the United States by *Lachnanthes*.

Economic plants and products: Genera such as *Anigozanthos* (kangaroo paw), *Conostylis*, and *Lachnanthes* (blood lily, redroot) are used as ornamentals.

Discussion: Haemodoraceae are considered monophyletic due to the presence of arylphenalenones; they are the only vascular plants to possess these pigments, which produce the orange-red to purple coloration characteristic of the roots and rhizomes of many genera (Simpson 1990, 1998a). Morphology supports recognition of two major clades: Haemodoroideae and Conostylidoideae (Simpson 1990). Monophyly of Haemodoroideae (e.g., *Haemodorum, Lachnanthes,* and *Xiphidium*) is supported by a reddish coloration in the roots and rhizomes, lack of sclereids in the placentas, and discoid seeds that are pubescent or marginally winged. Synapomorphies of Conostylidoideae (e.g., *Anigozanthos*) are multiseriate, branched hairs; pollen with rugulate (wrinkled) wall sculpturing and porate apertures; a nondeflexed style; and a base chromosome number of seven. The perianth of *Anigozanthos* has an adaxial slit; bilateral flowers have evolved within both tribes. Ovary position in the family has apparently undergone a shift from inferior to superior (Simpson 1990, 1993, 1998b).

The eastern North American *Lophiola* (Robertson 1976) resembles Haemodoraceae in its unifacial leaves, inflorescence of helicoid cymes, and densely hairy flowers, but differs in pollen ultrastructure, hair morphology, and stem anatomy. *Lophiola* lacks the arylphenalenones so characteristic of Haemodoraceae, as well as the UV-fluorescent, cell-wall-bound compounds that are present in Haemodoraceae (and in all other Commelinoid monocots). *Lophiola* is now placed within Nartheciaceae (Zomlefer 1997a,b,c, and this text).

The variously colored flowers of Haemodoraceae are usually pollinated by insects (especially bees and butterflies), but bird pollination is characteristic of *Anigozanthos*. Nectar usually provides the pollinator reward, but *Xiphidium* is pollinated by pollen-gathering bees. The often small, flattened, hairy, or winged seeds of many species probably are dispersed by wind.

Pontederiaceae Kunth
(Water Hyacinth Family)

Herbs, rhizomatous, floating to emergent aquatics; stems spongy. Hairs simple, only on reproductive parts. Leaves usually alternate and spiral, along stem or ± basal, ± differentiated into a petiole and blade, simple, entire, with parallel to palmate venation, sheathing at base; stipules lacking. Inflorescences determinate, but often appearing to be racemes or spikes, sometimes reduced to a single flower, terminal, but often appearing lateral, **associated with 2 bracts**. *Flowers* bisexual, radial to bilateral, *often tristylous. Tepals 6, showy,* variably **connate**, imbricate, *adaxial tepal of inner whorl often differentiated. Stamens usually 6;* **filaments adnate to the perianth tube,** *often unequal in length;* anthers opening by slits or pores; pollen grains with 1 or 2 furrows. Carpels 3, connate; ovary superior, with axile to occasionally intruded parietal placentation, sometimes with 2 locules sterile; stigma 1, capitate to 3-lobed. Ovules numerous to 1 in each locule. Nectaries often present in septa of ovary. *Fruit a loculicidal capsule or nut,* **surrounded by persistent basal portion of perianth tube** (Figure 9.28).

Floral formula: * or X, T-6-, A6, G③; capsule, nut

Distribution and ecology: Widespread in tropical and subtropical regions, with a few temperate species; plants of aquatic and wetland habitats.

Genera/species: 7/35. **Genera:** *Heteranthera* (12 spp.), *Eichhornia* (7), *Monochoria* (7), and *Pontederia* (6). The family is represented in the continental United States and/or Canada by *Eichhornia, Heteranthera,* and *Pontederia*.

Economic plants and products: *Pontederia* (pickerel weed) and *Eichhornia* (water hyacinth) are used as aquatic ornamentals; the latter is a very serious weed of still or slowly-flowing waters in the tropics and subtropics.

Discussion: Monophyly of Pontederiaceae has been supported by morphology (Eckenwalder and Barrett 1986) and molecular characters (Barrett and Graham 1997; Graham and Barrett 1995; Graham et al. 1998). *Pontederia* and *Eichhornia* form a clade on the basis of a long-lived, perennial habit, a curved inflorescence axis, and bilateral, tristylous flowers. Molecular data suggest that *Monochoria* belongs to the *Pontederia* + *Eichhornia* clade. *Pontederia* is distinct due to the apomorphies of a gynoecium with two locules abort-

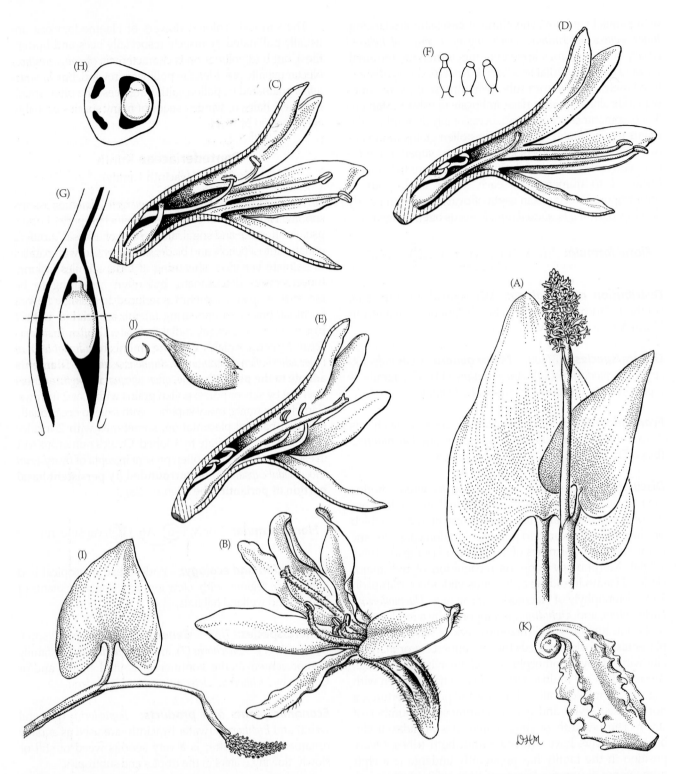

FIGURE 9.28 Pontederiaceae. *Pontederia cordata*: (A) leaf blade and portion of petiole behind flowering stem with leaf and bract subtending inflorescence (× 0.4); (B) flower of long-styled form, with style and three mid-length stamens exserted (× 4); (C) flower of short-styled form, in semidiagrammatic longitudinal section (e.g., hairs not shown) showing two of three adaxial, mid-length stamens and two of three abaxial, long stamens (× 4); (D) flower of mid-styled form showing two of three adaxial, short stamens and two of three abaxial, long stamens (× 4); (E) flower of long-styled form, showing two of three adaxial, short stamens and two of three abaxial, mid-length stamens (× 4); (F) glandular hairs of staminal filaments (× 72); (G) ovary in longitudinal section; note fertile locule with a single pendulous ovule; dashed line indicates position of cross-section in (H) (× 23); (H) ovary in cross-section, showing two aborted locules and one fertile locule (× 23); (I) terminal part of stem, with developing fruits (× 0.2); (J) fruit (× 4.5); (K) fruit enclosed in fleshy accrescent base of perianth (× 4.5). (From Rosatti 1987, *J. Arnold Arbor.* 68: p. 64.)

ing and the third containing only a single ovule, indehiscent fruits (nuts) surrounded by a ribbed to toothed persistent basal portion of the perianth, and relatively large seeds. *Eichhornia* is not monophyletic. *Heteranthera* is characterized by dimorphic stamens (fertile stamens and staminodes providing food for insects) with basifixed anthers (Eckenwalder and Barrett 1986).

Pontederiaceae is the only monocot family showing tristyly, which probably evolved just once within the group and then was lost several times. Only Oxalidaceae and Lythraceae also have tristylous species (Vuilleumier 1967).

The showy flowers of Pontederiaceae are open for only one day and are pollinated by various insects, especially bees, flies, and butterflies. Outcrossing is characteristic of heterostylous species, but self-pollination, as in *Heteranthera*, is also common. The warty nuts of *Pontederia* are water dispersed, as are the small seeds of capsular fruited species. Asexual reproduction by fragmentation of the rhizome system is common in some species.

Additional references: Cook 1998; Lowden 1973; Ornduff 1966; Price and Barrett 1982; Rosatti 1987; Strange et al. 2004.

Poales

Monophyly of Poales, as delimited in this text, is well supported by DNA sequence data (Chase et al. 1995a, 2000, 2006; Davis et al. 2004; Graham et al. 2006; Soltis et al. 2000). Possible morphological characters supporting this group of families include silica bodies in the epidermis, styles separate or connate but strongly branched, and the loss of raphide crystals. Wind pollination (with the loss of septal nectaries) has evolved several times within Poales and is characteristic of Typhaceae, Juncaceae, Cyperaceae, Restionaceae, and Poaceae. The order is often delimited very narrowly, including only Poaceae, Restoniaceae, and close relatives. However, Poales is here circumscribed broadly, including 17 families and about 19,500 species. Major families include **Typhaceae**, **Bromeliaceae**, **Eriocaulaceae**, **Xyridaceae**, **Juncaceae**, **Cyperaceae**, **Restionaceae**, and **Poaceae**.

Major clades within the Poales are illustrated in Figure 9.29. Typhaceae and Bromeliaceae are isolated and likely represent early divergent clades within the order. Eriocaulaceae and Xyridaceae are linked by their distinctive habit (i.e., rosette plants with dense cluster of flowers on a scape), perianth of a calyx and corolla, and ovules with

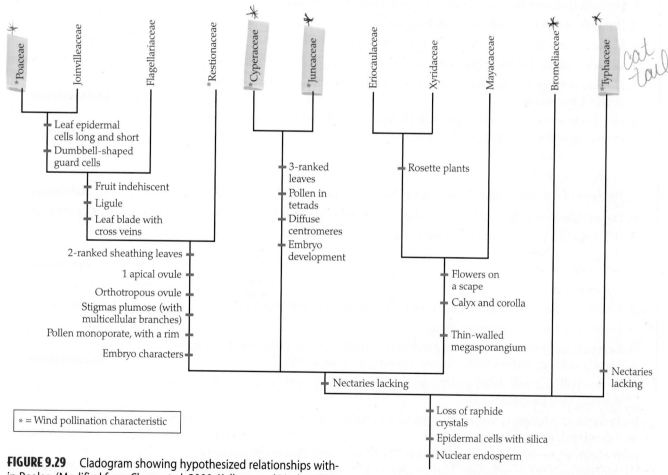

FIGURE 9.29 Cladogram showing hypothesized relationships within Poales. (Modified from Chase et al. 2000, Kellogg and Linder 1995, Soltis et al. 2000, 2005, and Stevens 2001, onward.)

thin-walled megasporangia (Dahlgren et al. 1985; Linder and Kellogg 1995). The remaining families mentioned above belong to either the sedge/rush clade or the core Poales.

Cyperaceae and Juncaceae (along with Thurniaceae) constitute the sedge/rush clade, which has been recognized formally in the past as the Juncales. This group is certainly monophyletic, based on morphology and *rbcL* sequences (Plunkett et al. 1995; Simpson 1995). Morphological synapomorphies include solid stems, 3-ranked leaves, pollen in tetrads (three of the grains reduced in Cyperaceae), chromosomes with diffuse centromeres, and details of embryo and pollen development. *Prionium* (sometimes placed in Juncaceae, but here considered in Thurniaceae) may be sister to the Cyperaceae/Juncaceae clade (Munro and Linder 1998). All evidence points to Cyperaceae being

monophyletic based on *rbcL* sequences (Muasya et al. 1998). Monophyly of Juncaceae is less clear, but is supported by ITS sequence data (Roalson 2005).

Members of the sedge/rush clade are wind-pollinated, superficially grasslike, and often confused with grasses. A common mnemonic is the following rhyme (origin uncertain): "Sedges have edges, and rushes are round, and grasses are hollow right to the ground." This rhyme refers to the sharply triangular stems of some (but not all) sedges (Cyperaceae) and the hollow stems of some (but not all) grasses (Poaceae). Rushes (Juncaceae) are indeed round-stemmed and solid, but the rhyme's characterization of sedges and grasses is an oversimplification.

The core Poales (or graminoid clade) includes Restionaceae, Flagellariaceae, Joinvilleaceae, and Poaceae, and a few other small families. Other than Poaceae, which are cos-

---

Key to Major Families of Poales

1. Leaves with water-absorbing peltate scales, occasionally merely stellat ....**Bromeliaceae**
1. Leaves lacking such hairs 2
2. Perianth differentiated into a calyx and usually colorful corolla 3
2. Perianth of tepals, or reduced to lacking 5
3. Leaves evenly distributed along stem; flowers solitary ....Mayacaceae
3. Leaves ± basal; flowers in heads or conelike spikes borne at apex of long scape 4
4. Flowers unisexual, in involucrate heads with numerous flowers open at one time; ovules 1 per locule ....**Eriocaulaceae**
4. Flowers bisexual, in conelike spikes, with only 1 or 2 flowers open at one time; ovules numerous on each placenta ....**Xyridaceae**
5. Leaves 3-ranked 6
5. Leaves 2-ranked 7
6. Perianth of 6 tepals; ovules 3–numerous; fruit a capsule ....**Juncaceae**
6. Perianth lacking or reduced to scales, bristles, or hairs; ovule 1; fruit an achene (nutlet) ....**Cyperaceae**
7. Inflorescence with numerous densely clustered flowers and appearing to be elongate/cylindrical spikes or globose clusters, the staminate flowers positioned above the carpellate ....**Typhaceae**
7. Inflorescence not as above 8
8. Leaves circinately inrolled in bud, with coiled, tendril-apex at maturity ....Flagellariaceae
8. Leaves not circinately inrolled at tip 9
9. Leaves usually with the blade very reduced, consisting merely of a sheath, a ligule usually lacking; anthers with 1 locule, opening by a single slit ....**Restionaceae**
9. Leaves with a ± well developed blade, a ligule usually present; anthers with 2 locules, opening by 2 longitudinal slits 10
10. Perianth of 6 tepals; flowers not in spikelets; gynoecium 3-carpellate; fruit a drupe, 1-3 seeded; embryo undifferentiated ....Joinvilleaceae
10. Perianth reduced, usually represented by 2 or 3 lodicules; flowers in spikelets; gynoecium 2- or 3-carpellate; fruit usually a caryopsis, 1 seeded; embryo grass-like ....**Poaceae**

mopolitan, these families are herbs of the Southern Hemisphere and especially the Pacific Ocean region. The largest family other than Poaceae in core Poales is Restionaceae, which grow mostly in South Africa and Australia. Monophyly of core Poales is supported by both morphological and DNA characters (Briggs et al. 2000; Chase et al. 2000, 2006; Dahlgren and Rasmussen 1983; Dahlgren et al. 1985; Graham et al. 2006; Kellogg and Linder 1995; Linder and Kellogg 1995; Michelangeli et al. 2003; Soltis et al. 2000; Stevenson et al. 2000); morphological synapomorphies include two-ranked leaves, each with a sheath around the stem, small flowers with plumose stigmas and one apical orthotropous ovule per carpel, pollen monoporate with rim around pore, and embryo characters (Kellogg and Linder 1995; Soreng and Davis 1998). The clade also shows deletions in the ORF 2280 region of the chloroplast genome. Evolutionary relationships within core Poales are shown in Figure 9.29.

Bromeliaceae A. L. de Jussieu
(Bromeliad Family)

Herbs, usually epiphytic; **solitary silica bodies** usually associated with epidermal cells. **Hairs water-absorbing peltate scales, occasionally merely stellate.** *Leaves* alternate and spiral, *often forming a tank-like basal rosette that holds water,* simple, entire to sharply serrate, with parallel venation, **containing water storage tissue** and air canals (often with stellate cells); sheathing at base; stipules lacking. Inflorescences indeterminate, terminal. *Flowers* usually bisexual, radial, **with perianth differentiated into a calyx and corolla,** *borne in the axils of often brightly colored bracts.* *Sepals 3,* distinct to connate, imbricate. *Petals 3,* distinct to connate, often with a pair of appendages at the base, imbricate. *Stamens 6;* filaments distinct to connate, sometimes adnate to petals; pollen grains monosulcate or disulcate, or with 2 to several pores. *Carpels 3, connate;* ovary superior to inferior, with axile placentation; stigmas 3, **usually spirally twisted.** Ovules numerous. Nectaries usually in septa of ovary. *Fruit a septicidal capsule or berry; seeds often winged or with tuft(s) of hairs* (Figure 9.30).

Floral formula:

$*$, K $\overline{③}$, C $\underline{\overline{③}}$, A $\underline{⑥}$, G $\overline{③}$; capsule, berry

Distribution and ecology: Tropical to warm temperate regions of the Americas (but one species of *Pitcairnia* in tropical Africa). An important group of epiphytes of moist montane forests; also occurring in xerophytic habitats.

Genera/species: 51/1520. **Major genera:** *Tillandsia* (450 spp.), *Pitcairnia* (250), *Vriesia* (200), *Aechmea* (150), *Puya* (150), and *Guzmania* (120). The family is represented in the continental United States by *Tillandsia, Catopsis,* and *Guzmania.*

Economic plants and products: The fruits of *Ananas comosus* (pineapple), a multiple fruit of fused berries associated with a fleshy, if somewhat fibrous, inflorescence axis, are important as an edible fruit. The dried stems and leaves of *Tillandsia usneoides* (Spanish moss) are used as stuffing material. Several genera, including *Aechmea, Billbergia, Bromelia, Guzmania, Neoregelia, Pitcairnia, Tillandsia,* and *Vriesia,* provide ornamentals.

Discussion: Monophyly of Bromeliaceae is supported by cpDNA restriction sites and *rbcL* sequence data (Chase et al. 1993, 1995a; Ranker et al. 1990), morphology, and possibly the base chromosome number of 25 (Dahlgren et al. 1985; Gilmartin and Brown 1987; Smith and Till 1998; Varadarajan and Gilmartin 1988). The family probably represents an early divergent clade within the order (Chase et al. 1995b; Dahlgren et al. 1985; Linder and Kellogg 1995); if basal, septal nectaries may be a plesiomorphy.

Within Bromeliaceae serrate leaf margins, superior ovaries, and capsular fruits with winged seeds are all probable plesiomorphies. The family is traditionally are divided into three subfamilies: "Pitcairnioideae," Bromelioideae, and Tillandsioideae. Tillandsioideae are considered monophyletic on the basis of their entire leaf margins, hair-tufted seeds, and distinctive peltate scales that have several rings of isodiametric cells in the center of the scale and a fringe of 32 or 64 radiating cells. This clade has usually superior ovaries and capsular fruits. Monophyly of Bromelioideae is supported by their inferior ovaries and berry fruits; "Pitcairnioideae" are a paraphyletic assemblage. Analyses of chloroplast DNA sequences support the monophyly of Tillandsioideae and Bromelioideae, and the paraphyly of Pitcairnioideae (Barfuss et al. 2005; Givnish et al. 2004; Terry et al. 1997a,b). The epiphytic habit evolved separately in Tillandsioideae and Bromelioideae.

Bromeliaceae show several adaptations to xerophytic and epiphytic conditions. The elongate, more or less concave leaves are typically clustered at the base of the plant (Plate 9.5I), and their expanded bases form a water-retaining tank. The leaf surface is covered with water-absorbing peltate scales; each scale has a uniseriate stalk (of living cells), while the radiating cells of the scale are dead at maturity. The dead cells expand when wetted, drawing water into (and under) the scale, where it is osmotically taken into the leaf by the stalk cells. Water loss is reduced by location of stomates in furrows and a thick cuticle. *Tillandsia usneoides* and *T. recurvata* grow happily on telephone wires. The plant's adventitious roots function mainly in holding it in place.

The showy flowers are pollinated by various insects, birds, or occasionally bats. The berries of members of the Bromelioideae are dispersed by birds or mammals, while the winged seeds of Pitcairnioideae and hair-tufted seeds of Tillandsioideae are wind-dispersed.

Additional references: Benzing 1980; Benzing et al. 1978; Brown and Gilmartin 1989; Smith and Wood 1975.

FIGURE 9.30   Bromeliaceae. (A, B) *Tillandsia recurvata:* (A) fruiting plant (× 0.4); (B) single stem with open capsule and seeds (× 0.75). (C–L) *T. usneoides:* (C) stem with flower and open fruit (× 0.75); (D) flower (× 2); (E) flower with two sepals, two petals, and five stamens removed to show gynoecium (× 3); (F) cross-section of ovary (× 22); (G) gynoecium in longitudinal section (× 22); (H) placenta and ovules (× 30); (I) open capsule with seeds (× 3); (J) seed with basal appendage of hairs (× 3); (K) seedling (× 6); (L) scale hair from leaf (× 75). (From Smith and Wood 1975, *J. Arnold Arbor.* 56: p. 386.)

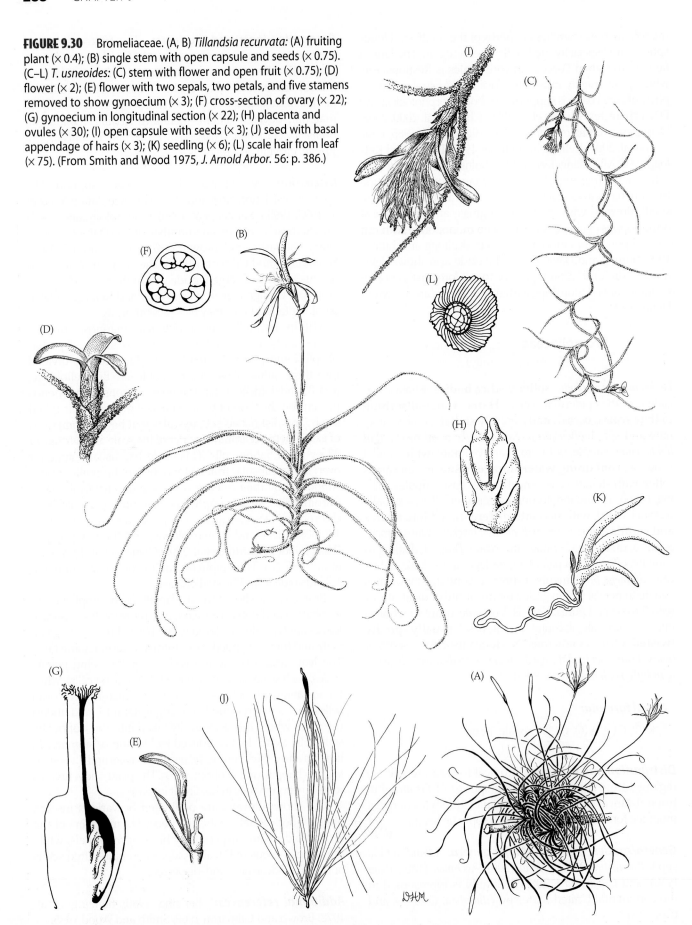

(A)

Poales: Poaceae, Panicoideae
Andropogon glomeratus var. *pumilis*:
plant in fruit

(B)

Poales: Juncaceae
Juncus dichotomous: fruits

(C)

Poales: Poaceae, Ehrhartoideae
Oryza sativa: spikelets

(D)

Poales: Typhaceae
Typha domingensis:
plant in bloom

(E)

Poales: Cyperaceae
Rhynchospora colorata:
plant in bloom

(F)

Poales: Poaceae, Chloridoideae
Uniola paniculata: spikelets

(G)

Poales: Restionaceae
Elegia capensis: stems with
sheathing leaves

(H)

Poales: Cyperaceae
Carex verrucosa: inflorescences

(I)

Poales: Bromeliaceae
Tillandsia hotteana: bromeliad
habit; plant in bloom

**PLATE 9.5 Monocots
Poales**

Typhaceae A. L. de Jussieu
(Cattail Family)

Rhizomatous herbs, aquatic or wetland, with leaves and stems distally floating or emergent. Hairs simple. *Leaves alternate, 2-ranked,* simple, *linear,* entire, with parallel venation, *often spongy, with air canals and partitions containing stellate cells,* sheathing at base; stipules lacking. *Inflorescences* determinate, terminal, highly modified *with numerous densely clustered flowers and appearing to be elongate/cylindrical spikes or globose clusters, the staminate flowers positioned above the carpellate,* often subtended by a linear bract. **Flowers unisexual (plants monoecious),** radial. **Tepals 1–6 and reduced**, *bractlike, numerous and bristle-like, or scale-like,* distinct. Stamens 1–8; filaments distinct or basally connate; anthers with connective sometimes expanded; pollen grains uniporate, in monads or tetrads. **Carpels** 3, connate, **usually only 1 functional**; ovary superior, with **apical placentation (and usually 1 locule)**, often on a stalk; stigma 1, extending along one side of style. **Ovule 1**. Nectaries lacking. *Fruit a drupe with dry-spongy "flesh" or an achene-like follicle;* seed or pit containing a pore, through which the embryo emerges (Figure 9.31).

Floral formula:

Staminate: *, T(1–∞), A1–8, G0

Carpellate: *, T3–∞, A0, G1; drupe, achene-like follicle

Distribution and ecology: Widely distributed, especially in the Northern Hemisphere; characteristic of aquatic and wetland habitats, especially marshes.

Genera/species: 2/28. **Genera:** *Sparganium* (15 spp.) and *Typha* (13). Both occur in the continental United States and Canada.

Economic plants and products: *Typha* (cattails; Plate 9.5D) and *Sparganium* (bur reeds) are occasionally used as ornamentals; the leaves of *Typha* are used as weaving material, and the starchy rhizomes, young staminate inflorescences, and pollen of both genera can be eaten.

Discussion: Cladistic analyses using morphology (Dahlgren et al. 1985; Linder and Kellogg 1995; Stevenson and Loconte 1995; Stevenson et al. 2000) strongly support the monophyly of Typhaceae (including Sparganiaceae). Analyses based on DNA sequences are more equivocal, with some supporting (Chase et al. 1993; Davis et al. 2004; Duvall et al. 1993) and others not supporting (Chase et al. 2000) the monophyly of the group. Analyses of DNA sequences and morphology (Chase et al. 1995b; Michelangeli et al. 2003) also supported the monophyly of Typhaceae (incl. Sparganiaceae). Typhaceae here are circumscribed broadly (including Sparganiaceae) based on morphological characters (see description).

Typhaceae are wind-pollinated. The persistent perianth bristles of *Typha* help in the wind dispersal of the achene-like fruits, which after dispersal split open, releasing the single seed. The dry-spongy drupes of *Sparganium* may be dispersed by birds, mammals, or water; the fruit can float because of its spongy outer layer.

Additional references: Thieret 1982; Thieret and Luken 1996.

Eriocaulaceae Martynov
(Pipewort Family)

Herbs with shortened, cormlike stems or rhizomes; stems with vascular bundles in 1 or 2 rings. Hairs simple and uniseriate, or T-shaped. *Leaves* alternate and usually spiral, *usually in basal rosettes, or in tufts along branching stems,* simple, *narrow and grasslike,* entire, with parallel venation, sheathing at base; stipules lacking. **Inflorescences** indeterminate, **forming a head subtended by an involucre of stiff papery bracts**, *terminal, on a long scape;* scapes 1 to several, basally enclosed by a sheathing bract. **Flowers unisexual** (plants usually monoecious), radial to bilateral, *individually inconspicuous, with the perianth differentiated into a calyx and corolla, and often fringed with hairs,* usually in the axil of a papery bract. *Sepals* 2 or 3, distinct or connate, usually valvate. *Petals* 2 or 3, distinct (in carpellate flowers) or *connate (in staminate flowers),* sometimes with nectar-producing glands near the apex, sometimes reduced (in carpellate flowers), usually valvate. Stamens 2–6, often unequal; filaments can be distinct or connate, **adnate to the petals**, sometimes arising from a stalk (formed by fused petals and filaments); anthers 1- or 2-locular; **pollen grains with an elongated and spiral germination furrow** (or, in some taxa, convoluted). Carpels 2 or 3, connate; *ovary superior, usually borne on a stalk, with axile placentation;* stigmas 2 or 3, minute. Ovules 1 in each locule, orthotropous, with a thin megasporangium. *Fruit a loculicidal capsule;* seed with testa ± longitudinally striated, reticulate, or "hairy."

FIGURE 9.31 Typhaceae. (A–J) *Typha latifolia:* (A) base of plant with rhizome and sheathing leaf bases (× 0.3); (B) inflorescence and apex of stem, staminate portion above, carpellate below (× 0.3); (C) three staminate flowers, the three filaments of each flower connate (× 15); (D) dehiscing anther, with tetrads of pollen grains (× 31); (E) tetrad of pollen grains (× 31); (F) cluster of four carpellate flowers (× 31); (G) cluster of three carpellate flowers with many trichomes omitted to show stalked ovaries (× 15); (H) ovary in longitudinal section to show single apical ovule (× 31); (I) infructescence shedding fruits (× 0.3); (J) mature fruit, a stalked, achenelike follicle, with numerous hairs on the stalk (× 3.75). (K–M) *T. angustifolia:* (K) inflorescence, showing gap between staminate and carpellate portions (× 0.3); (L) pollen grain (× 1250); (M) portion of carpellate inflorescence to show three flowers, simple and glandular hairs, and spatulate sterile flower (× 15). (From Thieret and Luken 1996, *Harvard Pap. Bot.* 1(8): p. 32.)

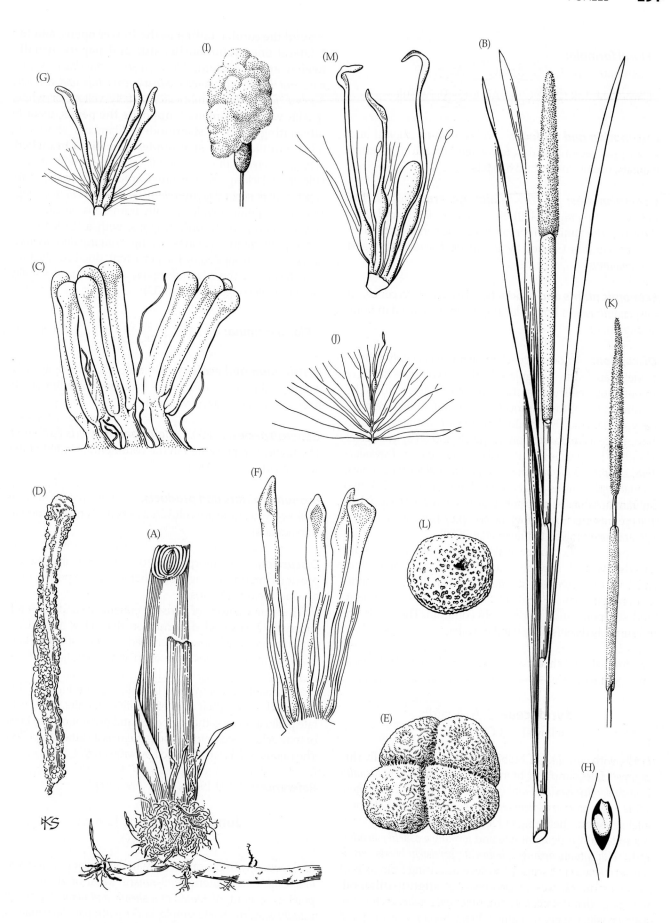

Floral formula:
Staminate: * or X, K (2–3), C (2–3), A2–6, G0
Carpellate: *, K (2–3), C2–3, A0, G (2–3); capsule

Distribution and ecology: Widespread in tropical and subtropical regions, with a few extending into temperate habitats; usually in wetland habitats.

Genera/species: 9/1175. **Major genera:** *Paepalanthus* (485 spp.), *Eriocaulon* (400), *Syngonanthus* (200), and *Leiothrix* (65). The family is represented in the continental United States and/or Canada by *Eriocaulon*, *Lachnocaulon*, and *Syngonanthus*.

Economic plants and products: Dried inflorescences of *Syngonanthus* and *Eriocaulon* (pipeworts) are used in floral arrangements.

Discussion: Eriocaulaceae are distinctive, clearly monophyletic, and easily recognized by their involucrate heads of minute flowers. They are therefore often referred to as the "Compositae of the monocots." Two subfamilies are typically recognized: Eriocauloideae (e.g., *Eriocaulon*), which have twice as many stamens as petals and an apical nectar gland on the petals, and Paepalanthoideae (e.g., *Paepalanthus*, *Leiothrix*, *Syngonanthus*, and *Lachnocaulon*), which have as many stamens as petals and lack nectar glands. The two subfamilies may be monophyletic (Unwin 2004). *Paepalanthus* represents a paraphyletic complex from which other genera have evolved (Giulietti et al. 2000).

Flowers of Eriocaulaceae, which have anthers and styles clearly exserted, may be wind-pollinated, although nectaries on the flowers of *Eriocaulon* suggest that insect pollination also occurs. Floral visitors appear to be infrequent, and self-pollination is probably common. The seeds are presumably dispersed by wind or water.

Additional references: Dahlgren et al. 1985; Kral 1966a, 1989; Stützel 1998.

Xyridaceae C. A. Agardh
(Yellow-Eyed Grass Family)

Herbs, with cormlike or bulblike stems, or occasionally rhizomes. Hairs simple or branched. *Leaves* alternate, *usually 2-ranked, often equitant and unifacial, flat to cylindrical, those of the upper portion of stem reduced*, simple, entire, with parallel venation, sheathing at base; stipules lacking. *Inflorescences* usually indeterminate and *forming a conelike head or spike, with spirally arranged, imbricate, persistent bracts, terminal, on a long scape*; scapes 1 to several, arising from axils of bracts or inner leaves. *Flowers* bisexual, **slightly bilateral**, *with perianth differentiated into a calyx and corolla*, each in the axil of a stiff, papery to leathery bract. Sepals 3, distinct and **dimorphic, the inner one membranous and wrapped**

around the corolla, falling as the flower opens, and the **2 lateral ones subopposite, stiff and papery, usually keeled, and persistent**. *Petals 3, distinct and clawed to connate, and then forming corolla with a narrow tube and distinctly 3-lobed, flaring limb*, imbricate, usually yellow or white, quickly wilting. **Stamens 3, opposite the petals, usually alternating with 3 staminodes**; filaments short and adnate to the petals; **staminodes apically 3-branched**, *densely covered with moniliform hairs*; pollen grains monosulcate or lacking an aperture. Carpels 3, connate; *ovary superior, with parietal to free-central or axile placentation*; stigmas 3, ± capitate. Ovules usually numerous on each placenta, anatropous to orthotropous, with a thick to thin megasporangium. Nectaries lacking. *Fruit usually a loculicidal capsule*, surrounded by the persistent, dried corolla tube and clasped by the 2 lateral sepals; seeds minute, usually longitudinally ridged (Figure 9.32).

Floral formula: X, K1+2, C (3), A3+3•, G (3); capsule

Distribution and ecology: Widespread in tropical and subtropical regions, with a few extending into temperate habitats; characteristic of wetlands.

Genera/species: 5/300. **Major genus:** *Xyris* (260 spp.). The family is represented in the continental United States and Canada only by *Xyris*.

Economic plants and products: A few species of *Xyris* (yellow-eyed grass) are cultivated as ornamentals, especially in aquaria.

Discussion: The morphology of the boat-shaped sepals provides important diagnostic characters for species, most of which belong to *Xyris*.

The showy flowers of *Xyris* are ephemeral, and the corollas usually are expanded for no more than a few hours. Usually only one or two flowers per head are open at once. Flowers of sympatric species often open at different times of day. Nectaries are consistently lacking, and pollination may be predominantly accomplished by pollen-gathering bees. The staminodes, with their tufts of moniliform hairs, may facilitate pollination by gathering pollen and presenting it to bees, or may delude bees into scraping them as if gathering pollen. The minute seeds are dispersed by wind and/or water.

References: Kral 1966b, 1983, 1992, 1998.

Juncaceae A. L. de Jussieu
(Rush Family)

Herbs, often with rhizomes; silica bodies absent; *stems round and solid*. *Leaves* alternate, 3-ranked, basal or along lower portion of stem, *composed of a sheath and blade, the sheath usually open*; the blade simple, entire, with parallel venation, linear, flat or cylindrical; ligule and stipules lacking. Inflores-

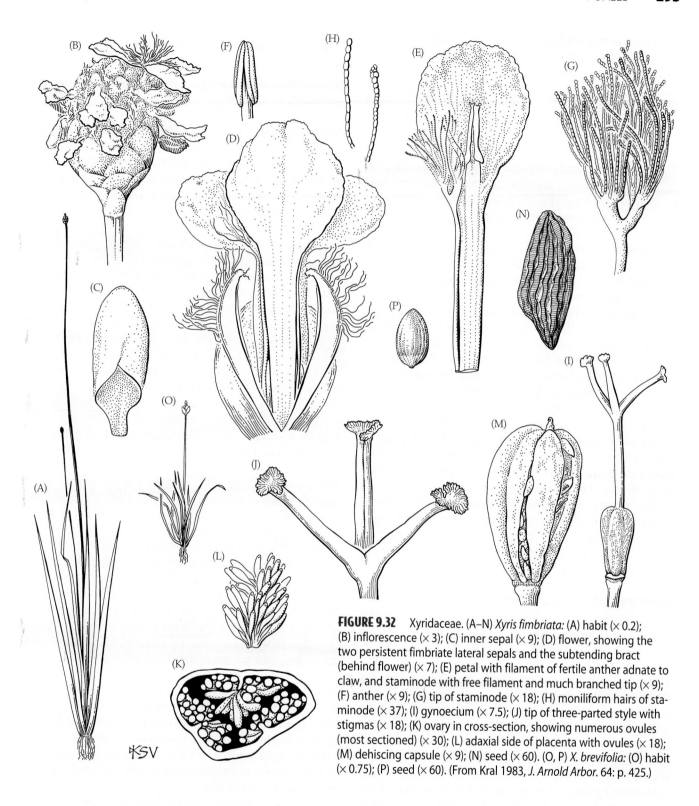

FIGURE 9.32 Xyridaceae. (A–N) *Xyris fimbriata:* (A) habit (× 0.2); (B) inflorescence (× 3); (C) inner sepal (× 9); (D) flower, showing the two persistent fimbriate lateral sepals and the subtending bract (behind flower) (× 7); (E) petal with filament of fertile anther adnate to claw, and staminode with free filament and much branched tip (× 9); (F) anther (× 9); (G) tip of staminode (× 18); (H) moniliform hairs of staminode (× 37); (I) gynoecium (× 7.5); (J) tip of three-parted style with stigmas (× 18); (K) ovary in cross-section, showing numerous ovules (most sectioned) (× 30); (L) adaxial side of placenta with ovules (× 18); (M) dehiscing capsule (× 9); (N) seed (× 60). (O, P) *X. brevifolia:* (O) habit (× 0.75); (P) seed (× 60). (From Kral 1983, *J. Arnold Arbor.* 64: p. 425.)

cences basically determinate, terminal, highly branched, but often condensed and headlike. *Flowers usually bisexual*, but occasionally unisexual (plants dioecious), radial, inconspicuous. *Tepals 6, distinct*, imbricate, generally dull colored (green, red-brown, black), but sometimes white or yellowish. *Stamens (3–)6*; filaments distinct; pollen monoporate, in obvious tetrads. *Carpels 3, connate; ovary superior*, with axile or parietal (occasionally basal) placentation; ovules numerous (rarely 3); stigmas 3, usually elongate. Nectaries lacking. *Fruit a loculicidal capsule*, with 3 to many seeds (Plate 9.5B).

Floral formula: *, T-6-, A(3–)6, G③; capsule

Distribution and ecology: Worldwide, mostly temperate and/or montane. Often in damp habitats, but with notable exceptions, such as the weedy *Juncus trifidus*.

Genera/species: 6/400. ***Major genera:*** *Juncus* (300 spp.), *Luzula* (80). Both occur in the United States and Canada.

Economic plants and products: *Juncus effusus* (soft rush) and *J. squarrosus* (heath rush) are used to produce split rushes for baskets and chair bottoms. A few *Juncus* and *Luzula* species are used as ornamentals.

Discussion: Monophyly of Juncaceae is supported by ITS sequences (Kristiansen et al. 2005; Roalson 2005). It is not clear which, if any, of the family characteristics are synapomorphic, most are generalized features of monocots. Recent phylogenetic analyses (Drábková et al. 2003; Roalson 2005) indicate that *Juncus* is not monophyletic, having given rise to both *Luzula* and a group of Andean cushionplants (*Oxychloe* and *Distichia*). The fact that *Juncus* is glabrous suggests that this condition may be a synapomorphy of the family (even if homoplasious).

Many members of this family look superficially like grasses, but the 3-ranked leaves, flowers with obvious tepals, and capsular fruit all make the distinction clear. In some species of *Juncus*, the large bract subtending the inflorescence is borne upright so that it looks like a continuation of the stem, and the inflorescence appears lateral.

The inconspicuous flowers of Juncaceae are predominantly wind pollinated; outcrossing is commonly brought about by protogyny, but some species are selfing. Dispersal of the small seeds occurs by wind, water, and external transport on animals.

Additional reference: Balslev 1998.

Cyperaceae A. L. de Jussieu
(Sedge Family)

Herbs, generally rhizomatous; **stems usually ± triangular in cross-section**, often leafless above the base. *Leaves alternate, 3-ranked*, **with conical silica bodies**; *composed of a sheath and blade*, **the sheath closed**, the blade simple, entire to minutely serrate, with parallel venation, linear, flattened; stipules lacking; *ligule generally lacking. Inflorescence a complex arrangement of small spikes (spikelets)*, often subtended by bracts. *Flowers bisexual or unisexual (plants then usually monoecious)*, *each subtended by a bract. Tepals lacking or reduced to usually 3–6 scales, bristles, or hairs.* **Stamens 1–3** (–6); filaments distinct; anthers not sagittate; **pollen** usually uniporate, **in pseudomonads (3 microspores degenerate and form part of the pollen wall)**. Carpels 2–3, connate; ovary superior, **with basal placentation**; **ovule 1**; styles 3, elongated. Nectaries lacking. **Fruit an achene (nutlet)**, often associated with persistent perianth bristles (Figure 9.33).

Floral formula: *, T-0–6-, A1–3 (–6), G③; achene

Distribution and ecology: Worldwide; often, but not exclusively, in damp sites.

Genera/species: 104/4500. ***Major genera:*** *Carex* (2000 spp.), *Cyperus* (600), *Fimbristylis* (300), *Scirpus* (300), *Rhynchospora* (200), *Scleria* (200), and *Eleocharis* (200). All of the above occur in North America; other common genera include *Cladium*, *Bulbostylis*, *Bulboschoenus*, *Eriophorum*, *Fuirena*, *Kyllinga*, *Schoenoplectus*, and *Trichophorum*.

Economic plants and products: *Cyperus papyrus* was used for making paper by ancient Egyptians and is also commonly planted as an ornamental. *Cyperus rotundus* (nut sedge) is an agricultural weed. *Cyperus esculentus* (nut sedge, chufa, tigernut) has edible underground storage organs, as do *Mariscus umbellatus*, *Scirpus tuberosus*, and *Eleocharis dulcis*; the last provides commercial "water chestnuts." Weaving materials are provided by stems and leaves of a few species of *Cyperus*, *Carex*, *Eleocharis*, *Lepironia*, and *Scirpus*. Roots of *Cyperus longus* (galingale) and *C. articulatus* are sweet-scented and used in perfumery. Roots of *Scirpus grossus* and *S. articulatus* are used in Hindu medicine. Various species of *Carex* (Plate 9.5H) are used as packing materials, hay, or straw.

Discussion: Cyperaceae contain unique conical silica bodies, which distinguish them from all other monocots. The family is apparently monophyletic (Muasya et al. 1998). In a comprehensive morphological study of the family, Bruhl (1995) recognized 2 subfamilies and 10 tribes. The tribes represent groups that appear in both phenetic and cladistic analyses of the family, although only 4 of the 10 have clear synapomorphies; phylogenetic relationships have also been assessed by a study of *rbcL* sequences (Muasya et al. 1998). The most common tribe in North America is the Cariceae, in which the spikelet prophyll forms a sac (the perigynium) enclosing the flower.

Like Juncaceae, Cyperaceae are often mistaken for grasses. They are distinguished by the more or less triangular stems, 3-ranked leaves, usual absence of a ligule, and closed sheaths. (The latter two characteristics occur in a few grasses, but not together.) Cyperaceae flowers are subtended by a single bract or, in *Carex*, bract plus prophyll (Plate 9.5H), whereas most grass flowers are associated with two bracts (lemma and palea).

Most Cyperaceae are wind pollinated, but insect pollination has evolved several times, e.g., *Hypolytrum*, some species of *Ascolepis* and *Rhynchospora*. Dispersal of the fruits is by water (due to corky fruit wall or associated structures, or perianth bristles holding air bubbles, e.g., *Cladium*, *Remirea*, *Eleocharis*, *Fuirena*), external transport (by means of a hooked inflorescence axis, as in, *Uncinia*, or muddy fruits sticking to feathers, fur, or clothing, as in *Fimbristylis*), birds (aril imitation, in *Scleria*), or wind (by action of elongate

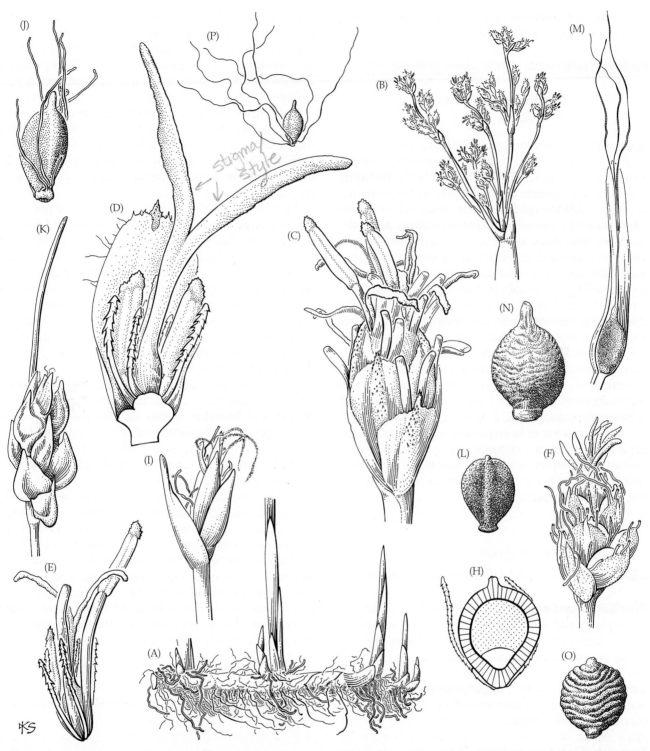

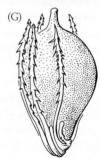

FIGURE 9.33 Cyperaceae. (A–H) *Scirpus* (*Schoenoplectus*) *tabernaemontani* (*S. validus*): (A) underwater rhizome (× 0.65); (B) apex of stem with inflorescence (× 1.35); (C) single spikelet with lower flowers past anthesis, upper ones with anthers visible and styles exserted (× 16); (D) flower and subtending bract removed from spikelet, view of adaxial surface, stigmas exserted, anthers still included, note barbed bristles (× 27); (E) flower showing different maturation of stamens (× 16); (F) spikelet at stage later than that in C, immature achenes below and flowers with receptive stigmas above (× 11); (G) mature achene with persistent bristles (× 16); (H) same in vertical section, fruit wall hatched, endosperm stippled, embryo unshaded, the seed coat too thin to show (× 16). (I–J) *Trichophorum cespitosum* (*Scirpus cespitosus*): (I) spikelet (× 11); (J) achene with bristles (× 16). (K–L) *Scirpus koilolepis*: (K) solitary spikelet, subtended by keeled bracts (× 16); (L) achene (× 16). (M–O) *S. erismaniae*: (M) basal flower in leaf axil (× 11); (N) achene from basal flower (× 16); (O) achene from spikelet borne on stem (× 16). (P) *S. cyperinus*: achene with elongate bristles (× 16). (From Tucker 1987, *J. Arnold Arbor*. 68: p. 372.)

bristles, as in *Eriophorum* or *Scirpus*). The perigynium of *Carex* often assists in water dispersal.

Additional references: Goetghebeur 1998; Tucker 1987.

Restionaceae R. Brown
(Restio Family)

Herbs, with rhizomes; *stems round, elliptic, or polyhedral in cross-section, and solid or hollow (in internodal region),* **anatomically distinctive, with "protective" cells lining substomatal chambers and 1 or 2 layers of chlorenchyma cells inside the epidermis, the chlorenchyma separated from the cortex by parenchymatous and sclerenchymatous rings**; silica bodies present. Hairs simple, sometimes flattened or peltate. *Leaves alternate, 2-ranked, usually with the blade very reduced and leaf consisting merely of an open sheath;* ligule usually lacking; stipules lacking. Inflorescences with flowers variously arranged, usually in spikelets. *Flowers usually unisexual (plants dioecious), often in dimorphic inflorescences,* radial, inconspicuous, each in the axil of a bract. *Tepals 3-6,* usually distinct, imbricate, generally dull colored or greenish. *Stamens (1-) 3,* sometimes represented by staminodes in carpellate flowers; filaments distinct to connate; *anthers with 1 locule, opening by a single slit;* pollen monoporate. Carpels 3, connate, ovary superior, with apical-axile placentation, and in some with only a single fertile locule; stigmas 3, elongate, often plumose. Ovules 1 per locule, orthotropous, with thin sporangium wall. Nectaries lacking. *Fruit a capsule, achene or nut,* often associated with a persistent perianth.

Floral formula:

Staminate: * T– 6 – , A ③, G0

Carpellate: * T– 6 – , A0, G③, capsule, achene, nut

Distribution and ecology: Distributed in the Southern Hemisphere and most diverse in Australia and South Africa; plants of low-nutrient soils, and frequently of seasonally arid and/or fire-prone habitats.

Genera/species: 53/485. ***Major genera:*** *Restio* (90 spp.), *Ischyrolepis* (50), and *Thamnochortus* (35). This clade does not occur in North America.

Economic plants and products: A few genera are of horticultural interest, e.g., *Chondropetalum* and *Elegia* (Plate 9.5G). *Thamnochortus* is used for thatch, and several genera provide important pasture plants.

Discussion: Monophyly of the family is supported by anatomical (Linder et al. 2000b) and molecular (Briggs et al. 2000) characters. Restionaceae are closely related to the Poaceae, Flagellariaceae, and especially Centrolepidaceae, as evidenced by their monoporate, spheroidal pollen grains and presence of silica bodies in parenchymatous tissues.

The latter family (which has much more reduced flowers) may be nested within Restionaceae (Linder et al. 2000b), but is treated here as their sister group, as supported by cpDNA sequence data (Briggs et al. 2000). The Restionaceae + Centrolepidaceae clade share a couple of remarkable embryological apomorphies (sporangium wall with vertically elongated cells, polar nuclei surrounded by large starch bodies) and both have 1-loculed anthers. Finally, molecular data supports the exclusion of two small genera, *Lyginia* and *Hopkinsia*, which are distinctive in having 2-loculate anthers, a plesiomorphic feature. They belong to the Australian Anarthriaceae.

Members of Restionaceae belong to one of two large clades, the first containing the African genera, and the second the Australian genera (Briggs et al. 2000; Eldenäs and Linder 2000; Linder 2000; Linder et al. 2000a,b). Members of the African clade are characterized by pollen grains with raised or swollen aperture margins and stems with cavities beneath the stomata lined with thick-walled ("protective") cells; Australian genera have the pollen exine margins attenuate towards the germination pore, and most have lost "protective" cells. Long-pubescent rhizomes support the monophyly of the Australian clade.

At one time only a few large genera were recognized within Restionaceae, e.g., *Restio*, which included both African and Australian species. At this time many, much smaller genera are delimited, and *Restio* is considered to be restricted to Africa (Briggs and Johnson 1998a,b; Linder 1985). Recent studies have shown that additional generic recircumscriptions are required.

The inconspicuous flowers of Restoniaceae are wind-pollinated, although some species are selfing or apomictic. Wind dispersal of the small seeds (or achenes or nutlets, in genera with indehiscent fruits) is common; some have modified tepals (wings, awns, hairs) that presumably aid dispersal. A few genera (e.g., *Willdenowia* and relatives) have fruits associated with fleshy structures, and may attract ants. *Alexgeorgea* is unusual because its carpellate flowers are borne underground (but with the long style/stigma lying on the surface); pollination of these species is by wind, but the mechanism of dispersal of the nutlets is unknown.

Additional references: Carlquist 1976b; Rudall and Linder 1988.

Poaceae Barnhart
(= Gramineae A. L. de Jussieu)
(Grass Family)

Herbs, often rhizomatous, but trees in tropical bamboos; *stems jointed, round to elliptical in cross-section, solid to hollow;* with silica bodies. *Leaves alternate, 2-ranked, consisting of sheath, ligule, and blade; sheaths tightly encircling the stem, the margins overlapping but not fused* or, occasionally, united to form a tube; ligule a membranous flange or fringe of hairs at adaxial apex of sheath; blades simple, usually linear, usu-

---

Key to Major Clades of Poaceae

1. Specialized cells, called arm cells, within the leaves of most members; stamens often more than 3 2
1. Arm and fusoid cells absent; stamens 3 or fewer 3
2. Stigmas 3; mostly trees ....Bambusoideae
2. Stigmas 2; herbs ....Ehrhartoideae
3. Spikelets compressed perpendicular to plane of arrangement of glumes and florets, not breaking at maturity into separate florets but falling as a unit, with 1 caryopsis-bearing floret ....Panicoideae
3. Spikelets not compressed, or compressed parallel to plane of arrangement of glumes and florets, breaking up at maturity above the glumes, mostly with more than 1 caryopsis-bearing floret 4
4. Veins in leaves separated by more than 4 mesophyll cells; bundle sheaths with few chloroplasts, appearing clear in cross section (C₃ leaf anatomy); bicellular microhairs present or lacking in leaf epidermis 5
4. Veins in leaves separated by 2–4 cells; bundle sheaths with numerous chloroplasts, appearing distinctly green in cross section (C₄ anatomy); bicellular microhairs present in leaf epidermis 6
5. Bicellular microhairs absent in leaf epidermis; subsidiary cells parallel-sided; plants variable in habit, but mostly less than 1 m tall ....Pooideae
5. Bicellular microhairs present in leaf epidermis; subsidiary cells dome-shaped, plants generally over 1 m tall ....Arundinoideae s.s.
6. Bicellular microhairs generally bulbous; awns, if present, unbranched ....Chloridoideae
6. Bicellular microhairs ± linear, awns divided into 3 parts ....Aristidoideae

---

ally with parallel venation, flat or sometimes rolled into a tube, continuous with the sheath or pseudopetiolate. Inflorescence a spike, panicle, cyme, or raceme of *spikelets*. *Spikelet composed of an axis bearing 2-ranked and closely overlapping basal bracts (glumes) and florets;* breaking up above the glumes or remaining intact at maturity; compressed parallel or perpendicular to the plane of arrangement of glumes and florets. *Glumes* usually 2, equal in size or unequal. **Florets** 1 to numerous per spikelet, made up of a bract (the *lemma*) subtending a flower and another bractlike structure (the *palea*, a prophyll) lying between the flower and the spikelet axis. Lemmas sometimes with 1 or more needle-like, straight or bent awns. Palea often translucent, smaller than, and partially enclosed by the lemma, often 2-keeled. Flowers small, bisexual or unisexual (plants then monoecious or dioecious), usually wind-pollinated, greatly reduced in size and number of floral parts. Lodicules (= perianth parts) mostly 2, translucent. Stamens (1–) 3 (–6, or numerous); *anthers usually sagittate;* pollen monoporate. Carpels 3, but often appearing as 2, connate; stigmas 2 (–3), plumose, papillae multicellular; ovary superior, with 1 locule and 1, subapical to nearly basal amphitropous or hemianatropous ovule, with a thin to thick megasporangium wall. **Fruit a single-seeded caryopsis (grain)** with fruit wall fused to the seed (or less often fruit wall free from the seed); often associated with parts of the spikelet for dispersal. **Embryo with a highly modified cotyledon (scutellum), lateral in position** (Figures 9.34 and 9.35).

Floral formula: *, T-2-, A(1–)3(–6), G(2–3); caryopsis

Note: Useful in identification are features of the spikelet, including size, plane of compression, presence or absence of glumes, number of florets, presence of sterile or incomplete florets, number of veins on glumes and lemmas, presence or absence of awns, and aggregation of spikelets in secondary inflorescences.

Distribution and ecology: Cosmopolitan; in desert to freshwater and marine habitats, and at all but the highest elevations. Native grasslands develop where there are periodic droughts, level to gently rolling topography, frequent fires, and in some instances grazing and certain soil conditions. Communities dominated by grasses, such as the North American prairie and plains, South American pampas, African veldt, and Eurasian steppes, account for about 24% of the Earth's vegetation. Woody bamboos play important roles in forest ecology in tropical and temperate Asia.

Genera/species: ca. 650/9700. Important genera are listed below under the major subgroups.

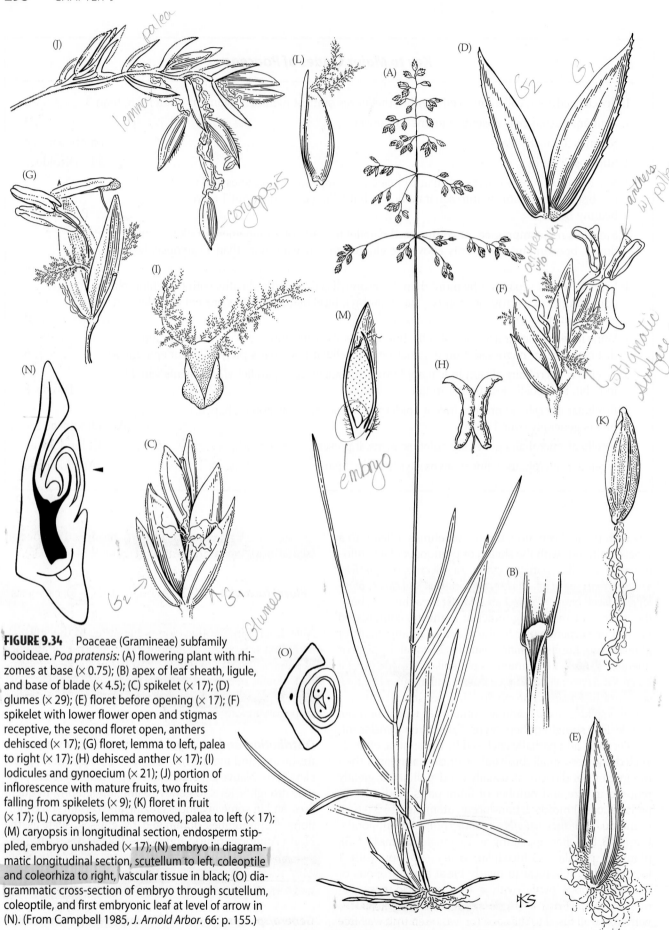

FIGURE 9.34 Poaceae (Gramineae) subfamily Pooideae. *Poa pratensis:* (A) flowering plant with rhizomes at base (× 0.75); (B) apex of leaf sheath, ligule, and base of blade (× 4.5); (C) spikelet (× 17); (D) glumes (× 29); (E) floret before opening (× 17); (F) spikelet with lower flower open and stigmas receptive, the second floret open, anthers dehisced (× 17); (G) floret, lemma to left, palea to right (× 17); (H) dehisced anther (× 17); (I) lodicules and gynoecium (× 21); (J) portion of inflorescence with mature fruits, two fruits falling from spikelets (× 9); (K) floret in fruit (× 17); (L) caryopsis, lemma removed, palea to left (× 17); (M) caryopsis in longitudinal section, endosperm stippled, embryo unshaded (× 17); (N) embryo in diagrammatic longitudinal section, scutellum to left, coleoptile and coleorhiza to right, vascular tissue in black; (O) diagrammatic cross-section of embryo through scutellum, coleoptile, and first embryonic leaf at level of arrow in (N). (From Campbell 1985, *J. Arnold Arbor.* 66: p. 155.)

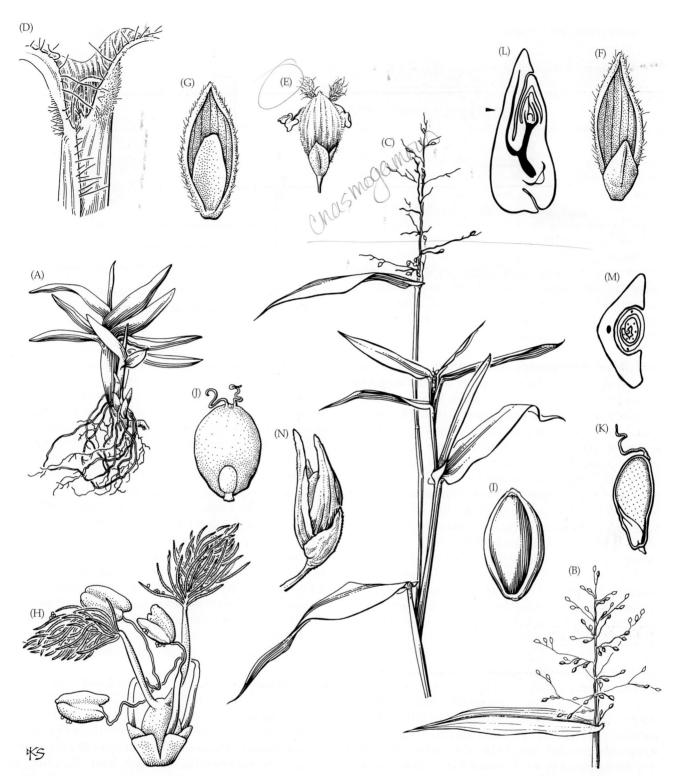

FIGURE 9.35 Poaceae (Gramineae) subfamily Panicoideae. (A–M) *Panicum* (*Dichanthelium*) *clandestinum*: (A) part of winter rosette (× 0.7); (B) inflorescence of chasmogamous spikelets (× 0.7); (C) upper part of plant, chasmogamous spikelets in fruit or shed from inflorescence, inflorescence of cleistogamous spikelets below (× 0.7); (D) upper part of leaf sheath, base of blade, and ligule (× 8); (E) chasmogamous spikelet at anthesis (× 8); (F) small first and larger second glume (× 14); (G) sterile lemma (pubescent) and sterile palea (× 14); (H) flower of cleistog- amous spikelet (× 27); (I) fertile lemma (behind) and palea enclos- ing mature caryopsis (× 14); (J) mature caryopsis (× 14); (K) longi- tudinal section of caryopsis, embryo to left, endosperm stippled (× 16); (L) embryo in diagrammatic longitudinal section, scutel- lum to left, coleoptile and coleorhiza to right, vascular tissue in black, note mesocotyl above trace to scutellum; (M) diagrammat- ic cross-section of embryo through scutellum, coleoptile, and first embryonic leaf, at level of arrow in (L). (N) *P. anceps:* spikelet in fruit (× 14). (From Campbell 1985, *J. Arnold Arbor.* 66: p. 172.)

Economic plants and products: The economic importance of grasses lies in their paramount role as food: about 70% of the world's farmland is planted in crop grasses, and over 50% of humanity's calories come from grasses. People have cultivated cereals for at least 10,000 years. From their earliest domestication, wheat (*Triticum aestivum*), barley (*Hordeum vulgare*), and oats (*Avena sativa*) in the Near East; sorghum (*Sorghum bicolor*) and pearl millet (*Pennisetum americanum*) in Africa; rice (*Oryza sativa*) in southeastern Asia,;and maize or corn (*Zea mays*) in Meso-America have made possible the rise of civilization. In terms of global production, the four most important crops are grasses: sugarcane (*Saccharum officinale*), wheat, rice, and maize. Barley and sorghum are in the top twelve.

Grasses are also used for livestock food, erosion control, turf production, and as a sugar source for the fermentation of alcoholic beverages, such as beer and whiskey. Bamboos are economically important in many tropical areas for their edible young shoots, fiber for paper, pulp for rayon, and strong stems for construction.

Subfamilial phylogenetic relationships: Grasses are easily recognized, and their monophyly has been supported by morphological and DNA characters. Recent molecular systematic studies based on *rbcL*, *ndhF*, *rpoC2*, ITS, granule-bound starch synthase I, and phytochrome B sequences agree with many phylogenetic relationships inferred from structural data (Grass Phylogeny Working Group, GPWG 2001) and support the recognition of 12 subfamilies. Anomochlooideae (native to Brazil), Pharoideae (Old and New World tropics), and Puelioideae (West Africa) are the three earliest diverging lineages, but together include only about 25 of the almost 10,000 species of the family. The remaining species fall into two large groups. One, the BEP clade, includes Bambusoideae s.s., Ehrhartoideae, and Pooideae; the PACCAD clade includes Panicoideae, Arundinoideae s.s., Chloridoideae, Centothecoideae, Aristidoideae, and Danthonioideae (Clark et al. 1995; Soreng and Davis 1998; Hilu et al. 1999; GPWG 2001). The PACCAD clade is supported by the embryological features of a long mesocotyl internode (see Figure 9.35L), and this clade emerges strongly in all molecular analyses. Support for the BEP clade is weak, and alternative arrangements of the three subfamilies are possible. All subfamilies recognized by GPWG (2001) are monophyletic, although only a few have morphological synapomorphies that characterize all members. More often they are supported by sets of character states that characterize large subgroups.

For each of the five major subfamilies—Bambusoideae s.s., Ehrhartoideae, Chloridoideae, Panicoideae, and Pooideae—geographic distribution, internal phylogenetic and systematic structure, and important genera are presented. Characters distinguishing the major subfamilies are given in the accompanying key; note that the best structural characters distinguishing subfamilies are anatomical, but morphological characters are used whenever possible. Phylogenetic relationships among these major subfamilies

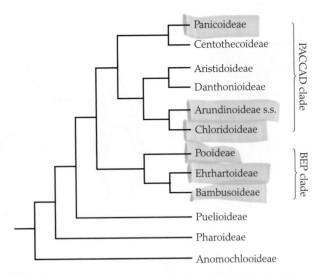

FIGURE 9.36 A phylogeny of Poaceae. (Modified from Clark et al. 1995 and Grass Phylogeny Working Group 2001.)

(along with some smaller clades) are shown in Figure 9.36 (GPWG 2001).

Anomochlooideae (including *Anomochloa* and *Streptochaeta*) were originally included in Bambusoideae, but are clearly only distantly related. The species are herbaceous and have morphologically unique inflorescences that are difficult to interpret, and certainly do not resemble the spikelets that are characteristic of other grasses. They are the sister group of the remaining members of the family (see Figure 9.36; Clark et al. 1995, 1996; Soreng and Davis 1998), thus suggesting that the grass spikelet probably arose after the Anomochlooideae originated.

Pharoideae and Puelioideae were, like Anomochlooideae, also traditionally included in Bambusoideae. They have conventional spikelets that are 1-flowered in Pharoideae and multiflowered in Puelioideae. The shift from 3 to 2 stigmas occurred after the origin of Pharoideae but before the origin of Puelioideae. The leaf blades of Pharoideae are turned upside down, with the abaxial surface uppermost.

Bambusoideae s.s. includes both woody and herbaceous plants, and are largely tropical in distribution. The herbaceous species form a clade; the woody bamboos are divided phylogenetically into a tropical clade and a temperate one. Woody bamboos, with stems up to 40 m in height, certainly do not resemble the turf in lawns. Flowering in many woody bamboos is also unusual, occurring in cycles of up to 120 years. Even though individual stems live for only one or a few decades, some form of "clock" directs stems to flower all at once throughout the range of a species. Important genera of woody bamboos are *Bambusa* (120 spp.), *Chusquea* (100), *Arundinaria* (50), *Sasa* (50), and *Phyllostachys* (45).

Ehrhartoideae include the Southern Hemisphere Ehrharteae, as well as the widespread Oryzeae. The latter are aquatic or wetland herbs. The most widely known Oryzeae are the commercially important Asian rice (*Oryza sativa*, Plate 9.5C) and North American wild rice (*Zizania aquatica*).

Pooideae are largely temperate in distribution, especially in the Northern Hemisphere. Important genera include several cereals (wheat, barley, oats; see above under economic plants and products) as well as rye (*Secale cereale*), turf grasses (e.g., bluegrasses, *Poa*, 500 spp.), fescues (*Festuca*, 450), pasture grasses (e.g., *Phleum*, *Dactylis*), and some weeds (e.g., *Agrostis*, 220, and *Poa*). Other important grasses of this subfamily are *Stipa* (300), *Calamagrostis* (270), *Bromus* (150), and *Elymus* (150).

Chloridoideae bear distinctive bicellular hairs on the leaf epidermis; these may be synapomorphic for a subset of the group. All but two chloridoids show C_4 photosynthesis. The subfamily is best developed in arid and semiarid tropical regions, where its C_4 photosynthesis is presumably advantageous. Centers of distribution in Africa and Australia suggest a Southern Hemisphere origin. Some important genera are *Eragrostis* (350 spp.), *Muhlenbergia* (160), *Sporobolus* (160), *Chloris* (55), *Spartina* (15), and *Eustachys* (10); the first three of these appear to be polyphyletic.

Panicoideae have long been recognized taxonomically because of their distinctive spikelets (see Key to Poaceae). The subfamily is primarily tropical or warm temperate and contains two large tribes, Andropogoneae and Paniceae, along with a number of smaller groups. Andropogoneae are relatively easy to recognize because the spikelets are often paired and grouped into a linear inflorescence (Plate 9.5A). Paniceae are not as homogeneous as Andropogoneae. The genus *Panicum* (in the broad sense) was once thought to be a large and heterogeneous group, but is now known to be demonstrably polyphyletic (Aliscioni et al. 2003). *Dichanthelium* clearly should be excluded from *Panicum*, as well as several other smaller genera. *Panicum* s.s. now includes ca. 150 species, all of which are C_4. Other important genera include *Paspalum* (330), *Andropogon* (100), *Setaria* (100), *Sorghum* (20), and *Zea* (4).

The other subfamilies in the PACCAD clade are structurally and genetically diverse. They range from small desert species (*Aristida*, 250 spp., the largest genus in the Aristidoideae) to giant wetland reeds (*Phragmites*, in the Arundinoideae s.s.).

Discussion: Poaceae rank behind Asteraceae, Orchidaceae, and Fabaceae in number of species, but rank first in global economic importance. Poaceae are unsurpassed among angiosperms in land surface area dominated. The family's monophyly is supported by phenotypic characters (reduced perianth, fruit a caryopsis, and embryo and pollen wall features) as well as DNA sequences. Similarities to sedges (Cyperaceae) in habit and spikelets represent convergent evolution. Sedges are more closely related to rushes (Juncaceae), and Poaceae belong to core Poales (see Figure 9.29).

Grasses have been successful ecologically and have diversified extensively due to several key adaptations. The grass spikelet (see Plate 9.5) protects the flowers while permitting pollination when the lodicules open the spikelet. Spikelets have various adaptations for fruit dispersal. Versatility in breeding systems, including inbreeding and agamospermy, helps make some grasses successful colonizers. C_3 and C_4 leaf anatomy adapt grasses to a wide range of habitats. Meristems are located at the bases of the internodes and sheaths. As a result, grasses tolerate fire and grazing better than many other plants. Development of grasslands during the Miocene epoch (25–5 mya) may have fostered the evolution of large herbivores, which in turn were an important food source and stimulus for the evolution of *Homo sapiens*.

The economic and ecological importance of the family has motivated considerable systematic study. Early in the nineteenth century, differences between the spikelets of pooids and panicoids led to division of the whole family into these two groups by Robert Brown. Early in the twentieth century, leaf epidermal characters and chromosome number led to separation of chloridoid grasses from the pooids. In the mid twentieth century, internal leaf anatomy and embryological features set up the recognition of 5 to 8 subfamilies. Differences in leaf anatomy are associated with different photosynthetic pathways. The C_3 pathway is more efficient in regions of cool and cold climate, whereas C_4 photosynthesis is advantageous in regions of high temperatures and low soil moisture. C_3 anatomy is the plesiomorphic state in the family; all C_4 species are in the PACCAD clade, although many C_3 taxa are included there as well.

Additional references: Aliscioni et al. 2003; Barker et al. 1995; Campbell 1985; Clark and Pohl 1996; Clark et al. 2000; Clark and Judziewicz 1996; Clayton and Renvoize 1986; Judziewicz et al. 1999; Kellogg and Linder 1995; Kellogg and Watson 1993; Soderstrom et al. 1987; Tucker 1996; Watson and Dallwitz 1992.

Zingiberales

Monophyly of Zingiberales is supported by morphology (Dahlgren and Rasmussen 1983; Kress 1990, 1995; Stevenson and Loconte 1995; Stevenson et al. 2000; Tomlinson 1962), DNA sequences (Chase et al. 1995b, 2000; Davis et al. 2004; Källersjö et al. 1998; Smith et al. 1993; Soltis et al. 1997, 2000). Putative synapomorphies include large herbs with vessels more or less limited to the roots; presence of silica cells in the bundle sheath; leaves clearly differentiated into a petiole and blade, with pinnate venation, often tearing between the secondary veins, with the blade rolled into a tube in bud, and the petiole (and midvein) with enlarged air canals, flowers bilateral (but, in more specialized taxa, lacking a plane of symmetry); pollen usually lacking an exine; ovary inferior; and the arillate seeds with perisperm (Figure 9.37A).

Phylogenetic relationships within core Zingiberales are fairly well understood due to careful morphological and cpDNA analyses (Dahlgren and Rasmussen 1983; Kress 1990, 1995; Kress et al. 2001; Tomlinson 1962, 1969a). Can-

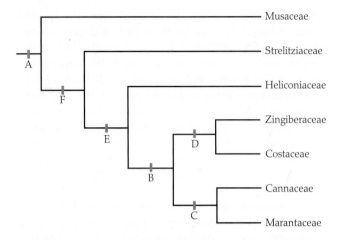

FIGURE 9.37 A phylogeny of Zingiberales, as discussed in the text. (Modified from Kress 1990, 1995, Kress et al. 2001, Soltis et al. 2000, 2005, Chase et al. 2000.)

naceae, Marantaceae, Zingiberaceae, and Costaceae form a clade based on reduction of the androecium to only a single functional stamen, presence of showy staminodes, seeds with more perisperm than endosperm, lack of raphide crystals in vegetative tissues, and leaves that are not easily torn (Figure 9.37B). Within this clade, Marantaceae and Cannaceae are hypothesized to be sister taxa, as evidenced by their flowers that lack a plane of symmetry, and androecium with only half of one stamen fertile, the other half of this stamen being expanded and staminodial (Figure 9.37C). Zingiberaceae and Costaceae constitute a clade that is supported by the unusual feature of the single functional stamen more or less grasping the style, a ligule at the apex of the leaf sheath, connate sepals, fused staminodes, and reduction of two of the three stigmas (Figure 9.37D). Heliconiaceae, Strelitziaceae, and Musaceae constitute a paraphyletic complex, and relationships among these families are still rather unclear. Heliconiaceae may be the sister taxon to the Cannaceae-Marantaceae-Zingiberaceae-Costaceae clade, as supported by the putative synapomorphies of a sterile outer median stamen, connate petals, and details of the root anatomy (Figure 9.37E). Strelitziaceae may be sister to the clade containing all of the above-listed families, all having 2-ranked leaves (Figure 9.37F). Musaceae (banana family) probably is sister group of the rest of the order; this family has retained the plesiomorphic condition of spirally arranged leaves.

Zingiberales contain 8 families and about 1980 species; major families include **Cannaceae**, **Marantaceae**, **Zingiberaceae**, Costaceae, Heliconiaceae, Strelitziaceae, and Musaceae.

Zingiberaceae Martinov
(Ginger Family)

Small to large, **spicy-aromatic herbs, scattered secretory cells containing ethereal oils, various terpenes, and**

phenyl-propanoid compounds. Hairs simple. *Leaves* alternate, *2-ranked*, simple, entire, usually petiolate, *with a well-developed blade, pinnate venation*, sheathing base, *and a ligule*; petiole with air canals, these separated into segments by diaphragms composed of stellate-shaped cells; stipules lacking. Inflorescences indeterminate, but comprised of determinate (cymose) units in the axils of usually conspicuous bracts. *Flowers* bisexual, *bilateral*, usually lasting for only 1 day. *Sepals 3, connate*, imbricate. *Petals 3, connate, with one lobe often larger than the others*, imbricate. *Stamen 1, grooved, grasping the style*; staminodes usually 4, **2 large, connate, and forming a lip-like structure (labellum)**, *and 2 smaller, these distinct or connate with the 2 larger staminodes*; pollen grains monosulcate or lacking apertures, exine very reduced. Carpels 3, connate; *ovary inferior*, with usually axile placentation; *style enveloped in groove between pollen sacs of the anther*; stigma 1, funnel-shaped. Ovules ± numerous. **Nectaries 2, positioned atop the ovary**. *Fruit a dry to fleshy capsule or berry*; seeds usually arillate; endosperm and perisperm present (Figure 9.38).

Floral formula:

X, K ③, C (2+1), A (2•+2•) + 1, G ③; fleshy capsule, berry

Distribution and ecology: Widespread in tropical regions; chiefly in shaded to semi-shaded forest understory habitats; occasionally in wetlands. Asexual reproduction occurs in some species of *Globba*.

Genera/species: 50/1000. ***Major genera:*** *Alpinia* (150 spp.), *Amomum* (120), *Zingiber* (90), *Globba* (70), *Curcuma* (60), *Kaempferia* (60), and *Hedychium* (50). The family is represented in the continental United States by *Alpinia*, *Curcuma*, *Hedychium*, and *Zingiber* (all rarely naturalized).

Economic plants and products: The family contains several important spices, including *Zingiber* (ginger), *Curcuma* (turmeric), and *Amomum* and *Elettaria* (cardamom). The rhizomes of several species of *Curcuma* are used as a starch source. *Alpinia* (shell ginger), *Curcuma* (hidden lily), *Hedychium* (garland lily), *Globba*, *Nicolaia* (torch ginger), *Renealmia*, and *Zingiber* (ginger) contain ornamental species.

Discussion: Monophyly of the Zingiberaceae has been supported by DNA (Smith et al. 1993; Kress 1995; Wood et al. 2000) and morphology (Kress 1990). The family is closely related to the Costaceae, which is often included within Zingiberaceae as a subfamily (see Rogers 1984).

Two early divergent lineages, *Tamija* and *Siphonochilus*, are distinctive because their lateral staminodes are well developed and fused to the labellum. This is the ancestral condition, also evident in the sister family, Costaceae. The remaining genera belong to two major clades: Alpinioideae (plane of the 2-ranked leaves perpendicular to rhizome) and Zingiberoideae (plane of the 2-ranked leaves parallel to rhizome) (Kress et al. 2002).

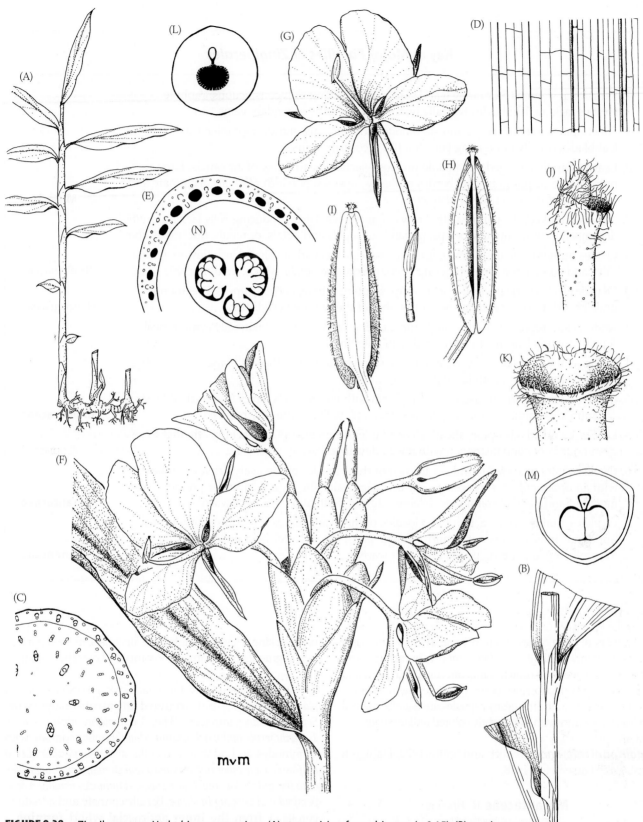

FIGURE 9.38   Zingiberaceae. *Hedychium coronarium:* (A) stem arising from rhizome (× 0.15); (B) portion of stem showing ligulate leaf bases (× 0.55); (C) stem, in cross-section (× 9); (D) leaf venation (× 9); (E) leaf sheath, in cross-section (× 9); (F) inflorescence (× 0.55); (G) flower, showing split, one-lobed calyx, narrow petals, two broad lateral staminodes, and liplike staminodes (with bifid apex) and single stamen grasping gynoecium (× 0.55); (H, I) anther, grasping style, adaxial and abaxial views (× 4.5); (J, K) stigma; (L) corolla tube, in cross-section above calyx (× 9); (M) cross-section of flower slightly above ovary, note two epigynous nectar glands and style (in groove in corolla tube) (× 9); (N) ovary in cross-section (× 9). (From Rogers 1984, *J. Arnold Arb.* 65: p. 24.)

Key to Major Families of Zingiberales

1. Functional stamens 5 or rarely 6 and staminodes lacking or inconspicuous; raphide crystals lacking; leaf blades typically tearing between the secondary veins 2

1. Functional stamens 1 or 1/2 and staminodes conspicuous, showy; raphide crystals present; leaf blades usually not tearing between the secondary veins 4

   2. Leaves spirally arranged, the petiole (in cross-section) with 1 row of air canals; latex-producing cells present; fruits berries or fleshy capsules; perianth of tepals, 5 connate and 1 member of the inner whorl distinct ....**Musaceae**

   2. Leaves 2-ranked, the petiole with 2 rows of air canals; latex-producing cells lacking; fruits dry capsules, or fleshy schizocarps; perianth of sepals and petals, or tepals, but not as above 3

      3. Ovules numerous in each locule; fruit a capsule; seeds with a colorful aril; flowers with a calyx and corolla, the sepals differing in color from the petals, which are dimorphic ....Strelitziaceae

      3. Ovules 1 in each locule; fruit a schizocarp, splitting into usually 3 drupelike segments; flowers with tepals, 5 connate and 1 member of the outer whorl distinct ....Heliconiaceae

   4. Androecium represented by a single functional stamen; flowers bilaterally symmetrical; sepals connate; leaf sheath associated with a ligule 5

   4. Androecium represented by 1/2 functional stamen; flowers without a plane of symmetry; sepals distinct; leaf sheath lacking a ligule 6

      5. Leaves 2-ranked, sheath usually open; plants with ethereal oils (spicy fragrant); at least 2 staminodes connate, forming a liplike structure; pollen exine very reduced ....**Zingiberaceae**

      5. Leaves 1-ranked and spiral, sheath closed (at least initially); plants lacking ethereal oils; staminodes not connate; pollen with a well developed exine ....Costaceae

      6. Ovule solitary in a single locule or in each of the 3 locules of the ovary; leaves petiolate, with an upper pulvinus; flowers in mirror-image pairs, the style held under pressure by modified and hooded staminode, released during pollination; fruit not warty ....**Marantaceae**

      6. Ovules ± numerous in each of the 3 locules of the ovary; leaves ± lacking a petiole and upper pulvinus; flowers not in mirror-image pairs, the style not held under pressure or forcefully moving during pollination; fruit warty ....**Cannaceae**

Flowers of Zingiberaceae are diverse in color and form and are mainly pollinated by bees, moths, butterflies, and birds. Many species are outcrossing, but self-pollination and vegetative reproduction also occur. Birds are the most common dispersal agents; the fleshy capsules are usually colorful and often contrast with the brightly colored arillate seeds.

Additional references: Burtt and Smith 1972; Dahlgren et al. 1985; Larsen et al. 1998.

Marantaceae R. Brown
(Prayer Plant Family)

Herbs, with erect stem and short, tuberlike, starchy rhizomes. Hairs simple and surrounded by inflated epidermal cells. *Leaves* alternate, *usually 2-ranked*, simple, entire, petiolate, **with an upper pulvinus**, *a well-developed blade* **that folds upward at night**, *and pinnate venation* **with sigmoid secondary veins and evenly spaced cross-veins**, sheathing at base, lacking a ligule; petiole with air canals, these separated into segments by diaphragms composed of stellate-shaped cells; stipules lacking. Inflorescences determinate, but often appearing indeterminate, terminal. *Flowers bisexual, lacking a plane of symmetry* **but arranged in mirror-image pairs**. *Sepals 3, distinct*, imbricate. *Petals 3, connate*, imbricate. *Stamen 1, partly fertile and partly staminodial;* filament connate with staminodes and adnate to corolla; *anther unilocular (i.e., a half-anther, the other half expanded and sterile), depositing pollen onto the style before the flower opens; staminodes usually 3 or 4, petaloid and varying in shape*, basally connate and adnate to corolla, **1 from the inner androecial whorl forming a hooded structure with 1 or 2 small appendages (i.e., the hooded or cucullate staminode that holds the curved style under tension before the flower is triggered by a floral visitor), the second from the inner androecial whorl forming a callose-thickened structure, "the callose staminode," which often serves as a landing platform for insect visitors and helps to brace the hooded**

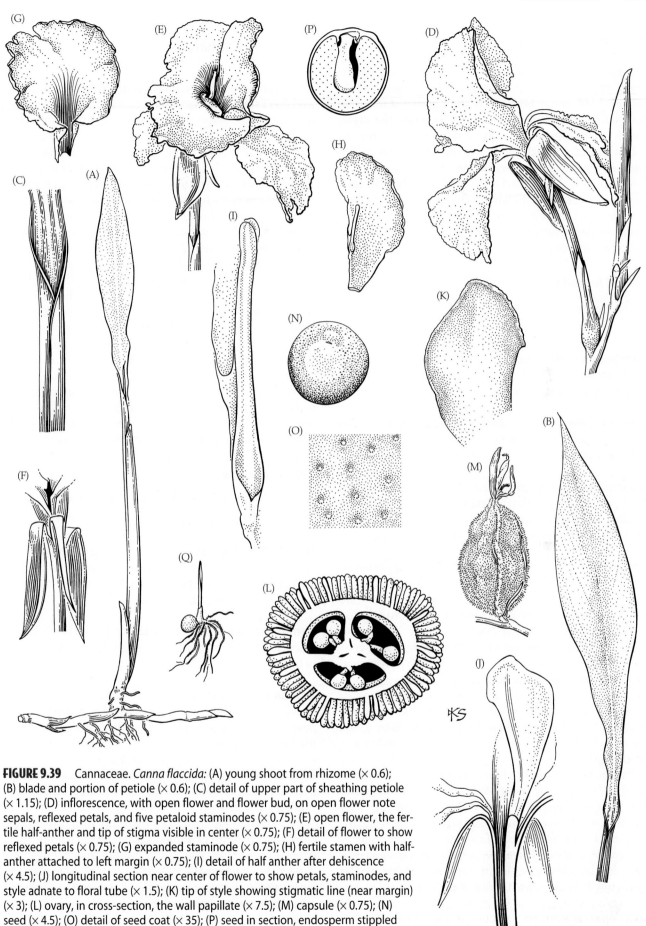

FIGURE 9.39 Cannaceae. *Canna flaccida:* (A) young shoot from rhizome (× 0.6); (B) blade and portion of petiole (× 0.6); (C) detail of upper part of sheathing petiole (× 1.15); (D) inflorescence, with open flower and flower bud, on open flower note sepals, reflexed petals, and five petaloid staminodes (× 0.75); (E) open flower, the fertile half-anther and tip of stigma visible in center (× 0.75); (F) detail of flower to show reflexed petals (× 0.75); (G) expanded staminode (× 0.75); (H) fertile stamen with half-anther attached to left margin (× 0.75); (I) detail of half anther after dehiscence (× 4.5); (J) longitudinal section near center of flower to show petals, staminodes, and style adnate to floral tube (× 1.5); (K) tip of style showing stigmatic line (near margin) (× 3); (L) ovary, in cross-section, the wall papillate (× 7.5); (M) capsule (× 0.75); (N) seed (× 4.5); (O) detail of seed coat (× 35); (P) seed in section, endosperm stippled (× 4.5); (Q) seedling (× 0.75). (From Rogers 1984, *J. Arnold Arb.* 65: p. 30.)

staminode, and 1 or 2 from the outer androecial whorl ± petal-like; pollen grains lacking apertures, exine very reduced. Carpels 3, connate; *ovary inferior*, with axile placentation, but 2 carpels often sterile and ± reduced; **style curved, held under tension by the hooded staminode that, when triggered by an insect, releases the style, which then elastically curves downward, scraping pollen from the insect's body and dusting it with pollen (held in a small cavity below the stigma)**; stigma 1, in depression between 3-lobed apex of style. **Ovules 1 in each locule, or solitary in the single functional carpel.** Nectaries in septa of ovary. *Fruit a loculicidal capsule or berry*; seeds usually arillate; embryo usually curved; endosperm and perisperm present.

Floral formula:

$, K3, C ③, A1• or 2•+2• + (½ + ½•, G ③; capsule, berry

Distribution and ecology: Widely distributed in tropical and subtropical regions. Most occur in tropical rain forest margins and clearings or in wetlands.

Genera/species: 30/450. **Major genus:** *Calathea* (250 spp.). The family is represented in the continental United States only by *Thalia* (native) and *Maranta* (rarely naturalized).

Economic plants and products: West Indian arrowroot starch is obtained from the rhizomes of *Maranta arundinacea*. *Calathea, Ctenanthe, Maranta* (arrowroot, prayer plant), and *Thalia* are cultivated because of their decorative leaves.

Discussion: Monophyly of Marantaceae is supported by DNA and morphology (Kress 1990, 1995; Smith et al. 1993). Phylogenetic relationships within the family are poorly understood, but have recently been investigated based on morphology and DNA sequences (Andersson and Chase 2001; Prince and Kress 2006). Traditionally, genera with one functional carpel (e.g., *Ischnosiphon, Maranta,* and *Thalia*) have been placed in the Maranteae, while genera with 3-locular ovaries (e.g., *Calathea* and *Marantochloa*) were placed in the Phrynieae. Both tribes, however, are polyphyletic. *Sarcophrynium* and relatives are sister to the rest of the family and have a single, bilobed hooded staminode appendage.

The complex flowers of Marantaceae are mainly bee-pollinated and outcrossing; nectar provides the pollinator reward. Arils associated with the seeds of Marantaceae are often brightly colored and contain deposits of lipids, suggesting dispersal by birds or ants. Fruits of *Thalia* are water dispersed.

Some Marantaceae have deep red abaxial leaf surfaces, a possible adaptation for efficient use of light in the shaded understory habitats where they commonly grow.

Additional references: Andersson 1981, 1998; Classen-Bockhoff 1991; Kennedy 2000; Rogers 1984.

Cannaceae A. L. de Jussieu
(Canna Family)

Rhizomatous herbs; **mucilage canals present in rhizome and erect stem**. Plants glabrous. Leaves alternate, 2-ranked to spiral, simple, entire, ± lacking a petiole, with a well-developed blade, the midvein of which possesses air canals, with pinnate venation, sheathing at base, lacking a ligule, pulvinus, and stipules. Inflorescences determinate or indeterminate, terminal, the main axis triangular in cross-section, with 3-ranked bracts, each bract usually associated with reduced, 2- or 1-flowered cymes. Flowers bisexual, lacking a plane of symmetry, often lasting only 1 day. Sepals 3, distinct, imbricate. Petals 3, connate, imbricate. Stamen 1; filament connate with staminodes and adnate to corolla; anther unilocular (i.e., a half-anther, the other half expanded and sterile); staminodes usually 3 or 4, petaloid, 1 larger than the others and recurved, all basally connate and adnate to corolla; pollen grains lacking apertures, exine very reduced. Carpels 3, connate; ovary inferior, **externally papillate**, with axile placentation; **style flattened and ± petaloid**; stigma 1, extending along one edge of the style. Ovules ± numerous in each locule. Nectaries in septa of ovary. **Fruit a warty capsule**, usually splitting irregularly by disintegration of the fruit wall; seeds spherical, black, associated with a tuft of hairs (modified aril); endosperm and perisperm present (Figure 9.39).

Floral formula:

$, K3, C ③, A1• or 2•+ 2• + (½ + ½•, G ③; warty capsule

Distribution and ecology: Occurring in tropical and subtropical regions of the Americas; some species naturalized in the Old World; plants of moist openings in tropical forests, along rivers, or in wetlands.

Genera/species: 1/19. **Genus:** *Canna*. The family is represented in the continental United States by a few species of *Canna*.

Economic plants and products: The rhizomes of *Canna edulis* (Queensland arrowroot) are used as a starch source. Several species and various hybrids (e.g., *C.× generalis*) are used as ornamentals.

Discussion: Monophyly of the Cannaceae has been supported by DNA and morphology (Kress 1990, 1995; Smith et al. 1993).

Pollen is deposited on the style before the flower opens, either directly onto or somewhat below the stigma. Most species are selfing. Pollination biology has been poorly studied, but nectar-gathering bees, butterflies, moths, and birds may be the most frequent pollinators. The seeds are often water dispersed and long-lived.

Additional references: Kubitzki 1998c; Rogers 1984.

EUDICOTS (TRICOLPATES)

This large group is considered to be monophyletic on the basis of tricolpate pollen (or conditions derived from this pollen type) as well as DNA nucleotide sequence characters (Chase et al. 1993; Hilu et al. 2003; Hoot et al. 1999; Judd and Olmstead 2004; Kim et al. 2004; Savolainen et al. 2000a,b; Soltis et al. 1997, 1998, 2000, 2003, 2005; Zanis et al. 2003). The clade is also characterized by cyclic flowers, that is, the parts are arranged in whorls, with the members of individual whorls alternating. The presence of differentiated outer and inner perianth members (i.e., a calyx and corolla) may be an additional, albeit homoplasious synapomorphy. The staminal filaments are usually slender, bearing well differentiated anthers. Most members have plastids of sieve elements with starch grains (S-type). The major clades of tricolpates are shown in Figures 9.3 and 9.4.

Ranunculales

Ranunculales are hypothesized to be monophyletic based on *rbcL, atpB, matK*, 18S and 26S rDNA sequences (Chase et al. 1993; Drinnan et al. 1994; Hilu et al. 2003; Hoot and Crane 1995; Hoot et al. 1999; Källersjö et al. 1998; Kim et al. 2004; Loconte et al. 1995; Savolainen et al. 2000a, b; Soltis et al. 1997, 1998, 2000). The presence of alkaloids of the berberine and morphine types and wood with reduced fiber-pit borders may also be synapomorphic. The order consists of 7 families and about 3490 species; major families include **Menispermaceae**, **Berberidaceae**, **Ranunculaceae**, and

Papaveraceae. These families traditionally have been associated because of their predominantly herbaceous habit; toothed to lobed or even compound leaves; the presence of alkaloids, typically of the benzyl-isoquinoline type; hypogynous flowers with the parts usually distinct and free, and often with many stamens; and seeds with tiny embryos and copious endosperm (Cronquist 1981; Thorne 1974, 1992). The group has often been associated with the magnoliid complex (i.e., Magnoliales, Laurales, and Canellales: Cronquist 1968, 1981, 1988; Dahlgren 1983; Takhtajan 1969, 1980, 1997; Thorne 1974, 1992), but morphology (Donoghue and Doyle 1989; Doyle et al. 1994) and DNA sequences (see Judd and Olmstead 2004) indicate that Ranunculales are sister to the remaining members of the tricolpate clade.

The woody Eupteleaceae and mainly herbaceous Papaveraceae are probably successively sister to the remaining families (Hoot and Crane 1995; Kim et al. 2004; Soltis et al. 2000; Thorne 1974) (Figure 9.40). Papaveraceae differs from the remaining members of the order in its syncarpous gynoecium, capsular fruits, quickly deciduous sepals, and presence of either laticifers and colored sap or specialized cells with clear, mucilaginous sap. These characters, along with arillate seeds, are probably synapomorphic for Papaveraceae (or evolved very early in the evolution of this family). Morphological synapomorphies of the clade comprising Berberidaceae, Menispermaceae and Ranunculaceae are not readily apparent; the monophyly of this group is supported, however, by *rbcL, atpB*, and 18S sequences (Chase et al. 1993; Drinnan et al. 1994; Hoot and Crane 1995; Hoot et al. 1999; Soltis et al. 2000). Various relation-

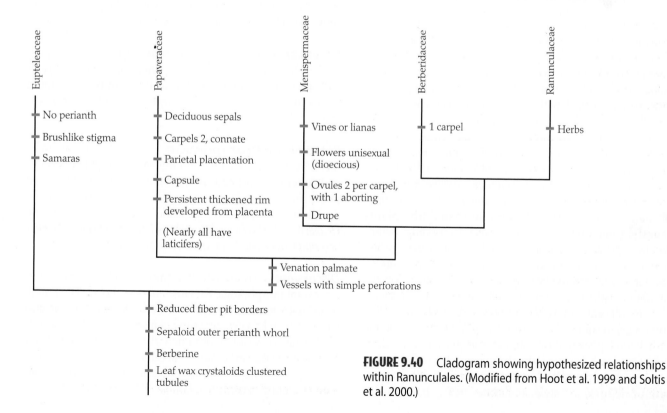

FIGURE 9.40 Cladogram showing hypothesized relationships within Ranunculales. (Modified from Hoot et al. 1999 and Soltis et al. 2000.)

Key to Major Families of Ranunculales

1. Gynoecium of 2 to many connate carpels, ovary with 1 locule and parietal placentas; fruits capsules; plants with clear, white, or colored sap . . . . **Papaveraceae**

1. Gynoecium of 1 to many distinct carpels, each ovary with lateral to ± basal placenta; fruits berries, follicles, or achenes; plants with clear sap (although wood may be yellow) or sap lacking 2

2. Flowers usually bisexual; plants mostly herbs or shrubs, rarely lianas; fruits achenes, follicles, or berries 3

2. Flowers usually unisexual; plants usually lianas; fruits drupes . . . . **Menispermaceae**

3. Carpels solitary; stamens usually as many or twice as many as and opposite the petals; anthers usually opening by 2 flaps; perianth 3-merous; fruits usually fleshy, berries . . . . **Berberidaceae**

3. Carpels usually numerous; stamens numerous, spirally arranged; anthers opening by 2 longitudinal slits; perianth usually not 3-merous; fruits usually dry, achenes or follicles . . . . **Ranunculaceae**

---

ships of these three families have been proposed (see Drinnan et al. 1994; Hoot and Crane 1995; Loconte and Estes 1989; Loconte and Stevenson 1991; Loconte et al. 1995). Berberidaceae apparently are sister to Ranunculaceae, as supported by nucleotide sequences (Hoot and Crane 1995; Hoot et al. 1999; Kim and Jansen 1995) and similarities in floral form (Endress 1995).

Additional reference: Endress and Igersheim 1999.

Menispermaceae A. L. de Jussieu
(Moonseed Family)

Twining vines or lianas, or less commonly shrubs or small trees; young stems with ring of vascular bundles separated by broad parenchymatous rays, *older stems often with anomalous secondary growth (successive rings of xylem and phloem) and flattened*; with various alkaloids (including the benzyl-isoquinoline type) and poisonous and very bitter sesquiterpenoids and diterpenoids. Hairs simple, sometimes glandular. *Leaves alternate and spiral, usually simple and entire, occasionally lobed or palmately compound, with usually palmate venation, often with an upper and lower pulvinus, often peltate or nearly so*; stipules lacking. Inflorescences usually determinate, axillary. **Flowers unisexual (the plants usually dioecious)**, radial, *with 3-merous perianth*. Sepals usually 6, distinct, imbricate or valvate. Petals usually 6, distinct or connate, imbricate. Stamens (3-) 6-numerous, sometimes represented by staminodes in carpellate flowers; filaments distinct or connate; pollen grains usually tricolpate, tricolporate, or variously porate. *Carpels (1-) 3-6 (-numerous)*, distinct, often on a gynophore; ovaries superior, with lateral placentation; stigma various, on very short style. **Ovules 2, but one aborting**, sometimes with a thin megasporangium wall. Nectaries lacking. **Fruit an aggregate of drupes**, the style sometimes nearly basal due to

asymmetrical growth, *with endocarp ± curved, variously sculptured, usually laterally compressed; seeds curved; embryo usually curved*; endosperm sometimes ruminate, sometimes lacking (Figure 9.41).

Floral formula:

Staminate: * K6, C⑥, A⟨6–∞⟩, G0

Carpellate: * K6, C⑥, A⟨6–∞⟩, G3–6, drupes

Distribution: Nearly pantropical, with a few genera extending into warm temperate regions; the group is characteristic of lowland tropical rain forests.

Genera/species: 71/450. **Major genera:** *Abuta* (35 spp.), *Cyclea* (30), *Stephania* (30), *Tinospora* (25), and *Cissampelos* (20). The family is represented in the continental United States and Canada by *Calycocarpum, Cocculus, Menispermum*, and *Cissampelos*.

Economic plants and products: Several species are used medicinally or as poisons, and the bark of members of several genera (including *Chondrodendron*; Plate 9.6A) is used in the preparation of curare, an arrow poison used by some South American Indian tribes, and medicinally as a muscle relaxant. A few species of *Cocculus* (coralbeads) and *Menispermum* (moonseed) are used as ornamentals.

Discussion: Monophyly of Menispermaceae is supported by several morphological characters (see above). The family is sister to Ranunculaceae and Berberidaceae (Hoot and Crane 1995). Lardizabalaceae are also vines or lianas (e.g., *Akebia*, an occasional ornamental), but they differ from Menispermaceae in having compound leaves, several ovules per carpel, and fleshy follicles or berries. Dioecy has evolved independently in these two families.

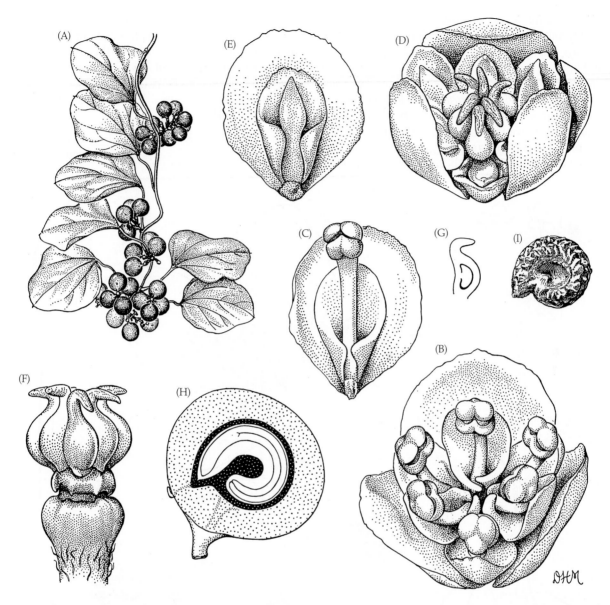

FIGURE 9.41 Menispermaceae. *Cocculus carolinus*: (A) pendent branch with fruit (× 0.6); (B) staminate flower, showing inner sepals, two series of petals, and stamens, the short outer sepals not visible (× 18); (C) inner sepal, clasping petal, and stamen (× 18); (D) carpellate flower, outer sepal in foreground removed to show inner sepals, two series of petals, staminodia, and carpels (× 18); (E) inner sepal, clasping petal, and staminode (× 18); (F) gynoecium, the carpels with reflexed stigmas (× 18); (G) vertical section of carpel (× 12); (H) vertical section of fruit, showing coiled embryo surrounded by endosperm (white), bony endocarp (dark), and fleshy fruit wall (× 4.8); (I) endocarp (× 3.6). (From Ernst 1964, *J. Arnold Arbor.* 45: p. 31.)

Relationships within Menispermaceae are unclear, but the clade is usually divided into 5–8 tribes on the basis of fruit and seed characters (endosperm present or absent, ruminate or homogeneous; embryo form; endocarp shape and sculpturing).

Pollination has not been studied but is probably by insects. The fleshy fruits are black, blue, purple, or red, small to quite large, and are presumably dispersed by birds and mammals.

Additional references: Ernst 1964; Kessler 1993b; Loconte et al. 1995; Mennega 1982; Thanikaimoni 1986.

Ranunculaceae A. L. de Jussieu
(Buttercup Family)

Herbs, shrubs, or occasionally vines; stems with vascular bundles often in several concentric rings or ± scattered; usually with alkaloids or ranunculin (a lactone glycoside); often with triterpenoid saponins. Hairs usually simple. *Leaves usually alternate and spiral,* occasionally opposite, *simple, sometimes lobed or dissected, to compound, usually serrate, dentate, or crenate,* with pinnate to occasionally palmate venation; *stipules usually lacking.* Inflorescences determinate, sometimes appearing indeterminate or reduced to a single flower, terminal. Flowers usually bisexual, radial to occasionally bilateral, with short to elongate receptacle. *Perianth parts usually not 3-merous. Tepals 4 to numerous, distinct,* and imbricate; *or perianth differentiated into calyx and corolla, then sepals usually 5, distinct, imbricate, and deciduous, and petals usually 5, distinct, imbricate, often with nectar-producing basal portion* or represented only by small nectar glands, probably derived from staminodes. *Stamens numerous;* filaments distinct; *anthers opening by longitudinal slits;* pollen grains tricolpate (or ± modified). *Carpels usually 5 to numerous,* occasionally reduced to 1, *usually distinct;* ovaries superior, with usually lateral placentation; stigmas punctate or extending along one side of the style. Ovules 1 to numerous per carpel. *Fruit usually an aggregate of follicles or achenes,* occasionally a berry (Figure 9.42).

Floral formula:

* or X, T-4–∞- or K5, C5, A∞, G<u>1–∞</u>; follicles, achenes, berries

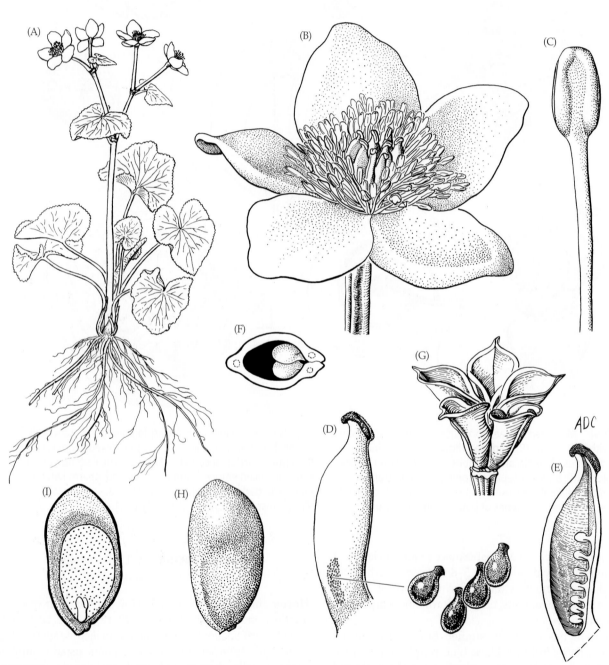

FIGURE 9.42 Ranunculaceae. Caltha palustris: (A) flowering plant (× 0.5); (B) flower (× 3); (C) stamen (× 12); (D) lateral view of carpel, with detail of nectar glands at base (× 10); (E) carpel in longitudinal section (× 10); (F) carpel in cross-section (× 15); (G) follicles from a five-carpellate flower (× 2); (H) seed (× 20); (I) seed in longitudinal section, note spongy seed coat, endosperm (stippled), and minute embryo (× 20). (From Wood 1974, *A student's atlas of flowering plants,* p. 29.)

(A)

Ranunculales: Menispermaceae
Chondrodendron tomentosum:
flowers

(B)

Ranunculales: Berberidaceae
Jeffersonia diphylla: young flower

(C)

Proteales: Platanaceae
Platanus occidentalis:
branch with inflorescences

(D)

Ranunculales: Papaveraceae
Argemone albiflora: flowers

A. albiflora: young fruit

(E)

Proteales: Proteaceae
Grevillea robusta: leaves and flowers

(F)

Ranunculales: Ranunculaceae
Ranunculus ficaria: flower

(G)

Ranunculales: Papaveraceae
Dicentra spectabilis: flowers

**PLATE 9.6 Eudicots:
Ranunculales and Proteales**

Distribution: Widespread, but especially characteristic of temperate and boreal regions of the Northern Hemisphere.

Genera/species: 47/2000. **Major genera:** *Ranunculus* (400 spp.), *Aconitum* (250), *Clematis* (250), *Delphinium* (250), *Anemone* (150), and *Thalictrum* (100). Some of the numerous genera in the continental United States and/or Canada (in addition to those listed above) are *Actaea, Aquilegia, Caltha, Coptis, Hydrastis, Isopyrum, Myosurus, Trollius,* and *Xanthorhiza.*

Economic plants and products: The family is chiefly important for its numerous ornamental herbs, such as *Anemone* (windflower, including *Hepatica*), *Aconitum* (monkshood), *Actaea* (baneberry, including *Cimicifuga*), *Aquilegia* (columbine), *Caltha* (marsh marigold), *Clematis* (virgin's bower), *Delphinium* (larkspur), *Helleborus* (hellebore), *Ranunculus* (buttercup), *Thalictrum* (meadow rue), and *Trollius* (globeflower). A number of genera are highly poisonous.

Discussion: The monophyly of this group is supported by morphology and nucleotide sequences (Drinnan et al. 1994; Hoot and Crane 1995; Hoot 1995; Jensen et al. 1995; Loconte and Estes 1989; Wang et al. 2005). *Hydrastis* has flowers with a 3-merous perianth, vessels with scalariform perforations, ovules with micropyle defined by two integuments, and fleshy follicles; cpDNA restriction sites and sequence data also support the position of this genus, along with *Glaucidium,* as sister to the rest of the family (Hoot 1995; Johansson and Jansen 1993). The remaining genera of Ranunculaceae may form a clade based on vessels with simple perforations, 4- or 5-merous perianth, and dry fruits. A large clade (Ranunculoideae) is supported by the synapomorphies of large chromosomes, stomates longer than 35 μm, a chromosome number of 8, and nucleotide characters (Hoot 1991, 1995). In addition, the alkaloid berberine has not been reported within this clade, and its loss may be synapomorphic. *Clematis, Ranunculus, Trautvetteria,* and *Anemone* probably form a subclade based on the synapomorphies of achenes, presence of ranunculin, and DNA sequence data (Hoot 1991, 1995; Johansson 1995; Johansson and Jansen 1993). A second subclade, including *Actaea, Caltha, Helleborus, Trollius, Nigella, Aconitum,* and *Delphinium,* may be diagnosed on the basis of petal-like tepals, but this grouping is not supported by nucleotide sequences (Hoot 1991, 1995). The remaining genera—*Xanthorhiza, Coptis, Aquilegia, Thalictrum,* and *Isopyrum*—constitute the "Thalictroideae," which are paraphyletic and comprise the early-diverging lineages within Ranunculaceae (along with *Hydrastis* and *Glaucidium*). These genera retain plesiomorphies—the presence of berberine, yellow creeping rhizomes, short, thin-walled hairs, and small chromosomes—found in Berberidaceae. Nucleotide sequences suggest that *Coptis* and *Xanthorhiza* form a clade that is sister to a clade containing the remaining genera of "Thalictroideae" + Ranunculoideae (Hoot 1995; Johansson and Jansen 1993; Wang et al. 2005).

Reduction in the number of ovules per carpel and the evolution of achenes have occurred several times within Ranunculaceae; petaloid staminodes also have evolved more than once, as suggested by their differing morphology, along with anatomical and developmental evidence (Hoot 1995). Woodiness probably is secondary within the family.

The wide range of floral structures within Ranunculaceae is associated with a diverse array of pollination syndromes. Most species are insect-pollinated, although some species of *Thalictrum* are wind-pollinated. *Anemone* and *Clematis* do not produce nectar and are pollinated by various pollen-gathering insects. In contrast, *Ranunculus, Delphinium,* and *Aquilegia* have modified nectar-secreting petals (sometimes spurred; Plate 9.6F), and their flowers are visited by nectar-gathering insects (mainly bees) or hummingbirds. *Caltha* has nectar glands at the base of the carpels and is also bee-pollinated. *Trollius europaeus* is exclusively pollinated by flies of the genus *Chiastochaeta* that enter the spherical flowers (which never open) to mate and feed on pollen and nectar; the females then lay their eggs (usually only one per flower) and the larvae feed on developing seeds.

Dispersal mechanisms vary widely. Achenes of *Clematis* have persistent, long, hairy styles, and are dispersed by wind, while those of *Ranunculus* may contain tubercules or hooked spines, leading to external transport by animals. The small seeds of follicular species may be dispersed by wind or water, and some (e.g., *Helleborus*) are secondarily dispersed by ants. The berries of some species of *Actaea* are mainly bird-dispersed.

Berberidaceae A. L. de Jussieu
(Barberry Family)

Herbs or shrubs; stems with vascular bundles sometimes ± scattered; usually with alkaloids, *the wood usually colored yellow by berberine* (an isoquinoline alkaloid). Hairs simple. *Leaves usually alternate and spiral* (opposite in *Podophyllum*), *simple, sometimes lobed or dissected, to compound,* sometimes reduced and unifoliolate (some species of *Berberis*), entire to serrate or spinose-serrate, sometimes reduced to spines, with pinnate to palmate venation; stipules lacking or present. Inflorescences various. Flowers bisexual, radial, with *perianth usually 3-merous.* Sepals usually 6, occasionally 4, distinct, imbricate. *Outer petals 6 (occasionally 4), distinct, imbricate, lacking nectar glands,* occasionally lacking; *inner petals (probably petal-like staminodes) usually 6, nectar-producing,* showy or represented only by small scales, sometimes lacking. *Stamens 4 to numerous, but most often 6, usually opposite petals;* filaments usually distinct; *anthers opening by 2 flaps that open from the base* (longitudinal slits in *Nandina* and *Podophyllum*); pollen grains tricolpate (or modified). **Carpel 1**; ovary superior, with lateral to basal placentation; stigma capitate to 3-lobed. Ovules usually numerous, sometimes reduced to 1. *Fruit usually a berry,* sometimes ± dehiscent; seeds often arillate (Figure 9.43).

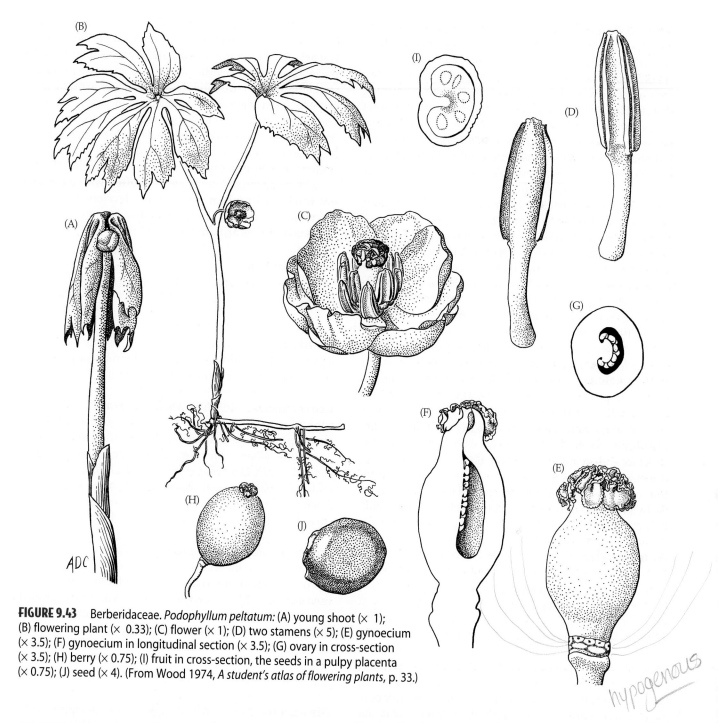

FIGURE 9.43 Berberidaceae. *Podophyllum peltatum:* (A) young shoot (× 1); (B) flowering plant (× 0.33); (C) flower (× 1); (D) two stamens (× 5); (E) gynoecium (× 3.5); (F) gynoecium in longitudinal section (× 3.5); (G) ovary in cross-section (× 3.5); (H) berry (× 0.75); (I) fruit in cross-section, the seeds in a pulpy placenta (× 0.75); (J) seed (× 4). (From Wood 1974, *A student's atlas of flowering plants,* p. 33.)

Floral formula:

*, K4–6, C4–6, A4–6• + 4–∞, G1; berry

Distribution: Widespread, especially in temperate regions of the Northern Hemisphere and the Andes of South America.

Genera/species: 15/650. **Major genus:** *Berberis* (600 spp.). *Achlys, Berberis, Caulophyllum, Diphylleia, Jeffersonia, Nandina, Podophyllum,* and *Vancouveria* occur in the continental United States and/or Canada.

Economic plants and products: Genera such as *Berberis* (barberry, Oregon grape) and *Nandina* (sacred bamboo) contain valuable ornamentals. Many are extremely poisonous.

Discussion: Monophyly of Berberidaceae is supported by morphology and DNA-based analyses (Hoot and Crane 1995; Loconte 1993; Loconte and Estes 1989; Loconte et al. 1994). *Nandina* (in Nandinoideae) is sister to the remainder of the family (Berberidoideae), which are united by their herbaceous habit and anthers that open by two flaps. Several clades can be recognized within Berberidoideae (Kim and Jansen 1998a; Loconte and Estes 1989; Meacham 1980). *Leontice, Gymnospermum,* and *Caulophyllum* are characterized by flabelliform nectar-secreting petals (staminodes), pollen with reticulate sculpturing, basal placentation, and a chromosome base number of 8. The remaining genera of Berberidoideae form a clade on the basis of carpels with several ovules. Within this clade, a group containing *Ranzania* and *Berberis* may be diagnosed by having inner petals with paired basal glands, sensitive stamens, and a base chromosome number of seven. *Berberis* has the apomorphies of secondary woodiness, spinose leaf teeth, irregular pollen apertures, basal placentation, and a large embryo; some species of this genus, often segregated as *Mahonia*, have pinnately compound leaves. *Podophyllum*, *Diphylleia*, and relatives form a clade having stems with scattered vascular bundles and simple leaves with the veins ending in the teeth. Finally, *Epimedium*, *Vancouveria*, *Jeffersonia*, and relatives may form a monophyletic group supported by leaves in a basal rosette and fruits that open by a horizontal slit. Chloroplast DNA restriction sites suggest, however, that the last group may be paraphyletic (Kim and Jansen 1995, 1998a).

The showy flowers of Berberidaceae are pollinated by various insects (mainly bees) that gather nectar or pollen. In *Berberis* the stamens are sensitive, and when touched by a bee as it probes for nectar, they spring inward, and the pollen-bearing flaps contact the insect's head. The often colorful berries of Berberidaceae typically are bird- or mammal-dispersed. Some have seeds with a hard, aril-like structure and are insect-dispersed. Fruits of *Vancouveria* are dispersed by wasps. *Caulophyllum* is distinct because the fruit ruptures early in development and the large, globose, fleshy blue seed develops in a completely exposed condition.

Additional references: Ernst 1964; Thorne 1974.

Papaveraceae A. L. de Jussieu
(Poppy Family)

Herbs to soft-wooded shrubs; stem with vascular bundles sometimes in several rings; *with laticifers present and plants with white, cream, yellow, orange, or red sap, or with specialized elongated secretory cells and sap then mucilaginous, clear;* sap with various alkaloids (including the benzyl-isoquinoline type). Hairs simple. *Leaves usually alternate and spiral, simple,* **but often lobed or dissected,** *entire to more commonly variously toothed, sometimes spinose,* with ± pinnate venation; *stipules lacking.* Inflorescences various. Flowers bisexual, radial (with numerous or only 2 main planes of symmetry) to bilateral. **Sepals usually 2** *or 3,* usually distinct, imbricate, usually **quickly deciduous,** large and surrounding bud to small and bractlike. **Petals** usually **4** *or 6,* sometimes numerous, distinct, imbricate *and often crumpled in bud and thus wrinkled when expanded; often the 2 (or 3) inner differentiated from the 2 (or 3) outer,* and sometimes with 1 or 2 of the outer petals with a prominent basal nectar spur or pouch and the 2 inner sticking together at apex, forming a cover over the stigmas. *Stamens numerous, to 6 that are ±*

connate in 2 groups of 3, rarely reduced to 4; filaments distinct to connate; pollen grains tricolporate to polyporate. **Carpels 2** to numerous, **connate; ovary** superior, **with parietal placentation**, the placentas sometimes intruded; stigma(s) distinct to connate, 1 or equaling number of carpels, often discoid and lobed, sometimes capitate. Ovules usually numerous, but sometimes reduced to 1 or 2, often amphitropous to campylotropous. Nectaries lacking, or sometimes one or more of the filaments with a basal nectar gland. **Fruit a capsule,** *opening variously, but often by apical pores, valves, or longitudinal slits,* **often with a persistent thickened replum** (or replums, developed from the placenta, but often lost), occasionally a nut or lomentlike; **seeds sometimes arillate** (Figure 9.44).

Floral formula:

* or X, K2–3, C4–6 (–∞), A 4–∞ or ③+③, G 2–∞ ;
capsule

Distribution: Widely distributed in mainly temperate regions; especially diverse in the Northern Hemisphere, but also in southern Africa and eastern Australia.

Genera/species: 40/770. **Major genera:** *Corydalis* (400 spp.), *Papaver* (100), *Fumaria* (50), and *Argemone* (30). Genera occurring in the continental United States and/or Canada include *Adlumia*, *Arctomecon*, *Argemone*, *Canbya*, *Chelidonium*, *Corydalis*, *Dendromecon*, *Dicentra*, *Eschscholzia*, *Fumaria*, *Glaucium*, *Hesperomecon*, *Meconella*, *Papaver*, *Platystemon*, *Romneya*, *Sanguinaria*, and *Stylophorum*. Delimitation of genera is often difficult (see Jork and Kadereit 1995).

Economic plants and products: *Papaver somniferum* (opium poppy) is the source of opium and derivatives such as morphine, heroin, and codeine; the seeds of this species are used as a spice. Many have showy flowers and are cultivated as ornamentals, such as species of *Argemone* (prickly poppy; Plate 9.6D), *Eschscholzia* (California poppy), *Papaver* (poppy), *Macleaya* (plume poppy), *Corydalis* (harlequin), *Sanguinaria* (bloodroot), and *Dicentra* (Dutchman's breeches, bleeding heart; Plate 9.6G). Most species are highly poisonous.

Discussion: Monophyly of Papaveraceae is supported by morphology and nucleotide sequence data (Drinnan et al. 1994; Hoot and Crane 1995; Kadereit et al. 1994, 1995; Loconte et al. 1995). The family is broadly circumscribed here, including Fumariaceae (see below).

Loconte et al. (1995) proposed (on the basis of morphological data) that subfamily Platystemonoideae is sister to the rest; *Platystemon* and relatives show numerous and only slightly fused carpels (with free stigmas) that separate from one another at maturity. In contrast, Kadereit et al. (1995) and Hoot et al. (1997) proposed (on the basis of morphological and nucleotide sequence data) that *Pteridophyllum* is sister to the remaining genera; this genus lacks secretory cells, while the remaining genera are united by the posses-

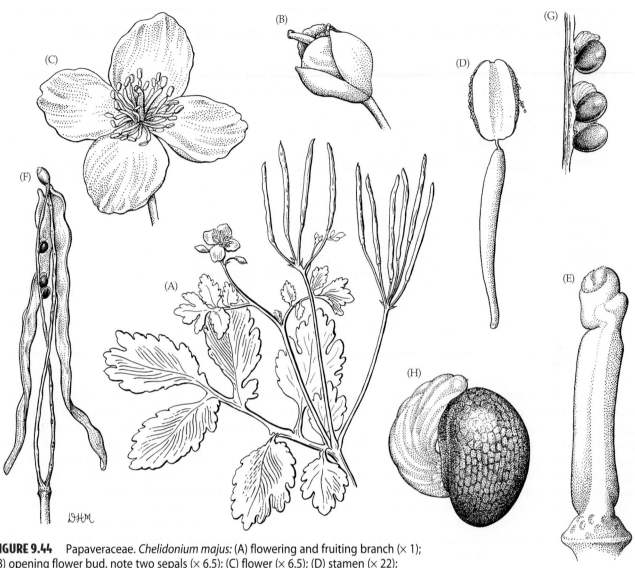

FIGURE 9.44 Papaveraceae. *Chelidonium majus:* (A) flowering and fruiting branch (× 1); (B) opening flower bud, note two sepals (× 6.5); (C) flower (× 6.5); (D) stamen (× 22); (E) gynoecium (× 22); (F) two-valved capsule, note persistent replum (× 4.5); (G) seeds attached to rim (× 9); (H) arillate seed (× 44). (From Ernst 1962, *J. Arnold Arbor.* 43: p. 325.)

sion of laticifers or idioblasts, a presumed synapomorphy. The differences between these hypotheses, the latter of which has more support, is largely on how rather similar topologies are rooted (see Blattner and Kadereit 1995; Judd et al. 1994; Schwarzbach and Kadereit 1995).

Within the laticifer-idioblast clade, the Fumarioideae (e.g., *Dicentra*, *Corydalis*, and *Fumaria*), often recognized (at least by many American systematists) as Fumariaceae (Cronquist 1981), are sister to the remaining subfamilies, comprising genera such as *Papaver*, *Argemone*, *Meconopsis*, *Eschscholzia*, *Chelidonium*, and *Sanguinaria* (Kadereit et al. 1995; Hoot et al. 1997). The monophyly of the Fumarioideae is supported by stamens fused into two groups of three (diadelphous) in which the central stamen of each group has two locules and the two lateral stamens have only a single locule, staminal nectaries, and flowers in which the two outer petals look very different from the two inner petals (which are more or less connate over the stigma). In addition, one or both of the outer petals are saccate or spurred. Nonfumarioid members of the laticifer-idioblast clade, Papaveraceae s.s., form a clade with the apomorphy of numerous stamens. Within this group, the Chelidonioideae (e.g., *Chelidonium*, *Sanguinaria*) are possibly sister to the rest. Both Fumarioideae and Chelidonioideae have elongated, consistently two carpellate fruits with a replum (Figure 9.44F). Papaveroideae (e.g., *Papaver*, *Argemone*, and *Meconopsis*) are more derived and are diagnosed by their many-carpellate flowers and multicellular-multiseriate hairs. *Platystemon* and relatives are placed within Papaveroideae by Kadereit et al. (1995). *Eschscholzia* (Eschcholzioideae) are distinct due to their unicellular hairs, polycolpate pollen, and explosive fruits.

Flowers of Papaveraceae attract bees, wasps, and flies as pollinators. Members of Papaveroideae and Chelido-

---

Key to Families of Proteales and some Phylogenetically Adjacent Eudicots

---

1. Plants aquatic with submerged rhizome; receptacle expanded with individual carpels sunken into its flattened apex; fruits nuts, embedded in an accrescent, spongy receptacle .... Nelumbonaceae

1. Plants of terrestrial habitats; receptacle not expanded; carpels free or connate; fruits achenes, follicles, or capsules 2

2. Vessels lacking; carpels with adaxial portion forming a nectar gland .... Trochodendraceae (incl. Tetracentraceae)

2. Vessels present; carpels lacking adaxial nectar gland 3

3. Carpels distinct, or carpel solitary; fruit not a capsule 4

3. Carpels at least partially connate; fruit a capsule .... Buxaceae

4. Perianth conspicuous, petaloid, and plants mainly animal pollinated; flowers usually bisexual; stamens 4, usually adnate to the perianth; leaves very often coriaceous; fruits usually follicles .... **Proteaceae**

4. Perianth minute, and plants wind pollinated; flowers unisexual; stamens 3–7, free from perianth; leaves not coriaceous; fruits achenes associated with hairlike bristles .... **Platanaceae**

---

nioideae, which have radially symmetrical flowers with an exposed androecium, do not produce nectar and are pollinated by pollen-gathering bees, flies, and beetles. The more specialized flowers of Fumarioideae have nectar spurs or sacs and are pollinated by bees. Bees probing for nectar depress the hooded inner petals, exposing the anthers, and the insect is dusted with pollen. *Bocconia* has inconspicuous flowers with pendulous stamens and is wind-pollinated. Both outcrossing and selfing may occur.

In many Papaveraceae, the small seeds are shaken out of the capsules (see Plate 9.6D) and thus passively dispersed (through the action of wind or contact with the plant), but in *Eschscholzia* the capsules dehisce explosively. The seeds are typically arillate, the arils (Figure 9.44H) being hard and oily, leading to dispersal by ants (several genera), or fleshy and brightly colored, leading to dispersal by birds (*Bocconia*). A fatty collar is produced around the base of the nuts of *Fumaria*, another adaptation for ant dispersal.

Additional references: Ernst 1962; Kadereit 1993; Lidén 1986, 1993; Ronse Decraene and Smets 1992; Thorne 1974.

Proteales and Other Tricolpates

Various families with reduced, wind-pollinated flowers that are often aggregated into dangling inflorescences (i.e., catkins or aments) have traditionally been considered to form a group of related families called the Hamamelidae (see Cronquist 1981, 1988; Takhtajan 1980; and the historical summary in Stern 1973). As shown by numerous lines of evidence, this assemblage clearly is not monophyletic (Thorne 1973a; Donoghue and Doyle 1989; Wolfe 1989; Crane and Blackmore 1989; Hufford and Crane 1989; Huf-

ford 1992; Chase et al. 1993; Manos et al. 1993). Some families traditionally placed in the Hamamelidae, such as Platanaceae, Trochodendraceae, and Buxaceae, are early divergent lineages of the eudicots; Proteaceae and Nelumbonaceae are probably also related. Morphology, *rbcL, atpB, matK,* 18S and 26S rDNA sequences support this position, with **Proteaceae**, **Platanaceae**, and Nelumbonaceae forming a clade, here called Proteales (Hilu et al. 2003; Hoot et al. 1999; Hufford 1992; Kim et al. 2004; Manos et al. 1993; Soltis et al. 1997, 2000, 2003). Gynoecia with one or two pendent ovules per carpel may be synapomorphic for Proteales. The Platanaceae + Proteaceae clade are supported by the putative synapomorphies of 4-merous flowers with stamens opposite the perianth parts and carpels with five vascular bundles, although the flowers of Platanaceae are highly reduced, making comparison difficult. These two families also have similar hair morphology (Carpenter et al. 2005). The families of Proteales, along with the phylogenetically adjacent Trochodendraceae and Buxaceae, are contrasted in the following key. Other hamamelid families with reduced flowers are placed in Saxifragales, Fagales or Malpighiales, of the core eudicots.

Platanaceae T. Lestiboudois
(Sycamore or Plane Tree Family)

Trees; **nodes multilacunar**; cyanides, *triterpenes*, and tannins present; *bark coming off in large, irregular sheets, leaving a patchy, light-colored, smooth surface.* **Hairs branched.** *Leaves alternate,* 2-ranked or spiral, *simple, usually palmately lobed and veined, ± coarsely toothed,* each tooth having a midvein that becomes attenuated toward a glandular apex, where it opens into a cavity; *petiole base expanded and* enclosing axillary bud; stipules present, usually large and conspicuous. **Inflorescences** indeterminate, a raceme of **globose heads** (Plate 9.6C), but sometimes reduced to a single head, pendulous, axillary. Flowers unisexual (**and in unisexual heads, plants monoecious**), radial, very reduced and inconspicuous. Sepals 3–7, distinct or slightly connate, minute. Petals 3–7, distinct, minute, usually lacking in carpellate flowers. Stamens 3–7; filaments very short; anthers with the connective prolonged into a peltate appendage; pollen grains tricolporate. Carpels usually 5–9, distinct; ovaries superior, **placentation apical**; stigma elongated and extending along adaxial side of recurved style. Ovule 2 per carpel, **but 1 aborting**, orthotropous. Nectaries lacking. **Fruits ± linear achenes, subtended by long bristles, in dense, globose clusters**.

Floral formula:

Staminate: *, K (3–7), C3–7, A3–7, G0

Carpellate: *, K (3–7), C0, A0, G5–9; achene

Distribution and ecology: Occurring in tropical to temperate regions of North America, south-central Europe, western Asia to Indochina; often along rivers or streams.

Genera/species: 1/7. **Genus:** Platanus.

Economic plants and products: Several species or hybrids of Platanus (plane tree, sycamore) are cultivated as ornamentals and provide timber.

Discussion: Platanus kerrii (of subg. Castaneophyllum) has evergreen leaves with pinnate venation and exposed axillary buds. It is sister to the remaining six species (of subg. Platanus), which have deciduous leaves with palmate venation and axillary buds covered by the enlarged petiole base (Feng et al. 2005). Inflorescence and achene characters are taxonomically useful in the genus (Nixon and Poole 2003). The inconspicuous and unisexual flowers of Platanus are wind-pollinated; flowering takes place in the early spring, just as the leaves are emerging. The globose cluster of achenes breaks apart in the fall, and the individual fruits, with their hair tufts, are dispersed by wind (and sometimes secondarily by water).

Additional references: Boothroyd 1930; Ernst 1963b; Hufford and Crane 1989; Kubitzki 1993d.

Proteaceae A. L. de Jussieu
(Protea Family)

Trees or shrubs; nodes trilacunar; tannins commonly present, sometimes cyanogenic. Roots not forming mycorrhizae, usually with clusters of specialized short lateral roots. Hairs simple or forked, 3-celled, with basal cells short and terminal cell elongate, often thick-walled. Leaves usually alternate and spiral, simple or pinnately compound, some-times lobed or deeply dissected, entire to serrate; stipules absent. Inflorescences determinate or indeterminate, terminal. Flowers usually bisexual, radial or bilateral, conspicuous. Tepals 4, distinct or more commonly connate, often deeply cleft on one side, or 3 tepals connate and 1 distinct, **valvate**. Stamens 4; filaments usually adnate to tepals; anthers usually with the connective prolonged and forming an appendage; pollen grains usually 3-porate or 3-colporate. **Carpel 1**, often on a stalk; ovary superior, with lateral placentation; stigma globose to elongated. Ovules 1 to numerous, anatropous to orthotropous. Fruits follicles, nuts, achenes, drupes, or samaras; seeds often winged; endosperm usually lacking.

Floral formula:

* or X, T(4), A4, G1; follicles, nut, achene, drupe

Distribution and ecology: Widespread in tropical and subtropical regions, especially southern Africa and Australia.

Genera/species: 80/1770. **Major genera:** Grevillea (200 spp.), Hakea (110), Protea (110), Helicia (80), Leucadendron (70), Banksia (50), and Leucospermum (40). The family is represented in North America only by a single introduced species of Grevillea (in Florida).

Economic plants and products: Banksia, Embothrium, Grevillea, Hakea, Protea, and Telopea are ornamentals. The seeds of Macadamia integrifolia (macadamia nuts) are edible. Many genera (Cardwellia, Euplassa, Grevillea, Roupala) yield highly decorative timbers.

Discussion: The family usually is divided into two subfamilies, Grevilleoideae (flowers in pairs) and Proteoideae (flowers single); the former has been divided by Johnson and Briggs (1975) into four subfamilies. The monophyly of Proteoideae (including genera such as Protea, Leucospermum, Leucadendron, and Conospermum) is supported by small chromosomes, ovules reduced to 1 or 2, and indehiscent, usually dry fruits. Grevilleoideae (e.g., Embothrium, Grevillea, Macadamia, Banksia, Hakea, Helicia, and Telopea) have their flowers in pairs, and their monophyly is supported by chloroplast DNA sequences (Hoot and Douglas 1998). Many have follicles and winged seeds, but indehiscent fruits with large, wingless seeds have evolved several times within the group. Bellendena, Placospermum, and Toronia may be sister to all other Proteaceae (Hoot and Douglas 1998), and Johnson and Briggs (1975) also noted their ancestral features.

Flowers of Proteaceae are pollinated by various insects, birds, and small mammals (marsupials and rodents). Species with follicles and winged seeds (e.g., Embothrium, Telopea, and Grevillea) are wind-dispersed, while those with dry to fleshy indehiscent fruits are dispersed by birds or mammals.

Additional reference: Rourke and Wiens 1977.

Core Eudicots (Core Tricolpates)

Monophyly of this large group (see Table 9.1 and Figure 9.7) is supported by *rbcL*, *atpB*, *matK*, and 18S sequences (Chase et al. 1993; Hilu et al. 2003; Hoot et al. 1999; Judd and Olmstead 2004; Savolainen et al. 2000a, b; Soltis et al. 1998, 2000, 2003; Zanis et al. 2003). *Gunnera* and *Myrothamnus* (Gunneraceae; which have 2-merous flowers) are well supported as sister to all other core eudicots. In contrast, the remaining core eudicots usually have 5-merous (or less commonly 4-merous) flowers; placentation is usually axile. It is noteworthy that phylogenetic analyses of angiosperm MADS-box genes show two gene clades within the core eudicots, i.e., euAP1 and euFUL, whereas non-core eudicot clades have only sequences similar to euFUL genes (Litt and Irish 2003). This indicates that a duplication event occurred in these floral genes in the common ancestor of the core eudicots. The euAP1 gene complex includes key regulators of floral development, which have been implicated in the specification of perianth identity.

Caryophyllales

Monophyly of the Caryophyllales (APG 1998, 2003; Stevens 2001 onwards) is supported by anther wall development, vessel-elements with simple perforations, cytochrome *c* amino acid sequences, and DNA sequence data. Their position, whether within the rosid complex, or sister to the asterid clade, is uncertain. The group includes two large clades, here recognized as the suborders Polygonineae and Caryophyl-

lineae (although they often have been treated as the orders Polygonales and Caryophyllales s.s., see Judd and Olmstead 2004), but Rhabdodendraceae are of problematic placement.

Most members of the Caryophyllineae belong to a well-supported clade, the core Caryophyllineae, as evidenced by numerous distinctive synapomorphies, such as sieve tubes of phloem with plastids with peripheral ring of proteinaceous filaments and often a central protein crystal (whereas most members of the tricolpate clade have sieve tube plastids with starch grains); betalains forming red to yellow pigments (but anthocyanins are found in Caryophyllaceae); loss of the *rpl2* intron in cpDNA; a single whorl of tepals; pollen with spinulose and tubuliferous/punctate exine; placentation free-central to basal; embryo curved; and presence of perisperm, with endosperm scanty or lacking (Behnke 1976, 1994; Behnke and Mabry 1994; Bittrich 1993b; Cuénoud 2003; Cuénoud et al. 2002; Downie and Palmer 1994a,b; Eckardt 1976; Hilu et al. 2003; Mabry 1973, 1976; Manhart and Rettig 1994; Rettig et al. 1992; Rodman 1990, 1994; Rodman et al. 1984). The monophyly of the group has been strongly supported by *rbcL*, *atpB*, *matK*, and 18S rDNA sequences (Chase et al. 1993; Källersjö et al. 1998; Savolainen et al. 2000a,b; Soltis et al. 1997, 2000, 2005). However, basal to the core Caryophyllineae are three small families, including the monotypic Simmondsiaceae, from the southwestern United States and adjacent Mexico. Their morphological characters, so far as known, differ in part from those of the core families, but stems with successive cambia (concentric rings of xylem and phloem or concentric rings of vascular bundles) and unilacunar nodes may be synapomorphic for the suborder (Figure 9.45). These families

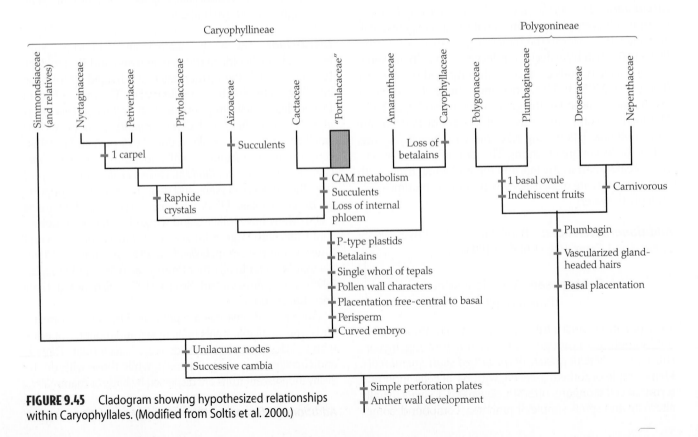

FIGURE 9.45 Cladogram showing hypothesized relationships within Caryophyllales. (Modified from Soltis et al. 2000.)

also typically have separate styles or well-developed style branches with elongated stigmas. The Caryophyllineae consists of 21 families and 8600 species; major families include **Aizoaceae**, **Caryophyllaceae**, **Phytolaccaceae**, Petiveriaceae, **Nyctaginaceae**, **Amaranthaceae** (including Chenopodiaceae), **Cactaceae**, and **Portulacaceae**.

Phylogenetic relationships within core Caryophyllineae have been much studied but are still poorly known (Downie et al. 1997; Downie and Palmer 1993a,b; Hilu et al. 2003; Manhart and Rettig 1994; Rettig et al. 1992; Rodman et al. 1984; Rodman 1990, 1994) (Figure 9.45). Family limits are unclear, and many critical taxa are poorly known. Caryophyllaceae have often been considered the sister family to the remaining members of the suborder because they possess anthocyanins (as red or yellow pigments), while the other members of the suborder have betalains (Ehrendorfer 1976; Mabry 1973, 1976; Mabry et al. 1972). If this hypothesis is true, then betalains are not a synapomorphy of the core Caryophyllineae, and instead diagnose a clade containing all members except Caryophyllaceae and Molluginaceae. Recent cladistic analyses (see references) cast doubt on this hypothesis, and suggest that Caryophyllaceae evolved from betalain-containing ancestors, and have lost betalains and reacquired anthocyanins, the two changes possibly being biochemically interrelated (Downie and Palmer 1994a,b). Phytolaccaceae, Petiveriaceae, Nyctaginaceae, and Aizoaceae

---

Key to Major Families of Caryophyllineae

1. Gynoecium of 1 carpel, with a single ovule 2

1. Gynoecium of 2 to numerous carpels, with 1 to numerous ovules 3

2. Inflorescence indeterminate; flowers 4-merous, tepals 4, distinct; stamens 4 to numerous; perianth not differentiated .... Petiveriaceae

2. Inflorescence determinate; flowers usually 5-merous, tepals 5, connate; stamens usually 5 to numerous; perianth differentiated, the distal portion ± conspicuous and withering, the proximal portion accrescent, persistent, associated with the fruit in dispersal .... **Nyctaginaceae**

3. Inflorescence indeterminate; carpels ± distinct to connate and ovary with axile placentation, with 1 ovule in each locule; fruit a berry, developing from superior ovary .... **Phytolaccaceae**

3. Inflorescence determinate; carpels connate and ovary with free-central, basal, parietal, or axile placentation, ovules 1 to numerous, and if axile, then ovules numerous in each locule; fruit a capsule, achene, utricle and developing from superior or inferior ovary, or a berry developing from ± inferior ovary 4

4. Flowers with hypanthium; ovary inferior to superior; stamens usually numerous 5

4. Flowers lacking a hypanthium; ovary superior; stamens few to numerous 6

5. Shoots ± equivalent; leaves succulent, never spiny, opposite or alternate; ovary superior to inferior, usually with axile placentation .... **Aizoaceae**

5. Shoots strongly dimorphic, the long shoots usually succulent and photosynthetic, the short shoots with a spine or cluster of spines (and sometimes also irritating hairs, glochids); stems succulent; leaves always alternate; ovary usually inferior, with basal or more commonly parietal placentation .... **Cactaceae**

6. Flowers with 2 bracteoles that appear to form a calyx, tepals petaloid; stamens often numerous, sometimes as few as five .... **"Portulacaceae"**

6. Flowers lacking 2 sepal-like bracteoles, with tepals greenish to petaloid, or with 5 "petals" and 5 tepals appearing to be a calyx; stamens 5–10 7

7. Flowers usually with 5 sepals and 5 "petals"; leaves opposite, ovary with free-central to basal placentation, ovules usually numerous (sometimes only 1); fruit usually a capsule (occasionally achene); anthocyanins present; sieve cell plastids with central protein crystal .... **Caryophyllaceae**

7. Flowers usually with 5 tepals; leaves alternate or opposite and nodes not swollen; ovary with basal placentation, ovule usually 1; fruit an achene or utricle; betalains present; sieve cell plastids lacking central protein crystal .... **Amaranthaceae**

likely form one clade, supported by the presence of raphide crystals and characteristics of the chloroplast genome, and Portulacaceae and Cactaceae form another clade, based on the presence of crassulacean acid metabolism (CAM), and the succulent habit. However, succulent leaves have also evolved in the Aizoaceae.

The second major clade of Caryophyllales is the suborder Polygonineae, which here is defined broadly, including **Polygonaceae**, Plumbaginaceae, Nepenthaceae, **Droseraceae**, and five other families, and comprising about 2050 species. The monophyly of Polygonineae is supported by *rbcL, atpB, matK*, and 18S sequences (Hilu et al. 2003; Soltis et al. 2000, 2003; Williams et al. 1994); supporting features include scattered secretory cells containing plumbagin, a naphthaquinone (but these have been lost in several clades, such as Plumbaginaceae subfamily Staticoideae and Polygonaceae); an indumentum of stalked, vascularized, gland-headed hairs, which are often mucilage-producing (lost in a few clades, such as Polygonaceae); basal placentation (but parietal in some Droseraceae, and axile in Nepenthaceae); and starchy endosperm.

Polygonineae may comprise two major clades (Figure 9.45). The first, comprising Polygonaceae and Plumbaginaceae, is supported by ovaries with a single basal ovule and usually indehiscent fruits (achenes or nutlets). The second, which includes Droseraceae and Nepenthaceae, is supported by molecular characters and probably the carnivorous habit as well. Most members of this clade also have circinate leaves and the corolla contorted in bud.

Some systematists have placed Nepenthaceae, Droseraceae, and Sarraceniaceae (Ericales) together on the basis of their carnivorous habit. The pitcherlike leaves of the Old World family Nepenthaceae are amazingly convergent with those of the New World Sarraceniaceae, and the Australian Cephalotaceae (Oxalidales).

Caryophyllaceae A. L. de Jussieu
(Carnation or Pink Family)

Usually herbs; stems sometimes with concentric rings of xylem and phloem; **anthocyanins present**; often with triterpenoid saponins. Hairs various. **Leaves opposite**, *simple, entire, often narrow, with pinnate venation, the secondary veins usually obscure and venation appearing ± parallel, the leaf pair often connected by a transverse nodal line, and nodes usually swollen*; stipules lacking or **present**. *Inflorescences determinate, sometimes reduced to a single flower, terminal*. Flowers usually bisexual, radial, sometimes with an androgynophore. *Tepals 4–5, distinct to connate*, imbricate, usually appearing to be sepals. *True petals lacking, but outer whorl of 4–5 stamens very often petal-like, here called "petals", these frequently bilobed, and sometimes differentiated into a long, thin, basal portion (claw) and an expanded apical portion (blade or limb) separated by appendaged joint*. Stamens 4–10; filaments distinct or slightly connate, sometimes adnate to "petals;" pollen grains tricolpate to polyporate. Carpels 2–5, connate; *ovary superior, with free-central*, occasionally axile or basal *placentation*; stigmas

minute to linear. Ovules usually numerous, occasionally few or only 1, ± campylotropous. Nectar produced by disk or staminal bases. **Fruit usually a loculicidal capsule**, *opening by valves or apical teeth* (Plate 9.7B), but sometimes a utricle; embryo usually curved; endosperm ± lacking, replaced by perisperm (Figure 9.46).

Floral formula:

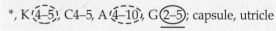

*, K (4–5), C4–5, A (4–10), G (2–5); capsule, utricle

Distribution and ecology: Widespread, but especially characteristic of temperate and warm temperate regions of the Northern Hemisphere, mostly of open habitats or disturbed sites.

Genera/species: 70/2200. ***Major genera:*** *Silene* (700 spp.), *Dianthus* (300), *Arenaria* (200), *Gypsophila* (150), *Minuartia* (150), *Stellaria* (150), *Paronychia* (110), and *Cerastium* (100). Numerous native and introduced genera occur in the continental United States and/or Canada; some of these, in addition to most of the above, include *Agrostemma, Drymaria, Geocarpon, Sagina, Saponaria, Spergula,* and *Stipulicida*.

Economic plants and products: The family is best known for ornamentals such as *Dianthus* (carnations, pinks), *Gypsophila* (baby's breath), *Saponaria* (soapwort), and *Silene* (catchfly, campion).

Discussion: Monophyly of Caryophyllaceae is supported by morphology, *rbcL*, and ORF2280 sequence characters. The family typically is divided into three subfamilies. The "Paronychioideae" (e.g., *Paronychia, Stipulicida, Spergula,* and *Spergularia*) are a paraphyletic, heterogeneous assemblage defined only by the presence of stipules (a likely plesiomorphy within Caryophyllaceae). Some members of this group have petals while others (e.g., *Paronychia*) lack them. The petaloid members of the "Paronychioideae" along with "Alsinoideae" and Caryophylloideae probably constitute a monophyletic group, which may be diagnosed by the presence of often bilobed petals, loss of stipules, and stamens twice the number of sepals (Fior et al. 2006; Lüders 1907). Caryophylloideae and "Alsinoideae" also differ from most "Paronychioideae" in their embryo development and their basally connate leaves. Within this clade, "Alsinoideae" (e.g., *Arenaria, Minuartia, Stellaria, Cerastium,* and *Sagina*) constitute a paraphyletic complex characterized by the symplesiomorphies of free sepals and nonjointed petals, while Caryophylloideae (e.g., *Silene, Saponaria, Dianthus,* and *Gypsophila*) form a clade, based on their connate sepals and usually clawed and jointed petals.

Flowers of Caryophyllaceae are pollinated by various insects (flies, bees, butterflies, and moths) that gather nectar. Protandry leads to outcrossing in most species, but many of the weedy species have inconspicuous flowers, a reduced number of stamens, and are self-pollinated. The small or winged seeds of most species are shaken from the

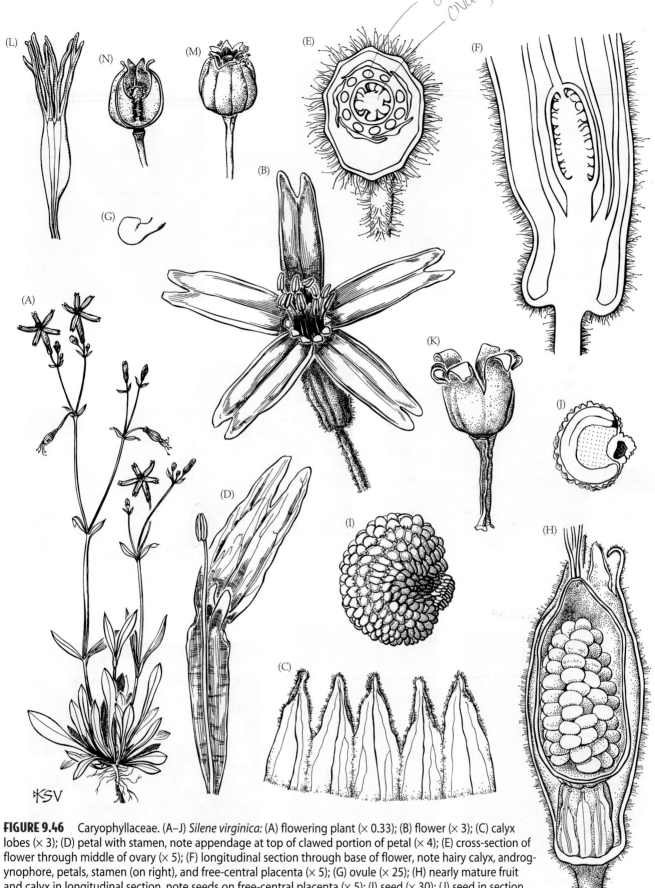

filaments of anther
Ovary

(L) (N) (M) (E) (F) (B) (G) (A) (K) (J) (D) (I) (C) (H)

KSV

FIGURE 9.46 Caryophyllaceae. (A–J) *Silene virginica:* (A) flowering plant (× 0.33); (B) flower (× 3); (C) calyx lobes (× 3); (D) petal with stamen, note appendage at top of clawed portion of petal (× 4); (E) cross-section of flower through middle of ovary (× 5); (F) longitudinal section through base of flower, note hairy calyx, androgynophore, petals, stamen (on right), and free-central placenta (× 5); (G) ovule (× 25); (H) nearly mature fruit and calyx in longitudinal section, note seeds on free-central placenta (× 5); (I) seed (× 30); (J) seed in section, note curved embryo and perisperm (stippled) (× 20). (K) *S. caroliniana:* capsule (× 3). (L) *S. ovata:* petal (× 3). (M–N) *S. antirrhina:* (M) capsule, surrounded by dried calyx (× 3); (N) capsule in longitudinal section (× 3). (From Wood 1974, *A student's atlas of flowering plants*, p. 27.)

(A)

Caryophyllales: Phytolaccaceae
Phytolacca americana: fruits

(B)

Caryophyllales: Caryophyllaceae
Agrostemma githago: capsule, seeds

(C)

Caryophyllales: Caryophyllaceae
Lychnis chalcedonica: flowers

(D)

Polygonales: Polygonaceae
Polygonum pensylvanicum: stem,
leaf, and ocrea

(E)

Caryophyllales: "Portulacaceae"
Portulaca 'sundial': face view
of flower

Back side of flower,
showing 2 sepals (= bracteoles)

(F)

Caryophyllales: Cactaceae
Consolea monililformis: plant in bloom

(G)

Caryophyllales: Amaranthaceae
Amaranthus hybridus:
flower and fruits

(H)

Caryophyllales:
Nyctaginaceae
Bougainvillea sp.: flowers

**PLATE 9.7 Eudicots:
Caryophyllales**

erect capsules by wind or passing animals. The dry utricles (associated with the persistent tepals) of *Paronychia* are probably also wind-dispersed. Sometimes the deciduous fruit cluster or the entire plant acts like a tumbleweed. Species such as *Sagina decumbens* have capsules that open widely only when wetted and have seeds that are dispersed by raindrops. Others are probably animal-dispersed, either by external transport or accidental ingestion as the entire plant is eaten.

References: Bittrich 1993a; Thomson 1942.

Phytolaccaceae R. Brown
(Pokeweed Family)

Usually *herbs*; stem with concentric rings of vascular bundles or alternating concentric rings of xylem and phloem; betalains present; triterpenoid saponins present; raphide crystals (calcium oxalate) usually present. Hairs usually simple. *Leaves alternate and spiral, simple, entire, with pinnate venation*; stipules lacking. **Inflorescences indeterminate (racemes or spikes)**, *terminal, but usually appearing lateral (and opposite the leaves)*. Flowers usually bisexual, radial. *Tepals usually 5 and distinct*, imbricate. *Stamens 10 to numerous*; filaments distinct to slightly connate; pollen grains tricolpate. *Carpels 3 to numerous, clearly to slightly connate*, or occasionally distinct; *ovary superior, usually with axile placentation; styles ± distinct; stigmas ± linear. Ovules 1 in each locule*, campylotropous. Nectar disk often present. **Fruit a berry**; embryo curved; endosperm lacking, replaced by perisperm (Figure 9.47).

Floral formula: *, T-5-, A⟨10–∞⟩, G③–∞; berry

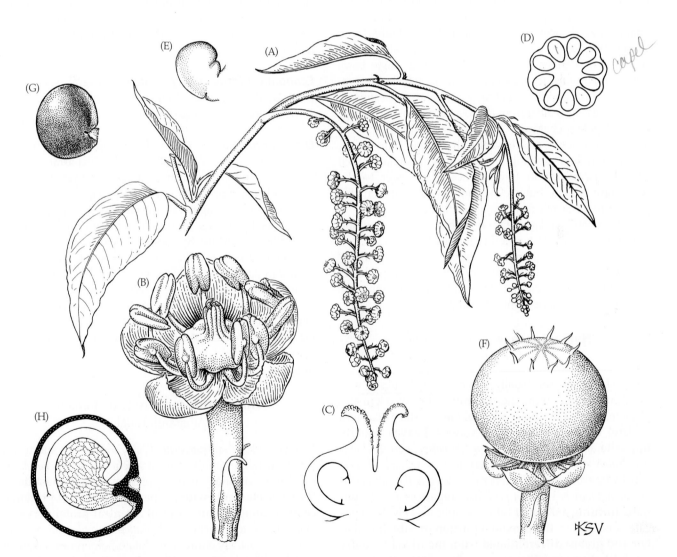

FIGURE 9.47 Phytolaccaceae. *Phytolacca americana* var. *americana:* (A) branch with flowers and immature fruits (× 0.7); (B) flower, stigmas not yet receptive (× 12); (C) gynoecium with receptive stigmas, in longitudinal section (× 20); (D) gynoecium in cross-section, with an ovule filling each locule (× 15); (E) ovule (× 1.5); (F) berry (× 4); (G) seed (× 7); (H) seed in section, note curved embryo and perisperm (large cells outlined with dots) (× 12). (From Rogers 1985, *J. Arnold Arbor.* 66: p. 12.)

Distribution and ecology: Widespread in tropical and warm temperate regions; characteristically early successional, sometimes with seeds that remain viable in the soil for several decades.

Genera/species: 4/30. ***Major genus:*** *Phytolacca* (25 spp.).

Economic plants and products: *Phytolacca* (pokeweed, pokeberry) is poisonous, containing a diverse array of chemicals, including single-chain to polymeric mitogens; ingestion or even contact with the sap should be avoided, but the young leaves are sometimes eaten after repeated boiling.

Discussion: Phytolaccaceae, as here circumscribed, are probably monophyletic. The family traditionally has been defined much more broadly (Rogers 1985; Rohwer 1993c), resulting in a polyphyletic assemblage that is difficult to characterize (Brown and Varadarajan 1985; Manhart and Rettig 1994; Rodman 1994). Taxa with four tepals, four (to numerous) stamens, a single carpel with a single basal ovule, and drupes, achenes, or samaras, may be best segregated as Petiveriaceae (e.g., *Petiveria*, *Rivina*, *Trichostigma*). Petiveriaceae also differ from Phytolaccaceae in embryological and pollen characters. *Stegnosperma*, which has capsules, arillate seeds, and flowers with petaloid staminodes, is placed in its own family. Brown and Varadarajan (1985) placed *Gisekia* within Phytolaccaceae, as sister to the remaining genera, but *rbcL* sequences suggest that this genus is more closely related to Petiveriaceae.

Flowers of Phytolaccaceae attract bees, wasps, flies, and butterflies. The red to purple-black fruits (Plate 9.7A) often contrast with the bright red inflorescence axis and pedicels, and are dispersed by birds. The long-lived seeds probably allow species of *Phytolacca* to take advantage of temporally separated disturbance events at a single site.

Nyctaginaceae A. L. de Jussieu
(Four O'Clock Family)

Herbs, shrubs, or trees, usually with concentric rings of vascular bundles or alternating concentric rings of xylem and phloem; betalains present; raphide crystals (calcium oxalate) usually present. Hairs various. **Leaves usually opposite,** *simple, entire, with pinnate venation;* stipules lacking. *Inflorescences determinate,* terminal or axillary. *Flowers* usually bisexual and radial, *often associated with conspicuous bracts that may be sepal- or petal-like.* **Tepals** usually 5, **connate, forming a distinct tube,** induplicate-valvate to plicate or contorted, **the proximal portion of tube (persistent and green) differentiated from the distal (usually petal-like) portion.** Stamens usually 5; filaments distinct or slightly connate; pollen grains tricolpate to polyporate. *Carpel 1; ovary superior (but often appearing inferior because of its close association with the lower, often constricted, portion of*

the perianth tube), with basal placentation; stigma capitate. *Ovule 1,* usually campylotropous. Nectar disk present. **Fruit an achene or nut, enclosed in persistent, leathery to fleshy, basal portion of perianth tube,** *which may have 5 lines of sticky, gland-headed or hooked hairs, and appearing drupaceous;* embryo usually curved; endosperm ± lacking, replaced by perisperm (Figure 9.48).

Floral formula:

*, T⑤, A⑤, G1; achene, nut (with accrescent tepals)

Distribution: Widespread in tropical and subtropical regions.

Genera/species: 31/350. ***Major genera:*** *Neea* (80 spp.), *Guapira* (60), *Mirabilis* (60), *Boerhavia* (40), *Pisonia* (40), and *Abronia* (20). All but *Neea*, along with *Acleisanthes, Allionia, Anulocaulis, Commicarpus, Cyphomeris, Nyctaginia, Okenia,* and *Tripterocalyx,* occur in the continental United States and/or Canada.

Economic plants and products: Several species of *Bougainvillea, Mirabilis* (four o'clock), and *Abronia* (sand verbena) are cultivated as ornamentals.

Discussion: Several tribes of Nyctaginaceae are recognized on the basis of variation in habit, leaf arrangement, bract formation, pubescence, stamen connation, stigma form, pollen morphology, and embryo shape.

Flowers of Nyctaginaceae attract a diversity of pollinators (bees, butterflies, moths, and birds); nectar is usually the pollinator reward. Species with fleshy, colorful, accrescent tepals usually are dispersed by birds. External transport occurs in species with sticky glands or hooked hairs on their accrescent tepals. Fruits of *Okenia* are pushed into the ground by rapid elongation of the pedicel. The colorful bracts of *Bougainvillea* become dry and papery in fruit, leading to wind dispersal of the associated achene (Plate 9.7H).

References: Bittrich and Kuhn 1993; Bogle 1974.

Amaranthaceae A. L. de Jussieu
(Amaranth Family)

Usually herbs or suffrutescent shrubs, sometimes succulent; usually with concentric rings of vascular bundles; betalains present; occasionally with C_4 photosynthesis; **plastids of the sieve elements with a ± peripheral ring of proteinaceous filaments, but without a central protein crystal.** Hairs simple to branched. *Leaves alternate and spiral, or opposite, simple,* usually entire or undulate, sometimes serrate or lobed, with pinnate venation, but veins often obscure, sometimes succulent; *stipules lacking;* nodes sometimes swollen. *Inflorescences determinate, terminal and axillary.* Flowers bisexual or, less commonly, unisexual (plants then

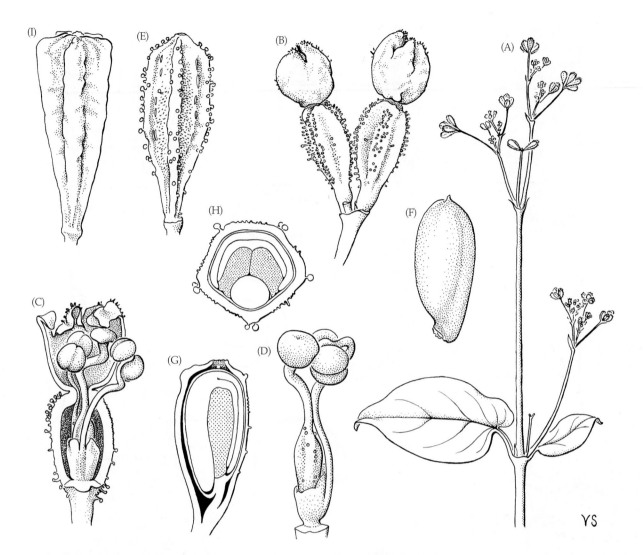

FIGURE 9.48 Nyctaginaceae. (A–H) *Boerhavia diffusa*: (A) flowering stem (× 1); (B) two flowers (× 17); (C) flower in longitudinal section (× 22); (D) gynoecium and one of four stamens (× 36); (E) accrescent perianth base, enclosing fruit (× 17); (F) achene (× 17); (G) perianth base and fruit in longitudinal section, the embryo white and perisperm stippled (× 17); (H) same, in cross-section, the cotyledons above and hypocotyl below (× 22). (I) *B. erecta*: accrescent perianth enclosing fruit (× 17). (From Bogle 1974, *J. Arnold Arbor.* 55: p. 24.)

monoecious to dioecious), radial, associated with fleshy to dry, papery bracts and/or bractlets, and often densely clustered. *Tepals usually 3–5, distinct to slightly connate, green and herbaceous or fleshy, to white (or reddish), dry, and papery, imbricate. Stamens 3–5, opposite tepals; filaments distinct, slightly to strongly connate; anthers 2- or only 1-locular;* **pollen grains 7-porate to polyporate, with pores scattered over the surface of the grain**. *Carpels usually 2 or 3, connate; ovary usually superior, with basal placentation; stigmas 1 to 3, elongate to capitate.* **Ovules 1** *to few*, usually campylotropous. Nectar disk or glands often present. *Fruit usually an achene, utricle, or a circumscissile capsule (pyxis), usually associated with persistent, fleshy to dry, perianth and/or bractlets; embryo curved to spirally twisted; endosperm ± lacking, replaced by perisperm* (Figure 9.49; see also Figure 4.47E).

Floral formula:

*, T(3–5), A(5), G(2–3); achene, utricle, 1-seeded capsule

Distribution and ecology: Cosmopolitan and especially characteristic of disturbed, arid, or saline habitats.

Genera/species: 169/2360. **Major genera:** *Atriplex* (250 spp.), *Gomphrena* (120), *Salsola* (120), *Alternanthera* (100), *Chenopodium* (100), *Ptilotus* (100), *Suaeda* (100), *Iresine* (80), *Amaranthus* (70), *Corispermum* (60), and *Celosia* (50). Numerous genera have native or naturalized species within the continental United States and/or Canada; some of

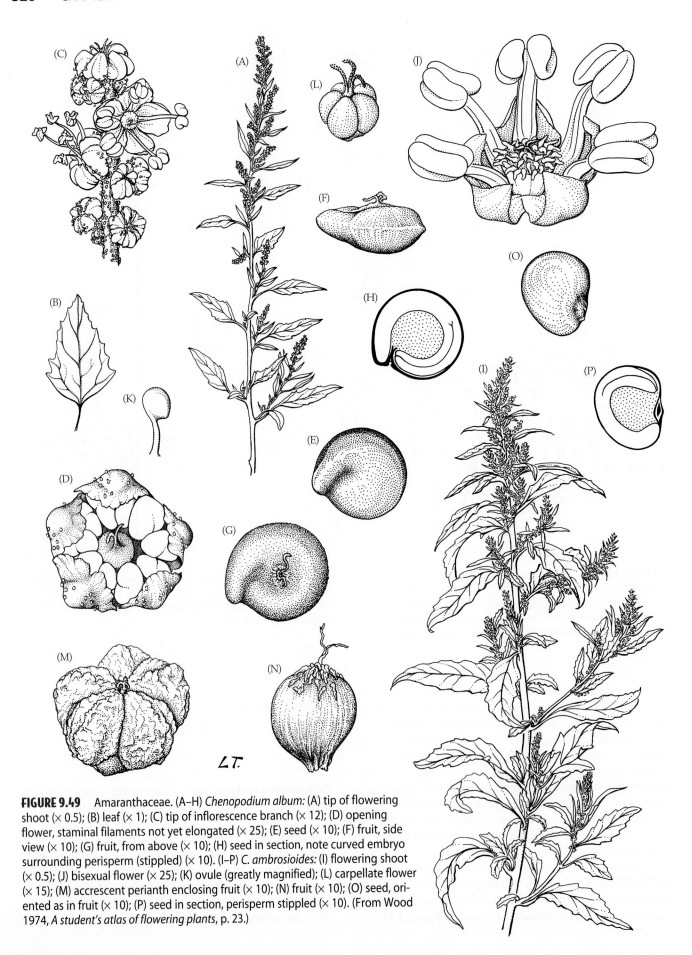

FIGURE 9.49 Amaranthaceae. (A–H) *Chenopodium album:* (A) tip of flowering shoot (× 0.5); (B) leaf (× 1); (C) tip of inflorescence branch (× 12); (D) opening flower, staminal filaments not yet elongated (× 25); (E) seed (× 10); (F) fruit, side view (× 10); (G) fruit, from above (× 10); (H) seed in section, note curved embryo surrounding perisperm (stippled) (× 10). (I–P) *C. ambrosioides:* (I) flowering shoot (× 0.5); (J) bisexual flower (× 25); (K) ovule (greatly magnified); (L) carpellate flower (× 15); (M) accrescent perianth enclosing fruit (× 10); (N) fruit (× 10); (O) seed, oriented as in fruit (× 10); (P) seed in section, perisperm stippled (× 10). (From Wood 1974, *A student's atlas of flowering plants,* p. 23.)

these include *Alternanthera, Amaranthus, Atriplex, Bluta-paron, Celosia, Chenopodium, Froelichia, Iresine, Gomphrena, Grayia, Monolepis, Nitrophila, Salicornia, Salsola,* and *Suaeda.*

Economic plants and products: The leaves and/or roots of a few species, such as *Beta vulgaris* (beet, Swiss chard), *Spinacia oleracea* (spinach), *Chenopodium* spp. (lamb's-quarters, goosefoot), and *Amaranthus* spp. (pigweed), are eaten. The seeds of several South American species of *Chenopodium* and *Amaranthus* (Plate 9.7G) are used to make flour. The family includes a few ornamentals, including *Celosia* (cockscomb), *Gomphrena* (globe amaranth), and *Iresine* (bloodleaf).

Discussion: Amaranthaceae are here broadly defined and include the Chenopodiaceae, which generally have been maintained as a separate family because of their usually distinct (vs. slightly to completely connate) stamens and greenish and membranous to fleshy (vs. white, white with green stripes, to pink or red, and dry/papery) tepals. The monophyly of Amaranthaceae, as broadly circumscribed, has been strongly supported by chloroplast DNA restriction sites, *rbcL* sequences, ORF2280 sequences, and morphology (Downie and Palmer 1994a,b; Downie et al. 1997; Kadereit et al. 2003; Manhart and Rettig 1994; Rodman 1990, 1994; Rodman et al. 1984). Separation of Chenopodiaceae from Amaranthaceae likely is arbitrary and results in a paraphyletic Chenopodiaceae (Downie et al. 1997; Kadereit et al. 2003; Müller and Borsch 2005; Rodman 1990, 1994). The genera *Polycnemum* and *Nitrophila* (Polycnemoideae)—traditionally included in Chenopodiaceae—are sister to all other Amaranthaceae s.l. The remaining genera fall into two large clades, corresponding to the traditional Amaranthaceae (with dry/papery tepals) and Chenopodiaceae (with greenish and membranous to fleshy tepals).

Currently recognized subfamilies and tribes stress ovule number, embryo shape, number of anther locules, pollen structure, and perianth form. Many of these groups are probably not monophyletic. Genera with papery tepals and bractlets and monadelphous stamens—*Celosia, Iresine, Froelichia, Alternanthera, Achyranthes, Blutaparon, Gomphrena,* and probably *Amaranthus,* form a clade. Within this group, genera with unilocular anthers—*Froelichia, Alternanthera, Blutaparon, Gomphrena,* and *Iresine*—likely constitute a subclade. Greenish, often fleshy tepals or bracts and separate stamens characterize *Atriplex, Chenopodium, Kochia, Salsola, Salicornia, Suaeda, Beta,* and *Spinacia,* a weakly supported clade. The spirally twisted embryos and reduced perisperm of *Salsola, Suaeda,* and relatives evolved independently. C_4 photosynthesis has also evolved several times within the family. Hybridization and polyploidy are common in some genera, leading to taxonomic difficulties at the species level.

The usually small, densely clustered flowers of Amaranthaceae are pollinated by wind or by various insects; selfing as well as outcrossing may occur. The small, dry fruits or seeds, which are typically associated with the accrescent and sometimes hairy perianth, usually are dispersed by wind or water. *Salsola* includes tumbleweeds. A few species form burrlike inflorescence units that are externally transported by animals. In *Amaranthus* and *Celosia* the small seeds tend to fall from the parent plant but germinate only when the site is again disturbed. Many seeds are accidentally eaten and dispersed by browsing animals.

Additional references: Blackwell 1977; Carolin 1983; Carolin et al. 1975; Judd and Ferguson 1999; Kühn et al. 1993; Robertson 1981; Townsend 1993.

Aizoaceae Martynov
(Stone Plant Family)

Succulent herbs; stem often with concentric rings of vascular bundles or alternating concentric rings of xylem and phloem; betalains present; alkaloids often present; raphide crystals (calcium oxalate) often present; usually with crassulacean acid metabolism (CAM) or C_4 photosynthesis; phytoferritin in phloem. Hairs various. *Leaves usually opposite, simple, usually entire and succulent (with clear cells in center of blade),* with venation pinnate, but veins ± obscure, **epidermis (and stem) with many large, bladderlike cells;** stipules usually lacking. Inflorescences determinate, sometimes reduced to a single flower, terminal or axillary. Flowers usually bisexual, radial, **with a hypanthium.** *Tepals usually 5,* ± connate, imbricate. *Stamens usually 5 to numerous, the outer often elongate petal-like staminodes;* filaments distinct to slightly connate; pollen grains usually tricolpate. Carpels usually 2–5, connate; *ovary superior to inferior,* **with usually axile placentation;** stigmas ± linear. Ovules 1 to numerous in each locule, ± anatropous to campylotropous. Nectar disk usually present. *Fruit usually a loculicidal, septicidal, or circumscissile capsule,* sometimes fleshy; seeds sometimes arillate; embryo curved; endosperm lacking, replaced by perisperm.

Floral formula: *, T⑤, A5-∞ + ∞•, G③–⑤; capsule

Distribution and ecology: Widespread in tropical and subtropical regions; primarily of coastal or arid habitats.

Genera/species: 127/2500. **Major genera:** *Conophytum* (290 spp.), *Delosperma* (150), *Lampranthus* (150), *Drosanthemum* (100), *Antimima* (60), *Lithops* (35), *Mesembryanthemum* (30), and *Carpobrotus* (30). Some genera occurring in the continental United States are *Carpobrotus, Cryophytum, Cypselea, Galenia, Sesuvium, Tetragonia,* and *Trianthema.*

Economic plants and products: The family contains numerous ornamentals, such as *Lampranthus, Dorotheanthus, Mesembryanthemum* (ice plant), *Ruschia,* and *Carpobrotus.* Some, such as *Lithops* (stone plants), are grown as curiosities. *Tetragonia* is used as a vegetable.

Discussion: The family appears to be monophyletic. Bittrich and Hartmann (1988) and Klak et al. (2003) have investigated infrafamilial relationships. The species with strongly succulent leaves, numerous stamens, the outermost of which are petal-like staminodes (i.e., Mesembryanthemoideae and Ruschioideae), form a monophyletic group (Hartmann 1993; Klak et al. 2003), which has explosively diversified in South Africa and Australia. These two large subfamilies are sister to the Aizooideae (*Galenia, Aizoon, Tetragonia*), which has only slightly succulent leaves and accessory lateral branches. All three subfamilies have explosively opening capsules (Bittrich 1990; Klak et al. 2003). *Sesuvium* and relatives (i.e., Sesuvioideae) probably form a clade on the basis of their circumscissile capsules and arillate seeds and are sister to the other members of the family. Generic limits have varied widely, with some systematists considering nearly all species of Mesembryanthemoideae within the large genus *Mesembryanthemum*.

Aizoaceae show several adaptations to extremely arid habitats. Bladderlike cells in the leaf (and stem) epidermis hold water, and the leaves themselves are succulent. Sometimes the plant is reduced to a single pair of opposite, nearly spherical leaves. In some genera (e.g., *Lithops*), only a clear "window" is exposed above the soil, with the chlorophyll-containing cells restricted to a thin layer along the sides or base of the nearly cylindrical leaves. These leaves are actually bracteoles, the adult plant evidently being a long-lived inflorescence. These and other bizarre features are probably adaptations to cope with intense sunlight.

The usually showy flowers attract bees, wasps, butterflies, flies, and beetles. The small seeds are usually dispersed by wind or water; in some genera the capsules open only when wetted.

Additional references: Bittrich and Struck 1989; Bogle 1970.

"Portulacaceae" A. L. de Jussieu
(Purslane Family)

Usually ± *succulent herbs*; mucilage cells common; betalains present; sometimes with crassulacean acid metabolism (CAM); phytoferritin present in phloem. Hairs usually simple. *Leaves opposite or alternate and spiral, simple*, entire, with pinnate venation, veins ± obscure; *stipules usually present, often scarious or tufts of short to elongate hairs*. Inflorescences determinate, sometimes appearing indeterminate or reduced to a single flower, terminal or axillary. Flowers usually bisexual and radial, *associated with 4 bracteoles, the inner 2 calyx-like. Tepals usually 4–6, occasionally numerous, petal-like*, distinct to slightly connate, imbricate. *Stamens usually 4–6, opposite tepals*, but sometimes fewer *or more numerous*; filaments distinct or slightly adnate to tepals; pollen grains tricolpate to polycolpate or polyporate. Carpels usually 2 or 3, connate; ovary superior to ± inferior, with ± free-central to basal placentation; stigmas usually linear. Ovules numerous to 1 per gynoecium, anatropous to campylotropous. Individ-

ual nectaries or a nectar disk. *Fruit usually a loculicidal or circumscissile capsule*; seeds sometimes arillate; embryo curved; endosperm lacking, replaced by perisperm (Figure 9.50).

Floral formula:

*, 2[bracts], T(4–6 (–∞)), A4–∞, G(2–3); capsule

Distribution: Widely distributed in tropical and temperate regions; especially diverse in western North America and the Andes of South America.

Genera/species: 20/450. **Major genera:** *Portulaca* (125 spp.), *Cistanthe* (35), *Phemeranthus* (30), *Claytonia* (30), *Lewisia* (16), and *Talinum* (15). Several genera are common in the continental United States and/or Canada, including those listed above and *Montia* and *Talinopsis*.

Economic plants and products: The leaves and young stems of *Portulaca oleracea* (purslane) are occasionally eaten. *Portulaca, Talinum* and *Phemeranthus* (flame flower), *Lewisia* (bitterroot), and *Calandrinia* (rock purslane) provide ornamentals.

Discussion: Monophyly of Portulacaceae has often been questioned, either in relation to the separation of two small families, Basellaceae and Didiereaceae (morphological data; Rodman 1990, 1994); the separation of the Cactaceae (DNA sequences; Edwards et al. 2005; Hershkovitz and Zimmer 1997), or the divergent placement of *Portulaca* and *Claytonia* (cpDNA restriction sites; Downie and Palmer 1994a,b). Cladistic analyses based on *ndhF* and *matK* sequences suggests very strongly that the family is paraphyletic, although many details regarding relationships within it have not yet been clarified (Applequist and Wallace 2001; Hilu et al. 2003). Portulacaceae may eventually be separable into three families—an *Anacampseros* + *Portulaca* + *Talinum* clade (incl. Cactaceae, diagnosed by fruit wall strongly differentiated into two layers and most with axillary hairs, see also Carolin 1987, 1993; Hershkovitz and Zimmer 1997; Nyffeler 2007), a *Portulacaria* + *Ceraria* clade (incl. Didiereaceae, mainly dioecious, woody plants with single-seeded, usually indehiscent fruits; see Stevens 2001 onwards), and a *Cistanthe* + *Claytonia* + *Montia* + *Lewisia* + *Phemeranthus* + *Calandrinia* clade (herbs with clasping leaves). Basellaceae may represent yet another lineage. More study of this problematic group is required; the traditional circumscription of the family is retained here on the basis of flowers closely associated with a pair (or occasionally more) of sepal-like bracteoles.

Several genera (e.g., *Claytonia, Lewisia, Montia, Portulaca*) show a wide range of chromosome numbers, suggesting a complex evolutionary history through development of polyploidy followed by the formation of aneuploids (see Chapter 4).

Flowers of Portulacaceae (Plate 9.7E) usually open only in full sunlight, and then only for a short time; bees, flies, bee-

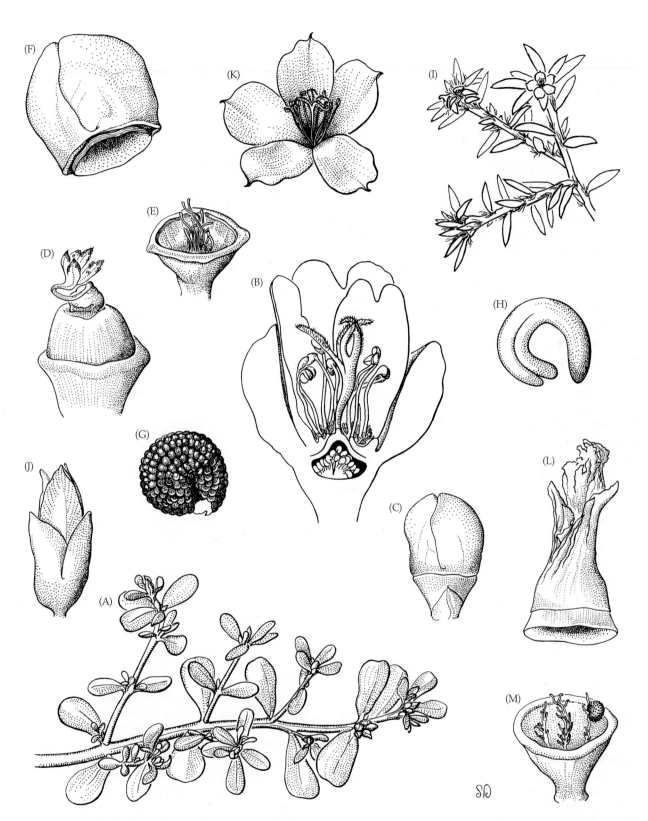

FIGURE 9.50 Portulacaceae. (A–H) *Portulaca oleracea:* (A) flowering and fruiting branch (× 0.75); (B) flower in longitudinal section (× 9); (C) nearly mature fruit, enclosed above by accrescent sepals (× 4.5); (D) same, with sepals removed (× 7); (E) base of circumscissile capsule after dehiscence, with basal funiculi (× 7); (F) upper part of fruit after dehiscence, with mature accrescent sepals (× 7); (G) seed (× 35); (H) curved embryo (× 35). (I–M) *P. pilosa:* (I) flowering and fruiting branch (× 0.5); (J) flower bud (× 12); (K) flower (× 6); (L) withered perianth adhering to upper part of circumscissile capsule (× 1.5); (M) base of capsule after dehiscence, with seed attached to placenta (× 1.5). (From Bogle 1969, *J. Arnold Arbor.* 50: p. 572.)

tles, and butterflies visit the flowers to gather nectar. Time and duration of flowering often differ among related species (reproductive isolation). The tiny seeds are dispersed by wind and water. Those with arils may be ant-dispersed.

Additional references: Bogle 1969; Nyananyo 1990.

Cactaceae A. L. de Jussieu
(Cactus Family)

Spiny *stem succulents (herbs to trees)*, **shoots differentiated**, *with usually succulent (and cylindrical, conical, globose, or flattened, often ridged or jointed)* **long shoots producing photosynthetic leaves** *(although these are usually reduced and quickly falling), or leaves lacking,* **and short shoots (areoles) producing a spine or spine cluster**, *and often irritating hairs (glochids);* with crassulacean acid metabolism (CAM); *stem epidermis usually with stomata;* phytoferritin present in phloem; often with alkaloids or triterpenoid saponins; betalains present. *Leaves of long shoots alternate and spiral, simple, entire, with pinnate or obscure venation, usually reduced to lacking;* **leaves of short shoots modified into spines**; stipules lacking. *Inflorescences* determinate, but usually *reduced to a single flower, terminal, but with flowers sunken into the apex of a modified branch (and thus appearing to be axillary).* Flowers usually bisexual, radial to slightly bilateral, **with short to elongate hypanthium**, *and almost always developmentally modified, with ovary surrounded by apex of modified stem, thus outer portion of ovary and hypanthium with spine-bearing areoles.* **Tepals numerous and spirally arranged**, *usually distinct, all petal-like, or these often intergrading with outer, sepal-like tepals,* imbricate. **Stamens numerous**; pollen grains tricolpate to polycolpate or polyporate. Carpels 3 to numerous, connate; *ovary almost always inferior*, but in some species of *Pereskia* only half-inferior or even superior, *almost always with parietal placentation*, but in *Pereskia* ± basal; stigmas 3 to numerous, elongated and radiating. Ovules numerous, usually campylotropous. Nectary a ring on inner surface of hypanthium. **Fruit a berry**, *outer portion with nodes and internodes, and usually with spines and/or glochids at areoles;* seed sometimes covered by a bony aril; embryo usually curved; endosperm lacking, but perisperm sometimes present (Figure 9.51; see also Figure 4.47F).

Floral formula: * or X, T-∞-, A∞, G—(3–∞)—; berry

Distribution and ecology: Mainly in North and South America, but *Rhipsalis* occurs in tropical Africa, and several species of *Opuntia* have been introduced in Africa, Australia, and India. Plants typically of deserts and other arid habitats, but sometimes epiphytic in tropical forests.

Genera/species: 100/1400. ***Major genera:*** *Mammillaria* (170 spp.), *Opuntia* (150), *Echinopsis* (70), *Cleistocactus* (50), *Echinocereus* (50), *Rhipsalis* (50), and *Cereus* (40, or many

more if various segregate genera are included). Several genera occur in the continental United States and/or Canada, including *Acanthocereus, Carnegia, Consolea, Cylindropuntia, Echinocactus, Echinocereus, Ferocactus, Grusonia, Harrisia, Lophophora, Mammillaria, Neolloydia, Opuntia, Pediocactus, Pilosocereus, Sclerocactus,* and *Thelocactus.*

Economic plants and products: The fruits of several species of *Opuntia* (prickly pear) are eaten. Nearly all genera are cultivated as ornamentals; some of the more common are *Opuntia, Carnegia* (giant saguaro), *Cereus* (hedge cactus, cereus), *Echinopsis* (sea-urchin cactus), *Epiphyllum* (orchid cactus), *Hylocereus* (night-blooming cereus), *Mammillaria* (pincushion cactus), *Melocactus* (Turk's-cap cactus), *Rhipsalis* (mistletoe cactus), and *Schlumbergera* (Christmas cactus). *Lophophora* contains mescaline alkaloids and is hallucinogenic.

Discussion: The family's monophyly is supported by numerous morphological characters, a 6 kb cpDNA inversion (R. Wallace, pers. comm.), and DNA sequence data (Edwards et al. 2005). *Pereskia* is paraphyletic and has given rise to all other Cactaceae (Edwards et al. 2005; Nyffeler 2002). *Pereskia* retains numerous plesiomorphic character states: nonsucculent stems, well-developed and persistent leaves, cymose inflorescences, several styles, and in at least some species superior ovaries with basal placentation. These character states are absent in all other members of the family. A group of eight species of *Pereskia* (=*Rhodocactus*) centered on the Caribbean basin are sister to all other cacti; they lack stem stomata. The rest of the family, including the remaining—mainly South American—species of *Pereskia,* form a clade and are characterized by stem stomata and delayed bark formation, leading to the evolution of stem-based photosynthesis (Edwards et al. 2005).

Most Cactaceae are placed in Opuntioideae and Cactoideae, and together these form a clade based on solitary flowers that are sunken into stem apices and inferior ovaries. (Inferior ovaries have independently evolved in a group of *Pereskia* species.) Opuntioideae (i.e., *Opuntia, Consolea,* and relatives) are monophyletic based on the synapomorphies of the presence of glochids (spinelike hairs) on the areoles (Plate 9.7F), the seed coat covered by a bony aril, and cpDNA characters. The monophyly of Cactoideae, a large complex containing more than three-fourths of the species of Cactaceae, is supported by the extreme reduction or complete loss of leaves, the characteristic hilum region of the seeds, and a deletion of the *rpoC1* intron in the chloroplast genome (Nyffeler 2002). Many species of Cactoideae have ribbed stems, another apomorphic feature. The family is taxonomically difficult, with problems in delimitation of both species and genera.

The showy flowers of Cactaceae are extremely variable in color and form and are visited by numerous different pollinators, including various insects (bees, flies, sphinx moths), birds, and bats, which may be attracted by either

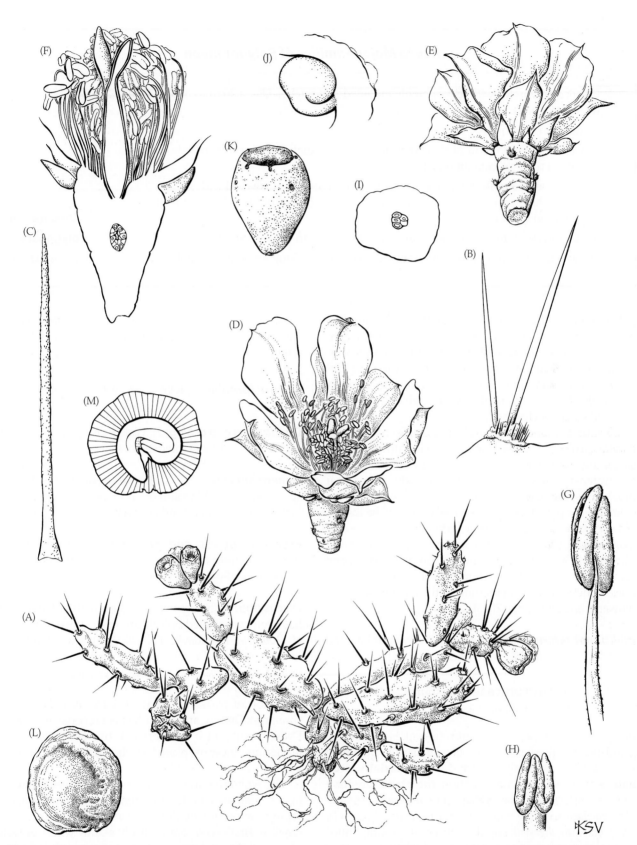

FIGURE 9.51 Cactaceae. *Opuntia pusilla:* (A) plant with immature fruits (× 0.5); (B) areole with spines and glochids (× 2); (C) single glochid (× 30); (D) flower (× 1); (E) underside of flower, sunken into apex of stem with areoles (× 1); (F) flower in longitudinal section (× 2); (G) stamen (× 14); (H) unexpanded stigmas (× 4); (I) ovary in cross-section (× 2); (J) ovule (greatly magnified); (K) berry (× 1); (L) seed, surrounded by bony aril (× 7); (M) seed in section, note curved embryo (× 7). (From Wood 1974, *A student's atlas of flowering plants*, p. 73.)

Key to Major Families of Polygonineae

1. Plants carnivorous, with leaves highly modified and pitcher-like, a snap trap, or covered with sticky hairs 2
1. Plants not carnivorous..3
2. Leaves modified to form pitcher traps; flowers unisexual; filaments completely connate, forming a tube; placentation axile ....Nepenthaceae
2. Leaves modified to form snap traps or variously shaped and covered with sticky, insect-catching hairs; flowers bisexual; filaments distinct or only slightly connate; placentation basal or parietal ....**Droseraceae**
3. Stipules present, fused around stem at the node (i.e., an ocrea); carpels 2 or 3 ....**Polygonaceae**
3. Stipules lacking, or if present then not as above (i.e., lacking an ocrea); carpels 5 ....Plumbaginaceae

---

pollen or nectar. They are predominantly outcrossing. The berries are dispersed by birds or mammals, but several genera (e.g., *Cereus*) are at least partially dispersed by ants, which are attracted to the fleshy funiculi of the seeds. Fruits of some species are spiny, leading to external transport by mammals; others have jointed stems, which may easily break off when touched.

Cacti show numerous specializations for survival in dry habitats. They have a specialized habit, with dimorphic stems and leaves; some stems are photosynthetic and contain water storage tissue, and others are reduced and bear protective spine-leaves. Their CAM metabolism allows the stomates to open at night (when less water will be lost) in order to take in carbon dioxide, which is then stored as malic acid and used in photosynthesis during the next day. They have a wide-spreading and shallow root system, and sometimes also a deep taproot, as well as stems with a thick cuticle and an epidermis with sunken stomates (Edwards and Donoghue 2006).

Additional references: Anderson 2001; Barthlott and Hunt 1993; Benson 1982; Boke 1964; Leins and Erbar 1994.

Droseraceae Salisbury
(Sundew Family)

Insectivorous herbs, sometimes suffrutescent. Hairs stalked, gland-headed, producing mucilage and usually containing xylem. *Leaves* usually alternate and spiral, **adaxially circinate,** simple, with obscure venation, **the blade sensitive** *and forming a snap trap or covered with conspicuous, tentacle-like, mucilage-secreting hairs; insects caught in a trap or by sticky hairs are digested;* stipules present or lacking. Inflorescences determinate, sometimes reduced to a solitary flower, terminal. Flowers bisexual, radial. Sepals usually 5, slightly connate, imbricate. *Petals usually 5, distinct*, convolute. Stamens usually 5, occasionally numerous; filaments distinct or slightly connate; **pollen grains triporate to multiporate**, released in tetrads. *Carpels usually 3, connate;*

ovary superior, with basal or parietal placentation; stigmas various. Ovules 3 to numerous. Fruit a loculicidal capsule (Figure 9.52).

Floral formula: *, K $\widehat{5}$, C5, A $\widehat{5}$, G $\underline{\textcircled{3}}$; capsule

Distribution and ecology: Widely distributed, commonly in wet, low-nutrient, acidic soils.

Genera/species: 3/109. **Major genus:** *Drosera* (107 spp.). *Drosera* and *Dionaea* occur in the continental United States; the former is found in Canada as well.

Economic plants and products: *Dionaea muscipula* (Venus's flytrap) and various species of *Drosera* (sundew) are occasionally cultivated as novelties.

Discussion: Phylogenetic pattern within Droseraceae has been investigated by Williams et al. (1994), Cameron et al. (2002), and Riradavia et al. (2003). The two small genera with snap-traps, *Dionaea* and *Aldrovanda*, are sister to the large genus *Drosera*, which has leaves covered with sticky, stalked, gland-headed hairs. *Drosera* shows the derived condition of parietal placentation. *Drosophyllum* is quite similar to *Drosera* and often has been considered within Droseraceae, but it has non-sensitive abaxially circinate leaves and now is placed in its own family (Kubitzki 2003a,b).

The white to purple flowers of Droseraceae are usually insect-pollinated; protandry often leads to outcrossing, but selfing can occur when the flowers close at the end of the day. The small seeds are probably mainly wind- or water-dispersed. Asexual reproduction commonly also occurs via the production of plantlets from inflorescences or detached leaves.

Dionaea is well known for its snap-trap leaves, which have hinged blades, each half of which is equipped with sensitive bristles (homologous to the glandular hairs of

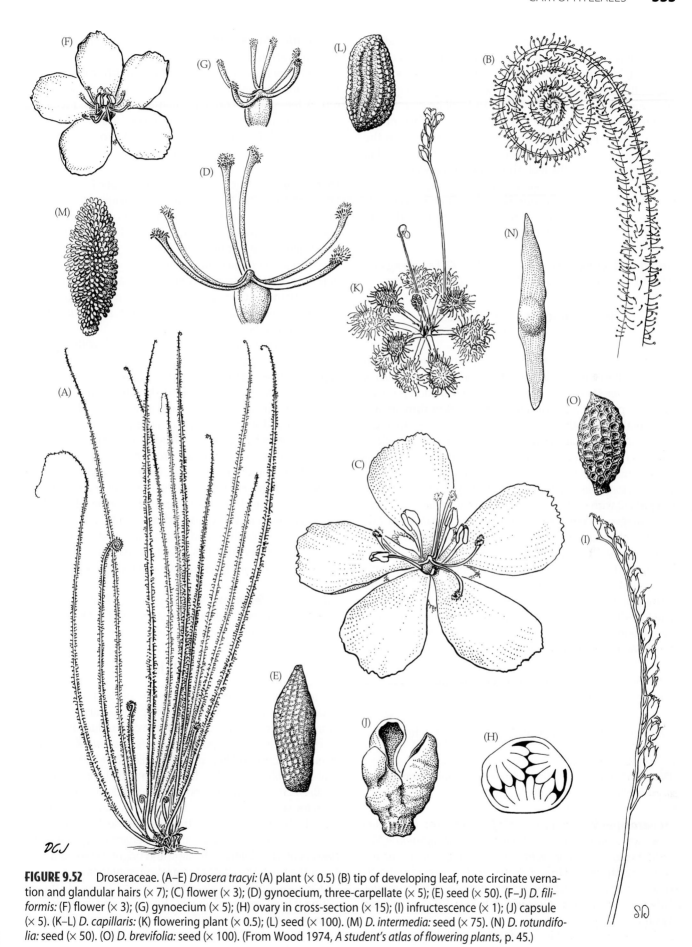

FIGURE 9.52 Droseraceae. (A–E) *Drosera tracyi:* (A) plant (× 0.5) (B) tip of developing leaf, note circinate vernation and glandular hairs (× 7); (C) flower (× 3); (D) gynoecium, three-carpellate (× 5); (E) seed (× 50). (F–J) *D. filiformis:* (F) flower (× 3); (G) gynoecium (× 5); (H) ovary in cross-section (× 15); (I) infructescence (× 1); (J) capsule
(× 5). (K–L) *D. capillaris:* (K) flowering plant (× 0.5); (L) seed (× 100). (M) *D. intermedia:* seed (× 75). (N) *D. rotundifolia:* seed (× 50). (O) *D. brevifolia:* seed (× 100). (From Wood 1974, *A student's atlas of flowering plants*, p. 45.)

Drosera). When these bristles are stimulated an action potential is generated, which is rapidly transmitted to the leaf joint, initiating closing of the blade. Thus, insects landing on the colorful leaf blades are likely to be entrapped by the overlapping marginal bristles, and will be digested by enzymes secreted by small glands on the surface of the blade. In *Drosera*, small insects become entangled in the mucilage secreted by tentacle-like, gland-headed hairs that cover the surface of the leaf. The hairs bend inward, pressing the captured insect against the leaf blade, which may also bend in order to surround the prey. Both movements involve cell elongation, so the number of times that an insect can be trapped is limited.

Additional references: Albert et al. 1992; Fagerberg and Allain 1991; Sibaoka 1991; Williams 1976; Wood 1960.

Polygonaceae A. L. de Jussieu
(Knotweed Family)

Herbs, shrubs, trees, or vines; nodes often swollen; usually with tannins; often with oxalic acid. Hairs various. *Leaves usually alternate, simple and spiral,* usually entire, venation pinnate; **stipules present and connate into an often thin sheath (or ocrea) around the stem** (lacking in *Eriogonum*). Inflorescences indeterminate or determinate, terminal or axillary. Flowers usually bisexual, sometimes unisexual (and plants then usually dioecious), radial. **Perianth** *of 6 tepals,* usually petaloid, *sometimes differentiated, with 3 sepals and 3 petals,* or **5 tepals** (2 with both margins on the outside in bud, 2 with both margins on the inside, and 1with a margin on the inside and the other on the outside, i.e., quincuncial aestivation), distinct to slightly connate, *persistent.* Stamens usually 5–9; filaments distinct to slightly connate; pollen grains usually tricolporate to multiporate. *Carpels usually 2 or 3, connate; ovary superior, with basal placentation;* stigmas punctate, capitate to ± dissected. *Ovule 1,* usually **orthotropous**. Nectary a disk around base of ovary, or paired glands associated with the filaments. *Fruit an achene or nutlet, often angled, and often associated with enlarged (fleshy or dry) perianth parts, these sometimes with various outgrowths;* embryo straight to curved (Figure 9.53).

Floral formula: *, T(5̲–6̲), A(6̲–9̲), G(2̲–3̲); achene

Distribution: Widely distributed; especially common in northern temperate regions.

Genera/species: 43/1100. **Major genera:** *Eriogonum* (250 spp., paraphyletic), *Rumex* (200), *Polygonum* (160, paraphyletic), and *Coccoloba* (120). Some genera occurring in the continental United States and/or Canada, in addition to those listed above, are *Antigonon, Chorizanthe, Fallopia, Nemacaulis, Oxytheca, Oxyria, Polygonella,* and *Stenogonum.*

Economic plants and products: Seeds of *Fagopyrum* (buckwheat) produce flour, while *Coccoloba* (sea grape) has edible fleshy fruits. The petioles of *Rheum* (rhubarb) are edible, as are the leaves of some species of *Rumex* (dock, sorrel). A few genera contain ornamental species, including *Antigonon* (coral vine) and *Coccoloba*. Many species of *Rumex* and *Polygonum* (possibly incl. *Persicaria* and *Polygonella*, knotweeds) are common weeds.

Discussion: Polygonaceae are monophyletic family (Lamb Frye and Kron 2003) and easily recognizable by their distinctive ocreas (Plate 9.7D; see Description); relationships within the group are in need of additional study. The species of *Eriogonum* are distinct in that they have lost ocreas; they have usually whorled or opposite leaves, and have obviously determinate inflorescences. Quincuncial aestivation characterizes many genera (e.g., *Polygonum, Coccoloba, Fagopyrum,* and *Antigonon*), while others (e.g., *Rumex, Eriogonum*) have six perianth parts.

Flowers of most Polygonaceae are small, with white to red, petaloid tepals and are pollinated by various insects, especially bees and flies. Flowers of *Rumex* are pendulous, greenish, with an expanded, divided stigma, and wind-pollinated. The fruits are often associated with persistent tepals that assist in dispersal by wind or water. In *Rumex* the tepals of the inner whorl expand and form membranous wings, while those of the outer whorl are winglike in *Triplaris* and *Ruprechtia*. Sometimes the achene itself is winged, as in *Rheum* and *Fagopyrum*. In *Polygonum virginianum* the achenes are forcibly tossed from the plant; additionally, the style is persistent with recurved hooked tips, leading to entanglement in hair or clothing and external transport. In *Coccoloba* the persistent tepals are fleshy and completely surround the achene; the fruit simulates a drupe and is dispersed by birds.

Additional references: Brandbyge 1993; Ronse Decraene and Akeroyd 1988; Ronse de Craene et al. 2004; Graham and Wood 1965.

Santalales

Santalales are apparently monophyletic based on the presence of polyacetylenes, roots lacking root hairs, 1-seeded, indehiscent fruits, and seeds with the coat reduced/crushed; monophyly is also supported by *rbcL, atpB, matK,* and 18S rDNA sequences (Hilu et al. 2003; Källersjö et al. 1998; Nickrent and Soltis 1995; Savolainen et al. 2000a,b; Soltis et al. 2000, 2003). Stamens are opposite the petals in most species. In many species, conventional roots are replaced by haustoria, which are complex in structure and development. The flowers vary from tiny and without a perianth (e.g., staminate flowers of *Misodendrum*) to large and brightly colored (many Loranthaceae). The ovary of many taxa is inferior.

Circumscription of this order and delimitation of families within it have been problematic. Recent molecular systematic work promises to clarify the situation substantially (see ref-

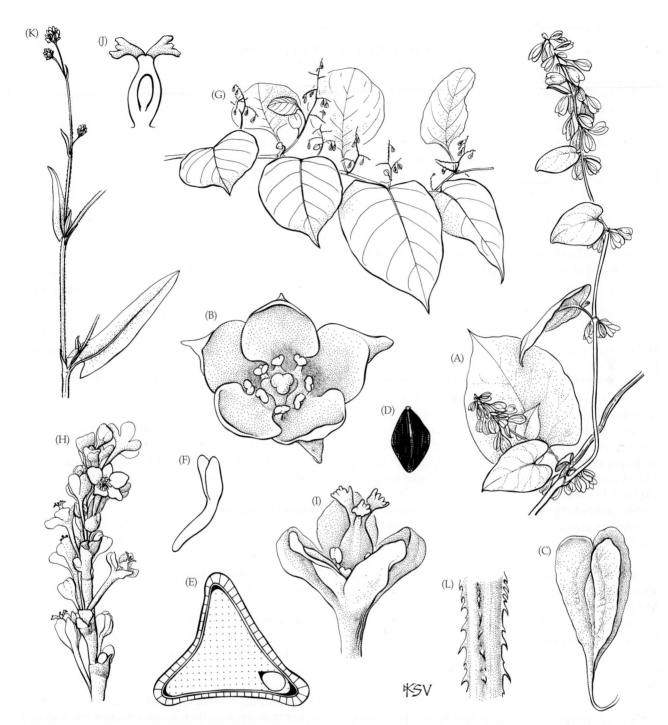

FIGURE 9.53 Polygonaceae. (A–F) *Polygonum scandens:* (A) fruiting branch (× 1); (B) flower (× 15); (C) accrescent perianth enclosing achene (× 4); (D) achene (× 5); (E) achene in cross-section, note embryo (lower right) and endosperm (stippled) (× × 20); (F) embryo (greatly magnified). (G–J) *P. cuspidatum:* (G) branchlet with fruits (× × 0.25); (H) tip of inflorescence (× 5); (I) flower (× 15); (J) gynoecium in longitudinal section, with basal, orthotropous ovule (× 15). (K–L) *P. sagittatum:* (K) flowering branch, with an ocrea at each node (× 1); (L) stem with retrorse prickles (× 4). (From Wood 1974, *A student's atlas of flowering plants*, p. 22.)

erences by Nickrent and the web site http://www.science. siu.edu/parasitic-plant/index.html). The number of families recognized is uncertain. Here we follow APG (2003) but the phylogeny of Nickrent and Malecot (http://www.science.siu. edu/parasitic-plants/Santalales.IPWC/Santalales.tree.JPEG)

should also be consulted. We recognize 8 families, including "Olacaceae," Misodendraceae, Schoepfiaceae, **Loranthaceae**, Opiliaceae, and **Santalaceae** (incl. Viscaceae). "Olacaceae" are paraphyletic, but relationships are too unclear to divide them. Viscaceae and Loranthaceae were

Key to Families of Santalales

1. Leaves alternate; flowers usually bisexual, sepals and petals present; calyx not reduced, with distinct or fused lobes; ovary with basal partitions; ovules with 1 or 2 integuments; seeds with a thin testa; root parasites or non-parasites ...."Olacaceae"

1. Leaves alternate or opposite; flowers bisexual or unisexual; calyx reduced or absent; ovary unilocular; ovules lacking integuments; seeds without a testa; always parasites 2

2. Ovary superior; tepals present only in carpellate flowers; fruit a nutlet with plumose staminodes; stem parasites of Nothofagus ....Misodendraceae

2. Ovary inferior, root or stem parasites 3

3. Sepals reduced (present as a calyculus); glandular tissue between ovary and stamens absent; fruit a one-seeded drupe, berry, or samara 4

3. Sepals absent; glandular tissue (a disk) between ovary and stamens usually present 5

4. Fruit viscid one-seeded berry or samara ....**Loranthaceae**

4. Fruit a nonviscid drupe ....Schoepfiaceae

5. Dried leaves finely tuberculate with cystoliths; all root parasites ....Opiliaceae

5. Dried leaves without cystoliths; root and stem parasites ....**Santalaceae**

---

historically combined as subfamilies of a single family (Loranthaceae), but more recent evidence supports their separation, with inclusion of Viscaceae within Santalaceae. Epiphytism has arisen more than once in the group. The Balanophoraceae, a strange group of obligate parasites, may also be placed here.

The best discussion of the biology of this group of plants remains the classic *Biology of parasitic flowering plants* (Kuijt 1969).

Additional references: Kuijt 1982; Nickrent 1996; Nickrent and Duff 1996; Nickrent et al. 1998; Wiens and Barlow 1971.

Loranthaceae A. L. de Jussieu
(Mistletoe Family)

Stem parasites, except for *Nuytsia, Gaiodendron,* and *Atkinsonia,* which are root parasites; *roots modified to form haustoria;* branches terete, compressed or quadrangular. Hairs simple. Leaves opposite or subopposite, simple, entire, pinnately veined, with or without petioles, lacking stipules. Flowers solitary or in various sorts of inflorescences with flowers borne singly or in groups of 3, with or without bracts and bracteoles; with or without pedicels; forming umbels, corymbs, racemes, spikes, or heads. Flowers usually bisexual, radial or bilateral. **Sepals reduced to form a rim or calyculus on top of ovary.** *Petals* (3–) 5–6 (–9), *distinct or connate,* valvate, erect to reflexed at flowering, *often bright red or yellow.* Stamens as many as and opposite the petals, often 3 long and 3 short (longer ones staminodial in *Dendropemon*); filaments adnate to petals; pollen 3-lobed.

Carpels 3–4, connate; ovary inferior, with basal placentation; stigma capitate or scarcely expanded, papillate. Ovules not differentiated and megagametophytes produced from 3 or 4 points on large placenta. *Fruit a berry with a single seed* or a samara (*Nuytsia*), *viscous;* seed lacking a testa.

Floral formula:

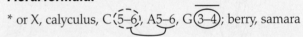

* or X, calyculus, C(5–6), A5–6, G(3–4); berry, samara

Distribution and ecology: Pantropical, but no single genus spans both Old and New Worlds.

Genera/species: 74/900.   **Major genera:** *Tapinanthus* (250 spp.), *Amyema* (100), *Psittacanthus* (50), *Struthanthus* (50), *Tapinanthus* (50), and *Phthirusa* (40). None occurs in temperate North America.

Economic plants and products: Mistletoes are a pest on timber trees because their haustoria cause irregularities in the wood structure.

Discussion: *Nuytsia,* a tree with root-haustoria, is sister to the rest of the family (Vidal-Russell and Nickrent 2005), most of which are stem (or epiphytic) parasites. Many genera have brightly colored corollas and are bird-pollinated (Plate 9.8E). The megagametophyte typically is "aggressive," growing out of the ovary and into the style and/or stigma, where fertilization occurs. The embryo is pushed back into the ovary by a long suspensor.

References: Calder and Bernhard 1983; Kuijt 1969, 1981.

(A)

Santalales: Santalaceae
Santalum paniculatum: fruits

(B)

Saxifragales: Crassulaceae
Kalanchoe daigremontiana:
leaf with plantlets

(C)

Saxifragales: Altingiaceae
Liquidambar styraciflua: staminate
and carpellate inflorescences

(D)

Saxifragales: Hamamelidaceae
Hamamelis mollis: inflorescences

(E)

Santalales: Loranthaceae
Dendrophthoe curvata: flowers

(F)

Santalales: Santalaceae
Arceuthobium pusillum: staminate
shoots (growing on *Picea*)

(G)

Saxifragales: Crassulaceae
Graptopetalum paraguayense: plant in bloom

(H)

Saxifragales: Saxifragaceae
Darmera peltata: flowers

PLATE 9.8 Eudicots:
Santalales and Saxifragales

Santalaceae R. Brown
(Sandalwood Family)

Root or stem parasites. Roots modified to form haustoria. Stems round, angled, or flattened, *jointed, breaking apart easily at the constricted nodes,* or not jointed, **with stomata transversely oriented**. Hairs simple. Leaves opposite or alternate and spiral, simple, often coriaceous or slightly succulent and shiny, entire, pinnately veined, with or without petioles, reduced to scales in some taxa, lacking stipules. Inflorescences variable, frequently spikes or racemes of 3-flowered cymes. *Flowers ± inconspicuous, bisexual or unisexual* (plants monoecious or dioecious), radial, with or without a hypanthium. Tepals 3–5, distinct to connate, erect or closed, valvate, greenish or drab. Stamens usually 3-5, opposite tepals, sometimes only unilocular or opening by an apical pore; pollen grains ± spherical. Carpels 3–4, connate; ovary usually inferior, with free-central to basal placentation; stigma punctate. Ovules with 1 integument and a thin-walled megasporangium, or not differentiated and 2 megagametophytes produced on the placenta. Fruit a nut or drupe, or a viscous berry with a single seed; seed lacking a testa (Figure 9.54; Plate 9.8A).

Floral formula:

Staminate: *, T–3–5–, A3–5, G0

Carpellate: *, T–3–5, A0, G(3–4); berry, drupe, nut

Distribution and ecology: Widely distributed, occurring from the tropics to temperate regions.

Genera/species: 44/950. **Major genera:** *Thesium* (325 spp.), *Phoradendron* (250), *Dendrophthora* (100), *Viscum* (130), and *Arceuthobium* (46). *Comandra, Geocaulon, Pyrularia, Nestronia, Buckleya,* and the stem parasites *Phoradendron* and *Arceuthobium* occur in the continental United States and/or Canada.

Economic plants and products: *Santalum* (sandalwood) is used for incense, and is the source of an aromatic oil used in cosmetics. Mistletoes—*Viscum album* in Europe and *Phoradendron leucarpon* in North America—are sold as decorations at Christmas time. Mistletoe species are forest parasites in many parts of the world, where they affect tree growth, vigor, fruiting, and wood quality. *Arceuthobium* (dwarf mistletoe) is a major pest on conifers in the western United States. The haustoria cause large knots, deformation in the wood, and "witches' brooms"—abnormally dense clusters of branches—which often make a tree completely useless for timber.

Discussion: The family is here broadly circumscribed, including the distinctive clade of stem (or epiphytic) parasites with stems that easily break apart at the constricted nodes (e.g., *Viscium, Phoradendron, Dendrophthora,* and relatives; often segregated as Viscaceae). The root-parasitic members of Santalaceae (e.g., *Santalum, Comandra, Buckleya*) comprise five or six clades, of uncertain relationship, but it is clear that they gave rise to the stem-parasitic clade.

The European mistletoe, *Viscum album,* is believed to be the Golden Bough of Aeneas. It was also the source of the spear that killed the Norse god Balder. The plant was central to Druid religious ceremonies and has long been a symbol of immortality.

It is significant that stem parasites have evolved from root parasites in both Santalaceae and Loranthaceae. In addition to the characters listed in the description, Santalaceae differ from Loranthaceae in their nonaggressive, usually *Allium* type (vs. aggressive and *Polygonum* type) gametophyte, green (vs. white) endosperm, and embryo with a short (vs. elongated) suspensor. Furthermore, the viscous part of the fruit (when present) arises from different tissues in the two families.

Reference: Kuijt 1982.

Saxifragales

Saxifragales are defined restrictively here, and include 14 families (e.g., **Saxifragaceae**, Iteaceae (e.g., *Itea*), Grossulariaceae (only *Ribes*), **Crassulaceae**, Haloragaceae, Cercidiphyllaceae, Paeoniaceae, **Hamamelidaceae**, and **Altingiaceae**, representing about 2470 species (Chase et al. 1993; Morgan and Soltis 1993; Soltis and Soltis 1997; Soltis et al. 1993, 1997). The group's monophyly is well supported by DNA sequences and morphology (Davis and Chase 2004; Fishbein et al. 2001; Hilu et al. 2003; Hoot et al. 1999; Hufford 1992; Savolainen et al. 2000b; Soltis and Soltis 1997; Soltis et al. 1997, 2000; Soltis and Hufford 2002). The order is clearly a member of the eudicot clade (Morgan and Soltis 1993; Soltis and Soltis 1997; Soltis et al. 1993, 1997, 2000, 2003a), and may be sister to the rosid clade. The order is morphologically similar to Rosales (and especially Rosaceae), but the major family, Saxifragaceae, can be readily separated from Rosaceae by the absence of stipules, fewer stamens, capsular fruits, and seeds possessing well-developed endosperm. The order is characterized by having the floral apex concave early in development and carpels that are free, at least apically; many also have flowers with a hypanthium. Members of this group have retained the plesiomorphies of 5-merous flowers with distinct parts.

Saxifragaceae A. L. de Jussieu
(Saxifrage Family)

Herbs; vessel elements with simple perforations; often with tannins, sometimes cyanogenic. Hairs often simple. Leaves usually alternate and spiral, sometimes in a basal rosette, simple to pinnately or palmately compound, entire to serrate or dentate, with venation pinnate to palmate; **stipules lacking** (or represented by expanded margins of the petiole base). Inflorescences determinate to indeterminate, usually terminal. *Flowers* bisexual to unisexual (plants then

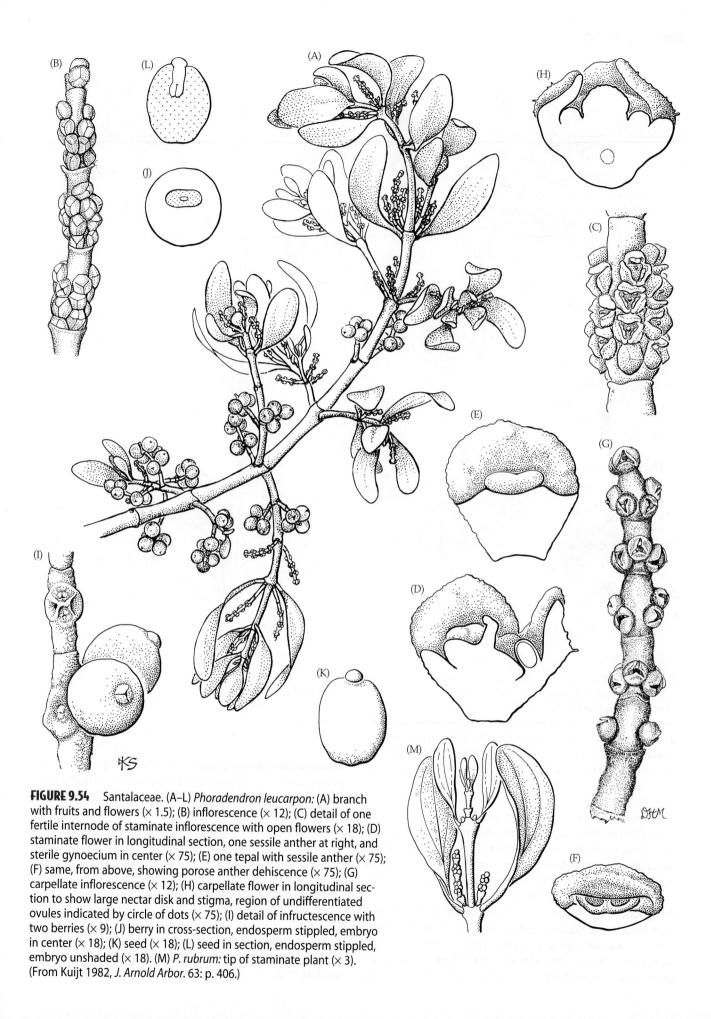

FIGURE 9.54 Santalaceae. (A–L) *Phoradendron leucarpon:* (A) branch with fruits and flowers (× 1.5); (B) inflorescence (× 12); (C) detail of one fertile internode of staminate inflorescence with open flowers (× 18); (D) staminate flower in longitudinal section, one sessile anther at right, and sterile gynoecium in center (× 75); (E) one tepal with sessile anther (× 75); (F) same, from above, showing porose anther dehiscence (× 75); (G) carpellate inflorescence (× 12); (H) carpellate flower in longitudinal section to show large nectar disk and stigma, region of undifferentiated ovules indicated by circle of dots (× 75); (I) detail of infructescence with two berries (× 9); (J) berry in cross-section, endosperm stippled, embryo in center (× 18); (K) seed (× 18); (L) seed in section, endosperm stippled, embryo unshaded (× 18). (M) *P. rubrum:* tip of staminate plant (× 3). (From Kuijt 1982, *J. Arnold Arbor.* 63: p. 406.)

petals

stamen

(C)

(J)

(B)

(A)

(F)

sepals

(H)

(E)

(G)

(D)

(I)

sepal

KS

FIGURE 9.55 Saxifragaceae. *Mitella diphylla*: (A) flowering plant (× 0.75); (B) detail of raceme (× 4); (C) flower (× 15); (D) flower in longitudinal section (× 17); (E) dehisced anther (× 35); (F) immature capsule (× 9); (G) top view of immature capsule (× 9) (H) floral cup and capsule in cross-section (× 9); (I) erect "splash cup" capsule after dehiscence (× 9); (J) seed (× 17). (From Spongberg 1972, *J. Arnold Arbor.* 53: p. 426.)

monoecious to ± dioecious), radial to bilateral, *with a variously developed hypanthium.* Sepals usually 4 or 5, distinct to connate. *Petals usually 4 or 5, distinct, often clawed,* sometimes variously dissected, imbricate or convolute, sometimes reduced or lacking. Stamens usually 3–10; pollen grains usually tricolpate or tricolporate. *Carpels 2 (–5), ± connate or less commonly distinct; ovary superior to inferior,* with axile or lateral placentation; stigmas separate, capitate. Ovules usually numerous on each placenta, with 1 or 2 integuments. Nectar disk often present around base of ovary. *Fruit a septicidal capsule or follicle* (Figure 9.55).

Floral formula:

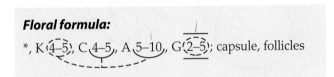

*, K (4–5), C (4–5), A 5–10, G (2–5); capsule, follicles

Distribution and ecology: Widely distributed in temperate and arctic regions, especially of the Northern Hemisphere, and often found in mountainous terrain.

Genera/species: 30/550. ***Major genera:*** *Saxifraga* (325 spp.), *Heuchera* (55), *Chrysosplenium* (55), *Mitella* (20), and *Astilbe* (20). In addition to the above listed genera, noteworthy genera in the colder regions of the continental United States and Canada include *Boykinia, Leptarrhena, Sullivantia, Tellima, Tolmiea,* and *Tiarella.*

Economic plants and products: *Saxifraga, Astilbe,* and a few other genera are cultivated in rock gardens or perennial borders.

Key to Major Families of Saxifragales

1. Plants woody and nonsucculent 2

1. Plants herbaceous and/or succulents 6

2. Leaves of long shoots opposite; fruits follicles; plants dioecious ....Cercidiphyllaceae

2. Leaves always alternate; fruits capsules or berries; plants with bisexual flowers,
   or if unisexual, then plants monoecious 3

3. Resin ducts in stems and leaves, the latter with sweet, resinous odor when crushed; flowers
   unisexual, carpellate flowers in a globular head, with gynoecia of adjacent flowers maturing to
   form a globose multiple fruit; each capsule surrounded by numerous minute lobes or scales ....**Altingiaceae**

3. Resin ducts absent, leaves without sweet odor; flowers bisexual, or if imperfect, carpellate
   flowers not as above; fruits not multiple, sometimes associated with a persistent calyx
   but not surrounded by numerous lobes or scales 4

4. Styles 2; hypanthium ± absent; petals often elongate and coiled; anthers usually
   opening by flaps; hairs ± stellate ....**Hamamelidaceae**

4. Style 1; hypanthium present, short to elongate; petals various but never coiled;
   anthers opening by slits; hairs not stellate 5

5. Leaves pinnately to palmately veined; flowers with a conspicuous hypanthium
   extending well above the inferior ovary; placentation parietal; fruits berries ....Grossulariaceae

5. Leaves pinnately veined; flowers with a short hypanthium and superior ovary;
   placentation axile; fruits capsules ....Iteaceae

6. Usually submerged to emergent aquatic herbs; flowers inconspicuous, solitary and
   axillary or in terminal spikes; ovules 1 per carpel; fruits nutlike or drupaceous,
   sometimes schizocarpic ....Haloragaceae

6. Terrestrial herbs or succulents; flowers conspicuous, usually in various determinate
   inflorescences; ovules few to numerous per carpel; fruits capsules or aggregates of follicles 7

7. Plants succulent; carpels as many as the petals, distinct or united only at base;
   each carpel subtended by a scale-like nectar gland; hypanthium usually lacking ....**Crassulaceae**

7. Plants not succulent; carpels usually fewer than the petals, usually ± connate; each carpel
   not associated with a nectar gland, although nectar disk sometimes present; hypanthium
   variously developed ....**Saxifragaceae**

Discussion: Saxifragaceae, as here circumscribed, are considered monophyletic on the basis of cpDNA restriction sites, *rbcL, matK,* and 18S sequences (Chase et al. 1993; Johnson and Soltis 1994; Morgan and Soltis 1993; Soltis and Soltis 1997; Soltis et al. 1993, 1997) and morphology. In addition, members of the family share an *rpl2* intron deletion.

Saxifragaceae traditionally have been broadly circumscribed, including both woody and herbaceous taxa with opposite or alternate leaves, and have been impossible to characterize (see Engler 1930; Schulze-Menz 1964). It is now clear that this broadly defined group is polyphyletic (Chase et al. 1993; Hufford 1992; Morgan and Soltis 1993; Soltis and Soltis 1997; Soltis et al. 1997). Some of the shrubby taxa previously associated with Saxifragaceae (e.g., *Itea* and *Ribes*) probably are closely related to the herbaceous saxifrages (Saxifragaceae, as here delimited), while others (e.g., *Hydrangea* and *Philadelphus*) are related to various groups within the asterid clade (see Hydrangeaceae). Even some herbaceous taxa are not closely related to Saxifragaceae, as recognized here; for example, *Parnassia* and *Lepuropetalon* (Parnassiaceae) are related to Celastraceae!

Until recently, relationships within Saxifragaceae s.s. have been poorly understood, probably as a result of the group's presumed rapid and recent diversification in cold temperate regions. Delimitation of some genera is problematic; many are monotypic, and some may not be monophyletic (e.g., *Saxifraga* and *Mitella*) (Soltis and Kuzoff 1995; Soltis et al. 1993, 1996, 2001). Hybridization sometimes causes taxonomic difficulties, and several intergeneric chloroplast capture events have been strongly supported (Soltis et al. 1991; Soltis and Kuzoff 1995). However, *matK, rbcL, trnL-F, psbA-trnH* nucleotide sequences and cpDNA restriction sites (Johnson and Soltis 1994; Soltis et al. 1991, 1995, 1996, 2001) support recognition of several monophyletic groups of genera within Saxifragaceae.

Flowers of Saxifragaceae (Plate 9.8H) are pollinated by various small, short-tongued insects (mainly flies and bees) that gather nectar and/or pollen. The numerous small seeds of *Saxifraga, Suksdorfia,* and *Boykinia* are borne on slender stems and probably scattered by wind or passing animals. In *Chrysosplenium* and *Mitella,* seeds are bounced from the erect, "splash-cup" capsule by raindrops. Vegetative reproduction by bulbils, vivipary, stolons, and rhizomes also occurs.

Additional references: Soltis and Hufford 2002; Spongberg 1972.

Crassulaceae J. St. Hilaire
(Stonecrop Family)

Succulent *herbs* to shrubs; stem often with cortical or medullary vascular bundles; **with crassulacean acid metabolism (CAM)**; tannins present; often with alkaloids, sometimes cyanogenic. Hairs simple, but plants more commonly glabrous and glaucous. **Leaves** alternate and spiral, opposite, or whorled, sometimes in a basal rosette, simple or rarely pinnately compound, entire to crenate, dentate or serrate, **succulent**, with pinnate venation, but veins often obscure; *stipules lacking.* Inflorescences determinate, sometimes reduced to a solitary flower, terminal or axillary. Flowers usually bisexual, radial, *lacking a hypanthium.* Sepals usually 4 or 5, distinct to connate. Petals usually 4 or 5, distinct to connate (and then forming a ± tubular corolla), imbricate. *Stamens 4–10;* filaments distinct to slightly connate, free or adnate to corolla; anthers opening by terminal pores; pollen grains tricolporate. *Carpels usually 4 or 5,* distinct *to slightly connate at base; ovaries superior,* with lateral placentation (or axile at base, if carpels fused); stigmas minute. **Each carpel subtended by a scale-like nectar-producing gland.** Ovules few to numerous in each carpel; the megasporangium wall thin or thick. *Fruit an aggregate of follicles,* rarely a capsule (Figure 9.56)

Floral formula:

*, K (4–5), C (4–5), A (4–10), G4–5; follicles

Distribution and ecology: Widespread from tropical to boreal regions; plants very often of arid habitats.

Genera/species: 35/1500. **Major genera:** *Sedum* (450), *Crassula* (300), *Echeveria* (150), and *Kalanchoe* (125). These, along with *Diamorpha, Dudleya, Graptopetalum, Lenophyllum,* and *Villadia,* occur in the continental United States and/or Canada.

Economic plants and products: *Sedum* (stonecrop), *Echeveria, Kalanchoe,* and *Sempervivum* (houseleek) are grown as ornamentals because of their distinctive succulent leaves.

Discussion: Analyses of *rbcL* and 18S rDNA sequences, along with morphology, support the monophyly of Crassu-

laceae (Chase et al. 1993; Morgan and Soltis 1993; Soltis and Soltis 1997; Soltis et al. 1997).

The family consists of two major clades: Crassuloideae (i.e., *Tillaea* and *Crassula*), diagnosed by their flowers with a single whorl of stamens and ovules with a thin megasporangium, and Sedoideae (e.g., *Aconium, Kalanchoe, Sedum, Echeveria, Villadia,* and *Sempervivum*) diagnosed by a usually ridged seed coat. Sedoideae have retained the plesiomorphic conditions of a thick megasporangium wall and stamens in two whorls. Within Sedoideae, the *Kalanchoe* clade may be sister to the rest. The *Kalanchoe* group has sympetalous flowers and opposite leaves, while the remaining genera (i.e., the *Sedum* clade) usually have separate petals and alternate leaves. The monophyly of the *Sedum* clade may be supported by its alternate leaves, assuming that the opposite and decussate condition of Crassuloideae and the *Kalanchoe* clade is plesiomorphic. Restriction sites and *matK* sequences strongly support the monophyly of Crassuloideae and Sedoideae (Mort et al. 2001; van Ham 1994; van Ham and t'Hart 1998). Several currently recognized genera (especially *Sedum*) are probably polyphyletic (Mort et al. 1998). Hybridization is frequent, and even occurs between members of different genera, which is hardly surprising given the non-monophyly of *Sedum* and some other genera.

The family is of physiological interest because it exhibits crassulacean acid metabolism (CAM), an adaptation for growth in arid habitats. The stomates open primarily at night and close during the day, thus reducing water loss. Carbon fixation occurs in the leaves at night, leading to the formation of malic acid. During the day, when the stomates are closed, the fixed carbon is reduced to carbohydrate. Other adaptations to arid conditions are the succulent leaves with abundant water storage tissue and the frequent occurrence of a thick waxy coating covering the epidermis (Plate 9.8G).

Flowers of most Crassulaceae are pollinated by a variety of insects, but some species of *Kalanchoe* have showy, sympetalous corollas and are bird-pollinated. The flowers are often protandrous, leading to outcrossing; but selfing may occur as well. The tiny seeds of Crassulaceae are probably dispersed by wind. Vegetative reproduction is common; some species of *Kalanchoe* produce numerous plantlets from adventitious buds associated with the leaf teeth (Plate 9.8B).

Additional reference: Spongberg 1978.

Hamamelidaceae R. Brown
(Witch Hazel Family)

Shrubs or trees; tannins often present. **Hairs stellate.** *Leaves alternate, often 2-ranked,* simple, entire to serrate, with pinnate or palmate venation; *stipules present, borne on the stem adjacent to the petiole base.* Inflorescences indeterminate, usually forming spikes, racemes, or heads; terminal or axillary. *Flowers*

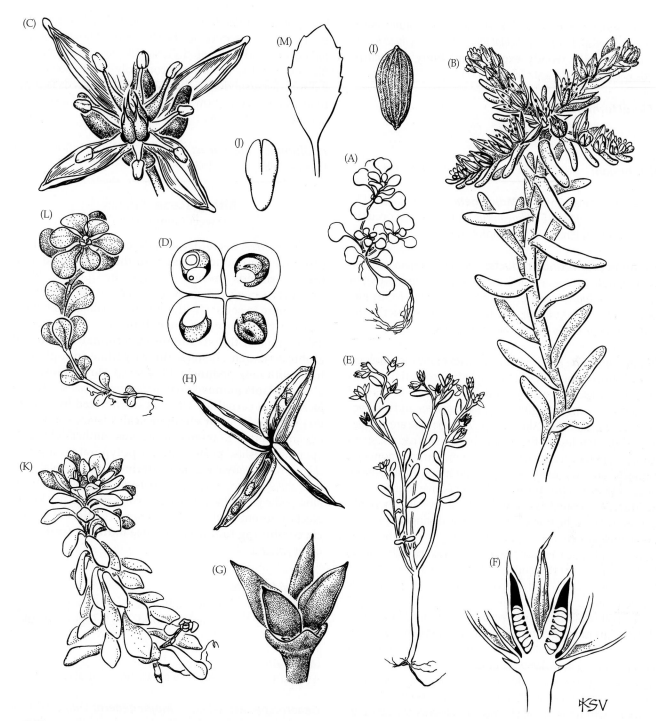

FIGURE 9.56 Crassulaceae. (A–D) *Sedum pulchellum:* (A) overwintering rosette (× 0.75); (B) flowering shoot (× 1.5); (C) flower (× 8); (D) cross-section through four carpels of gynoecium (× 30). (E–J) *S. pusillum:* (E) habit of mature plant (× 1.5); (F) immature follicles in longitudinal section, note nectaries (solid black) at base of carpels (× 9); (G) immature follicles (× 1.5); (H) mature, dehisced follicles (× 1.5); (I) seed (× 35); (J) embryo (× 35). (K) *S. glaucophyllum:* leafy shoot (× 1.5). (L) *S. ternatum,* leafy shoot (× 0.75). (M) *S. telephioides:* outline of leaf (× × 0.75). (From Spongberg 1978, *J. Arnold Arbor.* 59: p. 206.)

bisexual or unisexual (*and plants then monoecious), usually radial, showy to inconspicuous.* Sepals usually 4 or 5, distinct to connate, usually imbricate. *Petals usually 4 or 5, distinct, imbricate, valvate, or often circinately coiled in bud, sometimes lacking.* Stamens 4 or 5 and alternating with staminodes, or numerous; *anthers usually opening by 2 flaps;* pollen grains tricolpate to tricolporate. *Carpels 2, at least slightly connate;* **ovary half-inferior to inferior**, with axile placentation; *styles distinct, ± recurved, usually persistent;* stigmas 2, elongate along adaxial surface of style, or capitate. Ovules 1 to several per locule.

Nectar sometimes produced by staminodes or inner basal surface of petals. *Fruit a loculicidal to septicidal, woody to leathery capsule*, **with woody exocarp and bony endocarp**; seeds with thick hard testa (Figure 9.57)

Floral formula:

*, K (4–5), C4–5 or 0, A4–∞, G—②—; capsule

Distribution: Scattered in tropical to temperate regions.

Genera/species: 25/80. **Major genera:** *Corylopsis* (20 spp.) and *Distylium* (15). *Hamamelis* and *Fothergilla* occur in the continental United States and/or Canada.

Economic plants and products: Several genera provide ornamental shrubs or trees, including *Hamamelis* (witch hazel), *Corylopsis* (winter hazel), *Distylium, Fothergilla* (witch alder), *Loropetalum*, and *Rhodoleia*. An extract of the bark of *Hamamelis* is used as an astringent.

Discussion: The family is variable, with many morphologically distinct genera. Its monophyly has sometimes been questioned (Schwarzwalder and Dilcher 1991; Soltis and Soltis 1997; Soltis et al. 2000), and *Liquidambar* and relatives are here segregated as Altingiaceae. Altingiaceae are characterized by stipules on the petiole base, resin canals, spiral leaves, unisexual flowers that lack perianth parts, monoecy, elongate anthers, capsules tightly arranged in a globose head, and distinctive pollen grains that have four or more apertures. Persistent floral appendages (possibly perianth parts) surround the fruits. They are easily distinguished from Hamamelidaceae, which have stipules borne on the stem, often 2-ranked leaves, usually bisexual flowers with evident perianth parts, shorter anthers, fruits in spikes, racemes, or nonglobose heads, and usually tricolpate or tricolporate pollen grains.

Several subfamilies are recognized within Hamamelidaceae, but most species belong to the Hamamelidoideae, a probably monophyletic group (Endress 1989a,b 1993; Hufford and Crane 1989; Li et al. 1999). The Hamamelidoideae (e.g., *Corylopsis, Distylium, Fothergilla, Hamamelis,* and *Loropetalum*) may be united by the apomorphies of anthers with flaps, small stigmas, and leaves with pinnate venation with the secondary veins terminating in the teeth. Most also have 2-ranked leaves, carpels with only a single ovule, and ballistic seed dispersal. *Distylium, Fothergilla,* and relatives may form a clade characterized by loss of petals and aggregation of flowers into more or less capitate inflorescences (Endress 1989a). A possible clade containing *Hamamelis, Loropetalum,* and relatives has peculiar elongate petals (Plate 9.8D) that are circinately coiled in the bud (Figure 9.57C) (Endress 1989a), although it is not supported by ITS sequences (Li et al. 1999).

Both wind pollination and insect pollination (mainly by flies and bees) occur within the family (Endress 1977); either nectar or pollen is a pollinator reward. Flowers of the bird-pollinated *Rhodoleia* are aggregated and surrounded by showy inflorescence bracts. The reduced flowers of *Distylium* are wind-pollinated. *Fothergilla* has showy, inflated filaments and is probably secondarily insect-pollinated. Ballistic dispersal due to the ejection of the seeds is common. Many species grow along streams and are dispersed, at least in part, by water.

Additional references: Ernst 1963b; Hufford and Endress 1989; Tiffney 1986.

Altingiaceae Lindley
(Sweet Gum Family)

Shrubs or trees; **secretory canals containing aromatic resinous compounds present in the bark, wood, and leaves; iridoids present**; tannins often present. Hairs simple. *Leaves alternate and spiral*, simple, *often palmately lobed*, entire to serrate, with pinnate or palmate venation; *stipules present*, **borne on the petiole base**. Inflorescences indeterminate, **the staminate in terminal racemes of globose stamen clusters, the carpellate in a globose head with long peduncle** (Plate 9.8C). **Flowers unisexual (and plants monoecious)**, *radial, inconspicuous*. **Tepals lacking but carpellate flowers surrounded by numerous minute lobes or elongate scales** (sterile flowers or perianth parts). **Stamens numerous**; anthers elongate, opening by slits; **pollen grains polyporate**. *Carpels 2, slightly connate*; **ovary ± half-inferior**, with axile placentation; *styles distinct, ± recurved, persistent*; stigmas 2, elongate along adaxial surface of style. Ovules several per locule. Nectary absent. *Fruit a septicidal capsule*, **adjacent gynoecia combining to form a globose multiple fruit**; seeds often winged.

Floral formula:

Staminate: *, T-0-, A∞, G0
Carpellate: *, -∞-[minute lobes], A0, G—②—; capsule

Distribution: Temperate to tropical; Asia Minor, Southeast Asia, eastern North America south into Central America.

Genera/species: 1–3/12. **Major genera:** *Altingia* (7) and *Liquidambar* (4). Only *Liquidambar* occurs in the continental United States.

Economic plants and products: A fragrant gum is derived from several species of *Liquidambar* (sweet gum). Members of this genus are also used as ornamentals. Both *Altingia* and *Liquidambar* provide timber.

Discussion: The monophyly of Altingiaceae is supported by numerous morphological characters and DNA sequences (Ickert-Bond et al. 2005; Ickert-Bond and Wen 2006). The group is easily distinguished from Hamamelidaceae, in which it is often included (see discussion under that family).

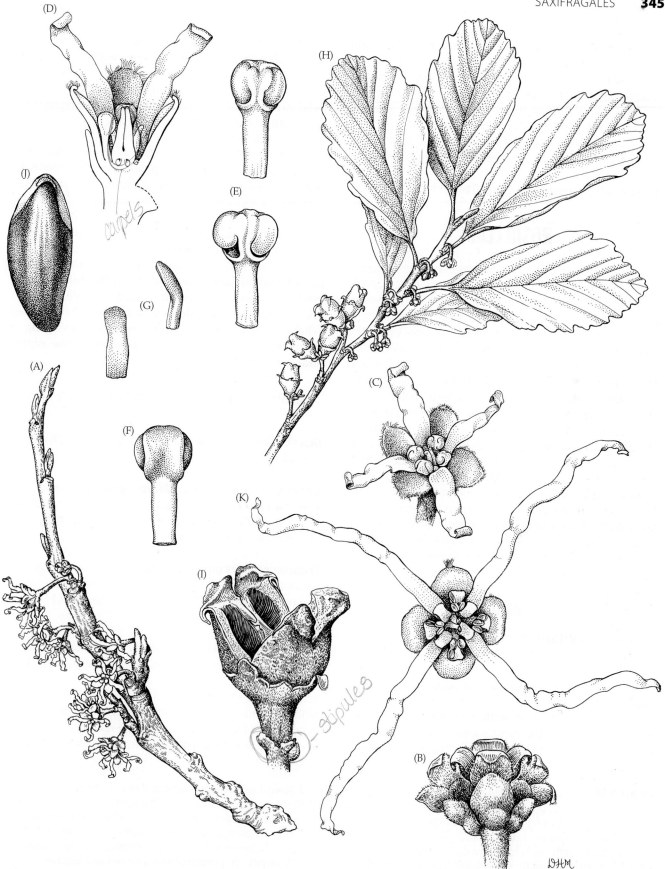

FIGURE 9.57 Hamamelidaceae. (A–J) *Hamamelis vernalis:* (A) branch in spring with flowers (× 1.5); (B) inflorescence of three flowers, petals removed to show sepals and bracts (× 6); (C) flower (× 6); (D) flower in longitudinal section, stamen at right removed to show staminode opposite petal (× 8); (E) two sta- mens, showing valves (× 22); (F) stamen, showing connective (× 22); (G) staminodes (× 22); (H) branch in fall with fruits and flower buds (× 0.75); (I) capsule (× 3); (J) seed (× 6). (K) *H. virgini- ana:* flower (× 8). (From Ernst 1963, *J. Arnold Arbor.* 44: p. 198.)

The western Asian species, *Liquidambar orientalis*, and the North American, *L. styraciflua*, form a clade, while *L. formosa* and *L. acalycina*, both of southeastern Asia, are related to a group of southeastern Asian species of *Altingia*; neither *Liquidambar* nor *Altingia* are monophyletic (Ickert-Bond et al. 2005; Ickert-Bond and Wen 2006).

Flowers of Altingiaceae are pollinated by the wind and the seeds, which are often winged, are wind-dispersed.

Additional references: Endress 1989a, 1993; Ernst 1963b.

ROSID CLADE

The monophyly of this rather heterogeneous grouping of orders has received support from analyses of *rbcL*, *atpB*, *matK*, and 18S rDNA sequences (see Judd and Olmstead 2004; Soltis et al. 2005). Most members of this group belong to one of two major subclades, which are here called the fabids (or eurosids I: Zygophyllales, Celastrales, Malpighiales, Oxalidales, Fabales, Rosales, Cucurbitales, and Fagales) and malvids (or eurosids II: Brassicales, Malvales, and Sapindales). The position of Myrtales is unclear (either fabid or malvid), and here it is unplaced and listed between these two major clades. Support for these two groups comes from recent phylogenetic analyses based on molecular characters (Angiosperm Phylogeny Group 1998, 2002; Hilu et al. 2003; Soltis et al. 2000). Rosales, Fabales, Cucurbitales, and Fagales likely form a clade, and it is noteworthy that some members of all of these orders have nitrogen-fixing nodules on their roots (usually inhabited by *Frankia*, but *Rhizobium* in Fabaceae; Soltis et al. 1995). The position of Saxifragales is still somewhat problematic; it also may belong in the rosid clade.

Vitales

Vitaceae A. L. de Jussieu
(Grape Family)

Usually lianas with leaf-opposed tendrils (modified inflorescences) that attach themselves by twining or by adhesive discs, or shrubs lacking tendrils, **the stem with ± swollen nodes; often with raphide sacs.** Hairs various. *Leaves alternate, spiral or 2-ranked, simple to palmately or pinnately compound, palmately to pinnately veined; stipules present. Inflorescences determinate,* terminal, but *usually appearing leaf-opposed due to growth of the axillary branch from the opposing leaf axil.* Flowers bisexual or unisexual (plants then polygamodioecious or monoecious), radial. *Sepals usually 4–6, ± connate,* **small,** often represented only by an obscurely toothed or lobed rim. *Petals usually 4–6, distinct or distally falsely connate* (due to interlocked papillae, as in *Vitis*) and at flowering deciduous as a cap, valvate. *Stamens usually 4–6,* **opposite the petals,** sometimes connate; pollen grains tricolporate. *Carpels 2, connate; ovary superior,* 2-loculate or 4-loculate (by secondary division) with axile placentation; stigma usually capitate. *Ovules 2 per locule.* Nectar disk prominent, usually forming a ring between the ovary and the stamens. **Fruit a berry;** *seeds 4, with a thin, translucent outer layer and a hard inner layer,* **with a cordlike raphe on the adaxial surface, extending from the hilum to the seed apex and onto the convex abaxial side, where it joins a round to linear, depressed to somewhat elevated "chalazal knot," and also with a deep groove varying in shape and length flanking both sides of the raphe; endosperm 3-lobed** (Figure 9.58).

Floral formula: *, K ④–⑤, C ④–⑤, A4–5, G②; berry

Distribution: Widely distributed, but most diverse in tropical and subtropical regions.

Genera/species: 14/725. ***Major genera:*** *Cissus* (300 spp.), *Vitis* (60), *Leea* (24), *Ampelopsis* (20), and *Parthenocissus* (15). All these, except *Leea*, occur in the continental United States and/or Canada.

Economic plants and products: Several species of *Vitis* (grapes) are of great commercial importance as sources of grapes, juice, wine, and raisins. *Parthenocissus* and *Leea* are used ornamentally.

Discussion: Vitaceae are easily recognized and are surely monophyletic (Ingrouille et al. 2002; Soejima and Wen 2006; Soltis et al. 2000). *Leea* is sister to the remaining genera, which are united by their liana habit and leaf-opposed tendrils. Genera of Vitaceae are distinguished on the basis of structure of the nectar disk, configuration of endosperm in cross-section, length of the style, and number and con-

FIGURE 9.58 Vitaceae. (A–R) *Vitis rotundifolia:* (A) portion of flowering vine (× 0.5); (B) staminate flower, note that petals have dropped off (× 8); (C) inflorescence opposite petiole (× 0.33); (D) flower bud with pseudoconnate petals (× 10); (E) opening bud of bisexual flower, with petals forming a cap (× 10); (F) bisexual flower, note pseudoconnate petals falling as a cap (× 10); (G) gynoecium in longitudinal section (× 20); (H) ovary in cross-section, showing four ovules (× 20); (I) branch with tendril, opposite a leaf, and infructescence (× 0.5); (J) berry in cross-section (× 4); (K) seed, with outer membranous seed coat attached (× 4); (L) seed, abaxial surface, note raphe and round, chalazal knot (× 4); (M, N) seeds, adaxial surface (× 4); (O) seed in cross-section, note three-lobed endosperm (stippled) (× 8); (P) seed in longitudinal section, endosperm stippled, embryo small, note chalazal knot (circular structure to left) (× 8); (Q, R) embryo (greatly magnified). (S–W) *V. vulpina:* (S) portion of vine with leaf and opposite infructescence (× 0.5); (T–V) seeds, abaxial surface (× 4); (W) seed, adaxial surface (× 4). (Original art prepared for the Generic Flora of the Southeastern U.S. Project. Used with permission.)

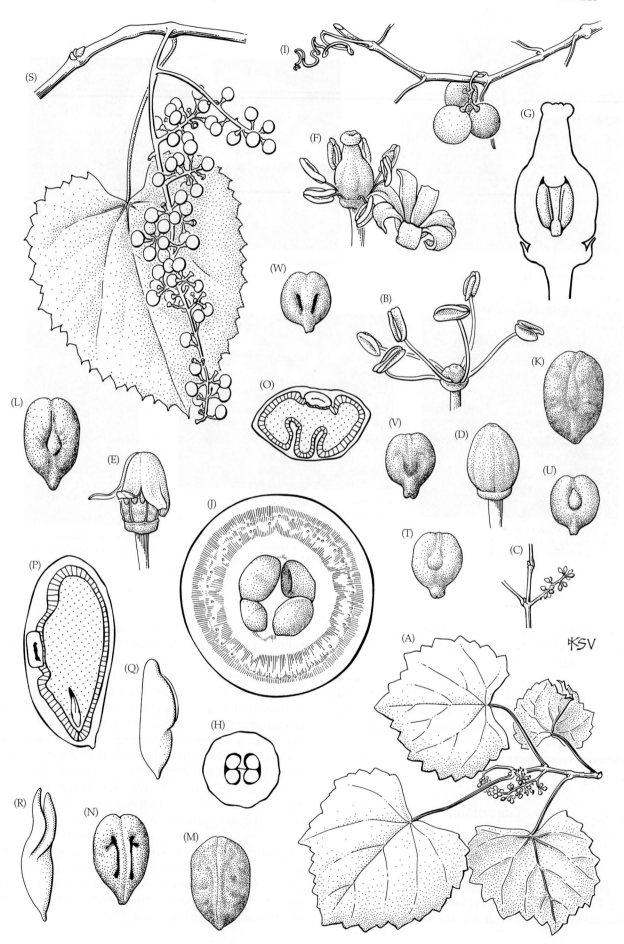

Zygophyllales: Zygophyllaceae
Guaiacum sanctum: flower

Oxalidales: Oxalidaceae
Oxalis pes-caprae: flowers

O. pes-caprae: dissected flower,
showing androecium and
gynoecium

Celastrales: Celastraceae
Euonymus americanus: branch
with capsules and arillate seeds

Celastrales: Celastraceae
Celastrus scandens: flowers

Geraniales: Geraniaceae
Geranium maculatum: fruit

**PLATE 9.9 Eudicots: Geraniales,
Zygophyllales, Oxalidales, and Celastrales**

nation of petals. Some are apparently non-monophyletic (Ingrouille et al. 2002), and there has been considerable parallel evolution in leaf shape and tendril and inflorescence form.

The small flowers, with their exposed nectaries, are visited by bees, wasps, flies, and beetles. Both selfing and outcrossing occur. The fleshy fruits are animal-dispersed.

Additional references: Brizicky 1965b; Gerrath et al. 2001.

Geraniales

Geraniaceae A. L. de Jussieu
(Geranium Family)

Usually herbs to subshrubs, with stems usually jointed at the nodes. Hairs simple, *often gland-headed, with aromatic oils. Leaves alternate and spiral, or opposite, simple and palmately lobed, dissected or compound,* ± serrate to entire, with ± palmate venation; stipules usually present. Inflorescences determinate, often umbel-like, sometimes reduced to a single flower, terminal or axillary. Flowers usually bisexual, radial or bilateral. *Sepals usually 5,* distinct or basally connate, occasionally the upper one prolonged backward into a nectar-producing spur that is an integral part of the pedicel. *Petals usually 5, distinct,* often emarginate, usually imbricate. Stamens (5–) 10–15; filaments distinct to slightly connate; pollen grains often tricolporate. Carpels usually 5, connate; *ovary superior,* ± *lobed,* with axile placentation and usually *with a prominent, elongate, persistent, terminal sterile column; style 1, stigmas 5, distinct,* ± *elongate. Ovules 2 in each locule,* anatropous to campylotropous. Nectar glands usually alternate with the petals (or lacking). *Fruit a schizocarp with 5 one-seeded segments that separate elastically from the persistent central column, and often opening to release the seed,* or a loculicidal capsule (*Hypseocharis*); embryo straight to curved; **endosperm scanty or lacking** (Figure 9.59).

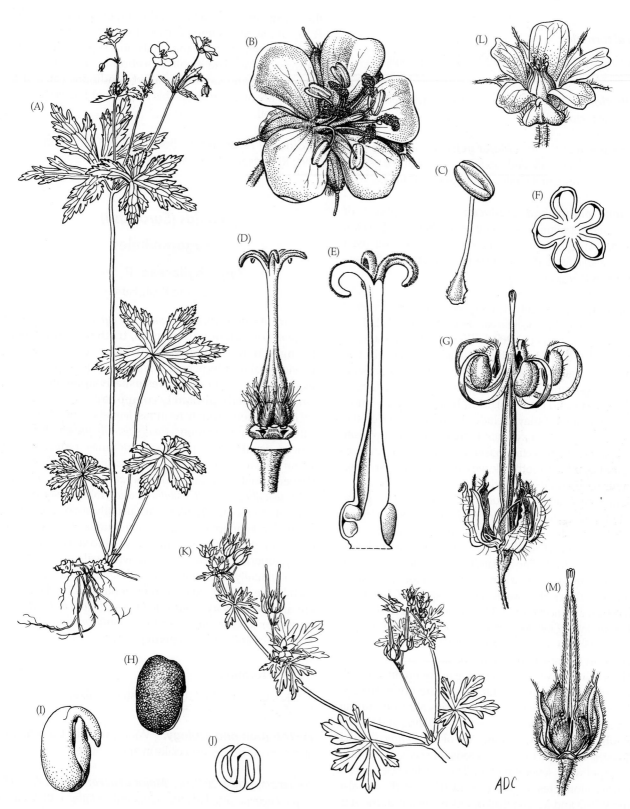

FIGURE 9.59 Geraniaceae. (A–J) *Geranium maculatum:* (A) flowering plant (× 0.5); (B) flower (× 2.25); (C) stamen of outer whorl (× 6); (D) gynoecium with nectar glands on receptacle (below pubescent ovary) (× 6); (E) gynoecium in longitudinal section (× 9); (F) ovary in cross-section (× 12); (G) dehisced fruit, showing segments attached to recurved hygroscopic awns (× 3.0); (H) seed (× 7.5); (I) embryo (× 7.5); (J) embryo in cross-section to show folding of cotyledons (× 7.5). (K–M) *G. carolinianum:* (K) branch with flowers and fruits (× 0.5); (L) flower with receptive stigmas, anthers mostly fallen (× 4.5); (M) mature fruit before dehiscence (× 3). (From Robertson 1972, *J. Arnold Arbor.* 53: p. 190.)

Floral formula:

* or X, K (5), C5, A (10–15), G (5); schizocarp, capsule

Distribution: Widespread, especially in temperate and subtropical regions.

Genera/species: 7/750. **Major genera:** *Geranium* (300 spp.), *Pelargonium* (250), and *Erodium* (75). All major genera occur in the continental United States and/or Canada.

Economic plants and products: Several species of *Pelargonium* (geranium), *Geranium* (cranesbill), and *Erodium* (storksbill) are grown as ornamentals. Geranium oil, used in perfumes, is distilled from the leaves and shoots of several species of *Pelargonium*.

Discussion: The family usually has been considered closely related to the Oxalidaceae, but DNA-based cladistic analyses (Savolainen et al. 2000b; Soltis et al. 2000) suggest that it is related to Melianthaceae, Vivianiaceae, and two other small families (APG 2003), with Geraniales possibly the sister group of Crossosomatales (Crossosomataceae, Staphyleaceae, and relatives). Morphological synapomorphies of Geraniales may include the outer whorl of stamens, or the single whorl when there is only one, opposite the sepals, the floral nectary positioned outside the androecium, and glandular-toothed leaves. In addition, their vessel elements have simple perforations.

Geraniaceae are a well-defined monophyletic group based on *rbcL* sequences (Price and Palmer 1993). *Hypseocharis* (usually placed in the Oxalidaceae) is sister to the remaining genera. This genus retains the plesiomorphic feature of capsular fruits. The remaining genera all have schizocarpic fruits, usually with a central stylar column (both apomorphies). Bilateral flowers and a spurred sepal are probably autapomorphies of *Pelargonium*, which is sister to a clade containing *Erodium*, *Geranium*, and relatives (Price and Palmer 1993). Note that the spur can be recognized only as an indistinct line and bump on the pedicel.

The showy flowers of Geraniaceae are pollinated by a wide variety of insects; nectar guides are often present, and nectar is the reward. Many species are protandrous and outcrossing, but some are self-pollinating and weedy. The schizocarps of most species of *Geranium* dehisce explosively and may throw the seeds (or seeds and associated carpel wall) several meters. The sterile upper part of the ovary elongates greatly in fruit (Plate 9.9E; see also Figure 9.59D and M. The fruit opens by means of the segments separating from the central column, each segment being composed of a globose basal portion (one of the ovary lobes) and an awnlike upper portion (narrow outer layer of the column). The segments curve upward and inward (Figure 9.59G), and the seeds may be forcibly ejected from the globose portion of the segment due to the drying and contraction of the carpel wall. In *Erodium*, *Pelargonium*, and some species of *Geranium*, the schizocarpic segments retain the seeds and separate from the central column. The awnlike portion of the segment is hygroscopic, and varying weather conditions cause it to coil and uncoil repeatedly. This movement forces the seed into the soil.

Additional references: Boesewinkel 1988; Rama Devi 1991; Robertson 1972a; Yeo 1984.

Fabids (Eurosids I)
Zygophyllales

Zygophyllaceae R. Brown
(Creosote Bush Family)

Trees, shrubs, or herbs, with stems often sympodial and jointed at the nodes; **xylem with vessels, tracheids, and fibers arranged in horizontally aligned tiers**; usually producing steroidal or triterpenoid saponins, sesquiterpenes, and alkaloids. Hairs various. **Leaves usually opposite**, *usually 2-ranked*, **usually pinnately compound**, *without a terminal leaflet*, often strongly resinous, leaflets entire, with pinnate to palmate venation; *stipules usually present*. Inflorescences usually determinate, sometimes reduced to a solitary flower, terminal, but often appearing to be lateral. Flowers usually bisexual, radial. *Sepals usually 5, ± distinct. Petals usually 5, distinct, usually clawed*, imbricate or convolute. Stamens 10–15; filaments usually associated with basal glands or appendages; pollen grains often tricolporate. Carpels usually 5, sometimes reduced to 2, connate; *ovary* superior, *ridged or winged*, with axile placentation; stigma usually 1, capitate to lobed. Ovules 1 to several in each locule, anatropous to orthotropous. Nectar disk present at base of ovary. *Fruit usually a septicidal or loculicidal capsule or a schizocarp, sometimes spiny or winged*; seeds sometimes arillate; endosperm present.

Floral formula:

*, K5, C5, A10–15, G (2–5); capsule, schizocarp

Distribution and ecology: Widespread in tropical and subtropical regions, especially in arid habitats.

Genera/species: 26/200. **Major genera:** *Zygophyllum* (80 spp.), *Fagonia* (40), *Balanites* (20), and *Tribulus* (20). Genera occurring in the continental United States include *Guaiacum*, *Kallstroemia*, *Larrea*, *Porlieria*, *Tribulus*, and *Zygophyllum*.

Economic plants and products: Species of *Guaiacum* (lignum vitae) provide a strong, hard, heavy, self-lubricating wood; the wood of *Bulnesia* is similar. The family also contains several ornamentals, including *Larrea* (creosote bush), *Guaiacum*, and *Tribulus* (caltrop).

Discussion: Zygophyllaceae are an isolated group, which are possibly sister to the remaining members of the Fabids. The family is considered to be monophyletic, based on morphological and DNA characters, after a few genera (e.g., *Nitraria, Malacocarpus,* and *Peganum*) are segregated as Nitrariaceae (Sapindales) (Gadek et al. 1996; Ronse Decraene et al. 1996). Analyses of *rbcL* sequence characters suggest that Zygophyllaceae are the sister group of Krameriaceae, a small, New World family of root parasites with bilateral resupinate flowers, petaloid calyx, two highly modified and oil-producing petals, and globose, spiny fruits (Cronquist 1981; Källersjö et al. 1998; Savolainen et al. 2000b; Soltis et al. 2000).

Within the Zygophyllaceae, analyses of *rbcL* and *trnL-F* sequences (Sheahan and Chase 2000) suggest that major clades include the Tribuloideae (e.g., *Tribulus, Kallstroemia, Balanites*), Larreoideae (e.g., *Bulnesia, Guaiacum, Larrea*), and Zygophylloideae (e.g., *Fagonia, Zygophyllum*).

The showy, nectar-producing flowers are pollinated by various insects (Plate 9.9A). Capsules of *Guaiacum* open to expose colorful arillate seeds and are probably bird-dispersed. The winged schizocarps of *Bulnesia* are dispersed by wind, while the spiny schizocarps of *Tribulus* are externally transported by animals.

Additional references: Porter 1972; Sheahan and Cutler 1993.

Oxalidales

Oxalidaceae R. Brown
(Wood Sorrel Family)

Herbs, often with bulblike tubers or fleshy rhizomes, to shrubs or trees; **with high levels of soluble and crystalline oxalates**. Hairs simple. *Leaves alternate and spiral, sometimes forming a basal rosette, palmately to pinnately compound, or reduced and trifoliate or unifoliate*, often with prominent pulvini and showing sleep movements, entire, often emarginate, with palmate to pinnate venation; stipules usually lacking. Inflorescences determinate, often umbel-like, sometimes reduced to a single flower, axillary. Flowers bisexual, radial, **heterostylous** (sometimes not in weedy species). *Sepals 5, distinct. Petals 5, distinct* or very slightly connate, usually convolute. *Stamens* usually 10; filaments connate basally, *with outer filaments shorter than inner*; pollen grains usually tricolpate or tricolporate. Carpels usually 5, connate; ovary superior, ± lobed, with axile placentation; styles usually 5, distinct; stigmas usually capitate or punctate. Ovules usually several in each locule, **with a thin megasporangium wall**. Nectar produced by base of filaments or glands alternating with petals. *Fruit a loculicidal capsule* or berry, *often lobed or angled; seeds sometimes arillate,*

the outer part of testa often elastically turning inside out and ejecting the seed (Figure 9.60).

Floral formula: *, K5, C5, A ⑩, G⑤; capsule, berry

Distribution: Widespread, especially in tropical and subtropical regions.

Genera/species: 6/770. **Major genera:** *Oxalis* (800 spp.) and *Biophytum* (70). Only *Oxalis* occurs in the continental United States and Canada.

Economic plants and products: *Averrhoa carambola* (carambola, star fruit) has edible fruits; numerous cultivars exist that vary greatly in acidity (oxalate content). The tubers of *Oxalis tuberosa* (oca) are eaten in Andean South America.

Discussion: Monophyly of Oxalidaceae is supported by morphological and DNA characters (Price and Palmer 1993).

Oxalidaceae usually have been considered closely related to Geraniaceae on the basis of their actinomorphic, 5-merous flowers with ten stamens and lobed, syncarpous gynoecium (Plate 9.9B). However, these similarities are symplesiomorphic. Separate styles nicely distinguish Oxalidaceae from Geraniaceae, but the feature is probably plesiomorphic. Phylogenetic studies based on *rbcL, atpB,* and 18S sequences indicate that Oxalidaceae are more closely related to Cunoniaceae and Cephalotaceae (and three other small families), here treated as the order Oxalidales, than to Geraniaceae (Price and Palmer 1993; Soltis et al. 2000). Cunoniaceae are pantropical trees and shrubs with opposite, pinnately compound, stipulate leaves; Cephalotaceae are Australian insectivorous herbs with a rosette of pitcherlike leaves. Oxalidales are morphologically heterogeneous, but the order's monophyly is supported by molecular analyses.

Heterostyly, both distyly and tristyly, is characteristic of Oxalidaceae. The family is characterized by outcrossing, but some weedy species are selfing. The showy, nectar-producing flowers of Oxalidaceae are pollinated by various insects. Most members of the family are self-dispersed by means of the explosive inversion of the smooth, turgid outer part of the testa.

Additional references: Denton 1973; Ornduff 1972; Robertson 1975.

Celastrales

Celastraceae R. Brown
(Bittersweet Family)

Trees, shrubs, or lianas climbing by twining stems or hook-like branches; often with tannins. Hairs simple to branched.

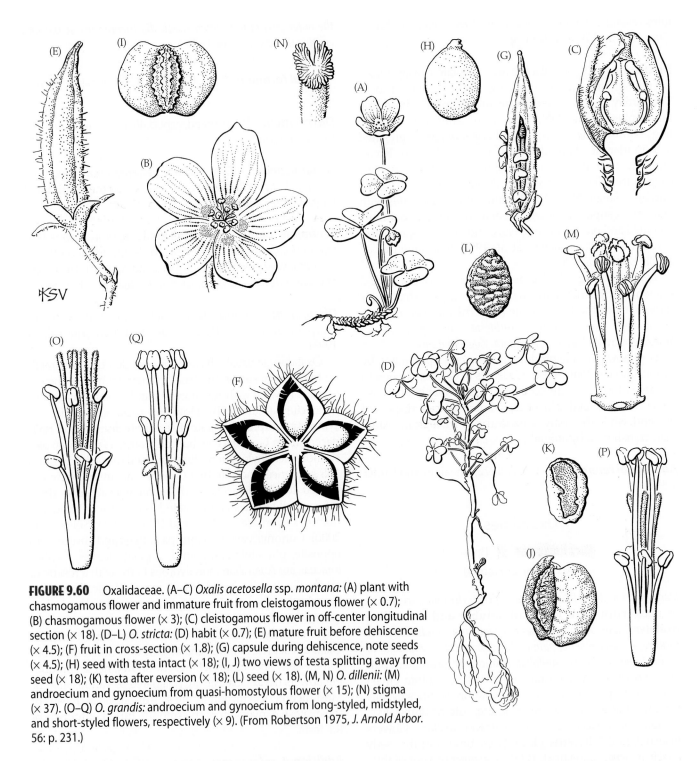

FIGURE 9.60 Oxalidaceae. (A–C) *Oxalis acetosella* ssp. *montana:* (A) plant with chasmogamous flower and immature fruit from cleistogamous flower (× 0.7); (B) chasmogamous flower (× 3); (C) cleistogamous flower in off-center longitudinal section (× 18). (D–L) *O. stricta:* (D) habit (× 0.7); (E) mature fruit before dehiscence (× 4.5); (F) fruit in cross-section (× 1.8); (G) capsule during dehiscence, note seeds (× 4.5); (H) seed with testa intact (× 18); (I, J) two views of testa splitting away from seed (× 18); (K) testa after eversion (× 18); (L) seed (× 18). (M, N) *O. dillenii:* (M) androecium and gynoecium from quasi-homostylous flower (× 15); (N) stigma (× 37). (O–Q) *O. grandis:* androecium and gynoecium from long-styled, midstyled, and short-styled flowers, respectively (× 9). (From Robertson 1975, *J. Arnold Arbor.* 56: p. 231.)

Leaves alternate and spiral or 2-ranked, or opposite, simple, entire to serrate, with pinnate venation; stipules present or lacking. Inflorescences usually determinate, terminal or axillary. Flowers bisexual or occasionally unisexual (plants monoecious to dioecious), radial, sometimes with a short hypanthium. *Sepals usually 4 or 5, distinct to slightly connate. Petals 4 or 5, distinct*, imbricate to occasionally valvate. **Stamens 3–5**, *alternating with petals;* filaments distinct to connate; pollen grains tricolporate or triporate, occasionally in tetrads or polyads. Carpels 2–5, connate; ovary superior to half-inferior, with axile placentation, **the locules dorsally bulged up, so septa apical**; stigma capitate to lobed. Ovules 2 to numerous per locule, sometimes with a thin megasporangium wall. **Conspicuous nectar disk present**, sometimes adnate to ovary. *Fruit a loculicidal capsule (sometimes strongly 3-lobed), schizocarp, drupe, or berry; seeds often with an orange to red aril or winged;* endosperm sometimes lacking.

Distribution: Widely distributed in tropical and subtropical regions, with a few genera extending into temperate areas.

Genera/species: 98/1221. **Major genera:** *Maytenus* (200 spp.), *Salacia* (200), and *Euonymus* (130). Noteworthy genera in the continental United States and/or Canada include *Celastrus*, *Crossopetalum*, *Euonymus*, *Hippocratea*, *Maytenus*, *Paxistima*, and *Schaefferia*.

Economic plants and products: *Celastrus* (bittersweet), *Euonymus*, and *Schaefferia* are used as ornamentals. The narcotic khat is derived from the leaves of *Catha*.

Discussion: Celastraceae are broadly circumscribed, including Hippocrateaceae, which are lianas with three stamens borne inside the disk, transverse anther dehiscence, berries or three-lobed capsules, and seeds that lack endosperm. The family is considered to be monophyletic on the basis of morphology and DNA sequences (Mathews and Endress 2005; Savolainen et al. 1994). Segregation of Hippocrateaceae would render Celastraceae paraphyletic (Clevinger and Panero 1998; Simmons and Hedin 1999; Simmons et al. 2000, 2001). Arils probably evolved twice in the family, in *Canotia* and in the common ancestor of all the remaining arillate genera (such as *Catha*, *Euonymus*, *Celastrus*, and *Maytenus*). In *Hippocratea* and relatives the aril has been modified into a basal wing, and in *Salacia* and relatives, into a mucilaginous pulp. The hippocrateoid group is likely monophyletic, as indicated by the loss of endosperm. The shrub *Brexia* (previously in Brexiaceae) is also included within Celastraceae.

The herbaceous genera *Parnassia* and *Lepuropetalon* (Parnassiaceae) are closely related to Celastraceae (Zhang and Simmons 2006).

The often inconspicuous, green to white flowers (Plate 9.9D) are pollinated by bees, flies, and beetles; nectar is the reward. The fruits are often brightly colored capsules that open to expose arillate seeds that are dispersed by birds (Plate 9.9C). The colorful drupes or berries of some species also are bird-dispersed. *Hippocratea* and relatives have winged seeds and are dispersed by wind.

Additional references: Brizicky 1964a; Hallé 1962.

Malpighiales

Monophyly of Malpighiales is indicated only by molecular data (Chase et al. 1993; Davis and Chase 2004; Hilu et al. 2003; Savolainen et al. 2000a,b; Soltis et al. 1998, 2000). The group is morphologically heterogeneous, but many have dry stigmas, a fibrous exotegmen, and toothed leaves, with teeth having a single vein running into a congested and often deciduous apex (i.e., teeth violoid, salicoid, or theoid). Several families are predominantly three-carpellate (e.g., Euphorbiaceae, Malpighiaceae, Passifloraceae, Violaceae). Groups such as Violaceae, Salicaceae, and Passifloraceae are distinct because of their parietal placentation (Cronquist 1981, 1988; Thorne 1992), and these families traditionally have been placed in the order "Violales."

Malpighiales contain 38 families and 16,000 species. Major families include Achariaceae, **Clusiaceae**, Chrysobalanaceae, **Euphorbiaceae, Hypericaceae, Malpighiaceae**, Ochnaceae, **Passifloraceae, Phyllanthaceae**, Picrodendraceae, Podostemonaceae, Rafflesiaceae, **Rhizophoraceae, Salicaceae**, and **Violaceae**.

Phylogenetic relationships within the order are problematic, but have recently been somewhat clarified by Chase et al. (2002), Davis and Chase (2004), and Davis et al. (2005). Salicaceae, Violaceae, Achariaceae, Turneraceae, and Passifloraceae are characterized by parietal placentation; most other Malpighiales have axile placentation. Achariaceae, Turneraceae, and Passifloraceae are related, and all share cyclopentenoid cyanogenic glucosides and cyclopentenyl fatty acids; Turneraceae and Passifloraceae (likely sister families) have flowers in which the stamens are not associated with the hypanthium, and arillate seeds. Rhizophoraceae and Erythroxylaceae may be sister families; they share tropane and pyrrolidine alkaloids, a terminal bud protected by stipules, and green embryos. Clusiaceae, Hypericaceae, and Podostemaceae have similar xanthone pigments and secretory tissues/cells.

Malpighiaceae A. L. de Jussieu
(Barbados Cherry Family)

Shrubs, trees, lianas, or occasionally perennial herbs. **Hairs various, but always 1-celled, usually ± attached such that the hair has 2 arms, and is T-, V-, or Y-shaped**, *the stalk often short and the arms straight to twisted. Leaves usually opposite, simple*, usually entire, occasionally lobed, with usually pinnate venation, *often with 2 or more glands on petiole or abaxially on blade*; stipules usually present. Inflorescences determinate, but often appearing indeterminate, terminal or axillary. *Flowers usually bisexual, usually ± bilateral.* **Sepals 5**, distinct to connate basally, **with 2 conspicuous, abaxial, oil-producing glands on all 5 sepals or on the 4 lateral sepals**, oil glands vestigial or lacking in a few Neotropical species and most Paleotropical species. *Petals 5, distinct*, **usually clawed**, often with fringed or toothed margins (Plate 9.10D), *the upper one usually slightly larger or smaller than the others and sometimes differentiated in color as well*, imbricate. Stamens usually 10; *filaments usually connate basally*; pollen grains usually 3–5-colporate or 4– to polyporate. *Carpels usually 3, connate*; ovary superior, with axile placentation; styles usually distinct; stigmas various. Ovules 1 in each locule; megagametophyte usually 16-nucleate.

Key to Major Families of Malpighiales

1. Plants aquatic, usually in rapidly flowing water; stems, roots, and leaves often not clearly distinguished .... Podostemonaceae

1. Plants terrestrial or in mangrove communities; stems, roots, and leaves clearly distinguished 2

2. Plants with white to colored latex or other exudate in secretory canals, or clear to black resins in secretory cavities (pellucid or black dots) 3

2. Plants lacking resins or latex, without canals or cavities 5

3. Leaves usually opposite or whorled, lacking stipules, with clear, black, or ± colored resins or exudate in canals or cavities; flowers bisexual or unisexual, styles usually not divided and number of stigmas equaling or fewer than the carpels; fruit a capsule, berry, or drupe 4

3. Leaves usually alternate, stipulate, with white (rarely colored) latex in laticifers; flowers always unisexual, styles forked to several times divided (and thus number of stigmas greater than number of carpels); fruit ± schizocarpic .... **Euphorbiaceae**

4. Secretory system usually of canals or canals and cavities; stems with colored sap; flowers bisexual or unisexual, with short to elongate styles, and usually ± expanded stigmas; seeds often arillate .... **Clusiaceae**

4. Secretory system usually of cavities; stems with clear sap; flowers bisexual, with elongate styles and minute stigmas; seeds not arillate .... **Hypericaceae**

5. Style gynobasic; gynoecium apparently a single carpel or the ovary deeply lobed and thus appearing to be of distinct carpels 6

5. Style terminal; gynoecium obviously syncarpous 7

6. Gynoecium apparently a single carpel, with a lateral style, fruit a single drupe .... Chrysobalanaceae

6. Gynoecium syncarpous, the ovary deeply lobed, developing into several drupe-like units contrasting in color with an expanded receptacle .... Ochnaceae

7. Placentation axile 8

7. Placentation parietal 12

8. Flowers unisexual; styles not divided to several times divided (and thus more numerous than the carpels) 9

8. Flowers bisexual; styles not secondarily divided 10

Nectaries lacking. *Fruit usually a samaroid schizocarp, schizocarp, drupe with usually 3-seeded, often ridged pit, or nutlike;* embryo straight to bent or coiled; endosperm ± lacking (Figure 9.61).

Floral formula:

X, K⑤, C1+4, A ⑩, G③; samaroid schizocarp, drupe, berry, nutlike

Distribution: More or less pantropical, but especially diverse in South America.

Genera/species: 66/1200. **Major genera:** *Byrsonima* (150 spp.), *Heteropterys* (120), *Banisteriopsis* (92), *Tetrapterys* (90), *Stigmaphyllon* (90), and *Bunchosia* (75). *Aspicarpa, Byrsonima, Galphimia, Janusia,* and *Malpighia* occur in the continental United States.

Economic plants and products: The drupes of *Malpighia emarginata* (Barbados cherry) are edible and contain large amounts of vitamin C. *Malpighia, Stigmaphyllon, Galphimia,* and *Byrsonima* provide ornamentals. *Banisteriopsis caapi* contains narcotic alkaloids.

Discussion: Monophyly of Malpighiaceae is not in doubt, being supported by both morphological and cpDNA sequence characters (Chase et al. 1993; Soltis et al. 2000). Recognition of two traditional subfamilies on the basis of unwinged versus winged fruits is surely artificial (Anderson 1977). Fleshy fruits, as found in *Byrsonima* and *Malpighia*, have evolved several times. The form of the style and stigma, pollen structure, and chromosome number are of phylogenetic significance within the family (Anderson 1977). Two subfamilies are recognized based on DNA sequence data (Davis et al. 2001): Byrsonimoideae (including *Byrsonima* and a few other genera) have a base chromosome num-

9. Each locule with a single ovule; seeds often arillate .... **Euphorbiaceae**

9. Each locule with two ovules; seeds not arillate .... **Phyllanthaceae**

10. Hairs T-, V-, or Y-shaped, the stalk often short and the arms straight to twisted; sepals often with paired, abaxial, oil-producing glands; petals usually clawed .... **Malpighiaceae**

10. Hairs simple; sepals lacking glands; petals not clawed 11

11. Leaves opposite; stipules interpetiolar; petals usually fringed or hairy and individually enclosing a single stamen or a staminal group; often plants of mangrove communities .... **Rhizophoraceae**

11. Leaves alternate; stipules lateral; petals not fringed or hairy and not enclosing a single stamen or group of stamens; plants not of mangrove habitats .... Ochnaceae

12. Flowers with a corona, consisting of 1 or more rows of filaments or scales, usually borne on a hypanthium; ovary usually on a short to long gynophore; often vines with tendrils; aril fleshy .... **Passifloraceae**

12. Flowers lacking a corona and usually a hypanthium as well; gynophore lacking; trees to herbs; always lacking tendrils; aril fleshy to hard and oily 13

13. Flowers usually bilateral; stamens usually 5, the filaments very short and stamens closely placed around style, connective prominent, some or all of the anthers with a glandlike or spurlike nectary on the back; style 1, often distally enlarged; stigma 1, sometimes lobed .... **Violaceae**

13. Flowers radial; stamens usually 10 to numerous, sometimes few and then flowers in catkins, the filaments elongate and stamens not closely placed around style, connective not prominent, anthers lacking nectar-producing tissue; styles several to 1, not apically enlarged; stigmas 2 to several 14

14. Leaves without salicoid teeth; petals much more numerous than sepals; nectar disk lacking; anthers usually linear; cyanogenic glycosides present .... Achariaceae

14. Leaves with salicoid teeth; petals lacking, or if present than equal in number to the sepals, or sepals and petals both lacking; nectar disk or glands present, but these sometimes lacking if flowers in catkins; anthers usually globose or ovoid; cyanogenic compounds lacking .... **Salicaceae**

ber of 6, while the remaining genera belong to the Malpighioideae and are characterized by $x = 10$.

In the Neotropics, the family is principally pollinated by oil-gathering anthophorid bees, while most Old World species are pollinated by pollen-gathering bees. Most Malpighiaceae are outcrossing, but selfing also occurs; some have been shown to be agamospermous. Winged fruits, as in *Stigmaphyllon* or *Tetrapterys*, are wind-dispersed; fleshy fruits, as seen in *Malpighia, Byrsonima,* or *Bunchosia,* are bird- and/or mammal-dispersed.

Additional references: Anderson 1979, 1990; Rao and Sarma 1992; Robertson 1972b; Vogel 1990.

Euphorbiaceae A. L. de Jussieu
(Spurge Family)

Trees, shrubs, herbs, or vines, sometimes succulent and cactuslike; internal phloem sometimes present; chemically diverse, with alkaloids, di- or triterpenoids, tannins, and cyanogenic glucosides; *often with laticifers* containing milky or colored latex; *usually poisonous.* Hairs simple to branched, stellate, or peltate. *Leaves usually alternate and spiral or 2-ranked,* sometimes opposite, simple, sometimes palmately lobed or compound, entire to serrate, with pinnate to palmate venation; *sometimes with paired nectar glands at base of blade or on petiole; stipules usually present. Inflorescences* determinate, but *often highly modified, sometimes forming false flowers* as in the cyathia of *Euphorbia,* terminal or axillary. **Flowers unisexual** (plants dioecious or monoecious), usually radial, showy to inconspicuous. Sepals usually 2–6, distinct to slightly connate. Petals usually 0–5, distinct to slightly connate, valvate or imbricate, often lacking. Stamens 1 to numerous; filaments distinct to connate; pollen grains often tricolporate or polyporate. Carpels usually 3, connate; ovary superior, usually 3-lobed, with axile placentation; **styles** *usually 3,* **each usually bifid, or several times divided**; stigmas various. **Ovules 1 in each locule.** Nectar disk usually present. **Fruit usually a schizocarp, with segments elastically dehiscent from a persistent central column**; seeds often arillate; embryo straight to curved (Figures 9.62 and 9.63).

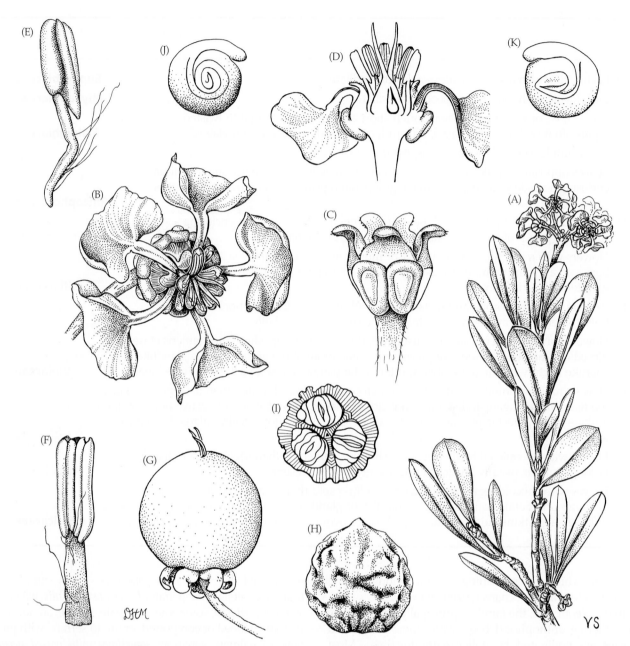

FIGURE 9.61 Malpighiaceae. *Byrsonima lucida:* (A) flowering branch (× 1.5); (B) flower (× 7.5) (C) calyx, showing glands on one sepal (glands on adjacent sepals removed) (× 11); (D) flower in longitudinal section (× 8); (E, F) stamens (× 22); (G) drupe (× 3); (H) pit (× 6); (I) pit in cross-section, showing three seeds and embryos (× 6); (J, K) opposite sides of the same embryo (× 9.5). (From Robertson 1972, *J. Arnold Arbor.* 53: p. 110.)

Floral formula:

Staminate: *, K 2–6, C 0–5, A 1–∞, G0

Carpellate: *, K 2–6, C 0–5, A0, G③; schizocarp

Distribution: Widespread, but most diverse in tropical regions.

Genera/species: 222/6100. **Major genera:** *Euphorbia* (2400 spp.), *Croton* (1300), *Acalypha* (400), *Macaranga* (250), *Manihot* (150), *Tragia* (150), *Jatropha* (150), *Mallotus* (120), *Sapium* (100), and *Dalechampia* (100). Numerous genera are represented in the continental United States and/or Canada; some of these include (in addition to most of the above) *Argythamnia, Bernardia, Cnidoscolus, Reverchonia, Sapium, Sebastiania,* and *Stillingia.*

Economic plants and products: *Hevea brasiliensis* (rubber tree) is the source of most natural rubber, and also is a timber source; *Aleurites moluccana* (candlenut tree) and *A. fordii* (tung tree) are sources of oils used in the manufacture of paints and varnishes; *Sapium sebiferum* (Chinese tallow tree) is a source of vegetable tallow and wax;

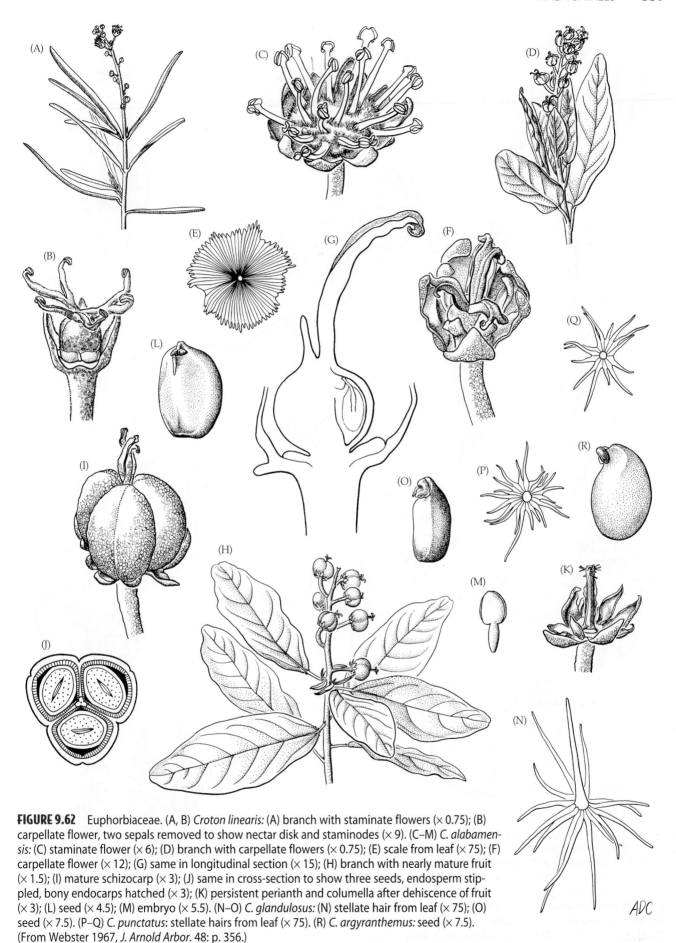

FIGURE 9.62 Euphorbiaceae. (A, B) *Croton linearis:* (A) branch with staminate flowers (× 0.75); (B) carpellate flower, two sepals removed to show nectar disk and staminodes (× 9). (C–M) *C. alabamensis:* (C) staminate flower (× 6); (D) branch with carpellate flowers (× 0.75); (E) scale from leaf (× 75); (F) carpellate flower (× 12); (G) same in longitudinal section (× 15); (H) branch with nearly mature fruit (× 1.5); (I) mature schizocarp (× 3); (J) same in cross-section to show three seeds, endosperm stippled, bony endocarps hatched (× 3); (K) persistent perianth and columella after dehiscence of fruit (× 3); (L) seed (× 4.5); (M) embryo (× 5.5). (N–O) *C. glandulosus:* (N) stellate hair from leaf (× 75); (O) seed (× 7.5). (P–Q) *C. punctatus:* stellate hairs from leaf (× 75). (R) *C. argyranthemus:* seed (× 7.5). (From Webster 1967, *J. Arnold Arbor.* 48: p. 356.)

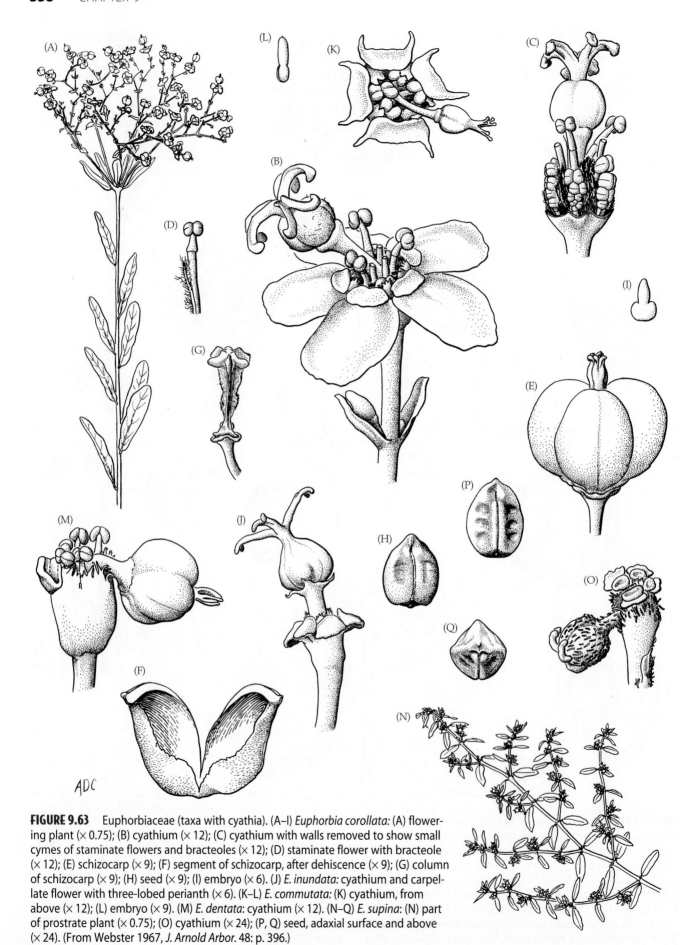

FIGURE 9.63 Euphorbiaceae (taxa with cyathia). (A–I) *Euphorbia corollata*: (A) flowering plant (× 0.75); (B) cyathium (× 12); (C) cyathium with walls removed to show small cymes of staminate flowers and bracteoles (× 12); (D) staminate flower with bracteole (× 12); (E) schizocarp (× 9); (F) segment of schizocarp, after dehiscence (× 9); (G) column of schizocarp (× 9); (H) seed (× 9); (I) embryo (× 6). (J) *E. inundata*: cyathium and carpellate flower with three-lobed perianth (× 6). (K–L) *E. commutata*: (K) cyathium, from above (× 12); (L) embryo (× 9). (M) *E. dentata*: cyathium (× 12). (N–Q) *E. supina*: (N) part of prostrate plant (× 0.75); (O) cyathium (× 24); (P, Q) seed, adaxial surface and above (× 24). (From Webster 1967, *J. Arnold Arbor.* 48: p. 396.)

Euphorbia spp. produce reduced hydrocarbon fuels. Many species are poisonous, and many have medicinal uses (see Rizk 1987). Species of *Euphorbia* and *Hippomane* have been used as arrow poisons, and members of several genera are used as fish poisons. Surprisingly, several species contain edible parts; the thick roots of *Manihot esculenta* (cassava, manioc, yuca) are an important starch source in tropical regions. The leaves of *Cnidoscolus chayamansa* (stinging nettle) are used as a vegetable. Finally, *Euphorbia* (poinsettia, crown-of-thorns, various succulents), *Jatropha*, *Codiaeum* (croton), *Acalypha* (chenille plant), and other genera provide ornamentals.

Discussion: The family has often been considered close to Malvales, although molecular data suggest that it is best placed in a broadly circumscribed Malpighiales (as in this text).

Euphorbiaceae are extremely diverse and have sometimes been divided into several families, or suggested to be polyphyletic. The monophyly of Euphorbiaceae, as broadly circumscribed, has not been supported in recent phylogenetic analyses based on DNA sequences (APG 2003; Chase et al. 2002; Davis and Chase 2004; Savolainen et al. 2000b; Soltis et al. 2000), and members of this "group" are here divided among four families: Euphorbiaceae s.s., Phyllanthaceae, Picrodendraceae, and Putranjivaceae. The mustard oil-containing *Drypetes* and *Putranjiva* are segregated as Putranjivaceae. The first three families are large, morphologically variable, and are discussed in more detail below. Evolutionary relationships within Euphorbiaceae, as broadly delimited, have been extensively studied by Webster (1967, 1987, 1994a,b), who recognized five subfamilies.

Phyllanthaceae (e.g., *Bischofia*, *Phyllanthus*, *Glochidion*, and *Antidesma*) are characterized by having two ovules per locule and nonarillate seeds (see treatment of this family). In contrast, Euphorbiaceae, as delimited in this text, have only a single ovule per locule and usually arillate seeds. Members of the Phyllanthaceae were considered as the subfamily Phyllanthoideae (of Euphorbiaceae) by Webster, who suggested that they represent the primitive group from which his other subfamilies were derived. Picrodendraceae (e.g., *Oldfieldia*, *Picrodendron*, and *Pseudanthus*; treated as Euphorbiaceae subfamily Oldfieldioideae by Webster) also have two ovules per locule. Spiny pollen supports the monophyly of Picrodendraceae; both Phyllanthaceae and Euphorbiaceae are characterized by nonspiny pollen. The monophyly of Euphorbiaceae, narrowly delimited, is supported by their having only one ovule per locule, unisexual flowers, ± divided styles, and schizocarpic fruits. Similar gynoecia with divided styles and schizocarpic fruits may have evolved independently in Phyllanthaceae. Picrodendraceae lacks divided styles and a nectar disk.

At least four subfamilies can be recognized in Euphorbiaceae s.s. (Wurdack et al. 2005). These include Acalyphoideae (*Acalypha*, *Alchornea*, *Tragia*, *Ricinus*, and relatives), which lack latex; and the Crotonoideae (*Croton*, *Manihot*, *Jatropha*, *Codiaeum*, *Aleurites*, *Cnidoscolus*, etc.)

and Euphorbioideae (*Hippomane*, *Hura*, *Euphorbia*, *Gymnanthes*, *Stillingia*, *Sapium*, etc.), both of which have latex. The Crotonoideae have distinctive polyporate pollen, often stellate, peltate, or branched hairs, and colored to white, noncaustic sap, while the Euphorbioideae have tricolporate pollen, simple hairs, and white, often caustic sap. Euphorbioideae contain the large tribe Euphorbieae (mainly *Euphorbia*, which includes *Poinsettia*, *Chamaesyce*, etc.). Members of this tribe are considered monophyletic on the basis of their inflorescences, which are cyathia (Plate 9.10F; Figure 9.63B). The carpellate flower is surrounded by numerous staminate flowers (each of which is reduced to a single stamen) within a cuplike structure formed from a highly reduced cymose inflorescence and associated bracts. One to five nectar glands, sometimes with petal-like appendages, are associated with the cuplike axis of each cyathium.

Most Euphorbiaceae are insect-pollinated (flies, bees, wasps, and butterflies), with nectar providing the floral attractant, yet some are probably pollinated by birds, bats, or other mammals. *Acalypha*, *Ricinus*, and *Alchornea*, among others, are wind-pollinated. Outcrossing is promoted by maturation of carpellate flowers before staminate ones. Most have elastic schizocarps. The large fruits of *Hura* or *Hevea* are able to explosively eject their seeds, shooting them several meters. Some are secondarily water-dispersed, while those with oily arils are sometimes secondarily dispersed by ants. Some taxa have fleshy arils (or indehiscent fleshy fruits) and are dispersed by birds.

Additional references: Levin 1986; Park and Elisens 2000; Rao 1971; Rizk 1987; Steinmann and Porter 2002; Sutter and Endress 1995.

Phyllanthaceae J. G. Agardh
(Phyllanthus Family)

Trees, shrubs, or herbs, sometimes with dimorphic (horizontal and erect) branches or phylloclades; chemically diverse, with alkaloids, triterpenoid saponins, and tannins; laticifers lacking; usually poisonous. Hairs simple. *Leaves alternate, often 2-ranked, usually simple,* entire to serrate, with pinnate venation; *stipules usually present.* Inflorescences determinate, axillary, occasionally reduced to a single flower. **Flowers unisexual** (*plants monoecious or dioecious), radial, ± inconspicuous.* Sepals usually 5, distinct or slightly connate. Petals usually 0–5, distinct to slightly connate, imbricate, often reduced or lacking. Stamens 3–8, occasionally fewer or more numerous; filaments distinct to connate; pollen grains tricolporate or polycolporate to polyporate, or inaperturate. *Carpels usually 3, connate; ovary superior, usually 3-lobed,* with axile placentation; *styles usually 3,* **each usually bifid**, occasionally entire; stigmas various. *Ovules 2 in each locule. Nectar disk usually present.* **Fruit usually a schizocarp, with segments elastically dehiscent from a persistent central column,** *sometimes a berry or drupe; seeds not arillate;* embryo straight to slightly curved.

(A)

Malpighiales: Clusiaceae
Clusia lanceolata: carpellate flower

C. lanceolata: staminate flower

(B)

Malpighiales: Rhizophoraceae
Rhizophora mangle: viviparous fruits

(C)

Malpighiales: Salicaceae
Casearia crassinervis: flowers

(D)

Malpighiales: Malpighiaceae
Mascagnia macroptera: flowers

(E)

Malpighiales: Euphorbiaceae, Crotonoideae
Jatropha interigma: carpellate flowers

(F)

Malpighiales: Euphorbiaceae, Euphorbioideae
Euphorbia cotinifolia: cyathium, developmental series

E. pulcherrima: cyathia

(G)

Malpighiales: Passifloraceae
Passiflora caerulea: flower

PLATE 9.10 Eudicots: Malpighiales

(H)

Malpighiales: Violaceae
Viola soraria: flower, longitudinal section

(I)

Malpighiales: Passifloraceae
Passiflora incarnata: immature fruits

(J)

Malpighiales: Hypericaceae
Hypericum tetrapetalum: flower

St. John's Wort

(K)

Malpighiales: Hypericaceae
Triadenum virginicum: leaf with
pellucid dots

(L)

Malpighiales: Salicaceae
Salix caroliniana: staminate catkins

S. caroliniana: carpellate catkin

**PLATE 9.10 Eudicots:
Malpighiales**

Floral formula:

Staminate: *, K⑤, C⟨0–5⟩, A⟨3–∞⟩, G0

Carpellate: *, K⑤, C⟨0–5⟩, A0, G③, schizocarp, drupe

Distribution: Widespread, but most diverse in tropical regions.

Genera/species: 55/1745. **Major genera:** *Phyllanthus* (1270 spp., incl. *Breynia, Glochidion*), *Antidesma* (150), and *Cleistanthus* (140). The group is represented in the continental United States and/or Canada by *Andrachne, Bischofia, Phyllanthus*, and *Savia*.

Economic plants and products: Several species are used medicinally, or have been used as fish-poisons. A few, such as *Bischofia* and *Phyllanthus*, provide ornamentals. A few species of *Antidesma* (bignay) and *Phyllanthus* (emblic, otaheite gooseberry) have edible drupes that are high in vitamin C.

Discussion: Phyllanthaceae can easily be distinguished from Euphorbiaceae, in which they are often included, by their gynoecia that have 2-ovulate locules (versus 1-ovulate locules). They also the lack the laticifers (with white or colored latex) and arillate seeds so characteristic of many Euphorbiaceae. They can be distinguished from the small, and mainly southern Hemisphere family, Picrodendraceae (e.g., *Austrobuxus, Oldfieldia, Pseudanthus, Tetracoccus*, and *Picrodendron*), which also have 2-ovulate locules, by the presence of a nectar disk and frequently also bifid styles in their flowers, and in the lack of the spiny pollen characteristic of this clade. Molecular studies have shown that Phyllanthaceae, Picrodendraceae, and Euphorbiaceae represent three distinct lineages, with Phyllanthaceae likely sister to Picrodendraceae (Chase et al. 2002; Davis and Chase 2004; Savolainen et al. 2000b; Soltis et al. 2000; Wurdack et al. 2004). Phyllanthaceae are presumed to be monophyletic based on their unisexual flowers, bifid styles and schizocarpic fruits, all characters that evidently evolved in parallel in Euphorbiaceae. The family is comprised of two major clades: *Phyllanthus, Savia, Cleistanthus, Andrachne,* and relatives (inflorescences fasciculate, tanniniferous epidermal cells lacking, fruits usually explosive, and plants monoecious) and *Antidesma, Bischofia, Hieronyma* and relatives (inflorescences elongate, tanniniferous epidermal cells present, fruits tardily or incompletely dehiscent, or indehiscent, and plants dioecious) (Samuel et al. 2005; Wurdack et al. 2004).

The small flowers of many Phyllanthaceae are pollinated by generalist insect visitors, mainly flies or small bees. Pollination mutualism involving moths of the genus *Epicephala* occur in many species of *Phyllanthus* (Kawakita and Kato 2004). Those with schizocarpic fruits have ballistic seed dispersal, while those with fleshy (and often colorful fruits) are mainly dispersed by birds.

Additional references: Webster 1967, 1994a,b.

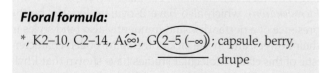

Clusiaceae Lindley
(= Guttiferae A. L. de Jussieu)
(Garcinia Family)

Trees, shrubs, or lianas; **with colored exudate** *in secretory canals or cavities.* Hairs simple, multicellular. *Leaves usually opposite or whorled, simple,* entire, with pinnate venation, *with pellucid or black dots and/or secretory canals;* stipules lacking, although paired glands may be present at the nodes. Inflorescences determinate, sometimes reduced to a single flower, usually terminal. Flowers bisexual to unisexual (then plants usually dioecious), radial. *Sepals usually 2–5,* and distinct. *Petals usually 4–5,* occasionally more numerous, *distinct, often asymmetrical,* imbricate or convolute. *Stamens usually numerous,* the innermost developing before the outermost, distinct or variously connate, often fascicled, sometimes with glands; pollen grains usually tricolporate. *Carpels usually 2–5,* sometimes numerous, connate; ovary superior, with usually axile placentation, or sometimes parietal; *stigmas peltate, lobed, or capitate, usually expanded in relation to the short to elongate style(s).* Ovules 1 to numerous per carpel, with a thin megasporangium. Nectaries usually lacking. *Fruit a variously dehiscent capsule, berry, or drupe; seeds arillate or not;* embryo straight, the cotyledons very large to small, and then the hypocotyl swollen; endosperm scanty or lacking.

Floral formula:

*, K2–10, C2–14, A⊛, G$\overline{(2-5\ (-\infty))}$; capsule, berry, drupe

Distribution: Mainly pantropical.

Genera/species: 27/1000. ***Major genera:*** *Garcinia* (200), *Calophyllum* (190), *Clusia* (160), and *Mammea* (70). *Clusia* occurs in the continental United States (southern Florida).

Economic plants and products: The fruits of *Garcinia mangostana* (mangosteen) and *Mammea americana* (mamey apple) are highly prized. Several genera provide timber.

Discussion: Clusiaceae are considered monophyletic on the basis of molecular and morphological synapomorphies (Chase et al. 2002; Gustaffson et al. 2002). The family is most closely related to Podostemaceae, an odd family of aquatics of rapidly flowing streams; Hypericaceae, a more temperate group with pellucid-dotted leaves and flowers with minute stigmas and slender styles; and Bonnetiaceae, largely New World, lacking exudate, but with serrulate leaves. It is of interest that Clusiaceae, Hypericaceae, and Podostemaceae all have canals, cavities, or individual cells with exudate, xanthone pigments, and ovules with a thin sporangium wall. Hypericaceae often are treated as a subfamily (Hypericoideae) within Clusiaceae, but a broadly circumscribed Clusiaceae are likely non-monophyletic as a result of the sister-group relationship of Hypericaceae and Podostemaceae. The traditional name for the family, Guttiferae, meaning "gum-bearing," refers to the colored resinous sap characteristic of the group.

Infrafamilial relationships have been considered by Stevens (personal communication) and Gustaffson et al. (2002). Kielmeyeroideae (e.g., *Calophyllum, Mammea,* and *Mesua*) are sister to the Clusioideae and are plants with secretory canals and sometimes black-dotted leaves; an androecium that is not obviously fasciculate; short to elongate, usually connate styles; an embryo with moderate-sized to huge cotyledons; and twigs with an exposed apical bud. Clusioideae (e.g., *Garcinia* and *Clusia*) are plants with secretory canals and have an androecium that is fasciculate or not, usually short, connate styles, an embryo usually with a very large hypocotyl, and an apical bud often protected by the petiole bases.

Clusia is noteworthy for its broad range in habit. Some species begin growth as epiphytes, growing like a strangling fig; others have numerous adventitious roots that assist in support of the stems.

The showy flowers with conspicuous stamens characteristic of Clusiaceae are pollinated mainly by bees and wasps (Plate 9.10A). Pollen is often the pollinator reward, but some taxa (e.g., *Clusia*) secrete terpenoid resins. Species with fleshy fruits (e.g., *Mammea, Calophyllum, Garcinia*) or capsules opening to expose colorful arillate seeds (e.g., *Clusia*) are usually dispersed by birds or mammals.

Additional references: Ramirez and Gomez 1978; Wood and Adams 1976.

Hypericaceae A. L. de Jussieu
(Saint-John's-Wort Family)

Trees, shrubs, or herbs; *with clear or black resinous sap, in secretory cavities.* Hairs simple, multicellular. *Leaves opposite or whorled, simple, entire,* with pinnate venation, *with pellucid or black dots (or lines;* Plate 9.10K); stipules lacking. Inflores-

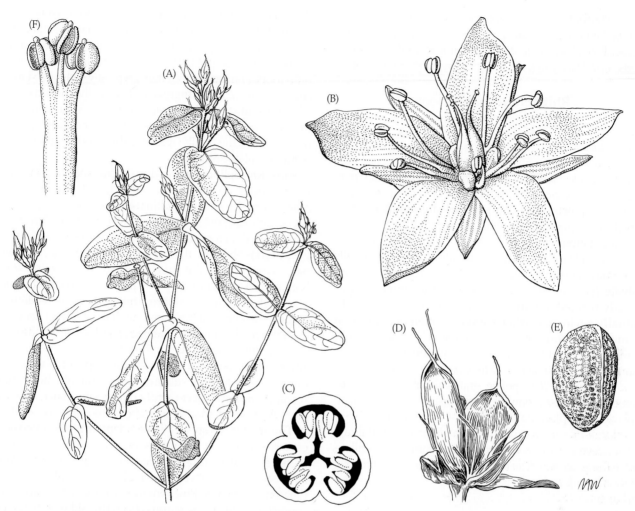

FIGURE 9.64   Hypericaceae. (A–E) *Triadenum virginianum:* (A) plant with fruits (× 0.75); (B) flower, note stamens in threes with alternating staminodes (× 7.5); (C) ovary in cross-section (× 22); (D) capsule (× 5.5); (E) seed (× 37). (F) *T. walteri:* fascicle of stamens, note apical glands (× 18). (From Wood and Adams 1976, *J. Arnold Arbor.* 57: p. 89.)

cences determinate, sometimes reduced to a single flower, usually terminal. Flowers bisexual, radial. *Sepals 4–5 and distinct or slightly connate. Petals usually 4–5, distinct, often asymmetrical,* imbricate or convolute. *Stamens numerous,* the innermost developing before the outermost, often fascicled; pollen grains usually tricolporate. *Carpels 3–5,* connate; *ovary superior,* with axile placentation, or sometimes parietal with deeply intruded placentas; **stigmas minute, punctate,** *on 3–5 elongate, distinct to basally connate styles.* Ovules numerous per carpel, with a thin megasporangium. Nectaries usually lacking. *Fruit a variously dehiscent capsule,* berry, or drupe; *seeds not arillate;* embryo straight, the cotyledons moderate sized; endosperm scanty or lacking (Figure 9.64).

Floral formula: *, K4–5, C4–5, A ⊙, G ③–⑤; capsule

Distribution: Widespread.

Genera/species: 9/540.   **Major genera:** *Hypericum* (360 spp.) and *Vismia* (55). *Hypericum* and *Triadenum* occur in the continental United States and/or Canada.

Economic plants and products: Some species of *Hypericum* are used medicinally, providing a popular treatment for depression. Species of *Hypericum* are used as ornamentals because of their showy flowers (Plate 9.10J).

Discussion: Hypericaceae are assumed to be monophyletic on the basis of preliminary molecular data (Gustaffson et al. 2002). These plants can be distinguished from the Clusiaceae by their more frequently shrubby to herbaceous habit, lack of colored exudate (with the leaves having only black and/or pellucid dots), and the elongate and often distinct styles, each with a minute stigma. Hypericaceae are well developed in temperate regions, in contrast to Clusiaceae. They have often been placed within an expanded Clusiaceae (Cronquist 1981; Wood and Adams 1976).

The showy flowers with conspicuous stamens characteristic of Hypericaceae are pollinated mainly by bees and wasps. Pollen is usually the pollinator reward. The small seeds, as in *Hypericum*, are dispersed by wind and/or water.

Rhizophoraceae Persoon
(Red Mangrove Family)

Trees or shrubs, often with prop roots or pneumatophores; tannins usually present. Hairs usually simple. *Leaves opposite, with adjacent leaf pairs usually diverging from each other at an angle less than or greater than 90°,* simple, serrate or crenate to entire, with pinnate venation; **stipules interpetiolar**, usually with colleters at the base of adaxial surface. Inflorescences usually determinate, axillary. Flowers usually bisexual, radial, often with a hypanthium. *Sepals usually 4 or 5,* occasionally numerous, usually slightly connate, *thick, fleshy or leathery,* **valvate**. *Petals usually 4 or 5,* occasionally numerous, *distinct,* **usually fringed or hairy**, convolute or infolded, and **individually enclosing either a single stamen or a group of stamens in bud**. Stamens usually 8–10, occasionally numerous; filaments distinct or basally connate, sometimes short or lacking; *anthers sometimes multilocular and opening by a longitudinal flap;* pollen grains usually tricolporate. Carpels usually 2–6, connate; ovary superior to inferior, usually with axile placentation; stigma ± lobed. Ovules 2–8 in each locule. Nectar disk often present. *Fruit a septicidal capsule or several to 1-seeded berry;* seeds sometimes arillate; *embryo large, often with an elongated hypocotyl and germinating while still inside the fruit* (Figure 9.65).

Floral formula:

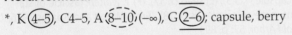

*, K (4–5), C4–5, A (8–10) (–∞), G (2–6); capsule, berry

Distribution and ecology: Pantropical; moist montane forests to tidally flooded wetlands (mangrove communities).

Genera/species: 12/84. **Major genera:** *Cassipourea* (55 spp.), *Rhizophora* (8), and *Bruguiera* (6). Only *Rhizophora* occurs in the continental United States.

Economic plants and products: *Rhizophora* is a source of tannin and charcoal. Mangrove swamps supply nutrition to marine communities and critical habitats for numerous marine organisms. They also stabilize shorelines and protect inland areas from wind and tides during tropical storms.

Discussion: The non-mangrove genera constitute a basal paraphyletic complex within Rhizophoraceae (Juncosa and Tomlinson 1988a,b; Schwarzbach and Ricklefs 2000; Toby and Raven 1988). In contrast, mangrove genera such as *Bruguiera* and *Rhizophora* form a well-supported clade, which can be diagnosed by their entire leaves, viviparous 1-seeded berries (Plate 9.10B), and specializations relating to

the structure and development of the seed or embryo. All mangrove species have anthers that open while still in the bud, depositing the pollen onto the hairy petals (another likely synapomorphy). Partially inferior ovaries, berries, prop roots, and lack of root hairs, typically considered to be characteristic of the mangrove taxa, actually evolved earlier in the evolution of the family, and also occur in tropical montane genera. *Rhizophora* is perhaps the most specialized genus in the family; it shows the unusual apomorphy of multiloculate anthers.

Most Rhizophoraceae produce nectar and are pollinated by butterflies, moths, various other insects, or birds. *Rhizophora* usually is wind-pollinated and does not produce nectar, although the flowers may be visited by pollen-gathering bees. Outcrossing is common. The mangrove species have fibrous "berries" that are one-seeded; the seed germinates while inside the fruit (viviparous), emerging from both seed coat and fruit up to 9 months before abscission. The cotyledons are fused into a tube; when the seedling is mature, the elongate hypocotyl (with its plumule) detaches from the cotyledons (and the berry) and falls from the tree. The seedling floats and survives in seawater. Eventually adventitious roots develop from the hypocotyl, which fix the seedling to the substrate. The initially horizontal seedlings become erect rapidly once rooted, such that the plumule is raised, snorkel-like, above the influence of tides.

The mangrove species have adapted to life in coastal estuary conditions by means of pneumatophores; prop roots; elongated, often curved and pointed, viviparous embryos; and the physiological blocking of salt uptake by the roots. Mangrove Rhizophoraceae have lost the ability to coppice (produce new sprouts from the old wood), which affects their ability to regenerate after major climatic or human-caused damage.

Additional references: Graham 1964b; Rabinowitz 1978; Tomlinson 1986; Tomlinson and Cox 2000; Tomlinson et al. 1979.

Violaceae Batsch
(Violet Family)

Herbs to shrubs or trees; often with saponins and/or alkaloids. Hairs often simple. *Leaves alternate and spiral or 2-ranked,* occasionally opposite, sometimes forming a basal rosette, simple, sometimes lobed, entire to serrate, with pinnate to palmate venation; *stipules present.* Inflorescences indeterminate, sometimes reduced to a solitary flower, usually axillary. Flowers usually bisexual, *usually slightly to strongly bilateral. Sepals 5,* usually distinct. *Petals 5,* distinct, imbricate to convolute, sometimes the abaxial with a spur. *Stamens usually 5,* **their edges touching to form a ring around the gynoecium, filaments very short**, distinct to slightly connate, **the 2 abaxial anthers or all anthers with dorsal glandlike or spurlike nectaries, connective often with triangular, membranous, apical appendage**, *shedding pollen inward,* **which is picked up by the modified style**; pollen grains

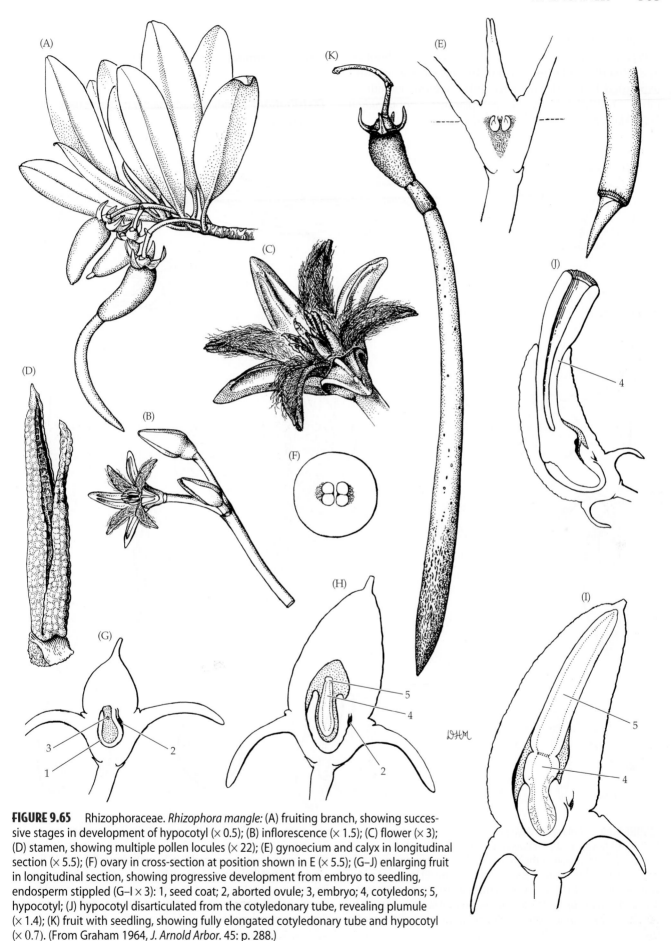

FIGURE 9.65 Rhizophoraceae. *Rhizophora mangle:* (A) fruiting branch, showing successive stages in development of hypocotyl (× 0.5); (B) inflorescence (× 1.5); (C) flower (× 3); (D) stamen, showing multiple pollen locules (× 22); (E) gynoecium and calyx in longitudinal section (× 5.5); (F) ovary in cross-section at position shown in E (× 5.5); (G–J) enlarging fruit in longitudinal section, showing progressive development from embryo to seedling, endosperm stippled (G–I × 3): 1, seed coat; 2, aborted ovule; 3, embryo; 4, cotyledons; 5, hypocotyl; (J) hypocotyl disarticulated from the cotyledonary tube, revealing plumule (× 1.4); (K) fruit with seedling, showing fully elongated cotyledonary tube and hypocotyl (× 0.7). (From Graham 1964, *J. Arnold Arbor.* 45: p. 288.)

usually tricolporate. *Carpels usually 3, connate; ovary superior, with parietal placentation;* **style 1, usually curved or hooked and distally enlarged or modified; stigma usually expanded but with small receptive region,** sometimes lobed. Ovules 1 to many on each placenta. *Fruit usually a loculicidal capsule; seeds often arillate* (Figure 9.66).

Floral formula: X, K5, C5, A⑤, G③; capsule

Distribution: Widespread; mainly herbaceous in temperate regions.

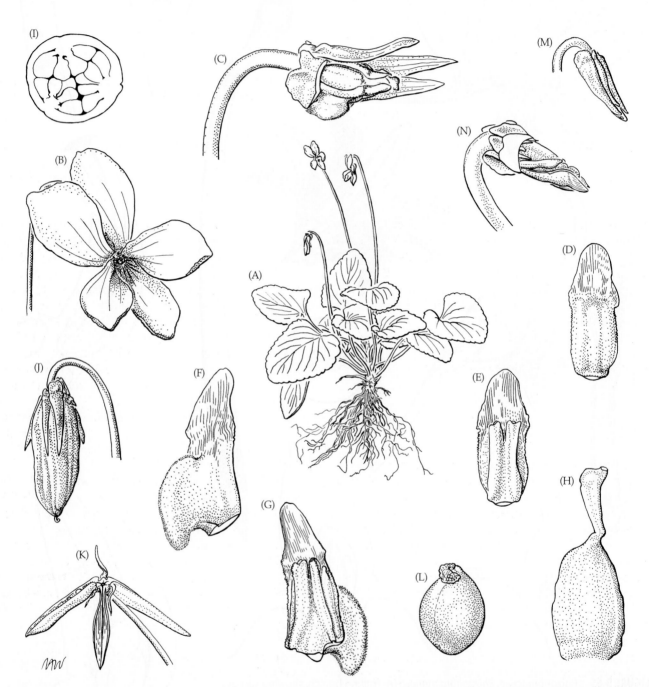

FIGURE 9.66 Violaceae. *Viola primulifolia:* (A) flowering plant (× 0.5); (B) flower (× 3); (C) flower, petals and two sepals removed, note sessile anthers closely placed around gynoecium (× 9); (D) lateral stamen, abaxial surface (× 16); (E) lateral stamen, adaxial surface (× 16); (F) lower stamen, abaxial surface, note nectar gland (× 16); (G) lower stamen, adaxial surface, note nectar gland, apical appendage, and open locule (× 16); (H) gynoeci- um, note unusual style with asymmetrically placed stigma (× 16); (I) ovary in cross-section (× 20); (J) nearly mature capsule (× 4); (K) capsule, seeds squeezed out by maturing valves (× 3); (L) seed (× 16); (M) cleistogamous flower (× 4); (N) cleistogamous flower, two sepals removed to expose reduced petals and two functional stamens (× 9). (From Wood 1974, *A student's atlas of flowering plants*, p. 71.)

Genera/species: 22/950. **Major genera:** *Viola* (500 spp.), *Rinorea* (300), and *Hybanthus* (100). *Viola* and *Hybanthus* occur within the continental United States and/or Canada.

Economic plants and products: Several species of *Viola* (violet, pansy) and *Hybanthus* (green-violet) are used as ornamentals. Rhizomes and roots of *Hybanthus ipecacuanha* have been used as a substitute for true ipecac (*Psychotria ipecacuanha*).

Discussion: Although Violaceae are clearly monophyletic, infrafamilial relationships are poorly understood. Most genera are placed in Rinoreae (e.g., *Rinorea*, *Gloeospermum*, and *Rinoreocarpus*), a group characterized by radial to slightly bilateral flowers, or Violeae (e.g., *Hybanthus* and *Viola*), which have strongly bilateral flowers (Plate 9.10H). Analyses of *rbcL* sequences suggest that neither is monophyletic (Hodges, personal communication).

All genera are insect-pollinated. The flowers are visited by various flies, bees, wasps, and butterflies, and are typically outcrossing. Nectar guides are often present on the petals, and nectar, the pollinator reward, is stored in a modified petal-spur. Inconspicuous cleistogamous flowers may also be produced. The fairly small seeds may fall passively from the capsule valves or may be forcibly ejected. In some species the seeds are secondarily dispersed by ants, which are attracted by an oily aril.

References: Brizicky 1961b; Gates 1943; Munzinger and Ballard 2003.

Passifloraceae A. L. de Jussieu ex Roussel
(Passionflower Family)

Vines or lianas with axillary tendrils (modified inflorescences), usually with anomalous secondary growth, occasionally shrubs or trees lacking tendrils; usually with cyanogenic glucosides having a cyclopentenoid ring system, often also with alkaloids. Hairs various. *Leaves alternate and spiral, usually simple, often lobed,* entire to serrate, *venation usually ± palmate, usually with nectaries on the petiole (and lamina);* stipules usually present. Inflorescences usually determinate, sometimes indeterminate or reduced to a single flower, axillary. Flowers usually bisexual, radial, usually with a cup-shaped to tubular hypanthium, often associated with conspicuous bracts. *Sepals usually 5,* distinct to slightly connate, *often petal-like. Petals usually 5 and distinct,* imbricate. **Complex corona borne on the apex and inner surface of the hypanthium**, *consisting of 1 to several rows of filaments, projections, or membranes. Stamens usually 5, often borne on a stalk along with gynoecium,* the filaments usually distinct; pollen grains tri- to 12-colporate. *Carpels usually 3, connate, ovary superior,* **and borne on a stalk** (often along with the androecium), *with parietal placentation;* stigmas usually 3, capitate to truncate. Ovules numerous on each placenta.

Nectar disk at base of hypanthium. *Fruit a loculicidal capsule or berry; seeds usually flattened, with a fleshy aril* (Figure 9.67).

Floral formula:

*, K $\underline{\textcircled{5}}$, C5, corona, A5, G$\underline{\textcircled{3}}$; capsule, berry

Distribution: Widespread in tropical to warm temperate regions.

Genera/species: 18/630. **Major genera:** *Passiflora* (400 spp.) and *Adenia* (100). Only *Passiflora* occurs in the continental United States.

Economic plants and products: Several species of *Passiflora* (passionflower) have edible fruits (Plate 9.10I); others are grown as ornamentals because of their showy flowers (Plate 9.10G).

Discussion: Monophyly of Passifloraceae is supported most strongly by the presence of a well-developed corona in the flowers (de Wilde 1971, 1974). "Paropsieae," which contains shrubs and trees that lack tendrils, probably represent a paraphyletic basal complex within the family. Passifloreae, in contrast, are clearly monophyletic, as evidenced by their viny habit, axillary tendrils, and specialized flowers.

Flowers of Passifloraceae may be white, green, red, blue, or purple, with the corona and perianth parts variously oriented and developed, and employ nectar as a pollinator reward. They attract a wide range of pollinators, including bees, wasps, moths, butterflies, birds, and bats. They also are the food plants of larvae of the butterfly genus *Heliconias*. Self-incompatibility is characteristic. Bird dispersal is common; the capsules or berries have fleshy, usually colorful arils surrounding the seeds.

Variation in corona structure is often taxonomically significant. The corolla includes all structures between the perianth and the androecium, and is composed of several series of short to elongate filamentous structures. The operculum, a delicate membrane, is positioned between the corona and the gynoecium, and covers the nectar disk. The corona and operculum are outgrowths of the hypanthium. The corona is usually colorful, attracting pollinators and guiding them to the nectar. The operculum helps to hold nectar within the base of the flower.

Additional references: Brizicky 1961a; Killip 1938; Muschner et al. 2003; Puri 1948.

Salicaceae Mirbel
(Willow Family)

Trees or shrubs; often with phenolic heterosides (salicin, populin), but usually lacking cyanogenic glycosides, usually with tannins. Hairs various. *Leaves deciduous, alternate, spiral*

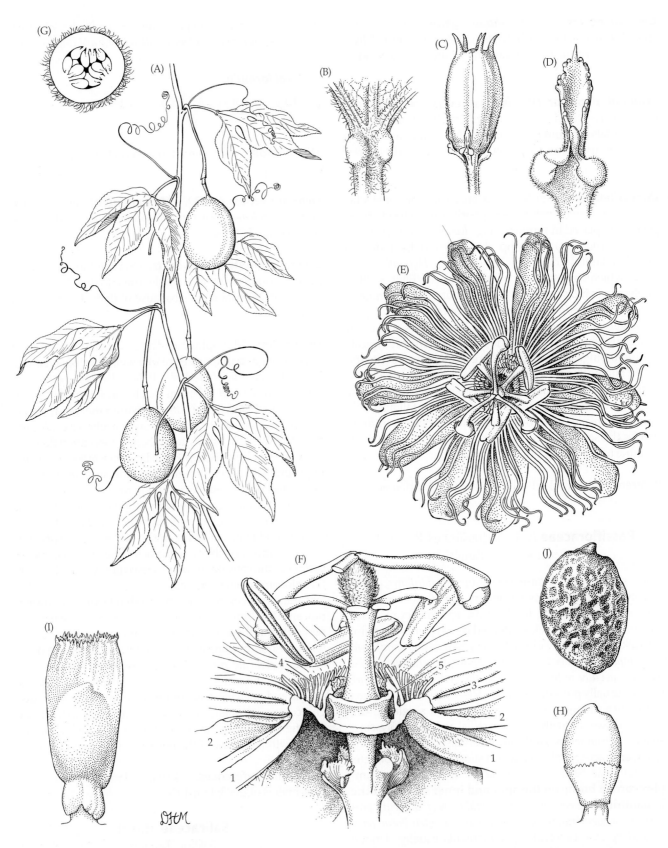

FIGURE 9.67 Passifloraceae. *Passiflora incarnata:* (A) vine with tendrils and fruits (× 0.3); (B) apex of petiole with nectar glands (× 4); (C) bud with bracts (× 1.5); (D) bract with glands (× 7); (E) flower (× 1.5); (F) flower, central portion in partial longitudinal section, two stamens, one style removed (× 3); 1, sepal; 2, petal; 3, outer corona; 4, inner corona; 5, operculum; (G) ovary in cross-section (× 8.5); (H) young seed with developing aril (× 4); (I) older seed with aril (× 4); (J) seed with fleshy portion of seed coat removed (× 5.5). (From Brizicky 1961, *J. Arnold Arbor.* 42: p. 213.)

or 2-ranked, simple, usually serrate to dentate, **the teeth salicoid (i.e., with vein expanding at the tooth apex and associated with spherical, glandular setae)** except in *Casearia,* with pinnate to palmate venation, occasionally with pellucid dots or lines; *stipules usually present.* Inflorescences determinate to indeterminate, various in form, *sometimes in erect to pendent catkins,* sometimes reduced to a solitary flower, terminal or axillary. *Flowers bisexual or unisexual (plants dioecious),* radial, *often reduced, subtended by a usually hairy bract in Salix and Populus.* Sepals usually 3–8, distinct to slightly connate, sometimes ± vestigial, forming a disk-shaped to cup-shaped structure in *Populus,* but absent in *Salix. Petals 3–8, distinct, or lacking.* Stamens 2 to numerous; filaments distinct to connate; pollen grains tricolpate, tricolporate, or lacking apertures. *Carpels usually 2–4, connate; ovary superior to half-inferior, with parietal placentation, the placentae sometimes deeply intruded and ovary then appearing axile;* stigmas usually 2–4, ± capitate to expanded and irregularly lobed. Ovules 1 to numerous on each placenta, sometimes with only 1 integument. Nectary a disk or of separate glands, sometimes lacking. *Fruit a loculicidal capsule, berry, or drupe; seeds often arillate or with basal tuft of hairs;* endosperm scanty or lacking (Figure 9.68).

Floral formula:

Staminate: *, K (3–8), C3–8 or 0, A2–∞, G0

Carpellate: *, K (3–8), C3–8 or 0, A0, G (2–4); capsule, berry, drupe

---

Distribution and ecology: Widespread, occurring from tropical to north temperate and arctic regions.

Genera/species: 58/1210. **Major genera:** *Salix* (450 spp.), *Casearia* (180), *Homalium* (180), *Xylosma* (85), *Populus* (35), and *Banara* (31). *Salix, Populus, Xylosma,* and *Flacourtia* occur in the continental United States and/or Canada.

Economic plants and products: *Salix* (willow) and *Populus* (poplar, cottonwood, aspen) provide lumber, wood pulp, and ornamentals. The bark of *Salix* was used medicinally due to the presence of salicylic acid, which reduces swelling and fever. Certain species of *Flacourtia* and *Dovyalis* are cultivated for their edible, fleshy fruits. Several genera provide useful ornamentals, e.g., *Oncoba, Casearia, Idesia,* and *Samyda.*

Discussion: Salicaceae are here circumscribed broadly, based on a phylogenetic analysis of *rbcL* sequences (Chase et al. 2002; Savolainen et al. 2000b; Soltis et al. 2000), and include not only *Salix* and *Populus,* but the numerous noncyanogenic members of "Flacourtiaceae". Monophyly of Salicaceae s.l. is well supported by DNA sequences, but the only morphological synapomorphy may be the distinctive salicoid tooth; even this is absent in *Casearia.* Such teeth are characteristic of a most other genera, although entire leaves

have occasionally evolved. The presence of salicin, imperfect apetalous flowers, and (of course) salicoid teeth in *Salix* and *Populus* are shared with *Idesia, Itoa, Poliothyrsis,* and several other genera of "Flacourtiaceae" (Boucher et al. 2003; Chase et al. 2002; Judd 1997b; Judd et al. 1994; Meeuse 1975). Molecular data support an especially close phylogenetic relationship of *Salix* and *Populus* to these genera, and also to the numerous noncyanogenic taxa traditionally placed in the "Flacourtiaceae" (e.g., *Azara, Banara, Dovyalis, Flacourtia, Homalium, Scolopia,* and *Xylosma*). In contrast, the cyanogenic "Flacourtiaceae" such as *Gynocardia, Kigellaria, Hydnocarpus,* and *Pangium* form a separate clade (Chase et al. 2002) within Malpighiales. Therefore, these cyanogenic genera have been segregated as the family Achariaceae. Members of this family also differ from Salicaceae in having many more petals than sepals, their linear (vs. globose or ovoid) anthers, and the lack of a nectar disk.

Phylogenetic relationships within Salicaceae are still poorly understood, but *Casearia* (Plate 9.10C) may be sister to the remaining genera. In addition, it is clear that *Populus* and *Salix* form a clade, which is well nested within an array of tropical genera with less reduced, and usually insect pollinated flowers.

Within the *Salix + Populus* clade, *Populus* shows many plesiomorphic characters, including several bud scales, two-integumented ovules, and a tendency toward palmate venation. The monophyly of both *Populus* and *Salix* is supported by molecular data (Chase et al. 2002). *Salix* has a single bud scale and flowers with the stipular nectar-producing structures.

Systematists of the Englerian tradition considered *Salix* and *Populus* (i.e., Salicaceae s.s.) to be primitive within angiosperms, and placed the family within the "Amentiferae," a group with inconspicuous flowers in pendulous catkins (aments; Plate 9.10L). Amentiferae include wind-pollinated families such as Platanaceae, Fagaceae, Betulaceae, and Juglandaceae, and are polyphyletic. The flowers of *Salix* and *Populus* may seem simple, but they are not primitive; they actually are highly reduced. It is clear that wind pollination has evolved numerous times within the angiosperms (Soltis et al. 2005; Thorne 1973a, 1992).

Flowers of Salicaceae vary from large and showy to tiny and inconspicuous. The common presence of a nectar disk, ring of nectar glands, or nectar-secreting petal-appendages and the exposed androecium and gynoecium suggest that various unspecialized insects are the predominant pollinators. However, some genera, such as *Populus,* have reduced unisexual flowers and are wind pollinated. The flowers of *Populus* are aggregated on dangling catkins produced in early spring, usually before the leaves emerge, and have a reduced perianth and expanded stigmas. *Salix* flowers have nectar glands and an odor, and they attract various insects, although wind pollination is probably important as well.

Fruit type varies greatly. Dispersal by birds or mammals is characteristic of taxa with drupes (*Flacourtia*), berries (*Banara, Xylosma, Idesia, Oncoba, Scolopia, Dovyalis*), or capsules opening to expose colorful arillate seeds (*Samyda,*

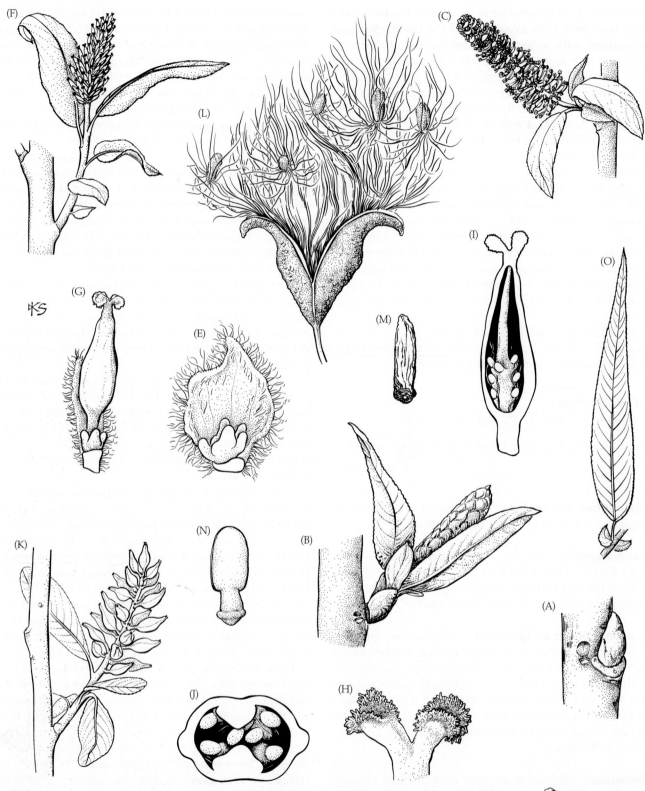

FIGURE 9.68   Salicaceae. *Salix caroliniana:* (A) winter bud (× 5); (B) expanding branchlet with young catkin (× 4); (C) staminate catkin (× 2); (D) staminate flower (× 10); (E) bract and nectar gland of staminate flower, stamens removed (× 20); (F) carpellate catkin (× 2); (G) carpellate flower, note lobed nectar gland at base (× 10); (H) stigmas (greatly magnified); (I) gynoecium in longitudinal section (× 20); (J) ovary in cross-section (× 40); (K) partly mature infructescence (× 2); (L) open capsule with escaping seeds (× 7); (M) seed, basal hairs removed (× 14); (N) embryo (greatly magnified); (O) leaf from rapidly growing summer shoot, note prominent stipules (× 1). (From Wood 1974, *A student's atlas of flowering plants*, p. 3.)

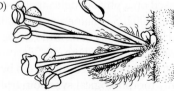

Casearia). Some genera (e.g., *Itoa, Poliothyrsis*) have capsules that open to release winged seeds and are wind dispersed. *Salix* and *Populus* have capsules that open, releasing tiny, hairy seeds that are dispersed by wind and/or water.

Species limits within *Salix* are often difficult as a result of hybridization.

Additional references: Argus 1974, 1986; Boucher et al. 2003; Brunsfeld et al. 1992; Fisher 1928; Tollsten and Kundsen 1992.

Fabales

Monophyly of the Fabales is supported by *rbcL, atpB,* and 18S sequences (Chase et al. 1993; Källersjö et al. 1998; Savolainen et al. 2000a,b; Soltis et al. 2000). Morphological synapomorphies may include vessel elements with single perforations, vestured pits, a large, green embryo, and the absence of ellagic acid. Fabales contain 4 families and about 18,860 species. Major families are **Fabaceae, Polygalaceae**, and Surianaceae.

Fabaceae Lindley
(= Leguminosae A. L. de Jussieu)
(Legume or Bean Family)

Herbs, shrubs, trees, or vines/lianas climbing by twining or tendrils; **with a high nitrogen metabolism and unusual amino acids,** *often with root nodules containing nitrogen-fixing bacteria (Rhizobium);* sometimes with secretory canals or cavities; tannins usually present; often with alkaloids; sometimes cyanogenic; sieve cell plastids with protein

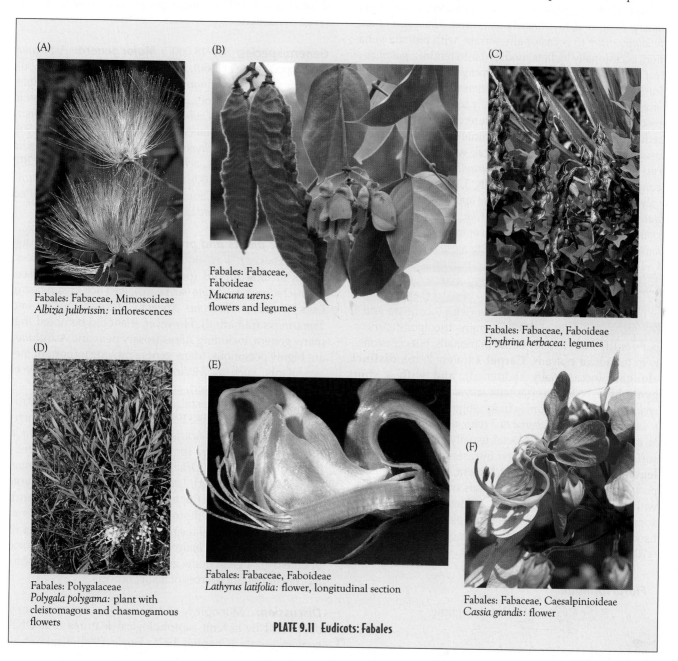

Fabales: Fabaceae, Mimosoideae
Albizia julibrissin: inflorescences

Fabales: Fabaceae, Faboideae
Mucuna urens:
flowers and legumes

Fabales: Fabaceae, Faboideae
Erythrina herbacea: legumes

Fabales: Polygalaceae
Polygala polygama: plant with cleistomagous and chasmogamous flowers

Fabales: Fabaceae, Faboideae
Lathyrus latifolia: flower, longitudinal section

Fabales: Fabaceae, Caesalpinioideae
Cassia grandis: flower

PLATE 9.11 Eudicots: Fabales

Key to Major Families of Fabales

---

1. Leaves usually compound, with stipules and well-developed pulvini; fruit often a legume .... **Fabaceae**
1. Leaves usually simple, lacking stipules, pulvini not prominent; fruit a capsule, samara, nut, drupe, or berry 2
   2. Carpels 1–5, distinct, styles gynobasic; flowers often radial; stamens usually opening by longitudinal slits .... Surianaceae
   2. Carpels usually 2–3, connate, style terminal; flowers ± bilateral; stamens usually opening by terminal pore .... **Polygalaceae**

---

crystals and usually also starch grains. Hairs various. *Leaves usually alternate, spiral to 2-ranked, pinnately (or twice pinnately) compound, to palmately compound, trifoliolate, or unifoliolate; entire* to occasionally serrate, with pinnate venation, occasionally leaflets modified into tendrils; *pulvinus of leaf and individual leaflets well developed, and leaf axis and leaflets usually showing sleep movements; stipules present, inconspicuous to leaflike, occasionally forming spines.* Inflorescences almost always indeterminate, sometimes reduced to a single flower, terminal or axillary. Flowers usually bisexual, radial to bilateral, **with a short**, *usually cup-shaped* **hypanthium**. *Sepals usually 5, distinct to more commonly connate. Petals usually 5, distinct or connate, valvate or imbricate, all alike, or the uppermost petal differentiated in size, shape, or coloration (i.e., forming a banner or standard), and positioned internally or externally in bud, the 2 lower petals often connate or sticking together and forming a keel, or widely flaring. Stamens* 1 to numerous, but *usually 10*, hidden by the perianth to long-exserted, and sometimes showy; filaments distinct to connate, then commonly monadelphous or diadelphous (with 9 connate and 1, the adaxial, ± distinct); pollen grains tricolporate, tricolpate, or triporate, usually borne in monads, but occasionally in tetrads or polyads. **Carpel 1** (rarely 2-16), **distinct, elongate** (occasionally shortened), **and with a short gynophore**; *ovary superior, with lateral placentation; style 1, arching upward*, sometimes hairy; stigma 1, small. *Ovules 1 to numerous per carpel, borne in 2 rows along an upper placenta, often campylotropous.* Nectar usually produced by inner surface of hypanthium or an intrastaminal disk. **Fruit a legume**, sometimes a samara, loment, follicle, indehiscent pod, achene, drupe, or berry; seeds often with hard coat with hourglass-shaped cells, sometimes arillate, and sometimes with a U-shaped line (pleurogram); embryo usually curved; endosperm often lacking (Figures 9.69, 9.70, 9.71; Table 9.2).

Floral formula:

X or *, K⑤, C⑤, A⑩–∞, G1; legume

Distribution and ecology: Nearly cosmopolitan; the third largest family of angiosperms, occurring in a wide range of habitats.

Genera/species: 630/18,000.   **Major genera:** *Astragalus* (2000 spp.), *Acacia* (1000), *Indigofera* (700), *Crotalaria* (600), *Mimosa* (500), *Desmodium* (400), *Tephrosia* (400), *Trifolium* (300), *Chamaecrista* (260), *Senna* (250), *Inga* (250), *Bauhinia* (250), *Adesmia* (230), *Dalbergia* (200), *Lupinus* (200), *Rhynchosia* (200), *Pithecellobium* (170), *Dalea* (150), *Lathyrus* (150), *Calliandra* (150), *Aeschynomene* (150), *Vicia* (140), *Albizia* (130), *Swartzia* (130), *Lonchocarpus* (130), *Caesalpinia* (120), *Lotus* (100), *Millettia* (100), and *Erythrina* (100). Over a hundred genera occur in Canada and/or the continental United States; some of these are listed in Table 9.2.

Economic plants and products: Fabaceae are second only to Poaceae in economic importance. Important food plants include *Arachis* (peanuts), *Cajanus* (pigeon peas), *Cicer* (chickpeas), *Glycine* (soybeans), *Inga* (ice-cream bean), *Lens* (lentils), *Phaseolus* (beans), *Pisum* (peas), and *Tamarindus* (tamarind). However, it should be noted that many genera, including *Abrus* (rosary pea) and *Astragalus*, are highly poisonous. Many genera provide important forage plants, such as *Medicago* (alfalfa), *Melilotus* (sweet clover), *Trifolium* (clover), and *Vicia* (vetch). Some species, when plowed into the soil, cause a large increase in nitrogen levels; this forms the basis of crop rotation. Ornamental species occur in *Acacia*, *Albizia* (mimosa), *Bauhinia* (orchid tree), *Calliandra* (powder puff), *Cassia*, *Cercis* (redbud), *Cytisus* (broom), *Delonix* (royal poinciana), *Erythrina*, *Gleditsia* (honey locust), *Laburnum* (goldenrain), *Lathyrus* (sweet pea), *Lupinus* (lupine), *Mimosa* (sensitive plant), *Parkinsonia* (Jerusalem thorn), *Robinia* (locust), and *Wisteria*. Commercial gums and resins are extracted from species of *Acacia* and *Hymenaea*; *Indigofera* (indigo) is used as a source of blue dye. A great many genera, e.g., *Dalbergia*, *Pterocarpus*, are important sources of luxury timbers.

Discussion: Monophyly of Fabaceae (or Leguminosae) is supported by several morphological features and DNA sequence data (Chappill 1994; Doyle 1994; Kajita et al.

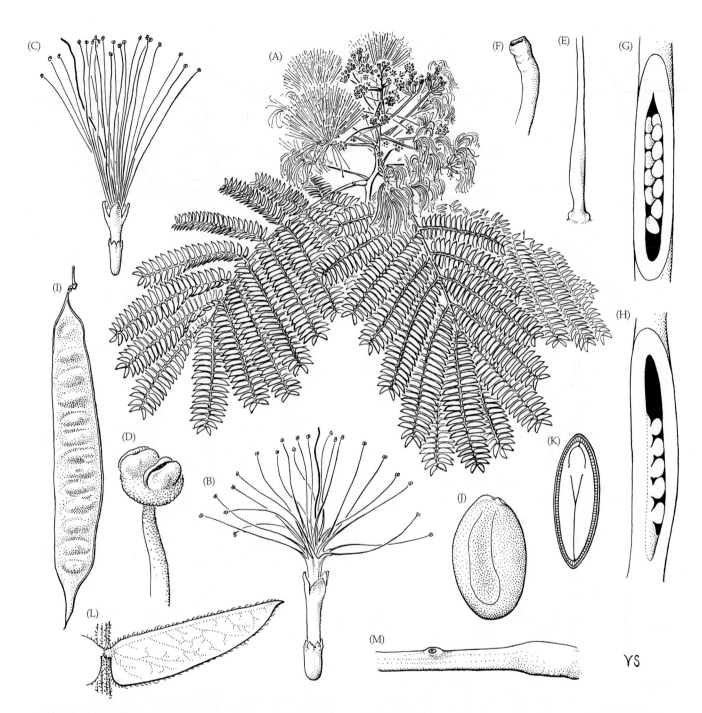

FIGURE 9.69 Fabaceae, subfamily Mimosoideae. *Albizia julibrissin:* (A) branch with flowers (× 0.4); (B) bisexual central flower of an individual inflorescence, showing elongated corolla tube and exserted staminal tube (× 2); (C) nonapical flower of an inflorescence, often functionally staminate (× 2); (D) anther (× 44); (E) ovary and lower part of style (× 9); (F) upper part of style and stigma (× 37); (G) ovary in longitudinal section from below, showing two rows of ovules (× 30); (H) ovary in longitudinal section, lateral view (× 30); (I) mature fruit (× 0.75); (J) seed, showing pleurogram (× 6); (K) seed in longitudinal section, showing large embryo (× 6); (L) leaflet (× 4.5); (M) base of petiole showing pulvinus and nectary (× 3). (From Elias 1974, *J. Arnold Arbor.* 55: p. 110.)

2001; Lavin et al. 2005). Nitrogen fixation occurs in numerous legumes, but is lacking in many early-diverging lineages; it is homoplasious and not synapomorphic for the family (Doyle et al. 1997).

Morphology (Hufford 1992) and DNA sequences (Chase et al. 1993; Savolainen et al. 2000b; Soltis et al. 2000; Kajita et al. 2001) clearly place the family within the rosid complex, close to the Polygalaceae and Surianaceae (in the Fabales, as here circumscribed; see Angiosperm Phylogeny Group 1998, 2003). Pinnately compound leaves, flowers with imbricate perianth parts and (at least sometimes) with a distinct nectar disk, along with similarities in woody

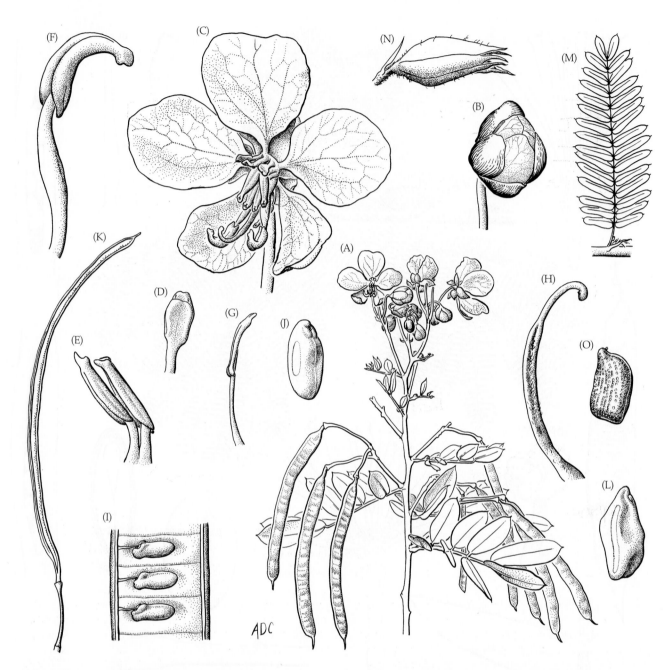

FIGURE 9.70 Fabaceae, subfamily "Caesalpinioideae." (A–J) *Senna bahamensis:* (A) flowering branch with immature fruits (× 0.75); (B) flower bud (× 3); (C) flower (× 2); (D) upper staminode (× 6); (E) functional lateral stamens (× 6); (F) functional lower stamen (× 6); (G) lower staminode (× 6); (H) gynoecium (× 6); (I) portion of immature fruit with developing seeds (× 4.5); (J) seed (× 6). (K–L) *S. obtusifolia:* (K) legume (× 0.75); (L) seed (× 6). (M–O) *Chamaecrista fasciculata:* (M) pinnately compound leaf with stipules (× 0.75); (N) flower bud (× 3); (O) seed (× 6). (From Robertson and Lee 1976, *J. Arnold Arbor.* 57: p. 38.)

anatomy and embryology have been used to suggest a relationship with Sapindales (Dickison 1981; Thorne 1992). Some botanists have considered Fabaceae to be closely related to Rosaceae because of the presence in both of stipulate leaves and hypanthia. Neither relationship is currently supported.

Three subgroups are generally recognized within Fabaceae: "Caesalpinioideae," Mimosoideae, and Faboideae (=Papilionoideae). In most classifications (Polhill et al. 1981)

these are considered subfamilies, but they are sometimes treated as separate families (Cronquist 1981; Dahlgren 1983). Diagnostic features for each subfamily are outlined in Table 9.2. Phylogenetic analyses of morphological characters (Chappill 1994; Tucker and Douglas 1994) and DNA sequences (Bruneau et al. 2001; Doyle 1987; Doyle and Luckow 2003; Doyle et al. 1997; Kajita et al. 2001; Lavin et al. 2005; Wojciechowski et al. 2004) show that "Caesalpinioideae" are paraphyletic, with some genera more closely

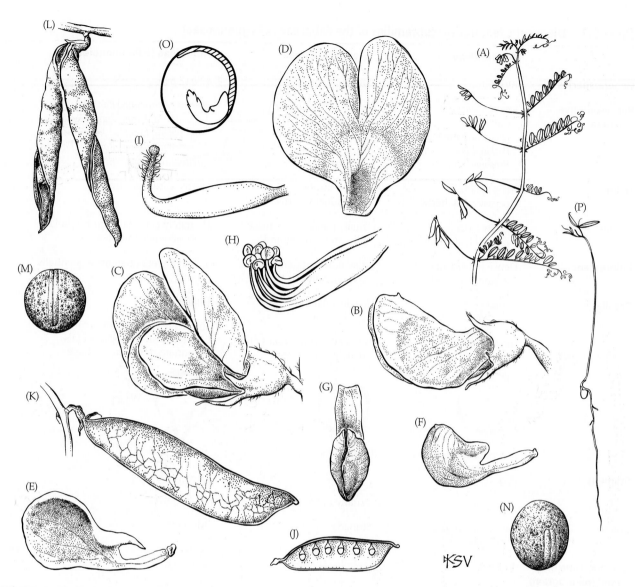

FIGURE 9.71 Fabaceae, subfamily Faboideae. *Vicia ludoviciana:* (A) tip of vine with flowers and fruits (× 0.3); (B) side view of flower bud (× 5); (C) flower (× 5); (D) banner petal (× 5); (E) inner surface of wing petal (× 5); (F) inner surface of keel petal (× 5); (G) keel seen from front (× 5); (H) androecium with nine stamens fused and one ± free (× 7); (I) gynoecium of one carpel (× 7); (J) young fruit, with one valve removed to show ovules (× 1.5); (K) mature legume (× 3); (L) dehisced legume (× 2.5); (M, N) seeds, note hilum half encircling seed (× 6); (O) seed in cross-section, note hilar region (dashed lines), and large embryo, cotyledon, and curved axis (× 8); (P) seedling. (From Wood 1974, *A student's atlas of flowering plants*, p. 60.)

related to Mimosoideae and others more closely related to Faboideae than they are to one another. A *Cercis + Bauhinia* clade, along with a clade containing *Hymenaea, Brownea, Amherstia,* among others, may be sister to the rest of the family. Within Faboideae, it is clear that the temperate herbaceous lines are more recent derivatives of tropical woody groups, although the number of origins of the herbaceous habit remains uncertain (Lavin, personal communication). Detailed cladistic analyses are now available for several tribes of Fabaceae. *Swartzia* and *Cladrastis* (and their relatives) probably represent basal clades of Faboideae.

Several genera of Fabaceae show interesting coevolutionary relationships with various species of ants. Extrafloral nectar glands are common in Mimosoideae and "Caesalpinioideae," and the modified stipules of some species of *Acacia* are inhabited by ants, which protect the plant from herbivory (McKey 1989).

Flowers of Fabaceae are extremely variable in size, form, coloration, and pollinators (Plate 9.11A,B,E,F). Nectar-gathering pollinators include bees, wasps, ants, butterflies, flies, beetles, birds, and bats, although bee pollination is most characteristic, especially of Faboideae (Arroyo 1981). The specialized bilaterally symmetrical flowers of this large group have a conspicuous banner (or standard) petal, which functions as a visual attractant, and two wings, which serve as a landing platform for visiting insects. When

TABLE 9.2 Diagnostic features for subfamilies of the Fabaceae (= Leguminosae).

	Mimosoideae	"Caesalpinioideae"	Faboideae (= Papilionoideae)
Genera/species	40/2500	150/2700	429/12,615 *[peanut]*
Representative genera	*Acacia, Albizia, Calliandra, Inga, Leucaena, Mimosa, Parkia, Pithecellobium, Prosopis*	*Bauhinia, Caesalpinia, Cassia, Chamaecrista, Cercis, Delonix, Gleditsia, Parkinsonia, Senna, Tamarindus* *[red bud] [green bean]*	*Arachis, Astragalus, Baptisia, Crotalaria, Desmodium, Glycine, Indigofera, Lupinus, Melilotus, Phaseolus, Pisum, Robinia, Tephrosia, Trifolium, Vicia, Wisteria* *[clover] [peas] [thorn tree]*
Habit	Trees to shrubs; occasionally herbs	Trees to shrubs; occasionally herbs	Herbs, shrubs, or trees
Leaves	Usually twice pinnately compound	Usually pinnately or twice pinnately compound	Pinnately compound to trifoliolate; occasionally unifoliolate
Inflorescence	Dense flowers opening ± simultaneously	± Lax flowers opening sequentially	± Lax flowers opening sequentially
Corolla	Radial *[only one in family]* Valvate — *[symmetrical]* — Individually not showy	Usually bilateral (some radial) — Imbricate, with upper petal usually innermost — *[assymetrical]* — Usually showy	Bilateral (most) — Imbricate, with upper petal outermost, two basal petals connate or coherent at apex — Banner / Wing / Keel — Showy
Stamens	10–∞ *[usually]* Showy	10–1 Usually not showy	10 or 9 + 1 *[monadelphous / diadelphous]* Not showy
Pollen	Monads, tetrads, polyads	Monads	Monads
U-shaped line in seed (pleurogram)	Present	Usually lacking	Lacking

the bee lands and probes for nectar, it depresses the keel petals, which enclose the stamens and carpel, causing the stamens and stigma to contact the insect's underside. Various other specialized pollen presentation mechanisms occur; the stigma and stamens of the flowers of *Genista* and *Medicago*, for example, are explosively presented. Outcrossing is favored by protandry.

Fabaceae show a diversity of dispersal mechanisms. A large number of species have elastically dehiscing legumes. The valves are hygroscopic, and pressure builds as the fruit dries; eventually the two valves become coiled and separate from each other, tossing out the seeds. Other species have strongly flattened legumes that twist in the wind as they open, passively tossing out the small seeds. *Cassia*, *Gleditsia*, and *Enterolobium* have more or less hard, indehiscent fruits in which the seeds are surrounded by a sweet to sour pulp. These fruits usually fall to the ground under the tree and are dispersed by mammals. Fruits of *Tamarindus* and *Inga* are mostly fleshy and form vertebrate-dispersed berries. Many species have fruits that open to expose either arillate (as in *Pithecellobium* and *Acacia*) or colorful seeds (as in *Abrus* and *Erythrina*; Plate 9.11C) and are mainly dispersed by birds—the latter pair by deceit as the seeds are not edible. Members of many genera have fruits adapted for wind dispersal; the fruits may be indehiscent, with one or a few seeds, and winged, as in *Hymenolobium*, *Tipuana*, *Peltophorum*, and *Pterocarpus*, or inflated and balloonlike, as in *Oxytropis* and *Crotalaria*. *Trifolium* and *Astragalus* have various accessory structures for wind dispersal. The fruits of *Piscidia* are loments that split into one to a few seeded, winged segments. The loments of *Desmodium* and *Aeschynomene*, which break transversely into one-seeded segments, are covered with hooked hairs, leading to external transport by various animals. A few species of *Acacia* have ant-dispersed seeds.

Additional references: Arroyo 1981; Augspurger 1989; Elias 1974; Herendeen et al. 1992; Isley 1998; Lewis et al.

2005; Oldeman 1989; Polhill 1981; Robertson and Lee 1976; Schrire 1989; Verma and Standley 1989; Weberling 1989.

Polygalaceae Hoffmannsegg & Link
(Milkwort Family)

Herbs to trees, or lianas; often producing triterpenoid saponins and methyl salicylate. Hairs simple. *Leaves alternate, spiral or 2-ranked, simple*, entire, with pinnate venation; stipules lacking, or paired glands or spines present. Inflorescences indeterminate, racemose to paniculate, sometimes reduced to a single flower, terminal or axillary. **Flowers** bisexual, **± bilateral**. *Sepals usually 5*, distinct to variably connate, *often only the 2 abaxial ones connate, often with the 2 lateral ones larger than the others and petaloid. Petals (5–) 3, with 2 adaxial and 1 abaxial, distinct, but often all adnate to staminal tube, the abaxial one often boat-shaped, sometimes also appendaged*, imbricate. *Stamens (4–) 8 (–10); filaments distinct or connate, adnate to petals; anthers opening by 1 or 2 ± apical pores*, or longitudinal slits; **pollen grains polycolporate**. Carpels 2 or 3 (–8), connate; ovary superior, with usually axile placentation, but sometimes pseudomonomerous; style often with one branch stigmatic and the other sterile, ending in a hair tuft; stigma capitate. Ovules 1 (2 to many) in each locule. Nectar disk sometimes present. *Fruit a loculicidal capsule, samara, drupe, berry, or nut;* seeds often with stiff hairs, sometimes arillate; endosperm present to lacking (Figure 9.72).

> **Floral formula:**
>
> X, K $\widehat{5}$, C3(–5), A $\widehat{4–10}$, G $\widehat{2–3}$; capsule
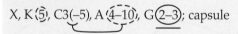

Distribution: Widespread; tropical to temperate.

Genera/species: 17/850. **Major genera:** *Polygala* (550 spp.), *Monnina* (125), and *Muraltia* (115). The first two occur in the continental United States and/or Canada.

Economic plants and products: A few species of *Polygala* and *Securidaca* are cultivated as ornamentals.

Discussion: Polygalaceae are hypothesized to be monophyletic on the basis of morphological and *trnL-F* characters. The group is composed of two major clades (Persson 2001). The first, including *Xanthophyllum*, is diagnosed by the accumulation of aluminum in the tissues and distinct staminal filaments. The second major clade, including most of the family, is supported by the boat-shaped lower petal, stamens opening by terminal pores, and curved or bent style. Most of the genera in this clade (*Polygala, Monnina, Muraltia,* and *Securidaca*) have several additional synapomorphies (e.g., flowers with three obvious petals, two carpels, and a bifid style with only one functional stigma) and constitute the Polygaleae, within which the large genus *Polygala* is nonmonophyletic.

Polygalaceae had been considered to be related to Malpighiaceae or Krameriaceae, and all three families have bilaterally symmetrical flowers (Cronquist 1981). Analyses of DNA sequences, however, place the family as the sister group of Fabaceae, in Fabales, as it is treated here. The flowers of most species of Polygalaceae do look remarkably similar to those of faboid legumes, yet the structures involved are not homologous. The "wings" of legume flowers are lateral petals, while in Polygalaceae they are lateral sepals. The "keel" in flowers of Faboideae is formed from two fused or closely associated petals, while the similar structure of Polygalaceae is a single petal.

The showy flowers of many Polygalaceae attract various bees and wasps and function similarly to the flowers of faboid legumes. Self-pollination is also well known, occurring by a curving of the style or by movement of the sterile, hair-tufted style branch. Some species produce cleistogamous flowers. The samaras of *Securidaca* are dispersed by wind. The loculicidal capsules of *Polygala* release seeds with lobed, aril-like structures that are dispersed short distances by ants. Fleshy-fruited species are vertebrate-dispersed.

Additional references: Eriksen 1993a,b; Miller 1971a; Verkerke 1985.

Rosales

Monophyly of Rosales receives strong support from analyses of DNA sequences (APG 1998, 2003; Hilu et al. 2003; Källersjö et al. 1998; Savolainen et al. 2000a,b; Soltis et al. 2000; Sytsma et al. 2002). The order is quite heterogeneous morphologically, but a reduction (or lack) of endosperm may be synapomorphic for these families. The presence of a hypanthium may also be synapomorphic, and this structure is seen in the families Rosaceae, Rhamnaceae, and some Ulmaceae; it has probably been lost in the more derived families Cannabaceae, Moraceae, and Urticaceae, which have very reduced flowers (Figure 9.73). Rosales also includes Elaeagnaceae (Russian olive).

Phylogenetic relationships within the order are still somewhat unclear, but Ulmaceae, Cannabaceae, Moraceae, and Urticaceae probably constitute a clade, which is diagnosed by globose to elongate cystoliths (concretions of calcium carbonate) within specialized cells (lithocysts), reduced, inconspicuous flowers with 5 or fewer stamens, and 2-carpellate, unilocular ovaries with a single, apical (to basal) ovule (Humphries and Blackmore 1989; Judd et al. 1994; Sytsma et al. 2002). The presence of urticoid teeth (which are nonglandular and have a slender median vein and converging lateral veins) and some characters of wood anatomy may also be synapomorphic. These families are often treated as the order Urticales (Cronquist 1981, 1988; Thorne 1992). They have sometimes been placed in the hamamelid complex (i.e., Platanaceae, Hamamelidaceae, Fagaceae, Betulaceae, etc.; see Cronquist 1981), but many systematists have considered them to be derived from Malvalean ancestors (Berg 1977, 1989; Dahlgren 1983; Thorne

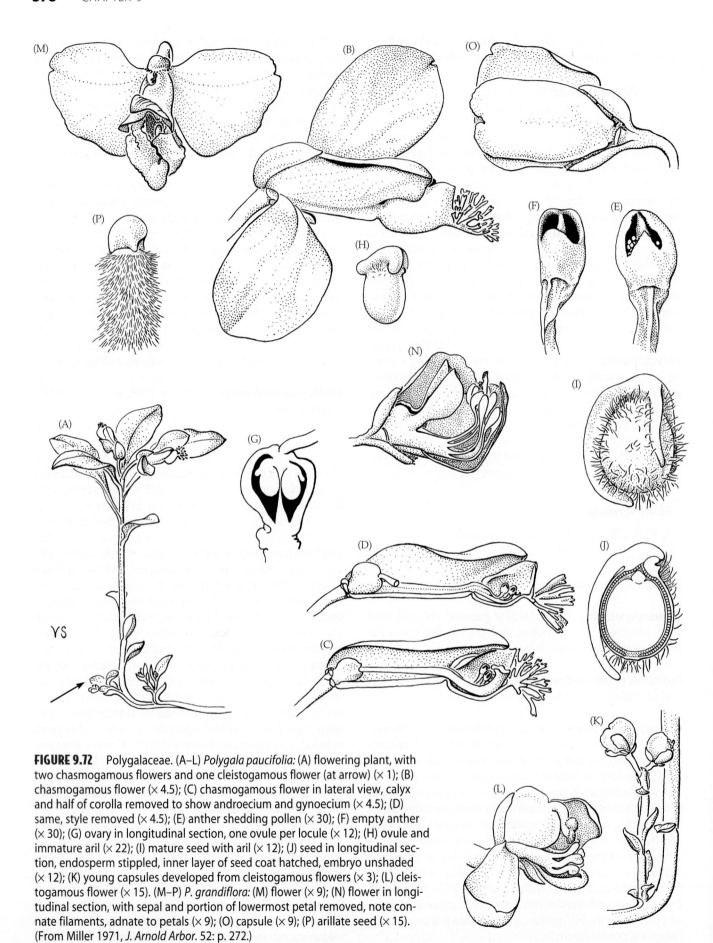

FIGURE 9.72 Polygalaceae. (A–L) *Polygala paucifolia*: (A) flowering plant, with two chasmogamous flowers and one cleistogamous flower (at arrow) (× 1); (B) chasmogamous flower (× 4.5); (C) chasmogamous flower in lateral view, calyx and half of corolla removed to show androecium and gynoecium (× 4.5); (D) same, style removed (× 4.5); (E) anther shedding pollen (× 30); (F) empty anther (× 30); (G) ovary in longitudinal section, one ovule per locule (× 12); (H) ovule and immature aril (× 22); (I) mature seed with aril (× 12); (J) seed in longitudinal section, endosperm stippled, inner layer of seed coat hatched, embryo unshaded (× 12); (K) young capsules developed from cleistogamous flowers (× 3); (L) cleistogamous flower (× 15). (M–P) *P. grandiflora*: (M) flower (× 9); (N) flower in longitudinal section, with sepal and portion of lowermost petal removed, note connate filaments, adnate to petals (× 9); (O) capsule (× 9); (P) arillate seed (× 15). (From Miller 1971, *J. Arnold Arbor.* 52: p. 272.)

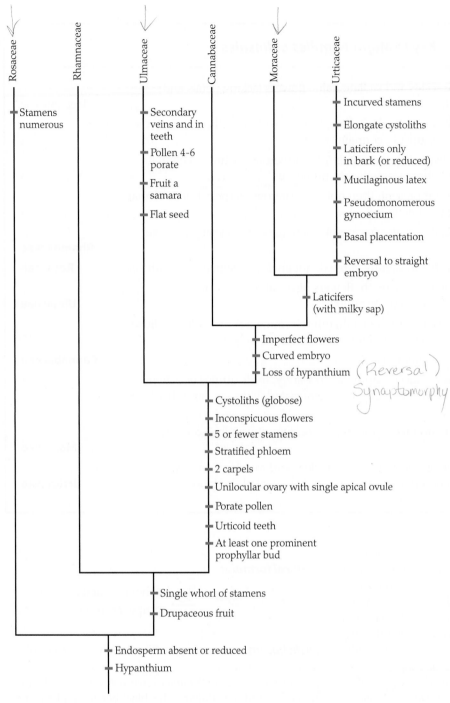

FIGURE 9.73 Cladogram showing hypothesized relationships within Rosales. This tree shows only structural features, but its topology is also supported by molecular data. (Adapted from Judd et al. 1994; Sytsma et al. 2002; and Datwyler and Weiblen 2004.)

tives) are distinguished by pattern of leaf venation and vernation; floral morphology, sexuality, and anatomy; fruit type; embryo shape; pollen form; wood anatomy; flavonoid chemistry; type of sieve cell plastids; and chromosome number (Grudzinskaja 1967; Omori and Terabayashi 1993; Terabayashi 1991). For most of these features the Cannabaceae are more similar to Urticaceae and Moraceae than Ulmaceae, some of these similarities probably representing additional synapomorphies of the Cannabaceae + Moraceae + Urticaceae clade. The Urticaceae + Moraceae clade is supported by DNA sequences and the presence of laticifers; within this clade, Moraceae are sister to Urticaceae, the latter being supported by laticifers restricted to the bark, a mucilaginous latex, pseudomonomerous gynoecia, basal placentation, and elongate cystoliths (but cystoliths lacking in *Cecropia* and relatives).

Rhamnaceae may be sister to the Ulmaceae-to-Urticaceae clade, as both groups show a reduction in the number of stamens (to a single whorl or less). The small family Elaeagnaceae (with distinctive, gold to silver peltate scales and lacking petals) may be sister to Rhamnaceae; both have basal ovules. Rosaceae may be the sister group of the remaining members of Rosales.

The Rosales contain 9 families and about 6300 species; major families include **Rosaceae, Rhamnaceae, Ulmaceae, Cannabaceae, Urticaceae**, and **Moraceae**.

Rosaceae A. L. de Jussieu
(Rose Family)

Herbs, shrubs, or trees (about three-quarters of the genera contain woody plants), often rhizomatous, infrequently climbing; thorns sometimes present; cyanogenic glycosides and sugar alcohol sorbitol present in some groups. Hairs

1992). Here they are considered to be closely related to Rosaceae and Rhamnaceae and therefore are placed within Rosales.

Ulmaceae are probably sister to the Cannabaceae + Moraceae + Urticaceae clade. This Cannabaceae + Moraceae + Urticaceae clade is diagnosed by consistently unisexual flowers and curved embryos. *Celtis, Trema,* and relatives, here placed in Cannabaceae (along with *Cannabis* and *Humulus*), were included within Ulmaceae by many systematists (Cronquist 1981), creating a paraphyletic assemblage (i.e., Ulmaceae s.l.). Ulmaceae and the woody taxa of Cannabaceae (e.g., *Celtis, Trema,* and rela-

Key to Major Families of Rosales

1. Indumentum of copious peltate scales and stellate hairs; flowers lacking petals and with a single carpel ....Elaeagnaceae
1. Indumentum various but usually not densely stellate; flowers with or without petals, the gynoecium with 1 to numerous, distinct to connate carpels 2
2. Flowers with nectar-producing hypanthium; carpels 1 to numerous, distinct to connate, and if connate, with axile placentation; stamens 4 to numerous 3
2. Flowers lacking a hypanthium, or if present, then not nectariferous; carpels 1 or 2, connate, with apical or basal placentation; stamens 1–5 4
3. Stamens 4 or 5, opposite the petals; petals concave or hooded, each enclosing a stamen at anthesis ....**Rhamnaceae**
3. Stamens 10 to numerous; petals flat to slightly cupped, not enclosing stamens at anthesis ....**Rosaceae**
4. Secondary veins running directly into the teeth; flowers bisexual or unisexual; embryo straight; fruits dry, usually samaras or nutlike ....**Ulmaceae**
4. Secondary veins not running directly into the teeth, often forming a series of loops; flowers consistently unisexual; embryo curved or straight; fruits drupes or achenes 5
5. Laticifers lacking; sap watery; fruits not multiple ....**Cannabaceae**
5. Laticifers present, but sometimes reduced; sap milky to watery; fruits drupes or achenes, usually densely clustered and multiple and often also associated with accessory structures; plants woody to herbaceous 6
6. Latex throughout plant, white; cystoliths ± globose; gynoecium of usually 2 carpels, with apical placentation ....**Moraceae**
6. Latex restricted to bark or essentially lacking, milky to clear and mucilaginous; cystoliths ± elongate or lacking; gynoecium pseudomonomerous, with ± basal placentation ....**Urticaceae**

typically simple, sometimes glandular, occasionally stellate; prickles sometimes present. Leaves usually alternate and spiral, simple to often palmately or pinnately compound, blade commonly with glandular-tipped teeth, with pinnate or palmate venation; *stipules usually present.* Inflorescences various. *Flowers often showy,* bisexual or infrequently unisexual (plants then dioecious or monoecious), usually radial, with a *hypanthium ranging from flat to cup-shaped or cylindrical and either free from or adnate to the carpels, often enlarging in fruit, with a nectar ring on the inside.* Sepals usually 5, sometimes alternating with epicalyx lobes. Petals usually 5, often clawed, imbricate, rarely absent. **Stamens usually numerous,** frequently 15 or more but sometimes 10 or fewer; filaments distinct or basally fused to nectar disk; pollen tricolporate. Carpels 1 to many, distinct or connate, sometimes adnate to hypanthium; ovary superior to inferior; styles same number as carpels; stigmas terminal; ovules 1, 2, or often more per carpel, basal, lateral, or apical (when carpels distinct) or ± axile placentation (when carpels connate). Fruit a follicle, achene (exposed or enclosed within the hypanthium, which is sometimes fleshy), pome, drupe, aggregate or accessory with drupelets, follicles, or achenes, rarely a capsule. Endosperm usually absent (Figures 9.74–9.77).

Floral formula:

$*$, K5, C5, A10–∞, G$\overline{(1-∞)}$; achene, druplet, follicles, drupe, pome

Distribution and ecology: The family is cosmopolitan and most abundant in the Northern Hemisphere. Some genera, such as *Lyonothamnus*, have very restricted distributions. In contrast, *Rubus*—the blackberries and raspberries—is native to 6 continents. Herbaceous members grow in temperate forest understory, salt marshes, freshwater marshes, arctic tundra, old fields, and roadsides. Woody members, such as *Rubus*, *Crataegus* (hawthorn), *Amelanchier*, and some species of *Prunus* (cherries), are prominent in early stages of succession. Tree species, such as *Prunus serotina* (black cherry), may be components of mature deciduous forests. Four genera—*Cercocarpus, Chamaebatiaria, Dryas,* and *Purshia* (all in the Dryadoideae as defined here)—fix nitrogen through a symbiosis with the actinomycete *Frankia.*

Genera/species: 90/3000; the major genera are listed in the discussions that follow.

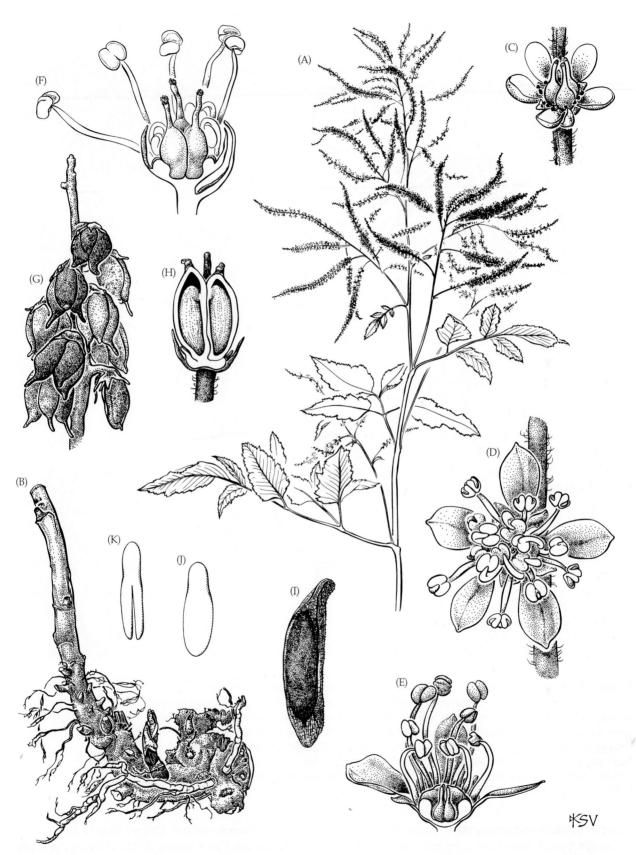

FIGURE 9.74 Rosaceae, subfamily Spiraeoideae, tribe Spiraeeae. *Aruncus dioicus:* (A) flowering stem of staminate plant (× 0.3); (B) portion of rhizome with stem base (× 0.75); (C) carpellate flower (× 17); (D) staminate flower (× 17); (E) staminate flower in longitudinal section, note nectar ring and rudimentary carpels (× 17); (F) bisexual flower (from "staminate" plant) in longitudinal section, petals removed (× 17); (G) fruits (× 7); (H) developing fruit in longitudinal section (× 17); (I) seed (× 30); (J) embryo (× 30); (K) same, side view (× 30). (From Robertson 1974, *J. Arnold Arbor.* 55: p. 326.)

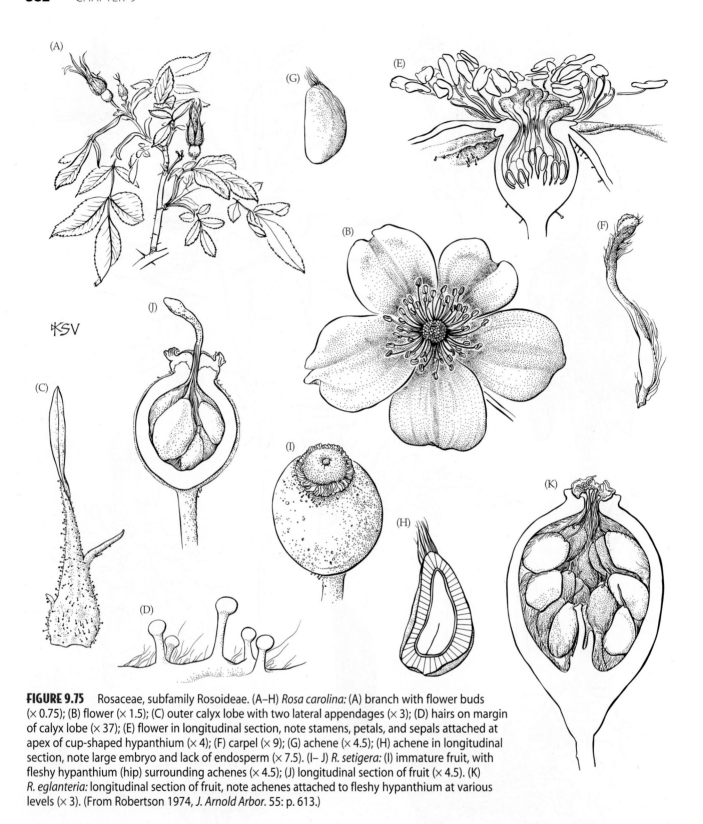

FIGURE 9.75 Rosaceae, subfamily Rosoideae. (A–H) *Rosa carolina:* (A) branch with flower buds
(× 0.75); (B) flower (× 1.5); (C) outer calyx lobe with two lateral appendages (× 3); (D) hairs on margin
of calyx lobe (× 37); (E) flower in longitudinal section, note stamens, petals, and sepals attached at
apex of cup-shaped hypanthium (× 4); (F) carpel (× 9); (G) achene (× 4.5); (H) achene in longitudinal
section, note large embryo and lack of endosperm (× 7.5). (I– J) *R. setigera:* (I) immature fruit, with
fleshy hypanthium (hip) surrounding achenes (× 4.5); (J) longitudinal section of fruit (× 4.5). (K)
R. eglanteria: longitudinal section of fruit, note achenes attached to fleshy hypanthium at various
levels (× 3). (From Robertson 1974, *J. Arnold Arbor.* 55: p. 613.)

Economic plants and products: The family is important
primarily for edible temperate-zone fruits and ornamentals.
Of first importance is *Malus domestica* (apple), a native of
the Old World, now with thousands of cultivars grown
throughout temperate regions. *Prunus* has a wide variety of
fruits: *P. dulcis* (almond), *P. armeniaca* (apricot), *P. avium*
(sweet cherry), *P. cerasus* (sour cherry), *P. persica* (peach),
and *P. domestica* (plum). Other edible fruits of commercial
interest are *Pyrus* (pear), *Rubus* (blackberry, loganberry,
raspberry), *Fragaria* (strawberry), *Cydonia* (quince), and *Eri-
obotrya* (loquat). Ornamental plants in the family include
herbs, such as *Alchemilla* (lady's mantle), *Geum* (avens), *Fil-
ipendula* (meadowsweet), and *Potentilla* (cinquefoil), and
many woody plants, notably *Amelanchier* (shadbush),

(A) Rosales: Rosaceae, Rosoideae
Fragaria × ananassa: accessory fruit

(B) Rosales: Urticaceae
Boehmeria cylindrica: plant in bloom

(C) Rosales: Rosaceae, Spiraeoideae, tribe Pyreae
Sorbus aucuparia: flowers

(D) Rosales: Rosaceae, Rosoideae
Rosa laevigata: leaves and flowers

(E) Rosales: Rosaceae, Rosoideae
Rosa rugosa: flower, longitudinal section

(F) Rosales: Cannabaceae
Trema micrantha: branch with fruits

(G) Rosales: Moraceae
Ficus citrifolia: branch with figs

(H) Rosales: Rosaceae, Spiraeoideae, tribe Pyreae
Crataegus crus-galli: pomes

(I) Rosales: Ulmaceae
Ulmus americana: fruit

(J) Rosales: Rhamnaceae
Rhamnus frangula: branch with fruits

PLATE 9.12 Eudicots: Rosales

Malus- originated in Asia, Himalayans
Rubus- only Rosaceae found on all continents

Key to Major Subgroups of Rosaceae

---

1. Carpels usually numerous; fruit an achene or drupelet; $x = 7$ (rarely 8); sorbitol, cyanogenic glycosides, and flavones absent; ellagic acid present ....Rosoideae

1. Carpels usually 1-5; fruit a drupe, follicle, pome, or capsule; $x = 8, 9, 15, 16,$ or 17; sorbitol, cyanogenic glycosides, and flavones usually present; ellagic acid absent 2

2. Plants with nitrogen-fixing root nodules; sorbitol present in trace amounts ....Dryadoideae

2. Plants not nitrogen-fixing; sorbitol present in significant amounts................................3 (Spiraeoideae)

3. Leaves opposite and fernlike ....Lyonothamnus

3. Leaves alternate (rarely opposite) and simple to compound, but not fernlike 4

4. $x = 8$ 5

4. $x = 9, 16, 16, 17$ (rarely 8); flowers with 2–5 carpels (rarely only 1); fruit a pome, achene, follicle, capsule, or drupe 6

5. Carpel 1; fruit a drupe ....Amygdaleae

5. Carpels 1–5; fruit a drupe, aggregate of drupelets, or capsule ....Osmaronieae

6. Ovules paired, basally inserted, anatropous, and with funicular obturators; host to *Phragmidium* and *Gymnosporangium* rusts; woody plants (*Gillenia* herbaceous); $x = 17$ (but 9 in *Gillenia*, 15 in *Vauquelinia*) ....Pyrodae

6. Ovules solitary, paired, or numerous in each carpel, variously inserted, and anatropous or otherwise but not paired, basally inserted, and anatropous; not hosts to *Phragmidium* and *Gymnosporangium* rusts; woody or herbaceous; $x = 9$ (rarely 8) 7

7. Stipules absent...Spiraeeae

7. Stipules present, although sometimes deciduous...8

8. Seed coat hard and shining ....Neillieae

8. Seed coat not hard and shining 9

9. Leaf epidermis with hard, wartlike projections; leaves simple ....Kerrieae

9. Leaf epidermis without hard, wartlike projections; leaves compound (but simple in *Adenostoma*) ....Sorbarieae

---

Chaenomeles (flowering quince), *Cotoneaster, Crataegus* (hawthorn), *Exochorda* (pearlbush), *Kerria, Malus* (especially crab apples), *Photinia, Physocarpus* (ninebark), *Prunus* (especially Japanese flowering cherries), *Pyracantha* (firethorn), *Rhodotypos* (jetbead), *Rosa* (rose), *Sorbus* (mountain ash, rowan), and *Spiraea* (bridal wreath). Roses, perhaps the most popular and widely grown garden flower in the world, are complex hybrids developed from about nine wild species. *Prunus serotina* produces a wood that is valued for furniture and cabinetry; several genera supply timber.

Discussion: Even though Rosaceae display considerable diversity in anatomy, vegetative features, and fruit morphology, the family has long been considered monophyletic (Potter et al. 2006). Numerous stamens and chloroplast DNA sequences (Evans 1999; Evans and Dickinson 1999a,b; Morgan et al. 1994; Potter et al. 2002), and the granule-bound starch synthase gene (*GBSSI*; Evans et al. 2000) strongly support the monophyly of Rosaceae. Alternate leaves with stip-ules plus showy flowers with radial symmetry, numerous stamens, and a hypanthium aid in recognition of the family. Saxifragaceae, Fabaceae, Crassulaceae, and other groups have been proposed as potential relatives of the family, but several genes from the chloroplast and nucleus place Rosaceae as sister to remaining members of Rosales.

Fruit type was the primary criterion in the traditional subdivision of Rosaceae into four subfamilies. For two of these subfamilies, Amygdaloideae (Figure 9.76) and Maloideae (Figure 9.77), fruit type was uniform. In pomes, the fruit type unique to Maloideae (Plate 9.12H), the hypanthium is fused to the ovary wall and often enlarges into an edible reward for fruit dispersers, as in apples and pears. Amygdaloideae (Prunoideae) bear drupes, and there is only one carpel in the flowers of the major genus, *Prunus*. Fruit type varies more in the other two subfamilies. Rosoideae (Figure 9.75) have achenes or drupelets. Achenes may be associated with a much-enlarged receptacle (*Fragaria*; Plate 9.12A) or enclosed in a more or less urn-shaped or cylindri-

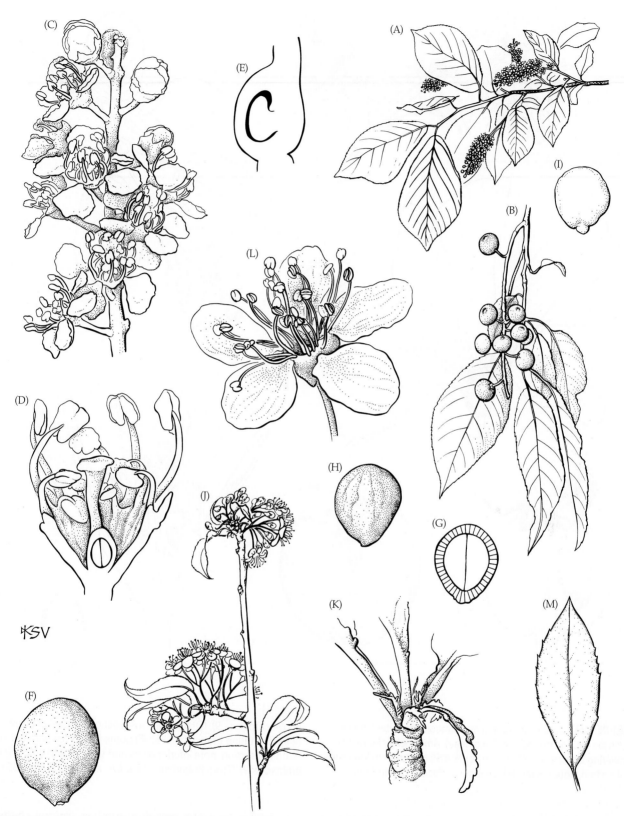

FIGURE 9.76 Rosaceae, subfamily Spiraeoideae, tribe Amygdaleae (the former Amygdaloideae). (A) *Prunus virginiana:* flowering branch (× 0.3). (B–I) *P. serotina:* (B) fruiting branch (× 0.3); (C) tip of raceme (× 4.5); (D) flower (petals removed), in longitudinal section, note hypanthium, solitary carpel with two ovules (× 15); (E) ovary in longitudinal section, at right angle to that of D (× 30); (F) pit from drupe (× 4.5); (G) pit in cross-section, the wall hatched, cotyledons unshaded (× 4.5); (H) seed (× 4.5); (I) embryo (× 4.5). (J–L) *P. pensylvanica:* (J) tip of branch with inflorescences (× 0.75); (K) shoot apex, showing leaf bases with glands and stipules (× 4.5); (L) flower (× 4.5). (M) *P. caroliniana:* spinulose-serrate leaf (× 0.75). (From Robertson 1974, *J. Arnold Arbor.* 55: p. 658.)

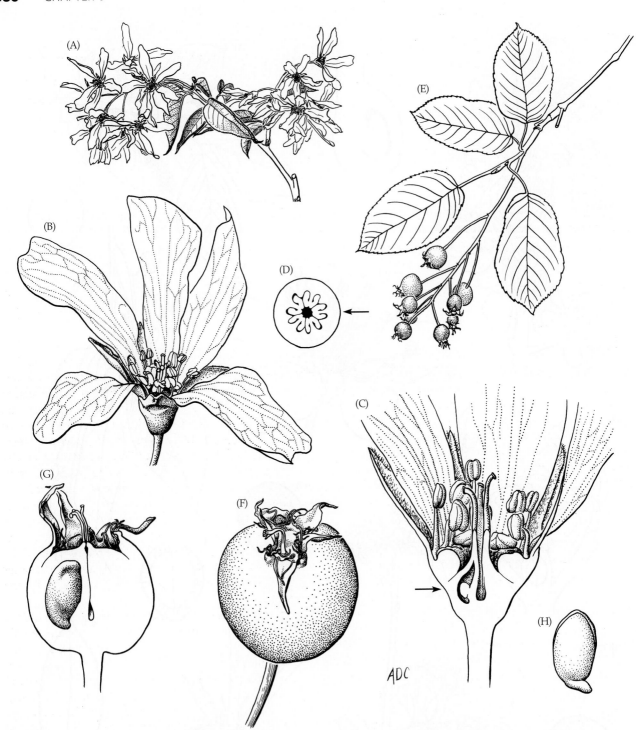

FIGURE 9.77 Rosaceae, subfamily Spiraeoideae, tribe Pyreae, subtribe Pyrinae (the former Maloideae). *Amelanchier laevis:* (A) flowering branch (× 0.75); (B) flower (× 4.5); (C) flower in longitudinal section, note nectar disk on inner surface of hypanthium, above carpels (× 9); (D) ovary in cross-section, at level of arrow in C (× 9); (E) fruiting branch (× 0.75); (F) pome (× 4.5); (G) pome in longitudinal section, note inferior ovary and seed (at left) (× 4.5); (H) embryo (× 9). (From Robertson 1974, *J. Arnold Arbor.* 55: p. 635.)

cal receptacle in many other genera. The fruit of *Rosa* is of this latter type, and the achenes together with the enclosing receptacle in this genus are called a hip. In *Rubus* the drupes are quite small, and therefore called drupelets, and aggregated. The fruit in the fourth traditional subfamily, "Spiraeoideae" (see Figure 9.74), is a follicle or capsule.

Neither base chromosome number, various chemical constituents, distribution of rust fungal parasites, nor chloroplast and nuclear DNA sequences agree with the subfamilies delimited on the basis of fruit type. These non-fruit characters are the basis of a new classification, in which the Rosoideae are defined more narrowly, Dryadoideae are rec-

TABLE 9.3   Subfamilies of Rosaceae

	Rosoideae	Dryadoideae	Spiraeoideae
Genera/species	28/1200–1900	4/31	57/1350
Base chromosome number	7 (8)	9	9, 17 (rarely 8, 15, 16)
Sorbitol	Absent	Present in trace amounts	Present in significant amounts
Cyanogenic glycosides	Absent	Present	Present
Flavones	Absent	Present	Present
Ellagic acid	Present	Absent	Absent
Leaves	Usually compound	Simple (compound in one genus)	Simple (compound in several genera)
Fruits	Achene, drupe; often aggregate	Achene or aggregate of achenes	Follicle, achene, capsule, of drupe, or pome; aggregate achenes or capsules
Symbiotic nitrogen fixation	Absent	Present	Absent

ognized as a subfamily, and Spiraeoideae are expanded to include the classical Amygdaloideae and Maloideae.

Members of the more narrowly defined Rosoideae are herbs or shrubs with usually compound leaves (a potential synapomorphy) and one to many carpels that are free from the hypanthium; a base chromosome number of 7 (8 in one genus); absence of the sugar alcohol sorbitol, cyanogenic glycosides, and flavones; and presence of ellagic acid. The subfamily contains about 25 genera and 1200–1900 species. Three tribes are recognized within Rosoideae. Sanguisorbeae contain a dozen small genera of herbs, shrubs, and trees. Potentilleae has about 10 genera, the best known being *Alchemilla* (250 spp.), *Fragaria* (15), and *Potentilla* (200–500). The largest genus of the third tribe, the Colurieae, is *Geum* (40 spp.). Genera of Rosoideae not included in tribes include *Rosa* (2050 spp.) and *Rubus* (400–740). *Alchemilla*, *Potentilla*, *Rosa*, and *Rubus* present great challenges due to the occurrence of hybridization, polyploidy, and/or agamospermy. Most molecular data place the Rosoideae as the sister to the remainder of Rosaceae (Table 9.3), but there is also some support for a sister relationship between Rosoideae and Dryadoideae.

The narrow definition of Rosoideae requires removal and redefinition of three tribes—Dryadeae, Kerrieae, and Sorbarieae—that differ from this subfamily in their base chromosome number of 9, chemical constituents, and DNA sequences. Dryadeae are elevated to a subfamily of four small genera of woody plants that mostly grow in western North America (one genus also occurs in eastern North America and Europe) and that are unique in the family in having the capacity for symbiotic nitrogen fixation (see "Distribution and Ecology," above). The other two tribes have been moved to the Spiraeoideae. The Kerrieae contain four small genera, the best known being two Asian, monotypic genera used as ornamentals, *Kerria* and

Rhodotypos. Kerrieae have been placed in a supertribe, Kerriodae, which also contain the tribe Osmaronieae. The latter has three genera that had been thought to be related to the cherries and relatives because of a base chromosome number of 8 and because two genera of Osmaronieae have drupes. Sorbarieae contain *Adenostoma* (3 species of western North America), *Chamaebatiaria* (2 species of the southwestern United States), *Sorbaria* (15 species of Asia), and *Spiraeanthus*, a monotypic genus with a restricted distribution in Asia. Another monotypic genus with a restricted distribution, *Lyonothamnus* is not closely related to any other group within the Spiraeoideae. *Lyonothamnus floribundus* is native to just three islands off the southern Californian coast, and it has opposite, fernlike leaves and other morphological features that are quite unusual in the Rosaceae.

Spiraeoideae also traditionally contain two other tribes, Neillieae and Spiraeeae, and have been considerably expanded by the addition of members of former subfamilies Amygdaloideae and Maloideae. Neillieae consist of two genera, *Neillia* (25 Asian species) and *Physocarpus* (15 eastern Asian and North American species). *Spiraea* (100 spp.), *Aruncus* (12), and about five other genera comprise the tribe Spiraeeae, which is monophyletic on the basis of molecular data and has the potential synapomorphy of loss of stipules.

Amygdaleae (formerly the Amygdaloideae) have traditionally comprised one or a few genera with drupaceous fruits and a base chromosome number of 8. The tribe contains one major genus, *Prunus*, with about 200 species mostly of the Northern Hemisphere (Figure 9.76).

The final group in Spiraeoideae is the supertribe Pyrodae (Figure 9.77), including the traditional subfamily Maloideae (i.e., plants with pomes) and four genera with capsular or follicular fruits—*Kageneckia*, *Lindleya*, *Gillenia*, and *Vauquelinia*. Ovules in Maloideae plus these four genera are paired in each carpel, basally inserted, and anatropous. In addition, rust fungi in the genus *Gymnosporangium* infect only plants with pomes (i.e., the traditional Maloideae) plus *Gillenia* and *Vauquelinia*. Within Pyrodae, the strongly supported clade of *Kageneckia, Lindleya, Vauquelinia*, and the traditional Maloideae, all of which are woody, and have a base chromosome number of 17 (x = 15 in *Vauquelinia*) is the tribe Pyreae, and the traditionally defined Maloideae becomes the subtribe Pyrinae, with about 30 genera and 1000 species.

Pyreae are remarkable for the role that hybridization has played in their evolution. The base chromosome number of 17 hints that the tribe could have arisen through allopolyploidization, and genetic studies confirm that members of the tribe are allopolyploids. Since about 1930, a favored hypothesis for origin of the subfamily has been an ancient hybridization between an amygdaloid (x = 8) and a spiraeoid (x = 9). Recent studies of GBSSI falsify this hypothesis and identify instead ancestors of the tribe closely related to *Gillenia*, a genus of herbaceous plants of the southeastern United States with x = 9. Aneuploidy could account for the Pyreae base number of 17 from 18.

Intergeneric hybridization is relatively common in Pyreae and has contributed to uncertainty about the generic limits of the tribe. There is the possibility that past gene flow is partly responsible for the lack of evolutionary divergence between the genera that has been observed in numerous genes and in wood anatomy. This lack of divergence is hard to explain given that several of the genera (e.g., *Amelanchier, Crataegus*, and *Photinia*) are relatively old, with fossils from 48–50 million years ago. Interspecific hybridization and agamospermy frequently occur in the larger genera, such as *Crataegus* (265 spp.), *Cotoneaster* (260), *Sorbus* (258), *Malus* (55), and *Amelanchier* (33), making recognition of species difficult. Other major genera are *Pyrus* (76) and *Photinia* (54).

Flowers of Rosaceae are unspecialized, with radial symmetry and flat or shallowly cup-shaped corollas, and are adapted primarily for generalist pollinators (Plate 9.12C, D, E). Smaller flowers are visited by flies and short-tongued bees, while species with larger flowers are pollinated by long-tongued bees, wasps, butterflies, moths, and beetles. There has been considerable diversification for fruit dispersal, with adaptations for wind dispersal (winged seeds, plumose styles, and winged hypanthia that enclose the seeds), animal attachment (hooked prickles or barbs on the outside of the dispersal unit, hooked styles), and ingestion (small oil bodies for ant dispersal, and fleshy fruit walls and/or hypanthia for consumption by birds, mammals, and reptiles).

Additional references: Alice et al. 2001; Bortiri et al. 2001; Campbell et al. 1995, 2006; Eriksson et al. 2003; Kalkman 2004; Lee and Wen. 2001; Phipps et al. 1990; Potter 2003; Potter et al. 2002; Robertson 1974; Robertson et al. 1991, 1992; Rohrer et al. 1991, 1994; Wolfe and Wehr 1988.

Rhamnaceae A. L. de Jussieu
(Buckthorn Family)

Trees, shrubs, often with thorns, or lianas with tendrils or twining stems, the axillary branches departing somewhat laterally; sometimes with nitrogen-fixing bacteria (*Frankia*) in root nodules; often with tannins. Hairs usually simple. *Leaves alternate and spiral*, or less commonly opposite, simple, entire to serrate, with strongly pinnate or palmate venation, tertiary veins often strongly ladder-like; *stipules present*, sometimes spinose. Inflorescences determinate, sometimes reduced to a single flower; axillary or terminal. *Flowers* usually bisexual, radial, *small, with disklike to cylindrical hypanthium*. Sepals usually 4 or 5, distinct, valvate. *Petals usually 4 or 5, distinct*, **usually ± concave or hooded and enclosing the anther at flowering**, *usually clawed. Stamens 4 or 5*, distinct, **opposite the petals; filaments adnate to base of petals**; pollen grains tricolporate. Carpels usually 2 or 3, connate; ovary superior to inferior, with axile placentation and usually 1 ovule attached at the base of each locule; stigmas usually ± capitate. Nectar-producing tissue on inner surface of hypanthium. **Fruit a dehiscent to indehiscent drupe (then ± schizocarpic) with 1 to several pits**, often with a conspicuous subbasal rim, rarely a samaroid schizocarp; endosperm present or lacking (Figure 9.78).

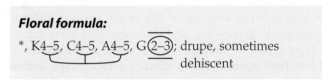

Floral formula:

*, K4–5, C4–5, A4–5, G(2–3); drupe, sometimes dehiscent

Distribution and ecology: Nearly cosmopolitan, but especially diverse in tropical regions; characteristic of limestone soils. *Ceanothus* and some other genera fix nitrogen.

Genera/species: 52/900. **Major genera:** *Phylica* (150 spp.), *Rhamnus* (100), *Zizyphus* (100), and *Ceanothus* (55), and *Gouania* (50). Noteworthy genera occurring in the continental United States and Canada are *Adolphia, Berchemia, Ceanothus, Colubrina, Gouania, Krugiodendron, Reynosia, Rhamnus, Sageretia*, and *Ziziphus*.

Economic plants and products: Fruits of *Ziziphus jujuba* (jujube) and the pedicels of *Hovenia dulcis* (raisin tree) are eaten. *Ceanothus, Colletia, Pomaderris, Rhamnus*, and *Phylica* provide ornamentals.

Discussion: Monophyly of Rhamnaceae is supported by morphology plus *rbcL* and *trnL-F* sequences (Richardson et al. 2000b). The family may be most closely related to Rosaceae, as supported by the presence of a well-developed hypanthium with a nectar-producing inner surface and stipulate leaves. However, DNA sequences suggest a closer relationship with Ulmaceae, Moraceae, and relatives.

Infrafamilial relationships have been studied by Richardson et al. (2000a,b); generic delimitation is often difficult.

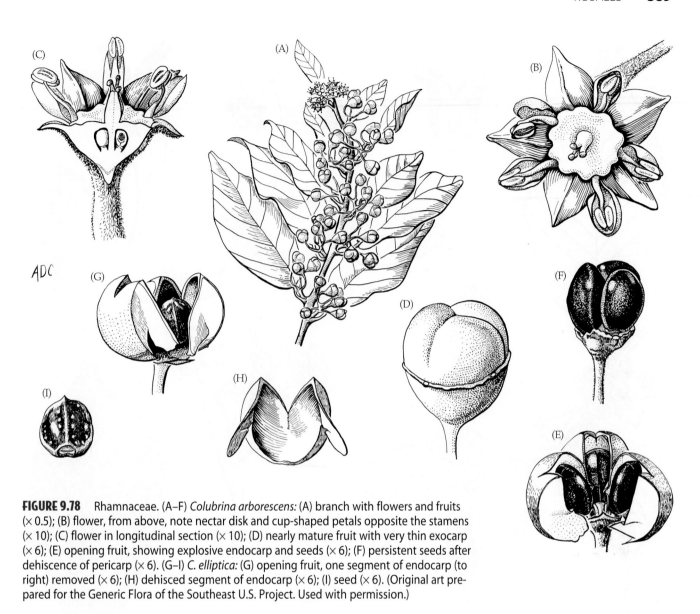

FIGURE 9.78 Rhamnaceae. (A–F) *Colubrina arborescens:* (A) branch with flowers and fruits (× 0.5); (B) flower, from above, note nectar disk and cup-shaped petals opposite the stamens (× 10); (C) flower in longitudinal section (× 10); (D) nearly mature fruit with very thin exocarp (× 6); (E) opening fruit, showing explosive endocarp and seeds (× 6); (F) persistent seeds after dehiscence of pericarp (× 6). (G–I) *C. elliptica:* (G) opening fruit, one segment of endocarp (to right) removed (× 6); (H) dehisced segment of endocarp (× 6); (I) seed (× 6). (Original art prepared for the Generic Flora of the Southeast U.S. Project. Used with permission.)

Extensive homoplasy is evident in ovary position, fruit type, and habit. Rhamnaceae may be composed of a few well-supported clades, for which no obvious morphological apomorphies are known. Major clades are the rhamnoids (e.g., *Rhamnus, Krugiodendron, Berchemia,* and *Reynosia*) and the ziziphoids (e.g., *Hovenia, Ziziphus, Ceanothus, Pomaderris, Phylica, Adolphia, Colubrina, Gouania,* and *Colletia*). Many rhamnoids have superior ovaries and drupaceous fruits, while ziziphoids usually have semi-inferior to inferior ovaries and schizocarpic fruits.

The usually inconspicuous flowers are visited by flies, bees, wasps, and beetles, and may be outcrossing or selfing. Seeds of fleshy-fruited genera such as *Rhamnus, Berchemia, Reynosia,* and *Krugiodendron* are usually dispersed by birds and/or mammals (Plate 9.12J). Those with dehiscent, more or less schizocarpic drupes (e.g., *Ceanothus* and *Colubrina*) typically eject their seeds. The samaroid schizocarps of *Gouania* are wind-dispersed.

Additional references: Brizicky 1964b; Medan and Schirarend 2004.

Ulmaceae Mirbel
(Elm Family)

Trees; growing by a central plagiotropic axis, which secondarily becomes erect, and the lateral branches similar, thus with a spreading aspect; often with tannins; cystoliths present; laticifers absent. *Hairs simple, often with mineralized cell walls. Leaves alternate and 2-ranked, simple, simply or doubly serrate,* **with pinnate venation, secondary veins ending in the teeth,** *blade with asymmetrical base;* stipules present. Inflorescences determinate, forming fascicles, axillary. Flowers bisexual or unisexual (plants then monoecious, dioecious, or polygamous), radial, *inconspicuous,* with hypanthium. Tepals 4–9, distinct to connate, usually imbricate. Stamens 4–9, opposite the tepals; *filaments* distinct, *erect in bud;* **pollen**

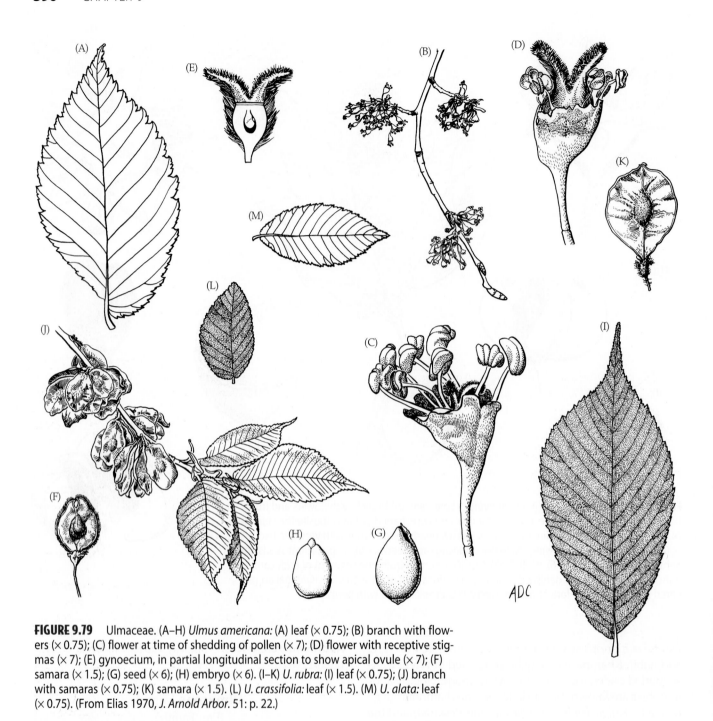

FIGURE 9.79 Ulmaceae. (A–H) *Ulmus americana:* (A) leaf (× 0.75); (B) branch with flowers (× 0.75); (C) flower at time of shedding of pollen (× 7); (D) flower with receptive stigmas (× 7); (E) gynoecium, in partial longitudinal section to show apical ovule (× 7); (F) samara (× 1.5); (G) seed (× 6); (H) embryo (× 6). (I–K) *U. rubra:* (I) leaf (× 0.75); (J) branch with samaras (× 0.75); (K) samara (× 1.5). (L) *U. crassifolia:* leaf (× 1.5). (M) *U. alata:* leaf (× 0.75). (From Elias 1970, *J. Arnold Arbor.* 51: p. 22.)

grains 4- to 6-porate. *Carpels 2, connate; ovary superior, with apical placentation* and usually 1 locule; stigmas 2, extending along adaxial side of styles. Ovule 1. **Fruit a samara or nutlet; seeds flat**; *embryo straight*; **endosperm of a single layer of cells and appearing absent** (Figure 9.79).

Floral formula: *, T(4–9), A4–9, G②; samara, nut

Distribution: Widely distributed, but most diverse in temperate regions of the Northern Hemisphere.

Genera/species: 6/40. **Major genus:** *Ulmus* (25). *Ulmus* and *Planera* occur in the continental United States and/or Canada.

Economic plants and products: *Ulmus* (elm) and *Zelkova* provide timber and important ornamentals.

Discussion: Ulmaceae are monophyletic based on numerous morphological synapomorphies (Zavada and Kim 1996) and easily distinguished from Cannabaceae (especially *Celtis* and relatives) by their leaves with secondary veins running directly into the teeth (vs. veins forming a series of loops); dry fruits, usually samaras (vs. fleshy fruits, i.e., drupes); 4–6-porate pollen with roughened exine (vs. 2–3-porate pollen with smooth exine); chemical makeup, with lignans, sesquiterpenes, and flavonols (vs. quebrachitol and glycoflavones); base chromosome number of 14 (vs. 10, 11); flowers bisexual or unisexual (vs. flowers consistently unisexual); seeds flat, with straight embryo (vs. seeds globular, with curved embryo); and styles with three vascular bundles (vs. a single bundle). In addition, *Ulmus* has plastids with protein crystals, while those of *Celtis* contain starch. Monophyly of Ulmaceae is also supported by DNA characters (Sytsma et al. 2002; Wiegrefe et al. 1998). Within Ulmaceae, *Ulmus, Planera, Hemiptelea,* and *Zelkova* may form a clade on the basis of their unusual leaf vernation, the blade being folded and held to one side of the axis.

Ampelocera and *Holoptelea* are sister to the remaining genera. *Ulmus, Planera,* and *Zelkova* form a clade; all have stipitate ovaries (Manchester and Tiffney 2001).

Two major clades are recognized within *Ulmus*: section *Oreoptelea,* which is diagnosed by elongate and articulated pedicels and ciliate-margined samaras; and section *Ulmus,* which has short, usually not obviously articulate pedicels and usually nonciliate-margined samaras. The monophyly of these two groups is strongly supported by cpDNA restriction sites (Wiegrefe et al. 1994).

The reduced flowers are wind-pollinated. The winged fruits of most species also are dispersed by wind. The warty, nutlike fruits of *Planera* are water-dispersed.

Additional references: Elias 1970; Grudzinskaja 1967; Manchester 1989; Omori and Terabayashi 1993; Terabayashi 1991; Todzia 1993; Ueda et al. 1997.

Cannabaceae Martynov
(Hemp or Hackberry Family)

Usually trees or shrubs, but also herbs (*Cannabis*) or vines (*Humulus*); cystoliths present; *laticifers absent* (but laticifer-like cells, with dark contents, present in *Cannabis* and *Humulus*). Hairs simple, often with mineralized cell walls, sometimes with gland-head containing aromatic substances or tetrahydrocannabinol. *Leaves alternate* (but opposite in *Humulus,* and opposite or alternate in *Cannabis*), and *usually 2-ranked, simple* (but palmately lobed in *Humulus* and palmately compound in *Cannabis*), *entire to serrate, with venation intermediate between pinnate and palmate,* usually with 3 main veins from the base, *or palmate,* with several major veins (*Humulus, Cannabis*), with blade with symmetrical to asymmetrical base; *stipules present.* Inflorescences determinate, sometimes fasciculate, racemelike, or reduced to a solitary flower, axillary. *Flowers unisexual* (plants monoecious or dioecious), radial, *inconspicuous.* Tepals usually 4 or 5, distinct to slightly connate, imbricate, reduced in carpellate flowers of some cultivars of *Cannabis.* Stamens 4 or 5 and opposite the tepals; filaments

distinct, free to slightly adnate to tepals, erect or incurved in bud; pollen grains 2–3-porate. *Carpels 2, connate; ovary superior, with apical placentation* and 1 locule; stigmas elongate and extending along one side of style, sometimes divided. Ovule 1. *Fruit a drupe* (Plate 9.12F), or achene (in *Humulus, Cannabis*); seeds globose; *embryo curved;* endosperm usually ± scanty (Figure 9.80).

Floral formula:
Staminate: *, T(4–5), A4–5, G0

Carpellate: *, T(4–5), A0, G②; drupe

Distribution: Widely distributed in tropical to temperate regions.

Genera/species: 11/180.   **Major genera:**   *Celtis* (100 spp.) and *Trema* (55). These, plus *Humulus* and *Cannabis,* occur in the continental United States and/or Canada.

Economic plants and products: *Celtis* (hackberry, sugarberry) provides timber and ornamental trees. The fruits are occasionally eaten. *Cannabis* is the source of fiber (hemp) and for psychotropic drugs (marijuana, hashish); *Humulus* is used in the flavoring of beer (providing bitter essential oils).

Discussion: *Celtis, Trema,* and other woody members of Cannabaceae at one time were included within Ulmaceae (as subfamily Celtidoideae); see discussions under Ulmaceae and Rosales for an outline of the characters differentiating them. Some species of *Aphananthe* have secondary veins terminating in the teeth (as in Ulmaceae); the remaining features of this genus, however, are typical of Cannabaceae, and this venation condition apparently evolved independently from that of Ulmaceae. The genus is not a "link" between Cannabaceae and Ulmaceae. Phylogenetic placement of *Cannabis* and *Humulus* has been argued, and the circumscription of Cannabaceae often has been restricted to these two distinctive genera (Cronquist 1981), but molecular characters place them firmly within the clade containing *Celtis, Trema,* and relatives (Song et al. 2001; Sytsma et al. 2002). Thus the clade formerly called Celtidaceae (or Ulmaceae subf. Celtidoideae) now must be called Cannabaceae s.l. (because *Cannabis* and *Humulus* are nested within it). *Cannabis* and *Humulus* are distinguished from *Celtis* and relatives by their herbaceous habit, leaves with secondary veins running directly into the teeth, and dioecy.

The inconspicuous flowers of Cannabaceae are wind-pollinated. The colorful drupaceous fruits of most genera have a sweet flesh and are adapted for dispersal by birds. Morphological synapomorphies of the group are unclear.

Additional references: Elias 1970; Grudzinskaja 1967; Omori and Terabayashi 1993; Terabayashi 1991; Ueda et al. 1997.

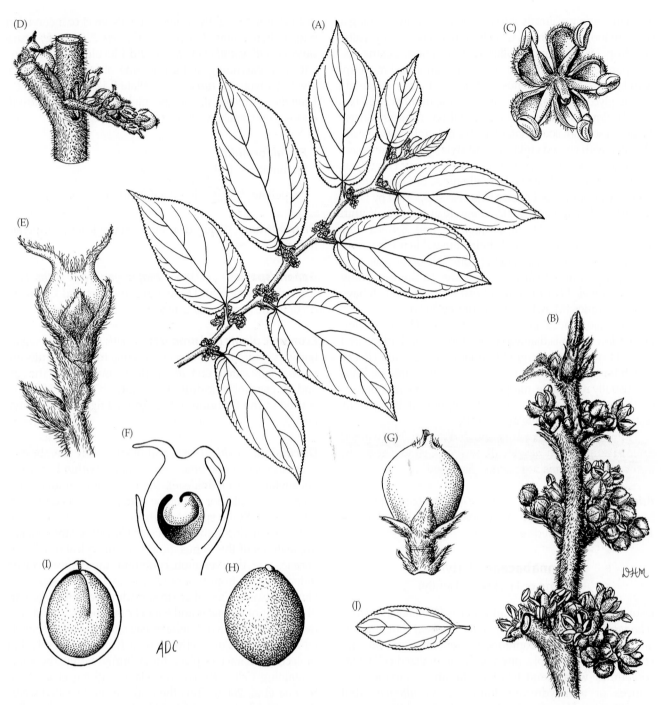

FIGURE 9.80 Cannabaceae. (A–I) *Trema micrantha:* (A) branch with staminate flowers (× 0.75); (B) same, leaves removed (× 3); (C) staminate flower with rudimentary gynoecium (× 7.5); (D) node with carpellate flowers (× 3); (E) carpellate flower (× 15); (F) same, in longitudinal section, showing solitary ovule (× 15); (G) drupe (× 7.5); (H) seed (× 15); (I) seed in partial longitudinal section to show embryo (× 15). (J) *T. lamarckiana:* leaf (× 0.75). (From Elias 1970, *J. Arnold Arbor.* 51: p. 38.)

Moraceae Gaudich.
(Mulberry or Fig Family)

Trees, shrubs, lianas, or rarely herbs; **with laticifers and milky sap, distributed in all parenchymatous tissues**; cystoliths present, usually globose; often with tannins. Hairs often simple, often with mineralized cell walls. *Leaves alternate, often 2-ranked but sometimes spiral, or opposite, usually simple,* sometimes lobed, *entire to serrate, with pinnate to palmate venation,* blade sometimes with cordate or asymmetrical base; *stipules usually present,* small to enlarged and leaving a circular scar on twig. *Inflorescences determinate, but sometimes appearing indeterminate,* axillary, *individual flowers usually congested and inflorescence axis often thickened and variously modified. Flowers unisexual* (plants monoecious), usually *radial, inconspicuous.* Tepals (0–) 4 or 5 (–8), distinct to connate, imbricate or valvate, often becoming fleshy and associated with mature fruits. *Stamens usually 1–5, opposite the tepals; filaments* distinct, *straight to incurved in bud*; anthers 2- or 1-locular; pollen grains usually 2–4- to multiporate. *Carpels 2, connate, sometimes with 1 carpel reduced*; ovary usually superior, *with apical placentation* and usually 1 locule; stigmas 2, extending along adaxial side of style, to capitate. Ovule 1, anatropous to campylotropous. *Fruit a drupe, dehiscent drupe, or achene, often closely clustered together and forming a multiple fruit; embryo curved to less commonly straight;* endosperm often lacking (Figure 9.81).

Floral formula:

Staminate: *, T(4–5), A1–5, G0

Carpellate: *, T(4–5), A0, G②; drupe; achene

Distribution: Widespread from tropical to temperate regions.

Genera/species: 53/1500. **Major genera:** *Ficus* (800 spp.) and *Dorstenia* (110). The family is represented in the continental United States and/or Canada by *Broussonetia*, *Fatoua*, *Ficus*, *Maclura*, and *Morus*.

Economic plants and products: Important edible fruits come from *Ficus* (figs; Plate 9.12G), *Morus* (mulberries), *Artocarpus* (jackfruit, breadfruit), and *Brosimum* (breadnut). Several genera provide useful timber. The leaves of a few species of *Morus* are used as food for silkworms. Finally, genera such as *Ficus*, *Maclura* (Osage orange), and *Dorstenia* provide ornamentals.

Discussion: Moraceae are circumscribed narrowly, including woody species with conspicuous milky latex distributed throughout the plant, gynoecia with usually two evident carpels and a more or less apical ovule, and usually curved embryos. *Cecropia* and relatives, which are phenetically intermediate between Moraceae and Urticaceae and are important early successional trees of tropical regions, are excluded. They belong in Urticaceae, as indicated by the restriction of laticifers to the bark, pseudomonomerous gynoecia, basal ovules, straight embryos, and cpDNA sequences (see Figure 9.78). Moraceae, as here circumscribed, are considered monophyletic on the basis of *rbcL* and *ndhF* sequences (Sytsma et al. 2002; Datwyler and Weiblen 2004). No morphological synapomorphies are known with certainty, although the presence of laticifers throughout all parts of the plant could be interpreted as synapomorphic.

Gynoecium reduction has occurred in several clades. It is common for the two styles to be slightly to strongly unequal in *Artocarpus*, *Dorstenia*, *Ficus*, and *Fatoua*. A complete loss of one of the two styles probably occurred in the ancestor of the Urticaceae clade.

Ficus shows an amazing array of growth forms. Some species begin life as epiphytes, but eventually become ground-rooted and strangle the tree on which they were

initially perched. Others have a broad spreading habit, with numerous adventitious roots functioning as supporting columns for the horizontally spreading branches.

The tiny flowers of Moraceae are often wind-pollinated, as in *Morus* and *Broussonetia*. The flowers of *Ficus*, in contrast, are pollinated by wasps that enter the syconium to lay eggs in the ovaries of the carpellate flowers. The wasp larvae feed on the ovary tissues of specialized flowers. Development of the larvae is timed with that of the staminate flowers, which open just as male and female wasps are emerging from their pupae. The wasps mate within the fig, and the females, carrying pollen, exit the developing inflorescence, and repeat their reproductive cycle. As the inflorescence continues to develop, the mature achenes are surrounded by the fleshy inflorescence axis. The evolution of figs apparently is well correlated with that of their fig-wasp pollinators (Weiblen 2000)

Moraceae exhibit a great diversity of inflorescence structures, although all based on a cymose theme, and of multiple fruits. In *Ficus* the tiny flowers are hidden on the inside of a hollow and cuplike inflorescence axis, with the achenes surrounded by a colorful, fleshy cup (syconium). In *Brosimum* the axis is also cuplike, but the staminate flowers are densely clustered on the outside, with a single carpellate flower in the center of the cup. The inflorescence axis of *Dorstenia* is discoid and flat-topped. In contrast, the inflorescence axes of *Artocarpus*, *Maclura*, and *Morus* are often elongate and covered by the densely packed flowers. The drupes of *Artocarpus* are densely aggregated. The massive infructescences of *Artocarpus* weigh up to 40 kg, and like those of *Morus* are associated with persistent, fleshy, accrescent tepals. The achenes of *Maclura* are also aggregated and associated with persistent fleshy tepals. Adaptive radiation, involving changes in the structure of the multiple fruits linked to changes in vertebrate dispersers (birds or mammals), appears to be characteristic of the more or less tropical, woody members of the family. The pits of *Dorstenia* are forcibly ejected from the infructescence; the fruits are explosive drupes!

Additional references: Bechtel 1921; Berg 1978; Faegri and van der Pijl 1980; Humphries and Blackmore 1989; Proctor and Yeo 1972; Rohwer 1993b.

Urticaceae A. L. de Jussieu
(Nettle Family)

Trees, shrubs, herbs or vines; with laticifers restricted to the bark and producing milky sap, or reduced and sap clear, mucilaginous; **cystoliths** *present,* **± elongate***, but lost in some; sometimes with tannins. Hairs often simple, often with mineralized cell walls, sometimes stinging. Leaves alternate and spiral or 2-ranked, or opposite, usually simple, sometimes lobed, entire to serrate, with pinnate to palmate venation,* blade sometimes with cordate or asymmetrical base; *stipules usually present. Inflorescences determinate, axillary, the individual flowers often congested*, sometimes reduced to a single flower. *Flowers unisexual* (plants monoecious to dioe-

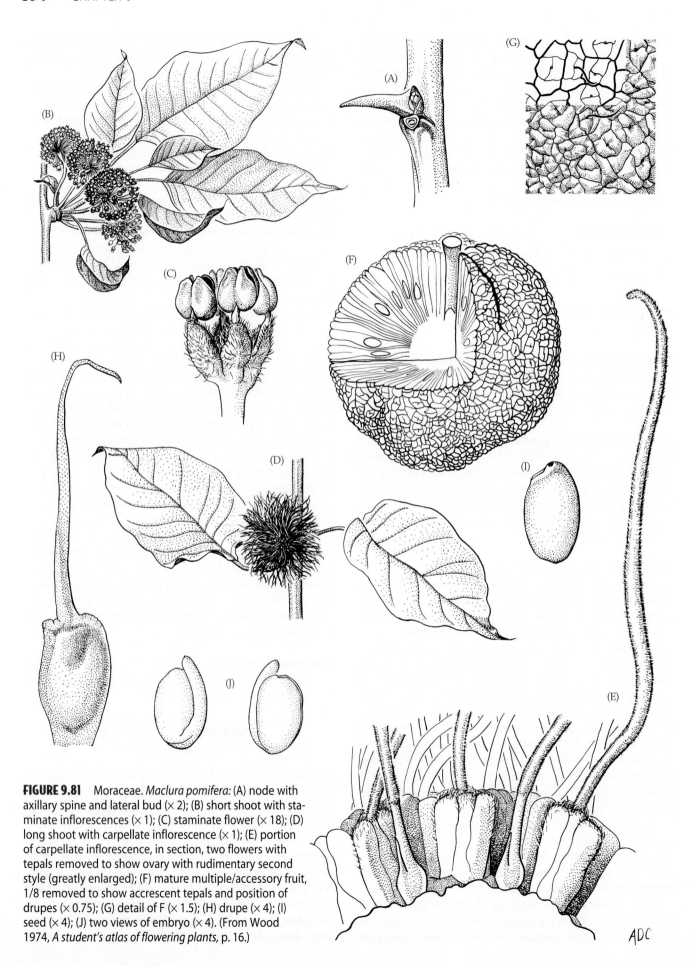

FIGURE 9.81 Moraceae. *Maclura pomifera:* (A) node with axillary spine and lateral bud (× 2); (B) short shoot with staminate inflorescences (× 1); (C) staminate flower (× 18); (D) long shoot with carpellate inflorescence (× 1); (E) portion of carpellate inflorescence, in section, two flowers with tepals removed to show ovary with rudimentary second style (greatly enlarged); (F) mature multiple/accessory fruit, 1/8 removed to show accrescent tepals and position of drupes (× 0.75); (G) detail of F (× 1.5); (H) drupe (× 4); (I) seed (× 4); (J) two views of embryo (× 4). (From Wood 1974, *A student's atlas of flowering plants,* p. 16.)

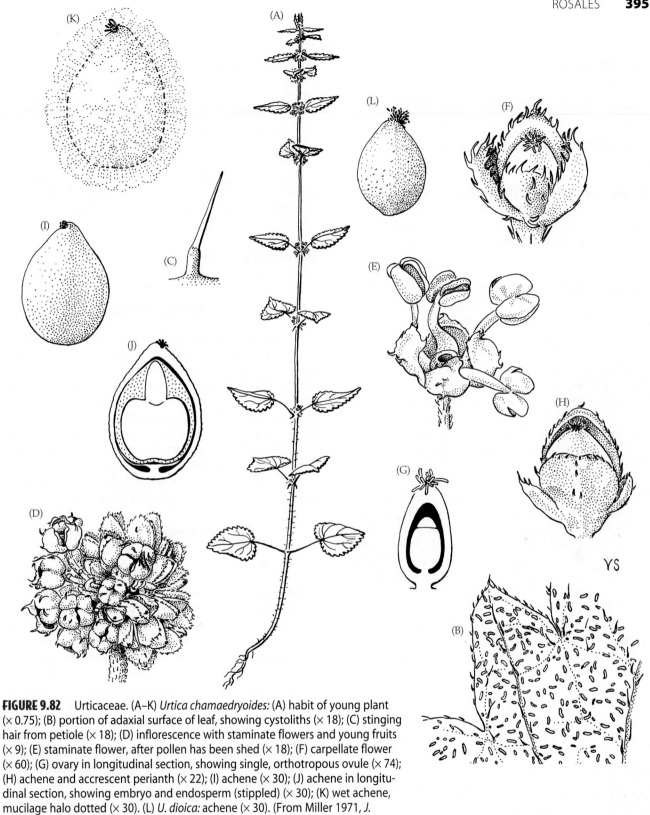

FIGURE 9.82 Urticaceae. (A–K) *Urtica chamaedryoides:* (A) habit of young plant (× 0.75); (B) portion of adaxial surface of leaf, showing cystoliths (× 18); (C) stinging hair from petiole (× 18); (D) inflorescence with staminate flowers and young fruits (× 9); (E) staminate flower, after pollen has been shed (× 18); (F) carpellate flower (× 60); (G) ovary in longitudinal section, showing single, orthotropous ovule (× 74); (H) achene and accrescent perianth (× 22); (I) achene (× 30); (J) achene in longitudinal section, showing embryo and endosperm (stippled) (× 30); (K) wet achene, mucilage halo dotted (× 30). (L) *U. dioica:* achene (× 30). (From Miller 1971, *J. Arnold Arbor.* 52: p. 48.)

cious), usually radial, *inconspicuous. Tepals* (2–) 4 (–6), distinct to connate, imbricate or valvate. *Stamens usually 2-5, opposite the tepals; filaments* distinct, **incurved in bud,** *and often elastically reflexed at anthesis,* but straight in *Cecropia* and relatives; anthers 2-locular; pollen grains usually 2- or 3- to multiporate. **Carpels apparently 1, but actually 2** **with 1 extremely reduced (pseudomonomerous)**; ovary superior, **with basal placentation** and 1 locule; stigma 1, extending along adaxial side of style, to capitate or punctate. Ovule 1, orthotropous. *Fruit an achene or small drupe, sometimes in a multiple fruit; embryo straight;* endosperm sometimes lacking (Figure 9.82).

Floral formula:
Staminate: *, T(4), A4–5, G0
Carpellate: *, T(4), A0, G①; drupe; achene

Distribution: Widespread from tropical to temperate regions.

Genera/species: 54/1160. **Major genera:** *Pilea* (400 spp.), *Elatostema* (200), *Boehmeria* (80), and *Cecropia* (75). The family is represented in the continental United States and/or Canada by *Boehmeria, Hesperocnide, Laportea, Parietaria,* and *Pilea.*

Economic plants and products: Fiber is extracted from *Boehmeria nivea* (ramie) and *Urtica dioica* (stinging nettle). *Pilea* (artillery plant) and *Soleirolia* (baby's tears) provide important ornamentals.

Discussion: Urticaceae are monophyletic (Sytsma et al. 2002) and circumscribed broadly, including woody to herbaceous species with more or less elongate cystoliths; laticifers restricted to the bark, or very reduced, usually producing a clear, mucilaginous sap; pseudomonomerous gynoecia with a more or less basal ovule; curved stamens; and straight embryos. Cecropiaceae (*Cecropia, Coussapoa, Poikilospermum,* and relatives; see Berg 1978; Cronquist 1981) are polyphyletic, forming at least two lineages that are nested within Urticaceae (Sytsma et al. 2002). The close phylogenetic relationship of these woody plants to the mainly herbaceous Urticaceae s.s. has long been apparent (Judd et al. 1994; Humphries and Blackmore 1989) and they are included here within an expanded Urticaceae. The family also has distinctive wood anatomy (Berg 1977, 1989; Friis 1989, 1993). Cystoliths have been lost in *Cecropia* and relatives, and their flowers have straight stamens (a possible reversal from the curved condition).

Aborted vascular bundles in gynoecia of *Laportea* and *Urtica* suggest that the unicarpellate ovary has been derived through abortion of a second carpel. The basal ovule of Urticaceae, likewise, has apparently been derived from an apical ovule, as in Moraceae. In *Boehmeria cylindrica* the vascular bundle supplying the ovule ascends the carpel wall for a short distance and then reverses direction to enter the ovule at the base of the ovary, suggesting a shift from apical to basal placentation in the common ancestor of Urticaceae (see Figure 9.78).

Distinctive stinging hairs occur in genera such as *Urtica, Laportea,* and *Urera.* Each hair consists of a single, long, narrow, tapering stinging cell with a saclike base embedded in a multicellular outgrowth. The hair is closed at the tip by a small bulb that easily breaks off, producing an extremely sharp point. Upon contact with human skin, the hair punctures the surface, and compression of the base forces fluid containing histamines and acetylcholines into the wound.

The process results in a reddening of the skin and painful itching and burning.

The tiny flowers of Urticaceae are usually wind-pollinated (Plate 9.12B). In many species, the inflexed stamens extend elastically at anthesis, causing the anthers to release their pollen in a sudden burst. This facilitates the movement of pollen by air currents, even in species of forest understory habitats.

The achenes of Urticaceae often are associated with fleshy to dry tepals. The fruits may be eaten by birds, or dispersed by external transport on hair or feathers, by wind, or by ballistic methods.

Additional references: Bechtel 1921; Bonsen and ter Welle 1983; Miller 1970, 1971b; Woodland 1989.

Cucurbitales

It is clear that *Cucurbitaceae, Begoniaceae,* and Datiscaceae are closely related, as they share the apomorphies of stems with separate vascular bundles, an inferior ovary, parietal placentae (often strongly intruded), usually forked stigmas, unisexual flowers, the distinctive cucurbitoid tooth type, and the presence of cucurbitacins (oxidized triterpenes). Analyses of DNA sequences and serological characters also support the monophyly of a group containing these three, along with four other small families (APG 2003; Hilu et al. 2003; Källersjö et al. 1998; Savolainen et al. 2000b; Soltis et al. 2000). This clade is here recognized as the order Cucurbitales.

Cucurbitaceae A. L. de Jussieu
(Cucurbit Family)

Herbaceous or soft-woody vines, **usually with spirally coiled and often branched tendrils, borne ± laterally at the nodes** (possibly modified shoots); vascular bundles usually bicollateral, often in 2 concentric rings; alkaloids and bitter tetra- and pentacyclic triterpenoid saponins usually present. Hairs simple, with calcified walls and a cystolith at the base. *Leaves alternate and spiral, usually simple, often palmately lobed, ± serrate, teeth cucurbitoid (i.e., with several veins entering tooth and ending at an expanded ± translucent glandular apex) with palmate venation;* **stipules lacking.** Inflorescences determinate, sometimes reduced to a single flower, axillary. *Flowers usually unisexual (plants monoecious or dioecious),* usually radial, **with short to elongate hypanthium,** often open for only a day. *Sepals usually 5,* usually connate, often reduced. **Petals** *usually 5,* **connate,** *bell-shaped, with a narrow tube and flaring lobes, or nearly flat, white, yellow to orange or red,* the lobes valvate or folded inward. **Stamens** 3–5, adnate to hypanthium, **variously connate and modified,** *usually appearing as 3 (or even appearing to be solitary due to complete connation of filaments and modification of anthers; filaments usually variously connate; anthers unilocular, but often appearing 2-locular or multilocular (due to connation),* **the locules usually bent to con-**

(A)

Cucurbitales: Cucurbitaceae
Cucumis melo: vine with flowers, fruits

(B)

Fagales: Betulaceae
Betula papyrifera: bark (left);
branch with staminate and carpellate
catkins (above)

(C)

Q. virginiana:
habit

Q. velutina:
staminate catkins

Fagales: Fagaceae
Quercus virginiana: acorns

(D)

Cucurbitales: Cucurbitaceae
Cucurbita pepo: gynoecium (left); androecium (right)

(E)

Cucurbitales: Begoniaceae
Begonia sp. 'Baby Wing Pink':
staminate and carpellate flowers

(F)

Fagales: Juglandaceae
Juglans cordiformis:
twig with fruits

J. nigra:
fruit dissected

PLATE 9.13 Eudicots: Cucurbitales and Fagales

voluted; pollen grains various, with 3 to many furrows and/or pores. *Carpels usually 3, connate; ovary half-inferior to inferior, with parietal placentation*, the placentae expanded and intruded; stigmas usually 3, each bilobed. Ovules usually numerous on each placenta. Nectaries various. *Fruit a berry, the rind often leathery to hard (then a pepo)*, or occasionally a fleshy to dry, variously dehiscing capsule; **seeds flattened, the seed coat with several layers**, the outermost sometimes fleshy; endosperm scanty or lacking (Figure 9.83; also see Figure 4.47D).

Floral formula:

Staminate: *, K⑤, C⑤, A⑤, G0

Carpellate: *, K5, C5, A0, G③; berry, capsule

Distribution: Widely distributed in the tropics and subtropics, with a few species occurring in temperate regions.

Genera/species: 118/825. **Major genera:** *Cayaponia* (60 spp.), *Momordica* (45), *Gurania* (40), and *Sicyos* (40). Noteworthy genera occurring in the continental United States and/or Canada are *Cayaponia, Cucumis, Cucurbita, Cyclanthera, Echinocystis, Ibervillea, Marah, Melothria, Momordica,* and *Sicyos*.

Economic plants and products: The family is very important for edible fruits and seeds, such as *Cucurbita* (pumpkins, winter and summer squashes, gourds), *Cucumis* (Plate 9.13A; cantaloupe, muskmelon, honeydew melon, cucumber), *Citrullus* (watermelon), *Benincasa* (wax gourd), and *Sechium* (chayote). The dried fruits of *Lagenaria* (bottle gourd) are used as containers, and the dried fruits of *Luffa* (loofah) are used as a vegetable sponge. Some, such as *Momordica* (balsam apple), are used medicinally.

Discussion: Cucurbitaceae are easily recognized and monophyletic. Two subfamilies are recognized (Jeffrey 1967, 1980, 1990a,b). "Zanonioideae," a small group characterized by separate styles, contains genera with numerous plesiomorphic features and probably is paraphyletic. Monophyly of Cucurbitoideae is supported by their completely connate styles and nrITS sequences (Jobst et al. 1998). This group is divided into several tribes based on characters such as ovule position and number, ornamentation of the pollen grains, shape of the hypanthium, fruit form, and characteristics of the androecium.

Flowers of Cucurbitaceae are showy and attract various insects, birds, and bats. Pollen and nectar are the rewards. Monoecy or dioecy leads to outcrossing. The highly modified androecium is similar in appearance to the style and stigma, and insects are fooled into visiting both staminate and carpellate flowers (Plate 9.13D). The berries of most cucurbits are dispersed by various animals. *Echinocystis* has capsules that open explosively; in *Momordica*, the capsules open to expose brightly colored and fleshy seeds, which are dispersed by birds.

Additional references: Chakravarty 1958; Robinson and Decker-Walters 1997.

Begoniaceae C. A. Agardh
(Begonia Family)

Herbs or soft-woody shrubs with jointed stems; tannins present; with large water-storage cells in leaf hypodermis. Hairs simple, often with a cystolith at the base. *Leaves alternate and 2-ranked, simple*, sometimes palmately lobed or palmate, usually ± *toothed, the teeth similar to those of Cucurbitaceae, with palmate venation*, **usually asymmetrical**; *stipules present, often large, persistent*. Inflorescences determinate, axillary. *Flowers unisexual (plants monoecious)*, radial. **Perianth of petaloid tepals**, 2–10, but usually 4 (in 2 whorls) in staminate and 5 (in a single whorl) in carpellate flowers, distinct, imbricate or valvate, those of inner and outer whorls often ± differentiated. *Stamens 4-numerous*, the filaments distinct or basally connate; the connective expanded; pollen grains tricolporate. *Carpels usually 3, connate; ovary ± inferior*, **with axile placentation, each placenta expanded**, variably divided (derived from intruded parietal, and this condition present in a few), *usually with 3 prominent wings; stigmas 3 or 6*, **elongated, twisted, yellow, strongly papillose**. Ovules numerous in each locule. Nectaries lacking. *Fruit a loculicidal capsule, winged*, occasionally a berry. Seeds tiny, with ring of elongated testa cells that rupture, forming a cap (operculate); endosperm scanty or lacking.

Floral formula:

Staminate: * or X, K2, C2, A∞, G0

Carpellate: * or X, T– 5 –, A∞, G③, capsule

Distribution: Widely distributed in the tropics and subtropics (excluding Australia), and especially characteristic of moist shaded understory habitats.

Genera/species: 2/920. **Major genus:** *Begonia* (919 spp.). The group is represented in the continental United States only by *Begonia* (naturalized). *Hillebrandia* is endemic to Hawaii.

Economic plants and products: The family contains numerous commercially important species and hybrids that are cultivated for their beautiful foliage and flowers (Plate 9.13E).

Discussion: Begoniaceae are fairly homogenous, with nearly all species belonging to the large genus *Begonia*, and the group's monophyly is supported by several morphological characters. *Hillebrandia sandwichensis* is sister to the remaining members of the family, which constitute the large and diverse genus *Begonia*, a clade characterized by winged ovaries (Clement et al. 2004; Forrest et al. 2005; Plana 2003; Swensen et al. 2001). Recognition of *Symbegonia* makes *Begonia* paraphyletic.

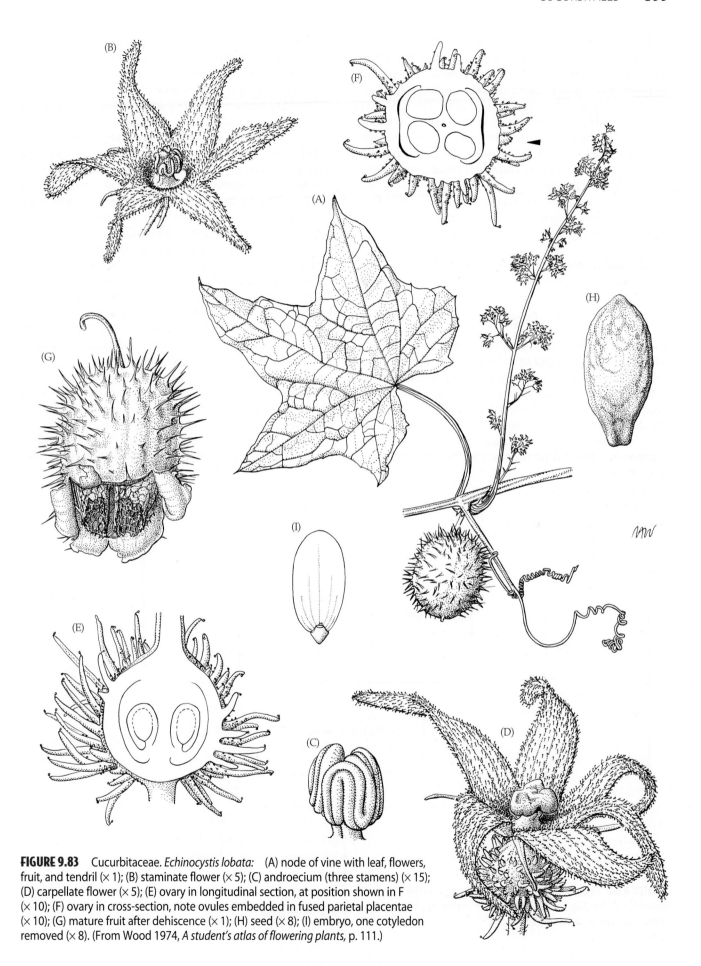

FIGURE 9.83 Cucurbitaceae. *Echinocystis lobata:* (A) node of vine with leaf, flowers, fruit, and tendril (× 1); (B) staminate flower (× 5); (C) androecium (three stamens) (× 15); (D) carpellate flower (× 5); (E) ovary in longitudinal section, at position shown in F (× 10); (F) ovary in cross-section, note ovules embedded in fused parietal placentae (× 10); (G) mature fruit after dehiscence (× 1); (H) seed (× 8); (I) embryo, one cotyledon removed (× 8). (From Wood 1974, *A student's atlas of flowering plants,* p. 111.)

The showy staminate flowers are visited by pollen-gathering bees. The bright yellow twisted stigmas of the carpellate flowers mimic the anthers of staminate flowers, and thus also attract pollen-gathering bees (resulting in pollination by deceit; see also Proctor et al. 1996). The tiny seeds are presumably dispersed by wind or water (rain wash), and the prominent (often asymmetrical) wings may assist in shaking seeds out of the capsule. The cymes produce more staminate flowers than carpellate flowers and the staminate flowers also open sooner, leading to outcrossing. Vegetative propagation is common; plantlets develop from small tubers that often occur in the leaf axils or from adventitious buds that form on detached leaves.

Additional references: de Lange and Bouman 1992; Smith and Wasshausen 1986.

Fagales

Fagales are monophyletic, based on their unisexual flowers with tepals very reduced or lacking, usually inferior ovary with one or two 1-integumented ovules per locule, pollen tube entering the ovule via the chalaza, absence of nectaries, and indehiscent 1-seeded fruits. They are trees or shrubs with tannins, alternate, stipulate leaves, typically wind-pollinated flowers aggregated in catkins, and seeds more or less lacking endosperm. They usually have gland headed and/or stellate hairs. The monophyly of Fagales is supported by chloroplast DNA restriction sites, *rbcL, atpB, matK,* and 18S sequences (Chase et al. 1993; Hilu et al. 2003; Källersjö et al. 1998; Manos et al. 1993; Savolainen et al. 2000a,b; Soltis et al. 2000), and morphology (Hufford 1992). The order consists of 8 families and about 1115 species; major families include **Fagaceae**, Nothofagaceae, **Betulaceae**, **Casuarinaceae**, **Myricaceae**, and **Juglandaceae**. Fagales are not closely related to the Hamamelidaceae, Platanaceae, or other "basal" tricolpates, and the Hamamelidae (as traditionally defined; see Cronquist 1981, 1988; Stern 1973) are polyphyletic (Chase et al. 1993; Hufford 1992; Manos et al. 1993; Meurer-Grimes 1995).

It is clear that either Fagaceae or, more probably, Nothofagaceae are sister to the remaining families of Fagales (Hufford 1992; Li et al. 2004; Manos et al. 1993, 2001; Manos and Steele 1997; Nixon 1989; Wolfe 1989). The Juglandaceae, Myricaceae, Casuarinaceae, and Betulaceae form a clade (core Fagales) on the basis of cpDNA features (restriction sites, *rbcL,* and *matK* nucleotide sequences) (Chase et al. 1993; Hufford 1992; Manos et al. 1993; Manos and Steele 1997; Nixon 1989; Wolfe 1989) (Figure 9.84). Members of core Fagales are characterized by more or less triporoporate pollen (pores with well-developed endoapertures) and ovules with multiple female gametophytes, and these may be an additional synapomorphies. This group of families represents the extant members of the Normapolles

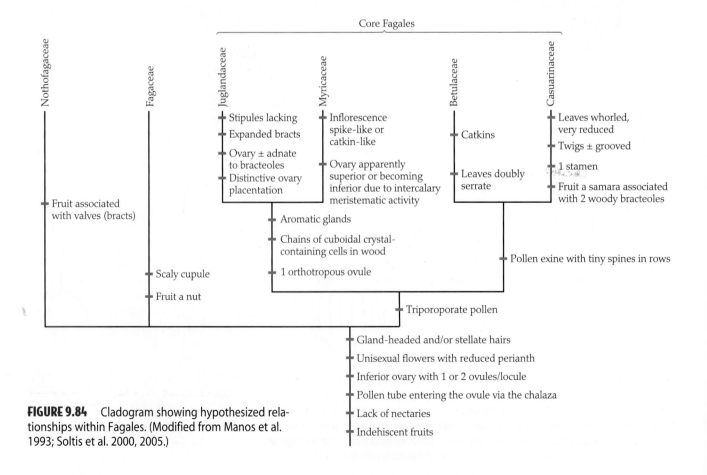

FIGURE 9.84 Cladogram showing hypothesized relationships within Fagales. (Modified from Manos et al. 1993; Soltis et al. 2000, 2005.)

Key to Major Families of Fagales

1. Fruits associated with a conspicuous cupule; carpels 2–6; pollen colpate or colporate 2

1. Fruits not associated with a cupule, although usually with variously developed bracts
and bracteoles; carpels usually 2; pollen porate 3

2. Carpels 3 (–12); ovules with 2 integuments; stipules narrowly triangular,
lacking colleters; leaves entire to serrate or lobed but never doubly serrate;
plants of the Northern Hemisphere ....**Fagaceae**

2. Carpels 2–3; ovules with 1 integument; stipules peltate, with colleters;
leaves entire to doubly serrate; plants of the Southern Hemisphere ....Nothofagaceae

3. Leaves pinnately compound ....**Juglandaceae**

3. Leaves simple 4

4. Leaves whorled, reduced to tiny scales; stems longitudinally ridged ....**Casuarinaceae**

4. Leaves alternate, with obvious blades; stems not ridged 5

5. Leaves entire to serrate; fruits achenes or drupes, usually with a covering of waxy
papillae; plants dioecious or occasionally monoecious; ovule 1, basal, orthotropous ....**Myricaceae**

5. Leaves ± doubly serrate; fruits achenes, samaras, or nuts; plants monoecious;
ovules 2 per locule at apex of incomplete septum, anatropous ....**Betulaceae**

---

complex, well represented in the fossil record (Kedves 1989). Juglandaceae and Rhoipteleaceae are united by the apomorphy of pinnately compound leaves, and Rhoipteleaceae are noteworthy in having the most Normapolles-like pollen of any living member of Fagales. Myricaceae shares with Juglandaceae peltate gland-headed hairs and a gynoecium with a single, orthotropous ovule, suggesting a closer relationship between these families than suggested by *matK* and *rbcL* sequences (Manos and Steele 1997). It is of interest that a recent analysis based on six DNA regions (Li et al. 2004) placed Myricaceae as sister to a Rhoipteleaceae + Juglandaceae clade.

Fagaceae Dumortier
(Beech or Oak Family)

Trees or shrubs; tannins present. *Hairs simple or stellate, often also glandular scales. Leaves usually alternate and spiral, simple, but often lobed, entire to serrate, with pinnate venation*; stipules present. *Inflorescences determinate, often erect and spicate, a dangling catkin, headlike cluster*, or even a solitary flower, terminal or axillary, the staminate and carpellate flowers on the same or different inflorescences. *Flowers unisexual (plants usually monoecious)*, radial, ± inconspicuous, the staminate flowers in reduced cymes and associated with a bract, *the carpellate flowers usually in groups of 1–3* and **associated with a scaly cupule**. Tepals usually 6, reduced and inconspicuous, distinct to slightly connate, imbricate. Stamens 4–numerous; filaments distinct; pollen grains tricolporate or tricolpate. *Carpels 3 (–12)*, connate; *ovary inferior*, with axile placentation; stigmas separate, porose, or expanded along upper side of style. Ovules 2 in each locule, but all except 1 aborting. Nectaries usually lacking. **Fruit a nut, closely associated with a spiny to scaly, usually 4-valved or nonvalved cupule**; endosperm lacking (Figure 9.85).

Floral formula:
Staminate: *, T-6-, A4–∞, G0

Carpellate: *, T-6-, A0, $\overline{G③}$; nut (with cupule)

Distribution: Widespread in tropical to temperate regions of the Northern Hemisphere.

Genera/species: 9/900. **Major genera:** *Quercus* (450 spp.), *Lithocarpus* (300), and *Castanopsis* (100). The family is represented in Canada and/or the continental United States by *Castanea*, *Chrysolepis*, *Fagus*, and *Quercus*.

Economic plants and products: The nuts of *Castanea* (chestnuts) are edible; those of *Quercus* (oak) and *Fagus* (beech) are also occasionally eaten. Cork is made from the bark of *Quercus suber*. *Quercus*, *Fagus*, and *Castanea* provide ornamental trees. The family is exceptionally important as a source of timber for construction, furniture, cabinetry, barrels, and other uses.

Discussion: Monophyly of Fagaceae is supported by morphology, cpDNA restriction sites (Manos et al. 1993), and nucleotide sequences (Li et al. 2004; Manos and Steele 1997). *Fagus* (Fagoideae) is sister to the remaining genera, which constitute the Quercoideae. *Trigonobalanus*, a group that may

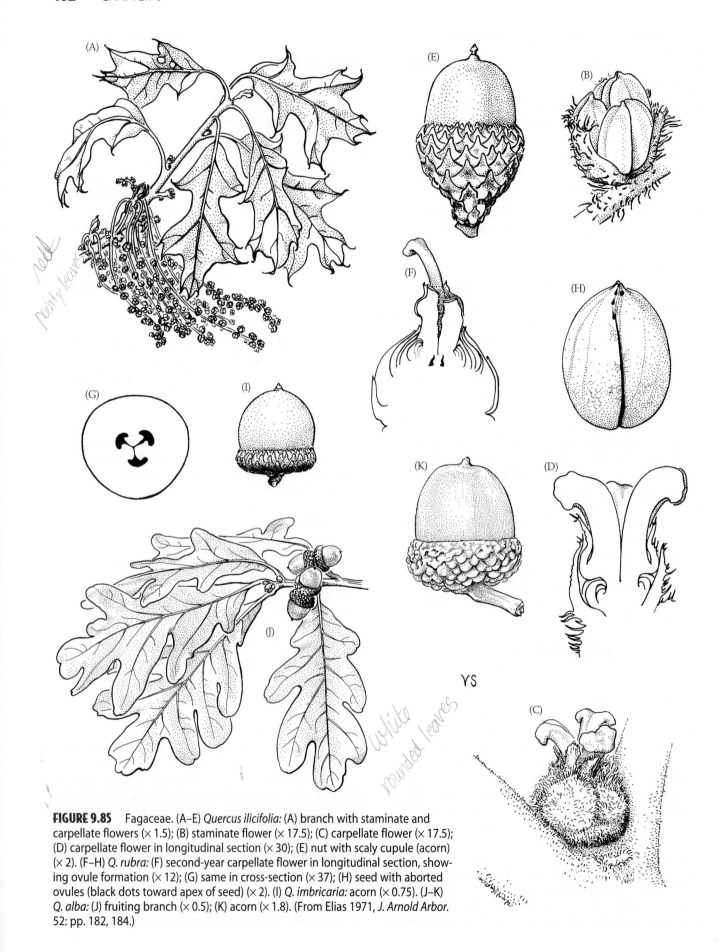

FIGURE 9.85 Fagaceae. (A–E) *Quercus ilicifolia:* (A) branch with staminate and carpellate flowers (× 1.5); (B) staminate flower (× 17.5); (C) carpellate flower (× 17.5); (D) carpellate flower in longitudinal section (× 30); (E) nut with scaly cupule (acorn) (× 2). (F–H) *Q. rubra:* (F) second-year carpellate flower in longitudinal section, showing ovule formation (× 12); (G) same in cross-section (× 37); (H) seed with aborted ovules (black dots toward apex of seed) (× 2). (I) *Q. imbricaria:* acorn (× 0.75). (J–K) *Q. alba:* (J) fruiting branch (× 0.5); (K) acorn (× 1.8). (From Elias 1971, *J. Arnold Arbor.* 52: pp. 182, 184.)

have retained epigeal cotyledons and ancestral inflorescence conditions (see Fig. 9.86), probably is sister to the other genera of Quercoideae. The single North American species of *Lithocarpus* groups with *Quercus*, *Castanea*, and *Castanopsis*, while all the other species of *Lithocarpus* group with *Chrysolepis* (Oh and Manos 2006). *Lithocarpus*, *Castanea*, *Castanopsis*, and *Lithocarpus* have retained numerous plesiomorphic morphological characters: bisexual inflorescences, flowers with a less reduced perianth, exserted stamens, and tiny stigmas.

Quercus is considered to be monophyletic on the basis of its fruits—a single nut surrounded by a nonvalved cupule (forming an acorn; Plate 9.13C). Its monophyly has also been supported by DNA sequences (Manos and Steele 1997; Manos et al. 2001). The genus has separate carpellate and staminate inflorescences, the latter forming lax catkins (Plate 9.13C). Two major phenetic groups can be recognized within *Quercus*: subgenus *Cyclobalanopsis* (those species with cyclically arranged cupule scales), and subgenus *Quercus* (those species with imbricate cupule scales). More study is needed to determine whether or not these represent clades. Subgenus *Quercus* is composed of three major groups. Red oaks (section *Lobatae*) are characterized by leaves often with bristle tips; flowers with elongate, linear-spatulate styles and usually retuse anthers; aborted ovules near the apex of the nut; and usually biennial fruits. White oaks (section *Quercus*) have leaves without bristle tips; flowers with short, abruptly dilated styles and usually apiculate anthers; aborted ovules at the base of the nut; and annual fruits. A third, smaller group, section *Protobalanus*, is similar to section *Quercus*, but differs in its biennial (vs. annual) fruits and clearly pubescent (vs. glabrous) inner fruit wall (Nixon et al. 1995). DNA analyses support the segregation of *Q. cerris*, *Q. suber*, and *Q. ilex*, and their relatives from section *Quercus*, and their recognition as section *Cerris* (Manos et al. 1999, 2001).

Variation in the cupule provides characters useful in generic recognition (Figure 9.86). The cupule is probably a cymose inflorescence in which the outer axes of the cyme are modified into cupule valves, which bear spines or scales (Brett 1964; Fey and Endress 1983; Nixon 1989; Okamoto 1989). The valves may be more or less evident due to fusion. The cupules of *Nothofagus* (Nothofagaceae) probably are

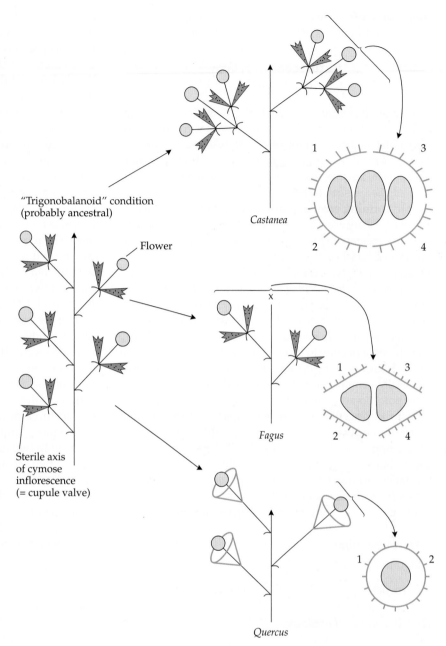

"Trigonobalanoid" condition (probably ancestral)

Flower

Castanea

Fagus

Sterile axis of cymose inflorescence (= cupule valve)

Quercus

FIGURE 9.86   Hypothesis of evolution of various cupule forms in Fagaceae.

not homologous with those of Fagaceae. They may be composed of clustered bracts and stipules.

The inconspicuous flowers of *Fagus* and *Quercus* are borne in unisexual inflorescences that develop in the spring just before or with the leaves, and are wind-pollinated. *Castanea* and *Castanopsis* bear staminate flowers that are conspicuous, give off a strong odor, and are pollinated by flies, beetles, and bees. Wind pollination may have evolved three times within the family. The large nuts of Fagaceae are dispersed by birds and mammals (especially rodents).

Castanea has been devastated in North America by a disease caused by the fungus *Endothea parasitica*.

Additional references: Abbe 1974; Elias 1971a; Kaul and Abbe 1984; Kubitzki 1993b; MacDonald 1979b.

Betulaceae S. F. Gray
(Birch Family)

Trees or shrubs; tannins present; bark smooth to scaly, sometimes exfoliating in thin layers, sometimes with prominent horizontal lenticels. *Hairs simple, gland-headed or peltate. Leaves alternate and spiral or 2-ranked, simple,* **doubly serrate***, with pinnate venation, the secondary veins running into a serration; stipules present.* Inflorescences determinate, appearing spicate, **forming erect or pendulous catkins** (aments), terminal or axillary, sometimes exposed during the winter, solitary or in racemose clusters, with conspicuous bracts, *the staminate and carpellate flowers in separate inflorescences. Flowers unisexual (plants monoecious), radial, inconspicuous, usually 2 or 3, forming a cymose unit in the axil of each inflorescence bract, also usually associated with variously fused second- and third-order bracteoles. Tepals (0–) 1–4 (–6), reduced,* ± distinct, sometimes lobed, slightly imbricate. *Stamens usually (1–) 4 (–6),* sometimes appearing to be more due to the close association of the 3 flowers of the cymose unit; filaments short, distinct to basally connate, sometimes divided; pollen grains (2–) 3–7-(poro)porate. *Carpels usually 2, connate; ovary inferior,* with axile placentation (but incompletely 2-locular, and ovules borne at apex of incomplete septum); stigmas 2, running along adaxial surface of styles. Ovules usually 2 per locule, but all except 1 aborting, usually with 1 integument. Nectaries lacking. *Fruit an achene, nut, or 2-winged samara, associated with variously fused and developed bract-bracteole complex;* endosperm present or lacking (Figure 9.87).

Floral formula:

Staminate: *, T-0–6-, A (1–4), G0

Carpellate: *, T-0–6-, A0, G (2); achene, samara, nut

Distribution and ecology: Widespread in temperate to boreal regions of the Northern Hemisphere, but *Alnus* extending into South America (Andes); species occur in early successional habitats or in wetlands, or as dominant forest trees. Nitrogen fixation occurs in specialized nodules (containing symbiotic bacteria) on the roots of *Alnus*.

Genera/species: 6/157. ***Major genera:*** *Betula* (60 spp.), *Alnus* (35), *Carpinus* (35), *Corylus* (15), and *Ostrya* (10). All of these occur in the continental United States and/or Canada.

Economic plants and products: The nuts of *Corylus* (filberts, hazelnuts) are edible. Several species of *Betula* (birch) and *Alnus* (alder) are important for timber or wood pulp; the latter is also important in land reclamation. *Betula, Corylus, Carpinus* (ironwood, hornbeam), and *Ostrya* (ironwood, hop hornbeam) provide valuable ornamentals.

Discussion: Betulaceae are considered monophyletic on the basis of having both their staminate and carpellate flowers in catkins (Plate 9.13B) and doubly serrate leaf margins (Crane 1989) and DNA sequences (Li et al. 2004). Betulaceae and the Neotropical genus *Ticodendron* (Ticodendraceae) are sister taxa. *Ticodendron* has leaves with almost all the teeth directly vascularized by secondary veins, stipules encircling the twigs, and drupes. Betulaceae comprise two major monophyletic groups: Betuloideae (including *Alnus* and *Betula*) and Coryloideae (including *Carpinus, Corylus, Ostrya,* and *Ostryopsis*) (Crane 1989; Chen et al. 1999). Betuloideae are hypothesized to form a clade on the basis of their flattened fruits, bracts and bracteoles fused into a scale, and carpellate flowers lacking a perianth. The monophyly of Coryloideae is supported by their staminate flowers lacking a perianth and carpellate flowers with the bracteoles connate and expanded. Nuclear ribosomal and *rbcL* sequences also support these two subfamilies (Chen et al. 1999; Forest et al. 2005).

The development, anatomy, and morphology of the highly modified flowers and inflorescences of Betulaceae have received extensive study (Abbe 1935, 1974) and are important in generic identification (Figure 9.88). All members of the family have both staminate and carpellate catkins with flowers borne in cymose units consisting of one to three flowers and one to seven bract/bracteoles. The cymose units of the carpellate catkin of *Alnus* are reduced to two flowers, and each cyme is associated with a bract, two secondary bracteoles, and two tertiary bracteoles (all of which are connate, persistent, and woody). The cymose units of the carpellate catkin of *Betula* are usually composed of three flowers; these are associated with a bract and two fused bracteoles, forming a deciduous three-lobed "bract." In *Carpinus*, the two nutlets (of each cymose unit) are associated with a small bract, and each nutlet is associated with an expanded secondary bracteole that is connate to the two adjacent tertiary bracteoles. In contrast, the achenes of *Ostrya* are completely surrounded by an expanded, bladderlike structure that develops from the fused secondary and two tertiary bracteoles. The nuts of *Corylus* are surrounded by the greatly expanded secondary bracteoles.

The inconspicuous flowers of Betulaceae are borne in usually erect carpellate catkins and pendulous staminate catkins. All members of the family bloom in the spring, before or just as the leaves emerge, and are wind-pollinated. The ovary and ovules are undeveloped at the time of pollination (as in Fagaceae). The small, two-winged samaras of *Betula* and *Alnus* are wind-dispersed. In some species of *Alnus* the wings are reduced, and these fruits, which float, are usually water-dispersed. The nutlets of *Carpinus* and *Ostrya* are associated with expanded and fused bracteoles, and both are wind- or water-dispersed. *Corylus* has large nuts that are dispersed by rodents.

Additional references: Furlow 1990; Kubitzki 1993a.

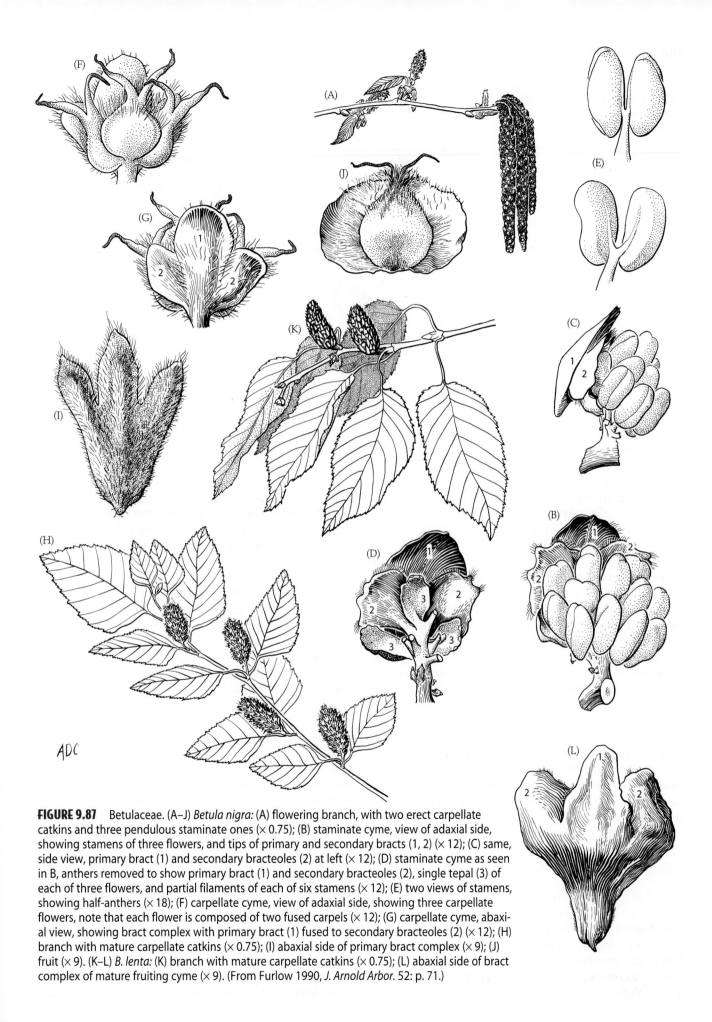

FIGURE 9.87 Betulaceae. (A–J) *Betula nigra:* (A) flowering branch, with two erect carpellate catkins and three pendulous staminate ones (× 0.75); (B) staminate cyme, view of adaxial side, showing stamens of three flowers, and tips of primary and secondary bracts (1, 2) (× 12); (C) same, side view, primary bract (1) and secondary bracteoles (2) at left (× 12); (D) staminate cyme as seen in B, anthers removed to show primary bract (1) and secondary bracteoles (2), single tepal (3) of each of three flowers, and partial filaments of each of six stamens (× 12); (E) two views of stamens, showing half-anthers (× 18); (F) carpellate cyme, view of adaxial side, showing three carpellate flowers, note that each flower is composed of two fused carpels (× 12); (G) carpellate cyme, abaxial view, showing bract complex with primary bract (1) fused to secondary bracteoles (2) (× 12); (H) branch with mature carpellate catkins (× 0.75); (I) abaxial side of primary bract complex (× 9); (J) fruit (× 9). (K–L) *B. lenta:* (K) branch with mature carpellate catkins (× 0.75); (L) abaxial side of bract complex of mature fruiting cyme (× 9). (From Furlow 1990, *J. Arnold Arbor.* 52: p. 71.)

FIGURE 9.88 Floral diagrams of the cymose units of the carpellate catkins of Betulaceae.

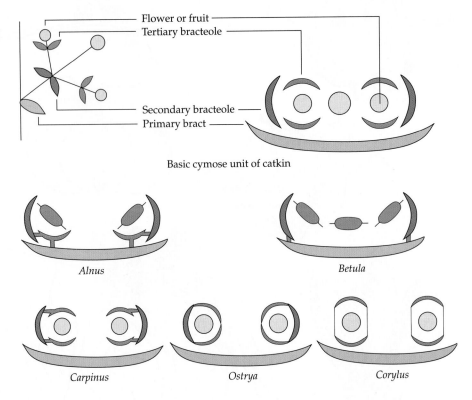

Basic cymose unit of catkin

Alnus

Betula

Carpinus

Ostrya

Corylus

Casuarinaceae R. Brown
(She-Oak Family)

Trees or shrubs with **slender, green, jointed, grooved twigs; roots with nodules containing nitrogen-fixing bacteria**; tannins present. Hairs simple or branched. **Leaves whorled, in groups of 4–20, simple, scale-like, ± connate, forming a toothed sheath at each node**; stipules lacking. *Inflorescences* indeterminate, terminal, *forming catkins at the tips of lateral branches. Flowers unisexual (plants monoecious or dioecious), radial, inconspicuous, solitary in the axil of each inflorescence bract, and associated with 2 bracteoles.* Tepals lacking. **Stamen 1**; pollen grains usually triporoporate. *Carpels 2, connate;* ovary presumably inferior, with axile placentation; stigmas 2, running along inner surface of styles. Ovules 2 in each locule, or sometimes lacking in 1 of the locules, and all but 1 aborting, orthotropous, **with 2 integuments**. Nectaries lacking. **Fruit a samara, associated with 2 woody bracteoles, in conelike catkins**; endosperm lacking.

Floral formula:
Staminate: *, T-0-, A1, G0
Carpellate: *, T-0-, A0, G$\overline{②}$; samara

Distribution and ecology: Widely distributed in southeastern Asia, Australia, and the islands of the southwestern Pacific, but naturalized in coastal habitats in subtropical and tropical regions of Africa and the Americas. Often plants of xeric habitats.

Genera/species: 4/96. **Major genera:** *Allocasuarina* (59 spp.), *Gymnostoma* (18), and *Casuarina* (17). The family is represented in the continental United States (Florida) by three introduced species of *Casuarina*.

Economic plants and products: Several species are important timber trees or ornamentals. *Casuarina* is a problem weed tree in Florida.

Discussion: Casuarinaceae are easily recognized, and are considered monophyletic (see characters listed in bold above). The traditional broadly defined *Casuarina* has now been divided into four genera based on characters such as nature of the stem grooves (shallow and open vs. deep and narrow), number of teeth per whorl, shape of the bracts in carpellate catkins, fruit color, and chromosome number (Johnson and Wilson 1989, 1993). *Gymnostoma* is sister to the remaining genera; it has branchlets with shallow and open furrows, while the other genera have deep and narrow furrows (concealing the stomata).

The flowers of Casuarinaceae are wind-pollinated, and the fruits are wind-dispersed.

Additional references: Rogers 1982; Torrey and Berg 1988.

Myricaceae Richard ex Kunth
(Bayberry Family)

Aromatic trees or shrubs; triterpenes and sesquiterpenes present; tannins present; **roots usually with nodules that contain nitrogen-fixing bacteria**. *Peltate scales with a glandular,*

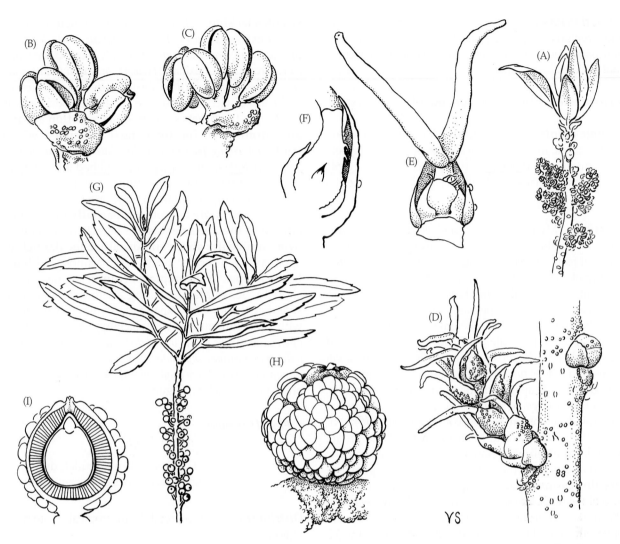

FIGURE 9.89 Myricaceae. (A–F) *Myrica pensylvanica:* (A) branch with staminate catkins (× 1.5); (B) staminate flower (× 14.5); (C) staminate flower, lateral view (× 14.5); (D) carpellate catkin (× 9); (E) carpellate flower with bracts (× 22); (F) carpellate flower in longitudinal section, showing basal ovule (× 30). (G–I) *M. cerifera:* (G) branch with fruits (× 0.75); (H) drupe (× 12); (I) fruit in longitudinal section, note waxy papillae, endocarp (indicated with numerous radiating lines), and embryo (× 12). (From Elias 1971, *J. Arnold Arbor.* 52: p. 310.)

usually golden-yellow, swollen head, containing various aromatic oils and/or resins. *Leaves alternate and spiral, simple* (deeply lobed in *Comptonia*), *entire to serrate, with pinnate venation;* stipules absent, or present (*Comptonia*). **Inflorescences** indeterminate, **often spike-like or catkin-like**, erect to ± pendulous, axillary, staminate and carpellate flowers usually in separate inflorescences. *Flowers unisexual* (*plants monoecious or dioecious*), *radial, inconspicuous, 1 in the axil of each inflorescence bract. Perianth lacking*, except in *Canacomyrica* where represented by 6 minute tepals at ovary apex, but flowers usually associated with bracts and bracteoles. *Stamens 2–9*, but appearing more numerous due to clustering of several flowers; pollen grains usually triporoporate. *Carpels 2, connate; ovary apparently superior* (due to loss of perianth; *Comptonia*), **becoming inferior due to intercalary meristematic activity around and/or beneath the gynoecium, forming a cuplike structure**, which raises the bracteoles up as part of the fruit wall (*Myrica gale* group), *or inferior even at the time of pollination, due to early intercalary activity* that forms a thick structure with (*Myrica,* most species) or without (*Canacomyrica*) papillae, with basal placentation; stigmas 2, elongated. Ovule 1 per gynoecium, orthotropous, with 1 integument. Nectaries lacking. *Fruit a drupe, covered either with waxy or fleshy papillae, or an achene,* not associated with conspicuous bracteoles (*Myrica,* most species; *Canacomyrica*), with 2 bracteoles fused to achene (*Myrica gale* group), or simply surrounding fruit (*Comptonia*); endosperm lacking, or nearly so (Figure 9.89).

Floral formula:

Staminate: *, T-0-, A1–9, G0

Carpellate: *, T-0-, A0, G$\overline{②}$; drupe, achene

Distribution and ecology: Widespread in temperate to tropical regions; often early successional or in wetlands; plants associated with nitrogen-fixing, filamentous bacteria in root nodules.

Genera/species: 3/40.   ***Major genus:*** *Myrica* (38 spp.).

Economic plants and products: Aromatic wax is extracted from the fruits of several species of *Myrica* (bayberry, wax myrtle, candleberry); a few species have edible fruits. Several species of *Myrica* are used as ornamental shrubs.

Discussion: Myricaceae are considered to be monophyletic on the basis of their numerous specialized morphological features (MacDonald 1974, 1977, 1979a, 1989) and molecular evidence (Herbert et al. 2006). *Canacomyrica,* an endemic to New Caledonia, is sister to the *Comptonia* + *Myrica* clade, in which the perianth has been lost. *Comptonia* may be sister to a broadly circumscribed *Myrica. Comptonia* has retained stipules, diffuse porous wood, and bracteoles that are free from the ovary. Its fruits are small achenes, and it does not show the intercalary development of a modified and cuplike portion of the inflorescence axis, which forms a large portion of the fruit wall in *Myrica.* Ovary/fruit development is most complex in *M. cerifera, M. pensylvanica, M. rubra,* and relatives; in this large group, waxy or fleshy papillae develop on the cuplike meristematic region that encircles the fruit. The result is a papillose drupe. (In *Canacomyrica,* the cuplike meristematic region and resulting drupe are smooth.) *Myrica gale* (and relatives) show the apomorphy of enlarged bracteoles that are strongly adnate to the fruit wall, and may be sister to the *M. cerifera* group. The distinctiveness of *M. cerifera* and relatives has led some to segregate these species as *Morella* (Wilbur 1994, 2001).

Abbe (1974) suggested that what appears to be the male flower (in the bract axil) is actually a cluster of staminate flowers, each of which consists of only a single stamen.

The reduced flowers of Myricaceae are wind-pollinated, and most species have separate staminate and carpellate catkins. The fleshy or waxy drupes of *Myrica cerifera* (and other similar species) are mainly bird-dispersed. The small fruits of *Myrica gale* are dispersed by water, the enlarged bracts acting as floats.

Additional references: Elias 1971b; Kubitzki 1993c; MacDonald 1978.

Juglandaceae A. P. de Candolle ex Perleb
(Walnut Family)

Aromatic trees; tannins present. Hairs various, often stellate, *and peltate scales with a glandular, swollen head containing various aromatic oils and/or resins. Leaves alternate and spiral,* occasionally opposite, **pinnately compound**, occasionally unifoliate, entire to serrate, with pinnate venation;

stipules lacking. *Inflorescences indeterminate, erect to pendulous spikes or panicles, staminate and carpellate flowers often in separate inflorescences, and staminate inflorescence usually a catkin,* terminal or axillary. *Flowers unisexual (plants monoecious to less commonly dioecious),* ± radial, inconspicuous, 1 in the axil of each inflorescence bract and associated with 2 bracteoles, **the bracts** *sometimes 3-lobed,* **often expanded and forming a wing (or wings) associated with the fruit**, *or forming part of a cuplike husk that surrounds the fruit. Tepals 0–4,* inconspicuous, modified into a stigmatic disk in *Carya.* Stamens 3–numerous; filaments short; pollen grains triporoporate to polyporoporate. *Carpels usually 2, connate; ovary inferior,* **and partially to completely adnate to the 2 bracteoles** *and often to the bract as well,* **ovary unilocular above and usually 2-locular below**, or apparently 4–8-locular due to false partitions, **with ovule borne at apex of incomplete septum**; stigmas usually 2, short to elongate and running along adaxial surface of style branches, often expanded. Ovule 1, orthotropous, with 1 integument. Nectaries lacking. *Fruit a nut or nutlet, often samara-like due to associated bracts and/or bracteoles, or a drupe, sometimes with outer husk splitting to release the bony pit;* embryo with *large cotyledons, these sometimes corrugated;* endosperm ± lacking (Figure 9.90).

Floral formula:

Staminate: *, T-4–0-, A3–∞, G0

Carpellate: *, T-4±0-, A0, G$\overline{②}$; nut; nutlet; drupe (sometimes with ± dehiscent husk)

Distribution: Widespread from tropical to temperate regions.

Genera/species: 8/59.   ***Major genera:*** *Carya* (25 spp.) and *Juglans* (20); both occur in the continental United States and/or Canada.

Economic plants and products: *Juglans regia* (walnut), *J. nigra* (black walnut), *Carya illinoensis* (pecan), and *C. ovata* (shagbark hickory) have edible nuts. *Juglans, Carya,* and *Engelhardia* are important as timber trees, and the first two genera, along with *Pterocarya* (wingnut), contain species that are used as ornamentals.

Discussion: Monophyly of Juglandaceae receives clear support from both morphology and DNA sequences (Li et al. 2004; Manchester 1987; Manning 1978; Smith and Doyle 1995). Phylogenetic relationships within Juglandaceae have been investigated in analyses based on morphology as well as molecular characters (Smith and Doyle 1995; Manos and Stone 2001), as well as more subjective considerations of morphological features (Manchester 1987; Manning 1978; Stone 1989, 1993). Engelhardioideae, including the tropical genera *Engelhardia, Alfaroa,* and *Oreomunnea,* are sister to the remaining genera. This group is characterized by three-lobed bracts and nuts with a fibrous

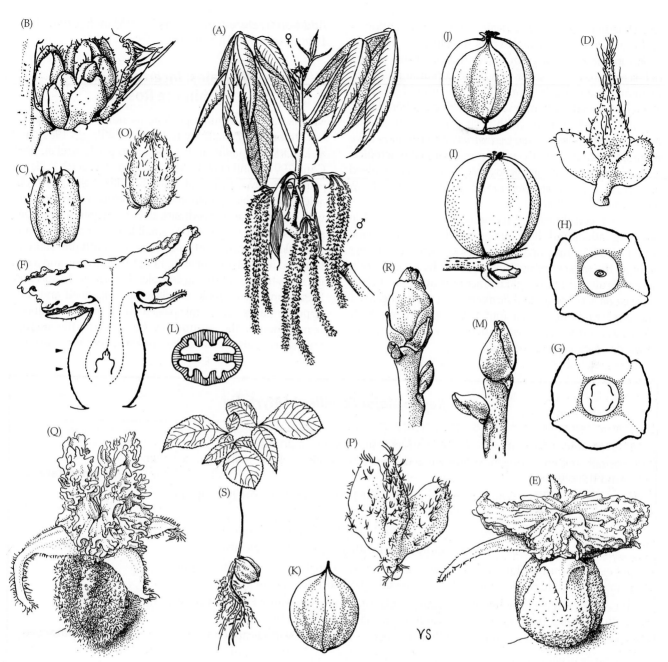

FIGURE 9.90 Juglandaceae. (A–M) *Carya ovata:* (A) flowering branch with staminate catkins and carpellate flowers (× 0.5); (B) staminate flower (× 14); (C) stamen (× 17); (D) bracts subtending staminate flower (× 14); (E) carpellate flower (× 8.5); (F) carpellate flower in longitudinal section, showing basal orthotropous ovule (× 8.5); (G) carpellate flower in cross-section, at level of lower arrow in F, showing 4-locular condition (× 8.5); (H) same, at level of upper arrow in F, showing ovule (× 8.5); (I) nut with dehiscent husk (× 1); (J) nut exposed by removal of two husk segments (× 1); (K) nut (× 1); (L) nut in cross-section, showing large embryo (unshaded) with corrugated cotyledons (× 1); (M) apical bud in winter condition (× 1.5). (N–S) *C. laciniosa:* (N) staminate flower (× 14); (O) stamen (× 17); (P) bracts subtending staminate flower (× 14); (Q) carpellate flower (× 8.5); (R) apical bud in winter condition (× 1.5); (S) seedling (× 0.25). (From Elias 1971, *J. Arnold Arbor.* 53: p. 37.)

shell. The remaining genera (i.e., *Platycarya, Carya, Cyclocarya, Pterocarya,* and *Juglans*—the Juglandoideae) form a well-supported clade; noteworthy synapomorphies of this group include buds with scales, odd-pinnately compound leaves with often serrate leaflets, bractlets that are com-pletely fused to the ovary, isodiametric sclereids of the nut shell, wood with growth rings, and vessel elements with exclusively simple perforations. *Platycarya* may be basal within Juglandoideae, with the remaining taxa linked by their drupaceous fruits, unisexual inflorescences, and elongate style branches. *Carya* is distinct due to its retention of highly modified, persistent tepals (forming a "stigmatic disk") in the carpellate flowers, and it has fruits with a dehiscent husk. It is sister to a clade containing *Juglans*, *Pterocarya*, and *Cyclocarya*, which are united by their stalked staminate catkins and twigs with chambered pith. *Juglans* has large drupes (with flesh derived from adnate bract and bracteoles), while *Cyclocarya* and *Pterocarya* have retained small fruits with conspicuous winged bracteoles. Large fruits (with usually fleshy cotyledons) have evolved separately in *Alfaroa*, *Carya*, and *Juglans* (Stone 1989; Manos and Stone 2001).

The reduced flowers of Juglandaceae are clustered in dangling catkins and show numerous adaptations for wind dispersal. Flowering occurs as, or soon after, the leaves emerge; the ovules are not developed at the time of pollination. Fruits with a winglike bract or bracteoles such as *Englehardia*, *Cyclocarya*, and *Pterocarya*, are wind-dispersed. The large fruits of *Alfaroa* (nuts), *Juglans* (drupes; Plate 9.13F), and *Carya* (nuts with a dehiscent outer husk) are adapted for dispersal by rodents.

Additional references: Elias 1972; Manning 1938, 1940, 1948; Stone 1973; Tiffney 1986.

Myrtales: *Incertae Sedis* within the Rosids

Myrtales are clearly monophyletic, as indicated by morphology, embryology, anatomy (Johnson and Briggs 1984), and *rbcL, matK, atpB, ndhF,* and 18S sequences (Chase et al. 1993; Conti 1994; Conti et al. 1996, 1997; Hilu et al. 2003; Källersjö et al. 1998; Savolainen et al. 2000a,b; Soltis et al. 2000). Probable morphological synapomorphies include vessel elements with vestured (i.e., fringed) pits, stems with internal phloem, stipules absent or present as small lateral or axillary structures, flowers with short to elongate hypanthium, stamens incurved in the bud (but straight in Onagraceae), and a single style, the carpels being completely connate. In addition, these plants have simple, usually entire-margined, and often opposite leaves. Taxonomic placement of Myrtales is uncertain, although the group may be sister to the malvids (eurosids II; Jansen et al. 2006). The order consists of 14 fami-

---

Key to Major Families of Myrtales

1. Ovary unilocular, with apical placentation; fruit 1-seeded, a drupe, usually flattened, ridged, and/or winged; hairs long, straight, sharp-pointed, unicellular, and thick-walled .... **Combretaceae**

1. Ovary multilocular with axile placentation; fruit 1–many-seeded, usually a capsule or berry; hairs various, but not as above 2

2. Leaves with pellucid dots containing ethereal oils, thus aromatic when crushed; stamens usually numerous, the anther often with an apical secretory cavity .... **Myrtaceae**

2. Leaves lacking pellucid dots, not aromatic; stamens equaling, or twice the number of petals to numerous, the anther lacking an apical secretory cavity 3

3. Leaves with 2–8 prominent secondary veins originating at or near base and converging toward apex, these usually connected by prominent tertiary veins oriented subperpendicular to midvein; anther connective, often with various appendages .... **Melastomataceae**

3. Leaves usually with pinnate venation; anther lacking appendages, but sometimes with conspicuous gland 4

4. Leaves containing large branched sclerids; stamen connective thickened, with a conspicuous circular to depressed, terpenoid-producing gland; seeds often few, large, the often large embryo with thick or convolute cotyledons; trees or shrubs; calyx ± imbricate .... Memecylaceae

4. Leaves lacking large branched sclereids; stamens without gland on connective; seeds several to numerous, small, the minute embryo with short cotyledons; herbs to trees; calyx valvate 5

5. Pollen associated with viscin threads; stamens arising from rim of hypanthium, straight in bud; flowers usually 4- or 2-merous; petals smooth, not crumpled in bud; ovary inferior; hypanthium elongate with sepals ± reflexed (except *Ludwigia*) .... **Onagraceae**

5. Pollen lacking viscin threads; stamens arising from inner surface of hypanthium, inflexed in bud; flowers usually 5-merous; petals wrinkled, crumpled in bud; ovary superior (except *Punica*); hypanthium short to elongate .... **Lythraceae**

(A)

Myrtales: Melastomataceae
Blakea wilsoniorum: flower

(B)

Myrtales: Myrtaceae
Callistemon rigidus: branch
with flowers

(C)

Myrtales: Onagraceae
Fuchsia pringsheimii: branch with flowers

(D)

Myrtales: Lythraceae
Cuphea cf. *ignea:* branch
with flowers

(E)

Myrtales: Myrtaceae
Myrcianthes fragrans: branch with fruits

(F)

Myrtales: Melastomataceae
Rhexia nuttalllii: fruits

(G)

Myrtales: Lythraceae
Lagerstroemia speciosa: fruits

(H)

Myrtales: Melastomataceae
Sagraea scalpta: branch with fruits

(I)

Myrtales: Combretaceae
Combretum aubletii: winged fruits

**PLATE 9.14 Eudicots:
Myrtales**

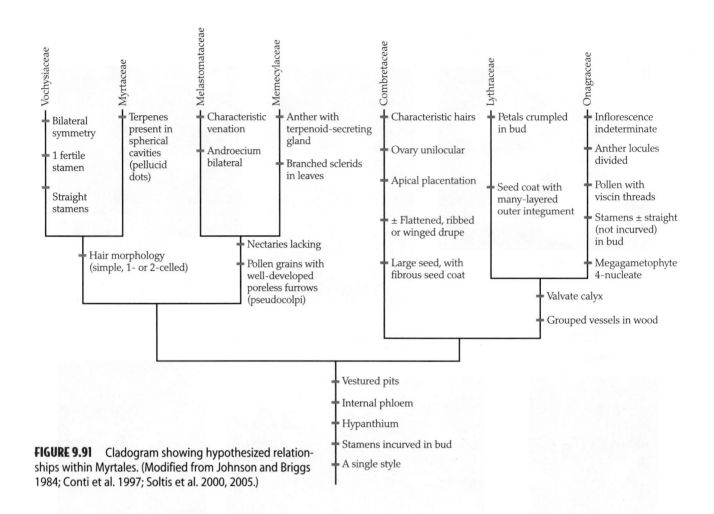

FIGURE 9.91 Cladogram showing hypothesized relationships within Myrtales. (Modified from Johnson and Briggs 1984; Conti et al. 1997; Soltis et al. 2000, 2005.)

lies and about 9000 species; major families include **Lythraceae**, **Onagraceae**, **Myrtaceae**, **Melastomataceae**, Memecylaceae, and **Combretaceae**. Vochysiaceae, a South American family with one sepal usually larger than the others and swollen or spurred at the base, and usually only one fertile stamen, positioned opposite the spurred sepal, also belongs here, as indicated by the presence of internal phloem, a hypanthium, and *rbcL* and *ndhF* sequences (Conti 1994a,b). It is closely related to Myrtaceae.

Phylogenetic relationships of families within the order have been studied by Johnson and Briggs (1984), who employed structural characters, and Conti (1994), Conti et al. (1996, 1997), and Sytsma et al. (2004), who used *rbcL* and *ndhF* sequences. Memecylaceae and Melastomataceae are sister taxa, whose pollen grains have well-developed poreless furrows (pseudocolpi) and lack floral nectaries. Onagraceae plus Lythraceae, with valvate calyx and grouped vessels in the wood, form another pair. Myrtaceae may be the sister group of the Melastomataceae + Memecylaceae clade. Relationships of Combretaceae are uncertain (Figure 9.91), but they may be associated with Onagraceae and Lythraceae.

Additional references: Baas et al. 2003; Dahlgren and Thorne 1984; Morley 1976; Patel et al. 1984; Tobe 1989; Weberling 1988a.

Lythraceae J. St.-Hilaire
(Loosestrife Family)

Trees, shrubs, or herbs. Hairs various, sometimes silicified. *Leaves opposite,* less often whorled or alternate and spiral, simple, entire, *with pinnate venation;* stipules typically reduced, commonly appearing as a row of minute hairs. Inflorescences various. *Flowers bisexual, commonly di- or tristylous,* radial to occasionally bilateral, *with well-developed hypanthium,* frequently associated with an epicalyx. *Sepals usually 4–8, distinct or slightly connate, valvate,* often rather thick. *Petals usually 4–8, distinct,* imbricate, **crumpled in bud and wrinkled at maturity,** occasionally lacking. *Stamens (4–) 8–16 (–numerous), usually attached well to slightly below summit of hypanthium, filaments unequal in length;* pollen usually tricolporate, sometimes with alternating poreless furrows. Carpels 2 to numerous, connate; *ovary superior* to rarely inferior, with axile placentation; stigma ± capitate. Ovules 2 to numerous in each locule. Nectaries often at base of hypanthium. *Fruit usually a dry, variously dehiscent capsule* (Plate 9.14G), occasionally a berry; *seeds usually flattened and/or winged,* **seed coat with many-layered outer integument,** sometimes with epidermal hairs that expand and become mucilaginous upon wetting; endosperm ± absent (Figure 9.92).

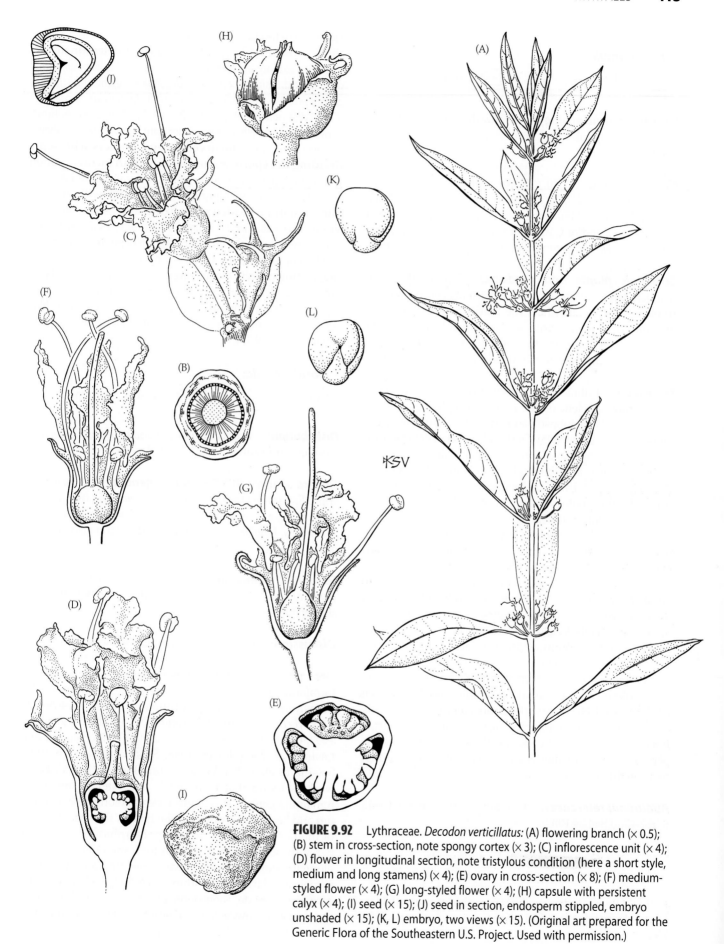

FIGURE 9.92 Lythraceae. *Decodon verticillatus:* (A) flowering branch (× 0.5); (B) stem in cross-section, note spongy cortex (× 3); (C) inflorescence unit (× 4); (D) flower in longitudinal section, note tristylous condition (here a short style, medium and long stamens) (× 4); (E) ovary in cross-section (× 8); (F) medium-styled flower (× 4); (G) long-styled flower (× 4); (H) capsule with persistent calyx (× 4); (I) seed (× 15); (J) seed in section, endosperm stippled, embryo unshaded (× 15); (K, L) embryo, two views (× 15). (Original art prepared for the Generic Flora of the Southeastern U.S. Project. Used with permission.)

Floral formula:

* or X, K(4–8), C4–8, A4–∞, G(2–∞); capsule

Distribution and ecology: Widely distributed, but most species are tropical. Many occur in aquatic or semiaquatic habitats.

Genera/species: 30/600. **Major genera:** *Cuphea* (275 spp.), *Diplusodon* (72), *Lagerstroemia* (56), *Nesaea* (50), *Rotala* (45), and *Lythrum* (35). Genera occurring in the continental United States and/or Canada are *Ammannia, Cuphea, Decodon, Didiplis, Heimia, Lythrum, Nesaea, Peplis, Rotala,* and *Trapa*.

Economic plants and products: *Cuphea* (Mexican heather), *Lagerstroemia* (crape myrtle), and *Lythrum* (loosestrife) provide ornamentals. The berries of *Punica* (pomegranates) contain numerous seeds with red, fleshy, and edible seed coats. *Lythrum salicaria* is a major exotic weed of wetlands in temperate North America.

Discussion: Lythraceae are easily recognized, and are considered monophyletic on the basis of morphological (Graham et al. 1993a,b; Johnson and Briggs 1984) and DNA features (Graham et al. 2005). The family is here circumscribed broadly, including *Punica* (often placed in Punicaceae), *Sonneratia* and *Duabanga* (often placed in Sonneratiaceae), and *Trapa* (Trapaceae). This broad familial circumscription is supported by morphology (Graham et al. 1993b) and DNA sequences (Conti 1994; Graham et al. 2005). *Sonneratia, Duabanga, Punica, Lagerstroemia, Trapa, Lawsonia,* and relatives may form a clade and are characterized by determinate inflorescences and wet stigmas, while most other genera have indeterminate inflorescences, dry stigmas, and a reduced number of carpels. *Decodon, Lythrum,* and *Pemphis* probably represent early divergent lineages. Herbaceousness has evolved in several specialized, temperate genera.

Heterostyly is common. Most species of the family are pollinated by bees, beetles, and flies, although bird pollination occurs in *Cuphea,* in which the hypanthium is sometimes colorful and spurred (Plate 9.14D), and *Sonneratia* is bat-pollinated. Nectar and pollen are employed as rewards. Cleistogamous or nearly cleistogamous flowers occur in *Peplis* and *Ammannia*. Seeds of Lythraceae are usually dispersed by wind or water. Some are buoyant due to spongy tissue in the outer seed coat.

Additional references: Ganders 1979; Graham 1964a; Stubbs and Slabas 1982.

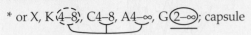 Onagraceae A. L. de Jussieu
(Evening Primrose Family)

Herbs to shrubs or occasionally trees; **raphides present**. Hairs simple. *Leaves* alternate and spiral, opposite, or whorled, *simple,* entire to toothed, sometimes lobed, *with pinnate venation;* stipules small to lacking. **Inflorescences indeterminate,** terminal, or axillary and solitary. *Flowers* usually bisexual, radial or bilateral, *usually with well-developed hypanthium that is clearly prolonged above ovary* (except in *Ludwigia*). *Sepals* (2–) 4 (–7), distinct, valvate. *Petals* (2–) 4 (–7), distinct, sometimes clawed, occasionally lacking, imbricate, convolute, or valvate. *Stamens* (1–) 8, **not incurved in bud, anthers with septa dividing the sporogenous tissue within locules; pollen grains** in monads, tetrads, or polyads, usually triporate, occasionally colpate, tricolporate, or biporate, **with unique paracrystalline beaded outer exine, and associated with viscin threads**. *Carpels usually 4,* connate; **ovary inferior,** usually with axile placentation; stigma capitate or clavate to 4-lobed or 4-branched. Ovules 1–numerous in each locule; **megagametophyte 4-nucleate** (i.e., *Oenothera*-type). Nectary usually near or at base of hypanthium. *Fruit a loculicidal capsule,* berry, or sometimes small, indehiscent, and nutlike; seeds sometimes winged or with a tuft of hairs; endosperm lacking (Figure 9.93).

Floral formula:

* or X, K4, C4, A4 or 8, G(4̄); capsule, berry, nut

Distribution: Widely distributed and especially diverse in western North America and South America.

Genera/species: 18/655. **Major genera:** *Oenothera* (192 spp.), *Epilobium* (183), *Fuchsia* (119), *Ludwigia* (85), *Clarkia* (66), and *Camissonia* (49). *Chamerion, Chylismia, Chylismiella, Circaea, Eulobus, Gayophytum,* and *Taraxia* also occur in North America.

Economic plants and products: *Fuchsia, Oenothera* (evening primrose), and *Clarkia* are ornamentals with showy flowers (Plate 9.13C).

Discussion: Monophyly of Onagraceae is supported by morphology, rDNA, and *rbcL* sequence data (Hoch et al. 1993; Johnson and Briggs 1984; Sytsma and Smith 1988). Infrafamilial relationships have been investigated by numerous authors using morphology, cpDNA restriction sites, *rbcL, ndhF,* and ITS sequences (Bult and Zimmer 1993; Conti et al. 1993; Crisci et al. 1990; Hoch et al. 1993; Levin et al. 2003; Raven 1988; Sytsma and Smith 1992; Sytsma et al. 1998a; Wagner and Hoch 2005). All these workers found that *Ludwigia* is the sister taxon to the remaining genera. This genus lacks an elongate hypanthium and has a nectary around the base of the style. The rest of the family is united by the apomorphies of deciduous perianth, consistently 4-merous (or 2-merous) flowers, an elongate and deciduous hypanthium, and a reduction of carpel vascularization. Within this group, *Fuchsia* and *Circaea* form a clade; their outer integument develops in an unusual way; *Lopezia* and *Megacorax* also form a clade. The

FIGURE 9.93 Onagraceae. *Oenothera macrocarpa:* (A) part of flowering plant (× 0.5); (B) upper part of flower (× 1); (C) upper part of flower to show insertion of stamens at apex of hypanthium (× 1.5); (D) pollen grains connected by viscin threads (greatly magnified); (E) stigmas (× 30); (F) ovary in longitudinal section, with base of hypanthium and base of style, note ovules (× 3); (G) fruit (× 1); (H) fruit in cross-section (× 1); (I) seed (× 30); (J) seed in longitudinal section, note large embryo (× 30); (K) embryo (× 30). (From Wood 1974, *A student's atlas of flowering plants*, p. 77.)

remaining genera, e.g., *Epilobium, Chamerion, Camissonia, Oenothera* (incl. *Gaura*), and *Clarkia*, form a clade diagnosed by loss of stipules and the occurrence of translocation heterozygosity (reciprocal interchange of parts of nonhomologous chromosomes; Dietrich et al. 1977). Problems of generic circumscription have recently been clarified in *Camissonia* and *Oenothera*.

Common pollinators of Onagraceae include bees, moths, flies, and birds; nectar is the reward. Species may be self-pollinated or outcrossing. Dispersal in capsular-fruited genera is typically by wind or water. The berries of *Fuchsia* are bird-dispersed, and hooked hairs on the small, nutlike fruits of *Circaea* facilitate external transport. The seeds of *Epilobium* have a tuft of elongate hairs, while those of *Hauya* are winged; both are wind-dispersed.

Additional references: Berry et al. 2004; Raven 1979; Skvarla et al. 1978; Tobe and Raven 1983.

Combretaceae R. Brown
(White Mangrove Family)

Trees, shrubs, or lianas, sometimes with erect, monopodial trunk supporting a series of horizontal, sympodial branches. Hairs various, **but some long, straight, sharp-pointed, unicellular, and very thick-walled, with a conical internal compartment at the base.** Leaves alternate and spiral, or opposite, **entire**, *with pinnate venation, often with domatia; petiole or base of blade often with 2 flask-shaped cavities, each containing a nectar gland;* stipules small or lacking. Inflorescences determinate, terminal or axillary. Flowers bisexual or unisexual (plants then monoecious, dioecious, or polygamous), usually radial, *with hypanthium slightly to conspicuously prolonged beyond the ovary. Sepals usually 4 or 5,* distinct or slightly connate, imbricate to valvate. *Petals usually 4 or 5,* distinct, imbricate or valvate, sometimes lacking. *Stamens 4–10;* filaments often long-exserted; pollen grains tricolporate or triporate, and often with poreless furrows. Carpels 2–5, connate; **ovary inferior and unilocular, with apical placentation**; stigma punctate to capitate. **Ovules few, pendulous on elongate funiculi from the tip of the locule.** Nectary usually a disk above ovary, often hairy. **Fruit a 1-seeded, ± flattened, ribbed and/or winged drupe; seed large, outer portion of seed coat fibrous;** embryo with usually folded or spirally twisted cotyledons; endosperm absent.

Floral formula:

*, K4–5, C4–5 or 0, A4–10, G$\overline{(4–5)}$; drupe (usually ribbed or winged)

Distribution and ecology: Pantropical; *Laguncularia, Lumnitzera,* and *Conocarpus* are mangroves or mangrove associates; others occur in tropical broad-leaved forests or savannas.

Genera/species: 20/600. **Major genera:** *Combretum* (250 spp.) and *Terminalia* (200). The family is represented in the continental United States by *Bucida, Conocarpus, Laguncularia,* and *Terminalia.*

Economic plants and products: *Terminalia* (tropical almond) has edible fruits; other genera are ornamentals with showy flowers, such as *Combretum* and *Quisqualis* (Rangoon creeper), or distinctive habit, such as *Terminalia, Bucida* (black olive), and *Conocarpus* (buttonwood). Many genera yield timber, especially *Terminalia.*

Discussion: Morphology (Dahlgren and Thorne 1984; Johnson and Briggs 1984) and *rbcL* sequence characters (Conti 1994) both support the monophyly of Combretaceae.

The flowers typically produce nectar and attract insects, birds, or small mammals; the hypanthium varies from green to brightly colored. Many have bisexual flowers, and outcrossing is promoted by protogyny. *Conocarpus erectus,* however, is dioecious, and its flowers are densely packed in

heads. In *Laguncularia racemosa,* some individuals bear staminate flowers, while others bear bisexual flowers that are very similar to the staminate ones. Flowers of *Terminalia* are also staminate or bisexual, but both occur on the same tree. The drupes of most genera are well adapted to dispersal by water due to their spongy mesocarp; some are distinctly winged and dispersed by wind (Plate 9.14I), others are fleshy and bird or mammal-dispersed.

Domatia occur frequently in the family, usually associated with the veins on the abaxial leaf surface. These are usually inhabited by mites, as in *Terminalia* and *Conocarpus,* which probably protect the plant from fungi or tiny herbivores.

Laguncularia racemosa, the white mangrove, shows adaptations to daily flooding by salt water. Leaves contain salt excretory glands, seeds germinate while still attached to the tree (vivipary; see Chapter 4), and pneumatophores (erectly growing roots containing aerenchyma tissue and functioning in gas exchange) may be present.

Additional references: Graham 1964b; Stace 1965.

Myrtaceae A. L. de Jussieu
(Myrtle Family)

Trees or shrubs, often with flaky bark; **terpenes present. Hairs simple, unicellular or bicellular.** *Leaves opposite, or alternate and spiral,* rarely whorled, **entire**, *usually with pinnate venation,* **with scattered pellucid dots (i.e., spherical secretory cavities containing terpenoids and other aromatic, spicy-resinous compounds)**; stipules minute or lacking. Inflorescences determinate, but sometimes appearing indeterminate, terminal or axillary, sometimes reduced to a single flower. Flowers usually bisexual, radial, with well-developed hypanthium. *Sepals usually 4 or 5,* distinct to connate, imbricate, sometimes fused into a circumscissilely or irregularly rupturing cap. *Petals usually 4 or 5,* distinct to connate, imbricate, sometimes fused into a cap (and also adnate to the sepals), sometimes lacking. *Stamens usually numerous,* developing from the outside of the flower inward, *distinct or basally connate into 4 or 5 fascicles;* anthers often with connective with an apical secretory cavity; pollen grains usually tricolporate, with furrows fused. Carpels usually 2–5, connate; *ovary usually inferior to ± half-inferior,* with axile placentation, or less commonly parietal with intruded placentae; stigma usually capitate. Ovules 2 to numerous in each locule, anatropous to campylotropous. Nectariferous tissue atop ovary or lining inner surface of hypanthium. *Fruit usually a 1- to many-seeded berry or loculicidal capsule,* rarely a nut. Embryo with small to large cotyledons, these sometimes connate or folded or twisted together; endosperm scanty or lacking (Figure 9.94).

Floral formula:

*, K$\widehat{(4–5)}$, C$\widehat{(4–5)}$, A∞, G$\overline{(2–5)}$; berry, capsule

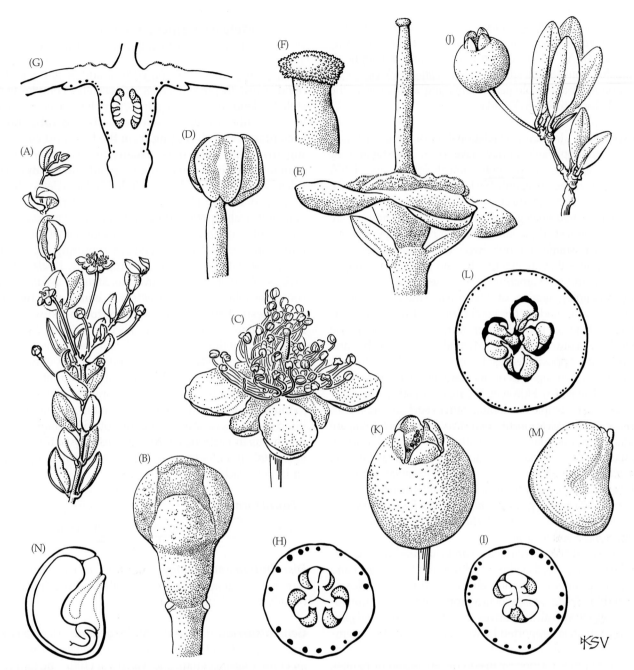

FIGURE 9.94   Myrtaceae. *Mosiera longipes:* (A) flowering branch (× 0.5); (B) flower bud (× 8); (C) flower (× 5); (D) stamen (× 25); (E) gynoecium and sepals (× 8); (F) stigma (× 25); (G) ovary in longitudinal section, note ovules and resin cavities (× 8); (H, I) two ovaries in cross-section, note resin cavities (× 12); (J) twig with berry (× 1.5); (K) berry (× 3); (L) berry in cross-section (× 5); (M) seed (greatly enlarged); (N) seed in section, note large embryo. (Original art prepared for the Generic Flora of the Southeastern U.S. Project. Used with permission.)

Distribution: Pantropical, in a wide variety of habitats; also diverse in warm temperate Australia.

Genera/species: 144/3100.   **Major genera:** *Eucalyptus* (500 spp.), *Eugenia* (400), *Myrcia* (300), *Syzygium* (300), *Melaleuca* (100), *Psidium* (100), and *Calyptranthes* (100). All of the above (except *Myrcia*), along with *Mosiera*, *Rhodomyrtus* and *Myrcianthes*, are native or naturalized in the continental United States.

Economic plants and products: *Eucalyptus* is an important source of timber. Many genera contain valuable ornamentals due to their showy sepals, petals, and/or stamens; these include *Myrtus* (myrtle), *Eucalyptus* (gum tree), *Callistemon* (bottlebrush), *Melaleuca* (bottlebrush), *Leptospermum*, and *Rhodomyrtus* (downy myrtle). Flower buds of *Syzygium aromaticum* are cloves, and fruits of *Pimenta dioica* are the source of allspice. Edible fruits are common, including *Psidium guajava* (guava), *Syzygium jambos* (rose apple), *S. malac-*

cense (Malay apple), *Myrciaria cauliflora* (jaboticaba), *Eugenia uniflora* (Surinam cherry), and *Acca sellowiana* (pineapple guava). An oil, used as a flavoring and antiseptic, is extracted from several species of *Eucalyptus*. *Melaleuca* is an economically and ecologically damaging weedy tree in the Everglades of southern Florida.

Discussion: Infrafamilial relationships of this distinct and clearly monophyletic family have been investigated by cladistic analyses using both morphological and DNA sequence characters (Conti 1994; Johnson and Briggs 1984; Lucas et al. 2005; Sytsma et al. 1998, 2004; Wilson et al. 1996, 2001, 2005). Two small genera, *Heteropyxis* and *Psiloxylon*, are sister to the rest of the family, having retained perigynous flowers and stamens in only two whorls. Both genera also have stamens that are erect in the bud (a possible synapomorphy), and are sometimes segregated from Myrtaceae. The remaining genera of Myrtaceae—the "core" Myrtaceae—are united by the apomorphies of numerous stamens and at least partly inferior ovaries. These genera have usually been divided into the "Myrtoideae" (with berries and consistently opposite leaves; Plate 9.14E) and "Leptospermoideae" (with capsules or nuts and alternate or opposite leaves; Plate 9.14B). However, "Leptospermoideae," a mainly Australian complex including genera such as *Eucalyptus*, *Leptospermum*, *Metrosideros*, *Callistemon*, and *Melaleuca*, are paraphyletic and mainly basal within the "core" Myrtaceae, while the "Myrtoideae", including genera such as *Eugenia*, *Psidium*, *Calyptranthes*, *Syzygium*, and *Myrcianthes*, are polyphyletic because berry fruits have evolved independently in the myrtoid clade (e.g., *Acca*, *Calyptranthes*, *Eugenia*, *Psidium*, *Mosiera*, *Myrcianthes*, and *Myrtus*) and the *Acmena* group (e.g., *Acmena* and *Syzygium*).

Genera of "core" Myrtaceae can be divided among several well-supported clades, in addition to the *Acmena* and myrtoid groups mentioned above. *Callistemon* is closely related to *Melaleuca* (Melaleuca group). *Eucalyptus, Corymbia, Angophora*, and relatives constitute the *Eucalyptus* group, which is supported by wood anatomy. Most members of this clade have evolved various kinds of perianth caps (calyptras, formed by fusion of sepals and/or petals), but *Angophora* has separate petals. Cladistic analyses have shown that *Eucalyptus*, as broadly delimited, is not monophyletic, and segregates, such as *Corymbia*, could be recognized (Steane et al. 2002). Embryo characters often are stressed in generic delimitations.

The sweet-scented flowers of Myrtaceae are pollinated by various insects, birds, or mammals; nectar is the reward. In genera such as *Eucalyptus, Melaleuca*, and *Callistemon*, the stamens are more conspicuous than the petals, giving a bottlebrush effect. Fleshy-fruited species are dispersed by birds and mammals; capsular species frequently have small, and sometimes winged, seeds and are wind- or water-dispersed.

Additional references: Johnson 1976; Schmid 1980; Wilson 1960d.

Melastomataceae A. L. de Jussieu
(Melastome or Meadow Beauty Family)

Trees, shrubs, vines, *or herbs*; cortical and/or pith vascular bundles usually present. Hairs various, often complex (e.g., stellate, dendritic, stalked glands, or peltate scales). *Leaves opposite*, entire to serrate, **with usually 2–8 subparallel secondary veins diverging from the base and converging toward the apex, these usually connected by prominent tertiary veins oriented ± perpendicular to midvein** (Plate 9.14H); stipules lacking. Inflorescences determinate, terminal or axillary. Flowers usually bisexual, radial to bilateral, with well-developed hypanthium. *Sepals usually 3–6* and slightly connate, imbricate to valvate, sometimes fused into a circumscissilely or irregularly rupturing cap, sometimes associated with external projections. *Petals usually 3–6*, usually distinct and convolute. *Stamens usually 6–12*, sometimes dimorphic; *filaments bent*, **commonly twisted at anthesis, bringing the anthers to one side of the flower**; *anthers sometimes unilocular, opening by apical pores*, or sometimes longitudinal slits, *the connective often thickened or appendaged at base*; endothecium ephemeral; pollen grains usually tricolporate, with 3 alternating poreless furrows. Carpels usually 2–10, connate; ovary superior to inferior, usually with axile placentation; stigma capitate, punctate, or occasionally slightly lobed. Ovules usually numerous in each locule. *Nectaries usually lacking. Fruit a loculicidal capsule or berry; seeds usually numerous and small;* endosperm lacking (Figure 9.95).

Floral formula:

X or *, K ③–⑥, C3–6, A6–12, G ②–⑩; capsule, berry

Distribution and ecology: Pantropical; especially common in tropical montane habitats; often light-loving shrubs of early successional habitats.

Genera/species: 150/3000. ***Major genera:*** *Miconia* (1000 spp.), *Medinilla* (300), *Tibouchina* (240), *Leandra* (175), *Sonerila* (150), *Clidemia* (100), and *Microlicia* (100). The family is represented in the continental United States only by *Rhexia* and *Tetrazygia*.

Economic plants and products: Some genera contain ornamentals with showy flowers or striking leaves, including *Dissotis*, *Medinilla*, *Rhexia* (meadow beauty), and *Tibouchina* (princess flower).

Discussion: Morphological features (Renner 1993; Johnson and Briggs 1984) and DNA sequences (Clausing et al. 2000; Clausing and Renner 2001; Conti 1994; Michelangeli et al. 2004; Renner 2004; Renner et al. 2001) support the monophyly of Melastomataceae. *Pternandra* probably is the sister group to the remaining genera (core Melastomataceae); this genus has retained an endothecium, while core

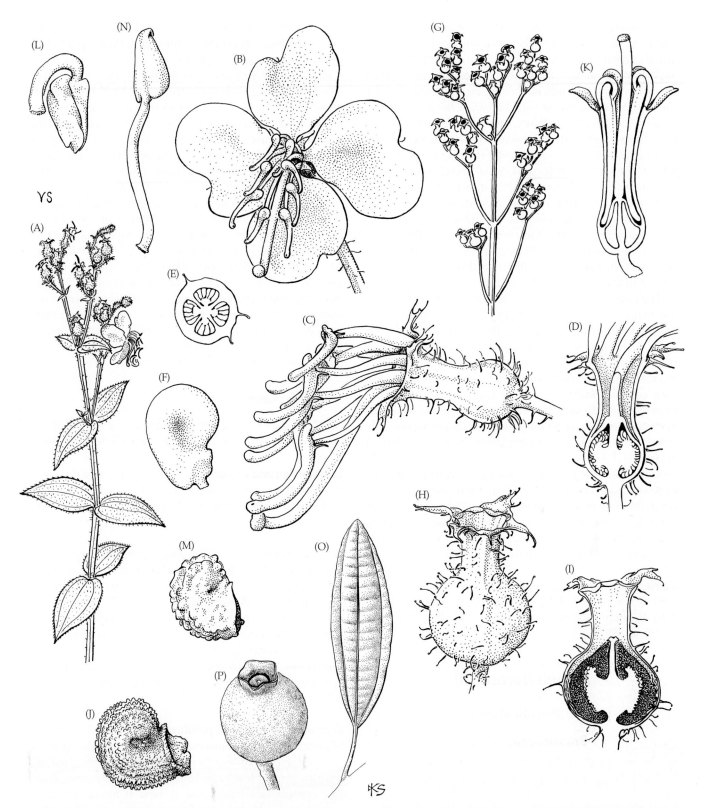

FIGURE 9.95 Melastomataceae. (A–J) *Rhexia virginica:* (A) tip of flowering plant (× 0.7); (B) flower (× 2.5); (C) same, side view with petals removed to show hypanthium and stamens (× 5.5); (D) hypanthium and gynoecium in longitudinal section (× 5.5); (E) ovary in cross-section (× 5.5); (F) ovule (× 65); (G) semidiagrammatic view of infructescence with leaves, bracts, and hairs omitted (× 0.7); (H) hypanthium enclosing mature capsule (× 5.5); (I) same, in longitudinal section, seeds removed (× 5.5); (J) seed (× 40). (K) *R. nashii:* flower bud in longitudinal section, petals removed to show inverted stamens (× 4). (L–M) *R. nuttallii:* (L) stamen from flower bud (× 4); (M) seed (× 40). (N) *R. petiolata:* stamen from open flower (× 16). (O–P) *Tetrazygia bicolor:* (O) leaf (× 0.7); (P) berry (× 16). (From Wurdack and Kral 1982, *J. Arnold Arbor.* 63: pp. 434, 438.)

Melastomataceae are united by the loss of an endothecium, and most of these genera also have anthers that open by apical pores (Clausing and Renner 2001). Members of core Melastomataceae are divided into several tribes. Most of these have capsular fruits, but berries are synapomorphic for the large tribe Miconieae (e.g., *Miconia, Clidemia, Leandra, Conostegia, Tetrazygia, Tococa,* and *Mecranium*). Berries evolved independently in *Medinilla* (of the Dissochaeteae/Sonerileae complex) and Blakeeae (*Blakea, Topobea;* Plate 9.14A). The Merianieae (e.g., *Meriania*) and Rhexieae (*Rhexia;* Plate 9.14F) are characterized by anthers with short connectives that are usually dorsally appendaged; the Rhexieae have the additional apomorphies of unilocular anthers and snail-shaped seeds. The Melastomateae (e.g., *Tibouchina, Melastoma, Monochaetum, Dissotis*) possess ovaries with a conic and often setose apex; this tribe also has snail-shaped seeds and may be closely related to *Rhexia.*

The flowers typically do not produce nectar, and are visited mainly by pollen-gathering bees, which vibrate or otherwise manipulate the anthers; the characteristic anther appendages may serve as a hold for the bee's legs. Most melastomes are facultative outcrossers, with the small stigma well separated from the anthers. The seeds of species with capsular fruits are usually wind- or raindrop-dispersed, while those with berries are dispersed by birds, mammals, turtles, or lizards; some seeds are secondarily ant-dispersed. Apomixis is common in species of disturbed habitats.

Some species (e.g., *Tococa*), have leaves with large pouches at the blade or petiole base that are inhabited by ants, while *Mecranium, Calycogonium, Blakea,* and several other genera have special structures (domatia-forming hair tufts, pouches, etc.) in the axils of the secondary veins that are inhabited by predatory or fungus-eating mites.

Additional references: Judd 1986, 1989; Judd and Skean 1991; Renner 1989b, 1990; Stein and Tobe 1989; Vliet et al. 1981; Whiffin 1972; Wilson 1950; Wurdack 1986; Wurdack and Kral 1982.

Malvids (Eurosids II)

Brassicales

Brassicaceae Burnett

(= Cruciferae A. L. de Jussieu)

(Crucifer, Mustard, or Caper Family)

Trees, shrubs, or herbs; producing glucosinolates (mustard oil glucosides) and with myrosin cells; often cyanogenic. Hairs diverse, simple to branched, stellate, or peltate. *Leaves usually alternate and spiral, sometimes in basal rosettes, simple, often pinnately dissected or lobed, or palmately or pinnately compound,* entire to serrate, with palmate or pinnate venation; stipules present or absent. *Inflorescences indeterminate,* occasionally reduced to a solitary flower, terminal or axil-

lary. *Flowers* usually bisexual, radial or bilateral, *often lacking subtending bracts;* **receptacle prolonged, forming an elongate or shortened gynophore** (or androgynophore). *Sepals 4, distinct. Petals 4, distinct, often forming a cross,* often with an elongate claw and abruptly spreading limb, imbricate or convolute. *Stamens (2–) 6,* or numerous, *all ± the same length or the 2 outer shorter than the 4 inner (tetradynamous);* **filaments elongate** to rather short, distinct, or connate in pairs; pollen grains usually tricolporate or tricolpate. **Carpels usually 2,** connate; ovary superior, *with parietal placentation, frequently with the placentas forming a thick rim (replum) around the fruit and often connected by a false septum (a thin partition lacking vascular tissue; Plate 9.15F) that divides the ovary into 2 chambers;* stigma capitate, sometimes bilobed. Ovules 1 to numerous on each placenta, anatropous to campylotropous. Nectar disk or gland usually present. *Fruit a berry or capsule, frequently with 2 valves breaking away from a replum and often additionally with a persistent septum (the fruit then a silique),* these short to elongate, globose to flattened; seeds with or without broad to narrow invagination, occasionally arillate; *embryo curved or folded;* endosperm scanty or absent (Figure 9.96).

> **Floral formula:**
>
> * or X, K4, C4, A(2–) 6–∞, G②; berry, capsule, silique-like capsule, silique

Distribution and ecology: Cosmopolitan, most diverse in the Mediterranean region, southwestern and central Asia, and western North America. Many species occur in early successional communities.

Genera/species: 356/4130. **Major genera:** *Capparis* (350 spp.), *Draba* (350), *Cleome* (200), *Erysimum* (180), *Cardamine* (170), *Lepidium* (170), *Alyssum* (150), *Physaria* (120, incl. *Lesquerella*), *Arabis* (70), *Heliophila* (70), *Thlaspi* (70), *Rorippa* (70), and *Hesperis* (60). Numerous genera occur in the continental United States and/or Canada; in addition to most of the above, noteworthy genera include *Barbarea, Brassica, Cakile, Caulanthus, Capsella, Cochlearia, Descurainia, Dimorphocarpa, Leavenworthia, Physaria, Platyspermum, Polanisia, Schoenocrambe, Stanleya, Streptanthus,* and *Warea.*

Economic plants and products: The family contains many important food plants, including both edible species, such as *Capparis spinosa* (capers), *Raphanus sativus* (radish), *Brassica oleracea* (cabbage, kale, broccoli, cauliflower, Brussels sprouts, kohlrabi), and *B. rapa* (Chinese cabbage, turnip), and sources of condiments, such as *Brassica juncea* (Chinese mustard), *B. nigra* (black mustard), *Sinapis alba* (white mustard), and *Armoracia rusticana* (horseradish). Table mustard is prepared from a mixture of the seeds of white mustard and either black mustard or Chinese mustard. Vegetable oil is extracted from the seeds of several species of *Brassica,* especially *B. napus* (canola, rapeseed oil). The family contains many ornamentals, such as *Cleome*

Canada's #1 crop

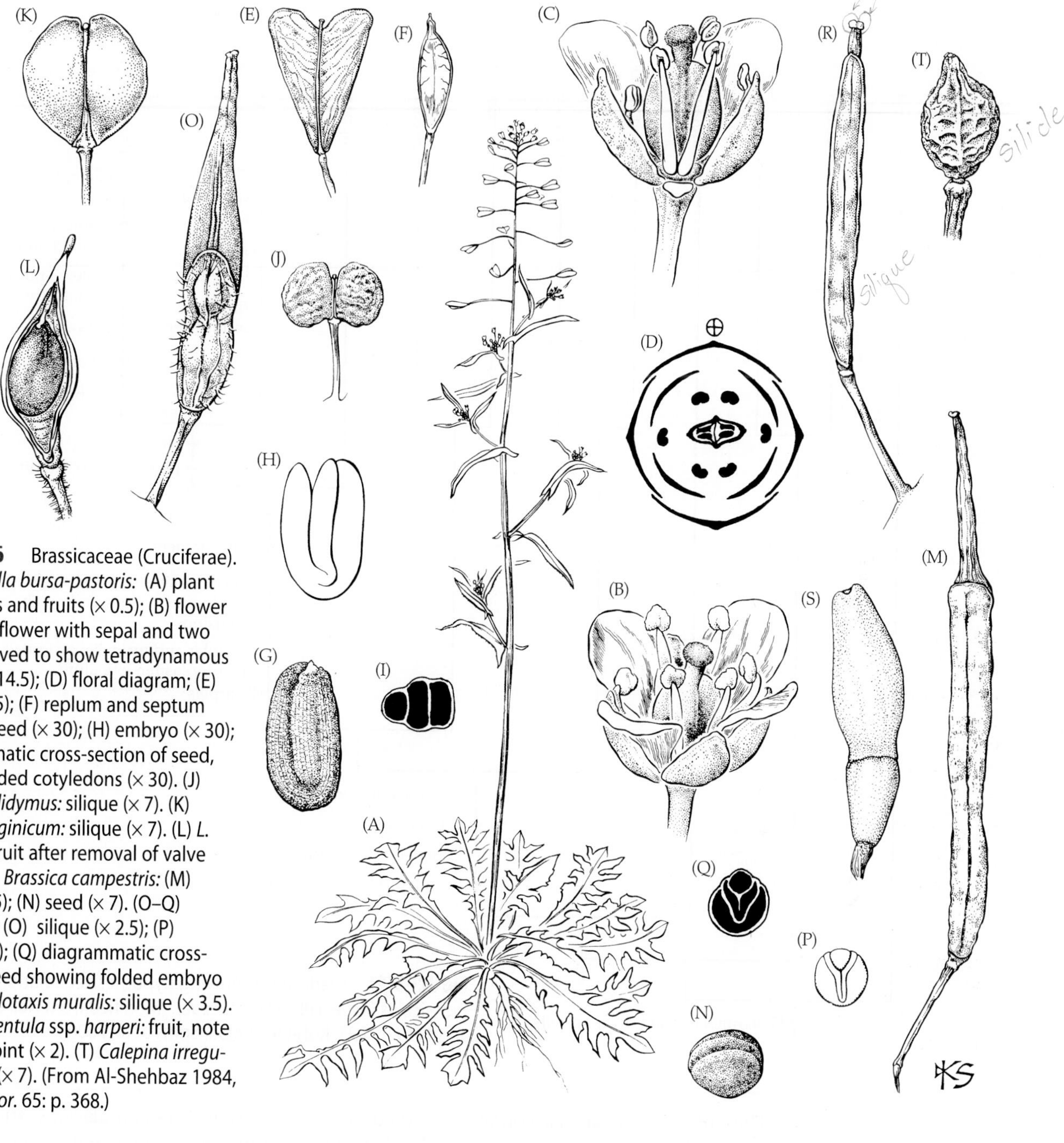

FIGURE 9.96 Brassicaceae (Cruciferae). (A–I) *Capsella bursa-pastoris:* (A) plant with flowers and fruits (× 0.5); (B) flower (× 14.5); (C) flower with sepal and two petals removed to show tetradynamous stamens (× 14.5); (D) floral diagram; (E) silique (× 3.5); (F) replum and septum (× 3.5); (G) seed (× 30); (H) embryo (× 30); (I) diagrammatic cross-section of seed, showing folded cotyledons (× 30). (J) *Coronopus didymus:* silique (× 7). (K) *Lepidium virginicum:* silique (× 7). (L) *L. campestre:* fruit after removal of valve (× 7). (M–N) *Brassica campestris:* (M) silique (× 2.5); (N) seed (× 7). (O–Q) *Sinapis alba:* (O) silique (× 2.5); (P) embryo (× 7); (Q) diagrammatic cross-section of seed showing folded embryo (× 7). (R) *Diplotaxis muralis:* silique (× 3.5). (S) *Cakile edentula* ssp. *harperi:* fruit, note transverse joint (× 2). (T) *Calepina irregularis:* silique (× 7). (From Al-Shehbaz 1984, *J. Arnold Arbor.* 65: p. 368.)

(spider flower), *Hesperis* (rocket, dame's violet), *Erysimum* (wallflower), *Iberis* (candytuft), *Lunaria* (honesty, money plant), *Lobularia* (sweet alyssum), *Aurinia* (golden alyssum), and *Arabis* (rock cress). Weedy taxa are also common, e.g., *Alliaria* (garlic mustard), *Capsella* (shepherd's purse), *Descurainia* (tansy mustard), and *Lepidium* (peppergrass). *Arabidopsis thaliana* (thale or mouse-ear cress), a Eurasian weed, is the most widely used vascular plant in molecular and experimental biology.

Discussion: Brassicaceae may also be called Cruciferae, meaning "cross-bearing," in reference to the cruciform arrangement of the petals (Plate 9.15D). The family is con-

sidered to be monophyletic on the basis of the elongate gynophore and elongate, exserted stamens, although both of these are shortened in most derived members of the family (Rodman 1991b; Judd et al. 1994). Additional synapomorphies include the type of glucosinolates, the structure of the endoplasmic reticulum (Rodman 1981; Jorgensen 1981), and *rbcL* sequences (Rodman et al. 1993).

Brassicaceae are the largest family of the Brassicales, an order of 15 families characterized by the presence of glucosinolates, which contain sulfur. When these compounds react with myrosinase (contained in specialized spherical myrosin cells) they release hot, pungent mustard oils. The presence of glucosinolates (and myrosin cells) is synapomorphic for

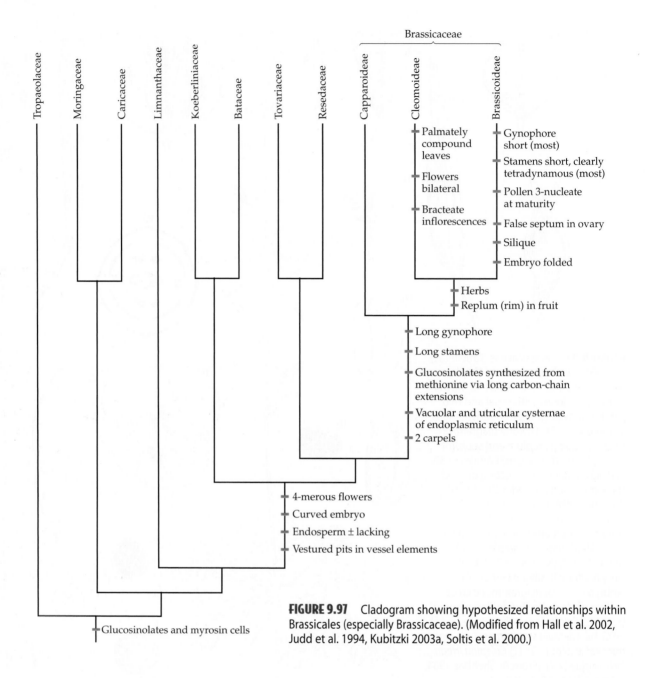

Brassicaceae

Capparoideae | *Cleomoideae* | *Brassicoideae*

- Palmately compound leaves
- Flowers bilateral
- Bracteate inflorescences

- Gynophore short (most)
- Stamens short, clearly tetradynamous (most)
- Pollen 3-nucleate at maturity
- False septum in ovary
- Silique
- Embryo folded

- Herbs
- Replum (rim) in fruit

- Long gynophore
- Long stamens
- Glucosinolates synthesized from methionine via long carbon-chain extensions
- Vacuolar and utricular cysternae of endoplasmic reticulum
- 2 carpels

- 4-merous flowers
- Curved embryo
- Endosperm ± lacking
- Vestured pits in vessel elements

- Glucosinolates and myrosin cells

FIGURE 9.97 Cladogram showing hypothesized relationships within Brassicales (especially Brassicaceae). (Modified from Hall et al. 2002, Judd et al. 1994, Kubitzki 2003a, Soltis et al. 2000.)

members of Brassicales. The only other taxa that contain these compounds are *Drypetes* and *Putranjiva* (see Putranjivaceae, and discussion for Euphorbiaceae), and mustard oils are thus hypothesized to have evolved twice (Rodman et al. 1998). Phylogenetic relationships within the order have been investigated by cladistic analyses of morphological and DNA characters (Rodman 1991b; Rodman et al. 1993, 1996; Judd et al. 1994; Källersjö et al. 1998; Soltis et al. 1997, 2000; Karol et al. 1999; Savolainen et al. 2000b) (Figure 9.97). Brassicaceae are members of a morphologically distinct subclade of Brassicales, which may be diagnosed by 4-merous flowers, seeds with curved or folded embryos and lacking or nearly lacking endosperm, vessels with vestured pits, and protein-rich, unspecialized to vacuolar cisternae of the endoplasmic reticulum. Thus, these "cruciferous" features are also present in Resedaceae, Tovariaceae, Koeberliniaceae, and Bataceae. Relationships among early divergent families, such as Moringaceae + Caricaceae and Tropaeolaceae + Akaniaceae are well supported. These families have retained characters such as 5-merous flowers, ovaries with axile placentation, and straight embryos.

Morphology (Judd et al. 1994; Rodman 1991b) and *rbcL* sequences (Rodman et al. 1993; Savolainen et al. 2000b; Soltis et al. 2000) suggest that *Capparis* and relatives (i.e., Capparoideae) form a basal paraphyletic complex within Brassicaceae, however recent more extensive molecular studies support the monophyly of Capparoideae (Hall et al. 2002), although the subfamily is without obvious morphological support. Cleomoideae (*Cleome* and relatives) and Brassicoideae (temperate zone mustards) form a monophyletic group based on the synapomorphies of an herbaceous habit, replum in the fruit, and DNA sequences (Figure 9.97). The monophyly of the Cleomoideae is supported by palmately compound leaves and bilaterally symmetrical flowers, while the monophyly of Brassicoideae is indicated by the presence of a false septum in the ovary and folded embryos (with a loss of the seed invagination). Members of Brassicoideae also differ from Cleomoideae in their strongly reticulate and colpate (vs. smooth to shallowly reticulate and colporate) pollen. Most Brassicoideae have short or obsolete gynophores and clearly tetradynamous stamens with short filaments, but a few, e.g., *Warea* and relatives, have elongate stamens and a distinct gynophore. Members of the clade here referred to as Brassicoideae are, in earlier textbooks, recognized at the family level (as Brassicaceae), and are distributed predominantly in the temperate zone, whereas the predominantly tropical Cleomoideae and Capparoideae are placed together in a paraphyletic "Capparaceae." A broadly defined Brassicaceae is adopted here, following APG (2003).

Genera of Brassicoideae often are difficult to distinguish, and have been placed in some ten poorly defined tribes. Generic and tribal delimitation in the past stressed fruit morphology, calyx aestivation, flower color and symmetry, stigma form, number of seeds per locule, type of embryo folding, and indumentum (Al-Shehbaz 1984; Rollins 1993). It is, therefore, easier to identify fruiting specimens than ones in flower. Al-Shehbaz et al. (2006) have begun a tribal reclassification following recent molecular findings; they recognize some 25 tribes; of these, Aethionemeae are sister to the rest of the Brassicoideae.

Flowers of Brassicaceae are frequently white, yellow, or pale to deep purple (Plate 9.15C,D) and are pollinated by bees, flies, butterflies, moths, and beetles, which gather nectar. Pollination by birds or bats occurs in some tropical *Capparis* (Plate 9.15G). Protogyny favors outcrossing, but many weedy species are selfing.

The small seeds may be explosively released from the siliques, as in *Cardamine*, but in most genera the valves of the silique (or capsule) merely fall away, exposing the seeds to the action of wind (or secondary dispersal by rain wash). In *Raphanus* and *Cakile* the corky fruits break transversely into one-seeded segments, which are then scattered by wind and/or water. Wings, bladders, or dust seeds facilitate dispersal by wind and have evolved several times within Brassicoideae. Fleshy fruits, as in *Capparis* (Plate 9.15G), are dispersed by mammals or birds.

Additional references: Al-Shehbaz 1985a,b, 1987, 1988a,b; Appel and Al-Shehbaz 2003; Ernst 1963a; Kers 2003; Kubitzki 2003; O'Kane and Al-Shehbaz 2003; Rodman 1991a; Sweeney and Price 2000; Vaughan et al. 1976; Warwick and Black 1993.

Malvales

Malvales are clearly monophyletic, as evidenced by their stratified phloem with fibrous and soft layers, wedge-shaped rays, mucilage (slime) canals and cavities, stellate hairs, connate sepals, malvoid leaf teeth (Judd and Manchester 1998), cyclopropenoid fatty acids, and *rbcL, atpB,* and 18S sequences (Alverson et al. 1998a; Bayer et al. 1999; Fay et al. 1998; Källersjö et al. 1998; Savolainen et al. 2000a,b; Soltis et al. 1998, 2000). The complex vascular system that occurs in the petioles may also be synapomorphic. Stamens are frequently numerous and develop centrifugally from only a few trunk vascular bundles (evidence of a secondary increase from originally only two whorls). The order has been variously circumscribed, but probably consists of 10 families and 3560 species; major families include

---

Key to Major Families of Malvales

1. Calyx and hypanthium colored and petaloid; internal phloem usually present ....Thymelaeaceae
1. Hypanthium absent and calyx usually not colorful; internal phloem lacking 2

2. Placentation parietal; mucilage and/or resin canals lacking; sepals dimorphic
   (i.e., 3 large and 2 small), but not winglike ....**Cistaceae**
2. Placentation axile; mucilage and/or resin canals present; sepals uniform or dimorphic
   but not as above, sometimes winglike 3

3. Nectaries of densely packed glandular hairs (usually on calyx); anthers not appendaged;
   venation pinnate to palmate; mucilage canals present and resin canals lacking;
   calyx lobes usually not winglike; fruit a loculicidal capsule, schizocarp, indehiscent pod,
   nut, berry, drupe, or aggregate of follicles ....**Malvaceae**
3. Nectaries not as above; anthers appendaged; venation usually pinnate; mucilage canals
   sometimes present and usually with resin canals; 2–5 calyx lobes becoming expanded
   and usually winglike in fruit; fruit usually a nut ....**Dipterocarpaceae**

(A) Malvales: Malvaceae
Thespesia populnea: flower

(B) Malvales: Malvaceae
Theobroma cacao: fruit

(C) Brassicales: Brassicaceae
Cleome domingensis: flowers

(D) Brassicales: Brassicaceae
Brassica napus: flowers

(E) Malvales: Cistaceae
Hudsonia tomentosa: habit; plant in bloom

(F) Brassicales: Brassicaceae
Sibara virginica: seeds, replum, and false septum of fruit

(G) Brassicales: Brassicaceae
Capparis flexuosa: flowers

C. flexuosa: fruits

(H) Malvales: Malvaceae
Dombeya × cayeuxii: flowers

PLATE 9.15 Brassicales and Malvales

Dipterocarpaceae, **Cistaceae**, **Malvaceae**, and Thymelaeaceae. Phylogenetic analyses of DNA sequences clearly place the order within the Malvids (Eurosid II), however, many systematists (e.g., Bessey 1915; Thorne 1992) had considered Malvales to be related to Urticales (here in Rosales, of the Fabid clade) based on the common occurrence of bands of fibers in the phloem, alternate leaves with often palmate venation, and stipules.

Malvaceae A. L. de Jussieu
(Mallow Family)

Trees, shrubs, lianas, or herbs; mucilage canals (and often also mucilage cavities) present. Hairs various, but usually stellate or peltate scales. Leaves usually alternate, spiral or 2-ranked, simple, often palmately lobed, or palmately compound, entire to serrate, the teeth malvoid (i.e., with major vein unexpanded, and end-

ing at the tooth apex), *with palmate or occasionally pinnate venation; stipules present*. Inflorescences indeterminate, mixed, or determinate, sometimes reduced to a single flower, axillary, **with determinate basic repeating unit and bearing 3 bracts, one of which is always sterile, while the others subtend lateral cymes or single flowers**. *Flowers* bisexual or unisexual, *usually radial, often associated with conspicuous bracts that form an epicalyx. Sepals usually 5, distinct to more commonly connate, valvate. Petals usually 5, distinct, imbricate,* convolute, or valvate, sometimes lacking. *Stamens 5 to numerous*, sometimes borne on a short to elongate androgynophore; *filaments distinct, basally connate and forming fascicles, but often strongly connate and forming a tube around the gynoecium (monadelphous); anthers 2-locular or unilocular (due to developmental modifications, the anther appearing as a half-anther)*, usually unappendaged; staminodes sometimes present, these sometimes elongate and alternating with stamens or groups of stamens; pollen grains usually tricolporate, triporate, to multiporate, sometimes distinctly spiny. *Carpels 2 to many, connate; ovary superior, placentation usually axile;* stigma(s) capitate or lobed. Ovules 1 to numerous in each locule, anatropous to campylotropous. **Nectaries composed of densely packed, multicellular, glandular hairs** *on sepals and sometimes on petals or androgynophore. Fruit usually a loculicidal capsule, schizocarp, nut, indehiscent pod, aggregate of follicles*, drupe, or berry; seeds sometimes with hairs or arillate, occasionally winged; embryo straight to curved; endosperm present, often with cyclopropenoid fatty acids (Figures 9.98 and 9.99).

Floral formula:

*, K ⑤, C5 or 0, A ⑤–∞, G ②–∞; capsule, schizocarp,

nut, indehiscent pod, follicles

Distribution: Cosmopolitan.

Genera/species: 204/2330.   **Major genera:** *Hibiscus* (300 spp.), *Sterculia* (250), *Dombeya* (250), *Sida* (200), *Pavonia* (200), *Grewia* (150), *Cola* (125), *Abutilon* (100), *Triumfetta* (100), *Bombax* (60), *Corchorus* (50), and *Tilia* (45). Noteworthy genera of the continental United States and Canada are *Abutilon, Callirhoe, Corchorus, Gossypium, Hibiscus, Kosteletzkya, Malva, Malvastrum, Malvaviscus, Modiola, Pavonia, Sida, Sidalcea, Sphaeralcea, Triumfetta, Tilia,* and *Urena*.

Economic plants and products: Important food plants include *Theobroma cacao* (chocolate, from seeds; Plate 9.15B), *Cola nitida* and *C. acuminata* (cola seeds), *Durio zibethinus* (durian, fruit), and *Hibiscus esculentus* (okra, fruit). A few genera yield valuable timber; balsa wood is from *Ochroma pyramidale*. Hairs associated with the seeds are used as stuffing material (such as kapok, from species of *Ceiba* and *Bombax*) or in fabrics (such as cotton, from species of *Gossypium*). The family contains numerous ornamentals, including *Tilia* (basswood, linden), *Fremontodendron* (flannelbush), *Dombeya, Grewia, Firmiana* (Chinese

parasol tree), *Ceiba, Abutilon, Althaea* (hollyhock), *Hibiscus* (hibiscus), *Pavonia, Malvaviscus* (turk's-cap), *Thespesia* (Portia tree), and *Malva* (mallow).

Discussion: The family is monophyletic and is circumscribed broadly; it usually has been divided into four families (i.e., Tiliaceae, Sterculiaceae, Bombacaceae, and Malvaceae s.s.). Traditional distinctions between these four families are arbitrary and inconsistent (Alverson et al. 1998, 1999 Baum et al. 1998; Bayer et al. 1999; Judd and Manchester 1998), and "Tiliaceae," "Sterculiaceae," and "Bombacaceae" are not monophyletic.

Infrafamilial relationships have been investigated by numerous systematists, and preliminary cladistic analyses based on morphology or DNA have been conducted by Judd and Manchester (1998), La Duke and Doebley (1995), Alverson et al. (1998, 1999), Bayer et al. (1999), and Baum et al. (2004). General phylogenetic relationships are summarized in Figure 9.100. Note that the genera traditionally placed in "Tiliaceae" and "Sterculiaceae" (e.g., *Tilia, Grewia, Triumfetta, Berrya, Theobroma, Byttneria, Sterculia,* and *Walthera*) have retained often numerous stamens with 2-locular anthers. The stamens may be free or connate, and connation appears to have evolved more than once. Evolutionary relationships within these plants are still rather unclear but several monophyletic groups can be discerned (see Figure 9.100)—some of which cannot be diagnosed morphologically. The clade including *Grewia, Leuhea, Corchorus, Triumfetta, Apeiba,* and relatives (Grewioideae) has lost calyx fusion. *Theobroma, Byttneria, Guazuma* and relatives (Byttnerioideae) may be monophyletic due to the shared possession of five elongated staminodes that alternate with the unusually shaped petals (often clawed and basally hooded). Within this group, *Waltheria* and *Melochia* form a clade supported by the reduction of the androecium to only five stamens. *Dombeya*, with its distinctive spiny pollen, also has elongated staminodes that alternate with the petals. The clade containing *Sterculia, Cola, Firmiana,* and relatives (Sterculioideae) is supported by the distinctive unisexual, apetalous flowers with an elongate gynophore and carpels that separate at maturity, forming a cluster of follicles.

The remaining members of the family constitute a monophyletic group, supported by adnation of the androecium to the corolla and the presence of usually half anthers. However, recent molecular evidence suggests that the underlying synapomorphy for this group may be transversely septate, 2-locular anthers that are strongly connate (von Balthazar et al. 2004), which have become developmentally modified in many taxa so as to appear unilocular (opening by a single slit). The genera traditionally placed in "Bombacaceae" form a nonmonophyletic group, although a subgroup of "Bombacaceae"—*Adansonia, Pachira, Ceiba, Bombax, Pseudobombax,* and relatives (Bombacoideae)—may be monophyletic because all have palmately compound leaves and flattened, triangular, nonspiny pollen. The largest group within the half-anther

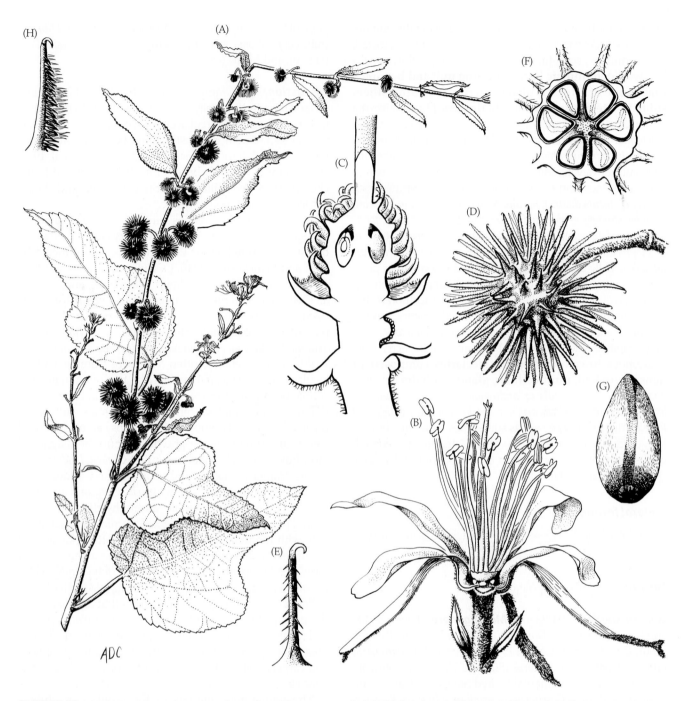

FIGURE 9.98 Malvaceae (representative of complex with separate stamens). (A–G) *Triumfetta semitriloba:* (A) stem with flowers and fruits (× 1.5); (B) flower with two sepals and one petal removed (×17); (C) flower in longitudinal section, with stamens removed (× 50); (D) fruit (× 8.5); (E) fruit prickle (×17); (F) fruit in cross-section, showing seeds (× 11); (G) seed (× 22). (H) *T. pentandra:* fruit prickle (× 17). (From Brizicky 1965, *J. Arnold Arbor.* 46: p. 301.)

clade includes the traditional Malvaceae (Malvoideae). Monophyly of Malvoideae is supported by globose spiny pollen, a staminal tube with five apical teeth, a well-developed epicalyx, and cpDNA restriction sites (La Duke and Doebley 1995). Molecular analyses associate several additional genera (e.g., *Quararibea, Matisia,* and *Fremontodendron*) with this clade. Within Malvoideae, the genera with loculicidal capsules and numerous seeds (e.g., *Hibiscus, Gossypium, Thespesia*) form a basal, paraphyletic complex

(see Figure 9.100). Loss of apical teeth on the staminal column and molecular characters diagnose Malveae (e.g., *Malva, Sida, Callirhoe, Abutilon,* and *Sphaeralcea*). These plants are also characterized by schizocarpic fruits, and only one or two ovules per carpel, but these characters have evolved in parallel in genera such as *Malvaviscus, Pavonia,* and *Urena* (closely related to *Hibiscus*); they have sterile carpels alternating with the fertile ones, making the number of styles twice the number of ovary locules. Many

FIGURE 9.99 Malvaceae (representative of monadelphous, spiny pollen clade). *Kosteletzkya virginica:* (A) flowering branch (× 0.5); (B) flower (× 1); (C) tip of staminal column with protruding styles, note half-stamens and teeth representing staminodes (× 10); (D) ovary in cross-section, each locule with one ovule (× 8); (E) capsule with calyx (× 4); (F) seed (× 7). (From Wood 1974, *A student's atlas of flowering plants*, p. 69.)

genera are nonmonophyletic (Pfeil et al. 2005; Tate et al. 2004).

Flowers of Malvaceae are diverse, and attract bees, wasps, ants, flies, moths, birds, and bats (Plate 9.15A,H). Nectar is the reward, and is usually produced on the inner surface of the connate sepals. Most species are outcrossing. Dispersal is also extremely varied. Capsular-fruited species have small seeds that are wind- or water-dispersed; they sometimes possess specialized dispersal structures such as wings or hairs. Many species with follicles (e.g., *Sterculia*) have seeds that contrast in color with the follicle wall, leading to dispersal by birds, but in *Firmiana* the follicle opens early in development and forms a dry, winglike structure for wind dispersal. Species with schizocarps are self-dispersed or adapted for external dispersal on birds or mammals. The large indehiscent pods of *Adansonia* contain an edible, somewhat sour, dry flesh, and are dispersed by large mammals. The nuts of *Tilia* are borne on a cyme that is usu-ally adnate to a conspicuous winglike bract; the entire infructescence and associated bract are dispersed by wind. Fleshy fruits or arillate seeds are usually dispersed by mammals or birds.

Additional references: Bayer 1998, 1999; Bayer and Kubitzki 2003; Brizicky 1965a, 1966a; Fryxell 1988; van Heel 1966; Whitlock et al. 2001.

Cistaceae A. L. de Jussieu
(Rockrose Family)

Shrubs or herbs, with tannins. *Hairs usually stellate* or peltate scales. Leaves opposite or alternate and spiral, simple, entire, usually pinnately veined, sometimes reduced, with a single vein; stipules present or absent. Inflorescences determinate, sometimes reduced to a single flower, terminal or axillary. Flowers bisexual, radial. *Sepals 5,* **the**

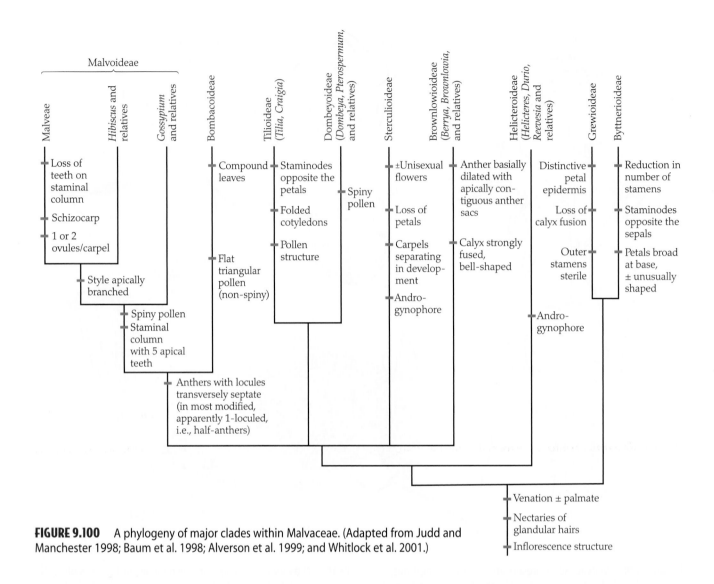

FIGURE 9.100 A phylogeny of major clades within Malvaceae. (Adapted from Judd and Manchester 1998; Baum et al. 1998; Alverson et al. 1999; and Whitlock et al. 2001.)

outer 2 distinctly narrower than the 3 inner, or only 3, distinct to connate. *Petals 5 (–3)*, *distinct*, **usually crumpled**, imbricate, usually convolute. *Stamens usually numerous; filaments distinct*; anthers 2-locular; pollen grains usually tricolporate. *Carpels usually 3*, connate; ovary superior, *placentation parietal*, the placentae often intruded; stigma punctate to capitate, often 3-lobed. Ovules usually 4 to numerous on each placenta, usually orthotropous. Nectar disk present. *Fruit a loculicidal capsule*; embryo variously curved or folded.

Floral formula: *, K (3+2), C5, A∞, G③; capsule

Distribution: Widely distributed in temperate regions, especially diverse in the Mediterranean region; mainly plants of open habitats and sandy or chalky soils.

Genera/species: 8/200. **Major genera:** *Helianthemum* (80 spp.) *Crocanthemum* (24), and *Lechea* (17). The last two occur in the United States and/or Canada, along with *Hudsonia* and *Cistus*.

Economic plants and products: Several genera, including *Cistus* (rockrose), *Helianthemum* (sunrose, rockrose), and *Hudsonia* (beach heather; Plate 9.15E), provide ornamentals.

Discussion: Monophyly of Cistaceae may be supported by the distinctive calyx. They may be most closely related to Dipterocarpaceae, and both families have an imbricate calyx, starchy endosperm, and similar seed coat anatomy (Kubitzki and Chase 2003). Phylogenetic relationships within Cistaceae are poorly understood, but *Fumana* and *Lechea* may be early divergent lineages. The New World genus *Crocanthemum* is very likely paraphyletic (including the specialized *Hudsonia*) and distinct from the Old World *Helianthemum* (Arrington and Kubitzki 2003).

The flowers of Cistaceae may be showy (often bright yellow), attracting various bees, flies, and beetles, or inconspicuous and predominantly selfing. The flowers open during bright sunlight and usually remain open for only a few hours. The small seeds are dispersed by wind or rainwash.

Additional references: Brizicky 1964c; Nandi 1998.

Dipterocarpaceae Blume
(Dipterocarp Family)

Trees; usually with branching resin canals in pith, wood, and bark, mucilage canals sometimes present in the cortex and pith; with tannins, *triterpenes and sesquiterpenes*. Hairs often fascicled, glandular, or peltate scales. *Leaves alternate, often 2-ranked, simple, entire, with pinnate venation, often with domatia; stipules present*. Inflorescences indeterminate or determinate, usually axillary. Flowers usually bisexual, radial, lacking an epicalyx. *Sepals 5, usually connate*, imbricate. *Petals 5, ± distinct*, imbricate and convolute. *Stamens (5-) 10-numerous*, occasionally on an androgynophore; filaments distinct or slightly connate; *anthers 2-locular*, **with terminal appendage developed from the connective**; pollen grains tricolpate or tricolporate. Carpels 2–4 (-5), connate; ovary superior, occasionally nearly inferior, with axile placentation; stigma capitate or lobed. Ovules 1–4 per locule. Nectary glandular. *Fruit usually a nut*, **the calyx persistent, with 2–5 calyx lobes becoming expanded** *and usually winglike*. Seeds lacking endosperm.

> **Floral formula:**
>
> * K ⑤, C5, A ⑩–∞, G ②–④, nut (with winged calyx)

Distribution and ecology: Especially diverse in tropical Asia and Indomalaysia, where they often dominate monsoonal forests and rainforests, but also occurring in Africa and northern South America.

Genera/species: 17/550. **Major genera:** *Shorea* (195 spp.), *Hopea* (100), *Dipterocarpus* (70), and *Vatica* (75). The family does not occur in North America.

Economic plants and products: Several genera, especially *Shorea, Hopea, Dipterocarpus*, and *Vatica*, provide valuable hardwood timber.

Discussion: The monophyly of Dipterocarpaceae (with the possible exception of *Pakaraimaea*) is supported by both morphological and molecular evidence (Dayanandan et al. 1999; Ashton 2003; Kubitzki and Chase 2003). The family is divided into three monophyletic groups, two of which (Monotoideae and Pakaraimoideae) are quite small (totaling about 35 species, taken together), and the third, the Dipterocarpoideae, contain the remaining species. Monotoideae (*Monotes, Marquesia, Pseudomonotes*; occurring in tropical Africa, Madagascar, and Colombia, South America) and Pakaraimoideae (*Pakaraimaea*; restricted to the Guayana highlands in northern South America) have stems lacking resin canals and have mucilage canals in the pith, tricolpate pollen, and equally expanded sepals. Monotoideae are distinguished from Pakaraimoideae in having petals longer (vs. shorter) than the sepals, flowers with an androgynophore (vs. this lacking), and differences in wood anatomy. The large and diverse Dipterocarpoideae are diagnosed by their branching resin canals, lack of mucilage canals in the pith, tricolpate pollen, and two or three calyx lobes conspicuously enlarged and wing-like. Phylogenetic relationships within the Dipterocarpoideae have been investigated by Dayanandan et al. (1999), Yulita et al. (2005), and Gamage et al. (2006).

The flowers are pollinated by a variety of insects. The modified, wing-like calyx lobes allow the nuts to be dispersed only fairly short distances by wind, largely because the fruits are quite heavy.

Additional references: Londoño et al. 1995; Maguire and Ashton 1977; Morton 1995.

Sapindales

Sapindales are clearly monophyletic, as indicated by the synapomorphies of pinnately compound leaves (occasionally becoming palmately compound, trifoliolate, or unifoliolate) and flowers with a distinct nectar disk. They are woody, with usually alternate and spiral exstipulate leaves and often rather small 4- or 5-merous flowers with imbricate perianth parts. The order's monophyly is strongly supported by analyses based on *rbcL, atpB*, and 18S sequence characters (Chase et al. 1993; Gadek et al. 1996; Hilu et al. 2003; Källersjö et al. 1998; Savolainen et al. 2000a,b; Soltis et al. 1998, 2000). The order consists of 9 families and about 5800 species; major families include **Anacardiaceae, Burseraceae, Meliaceae, Rutaceae, Sapindaceae**, and **Simaroubaceae**.

Based on DNA sequences (Gadek et al. 1996; Soltis et al. 2000), Anacardiaceae and Burseraceae clearly form a clade, which is also supported by the presence of resin canals and biflavonoids in the leaves; a Meliaceae + Rutaceae + Simaroubaceae clade is supported by DNA sequences and the presence of bitter triterpenoids (Figure 9.101). DNA-based studies indicate that Simaroubaceae, as traditionally circumscribed, is polyphyletic, and this family is here narrowly defined. Picramniaceae (usually included within Simaroubaceae) does not belong in the order (Fernando et al. 1995; Fernando and Quinn 1995; Soltis et al. 1998, 2000), and instead is a basal rosid.

dots on leaves (oils)

Rutaceae A. L. de Jussieu
(Citrus or Rue Family)

Usually trees or shrubs, sometimes with thorns, spines, or prickles; *usually with triterpenoid bitter substances*, alkaloids, and phenolic compounds; **with scattered secretory cavities (pellucid dots) containing aromatic ethereal oils**. Hairs various. *Leaves alternate and spiral*, or opposite, rarely whorled, *usually pinnately compound, or reduced and trifoliolate or unifoliolate*, occasionally palmately compound, *leaflets pellucid-dotted, especially near margin*, entire to crenate, with pinnate venation; stipules lacking. Inflorescences usually determinate, occasionally reduced to a single flower, terminal or axillary. Flowers bisexual or unisexual (plants then monoecious to dioecious), usually radial. *Sepals usually 4 or 5*, distinct to slightly connate basally. *Petals usually 4 or 5, distinct* or

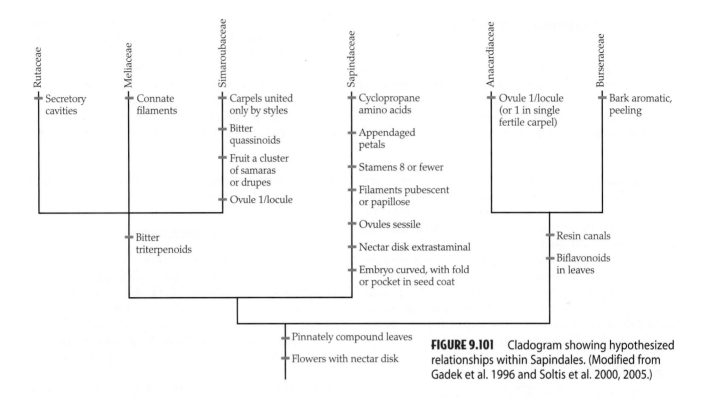

FIGURE 9.101 Cladogram showing hypothesized relationships within Sapindales. (Modified from Gadek et al. 1996 and Soltis et al. 2000, 2005.)

sometimes connate, usually imbricate. Stamens usually 8–10, sometimes numerous; filaments usually distinct but sometimes basally connate, glabrous or pubescent; pollen grains usually 3–6-colporate. *Carpels usually 4 or 5 to many, usually completely connate and with a single common style*, but occasionally with ovaries distinct and carpels only ± coherent by their styles; ovary superior, usually with axile placentation; stigmas various. Ovules 1 to several in each locule. *Nectar disk present, intrastaminal. Fruit a drupe, capsule, samara, cluster of follicles, or variously developed berry* (i.e., with mesocarp homogeneous or distinctly heterogeneous with a hard to leathery rind, with or without distinct partitions, and flesh or pulp derived from the ovary wall or from multicellular, juice-filled hairs); embryo straight to curved; endosperm present or lacking (Figure 9.102).

Floral formula:

*, K $\widehat{(4-5)}$, C $\widehat{(4-5)}$, A $\widehat{(4-\infty)}$, G $\underline{\widehat{(4-\infty)}}$; berry, drupe, samara,

cluster of follicles, capsule

Distribution: Nearly cosmopolitan, but mainly tropical and subtropical.

Genera/species: 155/930. **Major genera:** *Zanthoxylum* (200 spp.), *Agathosma* (180), and *Ruta* (60). Noteworthy genera in the continental United States and/or Canada include *Zanthoxylum, Amyris, Ptelea, Cneoridium, Poncirus*, and *Citrus*.

Economic plants and products: Several species of *Citrus* (oranges, tangerines, grapefruits, limes, lemons) are prized for

their edible fruits (Plate 9.16H). Fruits of *Fortunella* (kumquat) and *Casimiroa* (white zapote) are also eaten. *Ruta* (rue), *Zanthoxylum* (toothache tree; Plate 9.16B), *Citrus*, and *Casimiroa* are used medicinally. The family contains many ornamentals, including *Murraya* (orange-jasmine), *Phellodendron* (cork tree), *Poncirus* (trifoliate orange), *Severinia* (boxthorn), and *Triphasia* (limeberry). The resinous compounds characteristic of Rutaceae are flammable, and thus the wood of a few genera, such as *Amyris* (torchwood), is used for fuel and torches. *Zanthoxylum* and *Chloroxylon* yield furniture timbers.

Discussion: Rutaceae are broadly defined and their monophyly is evidenced by the presence of cavities containing ethereal oils, and appearing as pellucid dots in the mesophyll and other soft tissues, as well as *rbcL* and *atpB* sequences (Gadek et al. 1996; Morton et al. 2003; Chase et al. 1999). Traditionally recognized subfamilies are separated mainly by fruit type, carpel number, and extent of carpel connation. It is unlikely that any of these represent monophyletic groups (Morton et al. 1996; Chase et al. 1999) except for Aurantioideae (Citroideae, incl. *Citrus, Fortunella, Poncirus, Severinia, Atalantia, Aegle*, and relatives), a group characterized by globose berries and a base chromosome number of nine. Generic delimitations within this group are often problematic, especially those with berries with juice vesicles (e.g., *Citrus, Poncirus, Fortunella, Eremocitrus*, and *Microcitrus*). A recircumscribed Toddalioideae (incl. *Severinia, Zanthoxylum, Amyris, Phellodendron*, and relatives) may also be monophyletic, and these plants are characterized by fruits with a fibrous endocarp, usually forming follicles, capsules, or drupes with from one to several pits.

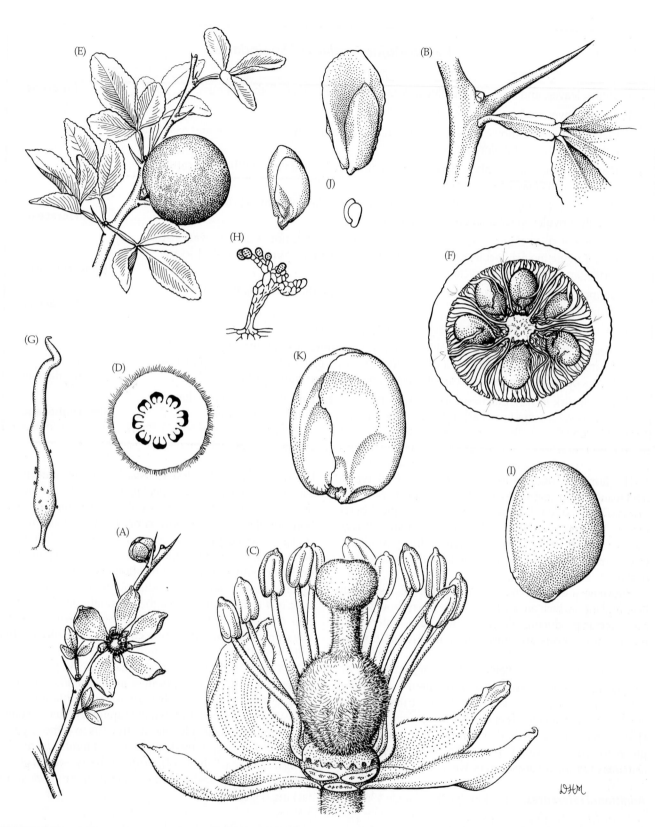

FIGURE 9.102 Rutaceae. *Poncirus trifoliata:* (A) flowering branch
(× 0.75); (B) portion of twig, showing winged petiole with base
of leaf blades, axillary thorn, and bud (× 1.5); (C) flower with two
petals, one sepal, and several stamens removed, showing stamens, nectar disk, and gynoecium (× 6); (D) ovary in cross-section (× 9); (E) fruiting branch (× 0.75); (F) cross-section of mature
berry, showing seeds embedded among pulp vesicles
(× 1.5); (G) pulp vesicle with minute multicellular lateral appendages (× 4.5); (H) appendage of pulp vesicle (greatly magnified);
(I) seed (× 4.5); (J, K) four of nine embryos from a single seed
(resulting from agamospermy) (× 6). (From Brizicky 1962,
J. Arnold Arbor. 43: p. 16.)

Key to Major Families of Sapindales

1. Stems, leaves, and fruits with pellucid dots containing aromatic ethereal oils ....**Rutaceae**

1. Stems, leaves, and fruits lacking pellucid dots 2

2. Plants strongly resinous, with vertical intercellular resin canals in the bark and associated with the phloem of major leaf veins 3

2. Plants lacking resin canals in bark and phloem of major leaf veins (although scattered secretory cells may be present) 4

3. Resins ± spicy-fragrant, not allergenic; bark smooth, often shredding; ovary with usually 2 ovules in each locule; placentation axile ....**Burseraceae**

3. Resins not spicy-fragrant, often allergenic; bark various, but not as above; ovary with only 1 ovule in each locule, and placentation axile, or more commonly only 1 locule fertile, and placentation apical ....**Anacardiaceae**

4. Stamens usually connate by their filaments, but if free, then fruit a capsule with winged seeds ....**Meliaceae**

4. Stamens distinct; fruit not a capsule with winged seeds 5

5. Nectar disk usually extrastaminal; ovules lacking a funiculus and broadly attached to an outgrowth of the placenta (obturator); stamens usually pubescent or papillose; leaves opposite or alternate; carpels persistently connate; bark not bitter ....**Sapindaceae**

5. Nectar disk intrastaminal; ovules with a funiculus; stamens glabrous; leaves alternate; carpels separating from each other after pollination; bark often very bitter ....**Simaroubaceae**

The family is notable for its variation in fruit types, which include a leathery-rinded berry ("hesperidium"; *Citrus, Poncirus*), hard-rinded berry (*Aegle*), typical berry with ± homogeneous flesh (*Triphasia*), samara (*Ptelea*), drupe (*Phellodendron, Amyris*), drupe with several pits (*Casimiroa*), cluster of follicles (*Zanthoxylum*), and capsule (*Ruta*).

Rutaceae are pollinated mainly by insects, especially bees and flies, which are attracted to the often showy, odor- and nectar-producing flowers. Most are outcrossing because the flowers are unisexual or, in bisexual flowers, because stigmas and anthers are separated physically or mature at different times. Selfing may occur as well. Asexual reproduction by agamospermy is frequent in *Zanthoxylum, Murraya, Poncirus*, and *Citrus*. Fleshy-fruited species are dispersed by mammals or birds. *Zanthoxylum*, with follicles of shiny black seeds that contrast in color with the green to red fruit wall, is also often bird-dispersed. The samaras of *Ptelea* are dispersed by wind.

Additional references: Brizicky 1962b; Swingle 1967.

Meliaceae A. L. de Jussieu
(Mahogany Family)

Trees or shrubs; commonly producing bitter triterpenoid compounds, usually with scattered secretory cells, *and inner bark often reddish*. Hairs usually simple, occasionally stel-

late or peltate scales. *Leaves usually alternate and spiral, once (or twice) pinnately compound*, occasionally trifoliolate or unifoliolate, leaflets usually entire, with pinnate venation; stipules lacking. Inflorescences usually determinate, axillary or less commonly terminal. *Flowers usually unisexual* (plants then monoecious, dioecious, or polygamous), but often with well-developed staminodes or pistillodes, radial. *Sepals usually 4 or 5*, distinct to ± connate. *Petals usually 4 or 5*, distinct or slightly connate basally, imbricate, convolute, or valvate. *Stamens usually 4–10, occasionally more numerous;* **filaments connate, forming a tube, with or without apical appendages**, but separate filaments in *Cedrela* (a reversal), glabrous or pubescent; pollen grains 2- to 5-colporate. *Carpels usually 2–6*, connate; ovary superior, usually with axile placentation; stigma variously shaped, but usually capitate. Ovules usually 2 to numerous in each locule, anatropous to orthotropous. *Nectar disk present, intrastaminal. Fruit a loculicidal or septifragal capsule* (Plate 9.16J), *drupe*, or berry; *seeds dry and winged* or with a fleshy seed coat; endosperm present or lacking (Figure 9.103).

Floral formula:

*, K (4–5), C4–5, A (4–10), G (2–6); capsule, drupe

Distribution: Widespread in tropical and subtropical regions.

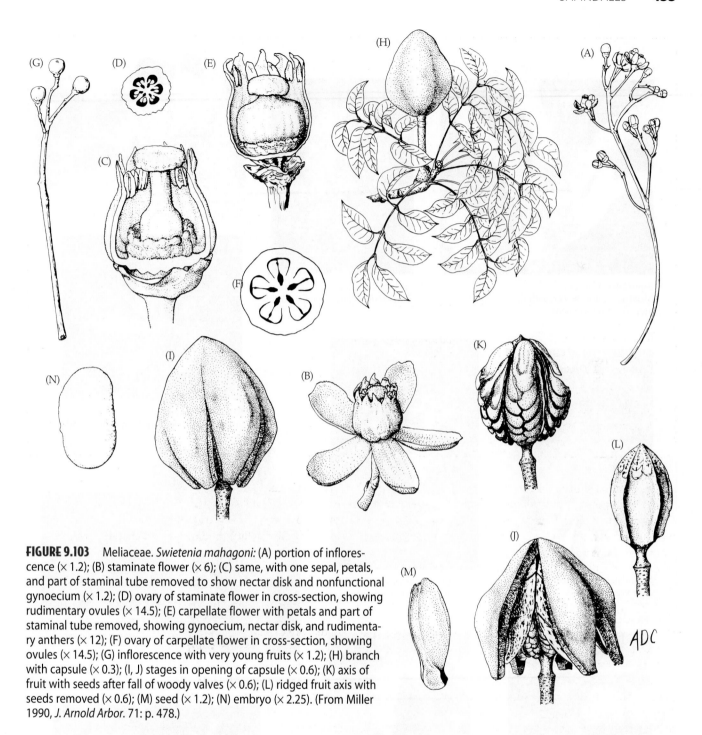

FIGURE 9.103 Meliaceae. *Swietenia mahagoni:* (A) portion of inflorescence (× 1.2); (B) staminate flower (× 6); (C) same, with one sepal, petals, and part of staminal tube removed to show nectar disk and nonfunctional gynoecium (× 1.2); (D) ovary of staminate flower in cross-section, showing rudimentary ovules (× 14.5); (E) carpellate flower with petals and part of staminal tube removed, showing gynoecium, nectar disk, and rudimentary anthers (× 12); (F) ovary of carpellate flower in cross-section, showing ovules (× 14.5); (G) inflorescence with very young fruits (× 1.2); (H) branch with capsule (× 0.3); (I, J) stages in opening of capsule (× 0.6); (K) axis of fruit with seeds after fall of woody valves (× 0.6); (L) ridged fruit axis with seeds removed (× 0.6); (M) seed (× 1.2); (N) embryo (× 2.25). (From Miller 1990, *J. Arnold Arbor.* 71: p. 478.)

Genera/species: 51/550. **Major genera:** *Aglaia* (100 spp.), *Trichilia* (66), *Turraea* (65), *Dysoxylum* (61), and *Guarea* (35). *Melia* and *Swietenia* occur in the continental United States.

Economic plants and products: The family is important chiefly as a source of valuable timber, which comes from *Swietenia* (mahogany), *Cedrela* (West Indian cedar), *Entandrophragma* (sipo, kosipo), and *Khaya* (African mahogany). *Azadirachta indica* (neem tree) is important medicinally and as a source of insecticides. *Melia azedarach* (chinaberry; Plate 9.16A) and *A. indica* are ornamental trees.

Discussion: Monophyly of Meliaceae is supported by analyses of DNA sequences (Gadek et al. 1996; Muellner et al. 2003) and morphology. All genera belong to either the Melioideae, with nonwinged seeds (in capsules, berries, or drupes), secondary xylem with one or two seriate rays, and naked buds; or to the Swietenioideae, with winged seeds (in capsules), secondary xylem with three to six seriate rays, and buds with scales (Pennington and Styles 1975). Swietenioideae, containing genera such as *Swietenia* and *Cedrela*, are monophyletic because of their distinctive flattened/winged seeds (in capsules) and scaly buds. *Cedrela* is dis-

(A)

Sapindales: Meliaceae
Melia azedarach: flower, with
monadelphous stamens

(B)

Sapindales: Rutaceae
Zanthoxylum clava-herculis: flowers

(C)

Sapindales: Sapindaceae
Blighia sapida: fruits with arillate seeds

(D)

Sapindales: Sapindaceae
Acer negundo: fruits

(E)

Sapindales: Sapindaceae
Koelreuteria paniculata: flowers

(F)

Sapindales: Simaroubaceae
Ailanthus altissima: fruits and compound leaf

(G)

B. simarouba:
staminate flowers

Sapindales: Burseraceae
Bursera simarouba: bark

(H)

Sapindales: Rutaceae
Citrus aurantiifolia: fruits and branch
with leaves

(I)

Sapindales: Anacardiaceae
Rhus glabra: leaves and fruits

(J)

Sapindales: Meliaceae
Cedrela toona: fruits

PLATE 9.16 Eudicots: Sapindales

tinct due to its separate filaments (a reversal) and erect petals. Melioideae, represented by genera such as *Trichilia, Guarea,* and *Melia,* are diverse morphologically. *Melia* and *Azadirachta* share drupaceous fruits, a likely synapomorphy. The leaves of *Guarea* (and the related *Chisocheton*) are unusual in that the apex is meristematic and continues to produce new leaflets for several years.

Bees and moths are the most important pollinators of the small, nectar-producing flowers of Meliaceae. Outcrossing is enforced by unisexuality. Genera with drupes or capsular fruits and colorful fleshy seeds (e.g., *Trichilia*) are mainly dispersed by birds and various mammals (including bats). Winged seeds, as in *Swietenia* and *Cedrela,* are wind-dispersed.

Additional reference: Miller 1990.

Simaroubaceae A. P. de Candolle
(Tree of Heaven Family)

Trees or shrubs, occasionally thorny; *scattered secretory cells often present in the leaves and bark, the pith conspicuous,* **with bitter triterpenoid compounds of the quassinoid type**. Hairs usually simple. *Leaves alternate and spiral, pinnately compound* to unifoliolate, leaflets entire to serrate, with pinnate venation (Plate 9.16F); stipules usually lacking. Inflorescences determinate, terminal or axillary; catkins in *Leitneria. Flowers unisexual* (plants monoecious or rarely dioecious), but staminodes and pistillodes often well developed, radial. *Sepals 4 or 5,* but minute or lacking in *Leitneria,* distinct to slightly connate. *Petals usually 5,* distinct, rarely lacking (*Leitneria*), imbricate or valvate. Stamens usually 10, but reduced to 4 in *Leitneria,* in which they appear to be numerous due to clustering of 3 reduced flowers; filaments distinct, often basally appendaged; pollen grains tricolporate. *Carpels usually 5, but only 1 in Leitneria,* **± united only by their styles**; *ovary superior, with axile placentation, but separating into individual carpels as the fruits develop;* stigma capitate to strongly lobed. **Ovules 1 per locule**. *Nectar disk present, intrastaminal,* but lost in *Leitneria.* **Fruit a cluster of samaras** (Plate 9.16F) **or ± dry to fleshy drupes**; endosperm ± lacking.

> **Floral formula:**
> Staminate: *, K (4–5), C5, A10, G⑤•
> Carpellate: *, K (4–5), C5, A10•, G⑤; cluster of
> samaras or drupes

Distribution: Widespread in tropical or subtropical regions, with a few genera extending into temperate habitats.

Genera/species: 21/100. **Major genera:** *Simaba* (30 spp.), *Ailanthus* (15), and *Castela* (12). The family is represented in the continental United States by *Castela, Simarouba, Leitneria,* and *Ailanthus.*

Economic plants and products: *Ailanthus* (tree of heaven) and *Simarouba* (paradise tree) are often cultivated as ornamentals. Several genera are used medicinally.

Discussion: Simaroubaceae are considered monophyletic on the basis of morphology and *rbcL* sequences (Fernando et al. 1995; Gadek et al. 1996). Anatomical data are consistent with the hypothesis that the reduced, wind-pollinated flowers of *Leitneria* evolved from flowers similar to those of other Simaroubaceae, and the placement of this genus is also supported by serological data.

Kirkia, Picramnia, Suriana, and *Alvaradoa* were previously included within Simaroubaceae, resulting in a heterogeneous, clearly polyphyletic assemblage (Fernando et al. 1993, 1995; Gadek et al. 1996; Fernando and Quinn 1995; Savolainen et al. 2000b). These genera should be treated in the families Kirkiaceae (*Kirkia*)—Sapindales; Picramniaceae (*Alvaradoa, Picramnia*)—basal rosid; and Surianaceae (*Suriana* and its relatives)—Fabales.

The flowers of Simaroubaceae are pollinated by various insects (especially bees) and birds. The flowers of *Leitneria* are wind-pollinated. Samaras (as in *Ailanthus*) are wind-dispersed (Plate 9.16F), while the drupes of *Simarouba* are bird-dispersed.

Additional references: Abbe and Earle 1940; Channell and Wood 1962; Cronquist 1944; Petersen and Fairbrothers 1983.

Anacardiaceae R. Brown
(Sumac or Poison Ivy Family)

Trees, shrubs, or lianas; *usually with tannins; well-developed vertical resin canals in the bark and associated with the larger veins of the leaves, often also in other parenchymatous tissues, the resin clear when fresh but drying black, often causing dermatitis.* Hairs various. *Leaves usually alternate and spiral, pinnately compound, but sometimes trifoliolate or unifoliolate,* leaflets entire to serrate, with pinnate venation; stipules ± lacking. Inflorescences determinate, terminal or axillary. *Flowers almost always unisexual (plants usually dioecious),* radial, small, often with well-developed staminodes or carpellodes. *Sepals usually 5,* distinct to slightly connate. *Petals usually 5,* distinct or slightly connate, ± imbricate. Stamens 5–10, occasionally more numerous or reduced to a single fertile stamen; filaments usually glabrous, usually distinct; pollen grains usually tricolporate or triporate. *Carpels typically 3,* sometimes 5, variously connate; ovary usually superior, sometimes all carpels fertile and gynoecium multilocular with axile placentation, *more commonly only 1 carpel fully developed and fertile (and others typically represented merely by their styles) and gynoecium ± asymmetrical and unilocular with apical placentation;* stigmas usually capitate. **Ovules 1 in each locule, or 1 in the single fertile carpel**. *Nectar disk present, usually intrastaminal. Fruit an often flattened asymmetrical drupe;* embryo curved to straight; endosperm scanty to lacking (Figure 9.104).

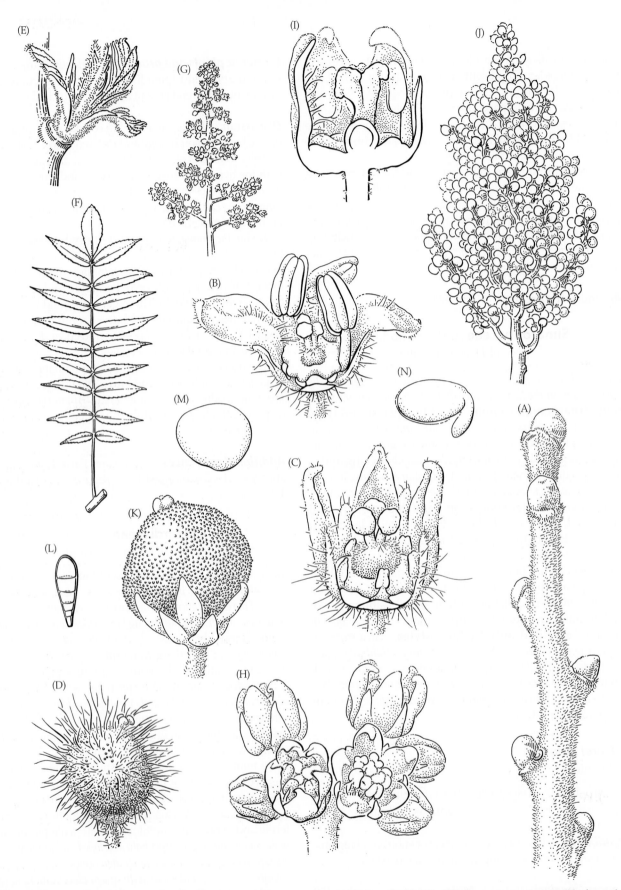

FIGURE 9.104 Anacardiaceae. (A–D) *Rhus typhina:* (A) twig in winter condition (× 1.25); (B) staminate flower (× 10); (C) carpellate flower (× 10); (D) drupe (× 7). (E–N) *R. glabra:* (E) bud, breaking dormancy (× 1.25); (F) leaf (× 0.2); (G) inflorescence (× 0.5); (H) cymose cluster of carpellate flowers (× 7); (I) carpellate flower in longitudinal section (× 10); (J) infructescence (× 0.8); (K) drupe (× 7); (L) glandular hair from fruit (greatly magnfied); (M) pit (× 10); (N) embryo (× 10). (Original art prepared for the Generic Flora of the Southeastern U.S. Project. Used with permission.)

Floral formula:

Staminate: *, K (5), C5, A5–10, G (3(–5))•

Carpellate: *, K (5), C5, A5–10•, G (3(–5)); drupe

Distribution: Mainly pantropical, with a few species in temperate regions.

Genera/species: 70/600. **Major genera:** *Rhus* (100 spp.), *Semecarpus* (50), *Lannea* (40), *Toxicodendron* (30), *Schinus* (30), and *Mangifera* (30). Noteworthy genera of the continental United States and/or Canada include *Cotinus*, *Metopium*, *Rhus*, *Schinus*, and *Toxicodendron*.

Economic plants and products: Fruits of *Mangifera indica* (mango) and *Spondias* (mombin, hog plum) are eaten, as are the roasted seeds of *Anacardium occidentale* (cashew) and *Pistacia vera* (pistachio). Fruits of several species of *Rhus* are used in drinks. A black lacquer is obtained from *Toxicodendron vernicifluum* (varnish tree). A few are ornamentals, including *Cotinus* (smoke tree), *Rhus* (sumac), and *Schinus* (Brazilian pepper). Some species, notably in *Astronium* and *Loxopterygium*, provide decorative timbers. Finally, the group is of medical significance because so many of its taxa, particularly *Toxicodendron* (poison ivy, poison oak, poison sumac) and *Metopium* (poison wood), promote dermatitis in susceptible individuals due to the phenolic compound 3-*n*-pentadecycatechol in the resin. It is worth noting that mangoes and cashews, even though edible, may still cause an allergic reaction.

Discussion: Anacardiaceae and Burseraceae both have resin canals, biflavones, and clearly form a clade based on their DNA sequences (Gadek et al. 1996; Savolainen et al. 2000b; Soltis et al. 2000). Anacardiaceae are tentatively considered to be monophyletic on the basis of a reduced number of ovules, other morphological features, and *rbcL* sequences (Gadek et al. 1996; Terrazas and Chase 1996).

The family is composed of two major subclades. Spondiadeae, which have retained many plesiomorphic features such as gynoecia with usually five carpels, multilocular ovaries, and fruits with a thick endocarp usually composed of lignified and irregularly oriented sclereids, may form a clade based on their septate fibers (Terrazas and Chase 1996). We note, however, that the group is often considered to be paraphyletic (Wannan and Quinn 1990, 1991). The remaining genera of the family form a large clade, many members of which have gynoecia with three (or fewer) carpels, unilocular ovaries with apical placentation, and fruits with an endocarp that is composed of discrete and regularly arranged layers of cells.

Rhus and *Toxicodendron* have often been confused, and some botanists have united these two genera (and several others). Fruits of *Rhus* are glandular-pubescent and red (Plate 9.16I), while those of *Toxicodendron* are glabrous and greenish to white. In addition, the resins of *Rhus* are not poisonous, while those of *Toxicodendron* cause a "poison ivy" rash. If combined, the resulting group would not be monophyletic.

The small, nectar-secreting flowers of Anacardiaceae are pollinated by various insects. Outcrossing is promoted by the more or less dioecious condition of members of this family. The large to small drupes are dispersed by various birds or mammals (including bats).

Additional references: Brizicky 1962a; Gillis 1971.

Burseraceae Kunth
(Frankincense Family)

Trees or occasionally shrubs; **bark aromatic, smooth and often peeling** (Plate 9.16G); *well-developed vertical resin canals in the bark and associated with the larger veins of the leaves, the resin clear, not allergenic.* Indumentum various. Leaves usually alternate and spiral, usually pinnately compound, but sometimes trifoliolate or unifoliolate, the leaflets entire to serrate, with pinnate venation; stipules ± lacking. Inflorescences usually determinate, axillary. *Flowers usually unisexual (and plants usually dioecious),* radial, small, and often with well-developed staminodes or carpellodes. *Sepals 4–5,* usually slightly connate, usually imbricate, usually deciduous. *Petals 4–5,* distinct, usually imbricate. *Stamens in 2 or 1 whorls, twice as many as or equaling the petals;* filaments usually distinct, usually glabrous; anthers 2-loculed, opening by longitudinal slits; pollen grains usually tricolporate. Carpels 3–5, connate; ovary superior, with axile placentation; style 1; stigma 1, capitate to lobed. *Ovules 2 in each locule,* anatropous to campylotropous. *Nectar disk present, usually intrastaminal. Fruit a drupe with 1–5 pits, often with dehiscent valves;* embryo straight to curved; endosperm ± lacking.

Floral formula:

Staminate: * K (4–5), C4–5, A4–10, G (3–5)•

Carpellate: * K (4–5), C4–5, A4–10•, G (3–5), drupe, dehiscent drupe

Distribution: Mainly pantropical; especially diverse in tropical America and Africa.

Genera/species: 17/500. **Major genera:** *Bursera* (100 spp.), *Commiphora* (100), *Protium* (80), and *Canarium* (75). The family is represented in the continental United States only by *Bursera*.

Economic plants and products: Frankincense comes from *Boswella carteri* and myrrh from *Commiphora habessinica* (and relatives). Some species of *Bursera* (gumbo-limbo) are used as ornamentals. Several genera provide timber species.

Discussion: Monophyly of Burseraceae is unclear, but has received preliminary support from a cladistic analysis of *rbcL* sequences (Gadek et al. 1995). The smooth bark is distinctive and may be synapomorphic. The small nectar-secreting flowers of Burseraceae are insect pollinated and the fleshy fruits are dispersed by birds (Plate 9.16G).

Additional reference: Brizicky 1962c.

Sapindaceae A. L. de Jussieu
(Soapberry Family)

Trees, shrubs, or lianas with tendrils; often with tannins, usually with triterpenoid saponins in secretory cells, **with a diverse array of cyclopropane amino acids**. Hairs various. *Leaves alternate and spiral, or opposite, pinnately or palmately compound, trifoliolate, or unifoliolate,* leaflets serrate or entire, with pinnate or palmate venation; stipules lacking or present. Inflorescences determinate, terminal or axillary. *Flowers usually unisexual* (plants monoecious, ± dioecious, or polygamous), radial to bilateral. *Sepals usually 4 or 5,* distinct or sometimes basally connate. *Petals usually 4 or 5,* sometimes lacking, distinct, *often clawed,* **with ± basal appendages on adaxial surface** (these lost in a few), imbricate. **Stamens 8 or fewer;** *filaments distinct,* **usually pubescent or papillose;** pollen grains usually tricolporate, the furrows sometimes fused with each other. **Carpels 2 or 3,** connate; ovary superior, usually with axile placentation; stigmas 2 or 3, minute to expanded. Ovules 1 or 2 in each locule, anatropous to orthotropous, **lacking a funiculus and broadly attached to a protruding portion of the placenta (the obturator). Nectar disk present, usually extrastaminal,** but sometimes with stamens borne upon it and ± intrastaminal. Fruit usually a loculicidal, septicidal, or septifragal capsule, arilloid berry, or schizocarp that splits into samara-like or sometimes drupelike segments, rarely a 1-seeded berry or drupe; *seeds often with an aril-like coat;* **embryo variously curved and with radicle separated from rest of embryo by deep fold or pocket in seed coat;** endosperm usually lacking (Figures 9.105 and 9.106).

Floral formula:

Staminate: *, or X, (4–5), 4–5, 4–8, 0 or (2–3)•

Carpellate: *, or X, (4–5), 4–5, 4–8•, (2–3); capsule, arilloid

berry, drupaceous or samaroid schizocarp

Distribution: Mainly tropical and subtropical, with a few genera most diverse in temperate regions.

Genera/species: 147/2215. **Major genera:** *Serjania* (220 spp.), *Paullinia* (150), *Acer* (110), and *Allophylus* (100). Noteworthy genera occurring in the continental United States and/or Canada are *Acer, Aesculus, Cardiospermum, Cupania, Dodonaea, Exothea, Hypelate, Koelreuteria, Sapindus,* and *Serjania.*

Economic plants and products: The family contains several members that are sources of important tropical fruits, such as *Euphoria* (longan), *Litchi* (lychee), and *Nephelium* (rambutan), in which the flesh is derived from a large aril. The arils of *Blighia* (akee; Plate 9.16C) are also eaten but are extremely poisonous if unripe. The seeds of *Melicoccus* (genip) have an edible fleshy coat, and those of *Paullinia cupana* (guarana) are used to make a caffeine-rich beverage. *Acer saccharum* (sugar maple) provides sugar and syrup. The fruits of species of *Sapindus* may be used as a natural soap due to the presence of saponins, and these poisonous compounds have led to the use of crushed branches and fruits of several genera as fish poisons. The group contains numerous ornamentals, including *Acer* (maple), *Aesculus* (horse chestnut), *Cardiospermum* (balloon vine), *Harpullia,* and *Koelreuteria* (goldenrain tree; Plate 9.16E). *Acer* and *Aesculus* also provide timber.

Discussion: Monophyly of Sapindaceae is supported by morphology and by DNA sequences (Gadek et al. 1996; Harrington et al. 2005; Judd et al. 1994; Soltis et al. 2000). The presence of hypoglycin, an unusual and toxic non-protein amino acid, may be another synapomorphic feature of the group. The group is defined broadly, including both Aceraceae (maples and relatives) and Hippocastanaceae (horse chestnuts and relatives). Morphological and molecular data suggest that if these are excluded, the remaining genera form a paraphyletic complex.

Infrafamilial phylogenetic relationships have been investigated using morphological data (Judd et al. 1994; Muller and Leenhouts 1976; Wolfe and Tanai 1987) and *rbcL* and *matK* sequences (Harrington et al. 2005). Four well-supported clades can be recognized. The first is the hippocastanoid clade (i.e., *Aesculus* and relatives), which is united by palmately compound leaves, petals with marginal appendages, usually seven stamens, and large, leathery capsules that have sclerotic inclusions in the pericarp and open to release a single large seed. The second clade includes *Acer* and *Dipteronia* (traditional Aceraceae) and is diagnosed by nonappendaged petals and more or less papillose stamens borne on a nectar disk. Molecular data suggest that these two clades are sister taxa, and both have opposite leaves and diffuse porous wood. The third is the dodonaeoid clade, which includes *Hypelate, Filicium, Harpullia, Dodonaea,* and relatives. These plants usually have two or more ovules per locule. The final, or sapindoid, clade is composed of the remaining genera, e.g., *Cupania, Cupaniopsis, Euphoria, Sapindus, Blighia, Litchi, Thouinia, Serjania, Allophylus,* and *Koelreuteria*. Nearly all of these genera have the ovules basally attached and reduced to one per carpel (but *Koelreuteria* has two ovules per locule, with only one reaching maturity). Within the sapindoids, an especially derived group (e.g., *Thouinia, Serjania, Cardiospermum, Allophylus, Paullinia, Bridgesia, Diatenopteryx*) form the thouinioid clade; they are characterized by samaroid-schizocarpic fruits and bilateral flowers with a unilateral nectar disk. Many (such as *Cardiospermum, Paullinia* and *Serjania*) are tendril-climbers with stipules. Morphology-based analyses suggest a rela-

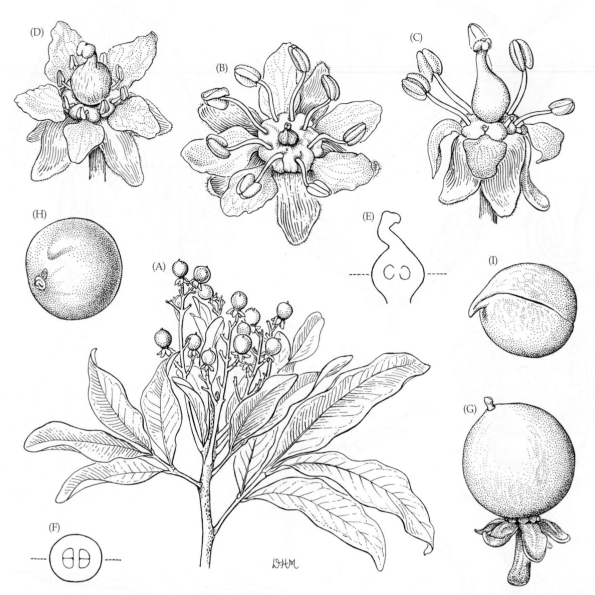

FIGURE 9.105 Sapindaceae (representative of dodonaeoid clade). *Exothea paniculata:* (A) branch with immature fruits (× 0.75); (B) staminate flower, showing nectar disk and pistillode (× 6); (C) bisexual flower, two stamens removed (× 6); (D) carpellate flower, showing staminodes (× 6); (E) gynoecium in longitu-dinal section, in plane marked by dashed line in F (× 9); (F) gynoecium in cross-section, in plane marked by dashed line in E (× 9); (G) fruit (× 3.5); (H) seed (× 3.5); (I) embryo (× 3.5). (From Brizicky 1963, *J. Arnold Arbor.* 44: p. 480.)

tionship between those genera with samaroid-schizocarps and *Acer* + *Dipteronia*, while DNA-based analyses suggest that samaroid fruits have evolved separately in the two clades.

The flowers of Sapindaceae vary from small and radially symmetrical to fairly large, showy, and bilaterally symmetrical; they are pollinated by birds and a wide variety of insects, which are rewarded by floral nectar. *Dodonaea* and some species of *Acer* are wind-pollinated. The dioecious condition enforces outcrossing. Dispersal varies widely. Many tropical groups, such as *Blighia*, *Harpullia*, and *Cupania*, have capsules contrasting in color with the often arillate seeds and are dispersed by birds or mammals. *Litchi*, *Nephelium*, and *Euphoria* have indehiscent fruits, each containing a single seed associated with a fleshy aril, and are also animal-dispersed. Others, such as *Koelreuteria* and *Dodonaea*, have dry inflated or winged capsules that are wind-dispersed. Wind dispersal characterizes the samaroid clade, in which the winged mericarps spin like a propeller as they move through the air (Plate 9.16D).

Additional references: Baas et al. 2003; Brizicky 1963; Hardin 1957; Umadevi and Daniel 1991; van der Pijl 1957.

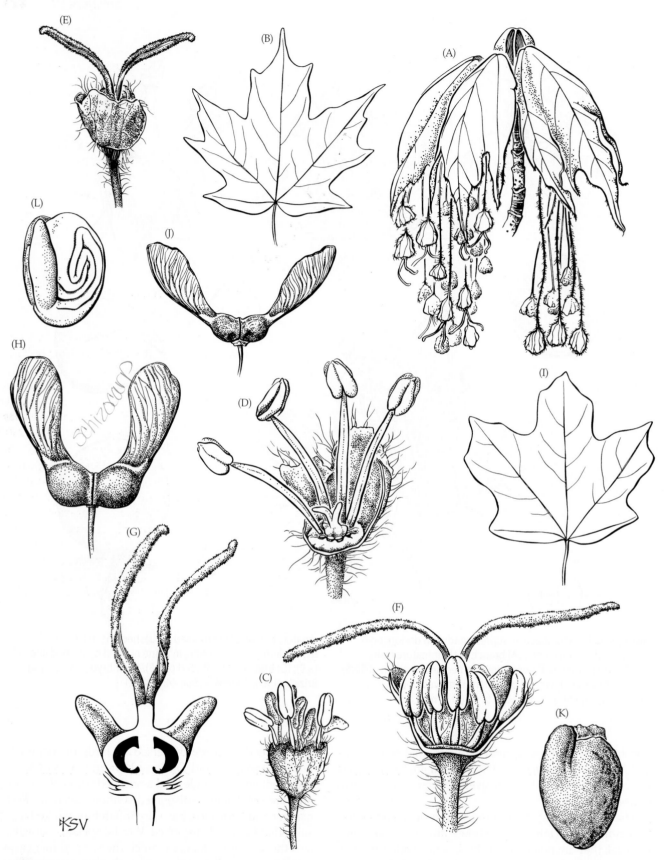

FIGURE 9.106 Sapindaceae (representatives of aceroid clade). (A–H) *Acer saccharum* ssp. *saccharum:* (A) branch with flowers and expanding leaves (× 1.5); (B) leaf (× 0.5); (C) staminate flower (× 6); (D) staminate flower in partial section, showing nectar disk and rudimentary gynoecium (× 9); (E) carpellate flower (× 6); (F) carpellate flower in partial section, note nonfunctional stamens (× 9); (G) gynoecium in partial longitudinal section (× 9); (H) samaroid schizocarp (× 1.5). (I–L) *A. saccharum* ssp. *floridanum:* (I) leaf (× 0.5); (J) samaroid schizocarp (× 1.5); (K) seed (× 6); (L) embryo (× 6). (From Wood 1974, *A student's atlas of flowering plants*, p. 68.)

ASTERID CLADE (SYMPETALAE)

This large and specialized subgroup of the tricolpate clade is hypothesized to be monophyletic on the basis of *rbcL, atpB, ndhF, matK,* and 18S rDNA sequence characters (Albach et al. 2001a,b; Bremer et al. 2002; Chase et al. 1993; Hilu et al. 2003; Olmstead et al. 2000; Savolainen et al. 2000a,b; Soltis et al. 1997, 2000, 2005) and probably ovules with only a single integument and a thin-walled megasporangium, although they are both homoplasious. Iridoids are widespread in the clade and could also be synapomorphic.

Cornales and Ericales may be successively sister to the rest of this clade. The core asterids, supported by the number of stamens equaling the number of petals, epipetalous stamens, an obviously sympetalous corolla, and DNA sequences, form the rest, although it should be noted that obviously sympetalous corollas and epipetalous stamens have also evolved in several members of the Ericales. The core asterids include two major clades, here called the **lamiids** (euasterids I: Garryales, Gentianales, Lamiales, and Solanales) and the **campanulids** (euasterids II: Aquifoliales, Apiales, Dipsacales, and Asterales) following the Angiosperm Phylogeny Group (1998, 2003), Judd and Olmstead (2004), and Soltis et al. (2005).

Cornales

The monophyly of Cornales is strongly supported by DNA sequence characters (Albach et al. 2001b; Bremer et al. 2002; Fan and Xiang 2003; Hilu et al. 2003; Olmstead et al. 2000; Savolainen et al. 2000b; Soltis et al. 2000, 2003a; Xiang et al. 1993, 1998, 2002), more or less inferior ovaries, usually reduced sepals, and an epigynous nectar disk. Many also have drupaceous fruits. The order, as here delimited, consists of perhaps six families, **Cornaceae** s.l. (including Nyssaceae and Alangiaceae), **Hydrangeaceae**, and **Loasaceae**, which together contain about 650 species (Hempel et al. 1995; Soltis et al. 1995; Xiang et al. 1998). Cornales may constitute the sister group to the rest of the asterid clade (Hempel et al. 1995; Olmstead et al. 1993; Xiang et al. 1993). Morphology also supports an asterid affinity, as these plants have ovules with a single integu-

ment, often a thin megasporangium, and iridoid compounds (Hufford 1992). Relationships within Cornales are still not well understood, and some botanists recognize Nyssaceae as distinct from Cornaceae (Xiang et al. 2002). Hydrangeaceae are probably sister to Loasaceae, and both share tuberculate trichomes with basal-cell pedestals.

Garryaceae and Vitaceae sometimes have been placed in Cornales. Garryaceae, a primarily western North American family of dioecious trees or shrubs with opposite, simple leaves, inconspicuous, 4-merous flowers borne in dangling catkins, and two-seeded berries, probably are more closely related to some families of the core asterid complex (Garryales, euasterids I). Vitaceae are an early divergent lineage of the rosid clade.

Hydrangeaceae Dumortier
(Hydrangea Family)

Shrubs, small trees, lianas or herbs; usually with tannins; often with iridoids, aluminum, and raphide crystals. Hairs usually simple. *Leaves usually opposite, simple,* but sometimes lobed, entire to serrate or dentate, with pinnate or palmate venation; *stipules lacking.* Inflorescences determinate, terminal or axillary. Flowers bisexual, radial, those at margin of inflorescence sometimes sterile and with enlarged petal-like sepals. *Sepals usually 4 or 5, connate,* the lobes often reduced. *Petals usually 4 or 5 and distinct,* imbricate, convolute, or valvate. *Stamens 8 or 10 to numerous;* filaments distinct or slightly connate pollen grains tricolpate or tricolporate. Carpels usually 2–5, connate; *ovary usually half-inferior to inferior,* **often ribbed**, with axile or deeply intruded parietal placentation; stigmas 2–5, usually elongate. Ovules usually several to numerous on each placenta, with 1 integument and a thin-walled megasporangium. *Nectar disk usually present atop ovary. Fruit usually a loculicidal or septicidal capsule;* seeds often winged.

Floral formula:

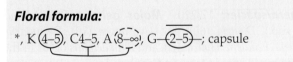

*, K (4–5), C4–5, A (8–∞), G (2–5); capsule

Distribution: Widespread, but especially characteristic of temperate to subtropical regions of the Northern Hemisphere.

Key to Families of Cornales

1. Placentation parietal; leaves with hooked or sticky hairs ....**Loasaceae**
1. Placentation axile or intrusive parietal; leaves with simple, Y- or T-shaped hairs 2
2. Fruit usually a septicidal or loculicidal capsule; ovules usually several to numerous in each locule ....**Hydrangeaceae**
2. Fruit a drupe; ovules solitary in each locule ....**Cornaceae**

(A) Cornales: Loasaceae
Eucnide sp.: flower

(B) Cornales: Loasaceae
Mentzelia sp.: Hairs
(scanning electron micrograph)

(C) Cornales: Cornaceae
Cornus nuttallii: inflorescences with
large bracts

(D) Cornales: Hydrangeaceae
Schizophragma hydrangeoides: inflorescences with
sterile marginal flowers

(E) Cornales: Cornaceae
Nyssa ogeche: drupes

(F) Cornales: Cornaceae
Cornus florida: inflorescence

PLATE 9.17 Eudicots: Cornales

Genera/species: 17/220. **Major genera:** *Philadelphus* (80 spp.), *Deutzia* (60), and *Hydrangea* (30). *Decumaria*, *Deutzia*, *Fendlera*, *Hydrangea*, and *Philadelphus* occur in the continental United States and/or Canada.

Economic plants and products: *Hydrangea* (hydrangea), *Decumaria* (climbing hydrangea), *Schizophragma* (climbing hydrangea), *Philadelphus* (mock orange), and *Deutzia* are frequently grown as ornamentals.

Discussion: Monophyly of Hydrangeaceae is supported by morphological and DNA characters (Albach et al. 2000b; Fan and Xiang 2003; Hempel et al. 1995; Hufford 1997; Morgan and Soltis 1993; Soltis et al. 1995; Xiang et al. 1993). This group has traditionally been included as a subfamily of Saxifragaceae (a basal tricolpate group), but cladistic analyses (Chase et al. 1993; Hempel et al. 1995; Hufford 1992; Morgan and Soltis 1993; Xiang et al. 1993) indicate that Hydrangeaceae are only distantly related to Saxifragaceae, and strongly support their placement near Cornaceae.

Fendlera and *Jamesia* (Jamesioideae) may be sister to the remaining genera of Hydrangeaceae (Hufford 1997; Hufford et al. 2001), here treated as Hydrangeoideae. The latter group is divided into two major clades: the Hydrangeae (*Hydrangea*, *Decumaria*, *Schizophragma*, and relatives) and the Philadelpheae (*Philadelphus*, *Deutzia*, *Fendlerella*, *Whipplea*, *Carpenteria*, and others) (Morgan and Soltis 1993; Soltis et al. 1995; Xiang et al. 1993; Hufford et al. 2001). Most members of the Hydrangeae have conspicuous sterile marginal flowers in the inflorescence (Plate 9.17D), valvate petals, and more or less loculicidal capsules. Members of the Philadelpheae lack sterile flowers and usually have imbricate petals and septicidal capsules. The genus *Hydrangea* definitely is nonmonophyletic.

Flowers of Hydrangeaceae may be large and showy, as in *Philadelphus*, or small (and then densely clustered and often associated with conspicuous sterile flowers), as in *Decumaria* or *Hydrangea*. The nectar produced in the epigynous floral disk rewards a variety of insect pollinators (butterflies, moths, flies, bees, wasps, and beetles). Outcrossing is promoted by protogyny, but self-pollination is also possible. The small seeds, which are often tailed or winged, are dispersed by wind.

Additional reference: Spongberg 1972.

Loasaceae A. L. de Jussieu
(Rock-Nettle Family)

Annual or perennial herbs or vines, occasionally shrubs, or trees; usually with iridoids. *Hairs stiff, silicified and often calcified, simple, with barbs or projections, these hooked or straight, often gland-headed and sticky, sometimes stinging, often associated with a cystolith. Leaves opposite, or alternate but with the first leaf-pair opposite, simple,* entire, serrate, and/or lobed, with palmate to pinnate venation; stipules lacking. Inflorescences determinate, terminal, sometimes reduced to a single flower. Flowers bisexual, radial. Sepals usually 5, distinct or basally connate. *Petals usually 5, distinct or connate, often concave, imbricate. Stamens* 2 to *numerous,* **often closely associated with or adnate to the corolla,** *often with staminodes,* and these variable in number, inconspicuous to petaloid, and sometimes nectariferous; filaments distinct or variably connate, sometimes adnate to corolla; pollen grains tricolporate. *Carpels usually 3-5, connate,* sometimes appearing to be a single carpel; *ovary inferior* (to nearly superior), usually *with parietal placentation,* the placentas sometimes deeply intruded; stigma linear to clubshaped. Ovules 1 to many, with 1 integument and a thin-walled megasporangium. *Fruit a variably dehiscent capsule,* often ribbed, less commonly an achene or nut; endosperm sometimes absent.

Floral formula:

*, K⑤, C⑤, A 2–∞ + 3•–∞• or 0•, G③–⑤; capsule

Distribution: Mainly in temperate and tropical regions of North and South America, but one genus (*Kissenia*) in Africa.

Genera/species: 14/280. **Major genera:** *Nasa* (105 spp.), *Loasa* (75), *Mentzelia* (60), and *Caiophora* (50). *Eucnide, Mentzenia,* and *Petalonyx* occur in the continental United States.

Economic plants and products: *Mentzelia* (blazing-star) and *Eucnide* (rock-nettle) are grown as ornamentals.

Discussion: Monophyly of Loasaceae is supported by the closely associated development of the corolla and androecium (but this unusual developmental pattern is lost in *Mentzelia, Petalonyx* and relatives). The family shows an amazing diversity of corolla and androecial form, with both centripetal and centrifugal androecium development. Recent molecular analyses (Hempel et al. 1995; Hufford et al. 2003) have confirmed the family's monophyly. Relation-

ships within the Loasaceae are still rather unclear, although *Eucnide* may be sister to the remaining genera (Hufford et al. 2003; but see also Weigend 2004).

The showy and remarkably variable flowers of Loasaceae (Plate 9.17A) are pollinated by bees, wasps, flies, butterflies, hawkmoths, hummingbirds, or rodents, and use nectar and/or pollen as a reward. The small seeds often lack any obvious dispersal mechanisms, but many are probably dispersed by wind or rain-wash. The capsules of some species of *Mentzelia* are tardily dehiscent and covered with hooked hairs (Plate 9.17B) and thus may be externally transported.

Additional references: Ernst and Thompson 1963; Hufford 2003.

Cornaceae Bercht. & Presl
(Dogwood Family)

Usually trees or shrubs; usually with iridoids. **Hairs** often calcified, **mesifixed,** *Y- or T-shaped. Leaves opposite, less commonly alternate and spiral, simple, usually entire,* but sometimes serrate, with pinnate to ± palmate venation, *secondary veins usually smoothly arching toward margin* or forming a series of loops; *stipules lacking.* Inflorescences determinate, terminal, sometimes associated with enlarged, showy bracts. Flowers bisexual or unisexual (plants then monoecious or dioecious), radial. *Sepals usually 4 or 5,* distinct or connate, *usually represented by small teeth,* sometimes lacking. *Petals usually 4 or 5,* ± *distinct,* imbricate or valvate. *Stamens 4–10;* filaments distinct; pollen grains usually tricolporate, **the apertures with an H-shaped thin region.** Carpels usually 2 or 3, connate, sometimes appearing to be a single carpel; ovary inferior, with axile placentation, **the axis lacking vascular bundles, and the ovules instead attached to vascular bundles that arch over the top of the each septum**; stigma usually capitate, lobed, or elongate. **Ovules 1 in each locule,** attached at apex, with 1 integument and a thin- to thick-walled megasporangium. *Nectar disk positioned atop ovary.* **Fruit a drupe, the pit 1- to few-seeded, ridged to winged, with thin regions** (i.e., germination valves, these sometimes obscure) (Figure 9.107).

Floral formula:

*, K④–⑤, C4–5, A4–10, G②–③; drupe

Distribution: Widespread; especially common in north temperate regions.

Genera/species: 7/110. **Major genera:** *Cornus* (60 spp.), *Mastixia* (20), and *Nyssa* (10). *Cornus* and *Nyssa* occur in the continental United States and Canada.

Economic plants and products: Ornamental trees and shrubs occur in *Nyssa* (tupelo), *Davidia* (dove tree), and *Cornus* (dogwood).

FIGURE 9.107 Cornaceae. (A–E) *Cornus amomum:* (A) flowering branch (× 0.75); (B) flower (× 9); (C) flower in longitudinal section, with petals and stamens removed (× 9); (D) drupe (× 3); (E) pit in lateral view and from above (× 6). (F–H) *C. florida:* (F) flowering branch (× 0.75); (G) flower (× 6); (H) pit from lateral view and from above (× 6). (I) *C. alternifolia:* pit in lateral view and from above (× 6). (From Ferguson 1966, *J. Arnold Arbor.* 47: p. 111.)

Discussion: Monophyly of Cornaceae, as defined broadly (i.e., including Nyssaceae and Alangiaceae), is weakly supported by morphology, a base chromosome number of 11, as well as *matK* and *rbcL* sequences (Xiang et al. 1998, 2002). Infrafamilial relationships have been investigated by Eyde (1988), Murrell (1993), Xiang et al. (1993, 1996, 1998, 2002, 2006), and Xiang and Murrell (1998). Two major clades can be recognized within Cornaceae: a nyssoid-mastixioid clade (*Nyssa, Camptotheca, Davidia, Mastixia,* and *Diplopanax*) with usually unisexual, 5-merous flowers; and a cornoid clade (*Cornus* and *Alangium*) with usually bisexual, 4-merous flowers.

Cornus has several specialized features, such as the number of stamens usually equaling that of the petals, ovules with the raphe positioned dorsally (i.e., apotropous), 4-merous flowers, and hairs coated with large crystals of calcium carbonate. Its monophyly is supported by morphology, cpDNA restriction sites, and DNA sequences. Combined evidence (morphology, restriction sites, *matK, rbcL,* 26S sequences: Fan and Xiang 2001; Xiang et al. 1993, 1996, 1998, 2006; Xiang and Murrell 1998) suggests that two major clades can be recognized within *Cornus*: blue-fruited dogwoods and red-fruited dogwoods. The latter contains the cornelian cherry clade (*C. mas* and relatives) and the big-bracted dogwood clade (*C. florida, C. kousa, C. nuttallii, C. canadensis,* and relatives; Plate 9.17C,F).

The flowers of Cornaceae typically produce nectar and attract bees, flies, and beetles, but wind pollination may occur, as in *Davidia*. The white, blue, blue-black, purple, or red drupes (Plate 9.17E), sometimes contrasting in color with the inflorescence axes, are dispersed by birds and mammals. The fruits of several species of *Nyssa* float well and are probably at least partly dispersed by water.

Species of *Cornus* can be identified in sterile condition by carefully tearing a leaf. The two halves will remain attached by delicate threads (unraveled spiral thickenings of the vessel elements), although some of the Hydrangeaceae also do this. The smoothly arcuate secondary veins are also diagnostic.

Additional references: Eyde 1966; Eyde and Xiang 1990; Ferguson 1966a; Kubitzki 2004a.

Ericales

Monophyly of Ericales has been strongly supported in analyses based on DNA sequences (Bremer et al. 2002; Chase et al. 1993; Hilu et al. 2003; Kron and Chase 1993; Morton et al. 1997a, 1998; Olmstead et al. 1993; Källersjö et al. 1998; Savolainen et al. 2000a,b; Soltis et al. 2000; Albach et al. 2001a,b). Ericales are here delimited broadly; these families usually have been divided into several smaller orders (e.g., Ebenales, Theales, Primulales, Polemoniales, Ericales s.s.). Morphological support for this group is weak, but a possible synapomorphy is the presence of theoid leaf teeth (i.e., a condition in which a single vein enters the tooth and ends in an opaque deciduous cap or gland; see Figure 4.13). Such teeth are found in at

least some members of most of the included families (Hickey and Wolfe 1975). In the Ericaceae themselves, the condition is somewhat modified, each tooth being associated with a multicellular, usually gland-headed hair. Placentas that protrude into the ovary locule/locules are a second possible synapomorphy (Nandi et al. 1998). Members of the Ericales usually can be distinguished from core asterids by their flowers with stamens usually twice the number of petals or numerous (vs. stamens equaling the number of petals or fewer). However, staminal reduction has occurred in Primulaceae, and increase in most Lecythidaceae, Actinidiaceae, and Theaceae.

Phylogenetic relationships within the order are not well understood, but recent work (Anderberg et al. 2002; Bremer et al. 2002; Schönenberger et al. 2005; Sytsma et al. 2006) suggests a number of groups. Sister to the rest of the order is a clade that includes Balsaminaceae. Fouquieriaceae plus Polemoniaceae may form a clade that is in turn sister to the remainder of the order. Relationships among many of the remaining families are problematic. Lecythidaceae are somewhat isolated, and Sapotaceae, along with Ebenaceae, may form a clade with Primulaceae s.l. Theaceae and Pentaphylacaceae (= Ternstroemiaceae) are not closely related; both were previously included in Theaceae s.l. The core Ericales include Actinidiaceae, Cyrillaceae, Clethraceae, Ericaceae, and Sarraceniaceae. They are clearly monophyletic (Anderberg 1992, 1993; Bayer et al. 1996; Judd and Kron 1993; Kron and Chase 1993); putative synapomorphies include anther inversion during development (so that the morphological base is apical), a usually hollow style that emerges from an apical depression in the ovary, and endosperm usually with haustoria at both ends (Anderberg 1992, 1993; Judd and Kron 1993; Kron 1996; Kron and Chase 1993). *Actinidia* and *Sarracenia* have many stamens, while *Actinidia* has incompletely fused carpels, so that the gynoecium has distinct styles. Cyrillaceae, Clethraceae, and Ericaceae may form a clade, which is supported by several embryological features. Morphological analyses support the sister group relationship of Clethraceae and Ericaceae (Anderberg 1993; Judd and Kron 1993), but molecular evidence suggests that Cyrillaceae may well be sister to Ericaceae (Anderberg et al. 2002).

Ericales includes 24 families and about 9450 species. Major families include Actinidiaceae, Balsaminaceae, Clethraceae, Cyrillaceae, **Ebenaceae, Ericaceae,** Fouquieriaceae, **Lecythidaceae,** Pentaphylacaceae, **Polemoniaceae, Primulaceae, Sarraceniaceae, Sapotaceae,** Styracaceae, Symplocaceae, and **Theaceae.**

Sapotaceae A. L. de Jussieu
(Sapodilla Family)

Trees or shrubs, sometimes with distinctly sympodial branches, or thorns; silica bodies often present; with tannins, often triterpenoids, and cyanogenic compounds; **with well-developed, elongate laticifers, the latex white. Hairs 2-branched, brownish, T-shaped, but one arm**

Key to Major Families of Ericales

1. Flowers resupinate, bilateral, with usually 3 sepals, the apparent lower one petaloid, often forming a nectar spur, with 5 petals, the seeming upper one distinct, concave and often sepal-like, the 4 others usually connate into 2 lateral pairs ....Balsaminaceae

1. Flowers not resupinate, radial or bilateral, with various numbers of distinct to connate sepals and petals, but not as above 2

2. Plants with laticifers and milky latex; hairs 2-branched, brownish, usually T-shaped; seeds with a large hilum ....**Sapotaceae**

2. Plants lacking laticifers; hairs not as above; seeds with a small hilum 3

3. Petals free 4

3. Petals connate 11

4. Plants carnivorous, with highly modified leaves that form pitcher-like traps ....**Sarraceniaceae**

4. Plants not carnivorous...5

5. Anthers inverting, or if apparently not inverting, then seed coat lacking 6

5. Anthers not inverting 9

6. Hairs usually stellate ....Clethraceae

6. Hairs not stellate 7

7. Stamens usually numerous; carpels usually incompletely fused, thus stigmas several; tissues with needle-like crystals of calcium oxalate ....Actinidiaceae

7. Stamens usually 2–10; carpels completely fused, thus a single stigma present; tissues lacking needle-like crystals 8

8. Seed coat present; pollen grains usually in tetrads; flowers pendulous or erect ....**Ericaceae**

8. Seed coat absent; pollen grains in monads; flowers ± erect ....Cyrillaceae

9. Ovary inferior or half-inferior; fruit usually opening by a circumscissile slit; stems with cortical vascular bundles ....**Lecythidaceae**

9. Ovary superior; fruit longitudinally dehiscent, or indehiscent; stems lacking cortical bundles 10

10. Stamens with filaments only slightly longer than the anthers; fruit fleshy and ± indehiscent; embryo curved ....Pentaphylacaceae

10. Stamens with filaments several times longer than anthers; fruit dry, longitudinally dehiscent; embryo straight ....**Theaceae**

11. Placentation free-central, with the placental axis expanded ....**Primulaceae**

11. Placentation ± axile 12

12. Anthers inverting either early or late in development; pollen grains usually in tetrahedral tetrads ..**Ericaceae**

12. Anthers not inverting; pollen grains in monads 13

13. Indumentum of either stellate hairs or peltate scales ....Styracaceae

13. Indumentum various, but not as above 14

14. Plants with long and short shoots, the leaves of the long shoots quickly deciduous, but with a portion of the petiole hardened and persisting as a spine ....Fouquieriaceae

14. Shoots ± monomorphic, the leaves not forming a spine 15

15. Ovary inferior or half inferior, fruit a drupe ....Symplocaceae

15. Ovary superior, fruit a berry or capsule 16

16. Shrubs or trees, with black or dark-colored naphthaquinones; stamens usually 8 to numerous; fruit a berry associated with an expanded calyx ....**Ebenaceae**

16. Usually herbs, lacking naphthaquinones; stamens usually 5; fruit a capsule; calyx not expanded ....**Polemoniaceae**

often ± reduced. *Leaves alternate and spiral*, sometimes distinctly clustered at shoot apices, *simple, entire*, with pinnate venation; stipules present or absent. Inflorescences determinate, *usually fasciculate*, sometimes reduced to a single flower, axillary. Flowers bisexual, radial. *Sepals 4–8, sometimes dimorphic*, distinct or connate basally. *Petals 4–8, connate, sometimes with paired petaloid appendages (outgrowths from lower portion of corolla lobes)*, imbricate. *Stamens 8–16*

and opposite the petals, usually alternating with staminodes; filaments and staminodes adnate to corolla; pollen grains usually 3- or 4-colporate. *Carpels 2 to numerous, connate; ovary superior*, with axile placentation; stigma capitate to slightly lobed. Ovules 1 in each locule, with 1 integument and a thin-walled megasporangium. **Fruit a berry; seeds usually with a hard shiny testa and large hilum;** endosperm sometimes absent (Figure 9.108).

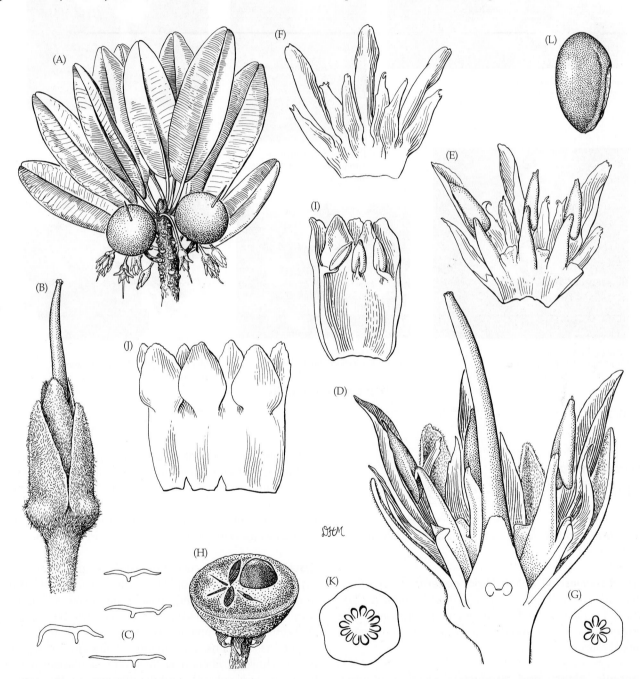

FIGURE 9.108 Sapotaceae. (A–H) *Manilkara jaimiqui* var. *emarginata:* (A) flowering and fruiting branch (× 0.75); (B) flower, after fall of corolla (× 4.5); (C) detached T-shaped hairs from calyx (× 75); (D) flower in longitudinal section, note staminodes alternating with stamens (× 7.5); (E) three corolla lobes from within, with appendages, stamens, and alternating staminodes (× 4.5); (F) three corolla lobes from without, to show dorsal appendages (× 4.5); (G) ovary in cross-section (× 6); (H) berry, the upper half removed to show single seed, locules with aborted ovules (× 1.5). (I–L) *M. zapota:* (I) portion of corolla from within, to show three corolla lobes, two petaloid staminodes alternating with stamens (× 4.5); (J) portion of corolla from without, to show three corolla lobes, tips of four staminodes (× 4.5); (K) ovary in cross-section (× 7.5); (L) seed, note elongate hilum (× 1.5). (From Wood and Channell 1960, *J. Arnold Arbor.* 41: p. 14.)

Ericales: Primulaceae
Primula sp.: flowers (heterostylous)

Ericales: Sarraceniaceae
Sarracenia leucophylla: leaf and flower

Ericales: Ericaceae
Vaccinium racemosum: leaves and fruits

Ericales: Lecythidaceae
Barringtonia racemosa: flowers

Ericales: Sapotaceae
Sideroxylon foetidissimum: flowers

Ericales: Sapotaceae
Manilkara zapota: branch, flowers, and fruit

PLATE 9.18 Eudicots: Ericales

Floral formula:

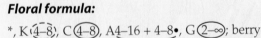

*, K(4–8), C(4–8), A4–16 + 4–8•, G(2–∞); berry

Distribution and ecology: Pantropical, especially in wet lowland forests.

Genera/species: 53/1100. **Major genera:** *Pouteria* (325 spp.), *Palaquium* (110), *Madhuca* (100), *Sideroxylon* (75), *Chrysophyllum* (70), and *Mimusops* (50). *Chrysophyllum, Sideroxylon, Manilkara,* and *Pouteria* occur in the continental United States.

Economic plants and products: *Manilkara zapota* (sapodilla; Plate 9.18F), *Pouteria mammosa* (mamey sapote), *Pouteria campechiana* (eggfruit), and *Chrysophyllum cainito* (star apple) provide delicious tropical fruits. Those of *Synsepalum dulcificum* affect the sense of taste; after eating even a portion of a single fruit, other food tastes sweet. Sev-

eral genera are important sources of latex, such as *Palaquium* (gutta-percha) and *Manilkara zapota* (chicle, for chewing gum). Many genera provide economically important timber; others are useful ornamentals, including *Chrysophyllum* (satinleaf), *Manilkara, Mimusops* (cherry mahogany), and *Sideroxylon* (buckthorn, ironwood, mastic; Plate 9.18E).

Discussion: Sapotaceae are easily recognized and are hypothesized to be monophyletic (Morton et al. 1997; Pennington 1991). The family usually has been considered closely related to Ebenaceae, a tropical family that also has alternate, entire, simple leaves, sympetalous flowers with epipetalous stamens, and superior ovaries developing into berries. However, Ebenaceae differ from Sapotaceae in their black or dark-colored naphthaquinones in the leaves, stems, and wood, absence of latex, dioecy, and lobed calyx that is persistent and expands as the fruit develops.

Infrafamilial and generic relationships have been investigated by Pennington (1991, 2004), using characters such as position of the corolla lobes relative to the stamens and

(G)

Ericales: Ebenaceae
Diospyros virginiana: fruits

(H)

Ericales: Ericaceae
Agarista populifolia: flowers

(I)

Ericales: Theaceae
Gordonia lasianthus: flower

(J)

Ericales: Ericaceae
Rhododendron viscosum: flowering plant

(K)

Ericales: Polemoniaceae
Ipomopsis rubra: flowering plant

(L)

Ericales: Primulaceae
Myrsine coriacea: fruits

PLATE 9.18 Eudicots: Ericales

one another, position of the staminodes relative to the ovary, shape of the corolla, presence of petal appendages, size of the anthers, and position of the hilum scar. It has been postulated that various genera having twice as many stamens as corolla lobes form a heterogeneous complex, which is probably paraphyletic and basal. The remaining genera have a number of stamens equaling the number of the corolla lobes, and probably constitute a clade. However *ndhF* sequences suggest that *Sarcosperma* (a genus with 5 stamens, 5 staminodes, elongate inflorescences, and opposite leaves) is sister to the remaining genera (Anderberg and Swenson 2003; Swenson and Anderberg 2005). *Mimusops, Manilkara,* and relatives (Sapotoideae) may form a clade based on their calyx of two differentiated whorls. A single calyx whorl characterizes *Chrysophyllum, Pouteria, Synsepalum,* and *Sideroxylon* (incl. *Bumelia, Dipholis,* and *Masticodendron*), all Chrysophylloideae.

The family is mainly insect-pollinated, although visitation by bats has been reported. Dispersal of the sweet berries is by various birds and mammals.

Additional reference: Wood and Channell 1960.

Ebenaceae Gürke
(Ebony Family)

Trees or shrubs; **with black or dark-colored naph-thaquinones (or related compounds) in nearly all tissues**; sometimes cyanogenic. Hairs simple, 2- or multi-branched, sometimes glandular. *Leaves alternate, usually 2-ranked, simple, entire,* with pinnate venation, **usually with nectar glands on abaxial surface**; stipules absent. Inflorescences determinate, often reduced to a single flower, axillary. *Flowers usually unisexual (plants dioecious),* radial. **Sepals 3–7, connate, usually persistent** and *variably enlarging as fruit develops.* **Petals 3–7, connate, ± urn-shaped**, with imbricate or valvate lobes, often contorted. *Stamens (3-) 6-numerous; filaments usually adnate to corolla;* anthers occasionally opening by apical pores; replaced by staminodes in carpellate flowers; pollen grains usually tricolporate. *Carpels usually 3–8, connate; ovary usually superior, with axile placentation,* the locules often secondarily divided; stigmas 3–8, capitate to slightly elongated. Ovules 1 or 2 in each locule, with a thin-walled megasporangium. **Fruit a berry**, astringent until very ripe; seeds fairly large, with thin testa (Plate 9.18G); endosperm sometimes ruminate.

Floral formula:

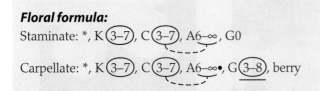

Staminate: *, K (3–7), C (3–7), A6–∞, G0

Carpellate: *, K (3–7), C (3–7), A6–∞•, G (3–8), berry

Distribution: Pantropical, with a few species extending into temperate regions.

Genera/species: 4/500.   **Major genus:** *Diospyros* (480 spp.), which is the only genus in the continental United States.

Economic plants and products: Several species of *Diospyros*, e.g., *D. digyna* (black sapote), *D. kaki* (Japanese persimmon), *D. lotus* (date plum), and *D. virginiana* (persimmon), have edible fruits; many species provide economically important timber (e.g., ebony).

Discussion: Ebenaceae are easily recognized and are hypothesized to be monophyletic based on their distinctive morphological characters and cpDNA sequences. *Lissocarpa* (glabrous plants, with inferior ovary, triporate pollen) is sister to the remaining genera (*Diospyros, Euclea,* and *Royena*: pubescent plants with superior ovary, tricolporate pollen) (Duangjai et al. 2006). The family is insect pollinated and dispersal of the berries is by both birds and mammals.

Ebenaceae are an especially important floristic element in the forests of Africa.

Additional references: Wallnöfer 2004; Wood and Channel 1960.

Primulaceae Batsch ex Borkh.
(Primrose Family)

Herbs, shrubs, trees, or lianas; often with benzoquinones, triterpenoid saponins and tannins; often **with secretory cavities containing yellow to red resinous materials, and then appearing as yellow-green, red, brown, or black dots to lines on leaves, stems, and floral parts**, sometimes with wandering fibers in the leaf. Hairs various, often elongate and septate or gland-headed. *Leaves* alternate and spiral, opposite, or whorled, often forming a basal rosette (when plants herbaceous), *simple*, entire to serrate, sometimes lobed, *with pinnate venation; stipules lacking.* Inflorescences determinate or indeterminate, sometimes reduced to a single flower, terminal or axillary. *Flowers bisexual* (occasionally unisexual, then plants ± dioecious), *usually radial*, sometimes heterostylous (Plate 9.18A). Sepals usually 4 or 5, distinct or connate. *Petals usually 4 or 5, connate,* imbri-

cate or convolute. **Stamens 4 or 5, opposite corolla lobes**, sometimes alternating with petaloid staminodes; filaments distinct or connate, adnate to corolla; anthers sometimes opening by apical pores; pollen grains tricolporate or 5–8 zonocolpate. *Carpels usually 3-5, connate; ovary usually superior* or occasionally half-inferior, **with free-central placentation, the central placental axis thick, ± globose, filling or nearly filling the locule**; stigma punctate or capitate, sometimes lobed. Ovules few to numerous, anatropous to campylotropous, with 1 or 2 integuments and a thin-walled megasporangium. Nectaries usually lacking. *Fruit a capsule*, opening by valves or circumscissile, *a berry*, with several seeds embedded in the fleshy placental axis, *or a drupe*, with a single, 1 to few-seeded pit; seeds often with a depressed hilum, occasionally arillate (Figure 9.109).

Floral formula:

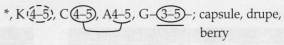

*, K (4–5), C (4–5), A4–5, G (3–5); capsule, drupe, berry

Distribution: Widespread, from temperate to tropical regions.

Genera/species: 57/2150.   **Major genera:** *Primula* (550 spp., incl. *Dodecantheon*), *Ardisia* (300), *Myrsine* (200), *Lysimachia* (200), *Maesa* (150), *Embelia* (130), *Androsace* (100), and *Clavija* (50). *Primula, Lysimachia, Androsace, Anagallis, Glaux, Hottonia, Samolus, Trientalis, Ardisia, Myrsine, Samolus,* and *Jacquinia* occur in continental United States and/or Canada.

Economic plants and products: *Primula* (primrose, shooting star), *Cyclamen* (sowbread), *Anagallis* (pimpernel), *Ardisia* (coralberry, marlberry), *Wallenia, Myrsine,* and *Jacquinia* are cultivated as ornamentals.

Discussion: It is clear that Primulaceae, as broadly circumscribed (i.e., including Theophrastaceae, Myrsinaceae, and Maesaceae), are monophyletic, as evidenced by their schizogenous secretory cavities or canals with yellow to red to black resinous material (lost in some), stamens equaling and opposite the corolla lobes; free-central placentation with a thick, more or less globose, central axis; and DNA sequences (Anderberg and Ståhl 1995; Bremer et al. 2002; Källersjö et al. 2000; Kron and Chase 1993; Olmstead et al. 1993). Relationships within this clade have been assessed by Anderberg and Ståhl (1995) and Anderberg et al. (1998). *Maesa*, a tropical genus of woody plants with the corolla lobes induplicate-valvate and a half-inferior ovary, likely is sister to all other primuloid genera, which are characterized by imbricate corolla lobes. Within this large clade, *Jacquinia, Clavija, Samolus,* and relatives (sometimes treated as Theophrastaceae), are the sister group of the remaining genera, and they have retained the outer whorl of stamens as well-developed staminodes; many have pseudoverticillate leaves with subepidermal fibers and distinctive berry fruits. The remaining genera form a monophyletic group, i.e., the *Primula + Lysimachia + Myrsine* clade, which is sup-

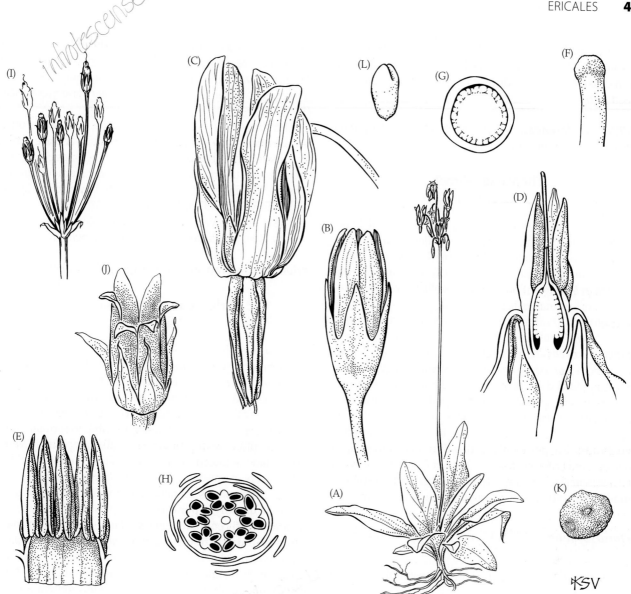

FIGURE 9.109 Primulaceae. *Primula (Dodecatheon) meadia:* (A) flowering plant (× 0.25); (B) flower bud (× 4); (C) flower, corolla lobes reflexed (× 4); (D) flower in longitudinal section (× 4); (E) androecial tube laid open, note that filaments are both connate, and adnate to corolla tube (× 4); (F) stigma (greatly magnified);

(G) ovary in cross-section, showing free-central placenta (× 13); (H) flower bud in cross-section (× 4); (I) infructescence (× 0.5); (J) capsule (× 4); (K) seed (× 15); (L) embryo (greatly magnified). (From Wood 1974, *A student's atlas of flowering plants,* p. 86.)

ported by two *ndhF* deletions, the loss of the outer staminal whorl, and perhaps capsular fruits.

The *Primula + Lysimachia + Myrsine* clade can be divided into two subclades. The first is Primuloideae, comprising *Primula, Androsace,* and relatives, which are herbaceous plants that lack resin cavities, with usually more or less campanulate corollas, capsular fruits, and a scapose inflorescence. The second is Myrsinoideae, which usually have resin cavities, and are either herbs with capsular fruits, such as *Coris, Ardisiandra,* and genera of Lysimachieae (e.g., *Anagallis, Cyclamen, Gaux, Lysimachia,* and *Trientalis*), or trees and shrubs with drupaceous fruits, such as *Ardisia, Myrsine, Wallenia,* and *Embelia* (Plate 9.18L). These drupaceous genera form a well-supported clade within Primulaceae s.l. (Anderberg and Ståhl 1995; Morton et al. 1997), supported

by DNA sequences, drupes with a single pit, and seeds with a depressed hilum. Myrsinaceae, as traditionally defined, was restricted to these woody, drupaceous genera, while the related capsular herbs were placed in a non-monophyletic Primulaceae s.s. (see Cronquist, 1981). Primulaceae are circumscribed broadly, anticipating changes in the APG system, and this group is both monophyletic and morphologically distinctive.

The showy to inconspicuous flowers of Primulaceae are pollinated by various insects. Heterostyly is common in *Primula,* but selfing also occurs. In many species outcrossing is favored by protogyny or the more or less dioecious condition. Recent work suggests that the species that used to be segregated as *Dodecatheon* are a buzz-pollinated lineage derived from within *Primula* (Mast et al. 2001; Trift et al.

2002). The small seeds are often wind- or water-dispersed, but some are dispersed by ants, which gather the oily arils. The red to purple-black drupes of some species are dispersed by birds.

Additional references: Caris and Smets 2004; Channell and Wood 1959; Kållersjö and Ståhl 2003.

Theaceae Mirbel
(Tea Family)

Trees or shrubs; usually with sclereids; tannins present. Hairs usually simple and unicellular. *Leaves alternate and spiral, sometimes 2-ranked, simple, toothed, the teeth theoid (i.e., with a deciduous glandular apex, see Figure 4.13), with pinnate venation; stipules lacking.* Inflorescences of solitary, axillary flowers. Flowers bisexual, radial, *the subtending bract(s) sometimes intergrading with calyx. Sepals usually 5, distinct or slightly connate basally, imbricate. Petals usually 5, distinct or very slightly connate basally,* imbricate, *slightly wrinkled along margin. Stamens numerous,* the ones nearest the gynoecium developing first, distinct or basally connate into a ring or 5 bundles that are opposite the petals; pollen grains tricolporate. *Carpels usually 3–5, connate; ovary superior, with axile placentation; stigma 1 and lobed, to 3–5 and capitate.* **Ovules 1 to few per locule,** with a thin-walled megasporangium. Nectariferous tissue at base of filaments or ovary. *Fruit a ± loculicidal capsule;* **± few seeds, often flattened or winged; embryo large**; endosperm present or absent.

Floral formula: *, K⑤, C5, A⑧, G③–⑤; capsule

Distribution: Widespread in temperate to tropical regions.

Genera/species: 9/300. **Major genera:** *Camellia* (100 spp.), *Gordonia* (60), *Pyrenaria* (40), *Polyspora* (35), *Laplacea* (30), and *Stewartia* (30). *Franklinia, Gordonia,* and *Stewartia* occur in the continental United States.

Economic plants and products: Tea is made from the leaves of *Camellia sinensis*. *Camellia* (camellia), *Gordonia* (loblolly bay), *Stewartia,* and *Franklinia* (Franklin tree) provide ornamentals.

Discussion: Theaceae, as traditionally circumscribed, probably are not monophyletic, a conclusion supported by DNA-based cladistic analyses (Morton et al. 1997; Prince and Parks 2001). Here, the family is restricted to the genera traditionally placed in Theoideae. Genera of former Ternstroemioideae (e.g., *Ternstroemia* and *Eurya*) are related to *Pentaphylax*, and are considered to represent a distinct family, the Pentaphylacaceae. The monophyly of both Theaceae and Pentaphylacaceae is supported by *rbcL* and *matK* sequences (Morton et al. 1997; Prince and Parks 2001). The morphological similarity between the flowers of some core

Ericales, such as Actinidiaceae, and Theaceae is striking. Theaceae are composed of three monophyletic tribes: Theeae (*Camellia, Polyspora, Laplacea,* and relatives), Gordonieae (*Gordonia, Franklinia,* and *Schima*), and Stewartieae (*Stewartia*).

The numerous stamens of Theaceae (and some other Ericales) develop centrifugally (Plate 9.18I) and are served by a limited number of vascular bundles, suggesting that they are evolutionarily derived from few stamens (in two whorls). This interpretation is supported by cladistic analyses of morphological and DNA characters.

The showy flowers of Theaceae are pollinated by various insects. Their seeds are wind- or water-dispersed.

Additional references: Keng 1962; Wood 1959b.

Ericaceae A. L. de Jussieu
(Heath Family)

Trees, shrubs, lianas, sometimes epiphytic, occasionally mycoparasitic herbs lacking chlorophyll, strongly associated with mycorrhizal fungi. Hairs simple, usually multicellular and unicellular mixed, sometimes dendritic, gland-headed, or peltate scales, but not stellate. *Leaves alternate and spiral, sometimes opposite or whorled, simple,* entire to serrate, sometimes revolute, with pinnate, ± parallel, or palmate venation, blade reduced in mycoparasites; stipules lacking. Inflorescences various. *Flowers usually bisexual,* rarely unisexual (then plants usually dioecious), radial to slightly bilateral, **usually ± pendulous.** Sepals usually 4 or 5, distinct to slightly connate. **Petals** *usually 4 or 5 and* **connate, often cylindrical to urn-shaped,** *with small to large, imbricate to valvate lobes, but sometimes ± bell-shaped or funnel-like,* occasionally distinct (a reversal). Perianth reduced to 2 or 3 sepals and petals, or 3 or 4 tepals in a few genera that are wind-pollinated. *Stamens 8–10,* but reduced to 2 or 3 in wind-pollinated species; filaments free or adnate to corolla, sometimes connate, **sometimes with paired projections (spurs) near or at junction with anther**; *anthers becoming inverted,* 2- or 1-locular, *usually opening by 2 apical pores,* **sometimes with 2 projections (awns)** *or with apex narrowed, forming a pair of tubules; pollen grains usually in tetrads,* usually tricolporate, sometimes associated with viscin threads. Carpels 2–10; ovary superior to inferior, usually with axile or deeply intruded parietal placentation; style 1, *hollow, internally fluted;* stigma capitate or slightly lobed. Ovules 1 to numerous per locule, with 1 integument and a thin-walled megasporangium. Nectariferous tissue around base or apex of ovary. *Fruit a septicidal or loculicidal capsule, berry, 1- or several-pitted drupe,* usually erectly held due to movement of pedicel; seed coat thin (Figure 9.110).

Floral formula:

* or X, K④–⑤, C④–⑤, A④–⑩, G②–⑩; capsule, berry, drupe

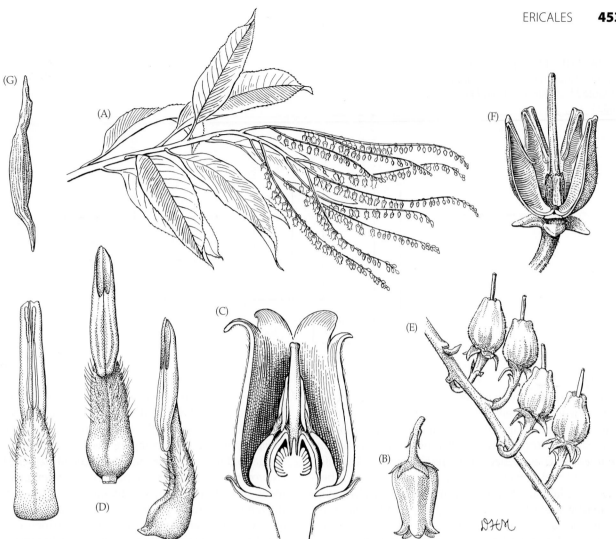

FIGURE 9.110 Ericaceae. (A–G) *Oxydendrum arboreum*: (A) flowering branch (× 0.4); (B) flower (× 3); (C) flower in longitudinal section (× 9); (D) outer, inner, and lateral views of stamens (× 18); (E) portion of raceme with immature fruits (× 3); (F) opened capsule, one valve removed, note deeply immersed style (× 6); (G) seed (× 15). (From Wood 1961, *J. Arnold Arbor.* 42: p. 57.)

Distribution and ecology: Cosmopolitan, but especially common in tropical montane habitats, southern Africa, eastern North America, eastern Asia, and Australia; usually light-loving shrubs of acid soils.

Genera/species: 124/4100. *blueberries* ***Major genera:*** *Erica* (860 spp.), *Rhododendron* (850), *Vaccinium* (740), *Agapetes* (400), *Leucopogon* (230), *Gaultheria* (130), *Cavendishia* (130), *Diplycosia* (100), *Arctostaphylos* (60), and *Epacris* (50). Noteworthy genera (in addition to most of the above) in the continental United States and/or Canada are *Agarista, Andromeda, Arbutus, Bejaria, Ceratiola, Chamaedaphne, Chimaphila, Corema, Empetrum, Eubotrys, Gaylussacia, Kalmia, Leucothoe, Lyonia, Monotropa, Monotropsis, Oxydendrum, Pieris, Pterospora,* and *Pyrola.*

Economic plants and products: The edible fruits of *Vaccinium* (blueberries, cranberries; Plate 9.18C) are economically important. The family contains many showy ornamentals, including *Arbutus* (madrone), *Calluna* (heather), *Erica* (heath), *Gaultheria* (wintergreen), *Kalmia* (mountain laurel), *Oxydendrum* (sourwood), *Pieris, Rhododendron* (azalea, rhododendron; Plate 9.18J), and *Leucothoe* (fetterbush). *Gaultheria procumbens* is the original source of oil of wintergreen (methyl salicylate).

Discussion: Ericaceae are broadly circumscribed, including five families (Empetraceae, Epacridaceae, Monotropaceae, Pyrolaceae, and Vacciniaceae) that are sometimes recognized separately. Recognition of these segregates would render Ericaceae s.s. paraphyletic. As circumscribed here, the group is monophyletic, based on evidence from morphology, *rbcL, matK,* and 18S rDNA sequences (Anderberg 1993; Chase et al. 1993; Judd and Kron 1993; Kron 1996; Kron and Chase 1993; Kron et al. 2002; Soltis et al. 1997). The eastern Asian genus *Enkianthus* is sister to the remaining genera, which form a clade based on their pollen shed as tetrahedral tetrads, lack of a fibrous endothecium (innermost layer of anther locule), and seed lacking a vascular bundle in the raphe (Anderberg 1993, 1994; Judd and Kron 1993; Kron 1996; Kron and Chase 1993; Kron et al. 2002). Several subclades can be discerned within this clade (Figure 9.111).

FIGURE 9.111 Cladogram showing hypothesized relationships within Ericaceae. (Adapted from Kron 1997 and Kron et al. 2002.)

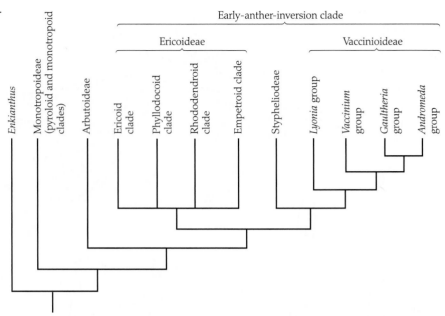

The Ericoideae are characterized by their erect to more or less horizontal flowers, loss of stamen appendages, and usually septicidal capsules, but all these show homoplasy. The rhododendroid clade, including *Rhododendron* and *Menziesia*, is characterized by protective bracts at the inflorescence base, septicidal capsules, usually showy, ± bell-shaped bilateral flowers, viscin threads, and nonappendaged anthers. A morphologically similar clade, the phyllodocoids, includes general such as *Phyllodoce* and *Kalmia*, and also has septicidal capsules; their flowers are radially symmetrical. The ericoid clade (*Erica* and *Calluna*) is diagnosed by a persistent corolla. All three groups are part of a clade that also includes the wind-pollinated empetroids. Empetroids—*Empetrum, Corema,* and *Ceratiola*—are clearly monophyletic, based on their strongly lobed and expanded stigma, reduced corolla, and drupaceous fruit.

Another major group within Ericaceae is the Vaccinioideae, including *Vaccinium, Gaylussacia, Cavendishia,* and relatives with inferior ovaries, plus *Lyonia, Pieris, Chamaedaphne, Leucothoe, Gaultheria, Andromeda,* and others, with superior ovaries. The genera with inferior ovaries form a monophyletic subgroup that is especially diverse in the montane tropics. Vaccinioideae, characterized by a base chromosome number of 12, are sister to Stphelioideae (e.g., *Epacris, Styphelia,* and *Leucopogon*). Stpheliodeae are monophyletic, as indicated by their monoloculate and epipetalous stamens in a single whorl, usually parallel-veined leaves, and usual lack of multicellular hairs.

Ericoideae, Vaccinioideae, and Stpheliodeae all show early developmental inversion of their anthers, and DNA sequences strongly indicate that together they constitute a clade (see Figure 9.111). In contrast, Monotropoideae and Arbutoideae (see below) represent earlier divergent clades within the family. They, along with *Enkianthus*, have late-inverting anthers.

Monotropoideae are related on the basis of their more or less herbaceous habit and reduced embryos (Anderberg 1993). Within this group, the monotropoids (e.g., *Monotropa, Monotropsis,* and *Pterospora*) are easily recognized by their specialized mycoparasitic habit, associated with the loss of chlorophyll. These genera represent an extreme development of the fungal symbiotic relationship characteristic of the family. The fungi parasitized by various monotropoids are mycorrhizal, and are also attached to the roots of various forest trees. *Monotropa* (Indian pipe) thus indirectly parasitizes these trees, with nutrients passing from tree to fungus to monotropoid. The Pyroloids (e.g., *Pyrola* and *Chimaphila*) are chlorophyllous herbs.

Arbutoideae (e.g., *Arctostaphylos* and *Arbutus*) are trees or shrubs with sympetalous, urceolate flowers, superior ovaries, and fleshy fruits (drupes, or berries with a fibrous inner fruit wall).

The pendulous, urn-shaped, cylindrical, or bell-shaped flowers of Ericaceae typically produce nectar and are visited by bees and wasps. The insect hangs onto the flower, probing for nectar, which is produced at the base of the corolla. In the process it brushes against the filaments or stamen appendages, causing pollen to fall onto its body. The stigma, which typically is centrally positioned at the narrowed corolla mouth, is situated so as to pick up pollen easily from visiting insects. In the montane tropics many species have tubular red flowers and are pollinated by birds. The viscin threads associated with the pollen of *Rhododendron* and related genera allow a large number of interconnected pollen tetrads to be simultaneously pulled out of an anther by pollinators. Capsular-fruited Ericaceae are largely wind-dispersed; most have small and/or winged seeds. The capsules are usually brought into an erect position before opening by movement of the pedicel. Species with berries (e.g., *Vaccinium, Cavendishia*) or drupes (e.g., *Arctostaphylos, Gaylussacia, Empetrum,* and *Styphelia*) are usually bird-dispersed.

Additional references: Bidartondo and Bruns 2001; Crayn and Quinn 2000; Crayn et al. 1998; Cullings and Bruns 1992; Hermann and Palser 2000; Hileman et al. 2001; Kron and Judd 1990; Kron and King 1996; Kron et al. 1999; Stevens et al. 2004a; Wallace 1975; Wood 1961.

Sarraceniaceae Dumortier
(Pitcher-Plant Family)

Carnivorous herbs or subshrubs; leaves alternate and spiral, **highly modified, forming pitcherlike traps**, *with ridge or laminar wing on adaxial side, and a flattened, relatively small, often hoodlike blade apically, inner surface often with retrorse hairs and glandular hairs,* sometimes in a basal rosette; stipules lacking. *Flowers large, ± pendulous, usually solitary on a scape,* bisexual, radial, *often associated with conspicuous bracts.* Sepals usually 5, distinct, often petaloid. Petals usually 5, distinct, imbricate. Stamens usually numerous; anthers sometimes inverting in development; pollen grains tricolporate to multicolporate. *Carpels 3 or 5, connate;* ovary superior, with axile or intruded parietal placentation; *style in Sarracenia expanded and peltate or umbrella-like with a small stigma under the tip of each of the 5 lobes;* stigmas truncate or minute. Ovules numerous, with 1 or 2 integuments and a thin-walled megasporangium. Nectaries absent. *Fruit a loculicidal capsule* (Figure 9.112).

Floral formula: *, K5, C5, A∞, G(3–5); capsule

Distribution and ecology: Limited to North America and northern South America; in acidic habitats.

Genera/species: 3/15. **Genera:** *Sarracenia* (8 spp.), *Heliamphora* (6), and *Darlingtonia* (1).

Economic plants and products: *Darlingtonia* and *Sarracenia* (pitcher plants) are grown as novelties.

Discussion: Sarraceniaceae are superficially similar to Nepenthaceae and Droseraceae because of the shared carnivorous habit. However, morphology and DNA characters suggest that Sarraceniaceae are not closely related to these families, which are in Caryophyllales, belonging instead within Ericales (Bayer et al. 1996; Hufford 1992; Kron and Chase 1993). The family may be the sister group of the carnivorous Roridulaceae. Morphology, *rbcL*, and ITS sequences (Bayer et al. 1996) support the monophyly of the family.

The tubular leaves of *Sarracenia* are variable, providing characters useful in specific delimitation. They may be erect or decumbent, and the apical hood usually prevents the entrance of rain (Plate 918B). These remarkable leaves are often conspicuously colored, have strong odors, and are provided with nectar glands. The leaves of several species have translucent, windowlike spots. Insects attracted by the coloration or odor may fall or crawl into the fluid-filled pitchers, where they are often trapped by downward-pointing hairs. If unable to escape, they die and are digested by enzymes in the fluid, apparently providing nitrogen that is often limiting in the acidic habitats occupied by pitcher plants.

Flowers are generally visited by various pollen-gathering bees and wasps. Dispersal of the small seeds is probably by wind and water.

Additional references: de Buhr 1975; Kubitzki 2004b; McDaniel 1971; Wood 1960; Renner 1989a.

Lecythidaceae A. Richard
(Brazil-nut Family)

Trees, or occasionally shrubs or lianas; **stems with cortical vascular bundles**; with triterpenoid saponins, often tannins, and sometimes mucilage canals. Hairs usually simple, glandular or nonglandular. *Leaves alternate and spiral, often clustered at the apices of the branches, sometimes 2-ranked, simple, entire or toothed, the teeth theoid* (with a deciduous glandular apex, see Figure 4.13), with pinnate venation; stipules present or lacking. Inflorescences indeterminate, terminal or axillary, sometimes reduced to a solitary axillary flower. *Flowers bisexual,* radial or *bilateral (due to unusual development of the androecium).* Sepals usually 4–6, distinct or connate, imbricate or valvate. *Petals usually 4–6, usually distinct,* imbricate, occasionally lacking. *Stamens usually numerous,* the ones nearest the gynoecium developing first, *usually connate, and in the more specialized genera the fused portion asymmetrical, produced on one side of flower, forming a flat structure that may be curved over the ovary, with some of the stamens often reduced and staminodial;* pollen grains tricolporate, sometimes with the colpi fused together. *Carpels 2–8, connate;* **ovary inferior or half-inferior**, rarely superior, with axile placentation; style capitate or lobed. Ovules 1 to many per locule, with a thin-walled megasporangium. Nectaries lacking, or nectar produced by staminodes. *Fruit a capsule, often very large and bony, usually opening by a circumscissile slit (with a cap), sometimes an indehiscent pod, drupe, or nut; seeds large, often with a fleshy aril, or aril flattened, forming a wing; embryo large, oily, often with a much-thickened hypocotyl; endosperm lacking or present (then usually ruminate).*

Floral formula:

X or *, K (4–6), C4–6, A ∞, G (2–6) capsule (opening by a cap)

Distribution and ecology: Widespread in the tropics, and especially diverse in moist to wet forests of South America.

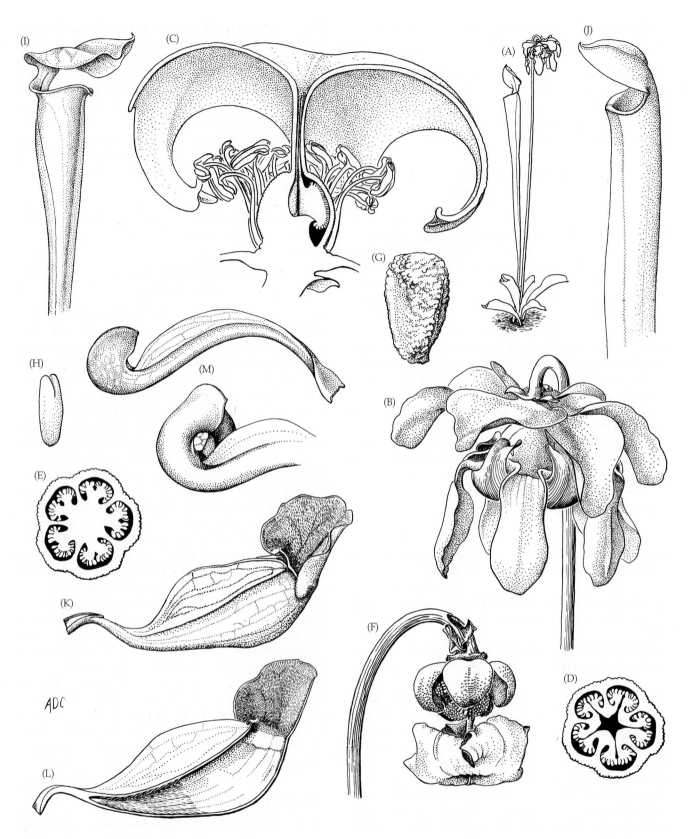

FIGURE 9.112 Sarraceniaceae. (A–H) *Sarracenia oreophila:* (A) flowering plant (× 7.5); (B) flower (× 1); (C) androecium and gynoecium in longitudinal section (× 2); (D) upper part of ovary in cross-section (× 3); (E) lower part of ovary in cross-section (× 3); (F) capsule, note persistent and expanded style (× 1); (G) seed (× 12); (H) embryo (× 12). (I) *S. flava:* upper part of leaf (× 0.5). (J) *S. rubra:* upper part of leaf (× 1). (K, L) *S. purpurea:* (K) leaf (× 0.5); (L) leaf in longitudinal section, note apical and basal portions with downward-pointing hairs (× 0.5). (M) *S. psittacina:* leaves (× 1). (From Wood 1974, *A student's atlas of flowering plants*, p. 44.)

Genera/species: 17/282. ***Major genera:*** *Eschweilera* (85), *Gustavia* (40), *Barringtonia* (40), and *Lecythis* (26). The family does not occur in temperate North America; *Barringtonia*, *Gustavia*, and *Couroupita* are occasionally cultivated in southern Florida.

Economic plants and products: *Bertholletia excelsa* provides edible seeds (Brazil nuts); the seeds of several species of *Lecythis* are also delicious. *Couroupita guianensis* (cannonball tree) and a few species of *Barringtonia* are used as ornamentals. Species of most genera provide useful timber.

Discussion: Both cpDNA nucleotide sequences and morphological characters (Morton et al. 1997a,b) support the monophyly of Lecythidaceae. The group consists of five major clades (which are here treated as subfamilies; see Morton et al. 1997b, 1998; Thorne 2001). *Napoleonaea* and *Crateranthus* (Napoleonaeoideae) are sister to the remaining members of the family, and form a clade diagnosed by extrorse anthers, an unusual androecium with an outer row of staminodes fused to form a radiating pseudocorolla, and the loss of petals. The next clade to diverge comprises *Asteranthos, Oubanguia, Scytopetalum*, and relatives (Scytopetaloideae); they are distinctive because they have seeds with ruminate endosperm. However, most genera and species belong to one of two subfamilies that are sister taxa. Barringtonioideae (*Barringtonia, Planchonia*, and relatives) are an Old World clade characterized by pollen with fused colpi and a reduction to only one seed per fruit. Lecythidoideae are limited to the neotropics and constitute the largest subfamily (containing, e.g., *Bertholletia, Couroupita, Eschweilera, Grias, Gustavia, Lecythis*). The monophyly of this group is supported by the haploid chromosome number of seventeen. Lecythidoideae contain all the genera having specialized bilaterally symmetrical flowers.

The large and showy flowers of Lecythidaceae usually are pollinated by various bees and wasps. The family shows an interesting evolutionary shift from species having radially symmetrical flowers with many stamens that use pollen as a pollinator reward to those with bilaterally symmetrical flowers with fewer fertile stamens (and many staminodes) that attract nectar-gathering euglossine bees. In some derived species, the nectar is produced by modified staminodes. Bat pollination is also known (in at least one species of *Lecythis* and in *Barringtonia*; Plate 9.18D). The seeds of Lecythidaceae are dispersed by a wide variety of animals (mammals, including rodents, monkeys, bats; various birds and fish), which are attracted by the arillate seeds or inner parts of the fruit wall, which are eaten. Water dispersal is common in some taxa that occur along rivers (e.g., *Allantoma*); a few (e.g., species of *Cariniana* and *Couratari*) are wind dispersed, and these have the aril modified into a wing.

Additional references: Mori and Prance 1990; Mori et al. 1978; Prance 1976; Prance and Mori 1978, 1979, 2004.

Polemoniaceae A. L. de Jussieu
(Phlox Family)

Herbs or occasionally shrubs, small trees, or lianas. Hairs various, often gland-headed. Leaves alternate and spiral, opposite, or whorled, simple, dissected, or pinnately compound, entire to serrate, with pinnate venation; stipules lacking. Inflorescences determinate, terminal, or sometimes solitary and axillary. *Flowers bisexual, radial, usually showy.* Sepals usually 5 and connate. *Petals usually 5, strongly connate, often forming a ± narrow tube, the lobes usually plicate and convolute.* **Stamens usually 5**; *filaments adnate to corolla tube*; pollen grains 4–many-colporate or porate. *Carpels 3,* connate; *ovary superior*, with axile placentation; stigmas on upper surface of style branches. Ovules 1 to numerous in each locule, with 1 integument and a thin-walled megasporangium. Nectar disk present. Fruit usually a loculicidal capsule; seed coat often mucilaginous when moistened (Figure 9.113).

Floral formula: *, K ⑤, C ⑤, A 5, G ③; capsule

Distribution: Widely distributed, but most diverse in temperate regions, especially western North America.

Genera/species: 18/380. ***Major genera:*** *Phlox* (70 spp.), *Gilia* (50), *Linanthus* (35), *Navarretia* (30), *Ipomopsis* (30), and *Polemonium* (25). Additional important genera in the United States and/or Canada are *Aliciella, Alophyllum, Collomia, Eriastrum, Giliastrum, Leptodactylon, Loeselia*, and *Microsteris*.

Economic plants and products: The family is best known for its ornamentals, including, *Gilia, Phlox*, and *Polemonium*, with showy, variously colored flowers.

Discussion: Polemoniaceae have been placed in Solanales because of their radially symmetrical flowers with sympetalous, plicate corollas (Plate 9.18H). However, DNA (Olmstead et al. 1993; Porter and Johnson 1998; Johnson et al. 1999) and morphology (Hufford 1992) indicate that the family belongs within Ericales. Analyses of nucleotide sequences indicate that the woody tropical genera of Polemoniaceae form a paraphyletic basal complex (now treated as Cobaeoideae, with uniform leaves, and Acanthogilioideae, with dimorphic leaves), while the herbaceous and more temperate genera such as *Ipomopsis, Linanthus, Navarretia, Polemonium, Phlox*, and *Gilia* constitute a monophyletic group (Johnson et al. 1996; Porter and Johnson 1998, 2000; Prather et al. 2000; Steele and Vilgalys 1994).

Flowers of Polemoniaceae are variously colored (sometimes even within a species, as in *Phlox drummondii*; see Kelly 1920) and shaped, attracting bees, flies, beetles, butterflies, and moths, as well as birds and bats. Most are out-

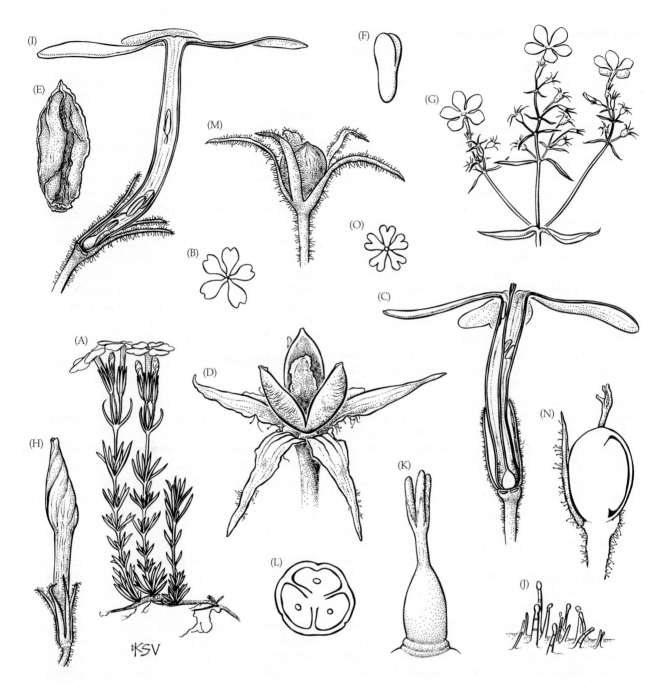

FIGURE 9.113 Polemoniaceae. (A–F) *Phlox nivalis:* (A) flowering plant (× 1); (B) flower, view from above (× 0.75); (C) flower in longitudinal section, note epipetalous stamens (× 4); (D) capsule with persistent calyx (× 4); (E) seed (× 15); (F) embryo (× 15). (G–N) *P. divaricata* var. *laphamii:* (G) inflorescence (× 0.5); (H) flower bud with convolute corolla (× 4); (I) flower in longitudinal section (× 4); (J) glandular hairs (greatly magnified); (K) gynoecium, note three stigmas (× 15); (L) ovary in cross-section with single ovule per locule (× 30); (M) immature fruit (× 4); (N) immature fruit in longitudinal section (× 5.5). (O) *P. divaricata* var. *divaricata:* flower, view from above (× 0.75). (Unpublished original art prepared for the Generic Flora of the Southeast U.S. series; used with permission.)

crossing due to protandry. Seed dispersal may be by external transport, facilitated by the mucilaginous seed coat, although movement of the small seeds by wind and/or water also occurs.

Additional references: Grant 1959; Grant and Grant 1965; Wilken 2004; Wilson 1960c.

CORE ASTERIDS

The remaining groups covered in this chapter form the clade of core asterids, which are divided into two subclades: lamiids (euasterids I) and campanulids (euasterids II) (see Figure 9.4). Core asterids have flowers in which the stamens are epipetalous and equaling (or less than) the corolla lobes in number, and a gynoecium usually of two fused carpels. Molecular data strongly supports the monophyly of this clade. Lamiids are generally characterized by opposite leaves, hypogynous flowers, and "early sympetaly" with a ring-shaped corolla primordium, while Campanulids typically have alternate leaves, epigynous flowers, and "late sympetaly" with distinct petal primordia (Bremer et al. 2001; Soltis et al. 2005).

Lamiids (Euasterids I)

Solanales

Solanales are tentatively taken as monophyletic, based on their characteristic radially symmetrical flowers with a plicate, sympetalous corolla. Members of this group have alternate, simple, exstipulate leaves and flowers in which the number of stamens equals the number of petals. They lack iridoids. The order consists of 6 families and about 7400 species; major families include **Solanaceae** (incl. Nolanaceae) and **Convolvulaceae** (incl. Cuscutaceae). The phylogenetic position of the **Boraginaceae** (incl. Hydrophyllaceae) is unclear, and this group, characterized by their scorpioid cymes, is included here merely for convenience. Analyses including Boraginaceae (e.g., Soltis et al. 2000; Albach et al. 2001b) have indicated a relationship with Lamiales, Lamiales + Solanales (Bremer et al 2002), Gentianales + Solanales + Lamiales (Hilu et al. 2003), or just Solanales (Chase et al. 1993; Olmstead et al. 2000); the last placement is also supported by their plicate corollas. DNA characters indicate that Solanales are most closely related to Gentianales and Lamiales (Bremer et al. 2002; Downie and Palmer 1992; Olmstead et al. 1992, 1993, 2000; Soltis et al. 2000). Solanaceae and Convolvulaceae are considered to be sister families, based on the anatomical synapomorphy of internal phloem, similar alkaloid chemistry, and several cpDNA characters (Figure 9.114).

Solanaceae A. L. de Jussieu
(Potato or Nightshade Family)

Herbs, shrubs, trees, or vines; internal phloem usually present; various alkaloids present. Hairs diverse, but often stellate or branched, sometimes with prickles. *Leaves alternate and spiral*, often in pairs, the members of a pair both on the same side of the stem, *simple, sometimes deeply lobed* or even pinnately compound, entire to serrate, with pinnate venation; stipules lacking. Inflorescences determinate, sometimes reduced to a single flower, terminal but usually appearing lateral. *Flowers usually bisexual and radial. Sepals usually 5, connate*, persistent, sometimes enlarging as fruit develops. *Petals usually 5, connate, often strongly so, and forming a wheel-shaped, bell-shaped, funnel-shaped, or tubular corolla, distinctly plicate (with fold lines), with marginal portions of each corolla lobe often folded inward, and often convolute*, sometimes imbricate or valvate. *Stamens usually 5; filaments adnate to corolla;* anthers usually 2-locular, opening by longitudinal slits or terminal pores, sometimes sticking to each other; pollen grains usually 3- to 5-colpate or colporate. *Carpels usually 2* (–5), **oriented obliquely to the median plane of the flower**, connate; *ovary superior*, entire to deeply lobed, *usually with axile placentation and 2 locules*; style terminal to gynobasic; stigma 2-lobed. *Ovules usually numerous in each locule*, but occasionally reduced to 1 per locule, with 1 integument and a thin-walled megasporangium. Nectar disk present or lacking. *Fruit usually a berry, septifragal capsule*, or schizocarp of nutlets; seeds often flattened (Figure 9.115).

Floral formula: *, K ⑤, C ⑤, A 5, G ②; berry, capsule

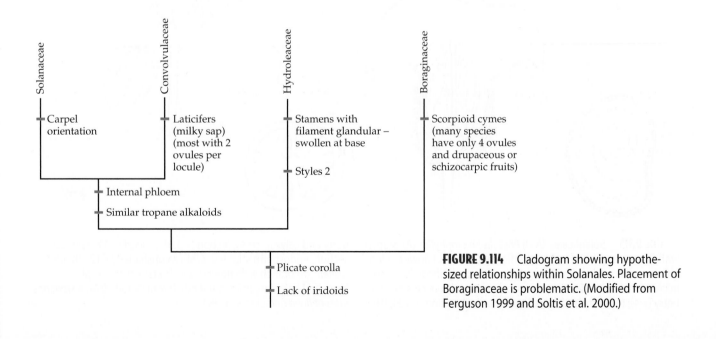

FIGURE 9.114 Cladogram showing hypothesized relationships within Solanales. Placement of Boraginaceae is problematic. (Modified from Ferguson 1999 and Soltis et al. 2000.)

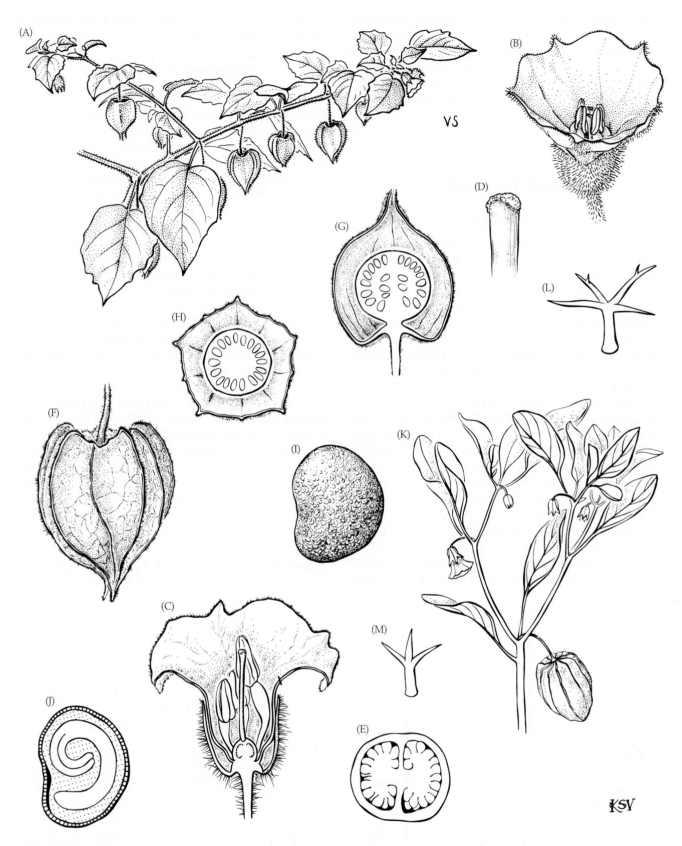

FIGURE 9.115   Solanaceae. (A–J) *Physalis heterophylla:* (A) branch with flowers and fruits (× 0.35); (B) flower (× 2.5); (C) flower in longitudinal section (× 5.5); (D) stigma (greatly magnified); (E) ovary in cross-section (× 22); (F) mature accrescent calyx enclosing berry (× 4); (G) berry and calyx in longitudinal section (× 1.5); (H) berry and calyx in cross-section (× 1.5); (I) seed (× 12); (J) seed in section, endosperm stippled, embryo unshaded (× 12). (K–M) *P. walteri:* (K) branch with flowers and fruits (× 0.35); (L, M) branched hairs (greatly magnified). (From Wood 1974, *A student's atlas of flowering plants,* p. 98.)

Key to Major Families of Solanales

1. Ovary with 4 ovules 2
1. Ovary with more than 4 ovules (or if only 4, then placentation parietal) 3

2. Stems lacking internal phloem and laticifers, sap not milky; inflorescence often distinctly scorpioid/helicoid; leaves with firm unicellular hairs that have a basal cystolith and often calcified or silicified walls and thus are rough to the touch; fruit a drupe with several pits or a schizocarp splitting into 4 nutlets .... **Boraginaceae**

2. Stems with internal phloem and usually also with laticifers (and sap milky); inflorescence and hairs not as above; fruit a capsule .... **Convolvulaceae**

3. Stems lacking internal phloem; inflorescence scorpioid; placentation parietal; flowers lacking a nectar disk .... **Boraginaceae**

3. Stems with internal phloem; inflorescence usually not scorpioid; placentation axile; flowers with a nectar disk .... **Solanaceae**

Distribution and ecology: Widespread, but most diverse in the Neotropics. Many species occur in disturbed habitats.

Genera/species: 102/2510. **Major genera:** *Solanum* (1400 spp.), *Lycianthes* (200), *Cestrum* (175), *Nicotiana* (100), *Physalis* (100), and *Lycium* (90). Numerous genera occur in the continental United States and/or Canada, some of which are *Capsicum, Datura, Solanum, Lycium, Nicotiana, Petunia,* and *Physalis*.

Economic plants and products: Most members of the family are poisonous due to the presence of tropane or steroid alkaloids. Solanaceae are the source of several pharmaceutical drugs, and some are powerful narcotics; some of these plants include *Nicotiana* (tobacco; Plate 9.19E), *Atropa* (belladona), and *Datura* (jimsonweed). Surprisingly, the family also provides edible fruits, such as cayenne, red, and green peppers (*Capsicum* spp., which contain the alkaloid capsaicin, providing a "hot" flavor), tomatoes (*Solanum lycopersicum*; Plate 9.19H), eggplants (*S. melongena*), tree tomatoes (*S. betacea*), and tomatillos (*Physalis ixocarpa*). The tubers of *Solanum tuberosum* (potatoes) are an important source of starch in the human diet. Many genera provide ornamentals, including *Brunfelsia* (lady-of-the-night, yesterday-today-and-tomorrow), *Cestrum* (night-blooming jessamine), *Datura* (angel's trumpet), *Petunia, Physalis* (ground-cherry), and *Solanum* (nightshade; Plate 9.19A).

Discussion: Solanaceae are considered monophyletic on the basis of morphological and cpDNA characters (Olmstead and Palmer 1991, 1992). Infrafamilial relationships have been investigated by D'Arcy (1979, 1991), using morphology, and by Olmstead and Palmer (1991, 1992, 1997), Olmstead and Sweere (1994), and Olmstead et al. (1995, 1999), based on a cladistic analysis of *rbcL* and *ndhF* sequences and cpDNA restriction site characters, and Martins and Barkman (2005), based on the nuclear gene salicylic acid methyltrans-

ferase, along with *rbcL* and *ndhF* sequences. The family has often been divided into two large subgroups, Cestroideae (e.g., *Brunfelsia, Petunia, Cestrum, Nicotiana,* and *Schizanthus*), defined by straight to slightly bent embryos and prismatic to subglobose seeds, and Solanoideae (e.g., *Solanum, Capsicum, Lycianthes, Datura, Physalis, Lycium, Atropa,* and *Mandragora*), with curved embryos and flattened, discoidal seeds. Members of the Cestroideae typically have capsules, while both berries and capsules occur in Solanoideae. *Nolana* and relatives, which are especially distinct due to their gynobasic style and deeply lobed ovary, are included in Solanaceae (e.g., as the small subfamily Nolanoideae, D'Arcy 1979, 1991; Thorne 1992) or within Solanoideae (Olmstead et al. 1999), but systematists (e.g., Cronquist 1981) used to place them in their own family. The cladistic analyses of Olmstead and Palmer (1992), Olmstead et al. (1995, 1999), and Martins and Barkman (2005) demonstrated that "Cestroideae" constitute a basal paraphyletic complex, which is divided into six subfamilies by Olmstead et al. (1995). Bilateral flowers have evolved several times within this complex. In contrast, monophyly of the Solanoideae is supported by cpDNA characters, discoidal seeds, and curved embryos. *Nolana* (Nolanoideae) nests within the Solanoideae; its recognition as family or subfamily has been due to over emphasis on its unusual multilobed ovary. Among former Cestroideae, genera such as *Nicotiana* and *Anthocercis* (now in Nicotianoideae) are sister to Solanoideae, forming a clade diagnosed by a base chromosome number of 12 (Martins and Barkman 2005; Olmstead et al. 1999). The monophyly of the large genus *Solanum* is supported only if *Lycopersicon* and *Cyphomandra* are included (Bohs and Olmstead 1997; Olmstead and Palmer 1997; Spooner et al. 1993, 2005). *Solanum* s.l. is diagnosed by its distinctive wheel-shaped and deeply lobed corollas, connivent anthers usually opening by apical pores, and cpDNA restriction site, *rbcL*, and *ndhF* characters. *Solanum*, and many other members of Solanoideae, have berries (an apomorphic feature).

Flowers of Solanaceae are usually showy and attract various bees, wasps, flies, butterflies, and moths. *Solanum* does not produce nectar and is pollinated by pollen-gathering bees and flies. Pollen is removed from the anthers by vibration or manipulation of the anthers. In contrast, *Cestrum* and *Datura* attract nectar-gathering insects, such as moths or butterflies. The brightly colored berries are dispersed by birds.

Additional references: Evans 1979; Roddick 1986, 1991.

Convolvulaceae A. L. de Jussieu
(Morning Glory Family)

Usually twining and climbing herbs, often rhizomatous, occasionally with little or no chlorophyll and parasitic; the roots often storing carbohydrates; internal phloem usually present; **usually with laticifers, often with milky sap**; sometimes with alkaloids. Hairs various, often 2-branched or simple. *Leaves alternate and spiral, simple*, sometimes lobed or compound, usually entire, with pinnate or sometimes palmate venation, occasionally reduced; *stipules lacking*. Inflorescences determinate, sometimes reduced to a solitary flower, terminal or axillary. *Flowers usually bisexual, radial. Sepals usually 5, usually distinct or only very slightly connate*, persistent. *Petals usually 5, strongly connate and forming funnel-like corolla, distinctly plicate (with fold lines), with marginal portions of each corolla lobe folded inward and the middle portion of each lobe valvately arranged in bud, and often also convolute with a clockwise twist*, but occasionally merely imbricate. *Stamens usually 5; filaments epipetalous*, often of unequal lengths; pollen grains usually tricolpate to multiporate. *Carpels 2, connate; ovary superior*, entire to deeply 2- or 4-lobed, *usually with axile placentation;* style(s) terminal to gynobasic; stigma(s) 1 or 2, capitate, lobed, or linear. *Ovules usually 2 in each locule*, with 1 integument and usually a thin-walled megasporangium. Nectar disk usually lobed. *Fruit usually a septifragal, circumscissile, or irregularly dehiscing capsule* (Plate 9.19D); embryo straight or curved, with folded cotyledons, or reduced (Figure 9.116; also see Figure 4.47G,H).

Floral formula: *, K5, C⑤, A⑤, G②; capsule

Distribution: Widely distributed; most diverse in tropical and subtropical regions.

Genera/species: 55/1930. **Major genera:** *Ipomoea* (600 spp.), *Convolvulus* (250), *Cuscuta* (150), and *Jacquemontia* (120). Numerous genera occur in the continental United States and/or Canada; noteworthy genera other than those listed above are *Bonamia*, *Calystegia*, *Dichondra*, *Evolvulus*, *Merremia*, and *Stylisma*.

Economic plants and products: *Ipomoea batatas* (sweet potato) is important for its edible roots. Many species are poisonous. *Ipomoea* (morning glory), *Jacquemontia*, *Porana* (Christmas vine), and *Dichondra* (ponyfoot) provide ornamentals.

Discussion: Monophyly of Convolvulaceae is supported by molecular characters and the presence of laticifers. *Humbertia* is probably sister to the remaining genera, which are united by a reduction to two ovules per carpel (Stefanovic et al. 2002, 2003). The group is divided into 12 tribes based on molecular data (sequences from the chloroplast, mitochondrial, and nuclear genomes) and morphology, e.g., presence or absence of the parasitic habit, leaf venation, number of styles and their length, stigma shape, fruit type, including dehiscence patterns, and pollen shape and surface features. Cuscutaceae are often segregated because of their numerous specializations associated with parasitism: reduction in chlorophyll content, scale-like leaves, haustoria, and reduced embryos. *Dichondra* and relatives are sometimes recognized as Dichondraceae due to their gynobasic styles. Recognition of either of these families makes Convolvulaceae s.s. paraphyletic (Neyland 2001; Stefanovic et al. 2002, 2003).

The flowers of Convolvulaceae are usually showy, and attract various insects (Plate 9.19B). They open for a day or less (often only a few hours), after which the corolla wilts. Some species of *Ipomoea* are pollinated by hummingbirds. Nectar is the reward. The seeds are relatively large, but are probably dispersed by wind, especially when retained in the rather papery and inflated capsule.

Additional references: Austin 1979; Allard 1947; Wilson 1960b.

Boraginaceae A. L. de Jussieu
(Borage Family)

Herbs or shrubs to trees, occasionally lianas or root-parasites; internal phloem lacking; often with alkaloids. Hairs various, but *often unicellular, with a basal cystolith and often calcified or silicified walls, and the plants rough to the touch*. *Leaves usually alternate and spiral, simple*, but sometimes deeply lobed or even compound, entire to serrate, with pinnate venation; *stipules lacking*. **Inflorescences determinate, usually forming scorpioid cymes (i.e., axes coiled, bearing flowers along the upper side and straightening as the flowers mature**; Plate 9.19F), usually terminal. *Flowers usually bisexual and radial*. Sepals usually 5, distinct to connate. *Petals usually 5, strongly connate and forming wheel-like, funnel-like, or tubular corolla, plicate (with fold lines)*, imbricate or convolute. *Stamens usually 5; filaments epipetalous*; pollen grains tricolporate or triporate to polycolpate or polycolporate. *Carpels usually 2, connate; ovary superior*, spherical to deeply 4-lobed, with axile placentation *and 4 locules (actual locule of each carpel becomes divided by development of a false partition) or parietal placentation and usually numerous ovules*; style(s) terminal or gynobasic; stigma(s) 1 and 2-lobed, 2, or 4, capitate to

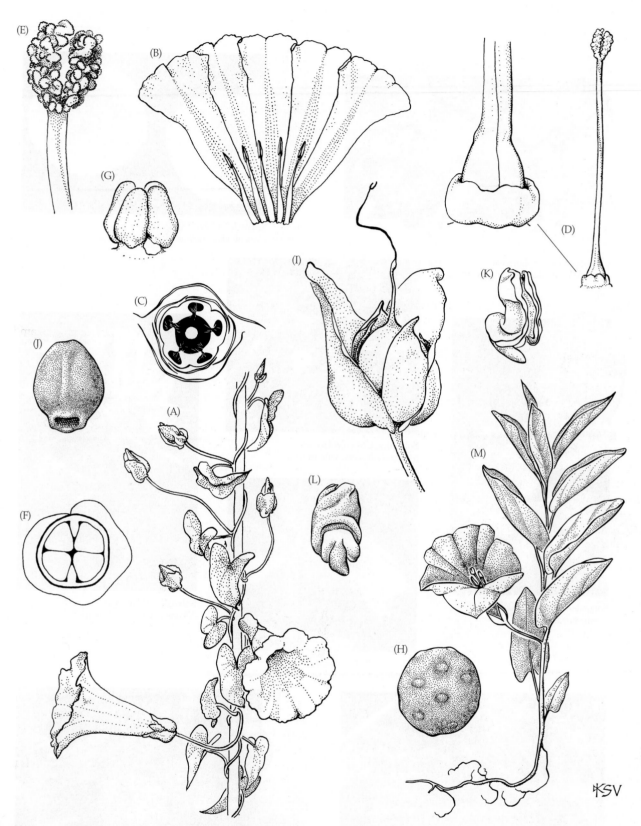

FIGURE 9.116 Convolvulaceae. (A–L) *Calystegia sepium:* (A) vine with flowers (× 0.5); (B) corolla opened out, note plications in corolla and epipetalous stamens (·× 1); (C) flower in cross-section just above ovary, note two large bracteoles, free sepals, corolla tube with adnate filaments, and style (× 2); (D) gynoecium and nectar disk (× 2; enlargement × 8); (E) apex of style and stigmas (× 7); (F) ovary and irregular nectar disk in cross-section, note incomplete septum and four ovules (× 8); (G) four ovules, removed from ovary (× 8); (H) pollen grain, pantoporate (greatly magnified); (I) capsule with persistent bracts and sepals (× 2); (J) seed, adaxial surface (× 5); (K, L) two views of an embryo, note folded cotyledons (greatly enlarged). (M) *C. spithamaea:* flowering plant. (From Wood 1974, *A student's atlas of flowering plants*, p. 92.)

(A)

Solanales: Solanaceae
Solanum dulcamara: flowers

(B)

Solanales: Convolvulaceae
Convolvulus arvensis: flowers

(C)

Solanales: Boraginaceae
Cordia lutea: flowers

(D)

Solanales: Convolvulaceae
Merremia dissecta: vine in fruit

(E)

Solanales: Solanaceae
Nicotiana tabacum: flowers

(F)

Solanales: Boraginaceae
Symphytum officinale: scorpioid inflorescence

(G)

Solanales: Boraginaceae
Cynoglossum amabile: 4-lobed fruits

**PLATE 9.19 Eudicots:
Solanales**

(H)

Solanales: Solanaceae
Solanum lycopersicon: fruits

truncate. Ovules 1 in each locule or numerous, with 1 integument and a thin-walled megasporangium. Nectar disk usually present around base of ovary. Fruit a drupe with 1 4-seeded, 2 2-seeded, or 4 1-seeded pit(s), a schizocarp of usually 4 1-seeded nutlets, or a loculicidal or irregularly dehiscent capsule; embryo straight to curved; endosperm present to lacking (Figure 9.117).

Floral formula:

*, K⑤, C⑤, A 5, G②; capsule, drupe, schizocarp of 2 or 4 nutlets

Distribution: Widely distributed in temperate and tropical regions.

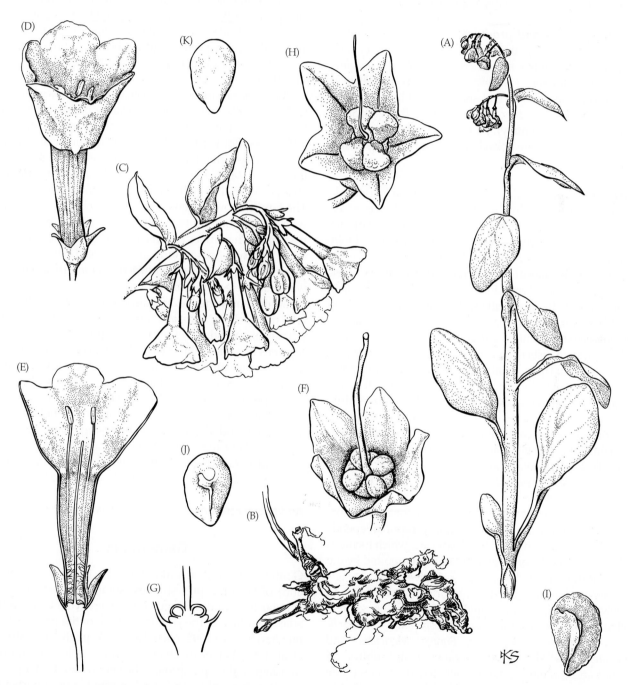

FIGURE 9.117 Boraginaceae. *Mertensia virginica:* (A) flowering plant (× 0.35); (B) root with two stem bases (upper left) and rhizome (lower left) (× 0.75); (C) inflorescence (× 1.5); (D) flower (× 3); (E) flower, in longitudinal section (× 3); (F) flower after corolla has fallen, to show gynoecium (× 7.5); (G) gynoecium in longitudinal section, note gynobasic style (× 7.5); (H) nutlets and accrescent calyx (× 4.5); (I) abaxial surface of nutlet (× 9); (J) seed (× 9); (K) embryo (× 9). (From Al-Shehbaz 1991, *J. Arnold Arbor. Suppl. Ser. 1:* p. 91.)

Genera/species: 134/2650. ***Major genera:*** *Cordia* (320 spp.), *Heliotropium* (260), *Tournefortia* (150), *Onosma* (150), *Cryptantha* (150), *Phacelia* (150), *Myosotis* (100), *Cynoglossum* (75), and *Ehretia* (75). Numerous genera occur in the continental United States and/or Canada, including *Amsinckia, Bourreria, Cordia, Cryptantha, Cynoglossum, Draperia, Ellisia, Eriodictyon, Hackelia, Heliotropium, Hydrophyllum, Lappula, Lithospermum, Mertensia, Myosotis, Nama, Nemophila, Onosmodium, Phacelia, Plagiobothrys,* and *Romanzoffia.*

Economic plants and products: *Heliotropium* (heliotrope), *Mertensia* (Virginia bluebells), *Myosotis* (forget-me-not), *Cordia* (geiger tree and others; Plate 9.19C), *Cynoglossum* (hound's-tongue), *Pulmonaria* (lungwort), and *Phacelia* provide ornamentals. Several species have been used as medicinal herbs, including *Borago officinalis* (borage), *Symphytum officinale* (comfrey), and *Lithospermum* spp. (puccoon); many species are poisonous. *Cordia* is important as a timber.

Discussion: The family is circumscribed broadly (including Lennoaceae, and Hydrophyllaceae, except *Hydrolea;* see Figure 9.114) and is considered to be monophyletic on the basis of its distinctive inflorescence form and DNA sequence data (Chase et al. 1993; Downie and Palmer 1992; Ferguson 1999; Långström and Chase 2002; Olmstead et al. 1992b, 1993, 2000; Olmstead and Ferguson 2001). Schizocarpic fruits with four nutlets, 4-locular ovaries, and a gynobasic style evolved independently within Boraginaceae, Verbenaceae, and Lamiaceae (Cantino 1982).

Most systematists recognize four large subfamilies (Al-Shehbaz 1991). The tropical Ehretioideae (e.g., *Ehretia* and *Bourreria*) have plesiomorphies such as a woody habit, terminal style divided into two branches, unlobed ovary, drupaceous fruit, flat cotyledons, and presence of endosperm. Cordioideae, which contains *Cordia* and two small genera, may constitute a related clade based on the four-branched style, plicate cotyledons, and lack of endosperm. This mainly tropical group has retained a woody habit, terminal styles, unlobed ovaries, and drupaceous fruits. The remaining two subfamilies both have schizocarpic fruits (with nutlets)—derived conditions that probably evolved independently. The monophyly of Heliotropioideae, which includes *Heliotropium, Tournefortia,* and *Argusia,* is supported by the often short, undivided style with a single stigma. Boraginoideae can be characterized by their gynobasic style and deeply lobed ovary; this large subfamily (106 genera, including *Hackelia, Plagiobothrys, Lithospermum, Onosmodium, Cynoglossum, Symphytum, Borago, Mertensia,* and *Myosotis*) is well developed in temperate and subtropical regions and contains numerous herbaceous species (Plate 9.19F). Members of all four subfamilies have gynoecia with only four ovules. Ehretioideae and Cordioideae often are segregated as "Ehretiaceae."

In addition, most genera traditionally placed in "Hydrophyllaceae" are closely related to the groups outlined above, and are here included within Boraginaceae. These plants have usually numerous ovules and capsular fruits (vs. gynoecia with only 4 ovules, developing into drupes or nutlets in the traditional Boraginaceae). Both numerous ovules and capsular fruits may be ancestral within the Boraginaceae/Hydrophyllaceae clade. *Codon* may be related most closely to Boraginoideae, while *Phacelia, Nama,* and relatives probably are basal and perhaps paraphyletic in a clade containing "Ehretioideae," Cordioideae, and Heliotropioideae. Finally, nucleotide sequences of the *ndhF* chloroplast gene suggest that *Pholisma* and relatives (Lennoaceae), an unusual group of root parasites, may also belong to Boraginaceae, being within the ehretioid complex (Olmstead and Ferguson 2001).

Boraginaceae are usually pollinated by bees, wasps, butterflies, and flies, which gather nectar or pollen, but moths, flies, beetles, bats, and birds have also been recorded as pollinators. Both self- and cross-pollination occur; distyly has been documented in species of several genera. The flowers of *Mertensia, Myosotis,* and *Cryptantha* rapidly change color after pollination, as a signal to pollinators. The family shows diverse dispersal mechanisms. Species with drupaceous fruits are usually dispersed by birds or mammals. The drupes of some coastal species of *Cordia* are corky and will float in (and are dispersed by) water. The corky nutlets of *Argusia* and *Mertensia maritima* are also water-dispersed. In some genera with nutlets, the calyx is persistent and winglike, and dispersal is by wind. The nutlets of *Hackelia, Lappula,* and *Cynoglossum* have appendages and stick to fur or clothing (Plate 9.19G), and *Myosotis* has hooks on its calyx. Many genera have an aril-like structure attached to the base of the nutlets that is hard and contains sugars, fats, and free amino acids; these are dispersed by ants. In *Heliotropium* and *Lithospermum* the nutlets are eaten by birds. Finally, the nutlets may be consumed and dispersed by browsing mammals. The small seeds of the capsular species are mainly wind- or water-dispersed. Again, some seeds have a hard, aril-like structure and may be at least partially dispersed by ants.

Additional references: Prior 1960; Wilson 1960c.

Gentianales

Gentianales are clearly monophyletic, based on the presence of vestured pits, stipules (sometimes reduced to a stipular line) and thick glandular hairs (colleters) on the adaxial surface of the stipules or base of the petiole (Bremer and Struwe 1992; Nicholas and Baijnath 1994; Struwe et al. 1994; Wagenitz 1959, 1992). The colleters produce mucilage and help to protect the shoot apex. Other possible synapomorphies include opposite leaves, a particular type of complex indole alkaloids, corollas that are convolute in bud (Bremer and Struwe 1992; Wagenitz 1992), *rbcL, matK, atpB, ndhF, 18S* sequences (Bremer et al. 1994, 2002; Chase et al. 1993; Olmstead et al. 1993; Endress et al.

Key to Major Families of Gentianales

1. Ovary inferior; internal phloem lacking; stipules ± interpetiolar; placentation axile ....**Rubiaceae**

1. Ovary or ovaries superior; internal phloem usually present; stipules various (i.e., adjacent to the petiole base, ± interpetiolar, often reduced or lacking, and plant then usually with a nodal line); placentation axile, or parietal, or marginal...2

2. Latex milky; style apically thickened, forming a stylar head; ovaries usually distinct, joined only by a common style or stylar head..**Apocynaceae**

2. Latex not milky; style not apically thickened and a stylar head lacking; ovaries usually connate 3

3. Stipules usually present, interpetiolar, these sometimes reduced to stipular lines at the node; placentation usually axile ....Loganiaceae

3. Stipules usually lacking; placentation parietal ....**Gentianaceae**

1996; Hilu et al. 2003; Källersjö et al. 1998; Backlund et al. 2000; Olmstead et al. 2000a; Savolainen et al. 2000a,b; Soltis et al. 2000; Albach et al. 2001a), and cpDNA restriction sites (Downie and Palmer 1992). The order consists of 5 families and about 14,200 species; families include **Gentianaceae, Rubiaceae**, and **Apocynaceae** (incl. Asclepiadaceae), Loganiaceae, and Gelsemiaceae. Gentianales are clearly within the asterid clade, as indicated by the single integument and a thin-walled megasporangium, and within the core asterid group because of their sympetaly and epipetalous stamens equaling the number of corolla lobes. Analyses of DNA sequences also strongly support the group's position within the asterid clade, and suggest

that the order is most closely related to the Solanales. Most members of both orders have internal phloem and radially symmetrical flowers with the number of stamens equaling the number of corolla lobes. Gentianales are easily distinguished by their usually opposite leaves, stipules or at least a stipular line, and colleters.

"Loganiaceae," as traditionally circumscribed, constitute a nonmonophyletic undefinable assemblage. Thus, *Fagraea* (Plate 9.20A) and *Potalia* should be transferred to Gentianaceae, and *Buddleja* and relatives constitute the Buddlejaceae (of the Lamiales). *Gelsemium* and *Mostuea* are taxonomically isolated and are best treated as Gelsemiaceae (Figure 9.118), but within Gentianales (Bremer and Struwe

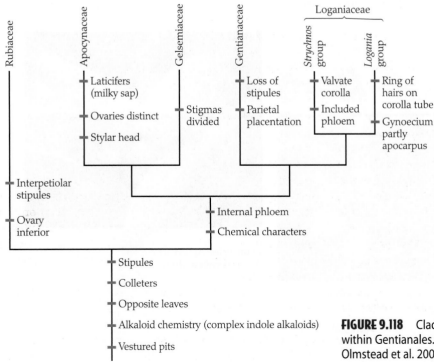

FIGURE 9.118 Cladogram showing hypothesized relationships within Gentianales. (Modified from Backlund et al. 2000, Olmstead et al. 2000a and Soltis et al. 2000.)

(A)

Gentianales: Gentianaceae
Fagraea ceilanica: leaves and flowers

(B)

Gentianales: Apocynaceae
Vinca minor: flower,
longitudinal section

(C)

Gentianales: Apocynaceae
Asclepias syriaca: flowers; latex

(D)

Gentianales: Rubiaceae
Coffea arabica: fruits

(E)

Gentianales: Apocynaceae
Nerium oleander: follicle with hairy seeds

(F)

Gentianales: Rubiaceae
Cubanola domingensis: flowers

(G)

Gentianales: Gentianaceae
Eustoma exaltatum: flowers

(H)

Gentianales: Apocynaceae
Asclepias syriaca: fruits

**PLATE 9.20 Eudicots:
Gentianales**

1992; Endress et al. 1996; Struwe et al. 1994; Backlund et al. 2000; Soltis et al. 2000). Most genera of Loganiaceae, however, probably do constitute a monophyletic group, containing two major subclades: the *Strychnos* group (*Strychnos* and *Spigelia*) and the *Logania* group (*Geniostoma, Labordia, Logania, Mitreola,* and *Mitrasacme*). The former is characterized by a valvate corolla and included phloem, while the latter has a ring of hairs at the mouth of the corolla tube and a partly apocarpous gynoecium (Struwe et al. 1994). Loganiaceae have often been considered to form a basal complex within Gentianales, but recent analyses (Backlund et al. 2000; Olmstead et al. 2000) support the recognition of two major clades within the order, the first comprising Rubiaceae and the second including the remaining families (Figure 9.118). In addition to the characters shown in Figure 9.118, these two groups also differ in the biosynthesis of their iridoid and indole alkaloids (Jensen 1992).

Additional references:　Wood 1983; Wood and Weaver 1982.

Rubiaceae A. L. de Jussieu
(Coffee or Madder Family)

Trees, shrubs, lianas, or herbs; lacking internal phloem; usually with iridoids, various alkaloids; raphide crystals common. Hairs various. *Leaves opposite or whorled, usually entire,* with pinnate venation; *stipules present,* **interpetiolar** *and usually connate,* occasionally leaflike, *with colleters on adaxial surface.* Inflorescences determinate, occasionally reduced to a single flower, terminal or axillary. *Flowers usually bisexual and radial, often heterostylous, frequently aggregated.* Sepals usually 4 or 5, connate, sometimes with colleters on adaxial surface. *Petals usually 4 or 5, connate,* forming a usually wheel-shaped to funnel-shaped corolla (Plate 9.20F), adaxial surface often pubescent, the lobes valvate, imbricate, or contorted. *Stamens usually 4 or 5; filaments usually adnate to corolla* and positioned within corolla tube or at its mouth, sometimes basally connate; anthers 2-locular, opening by longitudinal slits; pollen grains usually tricolporate. *Carpels usually 2 (-5), connate;* **ovary inferior,** with usually axile placentation; stigma(s) 1 or 2, linear, capitate, or lobed. Ovules 1 to numerous in each locule, with 1 integument and a thin-walled megasporangium. Nectar disk usually present above ovary. Fruit a loculicidal to septicidal capsule, berry, drupe, schizocarp, or indehiscent pod; seeds sometimes winged; embryo straight to curved; endosperm present or lacking (Figure 9.119).

Floral formula:

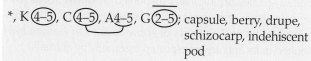

*, K④-⑤, C④-⑤, A4-5, G$\overline{②-⑤}$; capsule, berry, drupe, schizocarp, indehiscent pod

Distribution:　Cosmopolitan, but most diverse in tropical and subtropical regions.

Genera/species: 550/9000.　**Major genera:** *Psychotria* (1500 spp.), *Galium* (400), *Ixora* (400), *Pavetta* (400), *Hedyotis* (400), *Tarenna* (370), *Randia* (250), *Gardenia* (250), *Palicourea* (250), *Mussaenda* (200), *Borreria* (150), and *Rondeletia* (125). Some of the numerous genera in the continental United States and/or Canada are *Catesbaea, Cephalanthus, Chiococca, Diodia, Ernodea, Erithalis, Exostema, Galium, Genipa, Guettarda, Hamelia, Hedyotis* (with segregates such as *Houstonia, Oldenlandia, Pentodon, Stenaria*), *Mitchella, Morinda, Pentodon, Pinckneya, Psychotria, Randia, Richardia,* and *Spermacoce.*

Economic plants and products:　Coffee, a stimulating beverage containing caffeine, is made from the brewed seeds of *Coffea arabica* (Plate 9.20D) and *C. robusta.* Quinine, a drug used in treating malaria, comes from the bark of species of *Cinchona,* and ipecac, a drug used to induce vomiting, is derived from *Psychotria. Gardenia, Hamelia, Pentas, Randia, Rondeletia, Serissa, Hedyotis,* and *Ixora* provide ornamentals. *Rubia tinctoria* (madder) was long important as a red dye.

Discussion:　Rubiaceae are an easily recognized monophyletic group. Phylogenetic relationships within the family have been investigated by cladistic analyses of morphological and/or molecular characters (Andersson and Rova 1999; Andreasen and Bremer 2000; Bremer and Struwe 1992; Bremer and Jansen 1991; Bremer et al. 1995, 1999; Robbrecht and Manen 2006; Rova et al. 2002).

Cinchonoideae, which include *Gardenia, Casasia, Ixora, Coffea, Erithalis, Chiococca, Exostema, Hamelia, Randia, Cinchona, Cephalanthus, Pinckneya, Mussaenda, Portlandia, Catesbaea,* and *Rondeletia,* have traditionally been defined either on the basis of ovaries with numerous ovules or by the presence of endosperm, lack of raphide crystals, and seeds with a pitted-ridged coat (all likely symplesiomorphies). The monophyly of this group has not been supported by cladistic studies employing morphological or cpDNA restriction site characters, but receives support from analyses of *rbcL, ndhF,* or *rps16* intron sequences. However, a group of genera within Cinchonoideae, often segregated as Ixoroideae (e.g., *Ixora, Coffea, Gardenia, Mussaenda, Randia*), are clearly monophyletic. This clade is characterized by flowers with the corolla contorted (with a left twist) and a specialized pollination presentation mechanism. The anthers shed their pollen before the flower opens, and pollen is presented to pollinators on the hairy upper portion of the style; only later does the stigma become receptive. This mechanism is similar to the plunger pollination system of the Asterales.

Rubioideae, comprising the majority of the species, are usually defined by the combination of the presence of raphides and valvate corolla lobes. Some members of this clade have numerous ovules, while others show a reduction to a single ovule per locule. This subfamily probably is monophyletic; potential synapomorphies include the presence of raphides, seeds with a smooth coat, the usually herbaceous habit (presumably reversed in genera like *Psy-*

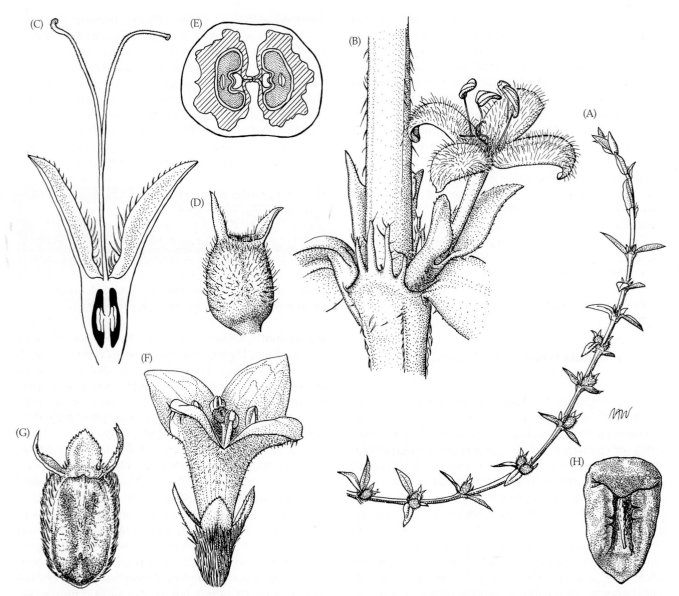

FIGURE 9.119 Rubiaceae. (A–E) *Diodia tetragona:* (A) flowering plant (× 0.4); (B) node with sessile axillary flower, note interpetiolar stipules (× 5); (C) gynoecium and calyx in longitudinal section, note inferior ovary (× 10); (D) nearly mature fruit (× 5); (E) dry drupe with 2 endocarps, in cross-section, endocarp hatched, endosperm stippled, embryo unshaded (× 10). (F–H) *D. teres:* (F) flower (× 5); (G) portion of schizocarp (× 5); (H) seed, adaxial surface (× 8). (From Wood 1974, *A student's atlas of flowering plants*, p. 106.)

chotria and *Palicourea*), and numerous cpDNA characters. Representative genera are *Pentas, Palicourea, Psychotria, Galium, Nertera, Hedyotis, Richardia, Diodia, Spermacoce, Pentodon, Morinda,* and *Mitchella.* Fleshy fruits and ovaries with a reduced number of ovules have evolved several times within Rubiaceae.

In *Galium* the stipules are expanded leaflike structures; the leaves, therefore, appear to be in whorls. The number of apparent leaves in a whorl varies depending upon the amount of fusion and splitting of the stipules.

The flowers of Rubiaceae are variable in color and shape, and may be pollinated by butterflies, moths, bees, flies, birds, or bats; wind pollination occurs in a few. Nectar is employed

as the pollinator reward. Outcrossing is promoted by protandry, a specialized pollen presentation mechanism (see above), or heterostyly. The Rubiaceae contain more species with heterostyly than any other angiosperm family, and may contain more than all other angiosperms combined. Fleshy-fruited species are typically dispersed by birds. Seeds from capsular fruits may be small or winged and are often dispersed by wind. The fruits of some species of *Galium* have hooked hairs and are transported externally by animals.

Additional references: Anderson 1973; Barrett and Richards 1990; Bremer 1996; Lersten 1975; Manen et al. 1994; Nilsson et al. 1990; Rogers 1987; Vuilleumier 1967.

Gentianaceae A. L. de Jussieu
(Gentian Family)

Herbs, sometimes mycoparasites (with reduced leaves and lacking chlorophyll), to shrubs or small trees; stems often winged; usually with internal phloem; usually with iridoids. Hairs often simple. *Leaves usually opposite, simple, usually entire, often ± sessile*, with pinnate venation; **stipules usually lacking**, *but colleters often present at base of adaxial surface of petiole*. Inflorescences determinate, sometimes reduced to a single flower, axillary or terminal. *Flowers usually bisexual, radial.* Sepals usually 4 or 5 and connate, often with colleters on adaxial surface. *Petals usually 4 or 5, connate*, forming a wheel-shaped, funnel-shaped, or bell-shaped corolla, the lobes sometimes fringed, often with nectar glands and/or scales on adaxial surface of the tube, usually convolute, sometimes plicate at sinuses. *Stamens usually 4 or 5; filaments adnate to corolla;* anthers occasionally opening by terminal pores; pollen grains usually tricolporate or triporate. *Carpels 2, connate; ovary superior,* **with parietal placentation, the placentas sometimes deeply intruded and 2-lobed**; stigma ± capitate to strongly 2-lobed, the lobes sometimes spirally twisted. Ovules usually numerous on each placenta, with 1 integument and a thin-walled megasporangium. Nectar-producing disk or glands present. *Fruit usually a septicidal capsule.*

Floral formula: *, K④–⑤, C④–⑤, A4–5, G②; capsule

Distribution: Widely distributed, but most diverse in temperate and subtropical regions and in the montane tropics.

Genera/species: 84/970. **Major genera:** *Gentiana* (300 spp.), *Gentianella* (125), *Sebaea* (100), *Swertia* (100), and *Halenia* (70). These genera (except *Sebaea*) occur in the continental United States and/or Canada, along with *Bartonia, Centaurium, Eustoma, Frasera, Sabatia,* and *Voyria*.

Economic plants and products: *Gentiana* (gentian), *Centaurium* (centaury), *Eustoma* (prairie gentian; Plate 9.20G), *Exacum* (German violet), and *Sabatia* (rose pink) provide ornamentals.

Discussion: Infrafamilial relationships have been investigated by Mészáros et al. (1996) and Thiv et al. (1999). It is clear that *Villarsia, Nymphoides, Lomatogonium, Obolaria,* and *Menyanthes* are only distantly related to Gentianaceae (Bremer et al. 1994; Chase et al. 1993; Downie and Palmer 1992; Michaels et al. 1993; Olmstead et al. 1993; Wood 1983; Wood and Weaver 1982). These genera are here placed in Menyanthaceae, a family of wetland or aquatic herbs with alternate and spiral or 2-ranked leaves and an induplicate-valvately folded corolla (see Asterales). The two families also differ chemically and anatomically.

The colorful flowers of Gentianaceae are pollinated mainly by bees and butterflies. Nectar is the reward. Most are outcrossing due to protogyny. The small seeds are probably wind- or water-dispersed.

Apocynaceae A. L. de Jussieu
(Milkweed Family)

Trees, shrubs, lianas, vines, and herbs, sometimes succulent and cactuslike; internal phloem present; **tissues with laticifers and sap usually milky**; with cardiac glycosides and various alkaloids; often with iridoids. Hairs various, often simple. *Leaves usually opposite*, sometimes alternate and spiral or 2-ranked or whorled, *entire*, with pinnate venation; *stipules reduced or lacking, colleters usually present at base of petiole.* Inflorescences determinate, but sometimes appearing indeterminate, occasionally reduced to a single flower, terminal or axillary, often not appearing strictly so. *Flowers usually radial.* Sepals usually 5, ± connate, sometimes reflexed, colleters often present at base of adaxial surface. *Petals usually 5, connate*, forming a wheel-shaped, bell-shaped, funnel-shaped, or tubular corolla, sometimes reflexed, often with coronal appendages or scales on inside or apex of tube, the lobes imbricate, valvate, or contorted. *Stamens usually 5;* filaments short, sometimes connate, always adnate to corolla; anthers often highly modified and 2-locular, *distinct or sticking together and forming a ring around the stylar head, free, or sticking to the stylar head by intertwining trichomes and sticky secretions (viscin), or strongly adnate to stylar head by solid parenchymatous tissue,* the connective often with apical appendages; *staminal outgrowths (corona) often present and petal-like, hood-shaped and horn-shaped* (Plate 9.20C); *pollen grains* usually tricolporate or di- or triporate, shed as monads, loosely sticking together by means of viscin, or as tetrads, also loosely sticking together, or strongly coherent and *forming hardened masses (pollinia),* then translators (structures connecting pollinia from anther sacs of adjacent anthers) present, often with a sticky gland (corpusculum). *Carpels usually 2,* **connate by styles and/or stigmas only and ovaries distinct**, fully connate in *Allamanda, Carissa,* and *Thevetia; ovaries superior,* with usually lateral or axile placentation; **apical portion of style expanded and highly modified, forming a head** (Plate 9.20B), **secreting viscin,** *usually abruptly enlarged and latitudinally differentiated into 3 zones* (i.e., specialized for pollen deposition, viscin secretion, and pollen reception), often 5-sided; stigmatic tissue often restricted to 5 small, lateral regions of the stylar head. Ovules 2 to numerous in each ovary, with 1 integument and a thin-walled (or secondarily thickened) megasporangium. Nectar glands or disk often present, or nectar secreted by stigmatic chambers (and held in the hollows of the staminal tube or coronal appendages). *Fruits often paired, each ovary usually developing into a fleshy or dry follicle, berry, or drupe,* the fruit surface smooth to covered with prickle-like outgrowths; *seeds flattened, often with a tuft of hairs;* embryo straight to bent (Figures 9.120 and 9.121).

Floral formula:

*, K⑤, C⑤, A5, G②; 1 or 2 follicles, berry, drupe

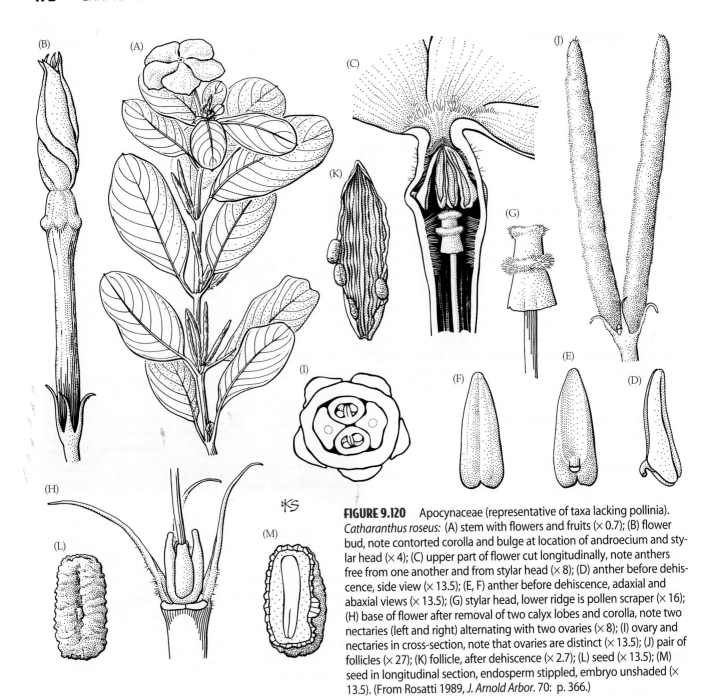

FIGURE 9.120 Apocynaceae (representative of taxa lacking pollinia). *Catharanthus roseus:* (A) stem with flowers and fruits (× 0.7); (B) flower bud, note contorted corolla and bulge at location of androecium and stylar head (× 4); (C) upper part of flower cut longitudinally, note anthers free from one another and from stylar head (× 8); (D) anther before dehiscence, side view (× 13.5); (E, F) anther before dehiscence, adaxial and abaxial views (× 13.5); (G) stylar head, lower ridge is pollen scraper (× 16); (H) base of flower after removal of two calyx lobes and corolla, note two nectaries (left and right) alternating with two ovaries (× 8); (I) ovary and nectaries in cross-section, note that ovaries are distinct (× 13.5); (J) pair of follicles (× 27); (K) follicle, after dehiscence (× 2.7); (L) seed (× 13.5); (M) seed in longitudinal section, endosperm stippled, embryo unshaded (× 13.5). (From Rosatti 1989, *J. Arnold Arbor.* 70: p. 366.)

Distribution: Widespread in tropical and subtropical regions, but with a few genera extending into temperate areas.

Genera/species: 355/3700. **Major genera:** *Asclepias* (230 spp.), *Tabernaemontana* (230), *Cynanchum* (200), *Ceropegia* (150), *Hoya* (150), *Matelea* (130), *Rauvolfia* (110), *Gonolobus* (100), *Secamone* (100), and *Mandevilla* (100). Noteworthy genera occurring in Canada and/or the continental United States include *Amsonia, Angadenia, Apocynum, Asclepias, Catharanthus, Cynanchum, Echites, Gonolobus, Macrosiphonia, Matelea, Morrenia, Pentalinon, Rhabdadenia, Sarcostemma,* and *Vallesia.*

Economic plants and products: Nearly all taxa are poisonous, and many have medicinal uses. *Catharanthus* (Madagascar periwinkle) provides antileukemia drugs, *Rauvolfia* provides a hypertension drug, and *Strophanthus* provides a drug used in treating heart ailments. Important ornamentals are *Allamanda, Amsonia* (bluestar), *Asclepias* (milkweed, butterfly weed; Plate 9.20H), *Carissa* (Natal plum), *Catharanthus* (Madagascar periwinkle), *Hoya* (wax plant), *Nerium* (oleander), *Plumeria* (frangipani), *Stapelia* (carrion flower), *Trachelospermum* (confederate jasmine), and *Vinca* (periwinkle). Species of some genera, especially *Aspidosperma,* yield timbers.

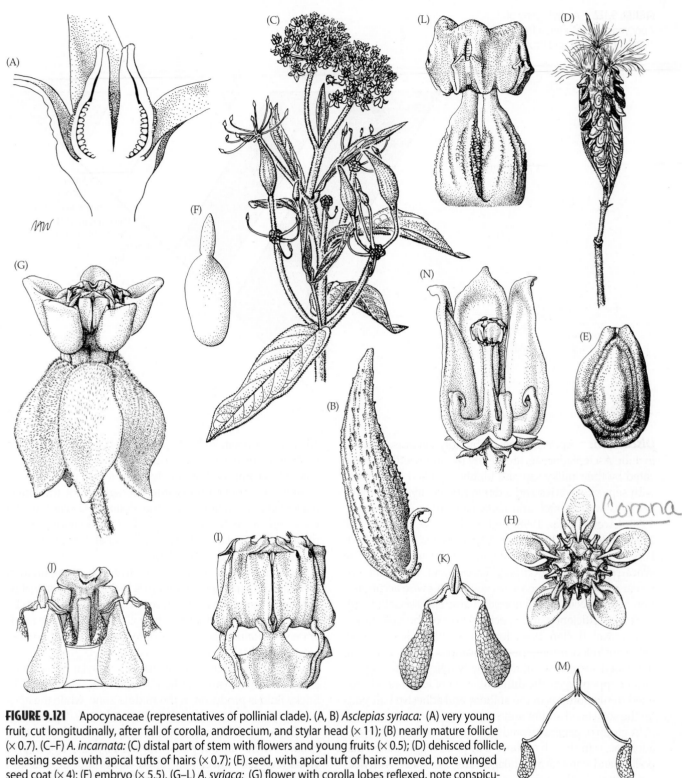

Corona

FIGURE 9.121 Apocynaceae (representatives of pollinial clade). (A, B) *Asclepias syriaca:* (A) very young fruit, cut longitudinally, after fall of corolla, androecium, and stylar head (× 11); (B) nearly mature follicle (× 0.7). (C–F) *A. incarnata:* (C) distal part of stem with flowers and young fruits (× 0.5); (D) dehisced follicle, releasing seeds with apical tufts of hairs (× 0.7); (E) seed, with apical tuft of hairs removed, note winged seed coat (× 4); (F) embryo (× 5.5). (G–L) *A. syriaca:* (G) flower with corolla lobes reflexed, note conspicuous, hood-shaped and horn-shaped staminal appendages (corona) (× 5); (H) flower from above, note hood-shaped and horn-shaped staminal appendages opposite anther flaps over stylar head and alternating with sticky glands (corpuscula) (× 5); (I) androecium (and, within, gynoecium) distal to level of adnation of corolla, with hood-shaped staminal appendages removed (× 12); (J) anther, adaxial side, with two pollinia, each connected by a translator arm, a corpusculum, and another translator arm to one pollinium from adjacent anther (× 12); (K) pair of pollinia (× 16); (L) gynoecium, the two carpels united only by stylar head (× 12). (M) *A. connivens:* pair of pollinia (× 12). (N) *A. pedicellata:* side view of flower with two corolla lobes removed (× 5). (From Rosatti 1989, *J. Arnold Arbor.* 70: pp. 492, 494.)

FIGURE 9.122   A phylogeny of Apocynaceae. (Modified from Judd et al. 1994 and Endress and Bruyns 2000.)

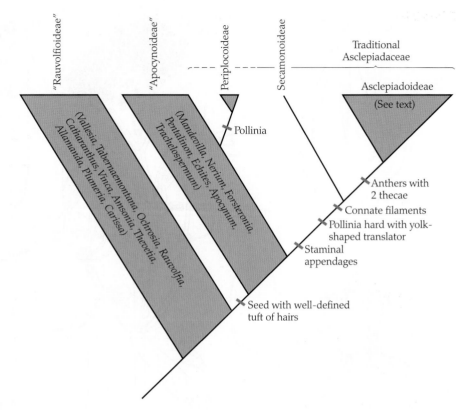

Discussion: Apocynaceae, as broadly circumscribed to include Asclepiadaceae, are clearly monophyletic, as indicated by their milky sap and highly modified gynoecium with separate ovaries and a differentiated stylar/stigmatic head, as well as *rbcL* and *matK* sequences (Chase et al. 1993; Civeyrel et al. 1998; Endress et al. 1996; Judd et al. 1994; Endress and Bruyns 2000; Endress and Stevens 2001; Potgieter and Albert 2001) (Figures 9.120 and 9.121). The syncarpous ovaries of *Carissa*, *Thevetia*, and *Allamanda* represent reversals. Apocynaceae exhibit a stepwise accumulation of androecial and gynoecial characters associated with increased efficiency and specialization of the pollination mechanism (Fallen 1986; Judd et al. 1994; Schick 1980, 1982). Included among these innovations are the enlarged, five-sided stylar head that is zonally differentiated, apical anther appendages, the differentiation of fertile and hardened sterile portions of the anthers, and adhesive hair pads on the style head and/or anthers.

A group of genera, treated here as the subfamily Asclepiadoideae, is further characterized by pollen agglutinated into pollinia and staminal appendages often associated with nectar storage. In these genera—*Asclepias*, *Calotropis*, *Cynanchum*, *Hoya*, *Matelea*, *Stapelia*, *Ceropegia*, *Gonolobus*, *Sarcostemma*, *Secamone*, and relatives—the pollinium develops by the agglutination of the pollen tetrads of a theca (i.e., pollen sac) into a hardened mass that is attached to a yoke-shaped translator with a gland; the filaments are also connate. All but *Secamone* and relatives (Secamonoideae) have the additional synapomorphy of anthers with only two thecae; these genera constitute the large group called Asclepi-

adoideae (Fishbein 2001); Figure 9.122. Pollinia have evolved independently in *Cryptostegia* and relatives (Periplocoideae), where it is composed of loosely agglutinated pollen tetrads resting on a cuplike translator. Recognition of the family Asclepiadaceae (comprising the pollinial species) would make the restrictively circumscribed Apocynaceae paraphyletic (Civeyrel et al. 1998; Endress et al. 1996; Judd et al. 1994; Sennblad and Bremer 1996).

Apocynaceae show a great diversity of pollination mechanisms; frequent pollinators include various nectar-gathering insects (butterflies, moths, bees, flies). The stylar head plays an important role in pollination. In many moderately specialized genera it is differentiated into three discrete horizontal zones (see above). The receptive tissue of the stigma is restricted to the base of the stylar head, under a collarlike extension (pollen scraper; see Figure 9.121). Sticky fluid is produced in the middle zone, while the apex of the stylar head presents the pollen, which is deposited by the inwardly dehiscing anthers that surround the stylar head. As the pollinator probes for nectar, pollen from a previously visited flower is scraped off its mouthparts and deposited in the stigmatic region. The mouthparts become sticky as they contact the sticky fluid produced in the middle region of the stylar head, then pick up pollen from the apical region. In the most specialized genera (e.g., *Asclepias*) the androecium and gynoecium are adnate and the pollen is fused into pollinia. Nectar accumulates in the elaborate outgrowths of the stamens, appendages called hoods and horns. In the process of probing for nectar in the corona, the insect's leg may become lodged in the space between

FIGURE 9.123 Cladogram showing hypothesized relationships within Lamiales. (Modified from Bremer et al. 2002, Olmstead et al. 2001, Oxelman et al. 2005, Wortley et al. 2005.)

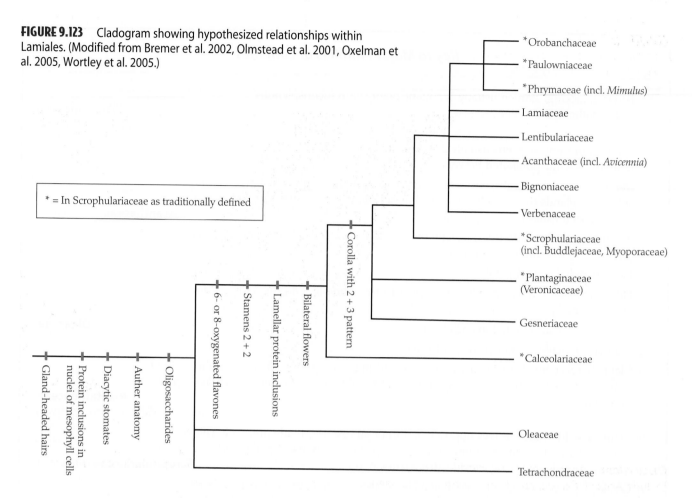

* = In Scrophulariaceae as traditionally defined

the anthers and make contact with the translator gland. When the insect pulls out its leg, it becomes stuck in the gland, pulling out the two pollinia. The pollinia are dislodged from the insect's leg when it slips into one of the stigmatic slits on the stylar head of another flower.

Most species of Apocynaceae have follicles with hair-tufted, flattened seeds (Plate 9.20E) that are dispersed by wind. In *Cameraria* the paired fruits are dry, distally winged, and dispersed by wind. The colorful (often blue or red) fleshy fruits of tropical genera, such as *Kopsia, Rauvolfia, Carissa,* and *Ochrosia,* are often mammal- or bird-dispersed. The seeds of *Tabernaemontana* are associated with a brightly colored, fleshy aril and dispersed by birds. Monkey dispersal has been reported for the arillate *Stemmadenia.*

Additional references: Endress et al. 1983; Kunze 1990, 1992; Rosatti 1989; Safwat 1962; Swarupanandan et al. 1996; Thomas and Dave 1991; Woodson 1954.

Lamiales

Lamiales (including Bignoniales or Scrophulariales) are clearly monophyletic. They are characterized by gland-headed hairs, oligosaccharides (which replace starch in carbohydrate storage), parenchyma tissue extending from anther connective into the locules, often diacytic stomates (i.e., stomates enclosed by one or more pairs of subsidiary cells whose common walls are at right angles to the guard cells), endosperm with a conspicuous micropylar haustorium, and protein inclusions in the nuclei of mesophyll cells (Dahlgren 1983; Judd et al. 1994; Wagenitz 1992; Yamazaki 1974), cpDNA restriction sites, and *rbcL, atpB, ndhF, matK,* and 18S sequences (Albach et al. 2001a; Bremer et al. 2002; Downie and Palmer 1992; Hilu et al. 2003; Källersjö et al, 1998; Lu 1990; Olmstead et al. 1992, 1993a,b, 2000a,b, 2001; Olmstead and Reeves 1995; Oxelman et al. 2005; Savolainen et al. 2000b; Soltis et al. 2000; Wagstaff and Olmstead 1997; Wortley et al. 2005).

Oleaceae and the small families Tetrachondraceae and Plocospermataceae are basally branching clades within the order. Oleaceae are characterized by actinomorphic flowers (usually of four petals) with only two stamens, and flowers of Tetrachondraceae (incl. *Polypremum*) and Plocospermataceae are also radial. Most other members of the order have two-lipped, bilaterally symmetrical flowers forming a 2 + 3 pattern, that is, with two upper petals and three lower petals forming a corolla with two opposing lobes (i.e., bilabiate), but secondarily radial and four-petaled in *Buddleja* (Figure 9.123). The remaining members of the order also have four stamens (two long and two short), although sometimes these have been reduced to two. This clade is also characterized by endosperm with haustoria at both micropylar and chalazal ends, lamellar protein inclusions in the nuclei, 6- or 8-oxygenated flavones, and shikimic-acid-derived anthraquinones.

Key to Major Families of Lamiales

1. Perianth lacking; stamen usually 1; aquatic plants with inconspicuous water-pollinated flowers .... Plantaginaceae
(in part, *Callitriche*)

1. Perianth present; stamens usually 2–4; terrestrial or aquatic plants with usually showy, animal- or wind-pollinated flowers 2

2. Mangrove trees or shrubs, with erect, pencil-like pneumatophores; leaves bearing subsessile glands that excrete a hypersaline solution that rapidly dries to form salt crystals; seed usually only 1 per fruit .... **Acanthaceae** (*Avicennia*)

2. Nonmangrove herbs to trees; lacking pneumatophores and salt-excreting glands; seeds (2–) 4 to numerous per fruit 3

3. Flowers radial, with 4 petals 4

3. Flowers bilateral, with 5 petals, the 2 upper forming one lip and the 3 lower forming an opposing lip 6

4. Stamens 2; ovules often 2 in each locule .... **Oleaceae**

4. Stamens 4; ovules ± numerous per locule 5

5. Plants herbaceous; leaves usually alternate and in basal rosettes, blade ± parallel-veined; flowers in spikes or heads, inconspicuous, wind pollinated; fruit a circumscissile capsule or achene .... **Plantaginaceae**
(in part; *Plantago* and relatives)

5. Plants usually woody, leaves opposite, not in basal rosettes; blade pinnately veined; flowers in various cymose inflorescences, conspicuous and insect pollinated; fruit a variously dehiscing capsule, drupe, or samara .... **Scrophulariaceae** (*Buddleja*)

6. Ovules 2 per carpel, each locule divided by a false partition, thus with 1 ovule per chamber of ovary; fruits drupes (with 1–4 pits) or schizocarps of usually 4 nutlets; style terminal to gynobasic; leaves opposite 7

6. Ovules usually numerous, but if reduced to 2 per carpel, then ovary locules 2 (i.e., not secondarily divided by false partition); fruits usually capsules, sometimes berries or drupes; style always terminal; leaves alternate or opposite 8

7. Inflorescences indeterminate, the lateral units individual flowers or appearing to be individual flowers; ovules directly attached to margins of false partitions; style consistently terminal; stigma conspicuous, lobed .... **Verbenaceae**

7. Inflorescences mixed, with ± indeterminate main axis and obviously determinate (cymosely branched) lateral units, these sometimes reduced and inflorescence then appearing pseudo-verticillate; ovules attached to false partitions very near the inrolled carpel margins; style terminal to gynobasic; stigma usually inconspicuous, at tips of stylar branches .... **Lamiaceae**

Within the two-lipped corolla clade, "Scrophulariaceae," as traditionally delimited, are polyphyletic, united only by symplesiomorphies (at this level of universality) such as flowers with two-lipped corollas, four stamens, and more or less globose capsules with small, endospermous seeds. Some members of this group are likely to be related more closely to members of various other families of Lamiales than they are to other scrophs (Figure 9.123). The family here is divided into three major groups, each of which is apparently monophyletic: Scrophulariaceae s.s., Orobanchaceae, and Plantaginaceae. Verbenaceae and Lamiaceae are often considered

sister taxa, and both have ovaries with four ovules, divided into four locules because of the development of a false septum, and aromatic, ethereal oils. Analyses of *rbcL* and *ndhF* sequences, however, support the hypothesis that these characters have evolved more than once within Lamiales (Olmstead and Reeves 1995; Wagstaff and Olmstead 1996; Oxelman et al. 1999), while the molecular analysis of Bremer et al. (2002) supports the sister-group relationship of Lamiaceae and Verbenaceae. Analyses based on *rbcL*, *ndhF*, and *rps*2 nucleotide sequences (Olmstead et al, 2000b) suggest that Orobanchaceae are most closely related to *Paulownia* and

8. Plants insectivorous, with sticky, mucilaginous hairs, or bladders; placentation free-central, anthers 2 ....**Lentibulariaceae**

8. Plants not insectivorous; placentation usually axile or parietal, anthers usually 4 9

9. Ovary superior to inferior; anthers connate or sticking together in pairs; placentation parietal ....**Gesneriaceae**

9. Ovary consistently superior; anthers usually distinct; placentation usually axile 10

10. Seeds with well-developed endosperm 11

10. Seeds with scanty or no endosperm 13

11. Plants hemiparasitic (and green) to holoparasitic (± lacking chlorophyll), forming haustorial connections to the roots of various host species; placentation axile or parietal ....**Orobanchaceae**

11. Plants autotrophic (and green), lacking haustoria; placentation axile 12

12. Anther locules confluent and anther opening by a single distal slit, the anther not sagittate; stamens 5, 4, or 2; flowers clearly to only slightly bilateral ....**Scrophulariaceae**
(in part)

12. Anther locules distinct, opening by 2 distinct longitudinal slits, or apical portion of anther sacs adnate and opening by a single U- or V-shaped slit, the anther ± sagittate; stamens 4 or 2; flowers ± clearly bilateral ....**Plantaginaceae**
(in part; numerous genera)

13. Fruits usually explosively dehiscent, seeds associated with an enlarged and specialized funiculus (retinaculum); often with cystoliths ....**Acanthaceae**

13. Fruits dehiscent, but not explosively so, or indehiscent; lacking retinacula; cystoliths lacking 14

14. Ovules ± numerous in each locule, borne on bilobed placentas; fruit usually an elongate capsule, never a drupe; leaves usually opposite, and compound, blades lacking pellucid dots; seeds usually flattened, winged ....**Bignoniaceae**

14. Ovules usually 1 per locule, borne on unmodified placentas; fruit a drupe; leaves usually alternate and spiral, simple, blades usually with pellucid dots; seeds not flattened, unwinged..**Scrophulariaceae**
(*Myoporum* and relatives)

Lamiaceae; Scrophulariaceae to Buddlejaceae and Myoporaceae (and the latter two groups are now included within a re-expanded Scrophulariaceae), and Verbenaceae to Bignoniaceae. Plantaginaceae, along with Gesneriaceae and Calceolariaceae, may be early divergent lineages within the two-lipped corolla clade (Figure 9.123). Current family limits are uncertain and the families presented in this text are distinguished on very slight differences. Some systematists have even suggested that all the families with bilaterally symmetrical flowers (i.e., with a 2 + 3 corolla pattern and usually 4 stamens) be combined, forming a single large family.

Lamiales surely belong to the core asterid clade, as indicated by DNA sequences, sympetalous corollas, ovules with a single integument, and thin-walled megasporangium. The order is probably most closely related to the Solanales and Gentianales (Downie and Palmer 1992; Hufford 1992; Olmstead et al. 1992a,b, 1993).

The order consists of about 22 families and 20,000 species; major families include **Oleaceae, Bignoniaceae, Scrophulariaceae, Orobanchaceae, Plantaginaceae**, Phyrmaceae, Calceolariaceae, **Gesneriaceae, Lentibulariaceae, Acanthaceae, Verbenaceae**, and **Lamiaceae** (Cantino 1990, 1992a,b; Cantino et al. 1992; Olmstead and Reeves 1995; Wagstaff and Olmstead 1996).

Oleaceae Hoffmannsegg and Link
(Olive Family)

Trees, shrubs, or lianas; sclereids often present; usually with phenolic glycosides; **buds 2 to several and superposed**; iridoids present. Hairs peltate scales. *Leaves usually opposite*, simple, pinnately compound, or trifoliolate, entire to serrate, with pinnate venation; *stipules lacking*. Inflorescences determinate, sometimes reduced to a single flower, termi-

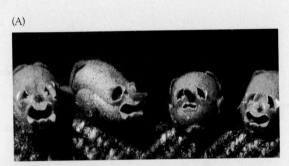

(A)

Lamiales: Plantaginaceae
Antirrhinum majus: fruits, opening by pores

(B)

Lamiales: Scrophulariaceae
Verbascum virgatum: flowers

(C)

Lamiales: Gesneriaceae
Columnea zebrina: plant in bloom

(D)

Lamiales: Plantaginaceae
Antirrhinum majus: flower, longitudinal section

(E)

Lamiales: Gesneriaceae
Aeschnanthus radicans:
connate anthers

(G)

Lamiales: Plantaginaceae
Plantago aristata: inflorescences

(F)

Lamiales: Bignoniaceae
Kigelia africana: flower

(H)

Lamiales: Oleaceae
Olea europaea: flowers

(I)

Lamiales: Orobanchaceae
Orobanche uniflora: plant
in bloom

PLATE 9.21 Eudicots: Lamiales

(J)

Lamiales: Verbenaceae
Lantana camara: flowers

(K)

Lamiales: Lamiaceae
Vitex agnus-castus: flowers

(L)

Lamiales: Acanthaceae
Pachystachys lutea: inflorescence
with colorful bracts

(M)

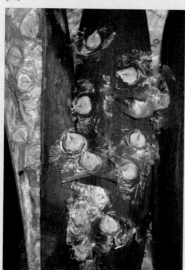

Lamiales: Bignoniaceae
Spathodea campanulata: fruit and winged seeds

(N)

Lamiales: Lentibulariaceae
Pinguicula caerulea: basal rosettes and flowers

(O)

 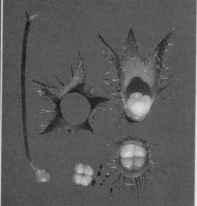

Lamiales: Lamiaceae
Stachys sp.: flowers

Stachys sp.: ovary dissection;
note 4 lobes

(P)

Lamiales: Lentibulariaceae
Utricularia inflata: leaves with bladders

PLATE 9.21 Eudicots: Lamiales

nal or axillary. *Flowers bisexual or sometimes unisexual* (plants then polygamous to ± dioecious), *radial. Sepals usually 4, connate, rarely lacking* (in some *Fraxinus*). *Petals usually 4,* but sometimes more numerous, *connate, imbricate or induplicate-valvate* (i.e., folded inward with central portion of lobes meeting), *rarely reduced or lacking* (in some *Fraxinus*). **Stamens 2**; *filaments adnate to corolla*; pollen grains tricolpate or tricolporate. *Carpels 2, connate; ovary superior, with axile placentation*; stigma 2-lobed or capitate. *Ovules usually 2 in each locule,* with 1 integument and a thin-walled megasporangium. Nectar disk often present. *Fruit a loculicidal or circumscissile capsule, samara, berry, or drupe, often 1-seeded*; endosperm present to absent (Figure 9.124).

Floral formula:

*, K④, C④, A2, G②; capsule, samara, drupe, berry

Distribution: Widely distributed in tropical to temperate regions.

Genera/species: 25/600.   **Major genera:** *Jasminum* (230 spp.), *Chionanthus* (90), *Fraxinus* (60), *Ligustrum* (35), *Noronhia* (35), *Syringa* (30), *Menodora* (25), *Olea* (20), *Forestiera* (15), and *Osmanthus* (15). All of the above except *Olea* and *Noronhia* are represented in the continental United States and/or Canada.

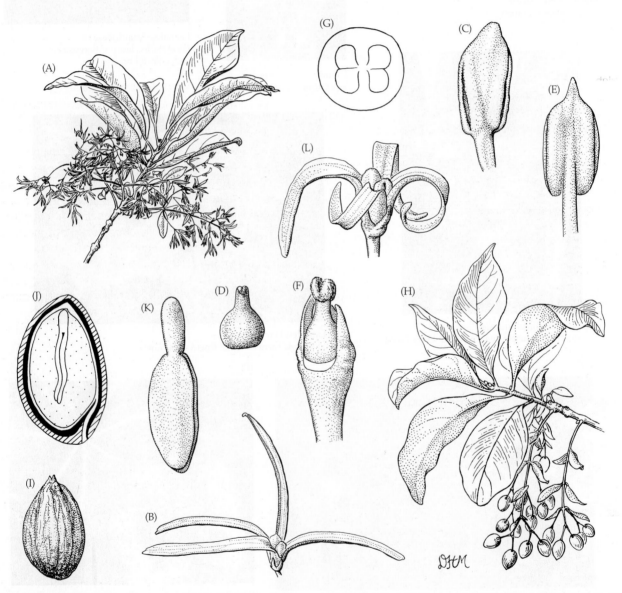

FIGURE 9.124   Oleaceae. (A–K) *Chionanthus virginicus:* (A) flowering branch (× 0.5); (B) staminate flower (× 3); (C) stamen from staminate flower (× 20); (D) nonfunctional gynoecium from staminate flower (× 3); (E) stamen from bisexual flower (× 20); (F) gynoecium from bisexual flower (× 15); (G) ovary in cross-section, with two ovules in each locule (× 30); (H) branch with drupes (× 0.5); (I) pit (× 2.5); (J) pit and seed in longitudinal section, endocarp wall hatched, endosperm stippled, embryo unshaded (× 4); (K) embryo (× 7.5). (L) *C. pygmaeus:* staminate flower, note two stamens (× 6). (From Wood 1974, *A student's atlas of flowering plants,* p. 87.)

Economic plants and products: *Olea europaea* (olive; Plate 9.21H) produces edible fruits and is a source of cooking oil. *Fraxinus* (ash) contains commercially important timber species. *Forsythia* (golden bells), *Jasminum* (jasmine), *Ligustrum* (privet), *Chionanthus* (fringe tree), *Osmanthus* (fragrant olive, wild olive), *Syringa* (lilac), and *Noronhia* (Madagascar olive) provide useful ornamentals.

Discussion: Oleaceae are considered monophyletic on the basis of *rps16* and *trnL-F* sequences and several morphological synapomorphies (Wallander and Albert 2000). Because this family is phenetically quite divergent from most other families of Lamiales, it sometimes has been placed in its own order (Oleales).

The family usually is divided into "Jasminoideae," characterized by ovules one to several, usually erect in each locule, and four to twelve petals, and Oleoideae, characterized by two pendulous ovules per locule and four petals. "Jasminoideae" are morphologically heterogeneous and paraphyletic (Kim and Jansen 1998b; Wallander and Albert 2000). Oleoideae are hypothesized to be monophyletic on the basis of the features cited above, the haploid chromosome number of 23, their flavone glycosides, and wood anatomy. Most Oleoideae have drupaceous fruits (e.g., *Chionanthus, Forestiera, Ligustrum, Noronhia, Olea,* and *Osmanthus*), but *Syringa* has loculicidal capsules, and *Fraxinus* has samaras. *Fraxinus* is also distinct due to its pinnately compound leaves (vs. simple in other Oleoideae) and reduced wind-pollinated flowers. Fruit type is variable within "Jasminoideae"; *Jasminum* has bilobed berries, *Fontanesia* has samaras, while *Forsythia* and *Menodora* have capsules.

The showy, variously colored, and often fragrant flowers of Oleaceae are pollinated by nectar-gathering bees, butterflies, and flies. The flowers of *Fraxinus* and *Forestiera* have a reduced perianth and are wind-pollinated. Both selfing and outcrossing occur. Fleshy-fruited taxa are dispersed by birds or mammals. The samaras of *Fraxinus* are wind-dispersed, as are the seeds of capsular species.

Additional references: Bigazzi 1989; Green 2004; Wilson and Wood 1959.

Gesneriaceae Richard and A. L. de Jussieu
(Gesneriad Family)

Herbs to shrubs, *often epiphytic*; often with phenolic glycosides; lacking iridoids. Hairs simple, often gland-headed. *Leaves usually opposite, usually simple*, entire to variously toothed, often softly hairy, with pinnate venation; *stipules usually lacking*. Inflorescences usually determinate, sometimes reduced to a single flower; axillary. Flowers bisexual, bilateral. Sepals 5, distinct to connate. *Petals 5, connate, the corolla usually 2-lipped*, the lobes imbricate. *Stamens 4, ± didynamous; filaments adnate to corolla;* **anthers sticking together in pairs or all together**; pollen grains usually tricolporate. *Carpels 2, connate; ovary superior to inferior,* **with parietal placentation**; stigma 2-lobed. Ovules numerous

on each placenta, with 1 integument and a thin-walled megasporangium. Nectar disk or glands usually present. *Fruit a loculicidal or septicidal capsule, sometimes fleshy, or a berry;* seeds small; endosperm present or absent.

Floral formula:

X, K⑤, C②₊③, A②₊₂, G②; capsule, berry

Distribution and ecology: Widely distributed in tropical regions.

Genera/species: 147/3500. **Major genera:** *Cyrtandra* (550 spp.), *Columnea* (270), *Besleria* (200), *Henckelia* (150), *Streptocarpus* (150, incl. *Saintpaulia*), *Aeschynanthus* (140), *Chirita* (80), *Alloplectus* (75), *Sinningia* (60), *Gesneria* (60). No species occur naturally in the continental United States or Canada.

Economic plants and products: The family is notable chiefly for its numerous ornamental herbs, including *Episcia, Columnea, Sinningia* (gloxinia), *Streptocarpus* (Cape primrose, African violet), and *Aeschynanthus* (Plate 9.21.E).

Discussion: Gesneriaceae are considered monophyletic based on both morphology and *ndhF* sequences (Smith 1996; Smith et al. 1997). Most genera belong to one of two large subfamilies. Gesnerioideae are monophyletic, as evidenced by their half-inferior to completely inferior ovaries and large pollen grains. The monophyly of Cyrtandroideae, which has superior ovaries, is supported by an unusual embryonic development pattern, in which one cotyledon becomes larger than the other after germination. Chromosome number, secondary compounds, and *ndhF* sequences (Smith et al. 1997) also lend support to these two groups. Within Gesnerioideae, genera such as *Asteranthera, Besleria, Napeanthus,* and *Sinningia* are early divergent lineages, while members of the Gloxinieae (*Kohleria, Gloxinia,* and relatives), Gesnerieae (*Gesneria, Sanango, Rhytidophyllum*) and Episcieae (e.g., *Codonanthe, Episcia, Alloplectus,* and *Columnea*) form a clade (Smith and Carroll 1997).

The showy flowers of Gesneriaceae show a diversity of pollination syndromes; major pollinators include bees, moths, butterflies, flies, bats, and birds. Nectar and pollen are used as rewards. Brightly colored calyces and even pigmented leaves sometimes assist in attracting pollinators. Species with fleshy fruits are usually bird-dispersed; the small seeds of capsular taxa are shaken from the capsule and may be dispersed by wind or water (rainwash).

Additional references: Burtt 1977; Harrison et al. 1999; Möller and Cronk 1997; Weber 2004; Wiehler 1983.

Plantaginaceae A. L. de Jussieu
(Snapdragon Family)

Herbs or less commonly shrubs, sometimes aquatic; autotrophic, lacking haustoria; often with phenolic glycosides and triterpenoid saponins, **acylated rhamnosyl iridoids**, and sometimes with cardiac glycosides. Hairs various, but usually simple, when glandular, the stalk elongate, *usually composed of 2 or more cells, and head ± globular to ellipsoidal,* **lacking vertical partitions**. *Leaves alternate and spiral, or opposite,* occasionally whorled, *simple,* entire to variously toothed, with pinnate venation, but ± parallel in *Plantago; stipules lacking.* Inflorescences various. *Flowers usually bisexual and bilateral, but ± radial in Plantago; reduced in Callitriche. Sepals usually 4 or 5, connate. Petals usually 5, or occasionally apparently 4 (due to fusion of 2 upper lobes), connate, the corolla 2-lipped,* sometimes with a basal nectar spur, the lower lip sometimes with a bulge obscuring the throat (personate), the lobes imbricate or valvate. *Stamens usually 4, didynamous,* sometimes reduced to 2, the fifth stamen sometimes present as a staminode (e.g., *Penstemon); filaments adnate to corolla; anthers 2-locular, locules distinct, opening by 2 longitudinal slits,* or apical portion of anther sacs sometimes adnate and opening by a single inverted, U- or V-shaped slit (i.e., sagittate), the pollen sacs divergent (anther sagittate); pollen grains often tricolporate. *Carpels 2, connate; ovary superior, with axile placentation, the placentas large, not divided;* stigma usually 2-lobed. *Ovules usually numerous in each locule,* with 1 integument and a thin-walled megasporangium. Nectar disk usually present, but lacking in *Plantago* and *Callitriche. Fruit usually a septicidal capsule, occasionally poricidal or circumscissile* (Plate 9.21A); seeds angular or winged (Figure 9.125).

Floral formula:

X or *, K (4–5), C (2+3) or (4), A2+2 or 2, G (2); capsule

Distribution: Nearly cosmopolitan, but most diverse in temperate areas.

Genera/species: 104/1820. **Major genera:** *Veronica* (450 spp.), *Penstemon* (250), *Plantago* (215), *Linaria* (120), *Antirrhinum* (40), *Limnophila* (35), *Gratiola* (20), *Scoparia* (20), and *Digitalis* (20). Noteworthy genera in the continental United States and/or Canada (in addition to most of those listed above) are *Bacopa, Callitriche, Chelone, Collinsia, Hippuris, Maurandya, Scoparia,* and *Veronicastrum.*

Economic plants and products: The family is well known for its ornamentals, such as *Angelonia, Antirrhinum* (snapdragon), *Digitalis* (foxglove), *Penstemon* (beardtongue), *Russelia* (firecracker plant), and *Veronica* (speedwell). Species of *Digitalis* are used medicinally.

Discussion: Circumscription of this group has long been ambiguous, and various genera have been considered phenetically transitional to other families of Lamiales. Two major problems relate to the placement of *Scrophularia* and *Verbascum,* and to the disposition of parasitic genera such as *Pedicularis, Agalinis,* and *Castilleja.*

Most of these genera have, until recently, been included in Scrophulariaceae, but *rbcL, ndhF,* and *rps2* sequences (Olmstead et al. 1992a, 1993, 2000a,b, 2001; Olmstead and Reeves 1995; Tank et al. 2006) strongly support the transfer of *Scrophularia* and *Verbascum* to the *Selago* clade. Morphological characters also support this transfer, that is, *Scrophularia* and *Verbascum* share with *Selago* and relatives the apomorphies of confluent anther locules, anthers opening by a single distal slit, and club-shaped (nonsagittate) anthers (Weberling 1989). Also, they all have glandular hairs with a flattened, discoidal head of two to numerous cells that are separated by vertical partitions (a likely plesiomorphy). The rules of nomenclature (see Appendix 1) require that the *Selago* clade be called Scrophulariaceae because this group contains *Scrophularia,* the type of the oldest available name. Members of the family formerly called Scrophulariaceae are here called Plantaginaceae, although the name Veronicaceae Durande has been used by Olmstead et al. (2001); see also Reveal et al. (1999).

Many parasitic herbs have been included in the Scrophulariaceae as traditionally defined. These plants maintain a physical connection to their host species by means of specially modified roots called haustoria. They range from hemiparasites, which are green and have chlorophyll, to holoparasites, which are white and lack chlorophyll. The parasitic genera are here included in Orobanchaceae. Both morphology and cpDNA sequences support this transfer (de Pamphilis et al. 1997; Tank et al. 2006).

Plantaginaceae, therefore, are circumscribed to exclude *Scrophularia, Verbascum,* and relatives, the parasitic genera, and the isolated genera *Paulownia* and *Schlegelia.* They include Callitrichaceae and Hippuridaceae, two families of specialized aquatic herbs with reduced flowers, and also a group of wind-pollinated herbs (Plantaginaceae s.s.; see the key, Cronquist 1981, and Rosatti 1984 for diagnostic features of *Plantago* and *Callitriche*).

Recently published data suggests that a few additional autotrophic genera should also be excluded. These include *Calceolaria,* which has an extremely saccate corolla (now placed in Calceolariaceae), *Leucophyllum,* which has secretory cavities in its leaves (now in an expanded Scrophulariaceae), *Mimulus,* which is related to *Phryma* and has a tubular toothed calyx and usually distinctive sensitive stigmas (now in Phrymaceae; Beardsley and Olmstead 2002; Olmstead et al. 2001), and *Lindernia, Torenia, Micranthemum* and relatives, which usually have square stems, pitted seeds, glandular hairs on the corolla, and geniculate anterior filaments with a basal swelling (now in Linderniaceae; Albach et al. 2005; Oxelman et al. 2005; Rahmanzadeh et al. 2005).

The monophyly of Plantaginaceae is supported by DNA characters (Albach et al. 2005; Olmstead et al. 1993, 2001; Olmstead and Reeves 1995; Oxelman et al. 2005; Wagstaff and Olmstead 1997). The link of *Plantago* to insect-pollinated members of the family is especially evident chemically, particularly in the presence of aucubin

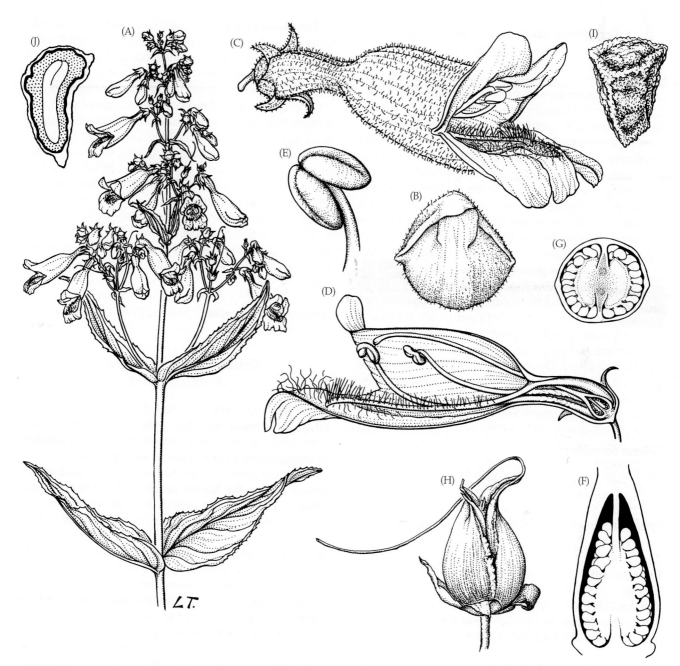

FIGURE 9.125 Plantaginaceae. *Penstemon canescens:* (A) flowering stem (× 1); (B) apex of flower bud (× 5); (C) flower (× 5); (D) flower in longitudinal section, note hairy staminode and stamens in two pairs (didynamous) (× 5); (E) anther (× 15); (F) ovary in longitudinal section (× 18); (G) ovary in cross-section, note numerous ovules (× 18); (H) capsule (× 4.5); (I) seed (× 36); (J) seed in longitudinal section, endosperm stippled, embryo unshaded (× 30). (From Wood 1974, *A student's atlas of flowering plants*, p. 99.

(an iridoid glycoside), mannitol (a sugar alcohol), and linoleic and oleic acids (as major fatty acids). The four-lobed corolla of *Plantago* probably has been derived through the complete fusion of the upper two corolla lobes of an ancestral five-lobed corolla, as is also seen in some species of *Veronica* and *Aragoa*. Morphological synapomorphies of Plantaginaceae are not readily apparent, although the early development of the androecium relative to the corolla may support the group's monophyly, as may hair morphology.

Flowers of Plantaginaceae come in a great variety of shapes, colors, and sizes, and are pollinated mainly by nectar-gathering bees, flies, and birds. Nectar guides are frequent. The reduced flowers of *Callitriche* are water pollinated, while those of *Plantago* are wind-pollinated (Plate 9.21G).

Additional references: Armstrong 1985; Boeshore 1920; Kamphy 1995; Miller 2001; Philbrick and Jansen 1991; Rahn 1996; Reeves and Olmstead 1998; Soekarjo 1992; Thieret 1967, 1971.

Scrophulariaceae A. L. de Jussieu
(Figwort Family)

Herbs or shrubs; autotrophic; with iridoids. *Hairs* usually simple, *when glandular, with a short to elongate stalk cell and a flattened, discoidal head composed of 2 or 4 to numerous cells that are separated by vertical partitions*, sometimes stellate, plumose, or peltate. Leaves alternate and spiral, or opposite, simple, entire to toothed, with pinnate venation, sometimes with scattered secretory cavities (pellucid dots); stipules lacking. Inflorescences indeterminate, terminal. *Flowers bisexual, bilateral to less commonly radial*. Sepals usually 3–5, connate. *Petals 4 or 5, connate, the corolla often ± 2-lipped, or with narrow tube and broadly flaring, imbricate lobes. Stamens 5, 4, or 2,* the fifth stamen sometimes represented by a staminode; *filaments adnate to corolla; anthers 2-locular,* **but the anther sacs usually confluent and opening by a single distal slit oriented at right angles to the filament or ± U-shaped,** anther base sagittate or not; pollen grains usually tricolporate. *Carpels 2, connate; ovary superior,* with axile placentation, the placentas undivided; stigma punctate, capitate to 2-lobed. Ovules numerous to 1 in each locule, with 1 integument and a thin- or thick-walled megasporangium. Nectar disk usually present. *Fruit a septicidal capsule, drupe, or schizocarp of achenes or druplets.*

Floral formula:

X, K④–⑤, C②+③, A5 or 2+2 or 2, G②; capsule, drupe, schizocarp

Distribution: Widely distributed from temperate to tropical regions.

Genera/species: 52/1680. **Major genera:** *Verbascum* (360 spp.), *Scrophularia* (250), *Eremophila* (210), *Selago* (190), *Buddleja* (125), *Manulea* (75), *Nemesia* (65), and *Sutera* (50). *Verbascum, Scrophularia, Leucophyllum, Bontia,* and *Buddleja* occur in the continental United States and Canada.

Economic plants and products: The family is of little economic importance; *Verbascum* (Plate 9.21B), *Buddleja,* and *Myoporum* are grown as ornamentals.

Discussion: Monophyly of Scrophulariaceae is clearly supported by morphology, *rbcL, ndhF,* and *rpo2* sequences (Olmstead et al. 1992a,b, 1993, 2001; Olmstead and Reeves 1995). *Verbascum* and *Scrophularia* are closely related, forming a clade on the basis of their distinctive seeds, hairy filaments, endosperm development, and cpDNA sequences (Olmstead and Reeves 1995; Thieret 1967; Olmstead et al. 2001). These genera probably form the sister clade to the Limoselleae (*Selago, Manulea, Sutera,* etc.). *Selago* and relatives form a clade based on their uniovulate locules and achenelike fruits, and are related to *Sutera, Manulea,* and relatives, which are phenetically intermediate between the *Verbascum + Scrophularia* clade and *Selago.*

The above listed genera are closely related to *Buddleja* (often treated as Buddlejaceae). *Buddleja* has radially symmetrical flowers with 4 corolla lobes and 4 stamens. The *Buddleja + Scrophularia + Selago,* etc. clade is, in turn, likely sister to a clade containing *Leucophyllum, Myoporum, Bontia, Eremophila,* and relatives (often recognized as Myoporaceae). That clade is characterized by leaf blades with pellucid dots and usually have only 1 ovule per locule and drupaceous fruits. *Leucophyllum* (likely sister to remaining members of the pellucid-dotted clade) has densely hairy leaves with only 2 apical pellucid dots and has multiovulate locules and capsular fruits (Lersten and Beaman 1998).

Flowers of Scrophulariaceae are pollinated by a variety of nectar-gathering insects. The seeds or achenes are dispersed by wind, and fleshy-fruited species often are bird dispersed.

Additional references: Hilliard 1994; Kornhall and Bremer 2004; Zona 1998.

Orobanchaceae Ventenant
(Broomrape Family)

Herbs; **hemiparasitic** (i.e., partial parasites, with chlorophyll) **to holoparasitic** (i.e., complete parasites, lacking chlorophyll), **with a single large or many small haustorial connections to the roots of host plants**; often with phenolic glycosides; usually with iridoids and **orobanchin, which causes the leaves (or the entire plant) to turn black in drying**. Hairs various, but usually simple, when glandular the stalks ± elongate, *usually composed of 2 or more cells, and the head ± globular to ellipsoidal,* usually **lacking vertical partitions**. *Leaves alternate and spiral, or opposite, simple, often pinnately lobed to dissected, sometimes reduced to scales,* entire to variously toothed, with pinnate venation; stipules lacking. **Inflorescences** usually **indeterminate**, sometimes reduced to a solitary flower, terminal or axillary. *Flowers bisexual, bilateral.* Sepals usually 5, connate. *Petals usually 5, connate, the corolla 2-lipped,* the lobes imbricate. *Stamens 4, didynamous,* the fifth stamen occasionally present as a staminode; *filaments adnate to corolla; anthers 2-locular, opening by 2 longitudinal slits,* 1 locule sometimes reduced or modified, pollen sacs divergent and anther sagittate; pollen grains often tricolporate. *Carpels 2, connate; ovary superior, with axile to parietal placentation, the placentas not divided (when placentation axile) or often divided (when placentation parietal);* stigma 2-lobed. Ovules usually numerous in each locule or on each placenta, with 1 integument and a thin-walled megasporangium. Nectar disk usually present around base of ovary. *Fruit a septicidal to loculicidal capsule;* seeds angular (Figure 9.126).

Floral formula: X, K⑤, C②+③, A2+2, G②; capsule

Distribution: Nearly cosmopolitan.

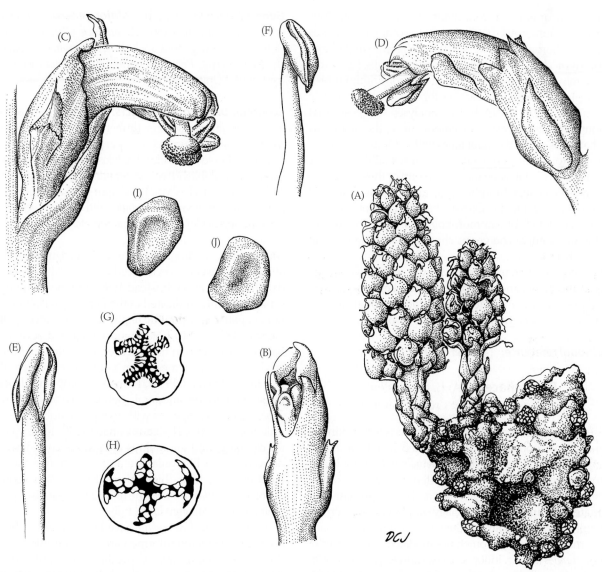

FIGURE 9.126 Orobanchaceae. *Conopholis americana:* (A) fruiting plant detached from root of *Quercus rubra*, which was to right, note scaly buds on "tuber" (× 0.75); (B) flower bud (× 6); (C, D) flowers (× 6); (E, F) stamens (× 12); (G) ovary in cross-section (× 9); (H) fruit in cross-section (× 3); (I, J) two seeds (× 18). (From Thieret 1971, *J. Arnold Arbor.* 52: p. 421.)

Genera/species: 65/1540. **Major genera:** *Pedicularis* (600 spp.), *Castilleja* (200), *Euphrasia* (200), *Orobanche* (100), *Buchnera* (100), *Agalinis* (60), *Striga* (50), and *Harveya* (40). Many genera occur in the continental United States and/or Canada; noteworthy genera (in addition to most of the above) include *Aureolaria, Conopholis, Epifagus, Melampyrum, Orthocarpus, Rhinanthus,* and *Seymeria.*

Economic plants and products: The family is of little economic importance, but a few taxa, such as *Striga* (witchweed), can severely damage crops.

Discussion: Orobanchaceae, as here circumscribed, are considered monophyletic on the basis of the hemi- to holoparasitic habit, hair morphology, and possibly racemose inflorescence. DNA-based cladistic analyses (Bennett and Mathews 2006; de Pamphilis and Young 1995; de Pamphilis

et al. 1997; Nickrent et al. 1998; Olmstead et al. 2001; Wolfe et al. 2005; Young et al. 1999) also support the group's monophyly. The hemiparasitic genera, such as *Aureolaria, Agalinis, Pedicularis, Castilleja,* and *Buchnera,* traditionally have been placed within a broadly defined Scrophulariaceae (including most members of Plantaginaceae and Scrophulariaceae, as here delimited), based on numerous symplesiomorphic features. Such taxa have been placed within the Scrophulariaceae subfam. Rhinanthoideae, along with autotrophic genera such as *Veronica.* In most classifications (Cronquist 1981), the holoparasitic taxa (e.g., *Orobanche, Epifagus, Conopholis,* and *Boschniakia*) are placed within a narrowly circumscribed family, and parietal placentas are stressed as a defining character. However, the traditional distinction between Scrophulariaceae and Orobanchaceae is difficult to accept in light of genera, such as *Harveya,* that are holoparasitic but have axile placentation. Orobanchaceae encompass a continuum from chlorophyllous plants with numerous root haustoria to complete parasites that lack chlorophyll (Plate 9.21I) and have a single large haustorium. *Orobanche, Epifagus,* and *Conopholis,* for example, merely represent the end point of a continuum representing increasing specialization for a parasitic existence (Boeshore 1920). DNA-based analyses suggest that although haustoria have evolved only once, the holoparasitic habit has evolved several times within Orobanchaceae.

The autotrophic genus *Lindenbergia* (usually treated as Scrophulariaceae) may be the sister group of these hemi- to holoparasitic genera. Morphological apomorphies linking it to them may include the abaxial lobes of the corolla outside the adaxial, details of hair morphology, and the indeterminate inflorescence. The genus should be included within Orobanchaceae.

Flowers of Orobanchaceae show a great diversity of shapes and colors and are pollinated by bees, wasps, flies, and birds. Bracts sometimes assist in pollinator attraction, as in *Castilleja.* Seeds are probably wind dispersed.

Additional reference: Thieret 1971.

Bignoniaceae A. L. de Jussieu
(Bignonia or Trumpet Creeper Family)

Trees, shrubs, or lianas; the lianas with a characteristic kind of anomalous secondary growth that results in a 4- or many-lobed or furrowed xylem cylinder; usually with iridoids and phenolic glycosides. Hairs various, often simple. *Leaves opposite or whorled,* occasionally alternate and spiral, **pinnately or palmately compound**, occasionally simple, entire to serrate, venation pinnate to palmate, the terminal and occasionally also some lateral leaflets sometimes modified into tendrils or hooks; *stipules lacking.* Inflorescences various. *Flowers bisexual, bilateral, usually large and showy.* Sepals 5, connate. *Petals 5, connate, the corolla ± 2-lipped,* the lobes usually imbricate. *Stamens usually 4, ± didynamous,* the fifth stamen sometimes represented by a small staminode, sometimes reduced to 2; *filaments adnate to corolla; anthers sagittate;* pollen grains diverse, sometimes in tetrads or polyads. *Carpels 2, connate; ovary superior, with usually axile placentation,* **the placentas divided (into 2 per locule)**; *stigma strongly 2-lobed, each lobe sensitive* (i.e., closing after contact with pollinator). Ovules numerous, with 1 integument and a thin-walled megasporangium. Nectar disk usually present. **Fruit a usually elongate**, *septicidal to loculicidal capsule,* occasionally a berry or indehiscent pod; **seeds usually flat, winged or fringed with hairs; endosperm lacking**; cotyledons deeply bilobed (Figure 9.127).

Floral formula: X, K $\circledS$, C $\widehat{(2+3)}$, A2+2, G $\underline{\circled{2}}$; capsule

Distribution: Widely distributed in tropical and subtropical regions, with a few species in temperate climates; most diverse in northern South America.

Genera/species: 104/860. ***Major genera:*** *Tabebuia* (100 spp.), *Adenocalymma* (80), *Arrabidaea* (70), and *Jacaranda* (40). Only a few genera, including *Bignonia, Catalpa, Campsis, Macfadyena, Pithecoctenium,* and *Tecoma,* occur in the continental United States.

Economic plants and products: Some species of *Catalpa* and *Tabebuia* are used for timber. *Spathodea* (African tulip tree), *Campsis* (trumpet creeper), *Bignonia* (cross vine), *Chilopsis* (flowering willow), *Clytostoma* (painted trumpet), *Crescentia* (calabash tree), *Jacaranda, Kigelia* (sausage tree), *Macfadyena* (cat-claw vine), *Podranea* (pink trumpet creeper), *Pyrostegia* (flame vine), *Tabebuia,* and *Tecoma* (yellow bells, Cape honeysuckle) provide ornamentals.

Discussion: Bignoniaceae are easily recognized and probably monophyletic, as evidenced by *rbcL* and *ndhF* sequence data (Spangler and Olmstead 1999) and the morphological synapomorphies indicated in the family description. Evolutionary patterns within the family have been investigated by Gentry (1974, 1980, 1990), who recognized several tribes based on variation in habit, leaves, placentation, fruits, and seeds, but several of theses may not be monophyletic (Spangler and Olmstead 1999). Pinnately compound leaves are considered to be ancestral, but *Crescentia, Catalpa,* and *Tabebuia* contain species with reduced, trifoliolate, or even unifoliolate leaves. The lianous habit (with leaves modified for climbing) and anomalous growth are considered to be derived within Bignoniaceae and characterize a group of Neotropical genera (Bignonieae; Plate 9.21F) (Lohmann 2006). *Crescentia, Parmentiera,* and *Kigelia* are distinct due to their indehiscent, often large, fruits.

Paulownia and *Schegelia* have been placed in Bignoniaceae or considered intermediate between this family and the Scrophulariaceae, as traditionally, but not here, defined. These genera lack the distinctive synapomorphies of Bignoniaceae (Armstrong 1985) and are here treated as members of separate small families (Spangler and Olmstead 1999). Their similarity to Plantaginaceae and Scrophulariaceae s.s. is probably symplesiomorphic.

Flowers of Bignoniaceae are variable in form, coloration, and time of anthesis. Pollinators are bees, wasps, butterflies, hawk moths, birds, and bats. Outcrossing is promoted by the sensitive stigma. The seeds of most species are dispersed by wind (Plate 9.21M).

Additional references: Dobbins 1971; Fischer et al. 2004b; Lohmann 2006a,b; Manning 2000.

Acanthaceae A. L. de Jussieu
(Acanthus Family)

Herbs or occasionally vines, shrubs, or trees; often with anomalous secondary growth; usually with phenolic glycosides; often with iridoids, alkaloids, and diterpenoids; *cystoliths of various forms often present.* Hairs usually simple, but sometimes branched, dendritic, or stellate. *Leaves*

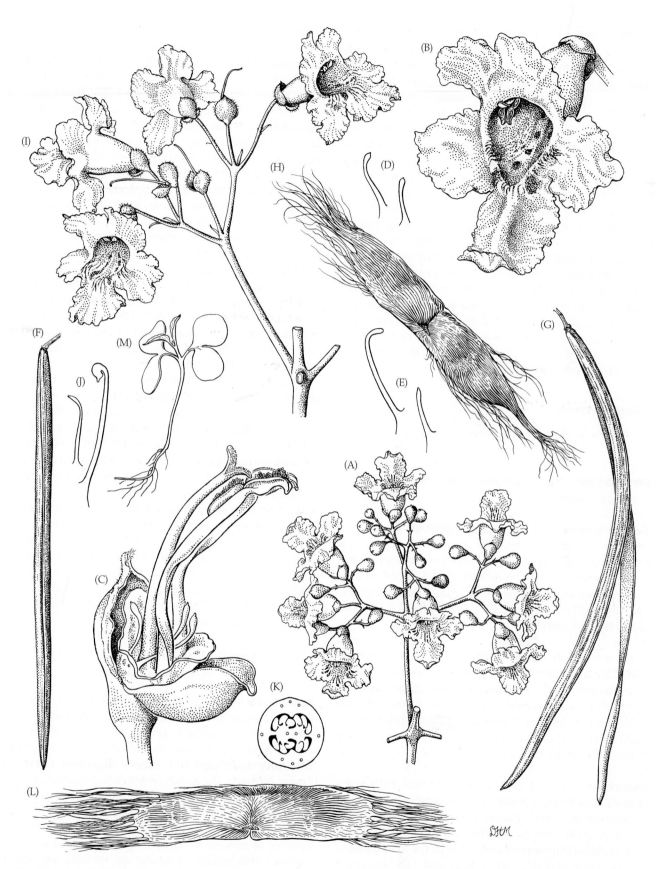

FIGURE 9.127 Bignoniaceae. (A–H) *Catalpa bignonioides:* (A) branch of cymose inflorescence (× 0.5); (B) flower (× 2); (C) flower, most of corolla removed, exposing gynoecium, two fertile stamens, three staminodes (× 4); (D, E) staminodes (× 4); (F) capsule (× 0.5); (G) open capsule (× 0.5); (H) seed (× 3). (I–M) *C. speciosa:* (I) inflorescence branch (× 0.7); (J) staminodes (× 4); (K) ovary in cross-section, note divided placentas (× 12); (L) seed (× 12); (M) seedling. (From Wood 1974, *A student's atlas of flowering plants*, p. 102.)

usually opposite, but alternate and spiral in Nelsonioideae, *simple*, sometimes lobed, entire to serrate or dentate, with pinnate venation; *stipules lacking*. Inflorescences various. *Flowers bisexual, bilateral, often associated with large, often colorful bracts and bractlets*. Sepals usually 4 or 5, connate. *Petals 5, connate, the corolla ± 2-lipped*, the lobes imbricate or convolute. *Stamens usually 4, didynamous, or sometimes only 2; filaments adnate to corolla; anthers* sometimes unilocular, often hairy, *often asymmetrical*, the pollen sacs sometimes widely separated on a modified connective; pollen grains various. *Carpels 2, connate; ovary superior, with usually axile placentation*; stigma funnel-shaped or 2-lobed, but 1 lobe sometimes reduced or lacking. *Ovules usually 2–10 in 2 rows in each locule*, rarely numerous (Nelsonioideae), anatropous to campylotropous, with 1 integument and a thin-walled megasporangium, *each ovule usually borne on a hook-shaped projection (retinaculum; i.e., a modified funiculus)*. Nectar disk present. *Fruit a loculicidal capsule, almost always explosively dehiscent*, but indehiscent and fleshy, with a single large seed in *Avicennia; seeds usually flat*, the coat sometimes mucilaginous; endosperm usually lacking (Figure 9.128).

Floral formula:

X, K ⑤, C (2+3), A2+2 or 2, G ②; capsule

Distribution: Widely distributed from tropical to warm temperate regions.

Genera/species: 202/3520. **Major genera:** *Justicia* (400 spp.), *Barleria* (250), *Strobilanthes* (250), *Ruellia* (200), *Thunbergia* (150), *Dicliptera* (150), and *Aphelandra* (150). Noteworthy genera in the continental United States include *Anisacanthus, Avicennia, Barleria, Carlowrightia, Dicliptera, Dyschoriste, Elytraria, Hygrophila, Justicia, Ruellia, Siphonoglossa,* and *Stenandrium*.

Economic plants and products: The family contains many ornamentals: *Aphelandra* (zebra plant), *Asystasia* (coromandel), *Barleria* (Philippine violet), *Justicia* (shrimp plant), *Eranthemum* (blue sage), *Fittonia* (mosaic plant), *Odontonema* (firespike), *Pachystachys* (yellow shrimp plant; Plate 9.21L), *Sanchezia, Thunbergia* (clock vine, blue-sky vine), and *Acanthus* (bear's-breech).

Discussion: Most genera of Acanthaceae form a well-defined monophyletic group, but some genera (here placed in Nelsonioideae and Thunbergioideae) have led some to doubt the group's naturalness (Bremekamp 1965). The family is broadly circumscribed and is considered monophyletic on the basis of *ndhF, rbcL, trnL-F*, and ITS sequences (Hedrén et al. 1995; Scotland et al. 1995; McDade et al. 2000). Morphological synapomorphies for Acanthaceae are not known.

Nelsonioideae (e.g., *Nelsonia, Elytraria*) may represent a paraphyletic basal complex within the family; this group has retained many plesiomorphic features, such as alternate and spiral leaves, numerous ovules that lack retinacula, and seeds with endosperm (Long 1970). *Thunbergia* and relatives (Thunbergioideae) and the remaining genera (i.e., Acanthoideae) may form a clade supported by DNA sequence characters, loss of endosperm, reduction in ovule number and possibly opposite leaves. Most members of the family belong to the large and clearly monophyletic Acanthoideae (Scotland 1990; McDade et al. 2000). Synapomorphies of this clade include retinacula, explosive capsules, and a particular pattern of imbricate corolla lobes. Two monophyletic subgroups are represented within Acanthoideae. The larger, supported by the presence of cystoliths in the leaves, includes *Sanchezia, Hygrophila, Ruellia, Eranthemum, Dyschoriste, Strobilanthes, Barleria, Pachystachys, Dicliptera, Odontonema, Justicia,* and *Fittonia*. The smaller subgroup, diagnosed by unilocular anthers, includes *Acanthus, Stenandrium,* and *Aphelandra* (Hedrén et al. 1995; Scotland et al. 1995; McDade and Moody 1999; McDade et al. 2000).

Interestingly, recent phylogenetic analyses based on molecular data support the inclusion of the mangrove genus *Avicennia* (Sanders 1997) within Acanthaceae (as subfamily Avicennioideae; Schwarzbach and McDade 2002), a placement also supported by articulated nodes, inflorescence structure, including flowers subtended by a bract and two bracteoles, a reduction in the number of ovules, and lack of endosperm.

Flowers of Acanthaceae, with their colorful corollas and often showy bracts, are pollinated by nectar-gathering bees, wasps, moths, butterflies, and birds. Protogyny leads to outcrossing, but some species of *Ruellia* produce cleistogamous flowers. The seeds are fairly large (compared to Plantaginaceae) and are flung some distance as the capsules dehisce explosively. The unusual fruits of *Avicennia* are dispersed by floating and tidal action.

Lentibulariaceae L. C. Richard
(Bladderwort Family)

Insectivorous herbs, aquatic or of wetlands, rooted in moist substrate to free-floating and lacking roots, sometimes epiphytic; vascular system often reduced; usually with iridoids. *Hairs sessile to stalked, gland-headed*, **some secreting mucilage and others digestive enzymes**. *Leaves alternate and spiral, or sometimes whorled, often in basal rosettes, simple, entire to finely divided,* **always highly modified**, *that is, flat and densely covered with sticky, mucilage-secreting and digestive hairs, and with margins inrolling in response to contact of glandular hairs with prey organism (Pinguicula), or tubular and spiraled, with downward pointing and digestive hairs and a basal chamber (Genlisea), or not obviously foliaceous, highly dissected, bearing prey-catching bladders, each with 2 sensitive valves forming a*

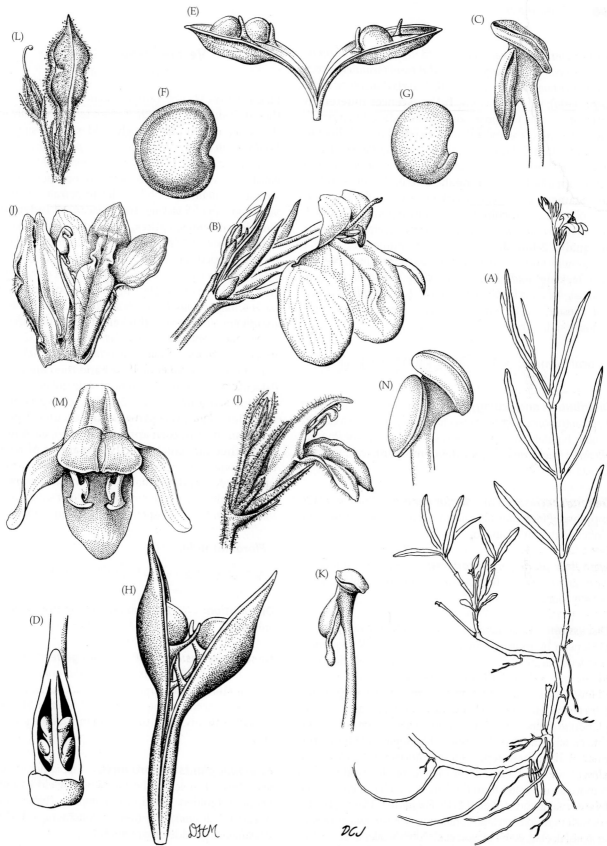

FIGURE 9.128 Acanthaceae. (A–G) *Justicia ovata* var. *angusta:*
(A) flowering plant (× 0.7); (B) flower and flower buds (× 4); (C)
anther (× 14); (D) nectar disk and ovary, with one side removed
to show four ovules (× 17); (E) capsule, one seed removed, note
retinacula (× 4); (F) moistened seed (× 8.5); (G) embryo (× 8.5). (H)
J. crassifolia: opening capsule with two seeds, four retinacula

(× 4). (I–L) *J. cooleyi:* (I) flower and bud (× 5.5); (J) corolla, opened
out, with one stamen (× 5.5); (K) anther (× 14); (L) partly mature
fruit and flower from which corolla has fallen (× 3). (M, N) *J. americana:* (M) corolla with stamens, from above (× 4); (N) anther
(× 14). (From Long 1970, *J. Arnold Arbor.* 51: p. 304.)

trapdoor entrance, which opens inward in response to a stimulus conveyed by 4 sensory hairs and then immediately closes again, and lined on the inside with branched digestive hairs (*Utricularia*); stipules lacking. **Inflorescences indeterminate**, on a scape, terminal, sometimes reduced to a single flower. Flowers bisexual, bilateral. Sepals 4 or 5, distinct to connate, often 2-lipped. *Petals 5, connate, the corolla 2-lipped, the lower lip usually with a nectar spur or sac,* and often with a bulge obscuring the throat, the lobes imbricate. **Stamens 2;** *filaments adnate to corolla; anthers unilocular;* pollen grains tricolporate to multicolporate. *Carpels 2, connate; ovary superior, with free-central placentation;* **stigma unequally 2-lobed**. Ovules usually numerous, with 1 integument and a thin-walled megasporangium. **Nectar disk lacking**, *and nectar produced by corolla nectar spur. Fruit usually a circumscissile, loculicidal, or irregularly dehiscent capsule;* seeds small, embryo ± undifferentiated; endosperm lacking.

Floral formula: X, K④–⑤, C②+③, A2, G②; capsule

Distribution and ecology: Widespread, occurring from tropical to boreal regions; insectivorous plants of wetlands, aquatic habitats, and moist forests. Some species of *Pinguicula* and *Utricularia* are epiphytes of moist tropical montane forests.

Genera/species: 3/310. ***Major genera:*** *Utricularia* (120 spp.) and *Pinguicula* (45). Both occur in the continental United States and Canada.

Economic plants and products: *Utricularia* (bladderwort) and *Pinguicula* (butterwort) are occasionally grown as novelty plants.

Discussion: Lentibulariaceae are clearly monophyletic, as evidenced by morphology and *rbcL* and *matK* sequences (Müller et al. 2000) although their exact phylogenetic placement within Lamiales is problematic. There are surprising differences in leaf form between *Pinguicula*, *Genlisea*, and *Utricularia*. However, the leaves of all three genera have glandular hairs that secrete both digestive enzymes and mucilage; a more or less unmodified leaf bearing gland-headed hairs may represent the ancestral condition. In *Utricularia* the hairs on the inside of the bladder also remove excess water from the lumen, which is then secreted by external glands (Plate 9.21P). *Pinguicula* (Plate 9.21N) is the sister group of a clade containing *Genlisea* and *Utricularia* (Cieslak et al. 2005; Jobson et al. 2003; Müller et al. 2000).

The showy (yellow, white, blue, purple) bilateral flowers are pollinated by nectar-gathering bees and wasps. In *Utricularia* the stigma is sometimes sensitive (closing when touched); outcrossing is characteristic.

Additional references: Fineran 1985; Taylor 1989; Wood and Godfrey 1957.

Verbenaceae J. St. Hilaire
(Verbena Family)

Herbs, lianas, shrubs, or trees, sometimes with prickles or thorns; *stems usually square in cross-section;* usually with iridoids; often with phenolic glycosides. *Hairs simple, gland-headed, with ethereal oils (including terpenoids),* and *nonglandular, these, when present, usually unicellular,* sometimes calcified or silicified. *Leaves opposite* or occasionally whorled, simple, sometimes lobed, entire to serrate, with pinnate venation; *stipules lacking*. **Inflorescences indeterminate, forming racemes, spikes, or heads**, terminal or axillary. Flowers bisexual, bilateral. *Sepals 5, connate,* the calyx tubular to bell-shaped, *persistent,* occasionally enlarged in fruit. *Petals 5 (but sometimes seemingly 4 by fusion of the upper pair), connate, the corolla weakly 2-lipped,* the lobes imbricate. *Stamens 4, didynamous; filaments adnate to corolla;* pollen grains usually tricolpate, **exine thickened near apertures**. *Carpels 2, connate; ovary superior, unlobed to ± 4-lobed, 2-locular but appearing 4-locular due to development of false septa,*but sometimes 1 carpel suppressed and then appearing only 2-locular, with axile placentation; style not apically divided, terminal; **stigma** usually 2-lobed, **conspicuous, with well-developed receptive tissue. Ovules 2 per carpel (i.e., usually 1 in each apparent locule), each marginally attached (attached directly to margins of false partitions)**, with 1 integument and a thin-walled megasporangium. Nectar disk usually present. *Fruit a drupe with 2 or 4 pits (single and 2-lobed in Lantana), or a schizocarp splitting into 2 or 4 nutlets;* endosperm lacking (Figure 9.129).

Floral formula:

X, K⑤, C②+③, A2+2, G②; drupe, 4 nutlets

Distribution: Widely distributed from tropical to temperate regions.

Genera/species: 35/1000. ***Major genera:*** *Verbena* (200 spp.), *Lippia* (200), *Lantana* (150), *Citharexylum* (130), *Glandularia* (100), *Stachytarpheta* (90), and *Duranta* (20). All of the above occur in the continental United States and/or Canada; other noteworthy genera include *Aloysia*, *Bouchea*, and *Priva*.

Economic plants and products: *Lippia* (lemon verbena) and *Priva* are used in herbal teas or yield essential oils. *Duranta* (golden dewdrop), *Lantana*, *Petraea* (queen's-wreath), *Stachytarpheta* (devil's-coachwhip), *Verbena*, and *Glandularia* provide ornamentals.

Discussion: Verbenaceae, as here circumscribed, are considered monophyletic on the basis of their morphology and *rbcL* sequences (Cantino 1992a,b; Chadwell et al. 1992; Judd et al. 1994; Olmstead et al. 1993; Wagstaff and Olmstead 1996). The family has been circumscribed much more broadly by most systematists (e.g., Cronquist 1981,

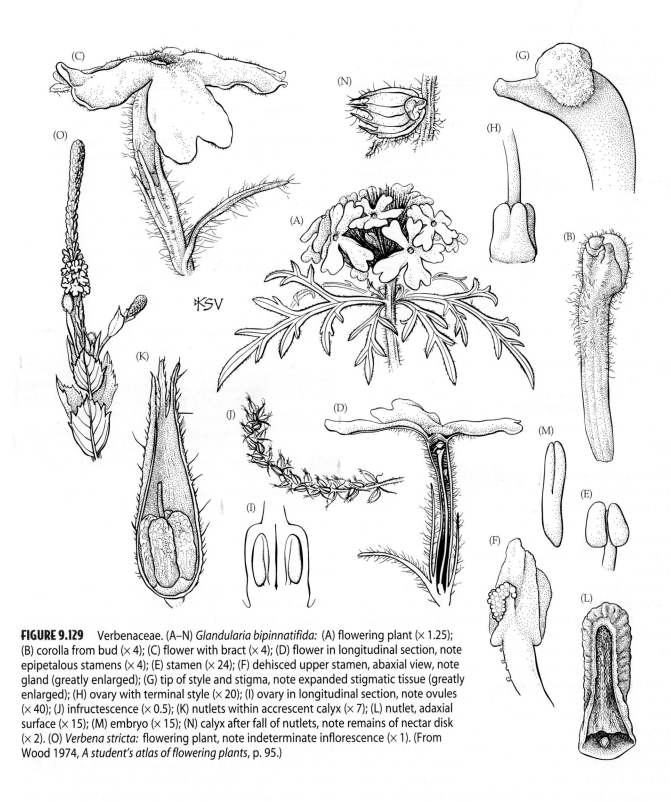

FIGURE 9.129 Verbenaceae. (A–N) *Glandularia bipinnatifida:* (A) flowering plant (× 1.25); (B) corolla from bud (× 4); (C) flower with bract (× 4); (D) flower in longitudinal section, note epipetalous stamens (× 4); (E) stamen (× 24); (F) dehisced upper stamen, abaxial view, note gland (greatly enlarged); (G) tip of style and stigma, note expanded stigmatic tissue (greatly enlarged); (H) ovary with terminal style (× 20); (I) ovary in longitudinal section, note ovules (× 40); (J) infructescence (× 0.5); (K) nutlets within accrescent calyx (× 7); (L) nutlet, adaxial surface (× 15); (M) embryo (× 15); (N) calyx after fall of nutlets, note remains of nectar disk (× 2). (O) *Verbena stricta:* flowering plant, note indeterminate inflorescence (× 1). (From Wood 1974, *A student's atlas of flowering plants*, p. 95.)

1988) and separated from Lamiaceae by the presence of a terminal (vs. gynobasic) style. Here we include only the traditional subfamily Verbenoideae (excluding the tribe Monochileae). As traditionally delimited, Verbenaceae are paraphyletic, while Lamiaceae are polyphyletic. In order to make Lamiaceae monophyletic, nearly two-thirds of the genera usually included within Verbenaceae (e.g., *Callicarpa, Clerodendrum, Vitex, Tectona*), are transferred to Lamiaceae (see Cantino 1992a,b; Cantino et al. 1992). As redefined, Verbenaceae can be distinguished from Lamiaceae by their indeterminate racemes, spikes, or heads (vs. inflorescences with an indeterminate main axis and cymosely branched lateral axes, these often reduced and forming pseudowhorls); ovules attached on the margins of false septa (vs. ovules attached on the sides of false septa); simple style with conspicuous, 2-lobed stigma (vs. usually apically forked style with inconspicuous stigmatic region at the tip of each style branch); pollen exine thickened near apertures (vs. not thickened); and nonglandular hairs exclusively unicellular (vs. multicellular, uniseriate). In addition, the flowers tend to be less strongly two-lipped. The style of Verbenaceae is exclusively terminal, while in Lamiaceae it varies from terminal to gynobasic.

The inclusion of *Petraea* within Verbenaceae is not supported by some analyses based on cpDNA data, but it is supported by others, and by a combination of morphology and cpDNA sequences; the genus may be sister to the rest of the family.

The showy flowers of Verbenaceae (Plate 9.21J) are pollinated by nectar-gathering bees, wasps, and flies. Outcrossing is favored by protandry. The colorful drupaceous fruits are dispersed by birds. The nutlets are associated with a persistent calyx, and are released over time by the action of wind or mechanical contact. They may be secondarily dispersed by water (rainwash) or eaten by birds. A few are externally transported by animals.

Additional references: Atkins 2004; Cantino 1990; Sanders 2001.

Lamiaceae Martynov

(= Labiatae A. L. de Jussieu)

(Mint Family)

Herbs, shrubs, or trees; *stems often square in cross-section; often with iridoids and phenolic glycosides. Hairs gland-headed, with ethereal oils (including terpenoids), and simple, nonglandular, these, when present, usually multicellular and uniseriate or a mixture of multicellular and unicellular. Leaves usually opposite,* occasionally whorled, simple, sometimes lobed or dissected, or pinnately or palmately compound, entire to serrate; *stipules lacking. Inflorescences with indeterminate main axis and determinate (cymosely branched) lateral axes, often congested into pseudowhorls,* terminal or axillary. Flowers bisexual, usually bilateral. *Sepals usually 5, connate,* calyx radial to bilateral, ± tubular, bell-shaped, or wheel-shaped, persistent, occasionally enlarged in fruit. *Petals usually 5, connate, usually 2-lipped,* the lobes imbricate. *Stamens 4, didynamous to ± equal, sometimes reduced to 2; filaments adnate to corolla;* pollen grains tricolpate or hexacolpate. *Carpels 2, connate; ovary superior, unlobed to deeply 4-lobed, 2-locular but appearing 4-locular due to development of false septa,* with axile placentation; *style usually apically divided, terminal to gynobasic; stigmas 2, tiny and inconspicuous at apices of style branches.* **Ovules 2 per carpel (i.e., one in each apparent locule), each laterally attached (attached on false septa very near the inrolled carpel margins)**, with 1 integument and a thin-walled megasporangium. Nectar disk often present. *Fruit a drupe with 1–4 pits, an indehiscent, 4-seeded pod, or a schizocarp splitting into 4 nutlets or 4 drupelets;* endosperm scanty or lacking (Figure 9.130).

Floral formula:

X, K⑤, C②+③, A2+2, G②; drupe, 4 nutlets

Distribution: Cosmopolitan.

Genera/species: 252/6800. **Major genera:** *Salvia* (800 spp.), *Hyptis* (400), *Clerodendrum* (400), *Thymus* (350), *Plectranthus* (300), *Scutellaria* (300), *Stachys* (300), *Nepeta* (250), *Vitex* (250), *Teucrium* (200), *Premna* (200), and *Callicarpa* (140). A large number of genera occur in the continental United States and/or Canada; noteworthy representatives (in addition to those listed above) include *Agastache, Ajuga, Collinsonia, Dicerandra, Dracocephalum, Glechoma, Hedeoma, Hyssopus, Lamium, Leonurus, Lycopus, Marrubium, Mentha, Monarda, Physostegia, Piloblephis, Prunella, Pycnanthemum, Pycnostachys, Satureja,* and *Trichostema.*

Economic plants and products: The family contains many species that are economically important either for their essential oils or for use as spices, including *Mentha* (peppermint, spearmint), *Lavandula* (lavender), *Marrubium* (horehound), *Nepeta* (catnip), *Ocimum* (basil), *Origanum* (oregano), *Rosmarinus* (rosemary), *Salvia* (sage), *Satureja* (savory), and *Thymus* (thyme). The tubers of a few species of *Stachys* are edible. *Tectona* (teak) is an important timber tree. Numerous genera provide ornamentals, including *Ajuga* (bugleweed), *Callicarpa* (beautyberry), *Clerodendrum, Plectranthus* (coleus), *Holmskioldia* (Chinese-hat plant), *Leonotis* (lion's ear), *Monarda* (bee balm), *Pycnanthemum* (mountain-mint), *Salvia, Scutellaria* (skullcap), and *Vitex* (monk's-pepper).

Discussion: Lamiaceae (or Labiatae), as here delimited, are considered monophyletic on the basis of laterally attached ovules, *rbcL,* and *ndhF* sequences (Cantino 1992a,b; Cantino et al. 1992; Junell 1934; Olmstead et al. 1993; Wagstaff and Olmstead 1996; Wagstaff et al. 1998). Phylogenetic relationships within Lamiales support a broad circumscription of the family, including many genera traditionally placed in Verbenaceae (see the discussion under Verbenaceae for a listing of the most useful distinguishing features). The ovary of Lamiaceae varies from rounded to deeply 4-lobed (vs. rounded to only moderately four-lobed in Verbenaceae), and the style is terminal to gynobasic (vs. consistently terminal in Verbenaceae). Most systematists have restricted Lamiaceae to those species with more or less gynobasic styles, a circumscription resulting in a polyphyletic assemblage because gynobasic styles have evolved more than once.

Several monophyletic groups can be distinguished within Lamiaceae (Cantino 1992a,b, Cantino et al. 1997; Wagstaff and Olmstead 1996; Wagstaff et al. 1998). Ajugoideae (or Teucrioideae) are considered monophyletic on the basis of their drupes with four pits, nonpersistent styles,

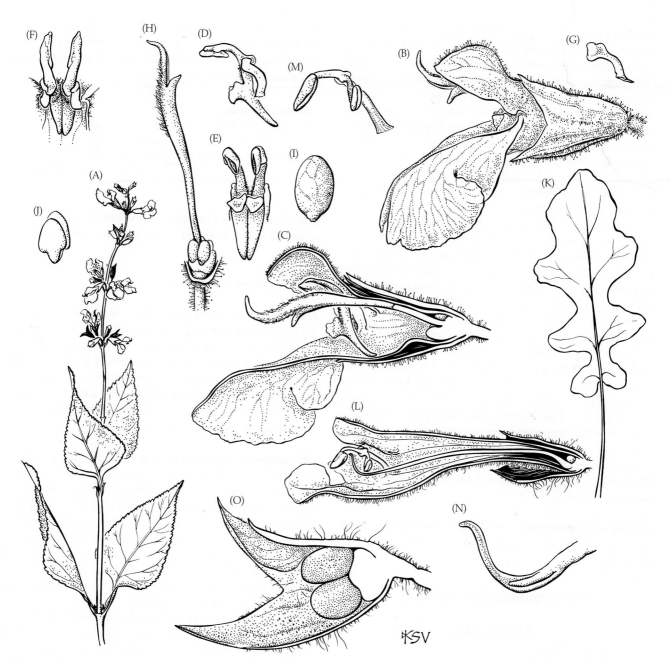

FIGURE 9.130 Lamiaceae (Labiatae). (A–J) *Salvia urticifolia:* (A) flowering plant, note reduced cymes forming pseudoverticillate inflorescence (× 0.5); (B) flower, side view (× 5); (C) flower, in longitudinal section (× 5); (D) stamen, side view, showing fertile anther half and sterile anther half (× 5); (E) stamens as seen in adaxial view (× 5); (F) stamens as seen from above and abaxially (× 5); (G) staminode (× 15); (H) gynoecium, note gynobasic style, lobed ovary and nectar disk (× 5); (I) nutlet (× 10); (J) embryo (× 10). (K–O) *S. lyrata:* (K) basal leaf (× 1); (L) flower in longitudinal section (× 5); (M) fertile stamen, both anther halves producing pollen (× 10); (N) tip of style with unequal style branches (greatly enlarged); (O) calyx in longitudinal section, with two (of four) nearly mature nutlets (× 10). (From Wood 1974, *A student's atlas of flowering plants,* p. 97.)

and pollen grains with branched to granular columns in the exine. This clade includes *Clerodendrum, Teucrium, Trichostema, Ajuga,* and relatives. *Scutellaria, Holmskioldia,* and relatives constitute the Scutellarioideae, a group with rounded calyx lobes and nutlets with tuberculate or plumose outgrowths. Lamioideae and Nepetoideae together may constitute a monophyletic group, and both have completely gynobasic styles (Cantino 1992a). Analyses of *rbcL* and *ndhF* sequences, however, place Lamioideae in a clade with Scutellarioideae and Ajugoideae, and Nepetoideae as sister to either "Chloanthoideae" or *Callicarpa* (Wagstaff and Olmstead 1996; Wagstaff et al. 1998). The monophyly of Nepetoideae, which includes *Glechoma, Dicerandra, Hyptis, Lycopus, Melissa, Mentha, Monarda,*

Nepeta, Ocimum, Origanum, Plectranthus, Prunella, Pycnan-themum, Pycnostachys, Salvia, Satureja, Lavandula, Thymus, and *Basilicum,* is supported by their hexacolpate pollen and DNA sequences; the other subfamilies have retained tricolpate pollen. Nepetoideae also lack endosperm and iridoids and have a high volatile terpenoid content. The monophyly of Lamioideae, which includes *Lamium, Prasium, Galeopsis, Leonotis, Physostegia, Marrubium, Pogostemon,* and *Stachys,* is strongly supported by DNA data. *Gmelina, Callicarpa, Vitex* (Plate 9.21K), *Premna, Cornutia,* and relatives are usually placed within the polyphyletic "Viticoideae;" *Tectona* and relatives have branched hairs, and have been treated as "Chloanthoideae" but are likely also polyphyletic.

The showy flowers of Lamiaceae are pollinated by bees, wasps, butterflies, moths, flies, beetles, and birds. The arching upper lip of the two-lipped corolla usually protects the stamens and stigma, while the lower lip provides a landing platform and is often showy (Plate 9.21O). The pollinator is dusted with pollen on its back or head as it probes for nectar. In *Ocimum* and relatives, however, the stamens lie close to the lower lip and deposit pollen on the underside of the pollinator. The stamens of *Salvia* are highly modified, the connective is expanded, forming a lever arm. Most species are protandrous, and outcrossing is common. Some species of *Lamium* have cleistogamous flowers. Species with drupaceous fruits are usually dispersed by birds or mammals. The nutlets of many species are merely shaken from the persistent calyx by the action of wind or disturbance of the plant. Nutlets may also be eaten by birds or dispersed by water.

Additional references: Abu-Asab and Cantino 1989; Cantino 1990; Cantino and Sanders 1986; Harley et al. 2004; Huck 1992.

Campanulids (Euasterids II)

Aquifoliales

Aquifoliaceae Bercht. & J. Presl
(Holly Family)

Trees or shrubs, bark ± smooth; sometimes with alkaloids, caffeine. Hairs simple. *Leaves alternate and spiral, simple,* entire to serrate, the teeth with deciduous glandular to spinose apex, with pinnate venation; **stipules minute**. Inflorescences determinate, axillary, but sometimes reduced to a solitary flower. *Flowers usually unisexual (plants dioecious), radial.* Sepals usually 4–6, slightly connate. *Petals usually 4–6, usually slightly connate,* imbricate. *Stamens usually 4–6 and slightly adnate to base of corolla;* pollen grains 3–4-colporate or porate; staminodes conspicuous in carpellate flowers. Carpels usually 4–6, connate; ovary superior, with axile placentation; **style very short or lacking**; *stigma capitate, discoid, or slightly lobed;* pistillode often conspicuous in staminate flowers. Ovules usually 1 in each locule, with 1 integument and a thin-walled megasporangium. Nectaries

lacking, or nectar secreted by papillose swellings on adaxial petal surface. *Fruit a colorful* (red, orange-red, purple-black, or pink) *drupe* (Plate 9.22G) **with a broad stigmatic zone and usually 4–6 pits** (Figure 9.131).

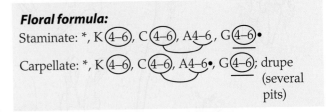

Floral formula:
Staminate: *, K (4–6), C (4–6), A4–6, G (4–6)•
Carpellate: *, K (4–6), C (4–6), A4–6•, G (4–6); drupe (several pits)

Distribution and ecology: Widespread, but especially common in the montane tropics; characteristic of acidic soils.

Genus/species: 1 (*Ilex*)/400.

Economic plants and products: Numerous species of *Ilex* (holly) are cultivated for their ornamental foliage and colorful fruits. The leaves of *I. paraguariensis* are brewed to make the high-caffeine beverage *maté.* The leaves of *I. vomitoria,* also high in caffeine, were used by native Americans of the southeastern U.S. coastal plain to make the stimulating "black drink." Hollywood is a fine timber, and is popular for inlay-work.

Discussion: Phylogenetic relationships within *Ilex* are obscure, although it is clear that recognition of *Nemopanthus* (on the basis of its slender, distinct petals) would create a paraphyletic *Ilex* (Gottlieb et al. 2005; Powell et al. 2000). Aquifoliaceae probably are most closely related to Helwingiaceae (*Helwingia*), a group of shrubs with alternate and spiral, pinnately veined leaves and epiphyllous inflorescences (Soltis et al. 2000; Albach et al. 2001a,b), and possibly also Phyllonomaceae (*Phyllonoma*), another epiphyllous group (Bremer et al. 2002).

Holly flowers are visited by various insects, especially bees. Outcrossing is characteristic due to dioecy (Plate 9.22D). Some species may be at least partially wind pollinated. The drupes of hollies are dispersed by birds. The tiny embryo matures after being dispersed; germination is slow, often requiring 1–3 years.

Additional references: Brizicky 1964a; Olmstead et al. 1993.

Apiales

Apiales clearly belong to the core asterid clade, as indicated by their ovules with a single integument and usually a thin megasporangium wall, sympetalous corollas (obvious in many Pittosporaceae), stamens in a single whorl (Judd et al. 1994; Judd 1996), and DNA characters (Downie and Palmer 1992; Källersjö et al. 1998; Olmstead et al. 1992a, 1993, 2000a; Plunkett et al. 1996a,b; Soltis et al. 2005). The order is monophyletic and closely related to Dipsacales and

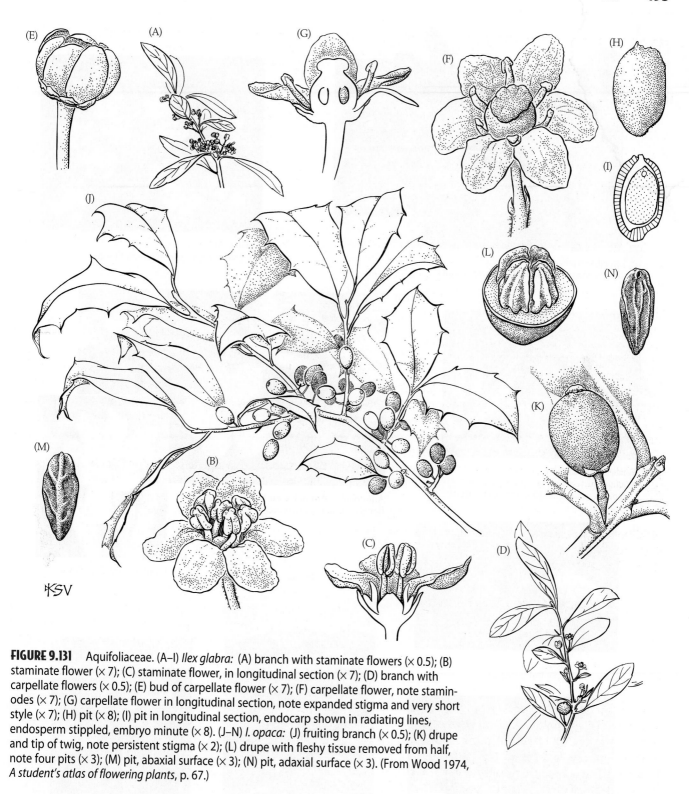

FIGURE 9.131 Aquifoliaceae. (A–I) *Ilex glabra:* (A) branch with staminate flowers (× 0.5); (B) staminate flower (× 7); (C) staminate flower, in longitudinal section (× 7); (D) branch with carpellate flowers (× 0.5); (E) bud of carpellate flower (× 7); (F) carpellate flower, note staminodes (× 7); (G) carpellate flower in longitudinal section, note expanded stigma and very short style (× 7); (H) pit (× 8); (I) pit in longitudinal section, endocarp shown in radiating lines, endosperm stippled, embryo minute (× 8). (J–N) *I. opaca:* (J) fruiting branch (× 0.5); (K) drupe and tip of twig, note persistent stigma (× 2); (L) drupe with fleshy tissue removed from half, note four pits (× 3); (M) pit, abaxial surface (× 3); (N) pit, adaxial surface (× 3). (From Wood 1974, *A student's atlas of flowering plants*, p. 67.)

Asterales (Hilu et al. 2003; Olmstead et al. 2000a; Soltis et al. 2000, 2005). Potential morphological synapomorphies of Apiales include corolla lobes well developed, stamens free from the corolla or nearly so, and only one or two ovules per carpel. We note that some members of all of these groups contain polyacetylenes. The order consists of 7 families and ca. 3780 species; major families are **Apiaceae, Araliaceae**, and Pittosporaceae.

Apiaceae Lindley
(= Umbelliferae A. L. de Jussieu)
(Carrot Family)

Usually herbs, aromatic; *stems usually hollow in internodal region; with secretory canals containing ethereal oils and resins, triterpenoid saponins, coumarins, falcarinone polyacetylenes, monoterpenes, and sesquiterpenes;* with umbelliferose (a trisaccharide) as carbohydrate storage product. Hairs various.

(A)

Apiales: Araliaceae
Didymopanax tremulum: branch with fruits

(B)

Apiales: Apiaceae
Oxypolis filiformis var. *filiformis*: flowers

(C)

Apiales: Apiaceae
*Anethum graveolens, Foeniculum vulgare,
Cuminum cyminum, Apium graveolens*: fruits

(D)

Aquifoliales: Aquifoliaceae
Ilex glabra: staminate flower

I. glabra: carpellate flower

(E)

Apiales: Apiaceae
Cicuta maculata: inflorescence

(F)

(G)

Aquifoliales: Aquifoliaceae
Ilex cassine: branch in fruit

Apiales: Myodocarpaceae
Myodocarpus fraxinifolius: plant
in bloom

(H)

Apiales: Araliaceae
Hydrocotyle umbellata: plant in bloom

PLATE 9.22 Eudicots: Aquifoliales and Apiales

Key to Major Families of Apiales

1. Ovary superior, with several ovules and usually parietal placentation, a stylopodium lacking; cells of seed coat each with a central depression that develops in the center of a projection; inflorescences not umbellate ....Pittosporaceae

1. Ovary inferior, with a single ovule per locule and usually with axile placentation, stylopodium present; cells of seed coat not as above; inflorescences usually umbellate or comprised of umbellate units 2

2. Usually herbs; stipules ± lacking; ovary 2-carpellate, developing into a schizocarp, usually with oil cavities and a carpophore ....**Apiaceae**

2. Usually trees or shrubs; stipules present; ovary 2-5-carpellate, usually developing into a globose berry or drupe (lacking oil cavities and a carpophore) ....**Araliaceae**

Leaves alternate and spiral, pinnately or palmately compound to simple, then often deeply dissected or lobed, entire to serrate, with pinnate to palmate venation; petioles ± sheathing; *stipules usually absent. Inflorescences determinate, forming simple or compound umbels*, sometimes condensed into a head, occasionally panicles or racemes of umbels, often subtended by an involucre of bracts, terminal. *Flowers* usually bisexual, usually radial, *small. Sepals usually 5, distinct, very reduced. Petals usually 5, distinct,* but developing from a ring primordium, *usually inflexed,* imbricate to valvate. *Stamens 5*; filaments distinct; pollen grains usually tricolporate. *Carpels 2, connate; ovary inferior,* usually with axile placentation; *styles ± swollen at base to form a nectar-secreting structure (stylopodium) atop ovary;* stigmas 2, tiny, capitate to truncate, or elongate. Ovules 2 in each locule, but only 1 fertile, with 1 integument and a thin-walled to less commonly thick-walled megasporangium. *Fruit a schizocarp, the 2 dry segments (mericarps)* usually **attached to an entire to deeply forked central stalk (carpophore)**; *globular to elongated oil canals (vittae) present in fruit-wall;* fruit surface smooth or ribbed, sometimes covered with hairs, scales, or bristles, sometimes flattened or winged; endosperm with petroselenic acid (Figure 9.132).

Floral formula: *, K5, C5, A5, G$\overline{②}$; schizocarp

Distribution: Widespread, from tropical to temperate regions.

Genera/species: 434/3780. **Major genera:** *Eryngium* (230 spp.), *Ferula* (150), *Peucedanum* (150), *Pimpinella* (150), *Bupleurum* (100), *Lomatium* (60), *Heracleum* (60), *Angelica* (50), *Sanicula* (40), and *Chaerophyllum* (40). Some of the numerous genera occurring in the continental United States and/or Canada are *Angelica, Apium, Carum, Centella, Chaerophyllum, Cicuta, Conioselinum, Daucus, Eryngium, Heracleum, Ligusticum, Lomatium, Osmorhiza, Oxypolis, Pastinaca, Ptilimnium, Sanicula, Sium, Spermolepis, Thaspium, Torilis,* and *Zizia.*

Economic plants and products: Apiaceae contain many food and spice plants: *Anethum* (dill), *Apium* (celery), *Carum* (caraway), *Coriandrum* (coriander), *Cuminum* (cumin), *Daucus* (carrot), *Foeniculum* (fennel), *Pastinaca* (parsnip), *Petroselinum* (parsley), and *Pimpinella* (anise). However, many are extremely poisonous, such as *Conium* (hemlock, which Socrates is said to have taken for suicide) and *Cicuta* (water hemlock).

Discussion: Monophyly of Apiaceae (Umbelliferae) is supported by DNA sequences (Chandler and Plunkett 2004; Olmstead et al. 1993; Plunkett et al. 1996a,b, 1997, 2004). Apiaceae are most closely related to Araliaceae, Pittosporaceae, and Myodocarpaceae and these four, along with three small families, constitute Apiales (Albach et al. 2001a,b; Judd and Olmstead 2004; Savolainen et al. 2000b; Soltis et al. 2000, 2005). Potential apomorphies of Apiaceae, Araliaceae, Myodocarpaceae, and Pittosporaceae include the distinctive resin/ethereal oil canals that are associated with the conducting tissues, a characteristic arrangement of lateral roots, the presence of falcarinone polyacetylenes, a tiny embryo, and reduced, bractlike leaves at the base of the shoots. Possible synapomorphies of Apiaceae, Myodocarpaceae, and Araliaceae include the presence of petroselenic acid in the seeds, umbelliferose as a carbohydrate storage product, umbellate inflorescences, and flowers with a stylopodium (Hegnauer 1971).

Apiaceae here are circumscribed ± narrowly, including herbaceous species with two-carpellate, usually dorsally flattened schizocarps with a carpophore and oil canals (vittae). Woody species with two- to five-carpellate, usually globose drupes (Plate 9.22A) that lack oil cavities, are usually considered as a separate family, Araliaceae, although some (e.g., Judd et al. 1994; Thorne 1973b) have combined these two clades into a single broadly circumscribed family. Although many members of Araliaceae and Apiaceae are distinct, the characters used to distinguish them are quite homoplasious (Plunkett et al. 1996a,b), and the genera

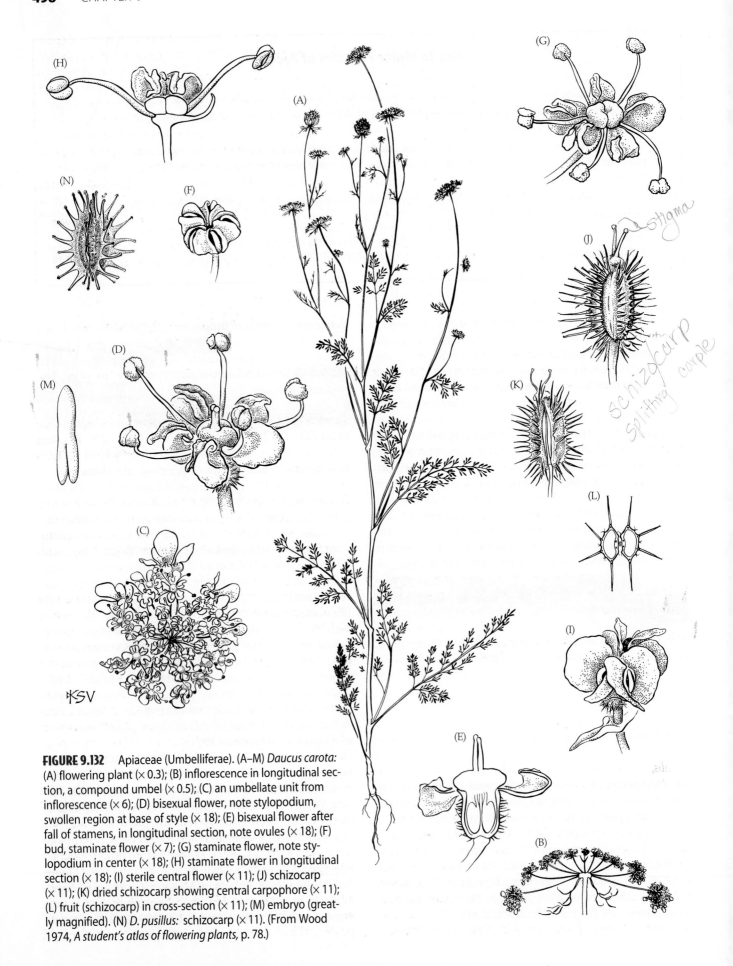

FIGURE 9.132 Apiaceae (Umbelliferae). (A–M) *Daucus carota:* (A) flowering plant (× 0.3); (B) inflorescence in longitudinal section, a compound umbel (× 0.5); (C) an umbellate unit from inflorescence (× 6); (D) bisexual flower, note stylopodium, swollen region at base of style (× 18); (E) bisexual flower after fall of stamens, in longitudinal section, note ovules (× 18); (F) bud, staminate flower (× 7); (G) staminate flower, note stylopodium in center (× 18); (H) staminate flower in longitudinal section (× 18); (I) sterile central flower (× 11); (J) schizocarp (× 11); (K) dried schizocarp showing central carpophore (× 11); (L) fruit (schizocarp) in cross-section (× 11); (M) embryo (greatly magnified). (N) *D. pusillus:* schizocarp (× 11). (From Wood 1974, *A student's atlas of flowering plants*, p. 78.)

Myodocarpus and *Delarbrea*, in particular, are problematic. These two genera (placed in the segregate family Myodocarpaceae) may represent remnants of a basal complex and are characterized by a woody habit, stipulate leaves, and two-carpellate, schizocarpic fruits. Herbaceousness and compound leaves undoubtedly have evolved several times within the derived lineages of Apiales.

Several clades can be discerned within Apiaceae although some are difficult to diagnose due to the lack of unambiguous morphological synapomorphies. Saniculoideae (e.g., *Sanicula* and *Eryngium*) have a stylopodium separated from the style by a narrow groove. Apioideae (e.g., *Coriandrum, Angelica, Apium, Chaerophyllum, Cicuta, Daucus,* and *Spermolepis*) have compound umbels and schizocarpic fruits with a more or less forked carpophore (see Downie et al. 1998, 2000). Saniculoideae + Apioideae form a clade that is supported by the herbaceous habit, lack of stipules, elongated oil canals (vittae) in the fruit, and DNA sequences. Genera like *Mackinlaya* and *Centella* are anomalous, e.g., *Mackinlaya* is woody, and in several features is similar to Araliaceae, while the related *Centella* is herbaceous and has schizocarpic fruits; both have stipules, a well-developed endocarp, and both lack vittae in their fruits. Such genera are placed in Apiaceae on the basis of DNA sequence characters, but are basal to Saniculoideae + Apioideae.

Genera traditionally placed in "Hydrocotyloideae" (such as *Centella* and *Hydrocotyle*) do not form a clade; some, such as *Centella*, show affinities to Apiaceae, and others, e.g., *Hydrocotyle*, to Araliaceae (Plunkett et al. 1996a,b, 1997, 2004).

The usually small flowers of Apiaceae are typically densely aggregated and pollinated by a wide range of small nectar-gathering flies, beetles, bees, and moths. Cross-pollination is favored by protandry. The dry, schizocarpic fruits are often wind-dispersed. The fruit segments of genera such as *Daucus* are covered by bristles, leading to dispersal by external transport.

Additional references: Baumann 1946; Constance 1971; Erbar 1991; Erbar and Leins 1988; Graham 1966; Hegnauer 1971; Jackson 1933; Kårehed 2003; Plunkett and Downie 1999; Sun et al. 2004.

Araliaceae A. L. de Jussieu
(Ginseng Family)

Shrubs, lianas, or trees, to occasionally herbs, *aromatic; with secretory canals containing ethereal oils and resins, triterpenoid saponins, coumarins, falcarinone polyacetylenes, monoterpenes, and sesquiterpenes;* with umbelliferose (a trisaccharide) as carbohydrate storage product. Hairs various, sometimes with prickles. *Leaves alternate and spiral, pinnately or palmately compound to simple, then sometimes dissected or lobed,* entire to serrate, with pinnate to palmate venation; petioles ± sheathing; *stipules usually present.* Inflorescences determinate, *forming simple umbels, and these usually arranged in racemes, spikes, or panicles. Flowers* usually bisex-ual but sometimes unisexual (plants then monoecious to dioecious), usually radial, *small. Sepals usually 5, distinct, very reduced. Petals usually 5, occasionally more, distinct,* but developing from a ring primordium, sometimes clearly connate, imbricate to valvate. *Stamens 5,* but occasionally numerous; filaments distinct; pollen grains usually tricolporate. *Carpels usually 2–5,* occasionally numerous, *connate; ovary inferior,* usually with axile placentation; *styles ± swollen at base to form a nectar-secreting structure (stylopodium) atop ovary;* stigmas usually 2–5, tiny, capitate to truncate, or elongate. Ovules 2 in each locule but only 1 fertile, with 1 integument and a thin-walled to less commonly thick-walled to thin-walled megasporangium. *Fruit a globose berry or drupe with 2–5 pits,* rarely a drupaceous schizocarp; endosperm with petroselenic acid (Figure 9.133).

Floral formula:

*, K5, C⑤, A5, G②–⑤ berry, drupe, drupaceous schizocarp

Distribution: Widespread, from tropical to temperate regions.

Genera/species: 43/1450. **Major genera:** *Schefflera* (600 spp.), *Polyscias* (200), *Oreopanax* (90), *Hydrocotyle* (80), and *Aralia* (68). Genera occurring in the continental United States and/or Canada include *Aralia, Hydrocotyle,* and *Panax.*

Economic plants and products: *Panax quinquefolia* and *P. ginseng* (ginseng) and various species of *Aralia* (wild sarsaparilla) are important medicinally. A few genera contain useful ornamentals, including *Hedera* (English ivy), and *Schefflera* (umbrella tree).

Discussion: Monophyly of Araliaceae is supported by DNA sequences (Chandler and Plunkett 2004; Plunkett et al. 1996a,b, 1997, 2004); the family is most closely related to Apiaceae, Myodocarpaceae, and Pittosporaceae (as discussed under Apiaceae). Araliaceae here are recognized in a circumscription that is slightly modified from that in traditional systems (Cronquist 1981), e.g., including *Hydrocotyle* and relatives (Plate 9.22H; formerly of the Apiaceae), but excluding *Myodocarpus* (Plate 9.22F) and *Delarbrea* (now in Myodocarpaceae), and *Mackinlaya* and *Stilbocarpa* (now in Apiaceae, and basal to Apioideae + Saniculoideae). Some genera are difficult to distinguish from those in Apiaceae, but the group is characterized by woody species with compound or simple, stipulate leaves and two- to five-carpellate, usually globose, berries or 2–5-pitted drupes that lack oil cavities and a carpophore, and thus is usually easily distinguished from the herbaceous, exstipulate, two-carpellate, and schizocarpic Apiaceae.

Several clades can be discerned within Araliaceae, although they are often difficult to diagnose, and some genera, e.g., *Schefflera* and *Polyscias*, are extremely non-monophyletic (Plunkett et al. 2004). *Hydrocotyle* and two related

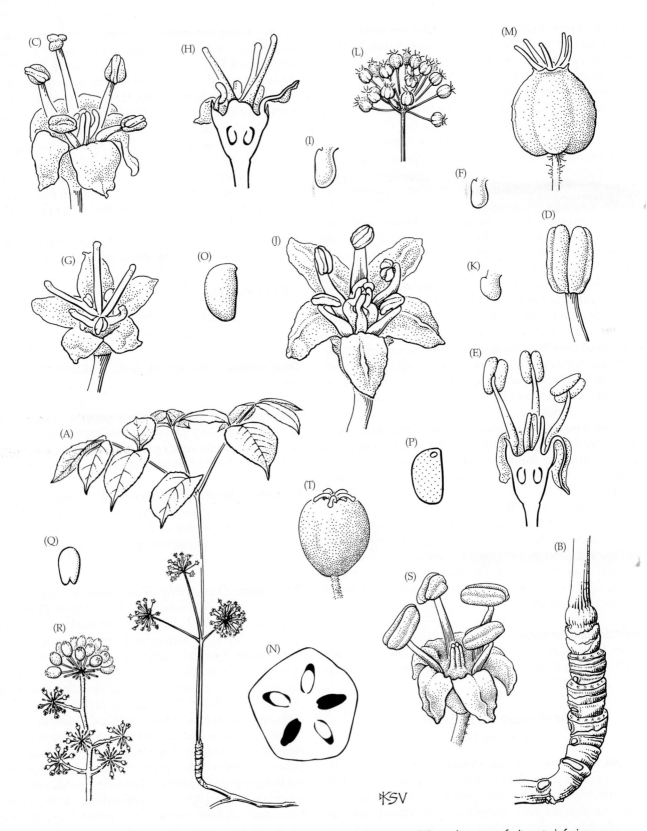

FIGURE 9.133 Araliaceae. (A–Q) *Aralia nudicaulis:* (A) flowering plant (× 0.3); (B) apex of rhizome (× 2); (C) staminate flower (× 8); (D) stamen (× 19); (E) staminate flower in longitudinal section (× 8); (F) rudimentary ovule (greatly magnified); (G) carpellate flower (× 8); (H) carpellate flower in longitudinal section (× 8); (I) functional ovule (greatly magnified); (J) staminate flower (× 8); (K) nonfunctional ovule (greatly magnified); (L) umbel of nearly mature fruits (× 1); (M) nearly mature fruit, note inferior ovary (× 5); (N) fruit in cross-section (× 5); (O) seed (× 8); (P) seed in section, endosperm stippled (× 8); (Q) embryo (greatly magnified). (R–T) *A. spinosa:* (R) portion of inflorescence (× 0.5); (S) flower (× 8); (T) berry (× 3). (Unpublished original art prepared for the Generic Flora of the Southeast U.S. series; used with permission; treatment of Araliaceae in 1965 *J. Arnold Arbor.* 47: 126–136.)

Key to Families of Dipsacales

---

1. Flowers ± bilateral, the style elongate with a capitate stigma; nectary composed of
   densely packed hairs on the inner surface of the corolla; pollen spiny ....**Caprifoliaceae**
1. Flowers radial, the style short with a lobed stigma; nectary glandular (atop ovary)
   or lacking; pollen non-spiny ....**Adoxaceae**

---

genera (Hydrocotyloideae) may be sister to the remaining genera (Aralioideae, including *Aralia, Dendropanax, Hedera, Oreopanax, Panax, Polycias, Schefflera, Tetrapanax,* and others).

The often small flowers of Apiaceae are typically densely aggregated and pollinated by a wide range of small nectar-gathering flies, beetles, bees, and moths. The drupaceous fruits usually are dispersed by birds.

Additional references: Erbar 1991; Erbar and Leins 1988; Graham 1966; Tingshuang et al. 2004; Wen et al. 2001.

Dipsacales

Dipsacales includes both **Caprifoliaceae** and **Adoxaceae**. The monophyly of this order is supported by opposite leaves, cellular endosperm development, anthers with a 3-

or 4-celled tapetum, *ndhF, rbcL, atpB, matK,* and 18S nucleotide sequences (Bell et al. 2001; Bremer et al. 2002; Donoghue 1983a,b; Donoghue et al. 1992, 2001; Hilu et al. 2003; Judd et al. 1994; Olmstead et al. 1993, 2000a; Savolainen et al. 2000b; Soltis et al. 2000; Albach et al. 2001a,b). Phylogenetic relationships within Dipsacales are summarized in Figure 9.134.

Caprifoliaceae A. L. de Jussieu
(Honeysuckle Family)

Herbs, shrubs, small trees, or lianas; often with phenolic glycosides, iridoids, and scattered secretory cells. Hairs various. *Leaves opposite, simple,* sometimes pinnately divided or compound, entire to serrate, with pinnate venation; stipules lacking. Inflorescences various. **Flowers** bisexual and **bilat-**

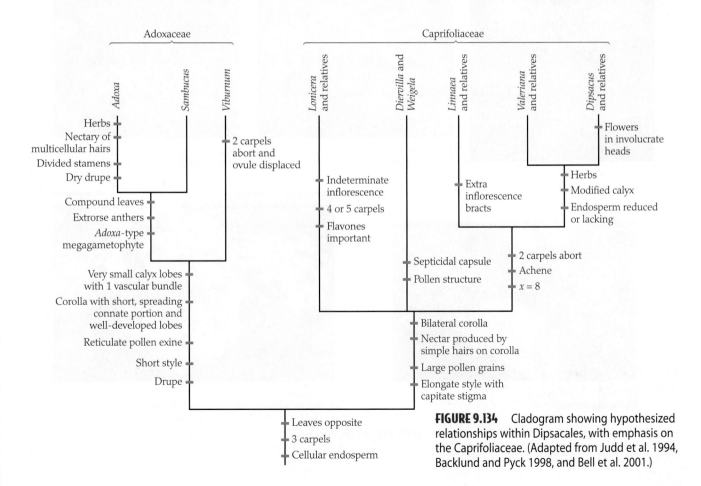

FIGURE 9.134 Cladogram showing hypothesized relationships within Dipsacales, with emphasis on the Caprifoliaceae. (Adapted from Judd et al. 1994, Backlund and Pyck 1998, and Bell et al. 2001.)

(A)

Dipsacales: Adoxaceae
Sambucus racemosa: branch with fruits

(B)

Dipsacales: Adoxaceae
Sambucus canadensis: flowers

(C)

Dipsacales: Caprifoliaceae
Dipsacus sylvestris: inflorescence

(D)

Dipsacales: Caprifoliaceae
Valeriana scandens: fruits

(E)

Dipsacales: Caprifoliaceae
Lonicera sempervirens: flowers

(F)

Dipsacales: Caprifoliaceae
Kolkwitzia amabilis: fruits

(G)

Dipsacales: Adoxaceae
Viburnum sargentii: inflorescences with sterile flowers

PLATE 9.23 Eudicots: Dipsacales

eral. Sepals usually 5, connate. *Petals usually 5, connate, often with 2 upper lobes and 3 lower lobes, or a single lower lobe and 4 upper ones*, the lobes imbricate or valvate. *Stamens (1–) 4 or 5;* filaments adnate to corolla; **pollen large, spiny**, usually tricolporate or triporate. *Carpels usually 2–5, connate*; ovary inferior, often elongate, with axile placentation, sometimes only 1 locule fertile; **style elongate; stigma capitate**. Ovules 1 to numerous in each locule, with 1 integument and a thin-walled megasporangium. **Nectar produced by closely packed glandular hairs on lower part of corolla tube**. Fruit a capsule, berry, drupe, or achene; endosperm present or lacking (Figure 9.135).

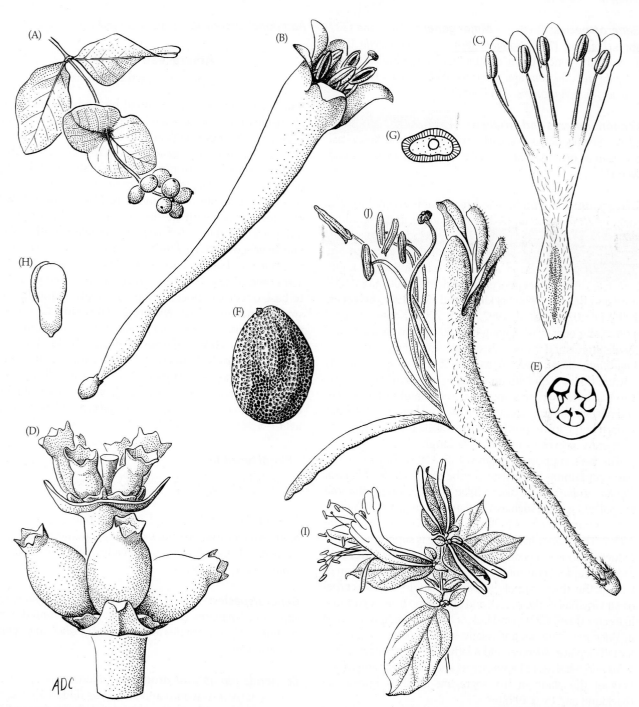

FIGURE 9.135 Caprifoliaceae. (A–H) *Lonicera sempervirens:* (A) fruiting branch (× 0.75); (B) flower (× 3); (C) corolla opened lengthwise to show attachment of stamens and distribution of hairs and nectar glands (× 2.3); (D) portion of inflorescence, corollas removed (× 12); (E) ovary in cross-section (× 15); (F) seed (× 9); (G) seed in cross-section, seed coat hatched, endosperm stippled, and embryo unshaded (× 9); (H) embryo (× 18). (I–J) *L. japonica:* (I) flowering branch (× 0.75); (J) flower (× 3). (From Ferguson 1966, *J. Arnold Arbor.* 47: p. 55.)

Floral formula:

X, K ⑤, C⑤, A4–5, G$\overline{(2-5)}$; drupe, berry, capsule,
                                        achene

Distribution: Widely distributed, especially in northern temperate regions.

Genera/species: 36/810.   **Major genera:** *Valeriana* (200 spp.), *Lonicera* (150), *Scabiosa* (80), and *Valerianella* (50). Noteworthy genera of the continental United States and/or Canada include *Lonicera, Valeriana, Valerianella, Dipsacus, Linnaea, Symphoricarpos,* and *Diervilla*.

Economic plants and products: *Lonicera* (honeysuckle), *Abelia, Symphoricarpos* (snowberry), *Weigela,* and *Kolkwitzia* are used as ornamentals. *Dipsacus* (teasel) is a widespread weed.

Discussion: Caprifoliaceae are circumscribed broadly, including Dipsacaceae and Valerianaceae, but excluding *Sambucus* and *Viburnum,* which are referred to Adoxaceae. As here delimited, the family is monophyletic, as evidenced by morphology (Judd et al. 1994), *rbcL, atpB, atpB, ndhF, trnL-F* and/or 18S sequences (Albach et al. 2001a,b; Backlund and Bremer 1998; Backlund and Pyck 1998; Bell et al. 2001; Donoghue et al. 1992, 2001, 2003; Källersjö et al. 1998; Soltis et al. 2001; Zhang et al. 2003). The family is easily distinguished from Adoxaceae (including *Sambucus, Viburnum, Adoxa,* and relatives) by its bilateral (vs. radial) flowers (Plate 9.23) with an elongate (vs. short) style, capitate (vs. lobed) stigma, spiny (vs. reticulate) pollen exine, and nectary composed of densely packed hairs on the inner surface of the lower portion of the corolla tube (vs. glandular nectary atop ovary or nectary lacking).

The best-supported group within Caprifoliaceae is the clade containing Linnaeeae, a tribe that includes *Linnaea, Dipelta, Abelia,* and *Kolkwitzia* (Plate 9.23F), along with more specialized genera such as *Valeriana* (often placed in Valerianaceae) and *Dipsacus* (often placed in Dipsacaceae; Plate 9.23C). The monophyly of the this clade is supported by the reduction to one nectary (hair cluster), four or fewer stamens, a haploid chromosome number of eight, abortion of two of the three carpels (leaving a single-ovuled carpel occupying half of the ovary), and an achene fruit. A pappus-like calyx (Plate 9.23D) and lack of endosperm have evolved in *Valeriana, Dipsacus,* and relatives. We note that some systematists place *Valeriana* and relatives in Valerianaceae, *Dipsacus* and relatives in Dipsacaceae, and the remaining members of the clade in the segregate family Linnaeaceae (Backlund and Pyck 1998).

Diervilla and *Weigela* constitute a monophyletic group, which is supported by their septicidal capsules and tectate pollen with poorly developed columellae; this clade is occasionally segregated as the Diervillaceae (Backlund and Pyck 1998).

Lonicera (Plate 9.23E), *Symphoricarpos,* and relatives probably form a clade, distinguished by indeterminate inflorescences and an often 4- or 5-carpellate gynoecium; Backlund and Pyck restrict the Caprifoliaceae to this group.

The showy flowers of Caprifoliaceae are pollinated by various nectar-gathering insects (mainly bees and wasps) and birds. The family shows a wide variety of dispersal syndromes.

Additional reference: Ferguson 1966b.

Adoxaceae   E. Meyer
(Moschatel or Elderberry Family)

Small trees, shrubs, or perennial herbs; with cyanogenic glycosides and iridoids. Hairs simple, stellate, or of peltate scales, glandular or nonglandular. *Leaves opposite, simple, trifoliolate, to pinnately compound,* entire or variously toothed, sometimes lobed, with pinnate to palmate venation; stipules present or lacking, sometimes glandular. Inflorescences determinate, often umbellate. *Flowers bisexual,* **radial,** a few species with sterile flowers around periphery of inflorescence. *Sepals 2–5, connate,* **reduced, with only a single vascular trace.** *Petals 4–5, connate, with usually short tube and well-developed,* imbricate or valvate *lobes. Stamens 5,* sometimes divided and appearing as 10; filaments adnate to base of corolla; pollen grains tricolpate or tricolporate, small to medium-sized, **with reticulate exine.** *Carpels 3–5, connate; ovary inferior or half-inferior,* with axile placentation; **style(s) short**; *stigma(s) capitate.* Ovules 1 in each locule, often only 1 functional, with 1 integument and a thin-walled megasporangium. *Nectar produced by glandular tissue atop ovary (Viburnum),* by cushion-like group of multicellular hairs *(Adoxa),* or lacking *(Sambucus).* **Fruit a drupe,** *with 1–5 pits.* (Figure 9.136).

Floral formula:

*, K ②-⑤, C ④-⑤, A 5, G $\overline{③-⑤}$, drupe

Distribution: Widespread in temperate regions of the Northern Hemisphere, but extending into the Africa, South America, Malesia, Australia, and New Zealand, especially in mountainous regions.

Genera/species: 5/245.   **Major genera:** *Viburnum* (220 spp.) and *Sambucus* (20). The family is represented in the continental United States and Canada by *Viburnum, Sambucus,* and *Adoxa.*

Economic plants and products: Numerous species of *Viburnum* (Arrow-wood) and *Sambucus* (Elder) are used as ornamentals, and the fruits are occasionally used for making jellies and wines. Several have medicinal uses.

Discussion: Monophyly of Adoxaceae is supported by both morphological and DNA sequence (*rbcL* and ITS)

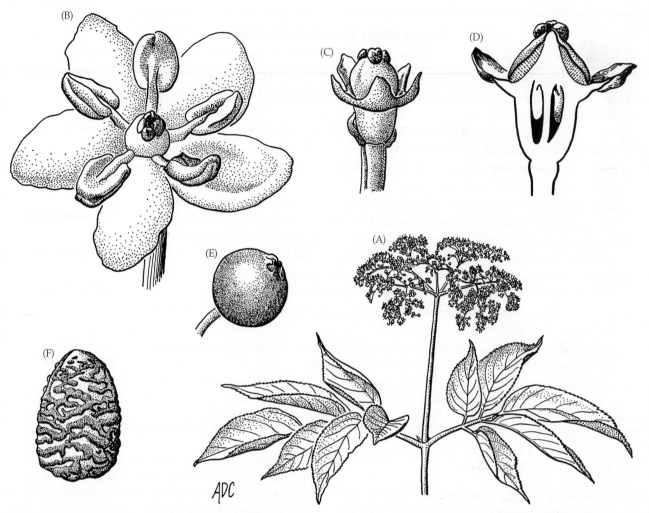

FIGURE 9.136 Adoxaceae. *Sambucus canadensis*: (A) flowering branch (× 0.4); (B) flower (× 9.6); (C) flower with corolla and stamens removed to show bracteoles and calyx lobes (× 9.6); (D) lon-gitudinal section of flower with corolla and stamens removed to show solitary, pendulous ovule in each locule (× 16); (E) fruit (× 4.8); (F) pit (× 14.7). (From Ferguson 1966, *J. Arnold Arbor.* 47: p. 38.)

characters (Donoghue et al. 1992, 2001; Eriksson and Donoghue 1997; Judd et al.1994). *Viburnum* (Opuloideae) is sister to *Sambucus* + *Adoxa* and relatives (Adoxoideae). The latter are linked based on their compound leaves, simple vessel perforations, extrorse anthers, and *Adoxa*-type megagametophyte development (i.e., developing from four megaspores, and eight-nucleate at maturity). Monophyly of the large and diverse genus *Viburnum* is supported by the unusual gynoecium development in which two carpels abort and the single functional ovule is displaced, developing in one of the sterile locules. *Adoxa*— and the closely related *Tetradoxa* and *Sinadoxa*— are especially distinctive, and easily distinguished from *Sambucus* by their herbaceous habit, nectary composed of multicellular hairs, divided stamens, and dry drupe (all synapomorphies). Morphological synapomorphies for *Sambucus* are uncertain, however, the small opening in the apical part of the endocarp is characteristic and may be synapomorphic; the production of sambunigrin (a cyano-genic glycoside) also may be a synapomorphy. Lack of nectaries in this genus has been suggested as a derived feature, but it could also represent the ancestral condition within Dipsacales.

It is noteworthy that rhizomatous herbs have evolved three times within Adoxaceae: in *Sambucus ebulus, S. adnata*, and relatives, in *S. gaudichaudiana*, and in the *Adoxa, Tetradoxa*, and *Sinadoxa* clade. The red-fruited species, e.g., *S. pubens* and *S. racemosa* (Plate 9.23A), form a derived monophyletic group; red fruits have evolved independently in *Viburnum* and *Sambucus*.

The flowers, although small, are presented in showy clusters (Plate 9.23B,G) and attract various insects, especially bees, wasps, and flies. The usually blue to red drupes are dispersed by birds.

Additional references: Donoghue 1980, 1981, 1983a,b, 1985; Ferguson 1966; Wilkinson 1948; Winkworth and Donoghue 2005.

Key to Major Families of Asterales

1. Stamens 2, extrorse, adnate to style by filaments, forming a column .... Stylidiaceae

1. Stamens 5, introrse, filaments not adnate to style, (although anthers often closely associated with style) 2

2. Ovary unilocular, with apical or basal ovule; flowers densely clustered in heads, which are surrounded by involucral bracts 3

2. Ovary multilocular, with axile placentation, or unilocular, with parietal placentation; flowers usually not in involucrate heads 4

3. Plants usually with resin canals or laticifers; ovary with basal ovule; anthers connate and filaments distinct; sepals forming a pappus of scales, bristles, or awns, or lacking .... **Asteraceae**

3. Plants with neither resin canals nor laticifers; ovary with apical ovule; anthers distinct and filaments connate; sepals not highly modified, of small lobes or teeth .... Calyceraceae

4. Ovary superior, with parietal placentation; flowers lacking a plunger or brush pollination mechanism, and style lacking pollen-gathering specializations .... Menyanthaceae

4. Ovary usually inferior or half-inferior, with axile placentation; flowers with a plunger pollination mechanism 5

5. Milky latex present; style with pollen-collecting hairs, these sometimes invaginating; flowers radial or bilateral; petals lacking marginal wings .... **Campanulaceae**

5. Milky latex lacking; styles with a pollen-collecting cup; flowers bilateral; petals with marginal wings .... Goodeniaceae

---

Asterales

Asterales are monophyletic, as evidenced by their valvate petals, storage of carbohydrates as the oligosaccharide inulin, and often plunger pollen presentation. The stamens are closely associated with one another (initially sticking together to completely connate) and form a tube around the style, with the anthers opening toward the inside. In plunger (or brush) pollination, pollen is pushed out of this tube, as out of a plunger, by specialized hairs on the style or by a specialized pollen-gathering cup. The style elongates to present pollen to floral visitors. Later the style branches diverge and the stigmas become receptive (Lammers 1992; Leins and Erbar 1990; Wagenitz 1977, 1992; Yeo 1993). The absence of a plunger mechanism in Menyanthaceae may be a reversal. The monophyly of the order also is supported strongly by cpDNA restriction sites, *rbcL, atpB, ndhF, matK,* and other chloroplast regions, and 18S sequences (Bremer et al. 2002; Chase et al. 1993; Cosner et al. 1994; Downie and Palmer 1992; Hilu et al. 2003; Källersjö et al. 1998; Lundberg and Bremer 2003; Michaels et al. 1993; Olmstead et al. 1992a, 1993, 2000a; Savolainen et al. 2000b; Soltis et al. 2000; Albach et al. 2001a,b). The order consists of 12 families and about 24,900 species; major families include **Campanulaceae** (incl. Lobeliaceae), Menyanthaceae, Goodeniaceae, Calyceraceae, Stylidiaceae, and **Asteraceae**. Analyses of DNA sequences suggest that the order is most closely related to Apiales and Dipsacales; all contain polyacetylenes.

Phylogenetic relationships of families within the order are still not entirely clear. Campanulaceae (possibly along with a few small families) is the sister group to a clade containing Menyanthaceae, Goodeniaceae, Calyceraceae, and Asteraceae. These two clades are frequently recognized as a pair of related orders—Campanulales and Asterales. Various embryological and chemical features (Lammers 1992) seem to be most useful in diagnosing these two groups, and the placement of a few families is questionable. The monophyletic group comprising Menyanthaceae, Goodeniaceae, Calyceraceae, and Asteraceae seems well supported; Menyanthaceae have superior ovaries. Anatomical and developmental studies indicate that flowers with inferior ovaries may have evolved twice within Asterales. The inferior ovary of Campanulaceae evolved by adnation of the ovary to a hypanthium. In contrast, epigyny in Goodeniaceae, and possibly also Calyceraceae and Asteraceae, was brought about by the adnation of the perianth to the ovary. These three families also share the apomorphy of a prominent and branched columella-layer in the pollen grains. Morphological analyses suggest that Calyceraceae are sister to Asteraceae; both share an unusual kind of corolla venation (possibly synapomorphic). Both families also have flowers densely clustered in heads that are surrounded by an involucre of bracts and reduced, unilocular, and uniovulate ovaries. However, all these are considered to be parallelisms (see Lammers 1992). On the other hand, *rbcL* sequences support the sister group relationship of Goodeniaceae and Calyceraceae.

(A)

Asterales: Campanulaceae
Triodanis perfoliata: flower

T. perfoliata: fruit

(B)

Asterales: Asteraceae
Helianthus annuus: radiate head

(C)

Asterales: Asteraceae
Hieracium aurantiacum: ligulate heads

(D)

Asterales: Asteraceae
Carphephorus paniculatus: discoid heads

(E)

Asterales: Asteraceae
Lactuca graminifolia: fruits

(F)

Asterales: Asteraceae
Helianthus angustifolius: disk flowers

(G)

Asterales: Campanulaceae
Lobelia cardinalis: flowers

PLATE 9.24 Eudicots: Asterales

Additional references: Bremer 1994; Brizicky 1966b.

Campanulaceae A. L. de Jussieu
(Bellflower or Lobelia Family)

Mostly herbs, but sometimes secondarily woody; plants storing carbohydrate as inulin (an oligosaccharide); **laticifers present with milky sap**; polyacetylenes present, but iridoids absent. Hairs usually simple, unicellular. *Leaves usually alternate and spiral, simple*, sometimes lobed, entire to serrate, with pinnate venation; stipules absent. Inflorescences various. *Flowers usually bisexual, radial to bilateral, with hypanthium*, sometimes twisting 180° in development (resupinate). Sepals usually 5, connate. *Petals usually 5, connate, forming a tubular or bell-shaped corolla, or 2-lipped to 1-lipped and then with a variously developed adaxial slit, the lobes valvate. Stamens usually 5; filaments distinct to distally connate, usually* **attached to disk at apex of ovary**; *anthers distinct but pressed together around the style or connate (syngenesious), forming a tube into which the pollen is shed, and the style then growing through this tube, picking up pollen with specialized hairs that later invaginate, or pushing it out, after which the stigmas become receptive (i.e., a plunger pollination mechanism)*; pollen grains with 3 to 12 apertures. *Carpels 2–5, connate; ovary usually inferior (or half-inferior), with usually axile placentation; style with pollen-collecting hairs near the apex; number of stigmas equaling number of carpels, globose to cylindrical*. Ovules usually numerous, with 1 integument and a thin-walled megasporangium. Nectar disk present above ovary. *Fruit a loculicidal or poricidal capsule (Plate 9.24A) or a berry (Figure 9.137)*.

Floral formula:

* or X, K $\underline{5}$, C $\underline{5}$, A $\underline{5}$, G $\overline{\underline{(2-5)}}$; capsule, berry

Distribution: Widely distributed in temperate and subtropical regions and in the montane tropics.

Genera/species: 65/2200. **Major genera:** *Lobelia* (400 spp.), *Campanula* (450), *Centropogon* (200), *Siphocampylus* (225), and *Wahlenbergia* (270). Genera occurring in Canada and/or the continental United States include *Campanula, Downingia, Githopsis, Heterocodon, Howellia, Jasione, Legenere, Lobelia, Nemacladus, Parishella, Porterella, Triodanis,* and *Wahlenbergia*.

Economic plants and products: *Campanula* (bellflower, bluebell), *Lobelia* (cardinal flower, lobelia; Plate 9.24G), and *Codonopsis* (bonnet bellflower) are used horticulturally.

Discussion: Monophyly of Campanulaceae, as here circumscribed, is supported by morphology and DNA sequences (Cosner et al. 1994; Lundberg and Bremer 2002). Three subfamilies—Campanuloideae, "Cyphioideae," and Lobelioideae (Thorne 1992)—are often distinguished and may be recognized at the family level (Lammers 1992). The Campanuloideae, with radially symmetrical flowers and nonconnate anthers, are considered monophyletic on the basis of invaginating hairs on the upper portion of the style. The Lobelioideae constitute a clade based on their connate anthers, resupinate flowers, and one- to two-lipped corollas with a variously developed slit in the upper lobe (developmentally adaxial, but abaxial when flower is resupinate). Analyses of *rbcL* and ITS sequence variation also support the monophyly of both subfamilies (Cosner et al. 1994; Eddie et al. 2003).

Generic delimitations are often problematic; *Campanula, Centropogon,* and *Lobelia* are certainly not monophyletic. The showy flowers of Campanulaceae attract a wide range of floral visitors, especially bees and birds. Outcrossing is favored by the brush/plunger secondary pollen presentation mechanism. Dispersal of the small seeds of capsular species is by wind; species with berries are bird-dispersed.

Additional references: Leins and Erbar 1990; Rosatti 1986; Shetler 1979.

Asteraceae Bercht. & J. Presl
(= Compositae Giseke)
(Aster or Composite Family)

Herbs, shrubs, or trees; storing carbohydrate as oligosaccharides, including inulin; *resin canals often present, laticifers often present, but one or the other of these sometimes lacking; polyacetylenes and terpenoid aromatic oils usually present;* **usually with sesquiterpene lactones (but lacking iridoids)**. Hairs various. *Leaves alternate and spiral, opposite,* or whorled, *simple, but sometimes deeply lobed or dissected,* entire to variously toothed, with usually pinnate or palmate venation; stipules lacking. **Flowers ± densely aggregated into indeterminate heads that are surrounded by an involucre of bracts (phyllaries)**; the heads arranged in a determinate secondary inflorescence, terminal or axillary. *Flowers bisexual or unisexual, sometimes sterile, radial or bilateral.* **Sepals highly modified, forming a pappus** *composed of 2–many, persistent, sometimes connate, scales, awns, or bristles that are capillary, hairy, minutely barbed, or plumose,* or sometimes lacking. *Petals 5, connate, forming a radial and tubular corolla (disk flower),* forming a bilateral and 2-lipped corolla (i.e., with usually 2 petals in upper lip and 3 petals in lower lip), *or forming a bilateral and 1-lipped corolla with upper lip ± lacking and lower lip elongated, ± 3-lobed (ray flower), or forming a bilateral, elongated, tonguelike corolla ending in 5 small teeth (ligulate flower); the heads with only disk flowers (discoid; Plate 9.24D), with disk flowers in center and ray flowers around the outside, the latter female or sterile (radiate; Plate 9.24B), or with only ligulate flowers (ligulate; Plate 9.24C), the corolla lobes valvate. Stamens usually 5; filaments distinct, adnate to corolla tube;* **anthers usually connate (syngenesious)**, *often with apical or basal appendages, forming a tube around the style into which the pollen is shed, and the style then growing through this tube, pushing out or picking up pollen (with variously developed hairs) and presenting it to floral visitors, after which the stigmas become receptive (i.e.,*

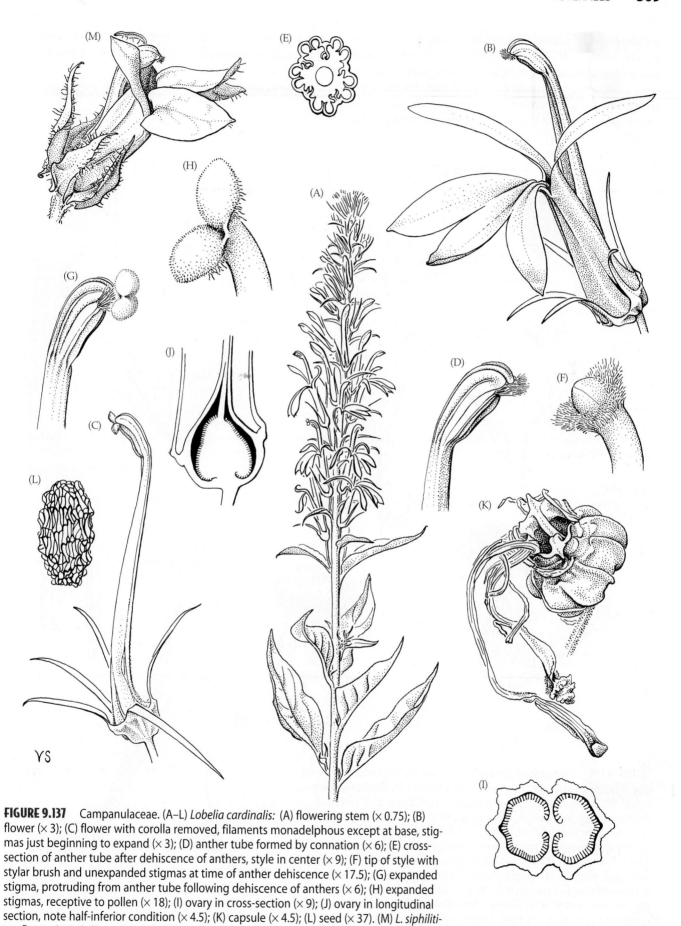

FIGURE 9.137 Campanulaceae. (A–L) *Lobelia cardinalis:* (A) flowering stem (× 0.75); (B) flower (× 3); (C) flower with corolla removed, filaments monadelphous except at base, stigmas just beginning to expand (× 3); (D) anther tube formed by connation (× 6); (E) cross-section of anther tube after dehiscence of anthers, style in center (× 9); (F) tip of style with stylar brush and unexpanded stigmas at time of anther dehiscence (× 17.5); (G) expanded stigma, protruding from anther tube following dehiscence of anthers (× 6); (H) expanded stigmas, receptive to pollen (× 18); (I) ovary in cross-section (× 9); (J) ovary in longitudinal section, note half-inferior condition (× 4.5); (K) capsule (× 4.5); (L) seed (× 37). (M) *L. siphilitica:* flower (× 4.5). (From Rosatti 1986, *J. Arnold Arbor.* 67: p. 67.)

with a plunger or brush pollination mechanism); pollen grains usually tricolporate. *Carpels 2, connate; ovary inferior,* **with basal placentation**; *style branches 2, with stigmatic tissue covering the inner surface or in 2 marginal lines.* **Ovule 1 per ovary**, with 1 integument and a thin megasporangium. Nectary at apex of ovary. *Fruit an achene,* **crowned by a persist-** ent pappus (Plate 9.24E), sometimes flattened, winged, or spiny; endosperm scanty or lacking (Figures 9.138–9.140).

Floral formula: * or X, K∞, C⑤, A⑤, G②; achene

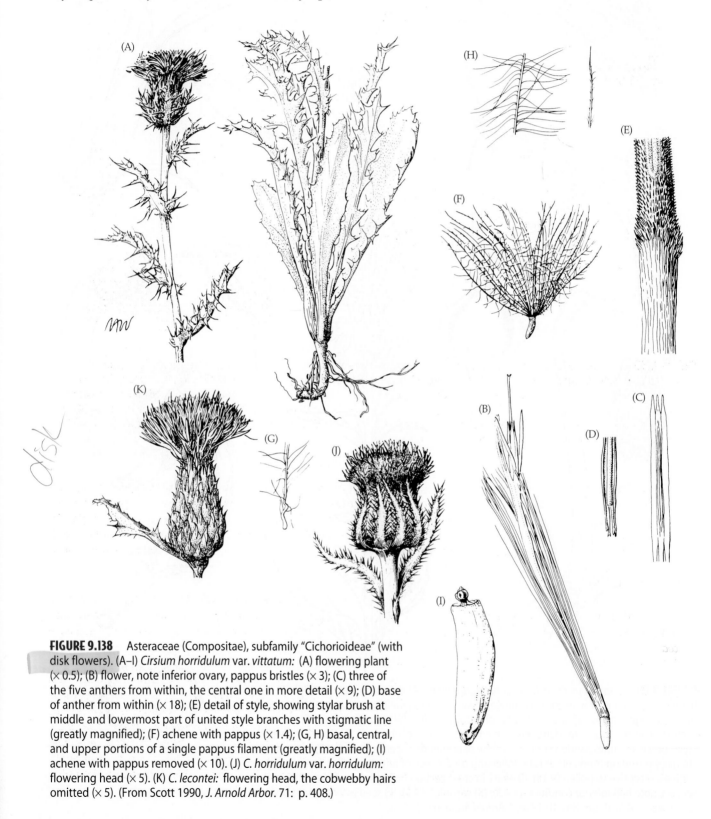

FIGURE 9.138   Asteraceae (Compositae), subfamily "Cichorioideae" (with disk flowers). (A–I) *Cirsium horridulum* var. *vittatum:* (A) flowering plant (× 0.5); (B) flower, note inferior ovary, pappus bristles (× 3); (C) three of the five anthers from within, the central one in more detail (× 9); (D) base of anther from within (× 18); (E) detail of style, showing stylar brush at middle and lowermost part of united style branches with stigmatic line (greatly magnified); (F) achene with pappus (× 1.4); (G, H) basal, central, and upper portions of a single pappus filament (greatly magnified); (I) achene with pappus removed (× 10). (J) *C. horridulum* var. *horridulum:* flowering head (× 5). (K) *C. lecontei:* flowering head, the cobwebby hairs omitted (× 5). (From Scott 1990, *J. Arnold Arbor.* 71: p. 408.)

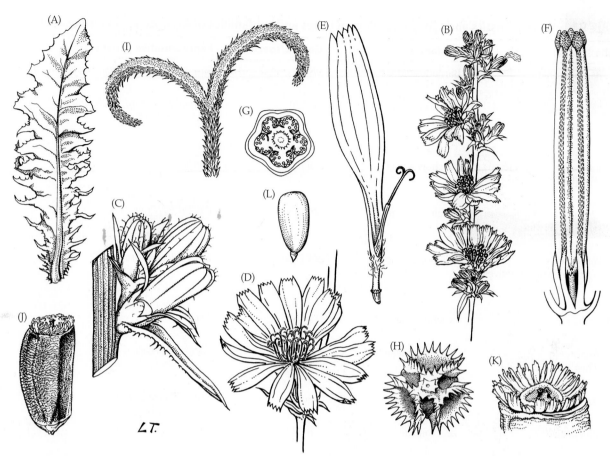

FIGURE 9.139 Asteraceae (Compositae), subfamily "Cichorioideae" (with ligulate flowers). *Cichorium intybus:* (A) basal leaf (× 0.3); (B) upper part of flowering stem (× 0.6); (C) cluster of heads in axil of bract (× 2.5); (D) head of flowers (× 1); (E) ligulate flower, note five corolla teeth (× 5); (F) androecium, from bud just before opening, note apical and basal appendages (× 9.5); (G) flower bud in cross-section, showing corolla, five coherent anthers, and style with pollen-collecting hairs (× 24); (H) pollen grain in polar view (greatly magnified); (I) style branches with stigmas (× 18); (J) achene (× 14); (K) detail of pappus scales (× 24); (L) seed (× 9.5). (From Vuilleumier 1973, *J. Arnold Arbor.* 54: p. 49.)

Distribution and ecology: Cosmopolitan, found especially in temperate or tropical montane regions and open and/or dry habitats.

Genera/species: 1535/23,000. **Major genera:** *Senecio* (1250 spp.), *Vernonia* (1000), *Cousinia* (650), *Eupatorium* (600), *Centaurea* (600), *Artemisia* (550), *Hieracium* (500), *Helichrysum* (500), *Baccharis* (400), *Mikania* (400), *Saussurea* (300), *Verbesina* (300), *Cirsium* (250), *Jurinea* (250), *Bidens* (200), *Crepis* (200), *Aster* (180, excluding segregates such as *Symphyotrichum, Sericocarpus*), *Gnaphalium* (150), *Tragopogon* (110), and *Solidago* (100). Generic limits are often problematic, and several of these large genera are frequently divided into numerous segregates (see Bremer 1994). Many genera occur in the continental United States and/or Canada (see various floras); especially important and/or widespread genera include *Acmella, Ageratina, Ambrosia, Antennaria, Arnoglossum, Artemisia, Baccharis, Balduina, Berlandiera, Bidens, Brickellia, Carphephorus, Centaurea, Chrysopsis, Cirsium, Conyza, Coreopsis, Crepis, Croptilon, Chaenactis, Chrysothamnus, Elephantopus, Ericameria, Erigeron, Eriophyllum, Eupatorium, Euthamia, Gaillardia, Gamochaeta, Gnaphalium, Haplopappus, Helenium, Helianthus, Hieracium, Iva, Krigia, Lactuca, Layia, Lessingia, Liatris, Lygodesmia, Packera, Pityopsis, Pluchea, Prenanthes, Rudbeckia, Senecio, Sericocarpus, Silphium, Solidago, Sonchus, Symphyotrichum, Taraxacum, Verbesina,* and *Vernonia.*

Economic plants and products: Food plants include *Cichorium* (endive, chicory), *Cynara* (artichoke), *Helianthus* (sunflower seeds and oil, Jerusalem artichoke), *Taraxacum* (dandelion greens), and *Lactuca* (lettuce). *Artemisia* (wormwood, tarragon) contains spice plants. *Tanacetum* (pyrethrum) and *Pulicaria* (fleabane) contain species with insecticidal properties. *Ambrosia* (ragweed) is important as a major cause of hay fever, and several genera are serious weeds. Finally, ornamentals come from *Calendula* (marigold), *Dendranthema, Argyranthemum, Leucanthemum* (chrysanthemum), *Dahlia, Tagetes* (French marigold), *Senecio, Spaghenticola, Gaillardia* (bandana daisy), *Helianthus* (sunflower), *Zinnia,* and many others.

TABLE 9.4 Characteristics and synapomorphies of tribes and subfamilies of Asteraceae (= Compositae).

Tribe	Subfamily	Genera/species	Representative genera
1. Barnadesieae	Barnadesioideae	9/92	*Barnadesia, Chuquiraga, Dasyphyllum*
2. "Mutisieae"	"Cichorioideae"	76/970	*Chaptalia, Gerbera, Gochnatia, Mutisia, Trixis*
3. Cardueae (= Cynareae) (Thistles)	"Cichorioideae"	83/2500	*Carduus, Centaurea, Cirsium, Cynara, Echinops*
4. Lactuceae (= Cichorieae) (Dandelions and relatives)	"Cichorioideae"	98/1550	*Cichorium, Crepis, Hieracium, Krigia, Lactuca, Pyrrhopappus, Sonchus, Taraxacum, Tragopogon, Youngia*
5. Vernonieae (Ironweeds and relatives)	"Cichorioideae"	98/1300	*Elephantopus, Vernonia*
6. Liabeae	"Cichorioideae"	14/160	*Liabum*
7. Arctoteae (Thistle-like plants)	"Cichorioideae"	16/200	*Arctotis, Berkheya*
8. Inuleae	Asteroideae	38/480	*Inula, Pulicaria, Telekia*
9. Plucheae	Asteroideae	28/220	*Pluchea, Sphaeranthus*
10. Gnaphalieae	Asteroideae	162/2000	*Anaphalis, Antennaria, Gamochaeta, Gnaphalium, Leontopodium*
11. Calenduleae	Asteroideae	8/110	*Calendula, Osteospermum*
12. Astereae	Asteroideae	174/2800	*Aster, Baccharis, Conyza, Erigeron, Haplopappus, Solidago, Symphyotrichum*
13. Anthemideae	Asteroideae	109/1740	*Achillea, Anthemis, Argyranthemum, Artemisia, Chrysanthemum, Leucanthemum, Seriphidium, Tanacetum*
14. Senecioneae	Asteroideae	120/3200	*Erechtites, Senecio*
15. "Helenieae"	Asteroideae	110/830	*Arnica, Flaveria, Gaillardia, Helenium, Pectis, Tagetes*
16. Heliantheae	Asteroideae	189/2500	*Ambrosia, Bidens, Calea, Coreopsis, Cosmos, Dahlia, Helianthus, Iva, Rudbeckia, Verbesina, Viguiera, Zinnia*
17. Eupatorieae	Asteroideae	170/2400	*Ageratum, Carphephorus, Eupatorium, Garberia, Liatris, Mikania*

17 tribes 3 subfamilies close to 25,000 species

* most advanced dicot

TABLE 9.4 (continued)

Flower types	Pappus types	Style branches	Major synapomorphies
Disk; 1+4 bilabiate	Usually bristles with long hairs	Inner surface stigmatic	Axillary spines; pubescence of long hairs on corolla, achene, pappus
Variable; 2+3 bilabiate, especially toward margin of head and/or disk	Usually bristles	Inner surface stigmatic	None
Disk (deeply lobed)	Usually bristles or scales	Inner surface stigmatic (often with fused style below style branches)	Leaves dissected; leaves and involucral bracts spine-tipped; ring of hairs
Ligulate	Usually bristles	Inner surface stigmatic	Ligulate flowers; copious milky latex
Usually disk (deeply lobed)	Usually bristles	Inner suface stigmatic (near base)	Anthers with glandular apical appendages; style branches long, slender, pilose, acute at apex; anatomical details of anthers (endothecial thickenings)
Usually disk (deeply lobed) and ray	Bristles or scales	Inner surface stigmatic	Leaves opposite; trinerved ray flowers
Usually disk (deeply lobed) and ray	Usually scales or short cup	Inner surface stigmatic	Ring of hairs below style branches; anther not caudate; ray flowers; often spiny
Disk (lobes short) and ray	Scales or capillary bristles	Marginal stigmatic lines	Marginal flowers filiform; elongate crystals in epidermis of achene
Usually disk (lobes short)	Scales or capillary bristles	Marginal stigmatic lines	Marginal flowers filiform; style branches with apically rounded, sweeping hairs along the branches and on upper part of shaft
Usually disk (lobes short), rarely with rays	Usually capillary bristles	Marginal stigmatic lines	Pollen grains with thick basal layer regularly perforated (gnaphaloid); upper level of prominent columellae; $x = 7$
Disk (lobes short) and ray	Lacking	Marginal stigmatic lines	Loss of pappus
Usually disk (lobes short) and ray (rays sometimes lacking)	Usually bristles or scales	Marginal stigmatic lines	Epidermal cells of ray flower corolla with median thickening in outer wall; style branches with triangular-subulate sterile appendages, i.e., asteroid style branches adaxially glabrous
Usually disk (lobes short) and ray (but ray flowers sometimes lost)	Scales, short cup, or lacking	Marginal stigmatic lines	Leaves pinnately dissected; involucral bracts with scarious margins; scaly or reduced pappus; truncate style branches; ray epidermal cells papillose
Disk (lobes short) and ray (or only disk)	Usually capillary bristles	Marginal stigmatic lines	Involucre usually uniseriate (single whorl of bracts); chemistry of sesquiterpene lactones, pyrrolizidine alkaloids
Disk (lobes short) and ray (these sometimes lost)	Scales, short cup, bristles	Marginal stigmatic lines	None, due to segregation of following two tribes—based on opposite leaves, carbonized achene wall, endothecium of short cells (bracts lacking on receptacle)
Usually disk (lobes short) and ray (these sometimes lacking)	Awns, scales, bristles, or lacking	Marginal stigmatic lines	Bracts on receptacle; black anthers
Disk (lobes short, or occasionally long)	Bristles	Marginal stigmatic lines	Loss of ray flowers; hairy style bases and style branches with very long sterile appendages; style branches glandular between stigmatic lines; anatomy of anther endothecium

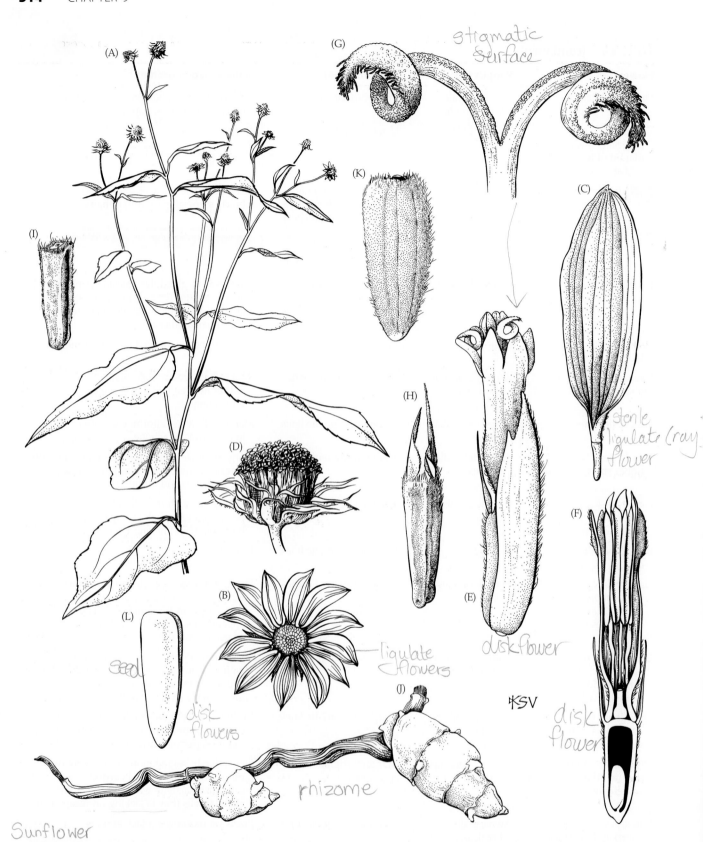

Sunflower

FIGURE 9.140 Asteraceae (Compositae), subfamily Asteroideae (with disk and ray flowers). (A–J) *Helianthus tuberosus:* (A) flowering plant (× 0.3); (B) head of flowers (× 0.5); (C) ray flower, sterile (× 2); (D) head, ray flowers removed, showing involucral bracts (phyllaries) (× 0.5); (E) disk flower and bract (× 10); (F) disk flower in longitudinal section, note anthers, pappus scales, and basal ovule (× 0.5); (G) style branches with their stigmatic lines (greatly magnified); (H) mature achene with pappus scales (× 10); (I) achene without pappus (× 8); (J) rhizome with overwintering tubers (× 1). (K–L) *H. annuus:* (K) achene (× 4.5); (L) embryo (× 4.5). (From Wood 1974, *A student's atlas of flowering plants,* p. 117.)

Discussion: Asteraceae (or Compositae) form an easily recognized and obviously monophyletic group; both morphological and molecular synapomorphies are numerous (K. Bremer 1987, 1994, 1996; Funk et al. 2005; Jansen et al. 1991, 1992; Karis 1993; Karis et al. 1992; Keeley and Jansen 1991; Kim et al. 1992). The family is divided into several tribes (Table 9.4), which are often arranged into three subfamilies (K. Bremer 1987, 1994; Bremer and Jansen 1992). The Barnadesioideae, a small South American group of mainly trees and shrubs, is the sister group to the remaining genera (Figure 9.141). This group lacks the chloroplast DNA inversion that characterizes the remaining species (Jansen and Palmer 1987). The remaining tribes are more or less equally divided into the "Cichorioideae" and the Asteroideae (K. Bremer 1987, 1994; Carlquist 1976; Thorne 1992). The former is paraphyletic, but is retained here because phylogenetic relationships within the complex are still incompletely known; it is often further divided (K. Bremer 1996; Funk et al. 2005; Panero and Funk 2002). "Cichorioideae" are characterized by style branches with the inner surface stigmatic. Their heads are usually discoid, except in the distinct tribe Lactuceae, which has ligulate heads. Both resin canals and laticifers occur within this subfamily, and the latex system is especially well developed in Lactuceae (Table 9.3). Lactuceae are phenetically distinct and have sometimes been placed in their own subfamily (Cronquist 1955, 1977, 1981), but are closely related to Vernonieae. The monophyletic Asteroideae can be diagnosed by the restriction of the stigmatic tissue to two marginal lines on each style branch; loss of laticifers; presence of ray flowers (and radiate heads, although these are lost in some genera); disk flowers with short lobes (but secondarily elongate in some), and DNA characters. Presumed synapomorphies of many commonly recognized tribes are outlined in Table 9.4 (see also Bremer 1994; Solbrig 1963). Morphological features that are taxonomically important at the tribal level include characteristics of the style branches, (e.g., location of stigmatic region, presence of hairs or sterile appendages, length and width, apex form); pappus form; corolla anatomy and shape; pollen morphology; anatomical and morphological features of the achenes; anatomy and form of the anthers; leaf arrangement; and presence or lack of nodal or marginal spines.

Crepis, Aster, Taraxacum, Tragopogon, Hieracium, and other genera are taxonomically difficult at the species level due to the combined action of hybridization, polyploidy, and agamospermy.

The tiny flowers of Asteraceae are not readily apparent (Plate 9.24F); the involucrate heads usually function as—and at first glance may even appear to be—a single flower. In radiate heads the ray flowers serve to attract pollinators and the disk flowers mature centripetally. Pollinators usually land on the ray flowers and deposit pollen from other plants on the stigmas of older, marginal disk flowers. The filaments of many Asteraceae respond to touch by contracting abruptly to force pollen out of the plunger mechanism and onto the pollinator's body. Corolla color is variable. Composite inflores-

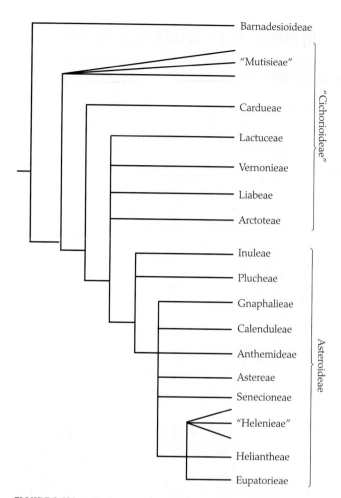

FIGURE 9.141   Cladogram showing hypothesized relationships within the Asteraceae. (Adapted from K. Bremer 1994.)

cences are generally outcrossing and attract a wide array of generalist pollinators (butterflies, bees, flies, and beetles), but pollination by solitary bees is especially common. Some genera have reduced flowers that are wind-pollinated (e.g., *Ambrosia* and *Baccharis*), and some have heads reduced to a single flower, but these reduced heads are then aggregated into compound heads (e.g., *Echinops*).

The achenes of most members of Asteraceae are dispersed by wind, with the pappus bristles functioning as a parachute (Plate 9.24E). The flattened and often winged fruit assists in wind dispersal. External transport on birds or mammals is facilitated by pappus modifications such as awns with retrorse barbs, fruit outgrowths, such as hooks or spines, or specialized involucral bracts.

Additional references: Anderberg 1991a,b,c; Arriagada and Miller 1997; Gustafsson 1996; Heywood et al. 1977; Jones 1982; Karis et al. 2001; Lane 1996; Leins and Erbar 1990; Michaels et al. 1993; Ownbey 1950; Rieseberg 1991; Scott 1990; Soltis and Soltis 1989; Urtubey and Stuessy 2001; Vuilleumier 1969, 1973; Wagenitz 1992; Xiaoping and Bremer 1993.

LITERATURE CITED AND SUGGESTED READINGS

In addition to the numerous books and scientific articles cited after each family, the following references will serve as additional sources of information on

the morphological variation, geographic distribution, economic importance, and evolutionary relationships of angiosperm families.

Suggested Readings

Angiosperm Phylogeny Group. 1998. An ordinal classification for the families of flowering plants. *Ann. Missouri Bot. Gard.* 85: 531–553.

Angiosperm Phylogeny Group. 2003. An update of the Phylogeny Group classification for the orders and families of flowering plants: APGII. *Bot. J. Linn. Soc.* 141: 399–436.

Bailey, L. H. and E. Z. Bailey. 1976. *Hortus third: a concise dictionary of plants cultivated in the United States and Canada.* Revised by the Bailey Hortorium. Macmillan, New York.

Benson, L. 1979. *Plant classification,* 2nd ed. D.C. Heath, Lexington, MA.

Cronquist, A. 1981. *An integrated system of classification of flowering plants.* Columbia University Press, New York.

Cronquist, A. 1988. *The evolution and classification of flowering plants,* 2nd ed. New York Botanical Garden, Bronx, NY.

Dahlgren, R. M. T. 1983. General aspects of angiosperm evolution and macrosystematics. *Nord. J. Bot.* 3: 119–149.

Dahlgren, R. M. T., H. T. Clifford and P. F. Yeo. 1985. *The families of the monocotyledons.* Springer-Verlag, Berlin.

Engler, A. and K. Prantl. 1887–1915. *Die natürlichen Pflanzenfamilien,* parts II–IV. G. Kreysing, Leipzig. [Also ed. 2, incomplete, 1924–.]

Heywood, V. H. (ed.). 1978. *Flowering plants of the world.* Mayflower Books, New York.

Heywood, V. H. (ed.). 2007. *Flowering plant families of the world.* Kew Publications, UK.

Hickey, M. and C. J. King. 1988. *100 families of flowering plants of the world,* 2nd ed. Cambridge University Press, Cambridge.

Hutchinson, J. 1973. *The families of flowering plants,* 3rd ed. Oxford University Press, Oxford.

Jones, S. B. and L. E. Luchsinger. 1986. *Plant systematics.* McGraw-Hill, New York.

Kubitzki, K. 1990–2004. *The families and genera of vascular plants.* 7 vols. Springer-Verlag, Berlin. [Additional volumes are planned; detailed taxonomic treatments of families and genera with descriptions, information on anatomy, embryology, chromosomes, pollen structure, pollination, dispersal, and phylogeny; keys to genera are provided.]

Lawrence, G. H. M. 1951. *Taxonomy of vascular plants.* Macmillan, New York.

Maas, P. J. M. and L. Y. T. Westra. 1993. *Neotropical plant families.* Koeltz Scientific Books, Konigstein.

Mabberley, D. J. 1997. *The plant-book,* 2nd ed. Cambridge University Press, Cambridge.

Melchior, H. (ed.) 1964. *A. Engler's Syllabus der Pflanzenfamilien,* 12th ed. Vol. 2. Gebruder Borntraeger, Berlin.

Rendle, A. B. 1925. *The classification of flowering plants.* 2 vols. Cambridge University Press, Cambridge.

Simpson, M. G. 2006. *Plant systematics.* Elsevier Academic Press, Oxford.

Soltis, D. E., P. S. Soltis, P. K. Endress, and M. W. Chase. 2005. *Phylogeny and evolution of angiosperms.* Sinauer Associates, Sunderland, MA.

Spears, P. 2006. *A tour of the flowering plants based on the classification system of the Angiosperm Phylogeny Group.* Missouri Bot. Gard. Press, St. Louis.

Spichiger, R. E., V. Savolainen, M. Figeat, and D. Jeanmonod. 2004. *Systematic botany of flowering plants: a new phylogenetic approach to angiosperms of the temperate and tropical regions.* Science Publ., Plymouth, UK.

Stevens, P. F. 2001 onward. Angiosperm Phylogeny Website. Version 2 Aug 2001. http://www.mobot.org/MOBOT/research/Apweb/.

Takhtajan, A. 1980. Outline of the classification of flowering plants (Magnoliophyta). *Bot. Rev.* 46: 225–359.

Takhtajan, A. 1997. *Diversity and classification of flowering plants.* Columbia University Press, New York.

Thorne, R. F. 1992. Classification and geography of the flowering plants. *Bot. Rev.* 58: 225–348.

Thorne, R. F. 2000. The classification and geography of the monocotyledon subclasses Alismatidae, Liliidae, and Commelinidae. In *Plant systematics for the 21st century,* B. Nordenstam, G. El-Ghazaly, M. Kassas, and T. C. Laurent (eds.), 75–124. Portland Press, London.

Thorne, R. F. 2001. The classification and geography of the flowering plants: dicotyledons of the class Angiospermae. *Bot. Rev.* 66: 441–647.

Van Balgooy, M. M. J. 1997. *Malesian seed plants,* vol. 1, Spot-characters. An aid for identification of families and genera. Rijksherbarium/Hortus Botanicus, Leiden.

Van Balgooy, M. M. J. 1998. *Malesian seed plants,* vol. 2. Portraits of tree families. Rijksherbarium/Hortus Botanicus, Leiden.

Van Balgooy, M. M. J. 2001. *Malesian seed plants,* vol. 3, Portraits of non-tree families. Rijksherbarium/Hortus Botanicus, Leiden.

Watson, L. and M. J. Dallwitz. 1997. The families of flowering plants: descriptions and illustrations. Website: <http://muse.bio.cornell.edu/delta/angio/www/index.htm>

Willis, J. C. 1973. *A dictionary of flowering plants and ferns,* 8th ed. Revised by H. K. Airy-Shaw. University Printing House, Cambridge.

Zomlefer, W. B. 1994. *Guide to flowering plant families.* University of North Carolina Press, Chapel Hill.

Literature Cited

Abbe, E. C. 1935. Studies in the phylogeny of the Betulaceae. I. Floral and inflorescence anatomy and morphology. *Bot. Gaz.* 97: 1–67.

Abbe, E. C. 1974. Flowers and inflorescences of "Amentiferae." *Bot. Rev.* 40: 159–261.

Abbe, E. C. and T. T. Earle. 1940. Inflorescence, floral anatomy, and morphology of *Leitneria floridana.* Bull. *Torrey Bot. Club* 67: 173–193.

Abu-Asab, M. S. and P. D. Cantino. 1989. Pollen morphology of *Trichostema* (Labiatae) and its systematic implications. *Syst. Bot.* 14: 359–369.

Adams, P. and R. K. Godfrey. 1961. Observations on the *Sagittaria subulata* complex. *Rhodora* 63: 247–266.

Agababian, V. S. 1972. Pollen morphology of the Magnoliaceae. *Grana* 12: 166–176.

Albach, D. C., P. S. Soltis, D. E. Soltis, and R. G. Olmstead. 2001a. Phylogenetic analysis of asterids based on sequences of four genes. *Ann. Missouri Bot. Gard.* 88: 162–212.

Albach, D. C., P. S. Soltis, and D. E. Soltis. 2001b. Patterns of embryological and biochemical evolution in the asterids. *Syst. Bot.* 26: 242–262.

Albach, D. C., D. E. Soltis, M. W. Chase, and P. S. Soltis. 2001c. Phylogenetic placement of the enigmatic angiosperm *Hydrostachys. Taxon* 50: 781–805.

Albach, D. C., H. M. Meudt, and B. Oxelman. 2005. Piecing together the "new" Plantaginaceae. *Amer. J. Bot.* 92: 297–315.

Albert, V. A., S. E. Williams, and M. W. Chase. 1992. Carnivorous plants: phylogeny and structural evolution. *Science* 257: 1491–1495.

Alice, L. A., T. Eriksson, B. Eriksen, and C. S. Campbell. 2001. Hybridization and gene flow between distantly related species of *Rubus* (Rosaceae): evidence from nuclear ribosomal DNA internal transcribed spacer region sequences. *Syst. Bot.* 26: 769–778.

Aliscioni, S. S., L. M. Glussani, F. O. Zuloaga, and E. A. Kellogg. 2003. A molecular phylogeny of *Panicum* (Poaceae: Paniceae). Tests of monophyly and phylogenetic placement

within the Panicoideae. *Amer. J. Bot.* 90: 796–821.

Allard, H. A. 1947. The direction of twist of the corolla in the bud, and the twining of the stems in Convolvulaceae and Dioscoreaceae. *Castanea* 12: 88–94.

Al-Shehbaz, I. A. 1984. The tribes of Cruciferae (Brassicaceae) in the southeastern United States. *J. Arnold Arbor.* 65: 343–373.

Al-Shehbaz, I. A. 1985a. The genera of Brassiceae (Cruciferae; Brassicaceae) in the southeastern United States. *J. Arnold Arbor.* 66: 279–351.

Al-Shehbaz, I. A. 1985b. The genera of Thelypodieae (Cruciferae; Brassicaceae) in the southeastern United States. *J. Arnold Arbor.* 66: 95–111.

Al-Shehbaz, I. A. 1987. The genera of Alysseae (Cruciferae; Brassicaceae) in the southeastern United States. *J. Arnold Arbor.* 68: 185–240.

Al-Shehbaz, I. A. 1988a. The genera of Anchonieae (Hesperideae) (Cruciferae; Brassicaceae) in the southeastern United States. *J. Arnold Arbor.* 69: 193–212.

Al-Shehbaz, I. A. 1988b. The genera of Arabideae (Cruciferae; Brassicaceae) in the southeastern United States. *J. Arnold Arbor.* 69: 85–166.

Al-Shehbaz, I. A. 1991. The genera of Boraginaceae in the southeastern United States. *J. Arnold Arbor.*, Suppl. 1: 1–169.

Al-Shehbaz, I. A, and B. G. Schubert. 1989. The Dioscoreaceae of the southeastern United States. *J. Arnold Arbor.* 70: 57–95.

Al-Shehbaz, I. A., M. A. Beilstein, and E. A. Kellogg. 2006. Systematics and phylogeny of the Brassicaceae (Cruciferae): an overview. *Plant Syst. Evol.* 259: 89–120

Alverson, W. S., K. G. Karol, D. A. Baum, M. W. Chase, S. M. Swensen, R. McCourt, and K. Sytsma. 1998. Circumscription of the Malvales and relationships to other Rosidae: evidence from *rbcL* sequence data. *Amer. J. Bot.* 85: 876–887.

Alverson, W. S., B. A. Whitlock, R. Nyffeler, and C. Bayer. 1999. Phylogeny of core Malvales: evidence from *ndhF* sequence data. *Amer. J. Bot.* 86: 1474–1486.

Ambrose, J. D. 1980. A re-evaluation of the Melanthioideae (Liliaceae) using numerical analyses. In *Petaloid monocotyledons,* Linnean Society Symposium Series no. 8, C. D. Brickell, D. F. Cutler and M. Gregory (eds.), 65–81. Academic Press, London.

Ambrose, J. D. 1985. *Lophiola,* familial affinity with the Liliaceae. *Taxon* 34: 140–150.

Anderberg, A. A. 1991a. Taxonomy and phylogeny of the tribe Plucheeae (Asteraceae). *Plant Syst. Evol.* 176: 145–177.

Anderberg, A. A. 1991b. Taxonomy and phylogeny of the tribe Inuleae (Asteraceae). *Plant Syst. Evol.* 176: 75–123.

Anderberg, A. A. 1991c. Taxonomy and phylogeny of the tribe Gnaphalieae (Asteraceae). *Opera Botanica* 104: 1–195.

Anderberg, A. A. 1992. The circumscription of the Ericales, and their cladistic relationships to other families of "higher" dicotyledons. *Syst. Bot.* 17: 660–675.

Anderberg, A. A. 1993. Cladistic relationships and major clades of the Ericales. *Plant Syst. Evol.* 184: 207–231.

Anderberg, A. A. 1994. Cladistic analysis of *Enkianthus* with notes on the early diversification of the Ericaceae. *Nordic J. Bot.* 14: 385–401.

Anderberg, A. A. and B. Ståhl. 1995. Phylogenetic interrelationships in the order Primulales, with special emphasis on the family circumscriptions. *Canad. J. Bot.* 73: 1699–1730.

Anderberg, A. A. and U. Swenson. 2003. Evolutionary lineages in Sapotaceae (Ericales): a cladistic analysis based on *ndhF* sequence data. *Int. J. Plant Sci.* 164: 763–773.

Anderberg, A. A., B. Ståhl, and M. Källersjö. 1998. Phylogenetic relationships in the Primulales inferred from *rbcL* sequence data. *Plant Syst. Evol.* 211: 93–102.

Anderberg, A. A., B. Ståhl, and M. Källersjö. 2000. Maesaceae, a new primuloid family in the order Ericales s. l. *Taxon* 49: 183–187.

Anderberg, A. A., C. Rydin, and M. Källersjö. 2002. Phylogenetic relationships in the order Ericales s. l.: analyses of molecular data from five genes from the plastid and mitochondrial genomes. *Amer. J. Bot.* 89: 677–689.

Anderson, F. 2001. *The cactus family.* Timber Press, Portland, Oregon.

Anderson, W. R. 1973. A morphological hypothesis for the origin of heterostyly in the Rubiaceae. *Taxon* 22: 537–542.

Anderson, W. R. 1977. Byrsonimoideae, a new subfamily of the Malpighiaceae. *Leandra* 7: 5–18.

Anderson, W. R. 1979. Floral conservation in neotropical Malpighiaceae. *Biotropica* 11: 219–223.

Anderson, W. R. 1990. The origin of the Malpighiaceae: the evidence from morphology. *Mem. New York Bot. Gard.* 64: 210–224.

Andersson, L. 1981. The neotropical genera of Marantaceae: circumscription and relationships. *Nordic J. Bot.* 1: 218–245.

Andersson, L. 1998. Marantaceae. In *The families and genera of vascular plants,* vol. 4, Monocotyledons: Alismatanae and Commelinanae (except Gramineae), K. Kubitzki (ed.), 278–293. Springer-Verlag, Berlin.

Andersson, L. and H. E. Rova. 1999. The *rps16* intron and the phylogeny of the Rubioideae (Rubiaceae). *Plant Syst. Evol.* 214: 161–189.

Andersson, L. and M. W. Chase. 2001. Phylogeny and classification of Marantaceae. *Bot. J. Linn. Soc.* 135: 275–287.

Andreasen, K. and B. Bremer. 2000. Combined phylogenetic analysis in the Rubiaceae: Ixoroideae. Morphology, nuclear, and chloroplast DNA data. *Amer. J. Bot.* 87: 1731–1748.

Angiosperm Phylogeny Group. 1998. An ordinal classification for the families of flowering plants. *Ann. Missouri Bot. Gard.* 85: 531–553.

Angiosperm Phylogeny Group. 2003. An update of the Phylogeny Group classification for the orders and families of flowering plants: APGII. *Bot. J. Linnaean Soc.* 141: 399–436.

Appel, O. and I. A. Al-Shebaz. 2003. Cruciferae. In *The families and genera of vascular plants,* vol. 5, Malvales, Capparales and non-betalain Caryophyllales. K. Kubitzki and C. Bayer (eds), 75–174. Springer-Verlag, Berlin.

Applequist, W. L. and R. S. Wallace. 2001. Phylogeny of the Portulacaceous cohort based on *ndhF* sequence data. *Syst. Bot.* 26: 406–419.

Argus, G. W. 1974. An experimental study of hybridization and pollination in *Salix* (willows). *Canad. J. Bot.* 52: 1613–1619.

Argus, G. W. 1986. The genus *Salix* (Salicaceae) in the southeastern United States. *Syst. Bot. Monogr.* 9: 1–170.

Armstrong, J. E. 1985. The delimitation of Bignoniaceae and Scrophulariaceae based on floral anatomy, and the placement of problem genera. *Amer. J. Bot.* 755–766.

Armstrong, J. E. and B. A. Drummond III. 1986. Floral biology of *Myristica fragrans* Houtt., the nutmeg of commerce. *Biotropica* 18: 32–38.

Arriagada, J. E. and N. G. Miller. 1997. The genera of Anthemideae (Compositae; Asteraceae) in the southeastern United States. *Harvard Pap. Bot.* No. 11: 1–46.

Arrington, J. M. and K. Kubitzki. 2003. Cistaceae. In *The families and genera of vascular plants,* Vol 5, Malvales, Capparales, and non-betalain Caryophyllales, K. Kubitzki and C. Bayer (eds.), 62–70. Springer-Verlag, Berlin.

Arroyo, M. T. K. 1981. Breeding systems and pollination biology in Leguminosae. In

Advances in legume systematics, part 2, R. M. Polhill and P. H. Raven (eds.), 723–769. Royal Botanic Gardens, Kew.

Ashton, P. S. 2003. Dipterocarpaceae. In *The families and genera of vascular plants,* vol. 5, Malvales, Capparales and non-betalain Caryophyllales, K. Kubitzki and C. Bayer (eds.), 182–197. Springer-Verlag, Berlin.

Asmussen, C. B. and M. W. Chase. 2001. Coding and noncoding plastid DNA in palm systematics. *Amer. J. Bot.* 88: 1103–1117.

Asmussen, C. B., W. J. Baker, and J. Dransfield. 2000. Phylogeny of the palm family (Arecaceae) based on *rps16* intron and *trnL-trnF* plastid DNA sequences. In *Monocots: systematics and evolution,* K. L. Wilson and D. A. Morrison (eds.), 525–535. Collingwood, Australia.

Asmussen, C. B., J. Dransfield, V. Deickmann, A. S. Barfod, J.-C. Pintaud, and W. J. Baker. 2006. A new subfamily classification of the palm family (Arecaceae): evidence from plastid DNA phylogeny. *Bot. J. Linnean Soc.* 151: 15–38.

Atkins, S. 2004. Verbenaceae. In *The families and genera of vascular plants,* vol. 7, Lamiales (except Acanthaceae, including Avicenniaceae), K. Kubitzki (ed.), 449–468. Springer-Verlag, Berlin.

Augspurger, C. K. 1989. Morphology and aerodynamics of wind-dispersed legumes. *Monogr. Syst. Bot. Missouri Bot. Gard.* 29: 451–466.

Austin, D. F. 1979. Studies of the Florida Convolvulaceae I. Key to genera. *Florida Scientist* 42: 214–216.

Azuma, H., J. G. García-Franco, V. Rico-Gray, and L. B. Thien. 2001. Molecular phylogeny of the Magnoliaceae: the biogeography of tropical and temperate disjunctions. *Amer. J. Bot.* 88: 2275–2285.

Baas, P., S. Jansen, and E. A. Wheeler. 2003. Ecological adaptations and deep phylogenetic splits – evidence and questions from the secondary xylem. In *Deep morphology: towards a renaissance of morphology in plant systematics,* T. F. Stuessy, V. Mayer, and E. Hörandl (eds.), 221–239. A. R. G. Gautner Verlag K. G., Ruggell.

Backlund, A. and B. Bremer. 1997. Phylogeny of the Asteridae s. str. based on *rbcL* sequences, with particular reference to the Dipsacales. *Plant Syst. Evol.* 207: 225–-254.

Backlund, A. and K. Bremer. 1998. To be or not to be: principles of classification and monotypic plant families. *Taxon* 47: 391–401.

Backlund, A. and N. Pyck. 1998. Diervillaceae and Linnaeaceae, two new families of caprifolioids. *Taxon* 47: 657–661.

Backlund, A., B. Oxelman, and B. Bremer. 2000. Phylogenetic relationships within the Gentianales based on *ndhF* and *rbcL* sequences, with particular reference to the Loganiaceae. *Amer. J. Bot.* 87: 1029–1043.

Bailey, I. W. and C. G. Nast. 1945. The comparative morphology of the Winteraceae. VII. Summary and conclusions. *J. Arnold Arbor.* 26: 37–47.

Bailey, I. W. and B. G. L. Swamy. 1948. *Amborella trichopoda* Baill., a new morphological type of vesselless dicotyledon. *J. Arnold Arbor.* 29: 245–254.

Baker, H. G. 1986. Yuccas and yucca moths: a historical commentary. *Ann. Missouri Bot. Gard.* 73: 556–564.

Baker, W. J., S. Zona, C. D. Heatubun, C. E. Lewis, R. A. Maturbongs, and M. V. Norup. 2006. *Dransfieldia* (Arecaceae): a new palm genus from western New Guinea. *Syst. Bot.* 31: 61–69.

Balslev, H. 1998. Juncaceae. In *The families and genera of vascular plants*, vol. 4, Monocotyledons: Alismatanae and Commelinanae (except Gramineae), K. Kubitzki (ed.), 252–260. Springer-Verlag, Berlin.

Barfuss, M. H. J., R. Samuel, W. Till, and T. F. Stuessy. 2005. Phylogenetic relationships in subfamily Tillandsioideae (Bromeliaceae) based on DNA sequence data from several plastid regions. *Amer. J. Bot.* 92: 337–351.

Barker, N. P., H. P. Linder and E. H. Harley. 1995. Polyphyly of Arundinoideae (Poaceae): evidence from *rbcL* sequence data. *Syst. Bot.* 20: 423–435.

Barkman, T. J., G. Chenery, J. R. McNeal, J. Luons-Weiler, W. J. Ellisens, G. Moore, A. D. Wolfe, and C. W. dePamphilis. 2000. Independent and combined analyses of sequences from all three genomic compartments converge on the root of flowering plant phylogeny. *Proc. Natl. Acad. Sci. USA* 97: 13166–13171.

Barrett, S. C. H. and S. W. Graham. 1997. Adaptive radiation in the aquatic plant family Pontederiaceae: insights from phylogenetic analyses. In *Molecular evolution and adaptive radiation*, T. J. Givnish and K. J. Sytsma (eds.), 225–238. Cambridge University Press, Cambridge.

Barrett, S. C. H. and J. H. Richards. 1990. Heterostyly in tropical plants. *Mem. New York Bot. Gard.* 55: 35–61.

Barthlott, W. and D. Fröhlich. 1983. Micromorphologie und Orientierungs muster epicuticularer Wachs-Kristalloide: ein neues systematisches Merkmal bei Monokotylen. *Plant Syst. Evol.* 142: 171–185.

Barthlott, W. and D. R. Hunt. 1993. Cactaceae. In *The families and genera of vascular plants.* vol. 2. Magnoliid, hamamelid and caryophyllid families, K. Kubitzki, J. G. Rohwer, and V. Bittrich (eds.), 161–197. Springer-Verlag, Berlin.

Baum, D. A., W. S. Alverson and R. Nyffeler. 1998. A durian by any other name: taxonomy and nomenclature of the core Malvales. *Harvard Pap. Bot.* 3: 315–330.

Baum, D. A., S. DeWitt, A. Yen, W. S. Alverson, R. Nyffeler, B. A. Whitlock, and R. L. Oldham. 2004. Phylogenetic relationships of Malvatheca (Bombacoideae and Malvoideae: Malvaceae sensu lato) as inferred from plastid DNA sequences. *Amer. J. Bot.* 91: 1863–1871.

Baumann, M. G. 1946. *Myodocarpus* und die Phylogenie der Umbelliferen-Frücht. *Ber. Schweiz. Bot. Ges.* 56: 13–112.

Bayer, C. 1998. Synflorescences of Malvaceae. *Nordic J. Bot.* 18: 335–338.

Bayer, C. 1999. The bicolor-unit: homology and transformation of an inflorescence structure unique to core Malvales. *Plant Syst. Evol.* 214: 187–198.

Bayer, C. and K. Kubitzki. 2003. Malvaceae. In *The families and genera of vascular plants*, vol. 5, Malvales, Capparales and non-betalain Caryophyllales. K. Kubitzki and C. Bayer (eds.), 225–310. Springer-Verlag, Berlin.

Bayer, C., M. F. Fay, A. Y. de Bruijn, V. Savolainen, C. M. Morton, K. Kubitzki, W. S. Alverson, and M. W. Chase. 1999. Support for an expanded family concept of Malvaceae within a re-circumscribed order Malvales: combined analysis of plastid *atpB* and *rbcL* DNA sequences. *Bot. J. Linn. Soc.* 129: 267–303.

Bayer, R. J., L. Hufford, and D. E. Soltis. 1996. Phylogenetic relationships in Sarraceniaceae based on *rbcL* and ITS sequences. *Syst. Bot.* 21: 121–134.

Beardsley, P. M. and R. G. Olmstead. 2002. Redefining Phrymaceae: the placement of *Mimulus*, tribe Mimuleae, and *Phryma*. *Amer. J. Bot.* 89: 1093–1102.

Bechtel, A. R. 1921. The floral anatomy of the Urticales. *Amer. J. Bot.* 8: 386–410.

Beck, C. B. (ed.) 1976. *Origin and early evolution of angiosperms*. Columbia University Press, New York.

Behnke, H.-D. 1976. Ultrastructure of sieve-element plastids in Caryophyllales (Centrospermae), evidence for the delimitation and classification of the order. *Plant Syst. Evol.* 126: 31–54.

Behnke, H.-D. 1994. Sieve-element plastids: their significance for the evolution and systematics of the order. In *Caryophyllales: evolution and systematics*, H.-D. Behnke and T. J. Mabry (eds.), 87–121. Springer-Verlag, Berlin.

Behnke, H.-D. and W. Barthlott. 1983. New evidence from the ultrastructural and micromorphological fields in angiosperm classification. *Nordic J. Bot.* 3: 43–66.

Behnke, H.-D. and T. J. Mabry (eds.). 1994. *Caryophyllales: evolution and systematics*. Springer-Verlag, Berlin.

Bell, C. D., E. J. Edwards, S.-T. Kim, and M. J. Donoghue. 2001. Dipsacales phylogeny based on chloroplast DNA sequences. *Harvard Pap. Bot.* 6: 481–499.

Bello, M. A., M. W. chase, R. G. Olmstead, N. Rønsted, and D. Albach. 2002. The páramo endemic *Aragoa* is the sister genus of *Plantago* (Plantaginaceae; Lamiales): evidence from plastid *rbcL* and nuclear ribosomal ITS sequence data. *Kew Bull.* 57: 585–597.

Bennett, J. R. and S. Mathews. 2006. Phylogeny of the parasitic plant family Orobanchaceae inferred from phytochrome A. *Amer. J. Bot.* 93: 1039–1051.

Bensel, C. R. and B. F. Palser. 1975. Floral anatomy in the Saxifragaceae sensu lato. III. Kirengeshomoideae, Hydrangeoideae, and Escallonioideae. *Amer. J. Bot.* 62: 676–687.

Benson, N. L. 1982. *The cacti of the United States and Canada*. Stanford University Press, Stanford.

Benzing, D. H. 1980. *The biology of the bromeliads*. Mad River Press, Eureka, CA.

Benzing, D. H., J. Seemann and A. Renfrow. 1978. The foliar epidermis in Tillandsioideae (Bromeliaceae) and its role in habitat selection. *Amer. J. Bot.* 65: 359–365.

Berg, C. C. 1977. Urticales, their differentiation and systematic position. *Plant Syst. Evol.*, Suppl. 1: 349–374.

Berg, C. C. 1978. Cecropiaceae, a new family of the Urticales. *Taxon* 27: 39–44.

Berg, C. C. 1989. Systematics and phylogeny of the Urticales. In *Evolution, systematics, and fossil history of the Hamamelidae*, vol. 2, "Higher" Hamamelidae. Systematics Association Special Vol. 40B, P. R. Crane and S. Blackmore (eds.), 193–200. Clarendon Press, Oxford.

Berg, R. Y. 1996. Development of ovule, embryo sac, and endosperm in *Dipterostemonx* and *Dichelostemma* (Alliaceae) relative to taxonomy. *Amer. J. Bot.* 83: 790–801.

Berry, P. E., W. J. Hahn, K. J. Sytsma, J. C. Hall, and A. Mast. 2004. Phylogenetic relationships and biogeography of *Fuchsia* (Onagraceae) based on noncoding nuclear and chloroplast DNA data. *Amer. J. Bot.* 91: 601–614.

Bessey, C. E. 1915. The phylogenetic taxonomy of flowering plants. *Ann. Missouri Bot. Gard.* 2: 109–164.

Bharatham, G. and E. A. Zimmer. 1995. Early branching events in monocotyledons: partial 18S ribosomal DNA sequence analysis. In *Monocotyledons: systematics and evolution*, P. J. Rudell, P. J. Cribb, D. F. Cutler and C. J. Humphries (eds.), 81–107. Royal Botanic Gardens, Kew.

Bidartondo, M. I. and T. D. Bruns. 2001. Extreme specificity in epiparasitic Monotropoideae (Ericaceae): widespread phylogenetic and geographical structure. *Mol. Ecol.* 10: 2285–2295.

Bigazzi, M. 1989. Ultrastructure of nuclear inclusions and the separation of Verbenaceae and Oleaceae (including *Nyctanthes*). *Plant Syst. Evol.* 163: 1–12.

Bittrich, V. 1993a. Caryophyllaceae. In *The families and genera of vascular plants*, vol. 2. Magnoliid, hamamelid and caryophyllid families, K. Kubitzki, J. G. Rohwer, and V. Bittrich (eds.), 206–236. Springer-Verlag, Berlin.

Bittrich, V. 1993b. Introduction to Centrospermae. In *The families and genera of vascular plants*, vol. 2, Magnoliid, hamamelid and caryophyllid families, K. Kubitzki, J. G. Rohwer, and V. Bittrich (eds.), 13–19. Springer-Verlag, Berlin.

Bittrich, V. 1990. Systematic studies in Aizoaceae. *Mittelungen aus dem Institut für Allgemeine Botanik Hannburg* 23b: 491– 507.

Bittrich, V. and H. Hartmann. 1988. The Aizoaceae: a new approach. *Bot. J. Linn. Soc.* 97: 239–254.

Bittrich, V. and U. Kühn. 1993. Nyctaginaceae. In *The families and genera of vascular plants*, vol. 2, Magnoliid, hamamelid and caryophyllid families, K. Kubitzki, J. G. Rohwer, and V. Bittrich (eds.), 473–486. Springer-Verlag, Berlin.

Bittrich, V. and M. Struck. 1989. What is primitive in Mesembryanthemaceae? *S. Afr. J. Bot.* 55: 321–331.

Blackwell, W. H. 1977. The subfamilies of the Chenopodiaceae. *Taxon* 26: 395–397.

Blattner, F. R. and J. W. Kadereit. 1995. Three intercontinental disjunctions in Papaveraceae subfamily Chelidonioideae: evidence from chloroplast DNA. *Plant Syst. Evol.* Suppl. 9: 147–157.

Bobrov, A. V. F. Ch., P. K. Endress, A. P. Melikian, M. S. Romanov, A. N. Sovokin, and A. Palmarola Bijerando. 2005. Fruit structure of *Amborella trichopoda* (Amborellaceae). *Bot. J. Linn. Soc.* 148: 265–274.

Boesewinkel, F. D. 1988. The seed structure and taxonomic relationships of *Hypseocharis* Remy. *Acta Bot. Neerl.* 37: 111–120.

Boeshore, I. 1920. The morphological continuity of Scrophulariaceae and Orobanchaceae. *Contr. Bot. Lab. Morris Arbor.* 5: 139–177.

Bogle, A. L. 1969. The genera of Portulacaceae and Basellaceae in the southeastern United States. *J. Arnold Arbor.* 50: 566–598.

Bogle, A. L. 1970. The genera of Molluginaceae and Aizoaceae in the southeastern United States. *J. Arnold Arbor.* 51: 431–462.

Bogle, A. L. 1974. The genera of Nyctaginaceae in the southeastern United States. *J. Arnold Arbor.* 55: 1–37.

Bogler, D. J. 1998. Nolinaceae. In *The families and genera of vascular plants,* vol 3. Flowering plants—Monocotyledons, K. Kubitzki (ed.), 392–397. Springer-Verlag, Berlin.

Bogler, D. J. and B. B. Simpson. 1995. A chloroplast DNA study of the Agavaceae. *Syst. Bot.* 20: 191–205.

Bogler, D. J. and B. B. Simpson. 1996. Phylogeny of Agavaceae based on ITS rDNA sequence variation. *Amer. J. Bot.* 83: 1225–1235.

Bogler, D. J., J. C. Pires, and J. Francisco-Ortega. 2006. Phylogeny of Agavaceae based on *ndhF, rbcL,* and ITS sequences: implications of molecular data for classification. *Aliso* 22: 313–328.

Bogner, J. and S. J. Mayo. 1998. Acoraceae. In *The families and genera of vascular plants,* vol. 4, Monocotyledons: Alismatanae and Commelinanae (except Gramineae), K. Kubitzki (ed.), 7–11. Springer-Verlag, Berlin.

Bogner, J. and D. H. Nicolson. 1991. A revised classification of Araceae with dichotomous keys. *Willdenowia* 21: 35–50.

Bohs, L. and R. G. Olmstead. 1997. Phylogenetic relationships in *Solanum* (Solanaceae) based on *ndhF* sequences. *Syst. Bot.* 22: 5–17.

Boke, N. H. 1964. The cactus gynoecium: a new interpretation. *Amer. J. Bot.* 51: 598–610.

Bonsen, K. and B. J. H. ter Welle. 1983. Comparative wood and leaf anatomy of the Cecropiaceae (Urticales). *Bull. Mus. Nat. Hist. Nat. Paris,* Sér. 4, 5 (sect. B, Adansonia, No. 2): 151–177.

Boothroyd, L. E. 1930. The morphology and anatomy of the inflorescence and flower of the Platanaceae. *Amer. J. Bot.* 17: 678–693.

Borstein, A. J. 1991. The Piperaceae in the southeastern United States. *J. Arnold Arbor.,* Suppl. 1: 349–366.

Bortiri, E., S.-H. Oh, J. Jiang, S. Baggett, A. Granger, C. Weeks, M. Buckingham, D. Potter, and D. E. Parfitt. 2001. Phylogeny and systematics of *Prunus* (Rosaceae) as determined by sequence analysis of ITS and chloroplast *trnL-trnF* spacer DNA. *Syst. Bot.* 26: 797–807.

Bos, J. 1998. Dracaenaceae. In *The families and genera of vascular plants,* vol. 3. Monocotyledons: Lilianae (except Orchidaceae), K. Kubitzki (ed.), 238–241. Springer-Verlag, Berlin.

Boucher, L. D., S. R. Manchester, and W. S. Judd. 2003. An extinct genus of Salicaceae based on twigs with attached flowers, fruits, and foliage from the Eocene Green River formation of Utah and Colorado, USA. *Amer. J. Bot.* 90: 1389–1399.

Bouman, F. 1995. Seed structure and systematics in Dioscoreales. In *Monocotyledons: systematics and evolution,* P. J. Rudall, P. J. Cribb, D. F. Cutler, and C. J. Humphries (eds.), 139–156. Royal Botanic Gardens, Kew.

Bowe, L. M., G. Coat, and C. W. dePamphilis. 2000. Phylogeny of seed plants based on all three genomic compartments: extant gymnosperms are monophyletic and Gnetales' closest relatives are conifers. *Proc. National Acad. Sci. USA* 97: 4092–4097.

Brandbyge, J. 1993. Polygonaceae. In *The families and genera of vascular plants,* vol. 2, Magnoliid, hamamelid and caryophyllid families, K. Kubitzki, J. G. Rohwer, and V. Bittrich (eds.), 531–544. Springer-Verlag, Berlin.

Bremekamp, C. E. B. 1965. Delimitation and subdivision of the Acanthaceae. *Bull. Bot. Surv. India* 7: 21–30.

Bremer, B. 1996. Phylogenetic studies within Rubiaceae and relationships to other families based on molecular data. *Opera Bot. Belg.* 7: 33–50.

Bremer, B. and R. K. Jansen. 1991. Comparative restriction site mapping of chloroplast DNA implies new phylogenetic relationships within the Rubiaceae. *Acta Bot. Neerl.* 15: 1–33.

Bremer, B. and L. Struwe. 1992. Phylogeny of the Rubiaceae and Loganiaceae: congruence or conflict between morphological and molecular data? *Amer. J. Bot.* 79: 1171–1184.

Bremer, B., K. Andreasen, and D. Olsson. 1995. Subfamilial and tribal relationships in the Rubiaceae based on *rbcL* sequence data. *Ann. Missouri Bot. Gard.* 82: 383–397.

Bremer, B., K. Bremer, N. Heidari, P. Erixon, R. G. Olmstead, A. A. Anderberg, M. Källersjö, and E. P. Barkhordarian. 2002. Phylogenetics of asterids based on 3 coding and 3 non-coding chloroplast DNA markers and the utility of non-coding DNA at higher taxonomic levels. *Mol. Phylo. Evol.* 24: 274–301.

Bremer, B., R. G. Olmstead, L. Struwe, and J. A. Sweere. 1994. *rbcL* sequences support exclusion of *Retzia, Desfontainia,* and *Nicodemia* from the Gentianales. *Plant Syst. Evol.* 190: 213–230.

Bremer, B., R. K. Jansen, B. Oxelman, M. Backlund, H. Lautz, and K.-J. Kim. 1999. More characters or more taxa for a robust phylogeny: case study from the coffee family (Rubiaceae). *Syst. Biol.* 48: 413–435.

Bremer, K. 1987. Tribal interrelationships of the Asteraceae. *Cladistics* 3: 210–253

Bremer, K. 1994. *Asteraceae: cladistics and classification.* Timber Press, Portland, OR.

Bremer, K. 1996. Major clades and grades of the Asteraceae. In *Compositae: systematics.* Proceedings of the International Compositae Conference, Kew, D. J. N. Hind and H. J. Beentje (eds.), 1–7. Royal Botanic Gardens, Kew.

Bremer, K. and R. K. Jansen. 1992. A new subfamily of the Asteraceae. *Ann. Missouri Bot. Gard.* 79: 414–415.

Brett, D. W. 1964. The inflorescence of *Fagus* and *Castanea* and the evolution of the cupules of the Fagaceae. *New Phytol.* 63: 96–117.

Briggs, B. G. and L. A. S. Johnson. 1998a. New genera and species of Australian Restionaceae (Poales). *Telopea* 7: 345–373.

Briggs, B. G. and L. A. S. Johnson. 1998b. A guide to a new classification of Restionaceae and allied families. In *Australian rushes: biology, identification and conservation of Restionaceae and allied families,* K. A. Meney and J. S. Pate (eds.), 25–56. University of Western Australia Press, Perth.

Briggs, B. G., A. D. Marchant, S. Gilmore, and C. L. Porter. 2000. A molecular phylogeny of Restionaceae and allies. In *Monocots: systematics and evolution,* K. L. Wilson and D. A. Morrison (eds.), 661–671. CSIRO, Collingwood, Australia.

Brizicky, G. K. 1961a. The genera of Turneraceae and Passifloraceae in the southeastern United States. *J. Arnold Arbor.* 42: 204–218.

Brizicky, G. K. 1961b. The genera of Violaceae in the southeastern United States. *J. Arnold Arbor.* 42: 321–333.

Brizicky, G. K. 1962a. The genera of Anacardiaceae in the southeastern United States. *J. Arnold Arbor.* 43: 359–375.

Brizicky, G. K. 1962b. The genera of Rutaceae in the southeastern United States. *J. Arnold Arbor.* 43: 1–22.

Brizicky, G. K. 1962c. The genera of Simaroubaceae and Burseraceae in the southeastern United States. *J. Arnold Arbor.* 43: 173–186.

Brizicky, G. K. 1963. The genera of Sapindaceae in the southeastern United States. *J. Arnold Arbor.* 44: 462–501.

Brizicky, G. K. 1964a. The genera of Celastrales in the southeastern United States. *J. Arnold Arbor.* 45: 206–234.

Brizicky, G. K. 1964b. The genera of Rhamnaceae in the southeastern United States. *J. Arnold Arbor.* 45: 439–463.

Brizicky, G. K. 1964c. The genera of Cistaceae in the southeastern United States. *J. Arnold Arbor.* 45: 346–357.

Brizicky, G. K. 1965a. The genera of Tiliaceae and Elaeocarpaceae in the southeastern United States. *J. Arnold Arbor.* 46: 286–307.

Brizicky, G. K. 1965b. The genera of Vitaceae in the southeastern United States. *J. Arnold Arbor.* 46: 48–67.

Brizicky, G. K. 1966a. The genera of Sterculiaceae in the southeastern United States. *J. Arnold Arbor.* 47: 60–74.

Brizicky, G. K. 1966b. The Goodeniaceae in the southeastern United States. *J. Arnold Arbor.* 47: 293–300.

Brown, G. K. and G. S. Varadarajan. 1985. Studies in Caryophyllales I: Re-evaluation of classification of Phytolaccaceae s. l. *Syst. Bot.* 10: 49–63.

Brown, G. K. and A. J. Gilmartin. 1989. Stigma types in the Bromeliaceae. A systematic study. *Syst. Bot.* 14: 110–132.

Bruhl, J. J. 1995. Sedge genera of the world: relationships and a new classification of the Cyperaceae. *Austr. Syst. Bot.* 8: 125–305.

Bruneau, A., F. Forest, P. S. Herendeen, B. B. Klitgaard, and G. P. Lewis. 2001. Phylogenetic relationships in the Caesalpinioideae (Leguminosae) as inferred from chloroplast *trnL* intron sequences. *Syst. Bot.* 26: 487–514.

Brunsfeld, S. J., D. E. Soltis and P. S. Soltis. 1992. Evolutionary patterns and processes in *Salix* sect. *Longifoliae:* evidence from chloroplast DNA. *Syst. Bot.* 17: 239–256.

Bult, C. J. and E. A. Zimmer. 1993. Nuclear ribosomal RNA sequences for inferring tribal relationships within Onagraceae. *Syst. Bot.* 18: 48–63.

Burger, W. C. 1977. The Piperales and the monocots: alternative hypothesis for the origin of the monocotyledon flower. *Bot. Rev.* 43: 345–393.

Burger, W. C. 1988. A new genus of Lauraceae from Costa Rica with comments on problems of generic and specific delimitation within the family. *Brittonia* 40: 275–282.

Burleigh, J. G. and S. Mathews. 2004. Phylogenetic signal in nucleotide data from seed plants: implications for resolving the seed plant tree of life. *Amer. J. Bot.* 91: 1599–1613.

Burns-Balogh, P. and V. A. Funk. 1986. A phylogenetic analysis of the Orchidaceae. *Smithsonian Contr. Bot.* 61: 1–79.

Burtt, B. L. 1977. Classification above the genus, as exemplified by Gesneriaceae, with parallels from other groups. *Plant. Syst. Evol.,* Suppl. 1: 97–109.

Butterworth, C. A. and R. S. Wallace. 2005. Molecular phylogenetics of the leafy cactus genus *Pereskia* (Cactaceae). Syst. Bot. 30: 800–808.

Buzgo, M., P. S. Soltis, and D. E. Soltis. 2004. Floral developmental morphology of *Amborella trichopoda* (Amborellaceae). *Int. J. Plant Sci.* 165: 925–947.

Caddick, L. R., P. J. Rudall, P. Wilkin, and M. W. Chase. 2000. Yams and their allies: systematics of Dioscoreales. In *Monocots: systematics and evolution*, K. L. Wilson and D. A. Morrison (eds.), 475–487. CSIRO, Collingwood, Australia.

Caddick, L. R., J. R. Rudall, P. Wilkin, T. A. J. Hedderson, and M. W. Chase. 2002a. Phylogenetics of Dioscoreales based on combined analyses of morphological and molecular data. *Bot. J. Linn. Soc.* 138: 123–144.

Caddick, L. R., P. Wilkin, P. J. Rudall, T. A. J. Hedderson, and M. W. Chase. 2002b. Yams reclassified: a recircumscription of Dioscoreaceae and Dioscoreales. *Taxon* 51: 103–114.

Calder, D. M. and P. Bernhard. 1983. *The biology of mistletoes.* Academic Press, New York.

Cameron, K. M. 2006. A comparison and combination of plastid *atpB* and *rbcL* gene sequences for inferring phylogenetic relationships within Orchidaceae. *Aliso* 22: 447–464.

Cameron, K. M. and M. W. Chase. 2000. Nuclear 18S rDNA sequences of Orchidaceae confirm the subfamilial status and circumscription of Vanilloideae. In *Monocots: systematics and evolution*, K. L. Wilson and D. A. Morrison (eds.), 457–464. CSIRO, Collingwood, Australia.

Cameron, K. M. and C.-X. Fu. 2006. A nuclear rDNA phylogeny of *Smilax* (Smilacaceae). *Aliso* 22: 598–605.

Cameron, K. M., M. W. Chase, W. M. Whitten, P. J. Kores, D. C. Jarrell, V. A. Albert, T. Tukawa, H. G. Hills and D. H. Goldman. 1999. A phylogenetic analysis of the Orchidaceae: evidence from *rbcL* nucleotide sequences. *Amer. J. Bot.* 86: 208–224.

Cameron, K. M., K. J. Wurdack, and R. W. Jobson. 2002. Molecular evidence for the common origin of snap-traps among carnivorous plants. *Amer. J. Bot.* 89: 1503–1509.

Campbell, C. S. 1985. The subfamilies and tribes of Gramineae (Poaceae) in the southeastern United States. *J. Arnold Arbor.* 66: 123–199.

Campbell, C. S., B. G. Baldwin, M. J. Donoghue and M. F. Wojciechowski. 1995. A phylogeny of the genera of Maloideae (Rosaceae): evidence from internal transcribed spacers of nuclear ribosomal DNA sequences and congruence with morphology. *Amer. J. Bot.* 82: 903–918.

Campbell, C. S., R. C. Evans, D. R. Morgan, T. A. Dickinson, and M. P. Arsenault. 2007. Phylogeny of subtribe Pyrinae (formerly the Maloideae, Rosaceae): limited resolution of a complex evolutionary histroy. *Plant Syst. Evol.* In press.

Canright, J. E. 1952. The comparative morphology and relationships of the Magnoliaceae. I. Trends in specialization in the stamens. *Amer. J. Bot.* 39: 484–497.

Canright, J. E. 1953. The comparative morphology and relationships of the Magnoliaceae. II. Significance of the pollen. *Phytomorphology* 3: 355–365.

Canright, J. E. 1960. The comparative morphology and relationships of the Magnoliaceae. III. Carpels. *Amer. J. Bot.* 47: 145–155.

Cantino, P. D. 1982. Affinities of the Lamiales: a cladistic analysis. *Syst. Bot.* 7: 237–248.

Cantino, P. D. 1990. The phylogenetic significance of stomata and trichomes in the Labiatae and Verbenaceae. *J. Arnold Arbor.* 71: 323–370.

Cantino, P. D. 1992a. Evidence for a polyphyletic origin of the Labiatae. *Ann. Missouri Bot. Gard.* 79: 361–379.

Cantino, P. D. 1992b. Toward a phylogenetic classification of the Labiatae. In *Advances in labiate science*, R. M. Harley and T. Reynolds (eds.), 27–32. Royal Botanic Gardens, Kew.

Cantino, P. D. and R. W. Sanders. 1986. Subfamilial classification of Labiatae. *Syst. Bot.* 11: 163–185.

Cantino, P. D., R. M. Harley, and S. J. Wagstaff. 1992. Genera of Labiatae: status and classification. In *Advances in labiate science*, R. M. Harley and T. Reynolds (eds.), 511–522. Royal Botanic Garden, Kew.

Caris, P. and E. F. Smets. 2004. A floral ontogenetic study on the sister group relationship between the genus *Samolus* (Primulaceae) and the Theophrastaceae. *Amer. J. Bot.* 91: 627–643.

Carlquist, S. 1976a. Tribal interrelationships and phylogeny of the Asteraceae. *Aliso* 8: 465–492.

Carlquist, S. 1976b. *Alexgeorgea*, a bizarre new genus of Restionaceae from Western Australia. *Austr. J. Bot.* 24: 281–295.

Carolin, R. C. 1983. The trichomes of the Chenopodiaceae and Amaranthaceae. *Bot. Jahrb. Syst.* 103: 451–466.

Carolin, R. C. 1987. A review of the family Portulacaceae. *Austr. J. Bot.* 35: 383–412.

Carolin, R. C. 1993. Portulacaceae. In *The families and genera of vascular plants*, vol. 2, Magnoliid, hamamelid and caryophyllid families, K. Kubitzki, J. G. Rohwer, and V. Bittrich (eds.), 554–555. Springer-Verlag, Berlin.

Carolin, R. C., S. W. L. Jacobs and M. Vesk. 1975. Leaf structure in Chenopodiaceae. *Bot. Jahrb. Syst.* 95: 226–255.

Carpenter, R. J., R. S. Hill, and G. J. Jordan. 2005. Leaf cuticular morphology links Platanaceae and Proteales. *Int. J. Plant Sci.* 166: 843–855.

Chadwell, T. B., S. J. Wagstaff, and P. D. Cantino. 1992. Pollen morphology of *Phryma* and some putative relatives. *Syst. Bot.* 17: 210–219.

Chakravarty, H. L. 1958. Morphology of the staminate flowers in the Cucurbitaceae with special reference to the evolution of the stamens. *Lloydia* 21: 49–87.

Chanderbali, A. S., H. van der Werff, and S. S. Renner. 2001. Phylogeny and historical biogeography of Lauraceae: evidence from the chloroplast and nuclear genomes. *Ann. Missouri Bot. Gard.* 88: 104–134.

Chandler, G. T. and G. M. Plunkett. 2004. Evolution in Apiales: Nuclear and chloroplast markers together in (almost) perfect harmony. *Bot. J. Linn. Soc.* 144: 123–147.

Channell, R. B. and C. E. Wood, Jr. 1959. The genera of the Primulales of the southeastern United States. *J. Arnold Arbor.* 40: 268–288.

Channell, R. B. and C. E. Wood, Jr. 1962. The Leitneriaceae in the southeastern United States. *J. Arnold Arbor.* 43: 435–438.

Chappill, J. A. 1994. Cladistic analysis of the Leguminosae: the development of an explicit hypothesis. In *Advances in legume systematics*, part 7, M. D. Crisp and J. J. Doyle (eds.), 1–9. Royal Botanic Gardens, Kew.

Chase, M. W. 2004. Monocot relationships: an overview. *Amer. J. Bot.* 91: 1645–1655.

Chase, M. W. and 41 others. 1993. Phylogenetics of seed plants: an analysis of nucleotide sequences from the plastid gene *rbcL. Ann. Missouri Bot. Gard.* 80: 528–580.

Chase, M. W. and 10 others. 1995a. Molecular systematics of Lilianae. In *Monocotyledons: systematics and evolution*, P. J. Rudall, P. J. Cribb, D. F. Cutler, and C. J. Humphries (eds.), 109–137. Royal Botanic Gardens, Kew.

Chase, M. W., D. W. Stevenson, P. Wilkin, and P. J. Rudall. 1995b. Monocot systematics: a combined analysis. In *Monocotyledons: systematics and evolution*, P. J. Rudall, P. J. Cribb, D. F. Cutler and C. J. Humphries (eds.), 109–137. Royal Botanic Gardens, Kew.

Chase, M. W., P. J. Rudall, and J. G. Conran. 1996. New circumscriptions and a new family of asparagoid lilies. Genera formerly included in Anthericaceae. *Kew Bull.* 57: 667–680.

Chase, M. W., C. M. Morton, and J. A. Kallunki. 1999. Phylogenetic relationships of Rutaceae: a cladistic analysis of the subfamilies using evidence from *rbcL* and *atpB* sequence variation. *Amer. J. Bot.* 86: 1191–1199.

Chase, M. W. and 12 others. 2000. In *Monocots: systematics and evolution*, K. L. Wilson and D. A. Morrison, (eds.), 3–16. CSIRO, Collingwood, Australia.

Chase, M. W., S. Zmarty, M. D. Lledó, K. J. Wurdack, S. M. Swensen, and M. F.. Fay. 2002. When in doubt, put it in Flacourtiaceae: a molecular phylogenetic analysis based on plastid *rbcL* DNA sequences. *Kew Bull.* 57: 141–181.

Chase, M. W. and 20 others. 2006. Multigene analyses of monocot relationships: a summary. *Aliso* 22: 63–75.

Chaw, S. M., A. Zharkikh, H. M. Sung, T. C. Lau, and W. H. Li. 1997. Molecular phylogeny of extant gymnosperms and seed plant evolution: analysis of nuclear 18S rRNA sequences. *Mol. Biol. Evol.* 14: 56–68.

Chen, Z.-O., S. R. Manchester, and H.-Y. Sun. 1999. Phylogeny and evolution of the Betulaceae as inferred from DNA sequences, morphology, and paleobotany. *Amer. J. Bot.* 86: 1168–1181.

Cieslak, T., J. S. Polepallli, A. White, K. Müller, T. Borsch, W. Barthlott, J. Steiger, A. Marchant, and L. Legendre. 2005. Phylogenetic analysis of *Pinguicula* (Lentibulariaceae): chloroplast DNA sequences and morphology support several geographically distinct radiations. *Amer. J. Bot.* 92: 1723–1736.

Civeyrel, L., A. LeThomas, K. Ferguson, and M. W. Chase. 1998. Critical reexamination of palynological characters used to delimit Asclepiadaceae in comparison to the molecular phylogeny obtained from plastid *matK* sequences. *Mol. Phylog. Evol.* 9: 517–527.

Clark, L. G., W. Zhang, and J. F. Wendel. 1995. A phylogeny of the grass family (Poaceae) based on *ndhF* sequence data. *Syst. Bot.* 20: 436–460.

Clark, L. G. and E. J. Judziewicz. 1996. The grass subfamilies Anomoclooideae and Pharoideae (Poaceae). *Taxon* 45: 641–645.

Clark, L. G. and R. W. Pohl. 1996. Agnes Chase's first book of grasses. Smithsonian Inst. Press, Washington, D.C.

Clark, L. G., M. Kobayashi, S. Mathews, R. E. Spangler, and E. A. Kellogg. 2000. The Puelioideae, a new subfamily of Poaceae. *Syst. Bot.* 25: 181–187.

Classen-Bockhoff, R. 1991. Untersuchungen zur Konstruktion des Bestäubungsapparates von *Thalia geniculata* (Marantaceen). *Bot. Acta* 104: 183–193.

Clausing, G., K. Meyer, and S. S. Renner. 2000. Correlations among fruit traits and evolution of different fruits within Melastomataceae. *Bot. J. Linn. Soc.* 133: 303–326.

Clausing, G. and S. S. Renner. 2001. Molecular phylogenetics of Melastomataceae and Memecylaceae: implications for character evolution. *Amer. J. Bot.* 88: 486–498.

Clayton, W. D. and S. A. Renvoize. 1986. Genera Graminum. Grasses of the world. *Kew Bull. Add. Ser.* 13.

Clement, W. L., M. C. Tebbitt, L. L. Forrest, J. E. Blair, L. Brouillet, T. Eriksson, and S. M. Swensen. 2004. Phylogenetic position and biogeography of *Hillebrandia sandwicensis* (Begoniaceae): a rare Hawaiian relict. *Amer. J. Bot.* 91: 905–917.

Clevinger, C. C. and J. L. Panero. 1998. Phylogenetic relationships of North American Celastraceae based on *ndhF* sequence data. *Amer. J. Bot.* 85(6) Suppl.: 120.

Cocucci, A. A. 2004. Oxalidaceae. In *The families and genera of vascular plants*, vol. 6, Celastrales, Oxalidales, Rosales, Cornales, Ericales, K. Kubitzki (ed.), 285–290. Springer-Verlag, Berlin.

Conran, J. G. 1998. Smilacaceae. In *The families and genera of vascular plants*, vol. 4, Monocotyledons: Alismatanae and Commelinanae (except Gramineae), K. Kubitzki (ed.), 417–422. Springer-Verlag, Berlin.

Conran, J. G. 1989. Cladistic analyses of some net-veined Liliiflorae. *Plant Syst. Evol.* 168: 123–141.

Conran, J. G. and M. N. Tamura. 1998. Convallariaceae. In *The families and genera of vascular plants*, vol. 3, Monocotyledons: Lilianae (except Orchidaceae), K. Kubitzki (ed.), 186–198. Springer-Verlag, Berlin.

Constance, L. 1971. History of the classification of Umbelliferae (Apiaceae). *Bot. J. Linn. Soc.* 64, Suppl. 1: 1–11.

Conti, E. 1994. Phylogenetic relationships of Onagraceae and Myrtales: evidence from *rbcL* sequence data. Ph.D. Dissertation, University of Wisconsin, Madison.

Conti, E., A. Fischback, and K. J. Sytsma. 1993. Tribal relationships in Onagraceae: implications from *rbcL* data. *Ann. Missouri Bot. Gard.* 80: 672–685.

Conti, E., A. Litt, and K. J. Sytsma. 1996. Circumscription of Myrtales and their relationship to other rosids: evidence from rbcL sequence data. *Amer. J. Bot.* 83: 221–233.

Conti, E., A. Litt, P. G. Wilson, S. A. Graham, B. G. Briggs, L. A. S. Johnson, and K. J. Sytsma. 1997. Interfamilial relationships in Myrtales: molecular phylogeny and patterns of morphological evolution. *Syst. Bot.* 22: 629–647.

Cook, C. D. K. 1982. Pollination mechanisms in the Hydrocharitaceae. In *Studies on aquatic vascular plants*, J. J. Symoens, S. S. Hooper, and P. Compère (eds.), 1–5. Royal Botanical Society of Belgium, Brussels.

Cook, C. D. K. 1998. Hydrocharitaceae and Pontederiaceae. In *The families and genera of vascular plants.*, vol 4, Monocotyledons: Alismatanae and Commelinanae (except Gramineae), K. Kubitzki (ed.), 234–248, 395–403. Springer-Verlag, Berlin.

Cosner, M. E., R. K. Jansen, and T. G. Lammers. 1994. Phylogenetic relationships in the Campanulales based on *rbcL* sequences. *Plant Syst. Evol.* 190: 79–95.

Cox, A. V., A. M. Pridgeon, V. A. Albert, and M. W. Chase. 1997. Phylogenetics of the slipper orchids (Cypripedioideae, Orchidaceae): nuclear rDNA ITS sequences. *Plant Syst. Evol.* 208: 197–223.

Cox, P. A. and P. B. Tomlinson. 1988. Pollination ecology of a seagrass, *Thalassia testudinum* (Hydrocharitaceae) in St. Croix. *Amer. J. Bot.* 75: 958–965.

Cox, P. A. and C. J. Humphries. 1993. Hydrophilous pollination and breeding system evolution in seagrasses: a phylogenetic approach to the evolutionary ecology of the Cymodoceaceae. *Bot. J. Linn. Soc.* 113: 217–226.

Crane, P. 1989. Early fossil history and evolution of the Betulaceae. In *Evolution, systematics, and fossil history of the Hamamelidae*, vol. 2, "Higher" Hamamelidae. Systematics Association Special Vol. 40B, P. R. Crane and S. Blackmore (eds.), 87–116. Clarendon Press, Oxford.

Crane, P. and S. Blackmore (eds.). 1989. *Evolution, systematics, and fossil history of the Hamamelidae*, vol. 1, Introduction and "lower" Hamamelidae, and vol. 2, "Higher" Hamamelidae. Systematics Association Special Vols. 40A and B. Clarendon Press, Oxford.

Crane, P. F., E. M. Friis, and K. R. Pedersen. 1995. The origin and early diversification of angiosperms. *Nature* 374: 27–33.

Crayn, D. M. and C. J. Quinn. 2000. The evolution of the *atpB-rbcL* intergenic spacer in the epacrids (Ericales) and its systematic and evolutionary implications. *Mol. Phylog. and Evol.* 16: 238–252.

Crayn, D. M., K. A. Kron, P. A. Gadek, and C. J. Quinn. 1998. Phylogenetics and evolution of epacrids: a molecular analysis using the plastid gene *rbcL* with a reappraisal of the position of *Lebetanthus*. *Austr. J. Bot.* 46: 187–200.

Crisci, J., E. A. Zimmer, P. C. Hoch, G. B. Johnson, C. Mudd, and N. S. Pan. 1990. Phylogenetic implications of ribosomal DNA restriction site variation in the plant family Onagraceae. *Ann. Missouri Bot. Gard.* 77: 523–538.

Croat, T. B. 1980. Flowering behavior of the neotropical genus *Anthurium* (Araceae). *Amer. J. Bot.* 67: 888–904.

Cronquist, A. 1944. Studies in the Simaroubaceae. IV. Resume of the American genera. *Brittonia* 5: 128–147.

Cronquist, A. 1955. Phylogeny and taxonomy of the Compositae. *Amer. Midl. Nat.* 53: 478–511.

Cronquist, A. 1968. *The evolution and classification of flowering plants*. Houghton Mifflin, Boston.

Cronquist, A. 1977. The Compositae revisited. *Brittonia* 19: 137–153.

Cronquist, A. 1981. *An integrated system of classification of flowering plants*. Columbia University Press, New York.

Cronquist, A. 1988. *The evolution and classification of flowering plants*, 2nd ed. New York Botanical Garden, Bronx.

Cuénoud, P. 2003. Introduction to expanded Caryophyllales. In *The families and genera of vascular plants*, vol. 5, Malvales, Capparales, and non-betalain Caryophyllales, K. Kubitzki and C. Bayer (eds.), 1–4. Springer-Verlag, Berlin.

Cuénoud, P., V. Savolainen, L. W. Chatrou, M. Powell, R. J. Grayer, and M. W. Chase. 2002. Molecular phylogenetics of Caryophyllales based on nuclear rDNA and plastid *rbcL*, *atpB*, and *matK* sequences. *Amer. J. Bot.* 89: 132–144.

Cullings, K. W. and T. D. Bruns. 1992. Phylogenetic origin of the Monotropoideae inferred from partial 28S ribosomal RNA gene sequences. *Canad. J. Bot.* 70: 1703–1708.

Dahlgren, G. 1989. The last Dalhgrenogram System of classification of the dicotyledons. In *Plant taxonomy, phytogeography, and related subjects*, K. Tan (ed.), 249–250. Edinburgh University Press, London.

Dahlgren, R. M. T. 1983. General aspects of angiosperm evolution and macrosystematics. *Nordic. J. Bot.* 3: 119–149.

Dahlgren, R. M. T. 1988. Rhizophoraceae and Anisophyllaceae: summary statement, relationships. *Ann. Missouri Bot. Gard.* 75: 1259–1277.

Dahlgren, R. M. T. and H. T. Clifford. 1982. *The monocotyledons: a comparative study*. Academic Press, London.

Dahlgren, R. M. T. and F. N. Rasmussen. 1983. Monocotyledon evolution: characters and phylogenetic estimation. *Evol. Biol.* 16: 255–395.

Dahlgren, R. M. T. and R. F. Thorne. 1984. The order Myrtales: circumscription, variation, and relationships. *Ann. Missouri Bot. Gard.* 71: 633–699.

Dahlgren, R. M. T., H. T. Clifford, and P. F. Yeo. 1985. *The families of the monocotyledons*. Springer-Verlag, Berlin.

D'Arcy, W. 1979. The classification of the Solanaceae. In *The biology and taxonomy of the Solanaceae*, J. G. Hawkes, R. N. Lester, and A. D. Skelding (eds.), 3–48. Academic Press, London.

D'Arcy, W. 1991. The Solanaceae since 1976, with a review of its biogeography. In *Solanaceae 3: taxonomy, chemistry, evolution*, J. G. Hawkes, R. N. Lester, M. Nee, and N. Estrada (eds.), 75–137. Royal Botanic Garden, Kew.

Datwyler, S. L. and G. D Weiblen. 2004. On the origin of the fig: phylogenetic relationships of Moraceae from *ndhF* sequences. *Amer. J. Bot.* 91: 767–777.

Davis, C. C. and M. W. Chase. 2004. Elatinaceae are sister to Malpighiaceae, Peridiscaceae belong to Saxifragales. *Amer. J. Bot.* 91: 262–273.

Davis, C. C., W. R. Anderson, and M. J. Donoghue. 2001. Phylogeny of Malpighiaceae: evidence from chloroplast *ndhF* and *trnL-F* nucleotide sequences. *Amer. J. Bot.* 88: 1830–1846.

Davis, C. C., C. O. Webb, K. J. Wurdack, C. A. Jaramillo, and M. J. Donoghue. 2005. Explosive radiation of Malpighiales supports a Mid-Cretaceous origin of modern tropical rain forests. *Amer. Naturalist* 165: E36–E65.

Davis, J. I. 1995. A phylogenetic structure for the Monocotyledons, as inferred from chloroplast DNA restriction site variation, and a comparison of measures of clade support. *Syst. Bot.* 20: 503–527.

Davis, J. I. and 12 others. 2004. A phylogeny of the monocots, as inferred from *rbcL* and *atpA* sequence variation, and a comparison of methods for calculating jackknife and bootstrap values. *Syst. Bot.* 29: 467–510.

Dayanandan, S., P. S. Ashton, S. M. Williams and R. B. Primack. 1999. Phylogeny of the tropical tree family Dipterocarpaceae based on nucleotide sequences of the chloroplast *rbcL* gene. *Amer. J. Bot.* 86: 1182–1190.

de Bruijn, A., V. A. Cox, and M. W. Chase. 1995. Molecular systematics of Asphodelaceae (Asparagales; Lilianae). *Amer. J. Bot.* 82 (6) Suppl.: 124.

de Buhr, L. E. 1975. Phylogenetic relationships of the Sarraceniaceae. *Taxon* 24: 297–306.

de Lange, A. and F. Bouman. 1992. Seed micromorphology of the genus *Begonia* in Africa: taxonomic and ecological implications. In *Studies in Begoniaceae III*, J. J. F. E. de Wilde (ed.), 1–82. Wageningen Agric. University Pap. Wageningen, The Nether-lands.

de Pamphilis, C. W. and N. D. Young. 1995. Evolution of parasitic Scrophulariaceae/ Orobanchaceae: evidence from sequences of chloroplast ribosomal protein gene *rps2* and a comparision with traditional classification schemes. *Amer. J. Bot.* 82(6) Suppl.: 126.

de Pamphilis, C. W., N. D. Young, and A. D. Wolfe. 1997. Evolution of plastid gene *rps2* in a lineage of hemiparasitic and holoparasitic plants: many losses of photosynthesis and complex patterns of rate variation. *Proc. Natl. Acad. Sci. USA* 94: 7362–7372.

de Wilde, W. J. J. O. 1971. The systematic position of the tribe Paropsieae, in particular the genus *Ancistrothyrsus*, and a key to the genera of Passifloraceae. *Blumea* 19: 99–104.

de Wilde, W. J. J. O. 1974. The genera of the tribe Passifloreae (Passifloraceae) with special reference to flower morphology. *Blumea* 22: 37–50.

den Hartog, C. 1975. Thoughts about the taxonomical relationships within the Lemnaceae. *Aquatic Bot.* 1: 407–416.

Denton, M. E. 1973. A monograph of *Oxalis*, section *Ionoxalis* (Oxalidaceae) in North America. *Publ. Mus. Michigan State University, Biol. Ser.* 4: 455–615.

Dickison, W. C. 1981. The evolutionary relationships of the Leguminosae. In *Advances in legume systematics*, part 1, R. M. Polhill and P. H. Raven (eds.), 35–54. Royal Botanic Gardens, Kew.

Dietrich, W., W. L. Wagner and P. H. Raven. 1997. Systematics of *Oenothera* section *Oenothera* subsection *Oenothera* (Onagraceae). *Syst. Bot. Monogr.* 50: 1–234.

Dilcher, D. L. 1989. The occurrence of fruits with affinities to Ceratophyllaceae in lower and mid-Cretaceous sediments. *Amer. J. Bot.* 76: 162.

Dobbins, D. R. 1971. Studies on the anomalous cambial activity in *Doxantha unguis-cati* (Bignoniaceae). II. A case of differential production of secondary tissues. *Amer. J. Bot.* 58: 697–705.

Donoghue, M. J. 1980. Flowering times in *Viburnum*. *Arnoldia* 40: 2–22.

Donoghue, M. J. 1981. Growth patterns in woody plants with examples from the genus *Viburnum*. *Arnoldia* 41: 2–23.

Donoghue, M. J. 1983a. A preliminary analysis of phylogenetic relationships in *Viburnum* (Caprifoliaceae s. l.). *Syst. Bot.* 8: 45–58.

Donoghue, M. J. 1983b. The phylogenetic relationships of *Viburnum*. In *Advances in cladistics*, vol. 2, N. I. Platnick and V. A. Funk (eds.), 143–166. Columbia University Press, New York.

Donoghue, M. J. 1985. Pollen diversity and exine evolution in *Viburnum* and the Caprifoliaceae sensu lato. *J. Arnold Arbor.* 66: 421–469.

Donoghue, M. J. and J. A. Doyle. 1989. Phylogenetic analysis of angiosperms and the relationships of Hamamelidae. In *Evolution, systematics, and fossil history of the Hamamelidae*, vol. 1, Introduction and "lower" Hamamelidae. Systematics Association Special Vol. 40A, P. R. Crane and S. Blackmore (eds.), 17–45. Clarendon Press, Oxford.

Donoghue, M. J., R. G. Olmstead, J. F. Smith and J. D. Palmer. 1992. Phylogenetic relationships of Dipsacales based on *rbcL* sequences. *Ann. Missouri Bot. Gard.* 79: 333–345.

Donoghue, M. J., T. Eriksson, P. A. Reeves, and R. G. Olmstead. 2001. Phylogeny and phylogenetic taxonomy of Dipsacales, with special reference to *Sinadoxa* and *Tetradoxa* (Adoxaceae). *Harvard Pap. Bot.* 6: 459–479.

Donoghue, M. J., C. D. Bell, and R. C. Winkworth. 2003. The evolution of reproductive characters in Dipsacales. *Int. J. Plant Sci.* 164 (Suppl.): S453–S464.

Donoghue, M. J., B. G. Baldwin, J. Li, and R. C. Winkworth. 2004. *Viburnum* phylogeny based on chloroplast *trnK* intron and nuclear ribosomal ITS DNA sequences. *Syst. Bot.* 29: 188–198.

Downie, S. R. and J. D. Palmer. 1992. Restriction site mapping of the chloroplast DNA inverted repeat: a molecular phylogeny of the Asteridae. *Ann. Missouri Bot. Gard.* 79: 238–266.

Downie, S. R. and J. D. Palmer. 1994a. A chloroplast DNA phylogeny of the Caryophyllales based on structural and inverted repeat restriction site variation. *Syst. Bot.* 19: 236–252.

Downie, S. R. and J. D. Palmer. 1994b. Phylogenetic relationships using restriction site variation of the chloroplast DNA inverted repeat. In *Caryophyllales: evolution and systematics*, H.-D. Behnke and T. J. Mabry (eds.), 223–233. Springer-Verlag, Berlin.

Downie, S. R., D. S. Katz-Downie and K.-J. Cho. 1997. Relationships in the Caryophyllales as suggested by phylogenetic analysis of partial chloroplast DNA ORF2280 homolog sequences. *Amer. J. Bot.* 84: 253–273.

Downie, S. R., S. Ramanath, D. S. Katz-Downie and E. Llanas. 1998. Molecular systematics of Apiaceae subfamily Apioideae: phylogenetic analysis of nuclear ribosomal DNA internal transcribed spacer and plastid *RPOC1* intron sequences. *Amer. J. Bot.* 85: 563–591.

Downie, S. R., D. S. Katz-Downie, and M. F. Watson. 2000. A phylogeny of the flowering plant family Apiaceae based on chloroplast DNA *rpl16* and *rpoC1* intron sequences: towards a suprageneric classification of subfamily Apioideae. *Amer. J. Bot.* 87: 273–292.

Doyle, J. A. 1996. Seed plant phylogeny and the relationships of the Gnetales. *Int. J. Plant Sci.* 157(6) (Suppl.): S3–S39.

Doyle, J. A. 1998. Phylogeny of vascular plants. *Annual Rev. Ecol. Syst.* 29: 567–599.

Doyle, J. A. and P. K. Endress. 2000. Morphological phylogenetic analyses of basal angiosperms: comparison and combination with molecular data. *Int. J. Plant Sci.* 161 (Suppl.): S121–S153.

Doyle, J. A. and A. LeThomas. 1994. Cladistic analysis and pollen evolution in Annonaceae. *Acta Bot. Gall.* 141: 149–170

Doyle, J. A. and A. Le Thomas. 1996. Phylogenetic analysis and character evolution in Annonaceae. *Bull. Mus. Natl. Hist. Nat., Paris, 4e Sér., 18, 1996, sect. B, Adansonia, No. 3–4: 279–334.

Doyle, J. A., C. L. Hotton and J. V. Ward. 1990. Early Cretaceous tetrads, zonasulculate pollen, and Winteraceae. II. Cladistic analysis and implications. *Amer. J. Bot.* 77: 1558–1568.

Doyle, J. A., M. J. Donoghue and E. A. Zimmer. 1994. Integration of morphological and ribosomal RNA data on the origin of angiosperms. *Ann. Missouri Bot. Gard.* 81: 419–450.

Doyle, J. A., H. Sauquet, T. Scharaschkin, and A. LeThomas. 2004. Phylogeny, molecular and fossil dating, and biogeographic history of Annonaceae and Myristicaceae (Magnoliales). *Int. J. Plant Sci.* 165(4) (Suppl): S55–S67.

Doyle, J. J. 1987. Variation at the DNA level: Uses and potential in legume systematics. In *Advances in legume systematics*, part 3, C. H. Stirton (ed.), 1–30. Royal Botanic Gardens, Kew.

Doyle, J. J. 1994. Phylogeny of the legume family: an approach to understanding the origins of nodulation. *Annu. Rev. Ecol. Syst.* 25: 325–349.

Doyle, J. J. and M. A. Luckow. 2003. The rest of the iceberg. Legume diversity and evolution in a phylogenetic context. *Plant Phys.* 131: 900–910.

Doyle, J. J., J. L. Doyle, J. A. Ballenger, E. E. Dickson, T. Kajita and H. Ohashi. 1997. A phylogeny of the chloroplast gene *rbcL* in the Leguminosae: taxonomic correlations and insights into evolution of nodulation. *Amer. J. Bot.* 84: 541–554.

Drábková, L., J. Kirschner, O. Seberg, G. Petersen, and C. Vlacek. 2003. Phylogeny of the Juncaceae based on *rbcL* sequences, with special emphasis on *Luzula* DC. and *Juncus* L. *Plant Syst. Evol.* 240: 133–147.

Dransfield, J. 1986. A guide to collecting palms. *Ann. Missouri Bot. Gard.* 73: 166–176.

Dransfield, J. and N. W. Uhl. 1998. Palmae. In *The families and genera of vascular plants.* vol. 4, Monocotyledons: Alismatanae and Commelinanae (except Gramineae), K. Kubitzki (ed.), 306–389. Springer-Verlag, Berlin.

Dransfield, J., N. Uhl, C. B. Asmussen, W. J. Baker, M. M. Harley, and C. E. Lewis. 2005. A new phylogenetic classification of the palm family, Arecaceae. *Kew Bull.* 60: 559–569.

Dressler, R. L. 1981. *The orchids: Natural history and classification.* Harvard University Press, Cambridge, MA.

Dressler, R. L. 1986. Recent advances in orchid phylogeny. *Lindleyana.* 1: 5–20.

Dressler, R. L. 1993. *Phylogeny and classification of the orchid family.* Dioscorides Press, Portland, OR.

Dressler, R. L. and M. W. Chase. 1995. Whence the orchids? In *Monocotyledons: systematics and evolution*, P. J. Rudall, P. J. Cribb, D. F. Cutler, and C. J. Humphries (eds.), 217–226. Royal Botanic Gardens, Kew.

Drinnan, A. N., P. R. Crane, and S. B. Hoot. 1994. Patterns of floral evolution in the early diversification of non-magnoliid dicotyledons (eudicots). *Plant Syst. Evol.* Suppl. 8: 93–122.

Duangjai, B. Wallnoefer, R. Samuel, J. Munzinger, and M. W. Chase. 2006. Phylogenetic relationships and infrafamilial classification of Ebenaceae s.l. based on six plastid markers. Abstract. *Botany 2006*: 218–219.

Duvall, M. R., M. T. Clegg, M. W. Chase, W. D. Clark, W. J. Kress, L. E. Eguiarte, J. F. Smith, B. S. Gaut, E. A. Zimmer, and G. H. Learns, Jr. 1993. Phylogenetic hypotheses for the monocotyledons constructed from *rbcL* sequence data. *Ann. Missouri Bot. Gard.* 80: 607–619.

Eckardt, T. 1976. Classical morphological features of Centrospermous families. *Plant Syst. Evol.* 126: 5–25.

Eckenwalder, J. E. and S. C. H. Barrett. 1986. Phylogenetic systematics of Pontederiaceae. *Syst. Bot.* 11: 373–391.

Eddie, W. M. M., T. Shulkina, J. Gaskin, R. C. Haberle, and R. K. Jansen. 2003. Phylogeny of Campanulaceae s. str. Inferred from ITS sequences of nuclear ribosomal DNA. *Ann. Missouri Bot. Gard.* 90: 554–575.

Edwards, E. J., R. Nyffeler, and M. J. Donoghue. 2005. Basal cactus phylogeny: implications of *Pereskia* (Cactaceae) paraphyly for the transition to the cactus life form. *Amer. J. Bot.* 92: 1177–1188.

Edwards, E. J. and M. J. Donoghue. 2006. *Pereskia* and the origin of the cactus life form. *Amer. Naturalist* 167: 777–793.

Ehrendorfer, F. 1976. Closing remarks: systematics and evolution of Centrospermous families. *Plant Syst. Evol.* 126: 99–105.

Eldenäs, P. A. and H. P. Linder. 2000. Congruence and complementarity of morphological and *trnL-trnF* sequence data and the phylogeny of the African Restionaceae. *Syst. Bot.* 25: 692–707.

Elias, T. S. 1974. The genera of Mimosoideae (Leguminosae) in the southeastern United States. *J. Arnold Arbor.* 55: 67–118.

Elias, T. S. 1970. The genera of Ulmaceae in the southeastern United States. *J. Arnold Arbor.* 51: 18–40.

Elias, T. S. 1971a. The genera of Fagaceae in the southeastern United States. *J. Arnold Arbor.* 52: 159–195.

Elias, T. S. 1971b. The genera of Myricaceae in the southeastern United States. *J. Arnold Arbor.* 52: 305–318.

Elias, T. S. 1972. The genera of Juglandaceae in the southeastern United States. *J. Arnold Arbor.* 53: 26–51.

Endress, M. E. and P. V. Bruyns. 2000. A revised classification of the Apocynaceae s. l. *Bot. Rev.* 66: 1–56.

Endress, M. E. and W. D. Stevens. 2001. The renaissance of the Apocynaceae s.l.: recent advances in systematics, phylogeny, and evolution: introduction. *Ann. Missouri Bot. Gard.* 88: 517–522.

Endress, M. E., B. Sennblad, S. Nilsson, L. Civeyrel, M. W. Chase, S. Huysmans, E. Grafsröm, and B. Bremer. 1996. A phylogenetic analysis of Apocynaceae s. s. and some related taxa in Gentianales: a multidisciplinary approach. *Opera Bot. Belg.* 7: 59–102.

Endress, P. K. 1977. Evolutionary trends in the Hamamelidales-Fagales group. *Plant Syst. Evol.*, Suppl. 1: 321–347.

Endress, P. K. 1986a. Floral structure, systematics, and phylogeny in Trochodendrales. *Ann. Missouri Bot. Gard.* 73: 297–324.

Endress, P. K. 1986b. Reproductive structures and phylogenetic significance of extant primitive angiosperms. *Plant Syst. Evol.* 152: 1–28.

Endress, P. K. 1989a. Phylogenetic relationships in the Hamamelidoideae. In *Evolution, systematics, and fossil history of the Hamamelidae*, vol. 1, Introduction and "lower" Hamamelidae. Systematics Association Special Vol. 40A, P. R. Crane and S. Blackmore (eds.), 227–248. Clarendon Press, Oxford.

Endress, P. K. 1989b. A suprageneric taxonomic classification of the Hamamelidaceae. *Taxon* 38: 371–376.

Endress, P. K. 1993. Hamamelidaceae. In *The families and genera of vascular plants*, vol. 2, Magnoliid, hamamelid and caryophyllid families, K. Kubitzki, J. G. Rohwer and V. Bittrich (eds.), 322–331. Springer-Verlag, Berlin.

Endress, P. K. 1994a. Evolutionary aspects of the floral structure in *Ceratophyllum. Plant Syst. Evol.* 8: 175–183.

Endress, P. K. 1994b. Shapes, sizes and evolutionary trends in stamens of Magnoliidae. *Bot. Jahrb. Syst.* 115: 429–460.

Endress, P. K. 1994c. *Diversity and evolutionary biology of tropical flowers.* Cambridge University Press, Cambridge.

Endress, P. K. 1995. Floral structure and evolution in Ranunculanae. *Plant Syst. Evol.* Suppl. 9: 47–61.

Endress, P. K. 2001. The flowers in extant basal angiosperms and inferences on ancestral flowers. *Int. J. Plant Sci.* 162: 1111–1140.

Endress, P. K. 2004a. L. A. S. Johnson Review No. 3. Structure and relationships of basal relictual angiosperms. *Austr. Syst. Bot.* 17: 343–366.

Endress, P. K. 2004b. Biologie und Evolution der Blüten basaler Blütenpflanzen. *Leopoldina* 49: 467–486.

Endress, P. K. and A. Igersheim. 1997. Gynoecium diversity and systematics of the Laurales. *Bot. J. Linn. Soc.* 125: 93–168.

Endress, P. K. and A. Igersheim. 1999. Gynoecium diversity and systematics of the basal eudicots. *Bot. J. Linn. Soc.* 130: 305–393.

Endress, P. K. and A. Igersheim. 2000a. Gynoecium structure and evolution in basal angiosperms. *Int. J. Plant Sci.* 161 (Suppl.): S211–S223.

Endress, P. K. and A. Igersheim. 2000b. The reproductive structures of the basal angiosperm *Amborella trichopoda* (Amborellaceae). *Int. J. Plant Sci.* 161 (Suppl.): S237–S248.

Endress, P. K., M. Jenny, and M. E. Fallen. 1983. Convergent elaboration of apocarpous gynoecia in higher advanced dicotyledons (Sapindales, Malvales, Gentianales). *Nordic J. Bot.* 3: 293–300.

Engler, A. 1930. Saxifragaceae. In *Die natürlichen Pflanzenfamilien*, 2nd ed., vol. 18a, A. Engler and K. Prantl (eds.), 74–226. Engelmann, Leipzig.

Erbar, C. 1991. Sympetalae—a systematic character? *Bot. Jahrb. Syst.* 112: 417–451.

Erbar, C. and P. Leins. 1988. Blutenentwicklungs-geschichtliche Studies an *Aralia* und *Hedera* (Araliaceae). *Flora* 180: 391–406.

Eriksen, B. 1993a. Floral anatomy and morphology in the Polygalaceae. *Plant Syst. Evol.* 186: 17–32.

Eriksen, B. 1993b. Phylogeny of the Polygalaceae and its taxonomic implications. *Plant Syst. Evol.* 186: 33–55.

Eriksson, T. and M. J. Donoghue. 1997. Phylogenetic relationships of *Sambucus* and *Adoxa* (Adoxoideae, Adoxaceae) based on nuclear ribosomal ITS sequences and preliminary morphological data. *Syst. Bot.* 22: 555–573.

Eriksson, T., M. S. Hibbs, A. D. Yoder, C. F. Delwiche, and M. J. Donoghue. 2003. The phylogeny of Rosoideae (Rosaceae) based on sequences of the internal transcribed spacer (ITS) of nuclear ribosomal DNA and the *trnL/F* region of chloroplast DNA. *Int. J. Plant Sci.* 164: 197–211.

Ernst, W. R. 1962. The genera of Papaveraceae and Fumariaceae in the southeastern United States. *J. Arnold Arbor.* 43: 315–343.

Ernst, W. R. 1963a. The genera of Capparaceae and Moringaceae in the southeastern United States. *J. Arnold Arbor.* 44: 81–95.

Ernst, W. R. 1963b. The genera of Hamamelidaceae and Platanaceae in the southeastern United States. *J. Arnold Arbor.* 44: 193–210.

Ernst, W. R. 1964. The genera of Berberidaceae, Lardizabalaceae, and Menispermaceae in the southeastern United States. *J. Arnold Arbor.* 45: 1–35.

Ernst, W. R. and H. J. Thompson. 1963. The Loasaceae in the southeastern United States. *J. Arnold Arbor.* 44: 138–142.

Evans, R. C. 1999. Molecular, morphological, and ontogenetic evaluation of relationships and evolution in the Rosaceae. Ph.D. dissertation, University of Toronto, Canada.

Evans, R. C. and T. A. Dickinson. 1999a. Floral ontogeny and morphology in subfamily Amygdaloideae. *Int. J. Plant Sci.* 160: 955–979.

Evans, R. C. and T. A. Dickinson. 1999b. Floral ontogeny and morphology in subfamily Spiraeoideae. *Int. J. Plant Sci.* 160: 981–1012.

Evans, R. C., L. A. Alice, C. S. Campbell, T. A. Dickinson, and E. A. Kellogg. 2000. The granule-bound starch synthase (GBSSI) gene in the Rosaceae: multiple loci and phylogenetic utility. *Mol. Phyl. Evol.* 17: 388–400.

Evans, T. M., R. B. Faden, M. G. Simpson, and K. J. Sytsma. 2000a. Phylogenetic relationships in the Commelinaceae. 1. A cladistic analysis of morphological data. *Syst. Bot.* 25: 668–691.

Evans, T. M., R. B. Faden, and K. J. Sytsma. 2000b. Homoplasy in the Commelinaceae: a comparison of different classes of morphological characters. In *Monocots: systematics and evolution*, K. L. Wilson and D. A. Morrison (eds.), 557–566. CSIRO, Collingwood, Australia.

Evans, T. M., K. J. Sytsma, R. B. Faden, and T. J. Givnish. 2003. Phylogenetic relationships in the Commelinaceae. II. A cladistic analysis of *rbcL* sequences and morphology. *Syst. Bot.* 28: 270–292.

Evans, W. C. 1979. Tropane alkaloids of the Solanaceae. In *The biology and taxonomy of the Solanaceae*, J. G. Hawkes, R. N. Lester and A. D. Skelding (eds.), 241–254. Academic Press, London.

Eyde, R. H. 1966. The Nyssaceae in the southeastern United States. *J. Arnold Arbor.* 47: 117–125.

Eyde, R. H. 1988. Comprehending *Cornus*: puzzles and progress in the systematics of the dogwoods. *Bot. Rev.* 54: 233–351.

Eyde, R. H. and Q. Xiang. 1990. Fossil mastixioid (Cornaceae) alive in eastern Asia. *Amer. J. Bot.* 77: 689–692.

Faden, R. B. 1998. Commelinaceae. In *The families and genera of vascular plants*, vol. 4, Monocotyledons: Alismatanae and Commelinanae (except Gramineae), K. Kubitzki (ed.), 109–128. Springer-Verlag, Berlin.

Faden, R. B. and D. R. Hunt. 1991. The classification of the Commelinaceae. *Taxon* 40: 19–31.

Faegri, K. and L. van der Pijl. 1980. *The principles of pollination ecology*. Pergamon Press, Oxford.

Fagerberg, W. R. and D. Allain. 1991. A quantitative study of tissue dynamics during closure in the traps of venus's flytrap *Dionaea muscipula* Ellis. *Amer. J. Bot.* 78: 647–657.

Fallen, M. E. 1986. Floral structure in the Apocynaceae: morphological, functional, and evolutionary aspects. *Bot. Jahrb. Syst.* 106: 245–286.

Fan, C.-Z. and Q.-Y. Xiang. 2001. Phylogenetic relationships within *Cornus* (Cornaceae) based on 26S rDNA sequences. *Amer. J. Bot.* 88: 1131–1138.

Fan, C.-Z. and Q.-Y. Xiang. 2003. Phylogenetic analyses of Cornales based on 26S rRNA and combined 26 rDNA-*matK-rbcL* sequence data. *Amer. J. Bot.* 90: 1357–1372.

Farmer, S. B. amd E. E. Schilling. 2002. Phylogenetic analyses of Trilliaceae based on morphological and molecular data. *Syst. Bot.* 27: 674–692.

Fay, M. F. and M. W. Chase. 1996. Resurrection of Themidaceae for the *Brodiaea* alliance, and recircumscription of Alliaceae, Amaryllidaceae, and Agapanthoideae. *Taxon* 45: 441–451

Fay, M. F., C. Bayer, W. S. Alverson, A. Y. De Bruijn, and M. W. Chase. 1998. Plastid *rbcL* sequence data indicate a close affinity between *Diegodendron* and *Bixa*. *Taxon* 47: 43–50.

Fay, M. F. and 11 others. 2000. Phylogenetic studies of Asparagales based on four plastid DNA regions. In *Monocots: systematics and evolution*, K. L. Wilson and D. A. Morrison (eds.), 360–371. CSIRO, Collingwood, Australia.

Fay, M. F., M. W. Chase, N. Rønsted, D. S. Devey, Y. Pillon, J. C. Pires, G. Petersen, O. Seberg, and J. I. Davis. 2006. Phylogenetics of Liliales: summarized evidence from combined analyses of five plastid and one mitochondrial loci. *Aliso* 22: 559–565.

Feild, T. S., M. A. Zwieniecki, and N. M. Holbrook. 2000. Winteraceae evolution: an ecophysiological perspective. *Ann. Missouri Bot. Gard.* 87: 323–334.

Feild, T. S., T. Brodribb, T. Jaffré, and N. M. Holbrook. 2001. Acclimation of leaf anatomy, photosynthetic light use, and xylem hydraulics to light in *Amborella trichopoda* (Amborellaceae*). Int. J. Plant Sci.* 162: 999–1008.

Feng, Y., S.-H. Oh, and P. S. Manos. 2005. Phylogeny and historical biogeography of the genus *Platanus* as infered from nuclear and chloroplast DNA. *Syst. Bot.* 30: 786–799.

Ferguson, D. M. 1999. Phylogenetic analysis and relationships in Hydrophyllaceae based on *ndhF* sequence data. *Syst. Bot.* 23: 253–268.

Ferguson, I. K. 1966a. The Cornaceae in the southeastern United States. *J. Arnold Arbor.* 47: 106–116.

Ferguson, I. K. 1966b. The genera of Caprifoliaceae in the southeastern United States. *J. Arnold Arbor.* 47: 33–59.

Fernando, E. S. and C. J. Quinn. 1995. Picramniaceae, a new family, and a recircumscription of Simaroubaceae. *Taxon* 44: 177–181.

Fernando, E. S., P. A. Gadek, and C. J. Quinn. 1995. Simaroubaceae, and artificial construct: evidence from *rbcL* sequence variation. *Amer. J. Bot.* 82: 92–103.

Fernando, E. S., P. A. Gadek, D. M. Crayn, and C. J. Quinn. 1993. Rosid affinities of Surianaceae: molecular evidence. *Mol. Phylog. Evol.* 2: 344–350.

Fey, B. S. and P. K. Endress. 1983. Development and morphological interpretation of the cupule in Fagaceae. *Flora, Morphol. Geobot. Oekophysiol.* 173: 451–468.

Fineran, B. A. 1985. Glandular trichomes in *Utricularia*: a review of their structure and function. *Israel J. Bot.* 34: 295–330.

Fior, S., P. O. Karis, G. Casazza, L. Minuto, and F. Sala. 2006. Molecular phylogeny of the Caryophyllaceae (Caryophyllales) inferred from chloroplast *matK* and nuclear rDNA ITS sequences. *Amer. J. Bot.* 93: 399–411.

Fischer, E., W. Barthlott, R. Seine, and I. Theisen. 2004a. Lentibulariaceae. In *The families and genera of vascular plants*, vol. 7, Lamiales (except Acanthaceae, including Avicen-

niaceae), K. Kubitzki (ed.), 276–282. Springer-Verlag, Berlin.

Fischer, E., I. Theisen, and L. G. Lohmann. 2004b. Bignoniaceae. In *The families and genera of vascular plants*, vol. 7, Lamiales (except Acanthaceae, including Avicenniaceae), K. Kubitzki (ed.), 9–38. Springer-Verlag, Berlin.

Fishbein, M. 2001. Evolutionary innovation and diversification in the flowers of Asclepiadaceae. *Ann. Missouri Bot. Gard.* 88: 603–623.

Fishbein, M., C. Hibsch-Jetter, D. E. Soltis, and L. Hufford. 2001. Phylogeny of Saxifragales (Angiosperms, Eudicots): Analysis of a rapid, ancient radiation. *Syst. Biol.* 50: 817–847.

Fisher, M. J. 1928. The morphology and anatomy of flowers of Salicaceae, I and II. *Amer. J. Bot.* 15: 307–326, 372–394.

Forest, F., V. Savolainen, M. W. Chase, R. Lupia, A. Bruneau, and P. R. Crane. 2005. Teasing apart molecular- versus fossil-based error estimates when dating phylogenetic trees: a case study in the birch family (Betulaceae). *Syst. Bot.* 30: 118–133.

Forrest, L. L., M. Hughes, and P. M. Hollingsworth. 2005. A phylogeny of *Begonia* using nuclear ribosomal sequence data and morphological characters. *Syst. Bot.* 30: 671–682.

French, J. C., M. G. Chung, and Y. K. Hur. 1995. Chloroplast DNA phylogeny of the Ariflorae. In *Monocotyledons: systematics and evolution*, P. J. Rudall, P. J. Cribb, D. F. Cutler, and C. J. Humphries (eds.), 255–275. Royal Botanic Gardens, Kew.

Freudenstein, J. V. and F. N. Rasmussen. 1999. What does morphology tell us about orchid relationships? A cladistic analysis. *Amer. J. Bot.* 86: 225–248.

Freudenstein, J. V., D. M. Senyo, and M. W. Chase. 2000. Mitochondrial DNA and relationships in the Orchidaceae. In *Monocots: systematics and evolution*, K. L. Wilson and D. A. Morrison (eds.), 421–429. CSIRO, Collingwood, Australia.

Freudenstein, J. V., C. van der Berg, D. H. Goldman, P. J. Kores, M. Molvray, and M. W. Chase. 2004. An expanded plastid DNA phylogeny of Orchidaceae and analysis of Jackknife branch support strategy. *Amer. J. Bot.* 91: 149–157.

Friedman, W. E. 2006. Embryological evidence for developmental lability during early angiosperm evolution. *Nature* 441: 337–340.

Friesen, N., R. M. Fritsch, and F. R. Blattner. 2006. Phylogeny and new intrageneric classification of *Allium* (Alliaceae) based on nuclear ribosomal DNA ITS sequences. *Aliso* 22: 372–395.

Friis, I. 1989. The Urticaceae: a systematic review. In *Evolution, systematics, and fossil history of the Hamamelidae*, vol. 2, "Higher" Hamamelidae. Systematics Association Special Vol. 40B, P. R. Crane and S. Blackmore (eds.), 285–308. Clarendon Press, Oxford.

Friis, I. 1993. Urticaceae. In *The families and genera of vascular plants*, vol. 2, Magnoliid, hamamelid and caryophyllid families, K. Kubitzki, J. G. Rohwer, and V. Bittrich (eds.), 612–630. Springer-Verlag, Berlin.

Fryxell, P. A. 1988. Malvaceae of Mexico. *Syst. Bot. Monogr.* 25: 1–522.

Fu, C.-X., H.-G. Kong, Y.-X. Qiu, and K. M. Cameron. 2005. Molecular phylogeny of the East Asian-North American disjunct *Smilax* sect. *Nemexia* (Smilacaceae). *Int. J. Plant Sci.* 166: 301–309.

Funk, V. A. and 11 others. 2005. Everywhere but Antarctica. Using a supertree to understand the diversity and distribution of the Compositae. *Biol. Skr.* 55: 343–374.

Furlow, J. J. 1990. The genera of Betulaceae in the southeastern United States. *J. Arnold Arbor.* 71: 1–67.

Gadek, P. A., E. S. Fernando, C. J. Quinn, S. B. Hoot, T. Terrazas, M. C. Sheahan, and M. W. Chase. 1996. Sapindales: molecular delimitation and infraordinal groups. *Amer. J. Bot.* 83: 802–811.

Gamage, D. T., M. P. de Silva, N. Inomata, T. Yamazaki, and A. E. Szmidt. 2006. Comprehensive molecular phylogeny of the subfamily Dipterocarpoideae (Dipterocarpaceae) based on chloroplast DNA sequences. *Genes Genet. Syst.* 81: 1–12.

Ganders, F. R. 1979. The biology of heterostyly. *New Zealand J. Bot.* 17: 607–635.

Gates, B. N. 1943. Carunculate seed dissemination by ants. *Rhodora* 45: 438–445.

Gentry, A. H. 1974. Coevolutionary patterns in Central American Bignoniaceae. *Ann. Missouri Bot. Gard.* 61: 728–759.

Gentry, A. H. 1980. *Bignoniaceae.* Part 1 (Crescentieae and Tourrettieae). *Flora Neotrop. Monogr.* 25: 1–130.

Gentry, A. H. 1990. Evolutionary patterns in neotropical Bignoniaceae. *Mem. New York Bot. Gard.* 55: 118–129.

George, A. S. 1998. *Proteus* in Australia: an overview of the correct state of taxonomy of the Australian Proteaceae. *Aust. Syst. Bot.* 11: 257–266.

Gerrath, J. M., V. Posluszny, and N. G. Dengler. 2001. Primary vascular patterns in the Vitaceae. *Int. J. Plant Sci.* 162: 729–745.

Gillis, W. T. 1971. The systematics and ecology of poison-ivy and the poison-oaks (*Toxicodendron*, Anacardiaceae). *Rhodora* 73: 72–159, 161–237, 370–443, 465–540.

Gilmartin, A. J. and G. K. Brown. 1987. Bromeliales, related monocots and resolution of relationships among Bromeliaceae subfamilies. *Syst. Bot.* 12: 493–500.

Giulietti, A. M. and 14 others. 2000. Multidisciplinary studies on neotropical Eriocaulaceae. In *Monocots: systematics and evolution*, K. L. Wilson and D. A. Morrison (eds.), 580–489. CSIRO, Collingwood, Australia.

Givnish, T. J., K. C. Millam, T. M. Evans, J. C. Hall, J. C. Pires, P. E. Berry, and K. J. Sytsma. 1994. Ancient vicariance or recent long-distance dispersal? Inferences about phylogeny and South American-African disjunctions in Rapateaceae and Bromeliaceae based on *ndhF* sequence data. *Int. J. Plant Sci.* 165 (4, Suppl.): S35–S54.

Givnish, T. J. and 16 others. 2005. Repeated evolution of net venation and fleshy fruits among monocots in shaded habitats confirms a priori predictions: evidence from an *ndhF* phylogeny. *Proc. Roy. Soc. London B* 272: 1481–1490.

Goetchebeur, P. 1998. Cyperaceae. In *The families and genera of vascular plants*, vol. 4, Monocotyledons: Alismatanae and Commelinanae (except Gramineae), K. Kubitzki (ed.), 141–190. Springer-Verlag, Berlin.

Goldblatt, P. 1990. Phylogeny and classification of Iridaceae. *Ann. Missouri Bot. Gard.* 77: 607–627.

Goldblatt, P. 1995. The status of R. Dahlgren's orders Liliales and Melanthiales. In *Monocotyledons: systematics and evolution*, P. J. Rudall, P. J. Cribb, D. F. Cutler and C. J. Humphries (eds.), 181–200. Royal Botanic Gardens, Kew.

Goldblatt, P., J. C. Manning, and P. Rudall. 1998. Iridaceae. In *The families and genera of vascular plants*, vol. 3, Monocotyledons: Lilianae (except Orchidaceae), K. Kubitzki (ed.), 295–333. Springer-Verlag, Berlin.

González, F. and P. Rudall. 2001. The questionable affinities of *Lactoris*: evidence from branching pattern, inflorescence morphology, and stipule development. *Amer. J. Bot.* 88: 2143–2150.

Gottlieb, A. M., G. C. Giberti, and L. Poggio. 2005. Molecular analyses of the genus *Ilex* (Aquifoliaceae) in southern South America, evidence from AFLP and ITS sequence data. *Amer. J. Bot.* 92: 352–369.

Gottsberger, G. 1977. Some aspects of beetle pollination in the evolution of flowering plants. *Plant Syst. Evol.*, 1: S211–S226.

Gottsberger, G. 1988. The reproductive biology of primitive angiosperms. *Taxon* 37: 630–643.

Gottsberger, G., I. Silberbauer-Gottsberger, and F. Ehrendorfer. 1980. Reproductive biology in the primitive relic angiosperm *Drimys brasiliensis* (Winteraceae). *Plant Syst. Evol.* 135: 11–39.

Graham, S. A. 1964a. The genera of Lythraceae in the southeastern United States. *J. Arnold Arbor.* 45: 235–250.

Graham, S. A. 1964b. The genera of Rhizophoraceae and Combretaceae in the southeastern United States. *J. Arnold Arbor.* 45: 285–301.

Graham, S. A. 1966. The genera of Araliaceae in the southeastern United States. *J. Arnold Arbor.* 47: 126–136.

Graham, S. A. and C. E. Wood, Jr. 1965. The genera of Polygonaceae in the southeastern United States. *J. Arnold Arbor.* 46: 91–121.

Graham, S. A., E. Conti, and K. Sytsma. 1993a. Phylogenetic analysis of the Lythraceae based on *rbcL* sequence divergence. *Amer. J. Bot.* Suppl. 80(6) Suppl.: 150.

Graham, S. A., J. V. Crisci, and P. C. Hoch. 1993b. Cladistic analysis of the Lythraceae sensu lato based on morphological characters. *Bot. J. Linn. Soc.* 113: 1–33.

Graham, S. A., J. Hall, K. Sytsma, and S.-H. Shi. 2005. Phylogenetic analysis of the Lythraceae based on four gene regions and morphology. *Int. J. Plant Sci.* 166: 995–1017.

Graham, S. W. and S. C. H. Barrett. 1995. Phylogenetic systematics of Pontederiales: implications for breeding-system evolution. In *Monocotyledons: systematics and evolution*, P. J. Rudall, P. J. Cribb, D. F. Cutler, and C. J. Humphries (eds.), 415–441. Royal Botanic Gardens, Kew.

Graham, S. W. and R. G. Olmstead. 2000. Utility of 17 chloroplast genes for inferring the phylogeny of the basal angiosperms. *Amer. J. Bot.* 87: 1712–1730.

Graham, S. W., J. R. Kohn, B. R. Morton, J. E. Eckenwalder, and S. C. H. Barrett. 1998. Phylogenetic congruence and discordance among one morphological and three molecular data sets from Pontederiaceae. *Syst. Biol.* 47: 545–567.

Graham, S. W. and 13 others. 2006. Robust inference of monocot deep phylogeny using an expanded multigene plastid data set. *Aliso* 22: 3–21.

Grant, V. 1959. Natural history of the *Phlox* family. In *Systematic botany*, vol. 1. Martinus Nijhoff, The Hague.

Grant, V. and K. A. Grant. 1965. *Flower pollination in the phlox family.* Columbia University Press, New York.

Grass Phylogeny Working group. 2001. Phylogeny and subfamilial classification of the grasses (Poaceae). *Ann. Missouri Bot. Gard.* 88: 373–457.

Grayum, M. H. 1987. A summary of evidence and arguments supporting the removal of *Acorus* from Araceae. *Taxon* 36: 723–729.

Grayum, M. H. 1990. Evolution and phylogeny of the Araceae. *Ann. Missouri Bot. Gard.* 77: 628–697.

Green, P. S. 2004. Oleaceae. In *The families and genera of vascular plants*, Vol. 7, Lamiales (except Acanthaceae, including Avicenniaceae), 296–306. Springer-Verlag, Berlin.

Grudzinskaja, I. A. 1967. Ulmaceae and reasons for distinguishing Celtidaceae as a separate family Celtidaceae Link. *Bot. Zhurn.* (Leningrad) 52: 1723–1749. [In Russian.]

Gustafsson, M. H. G. 1996. Phylogenetic hypotheses for Asteraceae relationships. In *Compositae: systematics*. Proceedings of the International Compositae Conference, Kew, D. J. N. Hind and H. J. Beentje (eds.), 9–19. Royal Botanic Gardens, Kew.

Gustafsson, M. H. G., V. Bittrich, and P. F. Stevens. 2002. Phylogeny of Clusiaceae based on *rbcL* sequences. *Int. J. Plant Sci.* 163: 1045–1054.

Hahn, W. J. 2002. A molecular phylogenetic study of the Palmae (Arecaceae) based on *atpB, rbcL*, and 18S nrDNA sequences. *Syst. Biol.* 51: 92–112.

Hall, J. C., K. J. Sytsma, and H. H. Iltis. 2002. Phylogeny of Capparaceae and Brassicaceae based on chloroplast sequence data. *Amer. J. Bot.* 89: 1826–1842.

Hallé, N. 1962. Monographie des Hippocratéacées. *Mémoires de L'Inst. Francais D'Afrique Noire* 64: 1–245.

Hambey, R. K. and E. A. Zimmer. 1992. Ribosomal RNA as a phylogenetic tool in plant systematics. In *Molecular systematics of plants*, P. S. Soltis, D. E. Soltis and J. J. Doyle (eds.), 50–91. Chapman and Hall, New York.

Hardin, J. W. 1957. A revision of the American Hippocastanaceae. *Brittonia* 9: 145–171.

Harley, M. M. and I. K. Ferguson. 1990. The role of the SEM in pollen morphology and plant systematics. In *Scanning electron microscope in taxonomy and functional morphology*. Systematics Association Special Vol. 41, D. Claugher (ed.), 45–68. Clarendon Press, Oxford.

Harley, R. M. and 12 others. 2004. Labiatae. In *The families and genera of vascular plants*, vol. 7, Lamiales (except Acanthaceae, including Avicenniaceae), K. Kubitzki (ed.), 167–275. Springer-Verlag, Berlin.

Harrington, M. G., K. J. Edwards, S. A. Johnson, M. W. Chase, and P. A. Gadek. 2005. Phylogenetic inference in Sapindaceae sensu lato using plastid *matK* and *rbcL* DNA sequences. *Syst. Bot.* 30: 366–382.

Harris, P. J. and R. D. Hartley. 1980. Phenolic constituents of the cell walls of monocotyledons. *Biochem. Syst. Ecol.* 8: 153–160.

Harrison, C. J., M. Möller, and C. B. Cronk. 1999. Evolution and development of floral diversity in *Streptocarpus* and *Saintpaulia*. *Ann. Bot.* 84: 49–60.

Hartmann, H. E. K. 1993. Aizoaceae. In *The families and genera of vascular plants*, vol. 2, Magnoliid, hamamelid and caryophyllid families, K. Kubitzki, J. G. Rohwer and V. Bittrich (eds.), 37–69. Springer-Verlag, Berlin.

Haynes, R. R. 1978. The Potamogetonaceae in the southeastern United States. *J. Arnold Arbor.* 59: 170–191.

Haynes, R. R. 1988. Reproductive biology of selected aquatic plants. *Ann. Missouri Bot. Gard.* 75: 805–810.

Haynes, R. R., D. H. Les, and L. B. Holm-Nielsen. 1998a. Alismataceae. In *The families and genera of vascular plants*, vol. 4, Monocotyledons: Alismatanae and Commelinanae (except Gramineae), K. Kubitzki (ed.), 11–21. Springer-Verlag, Berlin.

Haynes, R. R., D. H. Les, and L. B. Holm-Nielsen. 1998b. Potamogetonaceae. In *The families and genera of vascular plants*, vol. 4, Monocotyledons: Alismatanae and Commelinanae (except Gramineae), K. Kubitzki (ed.), 408–415. Springer-Verlag, Berlin.

Haynes, R. R. and L. B. Holm-Nielsen. 2001. The genera of Hydrocharitaceae in the southeastern United States. *Harvard Pap. Bot.* 5(2): 201–275.

Hedrén, M., M. W. Chase, and R. G. Olmstead. 1995. Relationships in the Acanthaceae and related families as suggested by cladistic analysis of *rbcL* nucleotide sequences. *Plant Syst. Evol.* 194: 93–109.

Hegnauer, R. 1971. Chemical patterns and relationships of Umbelliferae. In *The biology and chemistry of the Umbelliferae. Bot. J. Linn. Soc.* vol. 64, suppl. 1, V. H. Heywood (ed.), 267–277. Academic Press, London.

Hempel, A. L., P. A. Reeves, R. G. Olmstead, and R. K. Jansen. 1995. Implications of *rbcL* sequence data for higher order relationships of the Loasaceae and the anomalous aquatic plant *Hydrostachys* (Hydro-stachyaceae). *Plant Syst. Evol.* 194: 25–37.

Henderson, A. 1986. A review of pollination studies in the palms. *Bot. Rev.* 52: 221–259.

Henderson, A. 1995. *The palms of the Amazon.* Oxford University Press, New York.

Henderson, A., G. Galeano, and R. Bernal. 1995. *Field guide to the palms of the Americas.* Princeton University Press, Princeton, NJ.

Herbert, J., M. W. Chase, M. Möller, and R. J. Abbott. 2006. Nuclear and plastid DNA sequences confirm the placement of the enigmatic *Canacomyrica monticola* in Myricaceae. *Taxon* 55: 349–357.

Herendeen, P. S., W. L. Crepet, and D. L. Dilcher. 1992. The fossil history of the Leguminosae: phylogenetic and biogeographic implications. In *Advances in legume systematics*, part 4, P. S. Herendeen and D. L.

Dilcher (eds.), 303–316. Royal Botanic Gardens, Kew.

Hermann, P. M. and B. F. Palser. 2000. Stamen development in the Ericaceae. I. Anther wall, microsporogenesis, inversion, and appendages. *Amer. J. Bot.* 87: 934–957.

Hershkovitz, M. A. and E. A. Zimmer. 1997. On the evolutionary origins of the cacti. *Taxon* 46: 217–232.

Hesse, M. 2001. Pollen characters of *Amborella trichopoda* (Amborellaceae): a reinvestigation. *Int. J. Plant Sci.* 162: 201–208.

Heywood, V. H., J. B. Harborne, and B. L. Turner. 1977. An overture to the Compositae. In *The biology and chemistry of the Compositae*, vol. 1, V. H. Heywood, J. B. Harborne, and B. L. Turner (eds.), 1–20. Academic Press, London.

Hickey, L. J. and J. A. Wolfe. 1975. The bases of angiosperm phylogeny: vegetative morphology. *Ann. Missouri Bot. Gard.* 62: 538–589.

Hileman, L. C., M. C. Vasey, and V. T. Parker. 2001. Phylogeny and biogeography of the Arbutoideae (Ericaceae): implications for the Madrean-Tethyan hypothesis. *Syst. Bot.* 26: 131–143.

Hilliard, O. M. 1994. *The Manuleae, a tribe of the Scrophulariaceae.* Edinburgh University Press, Edinburgh.

Hilu, K. W., L. A. Alice, and H. Liang. 1999. Phylogeny of Poaceae inferred from *matK* sequences. *Ann. Missouri Bot. Gard.* 86: 835–851.

Hilu, K. W. and 15 others. 2003. Angiosperm phylogeny based on *matK* sequence information. *Amer. J. Bot.* 90: 1758–1776.

Hoch, P. C., J. V. Crisci, H. Tobe, and P. E. Berry. 1993. A cladistic analysis of the plant family Onagraceae. *Syst. Bot.* 18: 31–47.

Hoot, S. B. 1991. Phylogeny of the Ranunculaceae based on epidermal microcharacters and macromorphology. *Syst. Bot.* 16: 741–755.

Hoot, S. B. 1995. Phylogeny of the Ranunculaceae based on preliminary *atpB*, *rbcL* and 18S nuclear ribosomal DNA sequence data. *Plant Syst. Evol.* Suppl. 9: 241–251.

Hoot, S. B. and P. R. Crane. 1995. Interfamilial relationships in the Ranunculidae based on molecular systematics. *Plant Syst. Evol.* Suppl. 9: 119–131.

Hoot, S. B., A. W. Douglas. 1998. Phylogeny of the Proteaceae based on *atpB* and *atpB-rbcL* intergenic spacer region sequences. *Austr. Syst. Bot.* 11: 301–320.

Hoot, S. B., J. W. Kadereit, F. R. Blattner, K. B. Jork, A. E. Schwarzbach, and P. R. Crane. 1997. The phylogeny of the Papaveraceae s. l. based on four data sets: *atpB* and *rbcL* sequences, *trnK* restriction sites and morphological characters. *Syst. Bot.* 22: 575–590.

Hoot, S. B., S. Magallón, and P. R. Crane. 1999. Phylogeny of basal eudicots based on three molecular data sets: *atpB*, *rbcL*, and 18S nuclear ribosomal DNA sequences. *Ann. Missouri Bot. Gard.* 86: 1–32.

Howe, H. F. and G. A. Vande Kerckhove. 1980. Nutmeg dispersal by tropical birds. *Science* 210: 925–927.

Huber, H. 1977. The treatment of monocotyledons in an evolutionary system of classification. *Plant Syst. Evol.* 1: (Suppl.): S285–S298.

Huber, H. 1993. Aristolochiaceae. In *The families and genera of vascular plants*, vol. 2, Magnoliid, hamamelid and caryophyllid

families, K. Kubitzki, J. G. Rohwer, and V. Bittrich (eds.), 129–137. Springer-Verlag, Berlin.

Huck, R. B. 1992. Overview of pollination biology in the Lamiaceae. In *Advances in labiate science*, R. M. Harley and T. Reynolds (eds.), 167–181. Royal Botanic Gardens, Kew.

Hufford, L. 1992. Rosidae and their relationships to other nonmagnoliid dicotyledons: a phylogenetic analysis using morphological and chemical data. *Ann. Missouri Bot. Gard.* 79: 218–248.

Hufford, L. 1997. A phylogenetic analysis of Hydrangeaceae based on morphological data. *Int. J. Plant Sci.* 158: 652–672.

Hufford, L. 2001. Ontogeny and morphology of the fertile flowers of *Hydrangea* and allied genera of tribe Hydrangeae. *Bot. J. Linn. Soc.* 137: 139–187.

Hufford, L. 2003. Homology and dvelopmental transformation: models for the origins of the staminodes of Loasaceae subfam. Loasoideae. *Int. J. Plant Sci.* 164 (5, Suppl.): S409–S439.

Hufford, L. 2004. Hydrangeaceae. In *The families and genera of vascular plants*, vol. 6, Celastrales, Oxalidales, Rosales, Cornales, Ericales, K. Kubitzki (ed.), 202–215. Springer-Verlag, Berlin.

Hufford, L. and P. R. Crane. 1989. A preliminary phylogenetic analysis of the "lower" Hamamelidae. In *Evolution, systematics and fossil history of the Hamamelidae*, vol. 1, Introduction and "lower" Hamamelidae. Systematics Association Special Vol. 40A, P. R. Crane and S. Blackmore (eds.), 175–192. Clarendon Press, Oxford.

Hufford, L. D. and P. K. Endress. 1989. The diversity of anther structures and dehiscence patterns among Hamamelididae. *Bot. J. Linn. Soc.* 99: 301–346.

Hufford, L., M. L. Moody, and D. E. Soltis. 2001. A phylogenetic analysis of Hydrangeaceae based on sequences of the plastid gene *matK* and their combination with *rbcL* and morphological data. *Int. J. Plant Sci.* 162: 835–846.

Hufford, L. M. M. McMahon, A. M. Sherwood, G. Reeves, and M. W. Chase. 2003. The major clades of Loasaceae: phylogenetic analyses using the plastid *matK* and *trnL-trnF* regions. *Amer. J. Bot.* 1215–1228.

Humphries, C. J. and S. Blackmore. 1989. A review of the classification of the Moraceae. In *Evolution, systematics and fossil history of the Hamamelidae*, vol. 2, "Higher" Hamamelidae. Systematics Association Special Vol. 40B, P. R. Crane and S. Blackmore (eds.), 267–277. Clarendon Press, Oxford.

Hutchinson, J. 1934. *The families of flowering plants*, vol. 2, Monocotyledons. Macmillan, London.

Hutchinson, J. 1973. *The families of flowering plants*, 3rd ed. Clarendon Press, Oxford.

Ickert-Bond, S. M. and J. Wen. 2006. Phylogeny and biogeography of Altingiaceae: evidence from combined analysis of five non-coding chloroplast regions. *Mol. Phyl. Evol.* 39: 512–528.

Ickert-Bond, S. M., K. B. Pigg, and J. Wen. 2005. Comparative infructescence morphology in *Liquidambar* (Altingiaceae) and its evolutionary significance. *Amer. J. Bot.* 92: 1234–1255.

Igersheim, A., M. Buzgo, and P. K. Endress. 2001. Gynoecium diversity and systematics in basal monocots. *Bot. J. Linn. Soc.* 136: 1–65.

Igersheim, A. and P. K. Endress. 1997. Gynoecium diversity and systematics of the Magnolales and winteroids. *Bot. J. Linn. Soc.* 124: 213–271.

Igersheim, A. and P. K. Endress. 1998. Gynoecium diversity and systematics of the paleoherbs. *Bot. J. Linn. Soc.* 127: 289–370.

Iltis, H. H. 1999. Setchellanthaceae (Capparales), a new family for a relictual, glucosinolate-producing endemic of the Mexican deserts. *Taxon* 48: 257–275.

Ingrouille, M. J., M. W. Chase, M. F. Fay, D. Bowman, M. van der Bank, and A. D. E. Bruijn. 2002. Systematics of Vitaceae from the viewpoint of plastid *rbcL* DNA sequence data. *Bot. J. Linn. Soc.* 138: 421–432.

Isley, D. 1998. *Native and naturalized Leguminosae (Fabaceae) of the United States (exclusive of Alaska and Hawaii).* Brigham Young Univ. Press, Provo, UT.

Ito, M. 1986. Studies in the floral morphology and anatomy of Nymphaeales. III. Floral anatomy of *Brasenia schreberi* Gmel. and *Cabomba caroliniana* A. Gray. *Bot. Mag. Tokyo* 99: 169–184.

Ito, M. 1987. Phylogenetic systematics of the Nymphaeales. *Bot. Mag. Tokyo* 100: 17–36.

Jackson, G. A. 1933. A study of the carpophore of the Umbelliferae. *Amer. J. Bot.* 20: 121–144.

Jansen, R. K. and J. D. Palmer. 1987. A chloroplast DNA inversion marks an ancient evolutionary split in the sunflower family, Asteraceae. *Proceed. Nat. Acad. Sci. USA* 84: 5818–5822.

Jansen, R. K., H. J. Michaels and J. D. Palmer. 1991. Phylogeny and character evolution in the Asteraceae based on chloroplast DNA restriction site mapping. *Syst. Bot.* 16: 98–115.

Jansen, R. K., H. J. Michaels, R. S. Wallace, K.-J. Kim, S. C. Keeley, L. E. Watson, and J. D. Palmer. 1992. Chloroplast DNA variation in the Asteraceae: phylogenetic and evolutionary implications. In *Molecular systematics of plants*, P. S. Soltis, D. E. Soltis, and J. J. Doyle (eds.), 252–294. Chapman and Hall, New York.

Jansen, R. K. and 18 others. 2006. Phylogeny of angiosperms based on whole chloroplast genome sequences. Abstract. *Botany 2006*: 495.

Jaramillos, M. A. and P. S. Manos. 2001. Phylogeny and patterns of floral diversity in the genus *Piper* (Piperaceae*). Amer. J. Bot.* 88: 706–716.

Jaramillo, M. A., P. S. Manos, and E. A. Zimmer. 2004. Phylogenetic relationships of the perianthless Piperales: reconstructing the evolution of floral development. *Int. J. Plant Sci.* 165: 403–416.

Jeffrey, C. 1967. On the classification of the Cucurbitaceae. *Kew Bull.* 20: 417–426.

Jeffrey, C. 1980. A review of the Cucurbitaceae. *Bot. J. Linn. Soc.* 81: 233–247.

Jeffrey, C. 1990a. Appendix: an outline classification of the Cucurbitaceae. In *Biology and utilization of the Cucurbitaceae*, D. M. Bates, R. W. Robinson, and C. Jeffrey (eds.), 449–463. Cornell University Press, Ithaca, NY.

Jeffrey, C. 1990b. Systematics of the Cucurbitaceae: an overview. In *Biology and utilization of the Cucurbitaceae*, D. M. Bates, R. W. Robinson, and C. Jeffrey (eds.), 3–9. Cornell University Press, Ithaca, NY.

Jensen, S. R. 1992. Systematic implications of the distribution of iridoids and other chemical compounds in the Loganiaceae and other families of the Asteridae. *Ann. Missouri Bot. Gard.* 79: 284–302.

Jensen, U., S. B. Hoot, J. T. Johansson, and K. Kosuge. 1995. Systematics and phylogeny of the Ranunculaceae: a revised family concept on the basis of molecular data. *Plant Syst. Evol* 9 (Suppl.): S273–S280.

Jobson, R. W., J. Playford, K. C. Cameron, and V. A. Albert. 2003. Molecular phylogenetics of Lentibulariaceae inferred from plastid *rps16* intron and *trnL-F* DNA sequences: implications for character evolution and biogeography. *Syst. Bot.* 28: 157–171.

Jobst, J., K. King, and V. Hemleben. 1998. Molecular evolution of the internal transcribed spacers (ITS1 and ITS2) and phylogenetic relationships among species of the family Cucurbitaceae. *Mol. Phylog. Evol.* 9: 204–219.

Johansson, J. T. 1995. A revised chloroplast DNA phylogeny of Ranunculaceae. *Plant Syst. Evol.* 9: (Suppl.): S253–S261.

Johansson, J. T. and R. K. Jansen. 1993. Chloroplast DNA variation and phylogeny of the Ranunculaceae. *Plant Syst. Evol.* 187: 29–49.

Johnson, L. A. and D. E. Soltis. 1994. *matK* DNA sequences and phylogenetic reconstruction in Saxifragaceae s. str. *Syst. Bot.* 19: 143–156.

Johnson, L. A., J. L. Schultz, D. E. Soltis, and P. S. Soltis. 1996. Monophyly and generic relationships of Polemoniaceae based on *matK* sequences. *Amer. J. Bot.* 83: 1207–1224.

Johnson, L. A., D. E. Soltis, and P. S. Soltis. 1999. Phylogenetic relationships of Polemoniaceae inferred from 18S ribosomal DNA sequences. *Plant Syst. Evol.* 214: 65–89.

Johnson, L. A. S. 1957. A review of the family Oleaceae. *Contr. N. S. W. Nat. Herb.* 2: 397–418.

Johnson, L. A. S. 1976. Problems of species and genera in *Eucalyptus* (Myrtaceae). *Plant Syst. Evol.* 125: 155–167.

Johnson, L. A. S. and B. G. Briggs. 1975. On the Proteaceae: the evolution and classification of a Southern family. *J. Linn. Soc. Bot.* 70: 83–182.

Johnson, L. A. S. and B. G. Briggs. 1984. Myrtales and Myrtaceae: a phylogenetic analysis. *Ann. Missouri Bot. Gard.* 71: 700–756.

Johnson, L. A. S. and K. L. Wilson. 1989. Casuarinaceae: a synopsis. In *Evolution, systematics and fossil history of the Hamamelidae*, vol. 2, "Higher" Hamamelidae. Systematics Association Special Vol. 40B, P. R. Crane and S. Blackmore (eds.), 167–188. Clarendon Press, Oxford.

Johnson, L. A. S. and K. L. Wilson. 1993. Casuarinaceae. In *The families and genera of vascular plants*, vol. 2, Magnoliid, hamamelid and caryophyllid families, K. Kubitzki, J. G. Rohwer and V. Bittrich (eds.), 237–242. Springer-Verlag, Berlin.

Jones, S. B. 1982. The genera of Vernonieae (Compositae) in the southeastern United States. *J. Arnold Arbor.* 63: 489–507.

Jorgensen, L. B. 1981. Myrosin cells and dilated cisternae of the endoplasmic reticulum in the order Capparales. *Nordic J. Bot.* 1: 433–445.

Jork, K. B. and J. W. Kadereit. 1995. Molecular phylogeny of the Old World representatives of Papaveraceae subf. Papaveroideae with special emphasis on the genus *Meconopsis. Plant. Syst. Evol.* 9 (Suppl.): S171–S180.

Judd, W. S. 1986. Taxonomic studies in the Miconieae (Melastomataceae). I. Variation in inflorescence position. *Brittonia* 38: 238–242.

Judd, W. S. 1989. Taxonomic studies in the Miconieae (Melastomataceae). III. Cladistic analysis of axillary-flowered taxa. *Ann. Missouri Bot. Gard.* 76: 476–495.

Judd, W. S. 1996. The Pittosporaceae in the southeastern United States. *Harvard Pap. Bot.* No. 8: 15–26.

Judd, W. S. 1997a. The Asphodelaceae in the southeastern United States. *Harvard Pap. Bot.* No. 11: 109–123.

Judd, W. S. 1997b. The Flacourtiaceae in the southeastern United States. *Harvard Pap. Bot.* No. 10: 65–79.

Judd, W. S. 1998. The Smilacaceae in the southeastern United States. *Harvard Pap. Bot.* 3: 147–169.

Judd, W. S. 2000. The Hypoxidaceae in the southeastern United States. *Harvard Pap. Bot.* 5: 79–98.

Judd, W. S. 2001. The Asparagaceae in the southeastern United States. *Harvard Pap. Bot.* 6: 223–244.

Judd, W. S. 2003. The genera of Ruscaceae in the southeastern United States. *Harvard Pap. Bot.* 7: 93–149.

Judd, W. S. and T. K. Ferguson. 1999. The genera of Chenopodiaceae in the southeastern United States. *Harvard Pap. Bot.* 4: 365–416.

Judd, W. S. and K. A. Kron. 1993. Circumscription of Ericaceae (Ericales) as determined by preliminary cladistic analyses based on morphological, anatomical, and embryological features. *Brittonia* 45: 99–114.

Judd, W. S. and S. R. Manchester. 1998. Circumscription of Malvaceae (Malvales) as determined by a preliminary cladistic analysis employing morphological, palynological, and chemical characters. *Brittonia* 49: 384–405.

Judd, W. S. and R. G. Olmstead. 2004. A survey of tricolpate (eudicot) phylogenetic relationships. *Amer. J. Bot.* 91: 1627–1644.

Judd, W. S. and J. D. Skean, Jr. 1991. Taxonomic studies in the Miconieae (Melastomataceae). IV. Generic realignments among terminal-flowered taxa. *Bull. Florida Mus. Nat. Hist., Biol. Sci.* 36: 25–84.

Judd, W. S., W. L. Stern, and V. I. Cheadle. 1993. Phylogenetic position of *Apostasia* and *Neuwiedia* (Orchidaceae). *Bot. J. Linn. Soc.* 113: 87–94.

Judd, W. S., R. W. Sanders, and M. J. Donoghue. 1994. Angiosperm family pairs: preliminary cladistic analyses. *Harvard Pap. Bot.* No. 5: 1–51.

Judziewicz, E. J., L. G. Clark, X. Londoño, and M. J. Stern. 1999. *American bamboos.* Smithsonian Institution. Press, Washington, D. C.

Juncosa, A. M. and P. B. Tomlinson. 1988a. A historical and taxonomic synopsis of Rhizophoraceae and Anisophylleaceae. *Ann. Missouri Bot. Gard.* 75: 1278–1295.

Juncosa, A. M. and P. B. Tomlinson. 1988b. Systematic comparison and some biological characteristics of Rhizophoraceae and Anisophylleaceae. *Ann. Missouri Bot. Gard.* 75: 1296–1318.

Junell, S. 1934. Zur Gynäceummorphologie und Systematik der Verbenaceen und Labiaten. *Symb. Bot. Upsal.* 1(4): 1–219.

Kadereit, G., T. Borsch, K. Weising, and H. Freitag. 2003. Phylogeny of Amaranthaceae and Chenopodiaceae and the evolution of C₄ photosynthesis. *Int. J. Plant Sci.* 164: 959–986.

Kadereit, J. W. 1993. Papaveraceae. In *The families and genera of vascular plants*, vol. 2, Magnoliid, hamamelid and caryophyllid families, K. Kubitzki, J. G. Rohwer, and V. Bittrich (eds.), 494–506. Springer-Verlag, Berlin.

Kadereit, J. W., F. R. Blattner, K. B. Jork, and A. Schwarzbach. 1994. Phylogenetic analysis of the Papaveraceae s. l. (incl. Fumariaceae, Hypecoaceae and *Pteridophyllum*) based on morphological characters. *Bot. Jahrb. Syst.* 116: 361–390.

Kadereit, J. W., F. R. Blattner, K. B. Jork, and A. Schwarzback. 1995. The phylogeny of the Papaveraceae sensu lato: morphological, geographical and ecological implications. *Plant Syst. Evol.* Suppl. 9: 133–145.

Kajita, T., H. Ohashi, Y. Tateishi, C. D. Bailey, and J. J. Doyle. 2001. *rbcL* and legume phylogeny with particular reference to Phaseoleae, Millettieae, and allies. *Syst. Bot.* 26: 515–536.

Kalkman, C. 2004. Rosaceae. In *The families and genera of vascular plants*, vol. 6, Celastrales, Oxalidales, Rosales, Cornales, Ericales, K. Kubitzki (ed.), 343–389. Springer-Verlag, Berlin.

Källersjö, M. and B. Ståhl. 2003. Phylogeny of Theophrastaceae (Ericales s. lat.). *Int. J. Plant Sci.* 164: 579–591.

Källersjö, M., J. S. Farris, M. W. Chase, B. Bremer, M. F. Fay, C. J. Humphries, G. Petersen, O. Seberg, and K. Bremer. 1998. Simulatneous parsimony jackknife analysis of 2538 *rbcL* DNA sequences reveals support for major clades of green plants, land plants, and flowering plants. *Plant Syst. Evol.* 213: 259–287.

Källersjö, M., G. Bergqvist, and A. A. Anderberg. 2000. Generic realignment in primuloid families of the Ericales s. l.: a phylogenetic analysis based on DNA sequences from three chloroplast genes and morphology. *Amer. J. Bot.* 87: 1325–1341.

Kampny, C. M. 1995. Pollination and floral diversity in Scrophulariaceae. *Bot. Rev.* 61: 350–366.

Kårehed, J. 2003. The family Pennantiaceae and its relationships to Apiales. *Bot. J. Linn. Soc.* 141: 1–24.

Karis, P. O. 1993. Morphological phylogenetics of the Asteraceae-Asteroideae, with notes on character evolution. *Plant Syst. Evol.* 186: 69–93.

Karis, P. O., M. Källersjö, and K. Bremer. 1992. Phylogenetic analysis of the Cichorioideae (Asteraceae), with emphasis on the Mutisieae. *Ann. Missouri Bot. Gard.* 79: 416–427.

Karis, P. O., P. Eldenäs, and M. Källersjö. 2001. New evidence for the systematic position of *Gundelia* L. with notes on delimitation of Arctoteae (Asteraceae). *Novon* 50: 105–114.

Karol, K. G., Y. Suh, G. E. Schutz, and E. A. Zimmer. 2000. Molecular evidence for the phylogenetic position of *Takhtajania* in the Winteraceae: inference from nuclear ribosomal and chloroplast gene spacer sequences. *Ann. Missouri Bot. Gard.* 87: 414–432.

Karol, K. G., J. E. Rodman, E. Conti, and K. J. Sytsma. 1999. Nucleotide sequence of *rbcL* and phylogenetic relationships of *Setchellanthus caeruleus* (Setchellanthaceae). *Taxon* 48: 303–315.

Kato, H., S. Kawano, R. Terauchi, M. Ohara and F. H. Utech. 1995a. Evolutionary biology of *Trillium* and related genera (Trilliaceae). I. Restriction site mapping and variation of chloroplast DNA and its systematic implications. *Plant Species Biol.* 10: 17–29.

Kato, H., R. Terauchi, F. H. Utech, and S. Kawano. 1995b. Molecular systematics of the Trilliaceae sensu lato as inferred from *rbcL* sequence data. *Mol. Phylog. Evol.* 4: 184–193.

Kaul, R. B. 1968. Floral morphology and phylogeny in the Hydrocharitaceae. *Phytomorphology* 18: 13–35.

Kaul, R. B. 1970. Evolution and adaptation of the inflorescences in the Hydrocharitaceae. *Amer. J. Bot.* 57: 708–715.

Kaul, R. B. and E. C. Abbe. 1984. Inflorescence architecture and evolution in the Fagaceae. *J. Arnold Arbor.* 65: 375–401.

Kawakita, A. and M. Kato. 2004. Evolution of obligate pollination mutalism in New Caledonian *Phyllanthus* (Euphorbiaceae). *Amer. J. Bot.* 91: 410–415.

Kawano, S. and H. Kato. 1995. Evolutionary biology of *Trillium* and related genera (Trilliaceae). II. Cladistic analyses on gross morphological characters and phylogeny and evolution of the genus *Trillium*. *Plant Sp. Biol.* 10: 169–183.

Keating, R. C. 2004. Systematic occurrence of raphide crystals in Araceae. *Ann. Missouri Bot. Gard.* 91: 495–504.

Kedves, M. 1989. Evolution of the Normapolles complex. In *Evolution, systematics and fossil history of the Hamamelidae*, vol. 2, "Higher" Hamamelidae. Systematics Association Special Vol. 40B, P. R. Crane and S. Blackmore (eds.), 1–7. Clarendon Press, Oxford.

Keeley, S. C. and R. K. Jansen. 1991. Evidence from chloroplast DNA for the recognition of a new tribe, Tarchonantheae and the tribal placement of *Pluchea* (Asteraceae). *Syst. Bot.* 16: 173–181.

Kellogg, E. A. and H. P. Lindler. 1995. Phylogeny of Poales. In *Monocotyledons: systematics and evolution*, P. J. Rudall, P. J. Cribb, D. F. Cutler, and C. J. Humphries (eds.), 511–542. Royal Botanic Gardens, Kew.

Kellogg, E. A. and L. Watson. 1993. Phylogenetic studies of a large data set. I. Bambusoideae, Andropogonidae and Pooideae (Gramineae). *Bot. Rev.* 59: 273–343.

Kelly, L. M. 1997. A cladistic analysis of *Asarum* (Aristolochiaceae) and implications for the evolution of herkogamy. *Amer. J. Bot.* 84: 1752–1765.

Kelly, L. M. and F. González. 2003. Phylogenetic relationships in Aristolochiaceae. *Syst. Bot.* 28: 236–249.

Keng, H. 1962. Comparative morphological studies in Theaceae. *Univ. Calif. Publ. Bot.* 33: 219–384.

Keng, H. 1993. Illiciaceae. In *The families and genera of vascular plants*, vol. 2, Magnoliid, hamamelid and caryophyllid families, K. Kubitzki, J. G. Rohwer, and V. Bittrich (eds.), 344–347. Springer-Verlag, Berlin.

Kennedy, H. 2000. Diversification in pollination mechanisms in the Marantaceae. In *Monocots: systematics and evolution*, K. L. Wilson and D. A. Morrison (eds.), 335–343. CSIRO, Collingwood, Australia.

Kers, L. E. 2003. Capparaceae. In *The families and genera of vascular plants*, vol 5, Malvales, Capparales and non-betalain Caryophyllales, K. Kubitzki (ed.), 36–56. Springer-Verlag, Berlin.

Kessler, P. J. A. 1993a. Annonaceae. In *The families and genera of vascular plants*, vol. 2, Magnoliid, hamamelid and caryophyllid families, K. Kubitzki, J. G. Rohwer, and V. Bittrich (eds.), 93–129. Springer-Verlag, Berlin.

Kessler, P. J. A. 1993b. Menispermaceae. In *The families and genera of vascular plants*, vol. 2, Magnoliid, hamamelid and caryophyllid families, K. Kubitzki, J. G. Rohwer, and V. Bittrich (eds.), 402–418. Springer-Verlag, Berlin.

Killip, E. 1938. The American species of Passifloraceae. *Field Museum of Nat. Hist. Bot. Ser.* 19: 1–613.

Kim, J.-K. and R. K. Jansen. 1995. Phylogenetic implications of chloroplast DNA variation in the Berberidaceae. *Plant Syst, Evol. Suppl.* 9: 341–349.

Kim, J.-K. and R. K. Jansen. 1998a. Chloroplast DNA restriction site variation and phylogeny of the Berberidaceae. *Amer. J. Bot.* 85: 1766–1778.

Kim, J.-K. and R. K. Jansen. 1998b. Paraphyly of Jasminoideae and monophyly of Oleoideae in Oleaceae. *Amer. J. Bot.* 85(6) Suppl.: 139.

Kim, J.-K., R. K. Jansen, R. S. Wallace, H. J. Michaels, and J. D. Palmer. 1992. Phylogenetic implications of *rbcL* sequence variation in the Asteraceae. *Ann. Missouri Bot. Gard.* 79: 428–445.

Kim, S., C.-W. Park, Y.-D. Kim, and Y. Suh. 2001. Phylogenetic relationships in family Magnoliaceae inferred from *ndhF* sequences. *Amer. J. Bot.* 88: 717–728.

Kim, S., M.-J. You, V. A. Albert, J. S. Farris, P. S. Soltis, and D. E. Soltis. 2004. Phylogeny and diversification of B-function MADS-box genes in angiosperms: evolutionary and functional implications of a 260–million-year-old duplication. *Amer. J. Bot.* 91: 2102–2118.

Kim, Y.-D. and R. K. Jansen. 1998. Chloroplast DNA restriction site variation and phylogeny of the Berberidaceae. *Amer. J. Bot.* 85: 1766–1778.

Klak, C., A. Khunou, G. Reeves, and T. Hedderson. 2003. A phylogenetic hypothesis for the Aizoaceae (Caryophyllales) based on four plastid DNA regions. *Amer. J. Bot.* 90: 1433–1445.

Kocyan, A. and P. K. Endress. 2001. Floral structure and development in *Apostasia* and *Neuwiedia* (Apostasioideae) and their relationship to other Orchidaceae. *Int. J. Plant Sci.* 162: 847–867.

Kocyan, A., Y.-L. Qiu, P. K. Endress, and E. Conti. 2004. A phylogenetic analysis of Apostasioideae (Orchidaceae) based on ITS, *trnL-F* and *matK* sequences. *Plant Syst. Evol.* 247: 203–213.

Kornhall, P. and B. Bremer. 2004. New circumscription of the tribe Limoselleae (Scrophulariaceae) that includes the taxa of the tribe Manuleeae. *Bot. J. Linn. Soc.* 146: 353–467.

Kral, R. 1992. A treatment of American Xyridaceae exclusive of *Xyris*. *Ann. Missouri Bot. Gard.* 79: 819–885.

Kral, R. B. 1966a. Eriocaulaceae of continental North America north of Mexico. *Sida* 2: 285–332.

Kral, R. B. 1966b. *Xyris* (Xyridaceae) of the continental United States and Canada. *Sida* 2: 177–260.

Kral, R. B. 1983. The Xyridaceae in the southeastern United States. *J. Arnold Arbor.* 64: 421–429.

Kral, R. B. 1989. The genera of Eriocaulaceae in the southeastern United States. *J. Arnold Arbor.* 70: 131–142.

Kral, R. B. 1998. Xyridaceae. In *The families and genera of vascular plants*, vol. 4, Monocotyledons: Alismatanae and Commelinanae (except Gramineae), K. Kubitzki (ed.), 461–469. Springer-Verlag, Berlin.

Kress, W. J. 1990. The phylogeny and classification of the Zingiberales. *Ann. Missouri Bot. Gard.* 77: 698–721.

Kress, W. J. 1995. Phylogeny of the Zingiberanae: morphology and molecules. In *Monocotyledons: systematics and evolution*, P. J. Rudall, P. J. Cribb, D. F. Cutler and C. J. Humphries (eds.), 443–460. Royal Botanic Gardens, Kew.

Kress, W. J., L. M. Prince, W. J. Hahn, and E. A. Zimmer. 2001. Unraveling the evolutionary radiation of the families of the Zingiberales using morphological and molecular evidence. *Syst. Biol.* 50: 926–944.

Kress, W. J., L. M. Prince, and K. J. Williams. 2002. The phylogeny and a new classification of the gingers (Zingiberaceae): evidence from molecular data. *Amer. J. Bot.* 89: 1682–1619.

Kress, W. J., A. L. Liu, M. Neuman, and Q-J. Li. 2005. The molecular phylogeny of *Alpinia* (Zingiberaceae): a complex and polyphyletic genus of gingers. *Amer. J. Bot.* 92: 167–178.

Kristiansen, K. A., M. Cilieborg, L. Drábková, T. Jørgensen, G. Petersen, and O. Seberg. 2005. DNA taxonomy – the riddle of *Oxychloë* (Juncaceae). *Syst. Bot.* 30: 284–289.

Kron, K. A. 1996. Phylogenetic relationships of Empetraceae, Epacridaceae, Ericaceae, Monotropaceae and Pyrolaceae: evidence from nucleotide ribosomal 18S sequence data. *Ann. Bot.* 77: 293–303.

Kron, K. A. 1997. Phylogenetic relationships of Rhododendroideae (Ericaceae). *Amer. J. Bot.* 84: 973–980.

Kron, K. A. and M. W. Chase. 1993. Systematics of the Ericaceae, Empetraceae, Epacridaceae and related taxa based on *rbcL* sequence data. *Ann. Missouri Bot. Gard.* 80: 735–741.

Kron, K. A. and J. M. King. 1996. Cladistic relationships of *Kalmia*, *Leiophyllum* and *Loiseleuria* (Phyllodoceae, Ericaceae) based on nucleotide sequences from *rbcL* and nuclear ribosomal internal transcribed spacer regions (ITS). *Syst. Bot.* 21: 17–29.

Kron, K. A. and W. S. Judd. 1990. Phylogenetic relationships within the Rhodoreae (Ericaceae) with specific comments on the placement of *Ledum*. *Syst. Bot.* 15: 57–68.

Kron, K. A. and W. S. Judd. 1997. Systematics of the *Lyonia* group (Andromedeae, Ericaceae) and the use of species as terminals in higher-level cladistic analyses. *Syst. Bot.* 22: 479–492.

Kron, K. A., W. S. Judd, and D. M. Crayn. 1999. Phylogenetic analyses of Andromedeae (Ericaceae subfam. Vaccinioideae). *Amer. J. Bot.* 86: 1290–1300.

Kron, K. A., W. S. Judd, P. F. Stevens, D. M. Crayn, A. A. Anderberg, P. A. Gadek, C. J. Quinn, and J. L. Luteyn. 2002. A phylogenetic classification of Ericaceae: molecular and morphological evidence. *Bot. Rev.* 68: 335–423.

Kubitzki, K. 1993a. Betulaceae. In *The families and genera of vascular plants*, vol. 2, Magnoliid, hamamelid and caryophyllid families, K. Kubitzki, J. G. Rohwer, and V. Bittrich (eds.), 152–157. Springer-Verlag, Berlin.

Kubitzki, K. 1993b. Fagaceae. In *The families and genera of vascular plants*, vol. 2, Magnoliid, hamamelid and caryophyllid families, K. Kubitzki, J. G. Rohwer, and V. Bittrich (eds.), 301–309. Springer-Verlag, Berlin.

Kubitzki, K. 1993c. Myricaceae. In *The families and genera of vascular plants*, vol. 2, Magnoliid, hamamelid and caryophyllid families, K. Kubitzki, J. G. Rohwer, and V. Bittrich (eds.), 453–457. Springer-Verlag, Berlin.

Kubitzki, K. 1993d. Platanaceae. In *The families and genera of vascular plants*, vol. 2, Magnoliid, hamamelid and caryophyllid families, K. Kubitzki, J. G. Rohwer, and V. Bittrich (eds.), 521–522. Springer-Verlag, Berlin.

Kubitzki, K. (ed.) 1998a. *The families and genera of vascular plants*, vol. 3, Monocotyledons: Lilianae (except Orchidaceae). Springer-Verlag, Berlin.

Kubitzki, K. (ed.) 1998b. *The families and genera of vascular plants*, vol. 4, Monocotyledons: Alismatanae and Commelinanae (except Gramineae). Springer-Verlag, Berlin.

Kubitzki, K. 1998c. Cannaceae. In *The families and genera of vascular plants*, vol. 4, Monocotyledons: Alismatanae and Commelinanae (except Gramineae), K. Kubitzki (ed.), 103–106. Springer-Verlag, Berlin.

Kubitzki, K. 1998d. Typhaceae. In *The families and genera of vascular plants*, vol. 4, Monocotyledons: Alismatanae and Commelinanae (except Gramineae), K. Kubitzki (ed.), 457–461. Springer-Verlag, Berlin.

Kubitzk, K. 2003a. Introduction to Capparales. In *The families and genera of vascular plants*, Vol. 5, Malvales, Capparales, and non-betalain Caryophyllales, K. Kubitzki and C. Bayer (eds.), 7–10. Springer-Verlag, Berlin.

Kubitzki, K. 2003b. Droseraceae. In *The families and genera of vascular plants*, vol. 5, Malvales, Capparales, and non-betalain Caryophyllales, K. Kubitzki and C. Bayer (eds.), 198–202. Springer-Verlag, Berlin.

Kubitzki, K. 2004a. Cornaceae. In *The families and genera of vascular plants*, vol. 6, Celastrales, Oxalidales, Rosales, Cornales, Ericales, K. Kubitzki (ed.), 82–90. Springer-Verlag, Berlin.

Kubitzki, K. 2004b. Sarraceniaceae. In *The families and genera of vascular plants*, vol. 6, Celastrales, Oxalidales, Rosales, Cornales, Ericales, K. Kubitzki (ed.), 422–425. Springer-Verlag, Berlin.

Kubitzki, K. and M. W. Chase. 2003. Introduction to Malvales. In *The families and genera of vascular plants*, vol. 5, Malvales, Capparales, and non-betalain Caryophyllales, K. Kibutzki and C. Bayer (eds.), 12–16. Springer-Verlag, Berlin.

Kubitzki, K. and H. Kurz. 1984. Synchronized dichogamy and dioecy in neotropical Lauraceae. *Plant Syst. Evol.* 147: 253–266.

Kubitzki, K. and P. J. Rudall. 1998. Asparagaceae. In *The families and genera of vascular plants*, vol. 3, Monocotyledons: Lilianae (except Orchidaceae), K. Kubitzki (ed.), 125–129. Springer-Verlag, Berlin.

Kühn, U. and K. Kubitzki. 1993. Myristicaceae. In *The families and genera of vascular plants*, vol. 2, Magnoliid, hamamelid and caryophyllid families, K. Kubitzki, J. G. Rohwer, and V. Bittrich (eds.), 457–467. Springer-Verlag, Berlin.

Kühn, U., V. Bittrich, R. Carolin, H. Freitag, I. C. Hedge, P. Uotila, and P. G. Wilson. 1993. Chenopodiaceae. In *The families and genera of vascular plants*, vol. 2, Magnoliid, hamamelid and caryophyllid families, K. Kubitzki, J. G. Rohwer, and V. Bittrich (eds.), 253–281. Springer-Verlag, Berlin.

Kuijt, J. 1969. *The biology of parasitic flowering plants*. Univ. California Press, Berkeley.

Kuijt, J. 1981. Inflorescence morphology of Loranthaceae: an evolutionary synthesis. *Blumea* 27: 1–78.

Kuijt, J. 1982. The Viscaceae in the southeastern United States. *J. Arnold Arbor.* 63: 401–410.

Kunze, H. 1990. *Morphology and evolution of the corona in Asclepiadaceae and related families*. Akad. der Wiss. und der Literatur. Tropische und subtropische Pflanzenwelt 76. Steiner Verlag, Mainz-Stuttgart.

Kunze, H. 1992. Evolution of the translator in Periplocaceae and Asclepiadaceae. *Plant Syst. Evol.* 185: 99–122.

La Duke, J. C. and J. Doebley. 1995. A chloroplast DNA-based phylogeny of the Malvaceae. *Syst. Bot.* 20: 259–271.

Lamb Frye, A. S. and K. A. Kron. 2003. *rbcL* phylogeny and character evolution in Polygonaceae. *Syst. Bot.* 28: 326–332.

Lammers, T. G. 1992. Circumscription and phylogeny of the Campanulales. *Ann. Missouri Bot. Gard.* 79: 388–413.

Landolt, E. 1980. *Biosystematic investigation in the family of duckweeds (Lemnaceae)*, vol. 1, Key to the determination of taxa within the family of Lemnaceae. Veröff. Geobot. Inst. ETH Stiftung Rübel Zürich 70: 13–21.

Landolt, E. 1986. *Bio-systematic investigation in the family of duckweeds (Lemnaceae)*, vol. 2, The family of Lemnaceae, a monographic study, vol. 1. Veröff. Geobot. Inst. ETH Stiftung Rübel Zürich 71: 1–566.

Landolt, E. and R. Kandeler. 1987. *Biosystematic investigation in the family of duckweeds (Lemnaceae)*, vol. 4, The family of Lemnaceae, a monographic study, vol. 2. Veröff. Geobot. Inst. ETH Stiftung Rübel Zürich 95: 1–638.

Lane, M. A. 1996. Pollination biology of Compositae. In *Compositae: biology and utilization.* Proceedings of the International Compositae Conference, Kew, P. D. S. Caligari and D. J. N. Hind (eds.), 61–80. Royal Botanic Garden, Kew.

Långström, E. and M. W. Chase. 2002. Tribes of Boraginoideae (Boraginaceae) and placement of *Antiphytum, Echiochilon, Ogastemma* and *Sericostoma*: a phylogenetic analysis based on *atpB* plastid DNA sequence data. *Plant Syst. Evol.* 234: 137–153.

Larsen, K., J. M. Lock, H. Maas, and P. J. M. Maas. 1998. Zinbiberaceae. In *The families and genera of vascular plants*, vol. 4, Monocotyledons: Alismatanae and Commelinanae (except Gramineae), K. Kubitzki (ed.), 474–495. Springer-Verlag, Berlin.

Lavin, M., P. S. Herendeen, and M. F. Wojciechowski. 2005. Evolutionary rates analysis of Leguminosae implicates a rapid diversification of lineages during the Tertiary. *Syst. Biol.* 54: 475–594.

Lawrence, G. H. M. 1951. *Taxonomy of vascular plants.* Macmillan, New York.

Lee, S. and J. Wen. 2001. A phylogenetic analysis of *Prunus* and the Amygdaloideae (Rosaceae) using ITS sequences of nuclear ribosomal DNA. *Amer. J. Bot.* 88: 150–160.

Leins, P. and C. Erbar. 1990. On the mechanism of secondary pollen presentation in the Campanulales-Asterales complex. *Bot. Acta* 103: 87–92.

Leins, P. and C. Erbar. 1994. Putative origin and relationships of the order from the viewpoint of developmental flower morphology. In *Caryophyllales: evolution and systematics*, H.-D. Behnke and T. J. Mabry (eds.), 303–316. Springer-Verlag, Berlin.

Leins, P. and C. Erbar. 2003. Floral developmental features and molecular data in plant systematics. In *Deep morphology*, T. F. Stuessy, V. Mayer, and E. Hörandl (eds.), 81–105. A. R. G. Ganter Verlag K. G., Ruggell.

Lersten, N. R. 1975. Colleter types in Rubiaceae, especially in relation to the bacterial leaf nodule symbiosis. *Bot. J. Linn. Soc.* 71: 311–319.

Lersten, N. R. and J. M. Beaman. 1998. First report of oil cavities in Scrophulariaceae and reinvestigation of air spaces in leaves of *Leucophyllum frutescens. Amer. J. Bot.* 85: 1646–1649.

Les, D. H. 1988. The origin and affinities of the Ceratophyllaceae. *Taxon* 37: 326–435.

Les, D. H. 1989. The evolution of achene morphology in *Ceratophyllum* (Ceratophyllaceae). IV. Summary of proposed relationships and evolutionary trends. *Syst. Bot.* 14: 254–262.

Les, D. H. 1993. Ceratophyllaceae. In *The families and genera of vascular plants*, vol. 2, Magnoliid, hamamelid and caryophyllid families, K. Kubitzki, J. G. Rohwer, and V. Bittrich (eds.), 246–250. Springer-Verlag, Berlin.

Les, D. H. and D. J. Crawford. 1999. *Landoltia* (Lemnaceae), a new genus of duckweeds. *Novon* 9: 530–533.

Les, D. H. and R. R. Haynes. 1995. Systematics of subclass Alismatidae: A synthesis of approaches. In *Monocotyledons: systematics and evolution*, vol. 2, P. J. Rudall, P. J. Cribb, D. F. Cutler, and C. J. Humphries (eds.), 353–377. Royal Botanic Gardens, Kew.

Les, D. H., D. K. Garvin, and C. F. Wimpee. 1991. Molecular evolutionary history of ancient aquatic angiosperms. *Proc. Natl. Acad. Sci. USA* 88: 10119–10123.

Les, D. H., M. A. Cleland, and M. Waycott. 1997a. Phylogenetic studies in Alismatidae, II: Evolution of marine angiosperms (seagrasses) and hydrophily. *Syst. Bot.* 22: 443–463.

Les, D. H., E. Landolt, and D. J. Crawford. 1997b. Systematics of the Lemnaceae (duckweeds): inferences from micromolecular and morphological data. *Plant Syst. Evol.* 204: 161–177.

Les, D. H., E. L. Schneider, D. J. Padgett, P. S. Soltis, D. E. Soltis, and M. Zanis. 1999. Phylogeny, classification and floral evolution of water lilies (Nymphaeaceae; Nymphaeales): a synthesis of nonmolecular, *rbcL, matK*, and 18S rDNA data. *Syst. Bot.* 24: 28–46.

Les, D. H., D. J. Crawford, E. Landolt, J. D. Gabel, and R. T. Kimball. 2002. Phylogeny and systematics of Lemnaceae, the duckweed family. *Syst. Bot.* 27: 221–240.

Les, D. H., M. L. Moody, and C. L. Soros. 2006. A reappraisal of phylogenetic relationships in the monocotyledon family Hydrocharitaceae (Alismatidae). *Aliso* 22: 211–230.

Levin, G. A. 1986. Systematic foliar morphology of Phyllanthoideae (Euphorbiaceae). III. Cladistic analysis. *Syst. Bot.* 11: 515–530.

Levin, R. A., W. L. Wagner, P. C. Hoch, M. Nepokroeff, J. C. Pires, E. A. Zimmer, and K. J. Sytsma. 2003. Family-level relationships of Onagraceae based on chloroplast *rbcL* and *ndhF* data. *Amer. J. Bot.* 90: 107–115.

Lewis, G., B. Schrire, B. Mackinder, and M. Lock (eds.). 2005. *Legumes of the world. Royal Botanic Gardens*, Kew.

Li, J., A. L. Bogle, and A. S. Klein. 1999. Phylogenetic relationships of the Hamamelidaceae inferred from sequences of internal transcribed spacers (ITS) of nuclear ribosomal DNA. *Amer. J. Bot.* 86: 1027–1037.

Li, R.-Q., Z.-D. Chen, A.-M. Lu, D. E. Soltis, P. S. Soltis, and P. S. Manos. 2004. Phylogenetic relationships in Fagales based on DNA sequences from three genomes. *Int. J. Plant Sci.* 165: 311–324.

Lidén, M. 1986. Synopsis of Fumarioideae (Papaveraceae) with a monograph of the tribe Fumarieae. *Opera Bot.* 88: 5–129.

Lidén, M. 1993. Fumariaceae. In *The families and genera of vascular plants*, vol. 2, Magnoliid, hamamelid and caryophyllid families, K. Kubitzki, J. G. Rohwer and V. Bittrich (eds.), 310–318. Springer-Verlag, Berlin.

Linder, H. P. 1985. Conspectus of the African species of Restionaceae. *Bothalia* 15: 387–503.

Linder, H. P. 2000. Vicariance, climate change, anatomy and phylogeny of Restionaceae. *Bot. J. Linn. Soc.* 134: 159–177.

Linder, H. P. and E. A. Kellogg. 1995. Phylogenetic patterns in the commelinid clade. In *Monocotyledons: systematics and evolution*, P. J. Rudall, P. J. Cribb, D. F. Cutler, and C. J. Humphries (eds.), 473–496. Royal Botanic Gardens, Kew.

Linder, H. P., B. G. Briggs, and L. A. S. Johnson. 1998. Anarthriaceae, pp. 19–20, Ecdeiocoleaceae, 195–196, and Restionaceae, 425–44, In *The families and genera of vascular plants*, vol. 4, Monocotyledons: Alismatanae and Commelinanae (except Gramineae), K. Kubitzki (ed.). Springer-Verlag, Berlin.

Linder, H. P., B. G. Briggs, and L. A. S. Johnson. 2000a. Restionaceae. In *The families and genera of vascular plants*, vol. 4, Monocotyledons: Alismatanae and Commelinanae (except Gramineae), K. Kubitzki (ed.), 425–445. Springer-Verlag, Berlin.

Linder, H. P., B. G. Briggs, and L. A. S. Johnson. 2000b. Restionaceae: a morphological phylogeny. In *Monocots: systematics and evolution*, K. L. Wilson, and D. A. Morrison (eds.), 653–660. CSIRO Publ., Colling-wood, Australia.

Litt, A. and V. F. Irish. 2003. Duplication and diversification in the APETALA/FRUITFULL floral homeotic gene lineage: implications for the evolution of floral development. *Genetics* 165: 821–833.

Loconte, H. 1993. Berberidaceae. In *The families and genera of vascular plants*, vol. 2, Magnoliid, hamamelid and caryophyllid families, K. Kubitzki, J. G. Rohwer, and V. Bittrich (eds.), 147–152. Springer-Verlag, Berlin.

Loconte, H. and J. R. Estes. 1989. Phylogenetic systematics of Berberidaceae and Ranunculales (Magnoliidae). *Syst. Bot.* 14: 565–579.

Loconte, T. and D. W. Stevenson. 1991. Cladistics of the Magnoliidae. *Cladistics* 7: 267–296.

Loconte, H., L. M. Campbell, and D. W. Stevenson. 1995. Ordinal and familial relationships of ranunculid genera. *Plant Syst. Evol. Suppl.* 9: 99–118.

Lohmann, L. G. 2006a. Untangling the phylogeny of neotropical lianas (Bignonieae, Bignoniaceae). *Amer. J. Bot.* 93: 304–318.

Lohmann, L. G. 2006b. A new generic classification of Bignonieae (Bignoniaceae) based on molecular phylogenetic data and morphological synapomorphies. *Ann. Missouri Bot. Gard.* In press.

Long, R. W. 1970. The genera of Acanthaceae in the southeastern United States. *J. Arnold Arbor.* 51: 257–309.

Londoño, A. C., E. Alvarez, E. Forero, and C. M. Morton. 1995. A new genus and species of Dipterocarpaceae from the Neotropics. I. Introduction, taxonomy, ecology and distribution. *Brittonia* 47: 225–236.

Lowden, R. M. 1973. Revision of the genus *Pontederia* L. *Rhodora* 75: 426–483.

Lu, A.-M. 1990. A preliminary cladistic study of the families of the superorder Lamiiflorae. *Bot. J. Linn. Soc.* 103: 39–57.

Lucas, E. J., S. R. Belsham, E. M. N. Lughadha, D. A. Orlovich, C. M. Sakuragui, M. W. Chase, and P. G. Wilson. 2005. Phylogenetic patterns in the fleshy-fruited Myrtaceae: preliminary molecular evidence. *Plant Syst. Evol.* 251: 35–51.

Lüders, H. 1907. Systematische Untersuchungen über die Caryophyllaceen mit einfachem Diagramm. *Bot. Jahrb.* 40: 1–38.

Lundberg, J. and K. Bremer. 2003. A phylogenetic study of the order Asterales using one morphological and three molecular data sets. *Int. J. Plant Sci.* 164: 553–578.

Mabry, T. J. 1973. Is the order Centrospermae monophyletic? In *Chemistry in botanical classification*, G. Bendz and J. Santesson (eds.), 275–285. Academic Press, New York.

Mabry, T. J. 1976. Pigment dichotomy and DNA–DNA hybridization data for Centrospermous families. *Plant Syst. Evol.* 126: 79–94.

Mabry, T. J., L. Kilmer, and C. Chang. 1972. The betalains: structure, function and biogenesis and the plant order Centrospermae. In *Recent advances in phytochemistry*, vol. 5, Structural and functional aspects of phytochemistry, V. C. Runeckles and T. C. Tso (eds.), 105–134. Academic Press, London.

MacDonald, A. D. 1974. Floral development in *Comptonia peregrina* (Myricaceae). *Canad. J. Bot.* 52: 2165–2169.

MacDonald, A. D. 1977. Myricaceae: floral hypothesis for *Gale* and *Comptonia*. *Canad. J. Bot.* 55: 2636–2651.

MacDonald, A. D. 1978. Organogenesis of the male inflorescence and flowers of *Myrica esculenta*. *Canad. J. Bot.* 56: 2415–2423.

MacDonald, A. D. 1979a. Development of the female flower and gynecandrous partial inflorescence of *Myrica californica*. *Canad. J. Bot.* 57: 141–151.

MacDonald, A. D. 1979b. Inception of the cupule of *Quercus macrocarpa* and *Fagus grandifolia*. *Canad. J. Bot.* 57: 1777–1782.

MacDonald, A. D. 1989. The morphology and relationships of the Myricaceae. In *Evolution, systematics and fossil history of the Hamamelidae*, vol. 2, "Higher" Hamamelidae. Systematics Association Special Vol. 40B, P. R. Crane and S. Blackmore (eds.), 147–165. Clarendon Press, Oxford.

Maguire, B. and P. S. Ashton. 1977. Pakaramoideae, Dipterocarpaceae of the Western Hemisphere. II. Systematic, geographic, and phyletic considerations. *Taxon* 26: 359–368

Maheshwari, S. C. 1958. *Spirodela polyrrhiza*: link between the aroids and the duckweeds. *Nature* 181: 1745–1746.

Manchester, S. R. 1987. The fossil history of the Juglandaceae. *Monogr. Syst. Bot. Missouri Bot. Gard.* 21: 1–137.

Manchester, S. R. 1989. Systematics and fossil history of the Ulmaceae. In *Evolution, systematics and fossil history of the Hamamelidae*, vol. 2, "Higher" Hamamelidae. Systematics Association Special Vol. 40B, P. R. Crane and S. Blackmore (eds.), 221–251. Clarendon Press, Oxford.

Manchester, S. R. and B. H. Tiffney. 2001. Integration of paleobotanical and neobotanical data in the assessment of phytogeographic history of holarctic angiosperm clades. *Int. J. Plant Sci.* 162 (6, Suppl.): 519–527.

Manen, J. F., A. Natalil, and F. Ehrendorfer. 1994. Phylogeny of Rubiaceae–Rubieae inferred from the sequence of a cp-DNA intergene region. *Plant Syst. Evol.* 190: 195–211.

Manhart, J. R. and J. H. Rettig. 1994. Gene sequence data. In *Caryophyllales: evolution and systematics*, H.-D. Behnke and T. J. Mabry (eds.), 235–246. Springer-Verlag, Berlin.

Mann, L. K. 1959. The *Allium* inflorescence: some species of the section *Molium*. *Amer. J. Bot.* 46: 730–739.

Manning, J. C., P. Goldblatt, and M. F. Fay. 2004. A revised generic synopsis of Hyacintheaceae in sub-Saharan Africa, based on molecular evidence, including new combinations and the new tribe Pseudoprospereae. *Edinburgh J. Bot.* 60: 533–568.

Manning, S. D. 2000. The genera of Bignoniaceae in the southeastern United States. *Harvard Pap. Bot.* 5: 1–77.

Manning, W. E. 1938. The morphology of the flower of the Juglandaceae. I. The inflorescence. *Amer. J. Bot.* 25: 407–419.

Manning, W. E. 1940. The morphology of the flower of the Juglandaceae. II. The pistillate flowers and fruit. *Amer. J. Bot.* 27: 839–852.

Manning, W. E. 1948. The morphology of the flower of the Juglandaceae. III. The staminate flowers. *Amer. J. Bot.* 35: 606–621.

Manning, W. E. 1978. The classification within the Juglandaceae. *Ann. Missouri Bot. Gard.* 65: 1058–1087.

Manos, P. S. 1997. Systematics of *Nothofagus* (Nothofagaceae) based on rDNA spacer sequences: taxonomic congruence with morphology and plastid sequences. *Amer. J. Bot.* 84: 1137–1155.

Manos, P. S. and K. P. Steele. 1997. Phylogenetic analyses of "higher" Hamamelididae based on plastid sequence data. *Syst. Bot.* 84: 1407–1419.

Manos, P. S. and D. E. Stone. 2001. Evolution, phylogeny and systematics of the Juglandaceae. *Ann. Missouri Bot. Gard.* 88: 231–269.

Manos, P. S., K. C. Nixon and J. J. Doyle. 1993. Cladistic analysis of restriction site variation within the chloroplast DNA inverted repeat region of selected Hamamelididae. *Syst. Bot.* 18: 551–562.

Manos, P. S., J. J. Doyle, and K. C. Nixon. 1999. Phylogeny, biogeography, and processes of molecular differentiation in *Quercus* subgenus *Quercus* (Fagaceae). *Mol. Phylog. and Evol.* 12: 333–349.

Manos, P. S., Z.-K. Zhou, and C. H. Cannon. 2001. Systematics of Fagaceae: phylogenetic tests of reproductive trait evolution. *Int. J. Plant Sci.* 162: 1361–1379.

Martins, T. R. and T. J. Barkman. 2005. Reconstruction of Solanaceae phylogeny using the nuclear gene SAMT. *Syst. Bot.* 30: 435–447.

Mast, A. R. 1998. Molecular systematics of subtribe Banksiinae (*Banksia* and *Dryandra*; Proteaceae) based on cpDNA and nrDNA sequence data: implications for taxonomy and biogeography. *Austr. Syst. Bot.* 11: 321–342.

Mast, A. R., S. Kelso, A. J. Richards, D. L. Lang, D. M. S. Feller, and E. Conti. 2001. Phylogenetic relationships in *Primula* L. and related genera (Primulaceae) based on noncoding chloroplast DNA. *Int. J. Plant Sci.* 162: 1138–1400.

Mathews, S. 2006. The positions of *Ceratophyllum* and *Chloranthaceae* inferred from phytochrome data. Abstract. *Botany 2006*: 238.

Mathews, S. and M. J. Donoghue. 1999. The root of angiosperm phylogeny inferred from duplicate phytochrome genes. *Science* 286: 947–950.

Mathews, S. and M. J. Donoghue. 2000. Basal angiosperm phylogeny inferred from duplicate phytochromes A and C. *Int. J. Plant Sci.* 161(6) (Suppl.): S41–S55.

Matthews, M. L. and P. K. Endress. 2004. Comparative floral structure and systematics in Cucurbitales (Corynocarpaceae, Coriariaceae, Tetramelaceae, Datiscaceae, Begoniaceae, Cucurbitaceae, and Anisophylleaceae). *Bot. J. Linn. Soc.* 145: 129–185.

Matthews, M. L. and P. K. Endress. 2005. Comparative floral structure and systematics in Celastrales (Celastraceae, Parnassiaceae, Lepidobotryaceae). *Bot. J. Linn. Soc.* 149: 129–194.

Mayo, S. J., J. Bogner, and P. Boyce. 1995. The Arales. In *Monocotyledons: systematics and evolution*, P. J. Rudall, P. J. Cribb, D. F. Cutler, and C. J. Humphries (eds.), 277–286. Royal Botanic Gardens, Kew.

Mayo, S. J., J. Bogner, and P. C. Boyce. 1998. Araceae. In *The families and genera of vascular plants*, vol. 4, Monocotyledons: Alismatanae and Commelinanae (except Gramineae), K. Kubitzki (ed.), 26–74. Springer-Verlag, Berlin.

Mayr, E. M. and A. Weber. 2006. Calceolariaceae: floral development and systematic implications. *Amer. J. Bot.* 93: 327–343.

McDade, L. A. and M. L. Moody. 1999. Phylogenetic relationships among Acanthaceae: evidence from noncoding *trnL-trnF* chloroplast DNA sequences. *Amer. J. Bot.* 86: 70–80.

McDade, L. A., S. E. Masta, M. L. Moody, and E. Waters. 2000. Phylogenetic relationships among Acanthaceae: evidence from two genomes. *Syst. Bot.* 25: 106–121.

McDaniel, S. T. 1971. The genus *Sarracenia* (Sarraceniaceae). *Bull. Tall Timbers Res. Sta.* 9: 1–36.

McKelvey, S. D. and K. Sax. 1933. Taxonomic and cytological relationships of *Yucca* and *Agave*. *J. Arnold Arbor.* 14: 76–80.

McKey, D. 1989. Interactions between ants and leguminous plants. *Monogr. Syst. Bot. Missouri Bot. Gard.* 29: 673–718.

McPherson, M. A. and S. W. Graham. 2001. Inference of Asparagales phylogeny using a large chloroplast data set. Abstract. *Botany 2001*: 126.

Meacham, C. A. 1980. Phylogeny of the Berberidaceae with an evaluation of classifications. *Syst. Bot.* 5: 149–172.

Medan, D. and C. Schirarend. 2004. Rhamnaceae. In *The families and genera of vascular plants*, vol. 6, Celastrales, Oxalidales, Rosales, Cornales, Ericales, K. Kubitzki (ed.), 320–338. Springer-Verlag, Berlin.

Meerow, A. W. 1995. Towards a phylogeny of Amaryllidaceae. In *Monocotyledons: systematics and evolution*, P. J. Rudall, P. J. Cribb, D. F. Cutler, and C. J. Humphries (eds.), 169–179. Royal Botanic Gardens, Kew.

Meerow, A. W. and D. A. Suijman. 1998. Amaryllidaceae In *The families and genera of vascular plants*, vol. 3. Monocotyledons: Lilianae (except Orchidaceae), K. Kubitzki (ed.), 83–110. Springer-Verlag, Berlin.

Meerow, A. W. and D. A. Suijman. 2006. The never-ending story: multigene approaches to the phylogeny of Amaryllidaceae. *Aliso* 22: 355–366.

Meerow, A. W., M. F. Fay, C. L. Guy, Q.-B. Li, F. Q. Zaman, and M. W. Chase. 1999a. Systematics of Amaryllidaceae based on cladistic analysis of plastid *rbcL* and *trnL-F* sequence data. *Amer. J. Bot.* 86: 1325–1345.

Meerow, A. W., M. F. Fay, M. W. Chase, C. L. Guy, and Q.-B. Li. 1999b. The new phylogeny of the Amaryllidaceae. *Herbertia* 54: 180–203.

Meerow, A., M. F. Fay, M. W. Chase, C. l. Guy, Q.-B. Li, D. Snijman, and S.-L. Yang. 2000a. Phylogeny of Amaryllidaceae: molecules and morphology. In *Monocots: systematics and evolution*, K. L. Wilson and D. A. Morrison (eds.), 372–386. CSIRO, Collingwood, Australia.

Meerow, A. W., C. L. Guy, Q.-B. Li, and S.-L. Yang. 2000b. Phylogeny of the American Amaryllidaceae based on nrDNA ITS sequences. *Syst. Bot.* 25: 708–726.

Meeuse, A. D. J. 1975. Taxonomic relationships of Salicaceae and Flacourtiaceae. *Acta Bot. Neerl.* 24: 437–457.

Meeuse, B. J. D. and E. L. Schneider. 1979. *Nymphaea* revisited: a preliminary communication. *Israel J. Bot.* 28: 65–79.

Mennega, A. M. W. 1982. Stem structure of New World Menispermaceae. *J. Arnold Arbor.* 63: 145–171.

Mészáros, S., J. de Laet and E. Smets. 1996. Phylogeny of temperate Gentianaceae: a morphological approach. *Syst. Bot.* 21: 153–168.

Meurer-Grimes, B. 1995. New evidence for the systematic significance of acylated spermidines and flavonoids in pollen of higher Hamamelidae. *Brittonia* 47: 130–142.

Michaels, H. J., K. M. Scott, R. G. Olmstead, T. Szaro, R. K. Jansen, and J. D. Palmer. 1993. Interfamilial relationships of the Asteraceae: insights from *rbcL* sequence variation. *Ann. Missouri Bot. Gard.* 80: 742–751.

Michelangeli, F. A., J. I. Davis and D. W. Stevenson. 2003. Phylogenetic relationships among Poaceae and related families as inferred from morphology, inversions in the plastid genome, and sequence data from the mitochondrial and plastid genomes. *Amer. J. Bot.* 90: 93–106.

Michelangeli, F. A., D. S. Penneys, J. Giza, D. Soltis, M. H. Hils, and J. Dan Skean, Jr. 2004. Preliminary phylogeny of the tribe Miconieae (Melastomataceae) based on nrITS sequence data and its implications on inflorescence position. *Taxon* 53: 279–290.

Miller, N. G. 1970. The genera of Cannabaceae in the southeastern United States. *J. Arnold Arbor.* 51: 185–203.

Miller, N. G. 1971a. The genera of Polygalaceae in the southeastern United States. *J. Arnold Arbor.* 52: 267–284.

Miller, N. G. 1971b. The genera of Urticaceae in the southeastern United States. *J. Arnold Arbor.* 52: 40–68.

Miller, N. G. 1990. The genera of Meliaceae in the southeastern United States. *J. Arnold Arbor.* 71: 453–486.

Miller, N. G. 2001. The Callitrichaceae in the southeastern United States. *Harvard Pap. Bot.* 5(2): 277–301.

Möller, M. and Q. C. B. Cronk. 1997. Origin and relationships of *Saintpaulia* (Gesneriaceae) based on ribosomal DNA internal transcribed spacer (ITS) sequences. *Amer. J. Bot.* 84: 956–965.

Molvray, N., P. J. Kores, and M. W. Chase. 2000. Polyphyly of mycoheterotrophic orchids and functional influences on floral and molecular characters. In *Monocots: systematics and evolution*. K. L. Wilson and D. A. Morrison (eds.), 441–448. CSIRO, Collingwood, Australia.

Moore, H. E. 1973. The major groups of palms and their distribution. *Gentes Herb.* 11: 27–141.

Moore, H. E. and N. W. Uhl. 1982. The major trends of evolution in palms. *Bot. Rev.* 48: 1–69.

Morgan, D. R. and D. E. Soltis. 1993. Phylogenetic relationships among members of Saxifragaceae sensu lato based on *rbcL* sequence data. *Ann. Missouri Bot. Gard.* 80: 631–660.

Morgan, D. R., D. E. Soltis and K. R. Robertson. 1994. Systematic and evolutionary implications of *rbcL* sequence variation in Rosaceae. *Amer. J. Bot.* 81: 890–903.

Mori, S. A., G. T. Prance, and A. B. Bolten. 1978. Additional notes on the floral biology of neotropical Lecythidaceae. *Brittonia* 30: 113–130.

Mori, S. A. and G. T. Prance. 1990. Lecythidaceae. Part II. The zygomorphic-flowered New World genera (*Couroupita, Corythophora, Bertholletia, Couratari, Eschweilera,* and *Lecythis*). *Flora Neotrop. Monogr.* 21(II): 1–376.

Morley, T. 1976. Memecyleae (Melastomataceae). *Flora Neotropica Monographs* 15: 1–295.

Mort, M. E., D. E. Soltis and P. S. Soltis. 1998. Molecular systematics of Crassulaceae based on *matK* sequence data. *Amer. J. Bot.* 85(6) Suppl.: 145–146.

Mort, M. E., D. E. Soltis, P. S. Soltis, J. Francisco-Ortega, and A. Santos-Guerra. 2001. Phylogenetic relationships and evolution of Crassulaceae inferred from *matK* sequence data. *Amer. J. Bot.* 88: 76–91.

Morton, C. M. 1995. A new genus and species of Dipterocarpaceae from the Neotropics. II. Stem anatomy. *Brittonia* 47: 237–247.

Morton, C. M., M. W. Chase, and J. Kallunki. 1996. Evaluation of the six subfamilies of Rutaceae using evidence from *rbcL* sequence variation. *Amer. J. Bot.* 83(6) Suppl.: 180–181.

Morton, C. M., K. A. Kron, and M. W. Chase. 1997a. A molecular evaluation of the monophyly of the order Ebenales based upon *rbcL* sequence data. *Syst. Bot.* 21: 567–586.

Morton, C. M., S. A. Mori, G. T. Prance, K. G. Karol, and M. W. Chase. 1997b. Phylogenetic relationships of Lecythidaceae: a cladistic analysis using *rbcL* sequence and morphological data. *Amer. J. Bot.* 84: 530–549.

Morton, C. M., G. T. Prance, S. M. Mori, and L. G. Thorburn. 1998. Re-circumscription of the Lecythida-ceae. *Taxon* 47: 817–827.

Morton, C. M., M. Grant, and S. Blackmore. 2003. Phylogenetic relationships of the Aurantioideae inferred from chloroplast DNA sequence data. *Amer. J. Bot.* 90: 1463–1469.

Moseley, M. F., E. L. Schneider, and P. S. Williamson. 1993. Phylogenetic interpretations from selected floral vasculature characters in the Nymphaeaceae sensu lato. *Aquatic Bot.* 44: 325–342.

Muellner, A. N., R. Samuel, S. A. Johnson, M. Cheek, T. D. Pennington, and M. W. Chase. 2003. Molecular phylogenetics of Meliaceae (Sapindales) based on nuclear and plastid DNA sequences. *Amer. J. Bot.* 90: 471–480.

Muller, J. and P. W. Leenhouts. 1976. A general survey of pollen types in Sapindaceae in relation to taxonomy. In *The evolutionary significance of the exine*. Linn. Soc. Symp. Ser. No. 1, I. K. Ferguson and J. Muller (eds.), 407–455. Academic Press, London.

Müller, K. and T. Borsch. 2005. Phylogenetics of Amaranthaceae based on *matK/trnK* sequence data: evidence from parsimony, likelihood, and Bayesian analyses. *Ann. Missouri Bot. Gard.* 92: 66–102.

Müller, K., T. Borsch, L. Legendre, S. Porembski, and W. Barthlott. 2000. A phylogeny of Lentibulariaceae based on sequences of *matK* and adjacent noncoding regions. *Amer. J. Bot.* 87(6) (Suppl.): S145–S146.

Munro, S. L. and H. P. Linder. 1998. The phylogenetic position of *Prionium* (Juncaceae) within the order Juncales based on morphological and *rbcL* sequence data. *Syst. Bot.* 23: 43–55.

Munzinger, J. K. and H. E. Ballard, Jr. 2003. *Hekkingia* (Violaceae), a new arborescent violet genus from French Guiana, with a key to genera in the family. *Syst. Bot.* 28: 345–351.

Murrell, Z. E. 1993. Phylogenetic relationships in *Cornus* (Cornaceae). *Syst. Bot.* 18: 469–495.

Musaya, A. M., D. A. Simpson, M. W. Chase, and A. Culham. 1998. An assessment of suprageneric phylogeny in Cyperaceae using *rbcL* DNA sequences. *Plant Syst. Evol.* 211: 257–271.

Muschner, V. C., A. P. Lorenz, A. C. Cervil, S. L. Bonatto, T. T. Souza-Chies, F. M. Salzano, and L. B. Freitas. 2003. A first molecular phylogenetic analysis of *Passiflora* (Passifloraceae). *Amer. J. Bot.* 90: 1229–1238.

Nandi, O. I. 1998. Floral development and systematics of Cistaceae. *Plant Syst. Evol.* 212: 107–134.

Nandi, O. I., M. W. Chase, and P. L. Endress. 1998. A combined cladistic analysis of angiosperms using *rbcL* and non-molecular data sets. *Ann. Missouri Bot. Gard.* 85: 137–212.

Neyland, R. 2001. A phylogeny inferred from large ribosomal subunit (26S) rDNA sequences suggests that *Cuscuta* is a derived member of Convolvulaceae. *Brittonia* 53: 108–114.

Neyland, R. and L. E. Urbatsch. 1995. A terrestrial origin for the Orchidaceae suggested by a phylogeny inferred from *ndhF* chloroplast gene sequences. *Lindleyana* 10: 244–251.

Neyland, R. and L. E. Urbatsch. 1996. Phylogeny of subfamily Epidendroideae (Orchidaceae) inferred from *ndhF* chloroplast gene sequences. *Amer. J. Bot.* 83: 1195–1206.

Nicholas, A. and H. Baijnath. 1994. A consensus classification for the order Gentianales with additional details on the suborder Apocynineae. *Bot. Rev.* 60: 440–482.

Nickrent, D. L. and D. E. Soltis. 1995. A comparison of angiosperm phylogenies based upon complete 18S rDNA and *rbcL* sequences. *Ann. Missouri Bot. Gard.* 83: 208–234.

Nickrent, D. L. and R. J. Duff. 1996. Molecular studies of parasitic plants using ribosomal RNA. In *Advances in parasitic plant research*, M. T. Moreno, J. I. Cubero, D. Nerner, D. Joel, L. J. Musselman, and C. Parker (eds.), 28–52. Dirección General de Investigación Agraria, Córdoba, Spain.

Nickrent, D. L., R. J. Duff, A. E. Colwell, A. D. Wolfe, N. D. Young, K. E. Steinem, and C. W. de Pamphilis. 1998. Molecular phylogenetic and evolutionary studies of parasitic plants. In *Molecular systematics of plants*, vol. 2, DNA sequencing. D. Soltis, P. Soltis and J. Doyle (eds.), Chapter 8. Kluwer Academic, Boston.

Nickrent, D. L., A. Blarer, Y.-L. Qiu, D. E. Soltis, P. S. Soltis, and M. Zanis. 2002. Molecular data place Hydnoraceae with Aristolochiaceae. *Amer. J. Bot.* 89: 1809–1817.

Nilsson, L. A., E. Rabakonandrianina, B. Pettersson, and J. Ranaivo. 1990. "Ixoroid" secondary pollen presentation and pollination by small moths in the Malagasy treelet *Ixora platythyrsa* (Rubiaceae). *Plant Syst. Evol.* 170: 161–175.

Nixon, K. C. 1989. Origins of Fagaceae. In *Evolution, systematics and fossil history of the Hamamelidae*, vol. 2, "Higher" Hamamelidae. Systematics Association Special Vol. 40B, P. R. Crane, and S. Blackmore (eds.), 23–43. Clarendon Press, Oxford.

Nixon, K. C. and J. M. Poole. 2003. Revision of the Mexican and Guatemalan species of *Platanus* (Platanaceae). *Lundellia* 6: 103–137.

Nixon, K. C., R. J. Jensen, P. S. Manos, and C. H. Muller. 1995, Fagaceae. In *Flora North America*, vol. 3, 436–506.

Nooteboom, H. P. 1993. Magnoliaceae. In *The families and genera of vascular plants*, vol. 2, Magnoliid, hamamelid and caryophyllid families, K. Kubitzki, J. G. Rohwer, and V. Bittrich (eds.), 391–401. Springer-Verlag, Berlin.

Nordenstam, B. 1998. Colchicaceae. In *The families and genera of vascular plants*, vol. 3, Monocotyledons: Lilianae (except Orchidaceae), K. Kubitzki (ed.), 175–185. Springer-Verlag, Berlin.

Norman, E. M. and D. Clayton. 1986. Reproductive biology of two Florida pawpaws: *Asimina obovata* and *A. pygmaea* (Annonaceae). *Bull. Torrey Bot. Club* 113: 16–22.

Nyananyo, B. L. 1990. Tribal and generic relationships in the Portulacaceae (Centrospermae). *Feddes Repert.* 101: 237–241.

Nyffeler, R. 2002. Phylogenetic relationships in the cactus family (Cactaceae) based on evidence from *trnK/makK* and *trnL-trnF* sequences. *Amer. J. Bot.* 89: 312–326.

Nyffeler, R. 2007. The closest relatives of cacti: insights from phylogenetic analyses of chloroplast and mitochondrial sequences with special emphasis on relationships in the tribe Anacampseroteae. *Amer. J. Bot.* 94: 89–101.

Oh, S.-H. and P. S. Manos. 2006. Cups, nuts, and catkins: a phylogeny of Fagaceae based on CRABS CLAW sequences. Abstract. *Botany 2006:* 246.

O'Kane, S. L., Jr., I. A. Al-Shehbaz. 2003. Phylogenetic position and generic limits of *Arabidopsis* (Brassicaceae) based on sequences of nuclear ribosomal DNA. *Ann. Missouri Bot. Gard.* 90: 603–612.

Okomoto, M. 1989. New interpretation of the inflorescence of *Fagus* drawn from the developmental study of *Fagus crenata*, with description of an extremely monstrous cupule. *Amer. J. Bot.* 76: 14–22.

Oldeman, R. A. A. 1989. Biological implications of leguminous tree architecture. *Monogr. Syst. Bot. Missouri Bot. Gard.* 29: 17–34.

Olmstead, R. and D. Ferguson. 2001. A molecular phylogeny of the Boraginaceae/Hydrophyllaceae. Abstract. *Botany 2001:* 131.

Olmstead, R. and J. D. Palmer. 1991. Chloroplast DNA and systematics of the Solanaceae. In *Solanaceae 3: Taxonomy, chemistry, evolution*, J. G. Hawkes, R. N. Lester, M. Nee, and N. Estrada (eds.), 161–168. Royal Botanic Gardens, Kew.

Olmstead, R. G. and J. D. Palmer. 1992. A chloroplast DNA phylogeny of the Solanaceae: subfamilial relationships and character evolution. *Ann. Missouri Bot. Gard.* 79: 346–360.

Olmstead, R. G., H. J. Michaels, K. M. Scott, and J. D. Palmer. 1992a. Monophyly on the Asteridae and identification of their major lineages inferred from DNA sequences of *rbcL*. *Ann. Missouri Bot. Gard.* 79: 249–265.

Olmstead, R. G. and P. A. Reeves. 1995. Evidence for the polyphyly of the Scrophulariaceae based on chloroplast *rbcL* and *ndhF* sequences. *Ann. Missouri Bot. Gard.* 82: 176–193.

Olmstead, R. G., K. M. Scott, and J. D. Palmer. 1992b. A chloroplast DNA phylogeny for the Asteridae: implication for the Lamiales. In *Advances in labiate science*, R. M. Harley and T. Reynolds (eds.), 19–25. Royal Botanic Gardens, Kew.

Olmstead, R. G., B. Bremer, K. M. Scott, and J. D. Palmer. 1993. A parsimony analysis of the Asteridae sensu lato based on *rbcL* sequences. *Ann. Missouri Bot. Gard.* 80: 700–722.

Olmstead, R. G. and J. A. Sweere. 1994. Combining data in phylogenetic systematics: an empirical approach using three molecular data sets in the Solanaceae. *Syst. Biol.* 43: 467–481.

Olmstead, R. G., J. A. Sweere, R. E. Spangler, L. Bohs, and J. D. Palmer. 1995. Phylogeny and provisional classification of the Solanaceae based on chloroplast DNA. In *Solanaceae IV*, M. Nee, D. E. Symon, R. N. Lester, and J. P. Jessop (eds.), 111–137. Royal Botanic Gardens, Kew.

Olmstead, R. G. and J. D. Palmer. 1997. Implications for the phylogeny, classification and biogeography of *Solanum* from cpDNA restriction site variation. *Syst. Bot.* 22: 19–29.

Olmstead, R. G., K.-J. Kim, R. K. Jansen, and S. J. Wagstaff. 2000. The phylogeny of the Asteridae sensu lato based on chloroplast *ndhF* gene sequences. *Mol. Phylogenetics and Evol.* 16: 96–112.

Olmstead, R. G. and D. Ferguson. 2001. A molecular phylogeny of Boraginaceae/Hydrophyllaceae. Abstract. *Botany 2001:* 131.

Olmstead, R. G., J. A. Sweere, R. E. Spangler, L. Bohs, and J. D. Palmer. 1999. Phylogeny and provisional clasification of the Solanaceae based on chloroplast DNA. In *Solanaceae IV: advances in biology and utilization*. M. Nee, D. Symon, R. N. Lester, and J. P. Jessop (eds.), 111–137, Royal Botanic Gardens, Kew.

Olmstead, R. G., C. W. de Pamphilis, A. D. Wolfe, N. D. Young, W. J. Elisons, and P. A.

Reeves. 2001. Disintegration of the Scrophulariaceae. *Amer. J. Bot.* 88: 348–361.

Olmstead, R. and 17 others. 2000b. A synoptical classification of the Lamiales. Version 1.0 (in progress). Unpublished document.

Omori, Y. and S. Terabayashi. 1993. Gynoecial vascular anatomy and its systematic implications in Celtidaceae and Ulmaceae (Urticales). *J. Plant Res.* 106: 249–258.

Orgaard, M. 1991. The genus *Cabomba* (Cabombaceae): a taxonomic study. *Nordic J. Bot.* 11: 179–203.

Ornduff, R. 1966. The breeding system of *Pontederia cordata* L. *Bull. Torrey Bot. Club* 93: 407–416.

Ornduff, R. 1972. The breakdown of trimorphic incompatibility in *Oxalis* section *corniculatae*. *Evolution* 26: 52–65.

Osborn, J. M. and E. L. Schneider. 1988. Morphological studies of Nymphaeaceae sensu lato. XVI. The floral biology of *Brasenia schreberi*. *Ann. Missouri Bot. Gard.* 75: 778–794.

Osborn, J. M., T. N. Taylor, and E. L. Schneider. 1991. Pollen morphology and ultrastructure of the Cabombaceae: correlations with pollination biology. *Amer. J. Bot.* 78: 1367–1378.

Ownbey, M. 1950. Natural hybridization and amphiploidy in the genus *Tragopogon*. *Amer. J. Bot.* 37: 487–499.

Oxelman, B., M. Backlund, and B. Bremer. 1999. Relationships of the Buddlejaceae s. l. investigated using parsimony Jackknife and Branch Support Analysis of chloroplast *ndhF* and *rbcL* sequence data. *Syst. Bot.* 24: 164–182.

Oxelman, B., P. Kornhall, R. G. Olmstead, and B. Bremer. 2005. Further disintegration of Scrophulariaceae. *Taxon* 54: 411–425.

Padgett, D. J., D. H. Les, and G. E. Crow. 1999. Phylogenetic relationships in *Nuphar* (Nymphaeaceae): evidence from morphology, chloroplast DNA, and nuclear ribosomal DNA. *Amer. J. Bot.* 86: 1316–1324.

Panero, J. L. and V. A. Funk. 2002. Toward a phylogenetic classification for the Compositae (Asteraceae). *Proc. Biol. Soc. Washington* 115: 909–922.

Park, K. R. and W. S. Elisens. 2000. A phylogenetic study of tribe Euphorbieae (Euphorbiaceae). *Int. J. Plant Sci.* 161: 425–434.

Patel, V. C., J. J. Skvarla, and P. H. Raven. 1984. Pollen characters in relation to the delimitation of Myrtales. *Ann. Missouri Bot. Gard.* 71: 858–969.

Pennington, T. D. 1991. *The genera of Sapotaceae*. Royal Botanic Gardens, Kew and New York Botanical Gardens, Bronx.

Pennington, T. D. 2004. Sapotaceae. In *The families and genera of vascular plants*, vol. 6, Celastrales, Oxalidales, Rosales, Cornales, Ericales, K. Kubitzki (ed.). Springer-Verlag, Berlin.

Pennington, T. D. and B. T. Styles. 1975. A generic monograph of the Meliaceae. *Blumea* 22: 419–540.

Persson, C. 2001. Phylogenetic relationships in Polygalaceae based on plastid DNA sequences from the *trnL-F* region. *Taxon* 50: 763–779.

Petersen, F. P. and D. E. Fairbrothers. 1983. A serotaxonomic appraisal of *Amphipterygium* and *Leitneria*, two amentiferous taxa of Rutiflorae (Rosidae). *Syst. Bot.* 8: 134–148.

Pfeil, B. E. and M. D. Crisp. 2005. What to do with *Hibiscus*? A proposed nomenclatural resolution for a large and well known genus of Malvaceae and comments on paraphyly. *Austr. Syst. Bot.* 18: 49–60.

Pfosser, M. and F. Speta. 1999. Phylogenetics of Hyacinthaceae based on plastid DNA sequences. *Ann. Missouri Bot. Gard.* 86: 852–875.

Philbrick, C. T. and R. K. Jansen. 1991. Phylogenetic studies of North American *Callitriche* (Callitrichaceae) using chloroplast DNA restriction fragment analysis. *Syst. Bot.* 16: 478–491.

Philipson, W. R. 1993. Amborellaceae. In *The families and genera of vascular plants*, vol. 2, Magnoliid, hamamelid and caryophyllid families, K. Kubitzki, J. G. Rohwer, and V. Bittrich (eds.), 92–93. Springer-Verlag, Berlin.

Phipps, J. B., K. R. Robertson, P. G. Smith, and J. R. Rohrer. 1990. A checklist of the subfamily Maloideae (Rosaceae). *Canad. J. Bot.* 68: 2209–2269.

Pichon, M. 1946. Sur les Alismatacées et les Butomacées. *Not. Syst.* 12: 170–183.

Pires, J. C. and K. J. Sytsma. 2002. A phylogenetic evaluation of a biosystematic framework: *Brodiaea* and related petaloid monocots (Themidaceae). *Amer. J. Bot.* 89: 1342–1359.

Pires, J. C., I. J. Maureira, J. P. Rebman, G. A. Salazar, L. I. Cabrera, M. F. Fay, and M. W. Chase. 2004. Molecular data confirm the phylogenetic placement of the enigmatic *Hesperocallis* (Hesperocallidaceae) with *Agave. Madroño* 51: 307–311.

Pires, J. C., I. J. Maureira, T. J. Givnish, K. J. Sytsma, O. Seberg, G. Petersen, J. I. Davis, D. W. Stevenson, P. J. Rudall, M. F. Fay, and M. W. Chase. 2006. Phylogeny, genome size, and chromosome evolution of Asparagales. *Aliso* 22: 278–304.

Plana, V. 2003. Phylogenetic relationships of the Afro-Malagasy members of the large genus *Begonia* inferred from *trnL* intron sequences. *Syst. Bot.* 28: 693–704.

Plunkett, G. M. and S. R. Downie. 1999. Major lineages within Apiaceae subfamily Apioideae: comparison of chloroplast restriction site and DNA sequence data. *Amer. J. Bot.* 86: 1014–1026.

Plunkett, G. M., D. E. Soltis, P. S. Soltis, and R. E. Brooks. 1995. Phylogenetic relationships between Juncaceae and Cyperaceae: insights from *rbcL* sequence data. *Amer. J. Bot.* 82: 520–525.

Plunkett, G. M., D. E. Soltis, and P. S. Soltis. 1996a. Evolutionary patterns in Apiaceae: inferences based on *matK* sequence data. *Syst. Bot.* 21: 477–495.

Plunkett, G. M., D. E. Soltis, and P. S. Soltis. 1996b. Higher level relationships of Apiales (Apiaceae and Araliaceae) based on phylogenetic analysis of *rbcL* sequences. *Amer. J. Bot.* 83: 499–515.

Plunkett, G. M., D. E. Soltis, and P. S. Soltis. 1997. Clarification of the relationship between Apiaceae and Araliaceae based on *matK* and *rbcL* sequence data. *Amer. J. Bot.* 84: 567–580.

Plunkett, G. M., J. Wen, and P. P. Lowry, II. 2004. Infrafamilial classifications and characters in Araliaceae: insights from the phylogenetic analysis of nuclear (ITS) and plastid (*trnL-trnF*) sequence data. *Plant Syst. Evol.* 245: 1–39.

Polhill, R. M. 1981. Papilionoideae. In *Advances in legume systematics*, part 1, R. M. Polhill and P. H. Raven (eds.), 191–208. Royal Botanic Gardens, Kew.

Polhill, R. M., P. H. Raven, and C. H. Stirton. 1981. Evolution and systematics of the Leguminosae. In *Advances in legume systematics*, part 1, R. M. Polhill, and P. H. Raven (eds.), 1–26. Royal Botanic Gardens, Kew.

Porter, D. M. 1972. The genera of Zygo-phyllaceae in the southeastern United States. *J. Arnold Arbor.* 53: 531–552.

Porter, J. M. and L. A. Johnson. 1998. Phylogenetic relationships of Polemoniaceae: inferences from mitochrondrial *NAD1B* intron sequences. *Aliso* 17: 157–188.

Porter, J. M. and L. A. Johnson. 2000. A phylogenetic classification of Polemoniaceae. *Aliso* 19:55–91.

Potgieter, K. and V. A. Albert. 2001. Phylogenetic relationships within Apocynaceae s.l. based on *trnL* intron and *trnL-F* spacer sequences and propagule characters. *Ann. Missouri Bot. Gard.* 88: 523–549.

Potter, D. 2003. Molecular phylogenetic studies in Rosaceae. In *Plant genome: biodiversity and evolution*, vol. 1, Pt. A: Phanerogams, A. K. Sharma, A. Sharma (eds.), 319–351. Science Publishers, Enfield, NH.

Potter, D., F. Gao, P. E. Bortiri, S. Oh, and S. Baggett. 2002. Phylogenetic relationships in Rosaceae inferred from chloroplast *matK* and *trnL-trnF* nucleotide sequence data. *Plant Syst. Evol.* 231: 77–89.

Potter, D., T. Eriksson, R. C. Evans, S.-H. Oh, J. Smedmark, D. Morgan, M. Kerr, K. R. Robertson, M. Arsenault, T. A. Dickinson, and C. S. Campbell. 2006. Phylogeny and classification of Rosaceae. *Plant Syst. Evol.* In press.

Powell, M., V. Savolainen, P. Cuénoud, J.-F. Manen, and S. Andrews. 2000. The mountain holly (*Nemopanthus mucronatus*: Aquifoliaceae) revisited with molecular data. *Kew Bull.* 55: 341–347.

Prance, G. T. 1976. The pollination and androphore structure of some Amazonian Lecythidaceae. *Biotropica* 8: 235–241.

Prance, G. T. and S. A. Mori. 1978. Observations on the fruits and seeds of neotropical Lecythidaceae. *Brittonia* 30: 21–33.

Prance, G. T. and S. A. Mori. 1979. Lecythidaceae. Part. I. The actinomorphic-flowered New World Lecythidaceae (*Asteranthos, Gustavia, Grias, Allantoma*, and *Cariniana*). *Flora Neotrop. Monogr.* 21: 1–270.

Prance, G. T. and V. Plana. 1998. The American Proteaceae. *Austr. Syst. Bot.* 11: 287–299.

Prance, G. T. and S. A. Mori. 2004. Lecythidaceae. In *The families and genera of vascular plants*, Vol. 6, Celastrales, Oxalidales, Rosales, Cornales, Ericales, K. Kubitzki (ed.), 221–232. Springer-Verlag, Berlin.

Prather, L. A., C. J. Ferguson, and R. K. Jansen. Polemoniaceae phylogeny and classification: implications of sequence data from the chloroplast gene *ndhF. Amer. J. Bot.* 87: 1300–1308.

Price, R. A. and J. D. Palmer. 1993. Phylogenetic relationships of Geraniaceae and Geraniales from *rbcL* sequence comparisons. *Ann. Missouri Bot. Gard.* 80: 661–671.

Price, S. D. and S. C. H. Barrett. 1982. Tristyly in *Pontederia cordata* L. (Pontederiaceae). *Canad. J. Bot.* 60: 897–905.

Prince, L. M. and C. R. Parks. 2001. Phylogenetic relationships of Theaceae inferred from chloroplast DNA sequence data. *Amer. J. Bot.* 88: 2309–2320.

Prince, L. M. and W. J. Kress. 2006. Phylogenetic relationships and classification in Marantaceae: insights from plastid DNA sequence data. *Taxon* 55: 281–296.

Prior, P. V. 1960. Development of the helicoid and scorpioid cymes in *Myosotis laxa* Lehm. and *Mertensia virginica* L. *Proc. Iowa Acad. Sci.* 67: 76–81.

Proctor, M. and P. Yeo. 1972. *The pollination of flowers*. Taplinger, New York.

Proctor, M., P. Yeo, and A. Lack. 1996. *The natural history of pollination*. Timber Press, Portland, Oregon.

Prychid, C. J. and P. J. Rudall. 2000. Distribution of calcium oxalate crystals in monocotyledons. In *Monocots: systematics and evolution*, K. L. Wilson and D. A. Morrison (eds.), 159–162. CSIRO, Collingwood, Australia.

Puri, V. 1948. Studies in floral anatomy. V. On the structure and nature of the corona in certain species of the Passifloraceae. *J. Indian Bot. Soc.* 27: 130–149.

Qiu, Y.-L., M. W. Chase, D. H. Les, and C. R. Parks. 1993. Molecular phylogenetics of the Magnoliidae: cladistic analyses of nucleotide sequences of the plastid gene *rbcL. Ann. Missouri Bot. Gard.* 80: 587–606.

Qiu, Y.-L., M. W. Chase, and C. R. Parks. 1995. A chloroplast DNA phylogenetic study of the eastern Asia-eastern North America disjunct section *Rhytidospermum* of *Magnolia* (Magnoliaceae). *Amer. J. Bot.* 82: 1582–1588.

Qiu, Y.-L., J. Les, F. Bernascond-Quadroni, D. E. Soltis, P. S. Soltis, M. Zanis, E. A. Zimmer, Z. Chen, V. Savolainen, and M. W. Chase. 1999. The earliest angiosperms: evidence from mitochondria, plastid and nuclear genomes. *Nature* 402: 404–407.

Qiu, Y.-L., J. H. Lee, F. Bernasconi-Quadroni, D. E. Soltis, P. S. Soltis, M. Zanis, E. A. Zimmer, Z. D. Chen, V. Savolainen, and M. W. Chase. 2000. Phylogeny of basal angiosperms: analyses of 5 genes from 3 genomes. *Int. J. Plant Sci.* 161(6) (Suppl.): S3–S27.

Qiu, Y.-L. and 19 others. 2005. Phylogenetic analysis of basal angiosperms based on 9 plastid, mitochondrial, and nuclear genes. *Int. J. Plant Sci.* 166: 815–842.

Rabinowitz, D. 1978. Dispersal properties of mangrove propagules. *Biotropica* 10: 47–57.

Rahmanzadeh, R. K. Müller, E. Fisher, D. Bartels, and T. Borsch. 2005. The Linderniaceae and Gratiolaceae are further lineages distinct from the Scrophulariaceae (Lamiales). *Plant Biol.* 7: 67–78.

Rahn, K. 1996. A phylogenetic study of the Plantaginaceae. *Bot. J. Linn. Soc.* 120: 145–198.

Rahn, K. 1998. Alliaceae. In *The families and genera of vascular plants*, vol. 3. Monocotyledons: Lilianae (except Orchidaceae), K. Kubitzki (ed.), 70–78. Springer-Verlag, Berlin.

Rama Devi, D. 1991. Floral anatomy of *Hypseocharis* (Oxalidaceae) with a discussion on its systematic position. *Plant Syst. Evol.* 177: 161–164.

Ramivrez B., W. and L. D. Gomez P. 1978. Production of nectar and gums by flowers of *Monstera deliciosa* (Araceae) and some speices of *Clusia* (Guttiferae) collected by New World *Trigona* bees. *Brenesia* 14–15: 407–412.

Ranker, T. A., D. E. Soltis, and A. J. Gilmartin. 1990. Subfamilial phylogenetic relationships of the Bromeliaceae: evidence from chloroplast DNA restriction site variation. *Syst. Bot.* 15: 425–434.

Rao, C.V. 1971. Anatomy of the inflorescence of some Euphorbiaceae with a discussion on the phylogeny and evolution of the inflorescnece including the cyathium. *Bot. Not.* 124: 39–64.

Rao, S. R. S. and V. Sarma. 1992. Morphology of 2–armed trichomes in relation to taxonomy: Malpighiales. *Feddes Repert.* 103: 559–565.

Raven, P. H. 1979. A survey of reproductive biology in Onagraceae. *New Zealand J. Bot.* 17: 575–593.

Raven, P. H. 1988. Onagraceae as a model of plant evolution. In *Plant evolutionary biology: a symposium honoring G. Ledyard Stebbins*, L. D. Gottlieb and S. K. Jain (eds.), 85–107. Chapman and Hall, London.

Ray, T. S. 1987a. Diversity of shoot organization in the Araceae. *Amer. J. Bot.* 74: 1373–1387.

Ray, T. S. 1987b. Leaf types in the Araceae. *Amer. J. Bot.* 74: 1359–1372.

Reeves, P. A. and R. G. Olmstead. 1998. Evolution of novel morphological and reproductive traits in a clade containing *Antirrhinum majus* (Scrophulariaceae). *Amer. J. Bot.* 85: 1047–1056.

Reeves, G. , M. W. Chase, P. Goldblatt, P. Rudall, M. F. Fay, A. V. Cox, B. Lejeune, and T. Souza-Chies. 2001. Molecular systematics of Iridaceae: evidence from four plastid DNA regions. *Amer. J. Bot.* 88: 2074–2087.

Renner, S. S. 1989a. Floral biological observations on *Heliamphora tatei* (Sarraceniaceae) and other plants from Cerro de la Neblina in Venezuela. *Plant Syst. Evol.* 163: 21–29.

Renner, S. S. 1989b. A survey of reproductive biology in neotropical Melastomataceae and Memecylaceae. *Ann. Missouri Bot. Gard.* 76: 496–518.

Renner, S. S. 1990. Reproduction and evolution in some genera of neotropical Melastomataceae. *Mem. New York Bot. Gard.* 55: 143–152.

Renner, S. S. 1993. Phylogeny and classification of the Melastomataceae and Memecylaceae. *Nordic J. Bot.* 13: 519–540.

Renner, S. S. 1999. Circumscription and phylogeny of the Laurales: evidence from molecular and morphological data. *Amer. J. Bot.* 86: 1301–1315.

Renner, S. S. 2004. Bayesian analysis of combined chloroplast loci, using multiple calibrations, supports the recent arrival of Melastomataceae in Africa and Madagascar. *Amer. J. Bot.* 91: 1427–1435.

Renner, S. S. and A. Chanderbali. 2000. What is the relationshp among Hernandiaceae, Lauraceae, and Monimiaceae, and why is this question so difficult to answer? *Int. J. Plant Sci.* 161 (6 Suppl.) 161: S109–S119.

Renner, S. S., A. E. Schwarzbach, and L. Lohmann. 1997. Phylogenetic position and floral function of *Siparuna* (Siparunaceae: Laurales). *Int. J. Plant Sci.* 158 (Suppl.): S89–S98.

Renner, S. S., G. Clausing, and K. Meyer. 2001. Historical biogeography of Melastomataceae: the roles of Tertiary migration and long-distance dispersal. *Amer. J. Bot.* 88: 1290–1300.

Rettig, J. H., H. D. Wilson, and J. R. Manhart. 1992. Phylogeny of the Caryophyllales: gene sequence data. *Taxon* 41: 201–209.

Reveal, J. L., W. S. Judd, and R. Olmstead. 1999. (1405) Proposal to conserve the name Antirrhinaceae against Plantaginaceae (Magnoliophyta). *Taxon* 48: 182.

Richardson, J. E., M. F. Fay, Q. C. B. Cronk, and M. W. Chase. 2000a. Revision of the tribal classification of Rhamnaceae. *Kew Bull.* 55: 311–340.

Richardson, J. E., M. E. Fay, Q. C. B. Cronk, D. Bowman, and M. W. Chase. 2000b. A phylogenetic analysis of Rhamnaceae using *rbcL* and *trnL-F* plastid DNA sequences. *Amer. J. Bot.* 87: 1309–1324.

Rieseberg, L. H. 1991. Homoploid reticulate evolution in *Helianthus* (Asteraceae): evidence from ribosomal genes. *Amer. J. Bot.* 78: 1218–1237.

Rivadavia, F., K. Kondo, M. Kato, and M. Hasebe. 2003. Phylogeny of the sundews, *Drosera* (Droseraceae), based on chloroplast *rbcL* and nuclear 18S ribosomal DNA sequences. *Amer. J. Bot.* 90: 123–130.

Rizk, A.-F. M. 1987. The chemical constituents and economic plants of the Euphorbiaceae. *Bot. J. Linn. Soc.* 94: 293–326.

Roalson, E. H. 2005. Phylogenetic relationships in the Juncaceae inferred from nuclear ribosomal DNA internal transcribed spacer sequence data. *Int. J. Plant Sci.* 166: 397–413.

Robberecht, E. and J. F. Manen. 2006. The major evolutionary lineages of the coffee family (Rubiaceae, angiosperms): combined analysis (nDNA and cpDNA) to infer the position of *Coptosapelta* and *Luculia*, and supertree construction based on *rbcL*, *rps16*, *trnL-trnF* and *atpB-rbcL* data. A new classification in two subfamilies, Cinchonioideae and Rubioideae. *Syst. Geogr. Plants* 76: 85–146.

Roberts, M. L. and R. R. Haynes. 1983. Ballistic seed dispersal in *Illicium* (Illiciaceae). *Plant Syst. Evol.* 143: 227–232.

Robertson, K. R. 1972a. The genera of Geraniaceae in the southeastern United States. *J. Arnold Arbor.* 53: 182–201.

Robertson, K. R. 1972b. The Malpighiaceae in the southeastern United States. *J. Arnold Arbor.* 53: 101–112.

Robertson, K. R. 1974. The genera of Rosaceae in the southeastern United States. *J. Arnold Arbor.* 55: 303–332, 344–401, 600–662.

Robertson, K. R. 1975. The genera of Oxalidaceae in the southeastern United States. *J. Arnold Arbor.* 56: 223–239.

Robertson, K. R. 1976. The genera of Haemodoraceae in the southeastern United States. *J. Arnold Arbor.* 57: 205–216.

Robertson, K. R. 1981. The genera of Amaranthaceae in the southeastern United States. *J. Arnold Arbor.* 62: 267–313.

Robertson, K. R. and Y.-T. Lee. 1976. The genera of Caesalpinioideae (Leguminosae) in the southeastern United States. *J. Arnold Arbor.* 57: 1–53.

Robertson, K. R., J. B. Phipps, J. R. Rohrer, and P. G. Smith. 1991. A synopsis of genera of the Maloideae (Rosaceae). *Syst. Bot.* 16: 376–394.

Robertson, K. R., J. B. Phipps, and J. R. Rohrer. 1992. Summary of leaves in the genera of Maloideae (Rosaceae). *Ann. Missouri Bot. Gard.* 79: 81–94.

Robinson, R. W. and D. S. Decker-Walters. 1997. Cucurbits. *Crop production science in horticulture*, No. 6. CAB International, Wallingford, UK.

Roddick, J. G. 1986. Steroidal alkaloids of the Solanaceae. In *Solanaceae: biology and systematics*, W. G. D'Arcy, (ed.), 201–222. Columbia Univ. Press, New York.

Roddick, J. G. 1991. The importance of the Solanaceae in medicine and drug therapy. In *Solanaceae 3: Taxonomy, chemistry, evolution*, J. G. Hawkes, R. N. Lester, M. Nee and N. Estrada (eds.), 7–23. Royal Botanic Garden, Kew.

Rodman, J. E. 1981. Divergence, convergence, and parallelism in phytochemical characters: the glucosinolate-myrosinase system. In *Phytochemistry and angiosperm phylogeny*, D. A. Young and D. S. Seigler (eds.), 43–79. Praeger, New York.

Rodman, J. E. 1990. Centrospermae revisited, part 1. *Taxon* 39: 383–393.

Rodman, J. E. 1991a. A taxonomic analysis of glucosinolate-producing plants. I. Phenetics. *Syst. Bot.* 16: 598–618.

Rodman, J. E. 1991b. A taxonomic analysis of glucosinolate-producing plants. II. Cladistics. *Syst. Bot.* 16: 619–629.

Rodman, J. E. 1994. Cladistic and phenetic studies. In *Caryophyllales: evolution and systematics*, H.-D. Behnke and T. J. Mabry (eds.), 279–301. Springer-Verlag, Berlin.

Rodman, J. E., M. K. Oliver, R. R. Nakamura, J. U. McClammer, Jr., and A. H. Bledsoe. 1984. A taxonomic analysis and revised classification of Centrospermae. *Syst. Bot.* 9: 297–323.

Rodman, J. E., R. A. Price, K. Karol, E. Conti, K. J. Sytsma, and J. D. Palmer. 1993. Nucleotide sequences of the *rbcL* gene indicate monophyly of mustard oil plants. *Ann. Missouri Bot. Gard.* 80: 686–699.

Rodman, J. E., K. G. Karol, R. A. Price, and K. J. Sytsma. 1996. Molecules, morphology and Dahlgren's expanded order Capparales. *Syst. Bot.* 21: 289–307.

Rodman, J. E., P. S. Soltis, D. E. Soltis, K. J. Sytsma, and K. G. Karol. 1998. Parallel evolution of glucosinolate biosynthesis inferred from congruent nuclear and plastid gene phylogenies. *Amer. J. Bot.* 85: 997–1006.

Rogers, G. K. 1982. The Casuarinaceae in the southeastern United States. *J. Arnold Arbor.* 63: 357–373.

Rogers, G. K. 1983. The genera of Alismataceae in the southeastern United States. *J. Arnold Arbor.* 64: 383–420.

Rogers, G. K. 1984. The Zingiberales (Cannaceae, Marantaceae, and Zingiberaceae) in the southeastern United States. *J. Arnold Arbor.* 65: 5–55.

Rogers, G. K. 1985. The genera of Phytolaccaceae in the southeastern United States. *J. Arnold Arbor.* 66: 1–37.

Rogers, G. K. 1986. The genera of Loganiaceae in the southeastern United States. *J. Arnold Arbor.* 67: 143–185.

Rogers, G. K. 1987. The genera of Cinchonoideae (Rubiaceae) in the southeastern United States. *J. Arnold Arbor.* 68: 137–183.

Rohrer, J. R., K. R. Robertson, and J. B. Phipps. 1991. Variation in structure among fruits of Maloideae (Rosaceae). *Amer. J. Bot.* 78: 1617–1635.

Rohrer, J. R., K. R. Robertson, and J. B. Phipps. 1994. Floral morphology of Maloideae (Rosaceae) and its systematic relevance. *Amer. J. Bot.* 81: 574–581.

Rohwer, J. G. 1993a. Lauraceae. In *The families and genera of vascular plants*, vol. 2, Magnoliid, hamamelid and caryophyllid families, K. Kubitzki, J. G. Rohwer, and V. Bittrich (eds.), 366–391. Springer-Verlag, Berlin.

Rohwer, J. G. 1993b. Moraceae. In *The families and genera of vascular plants*, vol. 2, Magnoliid, hamamelid and caryophyllid families, K. Kubitzki, J. G. Rohwer, and V. Bittrich (eds.), 438–453. Springer-Verlag, Berlin.

Rohwer, J. G. 1993c. Phytolaccaceae. In *The families and genera of vascular plants*, vol. 2, Magnoliid, hamamelid and caryophyllid families, K. Kubitzki, J. G. Rohwer, and V. Bittrich (eds.), 506–515. Springer-Verlag, Berlin.

Rohwer, J. G. 1994. A note on the evolution of the stamens in the Laurales, with emphasis on the Lauraceae. *Bot. Acta* 107: 103–110.

Rohwer, J. G. 2000. Toward a phylogenetic classification of the Lauraceae: evidence from *matK* sequences. *Syst. Bot.* 25: 60–71.

Rohwer, J. G., and Rudolph, B. 2005. Jumping genera: the phylogenetic positions of *Cassytha*, *Hypodaphnis*, and *Neocinnamomum* (Lauraceae) based on different analyses of *trnK* intron sequences. *Ann. Missouri Bot. Gard.* 92: 153–178.

Rohwer, J. G., H. G. Richter, and H. van der Werff. 1991. Two new genera of neotropical Lauraceae and critical remarks on the generic delimitation. *Ann. Missouri Bot. Gard.* 78: 388–400.

Rollins, R. C. 1993. *The Cruciferae of continental North America*. Stanford University Press, Stanford.

Ronse Decraene, L. P. and J. R. Akeroyd. 1988. Generic limits in *Polygonum* and related genera (Polygonaceae) on the basis of floral characters. *Bot. J. Linn. Soc.* 98: 321–371.

Ronse Decraene, L. P. and E. Smets. 1992. An updated interpretation of the androecium of the Fumariaceae. *Canad. J. Bot.* 70: 1765–1776.

Ronse Decraene, L. P., J. de Laet, and E. F. Smets. 1996. Morphological studies in Zygophyllaceae. II. The floral development and vascular anatomy of *Peganum harmala*. *Amer. J. Bot.* 83: 201–215.

Ronse Decraene, L. P., P. S. Soltis, and D. E. Soltis. 2003. Evolution of floral structures in basal angiosperms. *Int. J. Plant Sci.* 164 (Suppl.): S329–S363.

Ronse Decraene, L. P., S.-P. Hong, and E. F. Smets. 2004. What is the taxonomic status of *Polygonella*? Evidence of floral morphology. *Ann. Missouri Bot. Gard.* 91: 320–345.

Rosatti, T. J. 1984. The Plantaginaceae in the southeastern United States. *J. Arnold Arbor.* 65: 533–562.

Rosatti, T. J. 1986. The genera of Sphenocleaceae and Campanulaceae in the southeastern United States. *J. Arnold Arbor.* 67: 1–64.

Rosatti, T. J. 1987. The genera of Pontederiaceae in the southeastern United States. *J. Arnold Arbor.* 68: 35–71.

Rosatti, T. J. 1989. The genera of suborder Apocynineae (Apocynaceae and Asclepiadaceae) in the southeastern United States. *J. Arnold Arbor.* 70: 307–401, 443–514.

Rourke, J. and D. Wiens. 1977. Convergent floral evolution in South African and Australian Proteaceae and its possible bearing on pollination by nonflying mammals. *Ann. Missouri Bot. Gard.* 64: 1–17.

Rova, J. H. E., P. G. Delprete, L. Andersson, and V. A. Albert. 2002. A *trnL-F* cpDNA sequence study of the Condamineeae-Rondeletieae-Sipaneeae complex with implications on the phylogeny of the Rubiaceae. *Amer. J. Bot.* 89: 145–159.

Rudall, P. 1994. Anatomy and systematics of Iridaceae. *Bot. J. Linn. Soc.* 114: 1–21.

Rudall, P. and D. F. Cutler. 1995. Asparagales: a reappraisal. In *Monocotyledons: systematics and evolution*, P. J. Rudall, P. J. Cribb, D. F. Cutler and C. J. Humphries (eds.), 157–168. Royal Botanic Gardens, Kew.

Rudall, P. and H. P. Linder. 1988. Megagametophyte and nucellus in Restionaceae and Flagellariaceae. *Amer. J. Bot.* 75: 1777–1786.

Rudall, P., C. A. Furness, M. W. Chase, and M. F. Fay. 1997a. Microsporogenesis and pollen sulcus type in Asparagales (Lilianae). *Canad. J. Bot.* 75: 408–430.

Rudall, P., M. W. Chase, J. G. Conran. 1997b. New Circumscriptions and a new family of asparagoid lilies: genera formerly included in Anthericaceae. *Kew Bull.* 51: 667–680.

Rudall, P., E. M. Engleman, L. Hanson, and M. W. Chase. 1998. Embryology, cytology and systematics of *Hemiphylacus*, *Asparagus* and *Anemarrhena* (Asparagales). *Plant Syst. Evol.* 211: 181–199.

Rudall, P., K. L. Stobart, W.-P. Hong, J. G. Conran, C. A. Furness, G. C. Kite, and M. W. Chase. 2000a. Consider the lilies: systematics of Liliales. In *Monocots: systematics and evolution*, K. L. Wilson and D. A. Morrison (eds.), 347–359. CSIRO, Collingwood, Australia.

Rudall, P., J. G. Conran, and M. W. Chase. 2000b. Systematics of Ruscaceae/Convallariaceae: a combined morphological and molecular investigation. *Bot. J. Linn. Soc.* 134: 73–92.

Rudall, P., J. C. Manning, and P. Goldblatt. 2003. Evolution of floral nectaries in Iridaceae. *Ann. Missouri Bot. Gard.* 90: 613–631.

Safwat, F. M. 1962. The floral morphology of *Secamone* and the evolution of the pollinating apparatus of Asclepiadaceae. *Ann. Missouri Bot. Gard.* 49: 95–119.

Sampson, F. B. 1993. Pollen morphology of the Amborellaceae and Hortoniaceae (Hoortonioideae: Monimiaceae). *Grana* 32: 154–162.

Samuel, R., H. Kathriarachchi, P. Hoffmann, M. H. J. Barfuss, K. J. Wurdack, C. C. Davis, and M. W. Chase. 2005. Molecular phylogenetics of Phyllanthaceae: evidence from plastid *matK* and nuclear *PHYC* sequences. *Amer. J. Bot.* 92: 132–141.

Sanders, R. W. 1997. The Avicenniaceae in the southeastern United States. *Harvard Pap. Bot.* No. 10: 81–92.

Sanders, R. W. 2001. The genera of Verbenaceae in the southeastern United States. *Harvard Pap. Bot. 5(2): 303–358.

Sauquet, H. 2003. Androecium diversity and evolution in Myristicaceae (Magnoliales),

with a description of a new Malagasy genus, *Doyleanthus* gen. nov. *Amer. J. Bot.* 90: 1293–1305.

Sauquet, H. and A. LeThomas. 2003. Pollen diversity and evolution in Myristicaceae (Magnoliales). *Int. J. Plant Sci.* 164: 613–628.

Sauquet, H., J. A. Doyle, T. Schoraschkin, T. Borsch, K. W. Hilu, L. W. Chatrou, and A. LeThomas. 2003. Phylogenetic analysis of Magnoliales and Myristicaceae based on multiple data sets: implications for character evolution. *Bot. J. Linn. Soc.* 142: 125–186.

Savolainen, V., J. F. Manen, E. Douzery, and R. Spichiger. 1994. Molecular phylogeny of families related to Celastrales based on *rbcL* 5′ flanking regions. *Mol. Phylog. Evol.* 3: 27–37.

Savolainen, V., M. W. Chase, S. B. Hoot, C. M. Morton, D. E. Soltis, C. Bayer, M. F. Fay, A.Y. de Bruijn, S. Sullivan, and Y.-L. Qiu. 2000a. Phylogenetics of flowering plants based upon a combined analysis of plastid *atpB* and *rbcL* gene sequences. *Syst. Biol.* 49: 306–362.

Savolainen, V. and 16 others. 2000b. Phylogeny of the eudicots: a nearly complete familial analysis based on *rbcL* gene sequences. *Kew Bull.* 55: 257–309.

Scharaschkin, T. and J. A. Doyle. 2006. Character evolution in *Anaxagorea* (Annonaceae). *Amer. J. Bot.* 93: 36–54.

Schick, B. 1980. Untersuchungen über Biotechnik der Apocynaceenblüte. I. Morphologie und Funktion des Narbenkopfes. *Flora, Morphol. Geobot. Oekophysiol.* 170: 394–432.

Schick, B. 1982. Untersuchungen über Biotechnik der Apocynaceenblüte. II. Bau und Funktion des Bestäubungsapparates. *Flora, Morphol. Geobot. Oekophysiol.* 172: 347–371.

Schmid, R. 1980. Comparative anatomy and morphology of *Psiloxylon* and *Heteropyxis* and the subfamilial and tribal classification of the Myrtaceae. *Taxon* 29: 559–595.

Schneider, E. L. and J. M. Jeter. 1982. Morphological studies of the Nymphaeaceae. XII. The floral biology of *Cabomba caroliniana*. *Amer. J. Bot.* 69: 1410–1419.

Schneider, E. L. and S. Carlquist. 1995. Vessels in the roots of *Barclaya rotundifolia*. *Amer. J. Bot.* 82: 1343–1349.

Schneider, E. L. and P. S. Williamson. 1993. Nymphaeaceae. In *The families and genera of vascular plants*, vol. 2, Magnoliid, hamamelid and caryophyllid families, K. Kubitzki, J. G. Rohwer and V. Bittrich (eds.), 486–493. Springer-Verlag, Berlin.

Schneider, E. L., S. Carlquist, K. Beamer and A. Kohn. 1995. Vessels in Nymphaeaceae: *Nuphar*, *Nymphaea* and *Ondinea*. *Int. J. Plant Sci.* 156: 857–862.

Schönenberger, J., A. A. Anderberg, and K. J. Sytsma. 2005. Molecular phylogenetics and patterns of floral evolution in the Ericales. *Int. J. Plant Sci.* 166: 265–288.

Schrire, B. D. 1989. A multidisciplinary approach to pollination biology in the Leguminosae. *Monogr. Syst. Bot. Missouri Bot. Gard.* 29: 183–242.

Schulze-Menz, G. K. 1964. Saxifragaceae. In *A. Engler's Syllabus der Pflanzenfamilien*, H. Melchior (ed.), 201–206. Gebruder Borntraeger, Berlin.

Schütze, P., H. Freitag, and K. Weising. 2003. An integrated molecular and morphological study of the subfamily Suaedoideae Ulbr. (Chenopodiaceae). *Plant Syst. Evol.* 164: 959–986.

Schwarzbach, A. E. and J. W. Kadereit. 1995. Rapid radiation of North American desert genera of the Papaveraceae: evidence from restriction site mapping of PCR-amplified chloroplast DNA fragments. *Plant Syst. Evol.* Suppl. 9: 159–170.

Schwarzbach, A. E. and R. E. Ricklefs. 2000. Systematic affinities of Rhizophoraceae and Anisophyllaceae and intergeneric relationships within Rhizophoraceae based on chloroplast DNA, nuclear ribosomal DNA, and morphology. *Amer. J. Bot.* 87: 547–564.

Schwarzbach, A. E. and L. A. McDade. 2002. Phylogenetic relationships of the mangrove family Avicenniaceae based on chloroplast and nuclear ribosomal DNA sequences. *Syst. Bot.* 27: 84–98.

Schwarzwalder, R. N. and D. L. Dilcher. 1991. Systematic placement of the Platanaceae in the Hamamelidae. *Ann. Missouri Bot. Gard.* 78: 962–969.

Scotland, R. W. 1990. Palynology and systematics of Acanthaceae. Ph.D. Thesis, University of Reading, England.

Scotland, R. W. J. A. Sweere, P. A. Reeves and R. G. Olmstead. 1995. Higher-level systematics of Acanthaceae determined by chloroplast DNA sequences. *Amer. J. Bot.* 82: 266–275.

Scott, R. W. 1990. The genera of Cardueae (Compositae; Asteraceae) in the southeastern United States. *J. Arnold Arbor.* 71: 391–451.

Semple, K. S. 1974. Pollination in Piperaceae. *Ann. Missouri Bot. Gard.* 61: 868–871.

Sennblad, B. and B. Bremer. 1996. The familial and subfamilial relationships of Apocynaceae and Asclepiadaceae evaluated with *rbcL* data. *Plant Syst. Evol.* 202: 153–175.

Sheahan, M. C. and D. F. Cutler. 1993. Contribution of vegetative anatomy to the systematics of the Zygophyllaceae R. Br. *Bot. J. Linn. Soc.* 113: 227–262.

Sheahan, M. C. and M. W. Chase. 2000. Phylogenetic relationships within Zygophyllaceae based on DNA sequences of three plastid regions, with special emphasis on Zygophylloideae. *Syst. Bot.* 25: 371–384.

Shetler, S. G. 1979. Pollen-collecting hairs of *Campanula* (Campanulaceae). I. Historical review. *Taxon* 28: 205–215.

Shinwari, Z. K., R. Terauchi, F. H. Htech, and S. Kawano. 1994. Recognition of the New World *Disporum* section *Prosartes* (Liliaceae) based on the sequence data of the *rbcL* gene. *Taxon* 43: 353–366.

Sibaoka, T. 1991. Rapid plant movements triggered by action potentials. *Bot. Mag. Tokyo.* 104: 73–95.

Silberbauer-Gottsberger, I., G. Gottsberger, and A. C. Webber. 2003. Morphological and functional flower characteristics of New and Old World Annonaceae with respect to their mode of pollination. *Taxon* 52: 701–718.

Simmons, M. P. 2004. Celastraceae. In *The families and genera of vascular plants*, vol. 6, Celastrales, Oxalidales, Rosales, Cornales, Ericales, K. Kubitzki (ed), 29–64. Springer-Verlag, Berlin.

Simmons, M. P. and J. P. Hedin. 1999. Relationships and morphological character change among genera of Celastraceae

sensu lato (including Hippocrateaceae). *Ann. Missouri Bot. Gard.* 86: 723–757.

Simmons, M. P., V. Savolainen, C. C. Clevinger, R. H. Archer, S. Mathews, and J. I. Davis. 2000. Phylogeny of the Celastraceae inferred from morphology and nuclear and plastid loci. *Amer. J. Bot.* 87 (6, suppl.): 156–157.

Simmons, M. P., C. C. Clevinger, V. Savolainen, R. H. Archer, S. Mathews, and J. J. Doyle. 2001. Phylogeny of the Celastraceae inferred from phytochrome B gene sequence and morphology. *Amer. J. Bot.* 88: 313–325.

Simpson, D. 1995. Relationships within Cyperales. In *Monocotyledons: systematics and evolution*, P. J. Rudall, P. J. Cribb, D. F. Cutler, and C. J. Humphries (eds.), 459–509. Royal Botanic Gardens, Kew.

Simpson, M. G. 1990. Phylogeny and classification of the Haemodoraceae. *Ann. Missouri Bot. Gard.* 77: 722–784.

Simpson, M. G. 1993. Septal nectary anatomy and phylogeny of the Haemodoraceae. *Syst. Bot.* 18: 593–613.

Simpson, M. G. 1998a. Haemodoraceae. In *The families and genera of vascular plants*, vol. 4, Monocotyledons: Alismatanae and Commelinanae (except Gramineae), K. Kubitzki (ed.), 212–222. Springer-Verlag, Berlin.

Simpson, M. G. 1998b. Reversal in ovary position from inferior to superior: evidence from floral ontogeny. *Int. J. Plant Sci.* 159: 466–479.

Skvarla, J. J., P. H. Raven, W. F. Chissoe, and M. Sharp. 1978. An ultrastructural study of viscin threads in Onagraceae pollen. *Pollen et Spores* 30: 5–143.

Smith, A. C. 1947. The families Illiciaceae and Schisandraceae. *Sargentia* 7: 1–224.

Smith, G. F. and B.-E. Van Wyk. 1991. Generic relationships in the Alooideae (Asphodelaceae). *Taxon* 40: 557–581.

Smith, G. F. and B.-E. Van Wyk. 1998. Asphodelaceae. In *The families and genera of vascular plants*, vol. 3, Monocotyledons: Lilianae (except Orchidaceae), K. Kubitzki (ed.), 130–140. Springer-Verlag, Berlin.

Smith, J. F. 1996. Tribal relationships within Gesneriaceae: a cladistic analysis of morphological data. *Syst. Bot.* 21: 497–513.

Smith, J. F. and J. J. Doyle. 1995. A cladistic analysis of chloroplast DNA restriction site variation and morphology from the genera of the Juglandaceae. *Amer. J. Bot.* 82: 1163–1172.

Smith, J. F. and C. L. Carroll. 1997. A cladistic analysis of the tribe Episcieae (Gesneriaceae) based on *ndhF* sequences: origin of morphological characters. *Syst. Bot.* 22: 713–724.

Smith, J. F., W. J. Kress, and E. A. Zimmer. 1993. Phylogenetic analysis of the Zingiberales based on *rbcL* sequences. *Ann. Missouri Bot. Gard.* 80: 620–630.

Smith, J. F., J. C. Wolfram, K. D. Brown, C. L. Carroll, and D. S. Denton. 1997. Tribal relationships in the Gesneriaceae: evidence from DNA sequences of the chloroplast gene *ndhF*. *Ann. Missouri Bot. Gard.* 84: 50–66.

Smith, L. B. and W. Till. 1998. Bromeliaceae. In *The families and genera of vascular plants*, vol. 4, Monocotyledons: Alismatanae and Commelinanae (except Gramineae), K.

Kubitzki (ed.), 74–99. Springer-Verlag, Berlin.

Smith, L. B. and D. C. Wasshausen. 1986. Begoniaceae, part I: illustrated key. *Smithsonian Contr. Bot.* 60: 1–129.

Smith, L. B. and C. E. Wood. 1975. The genera of Bromeliaceae in the southeastern United States. *J. Arnold Arbor.* 56: 375–397.

Soderstrom, T. R., K. W. Hilu, C. S. Campbell, and M. E. Barkworth (eds.). 1987. *Grass systematics and evolution*. Smithsonian Institution Press, Washington, DC.

Soejima, A. and J. Wen. 2006. Phylogenetic analysis of the grape family (Vitaceae) based on three chloroplast markers. *Amer. J. Bot.* 93: 278–287.

Soekarjo, R. 1992. General morphology in *Plantago*. In *Plantago: a multidisciplinary study*, P. J. C. Kuiper and M. Bos (eds.), 6–12. Springer-Verlag, Berlin.

Solbrig, O. T. 1963. The tribes of the Compositae in the southeastern United States. *J. Arnold Arbor.* 44: 436–461.

Soltis, D. E. and L. Hufford. 2002. Ovary position diversity in Saxifragaceae: clarifying the homology of epigyny. *Int. J. Plant Sci.* 163: 277–293.

Soltis, D. E. and R. K. Kuzoff. 1995. Discordance between nuclear and chloroplast phylogenies in the *Huchera* group (Saxifragaceae). *Evolution* 49: 727–742.

Soltis, D. E. and P. S. Soltis. 1997. Phylogenetic relationships in Saxifragaceae sensu lato: a comparison of topologies based on 18S rDNA and *rbcL* sequences. *Amer. J. Bot.* 84: 504–522.

Soltis, D. E. and P. S. Soltis. 1989. Allopolyploid speciation in *Tragopogon*: insights from chloroplast DNA. *Amer. J. Bot.* 76: 1119–1124.

Soltis, D. E., P. S. Soltis, T. G. Collier, and M. L. Edgerton. 1991. Chloroplast DNA variation within and among genera of the *Heuchera* group (Saxifragaceae): evidence for chloroplast transfer and paraphyly. *Amer. J. Bot.* 78: 1091–1112.

Soltis, D. E., D. R. Morgan, A. Grable, P. S. Soltis, and R. Kuzoff. 1993. Molecular systematics of Saxifragaceae sensu stricto. *Amer. J. Bot.* 80: 1056–1081.

Soltis, D. E., Q.-Y. Xiang, and L. Hufford. 1995. Relationships and evolution of Hydrangeaceae based on *rbcL* sequence data. *Amer. J. Bot.* 82: 504–514.

Soltis, D. E., P. S. Soltis, D. R. Morgan, S. M. Wensen, B. C. Mullin, J. M. Daud, and P. G. Martin. 1995. Chloroplast gene sequence data suggest a single origin of the predisposition for symbiotic nitrogen fixation in angiosperms. *Proc. Natl. Acad. Sci. USA* 92: 2647–2651.

Soltis, D. E., R. K. Kuzoff, E. Conti, R. Gornall, and K. Ferguson. 1996. *mat*K and *rbcL* gene sequence data indicate that *Saxifraga* (Saxifragaceae) is polyphyletic. *Amer. J. Bot.* 83: 371–382.

Soltis, D. E. and 15 others. 1997. Phylogenetic relationships among angiosperms inferred from 18S rDNA sequences. *Ann. Missouri Bot. Gard.* 84: 1–49.

Soltis, D. E., P. S. Soltis, M. E. Mort, M. W. Chase, V. Sarolainen, S. B. Hoot, and C. M. Morton. 1998. Inferring complex phylogenies using parsimony: an empirical approach using three large DNA data sets for angiosperms. *Syst. Biol.* 47: 32–42.

Soltis, D. E. and 15 others. 2000. Angiosperm phylogeny inferred from 18S rDNA, *rbcL*, and *atpB* sequences. *Bot. J. Linn. Soc.* 133: 381–461.

Soltis, D. E., R. K. Kuzoff, M. E. Mort, M. Zanis, M. Fishbein, L. Hufford, J. Koontz, and M. K. Arroyo. 2001. Elucidating deep-level phylogenetic relationships in Saxifraga-ceae using sequences for 6 chloroplast and nuclear DNA regions. *Ann. Missouri Bot. Gard.* 88: 669–693.

Soltis, D. E., A. E. Senters, M. J. Zanis, S. Kim, J. D. Thompson, P. S. Soltis, L. P. Ronse de Craene, P. K. Endress, and L. S. Farris. 2003a. Gunnerales are sister to other core eudicots: implications for the evolution of pentamery. *Amer. J. Bot.* 90: 461–470.

Soltis, D. E., P. S. Soltis, M. D. Bennett, and I. J. Leitch. 2003b. Evolution of genome size in the angiosperms. *Amer. J. Bot.* 90: 1596–1603.

Soltis, D. E., P. S. Soltis, P. K. Endress, and M. W. Chase. 2005. Phylogeny and evolution of angiosperms. Sinauer Associates, Sunderland, MA.

Soltis, P. S. and D. E. Soltis. 2004. The origin and diversification of angiosperms. *Amer. J. Bot.* 91: 1614–1626.

Soltis, P. S., D. E. Soltis, V. Savolainen, P. R. Crane, and T. G. Barraclough. 2002. Rate heterogeneity among lineages of tracheo-phytes: integration of molecular and fossil data and evidence for molecular living fossils. *Proc. Natl. Acad. Sci. USA* 99: 4430–4435.

Soltis, P. S., D. E. Soltis, M. W. Chase, P. K. Endress, and P. R. Crane. 2004. The diversi-fication of flowering plants. In *Assembling the tree of life*, J. Cracraft and M. J. Donoghue (eds.), 154–167. Oxford Univ. Press, Oxford.

Song, B.-H., X.-Q. Wang, F.-Z. Li, and D.-Y. Hong. 2001. Further evidence for paraphyly of the Celtidaceae from the chloroplast gene *matK*. *Plant Syst. Evol.* 228: 107–115.

Soreng, R. J. and J. I. Davis. 1998. Phylo-genet-ics and character evolution in the grass family (Poaceae): simultaneous analysis of morphological and chloroplast DNA restriction site character sets. *Bot. Rev.* 64: 1–85.

Soros, C. L. and D. H. Les. 2002. Phylogenetic relationships in the Alismataceae. Abstract. *Botany 2002*: 152.

Spangler, R. E. and R. G. Olmstead. 1999. Phylogenetic analysis of Bignoniaceae based on the cpDNA gene sequences *rbcL* and *ndhF*. *Ann. Missouri Bot. Gard.* 86: 33–46.

Speta, F. 1998. Hyacinthaceae. In *The families and genera of vascular plants*, vol. 3. Monocotyledons: Lilianae (except Orchidaceae), K. Kubitzki (ed.), 261–285. Springer-Verlag, Berlin.

Spongberg, S. A. 1972. The genera of Saxifragaceae in the southeastern United States. *J. Arnold Arbor.* 53: 409–498.

Spongberg, S. A. 1978. The genera of Crassulaceae in the southeastern United States. *J. Arnold Arbor.* 59: 197–248.

Spooner, D. M., G. A. Anderson, and R. K. Jansen. 1993. Chloroplast DNA evidence for the interrelationships of tomatoes, pota-toes, and pepinos (Solanaceae). *Amer. J. Bot.* 80: 676–688.

Spooner, D. M., I. E. Peralta, and S. Knapp. 2005. Comparison of AFLPs with other

markers for phylogenetic inference in wild tomatoes [*Solanum* L. section *Lycopersicon* (Mill.) Wettst.]. *Taxon* 54: 43–61.

Stace, C. A. 1965. Cuticular studies an an aid to plant taxonomy. *Bull. Brit. Mus. (Nat. Hist.), Bot.* 4: 1–78.

Ståhl, B. 1990. *Taxonomic studies in the Theophrastaceae*. Dissertation. Department of Systematic Botany, University of Goteborg, Sweden.

Steane, D. A., D. Nicolle, G. E. McKinnon, R. E. Vaillancourt, and B. M. Potts. 2002. Higher level relationships among the eucalypts are resolved by ITS-sequence data. *Austr. Syst. Bot.* 15: 49–62.

Steane, D. A., K. L. Wilson, and R. S. Hill. 2003. Using *matK* sequence data to unravel the phylogeny of Casuarinaceae. *Mol. Phylo. Evol.* 28: 47–59.

Steele, L. P. and R. Vilgalys. 1994. Phylogenetic analyses of Polemoniaceae using nucleotide sequences of the plastid gene *matK*. *Syst. Bot.* 19: 126–142.

Stefanoviæ, S., L. E. Krueger, and R. G. Olmstead. 2002. Monophyly of the Convolvulaceae and circumscription of their major lineages based on DNA sequences of multiple chloroplast loci. *Amer. J. Bot.* 89: 1510–1522.

Stefanoviæ, S., D. F. Austin, and R. G. Olmstead. 2003. Classification of Convol-vulaceae: a phylogenetic approach. *Syst. Bot.* 28: 791–806.

Stein, B. A. and H. Tobe. 1989. Floral nectaries in Melastomataceae and their systematic and evolutionary implications. *Ann. Missouri Bot. Gard.* 76: 519–531.

Steinmann, V. W. and J. M. Porter. 2002. Phylogenetic relationships in Euphorbieae (Euphorbiaceae) based on ITS and *ndhF* sequence data. *Ann. Missouri Bot. Gard.* 89: 453–490.

Stern, W. L. 1973. Development of the amentif-erous concept. *Brittonia* 25: 316–333.

Stern, W. L., M. W. Morris, W. S. Judd, A. M. Pridgeon, and R. L. Dressler. 1993. Comparative vegetative anatomy and sys-tematics of Spiranthoideae (Orchidaceae). *Bot. J. Linn. Soc.* 113: 161–197.

Stevens, P. F. and 14 others. 2004a. Ericaceae. In *The families and genera of vascular plants*, vol. 6, Celastrales, Oxalidales, Rosales, Cornales, Ericales, K. Kubitzki (ed.), 145–194. Springer-Verlag, Berlin.

Stevens, P. F., S. Dressler, and A. L. Weitzman. 2004b. Theaceae. In *The families and genera of vascular plants*, vol. 6, Celastrales, Oxalidales, Rosales, Cornales, Ericales, K. Kubitzki (ed.), pp. 463–471. Springer-Verlag, Berlin.

Stevenson, D. W. and H. Loconte. 1995. Cladistic analysis of monocot families. In *Monocotyledons: systematics and evolution*, P. J. Rudall, P. J. Cribb, D. F. Cutler, and C. J. Humphries (eds.), 543–578. Royal Botanic Gardens, Kew.

Stevenson, D. W., J. I. Davis, J. V. Freudenstein, C. R. Hardy, M. P. Simmons, and C. D. Specht. 2000. A phylogenetic analysis of the monocotyledons based on morphological and molecular character sets, with com-ments on the placement of *Acorus* and Hydatellaceae. In *Monocots: systematics and evolution*, K. L. Wilson and D. A. Morrison (eds.), 17–24. CSIRO, Collingwood, Australia.

Stockey, R. A., G. L. Hoffman, and G. W. Rothwell. 1997. The fossil monocot *Limnobiophyllum scutatum*: resolving the phylogeny of Lemnaceae. *Amer. J. Bot.* 84: 355–368.

Stone, D. E. 1973. Patterns in the evolution of amentiferous fruits. *Brittonia* 25: 371–384.

Stone, D. E. 1989. Biology and evolution of temperate and tropical Juglandaceae. In *Evolution, systematics and fossil history of the Hamamelidae*, vol. 2, "Higher" Hama-meli-dae. Systematics Association Special Vol. 40B, P. R. Crane and S. Blackmore (eds.), 117–145. Clarendon Press, Oxford.

Stone, D. E. 1993. Juglandaceae. In *The families and genera of vascular plants*, vol. 2, Magnoliid, hamamelid and caryophyllid fami-lies, K. Kubitzki, J. G. Rohwer, and V. Bittrich (eds.), 348–359. Springer-Verlag, Berlin.

Strange, A., P. J. Rudall, and C. J. Prychid. 2004. Comparative floral anatomy of Pontederi-aceae. *Bot. J. Linn. Soc.* 114: 395–408.

Struwe, L., V. A. Albert and B. Bremer. 1994. Cladistics and family level classification of the Gentianales. *Cladistics* 10: 175–206.

Struwe, L., M. Thiu, J. W. Kadereit, A. S.-R. Pepper, T. J. Motley, P. J. White, J. H. E. Rova, K. Potgieter, and V. A. Albert. 1998. *Saccifolium* (Saccifoliaceae), an endemic of Sierra de la Neblina on the Brazilian-Venezuelan border, is related to a temper-ate-alpine lineage of Gentianaceae. *Harvard Pap. Bot.* 3: 199–214.

Stubbs, J. M. and A. R. Slabas. 1982. Ultra-structural and biochemical characterization of the epidermal hairs of the seeds of *Cuphea procumbens*. *Planta* 155: 392–399.

Stützel, T. 1998. Eriocaulaceae. In *The families and genera of vascular plants*, vol. 4, Monocotyledons: Alismatanae and Com-melinanae (except Gramineae), K. Kubitzki (ed.), 197–207. Springer-Verlag, Berlin.

Suh, Y., L. B. Thien, H. E. Reeve, and E. A. Zimmer. 1993. Molecular evolution and phylogenetic implications of internal tran-scribed spacer sequences of ribosomal DNA in Winteraceae. *Amer. J. Bot.* 80: 1042–1055.

Sun, F.-J., S. R. Downie, and R. L. Hartman. 2004. An ITS-based phylogenetic analysis of the perennial, endemic Apiaceae sub-family Apioideae of western North America. *Syst. Bot.* 29: 419–431.

Sun, G., D. L. Dilcher, S. Zheng, and Z. Zhou. 1998. In search of the first flower: a Jurassic angiosperm, *Archaefructus*, from northeast China. *Science* 282: 1692–1695.

Sutter, D. and P. K. Endress. 1995. Aspects of gynoecial structure and macrosystematics in Euphorbiaceae. *Bot. Jahrb. Syst.* 116: 517–536.

Swarupanandan, K., J. K. Mangaly, T. K. Sonny, K. Kishorekumar, and S. Chand Basha. 1996. The subfamilial and tribal classifica-tion of the family Asclepiadaceae. *Bot. J. Linn. Soc.* 120: 327–369.

Sweeney, P. W. and R. A. Price. 2000. Polyphyly of the genus *Dentaria* (Brassicaceae): evi-dence from *trnL* intron and *ndhF* sequence data. *Syst. Bot.* 25: 468–478.

Swensen, S. M., W. L. Clement, L. L. Forrest, and M. C. Tebbit. 2001. *Hillebrandia sand-wichensis*: evolutionary relationships and biogeography. Abstract. *Botany 2001*: 95.

Swenson, U. and A. A. Anderberg. 2005. Phylogeny, character evolution, and classification of Sapotaceae (Ericales). *Cladistics* 21: 101–130.

Swingle, W. T. 1967. The botany of *Citrus* and its wild relatives [revised by P. C. Reece]. In *The Citrus industry*, vol. 1, History, world distribution, botany and varieties, W. Reuther, H. J. Webber and L. D. Batchelor (eds.), 190–430. Univ. of California Division of Agricultural Science, Berkeley.

Sytsma, K. J. and J. F. Smith. 1988. DNA and morphology: comparisons in the Onagraceae. *Ann. Missouri Bot. Gard.* 75: 1217–1237.

Sytsma, K. J. and J. F. Smith. 1992. Molecular systematics of Onagraceae: examples of *Clarkia* and *Fuchsia*. In *Molecular systematics of plants*, P. S. Soltis, D. E. Soltis and J. J. Doyle (eds.), 295–323. Chapman and Hall, New York.

Sytsma, K. J., D. A. Baum, A. Rodriguez, W. J. Hahn, L. Katinas, W. L. Wagner, and P. C. Hock. 1998a. An ITS phylogeny for Onagraceae: congruence with three molecular data sets. *Amer. J. Bot.* 85(6) Suppl.: 160.

Sytsma, K. J., M. L. Zjhra, M. Nepokroeff, C. J. Quinn, and P. G. Wilson. 1998b. Phylogenetic relationships, morphological evolution and biogeography in Myrtaceae based on *ndhF* sequence analysis. *Amer. J. Bot.* 85(6) Suppl.: S161.

Sytsma, K. J., J. Morawetz, J. C. Pires, M. Nepokroeff, E. Conti, M. Zjhra, J. C. Hall, and M. W. Chase. 2002. Urticalean rosids: circumscription, rosid ancestry, and phylogenetics based on *rbcL*, *trnLF*, and *ndhF* sequences. *Amer. J. Bot.* 89: 1531–1546.

Sytsma, K. J., A. Litt, M. L. Zjhra, C. Pires, M. Nepokroeff, E. Conti, J. Walker, and P. G. Wilson. 2004. Clades, clocks, and continents: historical and biogeographical analysis of Myrtaceae, Vochysiaceae, and relatives in the Southern Hemisphere. *Int. J. Plant Sci.* 165(4) (Suppl.): S85–S105.

Sytsma, K. J., J. B. Walker, J. Schönenberger, and A. A. Anderberg. 2006. Phylogenetics, biogeography, and radiation of Ericales. Abstract. *Botany 2006*: 71.

Takhtajan, A. 1969. *Flowering plants: origin and dispersal*. Smithsonian Institution Press, Washington, DC.

Takhtajan, A. 1980. Outline of the classification of flowering plants (Magnoliophyta). *Bot. Rev.* 46: 225–359.

Takhtajan, A. 1997. *Diversity and classification of flowering plants*. Columbia University Press, New York.

Tam, S.-M., P. C. Boyce, T. M. Upson, D. Barabé, A. Bruneau, F. Forest, and J. S. Parker. 2004. Intergeneric and infrafamilial phylogeny of subfamily Monsteroideae (Araceae) revealed by chloroplast *trnL-F* sequences. *Amer. J. Bot.* 91: 490–498.

Tamura, M. 1993. Ranunculaceae. In *The families and genera of vascular plants*, vol. 2, Magnoliid, hamamelid and caryophyllid families, K. Kubitzki, J. G. Rohwer and V. Bittrich (eds.), 563–583. Springer-Verlag, Berlin.

Tamura, M. N. 1998a. Calochortaceae. In *The families and genera of vascular plants,* vol 3. Monocotyledons. Lilianae (except Orchidaceae), K. Kubitzki (ed.), 164–172. Springer-Verlag, Berlin.

Tamura, M. N. 1998b. Liliaceae. In *The families and genera of vascular plants*, vol. 3. Monocotyledons: Lilianae (except Orchidaceae), K. Kubitzki (ed.), 343–353. Springer-Verlag, Berlin.

Tamura, M. N. 1998c. Melanthiaceae. In *The families and genera of vascular plants*, vol. 3. Monocotyledons: Lilianae (except Orchidaceae), K. Kubitzki (ed.), 369–380. Springer-Verlag, Berlin.

Tamura, M. N. 1998d. Trilliaceae. In *The families and genera of vascular plants*, vol. 3. Monocotyledons: Lilianae (except Orchidaceae), K. Kubitzki (ed.), 444–452. Springer-Verlag, Berlin.

Tank, D. C., P. M. Beardsley, S. A. Kelchner, and R. G. Olmstead. 2006. Review of the systematics of Scrophulariaceae s.l. and their current disposition. *Australian Syst. Bot.* 19: 289–307.

Tate, J. A., J. Fuertes Aguilar, S. J. Wagstaff, J. C. LaDuke, T. A. Bodo Slotta, and B. B. Simpson. 2005. Phylogenetic relationships within the tribe Malveae (Malvaceae, subfamily Malvoideae) as inferred from ITS sequence data. *Amer. J. Bot.* 92: 584–602.

Taylor, D. W. and L. J. Hickey. 1992. Phylogenetic evidence for the herbaceous origin of angiosperms. *Plant Syst. Evol.* 180: 137–156.

Taylor, D. W. and L. J. Hickey. 1996. Evidence for and implications of an herbaceous origin for angiosperms. In *Flowering plant origin, evolution and phylogeny*, D. W. Taylor and L. J. Hickey (eds.), 232–266. Chapman and Hall, New York.

Taylor, P. 1989. The genus *Utricularia*: a taxonomic monograph. *Kew Bull.*, Add. Ser. 14: 1–724.

Tebbs, M. C. 1993. Piperaceae. In *The families and genera of vascular plants*, vol. 2, Magnoliid, hamamelid and caryophyllid families, K. Kubitzki, J. G. Rohwer, and V. Bittrich (eds.), 516–522. Springer-Verlag, Berlin.

Terabayashi, S. 1991. Vernation patterns in Celtidaceae and Ulmaceae (Urticales) and their evolutionary and systematic implications. *Bot. Mag. Tokyo* 104: 1–13.

Terrazas, T. and M. Chase. 1996. A phylogenetic analysis of Anacardiaceae based on morphology and *rbcL* sequence data. *Amer. J. Bot.* 83(6) Suppl.: 197.

Terry, R. G., G. K. Brown, and R. G. Olmstead. 1997a. Examination of subfamilial phylogeny in Bromeliaceae using comparative sequencing of the plastid locus *ndhF*. *Amer. J. Bot.* 84: 664–670.

Terry, R. G., G. K. Brown, and R. G. Olmstead. 1997b. Phylogenetic relationships in subfamily Tillandsioideae (Bromeliaceae) using *ndhF* sequences. *Syst. Bot.* 22: 333–345

Thanikaimoni, G. 1986. Evolution of Menispermaceae. *Canad. J. Bot.* 64: 3130–3133.

Thien, L. B. 1974. Floral biology of *Magnolia*. *Amer. J. Bot.* 61: 1037–1045.

Thien, L. B. 1980. Patterns of pollination in the primitive angiosperms. *Biotropica* 12: 1–13.

Thien, L. B., D. A. White, and L. Y. Yatsu. 1983. The reproductive biology of a relic: *Illicium floridanum* Ellis. *Amer. J. Bot.* 70: 719–727.

Thien, L. B. and 12 others. 2003. The population structure and floral biology of *Amborella trichopoda* (Amborellaceae). *Ann. Missouri Bot. Gard.* 90: 466–490.

Thieret, J. W. 1982. The Sparganiaceae in the southeastern United States. *J. Arnold Arbor.* 63: 341–355.

Thieret, J. W. and J. O. Luken. 1996. The Typhaceae in the southeastern United States. *Harvard Pap. Bot.* No. 8: 27–56.

Thieret, W. 1967. Supraspecific classification in the Scrophulariaceae; a review. *Sida* 3: 87–106.

Thieret, W. 1971. The genera of Orobanchaceae in the southeastern United States. *J. Arnold Arbor.* 52: 404–434.

Thiv, M., L. Struwe, V. A. Albert, and J. W. Kadereit. 1999. The phylogenetic relationships of *Saccifolium bandeirae* (Gentianaceae) reconsidered. *Harvard Pap. Bot.* 4: 519–526.

Thomas, V. and Y. Dave. 1991. Comparative and phylogenetic significance of the colleters in the family Apocynaceae. *Feddes Repert.* 102: 177–182.

Thomson, B. F. 1942. The floral morphology of the Caryophyllaceae. *Amer. J. Bot.* 29: 333–349.

Thorne, R. F. 1973a. The "Amentiferae" or Hamamelidae as an artificial group: a summary statement. *Brittonia* 25: 395–405.

Thorne, R. F. 1973b. Inclusion of the Apiaceae (Umbelliferae) in the Araliaceae. *Notes Roy. Bot. Gard. Edinburgh* 32: 161–165.

Thorne, R. F. 1974. A phylogenetic classification of the Annoniflorae. *Aliso* 8: 147–209.

Thorne, R. F. 1976. A phylogenetic classification of the Angiospermae. *Evol. Biol.* 9: 35–106.

Thorne, R. F. 1983. Proposed new realignments in the angiosperms. *Nordic J. Bot.* 3: 85–117.

Thorne, R. F. 1992. Classification and geography of the flowering plants. *Bot. Rev.* 58: 225–348.

Thorne, R. F. 2001. The classification and geography of the flowering plants: dicotyledons of the class Angiospermae. *Bot. Rev.* 66: 441–647.

Tiffney, B. H. 1986. Fruit and seed dispersal and the evolution of the Hamamelidae. *Ann. Missouri Bot. Gard.* 73: 394–416.

Tingshuang. Y., P. P. Lowry, II, G. M. Plunkett, and J. Wen. 2004. Chromosomal evolution in Araliaceae and close relatives. *Taxon* 53: 987–1005.

Tobe, H. 1989. The embryology of angiosperms: its broad application to the systematic and evolutionary study. *Bot. Mag. Tokyo* 102: 351–367.

Tobe, H. and P. H. Raven. 1983. An embryological analysis of the Myrtales: its definition and characteristics. *Ann. Missouri Bot. Gard.* 70: 71–94.

Tobe, H. and P. H. Raven. 1988. Seed morphology and anatomy of Rhizophoraceae, inter- and infrafamilial relationships. *Ann. Missouri Bot. Gard.* 75: 1319–1342.

Todzia, C. A. 1993. Ulmaceae. In *The families and genera of vascular plants*, vol. 2, Magnoliid, hamamelid and caryophyllid families, K. Kubitzki, J. G. Rohwer, and V. Bittrich (eds.), 603–611. Springer-Verlag, Berlin.

Tollsten, L. and J. T. Kundsen. 1992. Floral scent in dioecious *Salix* (Salicaceae)—a cue determining the pollination system? *Plant Syst. Evol.* 182: 229–237.

Tomlinson, P. B. 1962. Phylogeny of the Scitamineae: morphological and anatomical considerations. *Evolution* 16: 192–213.

Tomlinson, P. B. 1969a. Commelinales—Zingiberales. In *Anatomy of the monocotyledons*, vol. 3, C. R. Metcalfe (ed.). Clarendon Press, Oxford.

Tomlinson, P. B. 1969b. On the morphology and anatomy of turtle grass, *Thalassia testudinum* (Hydrocharitaceae). *Bull. Mar. Sci.* 19: 286–305.

Tomlinson, P. B. 1982. Helobiae (Alismatidae), vol. 7. In *Anatomy of the monocotyledons*, C. R. Metcalfe (ed.). Oxford University Press, Oxford.

Tomlinson, P. B. 1986. *The botany of mangroves*. Cambridge University Press, Cambridge.

Tomlinson, P. B. 1990. *The structural biology of palms*. Clarendon Press, Oxford.

Tomlinson, P. B. and M. H. Zimmerman. 1969. Vascular anatomy of monocots with secondary growth. *J. Arnold Arbor.* 50: 159–179.

Tomlinson, P. B., R. B. Primack, and J. S. Bunt. 1979. Preliminary observations on floral biology in mangrove Rhizophoraceae. *Biotropica* 11: 256–277.

Tomlinson, P. B. and P. A. Cox. 2000. Systematic and functional anatomy of seedlings in mangrove Rhizophoraceae: vivipary explained. *Bot. J. Linn. Soc.* 134: 215–231.

Torrey, J. G. and R. H. Berg. 1988. Some morphological features for generic characterization among the Casuarinaceae. *Amer. J. Bot.* 75: 864–874.

Townsend, C. C. 1993. Amaranthaceae. In *The families and genera of vascular plants*, vol. 2, Magnoliid, hamamelid and caryophyllid families, K. Kubitzki, J. G. Rohwer, and V. Bittrich (eds.), 70–91. Springer-Verlag, Berlin.

Treutlein, J., G. F. Smith, B.-E. van Wyk, and M. Wink. 2003. Phylogenetic relationships in Asphodelaceae (subfamily Alloideae) inferred from chloroplast DNA sequences (*rbcL, matK*) and from genomic fingerprinting (ISSR). *Taxon* 52: 193–207.

Trift, I., M. Källersjö, and A. A. Anderberg. 2002. The monophyly of *Primula* (Primulaceae) evaluated by analysis of sequences from the chloroplast gene *rbcL*. *Syst. Bot.* 27: 396–407.

Tucker, G. C. 1987. The genera of Cyperaceae in the southeastern United States. *J. Arnold Arbor.* 68: 361–445.

Tucker, G. C. 1989. The genera of Commelinaceae in the southeastern United States. *J. Arnold Arbor.* 70: 97–130.

Tucker, G. C. 1996. The genera of Pooideae (Gramineae) in the southeastern United States. *Harvard Pap. Bot.* No. 9: 11–90.

Tucker, S. C., A. W. Douglas, and L. H.-X. Liang. 1993. Utility of ontogenetic and conventional characters in determining phylogenetic relationships of Saururaceae and Piperaceae (Piperales). *Syst. Bot.* 18: 614–641.

Tucker, S. C. and A. W. Douglas. 1994. Ontogenetic evidence and phylogenetic relationships among basal taxa of legumes. In *Advances in legume systematics*, part 6, I. K. Ferguson and S. Tucker (eds.), 11–32. Royal Botanic Gardens, Kew.

Ueda, K., K. Kosuge, and H. Tobe. 1997. A molecular phylogeny of Celtidaceae and Ulmaceae (Urticales) based on *rbcL* nucleotide sequences. *J. Plant Ses.* 110: 171–178.

Uhl, N. W. and J. Dransfield. 1987. *Genera palmarum*. L. H. Bailey Hortorium and International Palm Society, Ithaca, NY.

Uhl, N. W., J. Dransfield, J. I. Davis, M. A. Luckow, K. H. Hansen, and J. J. Doyle. 1995. Phylogenetic relationships among palms: cladistic analyses of morphological and chloroplast DNA restriction site variation. In *Monocotyledons: systematics and evolution*, P. J. Rudall, P. J. Cribb, D. F. Cutler, and C. J. Humphries (eds.), 623–661. Royal Botanic Gardens, Kew.

Umadevi, I. and M. Daniel. 1991. Chemosystematics of the Sapindaceae. *Feddes Repert.* 102: 607–612.

Unwin, M. M. 2004. *Molecular Systematics of the Eriocaulaceae Martinov*. Ph. D. Thesis, Miami University, Ohio.

Urtuey, E. and T. F. Stuessy. 2001. New hypotheses of phylogenetic relationships in Barnadesioideae (Asteraceae) based on morphology. *Taxon* 50: 1043–1066.

van Heusden, E. C. H. 1992. Flowers of Annonaceae: morphology, classification, and evolution. *Blumea*, Suppl. 7: 1–218.

van den Berg, C., D. H. Goldman, J. V. Freudenstein, A. M. Pridgeon, K. M. Cameron, and M. W. Chase. 2005. An overview of the phylogenetic relationships within Epidendropideae inferred from multiple DNA regions and recircumscription of Epidendreae and Arethuseae (Orchidaceae). *Amer. J. Bot.* 92: 613–624.

van der Pijl, L. 1957. On the arilloids of *Nephelium*, *Euphoria*, *Litchi* and *Aesculus* and the seeds of Sapindaceae in general. *Acta Bot. Neerl.* 6: 618–641.

van der Pijl, L. and C. H. Dodson. 1966. *Orchid flowers, their pollination and evolution*. University of Miami Press, Coral Gables, FL.

van der Werff, H. 1991. A key to the genera of Lauraceae in the New World. *Ann. Missouri Bot. Gard.* 78: 377–387.

van der Werff, H. and H. G. Richter. 1996. Towards an improved classification of Lauraceae. *Ann. Missouri Bot. Gard.* 83: 409–418.

van Ham, R. C. H. 1994. *Phylogenetic implications of chloroplast DNA variation in the Crassulaceae*. Dissertation. University of Utrecht, Belgium.

van Ham, R. C. H. and H. T. t'Hart. 1998. Phylogenetic relationships in the Crassulaceae inferred from chloroplast DNA restriction-site variation. *Amer. J. Bot.* 85: 123–134.

van Heel, W. A. 1966. Morphology of the androecium in Malvales. *Blumea* 13: 177–394.

Vander Wyk, R. and J. E. Canright. 1956. The anatomy and relationships of the Annonaceae. *Trop. Woods* 104: 1–24.

Varadarajan, G. S. and A. J. Gilmartin. 1988. Phylogenetic relationships of groups of genera within the subfamily Pitcairnioideae (Bromeliaceae). *Syst. Bot.* 13: 283–293.

Vaughan, J. G., J. R. Pjelan, and K. E. Denford. 1976. Seed studies in the Cruciferae. In *The biology and chemistry of the Cruciferae*, J. G. Vaughan, A. J. MacLeod, and B. M. G. Jones (eds.), 119–144. Academic Press, London.

Verhoek, S. 1998. Agavaceae. In *The families and genera of vascular plants*, vol. 3. Monocotyledons: Lilianae (except Orchidaceae), K. Kubitzki (ed.), 60–70. Springer-Verlag, Berlin.

Verkerke, W. 1985. Ovules and seeds of Polygalaceae. *J. Arnold Arbor.* 66: 353–394.

Verma, D.-P. S. and J. Standley. 1989. The legume–*Rhizobium* equation: coevolution of two genomes. *Monogr. Syst. Bot. Missouri Bot. Gard.* 29: 545–557.

Vidal-Russell, R. and D. Nickrent. 2005. A molecular phylogeny of the mistletoe family Loranthaceae. Abstract. *Botany 2005*: 131–132.

Vink, W. 1988. Taxonomy in Winteraceae. *Taxon* 37: 691–698.

Vink, W. 1993. Winteraceae. In *The families and genera of vascular plants*, vol. 2, Magnoliid, hamamelid and caryophyllid families, K. Kubitzki, J. G. Rohwer, and V. Bittrich (eds.), 630–638. Springer-Verlag, Berlin.

Vinnersten, A. and K. Bremer. 2001. Age and biogeography of major clades in Liliales. *Amer. J. Bot.* 88: 1695–1703.

Vinnersten, A. and G. Reeves. 2003. Phylogenetic relationships within Colchicaceae. *Amer. J. Bot.* 90: 1455–1462.

Vliet, G. J. C. M. van, J. Koek-Noorman, and B. J. H. ter Welle. 1981. Wood anatomy, classification and phylogeny of the Melastomataceae. *Blumea* 27: 463–473.

Vogel, S. 1990. History of the Malpighiaceae in the light of pollination ecology. *Mem. New York Bot. Gard.* 55: 130–142.

von Balthazar, M., W. S. Alverson, J. Schönenberger, and D. A. Baum. 2004. Comparative floral development and androecium structure in Malvoideae (Malvaceae s.l.). *Int. J. Plant Sci.* 165: 445–473.

Vuilleumier, B. S. 1967. The origin and evolutionary development of heterostyly in the angiosperms. *Evolution* 21: 210–226.

Vuilleumier, B. S. 1969. The genera of Senecioneae in the southeastern United States. *J. Arnold Arbor.* 50: 104–123.

Vuilleumier, B. S. 1973. The genera of Lactuceae (Compositae) in the southeastern United States. *J. Arnold Arbor.* 54: 42–93.

Wagenitz, G. 1959. Die systematische Stellung der Rubiaceae. *Bot. Jahr. Syst.* 79: 17–35.

Wagenitz, G. 1992. The Asteridae: evolution of a concept and its present status. *Ann. Missouri Bot. Gard.* 79: 209–217.

Wagenitz, G. 1977. New aspects of the systematics of Asteridae. *Plant Syst. Evol.* Suppl. 1: 375–395.

Wagner, W. L. and P. C. Hoch. 2005, onward. *Onagraceae. The evening primrose family website*. http://ravenel.si.edu/botany/ onagraceae/index.htm.

Wagstaff, S. J. and R. G. Olmstead. 1996. Phylogeny of Labiatae and Verbenaceae inferred from *rbcL* sequences. *Syst. Bot.* 22: 165–179.

Wagstaff, S. J., L. Hickerson, R. Spangler, P. A. Reeves, and R. G. Olmstead. 1998. Phylogeny in Labiatae s. l., inferred from cpDNA sequences. *Plant Syst. Evol.* 209: 265–274.

Walker, J. W. 1971. Pollen morphology, phytogeography and phylogeny of the Annonaceae. *Contr. Gray Herb.* 202: 1–131.

Wallace, G. D. 1975. Interrelationships of the subfamilies of the Ericaceae and the derivation of the Monotropoideae. *Bot. Not.* 128: 286–298.

Wallander, E. and V. A. Albert. 2000. Phylogeny and classification of Oleaceae based on *rbs16* and *trnL-F* sequence data. *Amer. J. Bot.* 87: 1827–1841.

Wallnöfer, B. 2004. Ebenaceae. In *The families and genera of vascular plants*, vol. 6, Celastrales, Oxalidales, Rosales, Cornales, Ericales, K. Kubitzki (ed.), 125–130. Springer-Verlag, Berlin.

Wang, W., R.-Q. Li, and Z.-D. Chen. 2005. Systematic position of Asteropyrum (Ranunculaceae) inferred from chloroplast and nuclear sequences. *Plant Syst. Evol.* 255: 41–54.

Wannan, B. S. and C. J. Quinn. 1990. Pericarp structure and generic affinities in the Anacardiaceae. *Bot. J. Linn. Soc.* 102: 225–252.

Wannan, B. S. and C. J. Quinn. 1991. Floral structure and evolution in the Anacardiaceae. *Bot. J. Linn. Soc.* 107: 349–385.

Watson, L. and M. J. Dallwitz. 1992. *The grass genera of the world*. C. A. B. International, Wallingford, UK.

Warwick, S. I. and L. D. Black. 1993. Molecular relationships in subtribe Brassicinae (Cruciferae, tribe Brassiceae). *Canad. J. Bot.* 71: 906–918.

Weber, A. 2004. Gesneriaceae. In The families and genera of vascular plants, vol. 7, Lamiales (except Acanthaceae, including Avicenniaceae), K. Kubitzki (ed.), 63–158. Springer-Verlag, Berlin.

Weberling, F. 1988a. The architecture of inflorescences in the Myrtales. *Ann. Missouri Bot. Gard.* 75: 226–310.

Weberling, F. 1988b. Inflorescence structure in primitive angiosperms. *Taxon* 37: 657–690.

Weberling, F. 1989. Structure and evolutionary tendencies of inflorescences in the Leguminosae. *Monogr. Syst. Bot. Missouri Bot. Gard.* 29: 35–58.

Webster, G. L. 1967. The genera of Euphorbiaceae in the southeastern United States. *J. Arnold Arbor.* 48: 303–430.

Webster, G. L. 1987. The saga of the spurges: a review of classification and relationships in the Euphorbiales. *Bot. J. Linn. Soc.* 94: 3–46.

Webster, G. L. 1994a. Classification of the Euphorbiaceae. *Ann. Missouri Bot. Gard.* 81: 3–32.

Webster, G. L. 1994b. Synopsis of the genera and suprageneric taxa of Euphorbiaceae. *Ann. Missouri Bot. Gard.* 81: 33–144.

Weigend, M. 2004. Loasaceae. In *The families and genera of vascular plants*, vol. 6, Celastrales, Oxalidales, Rosales, Cornales, Ericales, K. Kubitzki (ed.), 239–254. Springer-Verlag, Berlin.

Wen, J., G. M. Plunkett, A. D. Mitchell, and S. J. Wagstaff. 2001. The evolution of Araliaceae: a phylogenetic analysis based on ITS sequences of nuclear ribosomal DNA. *Syst. Bot.* 26: 144–167.

Wetsching, W. and M. Pfosser. 2003. The Scilla plumbea puzzle – present status of the genus *Scilla* sensu lato in southern Africa and description of *Spetaea lachenaliiflora*, a new genus and species of Massonieae (Hyacinthaceae). *Taxon* 52: 75–91.

Whiffin, T. 1972. Observations on some upper Amazonian formicarial Melastomataceae. *Sida* 5: 32–41.

White, D. A. and L. B. Thien. 1985. The pollination of *Illicium parviflorum* (Illiciaceae). *J. Elisha Mitchell Sci. Soc.* 101: 15–18.

Whitlock, B. A., C. Bayer, and D. A. Baum. 2001. Phylogenetic relationships and floral evolution of the Byttnerioideae ("Sterculiaceae" or Malvaceae s. l.) based on sequences of the chloroplast gene, *ndhF*. *Syst. Bot.* 26: 420–437.

Weiblen, G. D. 2000. Phylogenetic relationships of functionally dioecious Ficus (Moraceae) based on ribosomal DNA sequences and morphology. *Amer. J. Bot.* 87: 1342–1357.

Wiegrefe, S. J., K. J. Sytsma, and R. P. Guries. 1994. Phylogeny of elm (*Ulmus*, Ulmaceae): molecular evidence for a sectional classification. *Syst. Bot.* 19: 590–612.

Wiegrefe, S. J., K. J. Sytsma, and R. P. Gories. 1998. The Ulmaceae, one family or two? Evidence from chloroplast DNA restriction site mapping. *Plant Syst. Evol.* 210: 249–270.

Wiehler, H. 1983. A synopsis of the neotropical Gesneriaceae. *Selbyana* 6: 1–219.

Wiens, D. and B. A. Barlow. 1971. The cytogeography and relationships of the Viscaceous and Eremolepidaceous mistletoes. *Taxon* 20: 313–332.

Wiersema, J. H. 1988. Reproductive biology of *Nymphaea* (Nymphaeaceae). *Ann. Missouri Bot. Gard.* 75: 795–804.

Wilbur, R. L. 1994. The Myricaceae of the United States and Canada: genera, subgenera, and series. *Sida* 16: 93–107.

Wilbur, R. L. 2001. Five new combinations in the genus *Morella* (Myricaceae) for neotropical species. *Rhodora* 103: 120–122.

Wildinson, A. M. 1948. Floral anatomy and morphology of some species of the genus *Viburnum* of the Caprifoliaceae. *Amer. J. Bot.* 35: 455–465.

Wildman, W. C. and B. A. Pursey. 1968. Colchicine and related compounds. *In The alkaloids, chemistry and physiology 11*, R. H. F. Manske (ed.), 407–457. Academic Press, London.

Wilken, D. H. 2004. Polemoniaceae. In *The families and genera of vascular plants*, vol. 6, Celastrales, Oxalidales, Rosales, Cornales, Ericales, K. Kubitzki (ed.), 300–312. Springer-Verlag, Berlin.

Wilkin, P., P. Schols, M. W. Chase, K. Chayamarit, C. A. Furness, S. Huysmans, F. Rakotonasolo, E. Smets, and C. Thapyai. 2005. A plastid gene phylogeny of the yam genus, *Dioscorea*: roots, fruits, and Madagascar. *Syst. Bot.* 30: 736–749.

Williams, J. H., Jr. and W. E. Friedman. 2004. The 4-celled female gametophyte of *Illicium* (Illiciaceae; Austrobaileyales): implications for understanding the origin and early evolution of monocotyledons, eumagnoliids, and eudicots. *Amer. J. Bot.* 91: 332–351.

Williams, S. E. 1976. Comparative sensory plysiology of the Droseraceae: evolution of a plant sensory system. *Proc. Amer. Philos. Soc.* 120: 187–204.

Williams, S. E., V. A. Albert, and M. W. Chase. 1994. Relationships of Droseraceae: a cladistic analysis of *rbcL* sequence and morphological data. *Amer. J. Bot.* 81: 1027–1037.

Wilson, C. L. 1950. Vasculation of the stamen in the Melastomataceae, with some phyletic implications. *Amer. J. Bot.* 37: 431–444.

Wilson, K. A. 1960a. The genera of Arales in the southeastern United States. *J. Arnold Arbor.* 41: 47–72.

Wilson, K. A. 1960b. The genera of Convolvulaceae in the southeastern United States. *J. Arnold Arbor.* 41: 298–317.

Wilson, K. A. 1960c. The genera of Hydrophyllaceae and Polemoniaceae in the southeastern United States. *J. Arnold Arbor.* 41: 197–212.

Wilson, K. A. 1960d. The genera of Myrtaceae in the southeastern United States. *J. Arnold Arbor.* 41: 270–278.

Wilson, K. A. and C. E. Wood. 1959. The genera of Oleaceae in the southeastern United States. *J. Arnold Arbor.* 40: 369–384.

Wilson, P. G., P. A. Gadek, and C. J. Quinn. 1996. Phylogeny of Myrtaceae and its allies based on *matK* sequence data. *Amer. J. Bot.* 83(6) (Suppl.): S202.

Wilson, P. G., M. M. O'Brien, P. A. Gadek, and C. J. Quinn. 2001. Myrtaceae revisited: a reassessment of infrafamilial groups. *Amer. J. Bot.* 88: 2013–2025.

Wilson, P. G., M. M. O'Brien, M. M. Hestlewood, and C. J. Quinn. 2005. Relationships within Myrtaceae sensu lato based on *matK* phylogeny. *Plant Syst. Evol.* 251–3–19.

Wilson, T. K. and L. M. Maculans. 1967. The morphology of the Myristicaceae. I. Flowers of *Myristica fragrans* and M. *malabarica*. *Amer. J. Bot.* 54: 214–220.

Winkworth, R. C. and M. J. Donoghue. 2005. *Viburnum* phylogeny based on combined molecular data: implications for taxonomy and biogeography. *Amer. J. Bot.* 92: 653– 666.

Wojciechowski, M. F., M. Lavin, and M. J. Sanderson. 2004. A phylogeny of legumes (Leguminosae) based on analysis of the plastid *matK* gene resolves many well-supported subclades within the family. *Amer. J. Bot.* 91: 1846–1862.

Wolfe, A. D., C. P. Randle, L. Liu, and K. E. Steiner. 2005. Phylogeny and biogeography of Orobanchaceae. *Folia Geobot.* 40: 115–134.

Wolfe, J. A. 1989. Leaf-architectural analysis of the Hamamelididae. In *Evolution, systematics and fossil history of the Hamamelidae*, vol. 1, Introduction and "lower" Hamamelidae. Systematics Association Special Vol. 40A, P. R. Crane and S. Blackmore (eds.), 75–104. Clarendon Press, Oxford.

Wolfe, J. A. and T. Tanai. 1987. Systematics, phylogeny and distribution of *Acer* (maples) in the Cenozoic of North America. *J. Fac. Sci. Hokkaido Imperial University*, Ser. 4, Geology 22: 1–246.

Wolfe, J. A. and W. Wehr. 1988. Rosaceous *Chamaebatiaria*-like foliage from the Paleogene of western North America. *Aliso* 12: 177–200.

Wood, C. E. 1958. The genera of woody Ranales in the southeastern United States. *J. Arnold Arbor.* 39: 296–346.

Wood, C. E. 1959a. The genera of Nymphaeaceae and Ceratophyllaceae in the southeastern United States. *J. Arnold Arbor.* 40: 94–112.

Wood, C. E. 1959b. The genera of Theaceae of the southeastern United States. *J. Arnold Arbor.* 40: 413–419.

Wood, C. E. 1960. The genera of Sarraceniaceae and Droseraceae in the southeastern United States. *J. Arnold Arbor.* 41: 152–163.

Wood, C. E. 1961. The genera of Ericaceae in the southeastern United States. *J. Arnold Arbor.* 42: 10–80.

Wood, C. E. 1971. The Saururaceae in the southeastern United States. *J. Arnold Arbor.* 52: 479–485.

Wood, C. E. 1974. *A student's atlas of flowering plants: some dicotyledons of eastern North America.* Harper & Row, New York. Prepared as part of the Generic Flora of the Southeastern U.S. Project.

Wood, C. E. 1983. The genera of Menyanthaceae in the southeastern United States. *J. Arnold Arbor.* 64: 431–445.

Wood, C. E. and P. Adams. 1976. The genera of Guttiferae (Clusiaceae) in the southeastern United States. *J. Arnold Arbor.* 57: 74–90.

Wood, C. E. and R. B. Channell. 1960. The genera of Ebenales in the southeastern United States. *J. Arnold Arbor.* 41: 1–35.

Wood, C. E. and R. K. Godfrey. 1957. *Pinguicula* (Lentibulariaceae) in the Southeastern United States. *Rhodora* 59: 217– 230.

Wood, C. E. and R. E. Weaver, Jr. 1982. The genera of Gentianaceae in the southeastern United States. *J. Arnold Arbor.* 63: 441–487.

Wood, T. H., W. M. Whitten, and N. H. Williams. 2000. Phylogeny of *Hedychium* and related genera (Zingiberaceae) based on ITS sequence data. *Edinburgh J. Bot.* 57: 261–270.

Woodland, D. W. 1989. Biology of temperate Urticaceae (nettle) family. In *Evolution, systematics and fossil history of the Hamamelidae,* vol. 2, "Higher" Hamamelidae. Systematics Association Special Vol. 40B, P. R. Crane and S. Blackmore (eds.), 309–318. Clarendon Press, Oxford.

Woodson, R. E. 1954. The North American species of *Asclepias* L. *Ann. Missouri Bot. Gard.* 41: 1–211.

Wortley, A. H., P. J. Rudall, D. J. Harris, and R. W. Scotland. 2005. How much data are needed to resolve a difficult phylogeny? Case study in Lamiales. *Syst. Biol.* 54: 697–709.

Wurdack, J. J. 1986. Atlas of hairs for neotropical Melastomataceae. *Smithsonian Contr. Bot.* 63: 1–80.

Wurdack, J. J. and R. Kral. 1982. The genera of Melastomataceae in the southeastern United States. *J. Arnold Arbor.* 63: 429–439.

Wurdack, K. J., P. Hoffmann, R. Samuel, A. de Bruijn, M. van der Bank, and M. W. Chase. 2004. Molecular phylogenetic analysis of Phyllanthaceae (Phyllanthoideae pro parte, Euphorbiaceae sensu lato) using plastid *rbcL* DNA sequences. *Amer. J. Bot.* 91: 1882–1900.

Wurdack, K. J., P. Hoffman, and M. W. Chase. 2005. Molecular phylogenetic analysis of uniovulate Euphorbiaceae (Euphorbiaceae sensu lato) using plastid *rbcL* and *trnL-F* DNA sequences. *Amer. J. Bot.* 92: 1397–1420.

Xiang, Q. Y., D. E. Soltis, D. R. Moran, and P. S. Soltis. 1993. Phylogenetic relationships of *Cornus* L. sensu lato and putative relatives inferred from *rbcL* sequence data. *Ann. Missouri Bot. Gard.* 80: 723–734.

Xiang, Q. Y., S. J. Brunsfeld, D. E. Soltis and P. S. Soltis. 1996. Phylogenetic relationships in *Cornus* based on chloroplast DNA restriction sites: implications for biogeography and character evolution. *Syst. Bot.* 21: 515–534.

Xiang, Q. Y. and Z. Murrell. 1998. Relationships and biogeography of *Cornus* L. (Cornaceae) inferred from multiple molecular and morphological data sets. *Amer. J. Bot.* 85(6) Suppl.: 168.

Xiang, Q. Y., D. E. Soltis, and P. S. Soltis. 1998. Phylogenetic relationships of Cornaceae and close relatives inferred from *matK* and *rbcL* sequences. *Amer. J. Bot.* 85: 285–297.

Xiang, Q. Y., M. L. Moody, D. E. Soltis, C. Z. Fan, and P. S. Soltis. 2002. Relationships within Cornales and circumscription of Cornaceae: *matK* and *rbcL* sequence data and affects of outgroups and long branches. *Mol. Phylo. Evol.* 24: 35–57.

Xiang, Q.-Y., D. T. Thomas, W. Zhang, S. R. Manchester, and Z. Murrell. 2006. Species level phylogeny of the genus *Cornus* (Cornaceae) based on molecular and morphological evidence: implications for taxonomy and Tertiary intercontinental migration. *Taxon* 55: 9–30.

Xiaoping, Z. and K. Bremer. 1993. A cladistic analysis of the tribe Astereae (Asteraceae) with notes on their evolution and subtribal classification. *Plant Syst. Evol.* 184: 259–283.

Yamashita, J. and M. N. Tamura. 2000. Molecular phylogeny of the Convallariaceae (Asparagales). *In Monocots: systematics and evolution,* K. L. Wilson and D. A. Morrison (eds.), 387–400. CSIRO, Collingwood, Australia.

Yamazaki, T. 1974. A system of Gamopetalae based on embryology. *J. Fac. Sci. Univ. Tokyo,* Sect. 3, Bot. 11: 263–281.

Yeo, P. F. 1984. Fruit-discharge type in *Geranium* (Geraniaceae): its use in classification and its evolutionary implications. *Bot. J. Linn. Soc.* 89: 1–36.

Yeo, P. F. 1993. Secondary pollen presentation. *Plant Syst. Evol.* 6 (Suppl.): S1–S268.

Young, D. A. 1981. Are the angiosperms primitively vesselless? *Syst. Bot.* 6: 313–330.

Young, D. A. 1982. Leaf flavonoids of *Amborella trichopoda*. *Biochem. Syst. Evol.* 10: 21–22.

Young, N. D., K. E. Steiner, and C. W. de Pamphilis. 1999. The evolution of parasitism in Scrophulariaceae/Orobanchaceae: plastid gene sequences refute an evolutionary transition series. *Ann. Missouri Bot. Gard.* 86: 876–893.

Yulita, K. S., R. J. Bayer, and J. G. West. 2005. Molecular phylogenetic study of *Hopea* and *Shorea* (Dipterocarpaceae): evidence from the *trnL-trnF* and internal transcribed spacer regions. *Plant Species Biol.* 20: 167–182.

Zanis, M. J., D. E. Soltis, P. E. Soltis, S. Mathews, and M. J. Donoghue. 2002. The root of the angiosperms revisited. *Proc. National Acad. Sci. USA* 99: 6848–6853.

Zanis, M. J., P. S. Soltis, Y.-L. Qiu, E. Zimmer, and D. E. Soltis. 2003. Phylogenetic analysis and perianth evolution in basal angiosperms. *Ann. Missouri Bot. Gard.* 90: 129–150.

Zavada, M. S. and M. Kim. 1996. Phylogenetic analysis of Ulmaceae. *Plant Syst. Evol.* 200: 13–20.

Zhang, L.-B. and M. P. Simmons. 2006. Phylogeny and delimitation of the Celastrales inferred from nuclear and plastid genes. *Syst. Bot.* 31: 122–137.

Zhang, W.-H., Z.-D. Chen, J.-H. Li, H.-B. Chen, and Y.-C. Tang. 2003. Phylogeny of the Dipsacales s.l. based on chloroplast *trnL-F* and *ndhF* sequences. *Mol. Phylo. Evol.* 26: 176–189.

Zimmer, E. A., R. K. Hamby, M. L. Arnold, D. A. LeBlanc, and E. C. Theriot. 1989. Ribosomal RNA phylogenies and flowering plant evolution. In *The hierarchy of life,* B. Fernholm, K. Bremer, and H. Jornvall (eds.), 205–214. Elsevier, Amsterdam.

Zimmer, E. A., E. H. Roalson, L. E. Skog, J. K. Boggan, and A. Idnurm. 2002. Phylogenetic relationships in the Gesnerioideae (Gesneriaceae) based on nrDNA ITS and cpDNA *trnL-F* and *trnE-T* spacer region sequences. *Amer. J. Bot.* 89: 296–311.

Zomlefer, W. B. 1996. The Trilliaceae in the southeastern United States. *Harvard Pap. Bot.* No. 1(9): 91–120.

Zomlefer, W. B. 1997a. The genera of Melanthiaceae in the southeastern United States. *Harvard Pap. Bot.* 2: 133–177.

Zomlefer, W. B. 1997b. The genera of Nartheciaceae in the southeastern United States. *Harvard Pap. Bot.* 2: 195–211.

Zomlefer, W. B. 1997c. The genera of Tofieldiaceae in the southeastern United States. *Harvard Pap. Bot.* 2: 179–194.

Zomlefer, W. B. 1998. The genera of Hemerocallidaceae in the southeastern United States. *Harvard Pap. Bot.* 3: 113–145.

Zomlefer, W. B., N. H. Williams, W. M. Whitten, and W. S. Judd. 2001. Generic circumscription and relationships in the tribe Melanthieae (Liliales, Melanthiaceae), with emphasis on *Zigadenus*: evidence from ITS and *trnL-F* sequence data. *Amer. J. Bot.* 88: 1657–1669.

Zomlefer, W. B., W. M. Whitten, N. H. Williams, and W. S. Judd. 2003. An over-view of *Veratrum* s.l. (Liliales: Melanthiaceae) and an infrageneric phylogeny based on ITS sequence data. *Syst. Bot.* 28: 250–269.

Zomlefer, W. B., W. S. Judd, W. M. Whitten, and N. H. Williams. 2006a. A synopsis of Melanthiaceae (Liliales) with focus on character evolution in tribe Melanthieae. *Aliso* 22: 566–578.

Zomlefer, W. B., W. M. Whitten, N. H. Williams, and W. S. Judd. 2006b. Infrageneric phylogeny of *Schoenocaulon* (Liliales: Melanthiaceae) with clarification of cryptic species based on ITS sequence data and geographical distribution. *Amer. J. Bot.* 93: 1178–1192.

Zona, S. 1997. The genera of Palmae (Arecaceae) in the southeastern United States. *Harvard Pap. Bot.* No. 11: 71–107.

Zona, S. 1998. The Myoporaceae in the southeastern United States. *Harvard Pap. Bot.* 3: 171–179.

Zona, S. 2001. Starchy pollen in commelinoid monocots. *Ann. Bot. II* 87: 109–116.

Zona, S. and A. Henderson. 1989. A review of animal-mediated seed dispersal in palms. *Selbyana* 11: 6–21.

APPENDIX ONE
Botanical Nomenclature

Taxonomic groups require names if we are to efficiently communicate (or gain access to) information regarding their identity, phylogenetic relationships, and other aspects of their biology. The naming of plants is called **botanical nomenclature**. The principles and rules of botanical nomenclature have been developed and adapted by a series of international botanical congresses and are listed in the International Code of Botanical Nomenclature, or ICBN (McNeill et al. 2006). The major goal of the ICBN is to provide one correct name for each taxonomic group (or taxon) within a stable system of names (a classification).

Scientific Names

Systematic botanists (as well as other scientists) use Latinized scientific names. Each taxon—for example, a species, genus, or family—has one such name that is used throughout the world. The use of scientific names is essential for the efficient, accurate communication of information about plants on a worldwide basis. Common (or vernacular) names are not adequate for this purpose for a number of reasons. They are often limited to a single language or to a particular geographic region. Sometimes the same common name (e.g., "bluebell" or "cedar") is applied to several different taxa. Many species, especially if they are rare or of little economic importance, have no common names. Finally, common names are frequently misleading with respect to phylogenetic relationships; for example, poison oaks, species of *Toxicodendron* (Anacardiaceae), are not closely related to the true oaks of the genus *Quercus* (Fagaceae).

TABLE 1 Common specific epithets.

Specific epithet[a]	English meaning	Specific epithet	English meaning	Specific epithet	English meaning
acaulis	stemless	brevipes	short-footed	dentatus	toothed
acicularis	needlelike	brunneus	deep brown	didymus	twinned, in pairs
aduncus	hooked	bufonius	of toads	digitatus	fingered, palmate
aestivalis	summer	caeruleus	dark blue	discolor	having different colors
affinis	related	caesius	bluish gray		
agrestis	of the field	calvus	bald, hairless	dulcis	sweet
alatus	winged	calycinus	calyxlike	dumosus	bushy
albicans	whitish	campanulatus	bell-shaped	echinatus	prickly
albus	white	campestris	of the fields	edulis	edible
alpestris	alpine	candicans	white, hoary	effusus	loose-spreading
alpinus	alpine	capillaris	hairlike	elatior	taller
alternans	alternating	carinatus	keeled	elatus	tall
altissimus	very tall	caudatus	tailed	elegans	elegant
amabilis	lovely	cerifera	wax-bearing	ensifolius	sword-leaved
amarus	bitter	cernuus	drooping	eriocarpus	woolly-fruited
ambigens	ambiguous	chloranthus	green-flowered	esculentus	edible
amoenus	charming	chrysophyllus	golden-leaved	exiguus	little, slender
amplexicaulis	clasping	chrysostomus	golden-mouthed	fallax	deceptive
anceps	two-headed or two-edged	cinctus	girdled	farinosus	mealy
		clandestinus	concealed	fasciculatus	fascicled
angustatus	narrow	coarctatus	pressed together, narrowed	fastigiatus	erect and close together
angustifolius	narrow-leaved				
annotinus	year-old	coccineus	scarlet	filipes	threadlike stalks
annuus	annual	comatus	with hair	fistulosus	hollow, cylindrical
aphyllus	leafless	communis	gregarious	flabellatus	fanlike
apiculatus	tipped with a point	commutatus	changing	flagellaris	whiplike
appendiculatus	appendaged	comosus	bearded or tufted, long-haired	flavescens	yellowish
applanatus	flattened			flavus	yellow
arcuatus	bowlike	concinnus	neat	flexilis	flexible
arenarius	of sand	concolor	colored similarly	floribundus	free-flowering
areolatus	areolate, pitted	confertus	crowded	floridus	flowering
argenteus	silvery	confinis	bordered	fluitans	floating
argutus	sharp-toothed	conoideus	conelike	fluviatilis	of a river
argyreus	silvery	contortus	contorted	foetidus	bad-smelling
aridus	arid	corniculatus	with small horns	foliosus	leafy
aristatus	awned			formosus	beautiful
arundinaceus	reedlike	cornutus	horned	frondosus	leafy
arvensis	of cultivated fields	coronarius	used with garlands	fulgens	shining
asper	rough	crassifolius	thick-leaved	furcatus	forked
atratus	blackened	crassipes	thick-footed, i.e., with a stout pedicel, petiole, etc.	geniculatus	jointed
atropurpureus	dark purple			gracilis	slender
atrosanguineus	dark blood-red			gramineus	grassy
aureus	golden	crinitus	hairy	graveolens	heavy-scented
australis	southern	cristatus	crested	hebecarpus	pubescent-fruited
azureus	sky-blue	cuneiformis	wedge-shaped	hirtus	hairy
baccatus	berried	dasycarpus	shaggy-fruited	humifusus	sprawling
baculiformis	rod-shaped	dasystachys	shaggy-spiked	humilis	dwarf
bicolor	two-colored	debilis	weak	hyemalis	of winter
bidentata	two-toothed	decapetalus	ten-petaled	hyperboreus	far northern
biennis	biennial	decipiens	deceptive	hypogaeus	underground
bifidus	twice-cut	decorus	elegant	hypoglaucus	glaucous beneath
biflorus	two-flowered	decumbens	reclining	hystrix	bristly
borealis	northern	deflexus	bent downward	incanus	hoary
brachycarpus	short-fruited	demissus	low, weak	inermis	unarmed

[a] These epithets are adjectives, and are printed in their masculine form, but when used they will agree with the gender of the generic name.

TABLE 1   (continued)

Specific epithet[a]	English meaning	Specific epithet	English meaning	Specific epithet	English meaning
inodorus	without an odor	oliganthus	few-flowered	scandens	climbing
intumescens	swollen	oligocarpus	few-fruited	sclerophyllus	hard-leaved
junceus	rushlike	oligospermus	few-seeded	scoparius	broomlike
lactatus	milky	operculatus	with a lid	sensibilis	sensitive
lacustris	of lakes	orientalis	eastern	septentrionalis	northern
laevigatus	smooth	ornatus	adorned	serotinus	late-flowering
lanuginosus	woolly	orthocarpus	straight-fruited	serpens	creeping
latifolius	broad-leaved	ovatus	ovate	serpyllifolius	thyme-leaved
leptocladus	thin-stemmed	oxycanthus	sharp-spined	setaceus	bristlelike
leucanthus	white-flowered	paludosus	marsh-loving	speciosus	showy
linearis	linear	palustris	of swamps	spectabilis	spectacular
littoralis	of the seashore	parviflorus	small-flowered	squarrosus	with parts recurved
longipes	long-footed	parvifolius	small-leaved	stans	erect
lucidus	bright, clear	parvulus	very small	stellatus	star-shaped
lupulinus	hoplike	patens	spreading	stenophyllus	narrow-leaved
luteolus	yellowish	pauciflorus	few-flowered	strictus	strict, upright
macilentus	lean	pectinatus	comblike	tenellus	slender, soft, tender
macranthus	large-flowered	pedatus	like a foot	tenuis	slender, thin
macrocarpus	large-fruited	pentandrus	with five stamens	teres	terete
macrophyllus	large-leaved	peregrinus	exotic	ternatus	in threes
maculatus	spotted	perennans	perennial	tetrapterus	four-winged
maritimus	of the sea	plantagineus	plantainlike	thyrsiflorus	thyrse-flowered
medius	intermediate	platycarpus	broad-fruited	tinctorius	of dyes
megarrhizus	large-rooted	platycladus	broad-branched	tricoccus	three-lobed
micranthus	small-flowered	platyphyllus	broad-leaved	tridens	three-toothed
millefolius	many-leaved	polyanthus	many-flowered	trifidus	three-parted
mirabilis	wonderful	polystachyus	many-spiked	tripteris	three-winged
modestus	modest (e.g., rather small, with nodding or pink blushed flowers)	praecox	precocious	tristis	dull
		prasinus	grass-green	trivialis	common, ordinary
		procera	tall	umbrosus	shade-loving
mollis	soft	pulchellus	pretty, beautiful	uncinatus	hooked
moniliformis	constricted at regular intervals	pulcher	pretty, beautiful	undulatus	undulate
		pumilis	dwarf	uniflorus	one-flowered
monocephalus	single-headed	pungens	piercing	urceolatus	urn-shaped
monostachys	single-spiked	pycnanthus	densely flowered	urens	burning, stinging
montanus	of mountains	quadrifolius	four-leaved	ursinus	of bears
mutabilis	variable	quinquefolius	five-leaved	usitatissimus	most useful
nanus	dwarf	ramosus	branched	vaginatus	sheathed
natans	floating	repens	creeping	validus	strong
nemoralis	of groves	retroflexus	reflexed	velutinus	velvety
nictitans	blinking, nodding	riparius	of river banks	venosus	veiny
nigricans	black	rostratus	beaked	vernalis	vernal
nitens	shining	rubellus	reddish	vernus	of spring
nitidus	shining	rubiginosus	rusty	versicolor	variously colored
nivalis	snowy	rufus	red	vestitus	covered
niveus	snowy	rugosus	wrinkled	vimineus	of wickerwork
novae-angliae	of New England	rupestris	rock-loving	virens	green
noveboracensis	of New York	saccharinus	sugary, sweet	volubilis	twining
nudicaulis	naked-stemmed	sagittatus	arrowhead-shaped	vulgaris	common
nutans	nodding	salinus	salty	vulpinus	of foxes
obovatus	obovate	sanguineus	blood-red	xanthocarpus	yellow-fruited
occidentale	western	sativus	cultivated		
officinalis	official	saxatilis	found on the rocks		

Scientific names of species are **binomials**; that is, they are composed of two Latinized words (or names). The binomial system of nomenclature was first consistently used by Carolus Linnaeus in his *Species plantarum* (1753). The first word of a species name is a singular noun and is the name of the genus to which the plant is assigned. The second word is (1) an adjective modifying the generic name (and must agree in gender with the generic name), (2) a noun in apposition, or (3) a possessive noun; it is called the **specific epithet**.

Most specific epithets refer to a distinctive morphological, ecological, or chemical feature of the species. Some epithets refer to the species' geographic range; others honor the individual who first collected the plant or a scientist who contributed to the botanical knowledge of a particular geographic region or taxonomic group. Knowing the meanings of specific epithets aids in remembering scientific names. Table 1 lists some common specific epithets. Both the generic name and the specific epithet are italicized or underlined; the first letter of the generic name is capitalized, and the ICBN recommends that the specific epithet always be written in lowercase. Finally, the specific epithet may not exactly repeat the generic name, as in *Benzoin benzoin*; such names are called **tautonyms**.

The specific epithet is often followed by one or more **authorities**: the name (or names) of the person (or persons) who first described the species. These names may be abbreviated, and a list of approved abbreviations has been prepared (Brummitt and Powell 1992). For example, the scientific name, including authority, for the white oak is *Quercus alba* Linnaeus. The genus is *Quercus* (oak), the specific epithet is *alba* (an adjective meaning "white"), and the name of the authority is Linnaeus (which may be abbreviated "L."). Other examples include *Acer rubrum* L. (red maple), *Hibiscus coccineus* Walt. (scarlet rosemallow), *Pinus ponderosa* Dougl. (ponderosa pine), *Quercus virginiana* Mill. (live oak), *Rudbeckia laciniata* L. (cutleaf coneflower), and *Vaccinium corymbosum* L. (highbush blueberry).

Sometimes increased knowledge about a species' phylogenetic relationships results in a change in its name. Such name changes can interfere with the information retrieval function of names (and classifications), but they provide a more accurate reflection of hypothesized phylogenetic relationships, allowing our system of botanical nomenclature to be more predictive. In one such case, Thomas Walter described *Andromeda ferruginea* in 1788. Later Thomas Nuttall (1818) decided that this species (and its relatives) actually belongs to the genus *Lyonia* because it has capsules with distinctively thickened sutures (Figure 1). Thus the specific epithet *ferruginea* was transferred by Nuttall to the genus *Lyonia*. The resulting scientific name, with authorities, is *Lyonia ferruginea* (Walt.) Nutt., and that is the currently accepted name of the species. Note that the describing authority's name is placed in parentheses and is followed by the transferring authority's name. (Nuttall made this decision on the basis of his subjective evaluation

FIGURE 1 Capsules of *Lyonia ferruginea*. Note the thickened sutures, which prompted Thomas Nuttall to transfer this species from *Andromeda* to *Lyonia*.

of the pattern of character variation within these genera, but recent phylogenetic studies of this group of Ericaceae also suggest that *Andromeda*, as currently circumscribed, is not closely related to *Lyonia*.)

In a second case, German botanist G. C. Oeder described *Ledum groenlandicum* in 1771; he placed the species in the genus *Ledum* (at least in part) because it has radially symmetrical flowers with more or less distinct petals. A cladistic analysis of *Rhododendron* and related genera of the tribe Rhodoreae (Kron and Judd 1990) indicated that *Ledum groenlandicum* (and its close relatives) nests within the cladistic structure of *Rhododendron*, a genus in which nearly all species have sympetalous and slightly bilaterally symmetrical flowers. Therefore this species has been transferred to *Rhododendron* and given the name *Rhododendron groenlandicum* (Oeder) Kron & Judd. The new name reflects the close phylogenetic relationship of this species to others that also have multicellular peltate scales: the lepidote rhododendrons, such as *Rhododendron minus* and *R. lapponicum*.

Sometimes authority names are separated by the prepositions *ex* or *in*. Names separated by *ex* mean that the second author published a name that had been proposed (but was never published) by the first; names separated by *in* mean that the first author published the name in a book or article edited (or written, in part) by the second. Examples include *Gossypium tomentosum* Nutt. ex Seem., which may be shortened to *G. tomentosum* Seem., and *Viburnum ternatum* Rehder in Sargent, which may be shortened to *V. ternatum* Rehder. The ICBN allows that the author before the *ex* or after the *in* does not have to be listed.

Infraspecific taxa—**subspecies** or **varieties**—are sometimes recognized within variable species. Geographic races are often treated at the varietal or subspecific level, and in these cases varietal or subspecific epithets are provided,

TABLE 2 **Rank of taxonomic categories recognized by the ICBN.**

Category[a]	Standard suffix
Kingdom	-bionta
Phylum (or **Division**)	-phyta
Subphylum (or Subdivision)	-phytina
Class	-opsida
Subclass	-idae
Superorder	-anae
Order	-ales
Suborder	-ineae
Superfamily	-ariae
Family	-aceae
Subfamily	-oideae
Tribe	-eae
Subtribe	-inae
Genus	None; italicized, initial capital letter
Species	None; genus name plus specific epithet; italicized

[a]The seven major ranks are indicated in **boldface**.

such as *Lyonia ligustrina* (L.) DC. var. *ligustrina* and *L. ligustrina* (L.) DC. var. *foliosiflora* (Michx.) Fernald, or *Carpinus caroliniana* Walt. subsp. *caroliniana* and *C. caroliniana* Walt. subsp. *virginiana* (Marsh.) Furlow. The epithet of the variety or subspecies that contains the type specimen of the species (type specimens will be explained on page 549) repeats the species epithet but without naming any authority; this variety is often called **nominate**, or "typical."

Scientific names of higher taxa—genera and above—are **uninomials**; they are composed of a single word. The names of taxa above the rank of genus are Latinized plural nouns. The ending of the noun often represents the rank at which the taxon is placed. The ICBN recognizes seven major ranks (kingdom, phylum/division, class, order, family, genus, species), but allows others to be intercalated; such intercalation is often accomplished by the addition of the prefixes *super-* or *sub-*. The most commonly used ranks (from high to low), with their standard suffixes, are listed in Table 2.

It is common for scientists writing a monograph or flora treatment to assign every taxon to a higher-ranked taxon, but that is not required by the ICBN. If some genera within a family are placed in subfamilies, then it is conventional to place all the genera within that family in subfamilies, even if the result is multiple monotypic subfamilies or if it requires placing genera about which we know very little. This desire to place all taxa in the next higher taxon, however, has in the past led to the creation of "garbage can" taxa: groups created purely to hold miscellaneous taxa

about which little is known. Indeed, although the practice of assigning everything to a rank is pleasantly tidy, it is not necessary. Thus a genus can be placed in a family, but left with placement uncertain (*incertae sedis*) as to subfamily or tribe. In the APG (Angiosperm Phylogeny Group) classification of angiosperm families followed in this text, some families are unplaced with respect to order even if they are placed with respect to a higher-level group. For example, we have placed Buxaceae and Trochodendraceae in the eudicot clade, but have not placed them in an order.

Generic names do not have standardized endings. Generic names are capitalized and italicized (or underlined); names above the generic level are capitalized but usually not italicized (among American workers, although Europeans often italicize these names as well).

Family names are based on the name of the type genus for the family (type genera are explained on page 549)—for example, *Rosa* (Rosaceae), *Aster* (Asteraceae), *Erica* (Ericaceae), or *Cyperus* (Cyperaceae). The use of eight traditional family names, however, is specifically maintained by the ICBN. These names are Compositae (= Asteraceae), Cruciferae (= Brassicaceae), Gramineae (= Poaceae), Guttiferae (= Clusiaceae), Labiatae (= Lamiaceae), Leguminosae (= Fabaceae s.l.), Palmae (= Arecaceae), and Umbelliferae (= Apiaceae). (The names in parentheses are the newer names.) In addition, Papilionaceae (= Fabaceae s.s.), when this clade is regarded as a family, or Papilionoideae (= Faboideae), when regarded as a subfamily (as in this text), are allowed by the ICBN. The ICBN applies the principle of priority of publication to the ranks of family and below; for taxa above the rank of family, its rules are less restrictive. For this reason, there is appreciable variation in taxon names at ranks above the level of family.

An example of the taxonomic hierarchy as applied to *Acer rubrum* (red maple) is presented in Table 3. Keep in mind that taxa, particularly above the level of species, represent monophyletic groups and are, as such, considered to be the products of evolution. The ranks of the taxonomic hierarchy, on the other hand, are human constructs, having

TABLE 3 **An application of the classification hierarchy for *Acer rubrum* (red maple).**

Category	Taxon
Kingdom	Viridiplantae (green plants)
Phylum (or Division)	Embryophyta (embryophytes)
Subphylum (or Subdivision)	Tracheophytina (tracheophytes)
Class	Angiospermopsida (angiosperms)
Order	Sapindales
Family	Sapindaceae (soapberry family)
Genus	*Acer* (maple)
Species	*Acer rubrum* (red maple)

relative but not absolute meaning (see Chapter 2). It is thus more important to remember, for example, that the angiosperms are hypothesized to be a monophyletic group, supported by numerous synapomorphies, than to worry about what rank is most appropriate for this higher taxon.

Arguments against the Use of Ranks in Classification

As we have just seen, it is widely appreciated that taxonomic rank is arbitrary. Although a named group must be monophyletic, there is no way to determine a priori whether it should be named as a genus, family, or order—or some rank in between. If rank is arbitrary, then one logical step would be to eliminate ranks altogether. Taxa could be placed in named groups, but the groups would not be designated as genus, family, order, or other rank. However, groups assigned to orders, families, and genera are familiar and in common use, and acceptance of an entirely new sort of nomenclature is unlikely to happen rapidly or without protest. Nonetheless, an alternative system of nomenclature, known as the PhyloCode, is being developed.

The PhyloCode has been designed entirely outside the rules of the ICBN, which governs the use of Linnaean ranks and has long been used by all plant taxonomists. In other words, it is an alternative nomenclatural system rather than a revision of the existing one (see the PhyloCode Web site at www.ohiou.edu/phylocode). Under the PhyloCode, names of taxa refer only to monophyletic groups and are defined in one of three ways:

1. A **node-based definition** specifies the meaning of a taxon name by associating that name with a clade stemming from the immediate common ancestor of two (or more) designated descendants (Figure 2A).
2. A **stem-based definition** specifies the meaning of a taxon name by associating that name with a clade containing all organisms that share a more recent common ancestor with one designated descendant than with another (Figure 2B).
3. An **apomorphy-based definition** specifies the meaning of a taxon name by associating that name with a clade derived from the ancestor in which a particular apomorphic character arose (Figure 2C) (de Queiroz and Gauthier 1992).

The issue of ranking is particularly problematic with regard to species, and a full discussion of this subject would not fit into an introductory text. The system of binomial nomenclature (as outlined in the ICBN) requires that every species be placed in a genus, and that the genus name become the first element of the species name. For example, *Quercus virginiana*, live oak, is one member of the large genus *Quercus*. Such binomial constructions may appear to be philosophically problematic if Linnaean ranks, such as genus, are eliminated from the classification system; binomials, however, are simply a noun-plus-adjective combination in Latin, and as such they have no particular philosophical connotations (Stevens 2002, 2006). The problem is

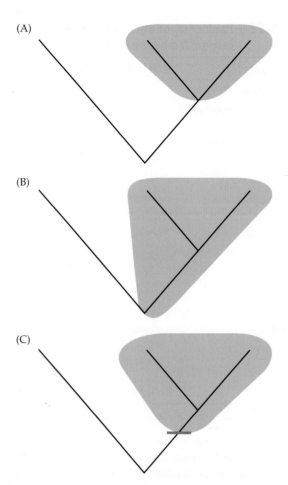

FIGURE 2 Three ways of defining taxon names under the PhyloCode. (A) Node-based definition. (B) Stem-based definition. (C) Apomorphy-based definition.

even worse if phylogenetic placement is uncertain; *incertae sedis* is not an option for a species. Currently, a systematist describing a new species does his or her best to assign it to the correct genus, but in some cases doing so may be overstating the evidence.

A solution to these problems would be to name species with a single word (a uninomial) rather than a binomial, and various ways in which uninomials could be used in naming species have been proposed (Cantino et al. 1999). Use of uninomials would increase the stability of species names, but would have several other less desirable consequences, such as the loss of information regarding phylogenetic placement from the species name. However, lack of consensus on how best to construct and implement uninomial species names has led to the decision to keep the governance of species names within the rank-based codes, but to develop conventions that will communicate the phylogenetic status of the generic portion of binomial species names (Laurin and Cantino 2007).

Ultimately, nomenclature should be thought of as part of language; words signify both particular things and relationships between things. The future success and acceptance of phylogenetic nomenclature may depend on the decision as to whether or not to use names that reflect as closely as possible the present binomial system and thus possess "historical" information and connotations readily grasped by most users of systematic information, including students (Sytsma and Pires 2001).

Pronunciation of Scientific Names

While scientific names may seem difficult to pronounce, most are easier to pronounce than unknown English words. In addition, it is difficult to pronounce scientific names "incorrectly" because there is no universally agreed upon system for pronouncing them!

Most North American botanists and horticulturists use "traditional English" pronunciation. In this system, most letters of the alphabet are pronounced as they would be in English. All vowels may be either long or short, as in English. However, it is important to remember that a Latin word has as many syllables as it has vowels or diphthongs (two vowels pronounced as one sound: *ea* as in meat, *eu* as in neutral). Every vowel or diphthong is pronounced, and there are no silent letters at the end of a word. However, when a word begins with *cn*, *gn*, *mn*, or *pt*, the first letter is silent. The letters *c* and *g* are usually hard—that is, pronounced like *k* or a hard *g*, respectively—but are soft—that is, pronounced like *s* or *j*, respectively—when followed by the letters/diphthongs *e*, *i*, *y*, *ae*, or *oe*. *Ch* is hard, and is pronounced like *k*. An *x* at the beginning of a word is pronounced like a *z*, while an *x* within a word is pronounced like *ks*. An *e* at the end of a word is long, while a final *a* is short. For *uu*, both letters are pronounced; the first is long and the second short.

Many Europeans pronounce Latin names using the "reformed academic" method, which attempts to approximate (insofar as we can tell!) the pronunciation of educated Romans. The pronunciations of Latin American scientists are also quite divergent from those of the traditional English method; they are much more similar to the pronunciations of most Europeans.

Nomenclatural Principles

1. **Botanical nomenclature is independent of zoological nomenclature.** Although the codes of botanical and zoological nomenclature are similar in their basic principles, there are many differences in detail. One result of the independence of the two codes is that a "plant" and an "animal" may have the same scientific name; *Cecropia* (showy moths and weedy tropical trees of the Urticaceae) and *Pieris* (cabbage butterflies and shrubs of the Ericaceae) are just two examples. Creating such duplications is strongly discouraged, however.

 Keep in mind that the botanical code applies not only to the green plant clade but also to other eukaryote clades—namely, stramenopiles, some alveolates (the dinoflagellates), rhodophytes (red algae), fungi, and vari-

ous other eukaryote clades (such as euglenoids and slime molds). Some "protist" groups, such as euglenoids and dinoflagellates, are sometimes considered "animals" and classified according to the nomenclatural rules of the zoological code. Therefore a few groups of organisms may have two names—one under the botanical code, and a second under the zoological code.

2. **The application of names to taxonomic groups is determined by means of nomenclatural types.** When a new species or infraspecific taxon is described, the author must designate a particular specimen as the nomenclatural type. This specimen, deposited in a particular herbarium where it will be available for study, is the **holotype** (also called the type specimen or name-bearing specimen). The name of the new species is therefore tied to this particular specimen, which illustrates what the author had in mind when he or she described the species. Duplicates of the holotype in other herbaria—that is, other portions of the same plant or other individuals of the same population, gathered at the same time and place as the holotype—are considered to be **isotypes**.

 The holotype and isotypes may be consulted by systematists seeking to clarify the application of particular scientific names. For example, biological investigations may support the existence of two species where only one was previously recognized. To which of these two species should the original species name be applied? The type concept provides an unambiguous answer by requiring that the original species name be applied to the species to which the type specimen belongs.

 The type specimen may not be at all "typical"; that is, it does not necessarily have the most common characters of the species to which it belongs. One should not disregard variation, or mistakenly think that all members of a species must possess certain "essential" or "key" features that are represented in the type specimen. The type is of no more importance than any other specimen in biological investigations, such as those determining specific delimitations or describing the pattern of variation; it is merely a nomenclatural aid. Once a systematist has determined, through systematic study, that recognition of certain species is justified (see Chapter 6 and Winston 1999), then type specimens are used in assigning names to those species.

 The type for the name of a genus is the type for the name of a particular species within that genus. This type is cited by the species' name; for example, *Lyonia ferruginea* is the type for the genus *Lyonia*. The type for the name of a family is the type for the name of a particular genus within that family (the type genus). As mentioned on page 547, the name of the family is based on the name of this genus; for example, *Aster* is the type genus of Asteraceae, and *Erica* is the type genus of Ericaceae.

3. **The nomenclature of taxonomic groups is based on priority of publication.** The correct name for a taxon is the earliest name that is in accordance with the rules of

nomenclature. Linnaeus's *Species plantarum* (published on 1 May 1753) is the starting point (for purposes of priority) for species names of vascular plants. Other plant groups may have other starting points. Later-published names for the same taxon are called **synonyms** and are not considered to be the correct names for the species. Names that duplicate names already in existence (for other species) are also not to be used and are called **homonyms**.

4. **Each taxon can bear only one correct name, except in specified cases.** Certain widely used names are not actually the earliest published in accordance with the rules, but to avoid unnecessary name changes, many of these have been (or are being) conserved—that is, allowed to be considered the correct name of a taxon— through the special action of botanical congresses. In addition, the eight families and one subfamily singled out on page 547 have more than one correct name.

5. **Scientific names are Latin or Latinized regardless of their derivation.** The use of Latin for scientific names originates from the use of Latin as the language of global communication in medieval scholarship. Botanical publications were frequently written in Latin even as late as the middle of the nineteenth century. The use of Latinized names greatly facilitates communication among plant systematists, who belong to diverse cultural and language groups.

6. **The rules of nomenclature are retroactive unless expressly limited.** The ICBN is a system of rules and conventions to be followed, but these do not necessarily have a biological basis.

Requirements for Naming a New Species

The ICBN outlines the steps necessary for describing a newly discovered taxon to ensure that its name is validly published; these steps are given here as they apply to species.

1. The species must be named. The name must be in Latin or Latinized, in a binomial format, and it must not duplicate any name already in existence.

2. The rank of the name must be clearly indicated.

3. A type specimen must be designated.

4. The species must be described in Latin, or described in another language and accompanied by a Latin **diagnosis** (a brief statement of the characters of the species or a comparison with a similar species), or linked to a reference in which there is a Latin description. A very useful reference for writing Latin descriptions and diagnoses is Stearn's *Botanical Latin* (1992).

5. All of this information must be **effectively published**; that is, it must be presented in a publication that is available to other botanists, such as a botanical journal or book. Publishing a new species in a seed catalogue, newspaper, e-mail message, or other ephemeral source does not qualify as effective publication, but the role of the World Wide Web in publication is under active discussion.

If all of these guidelines are followed, the species name is considered to be **validly published**. However, just because a name is validly published does not mean that it is the correct name for a particular species. For example, the name could be a synonym of an earlier, validly published name. The rules of nomenclature regarding valid publication are relatively straightforward. It is much more difficult to justify the delimitation of a species (as discussed in Chapter 6; see also Winston 1999) than it is to fulfill the technical requirements of the ICBN.

Hybrid Names

Hybrids between two species within the same genus may be designated by the alphabetical listing of both species names, separated by the multiplication sign (×); for example, *Verbascum lychnitis* × *V. nigrum*. Alternatively, a hybrid may be described and given its own epithet, which is preceded by the multiplication sign; for example, *Verbascum* ×*schiedeanum* (= *V. lychnitis* × *V. nigrum*). Such names require descriptions in Latin and the designation of type specimens, and they must be effectively published.

Hybrids between species in different genera may also be designated by the parental species names separated by a multiplication sign. Alternatively, they may be represented by a condensed generic formula along with a specific epithet. A condensed generic formula is composed of elements of the generic names and is preceded by a multiplication sign; for example, ×*Dialaeliocattleya* (an intergeneric hybrid of *Diacrium*, *Laelia*, and *Cattleya*). Such generic formulas require no Latin descriptions. These conventions are usually not applied to species originating through natural hybridization and polyploidy (see Chapter 6), especially when they reproduce sexually.

In addition, there exist so-called graft hybrids, which are chimeras (i.e., mixtures of cells of the parents). These plants can be designated by the parental species names separated by a plus sign (e.g., *Crataegus* + *Mespilus*) or have names of the form +*Crataegomespilus*.

Cultivated Plants

Plants produced in cultivation through hybridization, artificial selection, or other processes may receive additional epithets. The application of such names is covered by the International Code of Nomenclature for Cultivated Plants (ICNCP), which recognizes "cultivars" and "cultivar groups."

The term **cultivar** is a combination of the words *cultivated* and *variety*; these entities were often called "varieties" in the earlier literature. Cultivars should not be confused with *botanical* varieties, which usually represent naturally occurring geographic races or morphologically distinct populations (see Chapter 6). Cultivars are selections from any source—from the wild or from cultivation—that can be reliably reproduced by a prescribed method of propagation, either vegetatively (asexual) or by seed (sexual). Individuals within the same cultivar are usually genetically identical, but this is not in itself a requirement. Plants may be genetically diverse and still be considered within the same cultivar provided that no differences in the desired characters are perceptible (Brickell 2004). Cultivar groups are groups of cultivars that are united by a character of choice (e.g., early blooming yellow flowers). An example of a cultivar group is the Granny Smith apple. Cultivar epithets may be in any language, but epithets published after January 1, 1959, may not be in Latin. However, if a species is reduced to a cultivar, it will retain its epithet (e.g., *Mahonia japonica* becomes *Mahonia* 'Japonica'). Cultivar epithets may consist of more than one word, in which case each word is capitalized, and are written between single or double quotes. At one time the cultivar epithet was preceded with "cv.," but this has been disallowed since January 1, 1996. Cultivar names may be used after generic, specific, or infraspecific names. The following, for example, are equivalent designations of the same cultivar: *Citrullus* 'Crimson Sweet'; *Citrullus lanatus* 'Crimson Sweet' (see also Jeffrey 1977).

Summary

The principles and rules for constructing and using scientific names for plants are contained in the International Code of Botanical Nomenclature (ICBN). Scientific names are in Latin or Latinized. Species names are composed of two words (that is, they are binomials). The names of genera, families, and other higher taxa are uninomials (composed of only a single word), and are plural nouns.

The ICBN recognizes seven major ranks: kingdom, phylum/division, class, order, family, genus, and species. The ranks above genus have standardized suffixes. The ranks of the taxonomic hierarchy are human constructs, having relative but not absolute meaning, whereas the taxa (monophyletic groups of organisms) placed in these ranks are the products of evolution and, as such, represent real entities. For this reason, some taxonomists have advocated the elimination of ranks.

Botanical nomenclature is independent of zoological nomenclature, although the two codes are similar in their basic principles. The application of names to taxonomic groups is determined by means of holotypes (name-bearing specimens). The correct name for a taxonomic group is the earliest name that is in accordance with the rules of nomenclature; this principle is referred to as priority of publication. Each taxonomic group can bear only one correct name (except in a few specified cases). Special rules govern the naming of hybrids and plants produced in cultivation.

The ICBN outlines necessary steps for describing a new species. If all of these steps are followed, the species name is said to be validly published, and the name can compete for priority with other names. The earliest validly published name is the correct name for any particular species.

LITERATURE CITED AND SUGGESTED READINGS

Bailey, L. H. 1933. *How plants get their names.* Macmillan, New York. Reprinted 1963, Dover Publications, New York. [Historical development of the binomial system and the rules of nomenclature; information on the construction and pronunciation of scientific names; includes a list of common specific epithets.]

Brako, L., A. Y. Rossman and David F. Farr. 1995. *Scientific and common names of 7,000 vascular plants in the United States.* APS Press, American Phytopathological Society, St. Paul, MN. [Common name(s) for numerous vascular plants.]

Brickell, C. D. (ed.). 2004. *International code of nomenclature for cultivated plants.* Ed 7 (Acta Horticulturae 647; Regnum Vegetabile 144). International Society for Horticultural Science, Gent-Oostakken, Belgium.

Bridson, G. D. R. (ed.). 2004. *BPH-2. Periodicals with botanical content. Constituting a second edition of Botanico-Periodicum-Huntianum.* Hunt Institute for Botanical Documentation, Carnegie Mellon University, Pittsburgh, PA. [A listing of all periodical publications that regularly contain (or have contained) articles dealing with botany.]

Brummitt, R. K. 1992. *Vascular plant families and genera.* Royal Botanic Gardens, Kew, UK. [A listing of genera of vascular plants of the world, both alphabetically and by family, as recognized in the Kew Herbarium; now somewhat dated.]

Brummitt, R. K. and C. E. Powell. 1992. *Authors of plant names.* Royal Botanic Gardens, Kew, UK. [A list of the personal names of nearly 30,000 botanical authorities, along with standardized abbreviations.]

Cantino, P. D., H. N. Bryant, K. de Queiroz, M. J. Donoghue, T. Eriksson, D. M. Hillis and M. S.Y. Lee. 1999. Species names in phylogenetic nomenclature. *Syst. Biol.* 48: 790–807. [Various methods of using uninomial names for species.]

De Queiroz, K. and J. Gauthier. 1992. Phylogenetic taxonomy. *Annu. Rev. Ecol. Syst.* 23: 449–480.

Farr, E. R., J. A. S. Leussink and F. A. Stafleu (eds.). 1979. *Index nominum genericorum (plantarum).* 3 vols (Regnum Vegetabile, vols. 100–102). Bohn, Scheltema & Holkema, Utrecht, Netherlands. [A listing of all genera of "plants" and fungi, giving for each the authority, bibliographic citation, type species, and taxonomic placement.]

Greuter, W. 1993. *Family names in current use for vascular plants, bryophytes, and fungi* (Regnum Vegetabile, vol. 126). Published for the International Association for Plant Taxonomy by Koeltz Scientific Books, Königstein, Germany. [A listing of currently used family names, including authorities, literature citations, and type citations.]

Greuter, W., R. K. Brummitt, E. Farr, N. Kilian, P. M. Koirk and P. C. Silva (eds.). 1993. *Names in current use for extant plant genera* (Regnum Vegetabile, vol. 129). Published for the International Association for Plant Taxonomy by Koeltz Scientific Books, Königstein, Germany. [A listing of 28,041 generic names currently in use for extant algae, mosses, liverworts, vascular plants, and fungi; each listing includes the name, authority, literature citation, type citation, taxonomic disposition, and other nomenclatural information.]

Hooker, J. D. and B. D. Jackson. 1893–1990. *Index kewensis plantarum phanerogamarum.* 2 vols., 19 suppl. Clarendon Press, Oxford, and Royal Botanic Gardens, Kew. [A listing of all validly published names of vascular plants, including bibliographic citations.]

Jeffrey, C. 1977. *Biological nomenclature,* 2nd ed. Crane, Russak & Co., New York. [The legalistic complexities of the various nomenclatural codes presented in an easily understood manner; recommended for those interested in learning more about nomenclatural procedures.]

Kartesz, J. T. 1994. *A synonymized checklist of the vascular flora of the United States, Canada, and Greenland*, 2nd ed. 2 vols. Timber Press, Portland, OR. [A very useful listing of the vascular flora.]

Kartesz, J. T. and J. W. Thieret. 1991. Common names for vascular plants: Guidelines for use and application. *Sida* 14: 421–434. [Guidelines provided for the construction of common names for vascular plants.]

Kron, K. A. and W. S. Judd. 1990. Phylogenetic relationships within the Rhodoreae (Ericaceae) with specific comments on the placement of *Ledum*. *Syst. Bot.* 15: 57–68.

Laurin, M. and P. D. Cantino. 2007. Second meeting of the International Society for Phylogenetic Nomenclature: A report. *Zoologica Scripta* 36: 109–117.

Linnaeus, C. 1753. *Species plantarum*. 2 vols. Stockholm. [A listing of all species of plants then known, arranged according to Linnaeus's sexual system—that is, classes based on number and arrangement of the stamens. Brief descriptions are provided; first consistent use of the binomial system and the starting point for priority for vascular plants.]

Mabberley, D. J. 1997. *The plant-book*, ed. 2. Cambridge University Press, Cambridge. [A listing of all currently accepted generic and family names and commonly used English names of extant vascular plants; generic entries include authorities, family placement, approximate number of species, and geographic range; economically important plants and products are discussed briefly.]

McNeill, J. and 11 others (eds.). 2006. *International code of botanical nomenclature (Vienna Code)* (Regnum Vegetabile, vol. 146). A. R. G. Gantner Verlag, Ruggell, Lichtenstein.

McVaugh, R., R. Ross and F. A. Stafleu. 1968. *An annotated glossary of botanical nomenclature* (Regnum Vegetabile, vol. 56). International Bureau for Plant Taxonomy and Nomenclature of the International Association for Plant Taxonomy, Utrecht, Netherlands. [Clear definitions of nomenclatural terms.]

Nuttall, T. 1818. *Genera of North American plants*, vol. 1. Printed for the author by D. Heartt, Philadelphia.

Radcliffe-Smith, A. 1998. *Three-language list of botanical name components*. Royal Botanic Gardens, Kew, UK. [Greek, Latin, English words, word roots.]

Stafleu, F. A. and R. S. Cowan. 1976–1988. *Taxonomic literature*. 7 vols. (Regnum Vegetabile, vols. 94, 98, 105, 110, 112, 115, 116). Bohn, Scheltema & Holkema, Utrecht, Netherlands. [An indispensable guide to botanical publications, both books and journal articles that were separately printed, with dates, biographies, and bibliographies of individual authors, and indications of where the type specimens for taxa they described are likely to be found.]

Stafleu, F. A. and E. A. Mennega. 1992–1995. *Taxonomic literature*, suppl. 1–3 (Regnum Vegetabile, vols. 125, 130, 132). Koeltz Scientific Books, Königstein, Germany.

Stearn, W. T. 1992. *Botanical Latin*, 4th ed. Timber Press, Portland, OR. [A comprehensive handbook of botanical Latin, summarizing grammar and syntax; contains an illustrated guide to descriptive terminology in both English and Latin, along with an extensive vocabulary.]

Stevens, P. F. 2002. Why do we name organisms? Some reminders from the past. *Taxon* 51: 11–26.

Stevens, P. F. 2006. L. A. S. Johnson Review No. 5. An end to all things?—plants and their names. *Australian Syst. Bot.* 19: 1–19.

Sytsma, K. J. and J. C. Pires. 2001. Plant systematics in the next 50 years: Re-mapping the new frontier. *Taxon* 50: 713–732.

Weber, W. 1986. Pronunciation of scientific names. *Madroño* 33: 234–235.

Willis, J. C. 1973. *A dictionary of the flowering plants and ferns*, 8th ed., revised by H. K. Airy Shaw. Cambridge University Press, London. [A listing of all extant genera of vascular plants, giving authorities, family placement, approximate number of species, and geographic range; brief family descriptions are also provided.]

Winston, J. E. 1999. *Describing species: Practical taxonomic procedure for biologists*. Columbia University Press, New York. [A practical manual for the process of describing new species, including nomenclature, recognizing species, species and subspecies concepts, taxonomic literature, species descriptions, using museum collections, etymology, and key construction.]

APPENDIX TWO
Specimen Preparation and Identification

A collection of dried plant specimens is called an **herbarium** (plural **herbaria**). Such collections are essential for systematic research. Herbarium specimens form the basis for most of our understanding of the patterns of variation in nature. They document the morphological variability of populations, species, and higher taxa, their geographic distributions, and their ecological characteristics, including blooming and fruiting times. In addition, small portions of a specimen can be removed (with permission) to study palynology, ultrastructure, micromorphology, anatomy, and (if the specimen is of sufficient quality) nucleotide sequences.

Dried plant specimens also serve as **scientific vouchers** to document the presence of a species at a particular locality (in an environmental or floristic study) or the identity of a plant used in an experiment or from which a chromosome count, DNA sample, or chemical extract was obtained. Therefore, it is important to outline the steps involved in collecting, preserving, and identifying plants.

Collecting Plants

The plant collector must record certain facts while in the field. These data include (1) locality: country or state, county, or other local governmental unit, distance from roads or cities, and latitude/longitude or section-township-range; (2) date of collection; (3) habitat type, with associated species; (4) elevation, especially in mountainous regions; and (5) any information concerning the plant that will not be evident in the pressed and dried specimen, such as flower color, fragrance of flowers and/or leaves, habit, plant size, presence of colored or milky sap, bark characters, local abundance, and pollinators. The collector's name, along with the names of any individuals accompanying the collector, should be included with the field data. This information may be recorded in a field notebook (or collection book), with a tape recorder (for personal dictation), or on a portable computer for eventual transfer to a specimen label.

The plants collected should represent the variation seen in the population. Several specimens may be needed to document the observed pattern of variation adequately. Choose plants that are well developed and not diseased. If possible, the entire plant, including underground parts, should be collected; never collect just one flower or leaf! Do not collect specimens lacking flowers and fruits; such material will be difficult to identify.

Large herbs may have to be sharply folded or cut into pieces when they are collected. If so, the parts should be sequentially labeled, and if the plant is too large to keep all of the parts, the collector may decide to keep only certain portions, such as pieces from the base, middle, and apex of the plant. In such cases, care must be taken that no important information is lost. It is not necessary to preserve a portion of the root system when collecting trees or shrubs; the collector merely needs to cut off branches that document observed variation in vegetative and reproductive characters.

Certain kinds of plants require special treatment. Succulent plants or large, fleshy fruits can be cut longitudinally or in cross-section; preservation may be improved in these species if the plant material is initially killed by immersion in a preservative, such as ethyl alcohol, or in very hot water, or if the plant is first frozen. Floating or submerged aquatics can be floated in a tray of water and carefully lifted onto a piece of paper before further processing. Both staminate and carpellate flowers should be collected for dioecious or monoecious species; staminate and carpellate plants of dioecious species should normally be given separate specimen numbers.

Each plant collected should be assigned a specimen number, which allows information concerning a particular specimen to be kept separate from information on others. Most plant systematists simply begin with the number 1 and number consecutively throughout their careers, although other numbering systems are possible, such as using a new numbering sequence each year, or for each geographic region—for example, 07-1, 07-2 (for plants collected in 2007) or H-1, H-2 (for plants collected in Hispaniola).

Most systematic botanists consider the following items essential or at least very useful in the field: field notebook, field press or heavy-duty plastic bag, drying press, digging tools, pruning shears, newspapers, 10× hand lens, pocket knife, soft-leaded pencils, and maps. Other useful items include collecting jars or vials and liquid preservatives, camera and film, portable plant dryer, insect repellent, pole pruner, altimeter, compass, global positioning system unit, adhesive tape (for labeling specimens with numbers and for sealing specimen bags), and silica gel granules (to dry material for DNA studies). Figure 1 details the steps in creating an herbarium specimen.

FIGURE 1 Preparing and storing herbarium specimens. (A) Steps ▶ in making an herbarium specimen. (1) Find a plant in nature that is well developed and not diseased; record the geographic locality, habitat, and associated plants in your field notebook. Collect portions of the plant that will be useful in identification. (2) Fold the plant to fit into the press; then fold newspaper over the plant. (3) Tighten the plant press straps. (4) Place the plant press in a dryer (if available). (5) Mount the dried specimen. (6) Identify the specimen using floras and monographs in the herbarium library. (7) For the finished herbarium specimen, transfer the information from the field notebook to the specimen label. (B) Prepared herbarium specimens are stored in special herbarium cases (left), which may form part of compactor-storage system (right). (Photos by W. S. Judd and R. E. Judd; interior photos taken in the University of Florida Herbarium.)

Pressing and Drying Plants

Once a plant has been collected and its field data recorded, the specimen needs to be pressed and dried. It is important to press the plant before it wilts. A **plant press**, or drying press, consists of two 12″ × 18″ pieces of wood (the press panels); two press straps or ropes, which are used to tighten the press; blotters, which absorb moisture from the plant specimens; and corrugated ventilators (cardboard or aluminum corrugates), which allow air to pass through the press.

The plant specimen is placed in a folded piece of newspaper (or newsprint) and arranged so that upper and lower leaf surfaces, flowers, and fruits are visible. The plant material should be sufficient to more or less fill the sheet of newspaper; for tiny plants, several individuals should be gathered, and large plants will need to be folded sharply or cut into several pieces. The specimen number is written on the newspaper or on a tag attached to the plant. This number allows reference to the information in the field notebook. Large, hard fruits may have to be removed from the plant, carefully labeled, and processed separately from the rest of the specimen.

The press is assembled as follows: press frame, corrugated ventilator, blotter, newspaper (with plant specimen), blotter, corrugated ventilator, blotter, newspaper (with plant specimen), blotter, corrugated ventilator, and so on, finishing with the second press frame. The press is tightened with two straps or ropes. The blotter on one side of a specimen with medium-sized hard fruits, such as acorns or hickory nuts (in its folded piece of newspaper), may be replaced with a piece of foam rubber. The foam distributes the pressure evenly around the fruits, keeping leaves near the fruits from wrinkling. If foam is not available, wadded pieces of tissue paper placed around the fruits will also work. For some rigid plants, it may be best to open the press after a day and rearrange the specimens because they will be somewhat more flaccid.

It may be impractical to carry a large drying press to remote collecting localities. Therefore plant specimens are

(A) (1)

(2)

(3)

(4)

(5)

(6)

(7)

(B)

often carried in a **field press** (press frames with several corrugates and plenty of newspaper), a **vasculum** (plural **vascula**; a cylindrical metal container lined with moist newspaper), or a heavy-duty plastic bag to a place where they can be transferred to a drying press or processed in some other way.

Once in the plant press, the plants can be dried in several ways. The press may be placed in the sun or indoors in a dry place and the blotters changed every day, or it may be tied to the roof rack of a moving vehicle. However, it is preferable to use an artificial heat source. **Plant dryers** are boxlike or tablelike structures that suspend the press over a source of heat, provided by metal heat strips, forced-air space heaters, variable-control hot plates, field stoves, light bulbs, or fans. Too much heat (greater than 45°C) can discolor the specimens and may even cause the press and specimens to catch fire.

The press and dryer should be set up so as to allow the warm air to flow through the plant press and remove moisture. Large, hard fruits or cones should be removed from the specimens and dried separately. Succulent or delicate parts that may be best kept in a liquid preservative, such as alcohol or formalin, can be separated from the specimens and placed in a plastic container. Delicate flowers may be dipped in liquid preservative and should be pressed in a folded piece of tissue paper or nonabsorbent toilet paper. Bulky specimens, such as material with coriaceous leaves or thick, succulent stems, may be placed directly against a corrugate to speed drying. Aluminum corrugates dry material more quickly than cardboard corrugates and do not become dented with use. However, they are more expensive and heavier.

The plant press should be examined carefully as it dries and tightened when necessary. It may also be turned over if the heating is uneven. Most plant specimens will dry in 1 to 4 days, and the plant material usually will be rigid when dry.

When the specimens are dry, they should be kept within the folded newspapers and carefully bundled to protect them from breakage. If the moisture content of the air is high, the specimens should be placed (after cooling) in a tightly sealed plastic bag, to which silica gel granules may be added.

If drying facilities are not immediately available, the plant material can be temporarily preserved in alcohol (or other liquid preservatives). In this method, a bundle of specimens in folded newspapers is compressed by hand and tied. The tied bundle is placed in a heavy-duty plastic bag, and about half a liter of 60%–70% alcohol is sprinkled inside. The bag is then sealed with tape. The alcohol vapors inside the bag will preserve the plant specimens for several weeks, and they can be pressed and dried later. Note, however, that specimens treated in this way will usually be unsuitable for subsequent chemical (including DNA) analysis. A sealed bag of fresh plant specimens can also be safely kept in the refrigerator for a couple of days.

Mounting and Processing Herbarium Specimens

Once a plant specimen has been collected, pressed, dried, and identified, it is removed from the folded newspaper and mounted (with glue, linen adhesive strips, or thread) on a standard-sized (in the United States, 11.5″ × 16.5″) sheet of archive-quality paper. Valuable loose plant parts, such as seeds, fruits, or flower parts, are placed in a small folded packet, which is glued to the sheet. The specimen number and the information from the field notebook are transferred to a typed or computer-generated specimen label (Figure 2). The specimen label is glued to the sheet, usually in the lower right-hand corner. After the specimen has been mounted, the name of the herbarium is stamped on the sheet, documenting its ownership, and an accession number is often added for accurate record keeping (see Figure 1A, part 7).

Specimens may be filed in the herbarium alphabetically by family, genus, and species, or according to a particular system of classification such as that of A. Engler, A. Cronquist, or the Angiosperm Phylogeny Group (see Chapter 3 and the companion CD). Herbarium specimens will last indefinitely if handled with care. They are, however, susceptible to damage by insects, fungi, and fire, and they should be stored in specialized cabinets (herbarium cases) in a climate-controlled environment (see Figure 1B). (Valuable information regarding the curation and operation of herbaria can be found in Fosberg and Sachet 1965, Bridson and Forman 1998, and Metsger and Byers 1999.)

Conservation and the Law

Many plants that were once common are now rare, primarily due to habitat destruction, overexploitation by people, or competition with exotics (non-native plants introduced from other parts of the world). Collectors should be conservation-minded. Rare or uncommon plants usually should not be collected; photographing the plant is a good alternative in such circumstances.

Many laws protect native plant species, and they should be collected only with the proper permits. Permits are required for collecting in city, county, state, and national parks and forests. When collecting is done in a national or state park or in a foreign country, a complete set of specimens should be given to the appropriate authorities. Many states also have laws protecting specific endangered species, and the U.S. Endangered Species Act protects endangered and threatened species on a national level. No listed species should be collected without the proper permits. And, of course, one should never collect on private land without permission of the landowner.

Collecting in foreign countries often requires special governmental permission; local scientists should also be

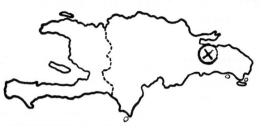

JARDIN BOTANICO NACIONAL "DR. RAFAEL M. MOSCOSO"
SANTO DOMINGO, REPUBLICA DOMINICANA

44523   Melastomataceae

Miconia racemosa(Aubl.)DC

Arbusto 1.5m. de alto; fr. verde, pero
púrpura al madurar; margen del bosque.

República Dominicana: Cordillera Oriental
(cima), Prov. El Seibo: 16.5km. al Sur de
Miches, en la carretera a El Seibo y Pedro
Sánchez; bosque nublado, alterado por fuego
en los últimos 10 años, Clusia rosea,
Alchornea latifolia, Inga fagifolia, Ormosia
krugii, Buchenavia capitata, Sloanea
berteriana, Miconia spp, con manantial.
18°55'N 69°09'Oeste, elev 540-560m.
27 jun. 1990
T. Zanoni, F. Jiménez

**FLORA OF EGLIN AIR FORCE BASE
PLANTS OF FLORIDA**

Aristida condensata Chapm.
< Poaceae >

WALTON COUNTY: T2N, R20W, S29 Eglin Air Force
Base. Mossy Head 7.5' Quad. North of Range C-62 and W. and
E. of Little Alaqua Creek. From intersection of RR 381 and
RR 385, ca. 0.3 mile W. of RR 385. *Pinus palustris*
dominated sandhill community with thick understory of
Quercus laevis. Dominant herbaceous plants include
Andropogon ternarius and *Schizachyrium scoparium*. Tall,
robust, herb. Leaves and inflorescences are tinged with
dark purple. Abundant.

coll. Brenda Herring # 1289   Oct. 7, 1997
    with Jeff McAdoo, Nancy Gobris, and Jim Sheehan
det. B.J.H.
Gift to FLAS from the Nature Conservancy / Brenda Herring, 23 Nov. 1998

FIGURE 2   Examples of labels from herbarium specimens.

Herbarium of the University of Florida
Gainesville, Florida U.S.A.
PLANTS OF DOMINICAN REPUBLIC

Miconia samanensis Urban

DISTRITO NACIONAL: Cordillera Central; Loma
Mariana Chica. ca. 6 km west of Villa Altagracia.
150-775 m elev. Disturbed moist montane forest at peak
(and ridge) of mountain, lower elevations with very
patchy and disturbed moist forest. Branches rigidly
ascending; new growth bright red. Shrub to 4 m tall;
occasional locally (but occurring just on peak and ridge,
ca. 750-775 m); petals pale pink, reflexed; stamens
white, actinomorphic; style white; fruits purple, turning
sky blue at maturity. Leaves with major veins impressed.

coll. Walter S. Judd # 6505   13 May 1992
    with James D. Skean, Jr.
det. W. S. Judd
Voucher-*Miconia* sect. *Chaenopleura* study.

Herbarium of the University of Florida
Gainesville, Florida, U.S.A.

**PLANTS OF: FLORIDA   COUNTY: ALACHUA
FLORA OF PAYNES PRAIRIE STATE PRESERVE**

Passiflora incarnata L.

S. of Gainesville. T10S, R20E, sec 25, SW ¼.
S. side of High Dike, ca. ½ mi W. of
pump station on Camp's Canal. Vine,
common. Sepals green. Corolla white,
outer corona white and violet striped
in concentric circles,   G'ville   Alachua
inner corona violet.   Sink
10 May 1981

coll. Janet C. Easterday #   460   I-75   441

---

contacted wherever possible. International trade in plant
material is covered in part by existing conservation legisla-
tion. The Convention on International Trade in Endangered
Species of Wild Fauna and Flora (CITES) regulates the
shipment of many plant groups, including Orchidaceae,
Cactaceae, Cycadaceae, Cyatheaceae, Nepenthaceae, Sarraceniaceae, and Zamiaceae. CITES-listed plants should
not be collected without the proper authorization.

Plant Identification

Plants may be identified after they have been collected,
pressed, and dried, but it is easier to identify fresh or liquid-
preserved material. Dried material may be dissected more
easily if it is first softened by being boiled in water containing

a wetting agent, such as aerosol OT (dioctyl sodium sulfos-
uccinate), or a detergent. Before identifying a plant specimen,
one should carefully observe certain features. A 10× hand
lens or even better, a good dissecting microscope will aid in
observation, especially of floral features and hair types. Dissection is easier with a pair of dissecting needles, sharp-
pointed forceps, and single-edged razor blades. A millimeter
rule is essential for measuring. Critical observations include
(1) habit; (2) leaf arrangement, shape, apex, base, margin,
and venation; (3) hair characters; (4) floral features; (5) placentation and ovule number; and (6) fruit type.

Keys

The systematist has several identification tools available,
the most important of which are dichotomous keys. A

dichotomous key presents the user with a series of choices between two mutually exclusive and parallel statements (leads of a **couplet**). If the user chooses correctly, he or she will be led to the name of the unknown object. The first dichotomous key to plants was published in 1778 by the French botanist Jean-Baptiste de Lamarck, and they have since become ubiquitous (Voss 1952). Dichotomous keys always have a flowchart structure, and they may be written in either an **indented** or a **bracketed** format (Figure 3). Bracketed keys are used throughout this book.

When using a dichotomous key, always read both leads of a couplet, do not guess on measurements, and look up any terms that you do not understand. Be sure to use a hand lens (or dissecting microscope) if necessary. Remember that living things are variable, so be sure to look at several leaves, flowers, or fruits in the process of observing the plant.

When constructing a key, keep in mind that the characters should be precisely defined, measurements should be used whenever possible (don't use terms such as *large* and *small*), characters that are constant within a taxon are more useful than those that are variable, and features that are available throughout the growing season or are easily observed are preferred over those that are ephemeral or difficult to see.

Couplets should start with a noun followed by adjectives, and both leads of a couplet should begin with the same word (see Figure 3). The two leads should be parallel; for example, if leaf shape is given in the first, a contrasting condition needs to be presented in the second. Do not use unqualified negative statements; instead, use positive statements wherever possible (e.g., "leaf apex acute" versus "leaf apex acuminate" is better than "leaf apex acute" versus "leaf apex not acute").

All parts of the key should be constructed dichotomously. Keys are most efficient when major couplets divide the entire set of included taxa into more or less equal-sized groups. A user is less likely to make an error if two or three characters are used per lead, but overly long leads lead to more mistakes.

A second kind of identification key is the **multi-access key**, or **polyclave**. This kind of key may be constructed through the use of punched cards, with one card per taxon and with the characters represented by holes around the margin of the card. If a taxon possesses a particular character, the hole for that character is notched. The taxon cards are sorted by use of a large needle. For example, if the cards are stacked and the user runs the needle through the character "leaves alternate," all the cards notched at that spot, which represent species possessing that character, will fall out. Characters may be selected (and the cards sorted) in any sequence, and sorting may be repeated until only one card remains. Alternatively, each card may represent a single character, with punched holes representing different taxa. Each taxon possessing a particular character will have its corresponding hole punched out on that character card. The user of such a key merely selects those cards representing characters evident on his or her specimen, then stacks

FIGURE 3 A flowchart, bracketed key, and indented key for five ▶ imaginary plants.

them and holds them up to the light. When only one hole shows through the stack, that taxon represents the identity of the unknown plant.

Both kinds of punched-card keys operate by a step-by-step elimination process, as does a dichotomous key. However, any selection of characters may be used, and in any order, which is why these keys are called *multi-access* keys. They are quite advantageous when the plant material being keyed is incomplete. Even if the information on the specimen is not sufficient to identify a single taxon, the user may be able to obtain a short list of possible identities for the unknown plant. Thus multi-access keys can be used as partial keys.

Polyclaves may also be published in written form. Each character is listed, followed by a list of all the taxa (usually represented by a numerical designation) possessing that character (Figure 4). To use the key, one takes the intersection of the sets of taxa specified by a series of characters. Note that a multi-access key, of necessity, contains much more information than a traditional dichotomous key. It contains the data necessary to construct many different dichotomous keys—thus the appropriateness of the term *polyclave*.

It is relatively easy, given the proper computer software, to convert multi-access keys to a computer-interactive format. Computer-assisted identification programs such as the DELTA (Descriptive Language for Taxonomy) or LUCID systems (Askevold and O'Brien 1994; Dallwitz 1993, 2000; Dallwitz et al. 2002; Pankhurst 1978, 1991; Watson and Dallwitz 1991, 1993) are sure to become more important in the future (Edwards and Morse 1995).

Floras and Monographs

Keys are usually presented in books together with plant descriptions, illustrations, distribution maps, and other biological information. It is important to read the appropriate taxon description after using a key; if a specimen has been properly keyed, the description of the identified species (or taxon) should match the plant in hand. The order of information in plant descriptions has been standardized as follows:

1. Habit (the plant as a whole)
2. Underground parts (roots, tubers, bulbs, and so on)
3. Vegetative buds
4. Stems
5. Leaves (including arrangement, structure, petiole, shape of blade, base of blade, margin of blade, apex of blade, and venation) and stipules
6. Inflorescences
7. Flower (including symmetry, sexual condition, calyx, corolla, androecium, gynoecium, placentation, ovules, and nectaries)

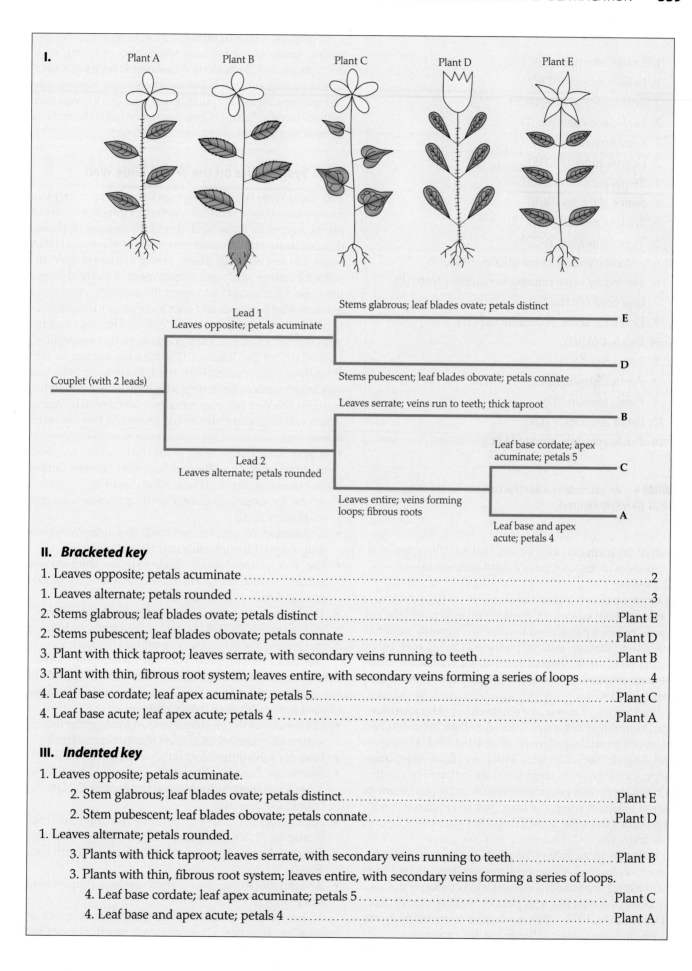

I.

Plant A   Plant B   Plant C   Plant D   Plant E

Stems glabrous; leaf blades ovate; petals distinct

E

Lead 1
Leaves opposite; petals acuminate

D

Stems pubescent; leaf blades obovate; petals connate

Couplet (with 2 leads)

Leaves serrate; veins run to teeth; thick taproot

B

Leaf base cordate; apex
acuminate; petals 5

C

Lead 2
Leaves alternate; petals rounded

Leaves entire; veins forming
loops; fibrous roots

A

Leaf base and apex
acute; petals 4

II. *Bracketed key*

1. Leaves opposite; petals acuminate 2

1. Leaves alternate; petals rounded 3

2. Stems glabrous; leaf blades ovate; petals distinct ....Plant E

2. Stems pubescent; leaf blades obovate; petals connate ....Plant D

3. Plant with thick taproot; leaves serrate, with secondary veins running to teeth ....Plant B

3. Plant with thin, fibrous root system; leaves entire, with secondary veins forming a series of loops 4

4. Leaf base cordate; leaf apex acuminate; petals 5 ....Plant C

4. Leaf base acute; leaf apex acute; petals 4 ....Plant A

III. *Indented key*

1. Leaves opposite; petals acuminate.

   2. Stem glabrous; leaf blades ovate; petals distinct ....Plant E

   2. Stem pubescent; leaf blades obovate; petals connate ....Plant D

1. Leaves alternate; petals rounded.

   3. Plants with thick taproot; leaves serrate, with secondary veins running to teeth ....Plant B

   3. Plants with thin, fibrous root system; leaves entire, with secondary veins forming a series of loops.

      4. Leaf base cordate; leaf apex acuminate; petals 5 ....Plant C

      4. Leaf base and apex acute; petals 4 ....Plant A

1. Leaves alternate (ABC)
1. Leaves opposite (DE)
2. Leaf blades ovate (ABCE)
2. Leaf blades obovate (D)
3. Leaves serrate (B)
3. Leaves entire (ACDE)
4. Stems pubescent (AD)
4. Stems glabrous (BCE)
5. Taproot present, thick (B)
5. Taproot lacking (ACDE)
6. Secondary veins forming loops (ACDE)
6. Secondary veins running to marginal teeth (B)
7. Leaf base cordate (C)
7. Leaf base acute or cuneate (ABDE)
8. Petals 4 (ABD)
8. Petals 5 (CE)
9. Petals distinct (ABCE)
9. Petals connate (D)
10. Petals acuminate (DE)
10. Petals apically rounded (ABC)

FIGURE 4 An example of a multi-access key for the imaginary plants shown in Figure 3.

8. Fruit (including dehiscence and fruit wall characteristics)
9. Seed (including coat, embryo, and endosperm)
10. Seedling
11. Chromosome number

A good general scheme for the description of any of the ten organs or plant parts listed here is the following: number and differentiation, relative position in relation to other organs, attachment, shape, size, color, surface texture, indumentum or detailed description of hairs, and anatomy (see Leenhouts 1968).

Illustrations, of course, are also useful in plant identification. Methods of botanical illustration have been outlined in several books (e.g., Zweifel 1961; West 1983; Holmgren and Angell 1986; Zomlefer 1994). In plant distribution maps, the range of the taxon is usually indicated by shading of the appropriate geographic region or by placement of dots at localities where the taxon has been collected; both approaches show the taxon's range at a glance (see Figures 4.44 and 6.27).

A **flora** is an account of the plants occurring in a particular area, including keys, descriptions, and illustrations. Floras may be local, applying to a relatively small region (e.g., Proctor 1984), or continental (e.g., Flora of North America Editorial Committee, 1993–2007). A selection of important North American floras is given at the end of this appendix.

In contrast, a **monograph** is a more comprehensive systematic study of a particular taxon, also including keys, descriptions, and illustrations. A **taxonomic revision** is similar to a monograph, but is less comprehensive. Monographs and revisions are usually published in botanical journals such as those listed in Table 1. Keys, descriptions, and illustrations are now also available on numerous Web sites.

Plant Systematics on the World Wide Web

The World Wide Web is an ever-growing resource for plant systematists. It now provides a wealth of information about plants, access to floras and checklists, images of plants, detailed treatments of particular groups of plants, and DNA sequences and other databases, as well as news of opportunities for further study and employment in the field. To get the same information or to reach the same Web site, there are many starting points and even more paths to the destination. This section lists a few useful Web sites for plant systematists that can serve as starting points for further exploration.

To start, try the Internet Directory for Botany, or IDB (http://www.botany.net/IDB). The IDB contains links to a very large number of botanical sites, which are listed alphabetically. The Web site also contains a useful search engine, which will help you find sites of interest. A few are listed below:

- A searchable database for plant chromosome numbers maintained by the Missouri Botanical Garden (http://mobot.mobot.org/W3T/Search/ipcn.html)
- A site for botanical glossaries (http://www.anbg.gov.au/glossary.html)
- A database of aquatic, wetland, and invasive plants (http://aquat1.ifas.ufl.edu/search80/NetAns2/)
- The International Plant Names Index (http://www.ipni.org), a database of the names and associated bibliographic details of all seed plants
- The Missouri Botanical Garden databases (http://www.mobot.org/MOBOT/Research/alldb.shtml), with floras and parts of floras from various parts of the world

The IDB links to many Web sites containing floristic checklists, floras, and taxonomic databases. A sample of these sites, with one per continent, shows the wide range of plant systematic work that is available:

- Conspectus of the Vascular Plants of Madagascar (http://www.mobot.org/MOBOT/Madagasc/welcome.html)
- Flora of China (http://flora.huh.harvard.edu/china/)
- Australian Flora & Vegetation Statistics (http://www.anbg.gov.au/flora/index.html) at the Australian National Botanic Gardens site
- Atlas Florae Europaeae (http://www.fmnh.helsinki.fi//map/afe/E_afe.htm?pageid=571&language=English)
- Flora of North America (http://hua.huh.harvard.edu/FNA/)
- Andean Botanical Information System (http://www.sacha.org/)

Major herbaria and botanical gardens have much of interest on their Web sites, including descriptions of the

TABLE 1   Some important journals containing articles in systematic botany.

Acta Botanica Mexicana	*Field Museum of Natural History, Botanical Series*
Aliso	*Harvard Papers in Botany (continuation of the Journal of*
American Fern Journal	*the Arnold Arboretum)*
American Journal of Botany	*International Journal of Plant Science (= Botanical Gazette)*
Annals of the Missouri Botanical Garden	*Journal of Plant Research (= Botanical Magazine, Tokyo)*
Australian Journal of Botany	*Journal of the Linnaean Society—Botany*
Australian Systematic Botany	*Kew Bulletin*
Austrobaileya	*Madroño*
Blumea	*New Zealand Journal of Botany*
Botanical Review	*Nordic Journal of Botany*
Botanische Jahrbücher	*Novon*
Bothalia	*Phytologia*
Brittonia	*Plant Systematics and Evolution*
Bulletin of the Torrey Botanical Club	*Rhodora*
Canadian Journal of Botany	*Sida*
Castanea	*South African Journal of Botany*
Cladistics	*Systematic Biology*
Contributions from the U.S. National Herbarium	*Systematic Botany*
Darwiniana	*Taxon*
Edinburgh Journal of Botany (= Notes from the Royal	
Botanical Garden, Edinburgh)	

history and nature of the institution, research at the institution, educational opportunities (such as special courses and graduate study), employment opportunities, and special events, as well as descriptions and in some cases searchable databases of collections, images, and links to other sites. Here are the Web addresses for several of the world's major herbaria and botanical gardens (sites for some other institutions are noted elsewhere in this section):

- Australian National Botanic Gardens, Canberra (http://www.anbg.gov.au/anbg/index.html)
- Fairchild Tropical Botanic Garden, Coral Gables, Florida (www.fairchildgarden.org/)
- Jardim Botânico do Rio de Janeiro, Brazil (http://www.jbrj.gov.br/)
- Missouri Botanical Garden, St. Louis (http://www.mobot.org/)
- Muséum National d'Histoire Naturelle, Paris (http://www.mnhn.fr/mnhn/pha/Collect.htm)
- New York Botanical Garden (http://www.nybg.org/)
- Royal Botanic Gardens, Kew, United Kingdom (http://www.rbgkew.org.uk/); a links page at this site (http://www/kew.org.uk/scihort/eblinks/botany.html) is very useful
- Royal Botanic Gardens, Sydney, Australia (http://www.rbgsyd.nsw.gov.au/)
- Department of Systematic Biology—Botany, National Museum of Natural History, Smithsonian Institution, Washington, DC (http://www.nmnh.si.edu/botany/)

Sites representing floras, such as the one listed above for the Flora of North America (FNA), present a wealth of information on the history, purpose, and scope of floristic projects as well as on the plants of the flora. The FNA site

lists the volumes that have been published to date. One can search for the families that have been treated in these published volumes and access descriptions of families and genera, as well as keys to genera and species within the treated families. The site also lists the families that will be treated in the other 26 volumes that are in preparation.

There are hundreds of regional and local floristic projects around the world, and some of them have rich Web sites. The Oregon Flora Project site (http://www.oregonflora.org/), for example, is making its checklist of the plants of Oregon available as an online, searchable database. Currently the Asteraceae portion of the database is available. This project is also developing an Oregon Plant Atlas, with a distribution map of each plant species that has been recorded from the state. These maps will be combinable with different base maps of the state, showing features such as landforms, ecoregions, mean annual precipitation, and vegetation types. Another useful site is the Atlas of Florida Vascular Plants (http://www.plantatlas.usf.edu/about.asp). This site has nomenclatural data and county distribution maps for every species recorded from Florida.

Also useful are Web sites for professional societies, such as the American Society of Plant Taxonomists (ASPT; http://www.aspt.net/), the Botanical Society of America (BSA; http://www.botany.org/), the International Association for Plant Taxonomy (IAPT; http://www.botanik.univie.ac.at/iapt/), the Society of Systematic Biologists (SSB; http://systbiol.org/), and the Willi Hennig Society (http://www.cladistics.org/). The BSA Web site has a page listing other Web sites of botanical interest (http://www. botany. org/bsa/newsite/education/wwwlinks.php), including the following categories: "Major Botany Links & Resources on the WWW"; "Jobs and Career Sites"; "How to Find Universities, People and Discussion Groups on the Web"; "Funding Sources, Graduate Fellowships"; "Electronic Journals and Newsletters"; and "Internet Resource Guides."

The best source of DNA sequences is GenBank (http://www.ncbi.nlm.nih.gov/GenBank), supported by the National Institutes of Health as part of the National Center for Biotechnology Information (NCBI). This site supports databases that contain about 13 billion nucleotide bases from about 100,000 species at this writing and are doubling about every 14 months. GenBank exchanges DNA sequence data on a daily basis with two other large databases: those of the European Molecular Biology Laboratory (EMBL) and the DNA Data Bank of Japan (DDBJ). One accesses sequence information at this site with the "Entrez" facility, a search-and-retrieval system that integrates information from the databases at NCBI.

Journals routinely require that, prior to publishing a paper, authors submit any DNA sequences (or amino acid sequences) included in that paper to GenBank. A unique alphanumeric code is assigned to each sequence submitted and is included in published papers using that DNA sequence. Plant systematists routinely search GenBank to find out which genes have been sequenced for a set of taxa of interest and to add previously published DNA sequences to data sets they are compiling. Students are encouraged to compile DNA sequence data for a group of plants that they can use to explore the group's phylogenetic relationships. If the data are the same as those, or a portion of those, in a published paper, comparisons can be made with the published results.

Some Web sites focus on phylogeny. For example, the Tree of Life Web Project (http://tolweb.org/tree/phylogeny.html) is a collection of about 2000 Web pages containing information about the diversity of life. These pages are authored by biologists from around the world. Each page contains information about one taxonomic group, and the pages are linked one to another in the form of the evolutionary tree of life, with the pages branching off from one group's page giving information about its subgroups. TreeBASE (http://www.treebase.org/treebase/index.html), a relational database of phylogenetic information, stores phylogenetic trees from published research papers and the data matrices used to generate them.

Detailed systematic treatments of some groups of plants covered in this text are available on the World Wide Web. Sites focused on a particular taxonomic group may contain discussions of work on the group, links to other Web sites with pertinent systematic information about the group, a list of collaborators, access to databases, and lists of publications. Here are four examples:

1. Phylogeny, Character Evolution, and Diversification of Extant Ferns (http://www.fieldmuseum.org/research_collections/botany/botany_sites/ferns/index.html)

2. The Parasitic Plant Connection (http://www.parasitic-plants.siu.edu/)

3. Neotropical Blueberries (http://www.nybg.org/bsci/res/lut2/), with information on Ericaceae

4. Systematics of Onagraceae (http://ravenel.si.edu/botany/onagraceae/index.htm)

Many Web sites contain images of plants, which may be useful for those learning to identify plant families, genera, and species. The Texas A&M University Bioinformatics Working Group Vascular Plant Image Gallery (http://www.csdl.tamu.edu/FLORA/gallery.htm), for example, is a broad collection of images of angiosperms. The Plant Systematics Web site, based at Cornell University (http://www.plantsystematics.org/), contains over 20,000 images, mainly of vascular plants (accessed by an image search engine or by order), as well as keys, cladograms, literature references, and software. The University of Wisconsin Virtual Foliage Home Page (http://botit.botany.wisc. edu/) contains images of angiosperms, gymnosperms, and ferns as well as microscopy views. The Plant Systematics Teaching Collection at this site contains over 4000 images of vascular plants organized taxonomically. It is used as a reference in much the same way as an herbarium would be used, and it represents a subset of a larger collection available on CD-ROM. The Land Plants Online site (http://www.science/siu.edu/landplants/index.html) is maintained at Southern Illinois University at Carbondale; it presents numerous images, life cycles, and cladograms of green plants, organized by major clade. Many of the flora pages and sites already noted have images of some of the plants in the flora. Images are also available for many of the plants in the Missouri Botanical Garden's W^{3}TROPICOS database (http://mobot.mobot. org/W3T/Search/vast.html).

The Web site of the herbarium of the Florida Museum of Natural History (http://www.flmnh.ufl.edu/herbarium/) contains useful supplementary documents (e.g., guidelines for preparation of plant specimens for deposit as vouchers, for annotation of herbarium specimens, and on the use of herbaria and herbarium specimens) and a downloadable program (Plabel) for generation of herbarium specimen labels.

The Web goes far beyond this textbook in many directions, giving the student access to searchable databases, floras, thousands of images of plants, and information about botanical gardens, herbaria, and employment opportunities. The Families of Flowering Plants home page (http://www.Biologie.uni-hamburg.de/b-online/delta/angio/index.htm) includes descriptions and representative illustrations for all families of flowering plants. The Angiosperm Phylogeny Web site (http://www.mobot.org/MOBOT/Research/APweb/welcome.html) also holds information on seed plant orders and families, and it is frequently updated. The current classification of plant orders and families at this site is an elaboration of the Angiosperm Phylogeny Group classification (see Chapter 9). These sites are particularly useful supplements to the information contained in this text.

TABLE 2 **Herbaria with 3 million or more specimens.**

Herbarium	Location	Number of specimens
Muséum National d'Histoire Naturelle	Paris, France	9,500,000
Royal Botanic Gardens	Kew, United Kingdom	7,000,000
New York Botanical Garden	New York, United States	7,000,000
Conservatoire et Jardin Botaniques	Geneva, Switzerland	6,000,000
Komarov Botanical Institute	St. Petersburg, Russia	5,770,000
Swedish Museum of Natural History	Stockholm, Sweden	5,600,000
Herbarium, Missouri Botanical Garden	St. Louis, MO, United States	5,400,000
Natural History Museum	London, United Kingdom	5,200,000
Harvard University	Cambridge, MA, United States	5,005,000
Naturhistorisches Museum	Vienna, Austria	5,000,000
Smithsonian Institution	Washington, DC, United States	4,368,000
Rijksherbarium	Leiden, Netherlands	4,100,000
Institute de Botanique	Montpellier, France	4,000,000
Université de Lyon	Villeurbane, France	4,000,000
Universitatis Florentinae	Museo Botanico Firenze, Italy	3,650,000
Friedrich-Schiller-Universität	Jena, Germany	3,000,000

Source: Based on data from Holmgren et al. 1990, updated with data from the on-line version maintained at http://sciweb.nybg.org/science2/IndexHerbariorum.asp and the institutions' Web sites.

Herbaria, Botanical Gardens, and Taxonomic Experts

Our understanding of the flora of North America is not yet complete, and many tropical regions have yet more incomplete floras. In addition, more than 95% of plant genera have not received recent monographic treatment. Our knowledge of the world's flora clearly is not as advanced as many believe. Thus herbaria, in addition to the uses outlined previously, serve an important identification function.

An herbarium is essentially a "library" of pressed and dried plant specimens. An unknown specimen can be compared with identified herbarium specimens and identified by the process of comparison. This process can be time-consuming, but in many cases it is the only way a specimen can be identified. Many herbaria operate plant identification and information services. An herbarium is therefore an essential resource for any taxonomic program, whether that program involves public service, teaching, or research. Major herbaria around the world are listed in Table 2.

Botanical gardens are also useful in plant identification and they have the advantage over herbaria of preserving plants in living condition. Finally, problematic collections may be sent to an expert on the systematics of a particular genus or family or on the flora of a particular region. (Lists of taxonomic experts may be found in Kiger et al. 1981 and Holmgren and Holmgren 1992.) Unfortunately, many genera and families have been studied by too few, or no, taxonomic experts.

LITERATURE CITED AND SUGGESTED READINGS

Major North American Floras (and other important references used in plant identification)

Abrams, L. 1923–1960. *Illustrated flora of the Pacific states: Washington, Oregon and California*. 4 vols. Stanford University Press, Stanford, CA. [Vol. 4 by L. Abrams and R. S. Ferris.]

Acevedo-Rodríguez, P. 1996. *Flora of St. John*. New York Botanical Garden, Bronx.

Acevedo-Rodríguez, P. 2005. Vines and climbing plants of Puerto Rico and the Virgin Islands. *Contr. U.S. Natl. Herbarium* 51: 1–483.

Acevedo-Rodríguez, P. 2005. Monocotyledons and gymnosperms of Puerto Rico and the Virgin Islands. *Contr. U.S. Natl. Herbarium* 52: 1–415.

Adams, C. D. 1972. *Flowering plants of Jamaica*. University of the West Indies, Mona, Jamaica.

Allen, P. H. 1956. *The rain forests of Golfo Dulce*. Stanford University Press, Stanford, CA.

Bailey, L. H. 1949. *Manual of cultivated plants*, rev. ed. Macmillan, New York.

Batson, W. T. 1984. *A guide to the genera of the plants of eastern North America*, 3rd ed.

University of South Carolina Press, Columbia.

Bisse, J. 1981. *Arboles de Cuba*. Ministerio de Cultura, Editorial Científico-Técnica, Havana, Cuba.

Calder, J. A. and R. L. Taylor. 1968. *Flora of the Queen Charlotte Islands*. 2 vols. (Canada Department of Agriculture, Research Branch, Monograph No. 4). Queen's Printer, Ottawa, Ontario.

Carter, J. L. 1997. *Trees and shrubs of New Mexico*. Johnson Books, Boulder, CO.

Clewell, A. F. 1985. *Guide to the vascular plants of the Florida Panhandle*. University Presses of Florida, Tallahassee.

Correll, D. S. and H. B. Correll. 1972. *Aquatic and wetland plants of southwestern United States*. U.S. Environmental Protection Agency, Washington, DC.

Correll, D. S. and H. B. Correll. 1982. *Flora of the Bahama Archipelago*. J. Cramer, Vaduz.

Correll, D. S and M. C. Johnston. 1970. *Manual of the vascular plants of Texas*. Texas Research Foundation, Renner, TX.

Croat, T. B. 1978. *Flora of Barro Colorado Island*. Stanford University Press, Stanford, CA.

Cronquist, A. 1980. *Vascular flora of the southeastern United States*. Vol. 1, *Asteraceae*. University of North Carolina Press, Chapel Hill.

Cronquist, A., A. H. Holmgren, N. H. Holmgren, J. L. Reveal and P. K. Holmgren. 1972–1997. *Intermountain Flora*. Vols. 1, 3A, 3B, 4, 5, and 6. New York Botanical Gardens, Bronx.

Crow, G. E. and C. B. Hellquist. 2000. *Aquatic and wetland plants of northeastern North America*. 2 vols. University of Wisconsin Press, Madison.

Cullen, J. 1997. *The identification of flowering plant families*, 4th ed. Cambridge University Press, Cambridge.

Diggs, G. M., Jr., B. L. Lipscomb and R. J. O'Kennon. 1999. *Shinners and Mahler's illustrated flora of north central Texas* (Sida, Botanical Miscellany, No. 16). Botanical Research Institute of Texas, Fort Worth.

Dorn, R. D. 1977. *Manual of vascular plants of Wyoming*. 2 vols. Garland, New York.

Duncan, W. H. and M. B. Duncan. 1988. *Trees of the southeastern United States*. University of Georgia Press, Athens.

Elias, T. S. 1980. *The complete trees of North America*. Van Nostrand Reinhold, New York.

Fassett, N. C. 1940. *A manual of aquatic plants*. McGraw-Hill, New York.

Fassett, N. C. 1976. *Spring flora of Wisconsin*, 4th ed. University of Wisconsin Press, Madison.

Fernald, M. L. 1950. *Gray's manual of botany*, 8th ed. American Book Co., New York.

Flora of North America Editorial Committee. 1993–2007. *Flora of North America north of Mexico*. Vols. 1–5, 19–26. Oxford University Press, New York.

Geesink, R., A. J. M. Leeuwenberg, C. E. Ridsdale and J. F. Veldkamp. 1981. *Thonner's analytical key to the families of flowering plants* (Leiden Botanical Series, vol. 5). Leiden University Press, The Hague, Netherlands.

Gleason, H. A. 1952. *The new Britton and Brown illustrated flora of the northeastern United States and adjacent Canada*. New York Botanical Garden, New York.

Gleason, H. A. and A. Cronquist. 1991. *Manual of vascular plants of the northeastern United States and adjacent Canada*, 2nd ed. New York Botanical Garden, Bronx.

Godfrey, R. K. 1988. *Trees, shrubs, and woody vines of northern Florida and adjacent Georgia and Alabama*. University of Georgia Press, Athens.

Godfrey, R. K. and J. W. Wooten. 1979–1981. *Aquatic and wetland plants of the southeastern United States*. 2 vols. University of Georgia Press, Athens.

Gould, F. W. 1951. *Grasses of southwestern United States* (University of Arizona, Biological Science Bulletin, No. 7). University of Arizona, Tucson.

Great Plains Flora Association. 1986. *Flora of the Great Plains*. University Press of Kansas, Lawrence.

Haines, A. D. and T. F. Vining. 1998. *Flora of Maine*. V. F. Thomas, Bar Harbor, ME.

Hammel, B. E., M. H. Grayum, C. Herrera and N. Zamora (eds.). 2003–2004. *Manual de plantas de Costa Rica*. Vols. 1–3. Missouri Botanical Garden Press, St. Louis, MO.

Hickman, J. C. (ed.). 1993. *The Jepson manual: Higher plants of California*. University of California Press, Berkeley.

Hinds, H. R. 1986. *Flora of New Brunswick*. Primrose Press, Fredericton, New Brunswick.

Hitchcock, C. L., A. Cronquist, M. Owenbey and J. W. Thompson. 1955–1969. *Vascular plants of the Pacific Northwest*. 5 vols. University of Washington Press, Seattle.

Holmgren, N. H. 1998. *The illustrated companion to Gleason and Cronquist's manual*. New York Botanical Garden, Bronx.

Howard, R. A. 1974–1989. *Flora of the Lesser Antilles*. 6 vols. Arnold Arboretum, Harvard University, Jamaica Plain, MA.

Hutchinson, J. 1968. *Key to the families of flowering plants of the world*. Clarendon Press, Oxford.

Isley, D. 1990. *Vascular flora of the southeastern United States*. Vol. 3, part 2, *Leguminosae*. University of North Carolina Press, Chapel Hill.

Isely, D. 1998. *Native and naturalized Leguminosae (Fabaceae) of the United States (exclusive of Alaska and Hawaii)*. Brigham Young University, Provo, UT.

Kearney, T. H. and R. H. Peebles. 1960. *Arizona flora*, 2nd ed. University of California Press, Berkeley.

Lellinger, D. B. 1985. *A field manual of the ferns and fern-allies of the United States and Canada*. Smithsonian Institution Press, Washington, DC.

Leon, Bro. and Bro. Alain. 1946–1962. *Flora de Cuba*. 5 vols. Cultural, Havana, Cuba.

Liogier, A. H. 1982–2000. *La flora de la Hispaniola*. 9 vols. University Central del Este, San Pedro de Macorís, Dominican Republic.

Liogier, A. H. 1985–1995. *Descriptive flora of Puerto Rico and adjacent islands*. Universidad de Puerto Rico, Río Piedras.

Little, E. L., Jr. and F. H. Wadsworth. 1964. *Common trees of Puerto Rico and the Virgin Islands*. Vol. 1 (Agriculture Handbook, No. 249). U.S. Department of Agriculture, Washington, DC.

Little, E. L., Jr., R. O. Woodbury and F. H. Wadsworth. 1974. *Trees of Puerto Rico and the Virgin Islands*. Vol. 2 (Agriculture Handbook, No. 449). U.S. Department of Agriculture, Washington, DC.

Llamas, K. A. 2003. *Tropical flowering plants*. Timber Press, Portland, OR.

Long, R. W. and O. Lakela. 1976. *A flora of tropical Florida*. Banyan Books, Miami, FL.

Looman, J. and K. F. Best. 1979. *Budd's flora of the Canadian Prairie Provinces* (Canada Department of Agriculture, Publication 1662). Research Branch, Agriculture Canada, Ottawa, Ontario.

Marie-Victorin, F. 1964. *Flore Laurentienne*, 2nd ed. Les Presses de l'Université de Montréal, Montreal, Quebec.

McVaugh, R. 1983–1993. *Flora Novo-Galiciana*. Vols. 5, 12–17. University of Michigan Herbarium, Ann Arbor.

Mohlenbrock, R. H. 1967–2001. *The illustrated flora of Illinois*. Southern Illinois University Press, Carbondale.

Moss, E. H. 1983. *Flora of Alberta*, 2nd ed. University of Toronto Press, Toronto, Ontario.

Munz, P. A. 1968. *Supplement to A California flora*. University California Press, Berkeley.

Munz, P. A. and D. D. Keck. 1959. *A California flora*. University of California Press, Berkeley.

Organization for Flora Neotropica. 1968 onward. *Flora Neotropica*. New York Botanical Garden, Bronx.

Pease, A. S. 1964. *Flora of northern New Hampshire*. New England Botanical Club, Cambridge, MA.

Proctor, G. R. 1984. *Flora of the Cayman Islands* (Kew Bulletin Additional Series, 11). H.M.S.O., London.

Proctor, G. R. 1985. *Ferns of Jamaica*. British Museum (Natural History), London.

Radford, A. E., H. E. Ahles and C. R. Bell. 1964. *Manual of the vascular flora of the Carolinas*. University of North Carolina Press, Chapel Hill.

Rehder, A. 2001. *Manual of cultivated trees and shrubs: Hardy in North America* (reprint of 2nd ed.). Blackburn, Caldwell, NJ.

Roland, A. E. and E. C. Smith. 1966–1969. *The flora of Nova Scotia* (revised ed., 2 parts). *Proc. Nova Scotian Inst. Sci.* 26(2): 3–224; 26(4): 277–743.

Sargent, C. S. 1922. *Manual of the trees of North America*. 2 vols. Houghton Mifflin, New York.

Scoggan, H. J. 1957. *Flora of Manitoba* (National Museum of Canada Bulletin, No. 140). Department of Northern Affairs and National Resources, Ottawa, Ontario.

Scoggan, H. J. 1978. *The flora of Canada*. 4 parts. National Museum of Natural Sciences, National Museums of Canada, Ottawa.

Seymour, F. C. 1982. *Flora of New England*, 2nd ed. (Phytologia Memoirs, 5). Moldenke, Plainfield, NJ.

Shreve, F. and I. L. Wiggins. 1964. *Vegetation and flora of the Sonoran Desert*. 2 vols. Stanford University Press, Stanford, CA.

Simpson, D. R. and D. Janos. 1974. *Punch card key to the families of dicotyledons of the Western Hemisphere south of the United States*. Field Museum of Natural History, Chicago.

Small, J. K. 1933. *Manual of the southeastern flora*. Author, New York.

Sosa, V. (chief ed.). 1978–2000. *Flora de Veracruz*. 118 fasc. INIREB, Xalapa, Mexico.

Standley, P. C. 1920–1926. *Trees and shrubs of Mexico*. 5 vols. U.S. Government Printing Office, Washington, DC.

Standley, P. C. 1937–1939. *Flora of Costa Rica*. 2 vols. (Botanical Series, Field Museum of Natural History, vol. 18). Field Museum of Natural History, Chicago.

Standley, P. C. and others. 1946–1977. *Flora of Guatemala*. 13 parts. Field Museum of Natural History, Chicago. *Fieldiana: Botany*, vol. 24.

Staples, G. W. and D. R. Herbst. 2005. A tropical garden flora. Bishop Museum Press, Honolulu, HI.

Stevens, W. D., C. U. Ulloa, A. Pool and O. M. Montiel (eds.). 2001. *Flora de Nicaragua*. 3 vols. Missouri Botanical Garden Press, St. Louis.

Steyermark, J. A. 1963. *Flora of Missouri*. Iowa State University Press, Ames.

Strausbaugh, P. D. S. and E. L. Core. 1971. *Flora of West Virginia*, 2nd ed. Seneca Books, Grantsville, WV.

Swink, F. 1990. *The key to the vascular flora of the northeastern United States and southeastern Canada*. Plantsmen's, Flossmoor, IL.

Tomlinson, P. B. 1980. *The biology of trees native to tropical Florida*. Author, Allston, MA.

Vines, R. A. 1960. *Trees, shrubs and woody vines of the Southwest*. University of Texas Press, Austin.

Voss, E. G. 1972–1997. *Michigan flora: A guide to the identification and occurrence of the native and naturalized seed plants of the state*. 3 vols. (Cranbrook Institute of Science Bulletin, 55, 59, 61). Cranbrook Institute of Science, Bloomfield Hills, MI.

Wagner, W. L., D. R. Herbst and S. H. Sohmer. 1990. *Manual of the flowering plants of Hawai'i*. 2 vols. University of Hawaii Press, Bishop Museum Press, Honolulu.

Weber, W. A. 1976. *Rocky Mountain flora*, 5th ed. Colorado Associated University Press, Boulder.

Welsh, S. L. 1974. *Anderson's flora of Alaska and adjacent parts of Canada*. Brigham Young University Press, Provo, UT.

Wiggins, I. L. 1980. *Flora of Baja California*. Stanford University Press, Stanford, CA.

Wood, C. E., Jr. and others. 1958 onward. *Generic flora of the southeastern United States*. J. Arnold Arbor.; Harvard Pap. Bot.

Woodson, P. E., Jr., R. W. Schery and others. 1943–1981. *Flora of Panama*. Ann. Missouri Bot. Gard. Vols. 30–67 passim.

Wunderlin, R. P. 1998. *Guide to the vascular plants of Florida*. University Press of Florida, Gainesville.

Wunderlin, R. P. and B. F. Hansen. 2000. *Flora of Florida*. Vol. 1. University Press of Florida, Gainesville.

Other Useful References

Askevold, I. S. and C. W. O'Brien. 1994. DELTA, an invaluable computer program for generation of taxonomic monographs. *Ann. Entomol. Soc. Am.* 87: 1–16.

Blanco, M. A., W. M. Whitten, D. S. Penneys, N. H. Williams, K. M. Neubig and L. Endara. 2006. A simple and safe method for rapid drying of plant specimens using forced-air space heaters. *Selbyana* 27: 83–87.

Bridson, D. and L. Forman. 1998. *The herbarium handbook*, 3rd. ed. Royal Botanic Gardens, Kew, UK.

Dallwitz, M. J. 1974. A flexible computer program for generating diagnostic keys. *Syst. Zool.* 23: 50–57.

Dallwitz, M. J. 1993. DELTA and INTKEY. In *Advances in computer methods for systematic biology: artificial intelligence, databases, computer vision*, R. Fortuner (ed.), 287–296. The Johns Hopkins University Press, Baltimore, MD.

Dallwitz, M. J. 2000 onward. A comparison of interactive identification programs. http://delta-intkey.com/www/comparison.htm.

Dallwitz, M. J., T. A. Paine and E. J. Zurcher. 1993. *User's guide to the DELTA system: A general system for processing taxonomic descriptions*, 4th ed. CSIRO, Division of Entomology, Canberra, Australia.

Dallwitz, M. J., T. A. Paine and E. J. Zurcher. 2002 onward. *Principles of interactive keys*. http://delta-intkey.com/www/interactivekeys.htm.

DeWolf, G. P., Jr. 1968. Notes on making an herbarium. *Arnoldia* 28: 69–111.

Dransfield, J. 1986. A guide to collecting palms. *Ann. Missouri Bot. Gard.* 73: 166–176.

Edwards, M. and D. R. Morse. 1995. The potential for computer-aided identification in biodiversity research. *Trends Ecol. Evol.* 10: 152–158.

Fosberg, F. R. and M.-H. Sachet. 1965. *Manual for tropical herbaria* (Regnum Vegetabile, vol. 39). International Bureau for Plant Taxonomy and Nomenclature, Utrecht, Netherlands.

Frodin, D. G. 2001. *Guide to standard floras of the world: An annotated, geographically arranged systematic bibliography of the principal floras, enumerations, checklists, and chorological atlases of different areas*, 2nd ed. Cambridge University Press, Cambridge.

Haynes, R. R. 1984. Techniques for collecting aquatic and marsh plants. *Ann. Missouri Bot. Gard.* 71: 229–231.

Kristiansen, K. A., M. Cilieborg, L. Drábková, T. Jørgensen, G. Petersen and O. Seberg. 2005. DNA taxonomy—the riddle of *Oxychloë* (Juncaceae). *Syst. Bot.* 30: 284–289.

Henderson, D. M. (ed.). 1983. *International directory of botanical gardens IV*, 4th ed. Koeltz Scientific Books, Königstein, Germany.

Holmgren, N. H. and B. Angell. 1986. *Botanical illustration: Preparation for publication*. New York Botanical Garden, Bronx.

Holmgren, P. K. and N. H. Holmgren. 1992. *Plant specialists index*. Koeltz Scientific Books, Königstein, Germany. [Index to specialists in the systematics of plants and fungi.]

Holmgren, P. K., N. H. Holmgren and L. C. Barnett. 1990. *Index Herbariorum, Part I: The herbaria of the world*, 8th ed. (Regnum Vegetabile, vol. 120). New York Botanical Garden, Bronx. [Also available on-line at the New York Botanical Garden website.]

Howard, R. A. 1969. The botanical garden—An unexploited source of information. *Boissiera* 14: 109–117.

Kiger, R. W., T. D. Jacobsen and R. M. Lilly. 1981. *International register of specialists and current research in plant systematics*. Hunt Institute for Botanical Documentation, Carnegie-Mellon University, Pittsburgh, PA.

Kudsen, J. W. 1972. *Collecting and preserving plants and animals*. Harper and Row, New York.

Lawrence, G. H. M. 1951. *Taxonomy of vascular plants*. Macmillan, New York. [Contains detailed chapters on plant identification, field and herbarium techniques, mono-graphs and revisions, and floristics; see pp. 223–283.]

Leenhouts, P. W. 1968. *A guide to the practice of herbarium taxonomy* (Regnum Vegetabile, vol. 58). International Bureau for Plant Taxonomy and Nomenclature, Utrecht, Netherlands.

Lot, A. and F. Chiang. 1986. *Manual de herbario*. Consejo Nacional de la Flora de México, Mexico.

Metsger, D. A. and S. C. Byers. 1999. *Managing the modern herbarium—An inter-disciplinary approach*. Society for the Preservation of Natural History Collections, Washington, DC.

Pankhurst, R. J. (ed.). 1975. *Biological identification with computers* (Systematics Association, Special Volume, No. 7). Academic Press, London.

Pankhurst, R. J. 1978. *Biological identification*. University Park Press, Baltimore, MD.

Pankhurst, R. J. 1991. *Practical taxonomic computing*. Cambridge University Press, Cambridge.

Pohl, R. W. 1965. Dissecting equipment and materials for the study of minute plant structures. *Rhodora* 67: 95–96.

Robertson, K. R. 1980. *Observing, photographing, and collecting plants* (Circular 55). Illinois Natural History Survey, Urbana.

Smith, C. E., Jr. 1971. *Preparing herbarium specimens of vascular plants* (Agriculture Information Bulletin No. 348). Agricultural Research Service, U.S. Department of Agriculture, Washington, DC.

Soderstrom, T. R. and S. M. Young. 1983. A guide to collecting bamboos. *Ann. Missouri Bot. Gard.* 70: 128–136.

Stern, M. J. and T. Eriksson. 1996. Symbioses in herbaria: Recommendations for more positive interactions between plant systematists and ecologists. *Taxon* 45: 49–58.

Tilling, S. M. 1984. Keys in biological identification: Their role and construction. *J. Biol. Educ.* 18: 293–304.

Voss, E. G. 1952. The history of keys and phylogenetic trees in systematic biology. *J. Sci. Lab. Denison Univ.* 43: 1–25.

Watson, L. and M. J. Dallwitz. 1991. The families of angiosperms: Automated descriptions, with interactive identification and information retrieval. *Aust. Syst. Bot.* 4: 681–695.

Watson, L. and M. J. Dallwitz. 1992. *Grass genera of the world*. C.A.B. International, Wallingford, UK.

Watson, L. and M. J. Dallwitz. 1993. *The families of flowering plants: Interactive identifications and information retrieval* (CD-ROM version 1.0 for MS-DOS). CSIRO, Melbourne, Australia.

West, K. 1983. *How to draw plants*. Herbert Press, London.

Zomlefer, W. B. 1994. *Guide to flowering plant families*. University of North Carolina Press, Chapel Hill.

Zweifel, F. W. 1961. *A handbook of biological illustration*. University of Chicago Press, Chicago.

GLOSSARY

Abaxial surface   Surface facing away from the apex of the axis.

Acaulescent plant   Plant that is apparently stemless; sometimes the stem is subterranean.

Accessory bud   Extra bud (or buds) produced on either side of or above or below the axillary bud.

Accessory fruit   Fruit (or a cluster of fruits) in which structures in addition to the matured gynoecium form a functional part of the fruit.

Achene   Fairly small, indehiscent, dry fruit with a thin and close-fitting wall surrounding, but free from, the single seed.

Acorn   Fruit of species belonging to the genus *Quercus*; a nut associated with a variously scaly cupule.

Actinomorphic   *See* Radial symmetry.

Acuminate apex   Apex with somewhat concave sides that taper to a sharp point (see Figure 4.10).

Acute apex   Apex ending in a point of less than 90 degrees, with the sides of the tapered apex more or less straight to slightly convex (see Figure 4.10).

Acute base   Base ending in a point of less than 90 degrees, with the sides of the tapered base slightly convex (see Figure 4.11).

Adaptation   Any change in an organism resulting from natural selection.

Adaxial surface   Surface facing toward the apex of the axis.

Adnate/adnation   Fusion of unlike parts, as stamens with the corolla.

Adventitious embryony   Development of the embryo directly from a somatic cell of the ovule without formation of a gametophyte.

Adventitious   Developing from any plant part other than the normal one, e.g., adventitious roots develop from stems or leaves, not from the embryonic root (radicle).

Aerial root   Root occurring above ground or water.

Agamic complex   Group of species with two to several taxa that reproduce asexually.

Agamospecies   Minimally differentiated individuals that reproduce asexually.

Agamospermy   Asexual seed production; not involving fertilization of egg by sperm.

Aggregate fruit   Fruit that develops from several separate carpels of a single flower.

Alkaloids   Structurally diverse, physiologically active secondary compounds; derived from various amino acids or from mevalonic acid by various biosynthetic pathways. (See pages 95–96.)

Allele   One of two or more molecular forms of a gene.

Allelopathy   Chemical inhibition of one organism by another.

Allopatric   Occurring in different geographical regions.

Allopatric speciation   Divergence of geographically isolated populations, leading to evolution of new species.

Allopolyploid speciation   Formation of a new species via hybridization followed by chromosome doubling in the hybrid.

Allopolyploidy   Presence in a cell of two or more genomes from different species.

Alternate leaves   Leaves borne one leaf per node along the stem (see Figure 4.4); such leaves may be spiral, 2-ranked, etc.

Ament   *See* Catkin.

Amoeboid   Having the form of an amoeba, i.e., with variable shape and various rounded extensions.

Amphiparacytic   Stoma with guard cells surrounded by two pairs of subsidiary cells, each parallel to the guard cells (see Figure 4.36).

Anagenesis   Evolutionary change within a single lineage over time.

Anatomy   Study of the internal structure (e.g., cells, tissues).

Anatropous ovule   Ovule that is inverted and fused to the funiculus (stalk) so that the micropyle (opening) is situated next to the funiculus (see Figure 4.40).

Androecium   Collective term for all the stamens in a flower (see Figure 4.16).

Androgynophore   Stalk bearing the androecium (stamens) and gynoecium (carpels).

Aneuploidy   Having a chromosome number that is not an even multiple of the base number.

Anisocytic   Stoma surrounded by three subsidiary cells, of which one is distinctly smaller (see Figure 4.36).

Anisophylly   When the two leaves of a pair, or alternating leaves of the same flush, are diverse in shape or size.

Annual   Plant growing from seed to maturity within a single growing season.

Annulus   Band or cluster of thick-walled, hygroscopic cells in the sporangium wall of most ferns; facilitates release of spores.

Anomalocidal capsule   Capsule that opens irregularly.

Anomocytic   Stoma lacking differentiated subsidiary cells (see Figure 4.36).

Anther   Pollen-bearing part of the stamen, borne at the top of the filament (see Figure 4.16).

Antheridium   Male reproductive structure producing and protecting the sperm in embryophytes (land plants).

Antheridogens   Compounds released by archegonia of embryophytes that stimulate neighboring gametophytes to produce antheridia.

Anthocyanins   Violet, blue, red, and yellow flavonoid pigments not containing nitrogen; widespread among angiosperms.

Antibody   One of a diverse array of antigen-binding recep-tors.

Antigen   Molecule that white blood cells recognize as foreign and that triggers an immune response.

Antipodals   Three cells of the female gametophyte clustered at the end opposite the egg and synergids.

Aperture of pollen   Thin, variously shaped region of the pollen wall through which the pollen tube emerges during germination.

Apetalous   Lacking petals.

Apex (pl. apices)/apical   Tip of a structure.

Apical meristem   Group of dividing cells at the growing apex of a stem or root.

Apical placentation   Ovule or ovules attached at the apex of the ovary (see Figure 4.22).

Apocarpous   Carpels separate.

Apomixis   Asexual reproduction, usually used for asexual seed production (see agamospermy) but also applied to the production of plantlets by vegetative propagules (e.g., gemmae).

Apomorphy   Derived character state; also called an apomorphic character (or character state).

Apopetalous   Petals separate.

Aposepalous   Sepals separate.

Apotepalous   Tepals separate.

Arachnoid   Having a cobwebby appearance.

Archegonium (pl. archegonia)   Female reproductive struc-ture producing and protecting the egg in embryophytes (land plants).

Areole   Spiny short-shoot characteristic of the Cactaceae.

Aril   A hard to juicy, often brightly colored outgrowth of the seed; usually developed from the funiculus or the outer integument.

Aromatic   Having a distinct, usually pleasant odor.

Articulate/articulation   Jointed.

Asexual reproduction   *See* Apomixis.

Asymmetrical flower   Flower lacking a plane of symmetry, i.e., neither radial nor bilateral.

Asymmetrical leaf   Leaf lacking a plane of symmetry; often wider on one side than the other, with widest point differ-ently positioned on each side of the blade, and with an oblique base.

Attenuate apex   Apex tapering gradually to a narrow tip (see Figure 4.10).

Autapomorphy   Derived character state restricted to a single terminal taxon (in a cladistic analysis).

Authority   In nomenclature, the citation following the taxon name and consisting of the name or names of individuals who described the taxon or transferred it from one rank to another.

Autopolyploidy   Presence in a cell of three or more chromo-some sets from the same species.

Awn   Narrow, bristlelike appendage, often at the tip of a structure such as a fruit or anther.

Axil   Region between the adaxial side of a leaf and the stem; literally, armpit.

Axile placentation   Ovules attached to the central axis of an ovary with two or more locules (see Figure 4.22).

Axillary branch   Branch that develops from an axillary bud.

Axillary bud   Bud located in the leaf axil (see Figure 4.2).

Backcrossing   Hybridization between F_1 hybrid individuals and members of either parental species.

Banner petal   Distinctive adaxial petal, usually the largest, of flowers, especially those of Fabaceae subfam. Faboideae; also called the **standard** or **flag petal**.

Bark   All tissues outside the vascular cambium in older trees, shrubs, or lianas.

Basal placentation   Ovule or ovules attached at the base of the ovary (see Figure 4.22).

Base number   Inferred ancestral haploid chromosome number of a taxon, usually prefixed by the letter "x", as in "$x = $".

Basifixed anther   Anther that is attached at its base to the filament.

Berry   Indehiscent, fleshy fruit with (one or) a few to many seeds; the flesh may be more or less homogeneous, or the outer part of the fruit may be firm, hard, or leathery.

Betalains   Nitrogenous red and yellow pigments; restricted to the Caryophyllales.

Bicollateral vascular bundle   Vascular bundle with phloem on two sides (abaxial and adaxial) of the xylem.

Biennial   Plant that lives for two years, growing vegetatively during the first and flowering/fruiting in the second.

Bifacial leaf   Leaf having anatomically distinct adaxial and abaxial surfaces.

Bijugate leaves   Opposite leaves in which those of adjacent nodes are rotated slightly less/more than 90 degrees.

Bilabiate   Two-lipped; typically applied to a calyx or corolla.

Bilateral symmetry   Divisible into equal halves by only one plane of symmetry; also called **zygomorphic** (see Figure 4.18).

Binary character   Character with only two character states.

Binomial nomenclature   System of naming developed by Linnaeus in which each species name is composed of two words   the genus name, a noun, and the specific epithet, an adjective or noun in apposition, e.g., *Acer rubrum*.

Biogeography   Scientific study of patterns in the geographic distribution of species and higher taxa.

Biological nomenclature   Rules for naming living organisms as outlined in the various nomenclatural codes.

Biradial symmetry   Divisible into equal halves by only two planes of symmetry; compare with radial symmetry.

Biseriate   With two rows or whorls of structures, as a flower with two perianth whorls.

Biseriate hair   Hair composed of two rows of cells.

Bisexual flower   Flower with both androecium (stamens) and gynoecium (carpels); also called **perfect flower**.

Bitegmic   With two integuments.

Bivalents   Pairs of homologous chromosomes in a cell during meiosis.

Blade   Flat, photosynthetic portion of a leaf, also called the lamina (see Figure 4.3), or leaflets, in a compound leaf (*see* Compound leaf).

Bole   *See* Trunk.

Bootstrap analysis   Statistical technique that uses resam-pling and replication of characters to assess how well the data set support a given tree topology.

Botanical nomenclature   Rules for naming plants as outlined in the *International Code of Botanical Nomenclature*.

Bracketed key   Key in which the two leads of a couplet (pair of contrasting choices) are placed on adjacent lines so that the leads of a couplet are not separated by intervening lines (compare, Indented key).

Bract   A reduced leaf, in the axil of which arises a flower or an inflorescence branch.

Bracteole   Prophyll of a flower, usually very small, borne below the flower on a pedicel; also called a bractlet.

Bremer support *See* Decay index.

Bud Small embryonic shoots, whether floral or vegetative, often protected by modified leaves (bud scales), stipules, or hairs.

Bud-scale Reduced, protective leaf associated with a bud.

Bud-scale scar Scar left after a bud-scale falls from the plant.

Bulb Short, erect, underground stem surrounded by thick, fleshy leaves or leaf bases.

C3 plants Plants having the photosynthetic pathway in which the first intermediate formed after fixation of carbon dioxide is the three-carbon compound phospho-glycerate.

C4 plants Plants having the photosynthetic pathway in which the first intermediate formed after fixation of carbon dioxide is the four-carbon compound oxaloacetate.

Calcium carbonate $CaCO_3$.

Calcium oxalate Oxalic acid (HO_2CCO_2H) associated with calcium, often occurring in plant tissues as raphide or druse crystals.

Callose Glucose polymer, e.g., deposited during development in sieve areas of sieve-tube elements.

Calyculus Reduced, rimlike calyx, as in members of Loranthaceae.

Calyx Collective term for all sepals of a flower (see Figure 4.16).

Campanulate Bell-shaped.

Campylotropous ovule Ovule that is curved so the micropyle is positioned near the funiculus (see Figure 4.40).

Canescent Gray, with dense, short hairs.

Capitate Enlarged at the tip, i.e., headlike.

Capitulum *See* Head.

Capsule Dry to rarely fleshy fruit from a two- to many-carpellate gynoecium that opens in various ways to release the seed or seeds.

Carbocyclic iridoids Iridoids that have two ring systems, one composed entirely of carbon (see page 98.)

Cardenolides Highly poisonous glucosides of a type of 23-carbon steroid.

Carnivorous Capturing animals and digesting them, as applied to plants.

Carpel Ovule-bearing unit(s) that make up the gynoecium.

Carpellate flower Flower with a gynoecium (carpel or carpels) but no functional androecium (stamens).

Carpellode Sterile carpel.

Carpophore Stalk supporting each portion of the dehiscing schizocarp of many members of Apiaceae.

Caruncle Rather firm outgrowth of a seed that develops from the outer integument (testa); in this text considered as a type of aril.

Caryopsis Small, indehiscent, dry fruit with a thin wall surrounding and more or less fused to a single seed.

Catkin Inflorescence consisting of a dense, elongated mass of inconspicuous, usually wind-pollinated flowers; also called an **ament**.

Caudate With a tail-like appendage.

Caulescent plant Plant with a distinct stem.

Cauliflory Flowers or inflorescences borne on older stems and/or trunks.

Cellular endosperm Division of the primary endosperm nucleus, and all subsequent nuclear divisions, immediately followed by formation of cell walls.

Centrifugal development Developing first at the center and then gradually toward the periphery.

Centripetal development Developing first at the periphery and then gradually toward the center.

Centromere Small, constricted region of a chromosome having attachment sites for spindle fibers during nuclear division.

Chalaza Indistinct region of an ovule (or seed) where the integuments are connected to the nucellus, at the end opposite from the micropyle.

Character Any feature of the organism, often used to refer to those showing group-diagnosing variation.

Character displacement Divergence in two related species in the zone where they co-occur, resulting from natural selection against individuals with intermediate phenotypes.

Character state polarization Determination of the direction of character state change, i.e., determining what state is ancestral and what state (or states) are derived.

Character state One of the various conditions (or values) of a character observed across a group of taxa.

Chartaceous Of papery texture.

Chasmogamous Condition of a flower that opens and is usually cross-pollinated. Contrast with Cleistogamous.

Chlorophyll Light-sensitive green pigment that absorbs violet-to-blue and also red light, and is essential for photosynthesis.

Chloroplast capture Transfer of chloroplasts from one species to another via hybridization and subsequent introgression.

Chloroplast genome Circular DNA strand of the chloroplast, an endosymbiont within the eukaryotic host cell.

Chromosome Complexly folded DNA molecule of the nuclear genome in eukaryotes with many associated histone proteins.

Circinate Rolled in a coil from the top downward, with the apex nearest the center of the coil (see Figure 4.14).

Circumscissile Structure opening by a line around its circumference, with the top usually coming off as a lid, e.g., a circumscissile capsule.

Circumscription Delimitation ("drawing a line around") of any feature or features, as for instance determining the limits or boundaries of a taxon.

Clade Monophyletic group, made up of an ancestor and all of its descendents.

Cladistics Taxonomic philosophy/method in which relationships of taxa are based on recency of common ancestry as assessed by the pattern of shared derived character states (synapomorphies).

Cladogenesis Origination of new clades (monophyletic groups) through the splitting of evolutionary lineages.

Cladogram Branching diagram that shows hypothesized phylogenetic (sister-group) relationships of a group of organisms.

Class As a concept, a group of individuals that all possess a particular defining characteristic. As a taxonomic rank, the category between the higher rank of division (or phylum) and the lower rank of order; in phylogenetic classifications, a monophyletic group of orders.

Classification Delimitation, ordering, and ranking of taxa; or the system of internested groups resulting from this process.

Clawed With distinctly narrowed basal portion, as in the clawed petals of Malpighiaceae.

Cleistogamous   Condition of a flower that never opens and is self-pollinated and self-fertilized. Contrast with Chasmogamous.

Cline   Gradual change in a single character across a geographical or ecological gradient.

Coevolution   Linked evolution of two or more closely interacting species.

Collateral vascular bundle   Vascular bundle with phloem on one side and xylem on the other.

Collection number   Number assigned to a preserved plant specimen, usually by its collector.

Colleter   Multicellular "hair" that produces a sticky, mucilaginous, or resinous secretion, usually occurring on the adaxial surface of a stipule, petiole base, or calyx lobe of certain plants, especially members of Gentianales.

Colpate   Pollen grain with long and grooved aperture(s), the colpus (i).

Columella   "Little column" in the pollen exine, separating the tectum from the foot layer of the nexine (see Figure 4.49).

Column   Structure formed by the fused style and stigma plus stamens of members of the Orchidaceae.

Combretaceous hair   Thick-walled, unicellular hair with a sharp pointed apex and a bulbous base that appears bicellular because of a conical or concave cellulose membrane in the cell wall.

Common name   Name of a plant in language of the culture or society where the plant occurs; also called the vernacular name.

Companion cell   Parenchymatous cell with a nucleus and dense cytoplasm that is positioned next to a sieve tube element and is closely involved in its metabolism.

Complete flower   Flower having a perianth, androecium, and gynoecium.

Compound leaf   Leaf with two or more blades (leaflets).

Compound umbel   Umbel of umbels, i.e., with the peduncle of each simple umbel arising from a common point.

Conducting tissues   Xylem (water-conducting) and phloem (sucrose-conducting) tissues of a vascular plant.

Conduplicate   Folded lengthwise with the adaxial surface within (see Figure 4.14).

Cone   Ovulate reproductive structure of the conifers, consisting of reduced branches borne on a rather short axis.

Connate/connation   Like parts fused, as the petals fused to form a flower with a sympetalous corolla.

Connective   Portion of the stamen connecting the two pollen sacs of an anther (see Figure 4.16).

Consistency index   Measure of the level of homoplasy of a character (or of all characters on a cladogram), equaling the minimum number of possible changes divided by the total number of actual changes on the tree (or the total tree length).

Contorted   Parts imbricate and whorled, each in turn overlapped by that on one side and overlapping that on the other, the whorl tightly rolled, appearing twisted.

Convergent evolution   Evolution of similar features from different ancestral conditions, resulting from similar selective pressures operating on distantly related, taxa, e.g., the evolution of stem succulence in both Cactaceae and Euphorbiaceae or carnivorous leaves in Sarraceniaceae and Nepenthaceae.

Convolute   Rolled, with the margins overlapping (see Figure 4.14).

Coralloid root   Short, thick, and much branched root, similar in appearance to some corals.

Cordate base   Base heart-shaped, with rounded lobes (see Figure 4.11).

Coriaceous   Thick, or of leathery texture.

Corm   Short, erect, underground, more or less fleshy stem covered with thin, dry leaves (or leaf bases).

Corolla   Collective term for all petals of a flower (see Figure 4.16).

Corona   Usually showy outgrowths of the perianth parts, stamens, or receptacle.

Cortex   Portion of a stem or root located between the pericycle (surrounding the primary vascular tissue) and the epidermis.

Cortical bundle   Vascular bundle occurring in the cortex of a stem.

Corymb   Raceme with the pedicels of the lowermost flowers elongated, bringing all flowers to more or less the same level, i.e., a flat-topped raceme (see Figure 4.30).

Costapalmate   More or less palmate leaf with an extension of the rachis, forming a midrib (or costa) in the blade, as in the leaves of many palms (Arecaceae).

Cotyledon   Leaf, or one of two or occasionally more leaves developed at the first node of the embryo (see Figure 4.43).

Couplet   A pair of contrasting choices in a dichotomous key.

Crenate margin   Margin with rounded teeth (see Figure 4.12).

Crossing over   Breakage and exchange of corresponding segments between a pair of homologous chromosomes during Prophase I of meiosis.

Cross-pollination   Transfer of pollen from the anther of flowers of one individual to the stigma of flowers of a second individual.

Cryptic species   Taxa that are reproductively isolated but are phenotypically very similar and hard to distinguish.

Crystal sand   Granular mass composed of very tiny crystals of calcium oxalate.

Cucullate   Hood-shaped.

Cultivar   Variant/variety of a plant produced in cultivation through hybridization, artificial selection, or any other process.

Cuneate base   Base that is wedge-shaped, triangular, and tapered to a point (see Figure 4.11).

Cupule   Any cup-shaped structure, such as the scaly to spiny "cup" associated with the nut of members of Fagaceae, or swollen receptacle and perianth parts usually associated with the drupe of members of Lauraceae.

Cuticle   Thin, waxy layer that covers the outer surface (epidermis) of most embryophytes.

Cyanogenesis   Process by which hydrogen cyanide is released from cyanogenic compounds.

Cyanogenic glycosides   Defensive compounds that are hydrolyzed to release hydrogen cyanide.

Cyathium   Cuplike cymose inflorescence unit of species of *Euphorbia* that mimics a single complete flower and consists of one to few nectar glands, several staminate flowers, and a single carpellate flower.

Cycasins   Poisonous compounds of cycads that cause neurological damage.

Cyclocytic   Stoma surrounded by subsidiary cells forming one or two narrow rings around the guard cells; the number of cells in each ring usually is four or more (see Figure 4.36).

Cyme   Determinate, compound inflorescence composed of repeating units of a pedicel bearing a terminal flower and below it, one or two bracteoles; each bracteole is associated with an axillary flower, and further bracteoles, and so on (see Figure 4.29).

Cystolith   Stalked concretion composed of calcium carbonate that projects into the center of a specialized cell or lithocyst.

Decay index   Number of extra steps required to find trees (cladograms) that do not link a particular group of taxa.

Deciduous leaves   Leaves that fall (are abscised) at the end of the growing season.

Decurrent base   Lamina base tapering gradually to the petiole (see Figure 4.11), also used for leaf base that extends down the stem.

Decussate leaves   Opposite leaves in which those of adjacent nodes are rotated 90 degrees from each other.

Dehiscence   Method or process of opening of a structure, such as a fruit, anther, or sporangium.

Dehiscent drupe   Fruit with a dry/fibrous to fleshy or leathery outer husk that eventually breaks apart or opens, exposing one or more nutlike pits, as in *Carya* (Juglandaceae).

Dendritic hair   Hair with a treelike branching pattern (see Figure 4.15).

Dendrogram   Treelike diagram showing relationships, whether phenetic, cladistic, or evolutionary taxonomic.

Dentate margin   With coarse teeth that are perpendicular to the margin (see Figure 4.12).

Denticidal capsule   Capsule that opens by a series of apical teeth.

Determinate inflorescence   Inflorescence in which the axis is converted into a flower, resulting in the cessation of growth of that axis.

Diacytic   Stoma enclosed by a pair of subsidiary cells whose common wall is at right angles to the guard cells (see Figure 4.36).

Diadelphous   Stamens with their filaments fused to form two groups.

Diagnosis   Brief statement of the attributes that allow recognition of a taxon and its separation from one or more others.

Dichotomous branching   Branching by division of the apical meristem or apical cell, forming two equal or unequal branches.

Dichotomous key   Key composed of pairs of contrasting statements (couplets).

Didynamous stamens   Four stamens, two long and two short.

Dimorphic   With two distinct forms of the same structure.

Dioecious   With staminate and carpellate flowers borne on separate plants.

Diploid   With two full sets of chromosomes in each cell.

Disk flower   Radially symmetrical flower of many members of the Asteraceae.

Dissected   Deeply divided into many narrow segments.

Distichous   *See* Two-ranked.

Distyly   Flowers of different individuals of the same species having two different style lengths, the stamen lengths often varying inversely; an outcrossing mechanism.

Diterpenoids   Twenty-carbon terpenoid compounds (see page 98).

Division   Taxonomic category between the higher rank of kingdom and the lower rank of class; also called phylum (pl. phyla); in a phylogenetic classification a division is a monophyletic group of classes.

Dollo parsimony   Parsimony assumption that allows only one evolutionary gain of a character state but numerous losses.

Domatium (pl. domatia)   Small cavity, pouch, or tuft of hairs on a leaf or bract that functions as a tiny home for mites, ants, or other small organisms.

Double fertilization   In angiosperms, fusion of one sperm with the egg to form the zygote, and fusion of the second sperm with the polar nuclei to form the nutritive tissue, endosperm.

Doubly serrate margin   Margin with coarse teeth bearing smaller teeth on their margins, and leaf margin thus with both small and large teeth (see Figure 4.12).

Drupe   Indehiscent, fleshy fruit in which the outer part is more or less soft (to occasionally leathery or fibrous) and the center contains one or more hard pits or stones consisting of a bony endocarp surrounding a seed or seeds.

Drupelet   Drupe developing from a single carpel and, as traditionally used, one portion of an aggregate fruit, as in *Rubus* (Rosaceae).

Druse   Spherical, compound crystal of calcium oxalate in which the numerous component crystals protrude from the surface giving the cluster a star-shaped appearance.

Ebracteate   Lacking bracts.

Ecotype   Distinct genetic entity of a species that is adapted to a particular set of ecological conditions.

Effective publication   Publication in a book or scientific journal as outlined in the *International Code of Botanical Nomenclature*.

Egg   Nonmotile, female gamete of embryophytes, often larger than the sperm, and formed in an archegonium or by a reduced female gametophyte.

Elaiosome   Hard, oily outgrowth (aril) of a seed that attracts ants.

Elater   Hygroscopic band attached to the spores of *Equisetum* that aid in dispersal.

Electrophoresis   Procedure using an electrical current to separate molecules (usually proteins) on the basis of their size and electrical charge.

Elliptic   Widest near the middle (see Figure 4.9).

Emarginate apex   Apex prominently notched (see Figure 4.10).

Embryo   Young sporophyte, from the first cell after fertilization (zygote) until germination of the seed (in seed plants) or emergence from protective gametophytic tissues (other embryophytes).

Embryo sac   Female gametophyte of the angiosperms (flowering plants).

Embryology   Study of the events leading to development of the male and female gametophytes, gamete development, syngamy, and early development of the embryonic sporophyte.

Endemic taxon   Taxon whose members are restricted to a particular geographic area.

Endexine   Lower part of the nexine, the innermost portion of the exine (see Figure 4.49).

Endocarp   Inner layer of the fruit wall or pericarp, as in the pit or stone of a peach or cherry (*Prunus*, Rosaceae).

Endosperm   Usually triploid nutritive tissue in the seed of most angiosperms, derived from the fusion of a sperm with the two polar nuclei of the female gametophyte.

Endothecium   Inner lining of the locule of an anther, usually fibrous.

Entire margin   With a smooth margin, lacking any teeth (see Figure 4.12).

Epicotyl   That portion of the embryonic stem above the cotyledons, and below the next leaves on the embryonic axis.

Epicuticular waxes   Wax deposits of variable form deposited on top of the cuticle on the surface of the plant.

Epidermis   Outermost layer of cells of the primary plant body, often with strongly thickened outer walls.

Epigynous   With perianth and stamens apparently borne upon the ovary (see Figure 4.21) and the ovary, therefore, inferior.

Epimatium   Modified ovuliferous scale that partly surrounds the ovule in members of the Podocarpaceae.

Epipetalous stamens   Stamens adnate (fused) to the corolla.

Epiphylly   Flower or inflorescence borne on a leaf.

Epiphyte   Plant growing upon another plant, which is used as a support.

Equitant leaves   Two-ranked leaves that are flattened in the plane of leaf insertion and with both surfaces morphologically and anatomically identical (i.e., unifacial), as in the leaves of *Iris* and relatives (Iridaceae).

Erose margin   Margin with irregular rounded to pointed teeth, as if gnawed (see Figure 4.12).

Essential oils   *See* Ethereal oils.

Ethereal oils   Aromatic, highly volatile, oily secondary plant products containing monoterpenoids and sesquiterpenoids (and other aromatic compounds), frequently in pellucid dots.

Eusporangium (pl. eusporangia)   Common sporangium form of embryophytes, sessile and with a wall composed of several cell layers.

Eustele   Vascular bundles of a stem forming a complete ring when viewed in cross-section (see Figure 4.39).

Even-pinnate leaf   Pinnately compound leaf with an even number of leaflets, i.e., lacking a terminal leaflet (see Figure 4.5).

Evergreen plant   Plant that is leafy throughout the year.

Evolution   Genetic change in a lineage (line of descent) over time.

Evolutionary taxonomy   Subjective approach to classification construction that attempts to reflect overall similarity, ecologically significant features, and evolutionary relationships.

Evolutionary tree   Branching diagram representing phylogenetic relationships (evolutionary history) of a group of taxa; in this text considered basically equivalent to a cladogram, and frequently shortened to "tree."

Exalbuminous seeds   Mature seeds lacking endosperm.

Exine   Outer layer of the two-layered wall of a pollen grain (see Figure 4.49).

Exocarp   Outer layer of the fruit wall or pericarp.

Exserted floral structures   Sticking out, as in stamens extending beyond the corolla.

Extrafloral nectary   Nectar-producing structure positioned on a non-floral structure, such as a leaf, stipule, stem, bract, or inflorescence axis.

Extrastaminal nectar disk   Nectar disk positioned between the staminal whorl (or whorls) and the perianth parts.

Extrorse anther   Anther that opens outwards, toward the periphery of the flower.

F₁ generation   First filial generation; offspring resulting from the cross of individuals representing two different species or distinct genetic entities.

False indusium   In ferns, a revolute leaf margin or marginal lobe that protects developing sporangia in a sorus.

Family   Taxonomic category between the higher rank of order and the lower rank of genus; in a phylogenetic classification, a monophyletic group of genera.

Fascicle   Reduced, axillary inflorescence (either determinate or indeterminate) with flowers in a bundle or cluster and no obvious inflorescence axis (see Figure 4.30).

Felsenstein zone   A combination of very fast and very slow evolutionary rates that will make accurate phylogeny reconstruction impossible; taxa with faster rates (long branches) will appear to be closely related.

Fertilization   Fusion of the sperm nucleus and the egg nucleus.

Fiber   Any long, narrow, thick-walled, and lignified cell.

Fibrous roots   Root system with roots of more or less equal thickness, and often well branched, the primary (tap) root absent or not obvious.

Field press   Device consisting of press frames, straps, corrugates, and newspapers used to press and transport plant specimens in the field.

Filament   Stalk of a stamen (see Figure 4.16).

Fissured xylem   Xylem broken up by the development of phloem or parenchyma tissue.

Fitch parsimony   Parsimony algorithm that treats all character states as unordered.

Fitness   An increase in adaptation to the environment, as brought about by genetic change; measured as reproductive success.

Flag petal   *See* Banner petal.

Flagellum (pl. flagella)   Tail-like motile structure of many eukaryotic cells, having a nine-plus-two array of microtubules.

Flavonoids   Phenolic compounds that usually occur in a ring system derived through cyclization of an intermediate from a cinnamic acid derivative and three malonyl Co-A molecules (see page 98.)

Fleshy root   Thick root, with tissues for storing water or carbohydrates.

Flora   Enumeration of the plants occurring in a particular geographical area, usually with keys, descriptions, illustrations, and distribution maps; also the plants occurring within a designated region.

Floral cup   *See* Hypanthium.

Floral diagram   Diagram summarizing floral symmetry and the number, fusion, and insertion of the floral parts.

Floral formula   Formula summarizing floral symmetry and the number, fusion, and insertion of floral parts.

Floral tube   *See* Hypanthium.

Floret   Very small flower, especially those of members of Asteraceae, Cyperaceae, and Poaceae; sometimes applied to the flowers plus associated bractlike structures.

Flower   Reproductive structure of the angiosperms, consisting of a determinate, modified shoot, the floral axis or receptacle, bearing modified leaves, the perianth parts, stamens, and/or carpels.

Follicle   Dry to rarely fleshy fruit derived from a single carpel that opens along a single (usually adaxial) longitudinal suture.

Foot layer (pollen)   Outer layer of the nexine, separated from the tectum by the columellae (see Figure 4.49).

Free-central placentation   Ovules attached to a freestanding column or central axis in the middle of a unilocular ovary (see Figure 4.22).

Frond   Leaf of a fern, usually large and divided or deeply lobed.

Fruit   Mature ovary with associated accessory parts, such as adnate hypanthium of *Malus* (apple, Rosaceae).

Funiculus (pl. funiculi)   Stalk of an ovule; also called the funicle.

Funnelform   Funnel-shaped.

Gametangium (pl.gametangia)   Gamete-producing structure, e.g., archegonium or antheridium.

Gamete   Haploid cell that fuses with another cell to form a zygote.

Gametophyte   Haploid, gamete-producing generation of the plant reproductive cycle.

Gemma (pl. gemmae)   Budlike structure or a cluster of cells that separate from the parent plant and grow, resulting in asexual propagation.

Gene   Stretch of DNA controlling a heritable trait.

Gene flow   Movement of alleles into and out of populations.

Gene trees   Phylogenetic history of a gene, based on its DNA sequence.

Genealogy   *See* Phylogeny.

Genome   All the DNA (genetic material) of the cells of a particular species; in photosynthetic eukaryotes comprised of the nuclear, chloroplast, and mitochondrial genomes.

Genotype   Genetic constitution of an individual.

Genus (pl. genera)   Taxonomic category between the higher rank of family and the lower rank of species; in a phylogenetic classification genera are monophyletic groups of species.

Geographical speciation   *See* Allopatric speciation.

Glabrous   Lacking hairs.

Glandular hair   Hair bearing a knoblike secretory swelling (head) at its apex (see Figure 4.15).

Glaucous   With a waxy covering, and thus often blue or white in appearance.

Globose   More or less spherical.

Glucosinolates   Mustard oil glucosides, hydrolyzed by myrosinases to yield pungent, hot-tasting mustard oils.

Glumes   Pair of bracts at the base of a grass spikelet.

Glycosides   Organic compounds that yield a sugar when hydrolyzed.

Grain   *See* Caryopsis.

Gynobasic style   Style that appears to be inserted at the base of the ovary.

Gynoecium   Collective term for all of the carpels of a flower (see Figure 4.16).

Gynophore   Stalk bearing a gynoecium.

Habit   General appearance of a plant, e.g., whether a tree, shrub, vine, herb, etc.

Hair   Epidermal outgrowth of diverse form, structure, and function (see Figure 4.15); often called a trichome.

Haploid   With one full set made up of one member each of the pairs of chromosomes in the nucleus.

Haustorium (pl. haustoria)   Specialized root of parasitic plants that penetrates another plant and absorbs water and nutrients.

Head   Compact determinate or indeterminate inflorescence with a very short, often disklike axis and usually sessile flowers (see Figures 4.29, 4.30).

Helicoid cyme   Coiled cyme in which the lateral branches all develop from the same side of the axis (see Figure 4.29).

Helobial endosperm   Endosperm development in which the primary endosperm nucleus divides, forming a transverse wall, creating a small chalazal cell and a much larger micropylar cell, followed by a sequence of free nuclear divisions in the micropylar and sometimes also the chalazal cell.

Hemiparasite   Parasitic plant that is green and able to produce at least some of its own carbohydrates but that has haustorial connections to other plants through which pass water as well as some carbohydrates.

Hemitropous ovule   Ovule half-inverted, so the funicle is attached near the middle with the micropyle at right angles to it.

Herbaceous   Not woody; the above-ground stem dying down at the end of the growing season.

Herbarium (pl. herbaria)   Place where plant collections are stored, typically as pressed and dried plants mounted to sheets of paper, identified, and provided with locality and habitat data, so they can be further studied.

Herb   Plant without a persistent above-ground woody stem, and either dying (in annuals) or dying back to the ground level (in herbaceous perennials) at the end of the growing season.

Heteromorphic   Of two or more distinct forms.

Heterophylly   Variation in form of the leaves produced during the life of the plant, e.g., sun/shade leaves or juvenile/adult leaves.

Heterosporous   Producing two types of spores; small spores (microspores) that give rise to male gametophytes, and usually large spores (megaspores) that give rise to female gametophytes.

Heterostyly   Flowers in different individuals of the same species having two or three different style lengths, the stamen lengths often varying inversely; an outcrossing mechanism.

Heterozygosity   Presence of two or more alleles (molecular forms of a particular gene) within an individual.

Hexaploid   With six full sets of chromosomes in a cell.

Hilum   Scar on a seed that indicates the point of attachment of the funiculus, the stalk of the ovule.

Hirsute   Having long, often stiff hairs.

Hispid   Having stiff or rough hairs; bristly.

Holoparasite   Plant that lacks chlorophyll and obtains all of its water and nutrients from other individuals (hosts) through haustorial connections.

Holotype   Sole specimen of a species or infraspecific taxon used as the type by the author of a name; or the one specimen designated by the author as the type or holotype. The name-bearing specimen.

Homology   Similarity due to inheritance of a feature from a common ancestor.

Homonym   Name identical in spelling to the name of another plant.

Homoplasy   Similarity due to parallelism or reversal of character states.

Homosporous   Producing only one type of spore, all the same size and producing bisexual gametophytes.

Homozygosity   Presence of only one allele (molecular form of a particular gene) within an individual.

Hybrid   Offspring of two organisms belonging to different species.

Hybrid speciation Production of a new species from two or more parental species via hybridization.

Hybrid swarm An array of individuals resulting from reproduction among F_1 hybrids and between hybrids and either parental species.

Hybridization Process of hybrid formation.

Hygroscopic tissues Tissues that swell, shrink, or change shape because of changes in moisture content.

Hypanthium Flat, cuplike, or tubular structure on which the sepals, petals, and stamens are borne (see Figure 4.21); usually formed from the fused bases of the perianth parts and stamens or from a modified receptacle; also called a floral cup or floral tube.

Hypocotyl Stem of an embryo below the cotyledon (or cotyledons).

Hypodermis (pl. hypodermes) Differentiated layer or layers of cells beneath the epidermis.

Hypogynous With perianth parts and stamens arising from below the ovary, and the ovary thus superior (see Figure 4.21).

Identification Determination of the name of an unknown plant.

Imbricate Overlapping, as of leaves or petals in bud, i.e., arranged like tiles or shingles on a roof, or scales of a fish.

Imperfect flower Flower lacking either an androecium (stamens) or a gynoecium (carpels); also called a unisexual flower.

Included Structure enclosed by others, as the stamens enclosed or hidden by the corolla, or scattered phloem strands surrounded by secondary xylem.

Incomplete flower Flower lacking one or more of the floral parts, i.e., missing either the perianth, androecium, and/or gynoecium.

Indehiscent Not opening.

Indehiscent pod Fairly dry fruit with few to many seeds that fails to open.

Indented key Key with each pair of contrasting choices (leads) that are equally indented and followed by the appropriate couplets, and thus the two leads are often widely separated in the key.

Independent assortment An outcome of random alignments of chromosomes at metaphase I of meiosis; each homologous chromosome and its partner are assorted into the different gametes independently of each other.

Indeterminate inflorescence Inflorescence in which the main axis produces only lateral flowers, branches, or groups of flowers, so that the lowermost or outermost flowers usually open first, with the main axis (and lateral axes, if present) often elongating as the flowers develop, without the production of terminal flowers.

Indeterminate thyrse Inflorescence with an indeterminate main axis and cymose lateral branches (see Figure 4.30).

Indumentum General term for the hairs (trichomes) on the surface of a plant.

Indusium (pl. indusia) Protective flap or outgrowth covering immature sori in ferns that develops from the abaxial surface of the leaf (or frond).

Inferior ovary Ovary that is positioned beneath the point of attachment of the other floral parts, which appear, therefore, to arise from its apex (see Figure 4.21).

Infructescence Mature inflorescence, with flowers replaced by fruits.

Insertion of floral parts Pattern of attachment of floral parts to the receptacle.

Intectate Exine of pollen grain lacking a tectum (outermost layer, separated from inner layers by small columns) (see Figure 4.49).

Integument Protective outer layer (or layers) surrounding the sporangium of an ovule, which will develop into the seed coat (see Figure 4.40).

Internal phloem Primary phloem in the form of strands or a continuous ring (as seen in cross-section) at the inner boundary of the xylem.

Internode Stem between two adjacent nodes (see Figure 4.2).

Interpetiolar stipules Stipules positioned at the node between the petiole bases of the opposing leaves.

Intine Inner layer of the two-layered pollen wall (see Figure 4.49).

Intrastaminal nectar disk Nectar disk positioned between the androecium (staminal whorl or whorls) and gynoecium.

Introgression Permanent incorporation of genes and/or organelles of one species into another through extensive hybridization and back-crossing.

Intron Non-coding portion of a gene transcript, excised before translation into protein.

Introrse anther Anther that opens inwards, toward the floral axis.

Intruded placenta (pl. placentae) Placenta that extends into the locule of the ovary (see Figure 4.22).

Involucre Series of bracts surrounding a flower or inflorescence.

Inulin Oligosaccharide containing about 30 fructose (and possibly also one or a few glucose) units.

Involute With margins rolled inward toward the adaxial surface (see Figure 4.14).

Iridoids Nine- or ten-carbon derivatives of a simple terpene unit.

Isolating mechanism Mechanism that prevents gene flow between closely related species.

Isotype Duplicate specimen of the holotype, being a part of the same individual (or gathering) that forms the holotype.

Karyotype Preparation of metaphase chromosomes sorted by length, centromere location, and other features.

Keel petals Two abaxial petals of the flowers of Fabaceae subfam. Faboideae; sometimes also used of similar-shaped petals of other families, e.g., Polygalaceae.

Key Series of contrasting choices used to identify an unknown organism by the process of elimination.

Kingdom Highest (most inclusive) taxonomic category in the Linnaean hierarchial system.

Kranz anatomy Foliar veins surrounded by conspicuously enlarged bundle sheath cells containing chloroplasts; associated with C4 photosynthesis.

Labellum "Lip" of an orchid perianth; the adaxial member of the inner whorl of perianth parts (but due to rotation of the flower, at anthesis usually placed as the lowermost perianth part), and differentiated from the other perianth parts in size, form, and/or coloration; or two (or four) large, petaloid, and connate staminodes forming a lip-like structure in gingers and relatives.

Lanate Woolly, with long intertwined, somewhat matted hairs.

Lanceolate   Narrowly ovate to narrowly elliptic; the term is ambiguous and so use is not recommended.

Laticifer   Cell or series of tubular cells, branched or unbranched, containing latex.

Latrorse anther   Anther that opens on the sides.

Leaf (pl. leaves)   Usually flat, determinate, photosynthetic part of a plant, borne on a branch or stem (axis) (see Figure 4.2), or more generally, in flowering plants, an appendicular structure of determinate growth usually subtending a bud or branch.

Leaf axil   Space in the angle between a leaf and the stem that bears it.

Leaf gap   Parenchymatous interruption left in the secondary vascular system of angiosperms by the departure of a vascular bundle(s) (or traces) to the leaves, hence, a trilacunar leaf gap, with three gaps (lacunae) left by the departure of three bundles to the leaf.

Leaflet   One of the blades of a compound leaf.

Leaf scar   Remains of a point attachment of a leaf.

Legume   Dry, more or less elongated fruit derived from a single carpel that opens, often explosively, along two longitudinal sutures; the most common fruit type of members of Fabaceae.

Lemma   Outermost or lower of the two bracts that surround the grass flower (floret).

Lenticel   Wartlike protuberance on the stem surface involved in gas exchange.

Leptosporangium   Stalked sporangium with a wall of only one cell layer (and in most ferns with a cluster or row of specialized hygroscopic cells, an annulus).

Liana   Woody climbing plant.

Ligulate flower   Flower with a straplike, usually five-toothed, corolla limb characteristic of many Asteraceae.

Ligule   Adaxial projection from the top of the leaf sheath, as in gingers (Zingiberaceae) or grasses (Poaceae).

Lineage   Ancestor-descendent sequence of populations.

Linear   Long and very narrow, i.e., very narrowly oblong.

Lithocyst   Enlarged cell containing a cystolith (a large, crystal-like concretion).

Lobate base   Base with two well developed rounded lobes (see Figure 4.11).

Lobed   Having large rounded projections along margin (see Figure 4.12).

Local speciation   Divergence of small, geographically isolated populations at the periphery of the range of the parental species, usually resulting from a combination of natural selection and genetic drift, leading to formation of a new species.

Locule   Compartment, cavity, or chamber within an ovary or anther.

Loculicidal capsule   Capsule that splits longitudinally between the septa (or placentae, if placentation is parietal) and into the locules (chambers) of the ovary.

Lodicules   Reduced perianth parts of members of Poaceae.

Loment   Dry, schizocarpic fruit derived from a single carpel that breaks transversely into one-seeded segments.

Long branch attraction   Lineages that evolve rapidly may contain large numbers of nucleotide similarities due to parallel evolution and may, therefore, group together in a cladistic analysis even though not each other's closest relatives.

Long shoot   Stem with long internodes; this term is applied only in plants in which internode length is clearly bimodal and both long and short shoots are present.

Mangrove   Tropical tree that grows in tidally flooded coastal banks or estuaries, showing various adaptations such as salt tolerance or secretion, vivipary, or the development of prop-roots; or the plant community occurring in such regions.

Marcescent   Withering, but persisting on the plant.

Marine   Of or found in the sea (but not fresh water).

Medullary bundle   Vascular bundle occurring in the pith of a stem or within a cylinder of vascular tissue or ring of vascular bundles within a petiole.

Megagametophyte   Female gametophyte, e.g., the embryo sac, of angiosperms.

Megaphyll   Large leaf, thought to have evolved by planation and webbing of a lateral branch system, characteristic of the ferns and seed plants, and forming leaf gaps in the central vascular system (or stele).

Megasporangium (pl. megasporangia)   Multicellular, spore-producing structure (sporangium) that bears megaspores.

Megaspore   Spore that germinates to produce a female gametophyte (megagametophyte).

Megasporocyte   Cell that undergoes meiosis and gives rise to megaspores.

Meiosis   Two-stage nuclear division process that reduces the chromosome number of a cell by half, to the haploid number, with each daughter nucleus receiving one of each type of chromosome, and followed by the production of spores.

Membranous   Thin, soft, and flexible, like a membrane; also called membranaceous.

Mericarp   One or few-seeded segment of a schizocarp.

-merous   The numerical plan of a flower, as in a *Lilium* flower, which is 3-merous, or a *Geranium* flower, which is 5-merous.

Microgametophyte   Male gametophyte.

Microhair   Minute, two-celled hair found on the leaves of most Poaceae.

Micromorphology   Study of the minute structures and surface features best seen with a scanning electron microscope.

Microphyll   Small leaf with a single vein, and not associated with a leaf gap in the central vascular system (stele); characteristic of the Lycopodiophytes.

Micropyle   Passage through the integument or integuments of an ovule through which the pollen grain or pollen tube usually enters (see Figure 4.40).

Microspecies   Minimally differentiated series of populations derived from uniparental reproduction (selfing or asexual means).

Microsporangium (pl. microsporangia)   Sporangium that produces microspores.

Microspore   Spore that germinates to produce a male gametophyte.

Microsporocyte   Cell that undergoes meiosis and gives rise to microspores or pollen.

Midvein   Central vein of a leaf or other organ (part), in a leaf often called the midrib.

Minimum spanning tree   Shortest path in multivariate space connecting an array of taxa (objects).

Mitochondrial genome   Genes contained on the circular DNA strand of the mitochondrion, an endosymbiont of the host eukaryotic cell.

Mitosis   Nuclear division that maintains the parental chromosome number for daughter cells, and the basis for growth in size and asexual reproduction in plants.

Monad   Generally, a single individual, free from others; specifically, a single pollen grain.

Monadelphous stamens   Stamens fused by their filaments in a single group, usually forming a tube.

Monocolpate   Pollen grain with a single, long, grooved aperture (see Figure 4.47).

Monoecious   Staminate and carpellate flowers separate, but borne on a single plant.

Monograph   Comprehensive systematic study of a particular taxonomic group, including keys, descriptions, illustrations, distribution maps, etc.

Monophyletic group   Group composed of an ancestor and all of its descendants; diagnosed by synapomorphies (shared derived characters), a clade.

Monopodial shoot   Shoot formed through the action of a single apical meristem.

Monoporate   Pollen grain with a single porelike aperture (see Figure 4.47).

Monosulcate   Pollen grain with a single, long and grooved aperture located at the pole (see Figure 4.47).

Monotelic   *See* Determinate inflorescence.

Morphology   Study of the form and structure (especially external) of living organisms.

Mucilage   Slimy or sticky fluids, often polysaccharides.

Mucronate apex   Apex terminated by a distinct, short and abrupt point or mucro (see Figure 4.10).

Multicellular hair   Hair composed of several cells.

Multilacunar node   Node with five or more leaf gaps (see Figure 4.34)

Multiple fruit   Fruit produced by the gynoecia of several closely clustered flowers.

Multiseriate hair   Hair composed of several rows of cells.

Multistate character   Character with three or more states.

Multivalent   Grouping (synapsis) of three or more chromosomes in a cell during meiosis.

Mustard oils   Hot, pungent compounds resulting from the hydrolyzation of glucosinolates in the mustards and their relatives (Brassicales).

Mutation   A heritable change in the molecular structure of DNA.

Mycorrhiza   Symbiotic association of various fungi with roots of vascular plants.

Myrosin cell   Cell containing the enzyme myrosinase, which hydrolyzes glucosinolates to produce mustard oils.

Naked bud   Bud not covered by scale leaves.

Native species   Species that occur naturally in an area.

Natural   Of classification or relationships, agreeing with a particular author's idea of what is natural; a useless, if much used, term.

Naturalized species   Species brought from another region by human action that maintain themselves in their new region, reproducing and spreading, without human intervention.

Nectar disk   Disklike or ringlike floral nectary.

Nectar spur   Hollow, slender, saclike projection of a floral part, usually a petal or sepal.

Nectary   Nectar-producing gland, often forming projections, lobes, or disklike structures.

Neospecies   Recently evolved species.

Nesting   Property of hierarchies, whereby smaller, less inclusive, groups are subsets, that is, are completely included within, larger, more inclusive, groups.

Nexine   Inner portion of the exine of pollen, composed of the endexine and foot layer (see Figure 4.49).

Node   Region of the stem where the leaf and bud are borne (see Figure 4.2).

Nominate   Infraspecific taxon that contains the type of a species; often called the "typical" infraspecific taxon.

Nonglandular hair   Hair that does not produce a secretion.

Nucellus   In angiosperms, another name for the sporangium wall within the ovule.

Nuclear endosperm   Successive divisions of the primary endosperm nucleus unaccompanied by cell wall formation (free nuclear divisions), leading to a multinucleate mass of protoplasm; sometimes later converted into cellular tissue.

Nuclear genome   DNA in the chromosomes within the nucleus of a eukaryotic cell.

Nucleotide   Small organic compound with a five-carbon sugar, nitrogen-containing base, and phosphate group; the structural unit of nucleic acids (DNA and RNA).

Nut   Fairly large, indehiscent, dry fruit with a thick bony wall surrounding a single seed.

Oblique   Having unequal sides or an asymmetrical base.

Oblong   With the sides nearly or quite parallel for most of their length (see Figure 4.9).

Obovate   Widest near the apex, i.e., the terminal half broader than the basal (see Figure 4.9).

Obtuse apex   Apex blunt, having an angle greater than 90 degrees, with straight to slightly convex sides (see Figure 4.10).

Obtuse base   Base blunt, having an angle greater than 90 degrees, with slightly convex sides (see Figure 4.11).

Ocrea (pl. ocreas, ocreae)   Nodal sheath or tube formed by the stipule(s), e.g., characteristic of most members of Polygonaceae.

Odd-pinnate   Pinnately compound leaf with an odd number of leaflets, i.e., a terminal leaflet is present (see Figure 4.5).

Oil gland   Gland producing oils, i.e., long-chain hydrocarbons.

Operational Taxonomic Unit (OTU)   Terminal taxon used in an analysis of relationships, especially in phenetic studies.

Operculate   Opening by a lid or cap.

Opposite   Pair of leaves or other structures borne along a stem, the members of which are positioned on opposing sides of the stem (see Figure 4.4).

Order   Taxonomic category between the higher rank of class and the lower rank of family; in a phylogenetic classification, a monophyletic group of families.

Ordering   Determining the sequence in which three or more states of a character originated.

Orthotropic   Growing erect.

Orthotropous ovule   Straight ovule with the micropyle at the apex and the funiculus at the base (see Figure 4.40).

OTU   *See* Operational Taxonomic Unit.

Ovary   Ovule-bearing part of a carpel (or several fused carpels).

Ovate   Widest near the base; i.e., with the broader end nearer the point of attachment (see Figure 4.9).

Ovule   Structure in seed plants comprised of the female gametophyte, the sporangium, one or two, rarely three, integuments, and a funiculus (stalk); after fertilization, it develops into the seed (See Figures 4.16, 4.17, 4.40).

Palea   Innermost or upper bract, of the two bracts surrounding the grass flower (floret); possibly a bracteole/prophyll.

Palmate venation   Three or more primary veins (or well-developed secondary veins) arising from at or near the base of the blade (see Figure 4.8).

Palmately compound leaf   Leaf with more than three leaflets attached to a common point, like the fingers of a hand (see Figure 4.5).

Palynology   Study of the form and structure of pollen and spores.

Panicle   Indeterminate inflorescence with two or more orders of branching, each axis bearing flowers or higher order axes (see Figure 4.30).

Panicle-like cyme   Determinate inflorescence in which the lateral branches produce several internodes before ending in a terminal flower (see Figure 4.29).

Pantoporate   Pollen grain with numerous porelike apertures.

Papillate   Covered with short, rounded, nipple-like bumps or projections (papillae).

Pappus   Highly modified calyx consisting of awns, scales, hairs, or fine bristles, as in members of Asteraceae.

Paracytic   Stoma surrounded by two subsidiary cells parallel to the guard cells (see Figure 4.36).

Parallel venation   Several to many parallel veins running the length of the blade.

Parallelism   Separate origination of the same character state in two (or more) organisms.

Paraphyletic group   Group containing a common ancestor and some, but not all, of its descendents; diagnosed by symplesiomorphies.

Paratetracytic   Stoma surrounded by four subsidiary cells, two of them being parallel to the guard cells and the other pair being polar and often smaller.

Parietal placentation   Ovules attached to the wall of the ovary (see Figure 4.22).

Parsimony   Acceptance of the simplest hypothesis (given the premises/assumptions) that explains the data.

Parthenogenesis   Development of an embryo from an unfertilized egg.

PCR   Polymerase chain reaction, a laboratory technique in which many copies of a DNA sequence can be replicated enzymatically.

Pedicel   Stalk of a single flower in an inflorescence (see Figure 4.16).

Peduncle   Stalk of a solitary, terminal flower or of an inflorescence (see Figure 4.28).

Pellucid dots   Translucent dots on the surface of a leaf, bract, petal, or other structure usually resulting from cavities or ethereal oil cells in leaf mesophyll or other parenchymatous tissues.

Peltate   Flat structures attached by their surfaces rather than the base or margin, e.g., like an umbrella.

Peltate scale   Shield-shaped or umbrella-like hair.

Pendulous   Hanging or drooping downward.

Peptide bonds   Bonds linking the amino acids in a protein.

Percurrent tertiary veins   Tertiary veins linking the secondaries, forming a ladderlike pattern (see Figure 4.6); often called scalariform.

Perennial   Plant that lives for three or more years and usually flowers and fruits repeatedly.

Perfect flower   Flower having both androecium (stamens) and gynoecium (carpels). Also known as **bisexual flower**.

Perforation plate   Area of the cell wall connecting two vessel elements.

Perianth   Collective term for calyx and corolla, or all tepals (when calyx and corolla are not distinguished), of a flower (see Figure 4.16).

Pericarp   Mature ovary wall of the fruit.

Pericytic   Stomate surrounded by a single subsidiary cell (see Figure 4.36).

Perigynium   Urn-shaped or saclike prophyll surrounding the carpellate flower of some Cyperaceae, e.g., *Carex*.

Perigynous   With perianth parts and stamens borne on an hypanthium that surrounds, but is not fused to, the superior ovary (see Figure 4.21).

Perisperm   Diploid nutritive tissue in the seeds of some angiosperms derived from the sporangium wall (nucellus).

Petal   Member of the inner perianth whorl, usually colorful and assisting in attracting pollinators (see Figure 4.16).

Petaloid   Functioning like a petal, whether sepal, tepal, staminode, or pistil (lode).

Petiole   Stalk of a leaf (see Figure 4.3).

Petiolule   Stalk of a leaflet.

Phenetics   Method of classifying organisms based on overall similarity, i.e., the sum of similarities and differences, with uniform rates of change assumed.

Phenogram   Treelike diagram representing phenetic relationships, i.e., with taxa clustered on the basis of overall similarity.

Phenotype   Observable trait or traits of an individual.

Phloem   Sucrose-conducting tissue of vascular plants, composed of sieve elements or cells, parenchyma cells, fibers, and sclereids.

Photosynthesis   Biochemical reactions occurring within chloroplasts that trap energy from the sun and convert it to chemical energy, followed by fixation of carbon dioxide and synthesis of glucose phosphates that become converted to sucrose.

Phyletic trend   Gradual change in a character on an evolutionary tree (cladogram).

Phyllaries   Involucral bracts of the heads of members of the Asteraceae.

Phylloclade   Flattened, leaflike stem.

Phylogenetic classification   Classification accurately reflecting hypothesized phylogenetic relationships among a group of taxa.

Phylogenetic tree   *See* Evolutionary tree.

Phylogeny   Evolutionary history of a group of organisms.

Phylum   Taxonomic category between the higher rank of kingdom and the lower rank of class; in plants usually called division.

Phytomelan   Opaque, black, carbonaceous substance that forms a crust on the testa of most Asparagales.

Pilose   Having scattered, long, slender, soft hairs.

Pin   Long-styled floral form in a heterostylous species.

Pinnate venation   Where secondary veins arise along the length of the single primary vein, like the teeth of a comb or the lateral units of a feather (see Figure 4.7).

Pinnately compound leaf   Compound leaf with leaflets more than three and attached along two sides of an axis (or rachis), i.e., featherlike (see Figure 4.5).

Pistil   Ovule-bearing part of a flower, formed from one or more carpels (see Figure 4.19).

Pistillode   Sterile pistil.

Pit   Central, hard portion of a drupe, composed of a seed or seeds surrounded by a bony endocarp (inner wall of the fruit); also called a stone or pyrene; also areas of the cell wall of vessels and tracheids through which water passes.

Pith   Soft tissue in the center of a stem, usually consisting of more or less isodiametric cells.

Placenta (pl. placentae)   Place or part of the ovary to which the ovules are attached.

Placentation   Arrangement of ovules within the ovary.

Plagiotropic   Growing horizontally.

Plane   Flat, as used of leaf margins (see Figure 4.12).

Plant   Any member of the green plant clade, i.e., an organism with a cellulose cell wall, chlorophyll *a* and *b*, chloroplast surrounded by two membranes, starch as carbohydrate storage product, and often having cells with two anterior, whiplash flagella; other photosynthetic eukaryotes are not plants.

Plant dryer   Box- or table-like device used to suspend a plant press over a source of heat.

Plant press   Device consisting of press frames, straps, corrugates, and blotting paper; used to press and dry a plant specimen.

Plasmodesmata   Pore and associated strand of cytoplasm that crosses the primary cell wall and connects two adjacent cells.

Plesiomorphy   Ancestral character state; also called a plesiomorphic character state.

Pleurogram   Fine, U-shaped or more or less circular line or groove in the seed coat of some Fabaceae.

Plicate   Folded, as in a fan (see Figure 4.14).

Plunger pollination   Secondary pollen presentation mechanism in which the modified style picks up and presents to the pollinators pollen from introrsely dehiscent anthers, as in flowers of members of Campanulaceae, Asteraceae, and a few other families.

Pneumatophore   Specialized root involved in gas exchange in some mangrove or swamp species.

Point mutation   Change of a single nucleotide in a strand of DNA.

Polar nuclei   Two nuclei located in the middle of the female gametophyte (embryo sac) of angiosperms, which fuse with a sperm nucleus to form the primary endosperm nucleus, and eventually the endosperm.

Pollen   Microspore containing a male gametophyte; in angiosperms germinating to form a pollen tube, which rapidly transports the sperm to the ovule (and egg).

Pollen sac   Chamber in the anther (of the stamen) that contains the pollen.

Pollen tube   Tube formed by the germinating pollen grain that carries the sperm to the ovule.

Pollination   Transference of pollen from the anther to the stigma of the same individual (selfing) or a different individual (outcrossing).

Pollination syndromes   Floral characteristics associated with pollination by various biotic or abiotic means, e.g., bird-pollination, bee-pollination, wind-pollination, etc.

Pollinium (pl. pollinia)   Mass of pollen grains transported as a unit, as in many Orchidaceae and Apocynaceae.

Polocytic   Stoma with guard cells partially surrounded by a single subsidiary cell (see Figure 4.36).

Polyacetylenes   Non-nitrogenous compounds formed from the linking of acetate units via fatty acids.

Polyad   Small cluster of pollen grains.

Polyclave   Multi-access key, as in a punched card key or computer interactive identification.

Polycolpate   Pollen grain with several long, grooved apertures (see Figure 4.47).

Polycolporate   Pollen grain with several long, grooved apertures, each with a central pore (see Figure 4.47).

Polygamous   With both bisexual and unisexual flowers on the same plant.

Polymerase chain reaction   *See* PCR.

Polypeptide   Polymer of amino acids.

Polyphyletic group   Group with two or more ancestors, but not including the true common ancestor of its members.

Polyploidy   With three or more complete sets of chromosomes in a cell.

Polyporate   Pollen grain with many porelike apertures.

Polysaccharide   Polymer of sugars.

Polytelic   *See* Indeterminate inflorescence.

Pome   Indehiscent, fleshy fruit in which the outer part is soft and the center contains papery to cartilaginous structures enclosing the seeds; characteristic fruit of apples, pears, quinces, and most other members of Rosaceae subfam. Spiraeoideae, tribe Pyreae.

Population   Individuals of a given locality that form a single interbreeding community.

Porate aperture   Round and porelike aperture.

Pore   Small opening, usually circular or elliptic.

Poricidal dehiscence   Opening by means of a pore(s) or flap(s), as in a poricidal capsule.

Prickle   Sharp-pointed hair, involving epidermal tissue, or emergence, involving epidermal and subepidermal tissues but not vascularized.

Primary growth   Growth of the plant body from activity of the stem and root apical meristems, or a primary thickening meristem (as in palms). Contrast with Secondary growth.

Primary metabolite   Molecules forming an essential part of the metabolic pathways of an organism.

Primary protein structure   Sequence of amino acids in a protein.

Primary vascular tissue   Xylem and phloem of the primary plant body, produced from and differentiated immediately behind the apical meristem.

Primary vein   Midvein of an appendicular structure (leaf, bract, petal, etc.) (venation pinnate), or one of several equally prominent veins arising from the base of an appendicular structure (venation palmate).

Primary xylem   Xylem of the primary plant body, produced from and differentiated immediately behind the apical meristem.

Primordium   Organ (plant part) or organized series of cells in their earliest stage of development.

Proleptic shoot   Axillary shoot that develops from a bud that shows a period of dormancy, often with bud scale scars at the base.

Prop roots   Adventitious roots usually arising from the basal portion of a trunk and assisting in support of the plant.

Prophyll   Basal leaf or leaves of axillary shoots, commonly two and lateral, but one and often two-keeled in monocots.

Protandry   Maturation of anthers, and shedding of pollen before stigmas become receptive (in a bisexual flower or monoecious individual).

Protein   Any of numerous organic molecules composed of amino acids linked together in chains.

Protogyny   Maturation of stigmas before pollen is shed (in a bisexual flower or monoecious individual).

Pseudobulb   Thickened, cormlike internode (or series of internodes) on the stem of many epiphytic orchids (Orchidaceae).

Pseudogene   A nonfunctional gene.

Pseudomonomerous   Falsely appearing to be comprised of a single carpel, due to extreme fusion and reduction of one or more carpels.

Pseudoterminal bud Axillary bud that has taken over the function of a terminal bud, in a sympodial shoot.

P-type plastids Plastids in sieve tube elements that accumulate proteins (or proteins and starch) in the form of crystalloids or filaments.

Ptyxis Way in which an individual leaf or perianth part is folded in bud.

Puberulent Having minute, short hairs.

Pubescent Covered with short, soft hairs; or more generally, as in this text, having hairs of any type.

Pulvinus (pl. pulvini) Swollen portion of the petiole (or petiolule) involved in movement, usually positioned at the petiole base, but sometimes at the apex (see Figure 4.3).

Pyrene *See* Pit.

Pyxis Circumscissile capsule.

Quincuncial aestivation Perianth parts overlapping as follows: two with both margins on the outside in bud, two with both margins on the inside, and one with a margin on the inside and the other on the outside.

Race Geographically defined population or aggregate of populations that differ morphologically from other races of the species; often treated at the taxonomic rank of subspecies or variety.

Raceme Simple, indeterminate inflorescence with a single axis bearing pedicellate flowers (see Figure 4.30).

Raceme-like cyme Reduced cyme with a single axis terminated by a flower and bearing pedicellate flowers (see Figure 4.29).

Rachis Main axis of a structure, such as a leaf or inflorescence.

Radial symmetry Divisible into equal halves by two or more planes of symmetry (see Figure 4.18); also called actinomorphic.

Radicle Embryonic root.

Rank Level in a taxonomic hierarchy of a category and of the taxa of that category.

Ranking Placement of a taxon in a category of the taxonomic hierarchy.

RAPD Random amplified polymorphic DNA.

Raphe Portion of the funiculus of an ovule that is fused to the integument; usually represented by a ridge.

Raphides Needle-shaped crystals occurring within a closely packed bundle.

Ray flower Bilaterally symmetrical, straplike, apically often three-lobed, carpellate or sterile flower occurring around the periphery of the head of many members of Asteraceae.

Ray *See* Wood ray.

Receptacle Floral axis that bears the flower parts (see Figure 4.16).

Recurved Curved backward.

Reduplicate Folded downward with the abaxial surface within (see Figure 4.14).

Reflexed Bent sharply backward or downward.

Regular symmetry *See* Radial symmetry.

Remane's criterion (pl. criteria) One of the criteria used in the determination of similarity (loosely called homology) between structures on different organisms.

Replum Persistent, thickened rim (modified placenta) of the fruit that bears the seeds in many Brassicaceae and Papaveraceae.

Reproductive isolating mechanisms Any heritable feature that prevents interbreeding between one or more genetically divergent populations.

Resins Sticky aromatic hydrocarbons that harden when oxidized.

Restriction enzyme One of a series of enzymes, derived from various bacteria, which cut DNA at a particular nucleotide sequence.

Resupinate Twisted 180 degrees; turned up-side down.

Retention index Measure applicable to either a single character or an entire set of characters on a cladogram, which reflects for a given tree topology how much of the variation displayed by the character reflects true synapomorphy; thus, a measure of phylogenetic informativeness of a character.

Reticulate tertiary veins Tertiary veins forming a netlike pattern (see Figure 4.6).

Retinaculum (pl. retinacula) Prominently thickened funiculus associated with a seed of most Acanthaceae, which functions in shooting out the seeds during capsule dehiscence.

Retuse apex Slightly notched apex (see Figure 4.10).

Reversal Derived character state changing back to the ancestral state.

Revolute Rolled toward the abaxial side (see Figure 4.12).

Rhizome Horizontal stem, often underground or on the surface of the ground, bearing scalelike leaves; often called a stolon (or runner) if above ground and having elongated internodes.

Ribosome Complex of ribosomal RNA and proteins that synthesizes protein from a messenger RNA template.

Rooting (of evolutionary trees) Determining the position of the root, i.e., converting a network into an evolutionary tree (or cladogram), and thus providing directionality to hypotheses of character state change.

Root Portion of the plant axis lacking nodes and leaves, usually branching irregularly and found below ground.

Rotate Wheel- or disk-shaped.

Rounded Round in outline; used to describe apices and bases (See Figures 4.10, 4.11).

Ruminate endosperm Endosperm with regular to irregular ingrowths of the seed coat.

Saccate Bag-shaped, bladder-like, or pouched.

Saccus (pl. sacci) Bladder or winglike appendage on the pollen grains of some conifers.

Sagittate base Arrowhead-shaped base (see Figure 4.11).

Salverform With a slender tube and abruptly flaring distal portion, or limb.

Samara Winged, indehiscent, dry fruit containing a single seed (or rarely two seeds).

Scabrous Rough.

Scalariform perforation plate A "plate" as in the cell wall of a vessel-element with several to many ladderlike bars with openings in between.

Scalelike leaf Small, flat leaf.

Scape Erect, leafless stem bearing an inflorescence or flower at its apex; usually composed of a single, elongated internode.

Schizocarp Dry to rarely fleshy fruit breaking into one-seeded (or few-seeded) segments (mericarps).

Scientific name Latin name of a taxon, in accordance with the rules outlined in the *International Code of Botanical Nomenclature* (or other nomenclatural codes).

Sclereid   Cell that is usually not strongly elongated in only one direction and that has a thick, usually lignified wall and is dead at maturity.

Scorpioid cyme   Coiled cyme in which the lateral branches (and flowers) develop alternately on opposite sides of the axis (see Figure 4.29).

Scutellum (pl. scutella)   Shield-shaped and specialized cotyledon of the embryo of members of Poaceae, to which the remaining portion of the embryo is laterally attached.

Seco-iridoids   Iridoid compounds that lack a carbocyclic ring.

Secondary growth   Growth resulting from the activity of the vascular cambium, producing xylem and phloem, and of the cork cambium, which produces bark, both of which result in an increase in stem diameter.

Secondary metabolites   Various organic compounds of ecological significance.

Secondary phloem   Phloem produced by the vascular cambium.

Secondary vascular tissue   Xylem and phloem produced by the vascular cambium.

Secondary vein   Vein that branches from a primary vein.

Secondary xylem   Xylem produced by the vascular cambium.

Secretory structures   Cavities, ducts, or other glandular structures that produce a secretion (nectar, oils, resins, volatile compounds, etc.)

Seed   Product of the ovule after fertilization, comprising the embryo, its nutritive tissue, and the seed coat.

Seed coat   Protective outer layer or layers of the seed, developed from the integument or integuments.

Self-compatibility   Ability of an individual's own pollen to effect pollination and self-fertilization.

Self-fertilization   Union of a sperm and egg from the same individual.

Self-incompatibility   Failure of an individual's own pollen to effect self-fertilization, typically due to its failure to germinate on the stigma (sporophytic incompatibility) or failure to grow through the style (gametophytic incompatibility).

Self-pollination   Transference of pollen from the anther to the stigma of the same individual.

Sensu lato (s. l.)   In the broad sense; with a wide or general interpretation.

Sensu stricto (s. s.)   In the strict or narrow sense; with a restricted interpretation.

Sepal   One member of the outer perianth whorl, when floral whorls are differentiated, usually green and protecting the inner flower parts in bud (see Figure 4.16).

Separate floral parts   Parts of the flower not fused.

Septal nectary   Nectar producing tissue located in the septa of an ovary, as in many monocots.

Septifragal capsule   Capsule with longitudinal valves formed by splitting across the ends of the septa (partitions), which remain attached to the center part of the fruit.

Septicidal capsule   Capsule with longitudinal valves formed by splitting down the middle of the septa (partitions between the locules).

Septum (pl. septa)   Partition or cross-wall, as in an ovary.

Sequencing convention   Rule stating that taxa forming a pectinate portion of a cladogram may be placed at the same taxonomic rank and arranged in their order of branching from the base.

Sericeous   Silky, with usually long, thin, appressed hairs.

Serology   Technique that assesses evolutionary relationships using an animal's immunological antibody-antigen reaction when exposed to a variety of foreign, including plant, proteins.

Serrate margin   Margin with saw-teeth, i.e., the teeth point forward (see Figure 4.12).

Sesquiterpene lactone   Type of terpenoid molecule, known primarily from Asteraceae.

Sessile   Lacking a stalk.

Sessile leaf   Leaf lacking a petiole.

Sexine   Outer portion of the exine of a pollen grain.

Shoot   Stem.

Short shoot   Stem with short internodes, other shoots having distinctly longer internodes; compare with Long shoot.

Shrub   Woody plant that is usually shorter than a tree and produces several stems or trunks from the base.

Sibling species   Taxa that are reproductively isolated without much phenotypic divergence; also called cryptic species.

Sieve cell   Long, slender, enucleate, sucrose-conducting cell of the phloem, which does not form a portion of a sieve tube.

Sieve tube element   Long, sucrose-conducting, enucleate cell of the phloem of flowering plants that forms part of the sieve tube.

Silica   Dioxide form of silicon; SiO_2.

Silique   Fruit derived from a two-carpellate gynoecium in which the two halves of the fruit split away from a persistent partition (around the rim of which the seeds are attached); sometimes restricted only to such fruits that are more than twice as long as wide (with shorter fruits then called silicles).

Simple fruit   Fruit developing from a single carpel or several fused carpels of a single flower.

Simple hair   Unbranched hair.

Simple leaf   Leaf with a single blade.

Sister groups   Two taxa that are each other's closest relatives.

S. l.   *See* Sensu lato

Sorus   Cluster of sporangia on the surface of a fern leaf; appearing as variously shaped brown patches on the abaxial surface.

Southern blot   Method of determining lengths of DNA fragments by transferring denatured fragments from an electrophoresis gel to a nylon membrane.

Spadix   Spike with a thickened, fleshy axis, as characteristic of members of the Araceae.

Spathe   Large bract surrounding or subtending an inflorescence, as the often showy inflorescence bract of members of the Araceae.

Speciation   Evolutionary process by which species are formed.

Species (pl. species)   A basic grouping of organisms. (See various definitions pertaining to particular species concepts in Chapter 6, pages 144–146.)

Species tree   A tree (cladogram) illustrating a hypothesized evolutionary history of an array of species.

Specific epithet   The second word of a binomial species name; usually an adjective that modifies the generic name, as *alba* in the species name *Quercus alba*.

Sperm   Motile, male gamete, often flagellated, formed in an antheridium or by a reduced male gametophyte.

Spike   Simple, indeterminate inflorescence with a single axis bearing sessile flowers (see Figure 4.30).

Spikelet   Small spike, as in the basic inflorescence units of members of Poaceae and Cyperaceae.

Spine   Reduced, sharp-pointed leaf or stipule, or sharp-pointed marginal tooth.

Sporangium (pl. sporangia)   Spore-bearing structure.

Spore   Reproductive cell resulting from meiotic cell divisions in a sporangium.

Sporocarp   Specialized reproductive leaves of members of Marsileaceae.

Sporophyll   Modified leaf that bears one or more sporangia.

Sporophyte   Diploid, spore-producing generation of the plant reproductive cycle.

Sporopollenin   Polymer with saturated and unsaturated hydrocarbons and phenolics that are cross-linked to each other; the most resistant biopolymer on Earth.

Spur   Variously shaped outgrowth of staminal filaments or connectives in Melastomataceae, Ericaceae, and a few other taxa; also used for tubular nectar-containing outgrowths of the calyx or corolla, the nectar spur.

S. s.   *See* Sensu stricto

Stamen   Pollen-bearing part of a flower, composed of a filament (stalk) and anther (pollen sacs, sporangia) (see Figure 4.16).

Staminate flower   Flower with an androecium (stamen or stamens) but not a functional gynoecium (carpel or carpels).

Staminode   Sterile stamen.

Standard petal   *See* Banner petal.

Starch   White, tasteless, solid carbohydrate composed of chains of glucose units; usually the carbohydrate storage compound of green plants.

Stellate hair   Hair with branches radiating outward like rays from a star.

Stem   The plant axis bearing leaves with axillary buds at the nodes separated by internodes; usually above ground.

Steroids   Hydrophobic molecules based on four connected carbon rings; often physiologically active.

Stigma   Part of the carpel (or several fused carpels) that receives and facilitates the germination of the pollen (see Figure 4.16).

Stipe   Petiole of a fern leaf; sometimes also used more generally to apply to any stalk, as in stipitate carpels.

Stipule scar   Mark left on the stem after a stipule has fallen off (abscised).

Stipule   One, usually of a pair of appendages located on either side of (or on) the petiole base; part of the leaf.

S-type plastid   Plastid in the sieve tube elements that accumulates starch.

Stolon   Rhizome having an elongated internode; see Rhizome.

Stoma (pl. stomata)   Controllable opening between two guard cells in the epidermis of the vascular plants (and mosses); also called a **stomate**.

Strigose   Having stiff hairs, all pointing in one direction.

Strobilus (pl. strobili)   Conelike cluster of modified, spore-bearing leaves inserted directly on an axis.

Style   More or less elongated part of the carpel (or of several fused carpels) between the stigma(s) and the ovary (see Figure 4.16), specialized for pollen tube growth.

Styloid   Large, elongated crystal with acute or square ends.

Stylopodium (pl. stylopodia)   Enlarged, nectariferous, basal portion of the style of the flowers of Apiaceae.

Subsidiary cell   Epidermal cell lying next to a guard cell and differing in size and/or shape from other epidermal cells.

Subspecies (pl. subspecies)   An infraspecific rank, usually applied to geographical races within a morphologically variable species.

Succulent   Fleshy and juicy.

Suffrutescent plant   Plant with woody base and herbaceous upper portion.

Sulcate   Pollen grain with a sulcus.

Sulcus   Long, grooved aperture positioned at the pole of a pollen grain.

Superior ovary   Ovary that arises above the point of insertion of the other flower parts (see Figure 4.21).

Superposed bud   Bud (or buds) located above or below the axillary bud.

Sylleptic shoot   Axillary shoot that elongates at the same time as the shoot on which it is borne, the branch thus often lacking basal bud-scale scars but with an elongated first internode.

Sympatric   Occurring together in the same geographical region.

Sympatric speciation   Speciation occurring without establishment of a geographical barrier; in plants, often involving hybridization and subsequent polyploidization.

Sympetalous flower   Flower with fused petals.

Symplesiomorphy   Shared ancestral character state.

Sympodial shoot   Shoot formed through activity of a series of axillary meristems.

Synandrous   Having the stamens fused.

Synapomorphy   Shared derived character state.

Syncarpous   Carpels fused, forming a compound pistil.

Syconium (pl. syconia)   Accessory and multiple fruit characteristic of figs (species of *Ficus*, Moraceae).

Syncolpate   Condition resulting from the fusion of the furrows (colpi) of a pollen grain.

Synergids   Two cells next to the egg in the female gametophyte of the angiosperms.

Syngameon   Most inclusive unit of interbreeding in a group of hybridizing species.

Syngenesious stamens   Stamens fused by their anthers.

Synonym   One of two or more names applied to the same taxon.

Synsepalous flower   Flower with fused sepals.

Syntepalous flower   Flower with fused tepals.

Systematics   The science of organismal diversity; frequently used in a sense roughly equivalent to taxonomy.

Systematist   A person trained in the discipline of systematics.

Tannins   Yellow, astringent, bitter phenolic compounds.

Tapetum   Innermost layer of the anther wall, which produces enzymes, hormones, and nutritive materials used during pollen formation.

Taproot   Major root of seed plants; usually enlarged and growing downward, sometimes more or less aborted.

Tautonym   Name of a species in which the specific epithet exactly repeats the generic name; not permissible in the botanical code.

Taxon (pl. taxa)   Group of organisms at any level in the taxonomic hierarchy.

Taxonomic category   Rank of the taxonomic hierarchy, e.g., kingdom, division, class, order, family, genus, and species.

Taxonomic revision   Taxonomic study similar to a monograph, but less comprehensive, and usually including keys, descriptions, illustrations, and distribution maps for a particular taxon.

Taxonomist   Person trained in the discipline of taxonomy.

Taxonomy   Theory and practice of grouping individuals into species, arranging species into larger groups, and giving those groups names, thus producing a classification.

Tectate exine   Exine with the outer layer, or tectum, supported by numerous small columns (columellae) (see Figure 4.49).

Tegmen (pl. tegmenta)   Inner part of the seed coat, developing from the inner integument.

Tendril   Elongated and twining structure (modified from an inflorescence, leaf, or stem) assisting in climbing.

Tepal   One of a series of perianth parts that are not differentiated into calyx and corolla.

Terminal   Occurring at the tip or apex of a structure.

Terminal bud   Bud at the apex of a stem (in a monopodial shoot).

Terminal style   Style arising from the apex of an ovary.

Terpenoids   Structurally diverse compounds formed by the union of five-carbon isopentenoid pyrophosphate units formed in the mevalonic acid pathway (see pages 97–98.)

Tertiary vein   Vein that branches from a secondary vein, typically forming either a net-like (reticulate) or a ladder-like (percurrent, scalariform) pattern in a leaf.

Testa (pl. testae)   Outer part of the seed coat, developing from the outer integument.

Tetracytic   Stoma surrounded by four subsidiary cells.

Tetrad   Group of four; usually applied to pollen grains united in a group of four that have not separated after meiosis.

Tetradynamous stamens   With four long and two short stamens.

Tetraploid   With four complete sets of chromosomes in a cell.

Thorn   Reduced, sharp-pointed stem.

Three-ranked leaves   Leaves borne in three planes along the stem.

Thrum   Short-styled floral form in a heterostylous species.

Tomentose   Having densely matted soft hairs.

Topology   Pattern of branching of an evolutionary tree or cladogram.

Tracheid   Imperforate, elongate, thick-walled, water-conducting cell with bordered pits, occurring in xylem and dead at maturity.

Tree   A woody plant with a single main trunk (or bole). (Also used as an abbreviation for "evolutionary tree," or cladogram.)

Trichome   *See* Hair.

Tricolpate   Pollen grain with three, long, grooved apertures (see Figure 4.47).

Tricolporate   Pollen grain with three, long, grooved apertures, each with a central pore (see Figure 4.47).

Trifoliolate leaf   Compound leaf with three leaflets; often called trifoliate.

Trigger flower   Flower with one or more moveable parts that cause the pollen to be forcibly deposited on the pollinator.

Trilacunar node   Node with three leaf gaps (see Figure 4.34).

Triploid   With three full sets of chromosomes in a cell.

Triporate   Pollen grain with three equatorial, porelike apertures (see Figure 4.48).

Tristichous   *See* Three-ranked.

Tristyly   Flowers of different individuals of the same species having three different style lengths, the stamen lengths often varying inversely; an outcrossing mechanism.

Triterpenes   Thirty-carbon terpenoid compounds. (See page 98.)

Truncate   Apex or base appearing as if cut off at the end, nearly straight across (see Figures 4.10, 4.11).

Trunk   Main stem of a tree below the branches.

T-shaped hairs   Branched hair in the shape of the letter T (see Figure 4.15).

Tuber   Swollen, fleshy portion of a rhizome or root involved in water and/or carbohydrate storage.

Tubule   Little tube, as on the anthers of species of *Vaccinium* (blueberries) and relatives (in the Ericaceae).

Twice-pinnately compound leaf   Leaf with two orders of pinnate axes and leaflets borne on second order axes (see Figure 4.5).

Twining   Spiraling around a support in order to climb.

Two-ranked leaves   Leaves borne along just two sides of a stem, i.e., all leaves in the same plane.

Type specimen   Name-bearing specimen of a species or infraspecific taxon.

Ultrastructure   Structures best observed using a transmission electron microscope.

Umbel   Determinate or indeterminate inflorescence in which all flowers have pedicels of equal or unequal length that arise from a single region at the apex of the inflorescence axis (see Figures 4.29, 4.30).

Unicellular hair   Hair composed of only a single cell (see Figure 4.15).

Unifacial leaf   Leaf that has both surfaces alike, i.e., not anatomically differentiated into an adaxial and abaxial surface.

Unifoliolate leaf   Compound leaf that through evolutionary reduction has only a single leaflet; typically distinguished from a simple leaf by the presence of a distinct joint or a pulvinus at the blade-petiole junction, e.g., *Citrus*, many *Berberis*.

Unilacunar node   Node with a single leaf gap (see Figure 4.34).

Uninformative character   Character that is not useful in constructing hypotheses of phylogenetic relationship.

Uninomial   Name consisting of one word, such as the name of a genus (e.g., *Erica* or *Acer*) or a family (Ericaceae or Sapindaceae).

Uniseriate   With a single series or whorl of structures, as a single whorl of bracts, or perianth parts.

Uniseriate hair   Hair composed of a single row of cells (see Figure 4.15).

Unisexual flower   Flower lacking either an androecium (stamens) or a gynoecium (carpels); see Imperfect.

Unitegmic   With one integument.

Univalent   A solitary chromosome for which there is no homologue in the cell, so that during meiosis it does not pair with another chromosome.

Urceolate   Urn-shaped.

Utricle   Small, indehiscent, dry fruit with a thin wall (bladderlike) that is loose and free from the single seed.

Valid publication   Publication of scientific names in accordance with criteria for valid publication outlined in the *International Code of Botanical Nomenclature* (see page 550).

Valvate   Meeting at the edges without overlapping, as sepals or petals when in the bud.

Valve   One of the segments of a dehiscent fruit, separating from other such units at maturity in order to release the seeds.

VNTR   Variable number tandem repeats; DNA regions that are repeated a large number of times throughout the genome.

Variety   An infraspecific rank below that of subspecies although used by some as equivalent to it; usually used for geographical races and well-marked ecotypes within morphologically variable species of plants.

Vascular cambium   Cylinder of dividing cells that produces secondary xylem toward the center and secondary phloem toward the outside of a stem (see Figure 4.39).

Vascular tissues   Conducting tissues, xylem and phloem.

Vasculum   Cylindrical metal container used to carry freshly collected plant specimens.

Vein   Vascular bundle, usually visible externally, as in a leaf.

Velutinous   Velvety.

Venation   Pattern of veins in a leaf or other plant part.

Venn diagram   In systematics, represented by a set of nested ovals, circles, or parentheses describing nested mono-phyletic groups.

Vernation   Way in which leaves or perianth parts are folded or arranged in the bud in relation to one another.

Versatile   Structure, such as an anther, that is attached at its midpoint.

Vessel   Series of cells that are joined to form a tubelike structure, the adjacent ends more or less breaking down, functioning in water transport in xylem.

Vessel element   One of the cellular components of a vessel.

Vestured pit   Pit in the xylem with the pit cavity wholly or partially lined with small projections from the cell wall.

Villous   Covered with long, fine, soft hairs.

Vine   Herbaceous climbing plant.

Viscid   Covered with a sticky substance.

Viscin   Elastic and/or somewhat sticky material often covering pollen grains.

Vitta (pl. vittae)   Aromatic oil or resin canal in the pericarp of the fruits of many Apiaceae.

Viviparous fruit   Fruit with the seeds germinating while fruit is still attached to the parent plant, as in many mangrove species.

Wagner parsimony   Type of parsimony used in a cladistic analysis in which all characters are treated as ordered.

Weighting   Assignment of differential importance to characters in a phenetic or cladistic analysis.

Whorled leaves   Three or more leaves at a single node.

Wing   Thin, often membranous extension or outgrowth of a structure. Also used for the lateral pair of petals in the flowers of Fabaceae subfam. Faboideae and for the lateral petaloid sepals of many Polygalaceae.

Wood ray   Ribbonlike aggregate of cells extending radially in the secondary xylem or wood.

Woody   Hard in texture (containing secondary xylem).

Xerophyte   Plant adapted to dry conditions.

Xylem   Water-conducting tissue of the vascular plants, composed mainly of tracheids and/or vessel elements and parenchyma cells.

Zonate apertures   Ring-shaped or band-shaped apertures of a pollen grain.

Zygomorphic symmetry   *See* Bilateral symmetry.

Zygote   First cell of a sporophyte; formed by the fusion of the sperm and egg at fertilization.

PHOTOGRAPHIC CREDITS

Photographs not listed here are credited in the accompanying figure caption.

CHAPTER 1

FIGURE 1.2: Blackberries, © AtWaG/istockphoto.com; raspberries, Christopher S. Campbell; cherries, © dirkr/istockphoto.com

FIGURE 1.8: Courtesy of Melissa Luckow

FIGURE 1.9: *Adansonia gibbosa,* © dfwalls/Alamy; *A. digitata,* Elizabeth A. Kellogg; *A. grandieri,* © Nick Garbutt/naturepl.com

CHAPTER 3

Photo of Arthur Cronquist courtesy of The LuEsther T. Mertz Library of The New York Botanical Garden, Bronx, New York

Photograph of Emil Hans Willi Hennig courtesy of Bernd Hennig

CHAPTER 4

FIGURE 4.24: © Michael and Patricia Fogden/Minden Pictures

CHAPTER 6

FIGURE 6.5: © Dr. Merlin D. Tuttle/Photo Researchers, Inc.

FIGURE 6.18A,B: Christopher S. Campbell

FIGURE 6.29: Christopher S. Campbell

CHAPTER 8

FIGURE 8.1A: Walter S. Judd

FIGURE 8.2A: Walter S. Judd

FIGURE 8.7A: © David Sieren/Visuals Unlimited

FIGURE 8.20: Walter S. Judd

FIGURE 8.22: David McIntyre

APPENDIX 1

FIGURE 1: Walter S. Judd

COLOR PLATES, CHAPTERS 8 AND 9

PLATE 8.1: Leptosporangiate Ferns
(A) David McIntyre; (B) J. Richard Abbott; (C) Kurt M. Neubig; (D, E) David McIntyre; (F) Walter S. Judd (habit), Barbara S. Carlsward (fertile leaf); (G) Walter S. Judd

PLATE 8.2: Coniferales
(A–C) Walter S. Judd; (D) J. Richard Abbott; (E, F) Walter S. Judd; (G) Christopher S. Campbell

PLATE 9.1: ANITA Grade
(A) Kenneth R. Robertson; (B) Peter K. Endress (both images); (C) Walter S. Judd (fruit & leaves), J. Richard Abbott (flowers); (D) Walter S. Judd

PLATE 9.2: Magnoliids
(A) Kenneth R. Robertson & Daniel L. Nickrent; (B) Kenneth R. Robertson; (C) Walter S. Judd; (D) J. Richard Abbott; (E) Reuben E. Judd; (F–I) Walter S. Judd; (J) Scott Zona

PLATE 9.3: Alismatales and Liliales
(A) Kenneth R. Robertson; (B) Kenneth R. Robertson & Daniel L. Nickrent; (C, D) Walter S. Judd; (E) J. Dan Skean, Jr.; (F–H) Walter S. Judd; (I) Kenneth R. Robertson & Daniel L. Nickrent

PLATE 9.4: Asparagales
(A) Kenneth R. Robertson & Daniel L. Nickrent; (B) Walter S. Judd (both images); (C) Walter S. Judd; (D) Walter S. Judd (both images); (E) J. Richard Abbott; (F) Walter S. Judd; (G) Kenneth R. Robertson & Daniel L. Nickrent; (H) J. Richard Abbott

PLATE 9.5: Poales
(A–F) Walter S. Judd; (G) Gretchen M. Ionta; (H) Walter S. Judd; (I) Reuben E. Judd

PLATE 9.6: Ranunculales and Proteales
(A) J. Richard Abbott; (B) Christopher S. Campbell; (C–G) Walter S. Judd

PLATE 9.7: Caryophyllales
(A) J. Richard Abbott; (B) Kenneth R. Robertson & Daniel L. Nickrent (both images); (C) Kenneth R. Robertson; (D) Walter S. Judd; (E) Kenneth R. Robertson (both images); (F) Walter S. Judd; (G) Kenneth R. Robertson; (H) Walter S. Judd

PLATE 9.8: Saxifragales and Santalales
(A, B) Daniel L. Nickrent; (C, D) Kenneth R. Robertson & Daniel L. Nickrent; (E) Daniel L. Nickrent; (F) Kenneth R. Robertson & Daniel L. Nickrent; (G, H) Barbara S. Carlsward

PLATE 9.9: Zygophyllales, Oxalidales, Celastrales, Geraniales
(A) Walter S. Judd; (B) Scott Zona (both images), (C) Walter S. Judd; (D, E) Kenneth R. Robertson & Daniel L. Nickrent

PLATE 9.10: Malpighiales
(A, B) Walter S. Judd; (C, D) J. Richard Abbott; (E) Walter S. Judd; (F) Kenneth R. Robertson & Daniel L. Nickrent (left image), Walter S. Judd (right image); (G) Daniel L. Nickrent; (H) Kenneth R. Robertson & Daniel L. Nickrent; (I) Walter S. Judd; (J) Barbara S. Carlsward; (K) J. Richard Abbott; (L) Walter S. Judd (left image), Barbara S. Carlsward (right image)

PLATE 9.11: Fabales
(A) Walter S. Judd; (B) Reuben E. Judd; (C) Walter S. Judd; (D) Daniel L. Nickrent; (E) Kenneth R. Robertson & Daniel L. Nickrent; (F) Walter S. Judd

PLATE 9.12: Rosales
(A–G) Walter S. Judd; (H) Kenneth R. Robertson; (I) Kenneth R. Robertson; (J) Walter S. Judd

PLATE 9.13: Cucurbitales and Fagales
(A, B): Walter S. Judd; (C) Kenneth R. Robertson (habit), Walter S. Judd (acorns, staminate catkins); (D) Walter S. Judd (both images); (E) J. Richard Abbott; (F) Walter S. Judd (both images)

PLATE 9.14: Myrtales
(A) Darin S. Penneys; (B, C) Walter S. Judd; (D) Kurt M. Neubig; (E–I) Walter S. Judd

PLATE 9.15: Brassicales and Malvales
(A) Walter S. Judd; (B) Daniel L. Nickrent; (C) Walter S. Judd; (D) Kenneth R. Robertson; (E) Walter S. Judd; (F) J. Richard Abbott; (G) J. Richard Abbott (left), Walter S. Judd (right); (H) Walter S. Judd

PLATE 9.16: Sapindales
(A) Kurt M. Neubig; (B) Walter S. Judd; (C) J. Dan Skean, Jr.; (D) Walter S. Judd; (E) Kenneth R. Robertson; (F) Walter S. Judd; (G) Walter S. Judd (left), Scott Zona (right); (H–J) Walter S. Judd

PLATE 9.17: Cornales
(A–C) Kenneth R. Robertson; (D, E) Walter S. Judd; (F) Kenneth R. Robertson & Daniel L. Nickrent

PLATE 9.18: Ericales
(A) Kenneth R. Robertson; (B) Scott Zona; (C–F) Walter S. Judd; (G) J. Richard Abbott; (H) Kurt M. Neubig; (I–K) Walter S. Judd; (L) J. Richard Abbott

PLATE 9.19: Solanales
All photos Walter S. Judd

PLATE 9.20: Gentianales
(A) Walter S. Judd; (B) Kenneth R. Robertson & Daniel L. Nickrent; (C, D) Walter S. Judd; (E) Barbara S. Carlsward; (F) Scott Zona; (G, H) Walter S. Judd

PLATE 9.21: Lamiales
(A) Margaret H. Stone; (B) Barbara S. Carlsward; (C) Margaret H. Stone; (D, E) Kenneth R. Robertson & Daniel L. Nickrent; (F) Walter S. Judd; (G) J. Richard Abbott; (H) Scott Zona; (I) J. Dan Skean, Jr.; (J–M) Walter S. Judd; (N) Barbara S. Carlsward; (O) Kenneth R. Robertson & Daniel L. Nickrent (both images); (P) J. Richard Abbott

PLATE 9.22: Aquifoliales and Apiales
(A) Walter S. Judd; (B) Scott Zona; (C) Kenneth R. Robertson; (D) Kurt M. Neubig (both images); (E) Walter S. Judd; (F) Scott Zona; (G, H) Walter S. Judd

PLATE 9.23: Dipsacales
(A) Michael J. Donoghue; (B) Walter S. Judd; (C) Kenneth R. Robertson & Daniel L. Nickrent; (D) J. Richard Abbott; (E, F) Walter S. Judd; (G) Michael J. Donoghue

PLATE 9.24: Asterales
(A) Kurt M. Neubig (both images); (B, C) Walter S. Judd; (D) J. Richard Abbott; (E) Walter S. Judd; (F) Kenneth R. Robertson & Daniel L. Nickrent; (G) Kenneth R. Robertson

TAXONOMIC INDEX

Boldface items refer to family discussions. Page numbers in *italics* refer to material in illustrations.

SUBJECT INDEX

Page numbers in *italic* refer to information in an illustration or table.

Abaxial surface, 56
Acaulescence, 55
Accelerated transformation, 32
Accessory buds, 56
Accessory structures, 75
ACCTRAN, 32
Achenes, 76, 77
Actinomorphic flowers, 64
Adanson, Michel, 46
Adaptation
    fitness and, 122
    in pollination syndromes, 8–9
Adaxial surface, 56
Adnation, 64, 65
Adventitious embryony, 89
Adventitious roots, 55
Aeneas, 338
Aerial roots, 55
Agamic complexes, 147
Agamospecies, 147
Agamospermous microspecies, 147
Agamospermy
    agamospecies, 147
    described, 89–90, 143
    reproduction isolation and, 130–131
Aggregate fruits, 75
Alkaloids, 95–96
Allergies, 215–216
Allopatric speciation, 126, 127, 149
Allopolyploidy
    overview of, 90, 91
    speciation and, 140–142
Alternate leaves, 56–57
Alternation of generations, 156
Amarolide, *98*
Ambiguity, 24, 31
Ament, 74
Amyotrophic lateral sclerosis/Parkinsonism dementia complex, 207
Anagenesis, 123
Analytic systems, 40
Anatomical characters
    crystals, 85
    features of, 81
    flower anatomy and development, 87
    leaves, 83–84
    nodes, 82–83
    phloem, 82
    secondary xylem, 81–82
    secretory structures, 84–85
    systematics and, 49
    xylem and phloem arrangements in the stem, 86–87

Anatropous ovule, 88, 175
Anderson, Edgar, 48, 137
Andrena (bee), *129*
Androecium (androecia), 61, 62
Androgynophore, 66
Aneuploidy, 90, 91, 124
Angiosperm Phylogeny Group, 226
    Web site, 562
Angiosperms
    "anthophyte hypothesis," 186
    chromosome numbers, 90
    crown clade, 176
    derivation of herbaceous lineages, 181
    embryology, *87*, 88
    families, *230–231*
    flowers, 175
    fruit morphology and dispersal, 180
    Gnetales and, 220
    growth form and, 180–181
    in historical classifications, 50
    insect pollination and, 180
    life cycle, *63*, 175
    meaning of name, 185
    monophyly, 225
    phylogenetic relationships to other groups, 24, *25*, 176–177, 178
    phylogenetic relationships within, 28, 178–180, 225–229
    pollen types, 95
    polyploidy, 140
    reproductive features, 173, 175
    secondary growth, 86
    secondary xylem, 81
    seeds, 89
    significance of, 2
    time of origin, 175–176
Animals
    dispersal of seeds and fruits, 80–81
    pollination and, 68–70
Annals of the Missouri Botanical Garden, 113, *561*
Annuals, 54
Annulus, 167, 190, 195
Anomocytic stomata, 84
Antheridium (antheridia), 87, 162
Antheridiogens, 190
Antherozoids, 190
Anthers, 62, 93
Anthocyanins, 96
Anthophyte hypothesis, 176, *178*, 186
Antipodal cells, 88, 175

Antophora erschowi (bee), *129*
Ants, dispersal of fruits and seeds, 81
Apertures, 94
Apical meristems, 54, 169
Apigeninidin, *96*
Apocarpous flowers, 64
Apomixis, 143
    See also Agamospermy
Apomorphy species concept, 145
Apomorphy-based definition of taxa, 548
Apopetalous flowers, 64
Aposepalous flowers, 64
Apotepalous flowers, 64
Arachnoid leaves, 60
Archegonium (archegonia), 87, 162
Arils, 79, *212*, 220
Arylphenalenones, 283
Asexual reproduction
    effects on heterozygosity, 143
    See also Agamospermy
Asymmetrical flowers, 64
atp1 gene, 113
ATP synthase, 113
atpA gene, 113
atpB gene, 113
Autapomorphy, 23
Authorities, botanical nomenclature and, 546
Autopolyploidy
    described, 90, 91
    speciation and, 142–143
Avocado, 242
Axile placentation, 67
Axillary branches/branching, 54, 168
Axillary buds, 56

Bacillus amyloliquefaciens, 115n
Backcrossing, 137
Bacteria, *155*
Bacterial Artificial Chromosomes (BACs), 107
Baculum, *95*
Balder, 338
*Bam*HI, 115
Bark, 219
Bartlett, H. H., 48
Basifixed flowers, 67
Bats
    dispersal of fruits and seeds, 81
    as pollinators, 9, 68, *69*, 122, *123*
Bay leaves, 242

Bayesian methods, 24
Bees
    behavioral isolation in plants and, 129
    as pollinators, 68, 69, 71
Beetles, *69*
Behavioral isolation, 129
Bentham, George, 41, 42, 45, 47
Benzylisoquinoline alkaloids, 96
Berries, 76, 77
Bessey, Charles Edwin, 41, 43, 50
Betalains, 96
Betanin, *96*
Betulin, 98
Bias, 26–27
Bicolateral vascular bundles, 86–87
Bidirectional introgression, 137
Biennials, 54
Bilateral symmetry, 64
Binary characters, 25
Binomials, 45, 546–547
Biogeography, 9
Bioinformatics, 107
Biological nomenclature, 6
Biological species concept (BSC), 144–145, 146
Biological systematics, 13
Biradial flowers, 64
Birds
    dispersal of fruits and seeds, 80
    as pollinators, *69*, 130
Bisexual flowers, 61–62
Bitegmic ovules, 175
Bivalents, 90
BMAA, 207
Bole, 54
Bootstrap analysis, 29
Botanical nomenclature, 543–551
    pronunciation, 549
    list of specific epithets, *544–545*
    rules of, 549–551
Botany, 40–41
Bracketed key, 558, *559*
Bracteate raceme, *73*
Bracteoles, 61
Bracts, 61
Branch-and-bound searches, 20
Breeding systems
    described, 143
    plant diversity and, 120
Bremer support, 29
Brown, Robert, 50, 301
Bud scales, 56
Buds, 54, 56
Bulbs, 55

605

TRACHEOPHYTES

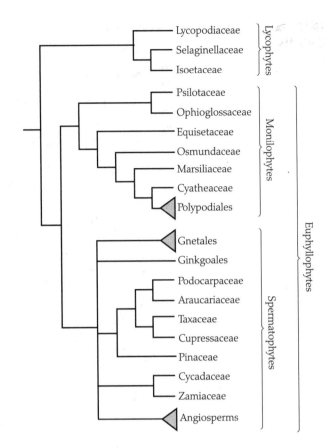

Lycopodiaceae ⎤
Selaginellaceae ⎥ Lycophytes
Isoetaceae ⎦

Psilotaceae ⎤
Ophioglossaceae ⎥
Equisetaceae ⎥
Osmundaceae ⎥ Monilophytes
Marsiliaceae ⎥
Cyatheaceae ⎥
Polypodiales ⎦

Gnetales ⎤
Ginkgoales ⎥
Podocarpaceae ⎥
Araucariaceae ⎥
Taxaceae ⎥
Cupressaceae ⎥ Spermatophytes
Pinaceae ⎥
Cycadaceae ⎥
Zamiaceae ⎥
Angiosperms ⎦

Euphyllophytes

ANGIOSPERMS

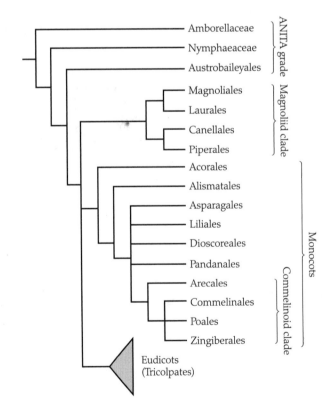

Amborellaceae ⎤
Nymphaeaceae ⎥ ANITA grade
Austrobaileyales ⎦

Magnoliales ⎤
Laurales ⎥ Magnoliid clade
Canellales ⎥
Piperales ⎦

Acorales ⎤
Alismatales ⎥
Asparagales ⎥
Liliales ⎥
Dioscoreales ⎥ Monocots
Pandanales ⎥
Arecales ⎤ ⎥
Commelinales ⎥ Commelinoid clade
Poales ⎥ ⎥
Zingiberales ⎦ ⎦

Eudicots
(Tricolpates)

Cladograms show simplified relationships of
the major groups of tracheophytes, detailing
angiosperms, as organized in this text.
See pages 185–187 (tracheophytes) and
225–231 (angiosperms).